STRENGTH OF MATERIALS

SECOND EDITION

R. SUBRAMANIAN
Formerly, Professor and Head
Civil Engineering Department
NITTTR, Chandigarh

OXFORD
UNIVERSITY PRESS

OXFORD
UNIVERSITY PRESS

Oxford University Press is a department of the University of Oxford. It furthers the University's objective of excellence in research, scholarship, and education by publishing worldwide. Oxford is a registered trademark of Oxford University Press in the UK and in certain other countries

Published in India by
Oxford University Press
YMCA Library Building, 1 Jai Singh Road, New Delhi 110001, India

First Edition published in 2005
Second Edition published in 2010
Eighth impression 2014

ISBN-13: 978-0-19-806110-6
ISBN-10: 0-19-806110-2

Typeset in Times Roman
by Archetype, New Delhi 110063
Printed in India by Raj Kamal Electric Press, Haryana

To
my wife, Usha, and daughter, Supriya,
for their
constant encouragement and inspiration

Preface to the Second Edition

Responding to the enthusiastic acceptance of the first edition of *Strength of Materials* in reputed universities and institutes, it gives me great pleasure to present this second edition to the engineering academic fraternity. Continuing on the lines of the first edition, the main emphasis of this text is on the understanding of the fundamental concepts and principles underlying the analysis and design of structures.

The book has been revised extensively and many new topics as well as solved and unsolved examples have been included in almost all chapters. I sincerely hope that the revised version of the book will further aid students in understanding the basic concepts and principles of this subject.

New to the Second Edition

The second edition improves upon the coverage of the original edition, making the text much more comprehensive and lucid. A special attempt has been made to cover all the standard cases of loading in analysing beams for shear force and bending moment or for slope and deflection, either through explicit explanations or through solved examples and illustrations. All solved examples have also been linked with the text to enable easy navigation.

Extended Chapter Material

Chapter 1 Definitions of basic relevant terms such as rigid body, deformable body, equilibrium, Newton's laws, principle of transmissibility, etc.

Chapter 2 Section modulus and its calculation

Chapter 3 Thermal stress in composite bars, shear stress and strain, stress concentration, residual stresses and fatigue, additional solved examples

Chapter 4 Determination of shear force and bending moment using singularity functions, additional solved examples

Chapter 5 Biaxial bending, additional solved examples of stress calculations

Chapter 6 Analysis of dams, walls, and chimneys, additional solved examples on structures subjected to lateral pressure

Chapter 7 Deflection in beams with unsymmetrical bending, additional solved examples

Chapter 8 Improved detailing of polar section modulus, torsional rigidity, torque capacity and design of shafts, additional solved examples

Chapter 9 Stress analysis of shafts subjected to torque, BM, and axial thrust, uniaxial and biaxial stress analysis, elementary discussion on three-dimensional stress analysis, additional solved examples

Chapter 10 Strain energy theorems, Castigliano's theorem and unit-load method for calculation of deflection in beams and frames, Maxwell–Betty theorem of reciprocal deflections

Chapter 11 Additional empirical methods for columns—straight line formula, Johnson's parabolic formula, IS code formula, Perry's formula, additional solved examples

Chapter 12 Wahl correction, Winkler–Bach formula, stresses in thick spherical shells, closed rings, and chain links, additional solved examples

Chapter 13 Determination of deflection in perfect trusses using Castigliano's theorem and the unit-load method, Williot–Mohr diagrams

Chapter 14 Introduction of force method and flexibility coefficients, analysis of indeterminate frames under loads, lack of fit, and temperature change using unit-load method and Castigliano's theorem

Chapter 15 Deformation in fixed beams, flexibility matrix method, stiffness method, stiffness matrix method using stiffness coefficients for the analysis of continuous beams. This chapter also includes the conceptual description of the moment-distribution method and its application to continuous beams.

Acknowledgements

I would like to gratefully acknowledge the feedback and suggestions given by various faculty members for the improvement of the book.

I am grateful to Oxford University Press for bringing out the revised edition in quick time and in a very elegant format. Comments and suggestions for the improvement of the book are welcome. Please send them to the publishers by logging on to their website www.oup.com or to the author at rsmani2k@yahoo.com.

R. Subramanian

Preface to the First Edition

The subject of strength of materials or mechanics of solids involves analytical methods for determining the strength, stiffness, and stability of the various load-carrying structural members. A thorough understanding of the underlying principles is useful to civil engineers and architects, and cuts broadly across all branches of engineering with several applications.

This book has been specially written for undergraduate students of engineering taking a first course on the subject. While there are many books available on this subject, I have written this book with a focus on the concepts and their engineering applications. During my tenure as a teacher, I have observed that a large number of students find it difficult to grasp the concepts and principles, and often omit several topics. I wanted to present the subject matter in an easy-to-comprehend form and explain the concepts and principles by applying them to a large number of examples, which, I hope, will help the students to internalize them. A strong grasp of these concepts and principles will help them to understand advanced topics with ease.

About the Book

I have attempted to make the presentation as lucid and comprehensible as possible. Going from the simple to the complex, the concrete to the abstract, and the known to the unknown are the cardinal tenets followed in arriving at the format and detailing of the book. There is lack of uniformity in the curriculum followed in different universities and engineering colleges. Therefore, during the preparation of the manuscript, I have attempted to give as comprehensive a coverage as possible of the topics, taking the various curricula in effect into consideration.

Content and Structure

A thorough grounding in mathematics and a basic course in engineering mechanics are prerequisites for a course in strength of materials. The book tries to build up on this foundation, developing the concepts and principles vital for advanced studies in structural analysis and design. The structure of the book is as follows.

Chapter 1 is devoted to a recapitulation of the relevant basic concepts of applied mechanics. Students should go through this chapter to give them the necessary background to understand the subject.

Chapter 2 deals with properties of sections such as moment of inertia (MOI), product of inertia, polar MOI, etc., which constantly find application in structural analysis. A thorough understanding of these topics is necessary for further work.

Chapter 3 deals with elementary concepts of stresses and strains, Hooke's law, elastic constants, stresses in compound sections, temperature stresses and some simple indeterminate problems, and mechanical properties of materials. This chapter is important for further studies in stress analysis.

Chapter 4 deals with an important and common structural element, the beam. The concepts of bending moment (BM) and shear force (SF) are introduced along with the differential relationships and methods of showing variations in BM and SF diagrams. The analysis of statically determinate rigid frames has also been covered.

Chapter 5 follows the concepts presented in Chapter 4 by introducing the bending equation along with methods for calculating bending and shearing stresses followed by their distribution. The concepts of shear centre and unsymmetrical bending have also been covered in this chapter.

Chapter 6 deals with the effect of combining direct and bending stresses, and their effects, which are important for analysing structures subjected to lateral pressure, such as chimneys, retaining walls, dams, etc.

Chapter 7 deals with the very important aspect of deformations in beams. The deformations have been calculated using different methods such as double integration, area-moment theorems, and conjugate beam. A thorough grounding of the concepts in this chapter is important for further analysis.

Chapter 8 deals with another important structural element—shafts. Torsion and torsional shear have been dealt with in detail.

Chapter 9 deals with analysis of plane stress, both analytically and graphically. The calculation of principal stresses and maximum shear stress has been explained in detail and illustrated with a number of examples.

Chapter 10 deals with strain energy, which has many applications in later studies. Strain energy due to uniaxial, bending, and shear stresses and due to torsion has been explained and illustrated. Suddenly applied and impact loads have also been covered.

Chapter 11 deals with another common structural element, the column, which also introduces the concept of stability. Euler's critical load method and the Rankine–Gordon method have been explained in detail.

Chapter 12 deals with some special structural elements such as springs, thin and thick pressure vessels, and bending of curved bars.

Chapter 13 deals with the truss or pin-jointed plane frame. The method of joints, method of sections, the relevant graphical methods, and the method of tension coefficients have been illustrated.

Chapter 14 is an introduction to indeterminate structural analysis. The basic concepts of flexibility and stiffness methods have been covered.

Chapter 15 deals with simple indeterminate structures such as fixed and continuous beams. It introduces the flexibility method of analysis and illustrates Clapeyron's theorem of three moments in detail.

Chapter 16 is intended as an introduction to advanced structural analysis. The method of moment distribution, credited to Hardy Cross, as applied to continuous beams and frames with no sway as well as side sway, has been introduced in this chapter.

To reinforce the theoretical concepts, the text is supplemented by a large number of worked-out examples with step-by-step solution procedures as well as a large number of review questions and exercise problems. The book also provides useful tables both in the text and in the appendices for ready reference.

Acknowledgements

I am grateful to the Bureau of Indian Standards for permission to reproduce a part of the steel tables from their publication. I am also grateful to the editorial team at Oxford University Press for the excellent job done in bringing this book out in a short time with a high degree of accuracy and precision. I am grateful as well to a large number of my teachers and students who have directly or indirectly helped me in this endeavour.

Despite best efforts by all concerned, it is possible that some errors might have crept into the final form in your hand. It will be appreciated if such errors are pointed out to me or to the publishers. In addition, suggestions from teachers and students regarding any additions in topics or subtopics required to make the book more useful are welcome. Such comments or suggestions will be taken care of in subsequent editions of the book.

R. Subramanian

Brief Contents

Detailed Contents

List of Symbols

A — area, constant
a — area, side of section
B — constant, width
b — width
C — constant
D, d — depth, diameter
E —Young's modulus of elasticity
e — eccentricity, shift of NA in curved bars
f — frequency
F — force, SF
G — modulus of rigidity
g — gravitational constant
H, h — height
I — MOI, second moment of area
J — polar MOI
K — bulk modulus, stiffness of springs
L, l — length, span
M — moment, BM
m — mass
N — speed, RPM
n — number of coils
P — force, concentrated load
p — pressure
R, r — radius
S — stiffness
T — torque, temperature
t — thickness, time, temperature
U — strain energy, principal axis
u, v — principal axis coordinates
V —SF, volume, principal axis
v volume
w — load intensity per m
x — variable, coordinate
X — coordinate axis
Y — coordinate axis
y — variable, coordinate
Z — section modulus, coordinate axis
z — coordinate
α — coefficient of thermal expansion, angle
β — angle, constant
γ — weight per unit volume, shear strain
ε — linear strain
ϕ — angle
θ — slope, angle, angle of twist
τ — shear stress
ν — Poisson's ratio
Δ — deflection, increment in function
δ — increment in function
σ — normal stress
@ — about, at the rate of
$\rightarrow +$ — positive to the right
$\uparrow +$ — positive upwards
↻ + — clockwise positive

CHAPTER **1**

Review of Basic Concepts

Learning Objectives

After going through this chapter, the reader will be able to

- apply the parallelogram law and triangle law to calculate the resultant of concurrent forces,
- state Varignon's theorem of moments,
- calculate the resultants of parallel and non-concurrent force systems using the principle of moments,
- use the graphical methods of the force polygon and funicular polygon to determine the resultant of planar force systems,
- calculate the centroid of a given area, and
- appreciate the importance of the concepts and principles of mechanics in engineering analysis and design.

1.1 INTRODUCTION

The methods and techniques used in the analysis and design of structures are based upon concepts and principles which the reader would have learnt in a first course in engineering mechanics. This chapter is intended as a review of such basic concepts and principles and as an introduction to structural engineering. This will ensure uniformity in symbols, units, and procedures in problem solving.

1.2 STRUCTURAL ENGINEERING

There are different types of structures built by man—buildings, bridges, culverts, water tanks, storage bins, roads, transmission towers, machines, etc. Such structures are built up of a number of structural elements joined together suitably.

In the case of any structure, two types of designs are involved—functional or architectural and structural. Functional or architectural design deals with aspects other than the strengths of the structure, like aesthetics, utility, orientation, general layout, etc. Once this important aspect of design is taken care of, structural designers take over. Their work involves analysis of the structure and its elements to find the forces and moments that they have to withstand and then design the dimensions of the elements and their interconnections. The former is the realm of structural analysis while the latter forms what is called structural design.

Structures are designed to withstand loads, i.e., forces and moments due to different causes. While some structures, like aircraft structures, machine foundations, etc. necessarily have to be designed for forces due to motion, many are designed considering them to be in equilibrium, or at rest. A body in uniform motion is also governed by the same laws as for a body at rest. In this book, we will deal with structures which are in equilibrium.

1.3 BASIC PRINCIPLES OF MECHANICS

Before taking up the principles, definition of the following terms must be clearly understood.

Matter is any substance that occupies space.

Particle in solid geometry is analogous to a point in plane geometry. A particle has mass but has no dimensions.

Body is matter that is bounded by a closed surface. Bodies can be classified as rigid and deformable.

A *rigid body* is a body that does not undergo any deformation, change in shape and size, on application of a force. All bodies in nature are deformable on application of a sufficiently large force. We would not have been able to fabricate many things that we use in day-to-day life if bodies were not deformable. But for the purpose of certain analysis, as in statics detailed in the next section, bodies are assumed to be rigid (strictly, we assume that deformations are too small that we can consider the body to be rigid).

A *deformable body* is one that undergoes deformations, change in size and shape, on application of a force. Materials have different degrees of deformations on application of forces. Rubber, for example, undergoes deformations on application of a small force. Steel, on the other hand, requires a very large amount of force to deform it. It is necessary to compute the deformations of bodies and it is done in many forms of analysis (see Chapter 3 on Simple Stresses and Strains and Chapter 7 on Deformations in Beams).

Inertia is an inherent property of matter by which it resists any change in its state.

Mass is a quantitative measure of inertia. Bodies undergo different degrees of deformations under the action of forces. Thus, steel has more inertia (mass) than aluminium as, under the action of the same force on two identical bodies of steel and aluminium, the aluminium body will undergo more change in motion than steel.

Space is a region that extends in all directions and contains all bodies. Position of a body in space is located by arbitrarily fixing reference axes and measuring its coordinates with respect to these axes. Rectangular coordinates use three mutually perpendicular axes while cylindrical coordinates use distances and angles to locate position in space.

Time is a measure of duration between successive events occurring. This is important in mechanics as bodies in motion change their position with time.

Equilibrium is a state of rest of a body. State of rest can be defined as a state in which the body does not change its position. Equilibrium equations also cover the case of a body moving with uniform velocity along a straight line.

Motion is a change of position of a body with respect to time. When an unbalanced force acts on a body, it has motion.

Scalar and vector quantities are physical quantities like mass, time, speed, velocity, force, etc. A scalar quantity is one that has only a magnitude as an attribute.

Two scalar quantities can be added using the conventional method of addition. Mass, time, length, etc. are scalar quantities. Vector quantities, on the other hand, have a direction in addition to magnitude. They cannot be added like scalar quantities and are manipulated using a different mathematical tool known as *vector algebra*. Force, moment, velocity, etc. are vector quantities.

The following basic principles of mechanics form the foundation on which the entire classical mechanics is based. They are in general proved by experience than mathematically.

1. ***Newton's laws of motion*** There are three Newton's laws of motion, which govern the motion of bodies under the action of forces.
   Newton's first law states that a particle or rigid body will remain in a state of rest or continue to move in a straight line with uniform velocity unless acted upon by an unbalanced force.
   Newton's second law states that a particle's rate of change of momentum is equal to the unbalanced force acting on it and takes place in the direction of the force.
   Newton's third law states that to every action there is an equal and opposite reaction.
2. ***Parallelogram law*** This law states that if two vector quantities are represented by two adjacent sides of a parallelogram, their sum is given by the diagonal of the parallelogram passing through the point of intersection of the two forces. This is discussed in detail in the next section.
3. ***Principle of transmissibility*** This principle states that the effect of a force acting on a rigid body does not change if the force is moved along its line of action to another point on the body.
4. ***Principle of superposition*** This principle states that if a number of forces act on a body, the total effect of all the forces is the summation of the effects of the individual forces. Thus, if a body is subjected to a number of forces, $F_1, F_2, F_3, \ldots$, the motion imparted to the body is the summation of the motion imparted to the body by the individual forces $F_1, F_2, F_3, \ldots$.
   The principle of superposition also means that the effect of a set of forces acting on a body does not change if another system of forces in equilibrium is added to, or subtracted from, it.
5. Two forces, acting on a body and keeping it in equilibrium, must be collinear, equal, and opposite.
6. ***Newton's law of gravitation*** It states that two bodies attract each other by a force given by Gm_1m_2/r_2, where G is the universal constant of gravitation, m_1, m_2 are the masses of the bodies, and r is the distance between them. Based on the law of gravitation, the earth attracts all material bodies with mass. The weight of a body is due to this attraction. Earth's gravity gives to a body of mass m a weight of gm, where $g = 9.81$ m/s^2 and $g = GM/R^2$, where G = Newton's universal gravitational constant, M is the mass of the earth, and R is the radius of the earth.
7. ***Law of conservation of energy*** It states that energy can neither be created nor destroyed but can change from one body to another or change form.

When a force acts on a mass and the body moves, it does work and the work is converted to kinetic energy of the body. Work and energy are scalar quantities and are easier to work with. Energy principles are used in many forms of analysis.

1.4 STATICS

Statics is a branch of mechanics that deals with forces and moments which are in equilibrium. The principles of statics are also applicable to bodies which are in motion but without acceleration.

1.4.1 Force

Force is a vector quantity and may be represented as shown in Fig. 1.1(a). The three attributes of force are shown here. AB shows the direction of the force, the arrow shows the sense (acting from A to B), and B is the point of application. The line *ab* shows the direction of the force and the arrow at the end shows the sense of the force as from *a* to *b* or $\overrightarrow{ab}$, and the point of application of the force is either at the beginning or at the end of line *ab*.

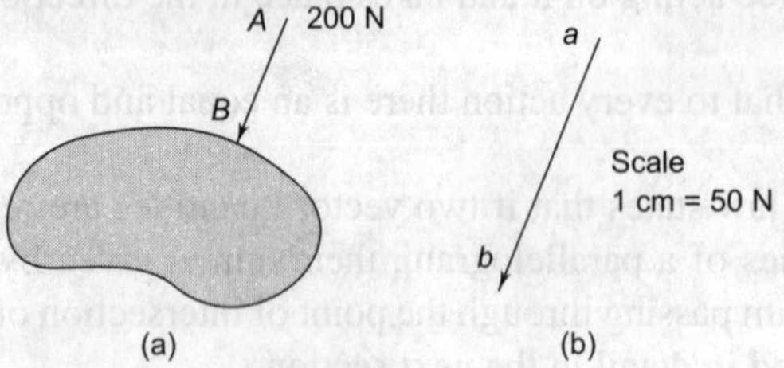

Fig. 1.1

Figure 1.1(b) shows a vectorial representation of the force $\overrightarrow{ab}$. In addition to the three attributes discussed above, the vectorial or graphic representation also includes the magnitude of the force represented by line *ab* to some scale. This representation is required for a graphic solution to problems.

In the SI (the abbreviation of the French form of The International System of Units) units, the unit of force is the newton. One newton is the force required to be applied to a mass of 1 kg to give it an acceleration of 1 m s^{-2}. Since force = mass × acceleration, force (newton) = kg m s^{-2}. Many multiples of this unit are used, e.g. 1 kilonewton = 1000 N, 1 meganewton (MN) = 1000 kN, 1 giganewton (GN) = 1000 MN, etc.

Force systems A number of forces acting on a body forms a force system. Force systems may be coplanar, in which case the lines of action of all the forces of the system lie in a plane. Such systems may also be spatial, when the lines of action do not lie in a plane (Fig. 1.2). Force systems can further be classified as concurrent, parallel and non-concurrent, non-parallel, as shown in Fig. 1.2.

Composition of forces Two given forces acting (concurrently) at a point can be combined into a single force, known as their resultant and having the same effect on the body as the two forces, by the parallelogram law or triangle law of forces. This is shown in Fig. 1.3, where F_1 and F_2 are the two forces and R is their resultant. Analytically, $R^2 = F_1^2 + F_2^2 - 2F_1F_2 \cos\theta$, where θ is the angle between the lines of action of the forces.

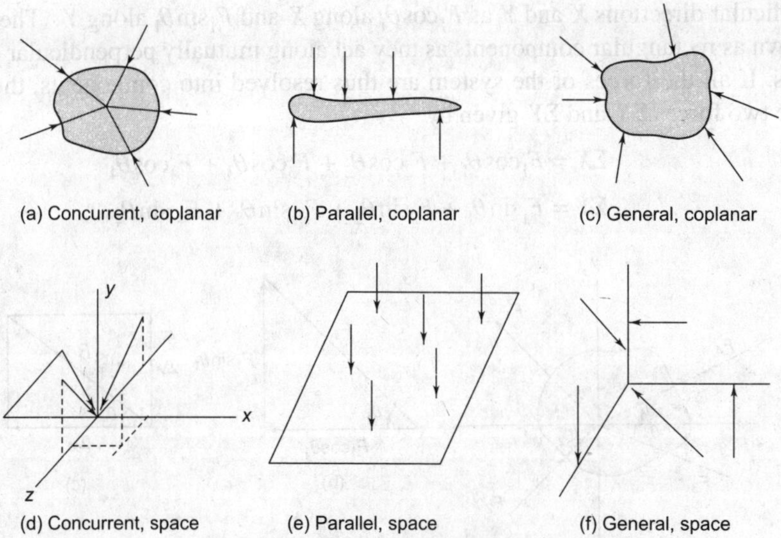

Fig. 1.2 Force systems

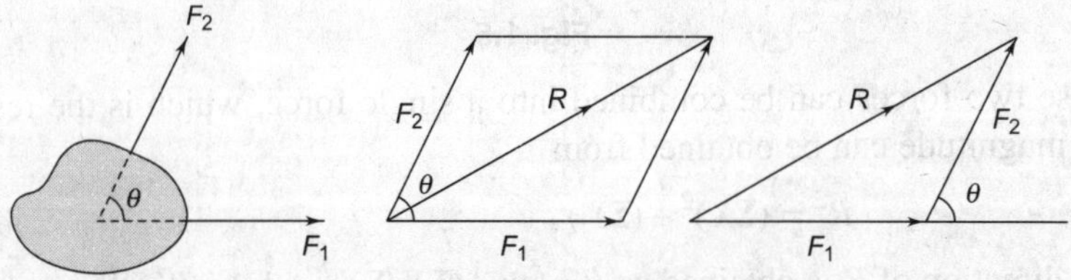

Fig. 1.3 Addition of two vectors

If more than two forces act at a point, the resultant is determined from the polygon of forces w1hich is obtained by the repeated application of the triangle law of forces (Fig. 1.4). The magnitude, line of action, and sense of the resultant are obtained from the force polygon, and the resultant passes through the point of concurrency.

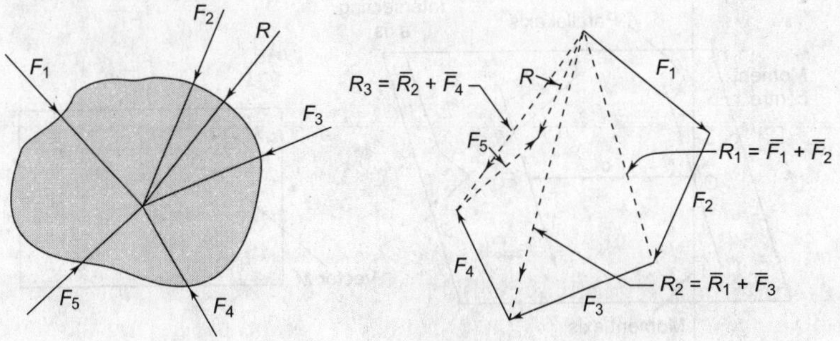

Fig. 1.4

To determine the resultant of more than two forces analytically, we use a combination of resolution and composition. In the concurrent force system shown in Fig. 1.5(a), any of the forces, say F_1, can be resolved into components along perpendicular directions X and Y as $F_1\cos\theta_1$ along X and $F_1\sin\theta_1$ along Y. These are known as rectangular components as they act along mutually perpendicular directions. If all the forces of the system are thus resolved into components, then we have two forces ΣX and ΣY given by

$$\Sigma X = F_1\cos\theta_1 + F_2\cos\theta_2 + F_3\cos\theta_3 + F_4\cos\theta_4$$

$$\Sigma Y = F_1\sin\theta_1 + F_2\sin\theta_2 + F_3\sin\theta_3 + F_4\sin\theta_4$$

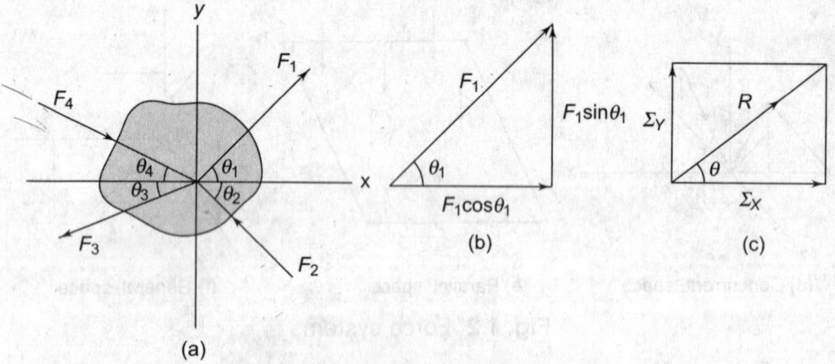

Fig. 1.5

These two forces can be combined into a single force, which is the resultant, whose magnitude can be obtained from

$$R^2 = (\Sigma X)^2 + (\Sigma Y)^2$$

The direction of R is obtained as $\theta = \tan^{-1}(\Sigma Y/\Sigma X)$, where θ is the angle made by R with the X-axis.

Moment of a force Moment is a vector quantity like force. As in Fig. 1.6, moment = force × distance = Fd about moment centre O which means moment about an axis passing through O and perpendicular to the plane containing the force. Moments are represented in diagrams using curved arrows about the moment centre, or by a line with double arrows using the right-hand screw notation.

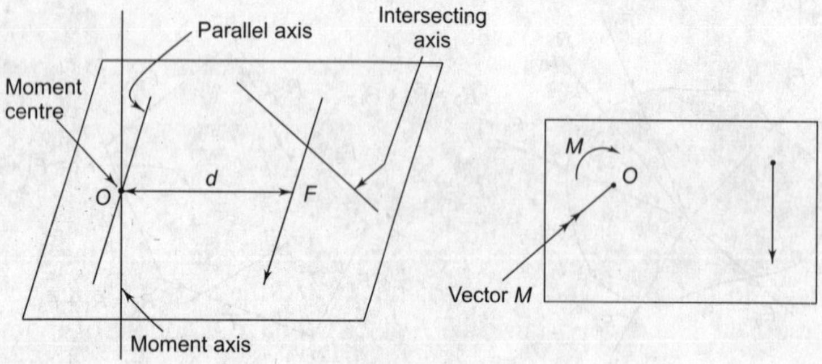

Fig. 1.6

Moment is a measure of the rotating effect of a force. The units of moment are the newton metre (Nm), kilonewton metre (kNm), etc. A force has no moment about an axis parallel to its line of action or intersecting it, as shown in Fig. 1.6.

Varignon's theorem or principle of moments states that *the algebraic sum of the moments of forces is equal to the moment of the resultant of the forces about the same axis*. Moments are given signs according to their nature or direction of rotation. Thus, the moment due to force F_1 is opposed to that due to F_2 about the moment centre O in Fig. 1.7.

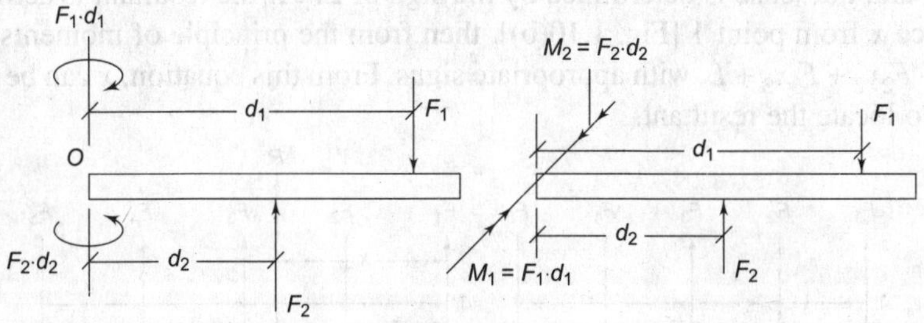

Fig. 1.7

A couple is a special case of a moment due to two equal and opposite forces acting at a distance (Fig. 1.8). The couple has the same moment Fd about any point in its plane.

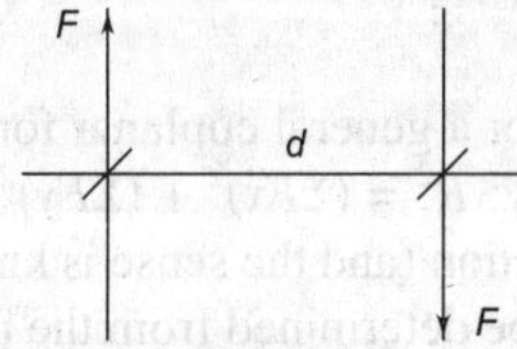

Couple Fd

Fig. 1.8

In rigid body mechanics, a force can be moved along its line of action without altering its effect. But a force can be translated parallel to its line of action only by adding a couple. Let us consider the case of the force F acting at point 1 to be translated to point 2 [Fig. 1.9(b)].

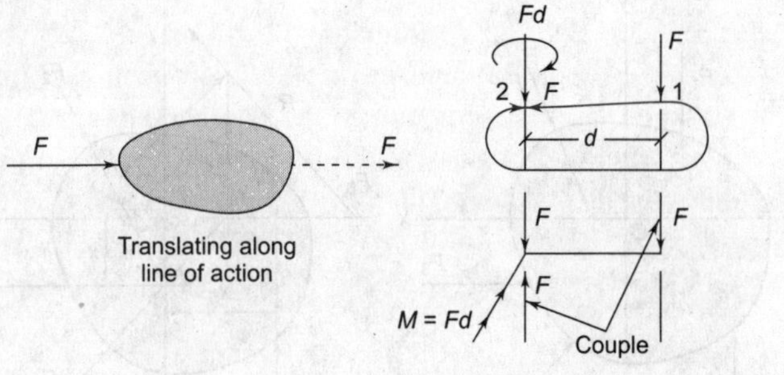

(a)

(b) Translating a force parallel to its line of action

Fig. 1.9

We add equal, opposite and collinear forces F at 2. Note that the addition of these forces do not affect the system of forces and their effect. Considering the given force F at 1 and opposite force F at 2, we have a couple of moment Fd. These two forces can be replaced by the couple Fd. The resultant system is shown in Fig. 1.9(b). The force F has been translated parallel to its line of action, which needs the addition of a couple Fd at 2.

The resultant of non-concurrent force systems In the case of non-concurrent forces, the resultant magnitude, direction, and sense can be obtained by resolution and composition or by the method of polygon of forces. But the location of the resultant in space is obtained by the principle of moments or by graphical method of the funicular polygon.

For the parallel force system shown in Fig. 1.10(a), resultant $R = \Sigma F =$ algebraic sum of the forces; the direction of the resultant is the same as that of the given forces and the sense is determined by the sign of ΣF. If the resultant is acting at a distance x from point 1 [Fig. 1.10(b)], then from the principle of moments, $Rx = F_1x_1 + F_2x_2 + F_3x_3 + L$, with appropriate signs. From this equation, x can be calculated to locate the resultant.

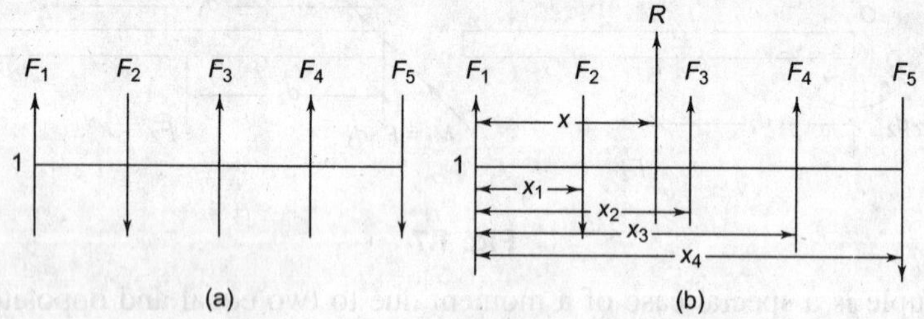

Fig. 1.10

For a general coplanar force system shown in Fig. 1.11(a), the same principles apply: $R^2 = (\Sigma Fx)^2 + (\Sigma Fy)^2$ gives the magnitude, $\theta = \tan^{-1} (\Sigma Fy/\Sigma Fx)$ gives the direction (and the sense is known from the signs of ΣFy and ΣFx), and its location can be determined from the principle of moments [Fig. 1.11(b)]. $Rd = F_1x_1 + F_2x_2 + F_3x_3 + \cdots$, where $x_1, x_2, x_3, \ldots$ are perpendicular distances of the forces from the moment centre O.

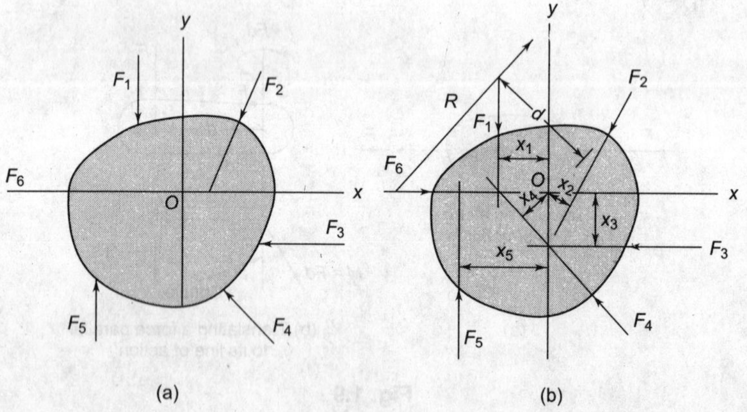

Fig. 1.11

Graphically, taking the force system shown in Fig. 1.12(a), the force polygon is drawn to a force scale in Fig. 1.12(b). Figure 1.12(a) is a space diagram which must be drawn to a linear scale for the graphical solution $\overline{ae}$ is the resultant in magnitude, direction, and sense, in Fig. 1.12(b). In the space diagram, forces are marked using Bow's notation. The letters A, B, etc. are assigned to spaces on either side of the force. Thus, AB is the force F_1, BC is the force F_2, etc.

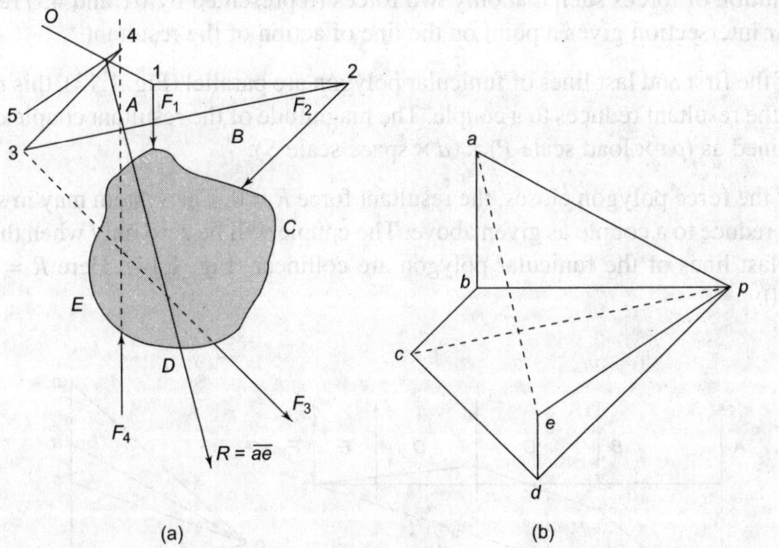

Fig. 1.12

We select a pole p (arbitrarily) and join pa, pb, etc. Draw lines parallel to pa, pb, etc. in the respective spaces A, B, etc. Note that marking of forces by Bow's notation in the space diagram is very important for drawing the funicular polygon. Where lines $\overline{01}$ and $\overline{45}$ of the funicular polygon meet is a point in the line of action of the resultant. The resultant R is shown in the space diagram.

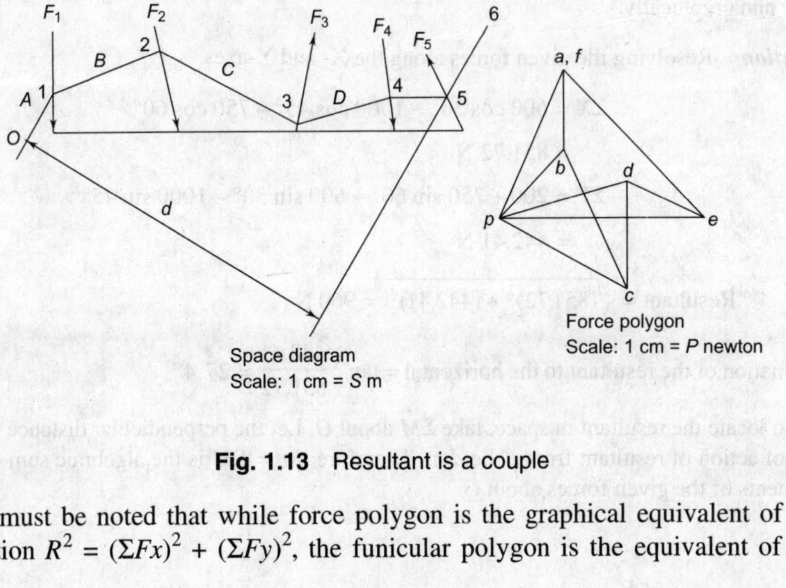

Fig. 1.13 Resultant is a couple

It must be noted that while force polygon is the graphical equivalent of the equation $R^2 = (\Sigma Fx)^2 + (\Sigma Fy)^2$, the funicular polygon is the equivalent of the principle of moments. In principle, the funicular polygon is an ingenious way of resolution of forces such that only two forces (represented by $\overline{01}$ and $\overline{45}$) remain. Their intersection gives a point on the line of action of the resultant.

If the first and last lines of funicular polygon are parallel (Fig. 1.13), this means that the resultant reduces to a couple. The magnitude of the resultant couple can be obtained as (pa × load scale P) × (d × space scale S).

If the force polygon closes, the resultant force $R = 0$. The system may in such a case reduce to a couple as given above. The couple will be zero only when the first and last lines of the funicular polygon are collinear (Fig. 1.14). Here $R = 0$ and $M = 0$.

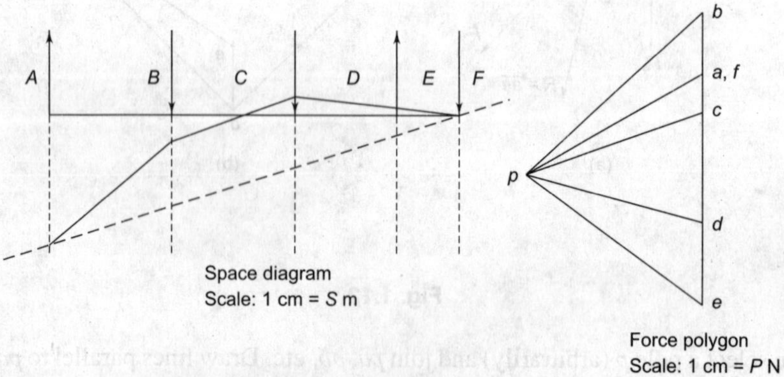

Fig. 1.14 Resultant force and couple zero

Example 1.1 Resultant of general coplanar forces

Determine the resultant of the four forces acting on a plate shown in Fig. 1.15(a), analytically and graphically.

Solution Resolving the given forces along the X- and Y-axes,

$$\Sigma X = 600 \cos 30° + 1000 \cos 45° - 750 \cos 60°$$

$$= 851.72 \text{ N}$$

$$\Sigma Y = 200 + 750 \sin 60° + 600 \sin 30° - 1000 \sin 45°$$

$$= 442.41 \text{ N}$$

$$\text{Resultant} = \sqrt{(851.72)^2 + (442.41)^2} = 960 \text{ N}$$

$$\text{Inclination of the resultant to the horizontal} = \tan^{-1}\frac{442.41}{851.72} = 27.4°$$

To locate the resultant in space, take ΣM about O. Let the perpendicular distance of the line of action of resultant from O be d m. Therefore, $Rd = \Sigma M$ is the algebraic sum of the moments of the given forces about O.

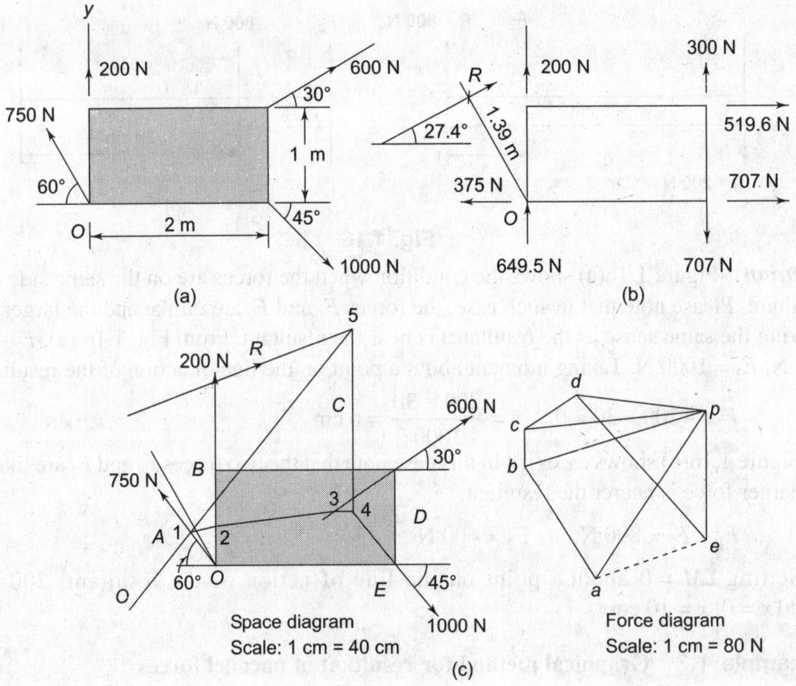

Fig. 1.15

In calculating ΣM, the forces 750 N and 200 N have no moment about O. To calculate the moments of the forces 600 N and 1000 N, resolve them into components as shown in Fig. 1.15(b). Taking clockwise moments as positive,

$$\Sigma M = 707 \times 2 - 300 \times 2 + 519.6 \times 1 = 1333.6 \text{ Nm}$$

$$d = \frac{\Sigma M}{R} = \frac{1333.6}{960} = 1.39 \text{ m from } O$$

Since the moment of the resultant is clockwise, the resultant is located as shown in Fig. 1.15(b).

Graphical solution Figure 1.15(c) gives the graphical solution. Note that the plate with forces is drawn to scale. Name the forces according to Bow's notation as shown. The force diagram is drawn to scale, giving the sides *a*, *b*, *c*, *d*, and *e*. Joining *ae* and (from *a* to *e*) gives the magnitude, sense and direction of the resultant.

To locate the resultant in space, we draw the funicular polygon. Take the pole *p* (arbitrarily) and join *pa*, *pb*, *pc*, *pd*, and *pe*. Draw lines parallel to *pa*, *pb*, etc. in paces *A*, *B*, etc. between the lines of action of forces. We get the funicular polygon 0-1, 1-2, 2-3, 3-4, 4-5. Extend the lines 0-1 and 4-5 to intersect at 5. 5 is a point in the line of action of the resultant and a line drawn parallel to $\overrightarrow{ae}$ through 5 locates the resultant in space. ■

Example 1.2 Resultant of two parallel forces

The resultant of two parallel forces F_1 and F_2 is a force of magnitude 800 N. If force F_1 = 200 N and is at a distance of 30 cm from the resultant, find the magnitude and location of force F_2 if the forces are (i) on the same side of the resultant and (ii) on either side of the resultant.

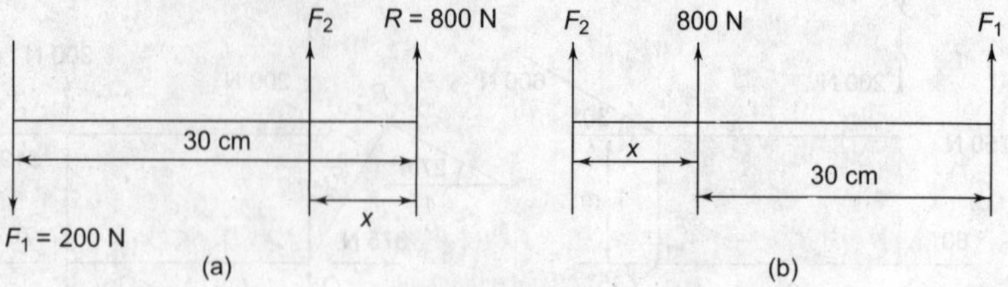

Fig. 1.16

Solution Figure 1.16(a) shows the condition when the forces are on the same side of the resultant. Please note that in such case, the forces F_1 and F_2 are unlike and the larger force (having the same sense as the resultant) is near the resultant. From Fig. 1.16 (a), $F_2 - F_1 = 800$ N; $F_2 = 1000$ N. Taking moment about a point on the line of action of the resultant,

$$F_2 x - 200 \times 30 = 0, \quad x = \frac{200 \times 30}{1000} = 6 \text{ cm}$$

Figure 1.16(b) shows case (ii). In this case note that the two forces F_1 and F_2 are like, and the larger force is nearer the resultant.

$$F_1 + F_2 = 800 \text{ N}, \qquad F_2 = 600 \text{ N}$$

Setting $\Sigma M = 0$ about a point on the line of action of the resultant, $200 \times 30 - 600\,x = 0$; $x = 10$ cm.

Example 1.3 Graphical method for resultant of parallel forces

Determine the resultant of the parallel force system shown in Fig. 1.17, analytically and graphically.

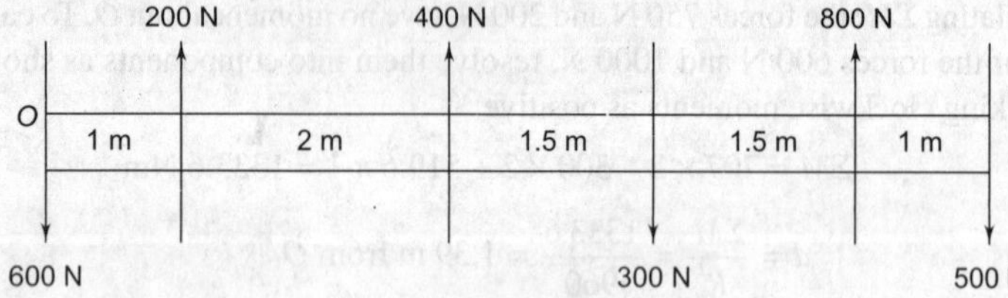

Fig. 1.17

Solution Analytically,

$$\text{Resultant} = \Sigma F$$

$$R = -600 + 200 + 400 - 300 + 800 - 500 = 0$$

Though the resultant force is zero, we have to check whether there is any resultant moment. Taking moments about O, a point on the line of action of the force of 600 N,

$$\Sigma M = +200 \times 1 + 400 \times 3 - 300 \times 4.5 + 800 \times 6 - 500 \times 7$$

$$= 1350 \text{ Nm anticlockwise}$$

The given force system has no resultant force but has a resultant couple.

Graphical solution The graphical solution is shown in Fig. 1.18. The force polygon is a straight line and closes because a and g coincide. The resultant force is thus zero. We select a pole p. Draw pa, pb, etc. and the funicular polygon 0-1-2-3-4-5-6-7. Lines 0-1 and 7-6 are parallel and do not meet. This shows that the force system reduces to a couple. The magnitude of the couple can be calculated as $M = (pa \times \text{load scale}) \times (d \times \text{space scale})$, where d is the distance between parallel lines 0-1 and 6-7.

$$M = (3.8 \times 0.5) \times (3.5 \times 200) = 1330 \text{ Nm}$$

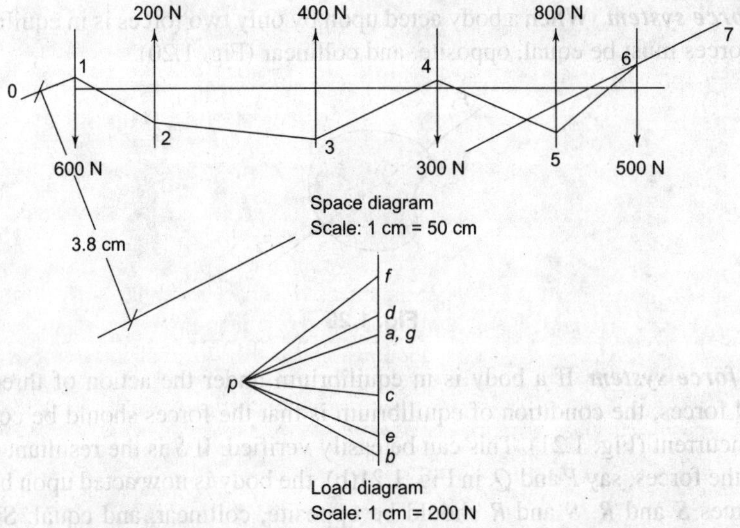

Fig. 1.18

1.5 EQUILIBRIUM

Equilibrium means a state of rest or motion without acceleration. Considering our discussions of coplanar force systems, a body subjected to a general coplanar force system may undergo translation, rotation, or both. As shown in Fig. 1.19, the possible displacements of a body are (i) translation Δd in the direction of the resultant force or in terms of its components Δx along the X-axis and Δy along Y-axis and (ii) a rotation in the plane due to a resultant couple.

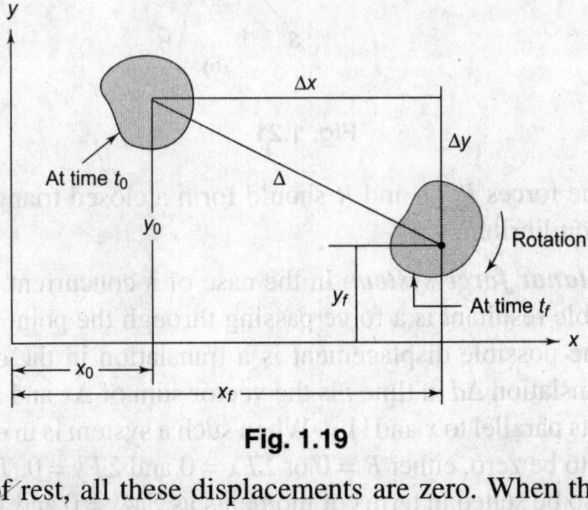

Fig. 1.19

In the state of rest, all these displacements are zero. When the body is not in equilibrium, it continuously changes its position with reference to the X- and Y-axes.

1.5.1 Conditions of Equilibrium

Depending upon the force system acting on the body and the condition that translation and rotation are zero in equilibrium, the following conditions of equilibrium can be derived.

Two-force system When a body acted upon by only two forces is in equilibrium, these forces must be equal, opposite, and collinear (Fig. 1.20).

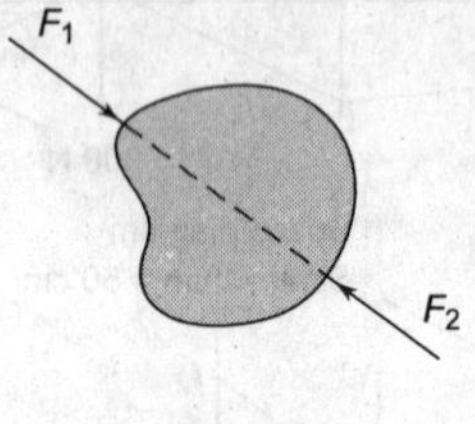

Fig. 1.20

Three-force system If a body is in equilibrium under the action of three non-parallel forces, the condition of equilibrium is that the forces should be coplanar and concurrent (Fig. 1.21). This can be easily verified. If S is the resultant of any two of the forces, say P and Q, in Fig. 1.21(b), the body is now acted upon by only two forces S and R. S and R should be opposite, collinear, and equal. Since S passes through the intersection of P and Q, R should pass through the same point. P, Q, and R are, therefore, concurrent. S lies in the same plane as P and Q, R should lie in the same plane. P, Q, and R, therefore, should be coplanar and concurrent for equilibrium.

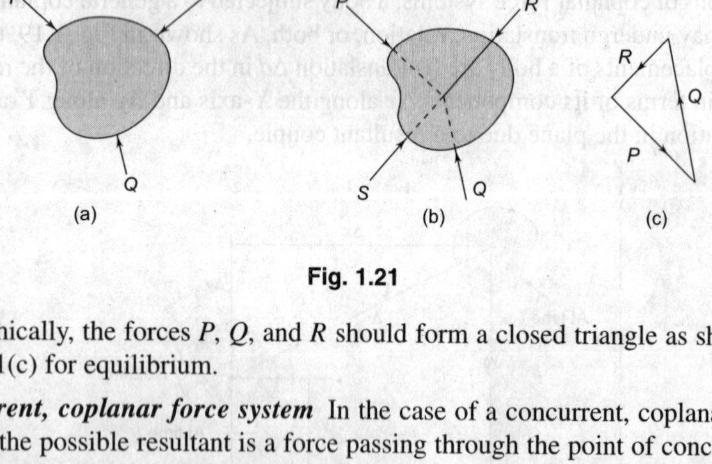

Fig. 1.21

Graphically, the forces P, Q, and R should form a closed triangle as shown in Fig. 1.21(c) for equilibrium.

Concurrent, coplanar force system In the case of a concurrent, coplanar force system, the possible resultant is a force passing through the point of concurrency O (Fig. 1.22). The possible displacement is a translation in the direction of the resultant. This translation Δd in time t is the vector sum of Δx and Δy, the component displacements parallel to x and11 y. When such a system is in equilibrium, for the displacement to be zero, either $R = 0$ or $\Sigma Fx = 0$ and $\Sigma Fy = 0$. The equilibrium conditions can also be stated in terms of moments as $\Sigma M_1 = 0$ and $\Sigma M_2 = 0$, where ΣM_1 and ΣM_2 are algebraic sums of moments about points 1 and 2, and points 1 and 2 are not collinear with the point of concurrency O [Fig. 1.22(c)]. This can be easily verified. If $\Sigma M_1 = 0$, either $R = 0$ or R passes through point 1. To eliminate the latter possibility, the moment about a second point is taken. If $\Sigma M_2 = 0$, then $R = 0$ provided 1 and 2 are not collinear with O.

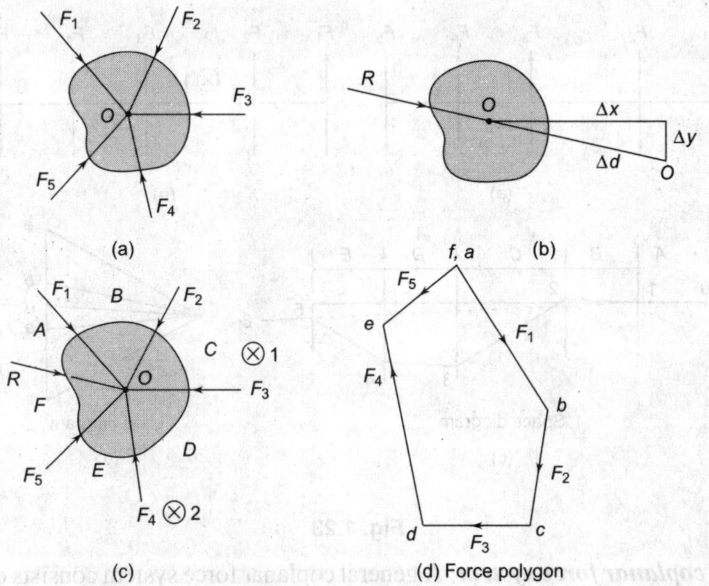

Fig. 1.22

Graphical condition of equilibrium The graphical condition for a concurrent, coplanar force system to be in equilibrium is that the force polygon drawn for such a force system closes. The first and last points coincide, which means that $R = 0$ [Fig. 1.22(d)].

Coplanar, parallel force system In the case of the parallel force system shown in Fig. 1.23(a), the resultant force, if any, is parallel to the given forces. Even if $R = 0$, the system may reduce to a couple M. The conditions of equilibrium for such a force system are that $R = 0$ and $\Sigma M = 0$ about any point in the plane of the forces. This ensures that the body neither translates nor rotates.

The conditions of equilibrium can also be stated in terms of moments as $\Sigma M_1 = 0$ and $\Sigma M_2 = 0$, where 1 and 2 are moment centres such that the line joining 1 and 2 is not parallel to the lines of action of forces. $\Sigma M_1 = 0$ ensures that there is no resultant couple and the resultant, if any, passes through point 1. $\Sigma M_2 = 0$ means that $R = 0$ since R cannot pass through points 1 and 2 [Fig. 1.23(b)].

Graphical conditions of equilibrium $R = 0$ requires that the force polygon closes. In the case of a parallel force system, the force polygon degenerates into a straight line as in Fig. 1.23(d) and closing of the polygon means that the first and last points a and f coincide. To ensure that there is no resultant couple, i.e., $\Sigma M = 0$, the funicular polygon should close. This means that the first and last lines of the funicular polygon are collinear. This is illustrated in Fig. 1.23(c), where the first line 0-1 and the last line 5-6 of the funicular polygon are collinear. The graphical conditions of equilibrium for a parallel force system are (i) the force polygon should close ($R = 0$) and (ii) the funicular polygon should close ($M = 0$).

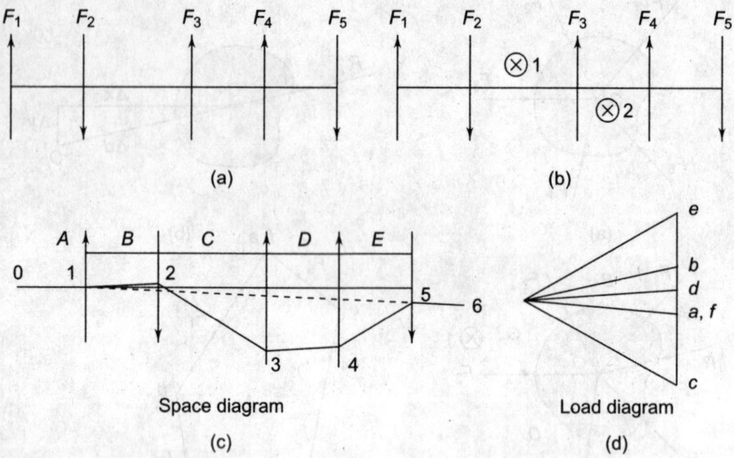

Fig. 1.23

General coplanar force system A general coplanar force system consists of forces which are neither concurrent nor parallel, as shown in Fig. 1.24(a). If a body acted upon by such a force system is in equilibrium, both linear displacement and rotation of the body are zero. The conditions of equilibrium are, therefore, $R = 0$ and $\Sigma M = 0$. The condition $R = 0$ can be expressed in terms of the rectangular components $\Sigma Fx = 0$ and $\Sigma Fy = 0$.

The equilibrium conditions can be expressed in terms of moments in many ways, as given below.

(i) $\Sigma Fx = 0, \Sigma M_1 = 0, \Sigma M_2 = 0$ with the condition that the line joining 1 and 2 is not parallel to the Y-axis, or

(ii) $\Sigma Fy = 0, \Sigma M_3 = 0, \Sigma M_4 = 0$ with the condition that the line joining 3 and 4 is not parallel to the X-axis, or

(iii) $\Sigma M_5 = 0, \Sigma M_6$, and $\Sigma M_7 = 0$ with the condition that 5, 6, and 7 are not collinear.

These conditions are shown in Fig. 1.24(b), and can be easily proved on arguments similar to the ones described for previous cases.

Graphical condition of equilibrium These conditions can be derived from the above. $R = 0$ requires that the forces polygon, *abcde*, closes (the first and last points coincide). $\Sigma M = 0$ requires that the funicular polygon closes, i.e. the first and last lines of the polygon are collinear. These conditions are shown in Fig. 1.24(c). The first line 0-1 and the last line 4-5 of the funicular polygon are collinear.

We have seen the conditions of equilibrium for coplanar force systems. In the case of spatial force systems, a third dimension z (in addition to x and y) will be added. Thus, the resultant force $R = 0$ means $\Sigma Fx = 0, \Sigma Fy = 0$, and $\Sigma Fz = 0$. The moment conditions will be in terms of the moments about the three coordinate axes, as $\Sigma M_x = 0, \Sigma M_y = 0, \Sigma M_z = 0$, as all the forces do not lie in a single plane.

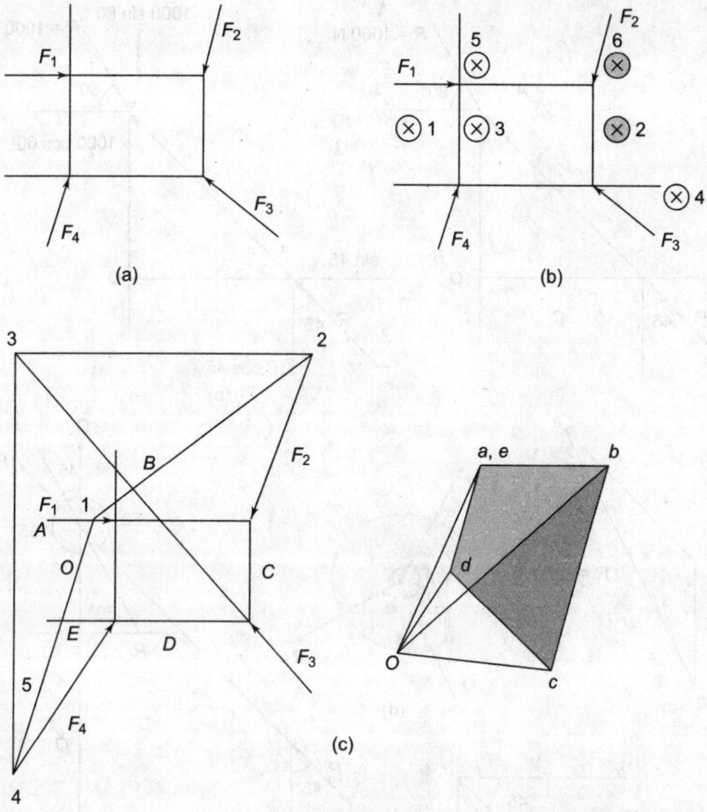

Fig. 1.24

The conditions of equilibrium given above are used to determine unknown forces in a given system if it is known that the system is in equilibrium. The following examples illustrate this application.

Example 1.4 Resultant of three coplanar forces

Three forces P, Q, and R keep a body in equilibrium, as shown in Fig. 1.25. Determine the magnitude of the forces P and Q.

Solution The equilibrium of a three (non-parallel)-force system is a very important concept in structural mechanics because it is possible to reduce a given force system into a three-force system and apply the conditions of equilibrium. The conditions of equilibrium require that the three forces should be concurrent and coplanar. There are many methods of solution.

Resolution For a concurrent force system, there are two conditions of equilibrium—$\Sigma Fx = 0$ and $\Sigma Fy = 0$. Resolving the forces along the X- and Y-axes shown in Fig. 1.25(b) and taking forces along the positive directions of these axes as positive,

$$\Sigma X = 0, \ \overrightarrow{+}, \ P \sin 45° - 1000 \cos 60° = 0, \ P = 707 \text{ N}$$

$$\Sigma Y = 0, \uparrow +, \ P \cos 45° + Q - 1000 \sin 60° = 0, \ Q = 366 \text{ N}$$

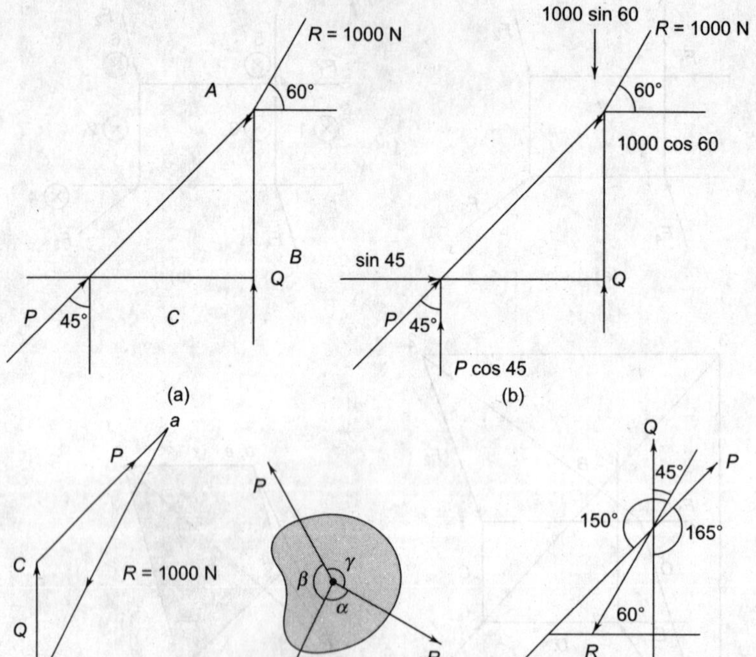

$$\frac{P}{\sin 30} = \frac{Q}{\sin 15} = \frac{R}{\sin 135}$$

$$\frac{P}{\sin(180-30)} = \frac{Q}{\sin(180-15)} = \frac{R}{\sin(180-135)}$$

Fig. 1.25

Please note that three forces have six quantities, involving magnitude and direction, for each force. Only two unknowns out of these can be determined from the conditions of equilibrium.

Graphical method Since the three forces are in equilibrium, the force polygon is a closed triangle. The graphical solution is shown in Fig. 1.25(c). We use Bow's notation to name the forces. Using a scale of 1 cm = 200 N, we lay out the force R = 1000 N as $\overrightarrow{ab}$. A vertical line is drawn from b to represent Q and a line is drawn parallel to the line of action of P from a. These intersect at c. The magnitude of P and Q can be scaled off from the force triangle. $P = (\overrightarrow{ca} \times 200) = 700$ N and $Q = (\overrightarrow{bc} \times 200) = 380$ N, which are the same as before within limits of graphical accuracy.

Lami's theorem Lami's theorem states that if three forces P, Q, and R keep a body in equilibrium, then $P/\sin\alpha = Q/\sin\gamma = R/\sin\beta$. This is shown in Fig. 1.25(d). Note that α is the angle between the lines of action of Q and R, γ the angle between P and R, etc. For using this theorem, we extend the lines of action of forces as shown in Fig. 1.25(e). The angles are 45° between P and Q, 150° between Q and R, and 165° between R and P. Therefore,

$$\frac{P}{\sin 150^\circ} = \frac{Q}{\sin 165^\circ} = \frac{R}{\sin 45^\circ}$$

This is equal to two equations,

$$\frac{P}{\sin 150^\circ} = \frac{1000}{\sin 45^\circ}, \quad P = 707 \text{ N}$$

$$\frac{Q}{\sin 165^\circ} = \frac{1000}{\sin 45^\circ}, \quad Q = 366 \text{ N}$$

Note that Lami's theorem can be directly derived from the force triangle of Fig. 1.25(c). This is shown in Fig. 1.25(f). ■

Example 1.5 General coplanar forces in equilibrium

The rod shown in Fig. 1.26(a) is in equilibrium under the action of the forces shown. Find the magnitudes of the forces P, Q, and R.

Solution The given force system is a general coplanar force system and has three conditions of equilibrium. $\Sigma Fx = 0$, $\Sigma Fy = 0$, and $\Sigma M = 0$. Three unknowns can be determined from these equations.

Analytical method Let us state the three equations of equilibrium.

$$\Sigma Fx = 0, \ \overset{\rightarrow}{+}, \ Q + 2000 \cos 30^\circ - 1000 \cos 60^\circ + R \cos 45^\circ = 0$$

$$\Sigma Fy = 0, \ \uparrow+, \ P + R \sin 45^\circ - 1000 \sin 60^\circ - 3000 - 2000 \sin 30^\circ = 0$$

$\Sigma M = 0$ about A 1 + (clockwise moments positive)

$$2000 \sin 30^\circ \times 1 + 3000 \times 2.5 + 1000 \sin 60^\circ \times 4.5 - R \sin 45^\circ \times 5 = 0$$

From the last equation, $R = 3384$ N. Substituting this value in the second equation, $P = 2473$ N. From the first equation, $Q = -3625$ N. The value of Q shows that Q must be acting towards the left for equilibrium. Therefore,

$$P = 2473 \text{ N} \uparrow, \ Q = 3625 \text{ N} \leftarrow, \ R = 3384 \text{ N} \uparrow$$

Graphical method Two graphical solutions are presented.

(a) Reducing to a three-force system The solution is shown in Fig. 1.26(b). We lay out the space diagram to scale, and determine the resultant of the three forces, 2000 N, 3000 N, and 1000 N. For this the force polygon *a-b-c-d* is drawn and $\overrightarrow{ad}$ is the resultant. The funicular polygon 0-1-2-3-4 is drawn by selecting a pole p and drawing pa, pb, pc, and pd. The line segments 0-1 and 3-4 intersect at 5. A line drawn parallel to $\overrightarrow{ad}$ through 5 locates the resultant.

(b) Considering a free body subject to all forces Assume that P and Q are combined to form a force T. The rod is now in equilibrium under the action of three forces $\overrightarrow{ad}$, R, and T. They must be concurrent and coplanar. $\overrightarrow{ad}$ and R intersect at 6. 6-L is the line of action of T,

where L is the point of intersection of P and Q. Knowing the directions of the three forces, the force triangle is drawn. A line is drawn parallel to R through d and parallel to T through a intersect at e. $\overrightarrow{ea} = T$ and $\overrightarrow{de} = R$. To determine P and Q, draw a vertical line through e and a horizontal line through a to intersect at f. $\overrightarrow{ef} = P$ and $\overrightarrow{fa} = Q$.

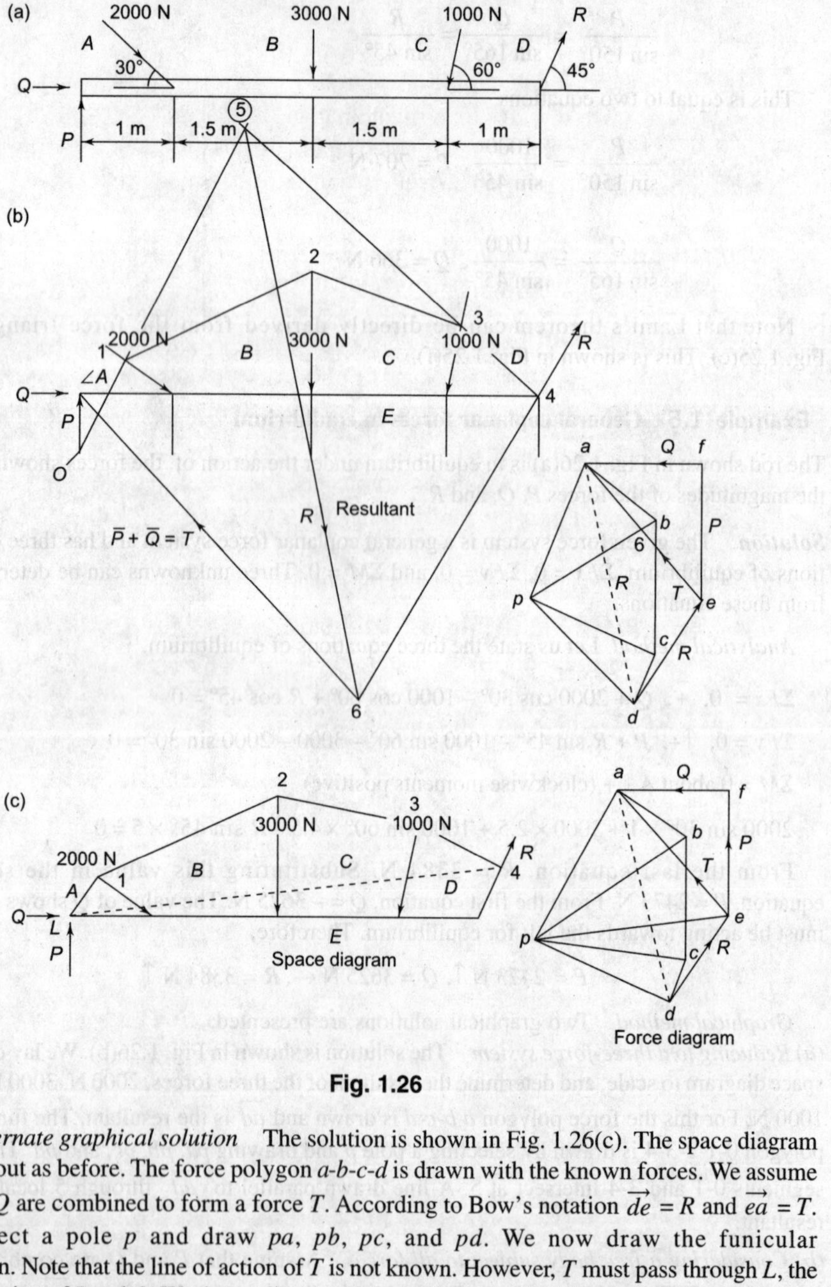

Fig. 1.26

(c) Alternate graphical solution The solution is shown in Fig. 1.26(c). The space diagram is laid out as before. The force polygon *a-b-c-d* is drawn with the known forces. We assume P and Q are combined to form a force T. According to Bow's notation $\overrightarrow{de} = R$ and $\overrightarrow{ea} = T$.

Select a pole p and draw pa, pb, pc, and pd. We now draw the funicular polygon. Note that the line of action of T is not known. However, T must pass through L, the intersection of P and Q. We draw a line parallel to pa in space A and start the line from L. The funicular polygon is continued till a line parallel to pd in space D intersects the line of action of R at 4. Close the funicular polygon by drawing 4-L. Draw a line parallel to 4-L through p, the pole. Point e must lie on this line. To locate e, draw a line parallel to the line of action of R from d to intersect the line parallel to 4-L at e. Join ea. $\overrightarrow{da} = R$ and $\overrightarrow{ea} = T$. $\overrightarrow{ea}$ can be resolved into horizontal and vertical components to get P and Q. In Fig. 1.26(c), $\overrightarrow{ef} = P$ and $\overrightarrow{fa} = Q$.

1.6 BODY CONSTRAINTS AND FREE BODY DIAGRAMS

We have seen analytical and graphical solutions to problems involving forces. The problems solved were direct, i.e., a mathematical model was presented for solution. In practice, problems in structural mechanics are quite complex, and the preparation of the mathematical model itself is quite difficult. To visualize a physical problem, to make suitable assumptions to simplify it, and then to prepare a mathematical model is the first step in structural analysis. The concepts of body constraints and free body diagrams are fundamental to this first step.

1.6.1 Body Constraints

A body constraint is a contrived support or force provided such that the body is in equilibrium. The constraints provide either reactive forces, couples, or both, depending upon the type of support.

Figure 1.27 shows the types of constraints generally assumed. While in a given situation, one may not exactly find parallel physical supports, one can carefully choose from those given in Fig. 1.27 the one that matches the most the physical condition obtaining. Let us briefly discuss the types of constraints that we come across.

Smooth surface It is difficult to find a surface which is perfectly smooth. A smooth surface prevents motion perpendicular to its surface but allows a translation parallel to the surface and rotation. Such a surface thus provides one reactive force perpendicular to the surface.

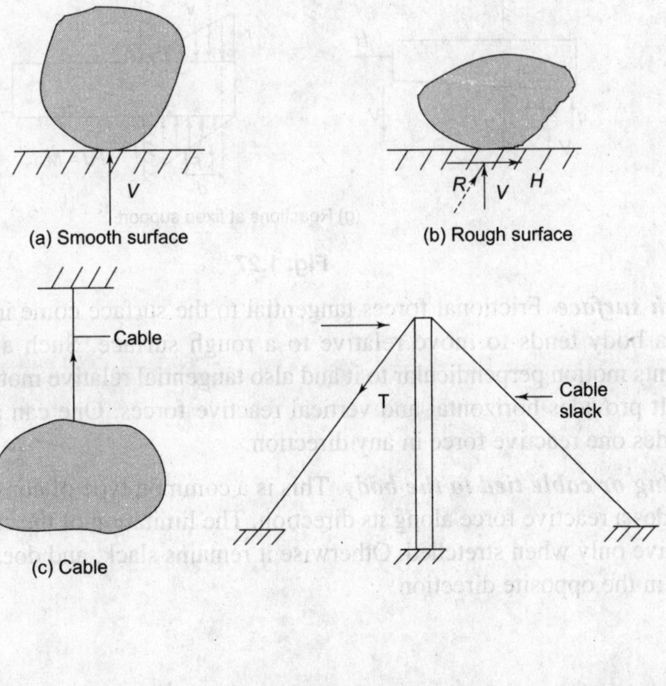

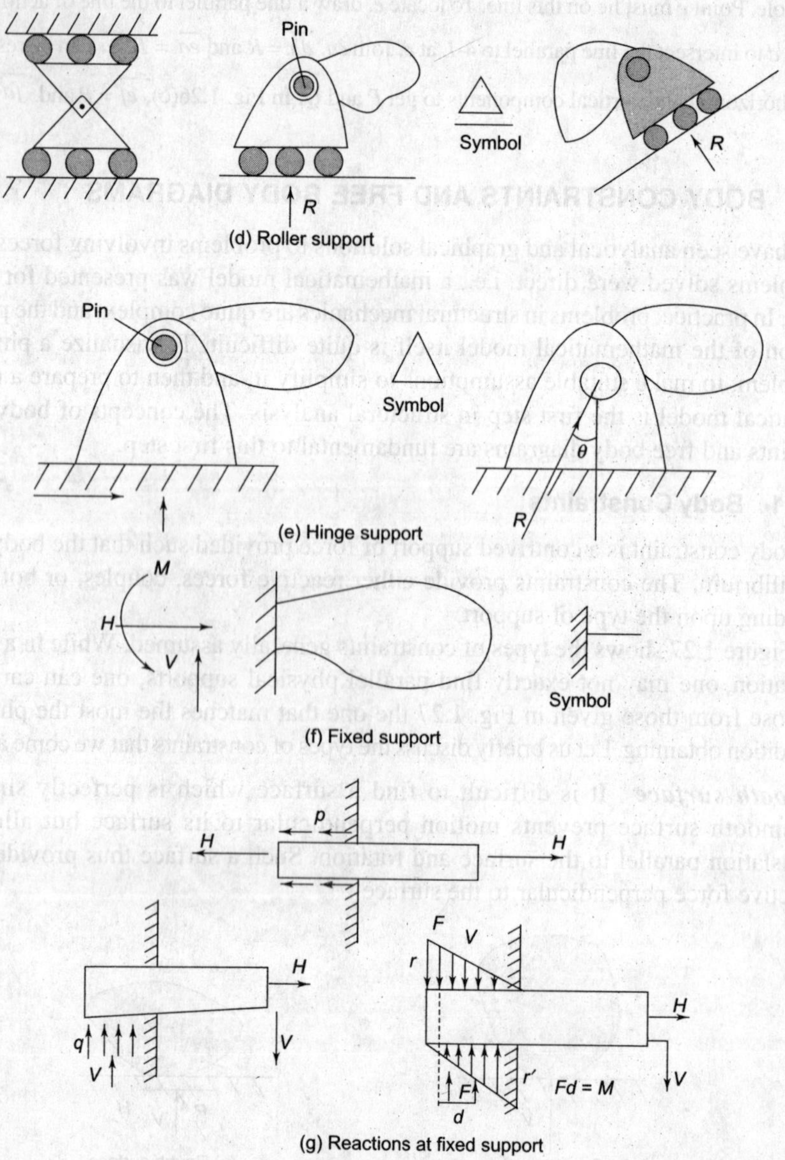

(d) Roller support

(e) Hinge support

(f) Fixed support

(g) Reactions at fixed support

Fig. 1.27

Rough surface Frictional forces tangential to the surface come into play whenever a body tends to move relative to a rough surface. Such a surface, thus, prevents motion perpendicular to it and also tangential relative motion due to friction. It provides horizontal and vertical reactive forces. One can also say that it provides one reactive force in any direction.

A string or cable tied to the body This is a common type of constraint. A cable provides a reactive force along its direction. The limitation of the cable is that it is effective only when stretched. Otherwise it remains slack, and does not provide a force in the opposite direction.

Roller support The roller support has rollers, which can move over a firm surface, and is attached to the body with a pin. By its very nature, the body can rotate about the pin, and also translate parallel to the surface on which the roller rests. However, it cannot move in a direction normal to this surface. The roller, thus, provides one reactive force perpendicular to the surface on which the rollers rest.

Hinged or pinned support This type of support is firmly attached to the ground and connected to the body through a pin connection. Thus, the body has the freedom to rotate but cannot translate horizontally and vertically or in any direction. The hinged support thus provides one reactive force in any direction or two reactive forces horizontally and vertically.

Fixed support This type of support does not allow the body to translate or rotate. The reactive forces provided by this support are shown in Fig. 1.27(g). Forces p provide the horizontal reactive force, forces q provide the vertical reactive forces, and forces r and r' provide a reactive couple. Such a support alone completely restrains a body acted upon by a general coplanar force system.

The equilibrium conditions can be used to determine the reactive forces provided by the constraints of a body. Depending upon the physical situation and nature of the constraints provided, the appropriate number of reactive forces or couples should be applied to the body.

1.6.2 Free Body Diagram

A number of external or active forces generally act upon a body in equilibrium. So do the reactive forces, couples, or both, provided by the body constraints. The equilibrium conditions are used to evaluate the reactive forces. To determine these conditions, the body is isolated and all the active and reactive forces acting on it are determined. When the isolated body is drawn with all the forces and reactive forces acting on it, such a diagram is known as a free body diagram. Note that in a complex structural system, the free body may be drawn for the whole system or a part of it. The free body diagram provides an excellent way of book-keeping in a structural problem by ensuring that all the forces and reactive forces are taken into account. The following problems illustrate the use of free body diagrams.

Example 1.6 Free body diagram

The drum shown in Fig. 1.28(a) weighs 800 N and is being rolled over a step 4 cm high. Determine the value of P required to roll it over the step if it is applied horizontally at the centre of the drum. Also determine the minimum value of P required to pull the drum if it can be inclined at any angle α.

Solution The free body diagram of the drum is shown in Fig. 1.28(b). Assuming the surfaces to be smooth, there is a reaction R at the floor, a reaction Q at the point of contact with the step, and the pull P in addition to the weight of the drum acting through its CG. Note that when the drum is about to be pulled over the step, it leaves contact with the ground and R becomes zero. There are then only three forces acting on the body, 800 N, P, and Q. The direction of Q will be such that it passes through the centre of the drum. In Fig. 1.28(b),

$$\sin\theta = \frac{16}{20} = 0.8, \quad \cos\theta = \frac{12}{20} = 0.6$$

$$\Sigma Fx = 0, \ \overset{\rightarrow}{+}, \ P - Q\cos\theta = 0$$

$$\Sigma Fy = 0, \ \uparrow +, \ Q\sin\theta - 800 = 0, \ Q = 1000 \text{ N}, \ P = 600 \text{ N}$$

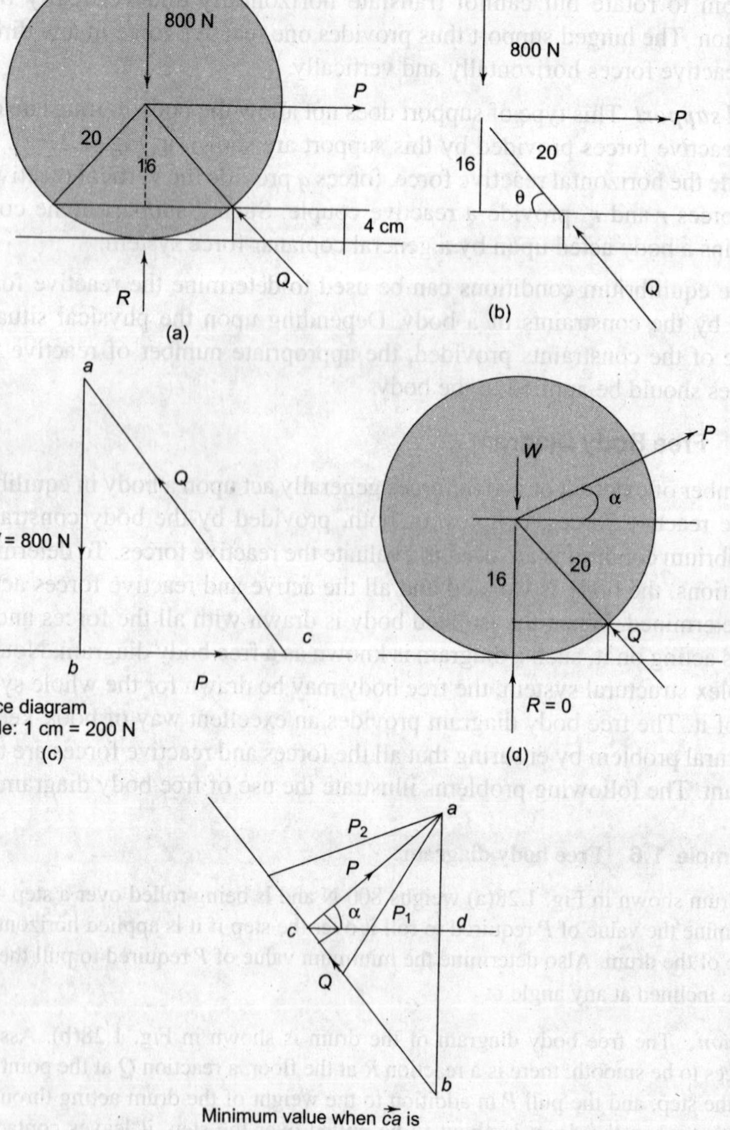

Fig. 1.28

Graphically, since the body is acted upon by a three-force system, the solution is found by drawing a force triangle shown in Fig. 1.28(c). $\overrightarrow{ab} = W = 800$ N to scale. Draw a line parallel to P through b and a line parallel to Q through a, intersecting at C. $\overrightarrow{bc} = P$ and $\overrightarrow{ca} = Q$ can be measured to scale.

To determine the minimum value of P inclined at an angle a, from Fig. 1.28(d), $P \cos\alpha = 0.6\ Q$; $W - P \sin \alpha = 0.8\ Q$. To eliminate Q, $4P \cos \alpha = 2.4\ Q$; $3W - 3P \sin\alpha = 2.4\ Q$. Subtracting,

$$4P \cos\alpha - 3W + 3P \sin\alpha = 0$$

$$P(4 \cos\alpha + 3 \sin\alpha) = 3W$$

For P to be minimum, $(4 \cos\alpha + 3 \sin\alpha)$ has to be maximum. Therefore,

$$\frac{d}{d\alpha}(4 \cos\alpha + 3 \sin\alpha) = 0$$

$$-4 \sin\alpha + 3 \cos\alpha = 0$$

$$\tan \alpha = 0.75, \quad \alpha = 36.87^\circ$$

$$P = \frac{3 \times 800}{4 \cos 36.87 + 3 \sin 36.87} = 480 \text{ N is the minimum value of } P.$$

Graphically, as in Fig. 1.28(e), we lay out $\overrightarrow{ab} = 800$ N to scale. Draw a line parallel to the line of action of Q from b. Any line drawn from a to intersect this line gives a value of P and its corresponding direction. The minimum value of P will be obtained when ac is at right angles to bc. Therefore, draw a perpendicular to bc from a. $\overrightarrow{ca}$ gives the minimum value of P and α can be measured from this triangle as $\overrightarrow{ca}$ and angle ad. ■

Example 1.7 Free body diagram

Two pipes 20 cm ϕ, weighing 1200 N and 16 cm ϕ, weighing 800 N, lie in a trench as shown in Fig. 1.29(a). Draw a free body diagram for the two pipes together and for each individual pipe. Assume the contacting surfaces as rough.

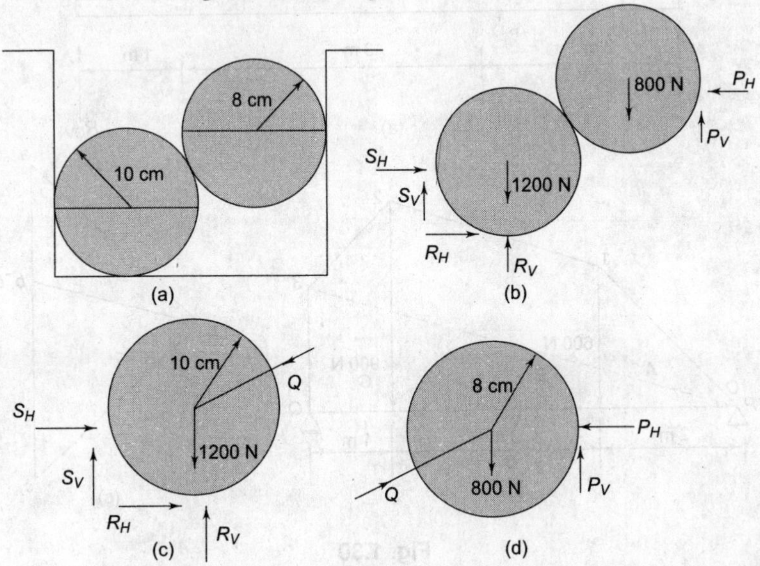

Fig. 1.29

Solution Figures 1.29(b), (c), and (d) show the free body diagram. Note that when the free body diagram of the two pipes together is drawn, the forces Q vanish. ■

Example 1.8 Reactive forces in a beam

The rod PQ shown in Fig. 1.30 is acted upon by two forces, one of 600 N and another of 900 N. It has a hinged support at P and a roller support at Q. Find the reactive forces at the supports.

Solution Selecting the X- and Y-directions as shown, from the free body diagram illustrated in Fig. 1.30(b),

$$\Sigma Fx = 0, \overset{\rightarrow}{+}, R_{PH} = 0$$

$$\Sigma Fy = 0, \uparrow +, R_{PV} - 600 - 900 + R_{QV} = 0$$

$$\Sigma M = 0 \text{ about } P +, \quad 600 \times 2 + 900 \times 5 - R_{QV} 6 = 0$$

From the last equation, $R_{QV} = 950$ N.

$$R_{PV} = 1500 - 950 = 550 \text{ N}$$

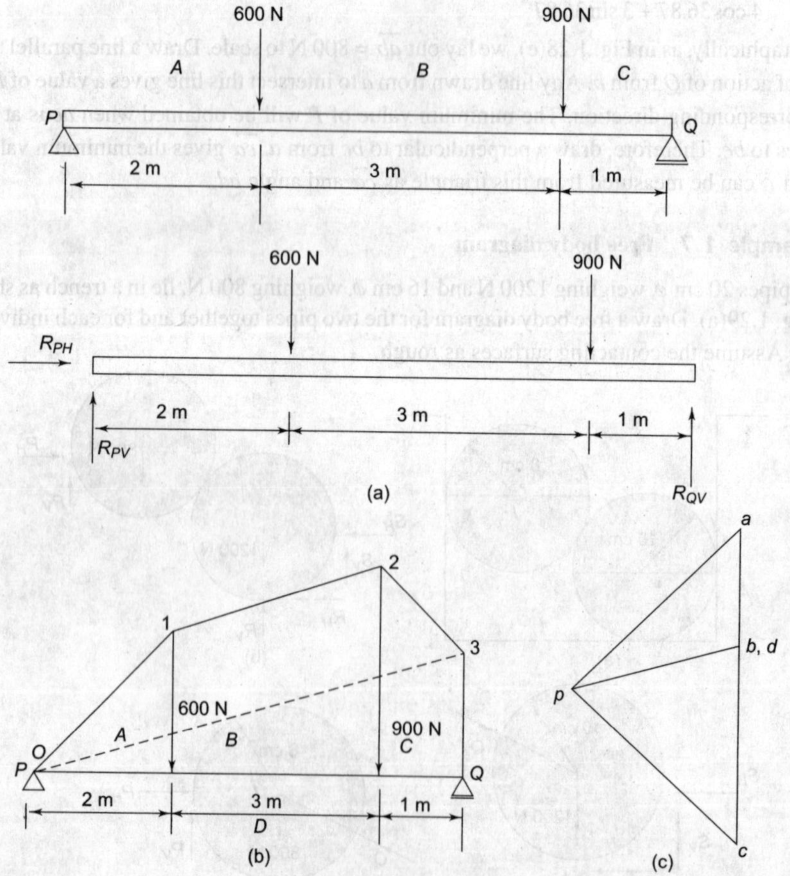

Fig. 1.30

Graphical Solution The graphical solution is shown in Fig. 1.30(c). We lay out the rod and the positions of the forces to a linear scale, and then draw the force polygon *a-b-c* to a load scale. Select a pole *p* and draw *pa*, *pb*, and *pc*. Then, draw lines parallel to these in the space diagram. Consider R_{PH} and R_{PV} as a combined force R_A whose direction is unknown but which passes through *P*. Start the funicular polygon from *P* to get *P*-1-2-3. The last segment 2-3 ends where it intersects reaction R_{QV} (vertical line through *Q*). Draw $\overrightarrow{P\text{-}3}$. Draw a line parallel to *P*-3 through *p* which intersects the vertical line through *c* at *d*. $\overrightarrow{cd} = R_{QV}$ and $\overrightarrow{da} = R_P$. Since R_P is vertical, $R_{PH} = 0$.

Example 1.9 Reactive forces in a frame

Determine the reactions at the supports of the structure carrying loads as shown in Fig. 1.31(a).

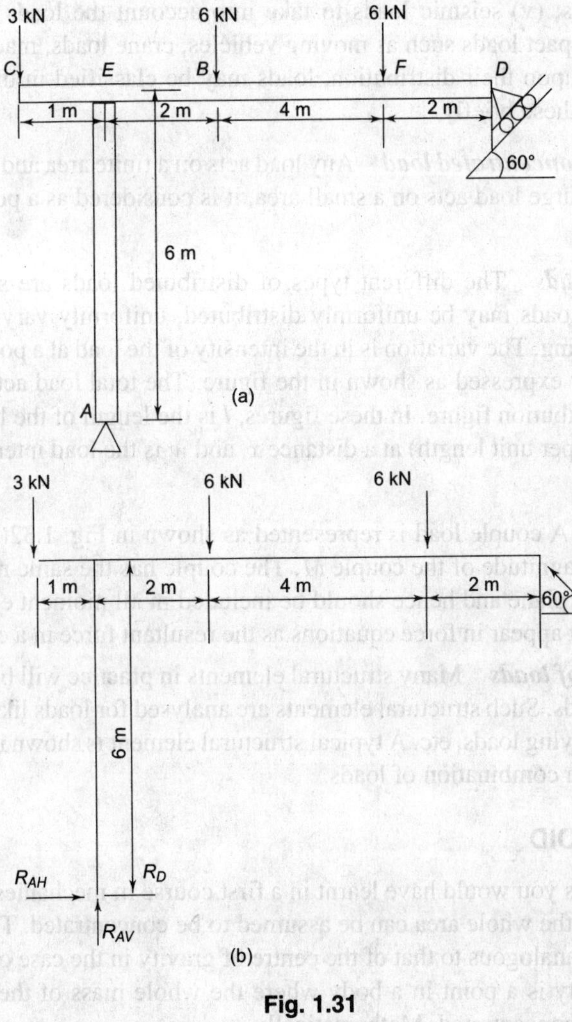

Fig. 1.31

Solution The free body diagram for the structure is shown in Fig. 1.31(b). The reactive components at A are R_{AH} and R_{AV}, and the reactive component at *D* is R_D, perpendicular to the plane of the roller and inclined at 60° to the horizontal.

Applying the conditions of equilibrium,

$$\Sigma Fx = 0, \overset{\rightarrow}{+}, \; R_{AH} - R_D \cos 60° = 0$$

$$\Sigma Fy = 0, \uparrow +, \; R_{AV} + R_D \sin 60° - 3 - 6 - 6 = 0$$

$$\Sigma M = 0 \text{ @ } A, \; -3 \times 1 + 6 \times 2 + 6 \times 6 - R_D \cos 60° \times 6 - R_D \sin 60° \times 8 = 0$$

From the last equation, $R_D = 4.53$ kN, $R_{AH} = 2.27$ kN, and $R_{AV} = 11.07$ kN. ■

1.7 LOADS ON STRUCTURES

Structures are subjected to different types of loads and due to different causes. Loads on structures can be classified into (i) dead loads, which include the self-weight of the structure and other fixed loads; (ii) live loads, which do not have a fixed position and can be placed anywhere for maximum effect; (iii) wind loads; (iv) snow loads; (v) seismic loads to take into account the loads due to earthquakes; (vi) impact loads such as moving vehicles, crane loads, machines, etc.

Depending upon their distribution, loads may be classified in different ways. Let us discuss these briefly.

Point load or concentrated load Any load acts on a finite area and not at a point. However, if a large load acts on a small area, it is considered as a point load [Fig. 1.32(a)].

Distributed loads The different types of distributed loads are shown in Fig. 1.32(b). Such loads may be uniformly distributed, uniformly varying, and non-uniformly varying. The variation is in the intensity of the load at a point and can be mathematically expressed as shown in the figure. The total load acts through the CG of the distribution figure. In these figures, l is the length of the load, w_x is the intensity (load per unit length) at a distance x, and w is the load intensity (known) at a distance l.

Couple load A couple load is represented as shown in Fig. 1.32(c), giving the location and magnitude of the couple M. The couple has the same moment about any point in its plane and hence should be included in all moment equations. The couple does not appear in force equations as the resultant force in a couple is zero.

Combination of loads Many structural elements in practice will be subjected to a variety of loads. Such structural elements are analysed for loads like self-weight, point loads, varying loads, etc. A typical structural element is shown in Fig. 1.32(d) acted upon by a combination of loads.

1.8 CENTROID

The centroid, as you would have learnt in a first course in mechanics, is a point in the area where the whole area can be assumed to be concentrated. The concept of the centroid is analogous to that of the centre of gravity in the case of a mass. The centre of gravity is a point in a body where the whole mass of the body can be assumed to be concentrated. Mathematically,

$$\bar{x} = \frac{\int (\delta M)\, x}{\int \delta M}$$

$$\bar{y} = z \text{and} \quad \bar{z} = \frac{\int (\delta M) z}{\int \delta M} \quad \text{[see Fig. 1.33(a)]}$$

Similarly,

$$\bar{x} = \frac{\int (\delta A) x}{\int \delta A} \quad \text{and} \quad \bar{y} = \frac{\int (\delta A) y}{\int \delta A}$$

for an area A [Fig. 1.33(b)].

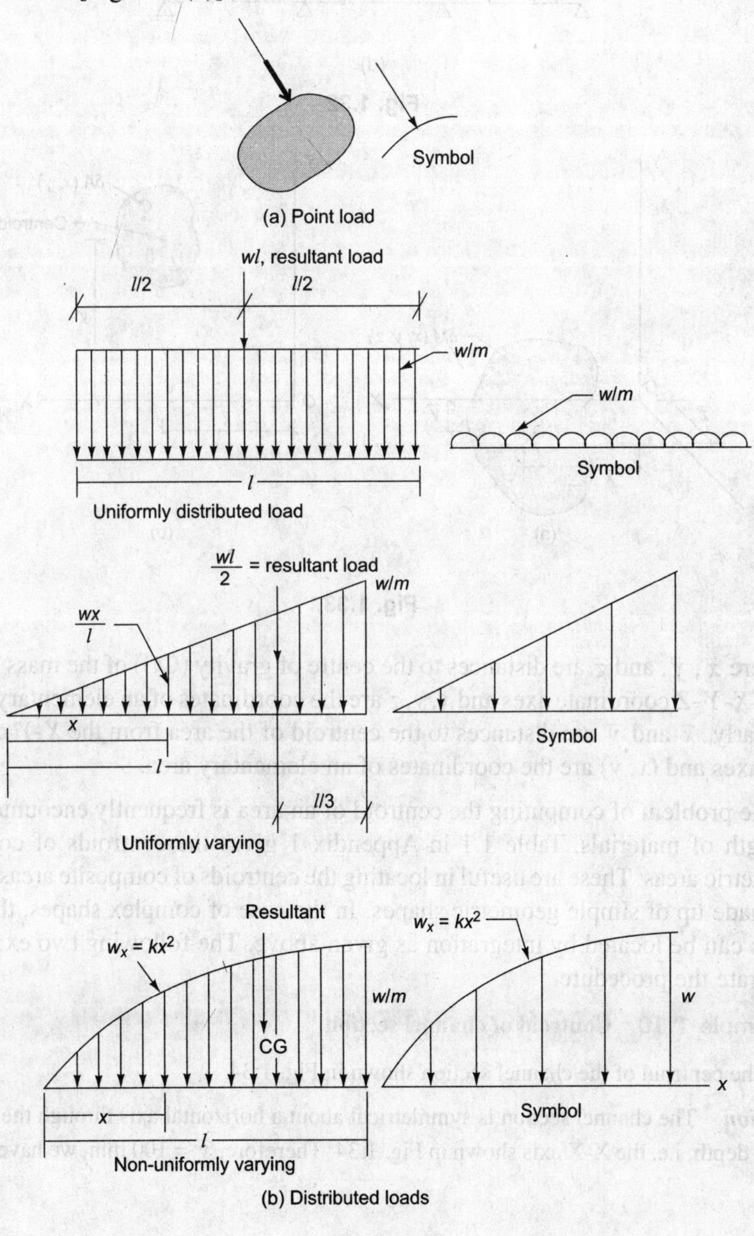

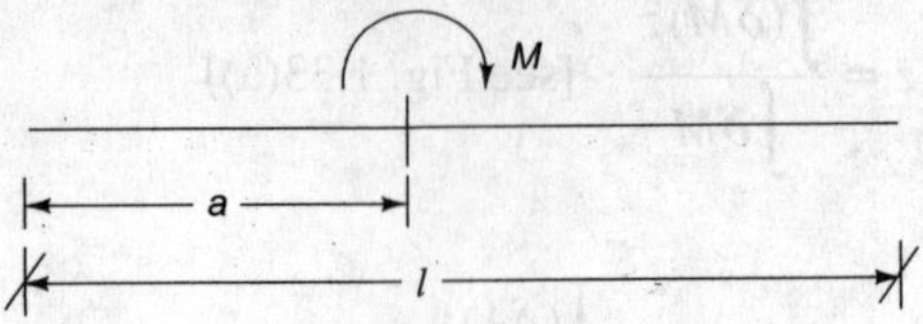

(c) Couple load

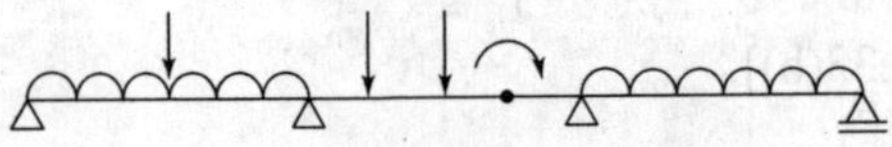

(d)

Fig. 1.32

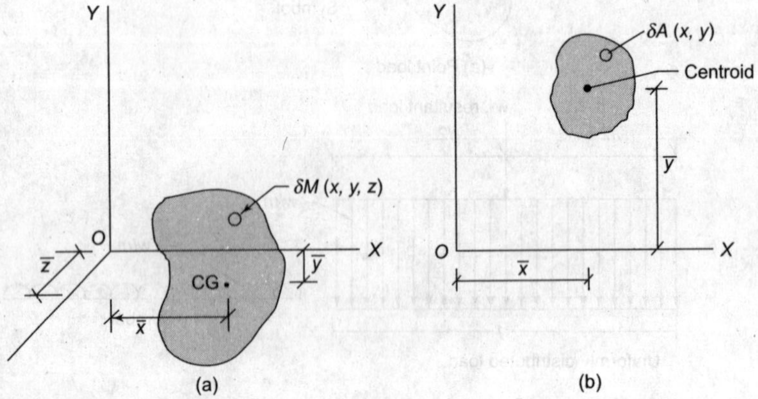

Fig. 1.33

Here $\bar{x}$, $\bar{y}$, and $\bar{z}$ are distances to the centre of gravity (CG) of the mass from a set of X-Y-Z coordinate axes and x, y, z are the coordinates of an elementary mass. Similarly, $\bar{x}$ and $\bar{y}$ are distances to the centroid of the area from the X-Y coordinate axes and (x, y) are the coordinates of an elementary area.

The problem of computing the centroid of an area is frequently encountered in strength of materials. Table 1.1 in Appendix 1 gives the centroids of common geometric areas. These are useful in locating the centroids of composite areas which are made up of simple geometric shapes. In the case of complex shapes, the centroids can be located by integration as given above. The following two examples illustrate the procedure.

Example 1.10 Controid of channel section

Find the centroid of the channel section shown in Fig. 1.34.

Solution The channel section is symmetrical about a horizontal axis through the middle of the depth, i.e. the X-X axis shown in Fig. 1.34. Therefore, $\bar{y}$ = 100 mm, we have to find $\bar{x}$.

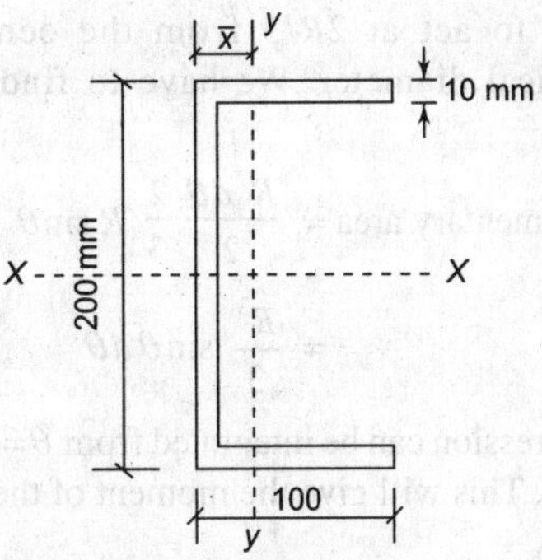

Fig. 1.34

From Fig. 1.34,

$$\bar{x} = \frac{2 \times 100 \times 10 \times 50 + 180 \times 10 \times 5}{2 \times 100 \times 10 + 180 \times 10}$$

$$= 28.68 \text{ mm}$$

Example 1.11 Centroid of semicircle and quarter circle areas

Find the centroid of a semicircular area and the quarter circle shown in Figures 1.35(a) and (b), respectively.

Solution Considering an elementary area subtending an angle $\delta\theta$ at the centre, the area of the hatched portion,

$$\text{Elementary area} = Rd\theta \frac{R}{2} = \frac{R^2 d\theta}{2}$$

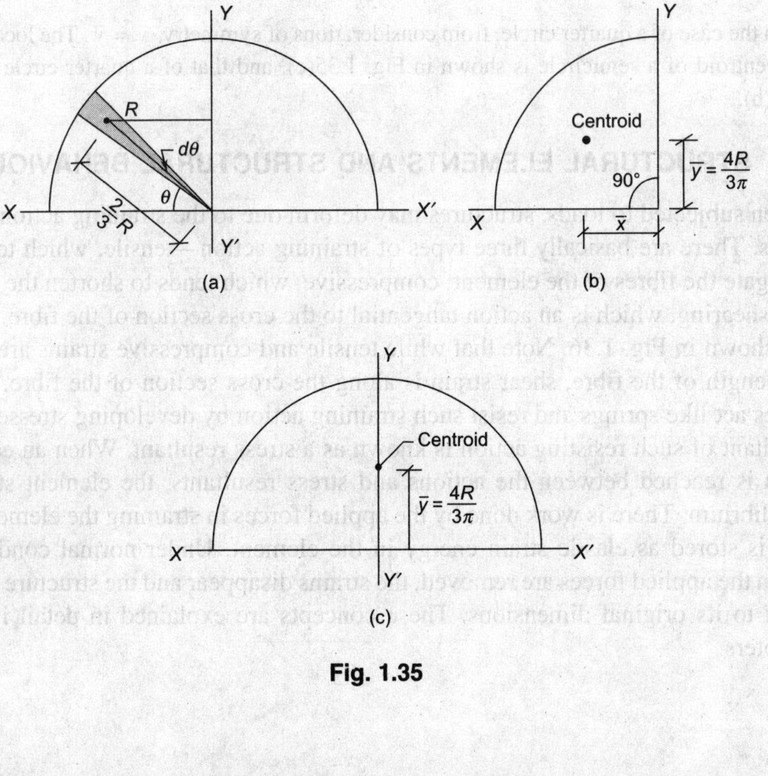

Fig. 1.35

This can be assumed to act at $2R/3$ from the centre O. The semicircle is symmetrical about a vertical diameter. We have to find $\bar{y}$ (the distance from the X-axis).

$$\text{Moment of elementary area} = \frac{R^2 d\theta}{2}\,\frac{2}{3}\,R\sin\theta$$

$$= \frac{R^3}{3}\sin\theta\, d\theta$$

For a semicircle, this expression can be integrated from $\theta = 0$ to $\theta = \pi$, and for the quarter circle from $\theta = 0$ to $\theta = \pi/2$. This will give the moment of the total area, which is equal to (area $\times\ \bar{y}$) for the areas.

Semicircle

$$\frac{\pi R^2}{2}\bar{y} = \int_0^{\pi}\frac{R^3}{3}\sin\theta\, d\theta = \frac{R^3}{3}\,[-\cos\theta]_0^{\pi}$$

$$= \frac{R^3}{3}(1+1) = \frac{2R^3}{3}$$

$$\bar{y} = \frac{2R^3 \times 2}{3\,\pi R^2} = \frac{4R}{3\pi}$$

Quarter circle

$$\frac{\pi R^2}{4}\bar{y} = \int_0^{\pi/2}\frac{R^3}{3}\sin\theta\, d\theta = \frac{R^3}{3}\,[-\cos\,\theta]_0^{\pi/2}$$

$$= \frac{R^3}{3}[-0+1] = \frac{R^3}{3}$$

$$\bar{y} = \frac{4R}{3\pi}$$

In the case of a quarter circle, from considerations of symmetry, $\bar{x} = \bar{y}$. The location of the centroid of a semicircle is shown in Fig. 1.35(c), and that of a quarter circle in Fig. 1.35(b).

1.9 STRUCTURAL ELEMENTS AND STRUCTURAL BEHAVIOUR

When subjected to loads, structures may deform due to the straining action of the loads. There are basically three types of straining action—tensile, which tends to elongate the fibres of the element; compressive, which tends to shorten the fibres; and shearing, which is an action tangential to the cross section of the fibre. These are shown in Fig. 1.36. Note that while tensile and compressive strains are along the length of the fibre, shear strain is along the cross section of the fibre. These fibres act like springs and resist such straining action by developing stresses. The resultant of such resisting action is known as a stress resultant. When an equilibrium is reached between the actions and stress resultants, the element stays in equilibrium. There is work done by the applied forces in straining the element and this is stored as elastic strain energy in the element. Under normal conditions, when the applied forces are removed, the strains disappear and the structure comes back to its original dimensions. These concepts are explained in detail in later chapters.

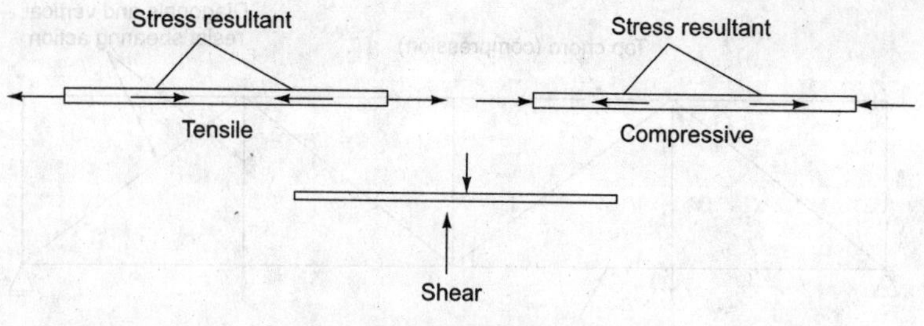

Fig. 1.36

There are different types of structures, consisting of different kinds of elements which exhibit different structural actions. Let us discuss them briefly.

Tension member A tension member can be represented as shown in Fig. 1.37(a). The action is equivalent to that of two forces tending to stretch the element, and is normal to the cross section, which may have any shape. The member increases in length due to the straining action, and develops stress resultants to oppose the applied forces as shown.

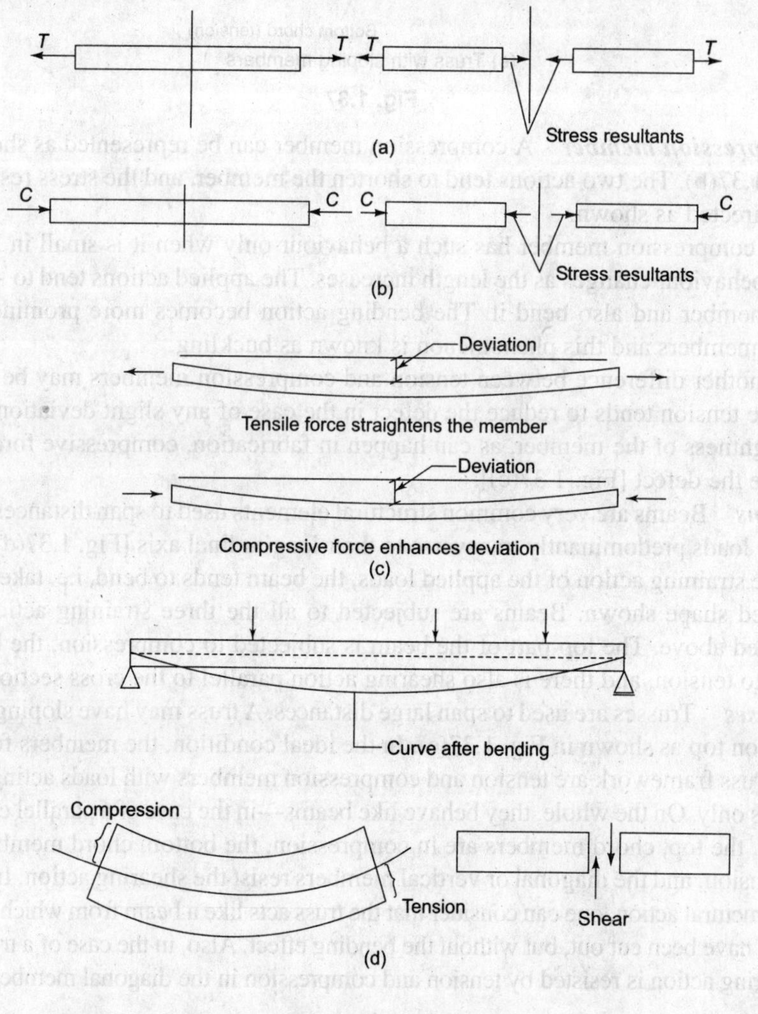

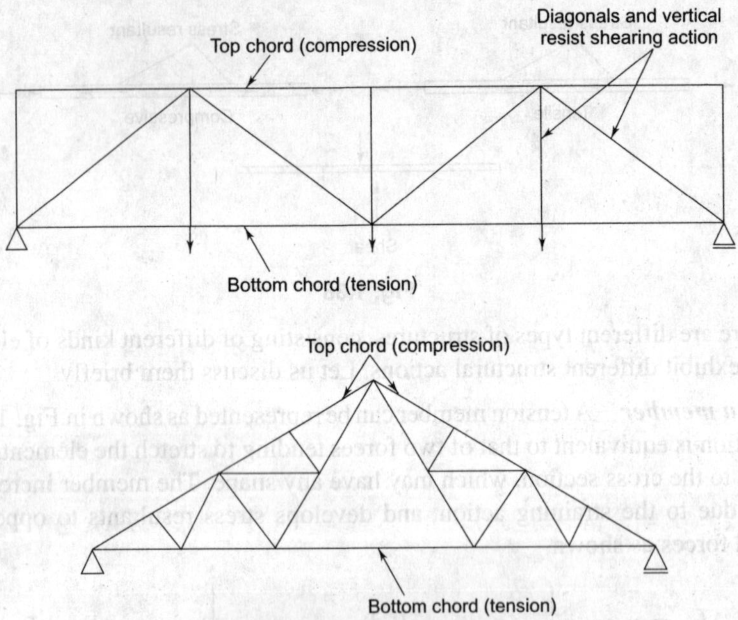

(e) Truss with sloping members

Fig. 1.37

Compression member A compression member can be represented as shown in Fig. 1.37(b). The two actions tend to shorten the member, and the stress resultants are directed as shown.

A compression member has such a behaviour only when it is small in length. The behaviour changes as the length increases. The applied actions tend to shorten the member and also bend it. The bending action becomes more prominent for long members and this phenomenon is known as buckling.

Another difference between tension and compression members may be noted. While tension tends to reduce the defect in the case of any slight deviation in the straightness of the member, as can happen in fabrication, compressive forces enhance the defect [Fig. 1.37(c)].

Beams Beams are very common structural elements used to span distances. They carry loads predominantly transverse to their longitudinal axis [Fig. 1.37(d)]. Due to the straining action of the applied loads, the beam tends to bend, i.e. take up the curved shape shown. Beams are subjected to all the three straining actions described above. The top part of the beam is subjected to compression, the bottom part to tension, and there is also shearing action parallel to the cross section.

Trusses Trusses are used to span large distances. A truss may have sloping members on top as shown in Fig. 1.37(e). In the ideal condition, the members forming the truss framework are tension and compression members with loads acting at the joints only. On the whole, they behave like beams—in the case of a parallel chorded truss, the top, chord members are in compression, the bottom chord members are in tension, and the diagonal or vertical members resist the shearing action. In terms of structural action, one can consider that the truss acts like a beam from which certain parts have been cut out, but without the bending effect. Also, in the case of a truss, the shearing action is resisted by tension and compression in the diagonal members.

Arches Arches are again elements used to span large distances. They basically carry load to supports by developing compressive stresses in them. In practice, some bending and shear are also developed. The supports for an arch need to be strong because the arch transfers the load at an inclination to the supports. Arches may be three-hinged, two-hinged, or fixed (Fig. 1.38). The load is transferred to the arch through spandrels or hangers to make them uniform.

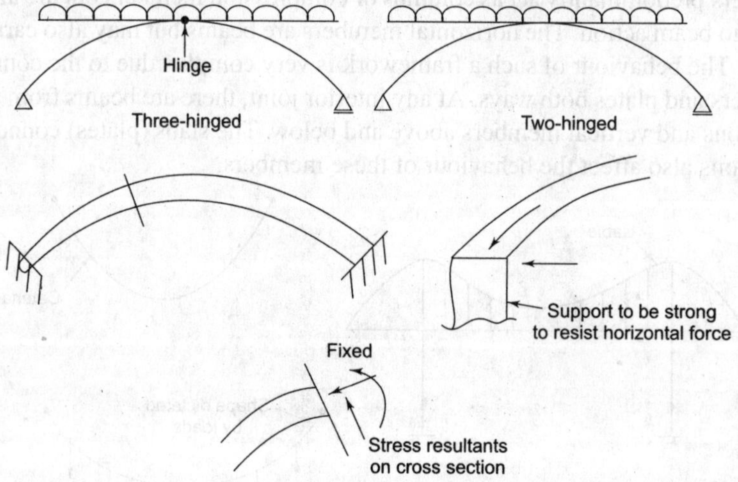

Fig. 1.38

If the arch has a funicular shape with reference to the load, the arch is under pure compression. For example, for a uniformly distributed load, the funicular shape is a parabola.

Plates A plate is a two-dimensional structure, and is generally subjected to loads perpendicular to its plane and supported along the edges or in many other ways. The behaviour of the plate element is very complex but can be simplified by assuming beam action of strips of the plate in two perpendicular directions (Fig. 1.39). The deflected shape is that of a saucer, with curvature in two directions. This is a very common element used as roof and floor slabs, bridge and culvert decks, etc.

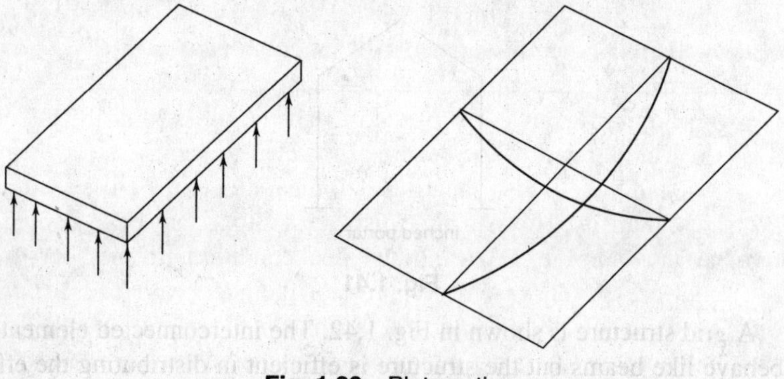

Fig. 1.39 Plate action

Cable A cable, as mentioned earlier, is effective only when stretched. It finds use in bridge structures as shown in Fig. 1.40, and is basically a tension member. A cable supported at ends hangs in a shape known as 'catenary' due to its own weight. Subjected to loads, the flexible cable takes up a shape dictated by the loads acting on it.

Rigid frames A rigid frame is a framework of elements, monolithically cast or otherwise rigidly jointed (unlike in a truss where the members are assumed to be pin-jointed). Such a framework is very common in multi-storeyed, multibay buildings. The top members may be inclined as a gabled portal (Fig. 1.41). The vertical members predominantly act as columns or compression members but are also subjected to beam action. The horizontal members are beams but may also carry axial forces. The behaviour of such a framework is very complex due to the connecting members and plates both ways. At any interior joint, there are beams from all four directions and vertical members above and below. The slabs (plates) connected to the beams also affect the behaviour of these members.

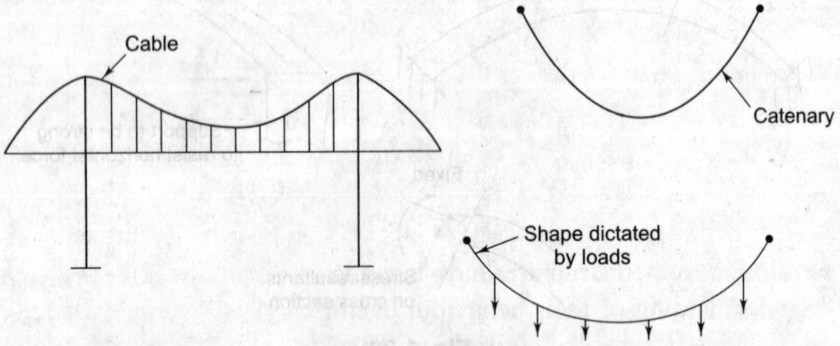

Fig. 1.40

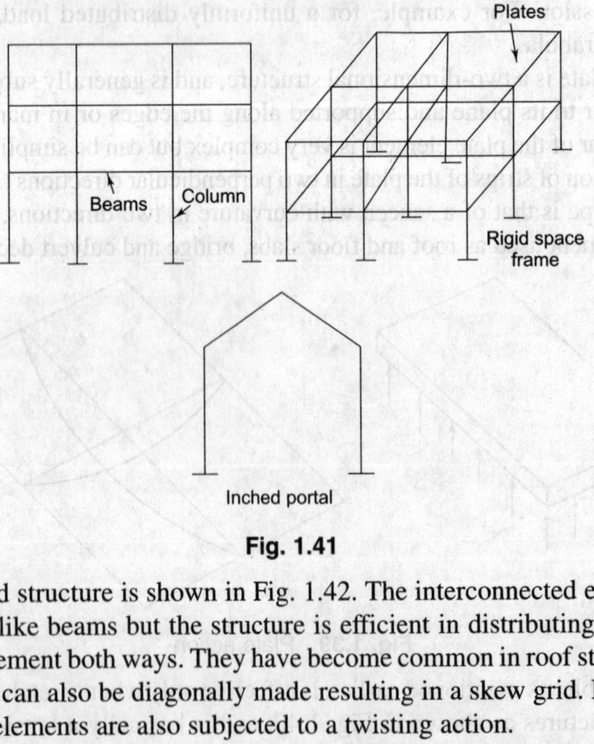

Fig. 1.41

Grids A grid structure is shown in Fig. 1.42. The interconnected elements both ways behave like beams but the structure is efficient in distributing the effect of loads on an element both ways. They have become common in roof structures. The grid structure can also be diagonally made resulting in a skew grid. In addition to bending, the elements are also subjected to a twisting action.

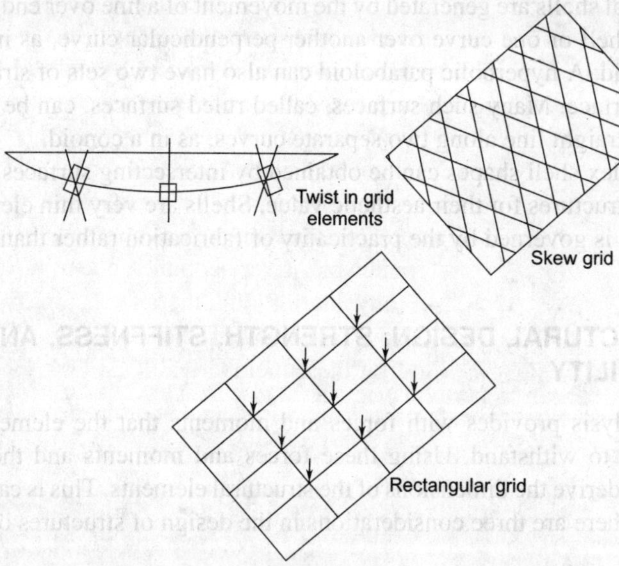

Fig. 1.42

Shell structures Shell structures have become very popular in recent times with a better understanding of their behaviour even though massive shells have been made since the early stages of civilization. There are innumerable forms of shells. Some of these are shown in Fig. 1.43.

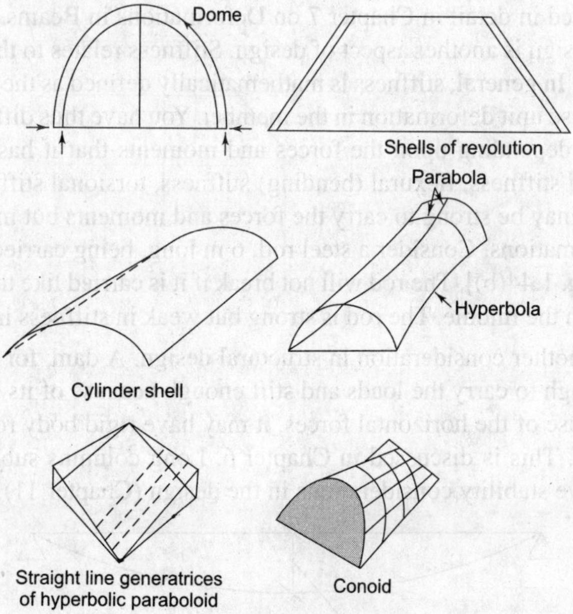

Fig. 1.43

Shells of revolution are those generated by the rotation of an arc about a line. Examples are different types of domes and conics. They are essentially thin elements, subjected to meridional and hoop stresses.

Translational shells are generated by the movement of a line over end arcs, as in a cylindrical shell or one curve over another perpendicular curve, as in a hyperbolic paraboloid. A hyperbolic paraboloid can also have two sets of straight lines lying on its surface. Many such surfaces, called ruled surfaces, can be generated by moving a straight line along two separate curves, as in a conoid.

More complex shell shapes can be obtained by intersecting surfaces, and have been used in structures for their aesthetic value. Shells are very thin elements and their thickness is governed by the practicality of fabrication rather than structural needs.

1.10 STRUCTURAL DESIGN: STRENGTH, STIFFNESS, AND STABILITY

Structural analysis provides with forces and moments that the elements of the structure have to withstand. Using these forces and moments and the material properties, we derive the dimensions of the structural elements. This is called *structural design*. There are three considerations in the design of structures or machine components.

Strength design is done to ensure that the stresses at any point in the element do not exceed the permissible value for the material. The structural element may be made of a single material like steel and timber. The material properties may be different in tension and compression as in timber. The element may also be made of two or more materials as in a composite element. Strength design ensures that in all cases, the stress in the element is within the appropriate permissible values. This is discussed in detail in Chapter 7 on Deformations in Beams.

Stiffness design is another aspect of design. Stiffness relates to the deformation of the member. In general, stiffness is mathematically defined as the force/moment required to cause unit deformation in the member. You have thus different stiffness for a member depending upon the forces and moments that it has to carry. You have thus axial stiffness, flexural (bending) stiffness, torsional stiffness, etc.

A member may be strong to carry the forces and moments but may have unacceptable deformations. Consider a steel rod, 6 m long, being carried by holding it at the ends [Fig. 1.44(b)]. The rod will not break if it is carried like that but will sag considerably in the middle. The rod is strong but weak in stiffness in this position. ***Stability*** is another consideration in structural design. A dam, for example, may be strong enough to carry the loads and stiff enough because of its size [Fig. 1.44 (c)]. But because of the horizontal forces, it may have rigid body rotation leading to overturning. This is discussed in Chapter 6. Long columns subjected to axial forces also have stability considerations in the design (Chapter 11).

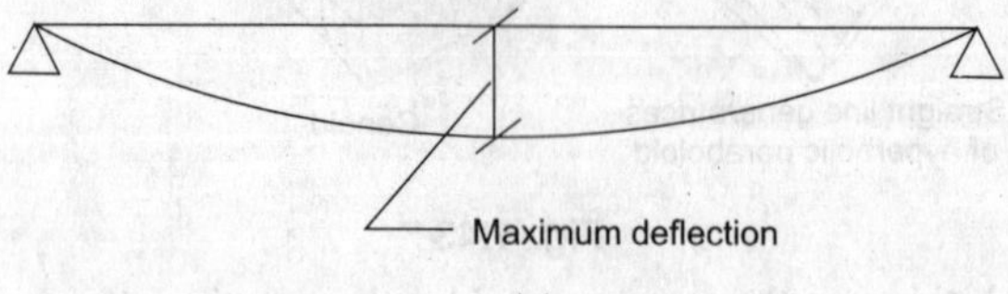

(a)

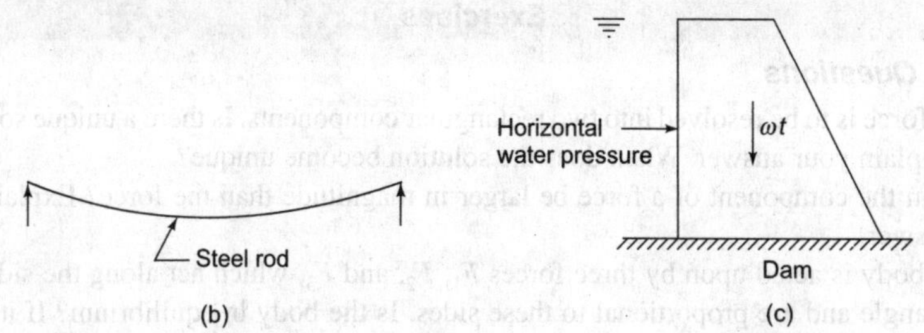

Fig. 1.44

1.11 SYMBOLS AND UNITS

A number of symbols are used in this book. They are explained wherever they are first encountered. A collection of such symbols is given at the beginning of the book. SI units will be used throughout in this work.

Summary

Structures, machine components, and other similar elements are designed to withstand forces and moments. Such designs, which belong to the realm of structural engineering or design theories, are based on the basic principles of statics and dynamics. Statics is the branch of mechanics dealing with forces and moments when the body is in equilibrium. The present discussion is limited to structures and components in equilibrium.

Force systems can be concurrent, parallel or non-concurrent. Resultants of force vectors are obtained by adding them according to the rules of vector algebra, which are different from those of scalar algebra. Two vectors are added according to the parallelogram law or triangle law. Force systems can also be coplanar or space systems. In all cases, forces and moments have to be added as vectors.

The resultant of two forces F_1 and F_2 is given by $R = (F_1^2 + F_2^2 - 2F_1F_2\cos\theta)^{1/2}$, where θ is the angle between the vectors. More than two forces can be added by repeated application of this principle. Forces can also be added by resolving them into rectangular components in two mutually perpendicular directions. The sum of such components, ΣFx and ΣFy, can then be combined into a single force $R = [(\Sigma Fx)^2 + (\Sigma Fy)^2]$.

Graphical methods use the force polygon and funicular polygon to find the resultant force and its location.

Body constraints are supports provided to a body to give forces and moments needed to keep the body in equilibrium. Free body diagrams are diagrams showing the body acted upon by the applied forces and moments, and the reactive forces and moments given by the body constraints. Once such a diagram is drawn, the reactive forces and moments can be determined by the equilibrium conditions, such as $\Sigma Fx = 0$, $\Sigma Fy = 0$, and $\Sigma M = 0$. Such conditions are used to calculate the reactive forces and moments, and analyse structures and machine components for design.

The concepts of centre of gravity and centroid of bodies and areas is also important in structural analysis. These also help us to locate the resultant of distributed forces.

Exercises

Review Questions

1. A force is to be resolved into two rectangular components. Is there a unique solution? Explain your answer. When does the solution become unique?
2. Can the component of a force be larger in magnitude than the force? Explain your answer.
3. A body is acted upon by three forces F_1, F_2, and F_3, which act along the sides of a triangle and are proportional to these sides. Is the body in equilibrium? If it is not, under what conditions will it be?
4. A body is acted by four forces F_1, F_2, F_3, and F_4, which act along the sides of a rectangle, and whose magnitudes are proportional to the respective sides. Is the body in equilibrium? If not, under what conditions the body will be in equilibrium?
5. Explain why it is not necessary to use the principle of moments to find the resultant of a concurrent force system.
6. If $\Sigma F = 0$, for a parallel force system, what can you say about the resultant of this system.
7. Can the funicular polygon close if the force polygon does not close? Explain your answer.
8. If the first and last lines of a funicular polygon are parallel, but not collinear, explain how you would calculate the resultant, with a diagram.
9. If, for a general coplanar force system, $\Sigma M_1 = 0$, $\Sigma M_2 = 0$, and $\Sigma M_3 = 0$, about three moment centres 1, 2, and 3, under what conditions will the force system be in equilibrium? Give reasons for your answer.

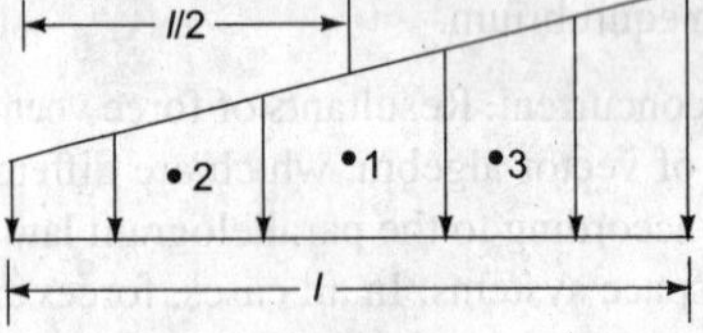

Fig. 1.45

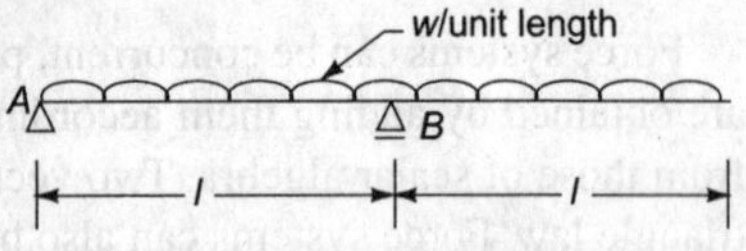

Fig. 1.46

10. For the load distribution shown in Fig. 1.45, will the resultant of the load be at 1, 2, or 3?
11. If a beam is loaded as shown in Fig. 1.46, what will be the reaction at A and B?

Problems

1. A force of 200 N is resolved into rectangular components. Find the magnitude and direction of the components if their magnitudes are in the ratio 1 : 2.
2. Resolve the force of 500 N (Fig. 1.47) into parallel components through (i) A and B and (ii) A and C.

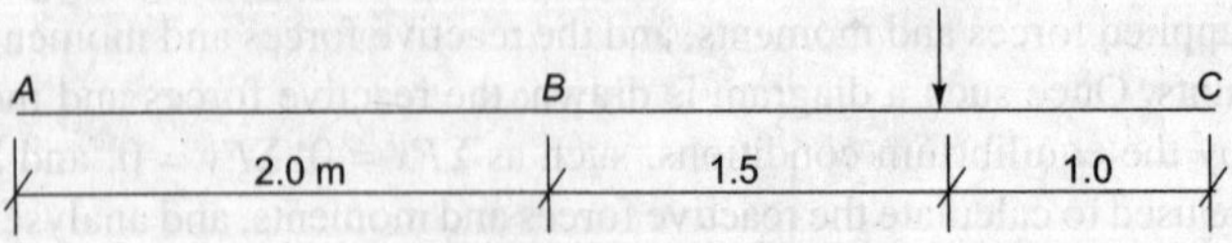

Fig. 1.47

3. A force is resolved into three components. If $\overline{DB}$ = 100 N, find the magnitudes of $\overrightarrow{AC}$ and $\overrightarrow{CD}$ for the two cases shown in Fig. 1.48. In Fig. 1.48 (a), $AB \parallel CD$.

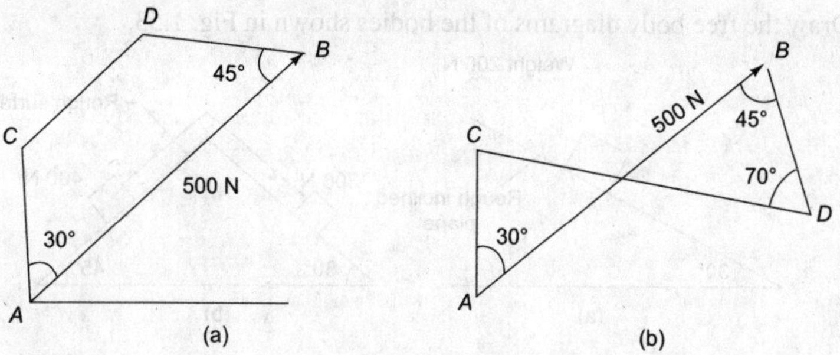

Fig. 1.48

4. A load of 1000 N is supported by two cables as shown in Fig. 1.49. Find the tension in these cables.

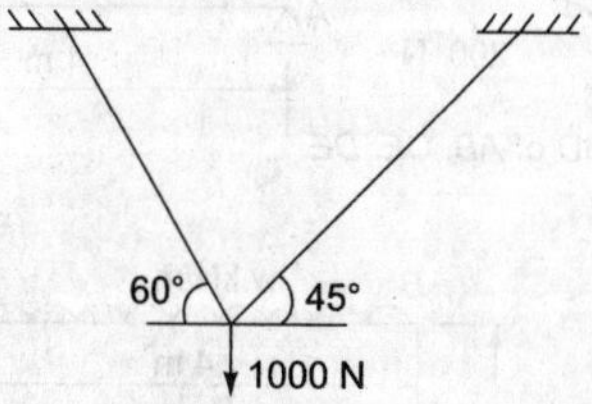

Fig. 1.49

5. Find the resultant of the parallel force system shown in Fig. 1.50, analytically and graphically.

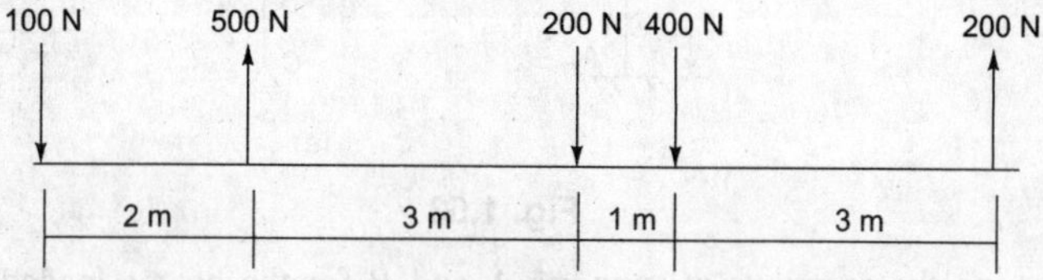

Fig. 1.50

6. Find the resultant of the parallel force system shown in Fig. 1.51, graphically.

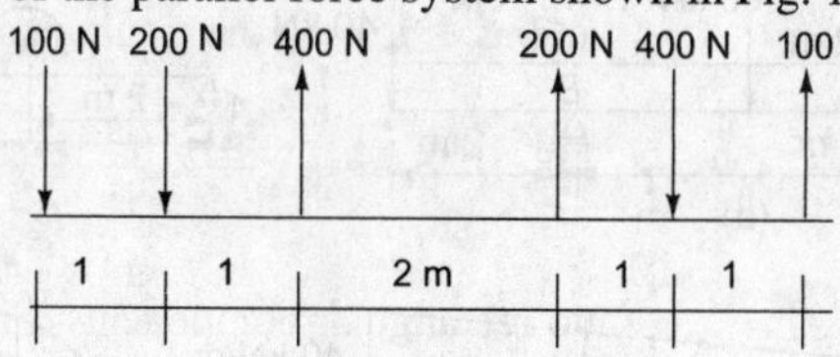

Fig. 1.51

7. Find the resultant of the three forces shown in Fig. 1.52.

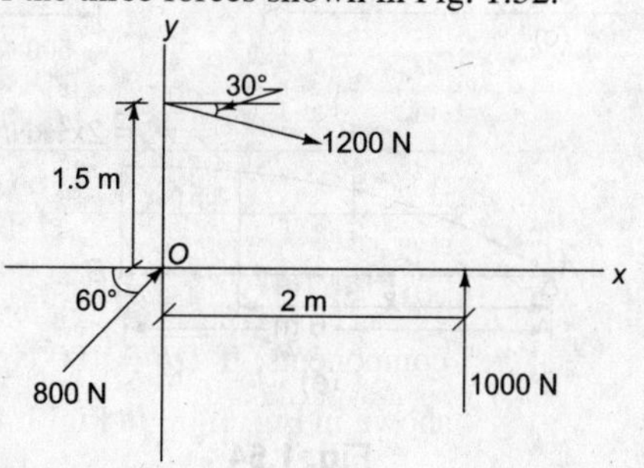

Fig. 1.52

8. Draw the free body diagrams of the bodies shown in Fig. 1.53.

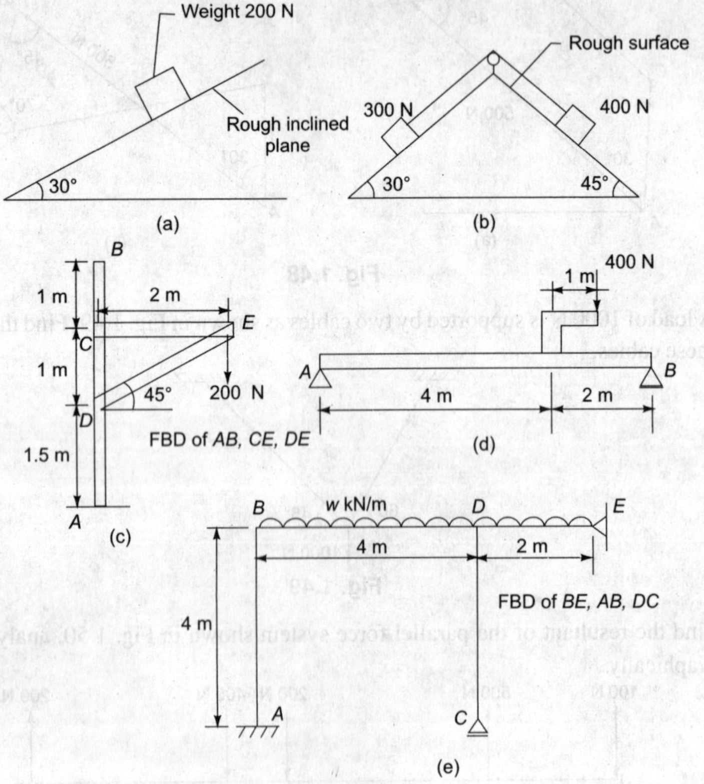

Fig. 1.53

9. Determine the reactions at supports A and B for the beams loaded as shown in Fig. 1.54.

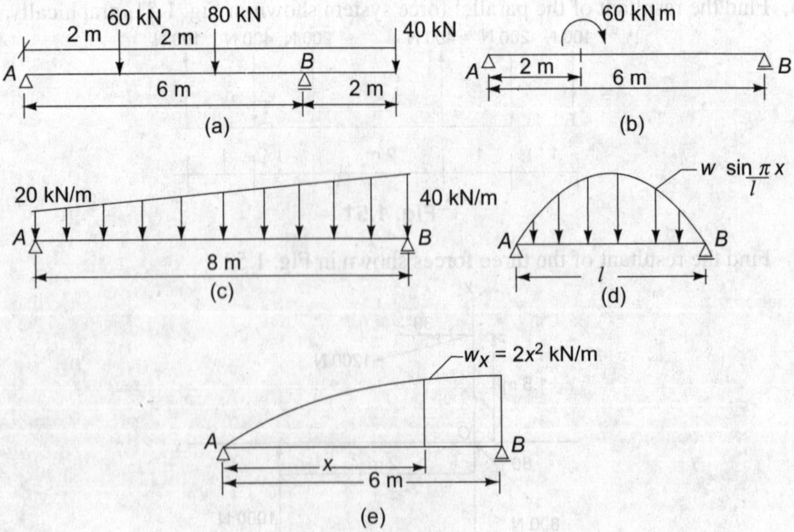

Fig. 1.54

10. For a beam loaded as shown in Fig. 1.55, if $x = l/2$, find the reactions at A and B. What is the value of x for which $R_B = 0$? What is the value of the reaction R_A in such case?

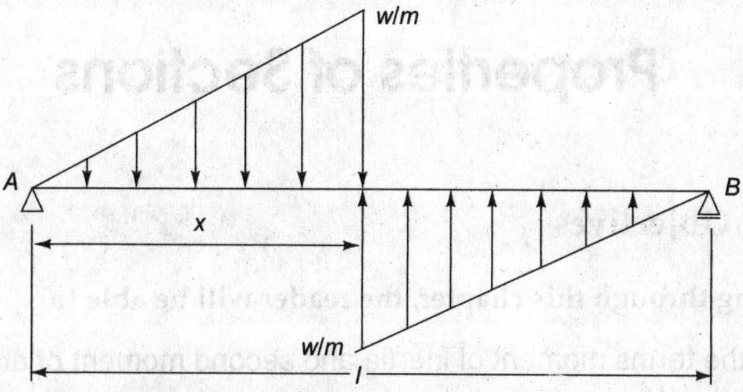

Fig. 1.55

11. Determine the reactions at supports A and B of the truss loaded as shown in Fig. 1.56.

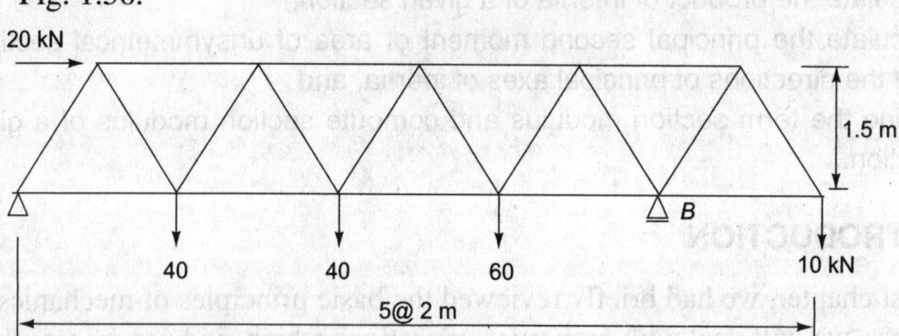

Fig. 1.56

12. Determine the reactions at A and B for the structures loaded as shown in Fig. 1.57.

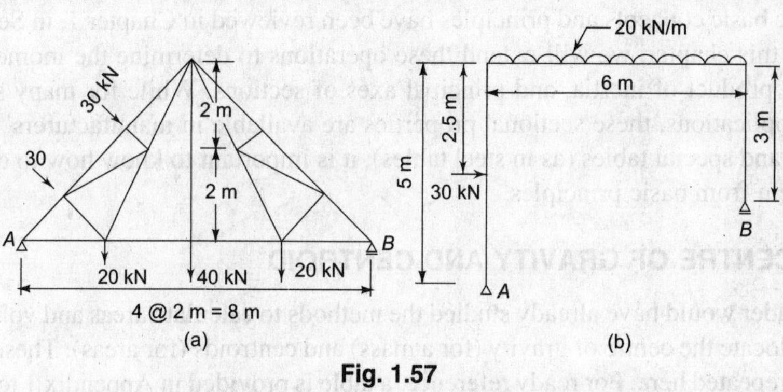

Fig. 1.57

CHAPTER 2

Properties of Sections

Learning Objectives

After going through this chapter, the reader will be able to

- define the terms moment of inertia and second moment of area,
- state and explain the parallel and perpendicular axes theorems,
- compute the second moment of area of a given section,
- calculate the radius of gyration of a given section,
- calculate the product of intertia of a given section,
- calculate the principal second moment of area of unsymmetrical sections and the directions of principal axes of inertia, and
- define the term section modulus and compute section modulus of a given section.

2.1 INTRODUCTION

In the first chapter, we had briefly reviewed the basic principles of mechanics. In this chapter, we will deal with properties of sections which find use repeatedly in many structural calculations. The reader would have already studied in detail two such properties, i.e., the centre of gravity and centroid. Computing areas of sections and locating their centroids is of fundamental importance in structural analysis. The basic concepts and principles have been reviewed in Chapter 1, in Section 1.7. In this chapter, we will extend these operations to determine the moment of inertia, product of inertia, and principal axes of sections. While for many structural applications, these sectional properties are available in manufacturers' catalogues and special tables (as in steel tables), it is important to know how to calculate them from basic principles.

2.2 CENTRE OF GRAVITY AND CENTROID

The reader would have already studied the methods to calculate areas and volumes and to locate the centre of gravity (for a mass) and centroids (for areas). These will not be repeated here. For ready reference, a table is provided in Appendix 1 to help calculate these for some standard sections.

2.3 MOMENT OF INERTIA

The reader would have already come across an expression (in physics or mechanics) $\int dM\, y^2$, particularly in connection with the rotation of solids about an axis. In Fig. 2.1, dM is an elementary mass and y is the distance of this elementary mass from an axis. This expression when evaluated gives a quantity which is called moment of inertia. It is usually denoted by the symbol I. Subscripts are used to indicate the axes about which the moments are taken. For example, I_{XX} represents the moment of inertia about the X-X axis.

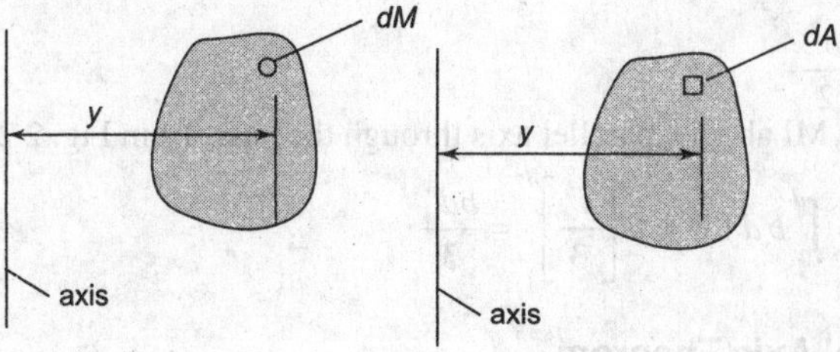

Fig. 2.1

A similar expression occurs in many structural calculations as $\int dA\, y^2$, where dA is an elementary area and y is the distance of dA from a given axis. The term moment of inertia does not apply to areas. The term used for this quantity is *second moment of area*. However, through long usage, these two terms are used to denote second moment of area.

This is a very important quantity for structural engineers, finding application in the design of beams, columns, etc.

2.4 COMPUTATION OF SECOND MOMENT OF AREA

The second moment of area is calculated from the integral $\int dA\, y^2$. The summation can be done by integration in the case of geometrical shapes like rectangle, triangle, circle, etc. Since the area is in L^2 units and y is in L units, the unit of moment of inertia will be in L^4 units, i.e., mm^4, m^4, etc. The following examples illustrate the basic procedure.

Example 2.1 Moment of inertia of a rectangle

Find the moment of inertia of a rectangle of width b and depth d about a horizontal axis through its centroid and about a parallel axis through its base (Fig. 2.2).

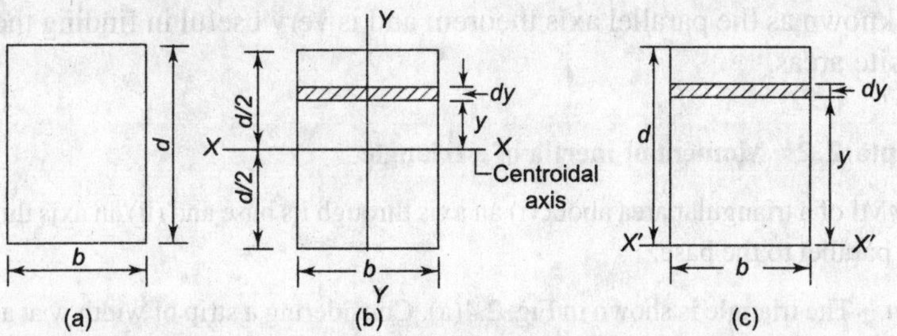

Fig. 2.2

Solution As shown in Fig. 2.2(b), consider an elementary strip of thickness d at a distance y from the centroid axis X-X.

$$\text{Area of the strip} = b\,dy$$
$$\text{MI about axis } X\text{-}X = b\,dy\,y^2$$

y has a range from $-d/2$ to $+d/2$. Therefore,

$$I_{XX} = \int_{-d/2}^{+d/2} b\,dy\,y^2 = b\left[\frac{y^3}{3}\right]_{-d/2}^{+d/2} = \frac{bd^3}{12}$$

Similarly, it can be shown that

$$I_{YY} = \frac{db^3}{12}$$

To determine the MI about a parallel axis through the base, from Fig. 2.2(c),

$$I_{X'-X'} = \int_0^d b\,dy\,y^2 = b\left[\frac{y^3}{3}\right]_0^d = \frac{bd^3}{3}$$

2.4.1 Parallel Axis Theorem

In Example 2.1, we found the MI about a parallel axis from the fundamental principles. A general theorem to find MI about parallel axes is stated as follows.

If I_{GG} is the MI of an area about an axis through its centroid, then the MI about any axis A-A at a distance d from GG and parallel to it is $I_{AA} = I_{GG} + A\,d^2$, where A is the area of the figure.

Considering Fig. 2.3, the MI of the elementary strip about the axis A-A is $dA(y + d)^2$. The MI of the whole area about the axis A-A is

$$I_{AA} = \int dA(y+d)^2 = \int dA(y^2 + d^2 + 2yd)$$

$$= \int dA\,y^2 + \int dA\,d^2 + \int dA \times 2yd$$

$$= I_{GG} + d^2 \int dA + 2d \int dA\,y$$

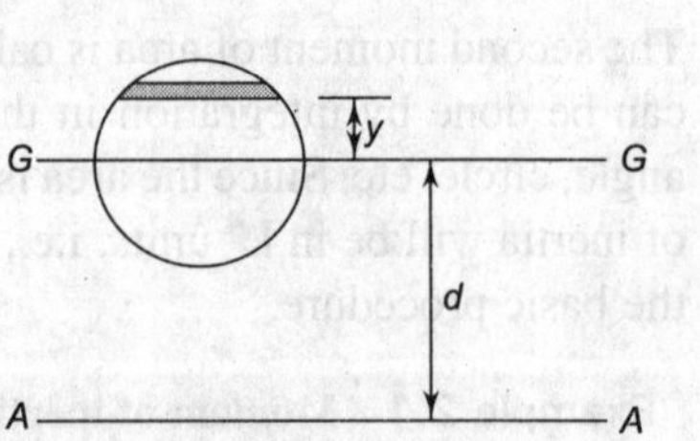

Fig. 2.3

$\int dA\,y = 0$

being the moment of the area about its centroid

$\int dA = A$

Therefore,

$$I_{AA} = I_{GG} + A\,d^2$$

This is known as the parallel axis theorem and is very useful in finding the MIs of composite areas.

Example 2.2 Moment of inertia of a triangle

Find the MI of a triangular area about (i) an axis through its base and (ii) an axis through the CG and parallel to the base.

Solution The triangle is shown in Fig. 2.4(a). Considering a strip of width w at a distance of y from vertex A, w = by/h, area = (by/h)dy, and I_{XX} = (by/h)dy(h–y)^2 for the elementary strip. For the whole area,

$$I_{XX} = \frac{b}{h}\int_0^h y\,dy\,(h-y)^2 = \frac{b}{h}\int_0^h y(h^2 + y^2 - 2hy)\,dy$$

$$= \frac{b}{h}\left[\frac{h^2y^2}{2} + \frac{y^4}{4} - 2h\frac{y^3}{3}\right]_0^h = \frac{bh^3}{12}$$

Considering Fig. 2.4(b), for the MI about axis *G-G*, at *h*/3 from the base,

$$I_{GG} = \int_0^h \frac{by}{h} dy \left(\frac{2}{3}h - y\right)^2 = \frac{b}{h}\int_0^h y\left(\frac{4}{9}h^2 + y^2 - \frac{4}{3}hy\right)dy$$

$$= \frac{b}{h}\left(\frac{4}{9}\frac{h^2y^2}{2} + \frac{y^4}{4} - \frac{4}{3}h\frac{y^3}{3}\right)_0^h$$

$$= \frac{bh^3}{36}$$

Knowing I_{XX}, I_{GG} could have been found by the parallel axis theorem as

$$I_{GG} = I_{XX} - Ad^2 = \frac{bh^3}{12} - \frac{bh}{2}\left(\frac{1}{3}h\right)^2 = \frac{bh^3}{12} - \frac{bh^3}{18} = \frac{bh^3}{36}$$

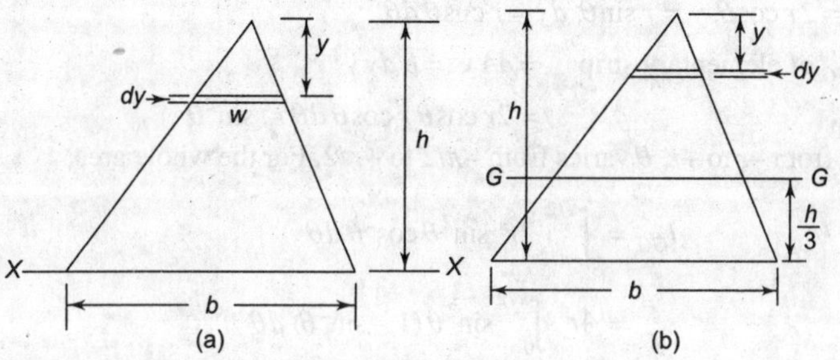

Fig. 2.4

2.4.2 Perpendicular Axes Theorem

We have seen how to determine the second moment of area about an axis lying in the plane of the area. In some structural calculations, we need to find the second moment of an area about an axis passing through the centroid and perpendicular to the plane containing the area. This is required in the case of torsion of members (Chapter 8). This is calculated using the perpendicular axes theorem.

The perpendicular axes theorem states that "*the second moment of an area about an axis perpendicular to the plane of the area through a point is equal to the sum of the second moment of areas about two mutually perpendicular axes through that point.*"

The second moment of area about an axis perpendicular to the plane of the area is known as polar moment of inertia.

2.4.3 Polar Moment of Inertia

The polar moment of inertia is defined as the MI about an axis perpendicular to the plane of the area. Polar MI is denoted by the symbol *J*.

Considering Fig. 2.5, $J_{ZZ} = \int_A dA\, r^2$. If *X* and *Y* are two axes in the plane of the area, mutually perpendicular and passing through *O*, then

$$J_{ZZ} = \int_A dAr^2 = \int_A dA(x^2 + y^2) = \int_A dA\, x^2 + \int dA\, y^2$$

$$= I_{XX} + I_{YY}$$

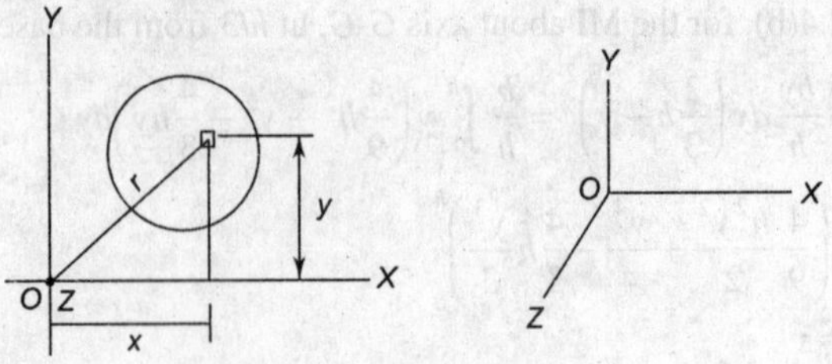

Fig. 2.5

Example 2.3 Moment of inertia of a circle

Determine the MI of a circular area about one of its diameters.

Solution Referring to Fig. 2.6, for the elementary strip parallel to the diametrical axis *G*-*G*, width $b = 2r\cos\theta$, $y = r\sin\theta$, $dy = r\cos\theta\, d\theta$,

$$I_{GG} \text{ of elementary strip} = dA\, y^2 = b\, dy\, y^2$$

$$= 2r\cos\theta\, r\cos\theta\, d\theta\, r^2 \sin^2\theta$$

As *y* varies from $-r$ to $+r$, θ varies from $-\pi/2$ to $+\pi/2$. For the whole area,

$$I_{GG} = \int_{-\pi/2}^{\pi/2} 2r^4 \sin^2\theta \cos^2\theta\, d\theta$$

$$= 4r^4 \int_0^{\pi/2} \sin^2\theta (1 - \sin^2\theta)\, d\theta$$

$$= 4r^4 \left[\int_0^{\pi/2} \sin^2\theta\, d\theta - \int_0^{\pi/2} \sin^4\theta\, d\theta\right]$$

$$= \frac{\pi r^4}{4}$$

Since $r = d/2$,

$$I_{GG} = \frac{\pi (d/2)^4}{4} = \frac{\pi d^4}{64}$$

This result can be easily obtained from $J_{ZZ} = I_{XX} + I_{YY}$. In Fig. 2.6(b), taking *X*- and *Y*-axes as shown, $I_{XX} = I_{YY}$ from considerations of symmetry for the elementary strip in the form of the ring shown, at a distance *y* from *O*.

$$J_{ZZ} = 2\pi y dy\, y^2$$

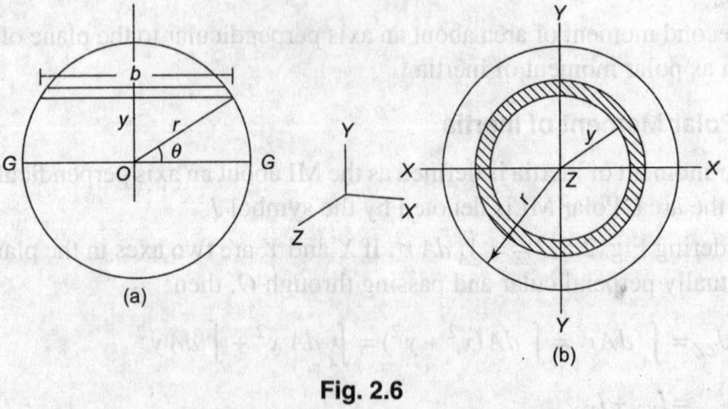

Fig. 2.6

For the whole area,

$$J_{ZZ} = \int_0^r 2\pi y^3 dy = \frac{\pi r^4}{2}$$

$$I_{XX} = I_{YY} = \frac{1}{2}\frac{\pi r^4}{2} = \frac{\pi r^4}{4}$$

Example 2.4 MI of a parabolic area

For the parabolic area shown in Fig. 2.7, find the MI with respect to the X- and Y-axes.

Solution In Fig. 2.7(b),

Width of elementary strip $= a - x = a - ky^2$

Area $= (a - ky^2)\,dy, \quad I_{XX} = (a - ky^2)\,dy\,y^2$

Integrating over the whole area,

$$I_{XX} = \int_0^b (a - ky^2)\,y^2 dy = \left(\frac{ay^3}{3} - \frac{ky^5}{5}\right)_0^b = \frac{ab^3}{3} - \frac{kb^5}{5}$$

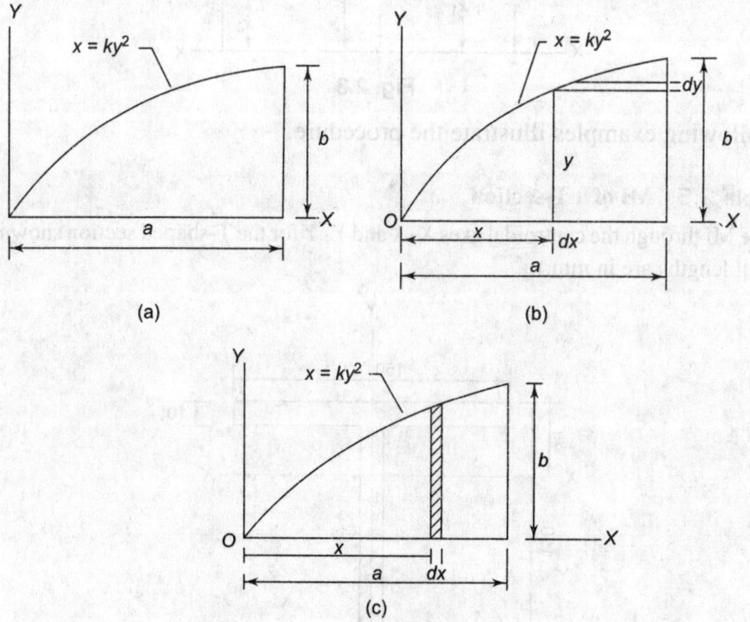

Fig. 2.7

Since $a = kb^2$, $b^2 = a/k$.

$$I_{XX} = \frac{ab^3}{3} - \frac{ab^3}{5} = \frac{2}{15}ab^3$$

To calculate I_{YY}, consider an elementary strip parallel to the Y-axis. For the elementary strip, area $= y\,dx = y \times 2ky\,dy$, $\quad I_{YY} = 2ky^2\,dy\,x^2$. Therefore,

$$I_{YY} = 2ky^2 dy k^2 y^4$$

Integrating over the whole area,

$$I_{YY} = \int_a^b 2k^3 ybdy = 2k^3\left(\frac{y^7}{7}\right)_0^b = \frac{2k^3b^7}{7} = \frac{2}{7}ba^3$$

2.4.4 Moment of Inertia of a Composite Area

The second moment of area composed of a number of simple areas can be found as the sum of the second moments of area of its parts. As the second moment of an area will never be negative, simple summation can be done. For example, considering the area shown in Fig. 2.8, the second moment of area about the $X X$-axes can be calculated as

$$I_{XX} = I_{G_1G_1} + A_1x_1^2 + I_{G_2G_2} + A_2x_2^2 + I_{G_3G_3} + A_3x_3^2$$

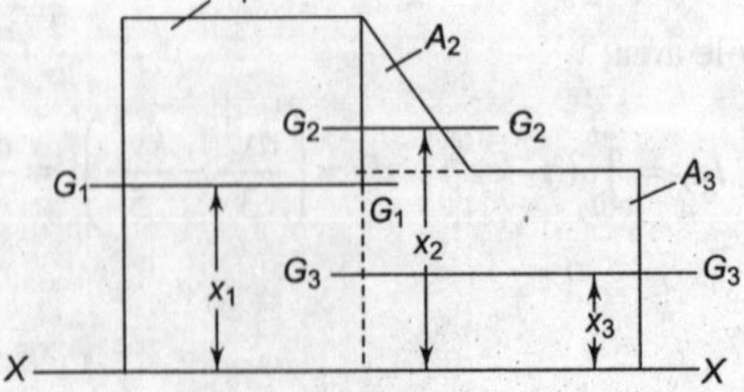

Fig. 2.8

The following examples illustrate the procedure.

Example 2.5 MI of a T-section

Find the MI through the centroidal axes X-X and Y-Y for the T-shaped section shown in Fig. 2.9. (All lengths are in mm.)

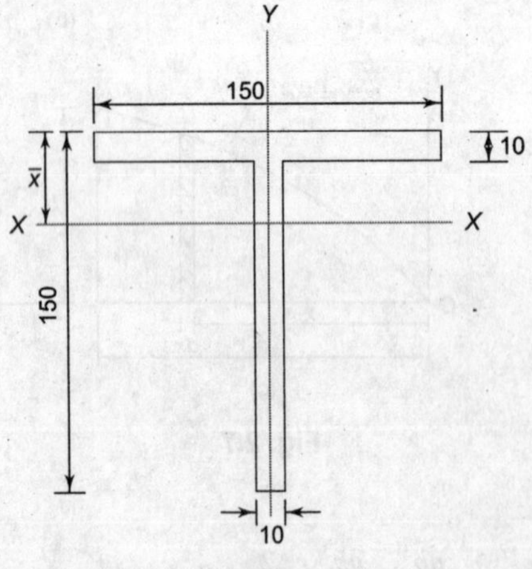

Fig. 2.9

Solution The section is symmetrical about the Y-axis. To locate the distance $\bar{x}$ (from the upper edge), take the moment about the top edge of the T-section of the areas:

$$(150 \times 10 + 140 \times 10)\ \bar{x} = 150 \times 10 \times 5 + 140 \times 10 \times 80$$

$$\bar{x} = 41.2 \text{ mm}$$

$$I_{XX} = 150 \times \frac{10^3}{12} + 150 \times 10\,(41.2 - 5)^2 + \frac{10 \times 140^3}{12} + 140 \times 10\,(108.8 - 70)^2$$

$$= 6.372 \times 10^6 \text{ mm}^4$$

$$I_{YY} = \frac{10 \times 150^3}{12} + \frac{140 \times 10^3}{12} = 2.8242 \times 10^6 \text{ mm}^4$$

Example 2.6 MI of an unequal angle section

Find the MI about the horizontal and vertical axes passing through the centroid of the unequal angle section shown in Fig. 2.10.

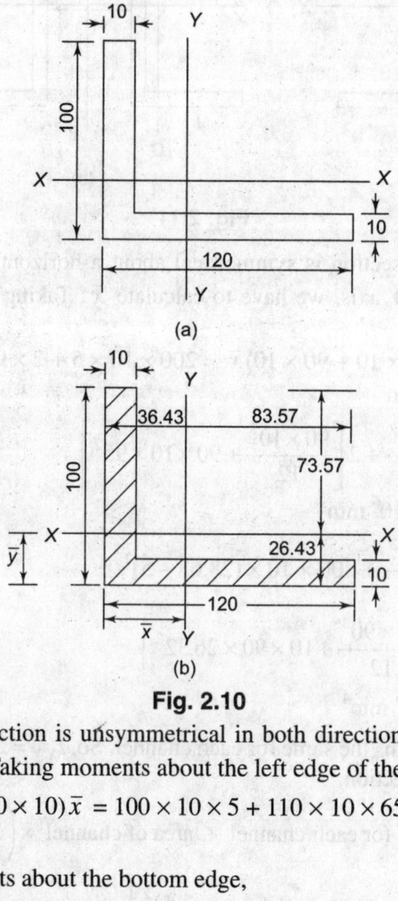

Fig. 2.10

Solution The angle section is unsymmetrical in both directions. Here $\bar{x}$ as well as $\bar{y}$ have to be determined. Taking moments about the left edge of the areas,

$$(100 \times 10 + 110 \times 10)\,\bar{x} = 100 \times 10 \times 5 + 110 \times 10 \times 65$$

$$\bar{x} = 36.43 \text{ mm}$$

Similarly, taking moments about the bottom edge,

$$(100 \times 10 + 110 \times 10)\,\bar{y} = 100 \times 10 \times 50 + 110 \times 10 \times 5$$

$$\bar{y} = 26.43 \text{ mm}$$

From the distances shown in Fig. 2.10(b),

$$I_{XX} = \frac{10 \times 100^3}{12} + 10 \times 100\,(73.57 - 50)^2 + \frac{110 \times 10^3}{12} + 110 \times 10\,(26.43 - 5)^2$$

$$= 1.9032 \times 10^6 \text{ mm}^4$$

$$I_{YY} = \frac{10 \times 100^3}{12} + 100 \times 10(36.43 - 5)^2 + \frac{110 \times 10^3}{12} + 10 \times 110\,(83.57 - 55)^2$$

$$= 3.0032 \times 10^6 \text{ mm}^4$$

Example 2.7 MI of a double channel section

Find the MI of the channel section shown in Fig. 2.11(a) about the *X-X* and *Y-Y* axes passing through its centroid. If two such sections are kept back to back as shown in Fig. 2.11(b), find the distance d such that $I_{XX} = I_{YY}$ for the compound section.

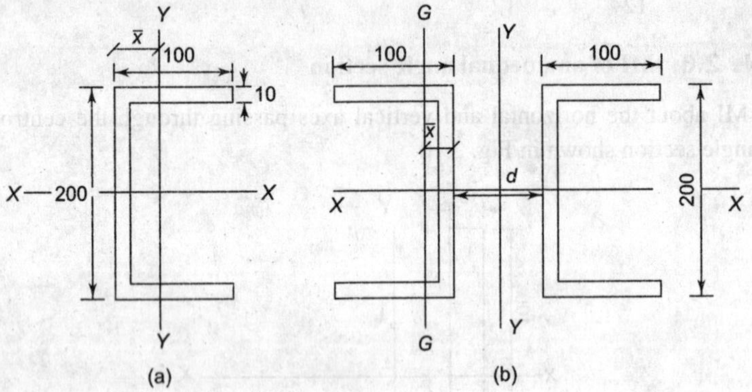

Fig. 2.11

Solution The channel section is symmetrical about a horizontal axis through its mid-height. To locate the *Y-Y* axis, we have to calculate $\bar{x}$. Taking moments about the left vertical edge,

$$(200 \times 10 + 90 \times 10 + 90 \times 10)\bar{x} = 200 \times 10 \times 5 + 2 \times 90 \times 10 \times 55$$

$$\bar{x} = 28.68 \text{ mm}$$

$$I_{XX} = \frac{10 \times 200^3}{12} + 2\left[\frac{90 \times 10^3}{12} + 90 \times 10 \times 95^2\right]$$

$$= 22.93 \times 10^6 \text{ mm}^4$$

$$I_{YY} = \frac{200 \times 10^3}{12} + 200 \times 10 \times (28.68 - 5)^2$$

$$+ 2\left[\frac{10 \times 90^3}{12} + 10 \times 90 \times 26.32^2\right]$$

$$= 3.6 \times 10^6 \text{ mm}^4$$

In Fig. 2.11(b), I_{XX} remains the same for each channel. So, $I_{XX} = 2 \times 22.93 \times 10^6 = 45.86 \times 10^6 \text{ mm}^4$ for the whole section.

$$I_{YY} = 2 \times \left[I_{GG} \text{ for each channel } + \text{ area of channel } \times \left(\bar{x} + \frac{d}{2}\right)^2\right]$$

$$= 2\left[3.6 \times 10^6 + 3800 \times \left(28.68 + \frac{d}{2}\right)^2\right]$$

$$= 7.2 \times 10^6 + 7600\left(28.68^2 + \frac{d^2}{4} + 28.68d\right)$$

This is equal to I_{XX}.

$$7.2 \times 10^6 + 6.2513 \times 10^6 + 7600\frac{d^2}{4} + 217{,}968d = 45.86 \times 10^6$$

$$1900d^2 + 217{,}968d - 32.41 \times 10^6 = 0$$

$$d = 85.3 \text{ mm}$$

Example 2.8 MI of a rectangular lamina with hole

From a rectangular plate, 100 mm × 200 mm, a circular portion is removed as shown in Fig. 2.12. Find the MI of the plate about an axis through its base.

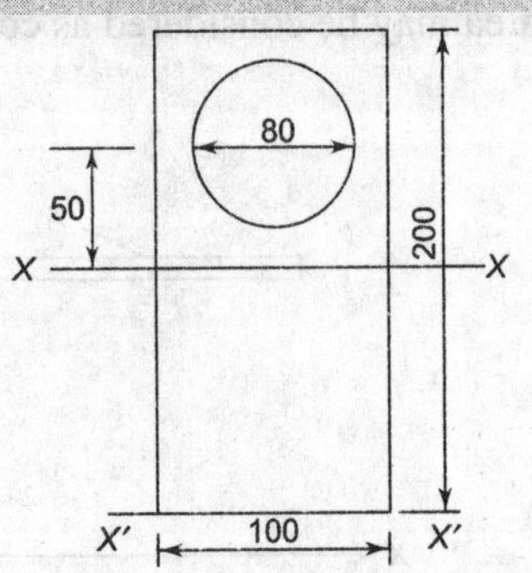

Fig. 2.12

Solution The circular portion can be taken as a negative area and its MI subtracted from that of the full rectangular plate.

$$I_{X'X'} = 100 \times \frac{200^3}{3} - \frac{\pi \times 80^4}{64} - \frac{\pi \times 80^2}{4} \times 150^2$$

$$= 1.5156 \times 10^8 \text{ mm}^4$$

Example 2.9 MI of a composite area

Find the MI of the composite area shown in Fig. 2.13(a) about the X-X and Y-Y axes.

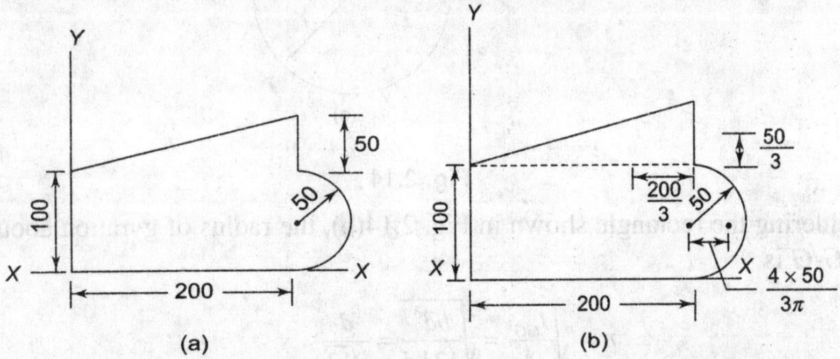

Fig. 2.13

Solution The distances of the centroids of each part of the area from the respective reference lines are shown in Fig. 2.13(b). I_{XX} and I_{YY} can be calculated as

$$I_{XX} = \frac{200 \times 100^3}{3} + \frac{\pi \times 50^4}{8} + \frac{\pi \times 50^2}{2} \times 50^2$$

$$+ \frac{200 \times 50^3}{36} + \frac{200 \times 50}{2} \times \left(\frac{350}{3}\right)^2$$

$$= 1.4768 \times 10^8 \text{ mm}^4$$

$$I_{YY} = \frac{100 \times 200^3}{3} + \frac{\pi \times 50^4}{8} + \frac{\pi \times 50^2}{2} \times (221.22)^2$$

$$+ \frac{50 \times 200^3}{36} + \frac{50 \times 200}{2}\left(\frac{400}{3}\right)^2$$

$$= 5.613 \times 10^8 \text{ mm}^4$$

2.4.5 Radius of Gyration

The radius of gyration is defined as $r = \sqrt{I/A}$, where I is the moment of inertia and A is the area of the section. This entity is frequently encountered in structural analysis and design.

The radius of gyration can be considered as the distance at which the whole area may be considered as concentrated as a strip, as shown in Fig. 2.14(a).

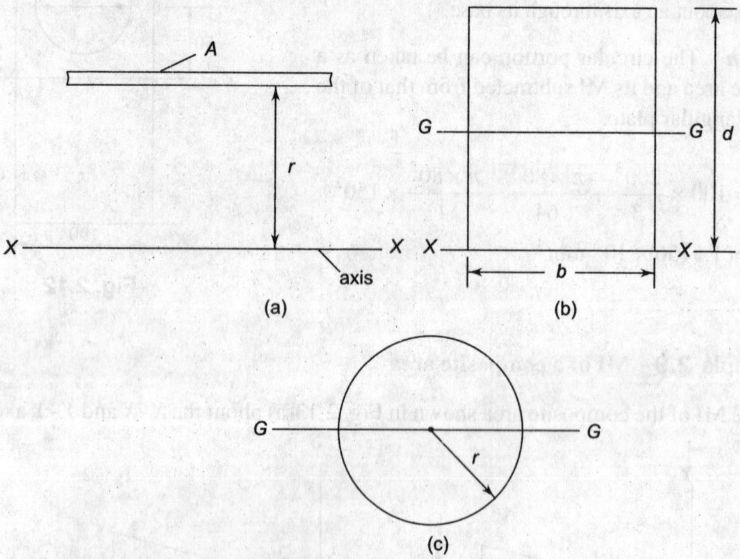

Fig. 2.14

Considering the rectangle shown in Fig. 2.14(b), the radius of gyration about the axis *G-G* is

$$r = \sqrt{\frac{I_{GG}}{A}} = \sqrt{\frac{bd^3}{12\,bd}} = \frac{d}{\sqrt{12}}$$

Similarly, for the circular area shown in Fig. 2.14(c),

$$R = \sqrt{\frac{I_{GG}}{A}} = \sqrt{\frac{\pi r^4}{4\pi r^2}} = \frac{r}{2}$$

In Fig. 2.14(b), $I_{XX} = I_{GG} + A\,h^2$, where $h = d/2$. From this, $A\,r_{XX}^2 = A\,r_{GG}^2 + A\,h^2$ or $r_{XX}^2 = r_{GG}^2 + h^2$. The polar radius of gyration, $r_z^2 = r_x^2 + r_y^2$.

2.5 SECTION MODULUS

Consider the beam shown in Fig. 2.15(b). When the beam is subjected to loads, it bends. The beam section is shown in Fig. 2.15(a). The centroid of the beam section can be found from the principles you might have studied in engineering mechanics. An axis through the centroid of the section and parallel to its base is called *neutral axis*. Note that this axis is perpendicular to the plane of the paper in Fig. 2.15(b). When the beam bends, each section of the beam rotates about the neutral axis. Because of this rotation, the MI (second moment of area) comes into effect in the bending of beams. This has been discussed in detail in the earlier sections.

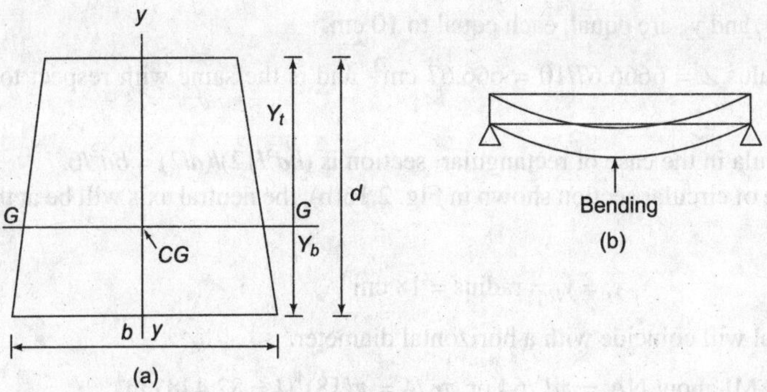

Fig. 2.15

The beam section is symmetrical about the Y-axis but can be unsymmetrical about the horizontal axis in the case of simple bending. The centroid, in such a case, will not be at the centre of the depth, d, but may be below or above. In the case shown in Fig. 2.15(a), the centroid is below the half depth $d/2$. The distances y_t and y_b thus are not equal. These distances are known as distances to the extreme fibres at top and bottom from the neutral axis (NA). The quantities I/y_t and I/y_b are known as moduli of section or section moduli of the beam section.

Section modulus is thus the MI of the section divided by the distance to the extreme fibre, either to the top or to the bottom, from the neutral axis. Section modulus is important because the least section modulus governs the design of the beam. Section modulus is usually denoted by the symbol Z. The unit of section modulus Z is L^3 being (L^4/L). The section modulus will have the units of mm^3, cm^3, m^3, etc.

We will illustrate the computation of section modulus through a number of examples. This will be useful while designing the beams.

Example 2.10 Section modulus

Compute the section modulus of the sections shown in Fig. 2.16.

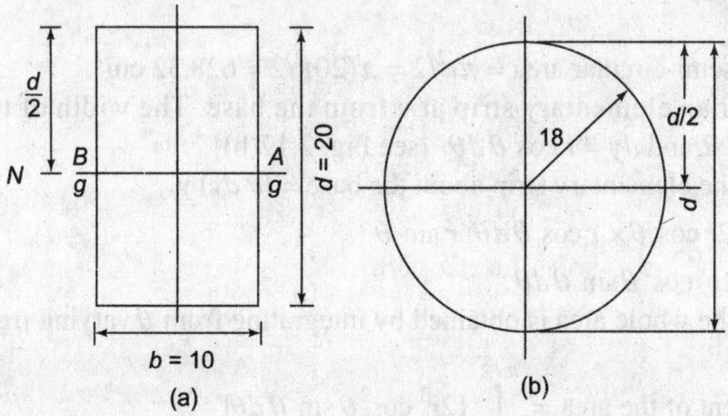

Fig. 2.16

Solution For the rectangular section shown in Fig. 2.16(a), the neutral axis will be at half depth as the section is symmetrical about that line. The neutral axis will be g-g at $d/2$ from the top fibre. The MI about the neutral axis is given by $bd^3/12$.

MI about neutral axis = $10 \times 20^3/12 = 6666.67\ cm^4$

In this case, y_t and y_b are equal, each equal to 10 cm.

Section modulus $Z = 6666.67/10 = 666.67$ cm^3 and is the same with respect to top and bottom fibres.

General formula in the case of rectangular section is $(bd^3/12)/(d/2) = bd^2/6$.

In the case of circular section shown in Fig. 2.16(b), the neutral axis will be at the centre of depth.

$$y_t = y_b = \text{radius} = 18 \text{ cm}$$

The neutral will coincide with a horizontal diameter.

$$\text{MI about NA} = \pi d^4/64 \text{ or } \pi r^4/4 = \pi(18)^4/4 = 82{,}448 \text{ cm}^4$$

Section modulus will be the same about top and bottom fibres.

$$\text{Section modulus} = 82{,}448/18 = 4580.44 \text{ cm}^3$$

The general formula for section modulus is $(\pi r^4/4)/r = \pi r^3/4 = \pi d^3/32$

Example 2.11 Section moduli of a semicircular section

A wooden semicircular log, of 20 cm radius, is used as a beam over a span of 2 m. Find the section moduli of the section.

Solution The section is shown in Fig. 2.17(a).

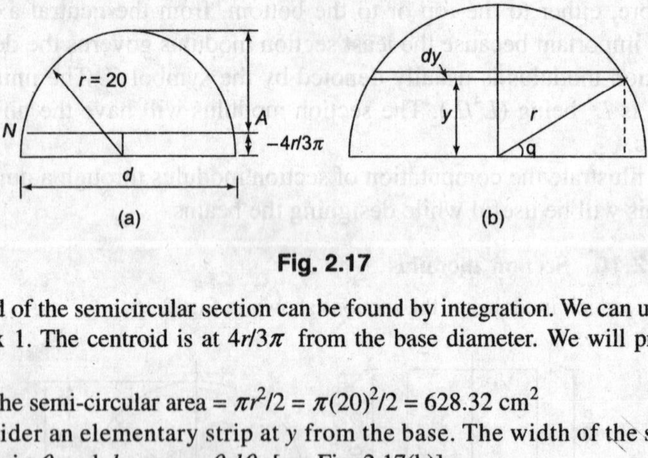

Fig. 2.17

The centroid of the semicircular section can be found by integration. We can use Table 1.1 of Appendix 1. The centroid is at $4r/3\pi$ from the base diameter. We will prove this by integration.

Area of the semi-circular area $= \pi r^2/2 = \pi(20)^2/2 = 628.32$ cm^2

We consider an elementary strip at y from the base. The width of the strip, $b = 2r \cos\theta$, $y = r\sin\theta$, and $dy = r\cos\theta\, d\theta$ [see Fig. 2.17(b)]

Moment of the elementary strip about the base = $(b\, dy)\, y$

$$\text{Moment} = (2r\cos\theta \times r\cos\theta\, d\theta)\, r\sin\theta$$

$$= 2r^3\cos^2\theta\sin\theta\, d\theta$$

Moment of the whole area is obtained by integrating from θ varying from 0 to $\pi/2$.

$$\text{Moment of the area} = \int_0^{\pi/2} [2r^3\cos^2\theta\sin\theta\, d\theta]$$

$$= 2r^3\,[-\cos^3\theta/3]_0^{\pi/2}$$

$$= 2r^3/3$$

$$\text{Area of the section} = \pi r^2/2$$

Distance of the centroid from the base y is given by

$$\hat{y} = (2r^3/3)/(\pi r^2/2)$$
$$= 4r/3\pi$$

Distance of the centroid from the base = $4 \times 20/3\pi = 8.49$ cm from the base
The neutral axis will be as shown in Fig. 2.17(b).
To find the moment of inertia, we find $(b\,dy)y^2$ and integrate the same from 0 to $\pi/2$.

Second moment of area of the elementary strip = $(2r\cos\theta \times r\cos\theta\, d\theta)(r\sin\theta)^2$
$= 2r^4\cos^2\theta\sin^2\theta\, d\theta$

$$\text{MI} = \int_0^{\pi/2} (2r^4\cos^2\theta\sin^2\theta)\, d\theta$$

$$= 2r^4 \int_0^{\pi/2} \sin^2\theta(1 - \sin^2\theta)\, d\theta$$

$$= 2r^4 \left[(\theta/2 - \sin 2\theta/4) + (1/4)\cos\theta\sin^3\theta - (3/4)\int \sin^2\theta\, d\theta \right]_0^{\pi/2}$$

$$= 0.11\, r^4$$
$$= 17{,}600 \text{ cm}^4$$
$$Z_t = 17{,}600/11.51 = 1529 \text{ cm}^3$$
$$Z_b = 17{,}600/8.49 = 2073 \text{ cm}^3$$

Example 2.12 Section moduli of a triangular section

Find the section moduli of an isosceles triangle. If two such triangles were joined base to base, what will be the section modulus of the figure? It is given that b = 30 cm and h = 18 cm.

Solution The isosceles triangle is shown in Fig. 2.18(a). The centroid of the triangle is at $h/3$ from the base. In this case, it is at 6 cm from the base.

MI of the triangle about the NA = $bh^3/36 = 30 \times 18^3/36 = 4860$ cm^4

As the section is unsymmetrical about the NA, there will be two section moduli.

(based on top edge) $Z_t = 4860/(2 \times 18/3) = 405$ cm^3

(based on bottom edge) $Z_b = 4860/6 = 810$ cm^3

If two such triangles were joined together as shown in Fig. 2.18(b),

MI about NA = $2 \times bh^3/12 = 2 \times 30 \times 18^3/12 = 29{,}160$ cm^4

Section modulus will be the same about top and bottom edges.

$Z = 29{,}160/18 = 1620$ cm^3

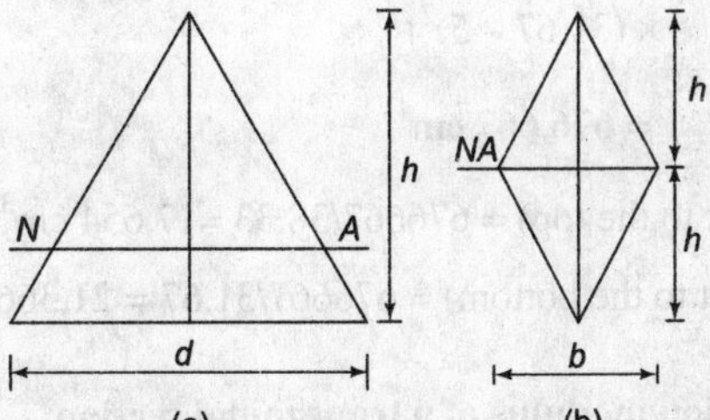

Fig. 2.18

Example 2.13 Section moduli of I-sections

Find the section modulus of the I-sections shown in Fig. 2.19.

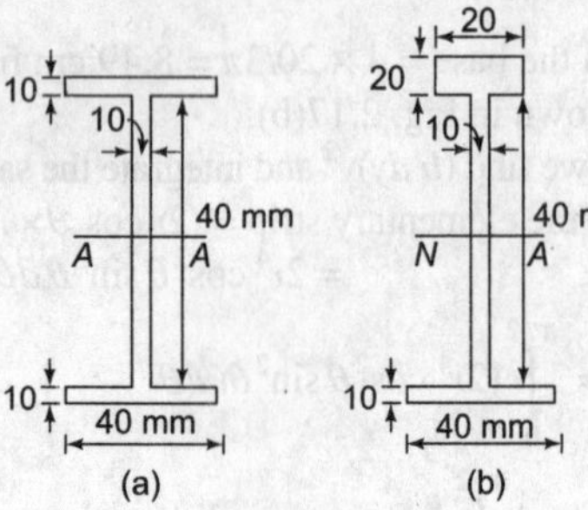

Fig. 2.19

Solution The I-section shown in Fig. 2.19(a) is symmetrical about the XX-axis (neutral axes). The section modulus about top and bottom edges is equal in this case.

The second moment of area of the section can be found as the sum of the second moments of area of the three rectangles. Flanges being equal, their second moments of area will also be equal.

$$MI = 2(40 \times 10^3/12 + 40 \times 10 \times 21^2) + 10 \times 40^3/12$$

We have used $I_{gg} + Ad^2$ in the case of flanges.

$$MI = 2[3333.33 + 176{,}400] + 53333.33 = 412{,}800 \text{ cm}^4$$

$$Z = 412{,}800/30 = 13{,}760 \text{ cm}^3$$

In the case of the I-section shown in Fig. 2.19(b), the flanges are not equal and hence the neutral axis will not be at half the depth. We first locate the centroid and then calculate MI about an axis passing through it.

Taking moments about the bottom edge,

$$\hat{y} = \frac{20\times 20\times 60 + 40\times 10\times 30 + 40\times 10\times 5}{20\times 20 + 40\times 10 + 40\times 10}$$

$$= 31.67 \text{ cm } (y_b)$$

Distance to the centroid from top edge = 38.33 cm (y_t)

$$\begin{aligned} \text{MI (about NA)} &= [20 \times 20^3/12 + 20 \times 20 \times (38.33 - 10)^2 + 10 \times 40^3/12 \\ &\quad + 10 \times 40 \times (31.67 - 30)^2 + 40 \times 10^3/12 + 40 \times 10 \\ &\quad \times (31.67 - 5)^2] \\ &= 676{,}667 \text{ cm}^4 \end{aligned}$$

Z_t (with respect to the top) = 676667/38.33 =17,654 cm^3

Z_b (with respect to the bottom) = 676667/31.67 = 21,366 cm^3 ■

Example 2.14 Section modulus of a trapezoidal section

Find the section modulus of the trapezium shown in Fig. 2.20.

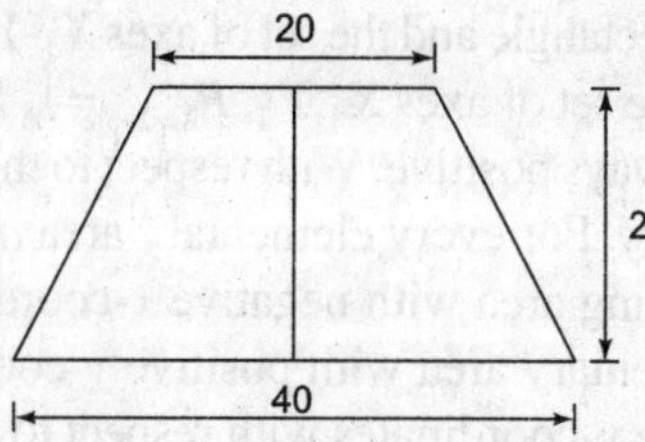

Fig. 2.20

Solution We first find the centroid of the section. Taking moments about the bottom edge, considering two triangles as shown,

$$\text{Area, } \bar{y} = [20 \times (20/2)\,(2 \times 20/3) + 40 \times (20/2)\,(20/3)] = 5333.33 \text{ cm}^3$$

$$\text{Area of trapezium} = [(40 + 20)/2] \times 20 = 600 \text{ cm}^2$$

$$\hat{y} = 5333.33/600 = 8.89 \text{ cm}$$

$$\text{Height of centroid from top} = 20 - 8.89 = 11.11 \text{ cm}$$

$$\text{MI about NA} = 40 \times 20^3/12 + (40 \times 20/2)\,(8.89 - 6.67)^2 + 20 \times 20^3/12$$
$$+ 20 \times (20/2)\,(11.11 - 6.67)^2$$
$$= 26{,}666.67 + 1971.36 + 13{,}333.33 + 3942.72$$
$$= 45{,}914 \text{ cm}^4$$

$$Z_t = 45{,}914/11.11 = 4132.68 \text{ cm}^3$$

$$Z_b = 45{,}914/8.89 = 5164.68 \text{ cm}^3$$

2.6 PRODUCT OF INERTIA

The product of inertia, usually denoted by the symbol P, is given by the mathematical expression $\int_A x\,y\,dA$. Though the concept does not find wide application, it is useful in dealing with unsymmetrical bending, finding principal axes of inertia, in advanced structural analysis, etc.

Consider Fig. 2.21. The product of inertia can be expressed as $P = \int_A x\,y\,dA$. The unit of P is the same as that of I, i.e., mm^4, m^4, etc. But while I is always positive, P can be positive or negative, depending upon the axes chosen. Note that P has to be with reference a set of coordinate axes and hence should be expressed as $P_{XY} = \int_A x\,y\,dA$.

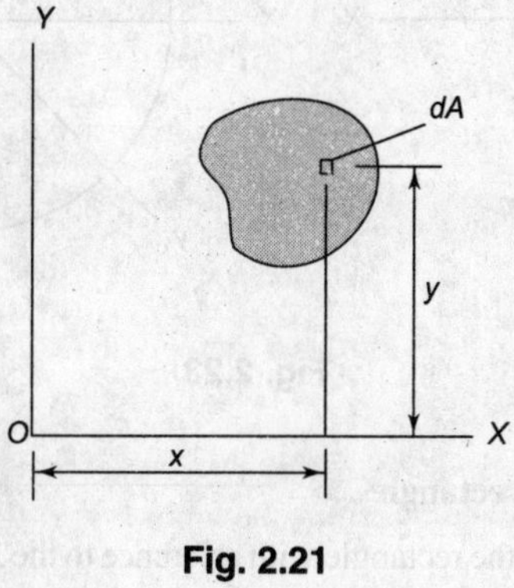

Fig. 2.21

Let us take the case of a rectangle and the set of axes X_1-Y_1, X_2-Y_2, X_3-Y_3, etc. as in Fig. 2.22. With respect to the set of axes X_1, Y_1, $P_{X_1Y_1} = \int_A X_1 Y_1 dA$. Since X_1, Y_1 are always positive, $P_{X_1Y_1}$ is always positive. With respect to the axes X_2, Y_2, note that X_2 and Y_2 are axes of symmetry. For every elementary area dA with positive x-coordinates, there is a corresponding area with negative x-coordinates with respect to the Y-axis, and for every elementary area with positive y-coordinates, there is a corresponding area with negative y-coordinates with respect to the X-axis. Thus, $P_{X_2Y_2} = \int_A x_2 y_2 dA = 0$.

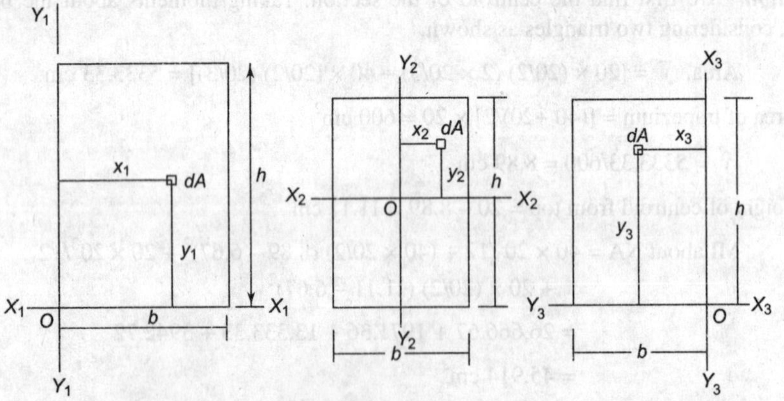

Fig. 2.22

With respect to axes X_3, Y_3, the area is so located that the y_3-coordinates are positive while the x_3-coordinates are negative. Therefore, $P_{X_3Y_3} = \int_A x_3 y_3 dA$ is negative.

An important property of product of inertia is that it is zero with respect to a set of axes if one or both of the axes are axes of symmetry.

Thus, in Fig. 2.23, $P_{XY} = 0$ for the T-section because the Y-axis is an axis of symmetry and $P_{XY} = 0$ for a circular area about the axes passing through its centre.

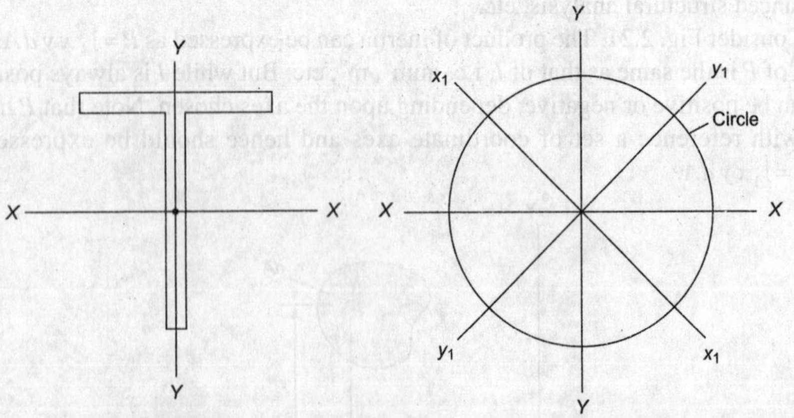

Fig. 2.23

Example 2.15 PI of a rectangle

Find the product of inertia of the rectangle with reference to the X, Y axes shown in Fig. 2.24.

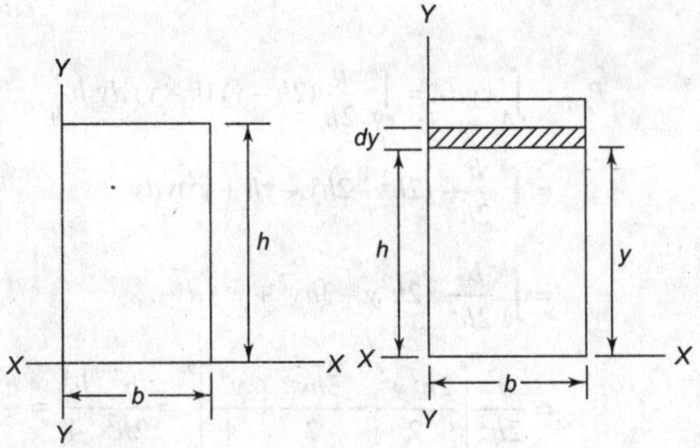

(a) (b)

Fig. 2.24

Solution Consider a strip parallel to the X-axis as shown in Fig. 2.24(b). Note that (x, y) are the coordinates of the centroid of the strip. From the figure, $dA = b\ dy$, $\bar{x} = b/2$, $\bar{y} = y$. Therefore,

$$P_{XY} = \int_A x\,y\ dA = \int_0^h \frac{b}{2}\ yb\ dy = \frac{b^2h^2}{4}$$

Example 2.16 PI of a right-angled triangle (general formula)

Find the product of inertia of the right-angled triangle shown in Fig. 2.25 with respect to the X, Y axes.

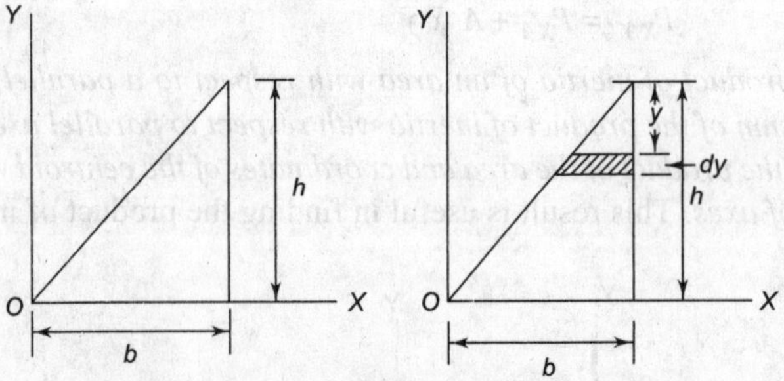

(a) (b)

Fig. 2.25

Solution The whole triangle is in the first quadrant. The x and y coordinates are positive throughout. Consider the elementary strip parallel to the base, as shown in Fig. 2.25(b).

$$\text{Width of the strip} = b\,\frac{y}{h}$$

$$\text{Area} = b\,\frac{y}{h}\,dy$$

$$x\text{-coordinate of centroid} \quad = b - \frac{1}{2}b\frac{y}{h}$$

$$= \frac{b}{2h}(2h - y)$$

$$y\text{-coordinate} = (h - y)$$

Therefore,

$$P_{XY} = \int_A xy dA = \int_0^h \frac{b}{2h}(2h-y)(h-y)\, dy\, b\frac{y}{h}$$

$$= \int_0^h \frac{b^2}{2h^2}(2h^2 - 2hy - yh + y^2)ydy$$

$$= \int_0^h \frac{b^2}{2h^2}(2h^2y - 3hy^2 + y^3)\, dy$$

$$= \frac{b^2}{2h^2}\left[\frac{2h^2y^2}{2} - \frac{3hy^3}{3} + \frac{y^4}{4}\right]_0^h = \frac{b^2}{2h^2}\frac{h^4}{4} = \frac{b^2h^2}{8}$$ ■

Transfer of axes for product of inertia In Fig. 2.26, G is the centroid of the area shown and P_{XY} is the product of inertia with respect to the axes X, Y through G. The product of inertia of this area with reference to axes X', Y' through point O, such that the coordinates of G (with respect to axes X', Y') are $(\bar{x}, \bar{y})$, is given by

$$P_{X'Y'} = \int_A (x + \bar{x})(y + \bar{y})\, dA$$

$$= \int_A xy\, dA + \int_A x\bar{y}\, dA + \int_A \bar{x} y\, dA + \int_A \bar{x}\,\bar{y}\, dA$$

Note that $(\bar{x}, \bar{y})$ are constants, being the coordinates of G. $\int_A \bar{y} x\, dA$ and $\int_A \bar{x} y\, dA$ are equal to zero, since they are the moments of the area about its centroid. $\int_A \bar{x}\,\bar{y}$, dA is $A\,\bar{x}\,\bar{y}$ and $\int_A xy\, dA = P_{XY}$. Therefore,

$$P_{X'Y'} = P_{XY} + A\,\bar{x}\,\bar{y}$$

That is, *the product of inertia of an area with respect to a parallel set of axes is equal to the sum of the product of inertia with respect to parallel axes through the centroid and the product of the area and coordinates of the centroid with respect to the new set of axes*. This result is useful in finding the product of inertia of composite areas.

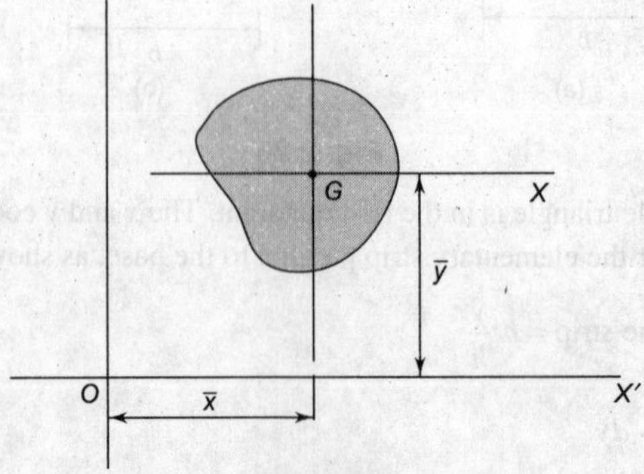

Fig. 2.26

Example 2.17 PI of an angle section

Find the product of inertia of the equal angle section shown, about the axes passing through its centroid and the axes passing through its edges (Fig. 2.27).

Solution We first determine the coordinates of the centroid of the section. $\bar{x} = \bar{y}$ as the section is an equal angle section.

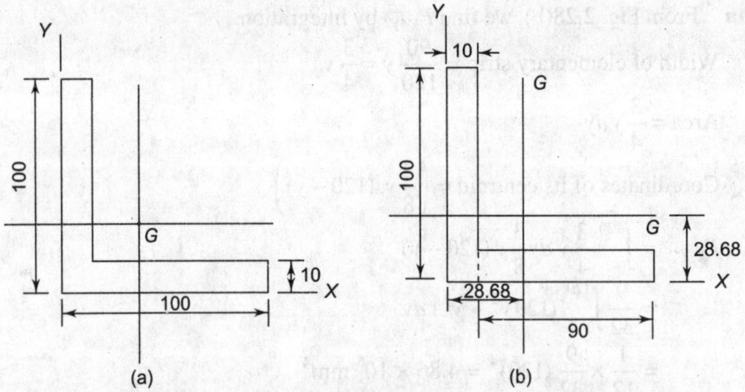

Fig. 2.27

$$(100 \times 10 + 90 \times 10)\ \bar{x} = 100 \times 10 \times 5 + 90 \times 10 \times 55,\ \bar{x} = 28.68 \text{ mm}$$

$$(100 \times 10 + 90 \times 10)\ \bar{y} = 100 \times 10 \times 50 + 90 \times 10 \times 5,\ \bar{y} = 28.68 \text{ mm}$$

The section consists of two rectangles, 100 × 10 and 90 × 10. The centroidal axes of both the rectangles are symmetrical and the product of inertia about their own centroidal axes is zero.

From Fig. 2.27(b),

$$P_{XY} = A_1 \bar{x}\,\bar{y} + A_2 \bar{x}\,\bar{y}$$

$$= 100 \times 10 \times 5 \times 50 + 90 \times 10 \times 55 \times 5$$

$$= 497.5 \times 10^3 \text{ mm}^4$$

$$P_{GG} = 100 \times 10 \times 23.68 \times (-21.32) + 900 \times 23.68 \times (-26.32)$$

$$= -1065.7 \times 10^3 \text{ mm}^4$$

Example 2.18 PI of a right-angled triangle

Find the product of inertia of the right-angled triangle shown in Fig. 2.28(a) about the X- and Y-axes passing through its centroid, and about the X'- and Y'-axes passing through its sides.

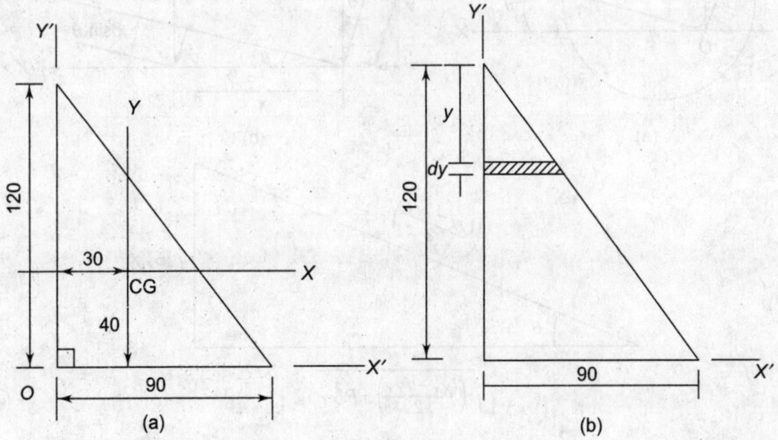

Fig. 2.28

Solution From Fig. 2.28(b), we find $P_{X'Y'}$ by integration.

Width of elementary strip $= \frac{90}{120} y = \frac{3}{4} y$

Area $= \frac{3}{4} y\,dy$

Coordinates of its centroid $= \left(\frac{3}{8} y, (120 - y)\right)$

$$P_{X'Y'} = \int_0^{120} \frac{3}{4} y\,dy \frac{3}{8} y\,(120 - y)$$
$$= \frac{9}{32} \int_0^{120} (120 y^2 - y^3)\,dy$$
$$= \frac{1}{12} \times \frac{9}{32} (120)^4 = 4.86 \times 10^6 \text{ mm}^4$$

P_{XY} (for centroidal axes) can be found from the transfer formula:

$$P_{XY'} = P_{XY} + A\,\bar{x}\,\bar{y}$$
$$P_{XY'} = 4.86 \times 10^6, \quad A = \frac{120 \times 90}{2} = 5400, \quad \bar{x} = 30, \quad \bar{y} = 40$$
$$P_{XY} = 4.86 \times 10^6 - 5400 \times 30 \times 40$$
$$= -1.62 \times 10^6 \text{ mm}^4$$

2.7 PRINCIPAL AXES FOR MI

We have seen that an area has a moment of inertia and a product of inertia about the axes passing through any point. Considering the area shown in Fig. 2.29(a), I_{XX}, I_{YY}, and P_{XY} can be calculated about the axes X-X and Y-Y passing through any point O.

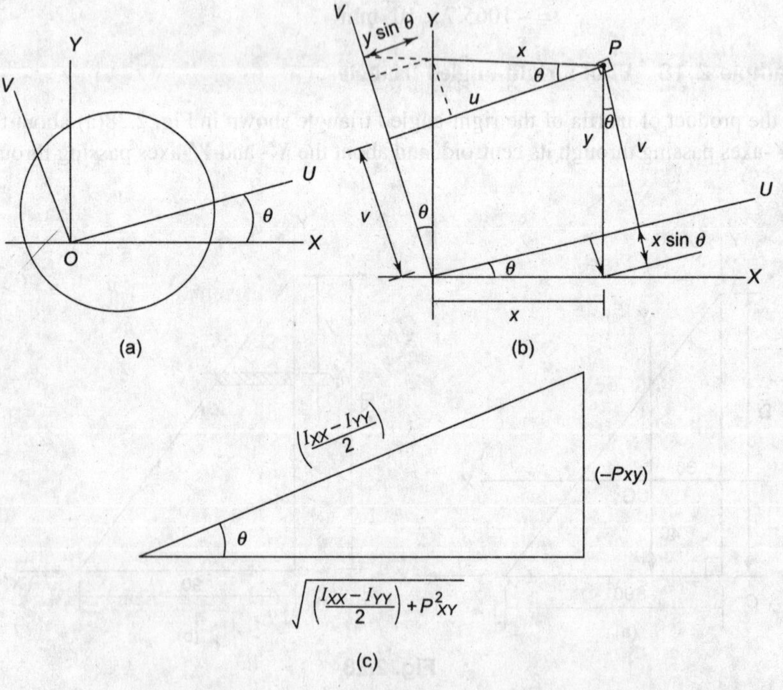

Fig. 2.29

If we consider any two mutually perpendicular axes U and V which are inclined at an angle θ to the axes X and Y and passing through O, it is possible to derive a relationship between I_{UU}, I_{VV}, and P_{UV} and the quantities I_{XX}, I_{YY}, and P_{XY}. Such a relationship has some important applications in structural mechanics.

From Fig. 2.29(b), we observe that point P has coordinates (x, y) with respect to the X, Y axes. The coordinates of point P with respect to axes U and V are u and v such that

$$v = y\cos\theta - x\sin\theta$$

$$u = y\sin\theta + x\cos\theta$$

I_{UU} and I_{VV} can be defined by the integrals $\int u^2 dA$ and $\int v^2 dA$. Thus,

$$I_{VV} = \int v^2 dA = \int (y\cos\theta - x\sin\theta)^2\, dA$$

$$= \int y^2\cos^2\theta\, dA + \int x^2\sin^2\theta\, dA - \int 2xy\sin\theta\cos\theta\, dA$$

$$\int y^2\cos^2\theta\, dA = I_{XX}\cos^2\theta \int x^2\sin^2\theta\, dA = I_{YY}\sin^2\theta$$

and

$$\int 2xy\sin\theta\cos\theta\, dA = P_{XY}\sin\theta\cos\theta$$

$$I_{VV} = I_{XX}\cos^2\theta + I_{YY}\sin^2\theta - 2P_{XY}\sin\theta\cos\theta$$

Similarly,

$$I_{UU} = \int u^2\, dA = \int (y\sin\theta + x\cos\theta)^2\, dA$$

$$= \int y^2\sin^2\theta\, dA + \int x^2\cos^2\theta\, dA + \int 2xy\sin\theta\cos\theta\, dA$$

$$= I_{XX}\sin^2\theta + I_{YY}\cos^2\theta + 2P_{XY}\sin\theta\cos\theta$$

Since $\cos^2\theta = \left(\frac{1+\cos 2\theta}{2}\right)$, $\sin^2\theta = \left(\frac{1-\cos 2\theta}{2}\right)$, and $2\sin\theta\cos\theta = \sin 2\theta$

$$I_{VV} = I_{XX}\left(\frac{1+\cos 2\theta}{2}\right) + I_{YY}\left(\frac{1-\cos 2\theta}{2}\right) - P_{XY}\sin 2\theta$$

$$= \frac{I_{XX}+I_{YY}}{2} + \frac{I_{XX}-I_{YY}}{2}\cos 2\theta - P_{XY}\sin 2\theta$$

$$I_{UU} = I_{XX}\left(\frac{1-\cos 2\theta}{2}\right) + I_{YY}\left(\frac{1+\cos 2\theta}{2}\right) + P_{XY}\sin 2\theta$$

$$= \left(\frac{I_{XX}+I_{YY}}{2}\right) - \left(\frac{I_{XX}-I_{YY}}{2}\right)\cos 2\theta + P_{XY}\sin 2\theta$$

Product of inertia $P_{UV} = \int uv\, dA$

$$= \int (y\cos\theta - x\sin\theta)(y\sin\theta + x\cos\theta)\, dA$$

$$= \int (y^2\sin\theta + \cos\theta + xy\cos^2\theta - xy\sin^2\theta - x^2\sin\theta\cos\theta)\, dA$$

$$= I_{XX}\sin\theta\cos\theta + P_{XY}\cos^2\theta - P_{XY}\sin^2\theta - I_{YY}\sin\theta\cos\theta$$

$$= \frac{I_{XX}-I_{YY}}{2}\sin 2\theta + P_{XY}\cos 2\theta \quad [\because \cos^2\theta - \sin^2\theta = \cos 2\theta]$$

As the angle θ changes, we get different orientations of the U- and V-axis. I_{UU}, I_{VV}, and P_{UV} change with values of θ. To get the maximum values of I_{UU} and I_{VV}, we differentiate with respect to θ and equate the results to zero to find the value of θ, and then substitute this value of θ in the expressions for I_{UU} and I_{VV}. $d(I_{UU})/d\theta = 0$ gives

$$\frac{I_{XX} - I_{YY}}{2}(-2\sin\theta) + P_{XY}(-2\cos 2\theta) = 0$$

which gives

$$\frac{\sin 2\theta}{\cos 2\theta} = \frac{-P_{XY}}{\left(\frac{I_{XX} - I_{YY}}{2}\right)}$$

$d(I_{VV})/d\theta = 0$ gives

$$\frac{I_{XX} - I_{YY}}{2} \times 2\sin 2\theta + P_{XY} \times 2\cos 2\theta = 0$$

Please note that the differentiation of I_{UU} and I_{VV} with respect to θ gives an expression which is the value of P_{XY}. This means that when the maximum or minimum value of I_{UU} and I_{VV} is obtained, P_{XY} is zero.

For the value of θ, we have the expression

$$\tan\theta = \frac{-P_{XY}}{\left(\frac{I_{XX} - I_{YY}}{2}\right)}$$

From Fig. 2.29(c), we get

$$\sin 2\theta = \frac{-P_{XY}}{\sqrt{\left(\frac{I_{XX} - I_{YY}}{2}\right)^2 + P_{XY}^2}}$$

$$\cos 2\theta = \frac{\left(\frac{I_{XX} - I_{YY}}{2}\right)}{\sqrt{\left(\frac{I_{XX} - I_{YY}}{2}\right)^2 + P_{XY}^2}}$$

These values can be substituted in the expressions for I_{VV} and I_{UU}. Substituting,

$$I_{VV} = \frac{I_{XX} + I_{YY}}{2} + \frac{\left(\frac{I_{XX} - I_{YY}}{2}\right)\left(\frac{I_{XX} - I_{YY}}{2}\right)}{\sqrt{\left(\frac{I_{XX} - I_{YY}}{2}\right)^2 + P_{XY}^2}} + \frac{P_{XY}(-P_{XY})}{\sqrt{\left(\frac{I_{XX} - I_{YY}}{2}\right)^2 + P_{XY}^2}}$$

$$= \frac{I_{XX} + I_{YY}}{2} - \sqrt{\left(\frac{I_{XX} - I_{YY}}{2}\right)^2 + P_{XY}^2}$$

$$I_{UU} = \frac{I_{XX} + I_{YY}}{2} - \frac{\left(\frac{I_{XX} - I_{YY}}{2}\right)\left(\frac{I_{XX} - I_{YY}}{2}\right)}{\sqrt{\left(\frac{I_{XX} - I_{YY}}{2}\right)^2 + P_{XY}^2}} + \frac{P_{XY}(-P_{XY})}{\sqrt{\left(\frac{I_{XX} - I_{YY}}{2}\right)^2 + P_{XY}^2}}$$

$$= \frac{I_{XX} + I_{YY}}{2} + \sqrt{\left(\frac{I_{XX} - I_{YY}}{2}\right)^2 + P_{XY}^2}$$

The maximum (or minimum) values of I are known as principal moments of inertia and the axes are known as principal axes. Also note that

$$I_{UU} + I_{VV} = I_{XX} + I_{YY} = J, \text{ the polar MI}$$

The principal moment of inertia can be expressed as

$$I_{1,2} = \frac{I_{XX} + I_{YY}}{2} \pm \sqrt{\left(\frac{I_{XX} - I_{YY}}{2}\right)^2 + P_{XY}^2}$$

The orientation of the principal axes is given by

$$\tan 2\theta = \frac{-P_{XY}}{\left(\frac{I_{XX} - I_{YY}}{2}\right)}$$

The product of inertia P_{XY} is zero about the principal axes.

The following points can easily be seen from the equations derived above.

(i) $I_{UU} + I_{VV} = I_{XX} + I_{YY} =$ constant (this constant is, of course, the polar MI J).

(ii) There will be a set of axes in any area, through its centroid, which are the principal axes. The MI is maximum about one of these and minimum about the other.

(iii) The product of inertia is zero about the principal axes.

(iv) Axes of symmetry of an area are principal axes. (The principal axes need not be the axes of symmetry.)

Example 2.19 Principal axes and principal MI of a triangle

For the right-angled triangle shown in Fig. 2.30(a), determine the principal axes through the centroid and the MI about these axes.

Solution Let X-X and Y-Y be two mutually perpendicular axes through the centroid. From Fig. 2.30(b), considering the elementary strip shown,

$$I_{XX} = \int_0^{120} \frac{3}{4} y\,dy\,(80 - y)^2 = \frac{3}{4}\int_0^{120} y\,(6400 + y^2 - 160y)dy$$

$$= \frac{3}{4}\left[\frac{6400y^2}{2} + \frac{y^4}{4} - \frac{160y^3}{3}\right]_0^{120}$$

$$= \frac{3}{4} \times 120 \times 120\left[3200 + \frac{120 \times 120}{4} - \frac{160 \times 120}{3}\right]$$

$$= 4.32 \times 10^6 \text{ mm}^4$$

$$I_{YY} = \int_0^{90} \frac{4}{3} x\,dx\,(60 - x)^2 = \frac{4}{3}\int_0^{90} x\,(3600 + x^2 - 120x)\,dx$$

$$= \frac{4}{3}\left[\frac{3600x^2}{2} + \frac{x^4}{4} - \frac{120x^3}{3}\right]_0^{90}$$

$$= \frac{4}{3} \times 90 \times 90 \left[1800 + \frac{90 \times 90}{4} - 40 \times 90\right]$$

$$= 2.43 \times 10^6 \text{ mm}^4$$

$P_{XY} = -1.62 \times 10^6$ mm^4, as already calculated in Example 2.18.

The MI about principal axes are $I_{1,2}$ and are given by

$$I_{1,2} = \frac{I_{XX} + I_{YY}}{2} \pm \sqrt{\left(\frac{I_{XX} - I_{YY}}{2}\right)^2 + P_{XY}^2}$$

$$= \frac{4.32 \times 10^6 + 2.43 \times 10^6}{2}$$

$$\pm \sqrt{\left(\frac{4.32 \times 10^6 - 2.43 \times 10^6}{2}\right)^2 + (-1.62 \times 10^6)^2}$$

$$= 5.25 \times 10^6 \text{ mm}^4 \quad \text{or} \quad 1.5 \times 10^6 \text{ mm}^4$$

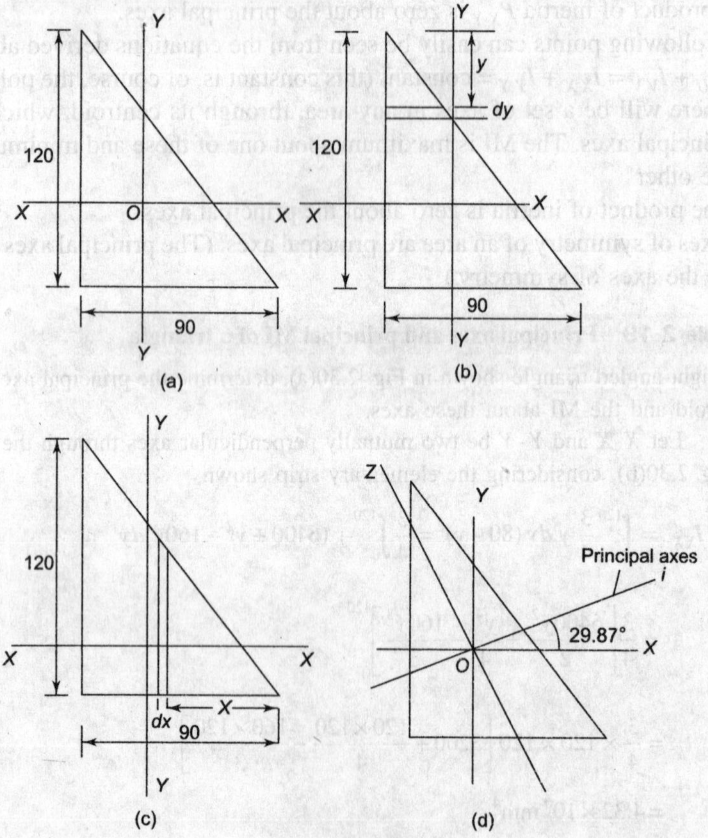

Fig. 2.30

The inclination of the principal axes is given by

$$\tan 2\alpha = \frac{-P_{XY}}{\left(\frac{I_X - I_Y}{2}\right)} = \frac{1.62 \times 10^6}{\left(\frac{4.32 - 2.43}{2}\right) \times 10^6} = 1.714$$

$$2\alpha = 59.74° \Rightarrow \alpha = 29.87°$$

The principal axes are shown in Fig. 2.30(d).

Example 2.20 Principal axes and principal MI for an angle section

Calculate the moment of inertia about the principal axes through the centroid of the angle section shown in Fig. 2.31(a).

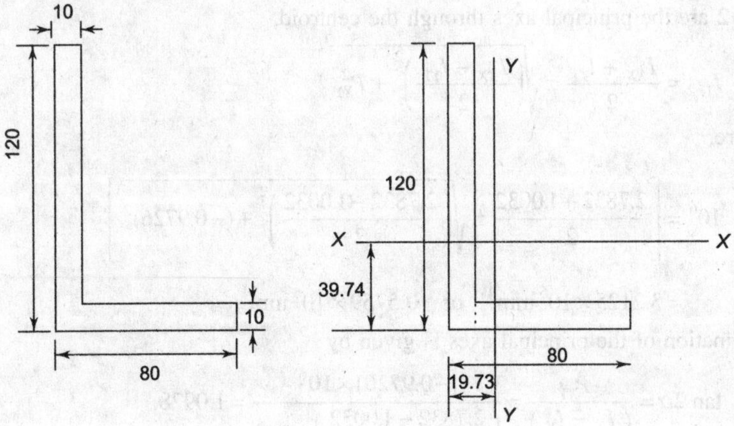

Fig. 2.31

Solution We first locate the position of the centroid:

$$(120 \times 10 + 70 \times 10)\bar{x} = 120 \times 10 \times 5 + 70 \times 10 \times 45$$

$$\bar{x} = 19.73 \text{ mm}$$

$$(120 \times 10 + 70 \times 10)\bar{y} = 120 \times 10 \times 60 + 70 \times 10 \times 5$$

$$\bar{y} = 39.74 \text{ mm}$$

I_{XX}, I_{YY}, and P_{XY} can be calculated from Fig. 2.31(b):

$$I_{XX} = \frac{10 \times 120^3}{12} + 1200\,(60 - 39.74)^2 + \frac{70 \times 10^3}{12} + 700 \times (39.74 - 5)^2$$

$$= 2.7832 \times 10^6 \text{ mm}^4$$

$$I_{YY} = \frac{120\times10^3}{12} + 1200\times14.73^2 + \frac{10\times70^3}{12} + 700\times25.27^2$$

$$= 1.0032\times10^6\ \text{mm}^4$$

$$P_{XY} = 1200\times14.73\times(-20.26) + 700\times34.74\times(-25.27)$$

$$= -0.9726\times10^6\ \text{mm}^4$$

If 1 and 2 are the principal axes through the centroid,

$$I_{1,2} = \frac{I_{XX}+I_{YY}}{2} \pm \sqrt{\left(\frac{I_{XX}-I_{YY}}{2}\right)^2 + P_{XY}^2}$$

Therefore,

$$10^6 = \left[\frac{2.7832+1.0032}{2} \pm \sqrt{\left(\frac{2.7832-1.0032}{2}\right)^2 + (-0.9726)^2}\right]$$

$$= 3.2125\times10^6\ \text{mm}^4 \quad \text{or} \quad 0.5759\times10^6\ \text{mm}^4$$

The inclination of the principal axes is given by

$$\tan 2\alpha = \frac{-P_{XY}}{\left(\dfrac{I_X - I_Y}{2}\right)} = \frac{(-0.9726)\times10^6}{\left(\dfrac{2.7832-1.0032}{2}\right)\times10^6} = 1.0928$$

$2\alpha = 47.54°$, $\alpha = 23.77°$. The principal axes are shown in Fig. 2.31(c).

2.8 MOHR'S CIRCLE FOR MI

There is an elegant graphical construction to help determine the principal MIs and the orientation of the principal axes. The concept involved is very similar to that of the principal stresses discussed later, in Chapter 9. The graphical construction of Mohr's circle is outlined below. Given I_{XX}, I_{YY}, and P_{XY} about the X-X and Y-Y axes, we have to find the principal MI and the orientation of the principal axes.

1. Draw a horizontal line which is the I-axis and a line perpendicular to it which is the P-axis (Fig. 2.32). You have thus an I-P coordinate system. The intersection point O is the origin.
2. From the origin, mark I_{XX} and I_{YY} to some scale.
3. If OB represents I_{YY} and OA represents I_{XX}, then $AB = (I_{XX} - I_{YY})$.
4. Bisect AB to obtain the point C. $OC = (I_{XX} + I_{YY})/2$.
5. Draw perpendicular lines at B and A and mark D and E such that $BD = P_{XY} = AE$. The sign of P_{XY} is associated with I_{XX} and the opposite sign with I_{YY}.
6. With CD (or CE) as the radius and the centre at C, draw a circle.
7. Since $AC = (I_{XX} + I_{YY})/2$ and $AE = P_{XY}$, $CE = \{[(I_{XX} - I_{YY})/2]^2 + P_{XY}^2\}^{1/2}$, which is the radius of the circle. This is known as Mohr's circle.
8. Let the circle intersect the I-axis at points 1 and 2.

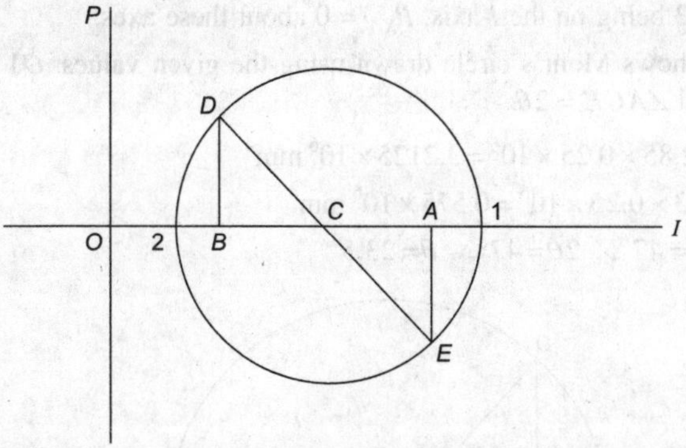

Fig. 2.32

9. The coordinates of points 1 and 2 give the principal moment of inertia. Note that $P_{XY} = 0$ at points 1 and 2. This statement can be easily proved:

$$O2 = OC - C2$$

$$= \frac{I_{XX} + I_{YY}}{2} - \sqrt{\left(\frac{I_{XX} - I_{YY}}{2}\right)^2 + P_{XY}^2}$$

as $OC = (I_{XX} + I_{YY})/2$ and $C2$ is the radius of Mohr's circle

$$= \sqrt{\left(\frac{I_{XX} - I_{YY}}{2}\right)^2 + P_{XY}{}^2}$$

Similarly,

$O1 = OC + C1$, $C1$ being the radius of the circle

$$= \frac{I_{XX} + I_{YY}}{2} + \sqrt{\left(\frac{I_{XX} - I_{YY}}{2}\right)^2 + P_{XY}^{\ 2}}$$

10. From the triangle CAE, $\tan 2\theta = \dfrac{-P_{XY}}{[(I_{XX} - I_{YY})/2]}$.
11. It can also be proved that the coordinates of any point in the circle represent the I and P about some axis. The diametrically opposite point gives I and P about an axis perpendicular to it. We will discuss these concepts in Chapter 9 in greater detail. With reference to principal stresses, which are applicable to principal moments of inertia as well.

The following example illustrates the procedure.

Example 2.21 Principal axes and principal MI for an angle section using Mohr's circle

For the angle section of Example 2.20, find the principal axes and inertias graphically using Mohr's circle.

Solution The scale chosen is 1 cm = 0.25×10^6 mm^4.

OI and OP are perpendicular axes.

$I_{XX} = 2.7832 \times 10^6$ mm^4, $I_{YY} = 1.0032 \times 10^6$ mm^4, $P_{XY} = -0.9726 \times 10^6$ mm^4.

In Mohr's circle, $OA = I_{XX}$, $OB = I_{YY}$, $AE = BD = P_{XY}$.

Join BD intersecting the I-axis at C. With C as the centre and CD as the radius, draw a circle intersecting the I-axis at points 1 and 2.

Points 1 and 2 being on the I-axis, $P_{XY} = 0$ about these axes.

Figure 2.33 shows Mohr's circle drawn using the given values. $O1$ and $O2$ give the principal MI, and $\angle ACE = 2\theta$.

$$O1 = 12.85 \times 0.25 \times 10^6 = 3.2125 \times 10^6 \text{ mm}^4$$

$$O2 = 2.3 \times 0.25 \times 10^6 = 0.575 \times 10^6 \text{ mm}^4$$

$$\angle ACE = 47°, \quad 2\theta = 47°, \quad \theta = 23.5°$$

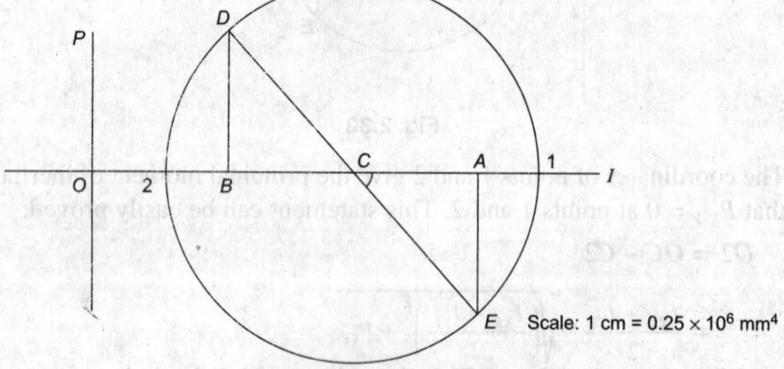

Fig. 2.33

2.9 GRAPHICAL CONSTRUCTION TO FIND MOMENTS OF INERTIA

A simple graphical procedure is available to find the MIs of irregular areas.

Consider the area shown in Fig. 2.34, whose MI is to be found about axis X-X. Draw a line X'-X' parallel to X-X on the other side as shown in the figure.

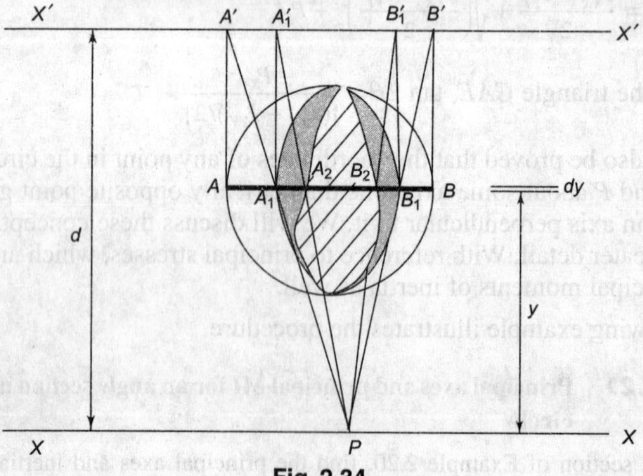

Fig. 2.34

Take a narrow strip AB of thickness dy in the area and project it on X'-X' as $A'B'$. Take a suitable point P on X-X and draw PA' and PB' such that they intersect AB at A_1 and B_1. The area of strip A_1B_1 can be proved to be the first moment of area of strip AB about P. By repeating the procedure with other strips, one can obtain the shaded figure. The area of this figure multiplied by d is equal to the moment of the area of the given figure about P.

Consider the strip A_1B_1 and repeat the procedure. Project A_1B_1 on X'-X' as $A_1'B_1'$ and draw PA_1' and PB_1'. Repeat the procedure with other strips of the shaded figure and obtain the hatched figure. The area of the hatched figure multiplied by d is the MI of the area about axis AB. This can be proved as follows.

$$\frac{A'B'}{d} = \frac{A_1B_1}{y}$$

because $A'B'P$ and A_1B_1P are similar triangles.

$$A'B'y = A_1B_1d$$

Since $A'B' = AB$,

$$A_1B_1d = ABy$$

Multiplying by dy,

$$A_1B_1d\,dy = ABy\,dy = (AB\,dy)y$$

where $AB\,dy$ is the area of the strip, which multiplied by y gives the moment of this area about P. A_1B_1dy is the area of the shaded strip, and this multiplied by d is thus equal to the first moment of area of the strip $AB\,dy$. This is true for all the elementary strips. The area of the shaded figure multiplied by d, the distance between X-X and X'-X', therefore, is equal to the first moment of area about P. The hatched figure is obtained by a similar procedure from the shaded figure. That is, the hatched figure is the first moment of the shaded figure about P. If A is the given area, A_1 is the area of the shaded figure, and A_2 is the area of the hatched figure, then

$A\bar{y} = A_1d$ as proved earlier

$\text{MI} = A\bar{y}\bar{y} = (A_1d)\bar{y}, \quad A_1\bar{y} = A_2d$ (on similar arguments)

$\text{MI} = A_2d \times d = A_2d^2$

The moment of inertia of the given area is equal to area A_2 multiplied by the square of distance d between X-X and X'-X'.

Summary

The moment of inertia (MI) or second moment of area is an important property of a section. It is calculated using the integral $\int y^2 dA$. The MIs of standard sections are listed in tables. The second moment of area is always a positive quantity.

The parallel axis theorem states that if I_{GG} is the second moment of area about an axis through the centroid, then the second moment of area about a parallel axis A-A is given by $I_{AA} = I_{GG} + Ad^2$, where d is the distance between the centroidal axis and the axis A-A.

The polar moment of inertia exists about an axis perpendicular to the area. $I_{ZZ} = I_{XX} + I_{YY}$, where X-X and Y-Y are two mutually perpendicular axes in the plane of the area and Z-Z is an axis perpendicular to the plane and passing through the intersection of the axes X-X and Y-Y.

The section modulus of a section is the moment of inertia divided by the distances to the extreme fibres of the section. Section modulus is useful in the design of beams.

The product of inertia $P_{XY} = \int xy\, dA$. It can be positive, negative, or zero. $P_{XY} = 0$ if either of the axes X-X or Y-Y is an axis of symmetry.

If X-X and Y-Y are the axes through the centroid of the area and X'-X' and Y'-Y' are another set of parallel axes in the plane of the area, then $P_{X'Y'} = P_{XY} + A\bar{x}\bar{y}$, where $\bar{x}, \bar{y}$ are the coordinates of the centroid of the area with respect to axes X', Y' and A is the area. This formula is used for calculating the product of inertia (PI) through the axes translated parallel to the centroidal axes.

If the axes are rotated by an angle θ, then the MI and PI about the rotated axes, U-U and V-V, can be found from

$$I_{UU} = \frac{I_{XX} + I_{YY}}{2} + \frac{I_{XX} - I_{YY}}{2} \cos 2\theta + P_{XY} \sin 2\theta$$

$$I_{VV} = \frac{I_{XX} + I_{YY}}{2} + \frac{I_{XX} - I_{YY}}{2} \cos 2\theta - P_{XY} \sin 2\theta$$

$$P_{UV} = \frac{I_{XX} - I_{YY}}{2} \sin 2\theta + P_{XY} \cos 2\theta$$

The principal axes of a section are those axes about which I_{UU} and I_{VV} are a maximum or minimum.

$$I_{(\text{max/min})} = \frac{I_{XX} + I_{YY}}{2} \pm \sqrt{\left(\frac{I_{XX} - I_{YY}}{2}\right)^2 + P_{XY}^2}$$

The angle θ for such a case is given by

$$\tan 2\theta = \frac{P_{XY}}{\left(\frac{I_{XX} - I_{YY}}{2}\right)}$$

The product of inertia about the principal axis is zero. The axis of symmetry of an area is the principal axis. The MI of an area can also be found by the graphical method.

Exercises

Review Questions

1. Can the MIs of an area about any axis in its plane be zero? Explain your answer.
2. Of the two equal triangles shown in Fig. 2.35, which has a greater MI about the Y-axis? Are their MIs about the X-axis equal?

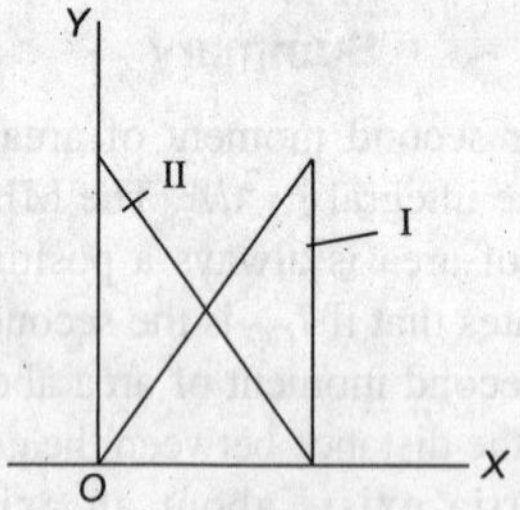

Fig. 2.35

3. The strength of a beam section is directly proportional to the MI of the section about its centroidal, horizontal axis (Fig. 2.36). Which of the two sections, of equal area, is preferable as a beam section? Why?

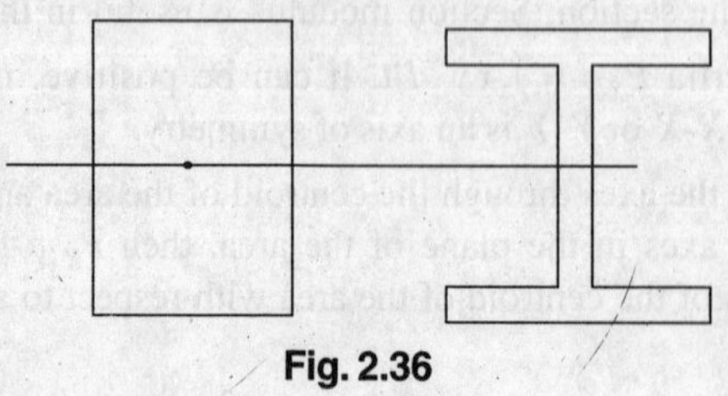

Fig. 2.36

4. Can the radius of gyration of any area be zero?
5. For which of the sections shown in Fig. 2.37, the product of inertia is not zero about the X-X and Y-Y axes?

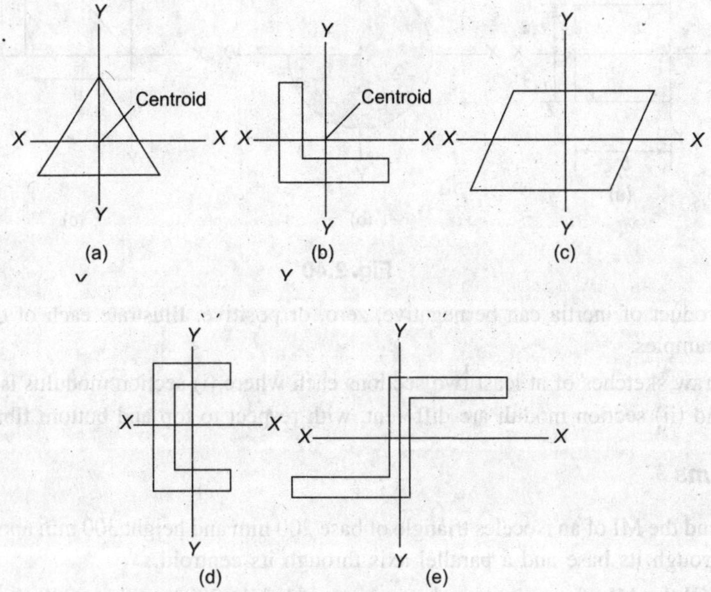

Fig. 2.37

6. In each case shown in Fig. 2.38, state, without calculation, whether the product of inertia is zero, negative, or positive.

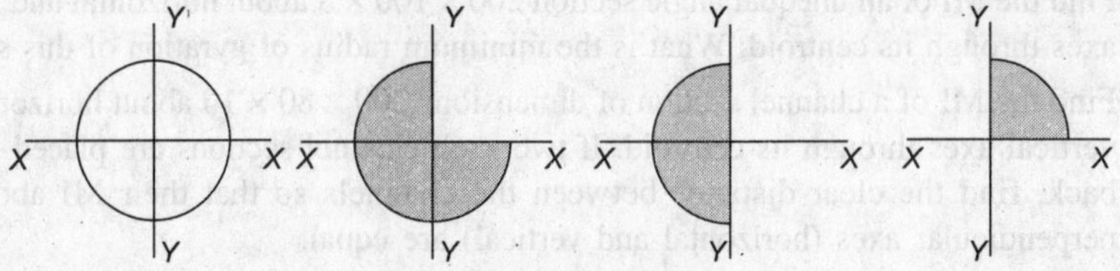

Fig. 2.38

7. In the shapes shown in Fig. 2.39, what are the MIs about the X-X axis?

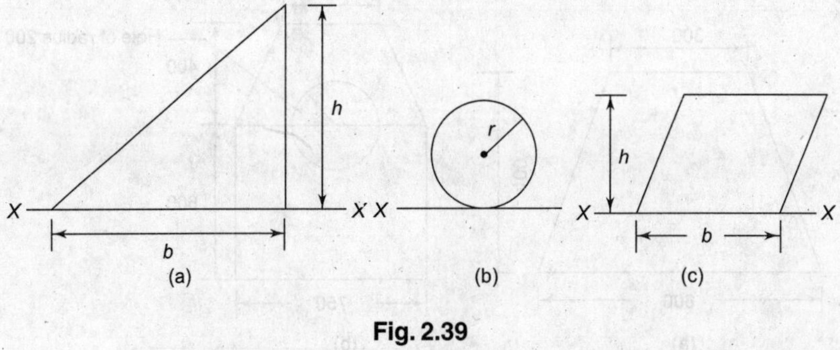

Fig. 2.39

8. In the cases shown in Fig. 2.40, about which axis the MI will be more—X-X or Y-Y? Why?

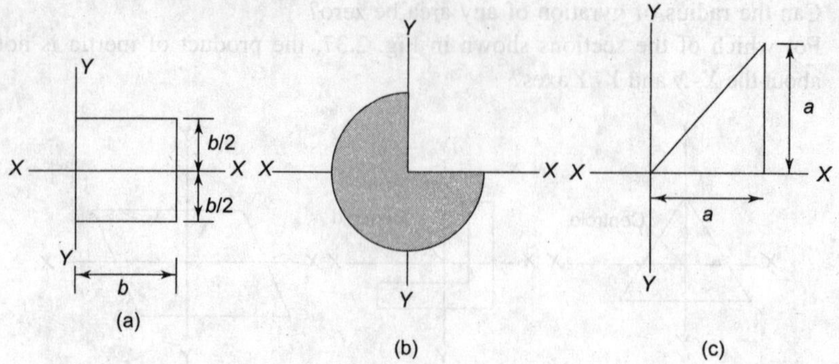

Fig. 2.40

9. Product of inertia can be negative, zero, or positive. Illustrate each of these with examples.
10. Draw sketches of at least two sections each where (i) section modulus is the same and (ii) section moduli are different, with respect to top and bottom fibres.

Problems

1. Find the MI of an isoceles triangle of base 200 mm and height 300 mm about an axis through its base and a parallel axis through its centroid.
2. Find the MI of a symmetrical trapezium of height 200 mm, base 200 mm, and top face 100 mm about an axis through its centroid.
3. Find the MI about the centroidal axes of (i) a hexagon of side 150 mm and (ii) an octagon of side 200 mm.
4. Find the MI of an unequal angle section $200 \times 100 \times 8$ about horizontal and vertical axes through its centroid. What is the minimum radius of gyration of this section?
5. Find the MI of a channel section of dimensions $200 \times 80 \times 10$ about horizontal and vertical axes through its centroid. If two such channel sections are placed back to back, find the clear distance between the channels so that their MI about two perpendicular axes (horizontal and vertical) are equal.
6. Find the MIs of the three sections shown in Fig. 2.41 about a horizontal axis through their centroids.

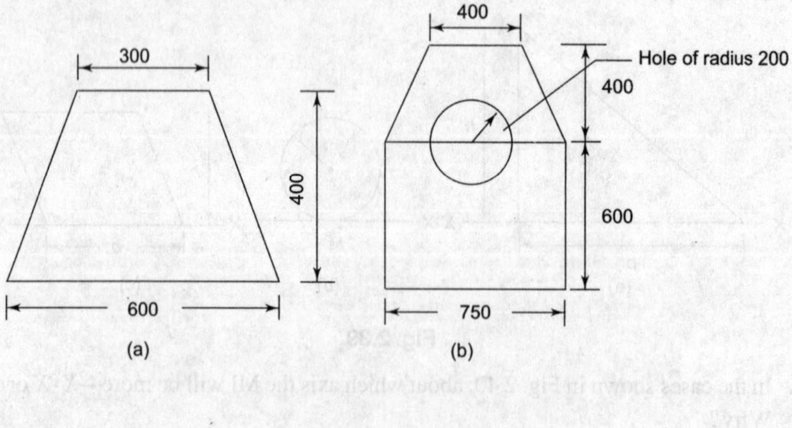

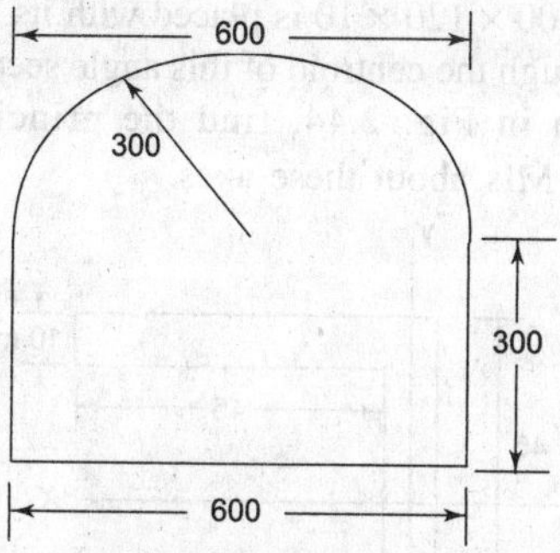

(c)

Fig. 2.41

7. Which of the following sections will have the largest radius of gyration, about an axis parallel to the base through the centroid or about a diameter, if they have the same area?
(i) A triangle of equal sides, (ii) a square, (iii) a circle, and (iv) a tube of thickness 0.05 times its outer diameter.
8. Find the MI about the X-X and Y-Y axes of the areas bound by the curves shown in Fig. 2.42.

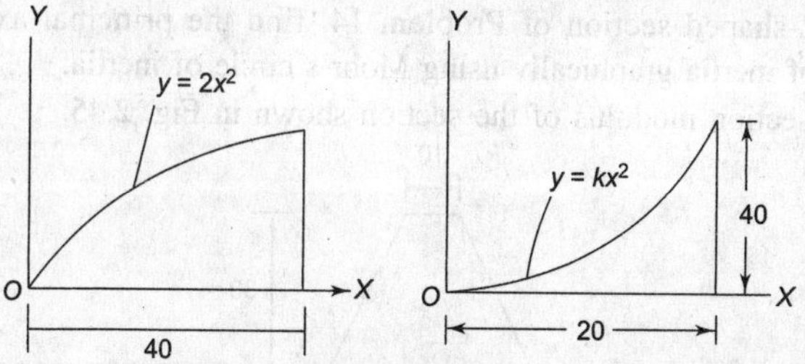

Fig. 2.42

9. Find the product of inertia of a trapezium if one of its sides is vertical and 200 mm long and the two parallel sides at right angles to it are 100 mm and 300 mm about the axis passing through the vertical side and the 300 mm base.
10. Find the product of inertia of the two triangles shown in Fig. 2.43 about the X-X and Y-Y axes.

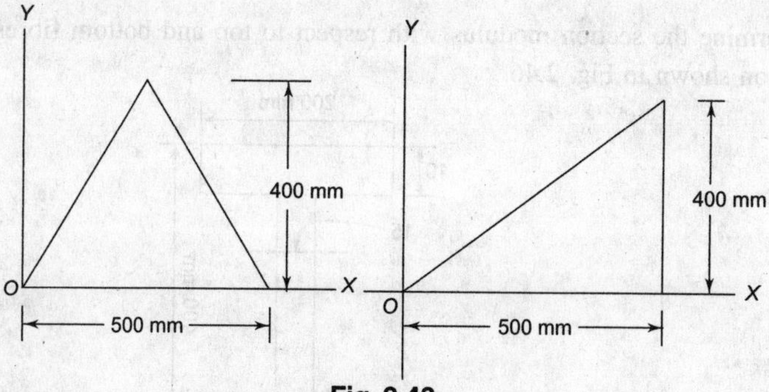

Fig. 2.43

11. Find the PI of a right-angled triangle about horizontal and vertical axes through its centroid. The base of the triangle is 250 mm long and its height is 300 mm.
12. Find the principal axes through the centroid of a right-angled triangle, of base 200 mm and height 300 mm. Find the principal MI about these axes.

13. An unequal angle section 300 × 120 × 10 is placed with its longer base horizontal. Find the principal axes through the centroid of this angle section, and principal MIs.
14. For the Z-section shown in Fig. 2.44, find the principal axes through its centroid and the principal MIs about these axes.

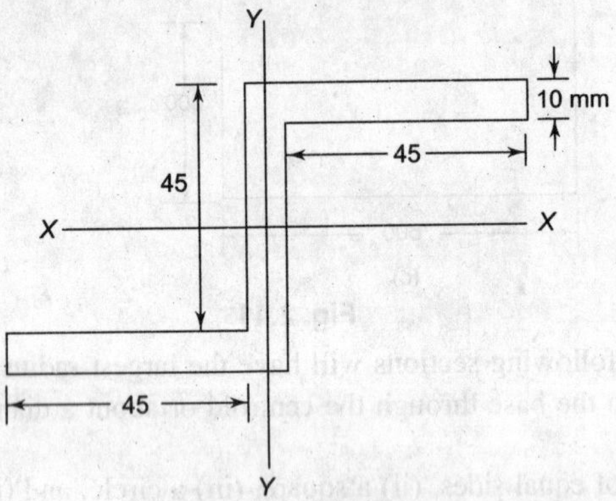

Fig. 2.44

15. For the angle section given in Problem 13, find the principal axes and inertias using Mohr's circle method.
16. For the Z-shaped section of Problem 14, find the principal axes and principal moment of inertia graphically using Mohr's circle of inertia.
17. Find the section modulus of the section shown in Fig. 2.45.

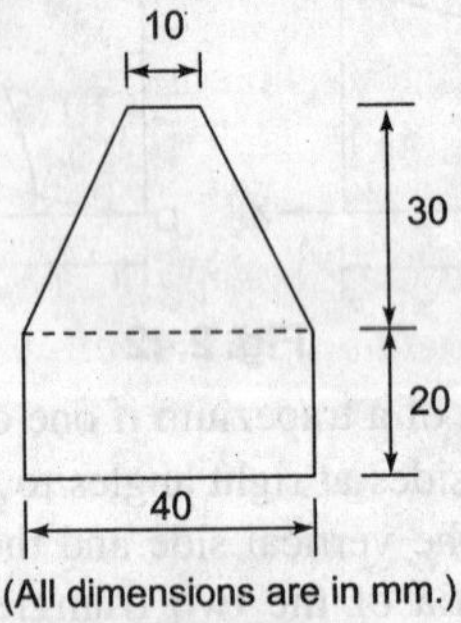

Fig. 2.45

18. Determine the section modulus with respect to top and bottom fibres for the T-section shown in Fig. 2.46.

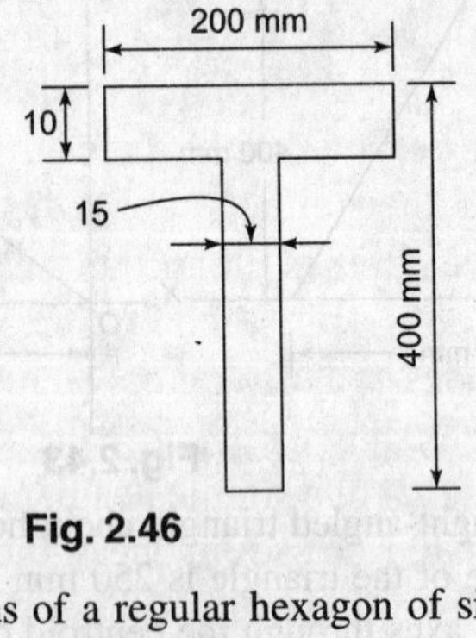

Fig. 2.46

19. Determine the section modulus of a regular hexagon of side 20 cm.
20. Compare the ratio of area to section modulus of a circular area of radius R and a hollow circular area of outer radius R and thickness $0.05R$.

CHAPTER 3

Simple Stresses and Strains

Learning Objectives

After going through this chapter, the reader will be able to

- classify stresses into various categories,
- state and apply Hooke's law,
- compute stress intensities caused by applied loads in simple and compound sections,
- compute stresses and strains due to temperature changes,
- calculate the reactive forces in simple indeterminate systems,
- derive and state the relations between elastic constants,
- define and describe the elastic properties of materials, and
- explain the terms stress concentration, residual stress, and fatigue.

3.1 INTRODUCTION

In Chapter 1, we discussed some basic concepts of mechanics. We considered the bodies to be rigid while discussing those concepts and principles. However, materials deform under the action of forces. In this chapter, we will discuss the basic nature of deformation of materials under load. From an engineering point of view, it is important to understand the mechanical properties of materials, i.e., the properties that describe the deformation of materials under the action of forces.

Tension, compression, and shear stresses and strains were briefly mentioned in Chapter 1. In actual practice, a structural element or a machine component is subjected to complex loading systems requiring stresses to be developed in all the principal directions X, Y, and Z, as shown in Fig. 3.1(a). Most structures or elements are three-dimensional; we consider them to be two-dimensional or one-dimensional systems where this does not significantly affect the behaviour of the system and leads to simpler computations.

Due to the presence of many applied loads and moments at a point in a stressed body, such stresses may exist in all the three directions, as in Fig. 3.1(b). By reducing the applied loads and their effects along the principal directions, we will have tensile or compressive stresses along the three principal directions at a point as shown in Fig. 3.1(c). σ_{XX}, σ_{YY}, and σ_{ZZ} are stresses along principal directions. Similarly, along the sides of a differential element, there may be tangential stresses in the three directions as shown (τ_{XY}, τ_{YZ}, etc.).

Some cases such as that of a simple beam, may be reduced to a two-dimensional system of forces. Then the stresses in one of the directions (say the Z-direction) are zero and you have a two-dimensional case [Fig. 3.1(d)]. The stress system at a point Q in the beam is shown in Fig. 3.1(d), where the Z-direction stresses are absent. Further, as in the case of a simple truss if we make certain assumptions, we can say that the members are subjected to simple tension or compression. Here we have a one-dimensional stress system, as shown in Fig. 3.1(e).

In this chapter, we will consider only uniaxial stresses and strains. We will study two-dimensional stress systems in Chapter 9.

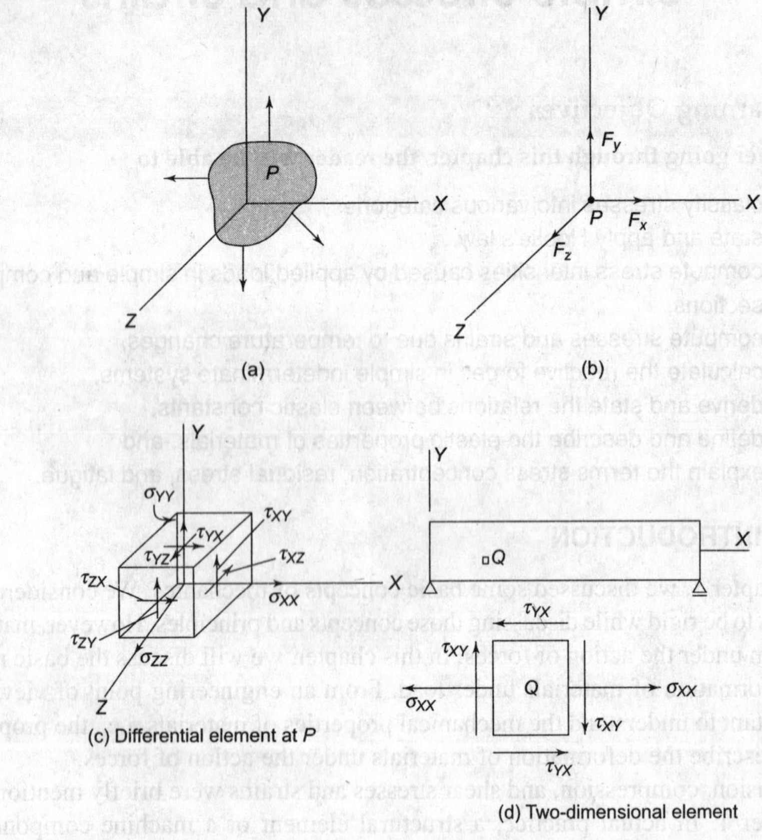

Fig. 3.1

3.2 STRESS AND STRAIN

Whenever a body is subjected to forces, the body deforms. The bodies in practice are not rigid. The body resists the deformation by developing stresses. The body reaches an equilibrium position when the internal stress resultant equals the external force. The deformation of the body and the stress can be related for any body depending upon its material properties.

3.2.1 Stress

Consider the effect of two axial forces *P* acting on a rod, as in Fig. 3.2, which tend to stretch or elongate the rod. It is a common experience that the rod increases in length due to these forces. The increase in length may be very evident if the rod is of, say, rubber, and may not be distinguishable if it is of steel, unless large forces are applied.

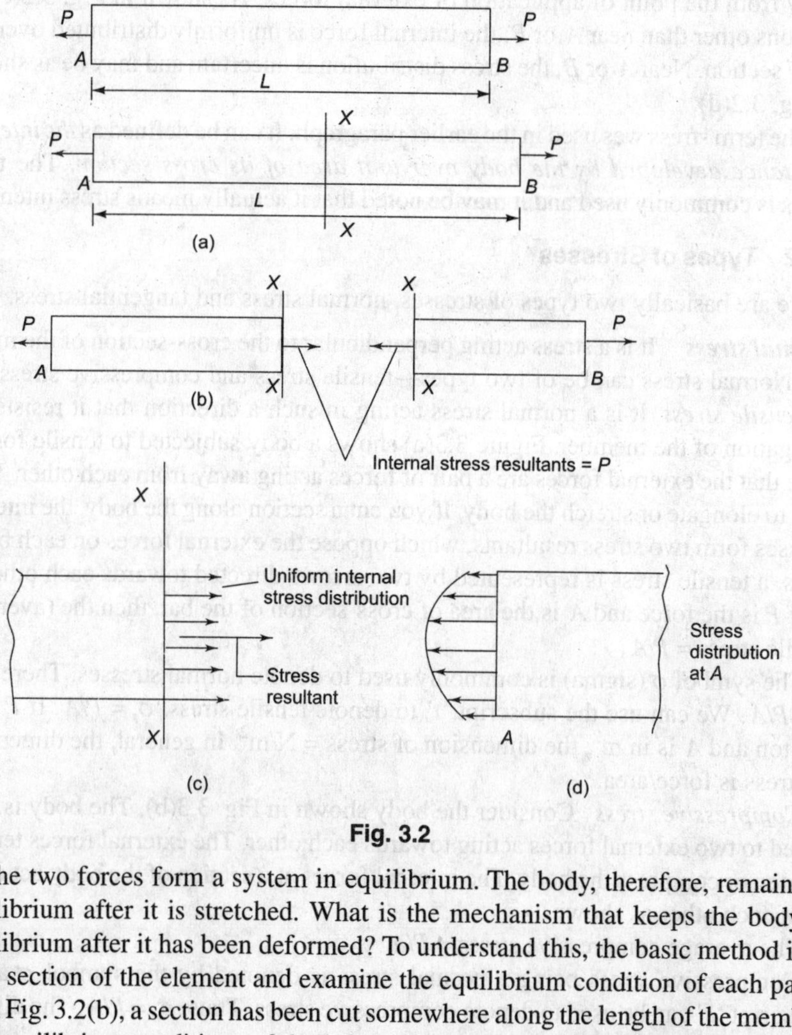

Fig. 3.2

The two forces form a system in equilibrium. The body, therefore, remains in equilibrium after it is stretched. What is the mechanism that keeps the body in equilibrium after it has been deformed? To understand this, the basic method is to cut a section of the element and examine the equilibrium condition of each part.

In Fig. 3.2(b), a section has been cut somewhere along the length of the member. The equilibrium conditions of the left and right parts require that forces equal to the external force act on the cut section. These forces are known as internal stress resultants.

A body tends to oppose the deformation due to the applied forces. While the applied forces have a deforming or straining action on the body, internal forces are developed to oppose such deforming action. The body thus remains in equilibrium in a strained state.

When the applied forces are removed, it is common experience that the body comes back to its original shape. Most engineering materials exhibit this property called elasticity even though there are those that do not come back to their original shape and size but remain in a deformed condition. These are discussed later in this chapter. Normally, the internal stress resultants become zero on removal of the applied forces, and in the process bring the body back to its original size.

How is the internal resisting force generated? According to St Venant's principle, the stress distribution is uniform over the cross section, at sections at a distance away from the point of application of external forces. As shown in Fig. 3.2(c), at sections other than near A or B, the internal force is uniformly distributed over the cross section. Near A or B, the stress distribution is uncertain and may be as shown in Fig. 3.2(d).

The term stress was used in the earlier paragraph. It can be defined as *the internal resistance developed by the body over unit area of its cross section.* The term stress is commonly used and it may be noted that it actually means stress intensity.

3.2.2 Types of Stresses

There are basically two types of stresses, normal stress and tangential stress.

Normal stress It is a stress acting perpendicular to the cross-section of the member. Normal stress can be of two types—tensile stress and compressive stress.

Tensile stress It is a normal stress acting in such a direction that it resists the elongation of the member. Figure 3.3(a) shows a body subjected to tensile forces. Note that the external forces are a pair of forces acting away from each other. They tend to elongate or stretch the body. If you cut a section along the body, the internal stresses form two stress resultants, which oppose the external forces on each body. Thus, a tensile stress is represented by two arrows directed towards each other.

If P is the force and A is the area of cross-section of the bar, then the (average) tensile stress = P/A.

The symbol σ (sigma) is commonly used to denote normal stresses. Therefore, $\sigma = P/A$. We can use the subscript 't' to denote tensile stress. $\sigma_t = P/A$. If P is in newton and A is in m^2, the dimension of stress = N/m^2. In general, the dimension of stress is force/area.

Compressive stress Consider the body shown in Fig. 3.3(b). The body is subjected to two external forces acting towards each other. The external forces tend to shorten or compress the body. The internal forces at a section of the body act away from each other as shown.

The average compressive stress = P/A

Compressive stress being a normal stress is denoted by the symbol σ and a subscript 'c' can be used to denote compressive stress. Thus, $\sigma_c = P/A$. The dimension of compressive stress is N/m^2.

Some normal stresses are known by special names. Consider a cylinder closed at the ends and subjected to internal pressure as shown in Fig. 3.3(c).

Longitudinal stress If we consider a section of the cylinder as shown in Fig. 3.3(c), the internal pressure acting on the ends causes a stretching of the fibres as shown. The forces acting on the ends require forces to be developed by stresses acting normal to the section as shown. These stresses are called *longitudinal stresses* as they act along the length of the cylinder. Note that they act normal to the section and are the same as normal stresses. The normal stress in this case has the same dimension as N/m^2 and is denoted by σ_l.

Hoop stress If we take a horizontal section of the cylinder, we see that equilibrium conditions require stresses as shown in Fig. 3.3(c). These are also normal stresses but as they act along the circumference of the cylinder, they are called *hoop* or *circumferential stress*. Hoop stress is denoted by σ_h. The dimension of the hoop stress is the same as that of other stresses, N/m^2.

Tangential stress Tangential or shear stress is shown in Fig. 3.3(d). If the external forces tend to shear the member along the cross section, internal forces developed act tangentially to the cross section. The average shear stress = P/A, where P is the shearing force and A is the area of cross section of the member. The unit of tangential stress is the same as all other stresses, N/m^2. Tangential stress is denoted by the symbol τ (tau). Therefore, $\tau = P/A$ (N/m^2)

In the case of all stresses, σ or τ = Force/Area = P/A. Therefore,

$$P = (\sigma \text{ or } \tau) \times \text{Area}$$

Units for stress The basic stress unit is N/m^2. This unit is also known as pascal (Pa). Depending upon the magnitude of the stress, we can use units like N/mm^2, N/m^2, kN/m^2, kN/mm^2, etc. Multiples of the basic unit (N/m^2 or Pa) can be used like megapascal (MPa), gigapascal (GPa), etc.

1 MPa = 10^6 Pa and 1 GPa = 10^9 Pa

Also note that MPa and N/mm^2 are numerically equal.

1 MPa = 10^6 N/m^2 = 10^6 N/10^6 mm^2 = 1 N/mm^2

The symbol σ (sigma) is commonly used to denote normal stresses. Therefore, $\sigma = P/A$. We can use the subscript 't' to denote tensile stress. $\sigma_t = P/A$. If P is in newton and A is in m^2, the dimension of stress = N/m^2. In general, the dimension of stress is force/area.

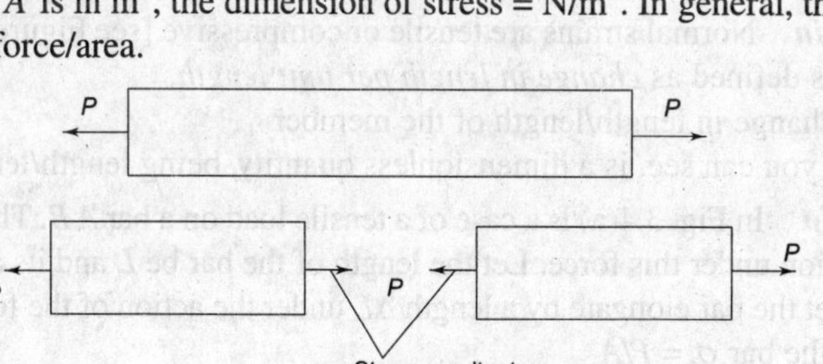

(a) Tensile stress

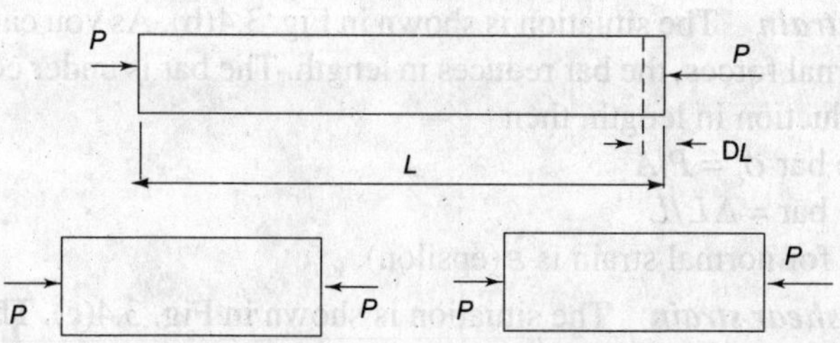

(b) Compressive stress

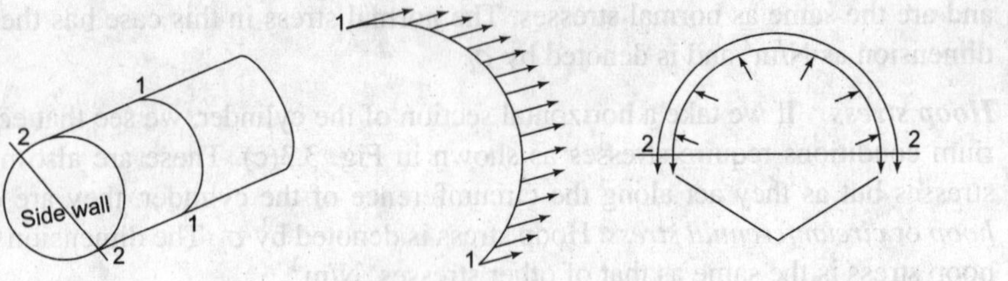

Cylinder under internal pressure | Longitudinal stress | Hoop stress

(c)

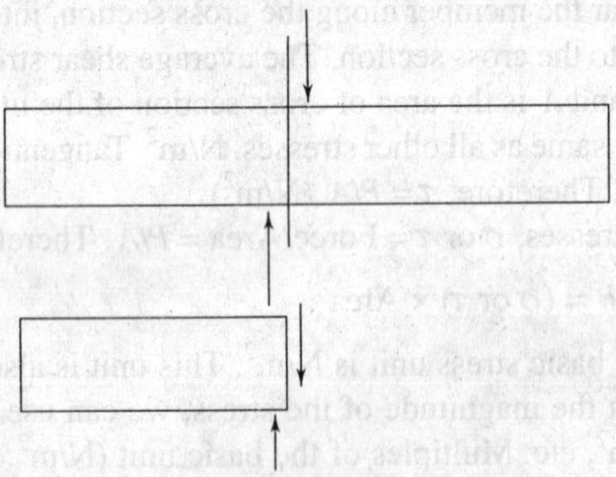

(d) Shearing stress resultant

Fig. 3.3

3.2.3 Strain

As the body is subjected to forces, the body deforms. Strain is a measure of the deformation of the body. Just as there are two types of stresses, normal and tangential, there are two types of strains, normal strain and tangential strain.

Normal strain Normal strains are tensile or compressive [see Figures 3.4(a) and (b)]. Strain is defined as *change in length per unit length.*

Strain = change in length/length of the member

Strain, as you can see, is a dimensionless quantity, being length/length.

Tensile strain In Fig. 3.4(a) is a case of a tensile load on a bar AB. The bar undergoes elongation under this force. Let the length of the bar be L and its area of cross section A. Let the bar elongate by a length ΔL under the action of the force P.

Stress in the bar $\sigma_t = P/A$

Strain in the bar = $\Delta L/L$

Compressive strain The situation is shown in Fig. 3.4(b). As you can easily see, due to the external forces, the bar reduces in length. The bar is under compression. If ΔL is the reduction in length, then

Stress in the bar $\sigma_c = P/A$

Strain in the bar = $\Delta L/L$

The symbol for normal strain is ε (epsilon).

Tangential or shear strain The situation is shown in Fig. 3.4(c). The force P is applied tangential to the top face of the block. The force tends to strain the bar as shown. The shear strain is the angle ϕ. From the triangle $\Delta ABB'$, we find that $\tan\phi = \Delta L/L$. As the shear deformations are generally small, $\tan\phi = \phi$. Thus, shear strain is given by $\phi = \Delta L/L$. Note that shear strain is a dimensionless quantity like other strains.

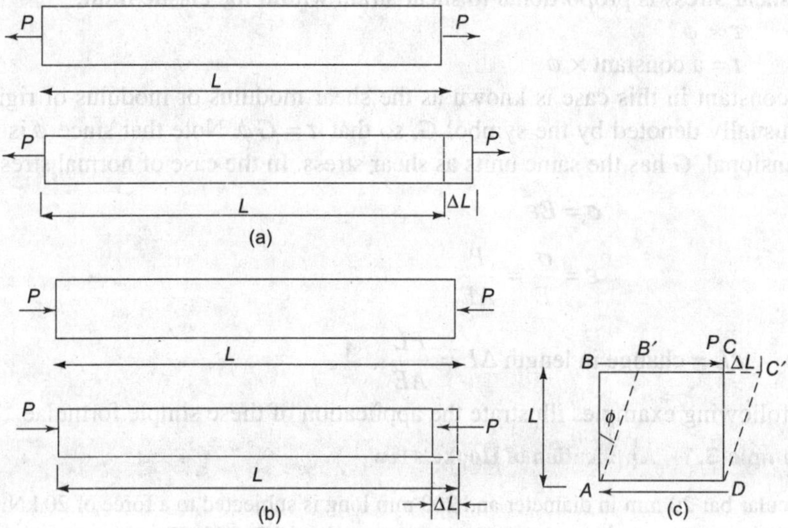

Fig. 3.4

3.2.4 Hooke's Law

An important relationship between stress and strain is given by Hooke's law, which states that stress is proportional to strain so long as the material behaves elastically. We have mentioned earlier that the property of a material to regain its shape on the removal of an applied force is known as *elasticity*. Some materials show elastic properties till they have attained considerable stress, while other materials exhibit such a property only for very small values of stress. The stress up to which the material behaves elastically is known as its *elastic limit*. Sir Robert Hooke first stated this property of a material in 1678 after noticing the linear relationship between applied forces and deformations. Thus, stress is proportional to strain within the elastic limit.

$$\text{Stress} \propto \text{strain}$$

$$\Rightarrow \quad \text{Stress} = \text{a constant} \times \text{strain}$$

In the case of normal stresses (tensile and compressive) this constant is denoted by the symbol E and is known as the *Young's modulus of elasticity*. Thus, $\sigma = E\varepsilon$ is the mathematical form of Hooke's law. E is a property of the material and may have a fairly constant value in some cases. But it can also vary depending upon the stress and may be different in compression and tension.

Note that since $\sigma = E\varepsilon$, and since ε is non-dimensional, the units of E must be the same as those of σ. Thus E can be expressed in N/mm^2, Pa, MPa, GPa etc.

In the case of shear stress and shear strain a similar relationship exists. Considering Fig. 3.4(c), shear strain is measured by the angular change between two lines which are originally at right angles as shown. It is given by

$$\phi = \frac{\Delta L}{L}$$

The shear stress is proportional to shear strain within the elastic limit.

$$\tau \propto \phi$$

$$\tau = \text{a constant} \times \phi$$

The constant in this case is known as the shear modulus or modulus of rigidity. It is usually denoted by the symbol G, so that $\tau = G\phi$. Note that since ϕ is non-dimensional, G has the same units as shear stress. In the case of normal stresses,

$$\sigma = E\varepsilon$$

$$\varepsilon = \frac{\sigma}{E} = \frac{P}{AE}$$

$$\varepsilon L = \text{change in length } \Delta L = \frac{PL}{AE}$$

The following examples illustrate the application of these simple formulae.

Example 3.1 Application of Hooke's law

A circular bar 20 mm in diameter and 200 mm long is subjected to a force of 20 kN. Find the stress, strain, and elongation in the bar if the value of $E = 80$ GPa.

Solution Area of the bar = $\pi \times 20^2/4 = 314.16$ mm^2

Stress $\sigma = P/A = 20 \times 10^3/314.16 = 63.66$ N/mm^2

$E = 80$ GPa = 80,000 MPa = 80,000 N/mm^2

Strain = $\sigma/E = 63.66/80{,}000 = 7.9 \times 10^{-4}$

Elongation = Strain × length = $7.9 \times 10^{-4} \times 200 = 0.16$ mm ■

Example 3.2 Young's modulus of elasticity

The ratio of Young's moduli of elasticity of two materials is 2.35. Find the ratio of the stresses and elongations in two bars of these materials if they are of the same length and same area and subjected to the same force P.

Solution Stress $\sigma = P/A$. Load P and area remaining the same, ratio of stresses = 1.0

Strain in one bar $\varepsilon_1 = \sigma/E_1$; strain in the other bar $\varepsilon_2 = \sigma/E_2$

Length being the same, the ratio of elongations will be the same as ratio of strains

Ratio of elongations = $\varepsilon_1/\varepsilon_2 = (\sigma/E_1)/(\sigma/E_2) = E_2/E_1 = 1/2.35 = 0.42$ ■

Example 3.3 Young's modulus of elasticity

A bar of cross-sectional area 314 mm^2 elongates by 0.8 mm over a length of 600 mm when subjected to a tensile force of 12000 N. Find the Young's modulus of elasticity of the material of the bar.

Solution Stress $\sigma = 12{,}000/314 = 38.22$ N/mm^2

Strain $\varepsilon = \sigma/E = 38.22/E$. Elongation = $\varepsilon L = 38.22\,L/E$

Elongation = 0.8 mm

Therefore, $38.22 \times 600/E = 0.8$ mm

$E = 38.22 \times 600/0.8 = 28665$ N/mm^2 = 28.67 GPa ■

Example 3.4 Young's modulus of elasticity of a circular pipe

A circular pipe of internal diameter 30 mm and thickness 4 mm is subjected to a force 30 kN and the elongation was measured as 1 mm. If the length of the pipe is 2 m, find the value of Young's modulus of elasticity and the stress in the pipe.

Solution Internal diameter of the pipe = 30 mm; thickness = 4 mm; external diameter = 38 mm; area of the pipe = $\pi(38^2 - 30^2)/4 = 427.26$ mm^2

Stress in the pipe material $\sigma = P/A = 30{,}000/427.26 = 70.2$ N/mm^2

Elongation = 1 mm; Length of pipe = 2 m = 2000 mm

Strain $\varepsilon = 1/2000 = 5 \times 10^{-4}$

E = stress/strain = $70.2/(5 \times 10^{-4}) = 140{,}400$ N/mm^2 = 140.4 GPa ■

Example 3.5 Stress in uniform bar

A uniform steel rod, 6 mm ϕ and 0.5 m long, is subjected to a tensile force of 3 kN. Find the stress in the bar and its elongation. E = 200 GPa.

Solution Tensile force = 3 kN = 3000 N

$$\text{Area of cross section} = \frac{\pi \times 6^2}{4} = 9\pi \,\text{mm}^2$$

$$\text{Stress} = \frac{3000}{9\pi} = 106 \text{ N/mm}^2$$

Change in length,

$$\Delta L = \frac{PL}{AE} = \frac{3000 \times (0.5 \times 1000)}{9\pi \times 200 \times 10^9/10^6}$$

(Note that 1 GPa = 10^9 Pa = 10^3 MPa = 10^3 N/mm^2.)

$$\Delta L = 0.265 \text{ mm}$$

■

Example 3.6 Determining stress and *E*

The length of an aluminium rod 10 mm ϕ and 400 mm long increases to 400.15 mm when subjected to a tensile force of 2 kN. Find the stress in the bar and the value of E for aluminium.

Solution $\text{Area of bar} = \dfrac{\pi \times 10^2}{4} = 25p \text{ mm}^2$

$$\text{Stress} = \frac{2 \times 1000}{25\pi} = 25.46 \text{ N/mm}^2$$

Change in length,

$$\Delta L = 400.15 - 400 = 0.15 \text{ mm}$$

$$\Delta L = \frac{PL}{AE}, \quad E = \frac{PL}{A\,\Delta L}$$

$$E \text{ for aluminium} = \frac{2000 \times 400}{25\pi \times 0.15} = 67{,}906 \text{ N/mm}^2 = 67.9 \text{ GPa}$$

■

Example 3.7 Uniform bar subjected to loads along length

A bar of uniform cross section 20 mm ϕ is subjected to loads as shown in Fig. 3.5(a). Find the total elongation of the bar and the maximum stress in the bar. E = 200 GPa. (All lengths are in mm.)

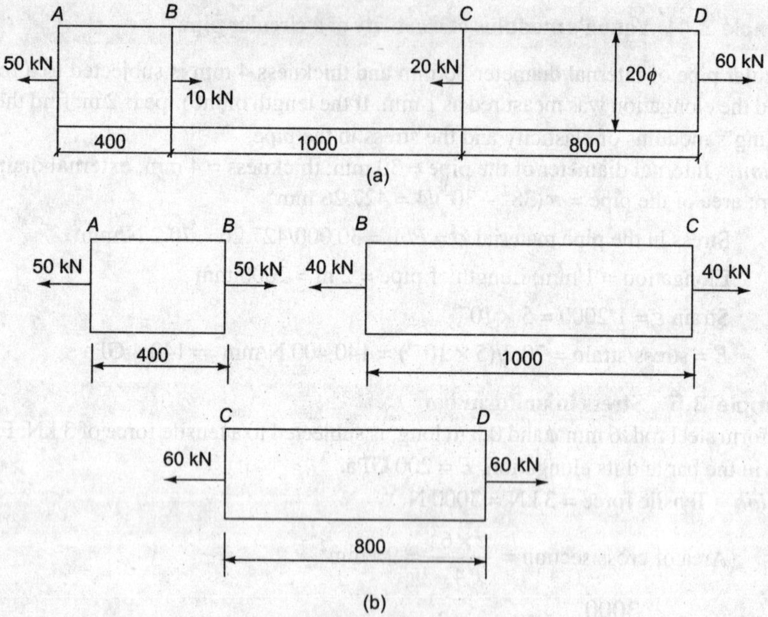

Fig. 3.5

Solution The total elongation will be the sum of the elongations of the individual sections AB, BC, and CD. The free body diagrams of each of these sections are shown in Fig. 3.5(b).

For each of the sections, the change in length is given by $\Delta L = PL/AE$, appropriate numerical values corresponding to the section being considered. For AB,

$$\Delta L_1 = \frac{50{,}000 \times 400}{100\pi \times 200{,}000} \qquad \left(A = \frac{\pi \times 20^2}{4} = 100\pi\right)$$

$$= 0.318 \text{ mm}$$

For BC,

$$\Delta L_2 = \frac{40{,}000 \times 1000}{100\pi \times 200{,}000} = 0.637 \text{ mm}$$

For CD,

$$\Delta L_3 = \frac{60{,}000 \times 800}{100\pi \times 200{,}000}$$

$$= 0.764 \text{ mm}$$

Total change in length $= \Delta L_1 + \Delta L_2 + \Delta L_3 = 0.318 + 0.637 + 0.764 = 1.719$ mm.

Since the cross sections of members are the same, the maximum stress will be in section CD, where the force is maximum.

$$\text{Maximum } \sigma = \frac{60{,}000}{100\pi} = 191 \text{ N/mm}^2$$

Example 3.8 Stress in a stepped bar

A bar has three sections of different diameters, 120 mm, 80 mm, and 100 mm, and is subjected to a load of 500 kN as shown in Fig. 3.6(a). Find the total elongation of the bar and the maximum stress in the material. $E = 200{,}000$ MPa.

Solution The total elongation will be the sum of the elongations of the three sections *AB*, *BC*, and *CD* shown in Fig. 3.6(b). Since the load on all the sections is the same,

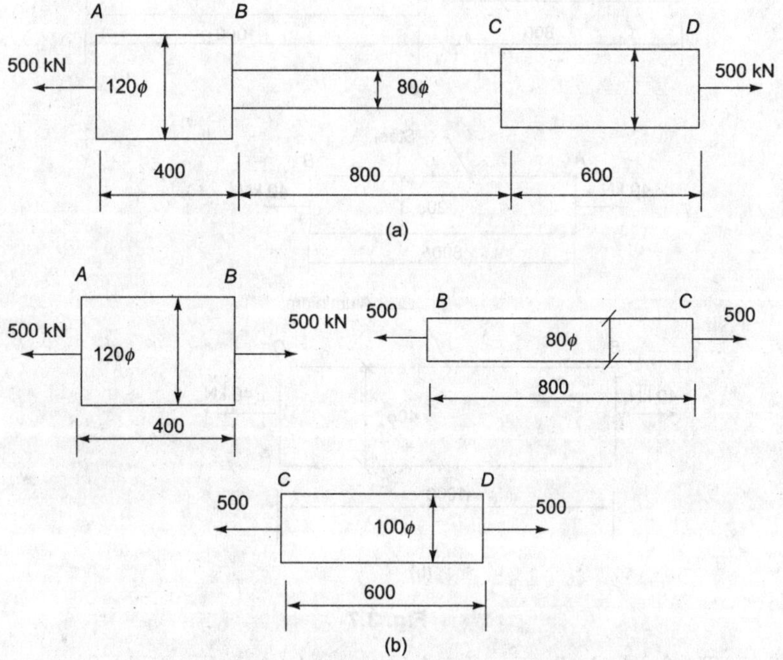

Fig. 3.6

$$\Delta L = \frac{P_1L_1}{A_1E_1} + \frac{P_2L_2}{A_2E_2} + \frac{P_3L_3}{A_3E_3} = \frac{P}{E}\left[\frac{L_1}{A_1} + \frac{L_2}{A_2} + \frac{L_3}{A_3}\right]$$

$$\Delta L = \frac{500{,}000}{200{,}000}\left[\frac{400}{(\pi/4)\times 120^2} + \frac{800}{(\pi/4)\times 80^2} + \frac{600}{(\pi/4)\times 100^2}\right]$$

$$= \frac{4}{\pi}\times\frac{5}{2}\left[\frac{4}{144} + \frac{8}{64} + \frac{6}{100}\right]$$

$$= 0.677 \text{ mm}$$

The maximum stress will be in section *BC*, which has the least area.

$$\sigma = \frac{500{,}000}{\pi/4\times 80^2} = 99.5 \text{ N/mm}^2$$

Example 3.9 Stress in a bar of different materials

A steel rod, 20 mm ϕ and 800 m long, is rigidly attached to an aluminium rod, 40 mm ϕ and 1 m long, as shown in Fig. 3.7. The combination is subjected to a tensile load of 40 kN. Find the stress in the materials and the total elongation of the bar. *E* for steel = 200 GPa, *E* for aluminium = 70 GPa.

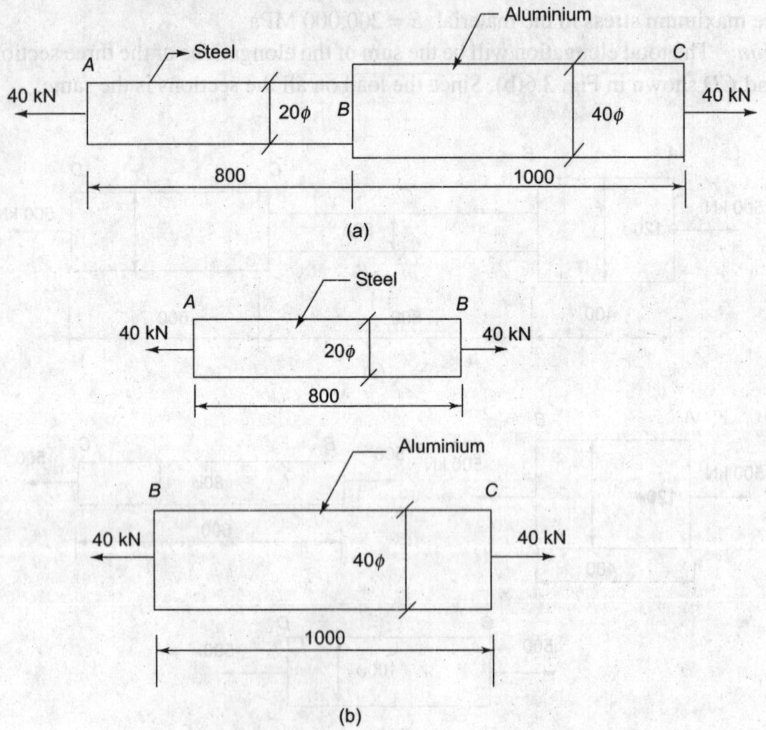

Fig. 3.7

Solution The free body diagram of each section is drawn in Fig. 3.7(b). Stress in steel,

$$\sigma_{\text{Al}} = \frac{40{,}000}{(\pi/4) \times 40^2} = 31.8 \text{ N/mm}^2$$

Change in length,

$$\Delta L = \frac{P_1 L_1}{A_1 E_1} + \frac{P_2 L_2}{A_2 E}$$

$$= 40{,}000 \left[\frac{800}{(\pi/4) \times 20^2 \times 200{,}000} + \frac{1000}{(\pi/4) \times 40^2 \times 70{,}000} \right]$$

$$= 0.5093 + 0.4547 = 0.964 \text{ mm}$$

Example 3.10 Stress in a piston

The piston of a steam engine is 60 mm ϕ and operates in a cylinder of diameter 400 mm. The piston rod is 1 m long. What is the maximum pressure that can be allowed in the cylinder, if the stress in the rod is limited to 80 N/mm^2? What will be the change in the length of the piston at this pressure? E = 200 GPa.

Solution The physical situation is represented in Fig. 3.8. If p is the maximum pressure permissible, the force on the rod is given by

$$P = p[(\pi/4) \times 400^2 - (\pi/4) \times 60^2]$$

$$\text{Stress in the rod} = \frac{P}{A} = \frac{p[(\pi/4) \times 400^2 - (\pi/4) \times 60^2]}{(\pi/4) \times 60^2}$$

This stress is limited to 80 N/mm^2.

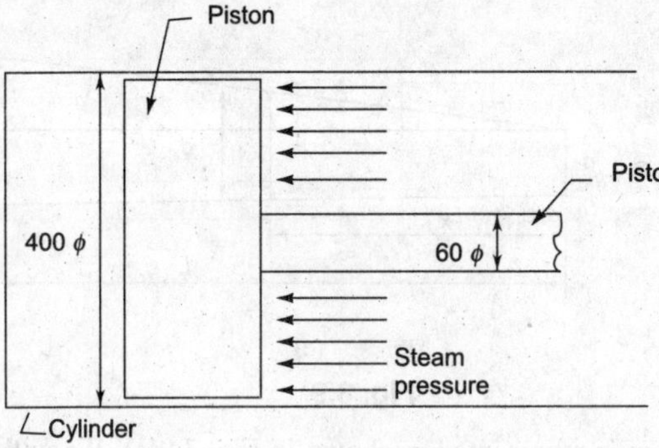

Fig. 3.8

Therefore,

$$80 = \frac{p(40{,}000\pi - 900\pi)}{900\pi}$$

$$p = 1.84 \text{ N/mm}^2$$

Change in length of the rod,

$$\Delta L = \frac{PL}{AE} = \frac{1.84 \times 39{,}100\pi \times 1000}{900\pi \times 200{,}000} = 0.3997 \text{ mm}$$

3.3 TAPERING SECTIONS

If a bar subjected to loads has a uniformly tapering section, the principles of analysis remain the same but it is necessary to arrive at the total elongation by integration. If the tapering is not uniform, one can follow the usual procedure to analyse the sections so long as the variation in the cross section can be expressed by a mathematical formula. The following examples illustrate this.

Example 3.11 Stress in a uniformly tapering round bar

A steel rod tapers in diameter from 18 mm at one end to 9 mm at the other over a length of 900 mm. Determine the elongation in the length of the bar if E = 200 GPa.

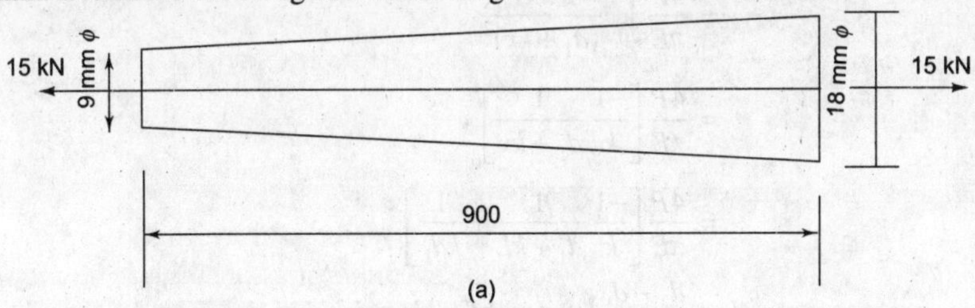

(a)

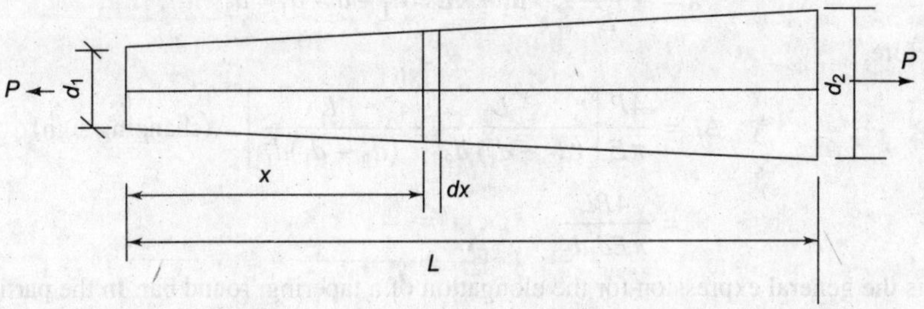

(b)

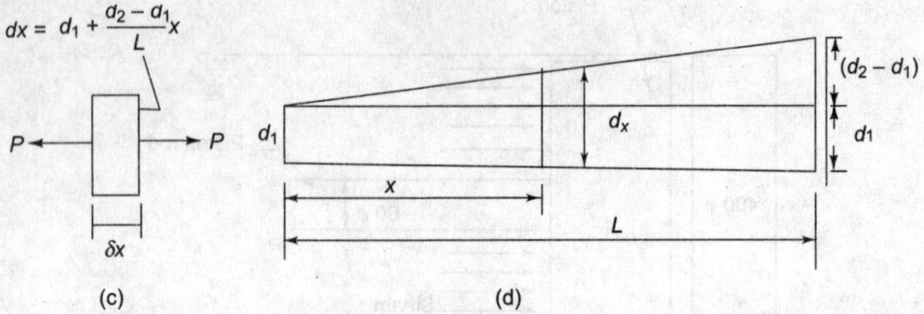

Fig. 3.9

Solution The problem is represented as shown in Fig. 3.9(a). We will derive a general formula for the case shown in Fig 3.9(b). Let d_1 be the diameter at the left end, d_2 at the right end, and L be the length of the member. We take a section at a distance x from the left end and consider an elementary length dx which is subjected to tensile forces P as shown in Fig. 3.9(c).

$$\Delta L \text{ for elementary length } dx = \frac{Pdx}{AE}$$

From a central cross section shown in Fig. 3.9(d), the diameter at x from the left end is given by

$$d_x = d_1 + \frac{d_2 - d_1}{L}x = d_1 + kx$$

where k is a constant $= (d_2 - d_1)/L$.

$$\text{Area at } x = \frac{\pi(d_1 + kx)^2}{4}$$

$$\Delta L \text{ for length } dx = \frac{Pdx}{\pi(d_1 + kx)^2 \, E/4}$$

For the whole bar, this expression is integrated from 0 to L. ΔL for the whole bar

$$= \int_0^L \frac{Pdx}{\pi/4\,(d_1 + kx)^2 E}$$

$$= \frac{4P}{\pi E}\int_0^L \frac{dx}{(d_1 + kx)^2}$$

$$= \frac{4P}{\pi E}\left[\frac{-1}{k}\,\frac{1}{d_1 + kx}\right]_0^L$$

$$= \frac{4P}{\pi E}\left[\frac{-1}{k}\,\frac{1}{d_1 + kL} + \frac{1}{kd_1}\right]$$

$$k = \frac{d_2 - d_1}{L}, \quad d_1 + kL = d_1 + d_2 - d_1 = d_2$$

Therefore,

$$\Delta L = \frac{4P}{\pi E}\left[\frac{-L}{(d_2 - d_1)\,d_2} + \frac{L}{(d_2 - d_1)d_1}\right] \quad \text{(changing sign)}$$

$$= \frac{4PL}{\pi E d_1 d_2}$$

This is the general expression for the elongation of a tapering, round bar. In the particular case, $d_1 = 9$ mm, $d_2 = 18$ mm, $P = 15$ kN $= 15{,}000$ N, $E = 200$ GPa $= 200{,}000$ N/mm^2, and

$L = 900$ mm. Therefore,

$$\Delta L = \frac{4 \times 15{,}000 \times 900}{\pi \times 200{,}000 \times 9 \times 18} = 0.53 \text{ mm}$$

Example 3.12 Stress in a uniformly tapering rectangular bar

A brass plate of uniform thickness 6 mm varies in width from 100 mm to 180 mm and is subjected to a load of 4 kN as shown in Fig. 3.10(a). Find the elongation of the bar. E for brass = 82 GPa.

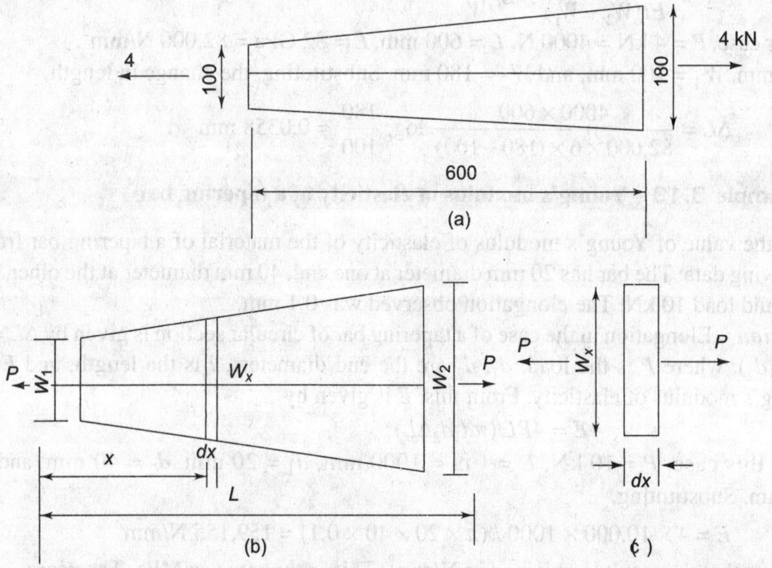

Fig. 3.10

Solution In this case again, let us derive a general formula for the case shown in Fig. 3.10(b). W_1 is the width at the left end, W_2 at the right end, and L is the length of the bar, which is subjected to a load P.

As in the previous case, we consider a section at a distance x from the left end and isolate a length dx as a free body, shown in Fig. 3.10(c).

$$\Delta L \text{ for the elementary length } dx = \frac{PL}{AE} = \frac{Pdx}{AE}$$

$$\text{The width } W_x \text{ at section of length } x = \frac{W_2 - W_1}{L} x + W_1$$

$$\text{Area of section of length } x = \left[W_1 + \frac{W_2 - W_1}{L} x \right] t$$

where t is the thickness of the member. Therefore,

$$\Delta L \text{ for length } dx = \frac{Pdx}{E[W_1 + (W_2 - W_1/L)x]t}$$

For the whole bar, we integrate this expression from 0 to L to find the elongation.

$$\Delta L \text{ for the bar} = \int_0^L \frac{Pdx}{Et(W_1 + kx)}$$

where $k = (W_2 - W_1)/L$. Therefore,

$$\Delta L = \frac{P}{Et} \int_0^L \frac{dx}{W_1 + kx}$$

$$= \frac{P}{Et}\frac{1}{k}[\log_e(w_1 + kx)]_0^L$$

$$= \frac{P}{Etk}\log_e\frac{W_1 + KL}{W_1}$$

$$= \frac{P}{Etk}\log_e\frac{W_2}{W_1}$$

$$= \frac{PL}{Et(W_2 - W_1)}\log_e\frac{W_2}{W_1}$$

In our case, $P = 4$ kN = 4000 N, $L = 600$ mm, $E = 82$ GPa = 82,000 N/mm^2, $t = 6$ mm, $W_1 = 100$ mm, and $W_2 = 180$ mm. Substituting, the change in length,

$$\Delta L = \frac{4000 \times 600}{82{,}000 \times 6 \times (180 - 100)}\log_e\frac{180}{100} = 0.0358 \text{ mm}$$

Example 3.13 Young's modulus of elasticity of a tapering bar

Find the value of Young's modulus of elasticity of the material of a tapering bar from the following data: The bar has 20 mm diameter at one end, 40 mm diameter at the other, length 1 m, and load 10 kN. The elongation observed was 0.1 mm.

Solution Elongation in the case of a tapering bar of circular section is given by $\Delta L = 4PL/(\pi E d_1 d_2)$, where P is the load, d_1, d_2 are the end diameters, l is the length, and E is the Young's modulus of elasticity. From this, E is given by

$$E = 4PL/(\pi d_1 d_2 \Delta L)$$

In this case, $P = 10$ kN, $L = 1$ m = 1000 mm, $d_1 = 20$ mm, $d_2 = 40$ mm, and $\Delta L =$ 0.1 mm. Substituting,

$$E = 4 \times 10{,}000 \times 1000 / (\pi \times 20 \times 40 \times 0.1) = 159{,}155 \text{ N/mm}^2$$

Note that the result is obtained in N/mm^2. This is the same as MPa. Therefore,

$$E = 159{,}155 \text{ MPa} = 159 \text{ GPa}$$

Example 3.14 Calculation of load on a tapering bar

A tapering bar of rectangular section, 20 mm wide at one end and 40 mm wide at the other, 8 mm thick, and 800 mm long, had an elongation of 0.08 mm under a load P. Find the load P if the modulus of elasticity of the material of the bar is 100 GPa.

Solution In the case of a tapering bar of rectangular section, the elongation is given by

$$\Delta L = PL \log_e(w_2/w_1)/[Et(w_2 - w_1)]$$

The load P is given by

$$P = \Delta L\, [Et(w_2 - w_1)] / [L \log_e(w_2/w_1)]$$

In this case, $\Delta L = 0.08$ mm, $E = 100$ GPa = 100,000 N/mm^2, $t = 8$ mm, $w_2 = 40$ mm, and $w_1 = 20$ mm. Substituting,

$$P = 0.08\,[\,100{,}000 \times 8\,(40 - 20)] / [800 \log_e(40/20)]$$
$$= 2308 \text{ N} = 2.3 \text{ kN}$$

3.4 DEFORMATION UNDER SELF-WEIGHT

When a rod or body is not subjected to any external force but is just hung so that the self-weight of the body acts as a force, it gets stretched. The weights acting at different levels vary, depending upon the length of the body below the section. The following examples illustrate the procedure for solving problems relating to such situations.

Example 3.15 Deformation under self-weight

A uniform rod of length L is hung from its top end. Determine the elongation of the bar. What is the maximum stress in the bar?

Solution The situation is represented in Fig. 3.11. L is the length of the bar, A its area of cross section, γ its density, and E the modulus of elasticity of the material. We isolate a length dx of the bar at a distance of x from the bottom end. The free body diagram of this elementary length is shown in Fig. 3.11(b).

Change in length of elementary length $\delta x = PL/AE$. Here, $P = A x \gamma$ and $L = \delta x$.

$$\Delta L \text{ for length } \delta x = \frac{Ax\gamma\,\delta x}{AE} = \frac{x\gamma\,\delta x}{E}$$

For the whole bar,

$$\Delta L = \int_0^L \frac{x\gamma\,dx}{E} = \frac{\gamma}{2E}(x^2)_0^L = \frac{\gamma L^2}{2E}$$

If W is the total weight of bar, then

$$W = AL\gamma$$

$$\Rightarrow \quad \gamma = \frac{W}{AL}$$

Then,

$$\Delta L = \frac{WL^2}{AL \times 2E} = \frac{WL}{2AE}$$

Note that this is half the elongation of a bar of the same dimensions subjected to a load W at the end.

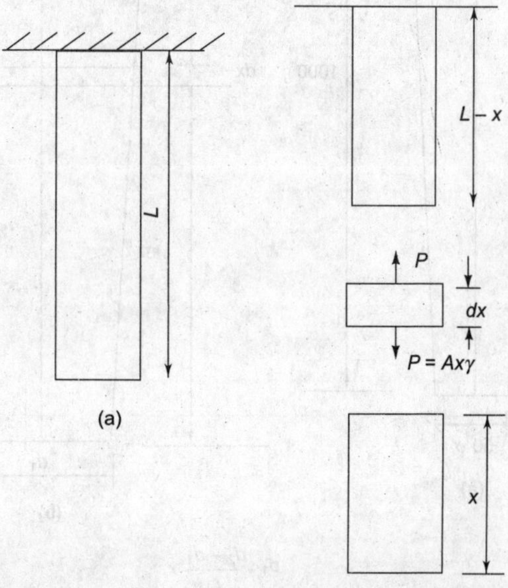

Fig. 3.11

Example 3.16 Elongation of a steel bar under self-weight

Determine the elongation of a steel bar of circular section under its own weight, if it is hung from the top, its diameter tapering from 100 mm at the top to 50 mm at the bottom over a length of 1 m. The unit weight of steel is 78.5 kN/m^3 and E = 200 GPa.

Solution The situation is represented in Fig. 3.12(a). We derive a general formula for this case by considering the general values shown in Fig. 3.12(b).

Consider an elementary length dx at a distance x from the bottom end. The diameter at x is given by

$$d_x = d_1 + \frac{d_2 - d_1}{L} x = d_1 + kx$$

where $k = (d_2 - d_1)/L$.

The free body diagram of the elementary length dx is given in Fig. 3.12(c). The distance y (the length remaining to make a full cone) is given by, from the principles of similar triangles,

$$\frac{d_1}{y} = \frac{d_2}{y+L}, \quad y = d_1(y+L) = d_2 y$$

$$y = \frac{d_1 L}{d_2 - d_1} = \frac{d_1}{(d_2 - d_1)/L} = \frac{d_1}{k}$$

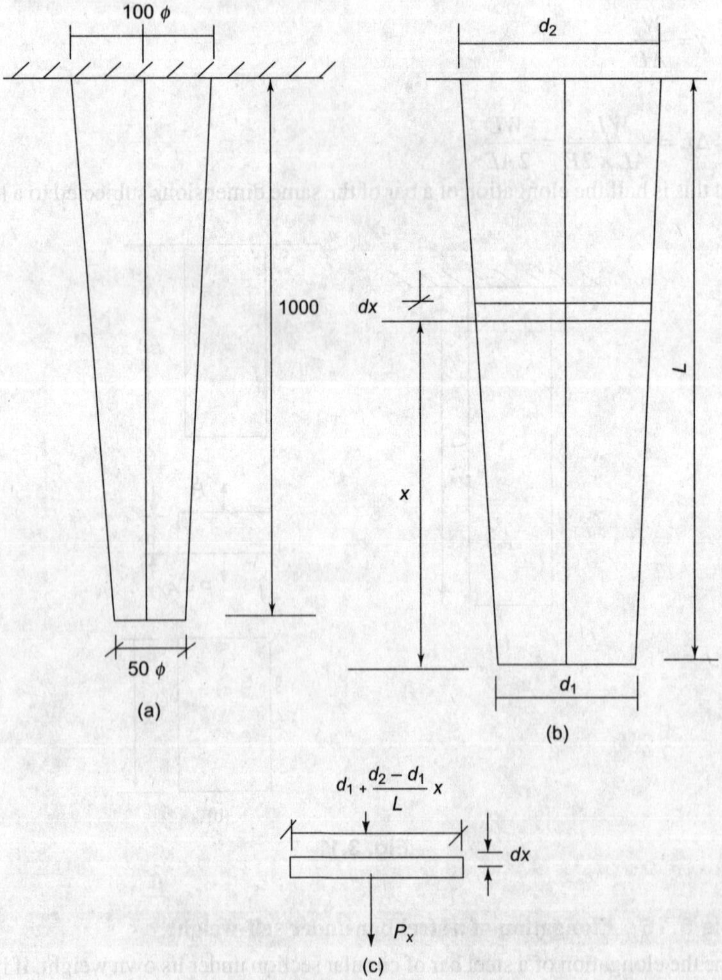

Fig. 3.12

The force P_x acting on the elementary length is due to the weight of the truncated cone below it. If γ is the weight density of the material, knowing that the volume of a cone is given by $\pi r^2 h/3$ or $\pi d^2 h/12$, we have

$$P_x = \frac{\gamma\pi}{12}[(d_1 + kx)^2 (y + x) - d_1^2 y]$$

$$= \frac{\gamma\pi}{12}\left[(d_1 + kx)^2\left(\frac{d_1}{k} + x\right) - d_1^2 \frac{d_1}{k}\right]$$

$$= \frac{\gamma\pi}{12k}[(d_1 + kx)^3 - d^3{}_1]$$

The increase in length for the elementary length δx is given by

$$\delta(\Delta L) = \frac{PL}{AE} = \frac{(\gamma\pi/12k)[(d_1 + kx)^3 - d_1^3]dx}{(\pi/4)(d_1 + kx)^2 E}$$

$$= \frac{\gamma}{3Ek}\left[(d_1 + kx) - \frac{d_1^3}{(d_1 + kx)^2}\right]$$

For the total elongation of the bar, integrate from 0 to L:

$$\Delta L = \frac{\gamma}{3Ek}\int_0^L \left[(d_1 + kx) - \frac{d_1^3}{(d_1 + kx)^2}\right]dx$$

$$= \frac{\gamma}{3Ek}\left[\frac{(d_1 + kx)^2}{2k} + \left(\frac{d_1^3}{k(d_1 + kx)}\right)\right]_0^L$$

$$= \frac{\gamma}{3Ek}\left[\frac{d_2^2}{2k} - \frac{d_1^2}{2k} + \frac{d_1^3}{kd_2} - \frac{d_1^3}{kd_1}\right]$$

$$= \frac{\gamma}{6Ek^2}\left[d_2^2 - d_1^2 + \frac{2d_1^3}{d_2} - 2d_1^2\right]$$

$$= \frac{\gamma L^2}{6E}\left[\frac{d_2^3 - d_1^2 d_2 + 2d_1^3 - 2d_1^2 d_2}{d_2(d_2 - d_1)^2}\right]$$

Therefore,

$$\Delta L = \frac{\gamma L^2}{6E}\left[\frac{d_2^3 - 3d_1^2 d_2 + 2d_1^3}{d_2(d_2 - d_1)^2}\right] \quad \text{[substituting } k = (d_2 - d_1)/L\text{]}$$

[*Note:* If $d_1 = 0$ and $d_2 = d$, for a cone hanging under its own weight, $\Delta L = \gamma L^2/6E$.]

In our case, $d_2 = 100$ mm, $d_1 = 50$ mm, $L = 1000$ mm, $E = 200{,}000$ N/mm^2, and $\gamma = 78.5 \times 10^{-6}$ N/mm^3. Substituting these values,

$$\Delta L = \left(\frac{78.5 \times 10^6 \times 1000 \times 1000}{6 \times 200{,}000}\right)\left[\frac{100^3 - 3 \times 50^2 \times 100 + 2 \times 50^3}{100(100 - 50)^2}\right]$$

$$= 0.000131 \text{ mm}$$

■

Example 3.17 Constant stress problem

Find the relation between the area of section and height y for the case shown in Fig. 3.13 for constant stress. The body is subjected to a force P at the lower end and self-weight. Unit weight of the bar material is γ. If the area at the lower end is A_b, find the area at the upper end.

Solution The situation is shown in Fig. 3.13. As the height y increases, the load acting on the section increases due to the self-weight. The area, therefore, has to change for keeping the stress constant.

Consider an element of depth dy at a height y from the free end. This element is subjected to the following forces.

(i) Upward force of $\sigma(A_y + dA)$
(ii) Downward force of σA_y
(iii) Downward force due to the self-weight of the element $\gamma A_y dy$

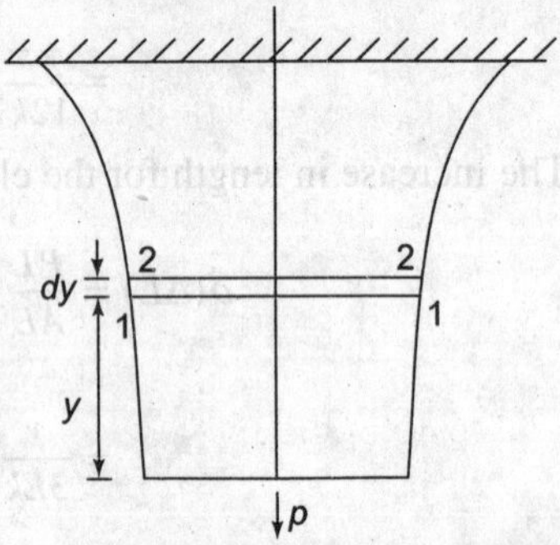

Fig. 3.13

The stress in the bar remains constant at P/A_b. Let the change in the area between 1-1 and 2-2 be dA. Area at section 1-1 is A_y and the area at 2-2 is $A_y + dA$.

For equilibrium of the element, the upward force must be equal to the sum of downward forces.

$$\sigma(A_y + dA) = \sigma A_y + \gamma A_y dy$$

$$\sigma dA = \gamma A_y dy \text{ or } dA/A_y = \gamma dy/\sigma$$

Integrating,

$$\int [dA/A_y] = \int [\gamma/\sigma] dy$$

$$\log_e A_y = \gamma y/\sigma + C$$

The constant of integration can be found from the condition that at $y = 0$, $A = A_b$.

$$C = \log_e A_b$$

Therefore,

$$\log_e A_y = \gamma y/\sigma + \log_e A_b$$

or $$\log_e [A_y/A_b] = \gamma y/\sigma$$

or $$A_y/A_b = e^{\gamma y/\sigma}$$

This is the general equation relating the area at height y to the base area.

$$A_y = A_b e^{\gamma y/\sigma}$$

Area at the top is obtained when $x = L$.

So, $$A_t = A_b e^{\gamma L/\sigma}$$

Example 3.18 Equation between cross-sectional area and height for constant stress

When a bar is hung from a point such that it is subjected to self-weight and the stress at different sections varies as the weight causing elongation varies. If the bar has to have constant stress along the length, the cross-section has to vary. Derive an equation relating the area of cross-section to the height above the base for a bar subjected to a force 100 kN at the bottom end and hanging freely if a constant stress of 50 N/mm^2 is desired. Unit weight of the material is 78.5 kN/m^3.

Solution The situation is shown in Fig. 3.14.

If the stress has to remain constant at 50 N/mm^2, the area at the bottom has to be 100,000/50 = 2000 mm^2.

We consider an elementary strip of thickness dy at a height y from the free end. From the free body diagram of the strip, we can write

$$\sigma(A_y + dA) = \sigma A_y + \gamma A_y\, dy,$$

where $\gamma = 78{,}500$ N/m^3 $= 0.0785$ N/mm^3

Reducing, we get

$$\sigma dA = \gamma A_y\, dy$$

or $$dA/A_y = (\gamma/\sigma)dy$$

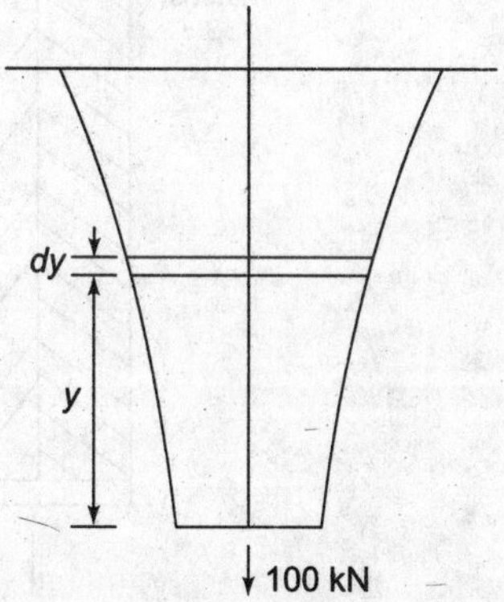

Fig. 3.14 Sectional variation for constant stress

Integrating,

$$\log_e A_y = (\gamma/\sigma)\, y + C$$

The constant can be found from the condition that at $y = 0$, $A_y = 2000$ mm^3.

$$C = \log_e(2000)$$

So,+ $$\log_e A_y = (\gamma/\sigma)y + \log_e 2000$$

or $$\log_e [A_y/2000] = (\gamma/\sigma)y$$

or $$A_y/2000 = e^{(\gamma/\sigma)y}$$

This is the general equation relating the area at any section to the base area.

$(\gamma/\sigma) = 0.0785/50 = 0.00157$

The area at the top where $y = 1$ m = 1000 mm is

$2000\, e^{0.00157 \times 1000} = 2000 \times e^{1.57} = 9613.3$ mm^2 ■

3.5 COMPOSITE SECTIONS

A composite section consists of two or more materials bonded together rigidly such that the straining action of the external load is shared by the materials forming the section. Figure 3.15 shows some composite sections in which the load P is shared by the two materials as they undergo deformations. In the problems we have considered so far, we have used only the conditions of static equilibrium and Hooke's law to find the deformations. In the case of composite sections, the conditions of static equilibrium are not enough to determine the forces in the members. Let us take a look at the composite section shown in Fig. 3.15. Such a section has two components of different materials which are constrained to deform together. This figure shows a composite section subjected to a load P. This load is shared by the two materials in a particular, unknown ratio, and it is not possible to find this ratio from the equilibrium conditions alone. If the loads shared by the materials are P_1 and P_2, then, from the conditions of equilibrium, $P_1 + P_2 = P$.

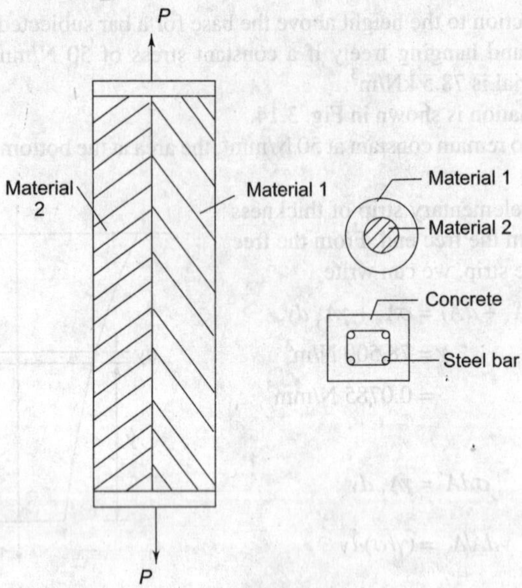

Fig. 3.15

We need one more equation relating P_1 and P_2 to find their values. This is obtained from what is known as a compatibility condition. The deformations of the two materials are equal (in this case the strains in the two materials are equal) since they have the same length. Since strain $\varepsilon = P/AE$, we can state $P_1/A_1E_1 = P_2/A_2E_2$ as the compatibility condition.

Modular ratio We have derived two conditions to solve problems of composite sections. (i) Equilibrium condition is $P_1 + P_2 = P$ and (ii) Compatibility condition is $P_1/(A_1E_1) = P_2/(A_2E_2)$. Note that P_1/A_1 is the stress σ_1 in the first material and P_2/A_2 is the stress σ_2 in the other material. Thus $\sigma_1/E_1 = \sigma_2/E_2$ is the compatibility condition for strains in the two materials. This gives,

$$\sigma_1 = \sigma_2\,(E_1/E_2) \text{ or } P_1 = P_2\,(A_1/A_2)\,(E_1/E_2)$$

The ratio E_1/E_2 is known as modular ratio (m) as it is the ratio of the modulus of elasticity of the two materials. Thus,

$$\sigma_1 = \sigma_2 m \text{ or } P_1 = P_2 m(A_1/A_2)$$

These two equations enable us to evaluate P_1 and P_2, and the stresses and strains in the two materials. The following examples illustrate the procedure.

Example 3.19 Stress in an RCC column

At short, reinforced cement concrete column 600 mm × 600 mm has eight steel rods of 25 mm ϕ as reinforcement. Find the stresses in steel and concrete, and the elastic shortening of the column if E = 200,000 N/mm^2 for steel and 10,000 N/mm^2 for concrete. Load on column = 3000 kN and length = 3 m.

Solution If P_c and P_s are the loads on concrete and steel, respectively, then $P_c + P_s = 3000 \times 1000$ N $= 3 \times 10^6$ N.

$$\text{Area of steel bars} = 8 \times \frac{\pi \times 25^2}{4} = 3927 \text{ mm}^2$$

Net area of concrete = $600 \times 600 - 3927 = 356{,}073\ \text{mm}^2$

The strains in the two materials are equal, i.e.,

$$\frac{P_c}{A_c E_c} = \frac{P_s}{A_s E_s}, \quad P_c = \frac{P_s A_c E_c}{A_s E_s}$$

So, $$P_c = P_s \times \frac{356{,}073 \times 10{,}000}{3927 \times 200{,}000} = 4.5336 P_s$$

Therefore,

$$4.5336 P_s + P_s = 3 \times 10^6\ \text{N}, \quad P_s = 542\ \text{kN}, \quad P_c = 2458\ \text{kN}$$

$$\text{Compressive stress in concrete} = \frac{2458 \times 10^3}{356{,}073} = 6.9\ \text{N/mm}^2$$

$$\text{Compressive stress in steel} = \frac{542 \times 10^3}{3927} = 138\ \text{N/mm}^2$$

Shortening,

$$\Delta L = \frac{PL}{AE} = \frac{6.9 \times 3000}{10{,}000} = 2.07\ \text{mm}$$

Check: $$\Delta L = \frac{138 \times 3000}{200{,}000} = 2.07\ \text{mm}$$

Check: $$\Delta L = \frac{138 \times 3000}{200{,}000} = 2.07\ \text{mm}$$

Example 3.20 Stress in a composite bar of steel and cast iron

A steel rod, 60 mm ϕ and 1 m long, is encased by a cast iron (CI) sleeve 8 mm thick and of internal diameter 60 mm. The assembly is subjected to a load of 40 kN. Find the stresses in the two materials and the elongation of the assembly. E for steel = 200 GPa and E for cast iron = 100 GPa.

Solution If the loads shared by steel and cast iron are P_s and P_c, then $P_s + P_c = 40$ kN (Fig. 3.16).

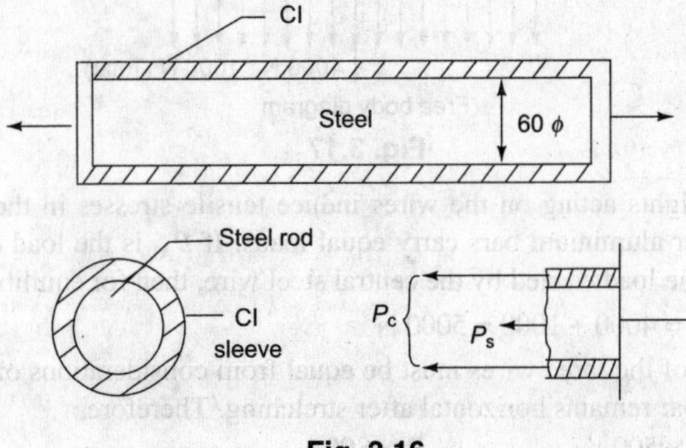

Fig. 3.16

$$\text{Area of cross section of steel rod} = \frac{\pi \times 60^2}{4} = 2827.4\ \text{mm}^2$$

$$\text{Area of cross section of cast iron} = \frac{\pi}{4}(76^2 - 60^2) = 1709\ \text{mm}^2$$

Since the strains in the two materials are equal,

$$\frac{P_s}{A_s E_s} = \frac{P_c}{A_c E_c}, \quad \frac{P_s}{2827.4 \times 200{,}000} = \frac{P_c}{1709 \times 100{,}000}, \quad P_s = 3.31P_c$$

$$3.31P_c + P_c = 40{,}000, \quad P_c = 9281 \text{ N}$$

$$P_s = 30{,}719 \text{ N}$$

$$\text{Stress in steel} = \frac{30{,}719}{2827.4} = 10.86 \text{ N/mm}^2$$

$$\text{Stress in cast iron} = \frac{9281}{1709} = 5.43 \text{ N/mm}^2$$

$$\text{Elongation of the assembly} = \frac{10.86 \times 1000}{200{,}000} = 0.0543 \text{ mm}$$

Example 3.21 Rigid rod hung with steel and aluminium rods

A rigid bar weighing 1000 N supports a vertical load of 4 kN at its midpoint, as shown in Fig. 3.17. The bar is attached to a fixed ceiling using three wires. The outer two wires, 500 mm long and 5 mm ϕ, are of aluminium and the middle wire is of steel, 800 mm long and 3 mm ϕ. Find the stresses in the wires if they are in the same vertical plane and are symmetrically attached to the bar. What is the distance by which the bar comes down? E for steel = 200 GPa and E for aluminium = 70 GPa.

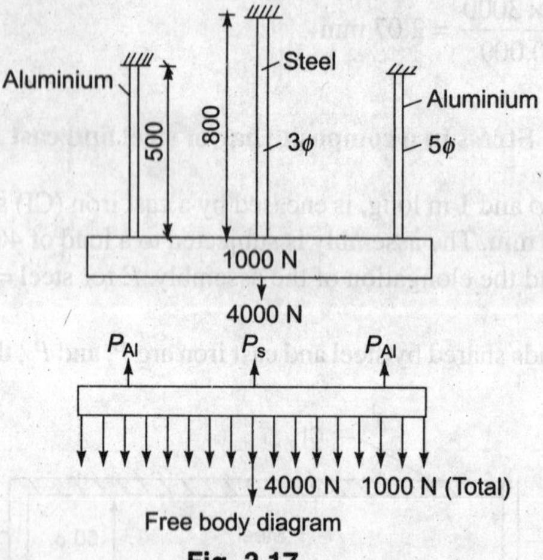

Fig. 3.17

Solution The weights acting on the wires induce tensile stresses in them. Because of symmetry, the outer aluminium bars carry equal loads. If P_{Al} is the load carried by each outer wire and P_s the load carried by the central steel wire, then for equilibrium,

$$P_s + 2P_{Al} = 4000 + 1000 = 5000 \text{ N}$$

The elongation of the three wires must be equal from considerations of symmetry and assuming that the bar remains horizontal after stretching. Therefore,

$$\frac{P_{Al} \times 500}{\pi \times (5^2/4) \times 70{,}000} = \frac{P_s \times 800}{\pi \times (3^2/4) \times 200{,}000}, \quad P_{Al} = P_s \times 1.555$$

Therefore,

$$P_s + 1.555\, P_s \times 2 = 5000, \quad P_s = 1216 \text{ N}, \quad P_{Al} = 1892 \text{ N}$$

$$\text{Stress in steel} = \frac{1216}{\pi \times (3^2/4)} = 172 \text{ N/mm}^2$$

$$\text{Stress in aluminium} = \frac{1892}{\pi \times (5^2/4)} = 96.4 \text{ N/mm}^2$$

$$\text{Elongation of the wires} = \frac{PL}{AE} = \frac{\sigma L}{E} = \frac{172 \times 800}{200{,}000} = 0.688$$

The bars come down by 0.688 mm. ■

Example 3.22 Stress in a composite section of aluminium and brass

The assembly shown in Fig. 3.18 consists of an aluminium tube through which a brass bolt has been passed between rigid plates. After the bolt has been tight-fitted initially, the nut is given quarter of a turn. The bolt is single-threaded with 2 mm pitch. Find the stress in the bolt and the tube. What is the tensile force that must be applied to the assembly so that the stress in the tube is zero? E for aluminium = 70 GPa and E for brass = 100 GPa.

Solution From Fig. 3.18, the aluminium tube is compressed while the brass bolt is subjected to tension by the turning of the nut. Since no external force is applied, $P_{Al} A_{Al} = p_b A_b$, where suffixes Al and b refer to aluminium and brass, respectively. Compression of the tube + extension of the bolt = $1/4 \times 2 = 0.5$ mm.

$$\sigma_{Al} A_{Al} = \sigma_b A_b \tag{1}$$

$$\frac{\sigma_{Al} L_{Al}}{E_{Al}} + \frac{\sigma_b L_b}{E_b} = 0.5 \tag{2}$$

σ_{Al} is the stress in the tube and σ_b is the stress in the brass bolt. From Eqn (1),

$$\sigma_{Al} = \frac{\sigma_b A_b}{A_{Al}}$$

Substituting in Eqn (2),

$$\frac{\sigma_b A_b}{A_{Al}} \frac{L_{Al}}{E_{Al}} + \frac{\sigma_b A_b}{A_{Al}} = 0.5$$

$$\sigma_b \left[\frac{A_b L_{Al}}{A_{Al} E_{Al}} + \frac{L_b}{A_b} \right] = 0.5$$

$$A_b = \frac{\pi}{4} \times 10^2 = 25\pi \text{ mm}^2, \quad L_{Al} = 350 \text{ mm}, \quad L_{Al} = 350 \text{ mm}$$

$$A_{Al} = \frac{\pi}{4}(18^2 - 15^2) = \frac{99\pi}{4} \text{ mm}^2, \quad L_b = 400 \text{ mm}$$

$$E_{Al} = 70{,}000 \text{ N/mm}^2, \quad E_b = 100{,}000 \text{ N/mm}^2$$

$$\sigma_b \left[\frac{25\pi \times 350}{(99/4)\pi 10{,}000} + \frac{400}{100{,}000} \right] = 0.5, \quad \sigma_b = 55.24 \text{ N/mm}^2 \text{ (tensile)}$$

$$\sigma_{Al} = \frac{55.24 \times 25\pi}{(99/4)\pi} = 55.8 \text{ N/mm}^2 \text{ (compressive)}$$

$$L_b = 400 \text{ mm}, \quad E_{Al} = 70000 \text{ N/mm}^2, \quad E_b = 100{,}000 \text{ N/mm}^2$$

Therefore,

$$p_b \left[\frac{25\pi \times 350}{(99/4)\pi \times 70{,}000} + \frac{400}{100{,}000} \right] = 0.5, \quad p_b = 55.24 \text{ N/mm}^2 \text{ (tensile)}$$

$$p_{Al} = \frac{55.24 \times 25\pi}{(99/4)\pi} = 55.8 \text{ N/mm}^2 \text{ (compressive)}$$

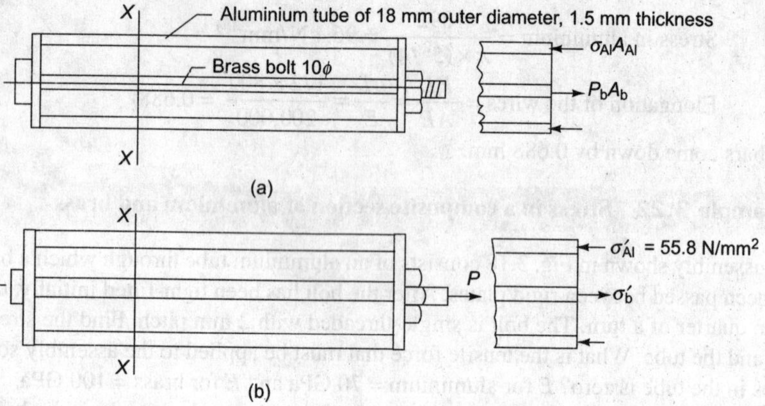

Fig. 3.18

In Fig. 3.18(b), tensile force P is applied such that the net stress in the tube is zero; σ_{Al} = 55.8 N/mm^2 (tensile). If σ'_b is the stress in the bolt due to force P, then $55.8 \times (99/4\pi) + \sigma'_b \times 25\pi = P$.

The elongation of the two due to the tensile force will be equal:

$$\frac{\sigma'_{Al} L_{Al}}{E_{Al}} = \frac{\sigma'_b L_b}{E_b}, \quad \sigma'_b = \frac{55.8 \times 350}{70{,}000} \times \frac{100{,}000}{400}$$

$$= 69.75 \text{ N/mm}^2$$

Stress in aluminium = – 55.8 + 55.8 = 0

Stress in brass = 69.75 + 55.24 = 125 N/mm^2

$$\text{Force } P = 69.75 \times 25\pi + 55.8 \times \frac{99}{4}\pi = 9817 \text{ N}$$

Alternate solution Instead of considering the tightening of the nut and the application of the force separately, they can be considered together and the net effect found since the stress in aluminium is zero at the end.

$$\sigma_b A_b = P, \quad \sigma_b \frac{L_b}{E_b} = 0.5$$

$$\sigma_b = \frac{0.5 \times 100{,}000}{400} = 125 \text{ N/mm}^2$$

Force $P = \sigma_b A_b = 125 \times 25\pi = 9817$ N ■

Example 3.23 Stress in a compound bar

A compound bar of brass and steel is fixed between rigid supports and loaded as shown in Fig. 3.19. Find the stresses in the two bars if load P = 30 kN, diameter of steel rod = 20 mm, diameter of brass rod = 30 mm, length of brass rod = 0.4 m, length of steel rod = 0.3 m, E for steel = 200 GPa, and E for brass = 105 GPa.

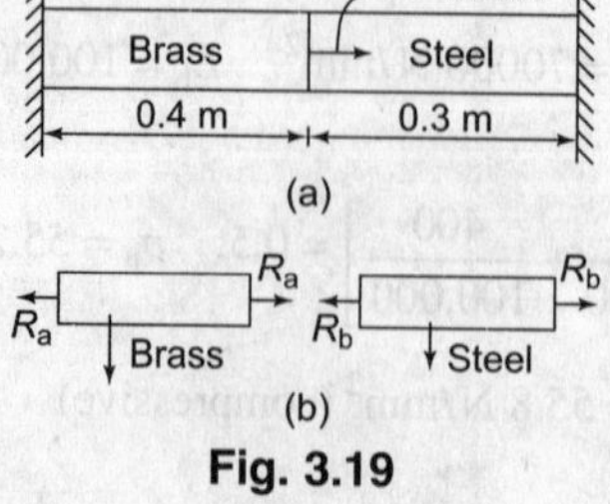

Fig. 3.19

Solution For brass rod, Area $A_b = \pi(30)^2/4 = 713.7\ \text{mm}^2$

For steel, Area $A_s = \pi(20)^2/4 = 314.2\ \text{mm}^2$

With the applied load, the free body diagrams of the two parts are shown in Fig. 3.19(b). Steel bar is under compression and brass is under tension. The compatibility condition requires that contraction of steel = extension of brass.

$$\frac{\sigma_s \times 400}{200{,}000} = \frac{\sigma_b \times 300}{105{,}000};\ \sigma_s = 1.43\sigma_b$$

Compression in steel + tension in brass = total force

$$317.2\ \sigma_s + 713.7\ \sigma_b = 30{,}000$$

or $\quad 317.2 \times 1.43\ \sigma_b + 713.7\ \sigma_b = 30{,}000$

so, $\quad \sigma_b = 25.7\ \text{N/mm}^2$ and $\sigma_s = 1.43 \times 25.7 = 36.75\ \text{N/mm}^2$

Example 3.24 Loads on a compound bar system

For the system shown in Fig. 3.20, find the distance x so that the rigid, weightless bar remains horizontal. For the bar of material 1, the properties are A_1, E_1, and L_1 and for material 2, the properties are A_2, E_2, and L_2.

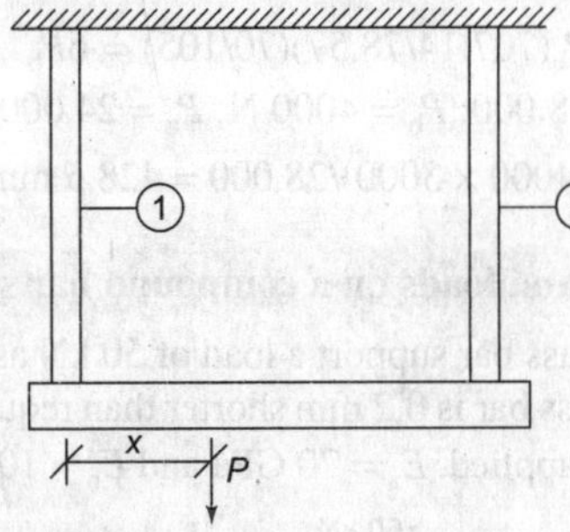

Fig. 3.20

Solution There are three unknowns in this problem–loads share of bars 1 and 2 and distance x. Three equations need to be formed to solve for the three unknowns. If P_1 and P_2 are the loads shared by bars 1 and 2, then,

$$P_1 + P_2 = P \tag{1}$$

Moment equation, $P_2 W = Px$ (2)

Taking moment @ A, as the bar remains horizontal, extension of the two bars will be equal, i.e.,

$$\frac{P_1 L_1}{A_1 E_1} = \frac{P_2 L_2}{A_2 E_2} \tag{3}$$

$$x = P_2 W / P$$

Example 3.25 Loads on a compound bar system

For the system shown in Fig. 3.21, find the value of x so that the rigid bar remains horizontal. For the aluminium bar, diameter = 30 mm, length = 1 m, and E = 70 GPa. For the brass bar, diameter = 10 mm, length = 1 m, and E = 105 GPa. Also P = 28 kN.

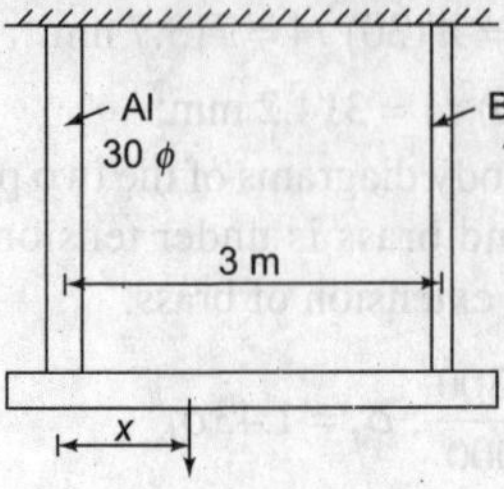

Fig. 3.21

Solution Area of aluminium bar = $\pi\, 30^2/4 = 707.14$ mm^2

Area of brass bar = $\pi 10^2/4 = 78.57$ mm^2

If P_a and P_b are the loads taken by the bars, $P_a + P_b = 28{,}000$

Taking moments about a point on the aluminium bar,

$28{,}000\, x = P_b \times 3000$

Also, as the bar remains horizontal, elongations of the two bars (in this case strains as lengths are equal) are the same.

$$\frac{P_a L_a}{A_a E_a} = \frac{P_b L_b}{A_b E_b}$$

$$P_a = P_b(707.14/78.57)(70/105) = 6P_b$$

$$6P_b + P_b = 28{,}000;\ P_b = 4000 \text{ N};\ P_a = 24{,}000 \text{ N}$$

$$x = (4000 \times 3000)/28{,}000 = 428.5 \text{ mm}$$

Example 3.26 Stresses and loads on a compound bar system

Two aluminium bars and a brass bar support a load of 50 kN as shown in Fig. 3.22. Due to an error in fabrication, the brass bar is 0.2 mm shorter than required. Find the stresses in the bars when a load of 50 kN is applied. E_a = 70 GPa and E_b = 105 GPa.

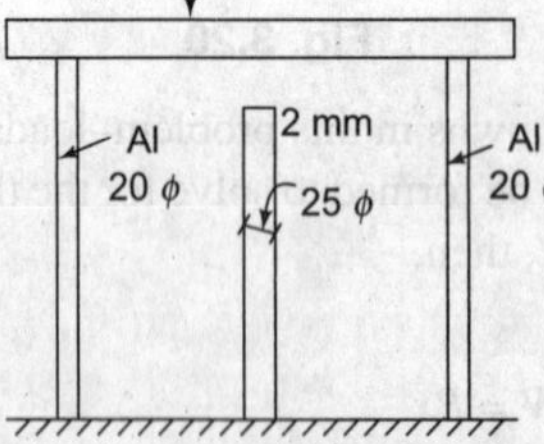

Fig. 3.22

Solution Due to the gap of 0.2 mm, initially the load will be taken by only the outer bars till their shortening is 0.2 mm. Due to symmetry, the stress in the two bars will be equal. If σ_0 is the stress in the two bars, then

Area of aluminium bar = $\pi \dfrac{(20)^2}{4}$ = 314.16 mm^2

Area of brass bar = $\pi \dfrac{(25)^2}{4}$ = 490.17 mm^2

$2\sigma_0 \times 1000/70{,}000 = 0.2$, $\sigma_0 = 7$ N/mm^2

The load taken by the two bars = $2 \times 7 \times 314.16 = 4404$ N

Balance load = $50{,}000 - 4404 = 45{,}596$ N

This load will be shared by all the three bars. If σ_a and σ_b are the stresses in the aluminium and brass bars, then

$$2 \times 314.16 \times \sigma_a + 490.87\sigma_b = 45{,}596$$

Also, strains in the two bars will be equal, i.e.,

$$\sigma_a \times 1000/70{,}000 = \sigma_b \times 1000/105{,}000, \quad \sigma_a = 0.6667\sigma_b$$

Substituting this value in the previous equation,

$$2 \times 314.16 \times 0.6667\sigma_b + 490.87\sigma_b = 45{,}596, \quad \sigma_b = 50.11 \text{ N/mm}^2$$

So, $\sigma_a = 0.6667 \times 50.11 = 33.41 \text{ N/mm}^2$

Stress in aluminium bar = 33.41 + 7 = 40.41 N/mm^2

Stress in brass bar = 50.11 N/mm^2

Example 3.27 Stress on a bar between rigid supports

A bar is fixed between rigid supports as shown in Fig. 3.23. If an axial load is applied at a distance *a* from the left support, find the stress in the bar.

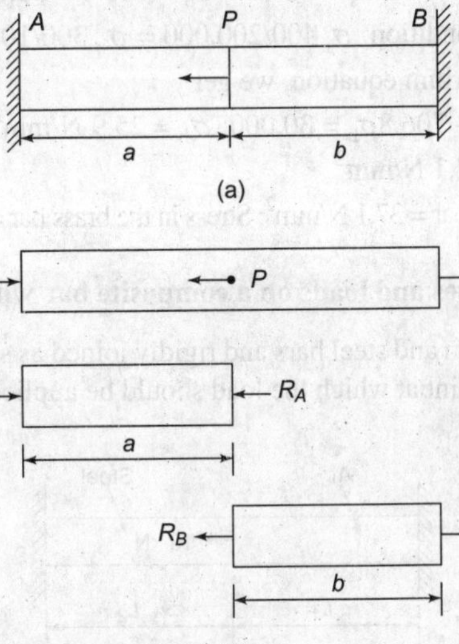

Fig. 3.23

Solution Let the reactions at the supports be R_a and R_b as shown. The free body diagrams of the two bars are as shown in Fig. 3.23(b). It is clear that the bar segment on the left side is under compression and the right segment is under tension. Also, the compression of the bar on the left is equal to the elongation of the bar on the right. Therefore, we have

$$R_a + R_b = P$$

Since A and E are the same for the two segments,

$$R_a a/AE = R_b b/AE$$

$$R_a a = R_b b; R_a = R_b(b/a)$$

Substituting in the first equation,

$$R_b\,(b/a) + R_b = P; R_b = PL/b; R_a = PL/a$$

Stress in the right segment = PL/Ab

Stress in the left segment = PL/Aa

Example 3.28 Stresses and loads on a composite bar

In the system shown in Fig. 3.24, find the stresses in steel and brass if the load P is 30 kN. Steel bar is of 20 mm diameter and the brass bar has a diameter of 30 mm. E for steel = 200 GPa and E for brass = 105 GPa.

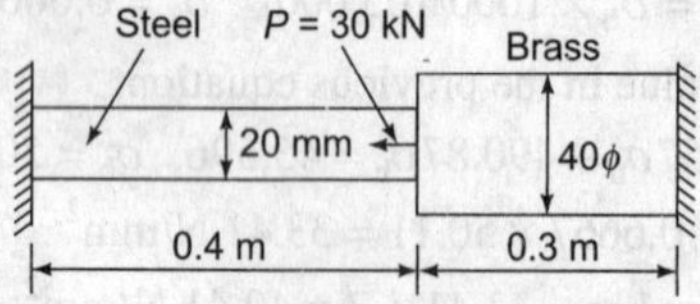

Fig. 3.24

Solution Area of steel bar = $\pi(20)^2/4 = 314.2$ mm^2

Area of brass bar = $\pi(30)^2/4 = 706.8$ mm^2

From the free body diagrams of steel and brass bars, we write

Equilibrium condition: $\sigma_s \times 314.2 + \sigma_b \times 706.8 = 30{,}000$

Compatibility condition: $\sigma_s\ 400/200{,}000 = \sigma_b\ 300/105{,}000$; $\sigma_s = 1.43\sigma_b$

Substituting in the equilibrium equation, we get

$1.43\sigma_b \times 314.2 + 706.8\sigma_b = 30{,}000$; $\sigma_b = 25.9$ N/mm^2

$\sigma_s = 1.43\sigma_b = 37.1$ N/mm^2

Stress in the steel bar = 37.1 N/mm^2; Stress in the brass bar = 25.9 N/mm^2

Example 3.29 Stresses and loads on a composite bar with a rigid joint

A bar is made of aluminium and steel bars and rigidly joined as shown. If a load of 10 kN is applied axially, find the point at which the load should be applied to have the bars carrying equal loads.

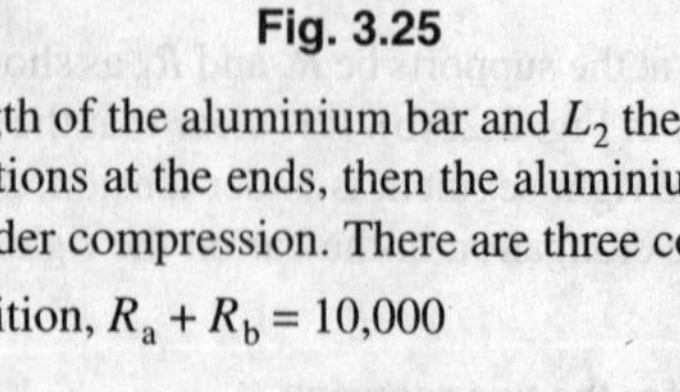

Fig. 3.25

Solution Let L_1 be the length of the aluminium bar and L_2 the length of the brass bar.

If R_a and R_b be the reactions at the ends, then the aluminium bar will be under tension and the steel bar will be under compression. There are three conditions available.

Equilibrium condition, $R_a + R_b = 10{,}000$

$R_a = R_b = 5000$ N

Compatibility condition, $R_a L_1/A\ 70{,}000 = R_b L_2/A\ 200{,}000$

$L_2 = 4000 - L_1$

$L_1/70{,}000 = L_2/200{,}000$ as $R_a = R_b = 5000$ N

$L_1/70000 = (4000 - L_1)/200{,}000$; $L_1 = 1037$ mm

Length of the aluminium bar = 1037 mm; length of the steel bar = 2963 mm

Load should be applied at 1037 mm from the left end.

3.6 STRESSES DUE TO TEMPERATURE CHANGE

When a body is subjected to temperature change, it elongates if temperature rises and shortens if the temperature decreases. Stresses are caused in the material if the natural expansion or contraction is prevented.

3.6.1 Effect of Temperature

It is common experience that materials expand on heating and contract on cooling. This is known as thermal expansion or contraction. If a rod of a material has a length L at any temperature t_0, then it increases to a length $L + \Delta L$ when heated to a temperature t_2 or decreases to a length $L - \Delta L$ if t_2 is less than t_0, as shown in Fig. 3.26. There is a change in length, ΔL, and the corresponding strain can be calculated as $\Delta L/L$. There will be no stress in the material if it is allowed to expand or contract freely, $L + \Delta L$ or $L - \Delta L$ being the normal length of the material at temperature t_2.

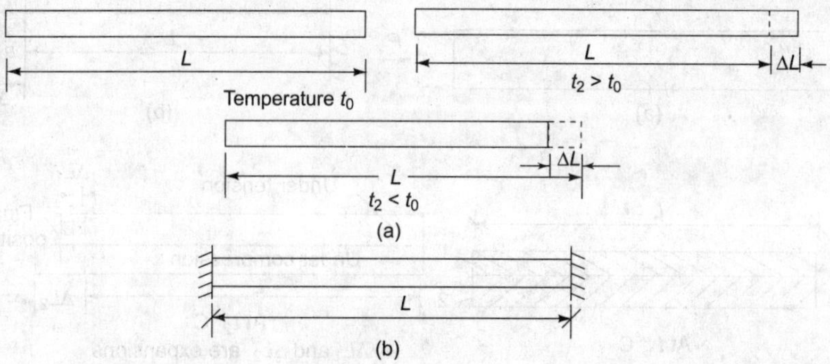

Fig. 3.26

However, if the material is constrained and is not fully or partly free to change its length, then stresses are developed in the material, called temperature (or thermal) stresses. As shown in Fig. 3.26(b), if the material is constrained fully, then there is no change in length (and no corresponding strain). The material is, however, stressed because it is not free to expand or contract to its normal length.

The problems of temperature stresses cannot be solved by equations of statics alone and in this sense are similar to the problems of composite sections dealt with previously.

3.6.2 Thermal Stress in Bars of Single Material

Consider the bar shown in Fig. 3.27(a). The length of the bar is L at temperature t_1°C and it is stress-free. If the bar is now heated to a temperature of t_2°C, it should increase in length but is, now, constrained. To find the stress in the bar, we assume that the constraint at B is removed. The bar is now free to expand to a length $L + \Delta L$, where

$$\Delta L = \text{length} \times \text{coefficient of linear expansion of the material} \times \text{change in temperature} = L\alpha t$$

where α is the coefficient of linear expansion and is the change in length per unit length per unit degree rise in temperature. Note that α is non-dimensional $t = (t_2 - t_1)$ is the change in temperature.

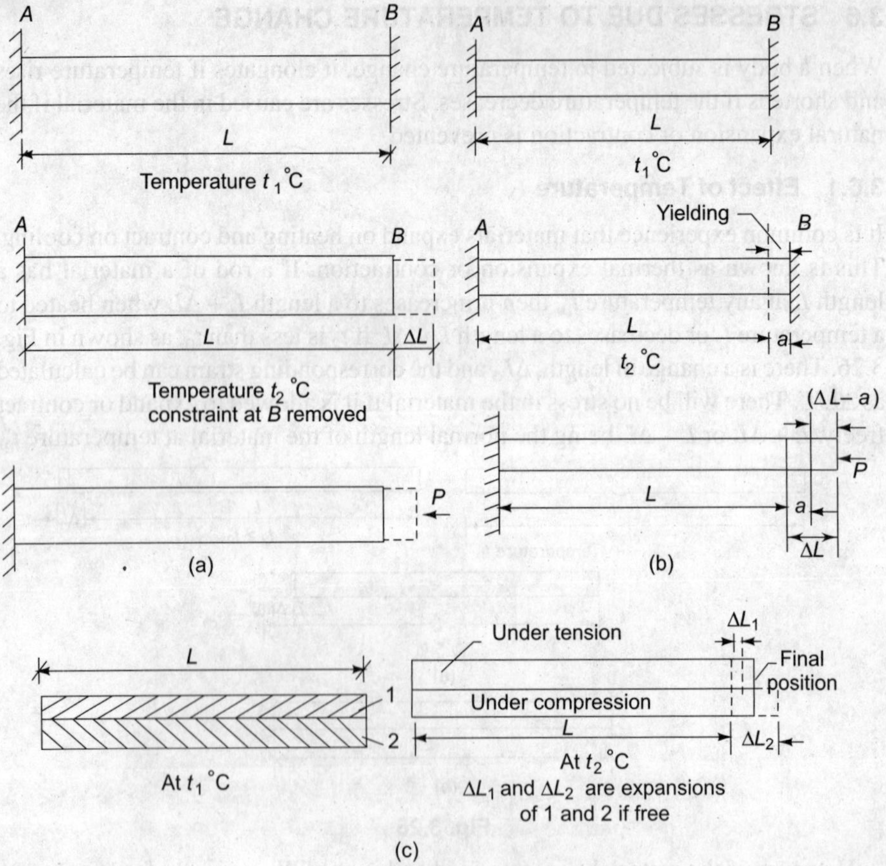

Fig. 3.27

In the situation shown in Fig. 3.27(a), the final length of the material is $L + \Delta L$. Therefore, we apply a compressive force P to reduce the length to L from $L + \Delta L$. Since $\Delta L = PL/AE$,

$$P = \frac{AE\Delta L}{L}$$

The stress due to temperature rise in the material will be equivalent to that due to compressive force P.

$$\text{Stress} = \frac{P}{A} = \frac{E\Delta L}{L}$$

Since $\Delta L/L = \alpha t$,

$$\text{Stress} = E\alpha t$$

Thermal stresses are produced due to constraints imposed on expansion or contraction. If, as shown in Fig. 3.27(b), the support acts as a constraint only partially and yields by an amount a, then the deformation prevented from taking place is only $(\Delta L - a)$ or $(L\alpha t - a)$. The force P will be correspondingly changed to $P = AE(\Delta L - a)/L$, and the corresponding stress can be calculated.

3.6.3 Thermal Stress in Composite Bars

The above discussion applies to bars of a single material only. When a bar consists of two materials rigidly joined together, thermal stresses are induced even without the bar being constrained. This is due to the difference in the thermal expansion characteristics of the materials.

As shown in Fig. 3.27(c), the bar is made up of two materials rigidly joined together. Coefficient of thermal expansion of material 2 is greater than that of 1, i.e., $\alpha_2 > \alpha_1$. The natural length of material 2 expanding independently due to temperature rise is $L + \Delta L_2$. The natural length of material 1 expanding independently under temperature rise is $L + \Delta L_1$. Now, $\Delta L_1 = L\alpha_1 T$ and $\Delta L_2 = L\alpha_2 T$, where T is rise in temperature.

The actual expansion of the two bars is ΔL, where $\Delta L_1 < \Delta L < \Delta L_2$. The component bar of material 1 is pulled and taken to the length ΔL ($>\Delta L_1$) and the material is thus under tension. The component bar of material 2 could expand only up to ΔL ($<\Delta L_2$) and is thus under compression. Note that the bars are stressed without any outside physical constraint. Each bar acts as a constraint on the other.

The procedure for solution to problems of thermal stresses is as follows.

(a) Temperature stresses in bodies of single material.

(i) Remove the constraint at one end and assume the body elongates or contracts freely with temperature variation.

(ii) Compute the extension/contraction of the body due to temperature. This change in length, ΔL, is equal to $L\alpha T$, where L is the length, α is the coefficient of linear expansion, and T is the rise or decrease in temperature = $T_f - T_i$, where T_f is the final temperature and T_i is the initial temperature.

(iii) If there is no yielding of support, apply a force P to bring the body back to its original length. The force required is obtained from $\Delta L = PL/AE$ as $P = AE\Delta L/L = AE\,(L\alpha T)/L$. Thus, $P = AE\alpha T$ and stress $\sigma = P/A = E\alpha T$, which is the stress due to temperature rise/decrease. The force P is compressive for a rise in temperature and tensile for a decrease in temperature.

(iv) If the support yields by an amount μ, then force P is required to create a change in length of $(\Delta L - \mu)$ only to bring the body to the original length.

Thus,

$$P = AE\,(L\alpha T - \mu)/L \text{ and stress } \sigma = E(L\alpha T - \mu)/L$$

(b) Composite sections: Let subscript 1 refer to material 1 and 2 to material 2. We proceed as follows. Composite sections need two conditions, an equilibrium condition and a compatibility condition to find the stresses in the two materials.

(i) If the two bars have the same length, then change in length $\Delta L_1 = L\alpha_1 T$ for the first material and $L\alpha_2 T$ for the second material, where $L_1 = L_2 = L$ and T is the rise or decrease in temperature and the bodies are free to expand independently of each other.

(ii) But, as the two bodies are constrained to move together, the change in length will be common. The final length will depend upon the values of

α_1 and α_2. If $\alpha_1 > \alpha_2$, the change in length ΔL will be $< L\alpha_1 T$ but more than $L\alpha_2 T$. Material 1 will be under compression as it is not able to expand to its natural length. Material 2 will be under tension as it has been pulled to a greater length. Reverse will be the case if $\alpha_2 > \alpha_1$.

(iii) If σ_1 and σ_2 are the stresses in the two materials, then the forces acting on them will be $P_1 = \sigma_1 A_1$ and $P_2 = \sigma_2 A_2$. If there is no external force, then these two forces are equal for equilibrium. $\sigma_1 A_1 = \sigma_2 A_2$.

(iv) We need one more equation to find these stresses. As the final change in length is common, the strains in the two materials are the same. $\varepsilon_1 = \sigma_1 L/E_1$ and $\varepsilon_2 = \sigma_2 L/E_2$. This gives $\sigma_1/E_1 = \sigma_2/E_2$ as the length L is the same.

(v) We find the stresses using these two equations.

Example 3.30 Thermal stress in a bar between rigid supports

A steel bar 2 m long is fixed between two supports. If the temperature of the bar is raised by 18°C, find the stress in the bar if the supports are rigid.

$$\alpha = 12 \times 10^{-6}/°\text{C/m and } E = 200 \text{ GPa}.$$

Solution The situation is shown in Fig. 3.28. The temperature rise causes the rod to expand but the rod is unable to expand because of the rigid supports.

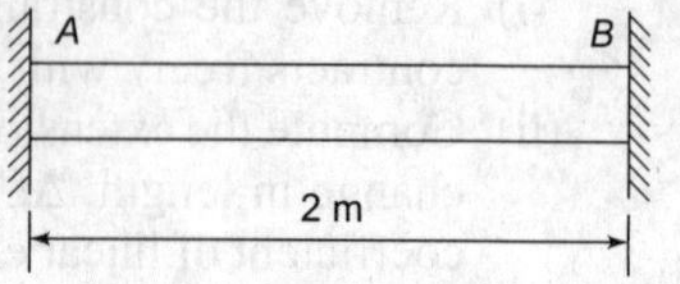

Fig. 3.28

Extension of the rod due to temperature rise $\Delta L = L\alpha T$.

In this case, $L = 2$ m = 2000 mm, $\alpha = 12 \times 10^{-6}$, and

$T = 18°$C

$\Delta L = 2000 \times 12 \times 10^{-6} \times 18 = 0.432$ mm. The force P to cause equal extension is given by

$P = AE\Delta L/L$ [from $\Delta L = PL/AE$] or $P/A = \sigma = E\Delta L/L$

Stress = $200{,}000 \times 0.432/2000 = 43.2$ N/mm^2. This stress is compressive.

Example 3.31 Temperature stress in a steel rod between rigid and yielding supports

A steel rod, 20 mm ϕ and 1.5 m long, is constrained between supports A and B as shown in Fig. 3.29. The material is stress-free at 27 °C. Determine the stress in the material when the temperature increases to 50 °C (a) if the supports are unyielding and (b) if the support at B yields by 0.1 mm. E for steel = 200 GPa and α for steel = $12 \times 10^{-6}/°$C.

Solution (a) *When the supports are unyielding* Remove the constraint at B and allow the material to expand. The change in length is given by

$$\Delta L = L\alpha t = 1500 \times 12 \times 10^{-6} \times (50 - 27)$$
$$= 0.414 \text{ mm}$$

The compressive force required to bring the material to its original length 1500 mm may be worked out as follows.

$$\frac{PL}{AE} = \Delta L$$

$$\Rightarrow \quad P = \frac{AE\Delta L}{L}$$

$$= \frac{(\pi/4) \times 20^2 \times 200{,}000 \times 0.414}{1500}$$

$$= 17{,}342 \text{ N}$$

$$\text{Stress } \sigma = \frac{P}{A} = \frac{17{,}432}{(\pi/4) \times 20^2} = 55.2 \text{ N/mm}^2 \text{ (compressive)}$$

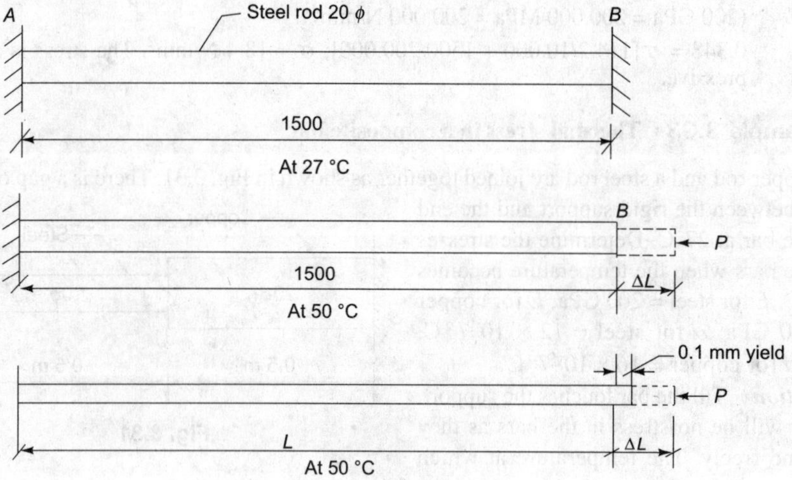

Fig. 3.29

(b) *When the support B yields by 0.1 mm* Change in length $\Delta L = 0.414$ mm as before. The final length of the member = 1500.1 mm. The compressive force P required is only for a change in length of (1500.414 – 1500.1) = 0.314 mm.

$$P = \frac{AE\,\Delta L}{L}, \quad \text{Stress} = \frac{P}{A} = \frac{E\,\Delta L}{L} = \frac{200{,}000 \times 0.314}{1500}$$

$$= 41.9 \text{ N/mm}^2 \text{ (compressive)}$$

Example 3.32 Thermal stress in a bar with rigid and spring supports

A steel bar 1.5 m long and 12 mm diameter is supported by a rigid support at one end and a spring of stiffness 10 kN/mm at the other. If the temperature of the bar is raised by 20°C, find the stress in the bar. $\alpha = 11.6 \times 10^{-6}$/°C and $E = 200$ GPa.

Solution The situation is shown in Fig. 3.30. Note that the spring's effect is similar to that of a yielding support.

If the rod were free, the expansion due to temperature would have been

$$\Delta L = L\alpha T$$

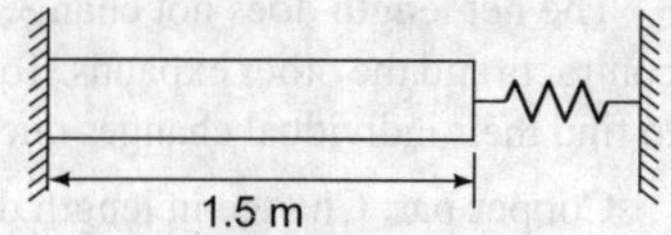

Fig. 3.30

Here L = 1.5 m = 1500 mm, $\alpha = 11.6 \times 10^{-6}$/°C, and $T = 20$°C

$$\Delta L = 1500 \times 11.6 \times 10^{-6} \times 20 = 0.348 \text{ mm}$$

This expansion is prevented but not fully because of the spring. If the spring is subjected to a force P as shown, the spring will exert an equal force on the bar and the stress in the bar due to temperature rise will be equal to that due to force P.

Due to force P, the spring will get compressed. This is similar to the yield of a support.

Compression of the spring = P/k = P/10,000 mm, where P is in newton and k = spring stiffness = 10 kN/mm = 10,000 N/mm.

Expansion of the bar prevented = 0.348 – P/10,000

Stress in the bar due to force $P = P/A = \sigma$, $P = A\sigma$

Compression in the bar due to $P = PL/AE = \sigma L/E$

Therefore, $0.348 - P/10{,}000 = \sigma L/E$

or $0.348 - A\sigma/10{,}000 = \sigma L/E$, $A = \pi \times 12^2/4 = 114.2$ mm^2

So, $0.348 - 114.2\sigma/10000 = \sigma \times 1500/200{,}000$

(200 GPa = 200,000 MPa = 200,000 N/mm^2)

or $0.348 = \sigma[114.2/10{,}000 + 1500/200{,}000]$; $\sigma = 18.4$ N/mm^2. The stress is compressive.

Example 3.33 Thermal stress in a composite bar

A copper rod and a steel rod are joined together as shown in Fig. 3.31. There is a gap of 0.1 mm between the rigid support and the end of the bar at 27°C. Determine the stresses in the bars when the temperature becomes 50°C. E for steel = 200 GPa, E for copper = 120 GPa; α for steel = 12×10^{-6}/°C, and α for copper = 16×10^{-6}/°C.

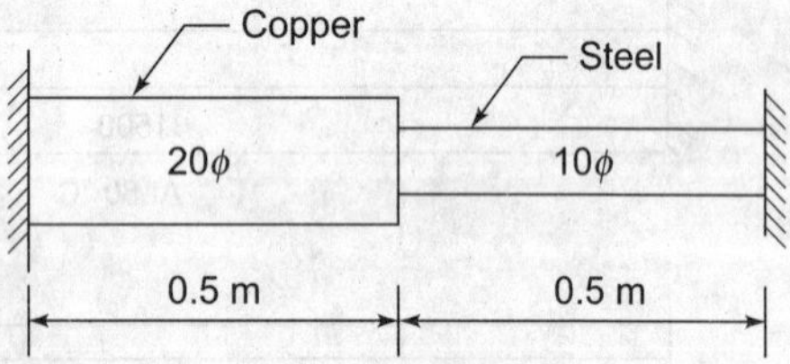

Fig. 3.31

Solution Till the bar touches the support, there will be no stress in the bars as they expand freely. The temperature at which this happens will be when the expansion of the copper and steel bars together equals 0.1 mm.

$$500 \times 16 \times 10^{-6} T + 500 \times 12 \times 10^{-6} T = 0.1;\ T = 7°\text{C}$$

Temperature is 27 + 7 = 34°C. The bars will be stressed due to any further rise in temperature. Temperature rise is 50 – 34 = 16°C.

Area of copper bar = $\pi(20)^2/4 = 314.2$ mm^2; area of steel bar = $\pi(10)^2/4 = 78.57$ mm^2

Total expansion of the bars = $500 \times 16\,[16 \times 10^{-6} + 12 \times 10^{-6}] = 0.224$ mm

This expansion is prevented. If P is the force, then

$$\frac{P \times 500}{314.3 \times 120{,}000} + \frac{P \times 500}{78.57 \times 200{,}000} = 0.224$$

$$P = 4978 \text{ N}$$

Stress in copper bar = 4978/314.2 = 15.84 N/mm^2

Stress in steel bar = 4978/78.57 = 63.36 N/mm^2

The net length does not change. But the copper and steel bars change in length, one bar contracts and the other expands. To find which bar extends and which bar contracts, we have to find their individual changes due to force and temperature rise.

Copper bar: Change in length due to temperature = $500 \times 16 \times 10^{-6} \times 16$
= 0.128 mm

Change in length due to force 4978 N = $4978 \times 500/(314.2 \times 120{,}000)$
= – 0.066 mm

Net change = 0.062 mm

Steel bar: Change due to temperature = $500 \times 12 \times 10^{-6} \times 16 = 0.096$ mm

Change due to force $P = 4978 \times 500/(78.57 \times 200{,}000) = -0.158$ mm

Net change = – 0.158 + 0.096 = – 0.062 mm

Example 3.34 Thermal stress in a bar with fixed supports

A bar, 20 mm diameter and fixed at A, is stretched with a force of 10 kN to bring to support B and fix it. Length $AB = 2$ m. Temperature is 27°C. Determine at what temperature will the stress become zero? What will be the stress in the bar if the temperature rises to (i) 40°C and (ii) 50°C?

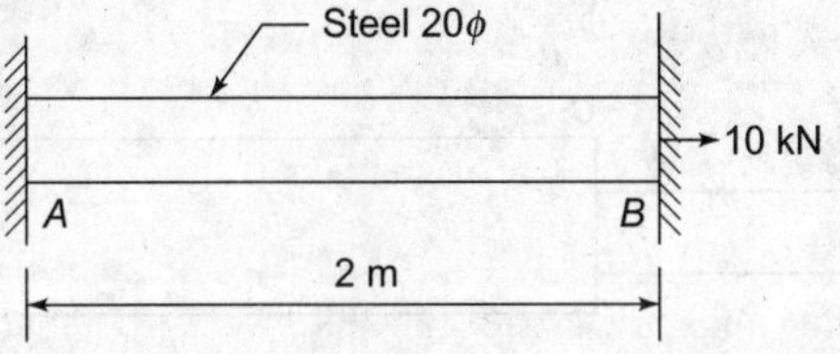

Fig. 3.32

Solution Area of the bar $= \pi(20)^2/4 = 314.2$ mm^2

Stress due to the force = 10,000/314.2 = 31.8 N/mm^2 (tensile)

Elongation due to this stress $= \sigma L/E = 31.8 \times 2000/200{,}000 = 0.318$ mm

A temperature rise will cause compressive stress in the bar. If T is the rise in temperature, this rise will cause an equal compressive stress to make the stress zero.

Stress due to temperature rise $T = E\alpha T = 200{,}000 \times 12 \times 10^{-6} \times T$

$200{,}000 \times 12 \times 10^{-6} \times T = 31.8; T = 13.25$°C

When the temperature reaches 27 + 13.25 = 40.25°C, the bar will be free of stress.

When the temperature is 40°C, temperature stress $= 200{,}000 \times 12 \times 10^{-6} \times 13$

$= 31.2$ N/mm^2

Net stress = 31.8 – 31.2 = 0.6 N/mm^2 (tensile)

At 50°C, temperature rise = 23°C

Thermal stress $= 200{,}000 \times 12 \times 10^{-6} \times 23 = 55.2$ N/mm^2

Net stress = 31.8 – 55.2 = –23.4 N/mm^2 (compressive)

Example 3.35 Temperature stresses in a stepped-section steel rod

The steel rod shown in Fig. 3.33 is in two parts. It has a diameter of 10 mm for a length of 1 m and of 20 mm for the remaining length of 1.5 m. If it is constrained between two supports A and B and is stress-free at 20 °C, find the stress in the material when it is heated to 70 °C. $E = 200$ GPa, $\alpha = 12 \times 10^{-6}$/°C.

Solution Assume that the support at B is removed and the material is allowed to expand. The change in length is given by

$$\Delta L = 2500 \times 12 \times 10^{-6} \times (70 - 20) = 1.5 \text{ mm}$$

If P is the equivalent compressive force required to bring the material back to its original length, then

$$\Delta L = \frac{PL_1}{A_1E_1} + \frac{PL_2}{A_2E_2}$$

the suffixes 1 and 2 referring to the parts AC and CB shown.

$L_1 = 1000$ mm, $A_1 = (\pi/4) \times 10^2 = 25\pi$ mm^2, $E_1 = 200$ GPa

$L_2 = 1500$ mm, $A_2 = (\pi/4) \times 20^2 = 100\pi$ mm^2, $E_2 = 200$ GPa

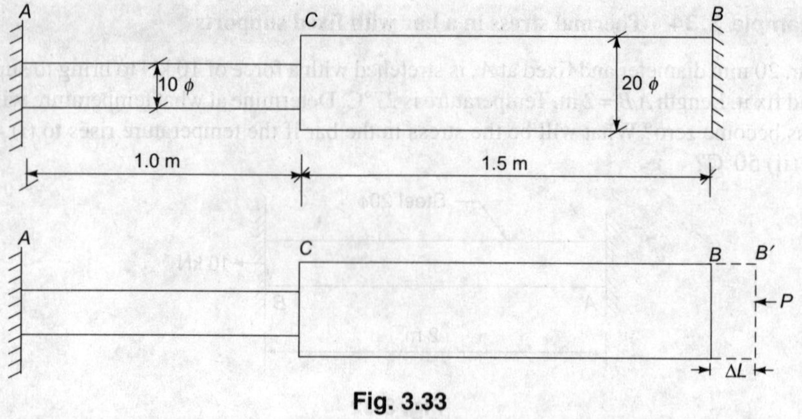

Fig. 3.33

Therefore,

$$1.5 = \frac{P \times 1000}{25\pi \times 200{,}000} + \frac{P \times 1500}{100\pi \times 200{,}000}$$

$$P = 17{,}136 \text{ N}$$

$$\text{Stress in } AC = \frac{17{,}136}{25\pi} = 218.18 \text{ N/mm}^2$$

$$\text{Stress in } CB = \frac{17{,}136}{100\pi} = 54.54 \text{ N/mm}^2$$

Note It must be clearly understood that while the total length of the member does not change, this is not the case with the individual parts AC and BC.

(i) Consider the part AC.

Change in length due to temperature $= 1000 \times 12 \times 10^{-6} \times 50 = +0.6$ mm

$$\text{Change in length due to force } P = \frac{-PL}{AE} = \frac{-17{,}136 \times 1000}{25\pi \times 200{,}000}$$

$$= -1.091 \text{ mm}$$

(The negative sign indicates a *decrease* in length.)

Net change in length = – 1.091 + 0.6 = – 0.491 mm (decrease)

(ii) Consider the part CB.

Change in length due to temperature rise $= 1500 \times 12 \times 10^{-6} \times 50$
$= 0.9$ mm

$$\text{Change in length due to } P = \frac{-17136 \times 1500}{100\pi \times 200{,}000} = -0.409$$

Net change in length = 0.9 – 0.409 = 0.491 mm (increase)

While the member AC decreases in length by 0.491 mm, CB increases in length by the same amount, the net change being zero for the whole member.

Example 3.36 Temperature stresses in steel and brass rods

If, in Example 3.35, the part CB was of brass, what would be the stress in the two materials? E for brass = 105 GPa, and α for brass $= 19 \times 10^{-6}/°\text{C}$.

Solution The total change in length would be due to expansion in the two materials and would be given by

$$\Delta L = L_1\alpha_1 t + L_2\alpha_2 t = 1000 \times 12 \times 10^{-6} \times 50 + 1500 \times 12 \times 10^{-6} \times 50$$
$$= 2.025 \text{ mm}$$

If P is the compressive force required as shown in Fig. 3.34, the two materials will contract by different lengths but to a total of 2.025 mm. Therefore,

$$\frac{P\times 1000}{25\pi\times 200{,}000}+\frac{P\times 1500}{100\pi\times 105{,}000}=2.025 \text{ mm}$$

$\Rightarrow \quad P = 18{,}555$ N

The stress in the two materials can now be calculated.

$$\text{Stress in steel} = \frac{18{,}555}{25\pi} = 236.25 \text{ N/mm}^2$$

$$\text{Stress in brass} = \frac{18{,}555}{100\pi} = 59.06 \text{ N/mm}^2$$

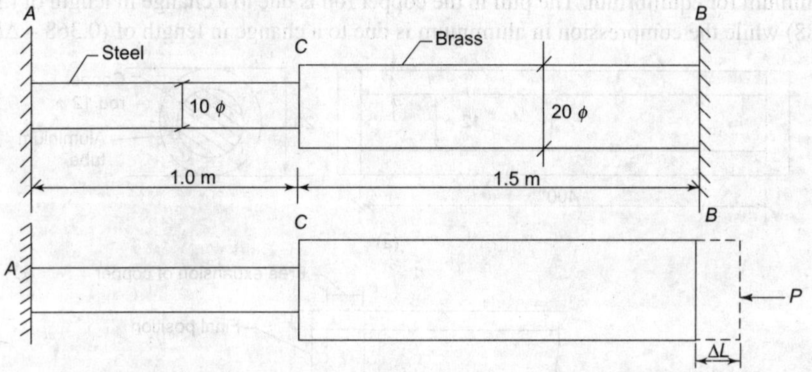

Fig. 3.34

With the application of force P, the total length remains the same, i.e., 2500 mm. However, the elongation in steel and brass individually is not zero, as given in the last example. Let us illustrate this.

Elongation of steel due to temperature rise $= 1000 \times 12 \times 10^{-6} \times 50$
$= 0.6$ mm

$$\text{Shortening of steel due to force } P = \frac{-236.25\times 1000}{2{,}000{,}000} = -1.1812$$

Net change in length = – 1.1812 + 0.6 = – 0.5812 (decrease)

Elongation of brass due to temperature rise $= 1500 \times 19 \times 10^{-6} \times 50$
$= 1.425$ mm

$$\text{Shortening brass due to force } P = \frac{59.06\times 1500}{105{,}000} = -0.8438 \text{ mm}$$

Net change in length for brass = 1.425 – 0.8435 = 0.5812 (increase)

There is a change in length in the two materials individually but they add up to zero.

Example 3.37 Temperature stresses in composite rod of copper and aluminium

A copper rod, 12 mm ϕ and 400 mm long, fits into an aluminium tube of external diameter 20 mm and thickness 4 mm of equal length. If the assembly is held together by a rigid plate at the end and is stress-free at 20 °C, find the stresses in the two materials when it is heated to 60 °C. For copper, $E = 120$ GPa and $\alpha = 18\times 10^{-6}$/°C. For aluminium, $E = 70$ GPa and $\alpha = 23\times 10^{-6}$/° C.

Solution The assembly is shown in Fig. 3.35. The aluminium tube tends to expand more but is held back by the copper rod. The tube is thus under compression. The copper rod, which expands less, is pulled by the tube and the rod is in tension. As shown in Fig. 3.35(b), the final length of the assembly will be more than what would have been the expansion of copper if it were free but less than the free expanded length of aluminium.

$$\text{Area of copper rod} = \frac{\pi}{4} \times 12^2 = 36\pi \text{ mm}^2$$

$$\text{Area of aluminium tube} = \frac{\pi}{4}(20^2 - 12^2) = 64\pi \text{ mm}^2$$

If the two materials were free to expand, the changes in length would have been

$$\Delta L \text{ for copper} = L\alpha t = 400 \times 18 \times 10^{-6} \times 40 = 0.288 \text{ mm}$$

$$\Delta L \text{ for aluminium} = 400 \times 23 \times 10^{-6} \times 40 = 0.368 \text{ mm}$$

If the final change in length is ΔL for the assembly, ΔL is more than 0.288 but less than 0.368. Since there is no external force, the tensile force in copper = compressive force in aluminium for equilibrium. The pull in the copper rod is due to a change in length of (ΔL – 0.288) while the compression in aluminium is due to a change in length of (0.368 – ΔL).

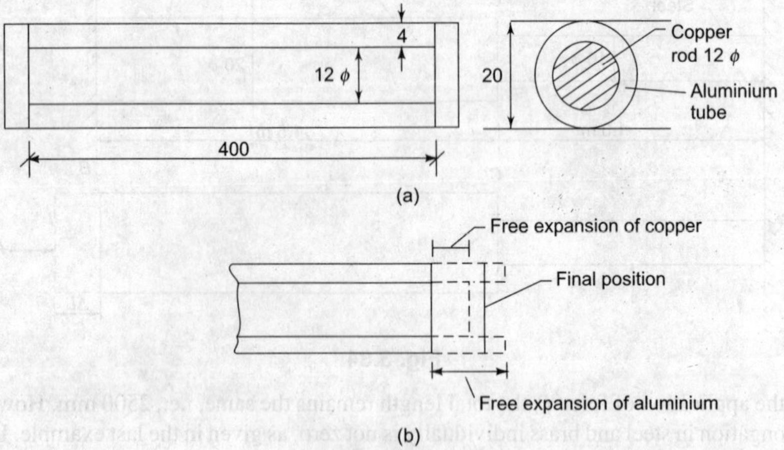

Fig. 3.35

$$\frac{PL}{AE} = \Delta$$

$$\Rightarrow \quad P = \frac{AE\Delta L}{L}$$

$$\frac{(\Delta L - 0.288) \times 36\pi \times 120{,}000}{400} = \frac{(0.368 - \Delta L) \times 64\pi \times 70{,}000}{400}$$

$$\Delta L = 0.3287 \text{ mm}$$

$$\text{Tensile stress in copper rod} = \frac{E\Delta L}{L} = \frac{120{,}000 \times (0.3287 - 0.288)}{400}$$

$$= 12.21 \text{ N/mm}^2$$

$$\text{Compressive stress in aluminium} = \frac{70{,}000\,(0.368 - 0.3287)}{400}$$

$$= 6.88 \text{ N/mm}^2$$

$$\text{Force } P = 12.21 \times 36\pi = 6.88 \times 64\pi = 1381 \text{ N}$$

Example 3.38 Temperature stresses in a tapering brass bar

A brass bar tapers from a diameter of 180 mm at one end to one of 90 mm at the other over a length of 900 mm. The bar is stress-free at a temperature of 27 °C. If the temperature drops to 0 °C, find the maximum normal stress on the bar if the ends are attached to rigid supports. $E = 100$ GPa and $\alpha = 18 \times 10^{-6}$/°C.

Solution We shall first derive a general expression for the case shown in Fig. 3.36(b). Assuming the left end is made free, the bar will decrease by a length given by

$$\Delta L = L\,\alpha t$$

This shortening is prevented by the supports which exert a tensile force P, which produces an expansion in length equal to ΔL. If we consider an elementary length dx at a distance x from the end A,

$$\text{Expansion of elementary length} = \frac{P\,dx}{AE}$$

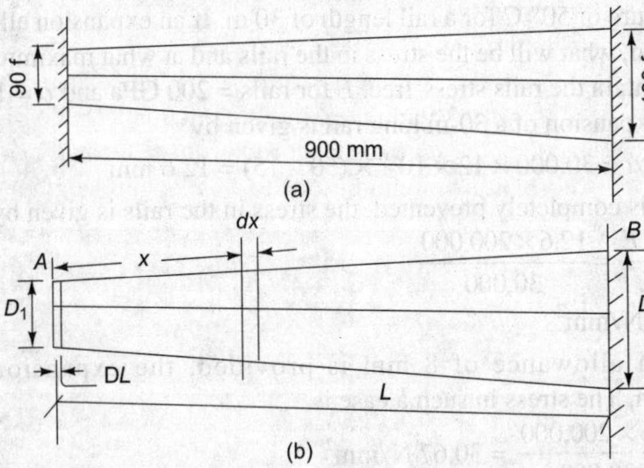

Fig. 3.36

For the whole bar,

$$\text{Expansion} = \int_0^L \frac{P\,dx}{A_x E}$$

The area at section x can be expressed in terms of end diameters as

$$D_X = \left[D_1 + \frac{(D_2 - D_1)x}{L}\right] = D_1 + kx, \quad A_x = \frac{\pi}{4}(D_1 + kx)^2$$

where $k = (D_2 - D_1)/L$. Therefore, expansion of the whole bar

$$= \int_0^L \frac{P\,dx}{(\pi/4)\,(D_1 + kx)^2\,E}$$

$$= \frac{4P}{\pi E}\int_0^L \frac{dx}{(D_1 + kx)^2}$$

$$= \frac{4P}{\pi E}\left(\frac{-1}{k}\right)\left[\frac{1}{D_2} - \frac{1}{D_1}\right]$$

$$= \frac{4P}{\pi E}\,\frac{L}{(D_2 - D_1)}\left[\frac{D_2 - D_1}{D_2 D_1}\right]$$

$$= \frac{4PL}{\pi E D_1 D_2}$$

This expansion is numerically equal to the shortening $\Delta L = L\,\alpha t$. Therefore, $L\alpha t = 4PL/\pi E D_1 D_2$ and $P = \alpha t \pi E D_1 D_2/4$. For the case in Fig. 3.36(a), substituting the values,

$$P = \frac{\pi \times 100{,}000 \times 90 \times 180 \times 18 \times 10^{-6} \times 27}{4}$$

$$= 618{,}360 \text{ N}$$

The maximum stress in the bar is at A.

$$\text{Maximum stress, } \sigma_{max} = \frac{618{,}360}{(\pi/4)\times 90^2} = 97.2 \text{ N/mm}^2$$

Example 3.39 Temperature stresses in steel rails

In a railway track, rails are placed and joined together such that they are stress-free at 15 °C. If no allowance is provided for expansion, find the maximum stress in the rails at peak summer temperature of 50° C for a rail length of 30 m. If an expansion allowance of 8 mm per rail is provided, what will be the stress in the rails and at what maximum temperature is it possible to maintain the rails stress-free. E for rails = 200 GPa and $\alpha = 12 \times 10^{-6}$/° C.

Solution The expansion of a 30-m long rail is given by

$$\Delta L = L\alpha t = 30{,}000 \times 12 \times 10^{-6} \times (50 - 15) = 12.6 \text{ mm}$$

If this expansion is completely prevented, the stress in the rails is given by

$$\sigma = \frac{\Delta LE}{L} = \frac{12.6 \times 200{,}000}{30{,}000}$$

$$= 84 \text{ N/mm}^2$$

If an expansion allowance of 8 mm is provided, the expansion prevented is 12.6 – 8 = 4.6 mm. The stress in such a case is

$$\sigma = \frac{4.6 \times 200{,}000}{30{,}000} = 30.67 \text{ N/mm}^2$$

If the rails are to remain unstressed,

$$L\alpha t = 8, \quad t = \frac{8}{L\alpha} = \frac{8}{30{,}000 \times 12 \times 10^{-6}} = 22.2 \text{ °C}$$

$$\text{Maximum temperature} = 15 + 22.2 = 37.2 \text{ °C}$$

3.7 SHEAR STRESS AND STRAIN

In the examples dealt with so far, the bodies have been subjected to axial or normal stresses. We have mentioned shear stress which is a tangential stress. Some situations where shear stress comes into play are shown in Fig. 3.37.

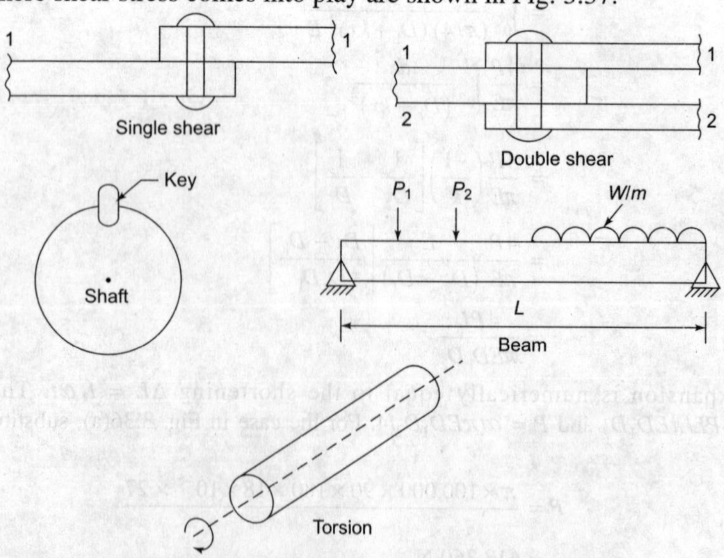

Fig. 3.37 Examples of shear stress

A rivet connecting two plates is subjected to shear stress across plane 1-1. If two plates are connected to a single plate as shown, shear stress exists across planes 1-1 and 2-2 and is known as double shear. A member subjected to torsion (Chapter 8) is also subjected to shear stress. A key inserted to prevent a shaft from rotating is also subjected to shear stress. A beam subjected to lateral loads also has shear stress. Shear stress acts tangential to the section and is known as tangential stress also.

The following examples illustrate the calculation of shear stress.

Example 3.40 Shear stress due to punching

A circular hole is punched in a metal plate by a punching machine. If the force required to punch a hole of 16 mm ϕ is 180 kN in the metal plate, which is 8 mm thick, find the maximum stress in the material.

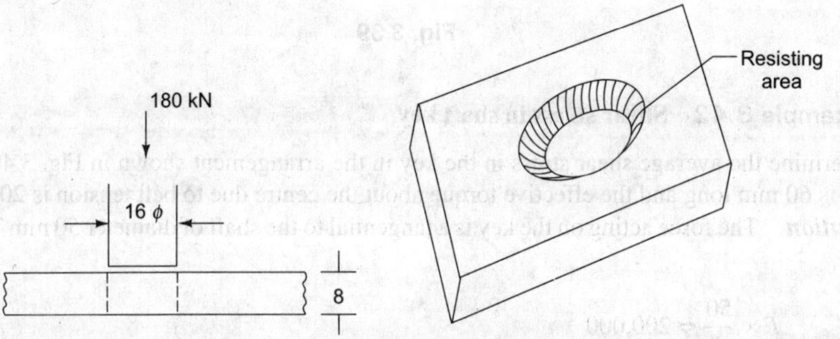

Fig. 3.38

Solution The punching machine exerts a shear force over an area which is cylindrical, as shown in Fig. 3.38. The surface area that is sheared is given by

$$A = \pi d t = \pi \times 16 \times 8 = 402 \text{ mm}^2$$

$$\text{Maximum stress (shear), } \tau = \frac{F}{A} = \frac{180{,}000}{402} = 448 \text{ N/mm}^2$$

Example 3.41 Shear stress in rivets

Determine the maximum shear stress in the rivets in the cases shown in Fig. 3.39. In Fig. 3.39(a), $P = 25$ kN and the diameter of the rivets = 20 mm. In Fig. 3.39(b), $P = 35$ kN and the diameter of the rivets = 16 mm.

Solution The tensile force applied on the plates produces a shearing action on the rivets. While in case (a), a single section of the rivet resists the shear, in case (b), two sections of the rivets resist the shearing action. Let τ be the shearing stress.

In case (a),

$$\tau \frac{\pi}{4} d^2 = T, \quad \tau = \frac{T}{(\pi/4) \times d^2} = \frac{25 \times 1000}{(\pi/4) \times 20^2} = 79.6 \text{ N/mm}^2$$

In case (b),

$$\tau \times 2 \frac{\pi}{4} \times d^2 = T, \quad \tau = \frac{T}{2 \times (\pi/4) \times d^2} = \frac{35{,}000}{2 \times (\pi/4) \times 16^2} = 87 \text{ N/mm}^2$$

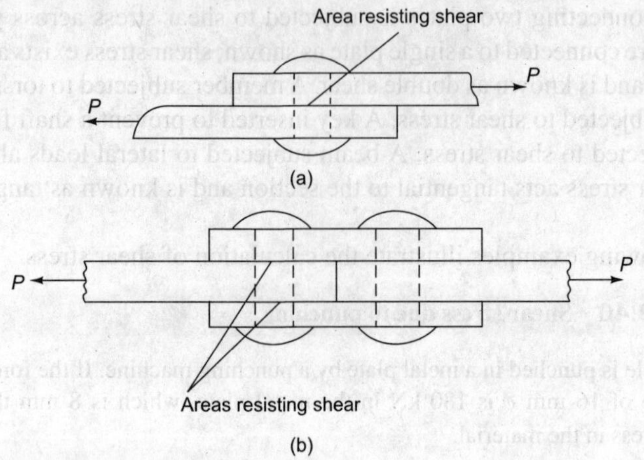

Fig. 3.39

Example 3.42 Shear stress in shaft key

Determine the average shear stress in the key in the arrangement shown in Fig. 3.40. The key is 60 mm long and the effective torque about the centre due to belt tension is 200 Nm.

Solution The force acting on the key is a tangential to the shaft of diameter 50 mm Therefore,

$$F \times \frac{50}{2} = 200{,}000$$

$$\Rightarrow \qquad F = 8000 \text{ N}$$

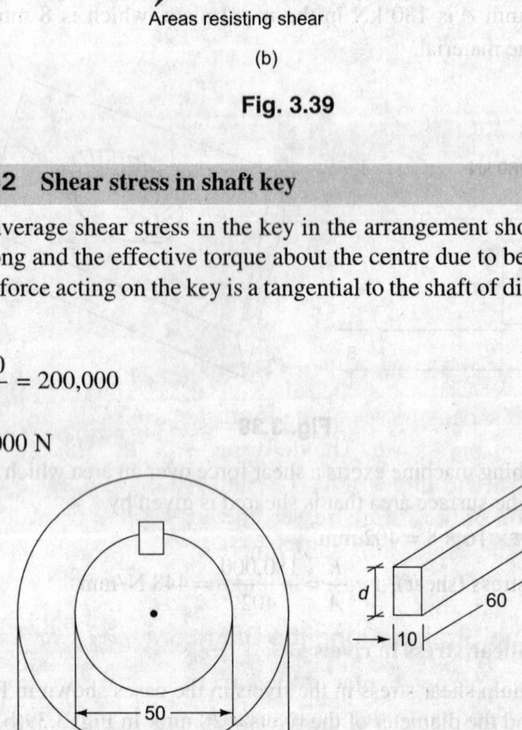

Fig. 3.40

This force acts as a shearing force on an area of 10×60 mm^2. Therefore, the average shearing stress is

$$\tau = \frac{8000}{60 \times 10} = 13.33 \text{ N/mm}^2$$

3.7.1 Complementary Shear Stress

We have already shown that if the shear stress is τ, shear strain $\phi = \Delta L/L = \tau/G$, where G is the shear modulus or modulus of rigidity.

Considering an elementary block of a material subjected to a shear stress τ along the sides $A\,B$ and CD, we immediately notice that while $\Sigma V = 0$ for the stresses shown (Fig. 3.41), $\Sigma M \neq 0$ as the two equal and opposite forces are non-collinear, and give rise to a couple. The equilibrium of the block requires that there be shearing forces on the parallel sides BC and AD in the directions shown in Fig. 3.41(c). It can be easily proved that the intensity of shear stress τ' on these two sides also must be equal to τ. For this, the couple due to shear stresses on AB and $CD = \tau \times AB \times 1 \times AD$, considering a unit length perpendicular to the paper. The couple due to shear stresses on BC and $AD = \tau' \times AD \times 1 \times AB$. Equating the two, $\tau = \tau'$. τ' is known as the complementary shear stress. The shear stress acting in any direction is accompanied by a shear stress acting in a perpendicular direction and of equal magnitude.

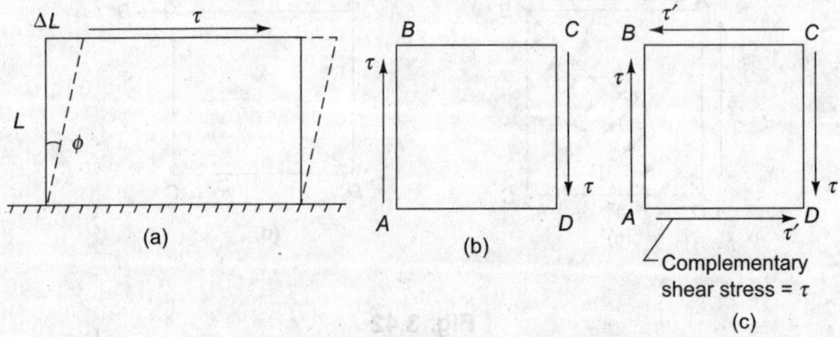

Fig. 3.41

3.7.2 Shear Strain and State of Pure Shear

A block subjected to shear stresses of the form shown in Fig. 3.42(a) is said to be subjected to a state of pure shear. The deformed shape of the material will be as shown in Fig. 3.42(b). The angles at A, B, C and D are originally right angles. Due to the straining action of the shear stresses, the angles at A and C decrease by the shear strain ϕ while the angles at B and D increase by the shear strain ϕ. It can be seen that the diagonal BD has decreased in length while AC has increased in length.

3.7.3 Stresses and Strains Along the Diagonals

Considering thickness perpendicular to the paper as unity, the shear forces acting along the sides will be equal to the shear stress multiplied by the area. Consider a block of equal sides, each side being equal to a.

Force on $AB = \tau \times a \times 1$, which will be same on all sides. The length of diagonals, BD or AC, is equal to $a\,(2)^{1/2}$. If we consider the equilibrium of the part ABD, the free body diagram is as shown in Fig. 3.42(c). Let the stresses on the diagonals be τ'. Resolving the forces perpendicular to the diagonal BD, we get

$$\tau' \times a\sqrt{2} \times 1 = 2 \times \tau \times a \times 1 \times \cos 45^\circ = 2\tau \frac{a}{\sqrt{2}} = \tau a\sqrt{2}$$

$$\tau' = \tau$$

The compressive stresses along the diagonals have the same magnitude as the stresses on the sides. The tensile stress along BD and the compressive stress along AC are both equal to τ.

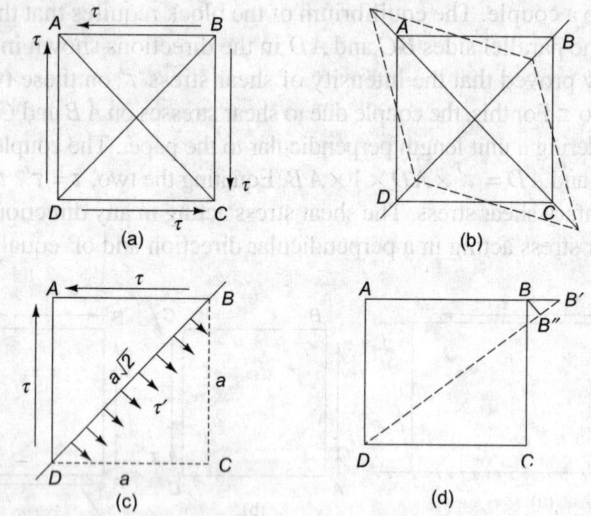

Fig. 3.42

The state of pure shear, with stress of magnitude τ, produces tensile and compressive stresses along the diagonals, also of magnitude τ.

Let us consider the distorted shape shown in Fig. 3.42(d). We draw a perpendicular from B to diagonal $B'D$ to intersect it at B''. $B'B''$ is considered as the strain of the diagonal since the angles ϕ are generally very small. $DB = DB''$ and $B'B' = BB' \cos 45°$.

$$\text{Strain along the diagonals} = \frac{B'B''}{BD} = \frac{BB' \cos 45°}{CD \sec 45°}$$

$$= \frac{1}{2}\frac{BB'}{CD} = \frac{1}{2}\phi$$

The linear strain on the diagonals is equal to half the shear strain.

3.8 LATERAL STRAIN AND POISSON'S RATIO

We have so far considered the effect of an axial force on a bar, as in Fig. 3.43, in terms of axial stresses and strains. When a bar is subjected to an axial tensile force as shown, the following effects are observed.

(a) There is an elongation of the bar in the axial direction given by $\Delta L = \sigma L/E$ where σ is the stress, L is the length, and E is Young's modulus of elasticity.

(b) Along two perpendicular directions, transverse to the axis, there is strain opposite to that along the axis (in this case compression).

(c) There is a change in the volume of the body.

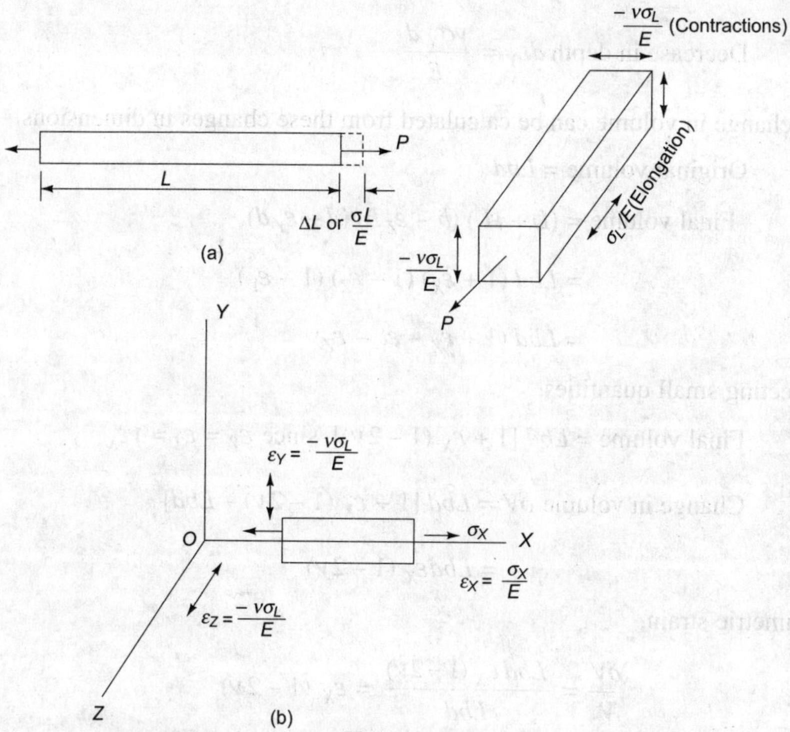

Fig. 3.43

3.8.1 Lateral Strains

The strains that are at right angles to the direction of load are called lateral strains. Also note that the lateral strains have a sense opposite to that of the longitudinal strains. Thus, if the load P is tensile and is taken positive, then the lateral strains are compressive and hence negative. Thus,

Longitudinal strain = + P/AE

Lateral strain in the other two cardinal directions = – (a constant). ε

3.8.2 Poisson's Ratio

In most materials, the ratio of lateral strain to longitudinal strain is fairly constant. This ratio is known as Poisson's ratio. Poisson's ratio is thus the ratio of lateral strain to longitudinal strain. It is denoted by the symbol ν or μ or 1/m. In Fig. 3.43(b), σ_X is the stress acting along the X-direction and the strain ε_X along this direction is σ_X/E. The strains in the lateral directions Y and Z are

$$\varepsilon_Y = -\nu\frac{\sigma_X}{E}, \quad \varepsilon_Z = -\nu\frac{\sigma_X}{E}$$

If the bar has length L and cross-sectional dimensions d and b, then

$$\text{Increase in length } L\varepsilon_X = \frac{\sigma_X L}{E}$$

$$\text{Decrease in width } b\varepsilon_Z = \frac{\nu\sigma_X b}{E}$$

Decrease in depth $d\varepsilon_Y = \dfrac{\nu\sigma_X d}{E}$

The change in volume can be calculated from these changes in dimensions.

Original volume = Lbd

$$\text{Final volume} = (L + \varepsilon L)(b - \varepsilon_Z b)(d - \varepsilon_Y d)$$

$$= Lbd(1 + \varepsilon_X)(1 - \varepsilon_Z)(1 - \varepsilon_Y)$$

$$= Lbd(1 + \varepsilon_X - \varepsilon_Y - \varepsilon_Z)$$

neglecting small quantities.

$$\text{Final volume} = Lbd[1 + \varepsilon_X(1 - 2\nu)], \text{ since } \varepsilon_Y = \varepsilon_Z = \nu\varepsilon_X$$

$$\text{Change in volume } \delta V = Lbd[1 + \varepsilon_X(1 - 2\nu) - Lbd]$$

$$= Lbd\varepsilon_X(1 - 2\nu)$$

Volumetric strain,

$$\frac{\delta V}{V} = \frac{Lbd\,\varepsilon_X(1-2\nu)}{Lbd} = \varepsilon_X(1 - 2\nu)$$

$$\varepsilon_X = \frac{\sigma_X}{E},\ \varepsilon_v = \varepsilon_X(1 - 2\nu)$$

Poisson's ratio = Lateral strain/Longitudinal strain = ν

3.8.3 Uniaxial, Biaxial, and Multi-axial Stresses

The stress conditions in a body have already been described in Section 1.2. Examples of the three stress conditions are shown in Fig. 3.44. A uniaxial stress condition exists [Fig. 3.44(a)] in members of a truss, which carry tensile or compressive forces. A biaxial stress condition exists [Fig. 3.44(b)] in the case of a cylinder under pressure where the longitudinal and hoop stresses form a biaxial stress condition.

Multiaxial stress condition is the most general form of stress where normal stresses act in all the three cardinal directions. A three-dimensional stress system also has shear stresses in all the three planes.

3.8.4 Multi-axial Stresses and Generalized Hooke's Law

Consider the cubic element shown in Fig. 3.44. Normal stresses σ_X, σ_Y, and σ_Z act along the principal directions X, Y, and Z. Let us assume that all the stresses are tensile. Each stress will result in a tensile strain along its direction and the opposite compressive strain along the other two directions. Thus, if ε_X, ε_Y, and ε_Z are the strains along the X, Y, and Z directions, then

$$\varepsilon_X = \frac{\sigma_X}{E} - \frac{\sigma_Y\nu}{E} - \frac{\sigma_Z\nu}{E}$$

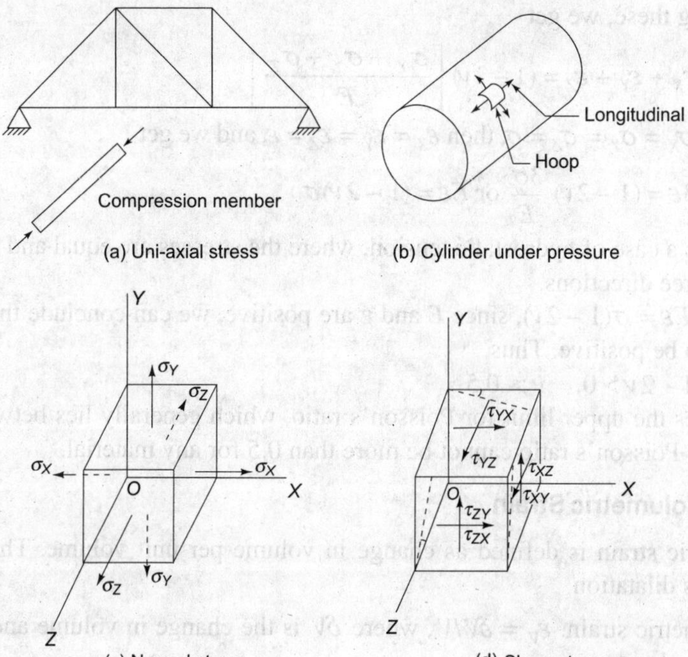

(a) Uni-axial stress (b) Cylinder under pressure

(c) Normal stress (d) Shear stresses

Fig. 3.44

$$\varepsilon_Y = \frac{\sigma_Y}{E} - \frac{\sigma_Z \nu}{E} - \frac{\sigma_X \nu}{E}$$

$$\varepsilon_Z = \frac{\sigma_Z}{E} - \frac{\sigma_X \nu}{E} - \frac{\sigma_Y \nu}{E}$$

These relations of strain in any principal direction in a multi-axial stress system comprise what is known as the generalized Hooke's law.

Similar relationships can be obtained from Fig. 3.44(d) for shear stresses and strains. Each shear stress, say τ_{XY}, is designated by two letters, X to show that it is on a plane whose normal is the X-axis and Y to show the direction of shear stress. The plane whose perpendicular is the X-axis has two shear stresses on it—τ_{XY} and τ_{XZ} and so on. The shear strains are also designated as ϕ_{XY} to show the plane in which it is measured. Thus,

$$\phi_{XY} = \frac{\tau_{XY}}{G}, \quad \phi_{YZ} = \frac{\tau_{YZ}}{G}, \quad \phi_{ZX} = \frac{\tau_{YZ}}{G}$$

These six relationships, three for normal stresses and strains and three for shear stresses and strains, give the most generalized form of Hooke's law.

Let us consider the first three relations of the generalized Hooke's law:

$$\varepsilon_X = \frac{\sigma_X}{E} - \frac{\nu\sigma_Y}{E} - \frac{\nu\sigma_Z}{E}$$

$$\varepsilon_Y = \frac{\sigma_Y}{E} - \frac{\nu\sigma_X}{E} - \frac{\nu\sigma_Z}{E}$$

$$\varepsilon_Z = \frac{\sigma_Z}{E} - \frac{\nu\sigma_X}{E} - \frac{\nu\sigma_Y}{E}$$

Adding these, we get

$$\varepsilon_X + \varepsilon_Y + \varepsilon_Z = (1 - 2\nu)\left[\frac{\sigma_X + \sigma_Y + \sigma_Z}{E}\right]$$

If $\sigma_X = \sigma_Y = \sigma_Z = \sigma$, then $\varepsilon_X = \varepsilon_Y = \varepsilon_Z = \varepsilon$, and we get

$$3\varepsilon = (1 - 2\nu)\frac{3\sigma}{E} \quad \text{or } E\varepsilon = (1 - 2\nu)\sigma$$

This is a case of hydrostatic tension, where the stresses are equal and tensile in all the three directions.

From $E\varepsilon = \sigma(1 - 2\nu)$, since E and ε are positive, we can conclude that $1 - 2\nu$ must also be positive. Thus,

$$1 - 2\nu > 0, \quad \nu \ngtr 0.5$$

This gives the upper limit for Poisson's ratio, which generally lies between 0.25 and 0.34. Poisson's ratio cannot be more than 0.5 for any material.

3.8.5 Volumetric Strain

Volumetric strain is defined as change in volume per unit volume. This is also known as dilatation.

Volumetric strain, $\varepsilon_V = \delta V / V$, where δV is the change in volume and ν is the original volume.

It can be shown that in the case of a multi-axial stress situation, volumetric strain is equal to the sum of the strains along the three mutually perpendicular directions.

3.8.6 Bulk Modulus

We have already come across two moduli of elasticity—E, the Young's modulus for normal stresses and strains and G, the shear modulus or modulus of rigidity.

Bulk modulus is defined in terms of stress and volumetric strain. When a body is subjected to stresses as shown in Fig. 3.45, and all the stresses are equal and like (tensile or compressive), the bulk modulus is given by $K = \sigma/\varepsilon_V$. Here $\sigma = \sigma_X = \sigma_Y = \sigma_Z$ and ε_V = volumetric strain.

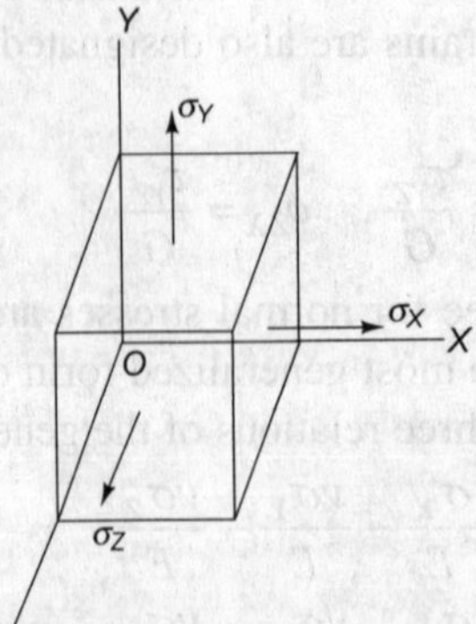

Fig. 3.45

3.9 RELATIONSHIP BETWEEN ELASTIC CONSTANTS

We need the three elastic constants, E, G, and K of a material and its Poisson's ratio ν, to be able to completely calculate its deformations. The relationship between these can be derived as follows.

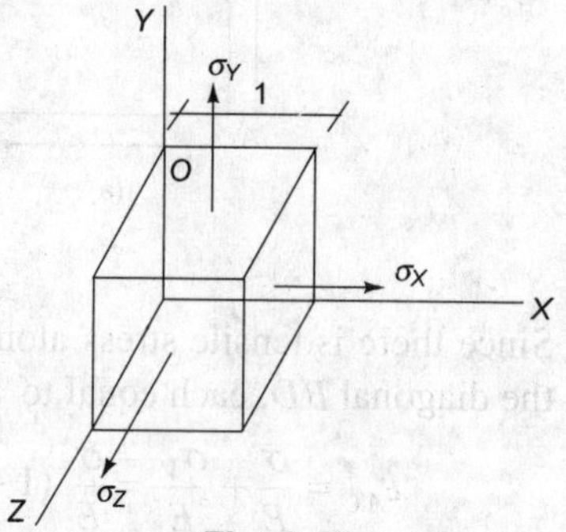

Fig. 3.46

Consider a cube of side 1 (unit length) subjected to tensile multi-axial stresses σ_X, σ_Y, and σ_Z, as shown in Fig. 3.46. Under the action of these stresses, the body deforms into a rectangular parallelepiped, and its volume is given by

$$V' = (1 + \varepsilon_X)(1 + \varepsilon_Y)(1 + \varepsilon_Z)$$

Original volume $V = 1 \times 1 \times 1 = 1$ cubic units and ε_X, ε_Y, and ε_Z are the strains in the respective directions.

$$V' = (1 + \varepsilon_X + \varepsilon_Y + \varepsilon_X \varepsilon_Y)(1 + \varepsilon_Z)$$

$$= 1 + \varepsilon_X + \varepsilon_Y + \varepsilon_X \varepsilon_Y + \varepsilon_Z + \varepsilon_X \varepsilon_Z + \varepsilon_Y \varepsilon_Z + \varepsilon_X \varepsilon_Y \varepsilon_Z$$

Note that ε_X, ε_Y, and ε_Z are small quantities and hence the terms $\varepsilon_X \varepsilon_Y$, $\varepsilon_X \varepsilon_Z$, $\varepsilon_Y \varepsilon_Z$, and $\varepsilon_X \varepsilon_Y \varepsilon_Z$ are smaller quantities which can be neglected.

$$V' = 1 + \varepsilon_X + \varepsilon_Y + \varepsilon_Z \quad \text{and} \quad \text{change in volume} = \varepsilon_X + \varepsilon_Y + \varepsilon_Z$$

Since $V = 1$,

$$\text{Volumetric strain, } \varepsilon_V = (\varepsilon_X + \varepsilon_Y + \varepsilon_Z)$$

Here ε_V is the change in volume per unit volume and is known as dilatation. Each strain can be expressed in terms of stresses as

$$\varepsilon_X = \frac{\sigma_X}{E} - \frac{\sigma_Y \nu}{E} - \frac{\sigma_Z \nu}{E} = \frac{\sigma_X}{E} - \frac{\nu}{E}(\sigma_X + \sigma_Z)$$

Similarly,

$$\varepsilon_Y = \frac{\sigma_Y}{E} - \frac{\nu}{E}(\sigma_Z + \sigma_X) \quad \text{and} \quad \varepsilon_Z = \frac{\sigma_Z}{E} - \frac{\nu}{E}(\sigma_X + \sigma_Y)$$

$$\varepsilon_X + \varepsilon_Y + \varepsilon_Z = \frac{\sigma_X + \sigma_Y + \sigma_Z}{E} - \frac{2\nu}{E}(\sigma_X + \sigma_Y + \sigma_Z) = \varepsilon_V$$

$$\varepsilon_V = \frac{(\sigma_X + \sigma_Y + \sigma_Z)}{E}(1 - 2\nu)$$

If $\sigma_X = \sigma_Y = \sigma_Z = \sigma$, then $\varepsilon_V = (3\sigma/E)(1 - 2\nu)$. In such a case, $K = \sigma/\varepsilon_V$ by definition. Therefore,

$$E = \frac{3\sigma}{\varepsilon_V}(1 - 2\nu) = 3K(1 - 2\nu), \quad K = \frac{E}{3(1 - 2\nu)}$$

This gives the relationship between Young's modulus E and bulk modulus K.

To derive a relationship between E and G, consider the case of pure shear shown in Fig. 3.47. This case has already been considered and we have found that there are compressive and tensile stresses along diagonals AC and BD, and that these stresses are equal to the shear stress τ. We have also found that the diagonal strains are each equal to half the shear strain, i.e., $\varepsilon_{AC} = \phi/2$. Therefore,

$$\varepsilon_{AC} = \frac{\phi}{2} = \frac{\tau}{2G}$$

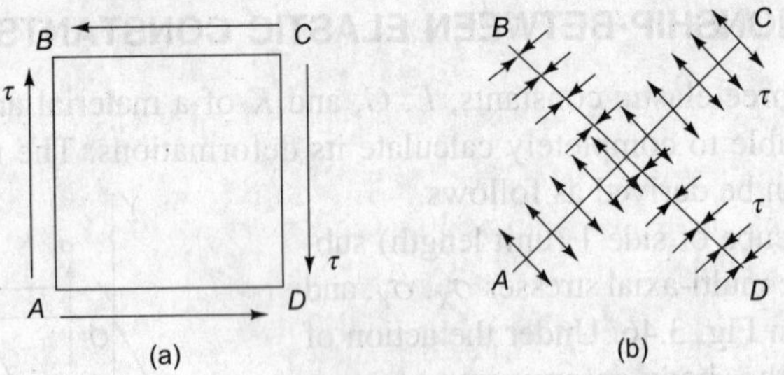

Fig. 3.47

Since there is tensile stress along the diagonal AC and compressive stress along the diagonal BD, each equal to τ in magnitude, from Fig. 3.47(b),

$$\varepsilon_{AC} = \frac{\sigma}{E} + \frac{\sigma_V}{E} = \frac{\sigma}{E}(1+\nu) = \frac{\tau}{E}(1+\nu).$$

Therefore,

$$\frac{\tau}{2G} = \frac{\tau}{E}(1+\nu)$$

$$\Rightarrow \quad E = 2G(1+\nu)$$

$$\Rightarrow \quad G = \frac{E}{2(1+\nu)}$$

This is the relationship between E and G.
Also, since $E = 3K(1-2\nu)$,

$$G = \frac{3K(1-2\nu)}{2(1+\nu)}$$

gives the relationship between G and K. Eliminating ν from these equations, one can get $E = 9KG/(3K+G)$.

Example 3.43 Change in dimensions of a steel bar due to axial stress

A steel bar 25 mm × 15 mm in cross section is 300 mm long and is subjected to a tensile force of 70 kN. Find the change in the dimensions of the bar and the change in volume. E = 200 GPa and $\nu = 0.3$.

Solution The situation is represented in Fig. 3.48(a). This is a case of uniaxial stress but will have an effect on the lateral dimensions.

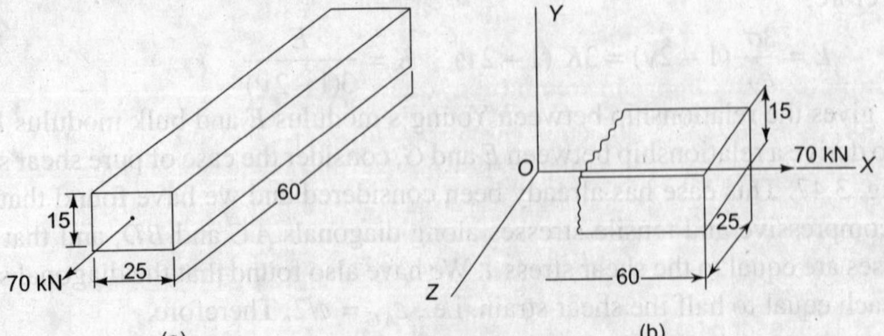

Fig. 3.48

In Fig. 3.48(b),

$$\text{Longitudinal strain, } \varepsilon_X = \frac{70{,}000}{25 \times 15 \times 200{,}000} = 0.00093$$

Lateral strain = $\nu\varepsilon_X = 0.00093 \times 0.3 = 0.00028$

Change in length = $0.00093 \times 300 = 0.279$ mm

Change in width = $0.00028 \times 25 = 0.007$ mm

Change in height = $0.00028 \times 15 = 0.0042$ mm

Volume of bar = $300 \times 25 \times 15 = 112{,}500$

Volumetric strain $\varepsilon_V = \varepsilon_X + \varepsilon_Y + \varepsilon_Z \quad = 0.00093 - 0.00028 - 0.00028$
$= 0.00037$

Change in volume = $V\varepsilon_V = 341.625$ mm^3 (increase)

Example 3.44 Changes in dimensions of a hollow circular tube

A steel tube of outside diameter 250 mm and thickness 10 mm is 2 m long, and carries a load of 1000 kN. Find the changes in length, outside diameter and thickness due to the tensile force if $E = 200$ GPa and $\nu = 0.33$.

Solution From Fig. 3.49, the strain along the length ε_X is given by

$$\varepsilon_X = \frac{P}{AE} = \frac{1000 \times 1000}{(\pi/4) \times (250^2 - 230^2)} = 0.00066$$

Lateral strain = $\varepsilon_X \nu = 0.00066 \times 0.33 = 0.00022$

Change in length = $\varepsilon_X L = 0.00066 \times 2000 = 0.165$ mm

Change in outside diameter = lateral strain × diameter = 0.00022×250
$= 0.055$ mm

Change in inside diameter = $0.00022 \times 230 = 0.0506$

$$\text{Change in thickness} = \frac{0.055 - 0.0506}{2} = 0.0022 \text{ mm}$$

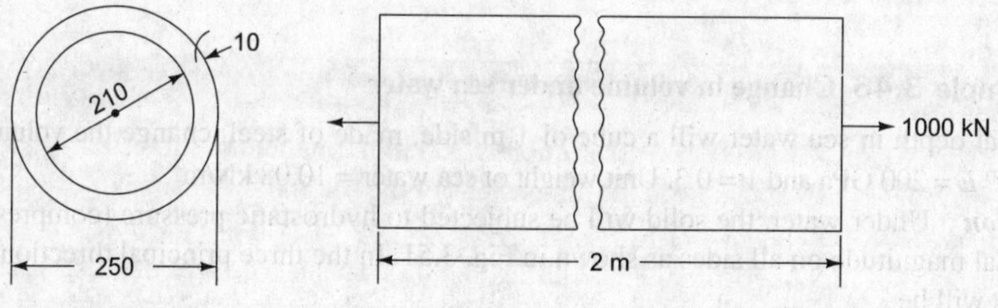

Fig. 3.49

Example 3.45 Change in volume of a square steel bar

A steel bar 20 mm square in section is subjected to an axial compressive load of 60 kN. Find the percentage change in volume if the bar is 500 mm long. What are the equal stresses that must be applied to the sides of the bar if the volumetric change is to be zero? $E = 200$ GPa and $\nu = 0.3$.

Solution From Fig. 3.50, taking compressive strains as negative,

$$\text{Longitudinal strain, } \varepsilon_X = \frac{-\sigma}{E} = \frac{-P}{AE} = \frac{-60 \times 1000}{20 \times 20 \times 200{,}000} = -0.00075$$

(compressive)

Lateral strains, $\varepsilon_Y = \varepsilon_Z = -\nu\varepsilon_X = 0.3 \times 0.00075 = 0.000225$ (tensile)

$\varepsilon_V = \varepsilon_X + \varepsilon_Y + \varepsilon_Z = -0.00075 + 0.000225 \times 2 = -0.3$

Volume, $V = 20 \times 20 \times 500 = 200{,}000\ \text{mm}^3$

Change in volume $= V\varepsilon_V = 200{,}000 \times -0.3 = -60\ \text{mm}^3$ (decrease)

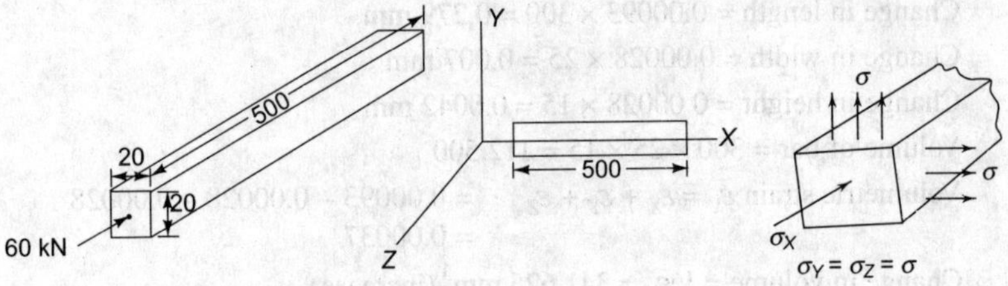

Fig. 3.50

To make the volume change zero, let tensile stresses of magnitude σ be applied in the Y- and Z-direction.

$$\varepsilon_X = \frac{\sigma_X}{E} - \frac{2\sigma\nu}{E}, \quad \varepsilon_Y = \varepsilon_Z = \frac{\sigma}{E} - \frac{\sigma\nu}{E} + \frac{\sigma_X\nu}{E}$$

$$\varepsilon_V = \varepsilon_X + \varepsilon_Y + \varepsilon_Z = -\frac{\sigma_X}{E} - \frac{2\sigma\nu}{E} + \frac{2\sigma}{E} - \frac{2\sigma\nu}{E} + \frac{2\sigma_X\nu}{E}$$

For no change in volume, $\varepsilon_V = 0$; that is,

$$-\sigma_X - 2\sigma\nu + 2\sigma - 2\sigma\nu + 2\sigma_X\nu = 0$$

$$\sigma_X = \frac{60 \times 1000}{20 \times 20} = 150\ \text{N/mm}^2, \quad \nu = 0.3$$

$$\sigma(2 - 4\nu) = \sigma_X(1 - 2\nu), \quad \sigma = \frac{150\,(1 - 0.6)}{(2 - 1.2)} = 75\ \text{N/mm}^2$$

Tensile stresses of 75 N/mm^2 have to be applied on the sides to make the volume change zero.

Example 3.46 Change in volume under sea water

At what depth in sea water will a cube of 1 m side, made of steel, change the volume by 0.05%? $E = 200$ GPa and $\nu = 0.3$. Unit weight of sea water = 10.08 kN/m^3.

Solution Under water, the solid will be subjected to hydrostatic pressure (compressive) of equal magnitude on all sides as shown in Fig. 3.51. In the three principal directions, the strains will be

$$\varepsilon_X = \frac{\sigma}{E}(1 - 2\nu) = \varepsilon_Y = \varepsilon_Z$$

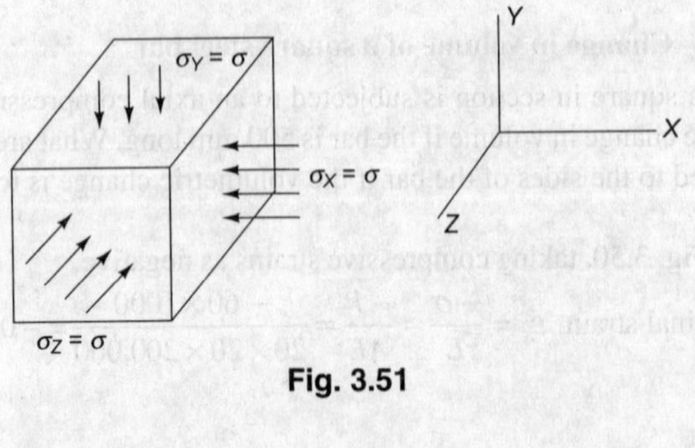

Fig. 3.51

Therefore,

$$\varepsilon_V = \varepsilon_X + \varepsilon_Y + \varepsilon_Z = \frac{3\sigma}{E}(1-2\nu)$$

$$\text{Change in volume} = \frac{0.05}{100} \times 1 = \varepsilon_V = \frac{3\sigma}{E}(1-2\nu)$$

$$\sigma = \frac{0.05}{100} \times \frac{E}{3(1-2\nu)} = \frac{0.05 \times 200{,}000}{100 \times 3(1-2\times 0.3)} = 83.33 \text{ N/mm}^2$$

Pressure at any depth, $h = wh = 10{,}080h$ N/m^2

Equating this to $\sigma = 83.33$ N/mm^2 = 83.33×10^6 N/m^2,

$$10{,}080h = 83.33 \times 10^6, \quad h = 8267 \text{ m}$$

Example 3.47 Changes in dimensions of a plate and that of a line etched on it

A steel plate of dimensions 300 × 400 × 10 mm is subjected to an axial stress of 150 N/mm^2 parallel to the 400 mm side as shown. Determine the percentage change in volume of the plate. What will be the change in the slope of a diagonal? If a circular area (80 mm dia.) is etched on the surface of the plate, how will this area change in shape and size? E = 200 GPa and $\nu = 0.3$.

Solution The situation is shown in Fig. 3.52.

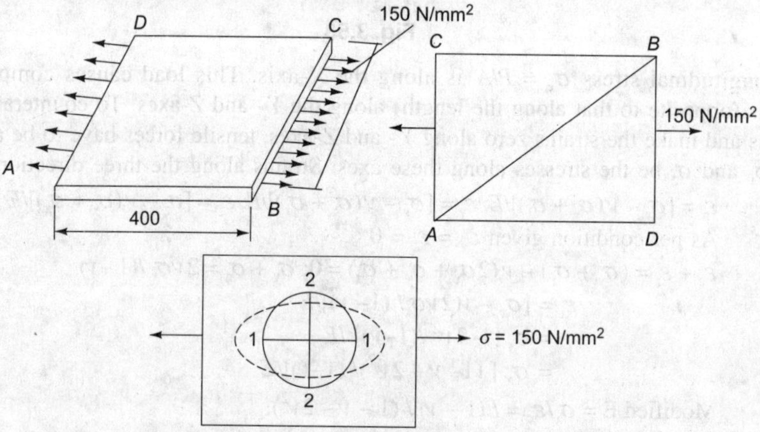

Fig. 3.52

$$\varepsilon_X = \frac{150}{200{,}000} = 0.00075, \quad \varepsilon_Y = \frac{-150}{200{,}000} \times 0.3 = -0.000225 = \varepsilon_Z$$

Volume of the plate = 300 × 400 × 10 = 1,200,000 mm^3

Change in volume = $\varepsilon_V V$

ε_V = volumetric strain = $\varepsilon_X + \varepsilon_Y + \varepsilon_Z = 0.0003$

$$\text{Percentage change in volume} = \frac{0.0003 \times 1{,}200{,}000 \times 100}{1{,}200{,}000} = 0.03$$

The diagonal AC has a slope given by $\tan^{-1}(3/4) = \tan^{-1}(0.75)$.

Change in width = 300 × (– 0.000225) = – 0.0675

Change in length = 400 × (+ 0.00075) = 0.3

$$\text{New slope of diagonal } AC = \tan^{-1}\left[\frac{(300-0.0675)}{400+0.3}\right] = \tan^{-1}(0.7492)$$

The circle etched on the plate will deform into an elliptical shape. The horizontal diameter increases in length while the vertical diameter decreases in length.

Length of diameter 1-1 $= 80 + 80 \times 0.00075$

$= 80.06$ mm

Length of diameter 2-2 $= 80 - 80 \times 0.000225 = 79.982$ mm

The de formed shape will be as shown.

Example 3.48 Longitudinal stress and compressive strains on a bar

If a bar having area of cross-section *A* is stretched by a force *P* and lateral strains in orthogonal directions is prevented by applying forces, find the modified value of *E*.

Solution The situation is shown in Fig. 3.53.

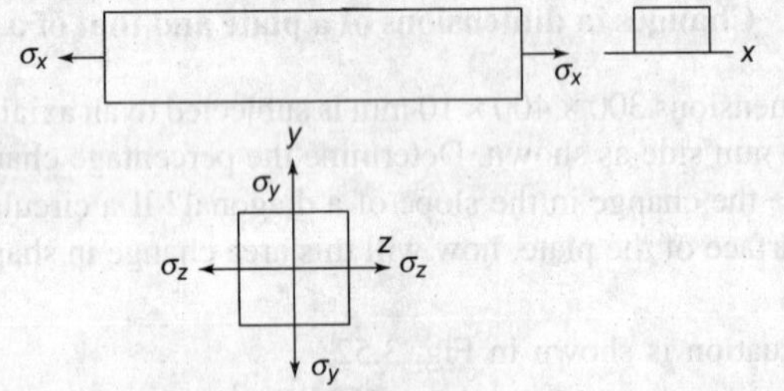

Fig. 3.53

Longitudinal stress $\sigma_x = P/A$ is along the *X*-axis. This load causes compressive strains (opposite to that along the length) along the *Y*- and *Z*-axes. To counteract these strains and make the strains zero along *Y*- and *Z*-axes, tensile forces have to be applied. Let σ_y and σ_z be the stresses along these axes. Strains along the three directions are

$\varepsilon_x = [\sigma_x - \nu(\sigma_y + \sigma_z)]/E$; $\varepsilon_y = [\sigma_y - \nu(\sigma_z + \sigma_x)]/E$; $\varepsilon_z = [\sigma_z - \nu(\sigma_y + \sigma_x)]/E$

As per condition given $\varepsilon_y = \varepsilon_z = 0$

$\varepsilon_y + \varepsilon_z = (\sigma_y + \sigma_z) - \nu(2\sigma_x + \sigma_y + \sigma_z) = 0$; $\sigma_y + \sigma_z = 2\nu\sigma_x/(1 - \nu)$

$\varepsilon_x = [\sigma_x - \nu(2\nu\sigma_x/(1 - \nu)]/E$

$= \sigma_x [1 - 2\nu^2/(1 - \nu)]/E$

$= \sigma_x [(1 - \nu - 2\nu^2)/(1 - \nu)]/E$

Modified E $= \sigma_x/\varepsilon_x = E(1 - \nu)/(1 - \nu - 2\nu^2)$

Example 3.49 Calculation of modulus of elasticity, modulus of rigidity, bulk modulus, and Poisson's ratio

A 25 mm diameter bar when subjected to a force of 40 kN has an extension of 0.08 mm on a gauge length of 200 mm. If the diametrical reduction is 0.003 mm, find the values of *E*, *G*, *K*, Poisson's ratio.

Solution Area of the bar $= p\,(25)^2/4 = 491$ mm^2

Stress $\sigma = 40{,}000/491 = 81.47$ N/mm^2

Strain ε = change in length /original length $= 0.08/200 = 4 \times 10^{-4}$

Modulus of elasticity $E = \sigma/\varepsilon = 81.47/(4 \times 10^{-4}) = 203$ GPa

Lateral strain ε (along diameter) $= 0.003/25 = \nu\varepsilon$

Poisson's ratio $\nu = 0.003/(25 \times 4 \times 10^{-4}) = 0.3$

Modulus of rigidity $G = E/[2(1 + \nu)] = 203/[2(1 + 0.3)] = 78.1$ GPa

Bulk modulus $K = E/[3(1 - 2\nu)] = 203/[3(1 - 2 \times 0.3)] = 169.2$ GPa

3.10 SOME INDETERMINATE PROBLEMS

We have already seen that in certain problems, as in composite sections or problems involving temperature stresses, the equations of static equilibrium are not enough to determine the unknowns. Such problems are classified as statically indeterminate. The additional equations needed for the solution are obtained by determining compatibility conditions based upon the deformation characteristics of the member. A simple example will illustrate the point.

Considering the member shown in Fig. 3.54, restrained at A and B, subjected to a force P as shown. We draw a free body diagram of the member. The end restraining forces are R_A and R_B.

$R_A + R_B = P$ is the static condition of equilibrium from $\Sigma H = 0$. There are two unknowns in this equation, which cannot be determined unless we formulate another equation relating these two.

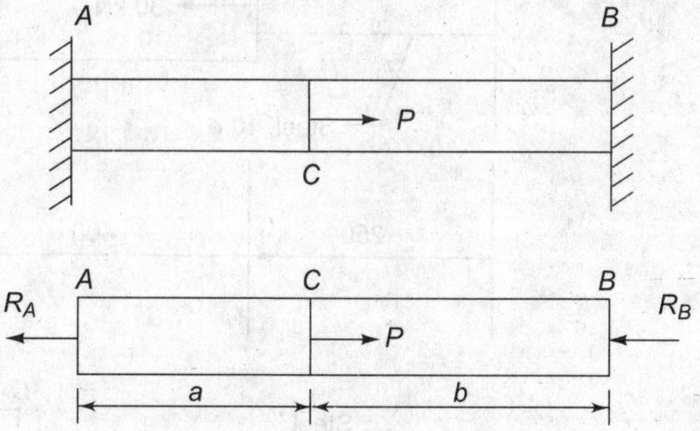

Fig. 3.54

Here we note that while AC is in tension, CB is in compression. The compatibility of deformation demands that the stretching of AC equals the shortening of CB. We can calculate these two quantities from what we have learnt so far.

$$\text{Change in length} = \frac{PL}{AE} = \frac{\text{force} \times \text{length}}{\text{area} \times E}$$

$$\text{Stretch of } AC = \frac{R_A AC}{AE} = \frac{R_A a}{AE}$$

$$\text{Shortening of } CB = \frac{R_B CB}{AE} = \frac{R_B b}{AE}$$

Equating the two,

$$\frac{R_A a}{AE} = \frac{R_B b}{AE}$$

This is the compatibility condition.

For a uniform member, A and E are constant. Therefore,

$$R_A a = R_B b \quad \text{or} \quad R_A = R_B \frac{b}{a}$$

Substituting,

$$R_B \frac{b}{a} + R_B = P \quad \text{or} \quad R_B = \frac{Pa}{a+b} \quad \text{and} \quad R_A = \frac{Pb}{a+b}$$

The following examples illustrate the procedure in a variety of cases.

Example 3.50 Indeterminate problem: reactions at the supports due to load in steel and aluminium bars

A steel rod, 10 mm ϕ, and an aluminium rod, 20 mm ϕ, are joined together fixed between supports as shown in Fig. 3.55. E for steel = 200 GPa and E for aluminium = 70 GPa. Find the reactions at the supports and the stresses in the metals.

Solution **If R_A and R_B are the reactions at the ends, then, from the conditions of equilibrium,**

$$R_A + R_B = 30 \text{ kN} = 30{,}000 \text{ N}$$

From the free body diagrams of AC and BC, the elongation of AC is equal to the shortening of BC.

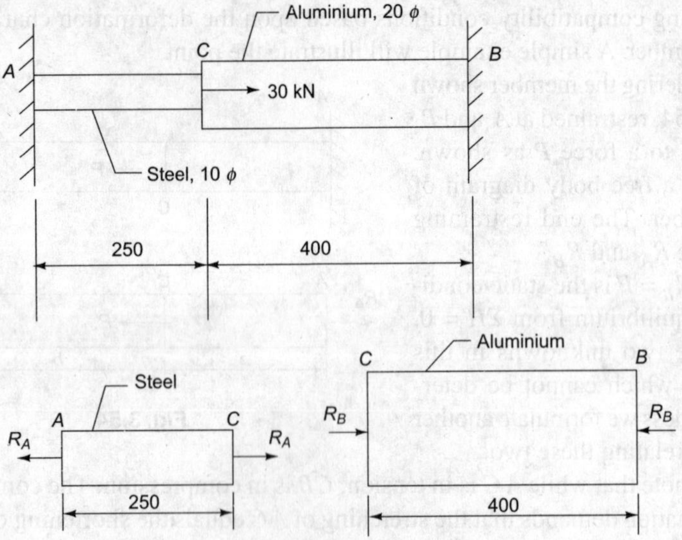

Fig. 3.55

$$\text{Elongation of } AC = \frac{PL}{AE} = \frac{R_A \times 250}{(\pi/4) \times 10^2 \times 200{,}000}$$

$$\text{Shortening of } BC = \frac{R_B \times 400}{(\pi/4) \times 20^2 \times 70{,}000}$$

Therefore,

$$\frac{R_A \times 250}{(\pi/4) \times 10^2 \times 200{,}000} = \frac{R_B \times 400}{(\pi/4) \times 20^2 \times 70{,}000}$$

according to the compatibility condition.

$$R_A = 1.143 R_B$$

$$1.143 R_B + R_B = 30{,}000, \quad R_B = 14{,}000 \text{ N}, \quad R_A = 16{,}000 \text{ N}$$

$$\text{Stress in steel} = \frac{16{,}000}{(\pi/4) \times 10^2} = 203.7 \text{ N/mm}^2$$

$$\text{Stress in aluminium} = \frac{14{,}000}{(\pi/4) \times 20^2} = 44.56 \text{ N/mm}^2$$

Example 3.51 Indeterminate problem in stress

In the arrangement shown in Fig. 3.56, the rigid bar is initially supported by the two outer steel rods 20 mm ϕ and 500 mm long. The central steel rod is 30 mm ϕ but 0.1 mm shorter than the other two bars. Determine the maximum value of P so that the central bar is not stressed. If this maximum load is doubled, what will be the stresses in the three bars? E = 200 GPa.

Solution **From considerations of symmetry, we can conclude that the outer two bars**

will carry equal loads. The central bar will not be stressed so long as the deformation of the outer bars is less than or equal to 0.1 mm.

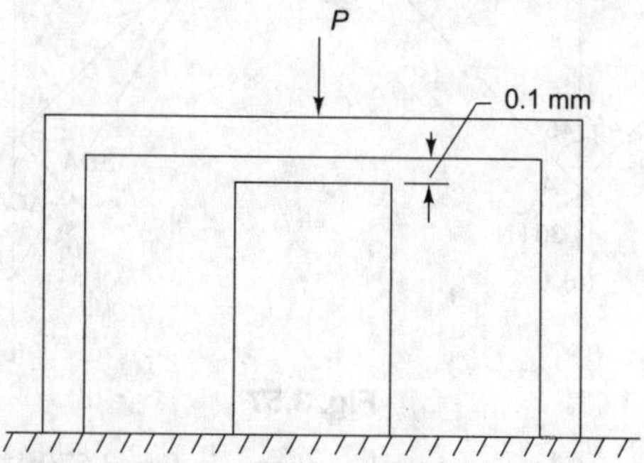

Fig. 3.56

If P is the load, the share of each outer bar is $P/2$.

$$\text{Elongation} = \frac{PL}{AE} = \frac{P \times 500}{2 \times (\pi/4) \times 20^2 \times 200{,}000}$$

For the maximum value of P (without stressing the central bar), this elongation is equal to 0.1 mm.

$$\frac{P \times 500}{2 \times (\pi/4) \times 20^2 \times 200{,}000} = 0.1 \text{ mm}$$

$$\Rightarrow \qquad P = 25 \text{ kN}$$

If this load is doubled to 50 kN, then the central bar is also stressed. Let P_1 and P_2 be the share of the loads of the outer and central bars. Then,

$$P_1 + P_1 + P_2 = (50 - 25) \text{ kN} = 25{,}000 \text{ N}$$

The elongation of all the bars will be equal. Therefore,

$$\frac{P_1 \times 499.9}{(\pi/4) \times 20^2 \times E} = \frac{P_2 \times 499.9}{(\pi/4) \times 30^2 \times E}, \; P_1 = P_2$$

$$\frac{4}{9} P_2 + \frac{4}{9} P_2 + P_2 = 25{,}000, \quad P_2 = 13{,}235 \text{ kN}$$

$$P_1 = 5882 \text{ N}$$

Total load taken by each outer bar = 12,500 + 5882 = 18,382 N

Stress in each outer bar = 58.5 N/mm^2

Stress in inner bar = 18.73 N/mm^2

Example 3.52 Indeterminate problem: three bars carrying a weight

In the framework shown in Fig. 3.57, determine the forces in each bar, if the central bar is of steel and the outer bars are of brass. The steel bar has a cross-sectional area of 200 mm^2 and the brass bar has an area of 300 mm^2. E = 200 GPa for steel and 100 GPa for brass. The length of the steel bar is 2 m.

Solution Note that the framework is statically indeterminate. The equation $\Sigma H = 0$ tells us that, because of symmetry, the forces in the two brass bars are equal. $\Sigma_V = 0$ gives us one equation,

$$2P_b \sin 45° + P_s = 30{,}000 \text{ N (static equilibrium)},$$

where P_b and P_s are the forces in the brass and steel bars, respectively.

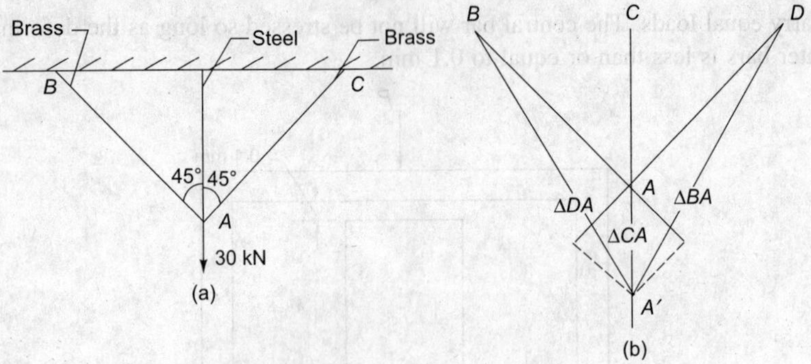

Fig. 3.57

The deformed shape of the framework is shown in Fig. 3.57(b). Note that the deformations are small compared to the lengths of bars. A moves down to A' and the angles made by the brass bars with the vertical are nearly equal to 45°. If ΔCA is the vertical downward movement of A, then ΔBA, the deformation in the bar BA is related to ΔCA by

$$\Delta BA = \Delta CA \cos 45^\circ$$

by the deformation condition.

$$\Delta BA = \frac{P_b \times 2000 \sec 45^\circ}{300 \times 100{,}000} \quad \text{and} \quad \Delta CA = \frac{P_s \times 2000}{200 \times 200{,}000}$$

Substituting,

$$\frac{P_b \times 2000\sqrt{2}}{300 \times 100{,}000} = \frac{P_s \times 2000\,(1/\sqrt{2})}{200 \times 200{,}000}$$

$$\Rightarrow \quad P_b = 0.375 P_s$$

$$2 \times 0.375 \times P_s \frac{1}{\sqrt{2}} + P_s = 30{,}000$$

$$\Rightarrow \quad P_s = 19{,}604 \text{ N}, \quad P_b = 7351 \text{ N}$$

Stress in steel,

$$\sigma_s = \frac{19{,}604}{200} = 98.0 \text{ N/mm}^2$$

Stress in brass,

$$\sigma_b = \frac{7351}{300} = 24.5 \text{ N/mm}^2$$

$$\text{Central vertical deflection} = \frac{19{,}604 \times 2000}{200 \times 200{,}000} = 0.9802 \text{ mm}$$

$$\text{Elongation of brass bar} = 0.9802 \times \frac{1}{\sqrt{2}} = 0.6931 \text{ mm}$$

Example 3.53 Indeterminate problem: stresses in brass and steel bars

In the mechanism shown in Fig. 3.58, find the stresses in the brass and steel bars. The brass bar has an area of 600 mm^2 and the steel bar has an area of 400 mm^2. E = 200 GPa for steel and 90 GPa for brass.

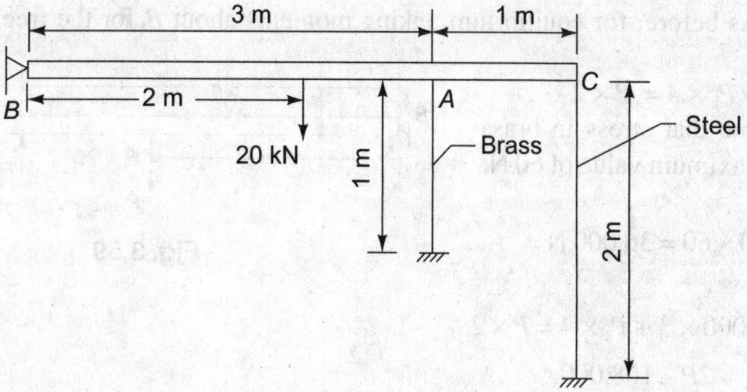

(a)

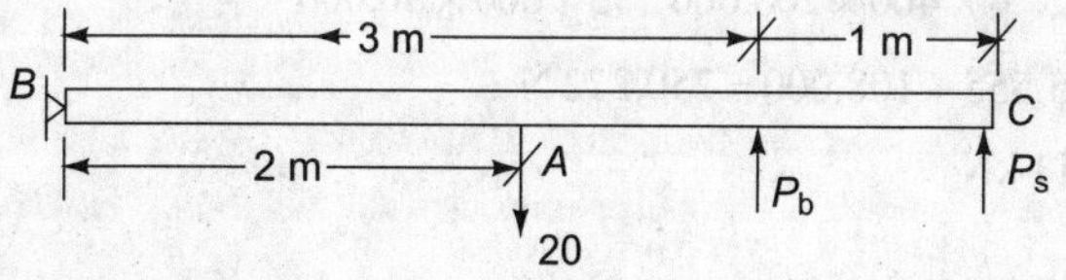

(b)

Fig. 3.58

Solution The free body diagram of the rigid bar is shown in Fig. 3.58(b). The bar is free to rotate about point B. If P_b and P_s are the forces in the bars, of brass and steel respectively, then taking moment about B for equilibrium,

$$\Sigma M = 0 \ @ \ B, \quad P_b \times 3 + P_s \times 4 - 20 \times 2 \times 1000 = 0 \qquad (1)$$

If Δs and Δb are the deformations (shortening as the bars are in compression), then from the principle of similar triangles,

$$\frac{\Delta s}{4} = \frac{\Delta b}{3} \text{ (deformation condition)}$$

$$\Delta s = \frac{P_s \times 2000}{400 \times 200{,}000}, \quad \Delta b = \frac{P_b \times 2000}{600 \times 90{,}000}$$

Therefore,

$$\frac{1}{4}\left(\frac{P_s \times 2000}{400 \times 200{,}000}\right) = \frac{1}{3}\left[\frac{P_b \times 1000}{600 \times 90{,}000}\right], \quad P_s = P_b \times 0.9876$$

$$P_b \times 3 + (P_b \times 0.9876) \times 4 = 40{,}000, \quad P_b = 5755 \text{ N}, \quad P_s = 5684 \text{ N}$$

Stress in the brass bar,

$$\sigma_b = \frac{5755}{600} = 9.59 \text{ N/mm}^2$$

Stress in the steel bar,

$$\sigma_s = \frac{5684}{400} = 14.21 \text{ N/mm}^2$$

Example 3.54 Indeterminate problem: load capacity of brass and steel bars

If in the mechanism shown in Fig. 3.59, the maximum stresses in the brass and steel bars are limited, respectively, to 60 N/mm^2 and 100 N/mm^2, what is the maximum value of the load at A? The other data remain the same as in Example 3.53.

Solution As before, for equilibrium, taking moments about B for the free body in Fig. 3.59.

$$P_b \times 3 + P_s \times 4 = P \times 2$$

Let us assume that stress in brass reaches its maximum value of 60 N/mm^2. Then,

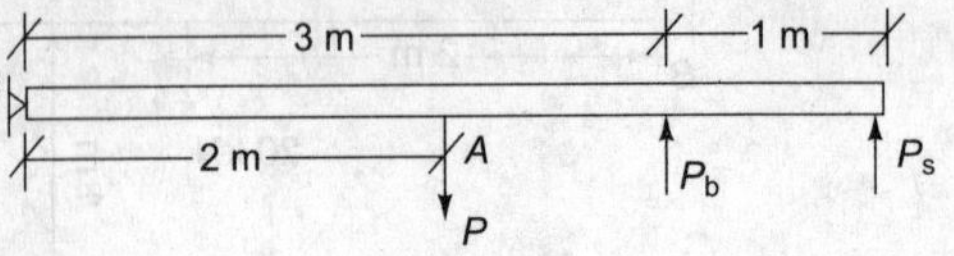

Fig. 3.59

$$P_b = 600 \times 60 = 36{,}000 \text{ N}$$

Therefore,

$$36{,}000 \times 3 + P_s \times 4 = P \times 2$$

$$4P_s = 2P - 108{,}000$$

$$\frac{\Delta s}{4} = \frac{\Delta b}{3}, \quad \frac{P_s \times 2000}{4 \times 400 \times 200{,}000} = \frac{36{,}000 \times 1000}{3 \times 600 \times 90{,}000}, \quad P_s = 35{,}555 \text{ N}$$

$$2P = 4 \times 35{,}555 + 108{,}000 = 250{,}222 \text{ N}$$

$$P = 125.111 \text{ kN}$$

Stress in steel,

$$\sigma_s = \frac{35{,}555}{400} = 88.9 \text{ N/mm}^2 < 100 \text{ N/mm}^2$$

Example 3.55 Indeterminate problem: stresses due to load and temperature

In the framework shown in Fig. 3.60(a), the rigid horizontal bar ABC is supported by a steel rod and a copper bar as shown. The mechanism is stress-free at 20°C. Find the stresses in the bars when the temperature rises to 50°C. For steel E = 200 GPa, $\alpha = 12 \times 10^{-6}$/°C, and for copper E = 120 GPa and $\alpha = 18 \times 10^{-6}$/°C.

Solution Figure 3.60(a) shows the framework. Due to the 100 kN load, both points B and C come down. Due to the rise in temperature, point B moves up while point C comes down. Let the net deformation be such that the rigid bar ABC rotates clockwise as shown in Fig. 3.60(b). Let P_s and P_c be the net forces in the steel and copper bars, respectively. Then for the steel bar, elongation due to temperature rise,

$$\Delta s_1 = 2000 \times 12 \times 10^{-6} \times (50 - 20)$$
$$= 0.72 \text{ mm}$$

Elongation due to temperature rise in the copper bar,

$$\Delta c_1 = 1000 \times 18 \times 10^{-6} \times 30$$
$$= 0.54 \text{ mm}$$

If Δs and Δc are the net downward deformations, then

$$\frac{\Delta s}{2} = \frac{\Delta c}{3}$$

by the deformation condition. Let Δs_1 and Δc_1 be the increase in lengths of the steel and brass bars due to temperature rise and Δs_2 and Δc_2 be the change in length (due to the applied force of 100 kN) due to P_s and P_b. Δs_1 and Δc_1 have been calculated above. Δs is equal to $\Delta s_2 - \Delta s_1$, where Δs_2 is the compressive strain due to P_s, i.e., $P_s \times 2000/(800 \times 200{,}000) = P_s/80{,}000$. Δc is equal to $\Delta c_2 + \Delta c_1$, where Δc_2 is the tensile strain due to P_c, i.e., $P_c \times 1000/(600 \times 120{,}000) = P_c/72{,}000$. Therefore,

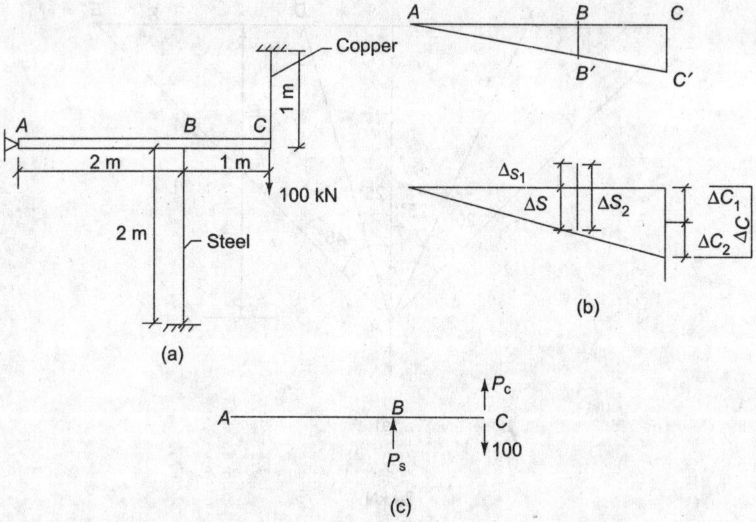

Fig. 3.60

$$\frac{(P_s/80{,}000)-0.72}{2}=\frac{0.54+(P_c/72{,}000)}{3}$$

or $$\frac{3P_s}{80{,}000}-2.16=1.08+\frac{2P_c}{36{,}000}$$

or $$\frac{3P_s}{80{,}000}-\frac{2P_c}{72{,}000}=3.24 \quad (1)$$

Also, taking moments about A, considering the free body diagram shown in Fig. 3.60(c).

$$P_s \times 2 + P_c \times 3 = 100{,}000 \times 3 \quad (2)$$

which is the condition for equilibrium. From (1),

$$\frac{3P_s}{80}-\frac{P_c}{36}=3240$$

From (2),

$$2P_s + 3P_c = 300{,}000$$

Solving,

$$P_s = 28{,}384 \text{ N}, \quad P_c = 81{,}077 \text{ N}$$

The stresses in the bars are

$$\sigma_s = \frac{28{,}384}{800} = 35.48 \text{ N/mm}^2, \quad \sigma_c = \frac{81{,}077}{600} = 135 \text{ N/mm}^2$$

Example 3.56 Indeterminate problem: four bars carrying a load

The framework of the four bars shown in Fig. 3.61(a) carries a load of 70 kN and is subjected to a temperature rise from a room temperature of 27 °C to 60 °C. Find the stresses in the bars and the distance by which point A comes down vertically. The bars AB and AE are of copper and AC and AD are of brass. In the unloaded condition, point A is 2 m below the support. For copper, E = 120 GPa and $\alpha = 18 \times 10^{-6}$/°C. The copper bars have an area of 300 mm^2. For brass E = 100 GPa and $\alpha = 20 \times 10^{-6}$/°C, and the bars have an area of 400 mm^2.

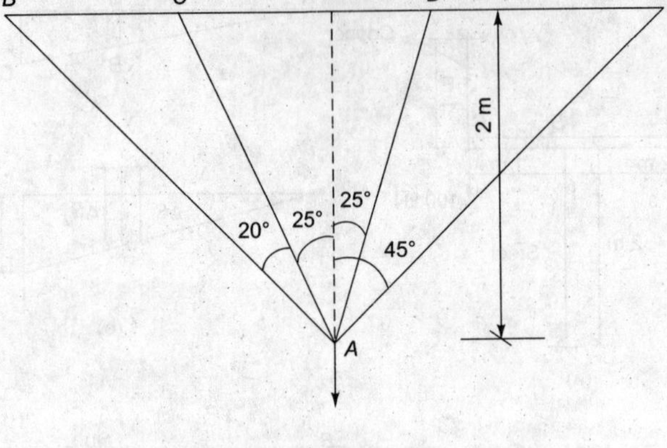

(a)

70 kN

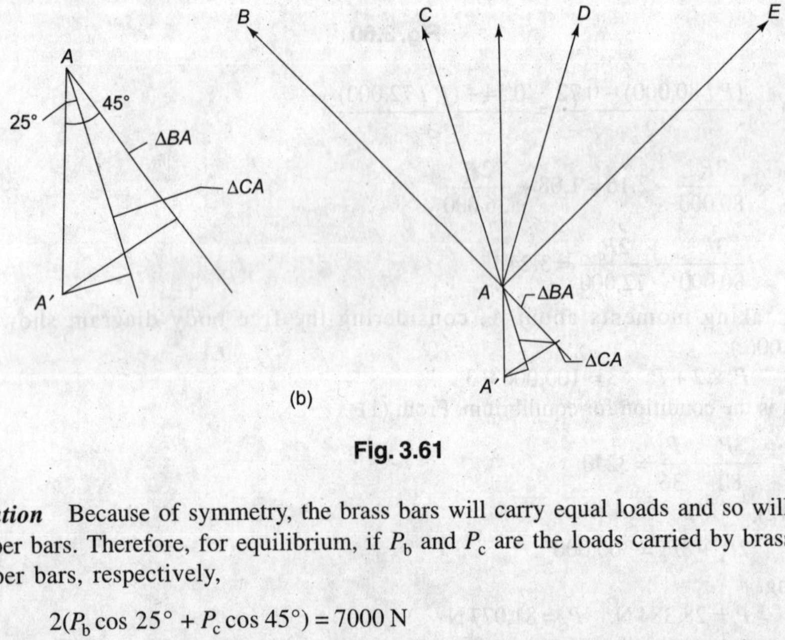

(b)

Fig. 3.61

Solution Because of symmetry, the brass bars will carry equal loads and so will the copper bars. Therefore, for equilibrium, if P_b and P_c are the loads carried by brass and copper bars, respectively,

$$2(P_b \cos 25° + P_c \cos 45°) = 7000 \text{ N} \tag{1}$$

If the point A comes down by an amount AA' vertically, then, because the deformations are small, from Fig. 3.61(b),

$$AA' = \frac{\Delta CA}{\cos 25°} = \frac{\Delta BA}{\cos 45°}$$

by the deformation condition. ΔCA and ΔBA are the deformations of bars CA and BA, respectively. The deformations in the bars are due to two causes.

Temperature rise

$$\Delta CAt = l\alpha t = \left(\frac{2000}{\cos 25°}\right) \times 20 \times 10^{-6} \times (60 - 27) = 1.456 \text{ mm}$$

$$\Delta BAt = l\alpha t = \left(\frac{2000°}{\cos 45°}\right) \times 18 \times 10^{-6} \times 33 = 1.68 \text{ mm}$$

Due to load

$$\Delta CAP = \frac{P_b \times 2207}{400 \times 100{,}000}, \quad \Delta BAP = \frac{P_c \times 2828}{300 \times 120{,}000}$$

$$\Delta CA = \Delta CAt + \Delta CAP = 1.456 + \frac{P_b \times 2207}{400 \times 100{,}000} = 1.456 + 55.175 \times 10^{-6} P_b$$

$$\Delta BA = \Delta BAt + \Delta BAP = 1.68 + \frac{P_c \times 2828}{300 \times 120{,}000} = 1.68 + 78.55 \times 10^{-6} P_c$$

$$\Delta CA = \Delta BA \times \frac{\cos 25°}{\cos 45°} = 1.2817\, \Delta BA$$

$$1.456 + 55.175 \times 10^{-6} P_b = 1.2817\,(1.68 + 78.55 \times 10^{-6} P_c)$$

$$= 2.1532 + 100.68 \times 10^{-6} P_c$$

$$P_b - 1.825 P_c = 12640 \qquad (2)$$

Solving Eqns (1) and (2),

$$P_c = 9972 \text{ N}, \quad P_b = 30{,}944 \text{ N}$$

Stress in the brass bar,

$$\sigma_b = 77.4 \text{ N/mm}^2$$

Stress in the copper bar,

$$\sigma_c = 33.2 \text{ N/mm}^2$$

$$AA' = \frac{\Delta CA}{\cos 25°} = \frac{1.456 + (77.4 \times 2207/100{,}000)}{\cos 25°} = 1.88 \text{ mm}$$

■

Example 3.57 Indeterminate problem: stresses in copper and brass bars in a framework

In the system shown in Fig. 3.62(a), a load of 100 kN is placed 3 m from A. The brass rod has a cross section of 30 × 30 mm and the copper rod 20 × 20 mm. E for brass is 85 GPa and for copper 120 GPa. Find the stresses in the bars. If the stress in brass is limited to 35 MPa, at what maximum distance from A can the 100 kN load be placed?

Solution If P_c and P_b are the stress resultants in the copper and brass bars, respectively, then from the free body diagram shown in Fig. 3.62(b), for equilibrium,

$$P_c \times 2 + P_b \times 5 - 100 \times 1000 \times 3 = 0$$

$$2P_c + 5P_b = 300{,}000$$

by the condition of equilibrium. If Δc and Δb are the deformations of the copper bar and brass bar, respectively, then from the deformation diagram of the rigid bar AC, the deformation condition is

$$\frac{\Delta c}{2} = \frac{\Delta b}{5}, \quad \Delta c = \frac{P_c \times 1500}{625 \times 120{,}000}, \quad \Delta b = \frac{P_b \times 3000}{900 \times 85{,}000}$$

(Area of copper bar = 25 × 25 = 625 mm^2, area of brass bar = 30 × 30 = 900 mm^2.)

$$\frac{1}{2} \times \frac{P_c \times 1500}{625 \times 120{,}000} = \frac{1}{5} \times \frac{P_b \times 3000}{900 \times 85{,}000}, \quad P_c = 0.7843 P_b$$

Substituting this in the equilibrium condition,

$2 \times 0.7843P_b + 5P_b = 300{,}000$

$\Rightarrow \quad P_b = 45{,}670$ N

$P_c = 35{,}820$ N

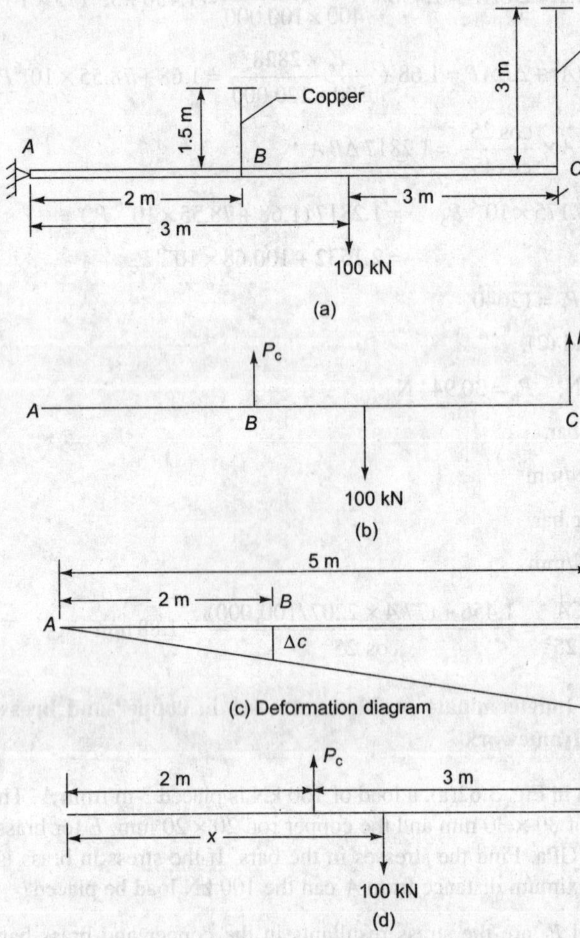

Fig. 3.62

$$\text{Stress in the copper bar} = \frac{35{,}820}{625} = 57.3 \text{ N/mm}^2$$

$$\text{Stress in the brass bar} = \frac{45{,}670}{900} = 50.7 \text{ N/mm}^2$$

Considering Fig. 3.62(d), the equilibrium condition is $P_b \times 5 + P_c \times 2 = 100{,}000x$.
The deformation condition remains the same, i.e.,

$P_c = 0.7843P_b$

$P_b = 35 \times 900 = 31{,}500$ N, $\quad P_c = 0.7843 \times 31{,}500 = 24{,}705$ N

$31{,}500 \times 5 + 24{,}705 \times 2 = 100{,}000x$

$\Rightarrow \quad x = 2.07$ m

Example 3.58 Indeterminate problem: stresses in cables carrying loads

A rigid bar ABC is hinged at A and supported by cables DC and DB, which are fixed to a point D, 2 m directly above A as shown in Fig. 3.63(a). Determine the maximum force in the cables if they are of the same material and have the same cross section.

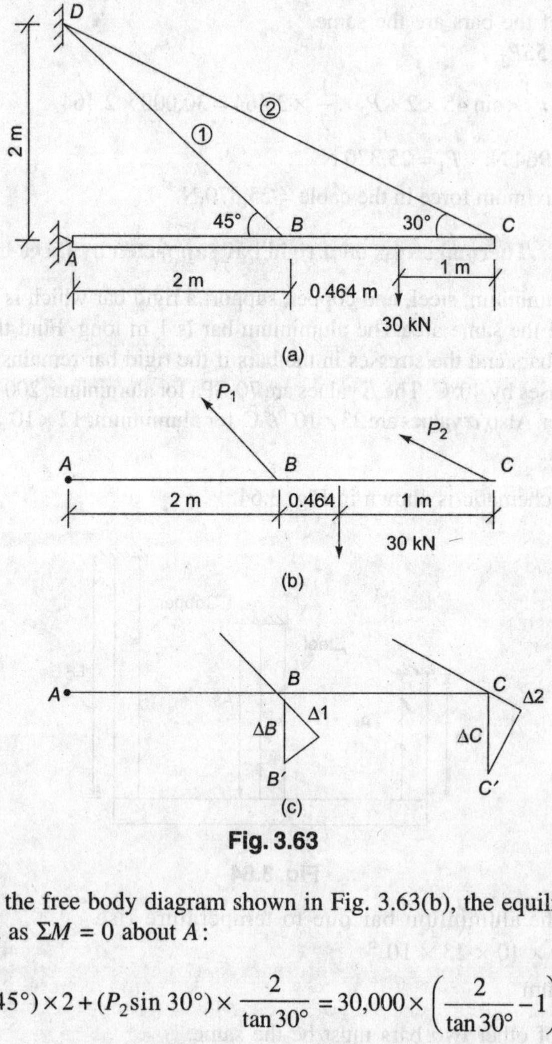

Fig. 3.63

Solution From the free body diagram shown in Fig. 3.63(b), the equilibrium condition can be expressed as $\Sigma M = 0$ about A:

$$(P_1 \sin 45°) \times 2 + (P_2 \sin 30°) \times \frac{2}{\tan 30°} = 30{,}000 \times \left(\frac{2}{\tan 30°} - 1\right)$$

From the deformation diagram shown in Fig. 3.63(c), since the deformations are small, we can infer that $\Delta B/2 = \Delta C/3.464$, where ΔB and ΔC are the vertical downward movements of points B and C. If Δ_1 and Δ_2 are the elongation of cables 1 and 2, respectively, then

$$\Delta B = \frac{\Delta_1}{\cos 45°} \quad \text{and} \quad \Delta C = \frac{\Delta_2}{\cos 60°}$$

$$\Delta_1 = \frac{\sigma_1 l_1}{E} \quad \text{and} \quad \Delta_2 = \frac{\sigma_2 l_2}{E}$$

where σ_1 and σ_2 are the stresses and l_1 and l_2 are the lengths of cables 1 and 2, respectively.

$$l_1 = 2\sqrt{2}\ \text{m} \quad \text{and} \quad l_2 = \frac{2}{\sin 30} = 4\ \text{m}$$

Therefore,

$$\frac{1}{2}\,\frac{\sigma_1 \times 2\sqrt{2}}{E \cos 45} = \frac{1}{3.464} \times \frac{\sigma_2 \times 4}{E \cos 60}, \quad \sigma_1 = 1.155\,\sigma_2$$

Since the areas of the bars are the same,

$$P_1 = 1.155 P_2$$

$$1.155 \times P_2 \times \sin 45 \times 2 + P_2 \times \frac{1}{2} \times 3.464 = 30{,}000 \times 2.464$$

$$P_2 = 21{,}964\ \text{N}, \quad P_1 = 25{,}370\ \text{N}$$

Therefore, the maximum force in the cable = 25,370 N.

Example 3.59 Thermal stress on a rigid bar supported by three bars

Three bars, of aluminium, steel, and copper, support a rigid bar which is centrally placed. All the bars have the same area.The aluminium bar is 1 m long. Find the lengths of the steel and copper bars and the stresses in the bars if the rigid bar remains horizontal when the temperature rises by 40°C. The E values are 70 GPa for aluminium, 200 GPa for steel, and 120 GPa for copper. Also α values are 23×10^{-6}/° C for aluminium, 12×10^{-6}/°C for steel, 16 $\times 10^{-6}$ for copper.

Solution The schematic is shown in Fig. 3.64.

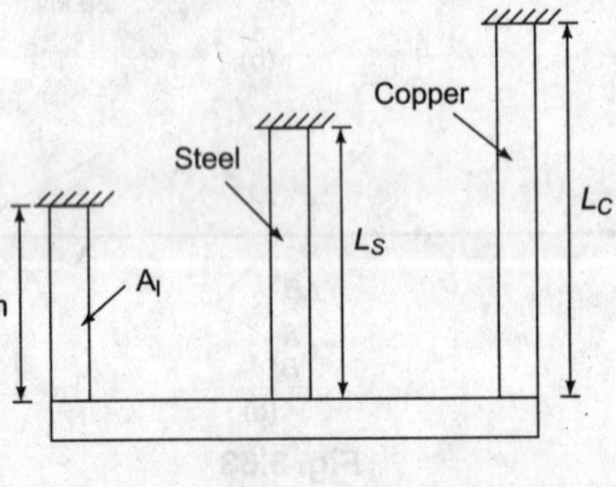

Fig. 3.64

Elongation of the aluminium bar due to temperature rise

$$= 1000 \times 40 \times 23 \times 10^{-6}$$
$$= 0.92\ \text{mm}$$

The elongation of other two bars must be the same.

For steel bar, $L_s \times 40 \times 12 \times 10^{-6} = 0.92$ mm; $L_s = 1917$ mm
Length of steel bar = 1.917 m
For copper bar, $L_c \times 40 \times 16 \times 10^{-6} = 0.92$; $L_c = 1438$ mm
Length of cmopper bar = 1.438 m

Stress in the aluminium bar, σ_a, is given by $\sigma_a L/E = 0.92$; $\sigma_a = 0.92 \times 70{,}000/1000 = 64.4$ N/mm^2

Stress in steel bar, $\sigma_s = 0.92 \times 200{,}000/1917 = 96$ N/mm^2
Stress in copper bar, $\sigma_c = 0.92 \times 120{,}000/1438 = 76.8$ N/mm^2

Example 3.60 Stresses on bars supporting a rigid bar

Three bars of the same material and cross sectional area support a rigid bar carrying 10 kN. The bars are of the same material and cross-section. Find the stresses in the bars and length of the middle bar if the rigid bar remains horizontal after the load is added.

Solution The situation is shown in Fig. 3.65.

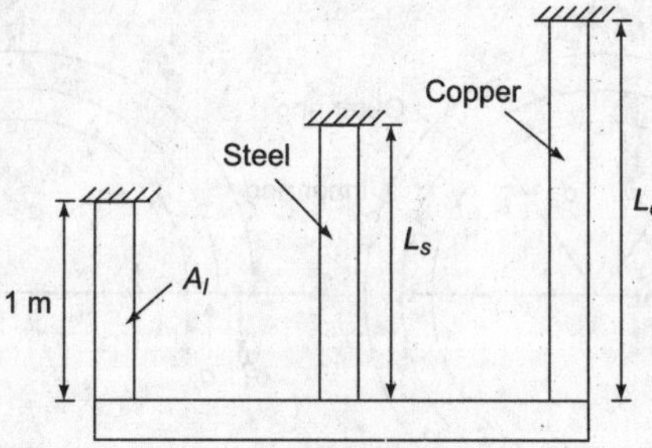

Fig. 3.65

We have four unknowns, the loads carried by the three bars and the length of the middle bar, L_2. We set up equations as follows.

Equilibrium equation, $P_1 + P_2 + P_3 = 10{,}000$ N ($\Sigma V = 0$)

Moment equation, $P_2 \times 3000 + P_3 \times 4000 = 10{,}000 \times 2000$ [Taking moment about 1]

Compatibility equation, $\dfrac{P_1 \times 1500}{AE} = \dfrac{P_2 L_2}{AE} = \dfrac{P_3 \times 2000}{AE}$

As AE is common, we have

$$P_1 \times 1500 = P_2 L_2 \text{ and } P_1 \times 1500 = P_3 \times 2000$$

These four equations can be used to solve for the unknowns.

$$P_2 = (P_1 \times 1500)/L_2;\ P_3 = (P_1 \times 1500)/2000 = 0.75P_1$$

Substituting in the first two equations,

$$P_1 + P_1 \times 1500/L_2 + 0.75P_1 = 10{,}000;\ 1.75\,P_1 + 1500\,P_1/L_2 = 10{,}000$$
$$3P_1 \times 1500/L_2 + 4P_1 \times 0.75 = 20{,}000;\ 4500P_1/L_2 + 3P_1 = 20{,}000$$
$$2[1.75P_1 + 1500\,P_1/L_2] = 20{,}000;\ 3.5P_1 + 3000\,P_1/L_2 = 20{,}000$$
$$3P_1 + 4500\,P_1/L_2 = 20{,}000$$

Solving,

$$P_1 = 4000 \text{ N},\ P_2 = 3000 \text{ N},\ P_3 = 3000 \text{ N, and } L_2 = 2000 \text{ mm}$$

3.11 STRESSES DUE TO SHRINK FIT

When a ring of one material is slipped on to a ring or wheel or another material, the inner diameter of the ring of the outer material is kept slightly smaller than the outer diameter of the inner ring. At normal temperature, it is not possible to fit one onto the other. The outer ring is heated so that its internal diameter becomes equal to or slightly more than outer diameter of the inner ring. The outer ring is then slipped on the inner ring. On cooling, the outer ring tries to shrink to its original size and exerts a pressure to cause compressive stress in the inner ring. The inner ring prevents the full recovery of size of the outer ring, causing tensile stresses in it. This is necessary so that the outer ring does not slip out at normal temperature.

Consider the situation shown in Fig. 3.66, σ_1 and σ_2 are the tensile and compressive stresses in the outer and inner rings after cooling d_1 and t_1 are the inner diameter and thickness of the outer ring and d_2 and t_2 are the outer diameter and thickness of the inner ring. Considering the free body diagram of a part of the rings, for equilibrium, $A_1\sigma_1 = A_2\sigma_2$, where A_1 and A_2 are the areas of outer and inner rings, respectively.

A second condition has to be evolved from the deformation condition of the rings. If the final diameter at the common surface is d, then d is less than d_2 but more than d_1.

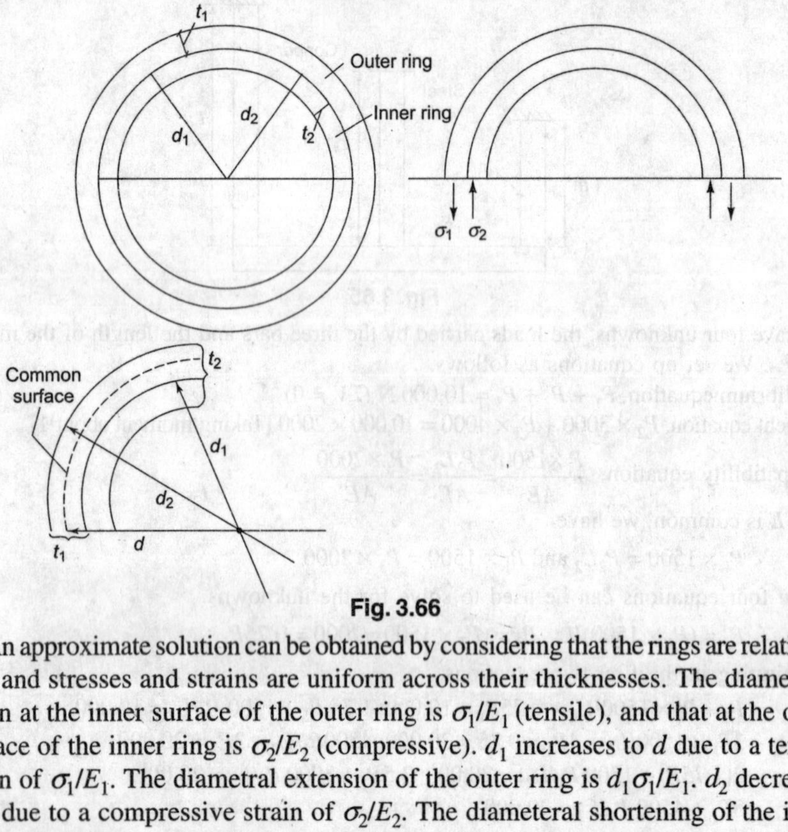

Fig. 3.66

An approximate solution can be obtained by considering that the rings are relatively thin and stresses and strains are uniform across their thicknesses. The diameteral strain at the inner surface of the outer ring is σ_1/E_1 (tensile), and that at the outer surface of the inner ring is σ_2/E_2 (compressive). d_1 increases to d due to a tensile strain of σ_1/E_1. The diametral extension of the outer ring is $d_1\sigma_1/E_1$. d_2 decreases to d due to a compressive strain of σ_2/E_2. The diameteral shortening of the inner ring is $d_2\sigma_2/E_2$.

$$d_1\frac{\sigma_1}{E_1} + d_2\frac{\sigma_2}{E_2} = d_2 - d_1$$

is the original difference in the diameters of the two rings. σ_1 and σ_2 are known as *hoop* or *circumferential* stresses, as these are along the circumferential direction. The following examples illustrate the principles involved.

Example 3.61 Stresses due to shrink fit

A steel ring of diameter 998 mm is to be slipped on to a wooden wheel of diameter 1 m. To what minimum temperature should the steel ring be heated to enable fitting? What will be the stresses in the two materials on cooling? E for steel is 200 GPa and for wood 8 GPa. The coefficient of expansion for steel is 12×10^{-6}/°C. cross-sectional dimensions are 20×5 mm for steel and 20×40 mm for wood.

Solution The steel ring has to be heated to a temperature T so that its diameter becomes at least equal to 1000 mm, i.e., an extension of 2 mm. Therefore,

$$998 \times 12 \times 10^{-6} \times T = 2 \text{ mm}$$

$\Rightarrow \quad T = 167°\ C$

On cooling, the steel ring will be in tension and wood under compression. If σ_s and σ_w are the stresses in steel and wood, then,

$$A_s\sigma_s = A_w\sigma_w, \quad 5\times 20\times \sigma_s = 40\times 20\times \sigma_w$$

$$\Rightarrow \quad \sigma_s = 8\sigma_w$$

Also, tensile elongation + compressive shortening = difference in diameters, i.e.,

$$998\times\frac{\sigma_s}{E_s} + 1000\times\frac{\sigma_w}{E_w} = 1000 - 998 = 2$$

$$\frac{998\times 8\sigma_w}{200{,}000} + 1000\times\frac{\sigma_w}{8000} = 2, \quad \sigma_t = 12\ \text{N/mm}^2$$

$$\sigma_s = 12\times 8 = 96\ \text{N/mm}^2$$

Example 3.62 Shrink fitting of a steel tube over a brass pipe

A steel pipe of internal diameter 799 mm and of thickness 2 mm has to be slipped on to a brass pipe of external diameter 800 mm and thickness 30 mm. What is the temperature to which the steel pipe should be heated for this to be possible? What are the stresses in the two materials on cooling? What is the maximum pressure that this compound pipe can withstand if the stress in steel is limited to 150 N/mm^2? E = 200 GPa for steel and 95 GPa for brass.

Solution The diameter of the steel pipe has to be increased from 799 to 800 mm. The temperature rise required is given by

$$l\alpha t = 1, \quad 799\times 12\times 10^{-6}\times t = 1$$

$$\Rightarrow \quad t = 104°\text{C}$$

The steel pipe is subjected to tension and the brass pipe to compression. Considering unit length of the pipe acting as a ring as shown in Fig. 3.67, for equilibrium,

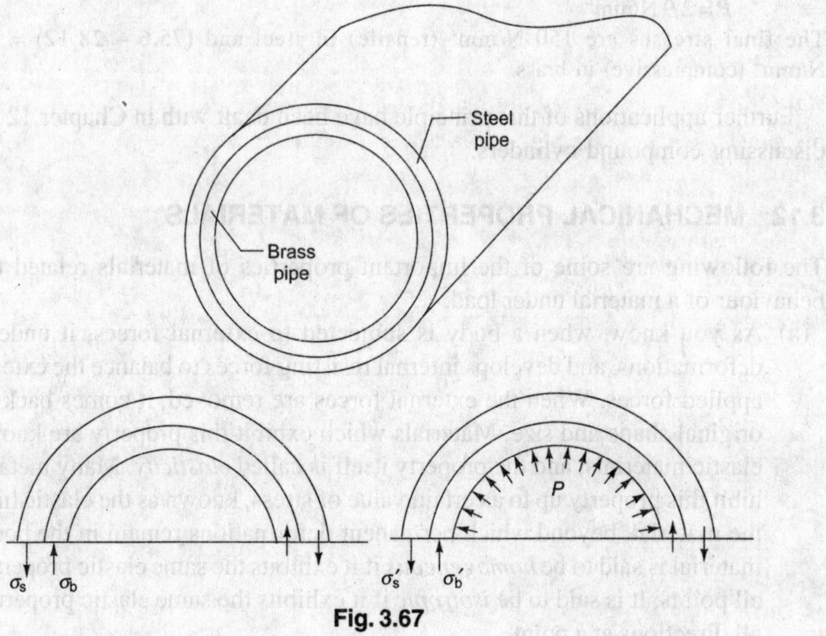

Fig. 3.67

Tensile force = compressive force

or $\quad A_s\sigma_s = A_b\,\sigma_b$

or $\quad 1 \times 25 \times \sigma_s = 1 \times 30 \times \sigma_b, \quad \sigma_b = \frac{5}{6}\sigma_s$

Also, if d_b and d_s are the diameters of the brass and steel pipes, respectively, then

$$d_b \frac{\sigma_b}{E_b} + d_s \frac{\sigma_s}{E_s} = 1$$

or $$\frac{800 \times (5/6)\sigma_s}{95{,}000} + \frac{799 \times \sigma_s}{200{,}000} = 1, \quad \frac{800 \times (5/6)\sigma_s}{95{,}000} + \frac{799 \times \sigma_s}{200{,}000} = 1$$

$\Rightarrow \quad \sigma_s = 90.8 \text{ N/mm}^2$ (tensile)

$\Rightarrow \quad \sigma_b = \frac{5}{6} \times 90.8 = 75.6 \text{ N/mm}^2$ (compressive)

When the pipe is subjected to internal pressure P, the steel and brass pipes are both subjected to tensile forces. Let us consider a half-circular segment of the pipe. From the free body diagram, the horizontal components of pressure P acting on the inner surface vanish. The net force due to pressure P is a vertical force acting on the projected area of $800L$ mm^2, where L mm is the length of the pipe. Considering unit length of the pipe,

$$2P \times 800 - 2\sigma'_s A_s - 2\sigma'_b A_b = 0$$

where σ'_s and σ'_b are the stresses due to internal pressure P. While the brass pipe is relieved of compressive stresses, the steel pipe will have additional tensile stress σ'_s. The maximum value of $\sigma'_s = 150 - 90.8 = 59.2 \text{ N/mm}^2$.

Again the strain at the common surface is equal. That is,

$$\frac{\sigma'_s}{E_s} = \frac{\sigma'_b}{E_b}, \quad \sigma'_s = \sigma'_b \frac{200{,}000}{95{,}000} = 2.105\sigma'_s$$

$\Rightarrow \quad \sigma'_b = \frac{59.2}{2.105} = 28.12 \text{ N/mm}^2$

$$2P \times 800 - 2 \times 59.2 \times 25 \times 1 - 2 \times 28.12 \times 30 \times 1 = 0$$

$\Rightarrow \quad P = 2.9 \text{ N/mm}^2$

The final stresses are 150 N/mm^2 (tensile) in steel and (75.6 – 28.12) = 47.48 N/mm^2 (compressive) in brass. ■

Further applications of this principle have been dealt with in Chapter 12 while discussing compound cylinders.

3.12 MECHANICAL PROPERTIES OF MATERIALS

The following are some of the important properties of materials related to the behaviour of a material under load.

(a) As you know, when a body is subjected to external forces, it undergoes deformations, and develops internal resisting forces to balance the externally applied forces. When the external forces are removed, it comes back to its original shape and size. Materials which exhibit this property are known as elastic materials, and the property itself is called *elasticity*. Many metals exhibit this property up to a certain value of stress, known as the elastic limit of the material, beyond which permanent deformations remain in the body. A material is said to be *homogeneous* if it exhibits the same elastic properties at all points. It is said to be *isotropic* if it exhibits the same elastic properties in all directions at a point.

(b) *Plasticity* is the property because of which a material subjected to forces undergoes deformations which do not disappear on the removal of external forces. Plasticity is important when a substance has to be moulded into components. Many materials become plastic at large values of stress or at high temperatures.

(c) *Ductility* is the property because of which it is possible to draw thin wires of a metal. Ductile materials can undergo large plastic deformations before breaking. Copper and mild steel exhibit this property.

(d) *Brittleness* is the tendency of a material to shatter on receiving a shock. This happens due to lack of ductility. The material does not have the capacity to undergo large deformations before failure. Glass and certain high-strength steels are brittle.

(e) *Toughness* is the property which enables a material to absorb large amounts of energy by undergoing large plastic deformations, particularly due to shock loading.

(f) *Resilience* is the ability of a material to recover its shape and size after deformation.

(g) *Hardness* is the resistance of a material to indentation, scratching, cutting or wear by abrasion.

(h) *Fatigue* is the phenomenon of a material failing under very little stress due to repeated cycles of loading. A fatigue failure is generally similar to brittle failure. This is important in case a component is likely to be subjected to cyclic or reversal loading, as in machine foundations or members subjected repeated dynamic loads.

(i) *Creep* is the property by which a material undergoes deformations at constant stress over a period of time. A body undergoes elastic deformation immediately on the application of a load. The stress remaining constant, it is found that deformations increase over a period of time due to creep.

3.13 STRESS-STRAIN DIAGRAM

In order to understand the behaviour of a material subjected to axial forces, it is a common practice to test a specimen of the material. A standard specimen of the material is prepared as per the relevant BIS specifications. For steel, standard specimen is of the form shown in Fig. 3.68. The specimen is tested in a universal testing machine, which can be set to a specified rate of loading. Extensometers are used to measure the extension of the specimen between gauge length. Most machines have a plotter attached to plot load vs extension diagram.

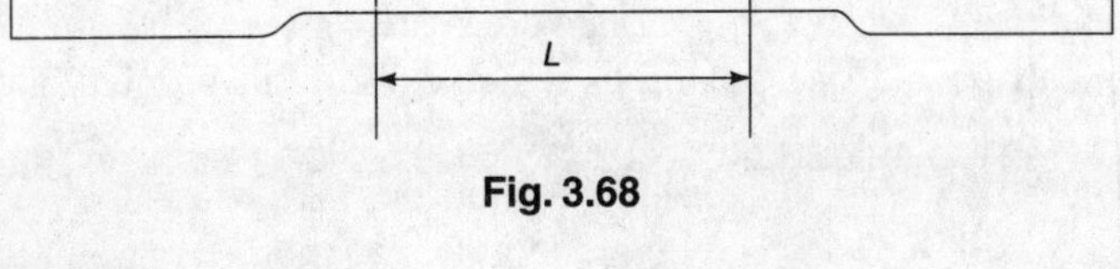

Fig. 3.68

3.13.1 Mild Steel

A mild steel specimen is generally subjected to a tensile test to obtain a P–ΔL diagram, also known as the load-deformation diagram. A standard specimen as shown in Fig. 3.68 is used. The elongation of the specimen is measured over a gauge length from which strains are calculated. Knowing the load and the area of the specimen, the stress at any point can be calculated. The stress-strain diagram can be plotted automatically by the machine if it is connected to a plotter.

A typical stress-strain diagram for mild steel is shown in Fig. 3.69. There are some important characteristics in this diagram that should be noted. The diagram is linear in the beginning and up to a point known as the proportional limit. The stress corresponding to this point is called proportional limit up to which stress is proportional to strain. The diagram curves off after this point up to the elastic limit, which is a point on the curve that limits the stress up to which full recovery of strain is noted. Thus, while stress is linearly proportional to strain up to the limit of proportionality, the material behaves elastically up to the elastic limit.

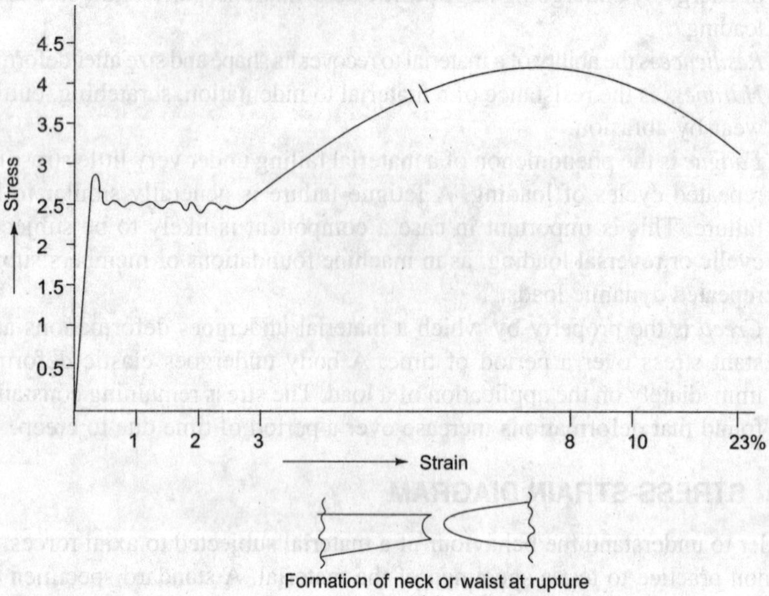

(a)

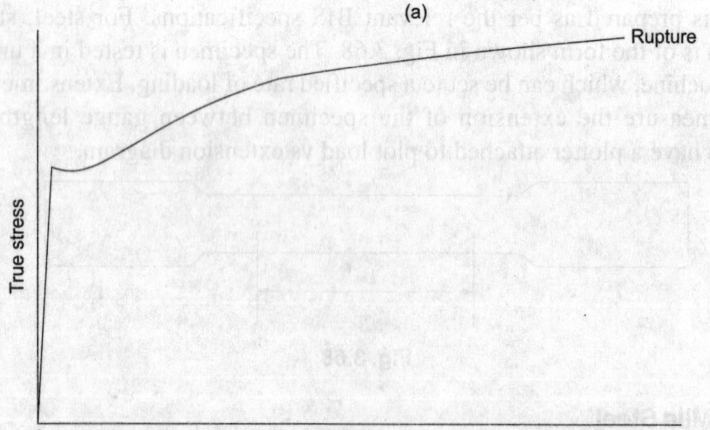

(b)

Fig. 3.69 Stress-strain diagram for mild steel

On further loading, the stress suddenly falls and the material suffers deformations more or less under constant stress. This point is called yield point and the stress corresponding to this is called yield stress. In a carefully conducted test, a distinction can be made between the upper yield point and a lower yield point. However, for engineering purposes, due to uncertainty in identifying clearly the two yield points, the stress corresponding to the lower yield point is taken as the yield stress.

The material undergoes large deformation after the yield stress is reached and this stage is known as the plastic stage. The structural change taking place in the material during this stage is known as strain hardening. At the end of this stage, stress and strain increase further till a maximum point, known as the ultimate stress, is reached. After this, the stress falls further with increasing strain until the specimen breaks at a point known as break point.

An examination of the ruptured specimen will show the formation of a neck of reduced area. The cupping of the specimen at the neck and the 45° planes along which the rupture takes place shows that the specimen actually fails in shear. The following are the salient points of the curve.

Limit of proportionality When a load test is done on a rod of any material, the load is increased slowly (as per some standard rate of loading). In the initial stages, stress is proportional to strain. The limit of stress up to which stress remains proportional to strain varies with the material. All forms of steel exhibit this property. The stress up to which stress is proportional to strain is known as *limit of proportionality*. In the case of steel, there is a well-pronounced limit which can be found from a tensile test. In the case of some other materials, this may be a small value or the material may not exhibit such a limit.

Elastic limit Elastic limit is the stress up to which the material recovers to its original length on removal of the load. Note the difference between the proportional and elastic limits. Up to the proportional limit, there is linearity relationship between stress and strain. Elastic property will be exhibited by the material in this stage also. Between the proportional and elastic limits, stress is not proportional to strain, but the material exhibits elastic properties. Strain becomes zero on removal of the load.

Yield point Yield point is the stress beyond which when the material is loaded, plastic deformations take place. At the same stress value, material undergoes large deformations. In the case of steel, it is possible to identify upper and lower yield points as well. Many materials do not exhibit a well-defined yield point. Yield point is a very important property of steel and the lower yield point is the maximum stress that is taken in any form of analysis.

Ultimate stress Ultimate stress is the maximum stress ordinate in a stress-strain diagram. The stress values in this case are calculated from the original area of cross section of the rod.

Breaking stress Breaking stress is the stress at which the specimen ruptures. The breaking stress appears as a lower value than ultimate stress if calculated on original area. Both ultimate stress and breaking stress are fictitious values due to reduction in the area of specimen at rupture section due to plastic deformation.

It should be noted that the ultimate stress as well as breaking stress as plotted on the diagram are fictitious values, and that the true stress and strain do increase continuously, as shown in Fig. 3.69(b). If A_o and L_o are the original area of cross-section and gauge length, then the stress is P/A_o and the strain is $\Delta L/L_o$ for plotting the diagram of Fig. 3.69(a). If we take into account the reduction in area due to necking and calculate the stress based upon the reduced area, we get the diagram of Fig. 3.69(b). In this, true strain is not calculated as $\Delta L/L_o$ but as the summation of elementary strains at each stage obtained by dividing the increase in length by the true length of the specimen between gauge marks.

Another point of interest on the stress-strain diagram is the per cent reduction in area. We have mentioned that due to high stress levels, a neck or waist is formed in the specimen. The reduction in area of the specimen at the neck divided by the original area A_o of the specimen expressed as a percentage is known as the per cent reduction in area. If A' is the area measured at the neck,

$$\text{Per cent reduction in area} = \frac{A_o - A'}{A_o} \times 100$$

This is a measure of the ductility of the specimen. The per cent reduction in area may be as high as 60 to 70 for mild steel. Per cent elongation (ductility) may also be expressed as per cent elongation, given by

$$\frac{(\text{Final length} - \text{original length}) \times 100}{\text{Original length}} = \frac{L_f - L_o}{L_o} \times 100$$

For mild steel, the per cent elongation is from 20–23%, and is an important characteristic of the material. Actually, ductility is commonly expressed as per cent elongation.

3.13.2 Other Materials

The stress-strain diagrams for some other materials are shown in Fig. 3.70. It should be noted that the stress diagram is non-linear from very low stresses as in the case of concrete and timber. Such diagrams are also different in tension and compression for materials such as timber and cast iron.

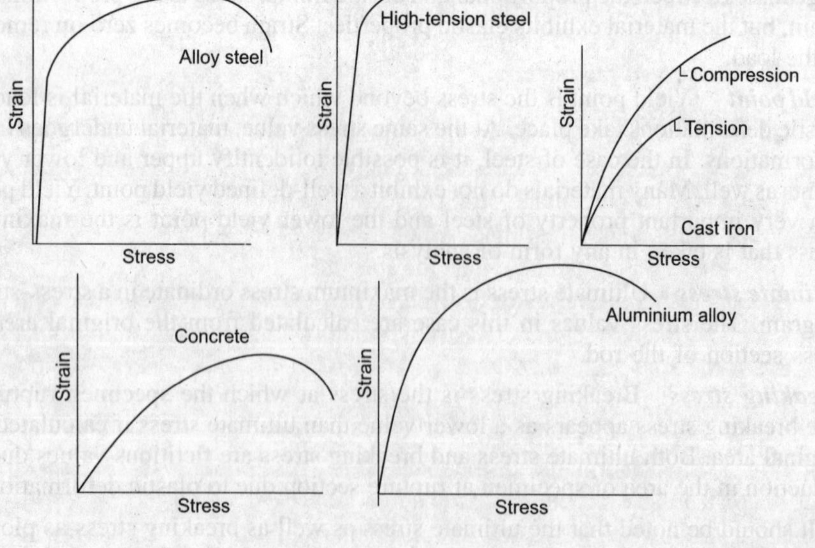

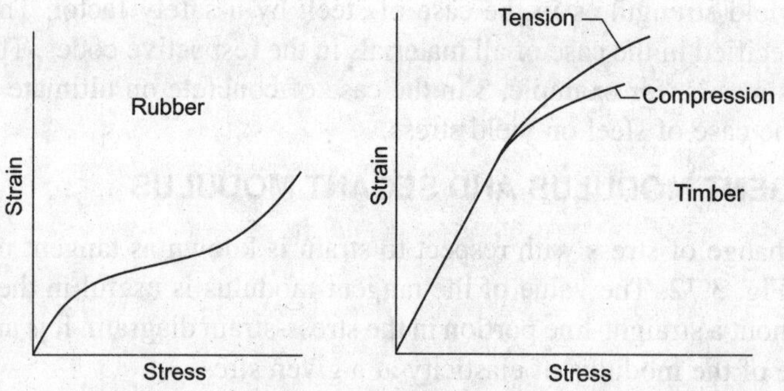

Fig. 3.70 Stress-strain diagrams for various materials

In the case of structural steels, the stress-strain diagram in tension and compression are identical. In compression, no neck or waist can be formed.

In the case of most structural steels, the initial stress-strain diagram is identical and hence the value of modulus of elasticity, *E*, for all structural steels is the same, being the slope of the linear portion of the diagram. However, high-strength steels have brittle characteristics and neither show a clear yield point nor have the large plastic deformations which are a characteristic of mild steel.

3.14 OBTAINING YIELD STRESS BY THE OFFSET METHOD

In the case of materials which do not show a clear yield point, the yield stress is obtained by the offset method shown in Fig. 3.71. By selecting a particular strain value of 0.2 per cent, a line is drawn from 0.002 on the strain axis parallel to the straight-line part of the stress-strain diagram. The stress corresponding to the point on the curve at which this line intersects the curve is taken as the yield stress of the material.

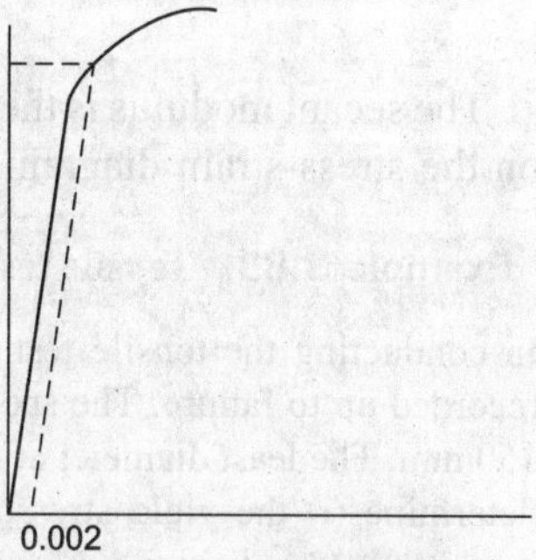

Fig. 3.71 The offset method

3.15 PROOF STRESS

To specify the yield stress of a material, many times a proof stress value is specified, indicating the percentage strain associated with it. Proof stress can be defined as the value of the stress up to which a specified permanent set in percentage of strain remains on unloading, when the specimen is unloaded. Thus, 0.1% proof stress means the stress value up to which when the specimen is loaded and then unloaded, a permanent set of 0.1% strain remains. This residual strain is specified along with the stress value as 0.1% proof stress.

3.16 WORKING STRESS AND FACTOR OF SAFETY

Many times, the performance of the material under working loads and conditions is of interest. In any structure, it is ensured that the stress in the material under normal conditions is much below the ultimate strength or yield strength. Such a stress is known as the working stress, and is obtained by dividing the ultimate strength, or yield strength as in the case of steel, by a safety factor. The safety factors are specified in the case of all materials in the respective codes. The factor of safety on stress is, for example, 3 in the case of concrete on ultimate strength and 1.68 in the case of steel on yield stress.

3.17 TANGENT MODULUS AND SECANT MODULUS

The rate of change of stress with respect to strain is known as tangent modulus, as shown in Fig. 3.72. The value of the tangent modulus is useful in the case of materials without a straight-line portion in the stress-strain diagram. It is an instantaneous value of the modulus of elasticity at a given stress.

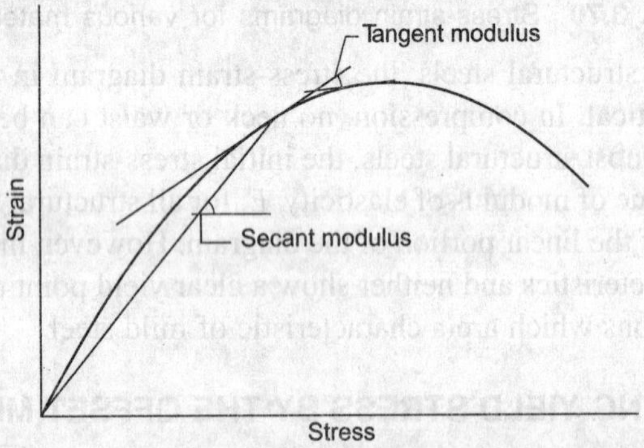

Fig. 3.72

The secant modulus is the slope of the line joining the origin to a specified point on the stress-strain diagram.

Example 3.63 Tensile test on a steel rod

In conducting the tensile test on a specimen of an alloy, the following test data were recorded up to failure. The specimen had a diameter of 14 mm and the gauge length was 150 mm. The least diameter at the waist was 10.8 mm. Draw the stress-strain diagram and determine (i) the yield stress, (ii) percentage elongation, (iii) percentage reduction in area, (iv) ultimate strength, (v) breaking stress, and (vi) true stress at rupture.

Load	*Elongation*	*Load*	*Elongation*
0	0	36,000	0.20
1,850	0.01	40,000	0.30
3,700	0.02	42,000	0.60
5,540	0.03	50,000	1.00
7,400	0.04	60,000	2.00
9,250	0.05	62,000	4.00
11,100	0.06	64,000	8.00
14,800	0.08	61,500	12.00
18,500	0.10	60,000	18.00
27,800	0.15	58,000	20.00
		54,000	Fracture

Solution The stress-strain diagram is shown in Fig. 3.73.

$$\text{Yield stress} = \frac{36{,}000}{(\pi/4)\times 14^2} = 234\ \text{N/mm}^2$$

$$\text{Percentage elongation} = \frac{20}{150}\times 100 = 13.33$$

$$\text{Percentage reduction in area} = \frac{14^2 - 10.8^2}{14^2}\times 100 = 40.5$$

$$\text{Ultimate strength} = \frac{64{,}000}{154} = 415\ \text{N/mm}^2$$

$$\text{Breaking stress} = \frac{54{,}000}{154} = 350\ \text{N/mm}^2$$

$$\text{True stress at rupture} = \frac{54{,}000}{(\pi/4)\times(10.8)^2} = 590\ \text{N/mm}^2$$

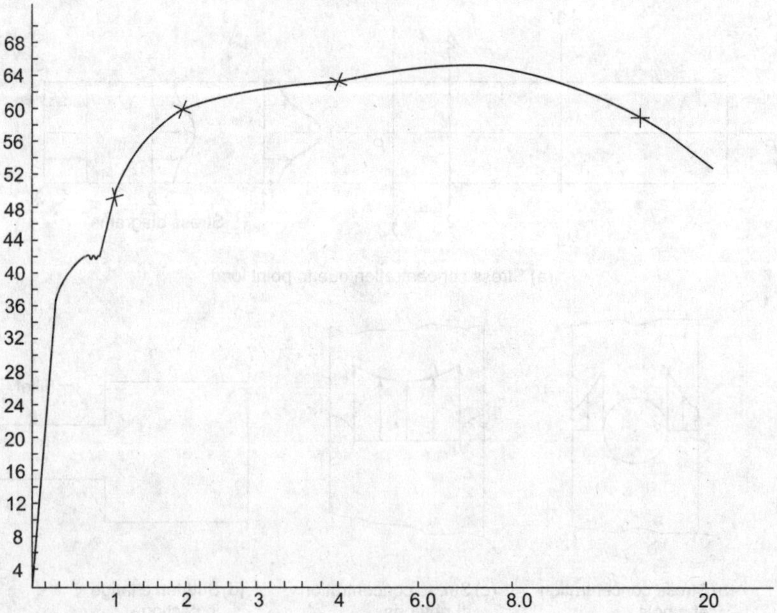

Fig. 3.73

3.18 STRESS CONCENTRATION

Consider the case of a bar loaded as shown in Fig. 3.74. While the stress at a section sufficiently away from the load is uniform, the stress at the section where the load is applied or near that end is not uniform. This principle is known as *St Venant's principle*. The local effect of the concentrated load is to increase the stress around the load point. This effect is called *stress concentration*.

Discontinuities in the section Any discontinuity in the material like a hole [Fig. 3.74(b)] or notches [Fig. 3.74(c)] in the section causes stress concentration as shown. A similar effect is seen in the case of stepped bars having sudden change of section [Fig. 3.74(d)].

The stress near the hole or notch, or the discontinuity is much higher than the average stress normally calculated as load/area.

Stress concentration factor is the ratio of the maximum stress to the average stress.

Stress concentration factor = maximum stress/average stress

St Venant's principle in simple terms states that the stress at a distance away from the load point is equal to the average stress.

In Fig. 3.74(a), it is generally assumed that the load P is uniformly distributed over the area of the member. But in practice, the load is a point load, concentrated at a point. Stress intensity is thus theoretically infinite at the point of the load. The load in fact cannot be a point load as all point loads are distributed over a small area.

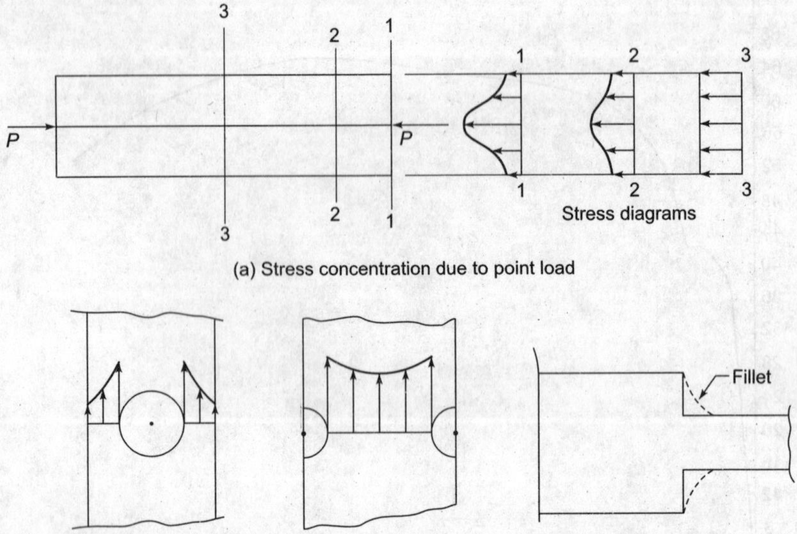

(a) Stress concentration due to point load

(b) Stress concentration at a hole

(c) Stress concentration at notches

(d) Sudden change in section

Fig. 3.74 Stress concentration

The stress distribution at or near the loaded point is such that there is a higher stress near the load. As we move away from the end sections, this higher stress goes on decreasing and at some distance away from the end section, the stress over the whole section becomes the average stress P/A. This distance is generally assumed to be the width of the member. It is also seen that even with higher stresses at some sections, the average stress remains the same.

In Fig. 3.74(b), the effect of a hole in the member (a rivet or bolt hole) is depicted. The stress is much higher at or near the edge of the hole than the average stress. The stress concentration factor can be as high as 3 in certain cases.

In Fig. 3.74(c), the effect of a notch on the side of the member is shown. Here again the stress near the notch can be very high compared to the average stress.

Another common example is shown in Fig. 3.74(d). This is a case of abrupt change in section. There is stress concentration near the change of section. This is taken care of by providing fillets, which change the section gradually.

Stress concentration factors are worked out using the theory of elasticity or experimentally by photoelasticity. The powerful numerical tool, finite element analysis, allows us to accurately determine the stress concentration at any form of discontinuity.

Stress concentration factors are listed in technical data tables (R.J. Roarke and W.C. Young, *Formulas for Stress and Strain*).

3.19 RESIDUAL STRESSES

Residual stresses can be considered in two cases. In the first instance, it is in the formation of materials themselves. Many manufacturing processes like rolling, forging, heat treatment, etc. create internal stresses. Such stresses are known as *residual stresses*. These stresses are induced without any external force being applied. The residual stresses are self-equilibrating stresses in the sense that they balance themselves.

Residual stresses can also occur in the case of loading beyond the plastic range. As shown in Fig. 3.75(a), a ductile material loaded beyond the elastic range and into the plastic range gets a permanent set. This is revealed while unloading which takes place elastically but with a permanent set (some strain remains). In an ideal elastic, perfectly plastic material, the curve is idealized as shown. Such a curve is assumed in many design practices by idealizing the actual curve. The unloading phenomenon is elastic with a permanent set [Fig. 3.75(b)].

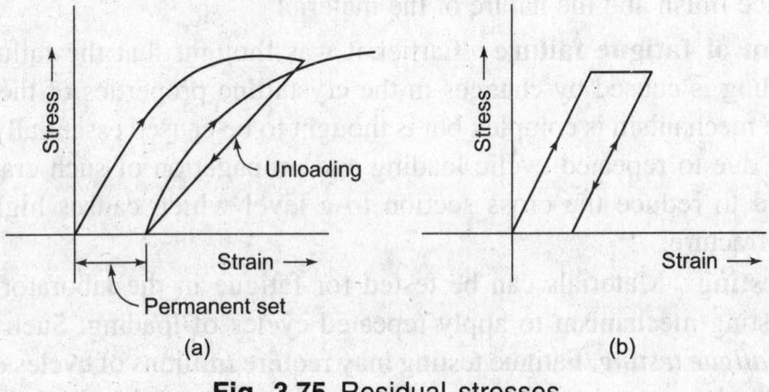

Fig. 3.75 Residual stresses

If the permanent is restricted in any way, internal stresses are created and are known as residual stresses. A detailed analysis of such stresses falls into the domain of inelastic behaviour (plastic analysis), which is beyond the scope of this book.

3.20 FATIGUE

Fatigue in materials is caused by cyclic loading. Cyclic loading means the loads that fluctuate between positive and negative values as shown in Fig. 3.76. Such loads are common in practice as in wind loads, earthquake loads, and moving machine components. Members subjected to such loads tend to fail at loads much smaller than the static loads they are capable of carrying.

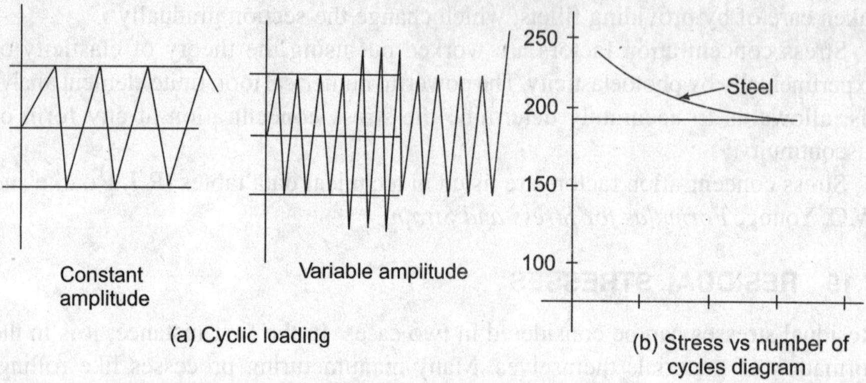

Fig. 3.76 Cyclic loading

Failure of materials due to cyclic loading is called *fatigue fracture*. Materials may fail after millions of cycles of loading or after a few cycles of loading. The number of cycles that the material can withstand before failure is called *fatigue life*. Fatigue in materials is caused by many factors like

(a) presence of stress concentration points on the inside or on the surface
(b) presence of any residual stress in the member due to manufacturing processes
(c) presence of manufacturing defects like cracks and other defects
(d) any environmental factors like chemical environment causing damage
(e) surface finish and the nature of the material

Mechanism of fatigue failure Earlier it was thought that the failure due to cyclic loading is caused by changes in the crystalline properties of the material. The failure mechanism is complex but is thought to be caused essentially by crack generation due to repeated cyclic loading and propagation of such cracks. Such cracks tend to reduce the cross section to a level which causes high stresses leading to fracture.

Fatigue testing Materials can be tested for fatigue in the laboratory using a suitable testing mechanism to apply repeated cycles of loading. Such testing is known as *fatigue testing*. Fatigue testing may require millions of cycles of loading and is a long-drawn testing procedure. Such tests can determine the fatigue life of a material used for beams or other structural members. An average fatigue life can be determined for a material using fatigue testing.

Stress vs number of cycles diagram With fatigue testing, one can draw stress vs number of cycles diagrams. The number of cycles being very large, a logarithmic scale is used for plotting the number of cycles. Such a semi-logarithmic graph may look like the one shown in Fig. 3.76(b). A tangent drawn to the graph gives a stress value called fatigue strength or endurance limit. The endurance limit gives the maximum stress value by a load that can be applied infinite number of times.

The general nature of cyclic loading is constant amplitude loading. Most of the machine parts are subjected to such loading. However, there can be situations where the amplitude may change when it is known as variable amplitude loading

[Fig. 3.76(c)]. Wind load on buildings and wave loading on marine structures and ships may cause such loading.

Summary

Due to complex loading systems on a body, a body may be subjected to complex stress systems with normal stresses and shear stresses existing along principal directions. Normal stresses are stresses acting normal to the cross section and shear stresses are those acting tangential to the cross section. Normal stresses are further categorised into tensile stresses causing elongation of the element and compressive stresses causing shortening of the element.

All materials deform under such stresses. The deformation due to normal stresses is governed by Hooke's law which states that stress is proportional to strain, within elastic limit. The constant of proportionality in the case of normal stresses and normal strains is known as Young's modulus of elasticity, E. Thus, the normal stress is $\sigma = E\varepsilon$, where ε is the strain. Strain is the change in length per unit length.

Due to shear stresses, materials again deform tangentially. The shear stress, τ, is proportional to the shear strain, ϕ. The constant of proportionality in this case is the shear modulus or modulus of rigidity G. Thus, $\tau = G\phi$ is the law governing shear stress and strain.

When a body is subjected to normal stress along one direction, it causes strains along the other two perpendicular directions known as lateral strains. The lateral strain bears a constant ratio to the normal strains called Poisson's ratio. Poisson's ratio is denoted by the letter ν. Thus, if ε_X is the normal strain along x-direction, $\varepsilon_Y = \varepsilon_Z = \varepsilon_X \nu$. The value of ν is between 0.24 to 0.34 for most materials and cannot exceed 0.5.

The generalised Hooke's law states that of σ_X, σ_Y, and σ_Z are the stresses acting along the principal directions X, Y, and Z, then

$$\varepsilon_X = \frac{\sigma_X}{E} - \frac{\sigma_Y \nu}{E} - \frac{\sigma_Z \nu}{E}, \quad \varepsilon_Y = \frac{\sigma_Y}{E} - \frac{\sigma_X \nu}{E} - \frac{\sigma_Z \nu}{E}$$

and

$$\varepsilon_z = \frac{\sigma_Z}{E} - \frac{\sigma_X \nu}{E} - \frac{\sigma_Y \nu}{E}$$

These three equations form what is known as Generalised Hooke's law. The three relations for shear stresses,

$$\phi_{XY} = \frac{\tau_{XY}}{G}, \quad \phi_{YZ} = \frac{\tau_{YZ}}{G} \quad \text{and} \quad \phi_{ZX} = \frac{\tau_{ZX}}{G}$$

also make up the total of stresses and strains relationships in a general case.

Normal strains and lateral strains result in a change in the volume of the body. The volumetric strain ε_V is the change in volume per unit volume. The bulk modulus $K = \sigma/\varepsilon_V$ where $\sigma = \sigma_X = \sigma_Y = \sigma_Z$. The three elastic constants, E, G, and K can be related to each other by the relationships, $E = 3K(1 - 2\nu) = 2G(1 + \nu)$.

Stresses and strains in the elements of a single material or of compound sections can be calculated from the above relationships. In the case of compound materials, a condition of compatibility of deformations is available for this purpose.

Materials when subjected to temperature changes deform, causing a change in length. When such deformations are restrained, stresses are developed in the materials. Due to temperature changes, change in length = $L\alpha t$, where L is the length, α = coefficient of linear expansion, and t = change in temperature. Compound sections of two or more materials are subjected to stresses even without a restraint because of the difference in the coefficients of expansion.

Important properties of materials are proportional limit, which is the stress up to which stress is proportional to strain; elastic limit, which is the stress up to which the material behaves elastically (meaning that the deformations disappear when the load is removed); yield stress, which is the stress beyond which the material undergoes plastic deformations. The stress-strain diagram for a material is drawn by conducting a tension/compression test on a standard piece of material to evaluate such properties.

Stress concentration is the increase in stress (above the average stress) at change of section, discontinuities like holes, etc. Residual stresses are stresses that remain in the material even after applied load is removed. Fatigue is a condition due to cyclic loading and the material loses strength and may fail at a much lower stress due to fatigue.

Exercises

Review Questions

1. Define the terms stress, strain, modulus of elasticity, modulus of rigidity, and bulk modulus.
2. Tensile and compressive stresses are called normal stresses. Why?
3. How is shear strain defined?
4. In each of the cases shown in Fig. 3.77, the steel and brass bars have the same length and cross-sectional area. If the ratio of modulii of elasticity of steel to brass ≈ 2, state, without calculation, which material will have a numerically larger stress in each case. What can you say about the strains?

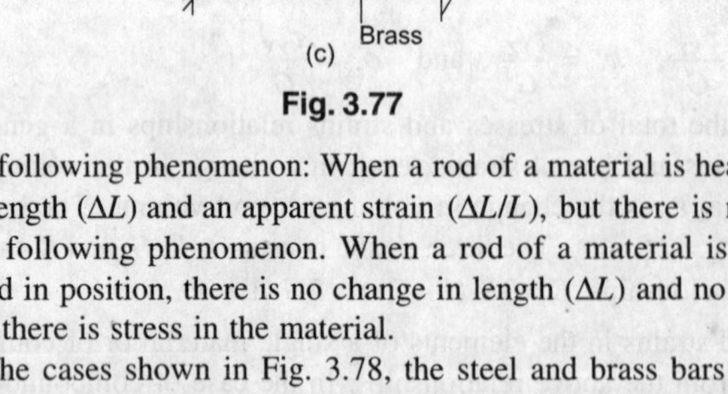

Fig. 3.77

5. Explain the following phenomenon: When a rod of a material is heated, there is an increase in length (ΔL) and an apparent strain ($\Delta L/L$), but there is no stress.
6. Explain the following phenomenon. When a rod of a material is heated and its ends are held in position, there is no change in length (ΔL) and no apparent strain (DL/L), but there is stress in the material.
7. In each of the cases shown in Fig. 3.78, the steel and brass bars have the same length and cross-sectional area, and are rigidly joined. If $E_s/E_b \approx 2$ and $\alpha_s = 12 \times 10^{-6}$/°C and $\alpha_b = 18 \times 10^{-6}$/°C, which material will have a larger stress due to a change in temperature?

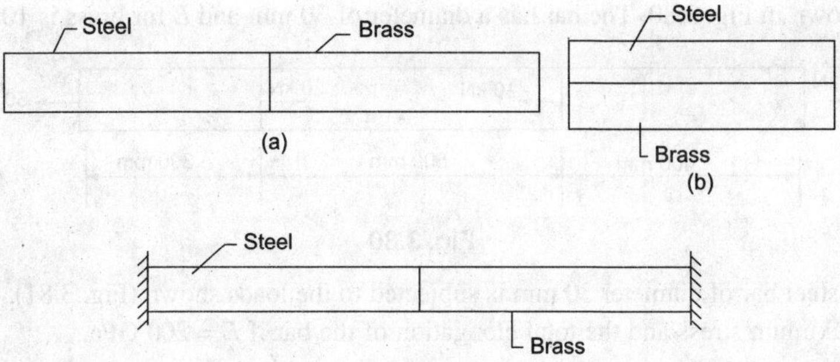

Fig. 3.78

8. If in a composite bar, the lengths are the same as shown in Fig. 3.79 and the ratio of modulii of elasticity is 2 : 1, what should be the ratio of the cross-sectional areas so that the stresses in the material are numerically the same.

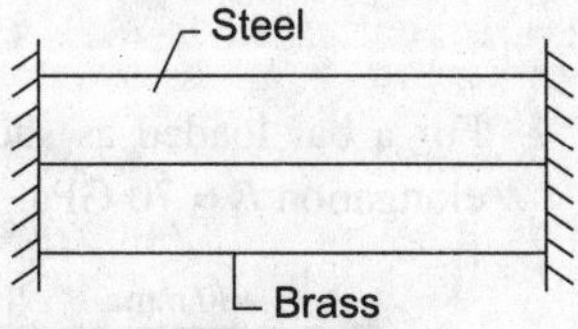

Fig. 3.79

9. Draw a typical stress-strain diagram for mild steel. Using the diagram, define proportional limit, elastic limit, yield stress, ultimate stress, and breaking stress.
10. What is indicated by percentage elongation and percentage reduction in area?
11. Explain the offset method of determining the yield stress.
12. Define and explain the term proof stress.
13. Draw typical stress-strain diagrams for high tensile steel, concrete, timber, and aluminium.
14. Explain the properties elasticity, plasticity, ductility, and brittleness.
15. Define the term Poisson's ratio. What is its maximum value?
16. State the relationship between E, G, and K.
17. Discuss the phenomenon of stress concentration and illustrate with examples.
18. Explain the term residual stress and the cause for such stress.
19. Explain the phenomenon of fatigue and the use of this concept.

Problems

1. A bar of an alloy material 20 mm ϕ was subjected to an axial tensile load of 25 kN. The elongation of a length of 400 mm was found to be 0.25 mm. Find the modulus of elasticity.
2. Find the elongation and stress in a steel bar, of diameter 16 mm and length 2 m when it is subjected to an axial load of 30 kN. E for steel = 200 GPa.
3. A hollow pipe of inside bore 50 mm and thickness 5 mm is used as a short column (3 m long), and carries an axial load of 28 kN. Find the stress in the material and the shortening in length if E = 100 GPa.
4. A masonry wall is 40 mm thick and 7 m high. The wall carries a UD load of 50 kN/m of its length in addition to its own weight. Find the maximum stress at the base if the masonry weighs 18 kN/m^3.
5. Find the maximum stress and total elongation of a brass bar subjected to the loads

shown in Fig. 3.80. The bar has a diameter of 30 mm and E for brass is 105 GPa.

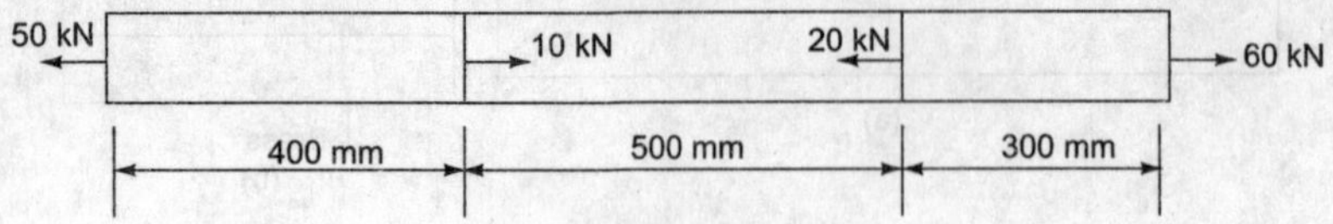

Fig. 3.80

6. A steel bar of diameter 30 mm is subjected to the loads shown (Fig. 3.81). Find the maximum stress and the total elongation of the bar if E = 200 GPa.

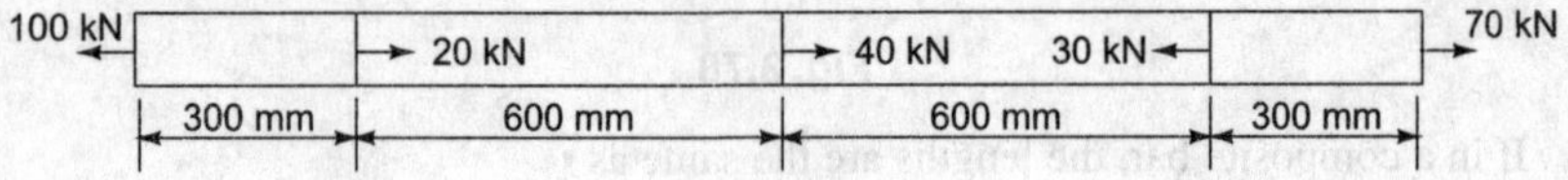

Fig. 3.81

7. For a bar loaded as shown in Fig. 3.82, find the maximum stress and the total elongation E = 70 GPa.

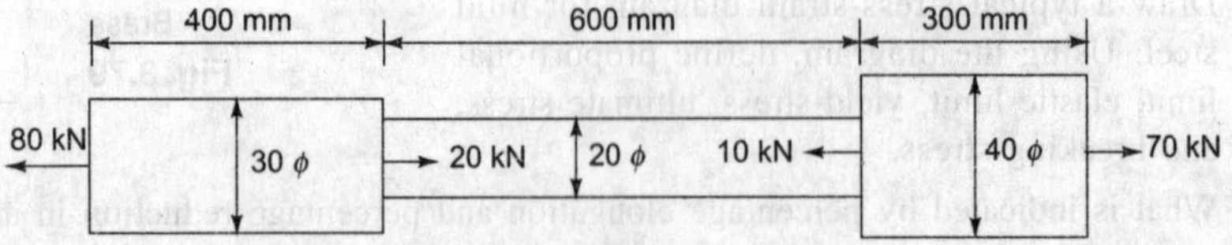

Fig. 3.82

8. A brass bar of varying sections is subjected to the loads shown in Fig. 3.83. Find the maximum stress and the total elongation if E = 100 GPa.

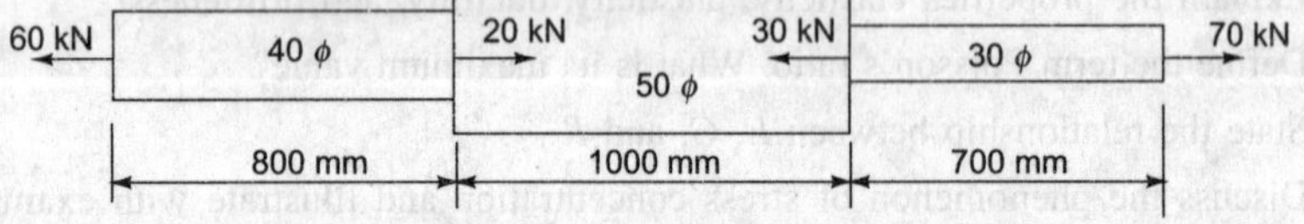

Fig. 3.83

9. Brass and steel bars of equal diameter are rigidly joined together and firmly fixed at one end, and subjected to a load of 80 kN as shown in Fig. 3.84. If the diameter of the bars is 40 mm, and E for brass and steel are 105 GPa and 200 GPa, respectively, find the stresses in the two materials and the total elongation.

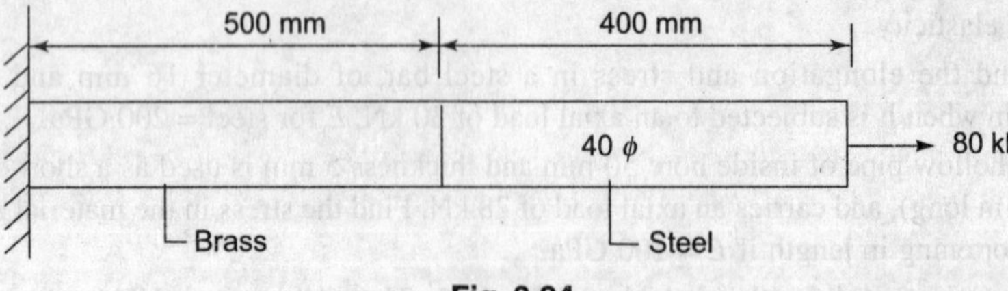

Fig. 3.84

10. A bar is made by rigidly connecting three bars of different materials and of different diameters as shown in Fig. 3.85. Find the stresses at the different sections of bar and its total elongation when it is subjected to a tensile load as shown. Take the values of E for steel, brass, and aluminium as 200 GPa, 100 GPa, and 70 GPa, respectively.

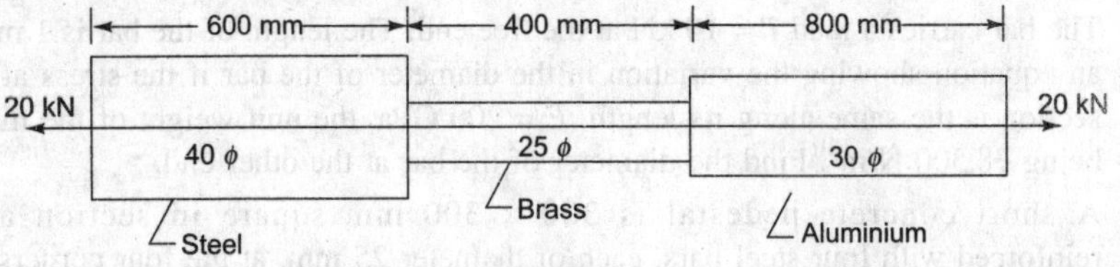

Fig. 3.85

11. A railway signal is operated by a 5 mm ϕ steel wire 500 m long. At the signal end, a movement of 200 mm is required for the proper orientation of the signal. A pull of 1.5 kN is applied at the signal-box end of the wire. What should be the movement so that the specified movement is available at the signal, E for steel = 200 GPa.
12. The steam pressure in a steam engine is 1.2 N/mm^2. If the piston is 350 mm in diameter and the piston rod is 50 mm in diameter, find the maximum stress and the elongation of the piston rod if it is 1m long. E = 200 GPa.
13. The lifting wire of a crane is 60 m long. Find the maximum load that can be lifted if the diameter of the wire is 10 mm, and the stress in the wire is limited to 100 N/mm^2. What is the extension of the wire under this load if E = 200 GPa?
14. As shown in Fig. 3.86, three wires are used to lift a load. The unstressed lengths of the wires are 24.195 m, 25 m, and 25.004 m. If the wires are 6 mm in diameter, find the maximum loads before each of the longer wires starts sharing the load. Find the stress in the longest wire if the load is 4 kN. Determine the stresses in the wires if the load is 2 kN. E = 200 GPa.
15. A rod has a diameter of 120 mm at one end and tapers uniformly to a diameter of 60 mm over a length of 400 mm. Find the elongation of the rod when it is subjected to a load of 200 kN. E = 200 GPa.
16. A brass plate of uniform thickness 5 mm is 600 mm long and varies in width from 70 mm at one end to 30 mm at the other. Find the load to be applied axially to this plate to cause an extension of 1 mm. E = 100 GPa.

Fig. 3.86

17. A rod, 500 mm long, of diameter 20 mm in the middle, varies uniformly to a diameter of 40 mm at the ends. If E = 200 GPa, find the extension of the rod when subjected to a tensile load of 40 kN.
18. A uniform steel rod of diameter 20 mm is connected to an aluminium rod of diameter 60 mm at one end. The aluminium rod tapers to a diameter of 20 mm at the other end. The steel rod is 0.6 m long and is connected rigidly to the 60 mm ϕ end of the aluminium rod which is 0.4 m long. If E = 200 GPa for steel and 70 GPa for aluminium, find the total extension of the rod under an axial tensile load of 30 kN.
19. A pillar, 2 m long, has a diameter of 200 mm at the top and tapers to a diameter of 800 mm at the base. It carries a load of 150 kN, uniformly distributed on its top face. Find the shortening of the pillar under this load, neglecting self-weight. E = 10 GPa.

20. A tapering steel rod, of diameter 100 mm at one end, and 40 mm at the other, and length 1 m, is hung freely by fixing it rigidly to a support at the larger end. Find the extension of the rod under its own weight. Steel weighs 78.5 kN/m^3 and E = 200 GPa. What is load P that should be applied from the free end so that the extension of the rod is zero?
21. A bar has a diameter of 20 mm at the free end and is hung freely from its larger end. The bar carries a load P = 10 kN at the free end. The length of the bar is 1 m. Find an equation showing the variation in the diameter of the bar if the stress at every section is the same along its length. E = 200 GPa, the unit weight of the material being 78,500 N/m^3. Find the diameter of the bar at the other end.
22. A short concrete pedestal is 300 × 300 mm square in section and is reinforced with four steel bars, each of diameter 25 mm, at the four corners. Find the maximum load that can be carried if the stress in concrete is limited to 4 N/mm^2 and that in steel to 130 N/mm^2. E = 200 GPa for steel and 15 GPa for concrete.
23. A 500 mm diameter reinforced concrete column has 8 bars of 20 mm diameter. The column is subjected to an axial thrust of 875 kN. Determine the stresses developed in concrete and steel. Assume $E_{steel} = 12E_{concrete}$.
24. A steel rod of diameter 20 mm and length 400 mm is covered by a rigidly connected brass shell of inside diameter 20 mm and outside diameter 40 mm. Find the stresses in the two materials and the deformation when the composite structure is subjected to a load of 100 kN. E = 200 GPa for steel and 100 GPa for brass.
25. Find the stresses in the steel bolt and brass tube in the assembly shown in Fig. 3.87 due to a quarter turn of the nut after they have been fitted fully in stress-free conditions. The bolt has a diameter of 8 mm and the pitch of the thread in the bolt is 2 mm. The brass tube has an inside diameter of 20 mm and a thickness of 3 mm. E = 200 GPa for steel and 105 GPa for brass.

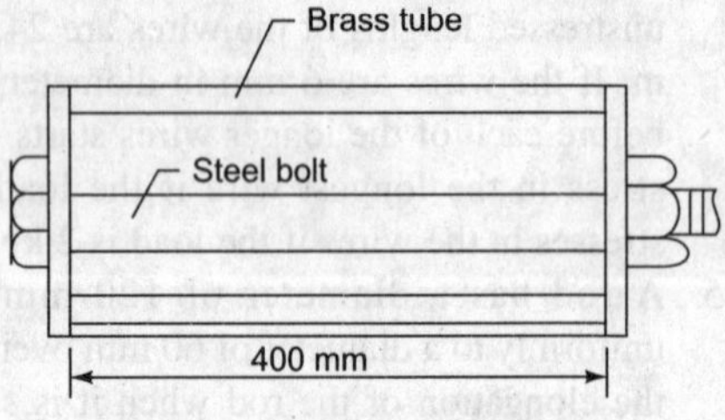

Fig. 3.87

26. A copper rod of diameter 20 mm is enclosed within a steel tube of 30 mm internal diameter and thickness 2 mm. The assembly is attached to rigid plates at the ends and is subjected to a tensile load of 28 kN. Find the stresses in the rod and the tube if E = 200 GPa for steel and 120 GPa for copper.
27. A wooden rod 300 mm square in section has two steel plates 300 mm × 3 mm rigidly attached to its opposite sides. Find the stresses in the two materials when the assembly is subjected to a compressive load of 100 kN. E = 200 GPa for steel and 10 GPa for wood.
28. A steel rod of diameter 10 mm and length 1400 mm is stress-free at 20 °C. Find the stress in the rod if the temperature rises to 70°C (i) if the ends are fully fixed and (ii) if one of the ends yields by 0.2 mm. $\alpha = 12 \times 10^{-6}$/°C and E = 200 GPa.
29. A bar tapers in diameter from 200 mm at one end to 80 mm at the other over a length of 900 mm. The bar fixed between supports is stress-free at 27°C. Find the maximum normal stress in the bar when the temperature rises to 100°C. E = 105 GPa and $\alpha = 20 \times 10^{-6}$/°C.

30. A steel plate, 10 mm thick, is 40 mm wide at one end and tapers uniformly to 10 mm at the other. The bar is held between rigid supports and is stress-free at 20°C. Find the maximum normal stress in the plate when the temperature rises by 60°C. The length of the bar is 1300 mm. $E = 200$ GPa and $\alpha = 12 \times 10^{-6}/°C$.
31. In the assembly shown in Fig. 3.88, the copper bar on the left has a diameter of 30 mm and the brass bar on the right one of 60 mm. The bars are stress-free at 25°C. Find the stresses and the change in length of the bars, when the temperature rises to 100°C if (i) the supports are unyielding and (ii) the right support yields by 0.2 mm. For copper, $E = 120$ GPa and $\alpha = 18 \times 10^{-6}/°C$, and for brass $E = 90$ GPa and $\alpha = 20 \times 10^{-6}/°C$.

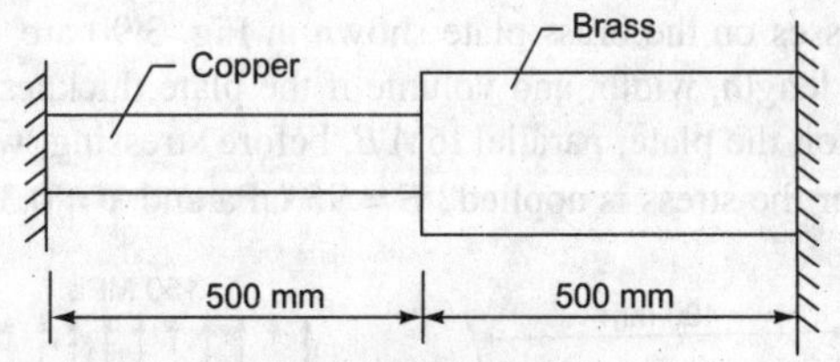

Fig. 3.88

32. A solid brass cylinder, 20 mm ϕ, is rigidly attached and surrounded by an aluminium tube of internal diameter 20 mm and thickness 5 mm. If the assembly is stress-free at 28°C, find the stresses in the two materials when the temperature rises to 90°C. For brass, $E = 90$ GPa and $\alpha = 20 \times 10^{-6}/°C$, and for aluminium $E = 70$ GPa and $\alpha = 23 \times 10^{-6}/°C$.
33. In the assembly shown in Fig. 3.89, the steel rod is rigidly attached to the brass bar and there is a gap of 0.5 mm when the temperature is 20°C, and the rods are stress-free. Find the stresses in the bars and the length of each bar when the temperature rises to 120°C. For steel, $E = 200$ GPa and $\alpha = 12 \times 10^{-6}/°C$. For brass, $E = 95$ GPa and $\alpha = 20 \times 10^{-6}/°C$.

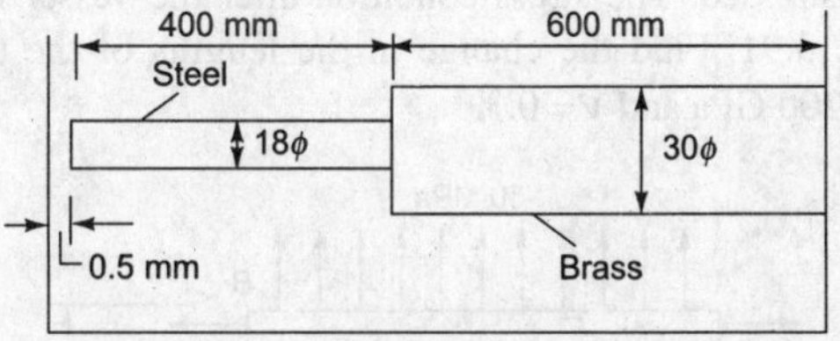

Fig. 3.89

34. A steel rod, 20 mm ϕ and 500 mm long, is stretched by a tensile force of 25 kN when the temperature is 20ºC. What will be the stress in the rod when the temperature drops to 0°C? At what temperature will the stress in the bar be zero? $E = 200$ GPa and $\alpha = 12 \times 10^{-6}/°C$.
35. Steel rails are laid at a temperature of 20°C. What should be the clearance between the rails if the maximum temperature expected is 55°C for no stress in the rails? If the actual clearance provided is half this value, what will be the stress in the rails at the maximum temperature? $E = 200$ GPa and $\alpha = 12 \times 10^{-6}/°C$.
36. Determine the force required to punch a hole of diameter 200 mm in a steel plate 8 mm thick, if the ultimate shear stress of steel is 300 MPa.
37. What is the average shear stress in an aluminium plate of thickness 5 mm while punching a hole of diameter 22 mm using a force of 70 kN?

38. A solid circular steel bar, 20 mm ϕ, is subjected to an axial tensile load of 60 kN. What is the decrease in the diameter of the bar? E = 200 GPa and ν = 0.3.
39. A square steel bar is 40 mm × 40 mm in cross section and is 400 mm long. It is subjected to an axial tensile force of 150 kN. Find the change in the length of the sides and the percentage change in volume. E = 200 GPa and ν = 0.3.
40. Show that the change in the diameter D of a solid cylinder, subjected to an axial load P is $4\,P\nu/\pi ED$, where n is Poisson's ratio and E is the modulus of elasticity.
41. On a steel plate of thickness 12 mm, size 300 mm square, a circle of diameter 150 mm is etched at the centre. If the plate is subjected to stresses as shown in Fig. 3.70, find the changes in the lengths of diameters parallel to the sides, and in thickness and volume. E = 200 GPa and ν = 0.33.
42. The axial stresses on the brass plate shown in Fig. 3.90 are 150 MPa. Determine the changes in length, width, and volume if the plate thickness is 10 mm. If a line $A'B'$ is etched on the plate, parallel to AB, before stressing, what will be the slope of this line after the stress is applied? E = 95 GPa and ν = 0.3.

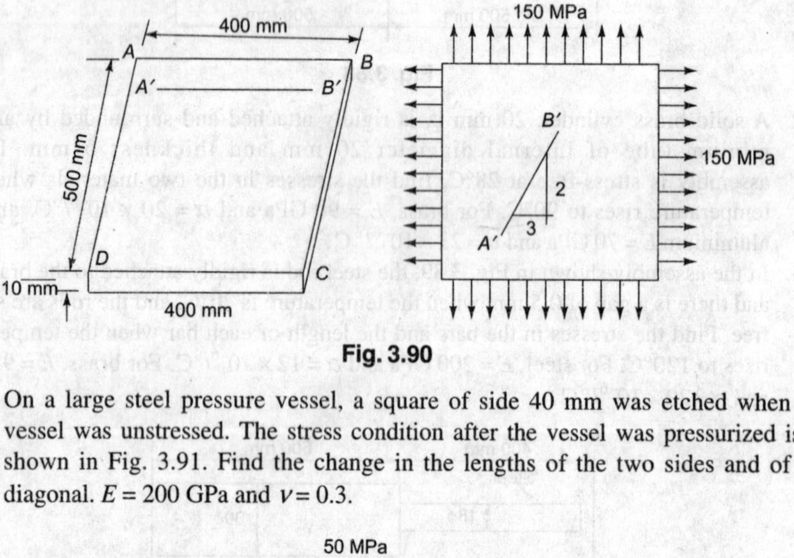

Fig. 3.90

43. On a large steel pressure vessel, a square of side 40 mm was etched when the vessel was unstressed. The stress condition after the vessel was pressurized is as shown in Fig. 3.91. Find the change in the lengths of the two sides and of the diagonal. E = 200 GPa and ν = 0.3.

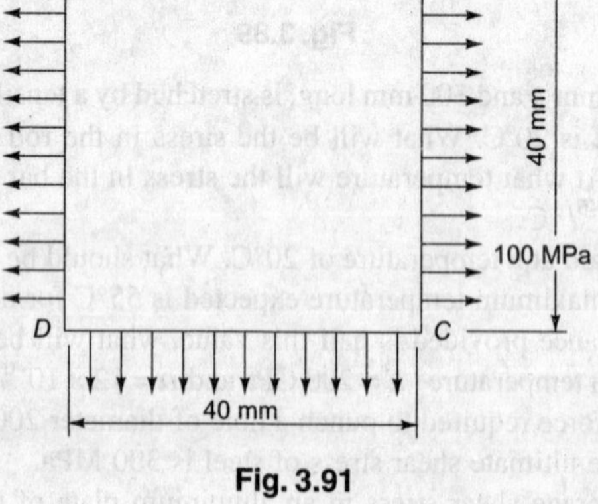

Fig. 3.91

44. At what depth under the sea will there be a 0.1% change in the volume of a brass cube of side a if $E = 100$ GPa and $\nu = 0.33$? Unit weight of sea water = 10.08 kN/m^3.
45. A solid aluminium sphere is lowered in sea water to a depth of 10 km. Find the change in the diameter and the volume of the sphere. What is the per cent increase in the density of the sphere? $E = 70$ GPa and $\nu = 0.33$. Unit weight of sea water = 10.08 kN/m^3.
46. If an element is subjected to stresses s in one direction only and the lateral strains are prevented in the other two directions, prove that the effective modulus of elasticity is $E(1-\nu)/[(1-2\nu)(1+\nu)]$.
47. On the surface of a cylinder, a square of side 20 mm is etched in a stress-free condition. If the shaft has a shear stress on its surface of 100 N/mm^2, find the change in the angle of the square. $G = 85$ GPa.
48. If, for a material, the modulus of elasticity is 140 GPa and the modulus of rigidity is 54 GPa, find the values of Poisson's ratio and bulk modulus.
49. A bar of certain material 60 mm × 60 mm in cross section is subjected to an axial pull of 180 kN. The extension over a length of 100 mm is 0.05 mm and decrease in each side is 0.00525 mm. Calculate modulus of elasticity, Poisson's ratio, modulus of rigidity, and bulk modulus.
50. A rigid bar ECD (Fig. 3.92) is kept in a horizontal position by two vertical rods, one of brass and the other of steel. The steel rod has a length of 3 m and a diameter of 20 mm. The brass rod has a length of 2 m and a diameter of 30 mm. Find the position of the load P so that the bar remains horizontal. $E = 200$ GPa for steel and 100 GPa for brass.

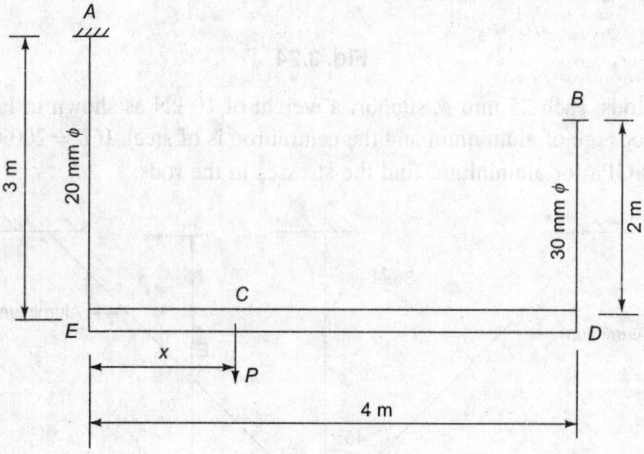

Fig. 3.92

51. A rigid bar of weight $W = 900$ kN is kept horizontal by the two rods, one of aluminium and the other of steel (Fig. 3.93). Find the ratio of the areas of the rods so that the bar remains horizontal. $E = 70$ GPa for aluminium and 200 GPa for steel.
52. A rigid bar of negligible weight is supported by two rods, one of copper and the other of brass (Fig. 3.94). With the 40 kN weight applied at the end, find the stresses in the bars. $E = 120$ GPa for copper and 105 GPa for brass.

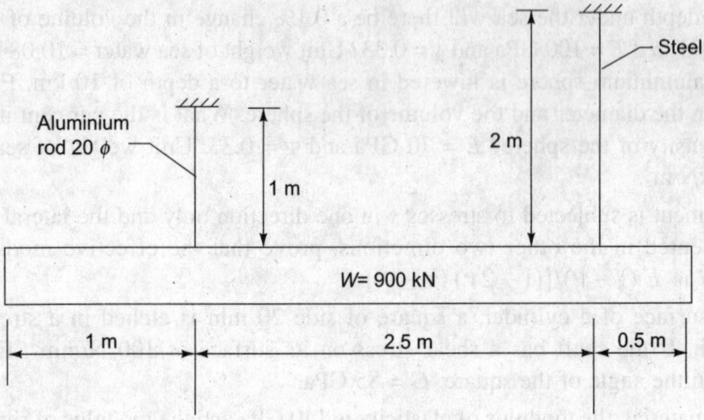

Fig. 3.93

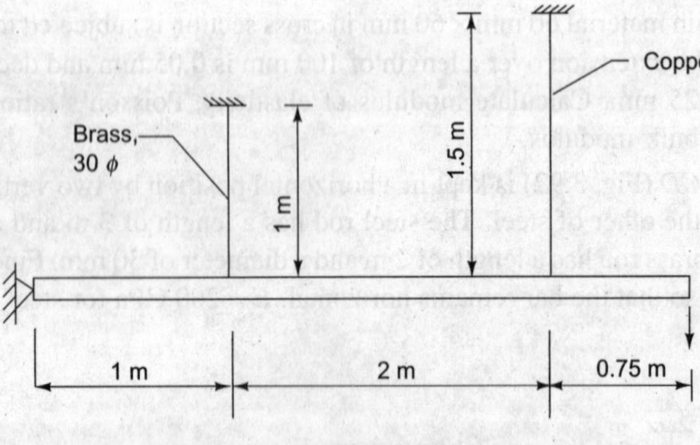

Fig. 3.94

53. Three rods, each 25 mm ϕ, support a weight of 10 kN as shown in Fig. 3.95. The outer rods are of aluminium and the central rod is of steel. If E = 200 GPa for steel and 70 GPa for aluminium, find the stresses in the rods.

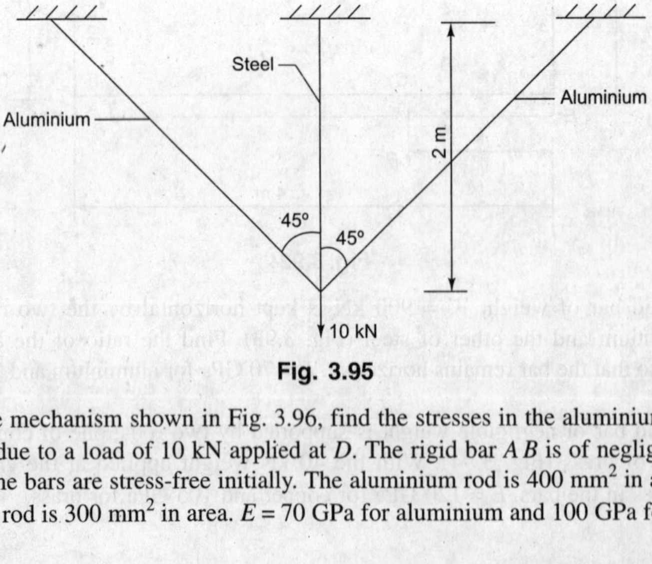

Fig. 3.95

54. In the mechanism shown in Fig. 3.96, find the stresses in the aluminium and brass rods due to a load of 10 kN applied at D. The rigid bar AB is of negligible weight and the bars are stress-free initially. The aluminium rod is 400 mm^2 in area and the brass rod is 300 mm^2 in area. E = 70 GPa for aluminium and 100 GPa for brass.

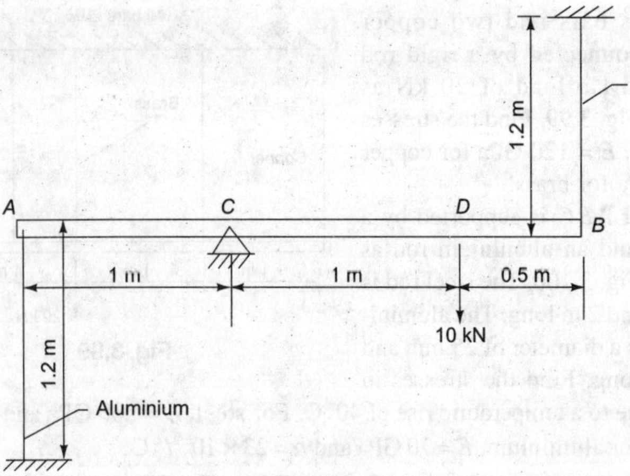

Fig. 3.96

55. A composite bar is made by attaching a copper rod of length 900 mm and diameter 30 mm and a brass rod of length 600 mm and diameter 45 mm (Fig. 3.97). A load of 20 kN is applied at 20°C. Find the stresses in the materials when the temperature rises to 50ºC. For copper, $E = 120$ GPa and $\alpha = 18 \times 10^{-6}$/°C. For brass, $E = 90$ GPa and $\alpha = 20 \times 10^{-6}$/ºC.

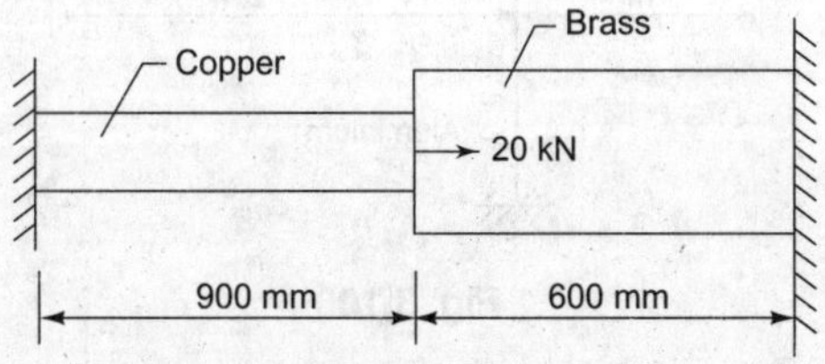

Fig. 3.97

56. A rigid bar of negligible weight is supported by three wires as shown in Fig. 3.98 and carries a load of 80 kN. After this load is applied at 20°C, the temperature is raised to 50°C. Find the stresses in the wires and the location of the load so that the rigid bar remains horizontal. For brass, $E = 95$ GPa and $\alpha = 20 \times 10^{-6}$/°C, for aluminium, $E = 70$ GPa and $\alpha = 23 \times 10^{-6}$/°C, and for copper, $E = 120$ GPa and $\alpha = 18 \times 10^{-6}$/°C.

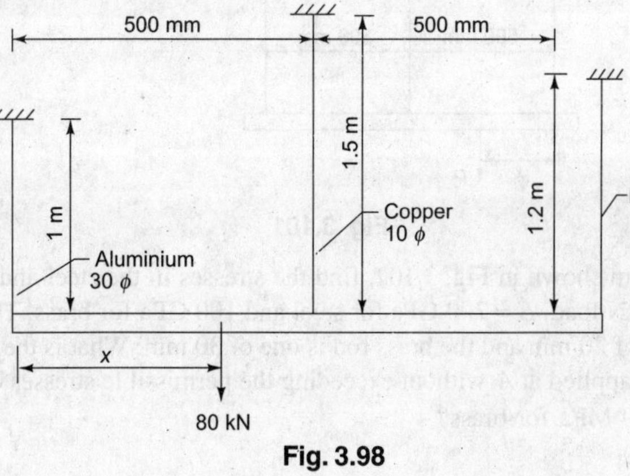

Fig. 3.98

57. Two brass bars and two copper bars are connected by a rigid rod and support a load of 20 kN as shown in Fig. 3.99. Find the stresses in the bars. $E = 120$ GPa for copper and 90 GPa for brass.

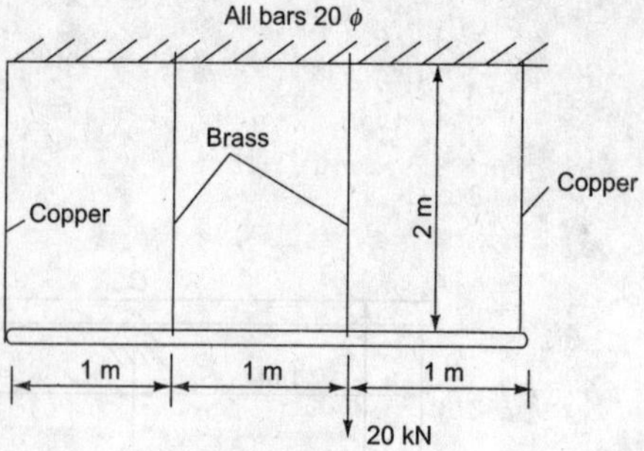

Fig. 3.99

58. A rigid bar ABC is supported by a steel rod and an aluminium rod as shown in Fig. 3.100. The steel rod is 20 mm ϕ and 2 m long. The aluminium rod has a diameter of 25 mm and is 1.5 m long. Find the stresses in the rods due to a temperature rise of 40°C. For steel, $E = 200$ GPa and $\alpha = 12 \times 10^{-6}$/°C, and for aluminium, $E = 70$ GPa and $\alpha = 23 \times 10^{-6}$/°C.

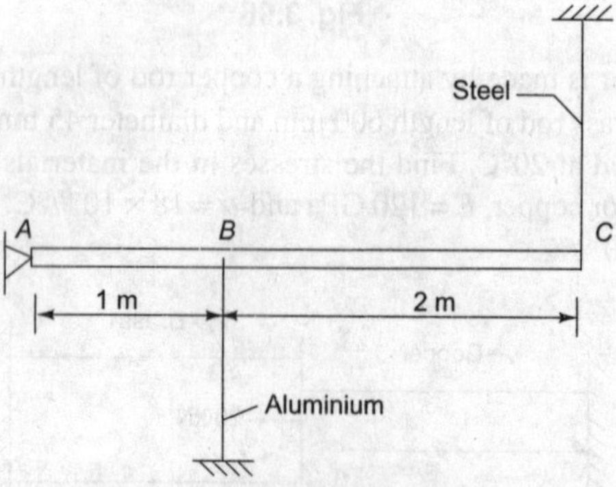

Fig. 3.100

59. A rigid bar is suspended by using three identical wires as shown in Fig. 3.101. Find the smallest value of x for which all the wires will remain in a stretched condition, and find the stresses in the wires if the load is equal to 20 kN.

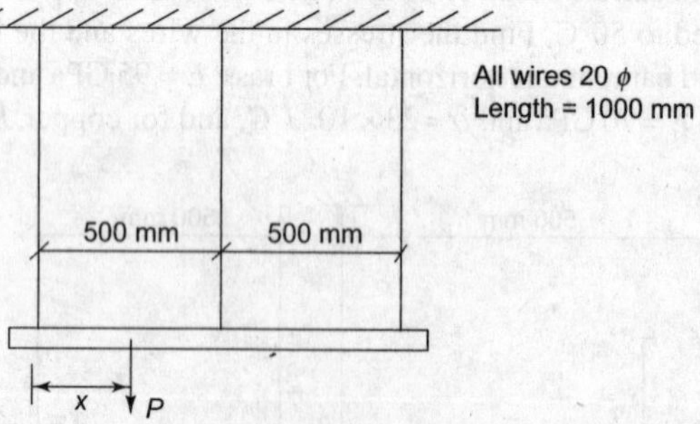

Fig. 3.101

60. In the system shown in Fig. 3.102, find the stresses in the steel and brass rods due to the 100 kN load. $E = 200$ GPa for steel and 100 GPa for brass. The steel rod has a diameter of 20 mm and the brass rod is one of 30 mm. What is the maximum load that can be applied at A without exceeding the permissible stresses of 150 MPa for steel and 60 MPa for brass?

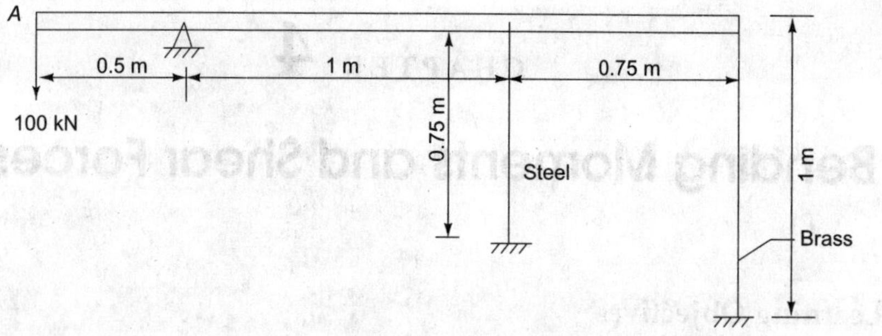

Fig. 3.102

61. A steel tyre 5 mm thick and 30 mm wide has a diameter 1 mm less than a wooden wheel of diameter 1 m. To what temperature must the steel tyre be heated so that it can be fitted on the wheel? What is the maximum hoop stress in the wheel? For steel, $E = 200$ GPa and $\alpha = 12 \times 10^{-6}$/°C and for wood, $E = 10$ GPa.
62. A copper ring of thickness 5 mm and width 30 mm has an internal diameter of 899 mm and is to be fitted on to a rigid brass ring of external diameter 900 mm. Calculate the temperature to which the copper ring should be heated so that it can be slipped on the brass ring. Calculate the stresses in the ring when it is cooled. Take $E = 120$ GPa and $\alpha = 18 \times 10^{-6/°}$C.
63. From the following data on the tension test on an alloy,
    (i) plot a stress-strain diagram,
    (ii) find the proportional limit, yield point, modulus of elasticity, ultimate strength, and rupture strength,
    (iii) find the yield strength at 0.2% offset, and
    (iv) find the percentage elongation.

The initial diameter of the specimen was 12.5 mm and the gauge length 50 mm. Final diameter at fracture is 10 mm.

Load (N)	***Elongation (mm)***	***Load (N)***	***Elongation (mm)***	***Load (N)***	***Elongation (mm)***
3000	0	35,000	0.057	53,000	1.875
6000	0.005	37,000	0.065	57,000	2.500
9000	0.010	40,000	0.080	61,500	3.750
12,000	0.015	41,500	0.095	64,000	5.00
15,000	0.020	42,500	0.145	67,500	7.50
18,000	0.025	42,500	0.170	68,500	9.50
21,000	0.03	42,000	0.200	65,500	11.250
24,000	0.035	41,000	0.250	57,500	12.5
27,000	0.040	41,000	0.500		
30,000	0.045	42,000	0.750		
32,500	0.050	47,000	1.25		

CHAPTER 4

Bending Moments and Shear Forces

Learning Objectives

After going through this chapter, the reader will be able to

- describe the structural behaviour of beams,
- determine the effects of external loads such as axial force, shear force, and bending moment at any section of the beam, using consistent sign conventions,
- derive and interpret the differential relationships between load intensity, shear force and bending moment,
- draw AF, SF, and BM diagrams for statically determinate beams and frames,
- draw shear force and BM diagrams for statically determinate beams using graphical procedure, and
- explain the singularity function approch to determine shear force and bending moments in beams.

4.1 INTRODUCTION

We discussed different types of structural elements in Chapter 1. In this chapter, we will analyse beams in detail. Beams are very common structural elements used to span distances. They carry loads predominantly transverse to their longitudinal axis. Determining the effect of external loads on beam cross sections is the first step in the design of beams. Bending moments and shear forces are the two effects of external loads on beam sections. In this chapter, we will limit ourselves to statically determinate beams and frames, discussing analytical as well as graphical methods.

4.1.1 Beams

Beams, in general, are very long compared to their cross-sectional dimensions. They are represented in diagrams by their centroidal, longitudinal axis and carry loads predominantly transverse to this axis (Fig. 4.1). The primary effect of the loads is to bend the beam as shown by the dotted line in Fig. 4.1. The changes in the geometry of the loads due to such bending are very small and are neglected. The beams we will discuss are planar, which means that the longitudinal axis, and the lines of action of the loads and reactive forces, lie in the same plane. The lon-

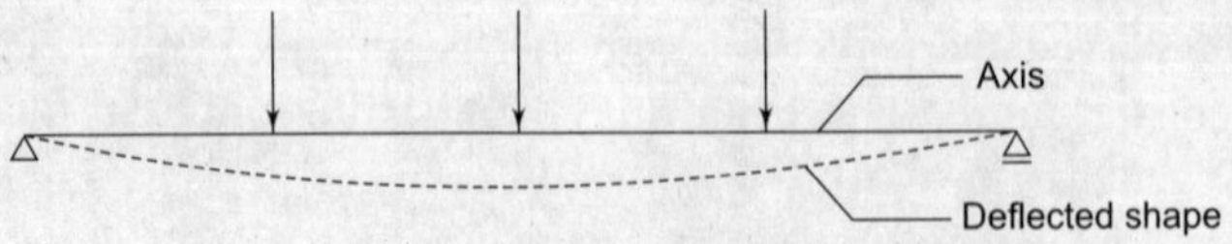

Fig. 4.1

gitudinal axes of the beams considered are straight lines and are generally horizontal though inclined beams are not uncommon.

The types of beams and frames we will deal with are shown in Fig. 4.2. Figure 4.2(a) shows *a simply supported beam* with a hinge at one end and a roller support at the other. Figure 4.2(b) shows a *cantilever* which has a fixed support at one end and is free at the other. Figure 4.2(c) shows an *overhanging beam* where one support of a simply supported beam is shifted to the interior. Figure 4.2(d) shows a *doubly overhung beam* with overhanging segments at both the ends. Figure 4.2(e) shows a *hinged beam* where a hinge is used to connect two segments of a beam. Figure 4.2(f) shows a *statically determinate rigid frame*. In all these figures, the number of reactive components provided by a support is shown near it.

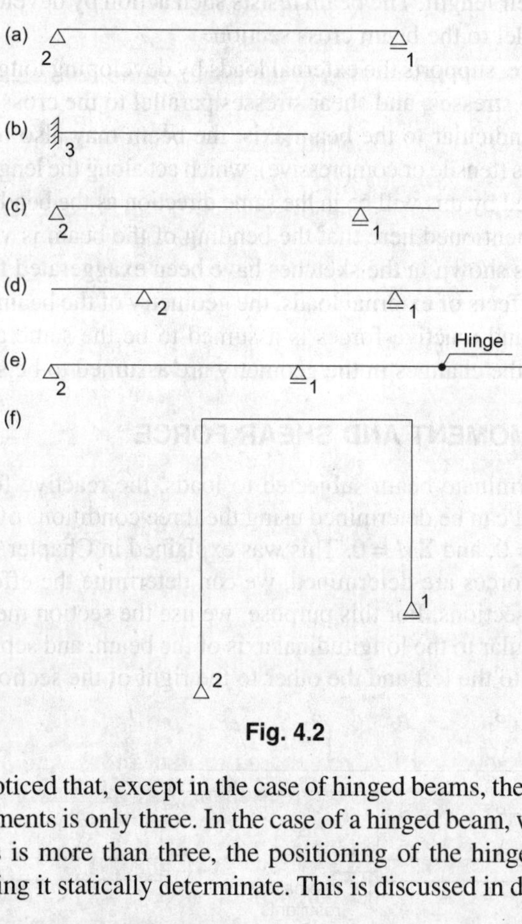

Fig. 4.2

It may be noticed that, except in the case of hinged beams, the number of reactive forces and moments is only three. In the case of a hinged beam, while the number of reactive forces is more than three, the positioning of the hinge provides an extra condition making it statically determinate. This is discussed in detail in Section 4.9.

4.1.2 Structural Action of a Beam

Under the action of loads, a beam gets deflected. This is known as bending or flexure. The curved shape of the beam after bending is known as the deflected shape or elastic curve. The beam may bend with convexity downwards or upwards. When the convexity is downwards, as in Fig. 4.3(a), it is called *sagging*, and when the convexity is upwards, as in Fig. 4.3(b), it is called *hogging*. It is easy to see that the fibres of the beam on the convex side are stretched, which means that they are in tension. The fibres on the concave side are compressed, meaning that they are in compression. Note that this stretching and compression are along the length of the beam.

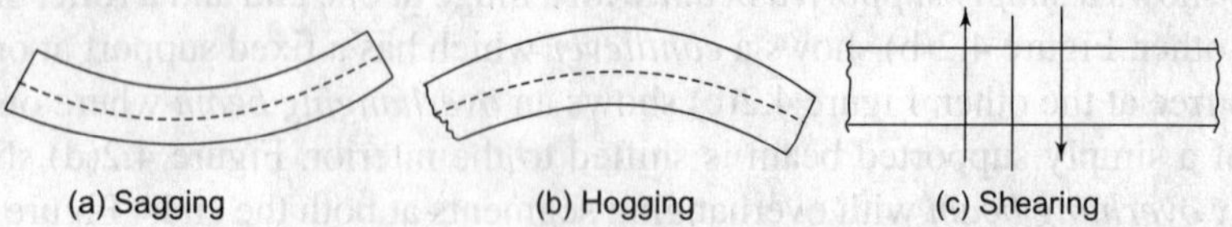

Fig. 4.3

In addition, as shown in Fig. 4.3(c), the external forces tend to shear the fibres perpendicular to their length. The beam resists such action by developing resisting shear stresses parallel to the beam cross section.

A beam, therefore, supports the external loads by developing longitudinal, compressive and tensile stresses, and shear stresses parallel to the cross section. If the loads are not perpendicular to the beam axis, the beam may also be subjected to axial forces or thrusts (tensile or compressive), which act along the length of the beam. The stresses produced by this will be in the same direction as the bending stresses.

It may also be mentioned here that the bending of the beam is very small, and the deflected shapes shown in the sketches have been exaggerated for illustration. In calculating the effects of external loads, the geometry of the beam axis and lines of action of loads and reactive forces is assumed to be the same as in the beam before bending, as the changes in the geometry are assumed to be small.

4.2 BENDING MOMENT AND SHEAR FORCE

In a statically determinate beam subjected to loads, the reactive forces and moments at the support can be determined using the three conditions of static equilibrium, $\Sigma H = 0$, $\Sigma V = 0$, and $\Sigma M = 0$. This was explained in Chapter 1.

Once the reactive forces are determined, we can determine the effect of external loads on the beam sections. For this purpose, we use the section method. We pass a section perpendicular to the longitudinal axis of the beam, and separate the beam into two parts, one to the left and the other to the right of the section (Fig. 4.4).

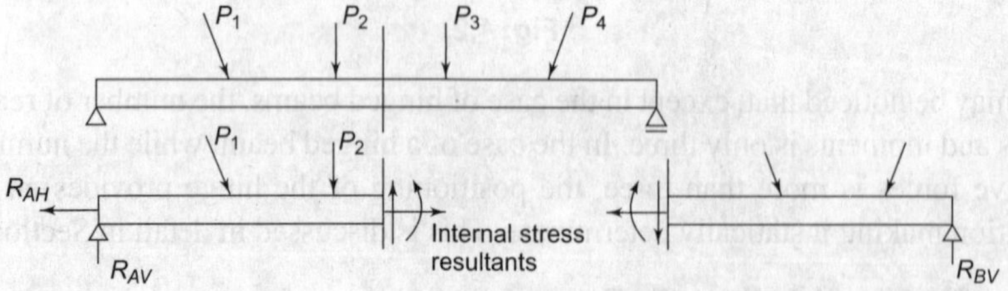

Fig. 4.4

The basic principle in analysing either part is that since the beam as a whole is in equilibrium, so should each part be. Either of the free bodies obtained, by passing a section as in Fig. 4.4, is in equilibrium due to the loads and reactive forces acting on it and the forces and moments at the section due to the continuity of the beam with the other part. Such forces and moments at the section are known as internal stress resultants.

Consider for illustration the beam loaded as shown in Fig. 4.5. There is a hinged support at A and a roller support at B. The reactive forces are R_{AH} (horizontal) and R_{AV} (vertical) at A and R_{BV} (vertical) at B. The beam section under consideration is at 5 m from A as shown by section 1-1 [Fig. 4.5(a)].

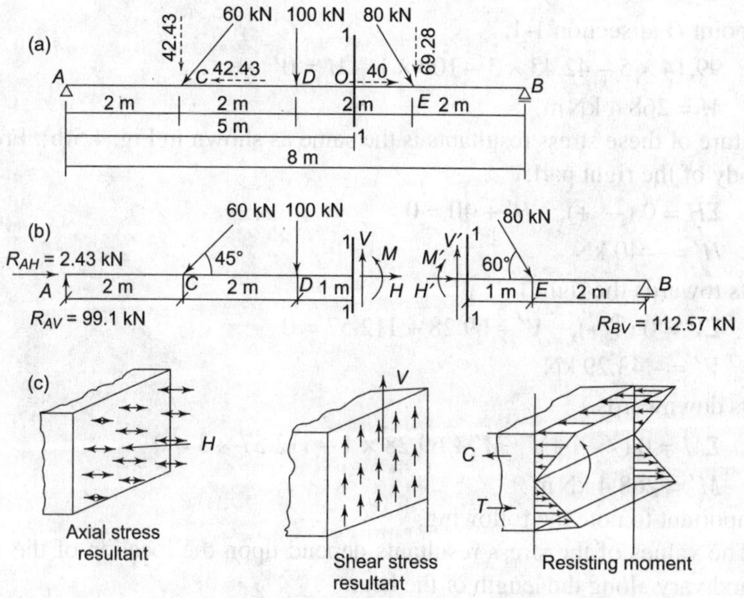

Fig. 4.5

The reactive forces can be determined using the conditions of equilibrium:

$$\Sigma H = 0\ (\rightarrow +),\quad R_{AH} - 42.43 + 40 = 0$$

$$\Rightarrow \quad R_{AH} = 2.43\text{ kN} \rightarrow$$

$$\Sigma V = 0\ (\uparrow +),\quad R_{AV} - 42.43 - 100 - 69.28 + R_{BV} = 0,\quad R_{AV} + R_{BV} = 211.71$$

$$\Sigma M = 0 \text{ about } A\ (\curvearrowright +),\quad 42.43 \times 2 + 100 \times 4 + 69.28 \times 6 - R_{BV} \times 8 = 0$$

$$R_{BV} = 112.57\text{ kN}$$

$$R_{AV} = 211.71 - 112.57 = 99.14\text{ kN}$$

After determining the reactive forces, we separate the free bodies to the left and right obtained by section 1-1. The free bodies, in general, are acted upon by coplanar force systems. The equilibrium of the free body requires that the three conditions of static equilibrium, $\Sigma H = 0$, $\Sigma V = 0$, and $\Sigma M = 0$, be satisfied by forces and moments acting on the body. In general, therefore, a horizontal force, a vertical force, and a couple may be acting at the section to satisfy the conditions of equilibrium. These are indicated as H, V, and M on the free body on the left and H', V', and M' on the free body on the right, as shown in Fig. 4.5(b). In Fig. 4.5(c), how these internal stress resultants are developed is illustrated. H is due to an axial stress acting over the section. V is due to tangential stresses. M is a couple developed due to the longitudinal compressive and tensile stresses.

The conditions of static equilibrium can be applied to the body to get the values of the internal stress resultants. From the free body of the left part,

$$\Sigma H = 0\ (\rightarrow +),\quad 2.43 - 42.43 + H = 0$$

$\Rightarrow \quad H = 40$ kN

$\Sigma V = 0\ (\uparrow +), \quad 99.14 - 42.43 - 100 + V = 0$

$\Rightarrow \quad V = 43.29$ kN

$\Sigma M = 0\ (\curvearrowright +)$

about point O at section 1-1.

$99.14 \times 5 - 42.43 \times 3 - 100 \times 1 - M = 0$

$\Rightarrow \quad M = 268.4$ kN m

The nature of these stress resultants is the same as shown in Fig. 4.5(b). From the free body of the right part,

$\Sigma H = 0\ (\rightarrow +), \quad H' + 40 = 0$

$\Rightarrow \quad H' = -40$ kN

(H' acts towards the right.)

$\Sigma V = 0\ (\uparrow +), \quad V' - 69.28 + 112.57 = 0$

$\Rightarrow \quad V' = -43.29$ kN

(V' acts downwards.)

$\Sigma M = 0\ (\curvearrowright +), \quad M' + 69.28 \times 1 - 112.57 \times 3 = 0$

$\Rightarrow \quad M' = 268.4$ kN m

It is important to note the following.

(i) The values of the stress resultants depend upon the location of the section and vary along the length of the beam.

(ii) H, V, and M and H', V', and M' are equal in magnitude but opposite in direction.

Considering the free body on the left, for example, since H, V, and M are the internal stress resultants, the effects of the loads at the section must be equal and opposite to these. The effects of external loads can be defined as follows.

Axial force (thrust) The axial force (AF) at any section of a beam is equal to the algebraic sum of the forces parallel to the axis of the beam either to the left or the right of the section.

Shear force The shear force (SF) at any section of a beam is equal to the algebraic sum of the forces perpendicular to the axis of the beam either to the left or the right of the section.

Bending moment The bending moment (BM) at any section of a beam is equal to the algebraic sum of the moments of all the forces, either to the left or to the right of the section, about the section. In the case of BM, it must be understood that when we say moment about a section, we mean that the moment centre is at that section. As explained in Chapter 1, the moment axis is an axis perpendicular to the plane of the paper.

Considering the same beam discussed earlier, reproduced in Fig. 4.6, let us calculate the three quantities defined above.

AF at section 1-1 $= 2.43 - 42.43 = 40$ kN $\leftarrow$

SF at section 1-1 $= 99.14 - 42.43 - 100 = 43.29$ kN $\downarrow$

BM at section 1-1 $= 99.14 \times 5 - 2.43 \times 3 - 100 \times 1$

$= 268.4 \text{ kN m}$ ↻

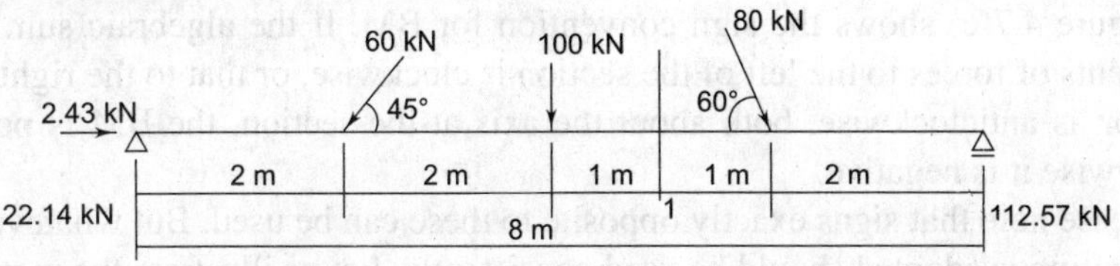

Fig. 4.6

Note that these values are exactly equal and opposite to those found earlier as internal stress resultants. Let us calculate these values for the same section from the right side.

$$\text{AF} = 40 \text{ kN} \rightarrow$$

$$\text{SF} = 112.57 - 69.28 = 43.29 \text{ kN} \uparrow$$

$$\text{BM} = 112.57 \times 3 - 69.28 \times 1 = 268.4 \text{ kN m (counterclockwise)}$$

An important point should be noted here. The values calculated from the left of the section are exactly equal and opposite to those calculated from the right. That this should be so can be easily explained. Since $\Sigma H = 0$, $\Sigma V = 0$, and $\Sigma M = 0$ for the whole beam, the values calculated from the left and those from the right when added are equal to zero. Therefore, at any section of the beam, the values calculated from the left are equal and opposite to those calculated from the right. This is directly related to the sign convention adopted for these quantities.

4.3 SIGN CONVENTION

The sign conventions for AF, SF, and BM that will be adopted in this book are shown in Fig. 4.7.

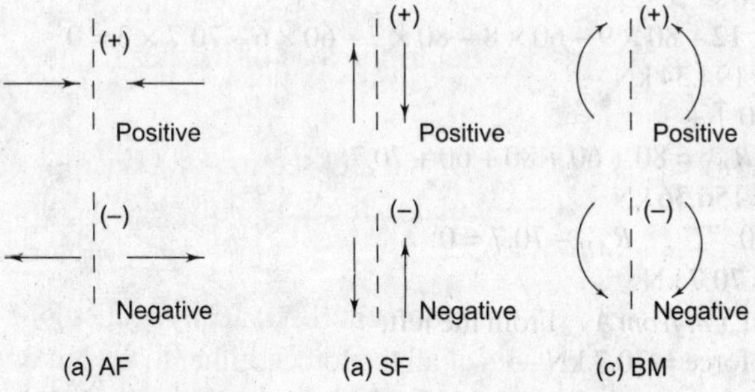

Fig. 4.7 Sign conventions for AF, SF, and BM

Figure 4.7(a) shows the sign convention for AF. If the algebraic sum of the forces parallel to the longitudinal axes calculated from the left is towards the right, or that calculated from the right is towards the left, it is taken as positive. Otherwise it is negative. You will note that this sign convention means that forces causing compression are positive and those causing tension are negative.

Figure 4.7(b) shows the sign convention for SF. If the algebraic sum of the forces perpendicular to the axis of the beam calculated from the left is upwards, or that calculated from the right is downwards, the SF is positive. Otherwise, the SF is negative.

Figure 4.7(c) shows the sign convention for BM. If the algebraic sum of the moments of forces to the left of the section is clockwise, or that to the right of the section is anticlockwise, both about the axis at the section, the BM is positive. Otherwise it is negative.

Please note that signs exactly opposite to these can be used. But whatever sign convention is adopted should be used consistently. Let us illustrate the method of calculating the quantities and sign conventions with examples.

Example 4.1 BM at sections of an SS beam

For the beam loaded as shown in Fig. 4.8, calculate the AF, SF, and BM values at 3 m, 5 m, and 10 m from A.

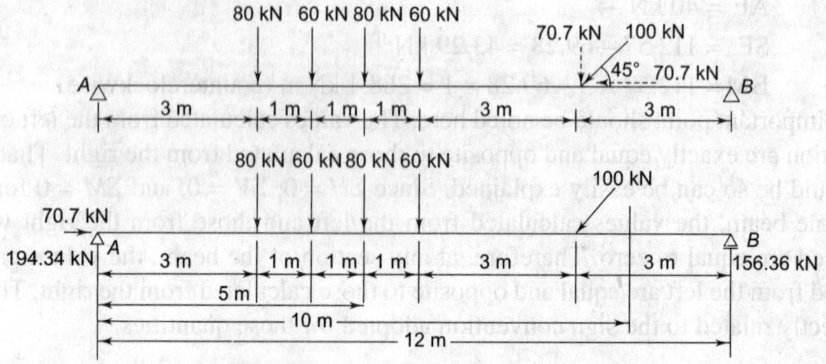

Fig. 4.8

Solution The horizontal and vertical components of the 100 kN load are equal to 70.7 kN each. Calculating the reactions,

$$\Sigma M = 0 \text{ @ } B \curvearrowright +$$
$$R_{AV} \times 12 - 80 \times 9 - 60 \times 8 - 80 \times 7 - 60 \times 6 - 70.7 \times 3 = 0$$
$$\Rightarrow \quad R_{AV} = 194.34 \text{ kN}$$
$$\Sigma V = 0 \uparrow +$$
$$R_{AV} + R_{BV} = 80 + 60 + 80 + 60 + 70.7$$
$$\Rightarrow \quad R_{BV} = 156.36 \text{ kN}$$
$$\Sigma H = 0 \xrightarrow{+}, \quad R_{AH} - 70.7 = 0$$
$$\Rightarrow \quad R_{AH} = 70.7 \text{ kN}$$

(i) *Section at 3 m from A* From the left,

Axial force = 70.7 kN →

Shear force calculations need some clarifications. Since the section being considered is at a point where a concentrated load is acting, should this load be included in the calculation? The solution is to calculate two values, one just to the left of the 80 kN load and another just to the right of this load. To put it in terms of computation, calculate two SF values at a concentrated load, one excluding this load and one including it. Therefore, SF = 194.34 kN ↑ just to the left of the section and 114.34 kN ↑ just to the right of the section.

$$\text{BM} = 194.34 \times 3 = 583.02 \text{ kN m} \curvearrowright$$

(ii) *Section at 5 m from A* Calculating from the left,

AF = 70.7 kN →

Shear force = 194.34 – 80 – 60 = 54.34 kN ↑ just to the left and 194.34 – 80 – 60 – 80 = – 25.66 kN ↓ just to the right.

Bending moment = 194.34 × 5 – 80 × 2 – 60 × 1 = 751.7 kN m ↷

(iii) *Section at 10 m from A* Calculating from the left,

AF = 70.7 – 70.7 = 0

SF = 194.34 – 80 – 60 – 80 – 60 – 70.7 = – 156.36 kN ↓

BM = 194.34 × 10 – 80 × 7 – 60 × 6 – 80 × 5 – 60 × 4 – 70.7 × 1
= 312.72 kN m ↷

Please note that in the last case it is much easier to calculate from the right, which directly gives

AF = 0

SF = – 156.36 kN (upward from the right)

BM = 156.36 × 2 = 312.72 kN m ↶

Example 4.2 BM and SF at quarter points of an SS beam

A beam 8 m long carries a uniformly distributed load of 20 kN/m over the whole span. Calculate the AF, SF, and BM at quarter-span points and at the middle of the span.

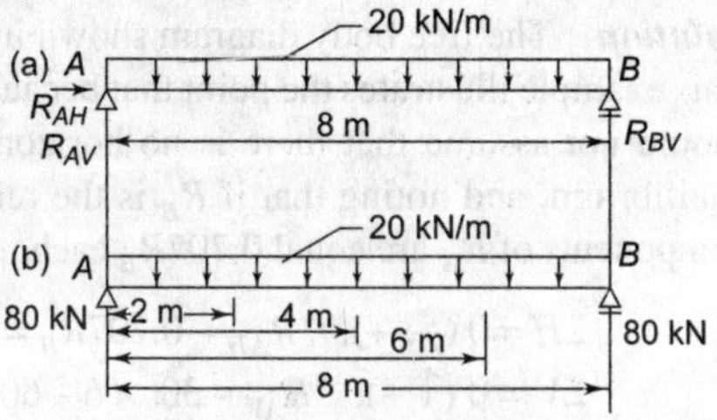

Fig. 4.9

Solution The beam with the loading is shown in Fig. 4.9(a). To calculate the reactions [Fig. 4.9(b)],

$\Sigma H = 0, \quad R_{AH} = 0$

$\Sigma M = 0$ @ B ↷+, $R_{AV} \times 8 - 8 \times 20 \times 4 = 0$

⇒ $R_{AV} = 80$ kN ↑

$\Sigma V = 0$ ↑ +

$R_{AV} + R_{BV} - 8 \times 20 = 0$

⇒ $R_{BV} = 80$ kN ↑

(As the beam is symmetrically loaded, one can easily deduce that both the reactions are equal to half of the load on the beam.) The three reactions are shown in Fig. 4.9(a). Axial force is zero at all sections.

Section at 2 m from A From the left,

SF = 80 – 20 × 2 = 40 kN ↑

BM = 80 × 2 – 20 × 2 × 1 = 120 kN m ↷

Section at 4 m from A Calculating from the left,

SF = 80 – 20 × 4 = 0

BM = 80 × 4 – 20 × 4 × 2 = 160 kN m ↷

Section at 6 m from A Calculating from the left,

SF = 80 – 20 × 6 = – 40 kN ↓

BM = 80 × 6 – 20 × 6 × 3 = 120 kN m ↷

Note that it is easier in such a case to write a general equation for a section at x m from A.

SF at x m = $80 - 20x$

BM at x m = $80x - 20x \times x/2$

By setting $x = 2$, $x = 4$, and $x = 6$ m in these equations, the same values as calculated before are obtained.

Example 4.3 SS beam with inclined roller support

A beam has a span of 6 m and is supported as shown in Fig. 4.10(a). The plane of the roller support at B is inclined at 45° to the horizontal. The loading on the beam consists of a uniformly distributed load of 30 kN/m and two concentrated loads of 60 kN each at one-third points. Calculate the AF, SF, and BM at one-third points and at the middle of the span.

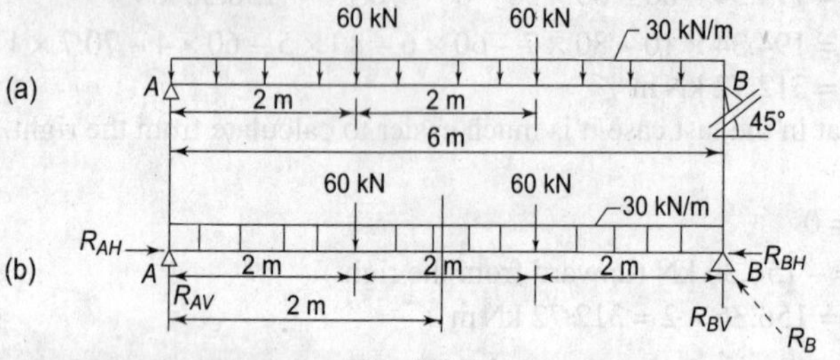

Fig. 4.10

Solution The free body diagram shown in Fig. 4.10(b) is used to calculate the reactions. This example illustrates the point that because there are only vertical loads on the span, one should not assume that there is no horizontal reaction at A. Applying the conditions of equilibrium, and noting that if R_B is the reaction at the roller, the horizontal and vertical components of R_B are equal $0.707R_B$ each,

$$\Sigma H = 0\ (\rightarrow +),\quad R_{AH} - 0.707R_B = 0 \qquad \text{(i)}$$

$$\Sigma V = 0\ (\uparrow +),\quad R_{AV} - 30\times 6 - 60 - 60 + R_B \times 0.707 = 0$$

$$R_{AV} + 0.707R_B = 300 \qquad \text{(ii)}$$

$$\Sigma M = 0 \text{ about } B \curvearrowright +,\quad R_{AV}\times 6 - 30\times 6\times 3 - 60\times 4 - 60\times 2 = 0$$

$$R_{AV} = 150 \text{ kN} \uparrow$$

From (ii),

$$R_B = 150\sqrt{2} = 212.1 \text{ kN}$$

From (i),

$$R_{AH} = 150 \text{ kN} \rightarrow$$

Section at 2 m from A Calculating from the left of the section,

$$\text{AF} = 150 \text{ kN} \rightarrow$$

$$\text{SF} = 150 - 30\times 2 = 90 \text{ kN} \uparrow \text{ just to the left}$$

and

$$150 - 30\times 2 - 60 = 30 \text{ kN} \uparrow \text{ just to the right}$$

$$\text{BM} = 150\times 2 - 30\times 2\times 1 = 240 \text{ kN m} \curvearrowright$$

Section at 3 m from A Calculating from the left,

$$\text{AF} = 150 \text{ kN} \rightarrow$$

$$\text{SF} = 150 - 30\times 3 - 60 = 0$$

$$\text{BM} = 150\times 3 - 30\times 3\times \frac{3}{2} - 60\times 1 = 255 \text{ kN m} \curvearrowright$$

Section at 4 m from A Calculating from the left,

$$\text{AF} = 150 \text{ kN} \rightarrow$$

$$\text{SF} = 150 - 30\times 4 - 60 = -30 \text{ kN} \downarrow \text{ just to the left}$$

$$\text{SF} = 150 - 30\times 4 - 60 - 60 = -90 \text{ kN} \downarrow \text{ just to the right}$$

$$\text{BM} = 150\times 4 - 30\times 4\times 2 - 60\times 2 = 240 \text{ kN m} \curvearrowright$$

Note that, where convenient, calculations can be done from the right, taking care of the sign convention.

4.4 BENDING MOMENT AND SHEAR FORCE DIAGRAMS

We have seen from the examples just discussed that the quantities AF, SF, and BM vary along the length of the beam. Diagrams showing such variation are drawn to help in the design. While the section is designed to resist the maximum BM and SF, there are many cases where the section can be reduced. For example, the reinforcement in an RCC beam and steel plates in a built-up beam can be curtailed at sections where the BM values are small enough to cut off the bars or plates (Fig. 4.11).

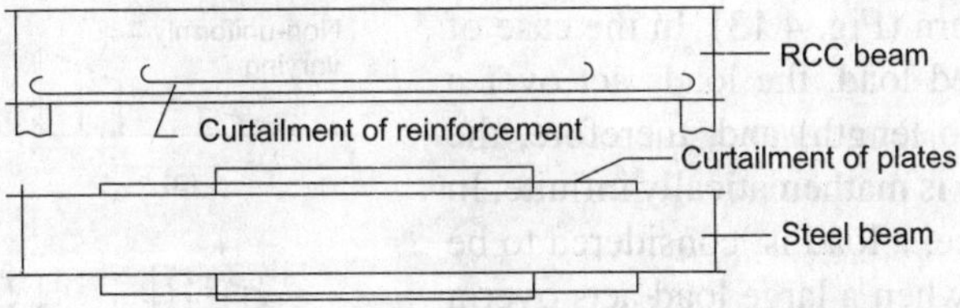

Fig. 4.11

These diagrams are drawn by plotting x, the distance along the length of the beam, against the AF, SF, or BM (Fig. 4.12). x-AF, x-SF, and x-BM relationships can be obtained as mathematical equations or by calculating these quantities at a number of sections corresponding to different values of x.

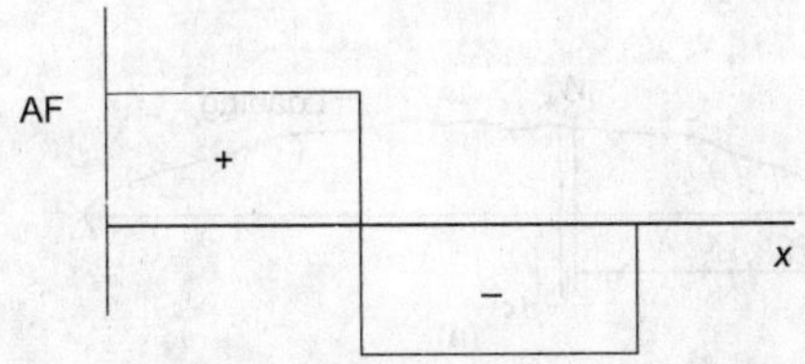

(a) A plot of AF against x, or the AF diagram

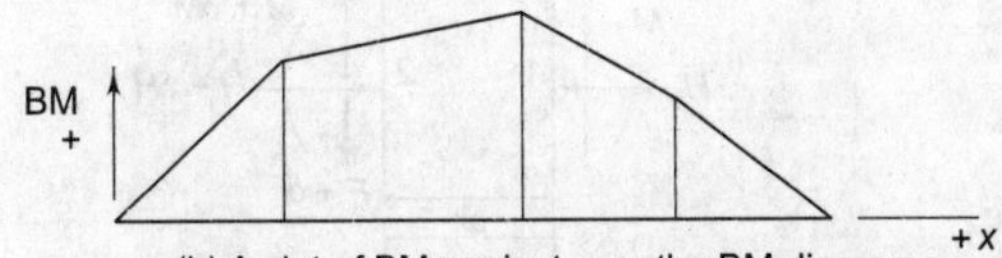

(b) A plot of BM against x, or the BM diagram

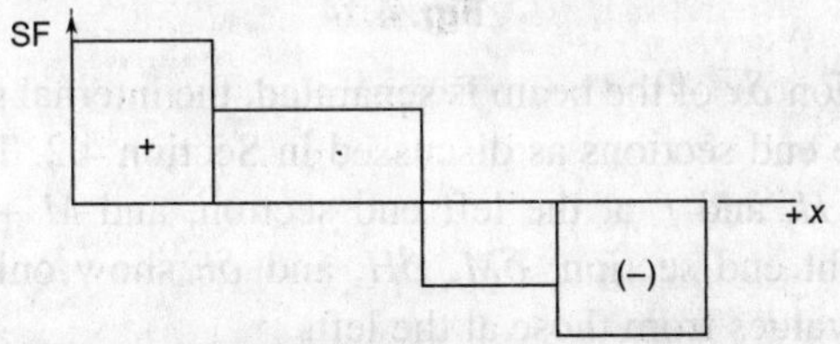

(c) A plot of SF against x, or the SF diagram

Fig. 4.12

The load on the beam, SF, and BM have mathematical relationships helpful in studying the variation of these quantities. Let us take up these relationships first before illustrating the SF and BM diagrams with examples.

4.5 DIFFERENTIAL RELATIONSHIP BETWEEN LOAD INTENSITY, SF, AND BM

In the case of beams, the term *load intensity* means load per unit length of the beam. The load intensity may be constant as in the case of a uniformly distributed (UD) load, uniformly varying as in the case of a triangular load, or varying in any other non-uniform pattern (Fig. 4.13). In the case of a concentrated load, the loads act over a point (of zero length) and, therefore, the load intensity is mathematically infinite. In actual practice, a load is considered to be a point load when a large load acts over a small length. The load intensity is positive if the load acts upwards and negative if it acts downwards [Fig. 4.13(c)]

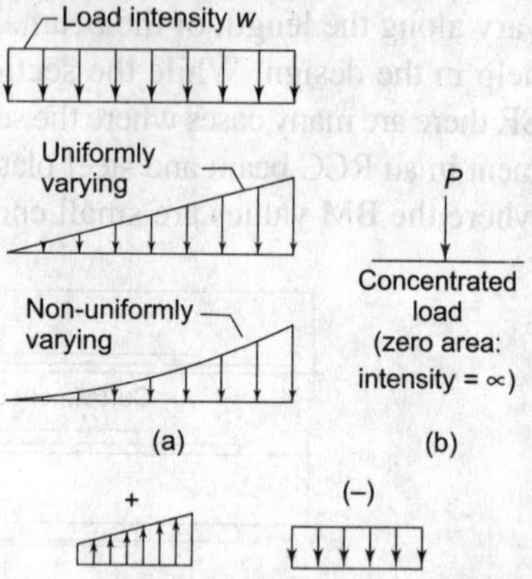

Fig. 4.13

Consider a beam loaded, as shown in Fig. 4.14(a). We consider an elemental length dx of the beam at a distance x from the left-hand support A. As the length δx is very small, the load intensity can be considered to be uniform over this length and is equal to W_x. For the sake of clarity, this length δx of the beam is shown exaggerated in Fig. 4.14(b).

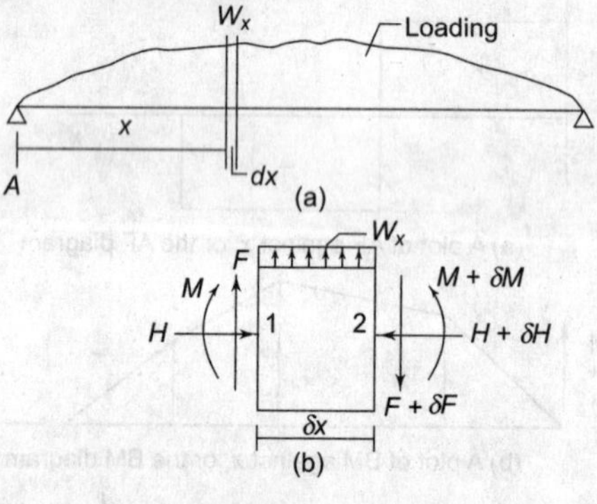

Fig. 4.14

When this section δx of the beam is separated, the internal stress resultants will appear at both the end sections as discussed in Section 4.2. These internal stress resultants are M, H, and F at the left end section, and $M + \delta M$, $H + \delta H$, and $F + \delta F$ at the right end section. δM, δH, and δF show only the increment or decrement in the values from those at the left.

We state the equations of equilibrium $\Sigma V = 0$ and $\Sigma M = 0$ for this free body:

$$\Sigma V = 0 \uparrow +, \quad F + W_x \, \delta x - F - \delta F = 0$$

$$\Rightarrow \quad \frac{\delta F}{\delta x} = W_x$$

As $\delta x \to 0$,

$$\frac{dF}{dx} = W_x$$

$$\Sigma M = 0 \curvearrowright +, \quad M + F\,\delta x + W_x\,\delta x\,\frac{\delta x}{2} - M - \delta M = 0$$

taking moments about point 2. $W_x\ \delta x \times \delta x/2$ can be neglected, being a small quantity of second order. This gives us $\delta M/\delta x = F$. As $\delta x \to 0$, this becomes $dM/dx = F$.

These two relationships are very useful in deciding the variation of SF and BM over the length of the beam and greatly facilitates the drawing of SF and BM diagrams.

4.5.1 Interpretation of Differential Relationships

The relationships $dM/dx = F$ and $dF/dx = W_x$ provide us with two very useful interpretations facilitating the computations and drawing of SF and BM diagrams.

Interpretation as slope of diagrams From our knowledge of differential calculus, we know that dM/dx gives the slope of the curve at the point on the curve corresponding to (x, M). This means that dM/dx gives the slope of the BM diagrams [Fig. 4.15(a)]. Similarly, dF/dx gives the slope of the SF diagram. Therefore, $dM/dx = F$ means that the slope of the BM diagram is equal to the SF at the section. While the exact value of the slope is not of interest to us, the variations in the slope enable us to draw the BM diagram. The variation in the slope of the BM diagram is the variation in the SF along the length of the beam. The same considerations apply to the equation $dF/dx = W_x$. The slope of the SF diagram is equal to the load intensity at a section. Here again, instead of the actual values of slope, the variation in the slope of the diagram is of interest, and these variations follow the variations of load intensity along the length of the beam [Fig. 4.15(b)].

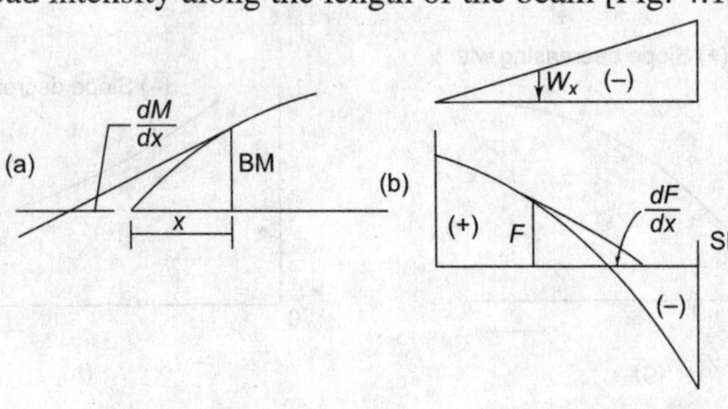

Fig. 4.15

Table 4.1 gives many types of curves and straight lines showing different variations of slope. The origin is at the left, x positive to the right and BM or SF positive upwards. The reader may study these diagrams carefully to understand what they represent. In the illustrative problems given later, these have been discussed wherever necessary.

The term $dM/dx = F$ has one more important significance. For the BM diagram, drawn based upon a relationship between x and M, the maximum value of M is obtained when $dM/dx = 0$. This is also the condition for minimum value. Therefore, BM has a maximum value where the SF is zero or changes sign. This is important in locating the point of maximum BM.

Table 4.1 Different shapes of diagrams and their slopes

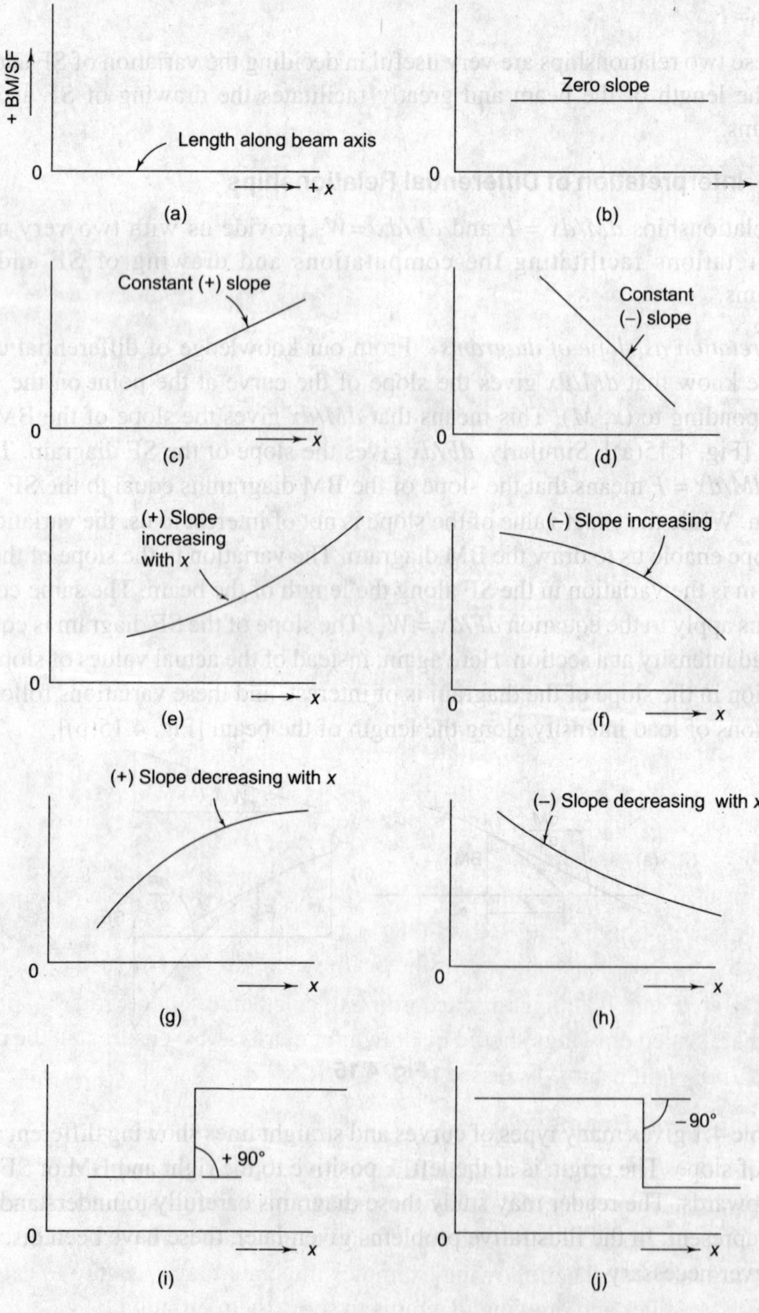

Interpretation as change in value The differential relationship can be expressed as $dM = F\,dx$ and $dF = W_x\,dx$. In Figs [4.16(a) and (b)], the meaning of the form is illustrated. dM is the change in BM between two sections and $F\,dx$ is the area of SF diagram between the sections. A similar interpretation applies to the second equation. The change in SF is equal to the area of the load diagram. It may be mentioned here that in using this interpretation, care should be taken while considering concentrated loads for SF diagrams and couple loadings for BM diagrams. This interpretation does not apply to such cases but change in SF is equal to the concentrated load, and change in BM is equal to the couple at the same section. This principle will be further described in the following examples.

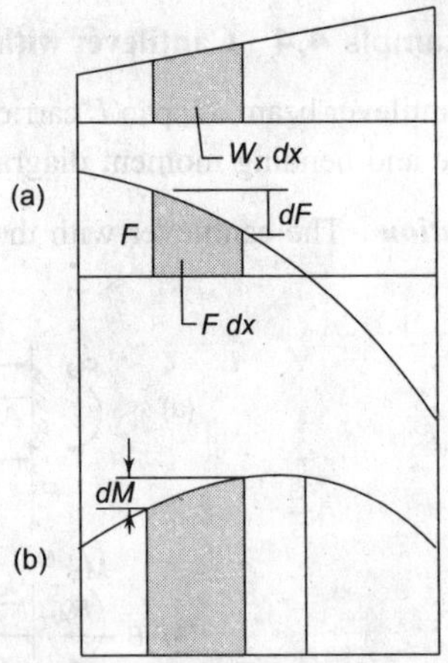

Fig. 4.16 Different shapes of diagrams and their slopes

4.6 STANDARD CASES

Table 4.2 gives the bending moments, shear forces, and their maximum values for some standard case of loading, which are frequently encountered in practice. We now consider a few standard cases of loading in the case of cantilever, simply-supported and overhanging beams. We illustrate the method to calculate salient values of BM and SF, their maximum values and show variation through diagrams. The general procedure is as follows.

(i) Draw a free body diagram of the beam indicating the applied loads and reactive components.

(ii) Calculate the reactions applying the three conditions of equilibrium, $\Sigma H = 0$, $\Sigma V = 0$, and $\Sigma M = 0$ to the free body. In a statically determinate beam, there are only three reaction components and these can be determined from these three equations.

(iii) Write equation for SF and BM. The equation will tell us whether the diagram will be a straight line, second degree or third degree curve. The equations can vary between segments of the beam where the load intensity changes.

(iv) Draw SF and BM diagrams according to calculated salient values. Approximate scaled drawings should be drawn for clarity. Also ensure that the differential relationships discussed in Section 4.5 are followed.

4.6.1 Cantilever Beams

A cantilever, as has been illustrated earlier, has a fixed support at one end, the other end remaining free. This is a stable determinate beam because the fixed support provides three reaction components, a vertical force, a horizontal force, and a restraining moment. The following examples illustrate the procedure to calculate SF and BM values and drawing diagrams to show their variation.

Example 4.4 Cantilever with point load

A cantilever beam of span L carries a point load at 'a' from the fixed end. Draw the shear force and bending moment diagrams for the beam.

Solution The cantilever with the point load P is shown in Fig. 4.17.

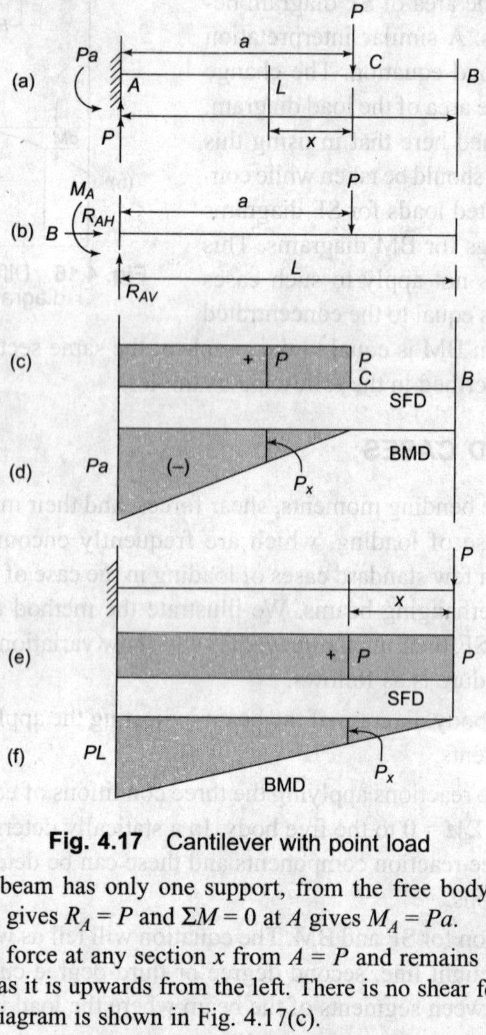

Fig. 4.17 Cantilever with point load

Reactions As the beam has only one support, from the free body diagram shown in Fig. 2.17(b), $\Sigma V = 0$, gives $R_A = P$ and $\Sigma M = 0$ at A gives $M_A = Pa$.

Shear force Shear force at any section x from $A = P$ and remains constant between A and C. It is positive as it is upwards from the left. There is no shear force between C and B. The shear force diagram is shown in Fig. 4.17(c).

Bending moment It is easier to calculate from the right. At any distance x from C, $M = -Px$ (negative sign because it is clockwise from the right); This is a straight line relationship with maximum value at $A = Pa$. The bending moment is zero under the load P and maximum at $A = Pa$. The bending moment diagram is shown in Fig. 4.17(d). Note that there is no bending moment between C and the free end B.

Load at free end If the load were acting at the free end B, the shear force and bending moment diagrams will be as shown in Fig. 4.17(e) and(f).

Maximum BM = PL; Maximum SF = P

Table 4.2 Standard cases of statically determinate beams

	Beam and loading	*Reactions*	*SF*	*BM*
1.	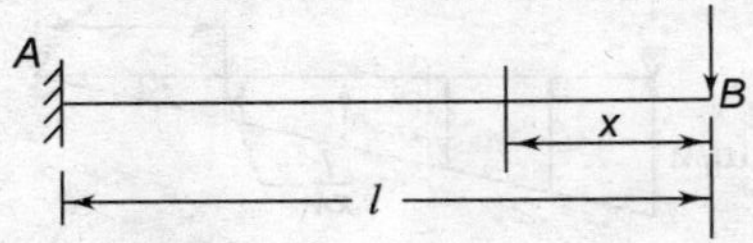	$R_{AV} = P\uparrow$ $M_A = Pl$ ↺	$+P$ throughout	BM at $x = -Px$ Max. BM $= -Pl$ at A
2.	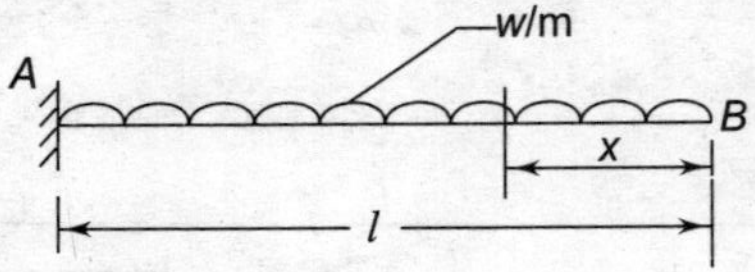	$R_{AV} = wl\uparrow$ $M_A = \frac{wl^2}{2}$ ↺	SF at $x = wx$ Max. SF $= wl$ at A	BM at $x = \frac{-wx^2}{2}$ Max. BM $= \frac{-wl^2}{2}$ at A
3.	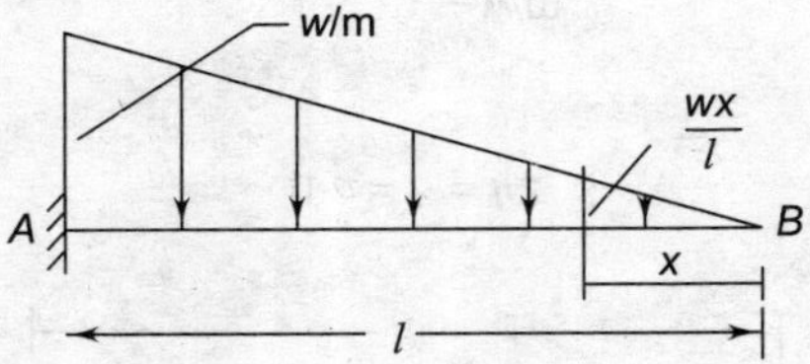	$R_{AV} = \frac{wl}{2}\uparrow$ $M_A = \frac{wl^2}{6}$ ↺	SF at $x = \frac{wx^2}{2l}$ Max. SF $= \frac{wl}{2}$ at A	BM at $x = \frac{-wx^3}{6l}$ Max. BM $= \frac{wl^2}{6}$ at A
4.	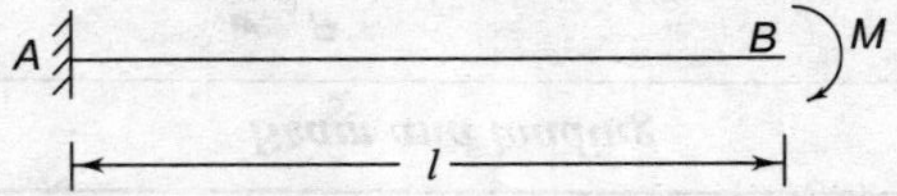	$M_A = M$ ↺	SF is zero throughout	BM $= -M$ throughout

(contd)

Table 4.2 *(contd)*

	Beam and loading	Reactions	SF	BM
5.	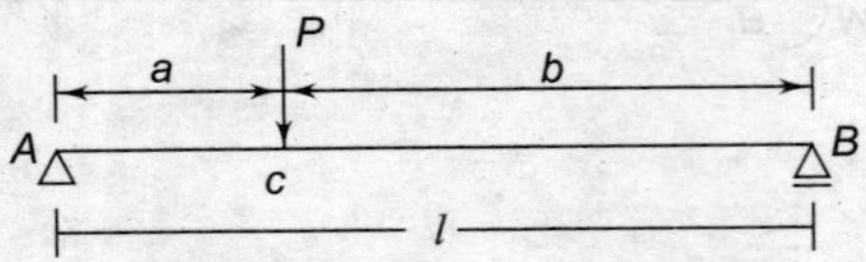	$R_{AV} = \frac{Pb}{l}$ $R_{BV} = \frac{Pa}{l}$	SF at $A = \frac{Pb}{l}$ SF at B $= \frac{-Pa}{l}$	Max. BM $= \frac{Pab}{l}$ at c
	If $a = b = l/2$	$R_{AV} = R_{BV} = \frac{P}{2}$	SF at $A = \frac{P}{2}$ SF at $B = \frac{-P}{2}$	Max. BM $= \frac{Pl}{4}$ at c
6.	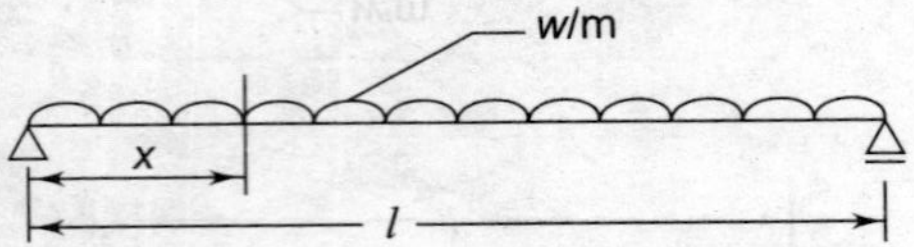	$R_{AV} = R_{BV} = \frac{wl}{2}$	SF at $x = \frac{wl}{2} - wx$ SF at $A = \frac{wl}{2}$ SF at $B = \frac{-wl}{2}$	BM at $x = \frac{wl}{2}x - \frac{wx^2}{2}$ Max. BM $= \frac{wl^2}{8}$ at $x = l/2$
7.	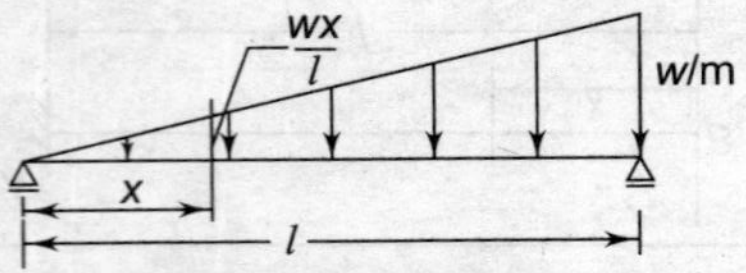	$R_{AV} = \frac{wl}{6}$ $R_{BV} = \frac{wl}{3}$	SF at $x = \frac{wl}{6} - \frac{wx^2}{2l}$ Max. SF $= \frac{-wl}{3}$ at B	BM at $x = \frac{wl}{6}x - \frac{wx^3}{6l}$ Max. BM $= \frac{wl^2}{9\sqrt{3}} = 0.06415wl^2$ at $x = \frac{l}{\sqrt{3}}$

Table 4.2 *(contd)*

S.No.	***Beam and loading***	***Reactions***	***SF***	***BM***
8.	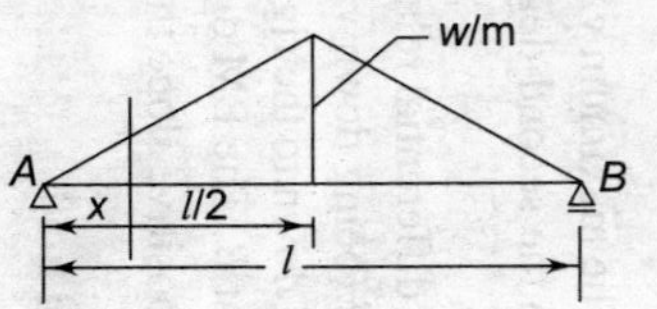	$R_{AV} = R_{BV} = \frac{wl}{4}$	SF at x ($x \le l/2$) $= \frac{wl}{4} - \frac{wx^2}{l}$ Max. SF $= \frac{wl}{4}$ at A and $\frac{-wl}{4}$ at B	BM at x ($x \le l/2$) $=$ $\frac{wl}{4}x - \frac{wx^3}{3l}$ Max. BM $= \frac{wl^2}{12}$ at mid-span
9.	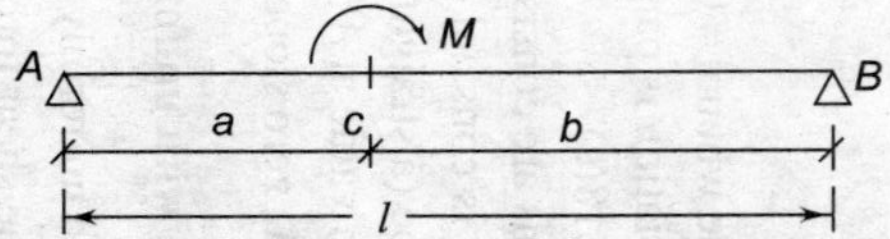	$R_{AV} = \frac{M}{l}\downarrow$ $R_{BV} = \frac{M}{l}\uparrow$	SF $= -\frac{M}{l}$ throughout	Max. BM $= \frac{-Ma}{l}$ or $+\frac{Mb}{l}$ at c
10.	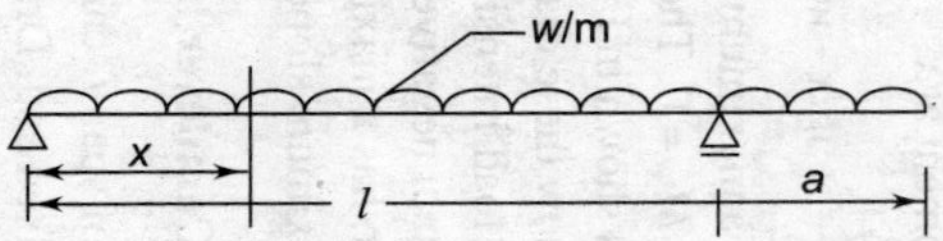	$R_{AV} = \frac{w(l^2 - a^2)}{2l}$ $R_{BV} = \frac{w(l + a)^2}{2l}$	SF at $x = R_{AV} - wx$	BM at $x = R_{AV}x - \frac{wx^2}{2}$ Max. BM $= \frac{w(l^2 - a^2)}{8l^2}$ $x = \frac{l^2 - a^2}{2l}$ and $\frac{-wa^2}{2}$ at B

Example 4.5 Cantilever with UD load

Figure 4.18(a) shows a cantilever with a uniformly distributed load. Figure 4.18(b) is the free body diagram for the beam. The reactive forces and moments are calculated first. $R_{AH} = 0$.

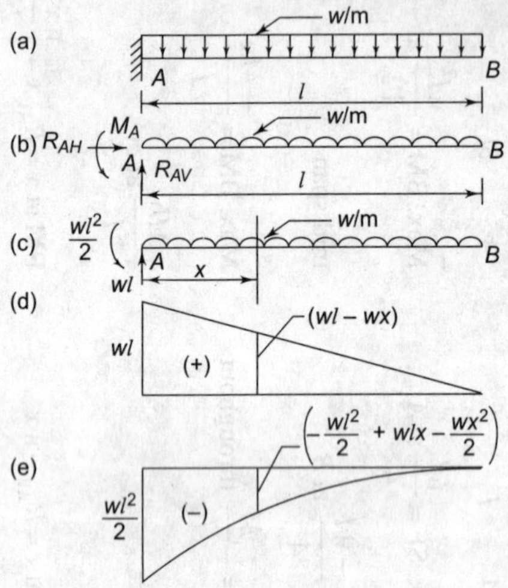

Fig. 4.18

$$\Sigma V = 0 \uparrow +, \quad R_{AV} - wl = 0$$

$$\Rightarrow \quad R_{AV} = wl$$

$$\Sigma M = 0 \text{ @ } B \curvearrowright +, \quad -M_A + R_{AV}l - wl\frac{l}{2} = 0$$

$$\Rightarrow \quad M_A = +\frac{wl^2}{2} \quad \text{(counterclockwise as assumed)}$$

Please note that M_A is a hogging moment.

General equations can be written for the SF and BM. In Fig. 4.18(c), the SF at $x = wl - wx$. When $x = 0$, SF $= wl$ at A. When $x = l$, the SF is zero at B. The variation is linear. The SF is positive throughout. The SF diagram is shown in Fig. 4.18(d).

The bending moment at x,

$$M_x = -\frac{wl^2}{2} + wlx - wx\frac{x}{2}$$

Notice that this has a maximum value when $x = 0$, i.e., at A. The maximum value of $M_x = -wl^2/2$. When $x = l$, $M_x = 0$. The variation is parabolic due to the second-degree equation. The BM diagram is shown in Fig. 4.18(e).

You must note how these diagrams are consistent with the differential relationships discussed earlier. The load intensity w is constant and negative (being downwards). The SF diagram has a constant negative slope (a straight line slanting down to the right). The SF is positive throughout, has a maximum value at A, and is zero at B. The BM diagram, correspondingly, has maximum slope at A, zero slope at B, and a positive slope throughout. ■

Example 4.6 Cantilever beam with uniformly varying load

A cantilever beam of span L carries a uniformly varying load varying from zero at the free end to w/m at the fixed end. Draw the shear force and bending moment diagrams.

Solution The beam with the load is shown in Fig. 4.19.

From the free body diagram shown in Fig. 4.19(b), total load on the beam = area of the triangle $= wL/2$. Reaction $R_A = wL/2$ from $\Sigma V = 0$. Taking moments about A,

$$-M_A + (wL/2)\,L/3 = 0;\ M_A = wL^2/6$$

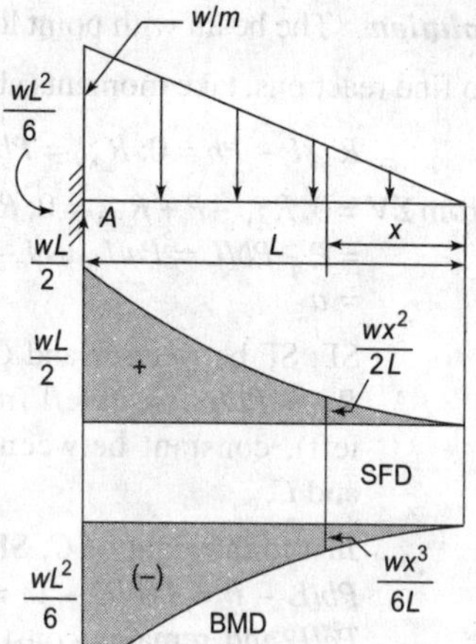

Fig. 4.19

Calculating from the right, SF at x from $B = (wx/L)\,x/2$. This is positive as it is ↓ from the right. This is a parabolic equation. SF is zero at B ($x = 0$) and $wL/2$ at A. The shear force diagram is a parabola and will be as shown. As the load intensity at B is zero, the diagram will have zero slope at B. The load intensity is negative (being downwards) and hence the slope of the SF diagram is also negative.

Calculating from the right, BM at section $x = -(wx^2/2L)\,x/3 = -wx^3/6L$. This is a third degree equation with zero value at B ($x = 0$) and $-wLxL/6$ at A. The BM is negative throughout, as it is clockwise from the right.

The BM diagram will be as shown in Fig. 4.19. As the SF is positive throughout, the slope of BM diagram is positive, with zero slope at B and maximum slope at A. ■

Example 4.7 Cantilever beam with couple load

A cantilever beam of span L carries a clockwise moment M at the free end. Draw the SF and BM diagrams.

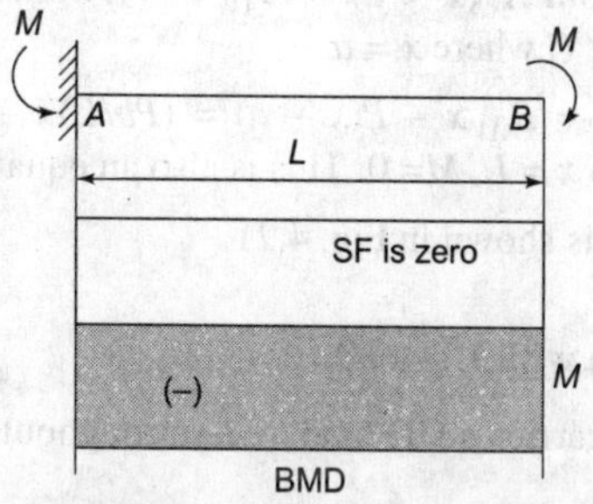

Fig. 4.20 Cantilever with couple load

Solution As there is no vertical load on the beam, with only one support, $R_{AV} = 0$; the reaction at A is a restraining couple $M_A = M$.

As there is no vertical force on the beam, shear force is zero throughout.

BM at x from $B = -M$ and remains constant throughout. The BM diagram is thus a rectangle of side M. The BM diagram is shown in Fig. 4.20. ■

4.6.2 Simply Supported Beams

A simply supported beam is a beam having a hinge support at one end and a roller support at the other. Hinge support can have two reaction components, vertical and horizontal, and the roller support has one reaction perpendicular to the plane of the roller. The beam so supported is stable and statically determinate. The following examples illustrate the calculation of values and drawing of SF and BM diagrams for the beam.

Example 4.8 SS beam with point load

A simply supported beam of span L carries a point load at a distance a from the left support. Draw the SF and BM diagrams indicating salient values.

Solution The beam with point load is shown in Fig. 4.21.

To find reactions, take moments about B;

$R_{AV}L - Pb = 0; R_{AV} = Pb/L$

From $\Sigma V = 0, R_{AV} - P + R_{BV} = 0; R_{BV} = P - Pb/L = Pa/L$, as $L - b = a$

SF: SF between A and $C = R_{AV} = Pb/L$ (positive ↑ from left), constant between A and C.

Just to the right of C, SF $= Pb/L - P = P(b/L - 1) = Pa/L$ and remains constant between C and B. This can also be calculated from the right as SF at x from $B = -Pa/L$ (negative being ↑ from right).

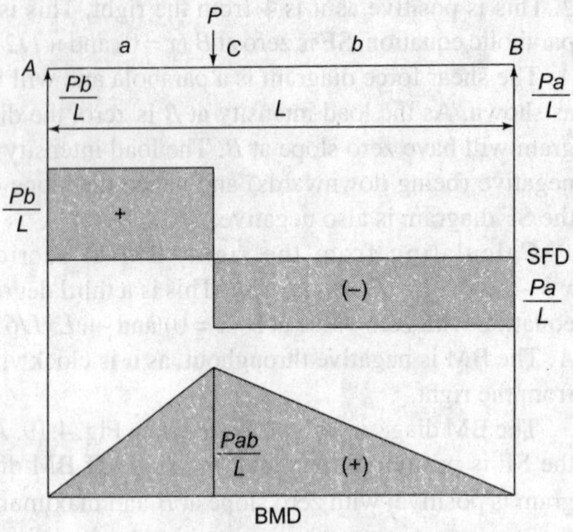

Fig. 4.21 SS beam with point load

The SF diagram is shown in Fig. 4.21.

BM: BM at x from A $(x < a) = R_{AV}\,x = (Pb/L)x$; This is zero at A where $x = 0$ and Pab/L at C where $x = a$.

BM at x $(x > a) = R_{AV}x - P(x - a) = (Pb/L)\,x - P(x - a)$; When $x = a$, $M = Pab/L$ and when $x = L$, $M = 0$. This is also an equation of a straight line.

The BM diagram is shown in Fig. 4.21.

Example 4.9 SS beam with UD load

A simply-supported beam carries a UD load w/m throughout its length. Draw the SF and BM diagrams.

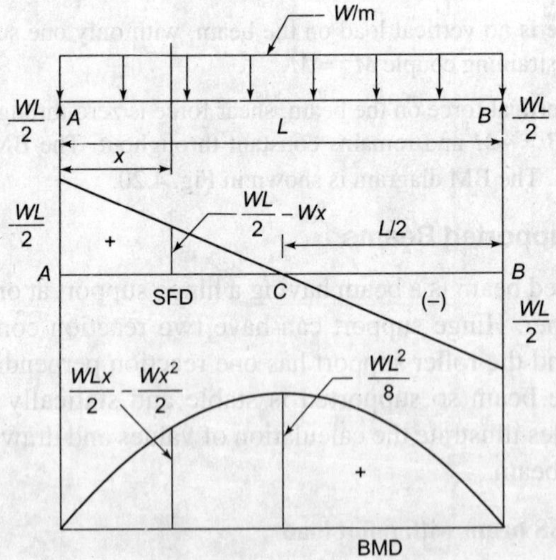

Fig. 4.22 SS beam with UD load

Solution The beam with the loading is shown in Fig. 4.22. As A and B are simple supports, there will be vertical reactions at A and B. By symmetry, the two reactions will be equal. $R_{AV} = R_{BV} = wL/2$. Otherwise, taking moments @ B,

$-wL\,L/2 + R_{AV}L = 0; R_{AV} = wL/2$

From $\Sigma V = 0, -wL + R_{AV} + R_{BV} = 0; R_{BV} = wL/2$

Shear Force at $A = +wL/2$, positive as it is upwards from the left.

At any distance x from A, $SF = wL/2 - wx$; this is a straight line equation.

$x = 0$, SF at $A = wL/2$. $x = L$, SF at $B = -wL/2$

SF becomes zero at, $wL/2 - wx = 0; x = L/2$

As the load intensity is negative throughout, the SF diagram will have negative slope throughout. The SF diagram will be as shown in Fig. 4.22.

BM is zero at A and B.

At any distance x from A,

BM at $x = (wL/2)x - wx(x/2) = (wL/2)x - wx^2/2$

This is an equation of a parabola. We find the x for maximum BM, by setting $dM/dx = 0$; $dM/dx = wL/2 - wx = 0$ (note that this is the SF equation); $x = L/2$. The maximum BM will be at the centre. Maximum BM is obtained by putting $x = L/2$ in the BM equation.

Maximum BM $= (wL/2)\,L/2 - wL^2/8 = wL^2/8$. The BM diagram is shown in Fig. 4.22.

Note that the SF is positive from A to C and decreases; BM diagram has positive decreasing slope. From C to B, the SF is negative and increasing, the BM diagram has negative increasing slope.

Example 4.10 SS beam with uniformly varying load

Figure 4.23(a) shows a simply supported beam with a uniformly varying load. Figure 4.23(b) shows the free body diagram. $R_{AH} = 0$ as there is no other horizontal force on the beam.

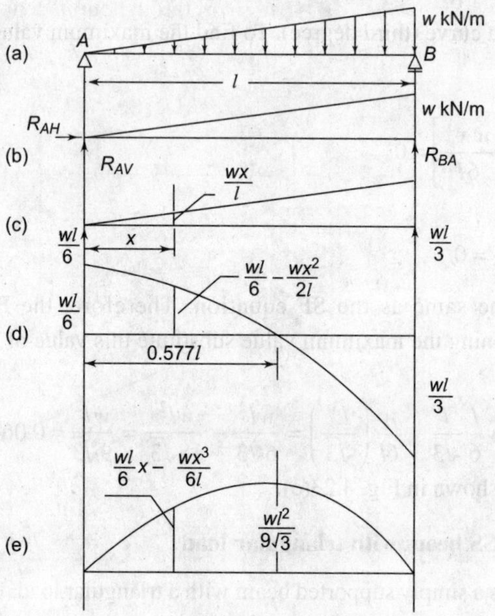

Fig. 4.23

$$\Sigma M = 0 \text{ @ } B \curvearrowright +, \quad R_{AV} l - w\frac{l}{2}\frac{l}{3} = 0$$

$$\Rightarrow \quad R_{AV} = \frac{wl}{6}$$

$$\Sigma V = 0 \uparrow +, \quad R_{AV} + R_{BV} = w\frac{l}{2}$$

$$\Rightarrow \quad R_{BV} = \frac{wl}{3}$$

General equations for the SF and BM can be written for a section x m from A. [Fig. 4.23(c)].
Shear force at x,

$$F_x = \frac{wl}{6} - w\frac{x}{l}\frac{x}{2}$$

(load intensity at $x = wx/l$).

$$\text{Shear force at } x = 0 = \frac{wl}{3}$$

$$\text{Shear force at } B\ (x = l) = \frac{wl}{6} - \frac{wl}{2} = -\frac{wl}{3}$$

The SF diagram is shown in Fig. 4.23(d). To find the location of zero shear force, set $F_x = 0$.

$$\frac{wl}{6} - \frac{wx^2}{2l} = 0, \quad x^2 = \frac{l^2}{3}$$

$$\Rightarrow \quad x = \frac{l}{\sqrt{3}} = 0.577l$$

The BM at x,

$$M_x = \frac{wl}{6}x - w\frac{x}{l}\frac{x}{2}\frac{x}{3} = \frac{wlx}{6} - \frac{wx^3}{6l}$$

M_x is zero at $x = 0$, and at $x = l$,

$$M_x = \frac{wlx}{6} - \frac{wx^3}{6l} = 0$$

The BM diagram is a curve (third degree). To find the maximum value of the BM,

$$\frac{dM}{dx} = 0$$

$$\frac{d}{dx}\left[\frac{wlx}{6} - \frac{wx^3}{6l}\right] = 0$$

That is,

$$\frac{wl}{6} - \frac{wx^2}{2l} = 0$$

Note that this is the same as the SF equation. Therefore, the BM is maximum at $x = 0.577l$. For obtaining the maximum value substitute this value of x in the equation for M_x:

$$M_{x(\max)} = w\frac{l}{6}\frac{l}{\sqrt{3}} - \frac{w}{6l}\left(\frac{l^3}{\sqrt{3}}\right) = \frac{wl^2}{6\sqrt{3}} - \frac{wl^2}{18\sqrt{3}} = \frac{wl^2}{9\sqrt{3}} = 0.06415wl^2$$

The BM diagram is shown in Fig. 4.23(e). ■

Example 4.11 SS beam with triangular load

Figure 4.24(a) shows a simply supported beam with a triangular load. Draw the SF and BM diagrams for the beam.

Solution Due to symmetry of loading, the reactions at A and B are equal. $R_{AH} = 0$.

$$\Sigma M_B = 0 \;@\; B \curvearrowright +$$

$$R_{AV}\, l - w\frac{l}{2}\,(l/2) = 0$$

$$\Rightarrow \quad R_{AV} = \frac{wl}{4} \uparrow$$

$$\Sigma V = 0 \uparrow +$$

$$R_{AV} - \frac{wl}{2} + R_{BV} = 0$$

$$\Rightarrow \quad R_{BV} = \frac{wl}{4} \uparrow$$

The general equations for SF and BM valid for the whole beam cannot be stated, as the load diagram cannot be represented by a single equation. A general equation is therefore stated for the left half from $x = 0$ to $x = l/2$. Referring to Fig. 4.24(b),

$$w_x = w\frac{x}{(l/2)} = \frac{w \times 2x}{l}$$

$$\text{Shear force at } x = \frac{wl}{4} - w\frac{2x}{l}\frac{x}{2} = \frac{wl}{4} - \frac{wx^2}{l}$$

The SF is zero at the middle of span at $x = l/2$. At A, $x = 0$ and SF = $wl/4$.

$$\text{BM at } x = \frac{wl}{4}x - w\frac{2x}{l}\frac{x}{2}\frac{x}{3} = \frac{wlx}{4} - \frac{wx^3}{3L}$$

The BM is maximum where the SF is zero, i.e., at $x = l/2$.

$$M_{\max} = \frac{wl}{4}\frac{l}{2} - \frac{w}{3l}\frac{l^3}{8} = \frac{wl^2}{12}$$

For the other half of the span, equations can be easily formulated from the right side. For $x = l/2$ to $x = l$ [Fig. 4.24(b)],

$$\text{SF at } x = \frac{wl}{4} + \frac{w(l-x)}{l/2}\frac{(l-x)}{2} = \frac{(l-x)}{3}\frac{wl}{4} + \frac{w(l-x)^2}{l}$$

$$\text{BM at } x = \frac{wl}{4}x - \frac{w(l-x)}{l}\frac{(l-x)}{2}\frac{(l-x)}{3} = \frac{wl}{4}x - \frac{w(l-x)^3}{l}$$

The maximum value of BM is again at $x = l/2$ and the same value as obtained from the left side is obtained. The SF and BM diagrams are shown in Figures 4.24(c) and (d). It must be noted how these diagrams satisfy the differential relationships derived earlier.

For the SF diagram, from $x = 0$ to $x = l/2$, the load intensity increases uniformly from 0 to w while from $x = l/2$ to $x = l$ it uniformly decreases from w to 0. In the SF diagram, confirm that the slope is zero at A, increases up to $l/2$ and then decreases from $l/2$ to l, becoming zero at B. The slope is negative throughout because the load is negative (acting downwards).

The slope of the BM diagram follows the magnitude of SF From $x = 0$ to $x = l/2$, the SF is positive and decreases from $wl/4$ to zero. The slope of the BM diagram is positive from $x = 0$ to $x = l/2$ and decreases from $x = 0$ to $x = l/2$. From $x = l/2$ to $x = l$, note that the SF is negative and

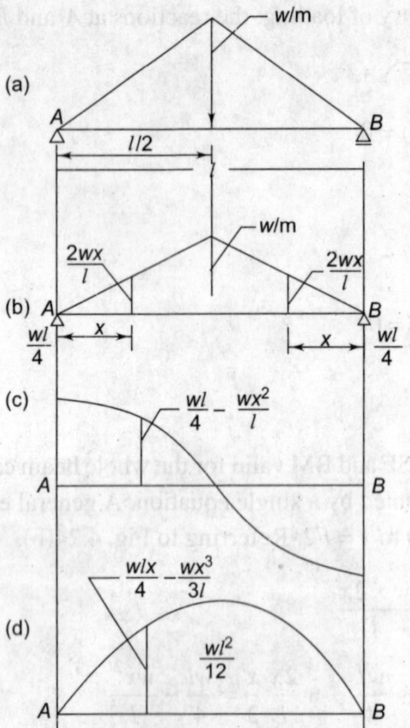

Fig. 4.24

increases from zero to $-wl/4$. The slope of the BM diagram is correspondingly negative and increasing as we go towards B.

Example 4.12 SS beam with couple load

Figure 4.25(a) shows a simply supported beam with a couple load.

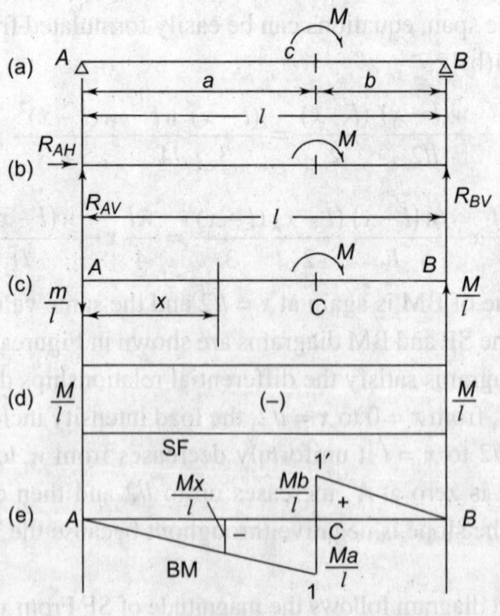

Fig. 4.25

A couple can be balanced only by another. Therefore, the reactive forces at A and B will be equal and opposite so that they form a couple. From Fig. 4.25(b),

$$\Sigma M = 0 \ @ \ B \ (+), \quad R_{AV} l + M = 0$$

$$\Rightarrow \quad R_{AV} = -\frac{M}{l} \text{ (acts downwards)}$$

$R_{BV} = M/l$ acting upwards. We can formulate separate equations for the segments AC and CB. When the section lies in segment AC ($x \le a$) [Fig. 4.25(c)], SF at $x = -M/l$ and is constant.

$$\text{BM at } x = \frac{M}{l} x$$

At $x = a$, the BM will have two values just to the left and right of the point at which the couple is acting.

BM (at $x = a$) = – $(Ma)/L$ just to the left of the section

$$\text{BM} = -\frac{M}{l} a + M = \frac{M(l-a)}{l}$$

just to the right of C. For the segment CB, we may compute from the right. For $a \le x \le l$, the SF at $x = -M/l$ (acting upwards from the right) and is constant.

BM at $x = (M/l)(l - x)$. This is zero at $x = l$ and $M(l-a)/l$ at $x = a$. The SF and BM diagrams are shown in Figures 4.25(d) and 4.25(e), respectively. The load intensity on the beam is zero. The slope of the SF diagram is, therefore, zero. The SF is constant and negative throughout the length. The BM diagram is, therefore, a straight line with negative slope. Note that the two segments $A1$ and $1'B$ are parallel. ■

4.6.3 Overhanging Beams

Overhanging beams have overhanging lengths at one end or both ends. The procedure for solution is similar to that of simply supported beams. The following example illustrates the general procedure.

Example 4.13 Overhanging beam with UD load

An overhanging beam of span L between supports and overhang length a is subjected to a UD load of w/m over the whole length. Draw SF and BM diagrams indicating salient values.

Solution The beam with the loading is shown in Fig. 4.26.

The reactions can be calculated as, by setting $\Sigma M = 0$ @ A,

$w(L + a)(L + a)/2 - R_{BV}L = 0$; $R_{BV} = w(L + a)^2/2L$. From $\Sigma V = 0$,

$R_{AV} = w(L + a) - w(L + a)^2/2L = w(L + a)[1 - (L + a)/2L] = w(L^2 - a^2)/2L$.

SF at x from $A = w(L^2 - a^2)/2L - wx$ for $x \le L$

This is equal to $w(L^2 - a^2)/2L$ at A and $w(L^2 - a^2)/2L - wL$ at B.

SF just to the right of $B = +wa$ (can be calculated easily from the right)

SF becomes zero at $w(L^2 - a^2)2L - wx = 0$; $x = [(L^2 - a^2)/2L]$

Bending moment is zero at A.

$$\text{BM at } B = \frac{w(L^2 - a^2)}{2L} L - \frac{wL^2}{2}$$

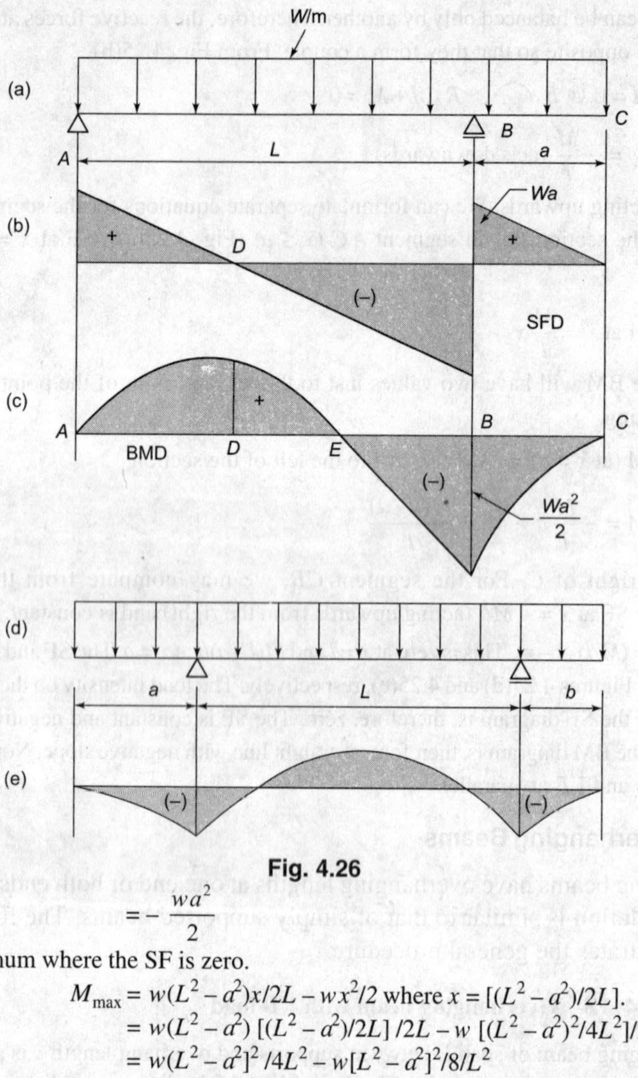

Fig. 4.26

$$= -\frac{wa^2}{2}$$

BM is maximum where the SF is zero.

$$M_{max} = w(L^2 - a^2)x/2L - wx^2/2 \text{ where } x = [(L^2 - a^2)/2L].$$
$$= w(L^2 - a^2)\,[(L^2 - a^2)/2L]\,/2L - w\,[(L^2 - a^2)^2/4L^2]/2$$
$$= w(L^2 - a^2]^2\,/4L^2 - w[L^2 - a^2]^2/8/L^2$$
$$= w(L^2 - a^2)^2/8L^2$$

The SF and BM diagrams are shown in Fig. 4.26.

Point of contraflexure From the bending moment diagram, it can be noted that there is a point *E*, where the BM is zero. Such a point is known as *point of contraflexure*. At the point of contraflexure, bending moment changes sign and passes through zero. In this case, BM changes from positive to negative. Such points normally do not exist in SS and cantilever beams. In an SS beam, BM is zero at the end supports. In a cantilever, BM is zero at the free end. In overhanging beams, such points do exist where the BM changes sign and passes through zero. In a doubly overhanging beam, two points of contraflexure exist as shown in Figures 4.26(d) and (e). ■

We will now discuss some more examples to illustrate the concepts and principles we have just outlined. While going through the examples, the reader should try to relate these concepts and principles to the solutions given. The AF, SF, and BM are generally calculated only at salient points and, the diagrams have been drawn knowing the nature of variation in these quantities.

Example 4.14 SS beam with UD load, point load, and couple load

Draw the AF, SF, and BM diagrams for the beam loaded as shown in Fig. 4.27(a).

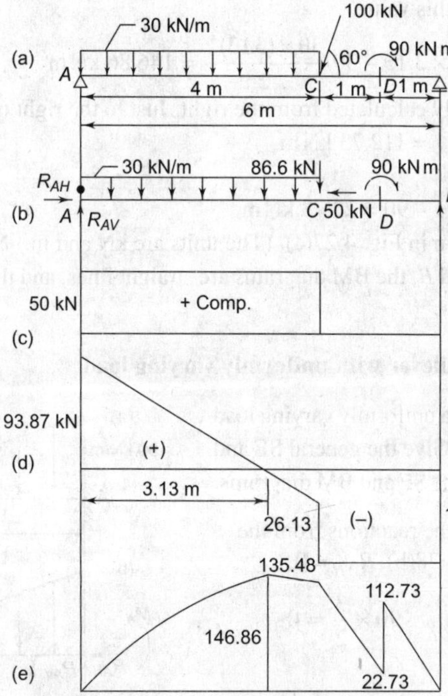

Fig. 4.27

Solution We start by calculating the reaction at A and B from the free body diagram shown in Fig. 4.27(b):

$$\Sigma H = 0 \rightarrow +, \quad R_{AH} - 100 \cos 60° = 0$$

$$\Rightarrow \quad R_{AH} = 50 \text{ kN} \rightarrow$$

$$\Sigma M = 0 \text{ @ } B \curvearrowright +, \quad R_{AV} \times 6 - 30 \times 4 \times 4 - 100 \sin 60° \times 2 + 90 = 0$$

$$\Rightarrow \quad R_{AV} = 93.87 \text{ kN} \uparrow$$

$$\Sigma V = 0 \uparrow +, \quad R_{AV} + R_{BV} - 30 \times 4 - 86.6 = 0$$

$$\Rightarrow \quad R_{BV} = 112.73 \text{ kN} \uparrow$$

Axial force (thrust) There is no axial force between C and B. The axial force remains constant between A and C. It is compressive and is equal to 50 kN. The axial force diagram is shown in Fig. 4.27(c). (The units are kN and m.)

Shear force From the left end, for $x \leq 4$ m,

$$\text{SF} = 93.87 - 30x$$

It is equal to 93.87 kN at $x = 0$ and – 26.13 kN at $x = 4$. The SF changes sign between $x = 0$ and $x = 4$. To locate the point of zero SF,

$$93.87 - 30x = 0$$

$$\Rightarrow \quad x = 3.13 \text{ m}$$

Just to the right of C, SF = 93.87 – 30 × 4 – 86.6 = – 112.73 kN and remains constant between C and B. Please note that the couple has no effect on SF calculations. The SF diagram is shown in Fig. 4.27(d). (The units are kN and m.)

Bending moment Calculating from the left, for $x \le 4$ m, BM = $93.87x - 30x^2/2$. This is equal to zero at $x = 0$ and + 135.48 kN m at $x = 4$. The maximum BM is at $x = 3.13$ m. Substituting this value,

$$\text{BM}_{\max} = 93.87 \times 3.13 - 30\,\frac{30 \times (3.13)^2}{2} = 146.86 \text{ kN m}$$

The BM at D can be easily calculated from the right. Just to the right of D,

$$\text{BM} = 112.73 \times 1 = 112.73 \text{ kN m}$$

Just to the left of D,

$$\text{BM} = 112.73 \times 1 - 90 = 22.73 \text{ kN m}$$

The BM diagram is shown in Fig. 4.27(e). (The units are kN and m.) Note that between C and D, and between D and B, the BM diagrams are straight lines, and that they are parallel.

Example 4.15 Cantilever with uniformly varying load

A cantilever beam carries a uniformly varying load as shown in Fig. 4.28(a). Give the general SF and BM equations and draw the SF and BM diagrams.

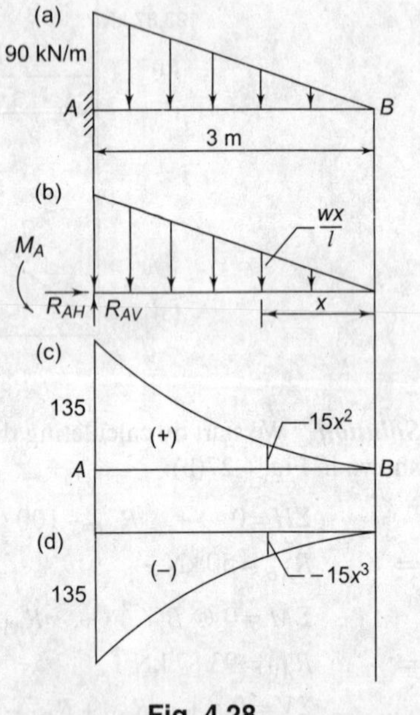

Fig. 4.28

Solution We calculate the reactions from the free body shown in Fig. 4.28(b). $R_{AH} = 0$.

$$\Sigma V = 0 \uparrow +, \quad R_{AV} - 90 \times \frac{3}{2} = 0$$

$$\Rightarrow \quad R_{AV} = 135 \text{ kN} \uparrow$$

$$\Sigma M = 0 \text{ @ B} \curvearrowright +$$

$$R_{AV} \times 3 - M_A - 90 \times \frac{3}{2} \times 2 = 0$$

$$\Rightarrow \quad M_A = 135 \text{ kN m} \smile$$

The general equations for SF and BM can now be stated. They are easier to formulate from the right side. The load intensity at x m from B is $30x$.

$$\text{SF at } x = 30x\,\frac{x}{2} = 15x^2$$

is positive, being downwards from the right. This is equal to zero at B ($x = 0$) and is equal to 135 kN at A ($x = 3$). The SF diagram is shown in Fig. 4.28(c).

$$\text{BM at } x = -30x\,\frac{x}{2}\,\frac{x}{3}$$

is negative, being clockwise from the right. It is equal to zero at B and is equal to –135 kN m at A. The BM diagram is shown in Fig. 4.28(d).

The load intensity is maximum at A, zero at B, and is negative. The slope of the SF diagram is negative, maximum at A, and zero at B. Similarly, the SF diagram is positive throughout with maximum value at A and zero at B. The slope of the BM diagram is, therefore, maximum at A, zero at B and positive.

Example 4.16 SS beam with multiple loads

Draw the AF, SF, and BM diagrams for the beam loaded as shown in Fig. 4.29(a). (The units are m, kN, and kNm.)

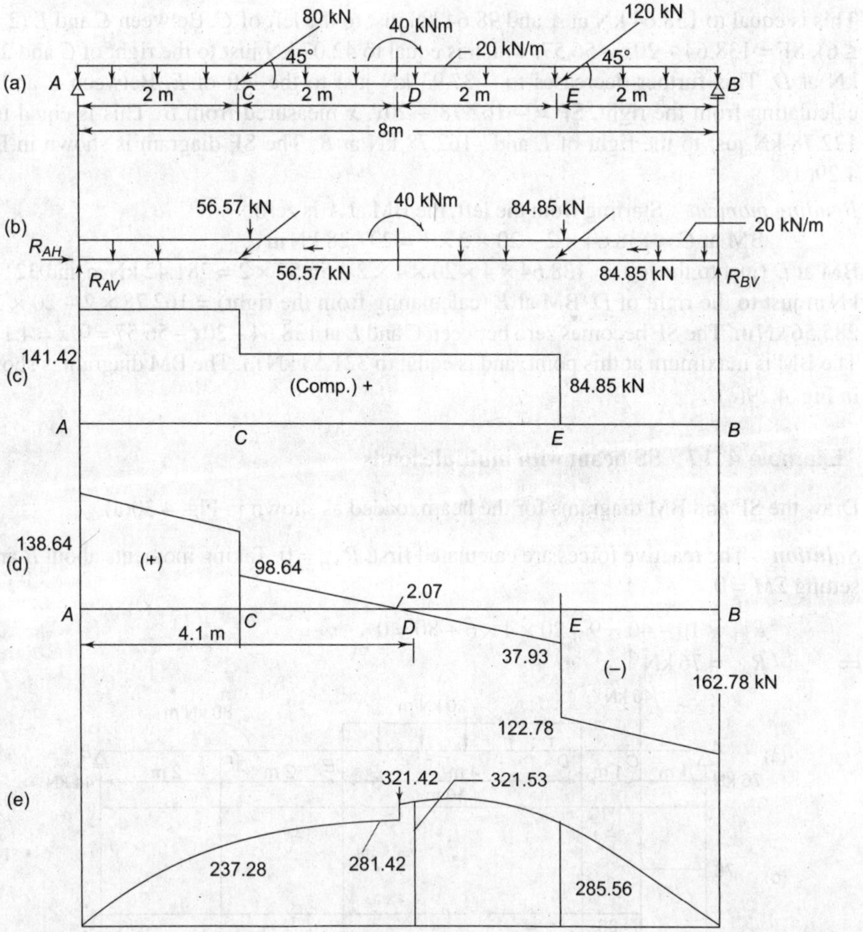

Fig. 4.29

Solution The vertical and horizontal components of the 80 kN load are each equal to 56.57 kN and those of the 120 kN load are 84.85 kN each. The free body diagram of the beam is shown in Fig. 4.29(b).

$$\Sigma H = 0 \xrightarrow[+]{}, \quad R_{AH} - 56.57 - 84.85 = 0$$

$$\Rightarrow \quad R_{AH} = 141.42 \text{ kN} \rightarrow$$

$$\Sigma M = 0 \text{ @ } B \curvearrowright +, \quad R_{AV} \times 8 - 56.57 \times 6 + 40 - 120 \times \sin 45° \times 2 - 20 \times 8 \times 4 = 0$$

$$\Rightarrow \quad R_{AV} = 138.64 \text{ kN} \uparrow$$

$$\Sigma V = 0 \uparrow +, \quad R_{AV} + R_{BV} - 56.57 - 84.85 - 20 \times 8 = 0$$

$$\Rightarrow \quad R_{BV} = 162.78 \text{ kN} \uparrow$$

Axial force (thrust) Computing from the left side, the AF is constant between *A* and just to the left of *C*, and is compressive. Its magnitude is 141.42 kN. Just to the right of *C*, the AF is equal to 141.42 – 56.57 = 84.85 kN and is compressive, and remains constant up to *E*. It becomes zero just to the right of *E*. The AF diagram is shown in Fig. 4.29(c).

Shear force Starting from the left, for $x \le 2$,

$$\text{SF} = 138.64 - 20x$$

This is equal to 138.64 kN at *A* and 98.64 kN just to the left of *C*. Between *C* and *E* ($2 \le x \le 6$), SF = $138.64 - 20x - 56.57$, which is equal to 42.07 kN just to the right of *C* and 2.07 kN at *D*. This further decreases to – 37.93 kN just to the left of *E*. Between *E* and *B*, calculating from the right, SF = $-162.78 + 20x$, *x* measured from *B*. This is equal to – 122.78 kN just to the right of *E* and –162.78 kN at *B*. The SF diagram is shown in Fig. 4.29(d).

Bending moment Starting from the left, the BM at *A* is zero.

$$\text{BM at } C = 138.64 \times 2 - 20 \times 2 \times 1 = 237.28 \text{ kN m}$$

BM at *D* (just to the left) = $138.64 \times 4 \times 20 \times 4 \times 2 - 56.57 \times 2 = 281.42$ kN m and 321.42 kN m just to the right of *D*. BM at *E* (calculating from the right) = $162.78 \times 2 - 20 \times 2 = 285.56$ kN m. The SF becomes zero between *C* and *E* at $138.64 - 20x - 56.57 = 0$, $x = 4.1$ m. The BM is maximum at this point, and is equal to 321.53 kN m. The BM diagram is shown in Fig. 4.29(e).

Example 4.17 SS beam with multiple loads

Draw the SF and BM diagrams for the beam loaded as shown in Fig. 4.30(a).

Solution The reactive forces are calculated first. $R_{AH} = 0$. Taking moments about *B* and setting $\Sigma M = 0$,

$$R_{AV} \times 10 - 40 \times 9 - 20 \times 4 \times 6 + 80 = 0$$

$$\Rightarrow \quad R_{AV} = 76 \text{ kN} \uparrow$$

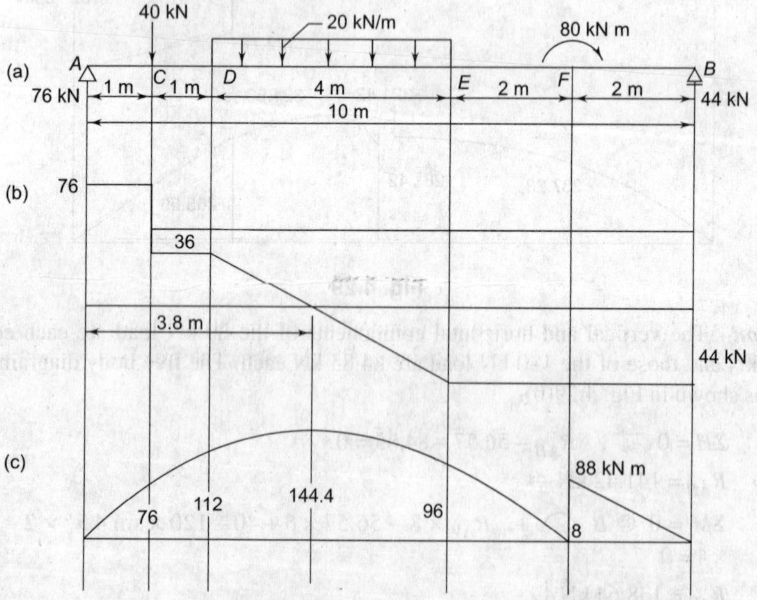

Fig. 4.30

From $\Sigma V = 0$,

$$R_{BV} = 44 \text{ kN} \uparrow$$

The axial forces are zero in the beam. The shear forces at salient points are as follows. SF at *A* (just to the right) = 76 kN. SF at *C* (just to the right) = 76 – 40 = 36 kN. Between *D* and *E*, SF =

$76 - 40 - 20x,\ 2 \le x \le 6$. This implies an SF of 36 kN at D and – 44 kN at E. The SF remains constant between E and B. The SF becomes zero between D and E at $76 - 40 - 20x = 0$, i.e., at $x = 1.8$ m from D. The SF diagram is shown in Fig. 4.30(b).

$$\text{BM at } A = 0$$

$$\text{BM at } C = 76 \times 1 = 76 \text{ kN m}$$

$$\text{BM at } D = 76 \times 2 - 40 \times 1 = 112 \text{ kN m}$$

$$\text{BM at } E = 76 \times 6 - 0 \times 5 - 20 \times 4 \times 2 = 96 \text{ kN m}$$

Between D and E, the BM varies according to the equation

$$76x - 40\,(x - 1) - 20\,\frac{(x-2)^2}{2}$$

(x measured from A). The maximum value is at $x = 3.8$, where the SF is zero.

$$M_{\max} = 76 \times 3.8 - 40 \times 2.8 - 20 \times \frac{3.8 \times 3.8}{2} = 144.4 \text{ kN m}$$

BM at $F = 44 \times 2 = 88$ kN m just to the right of F and 8 kN m just to the left of F. The BM diagram is shown in Fig. 4.30(c).

Example 4.18 Overhanging beam with many loads

A beam 10 m long is supported at its left end and at a point 8 m from the left end. The beam carries loading as shown in Fig. 4.31(a). Draw the SF and BM diagrams.

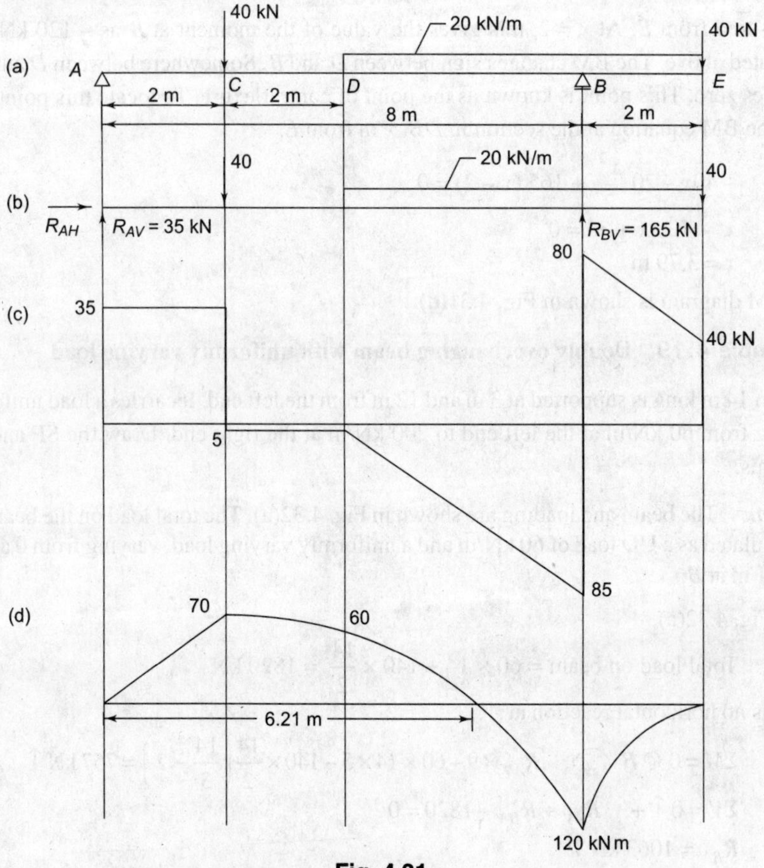

Fig. 4.31

Solution We start by calculating the reactions:

$R_{AH} = 0$ [Fig. 4.31(b)]

$\Sigma M = 0$ @ B ↷ +

$R_{AV} \times 8 - 40 \times 6 - 20 \times 6 \times 1 + 40 \times 2 = 0$

$\Rightarrow$ $R_{AV} = 35$ kN ↑

$\Sigma V = 0$ ↑ +, $R_{AV} + R_{BV} - 40 - 120 - 40 = 0$

$\Rightarrow$ $R_{BV} = 165$ kN ↑

Shear force The SF remains constant in segment AC and is equal to + 35 kN. Just to the right of C, SF = 35 – 40 = – 5 kN. This value remains constant in segment CD. From D to B, SF = 35 – 40 – 20x (x/2), measured from D. At B (x = 4 m), SF = – 85 kN.

In the segment BE, SF = 40 + 20x, calculating from the right. This yields SF = 40 kN at E and 80 kN at B. The SF diagram is shown in Fig. 4.31(c).

Bending moment Calculating from the left,

BM at $C = 35 \times 2 = 70$ kN m

BM at $D = 35 \times 4 - 40 \times 2 = 60$ kN m

BM at $B = 35 \times 8 - 40 \times 6 - 20 \times 4 \times 2 = -120$ kN m

This can be easily calculated from the right.

$$\text{BM in segment } EB = -40x - \frac{20\,x^2}{2}$$

x measured from E. At x = 2, this gives the value of the moment at B as – 120 kN m, as calculated above. The BM changes sign between D and B. Somewhere between D and B, it becomes zero. This point is known as the point of contraflexure. To locate this point, consider the BM equation at the section in DB, x m from E.

$$-40x - 20\frac{x^2}{2} + 165(x-2) = 0$$

$$x^2 - 12.5x + 33 = 0$$

$\Rightarrow$ $x = 3.79$ m

The BM diagram is shown in Fig. 4.31(d).

Example 4.19 Doubly overhanging beam with uniformly varying load

A beam 14 m long is supported at 3 m and 12 m from the left end. It carries a load uniformly varying from 60 kN/m at the left end to 200 kN/m at the right end. Draw the SF and BM diagrams.

Solution The beam and loading are shown in Fig. 4.32(a). The total load on the beam can be calculated as a UD load of 60 kN/m and a uniformly varying load, varying from 0 at A to 140 kN/m at B.

From Fig. 4.32(a),

$$\text{Total load on beam} = 60 \times 14 + 140 \times \frac{14}{2} = 1820 \text{ kN}$$

There is no horizontal reaction at A.

$$\Sigma M = 0 \text{ @ } B \curvearrowright +, \quad R_{AV} \times 9 - 60 \times 14 \times 5 - 140 \times \frac{14}{2}\left(\frac{14}{3} - 2\right) = 757 \text{ kN} \uparrow$$

$\Sigma V = 0$ ↑ +, $R_{AV} + R_{BV} - 1820 = 0$

$\Rightarrow$ $R_{BV} = 1063$ kN ↑

Shear force Calculating from the left, in the segment CA, SF = $-60x - 10x(x/2)$. The values are zero at C and -225 kN just to the right of A. Just to the left of A, the SF = $(-225 + 757) = 532$ kN and positive.

In the segment AB, SF = $-60x - 10x^2/2 + 757$. At B, $x = 12$ and the SF is equal to -683 kN (just to the left). Just to the right of B, SF = $+380$ kN. In the segment BD, SF = $-60x - 10x^2/2 + 757 + 1063$. This is zero at D and is positive for values of x from 12 m to 14 m.

The SF changes sign between A and B. To locate this point, $-60x - 10x^2/2 + 757 = 0$. This gives $x = 7.69$ m from C. The SF diagram is shown in Fig. 4.32(b). Note that the SF diagram is parabolic due to the varying load.

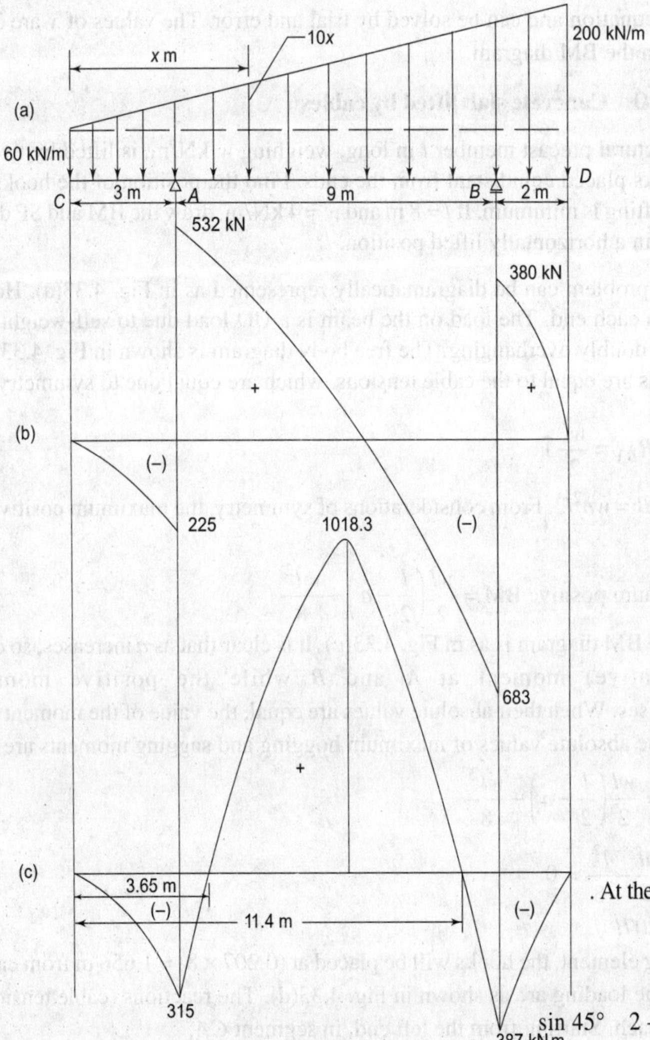

. At the point of

sin 45° 2 – 20 8

Fig. 4.32

Bending moment In segment CA,

$$\text{BM} = -60\,\frac{x^2}{2} - 10x\,\frac{x}{2}\,\frac{x}{3}$$

The BM at A is – 315 kN m. In segment AB,

$$\text{BM} = -60\frac{x^2}{2} - \frac{10x^3}{6} + 757(x-3)$$

At B, $x = 12$ and BM = – 387 kN m. BM_{max} (at $x = 7.69$, where SF is zero) = $-60 \times (7.69)^2/2 - 10 \times (7.69)^3/6 + 757\ (7.69 - 3)$ = 1018.3 kN m. The BM diagram is shown in Fig. 4.32(d). There are two points of contraflexure (point of zero BM) between A and B. To locate these, consider the equation

$$-\frac{60x^2}{2} - 10 \times \frac{x^3}{6} + 757\ (x-3) = 0$$

This is a cubic equation and can be solved by trial and error. The values of x are 3.65 and 11.4 as shown in the BM diagram. ■

Example 4.20 Concrete slab lifted by cables

A concrete structural precast member l m long, weighing w kN/m, is lifted by two cables attached to hooks placed equidistant from the ends. Find the position of the hooks so that the BM due to lifting is minimum. If $l = 8$ m and $w = 4$ kN/m, draw the BM and SF diagrams for the element in a horizontally lifted position.

Solution The problem can be diagramatically represented as in Fig. 4.33(a). Hooks are placed at a from each end. The load on the beam is a UD load due to self-weight. Notice that the beam is doubly overhanging. The free body diagram is shown in Fig. 4.33(b). The support reactions are equal to the cable tensions, which are equal due to symmetry.

$$R_{AV} = R_{BV} = \frac{wl}{2} \uparrow$$

The BM at A or B = $wa^2/2$. From considerations of symmetry, the maximum positive BM is at mid-span.

$$\text{Maximum positive BM} = \frac{wl}{2}\left(\frac{l}{2} - a\right) - \frac{wl^2}{8}$$

The shape of the BM diagram is as in Fig. 4.33(c). It is clear that as a increases, so does the hogging (negative) moment at A and B while the positive moment at mid-span decreases. When their absolute values are equal, the value of the moment is minimum. That is, the absolute values of maximum hogging and sagging moments are equal.

$$\frac{wa^2}{2} = \frac{wl}{2}\left(\frac{l}{2} - a\right) - \frac{wl^2}{8}$$

$$\frac{a^2}{2} + \frac{al}{2} - \frac{l^2}{8} = 0$$

$$\Rightarrow \qquad a = 0.207l$$

For the 8-m-long element, the hooks will be placed at (0.207 × 8) = 1.656 m from each end. The beam and the loading are as shown in Fig. 4.33(d). The reactions (cable tensions) are equal to 16 kN each. Starting from the left end, in segment CA,

$$\text{SF} = -4x\ (0 \le x \le 1.656)$$

SF at A = – 6.624 kN (just to the left) and $-4 \times 1.656 + 16 = 9.376$ kN (just to the right). The SF diagram will be the same on the other half as shown in the figure.

$$\text{BM at } A = \frac{-4x^2}{2} = -4 \times \frac{1.656^2}{2} = 5.48 \text{ kN m}$$

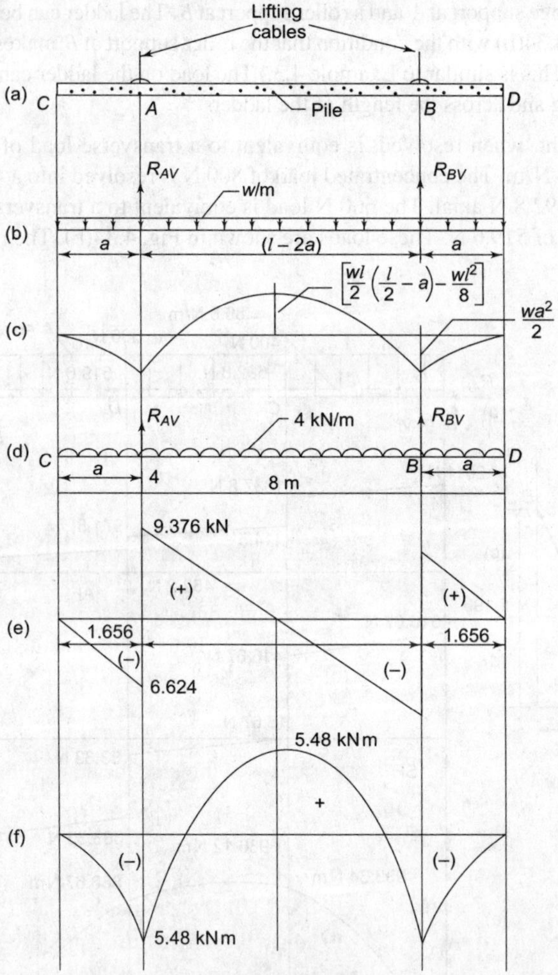

Fig. 4.33

The SF and BM diagrams are shown in Figures 4.33(e) and 4.33(f). To verify,

$$\text{BM at mid-span} = 16 \times (4 - 1.656) - \frac{4.4^2}{2} = 5.5 \text{ kN m}$$

4.7 INCLINED BEAMS

The beams illustrated so far have all been horizontal. Inclined beams are not uncommon. Staircase beams, ladders, etc. are examples of such beams. No new principles are involved. Remember that only loads transverse to the axis cause shear force and bending moment. Let us illustrate this with an example.

Example 4.21 Inclined ladder

A ladder is kept inclined at an angle of 60° against a smooth wall and rests on a rough floor. In addition to its self-weight of 100 N/m, the ladder carries the weights of two persons weighing 800 N at 2 m and 600 N at 4 m from the ground along the length of the ladder. Draw the AF, SF, and BM diagrams for the ladder.

Solution The physical representation of the problem is as shown in Fig. 4.34(a). It has the equivalent of a hinge support at A and a roller support at B. The ladder can be laid horizontally as shown in Fig. 4.34(b) with the condition that the roller support at B makes an angle of 60° with the ladder. (This is similar to Example 4.3.) The load on the ladder can be resolved into components along and across the length of the ladder.

The self-weight, when resolved, is equivalent to a transverse load of 50 N/m and an axial load of 86.6 N/m. The concentrated load of 800 N is resolved into a 400 N transverse component and 692.8 N axial. The 600 N load is equivalent to a transverse load of 300 N and an axial load of 519.6 N. These loads are shown in Fig. 4.34(b). The reaction can now be calculated.

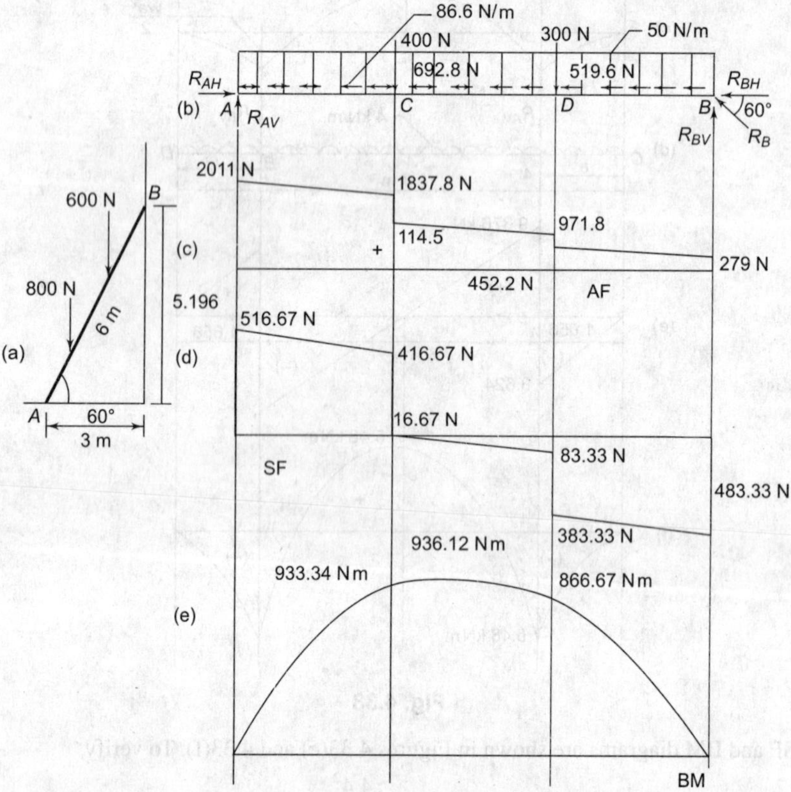

Fig. 4.34

$$\Sigma H = 0, \quad R_{AH} - 86.6 \times 6 - 692.8 - 519.76 - R_{BH} = 0, \quad R_{AH} - R_{BH} = 1732$$

$$\Sigma M = 0 \ @ \ A \curvearrowleft +, \quad R_{BV} \times 6 + 50 \times 6 \times 3 + 300 \times 4 + 400 \times 2 = 0$$

$$\Rightarrow \quad R_{BV} = 483.33 \text{ N}$$

$$R_{BV} = R_B \sin 60°$$

Therefore,

$$R_B = 558.10 \text{ N}$$

$$R_{BH} = R_B \cos 60° = 279 \text{ N}$$

$$R_{AH} = 1732 + R_{BH} = 2011 \text{ N}$$

$$\Sigma V = 0 \uparrow +, \quad R_{AV} + R_{BV} - 50 \times 6 - 400 - 300 = 0$$

$$\Rightarrow \quad R_{AV} = 516.67 \text{ N}$$

Axial force The AF is compressive throughout and hence positive.

At A, AF = 2011 N. Just to the left of C, AF = 2011 – 86.6 × 2 = 1837.8 N. Just to the right of C, AF = 1145 N. It decreases to 1145 – 86.6 × 2 – 971.8 N to the left of D and to 452.2 N to the right of D. This decreases to 452.2 – 86.6 × 2 = 279 N to the left of B. The AF diagram is shown in Fig. 4.34(c).

Shear force Starting from the left, the SFs are as follows:

SF at A = 516.67 N

At C (just to the left),

SF = 416.67 N

At C (just to the right),

SF = 16.67 N

At D (just to the left),

SF = 16.67 – 50 × 2 = – 83.33 N

At D (just to the right),

SF = – 383.33 N

At B,

SF = – (383.33 + 50 × 2) = – 483.33 N

The SF diagram is shown in Fig. 4.34(d).

Bending moment The BMs calculated at different sections are as follows.

At A,

BM = 0

At C,

BM = 516.67 × 2 – 50 × 2 = 933.34 N m

At D,

BM = 483.33 × 2 – 50 × 2 × 1 = 866.66 N m

from the right. The SF becomes zero at 16.67/50 = 0.333 m from C to the right. Maximum BM at this point = 516.67 × 2.333 – 50 × 2.333^2/2 – 400 × 0.333 = 936.12 N m. The BM diagram is shown in Fig. 4.34(e).

4.8 HINGED BEAMS

We mentioned hinged beams in the beginning of this chapter. An otherwise statically indeterminate beam can be made determinate by the provision of hinges. Consider the beam shown in Fig. 4.35(a). It has four reactive components and would have been statically indeterminate but for the hinge at D. The provision of the hinge means freedom of rotation about it for the beam segments connected, i.e., AC and CD. This has the advantage that disturbances like support settlement get localized to the segment where it happens, due to the freedom of rotation about the hinge.

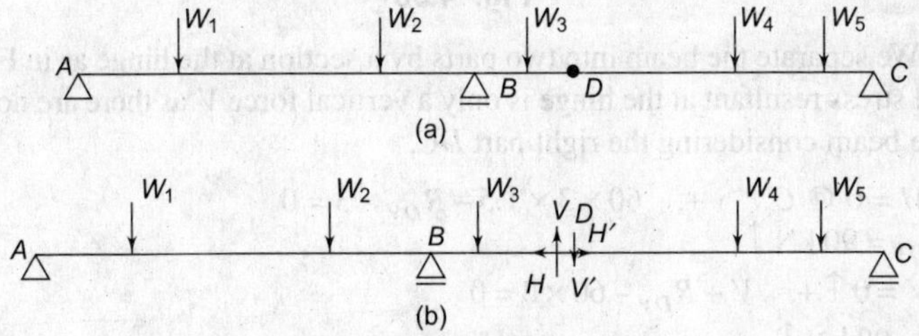

Fig. 4.35

The provision of a hinge means that the BM about the hinge is zero. If we separate the beam at the hinge into two parts, the internal stress resultants at the section will have only horizontal and vertical forces. The two separated parts [Fig. 4.35(b)] are statically determinate. This is the method employed for analysis. Let us illustrate the procedure with an example.

Example 4.22 Two-span hinged beam

Analyse the hinged beam shown in Fig. 4.36(a) and draw the required SF and BM diagrams.

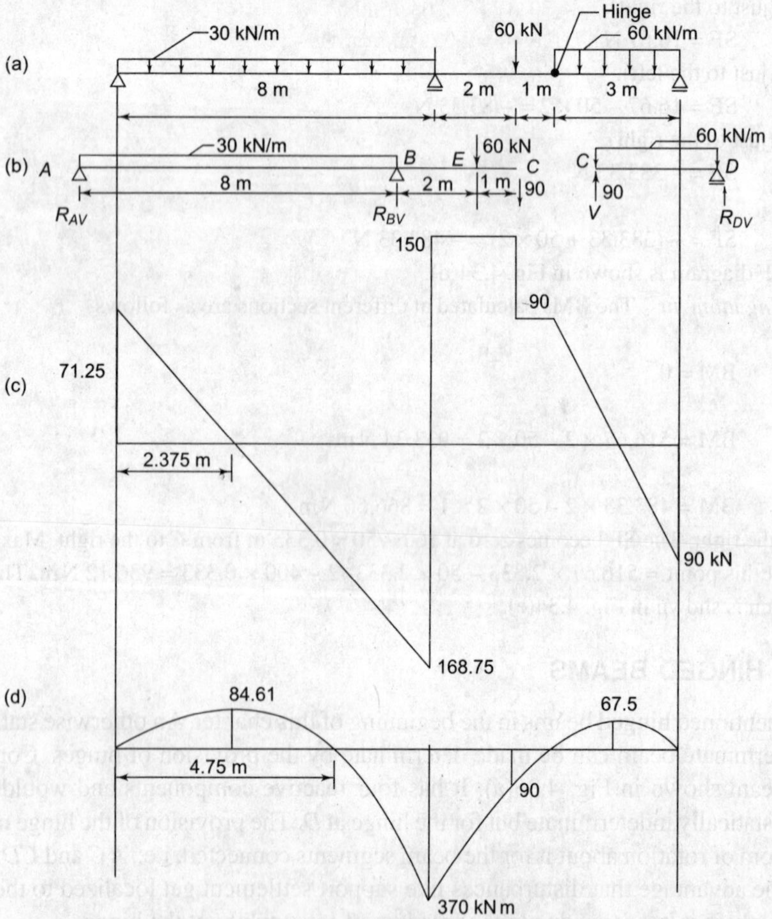

Fig. 4.36

Solution We separate the beam into two parts by a section at the hinge as in Fig. 4.30(b). The internal stress resultant at the hinge is only a vertical force V as there are no horizontal forces in the beam considering the right part DC,

$$\Sigma M = 0 \text{ @ } C \curvearrowright +, \quad 60 \times 3 \times 1.5 - R_{DV} \times 3 = 0$$

$$\Rightarrow \quad R_{DV} = 90 \text{ kN} \uparrow$$

$$\Sigma V = 0 \uparrow +, \quad V + R_{DV} - 60 \times 3 = 0$$

$$\Rightarrow \quad V = 90 \text{ kN} \uparrow$$

Considering the left part, there is a downward force of 90 kN acting at C. This part is a statically determinate overhanging beam and can be analysed as such.

$\Sigma M_A = 0 \curvearrowright + @ A$

$30 \times 8 \times 4 - R_{BV} \times 8 + 60 \times 10 + 90 \times 11 = 0$

$R_{BV} = 318.75 \text{ kN} \uparrow$

$\Sigma V = 0 \uparrow +$

$R_{AV} + R_{BV} - 30 \times 8 - 60 - 90 = 0$

$\Rightarrow \quad R_{AV} = 71.25 \text{ kN} \uparrow$

For the left part,

SF at A = 71.25 kN

SF at B = 71.25 – 30 × 8 = – 168.75 kN

The SF becomes zero at 71.25/30 = 2.375 m from A.

SF at B (just to the right) = 71.25 – 30 × 8 + 318.75 = 150 kN.

This reduces to 90 kN just to the right of E.

BM in $AB = 71.25x - 30x^2/2$.

This is equal to – 370 kN m at B (for x = 8 m). It is maximum at x = 2.375, where SF is zero and is equal to 84.61 kN m (positive).

BM at E = – 90 × 1 = – 90 kN m

from the right.

BM at C = 0

For the right part,

SF at C = 90 kN

and becomes zero at 90/60 = 1.5 m. SF = – 90 kN at D.

BM in CD at x m from $C = 90x - 60x^2/2$. It has a maximum value at x = 1.5 m and $M_{max} = 90 \times 1.5 - 30 \times 1.5^2 = 67.5$ kN m (positive). The SF and BM diagrams can be combined as in Figures 4.36(c) and (d).

4.9 STATICALLY DETERMINATE RIGID FRAMES

A rigid frame is a structural framework of members connected with rigid joints. In the case of a rigid joint, when the frame is subjected to loads and deforms, the angle between the members remains constant. To ensure this, the members are subjected to moments at the joints. A rigid frame is shown in Fig. 4.37. While the member BC primarily acts as a beam, AB and CD mainly act as axially loaded members. While multibay, multistorey frames are more common as in building frames, they are highly indeterminate structures. We will now illustrate the analysis of simple, statically determinate frames with examples.

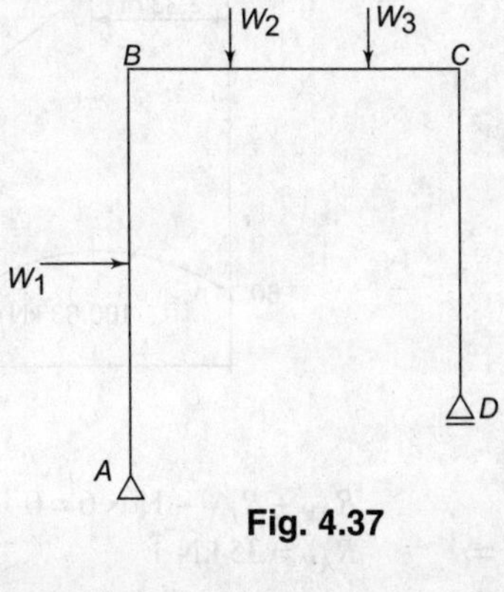

Fig. 4.37

Example 4.23 Statically determinate rigid frame

Analyse the rigid frame shown in Fig. 4.38(a) and draw the required SF and BM diagrams.

Solution We start by calculating reactions from the free body shown in Fig. 4.38(b).

$\Sigma M = 0 \; @ A \curvearrowright +$

$30 \times 2 + 15 \times 6 \times 3 - R_{DV} \times 6 = 0$

$\Rightarrow \quad R_{DV} = 55 \text{ kN} \uparrow$

$\Sigma H = 0 \rightarrow +$

$-R_{AH} + 30 = 0$

$\Rightarrow \quad R_{AH} = 30 \text{ kN} \leftarrow$

$\Sigma V = 0 \uparrow +$

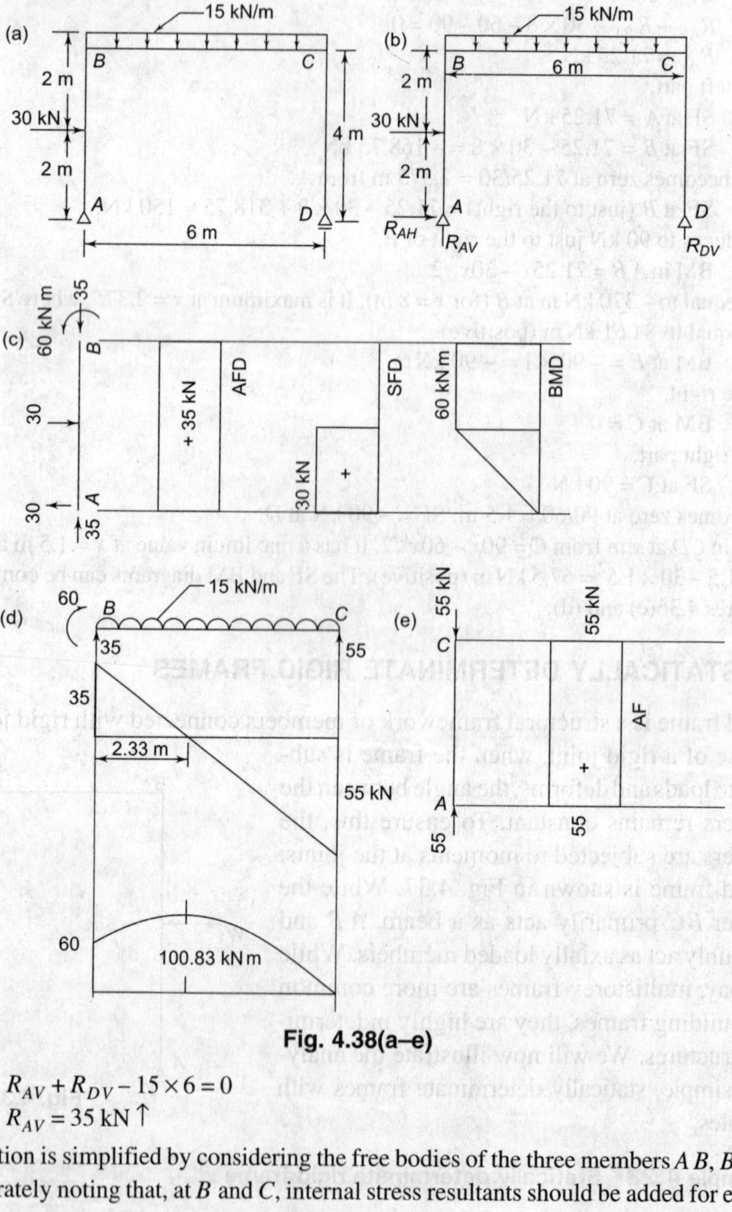

Fig. 4.38(a–e)

$R_{AV} + R_{DV} - 15 \times 6 = 0$

$\Rightarrow \quad R_{AV} = 35 \text{ kN} \uparrow$

The solution is simplified by considering the free bodies of the three members AB, BC, and CD separately noting that, at B and C, internal stress resultants should be added for equilibrium.

Member AB The free body diagram is shown in Fig. 4.38(c). V, M, and H are the internal stress resultants at B. Considering the equilibrium of the free body, $\Sigma H = 0$ gives $30 - 30 + H = 0$; $H = 0$. $\Sigma V = 0$ gives $V = 35$ kN $\downarrow$, $\Sigma M = 0$ @ B gives $30 \times 4 - 30 \times 2 + M = 0$ and $M = -60$ kNm (↶). The AF, SF, and BM diagrams for the member are shown in Fig. 4.38(c).

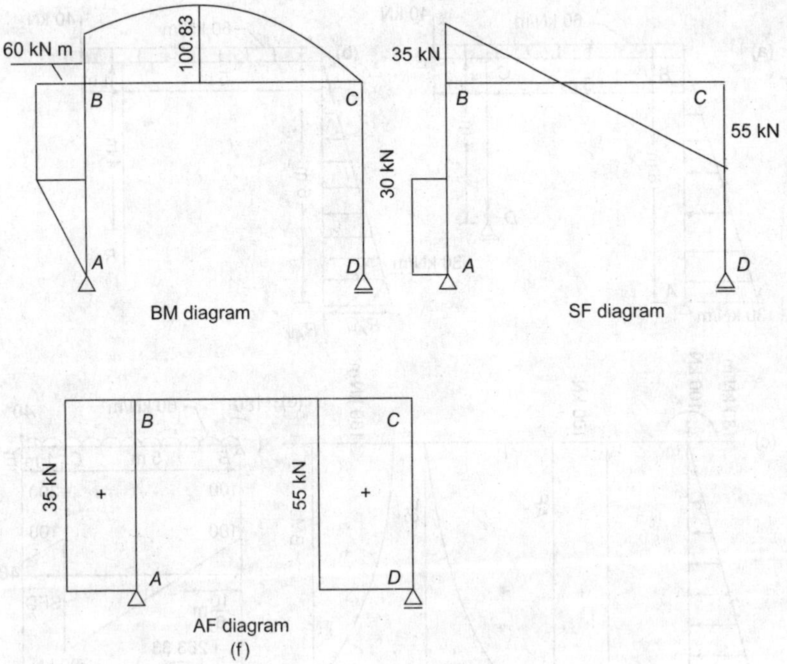

Fig. 4.38(f)

Member BC The free body of the member is shown in Fig. 4.38(d). Please note that the internal stress resultants at B are exactly equal and opposite to those at B in AB. The internal stress resultants at C can be evaluated from the free body.

$\Sigma V = 0 \uparrow +$ gives

$35 - 15 \times 6 + V = 0$

$\Rightarrow \quad V = 55 \text{ kN} \uparrow$

$\Sigma M = 0$ @ C gives

$60 + 35 \times 6 - 15 \times 6 \times 3 - M = 0$

$\Rightarrow \quad M = 0$

The AF, SF, and BM diagrams are shown in Fig. 4.38(d) for BC. Note that the SF is zero at 35/15 = 2.33 m from B and the maximum BM at this point is $60 + 35 \times 2.33 - 15 \times 2.33^2/2 = 100.83$ kN m.

Member CD In the free body diagram shown in Fig. 4.38(e), the internal stress resultant at C in CD is a force of 55 kN $\downarrow$ only, and the member has no transverse loading and acts purely as a compression member. The AF diagram is shown in Fig. 4.38(e). If the diagram for the whole frame is to be drawn, separate diagrams for BM, SF and AF for the frame can be drawn, as shown in Fig. 4.38(f).

Example 4.24 Rigid frame with unequal verticals

Analyse the rigid frame shown in Fig. 4.39(a) and draw the AF, SF, and BM diagrams.

Solution The reactions are obtained from the free body shown in Fig. 4.39(b). The total triangular load acting to the right on $AB = 30 \times 6/2 = 90$ kN.

$\Sigma H = 0 \rightarrow +, \quad 90 + R_{AH} = 0$

$\Rightarrow \quad R_{AH} = -90$ kN (acts to the left)

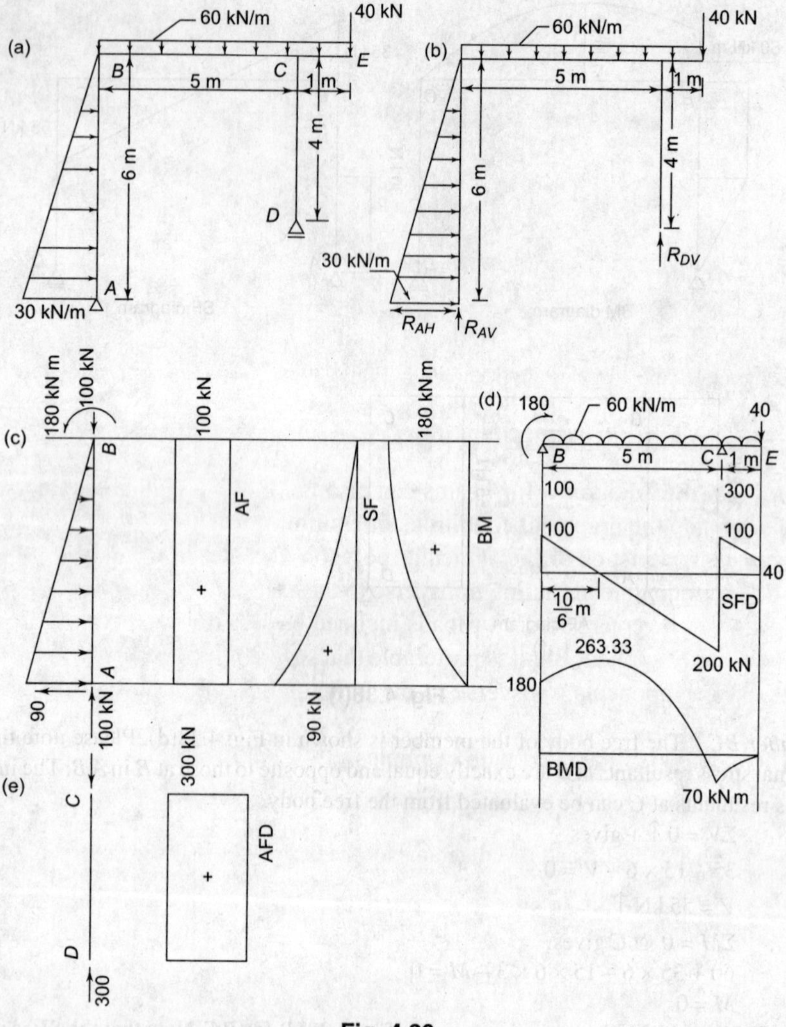

Fig. 4.39

$$\Sigma M = 0 \ @\ A \curvearrowleft +, \quad 30 \times \frac{6}{2} \times 2 + 60 \times 6 \times 3 + 40 \times 6 - R_{DV} \times 5 = 0$$

$$\Rightarrow \quad R_{DV} = 300 \text{ kN} \uparrow$$

$$\Sigma V = 0 \uparrow +, \quad R_{AV} + R_{DV} - 60 \times 6 - 40 = 0$$

$$\Rightarrow \quad R_{AV} = 100 \text{ kN} \uparrow$$

Member AB The free body is shown in Fig. 4.39(c). The internal stress resultants at B are as follows:

$$\Sigma H = 0 \ \overset{\longrightarrow}{+}, \quad -90 + 90 + H = 0$$

$$\Rightarrow \quad H = 0$$

$$\Sigma V = 0 \uparrow +, \quad 100 + V = 0$$

$$\Rightarrow \quad V = -100 \text{ kN} \downarrow$$

$$\Sigma M = 0 \curvearrowleft + \ @\ B, \quad 90 \times 6 - 90 \times 4 + M = 0$$

$$\Rightarrow \quad M = -180 \text{ kNm} \curvearrowleft$$

The AF, SF, and BM diagrams are shown in Fig. 4.39(c).

Member BCE The free body diagram is shown in Fig. 4.39(d). The internal stress resultants at B are a force of 100 kN ↷ and a couple of 180 kN m ↷ (opposite to those at B in AB). There is no axial force in this member.

$$\Sigma M = 0 \text{ @ } B \; \curvearrowright +, \quad 180 + 60 \times 6 \times 3 + 40 \times 6 - R_{CV} \times 5 = 0$$

$$\Rightarrow \quad R_{CV} = 300 \text{ kN} \uparrow$$

(Since there is no transverse force on CD, there will no internal moment at C.) The SF and BM diagrams are shown in Fig. 4.39(d).

Member CD The free body diagram is shown in Fig. 4.39(e). The internal stress resultant at C is a force of 300 kN ↓. The member is in pure compression and there is no SF and BM in the member. The AF diagram is shown in Fig. 4.39(e). ■

4.10 GRAPHICAL METHOD FOR DRAWING SF AND BM DIAGRAMS

The SF and BM diagrams for beams can also be drawn graphically. The method involves using the graphical conditions of equilibrium of a coplanar force system and the characteristics of the funicular polygon outlined in Chapter 1. Since SF and BM are due to loads acting transverse to the longitudinal axis only, only gravity loads will be considered. For loads inclined to the axis of a beam, the method can be directly applied. But it is preferable that such loads be resolved into components, and components transverse to the longitudinal axis are used for graphical analysis.

The method is illustrated with examples.

Example 4.25 SS beam: graphical method

Draw graphically the SF and BM diagrams for the beam loaded as shown in Fig. 4.40(a).

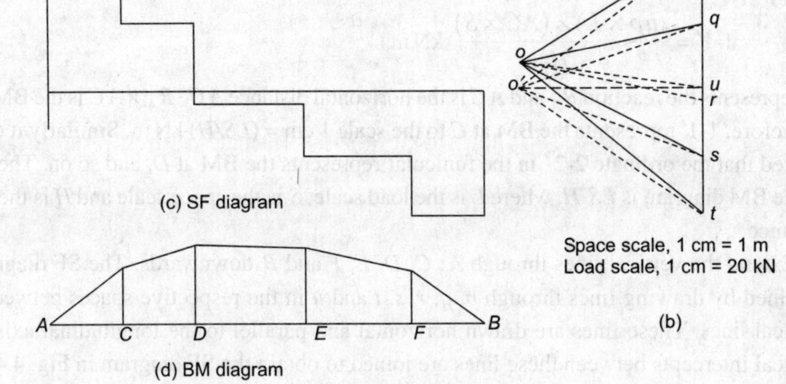

Fig. 4.40

Solution The beam should be drawn to scale and the load positions marked on the beam according to the scale selected. This is the space scale (1 cm = 1 m) and will be needed in calculations. The scale chosen is 1 cm = 1 m.

The forces are named according Bow's notation. The letters *P*, *Q*, *R*, *S*, *T*, and *U* are used. *tu* is the reaction at *B* and *up* is the reaction at *A*. Draw the force diagram to scale. The scale used is 1 cm = 20 kN. The force diagram is a straight line since the loads are parallel. *pqrst* is the force diagram for loads. The point *u* has to be located on this line to determine the reactions. To do this, we draw the funicular polygon. While the funicular polygon can be started anywhere on a vertical line through *A* (R_A being vertical), we illustrate by starting the funicular polygon from point *A* on the beam. This is particularly needed if there are inclined loads on the beam and we are not using resolved components of loads. In such a case, point *A* is on the line of action of R_A, whatever be the direction of R_A.

We select a pole *o* on either side of the load line *pqrst* and draw *op*, *oq*, *or*, *os*, and *ot*. The funicular polygon is drawn by drawing a line parallel to *op* starting from *A* till it intersects the line of action of the 40 kN load at 1 m from *A*. This intersection point is named 1. Similarly lines are drawn parallel to *oq*, *or*, *os*, and *ot* in the respective spaces. The end of the line parallel to *ot* should be on the reaction line at *B*. Since this is known to be vertical due to the roller at *B*, we draw a vertical line through *B* to obtain the point 5 on funicular polygon. Since the beam is in equilibrium, the funicular polygon should close. We close the funicular polygon by joining *A* to point 5. A line drawn parallel to *A*-5 through *o*, intersects the load line at *u*. *tu* and *up* are the reactions at *B* and *A* respectively to the scale chosen for the load diagram. The reactions can thus be found.

We can prove that the funicular polygon drawn represents the BM diagram for the beam to some scale. Name the points at which the closing line *A*-5 intersects the load lines as 1′, 2′, 3′, ... as shown. 1-1′ is the intercept of the funicular polygon at *C*.

Consider the triangles *A*-1-1′ and *pou*. These two triangles are similar because *A*-1 is parallel to *po*, *A* 1′ is parallel to *ou* and 1-1′ is parallel to *up*. Therefore, the corresponding sides and altitudes of these triangles will be in the same proportion. The distance of the pole *o* from the load line is known as the pole distance, and is shown in Fig. 4.40(b) as *H*. We can thus state that

$$\frac{1\text{-}1'}{up} = \frac{AC}{H} \Rightarrow 1\text{-}1' = \frac{up \times AC}{H}$$

If 1 cm = *S* m is the space scale and 1 cm = *L* kN is the load scale, then,

$$1\text{-}1' = \frac{(up \times L) \times (AC \times S)}{H} \text{ kN m}$$

up represents the reaction R_A and *A C* is the horizontal distance $AC \cdot R_A \times AC$ is the BM at *C*. Therefore, 1-1′ represents the BM at *C* to the scale 1 cm = (*LS/H*) kN m. Similarly it can be proved that the ordinate 2-2′ in the funicular represents the BM at *D*, and so on. The scale of the BM diagram is *LS/H*, where *L* is the load scale, *S* is the space scale and *H* is the polar distance.

Extend the vertical lines through *A*, *C*, *D*, *E*, *F* and *B* downwards. The SF diagram is obtained by drawing lines through *p*, *q*, *r*, *s*, *t* and *u* in the respective spaces between the vertical lines. These lines are drawn horizontal and parallel to the longitudinal axis. The vertical intercepts between these lines are joined to obtain the SF diagram in Fig. 4.40(c).

The following points may be noted with reference to the graphical procedure we have discussed.

(i) Since the BM diagram for the beam is positive, it is possible, if required, to ensure that the closing line is below the funicular polygon. To do this, the pole should be chosen on the left of the load line.

(ii) The closing line, which is the base of the funicular polygon, is inclined. If we desire to make it horizontal, we can do so by shifting the pole in a horizontal line with *u*. The pole was chosen earlier arbitrarily. Whichever pole is chosen, the position of *u* cannot change since the reactions remain the same. Selecting the pole *o′* on a horizontal line with *u* and redrawing the funicular polygon will give the BM diagram on a horizontal line. Pole *o′* can be joined to *p*, *q*, *r*, *s*, etc. and a new funicular polygon with a horizontal closing line drawn. Both these modifications have been made in Fig. 4.40(d).

(iii) The ordinate in the funicular polygon depends upon distance *H*. The larger the value of *H*, the smaller will be the ordinate.

This procedure of drawing the funicular polygon can be applied to beams subjected to UD loads as well. The procedure is to divide the load into an appropriate number of segments and assume that the load in each segment acts at its middle. The procedure then is exactly the same as in Example 4.25. The only modification is to round off the angularities in the diagram and get a smooth curve or straight line as required. The larger the number of segments, the better will be the approximation to the actual curve. This is illustrated with an example.

Example 4.26 SS beam with UD load: graphical method

Draw graphically the SF and BM diagrams for a beam 12 m long and subjected to a uniformly distributed load of 30 kN/m.

Solution The beam is shown in Fig. 4.41(a), drawn to a space scale of 1 cm = 1 m. The beam is divided into six segments, 2 m each, and the load on the segments is set as a point load in the middle. There are six loads of 60 kN each.

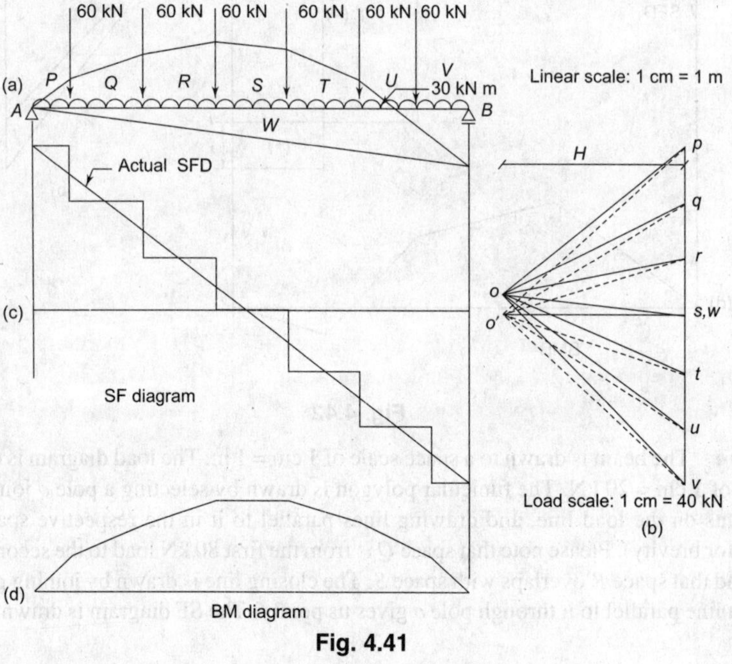

Fig. 4.41

The load line is drawn to a scale of 1 cm = 40 kN [Fig. 4.41(b)]. The reactions are obtained by drawing the funicular polygon.

The SF diagram is obtained as before by drawing horizontal lines through points *a*, *b*, *c*, ... in the respective spaces. Note that the line through *W* is drawn for the whole length, as space *W* lies between the reactions. The stepped SF diagram can be changed by just joining the points on the reaction lines by a line, which is the actual SF diagram [Fig. 4.41(c)].

For the sake of illustration, the BM diagram is redrawn on a horizontal base by shifting the pole on a horizontal line with *W* in Fig. 4.41(d). The new pole *o* is joined to *p*, *q*, *r*, *s*, etc. and a new funicular polygon drawn as in Fig. 4.41(d). This diagram has a horizontal base as in the conventional BM diagram. The sharp edges of the diagram can be rounded off by inscribing tangential curves to the line segments meeting at the points.

Example 4.27 Overhanging beam: graphical method

An overhanging beam is loaded as shown in Fig. 4.42(a). Draw graphically the SF and BM diagrams.

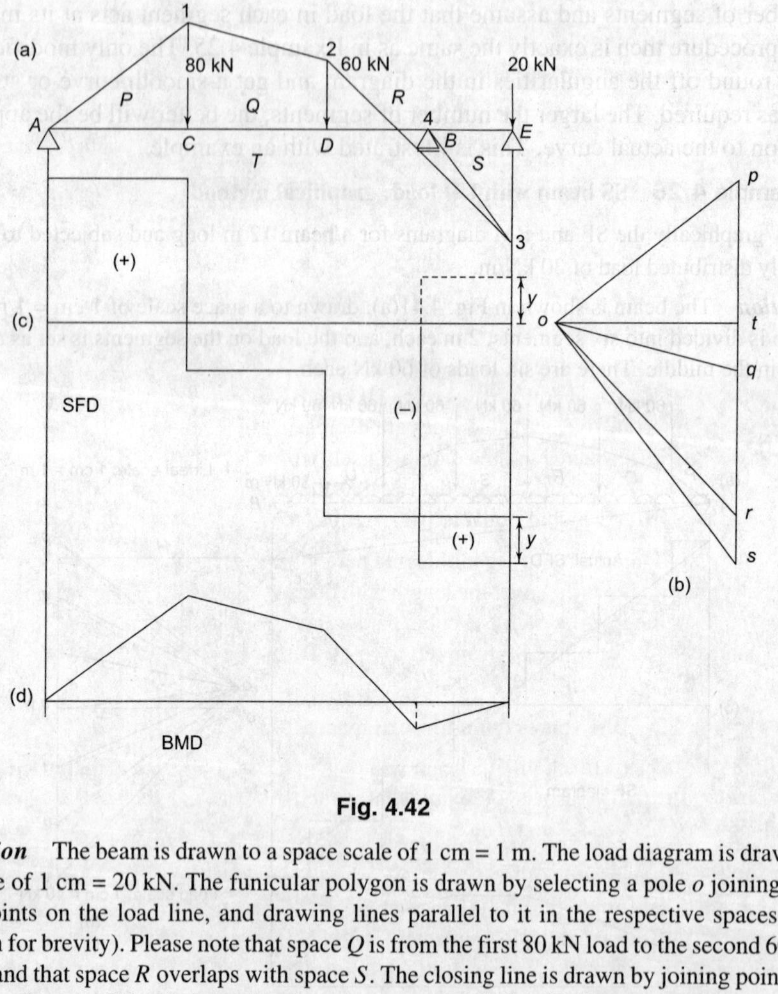

Fig. 4.42

Solution The beam is drawn to a space scale of 1 cm = 1 m. The load diagram is drawn to a scale of 1 cm = 20 kN. The funicular polygon is drawn by selecting a pole *o* joining it to the points on the load line, and drawing lines parallel to it in the respective spaces (not shown for brevity). Please note that space *Q* is from the first 80 kN load to the second 60 kN load, and that space *R* overlaps with space *S*. The closing line is drawn by joining points *A*, 4, and a line parallel to it through pole *o* gives us point *f*. The SF diagram is drawn as usual by drawing horizontal lines through points on the load line in the respective spaces as in Fig. 4.42(c). Figure 4.42(d) shows the BM diagram drawn with the pole in horizontal line with *f*. The funicular polygon is redrawn to make the line *A*-4 horizontal by selecting a new pole o_1 horizontally in line with *f*. A minor modification as shown by dotted lines in the SF and BM diagrams for blocks in space *E* gives us the conventional SF and BM diagrams.

We conclude with a few examples illustrating the use of differential relationships between BM, SF, and load intensity.

Example 4.28 Load and BM diagram from shear force diagram

Shear force diagrams on a 4 m length of two beams are shown in Figures 4.43(a) and (b). Draw the shapes of the load intensity and BM diagrams for the part of the beam shown.

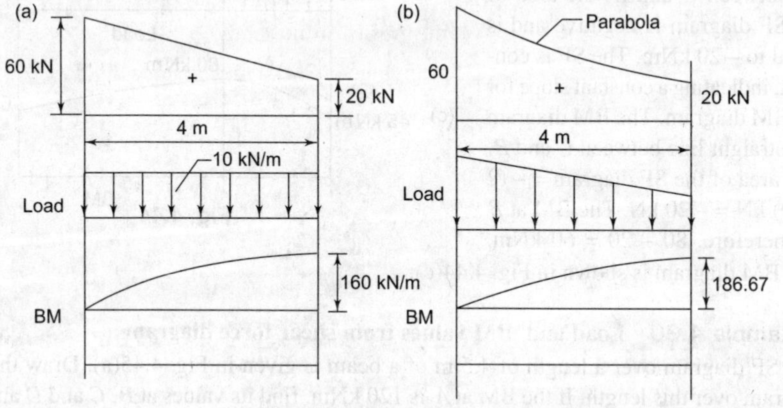

Fig. 4.43

Solution *Case (a)* The SF diagram is a straight line with a negative slope. This shows that the load intensity is uniform. If the load intensity is *w*/*m*, then $4w = 60 - 20$ (from $dF = w\,dx$) $\Rightarrow w = 10$ kN/m. Therefore, there is a UD load 10 kN/m over this part of the beam. The SF diagram is positive, and the values are decreasing. The slope of the BM diagram is thus positive and decreasing and the diagram itself will be as shown. The other deduction that can be made from the available data is that the difference in BM = area of the SF diagram ($dM = F\,dx$), which is equal to $[(60 + 20)/2] \times 4 = 160$ kNm.

Case (b) The SF diagram is a second-degree parabolic curve with decreasing slope. This shows that the load is varying over the length with decreasing intensity. The load diagram is as shown. Also, the BM diagram is a cubic parabolic curve with decreasing slope as given by the decreasing value of the SF from left to right. The slope of the BM diagram is shown in Fig. 4.43(b).

Example 4.29 BM values from force diagram

The SF diagram on a 3 m length of a beam is shown in Fig. 4.44(a). Draw the load diagram for the beam. If the BM at *A* is 45 kNm, what is the BM at *C* and *B*? Show the nature of variation of BM between *A* and *B*.

Solution Between *A* and *C*, the SF drops by $40 - 30 = 10$ kN. The SF diagram is a straight line, showing a uniformly distributed load in the segment *AC*. A change of 10 kN over a length of 1 m indicates that the load intensity is 10 kN/m. At *C*, the SF drops from + 30 kN to – 10 kN at *C*, indicating a concentrated load of 40 kN at *C*. The SF remains constant between *C* and *B*, indicating that there is no load in the segment *CB*. The load diagram is shown in Fig. 4.44(b).

The BM at A is 45 kNm. Since $dM = F\,dx$, the BM at C = BM at A + area of SF diagram between A and C:

$$\text{BM at } C = 45 + \frac{40 + 30}{2} \times 1 = 80$$

kN ms

The variation of BM is parabolic between A and C, with decreasing positive slope.

Between C and B, the area of the SF diagram is negative and is equal to – 20 kNm. The SF is constant, indicating a constant slope for the BM diagram. The BM diagram is a straight line between C and B. The area of the SF diagram = – (2 × 10) kN = – 20 kN. The BM at B is, therefore, 80 – 20 = 60 kNm. The BM diagram is shown in Fig. 4.44(c).

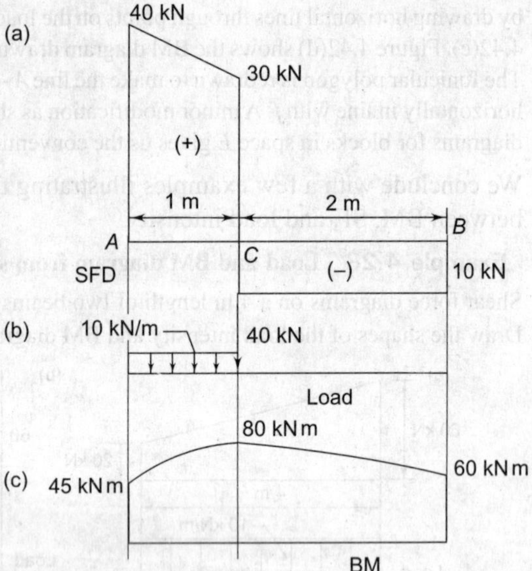

Fig. 4.44

Example 4.30 Load and BM values from shear force diagram

The SF diagram over a length of 4.5 m of a beam is given in Fig. 4.45(a). Draw the load diagram over this length. If the BM at A is 120 kNm, find its values at B, C and D and also show its variation.

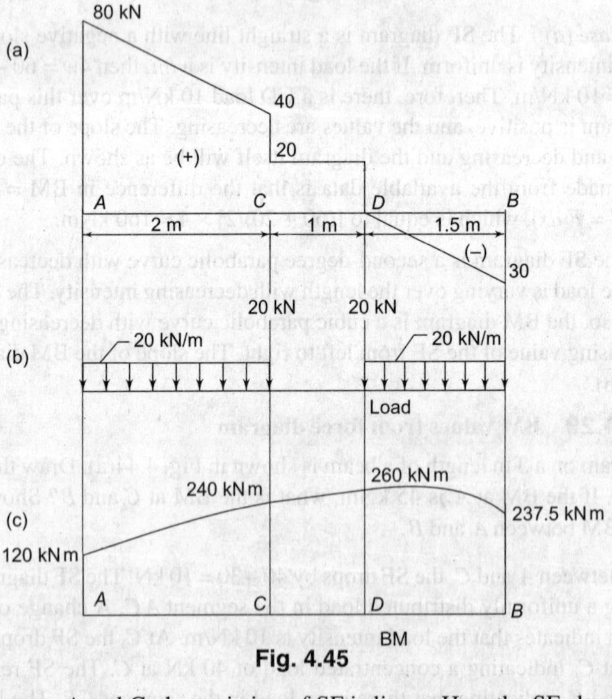

Fig. 4.45

Solution Between A and C, the variation of SF is linear. Here the SF changes by 40 kN over a length of 2 m. Therefore, this segment has a UD load of 40/2 = 20 kN/m. At C, the SF changes from 40 to 20 kN, indicating a concentrated load of 20 kN at C. The SF remains constant between C and D, there is no load over the length CD. At D, the SF drops by 20 kN, indicating a point load of 20 kN at D. Between D and B, the SF drops linearly to – 30 kN over a length of 1.5 m. The UD load over this length has a magnitude of 30/1.5 = 20 kN/m. The load diagram is shown in Fig. 4.45(b).

The BM values can be calculated from the area of the SF diagram from the relation $dM = F\,dx$:

$$\text{BM at } A = 120 \text{ kNm}$$

$$\text{BM at } C = 120 + \frac{80 + 40}{2} \times 2 = 240 \text{ kN m}$$

$$\text{BM at } D = 240 + 20 \times 1 = 260 \text{ kN m}$$

$$\text{BM at B} = 260 - 30 \times \frac{1.5}{2} = 237.5 \text{ kN m}$$

The variation of BM is shown in Fig. 4.45(c).

Example 4.31 Load and BM diagram from shear force diagram for SS beam

The SF diagram of a simply supported beam of 10 m span is shown in Fig. 4.46(a). Draw the load and BM diagrams for the beam.

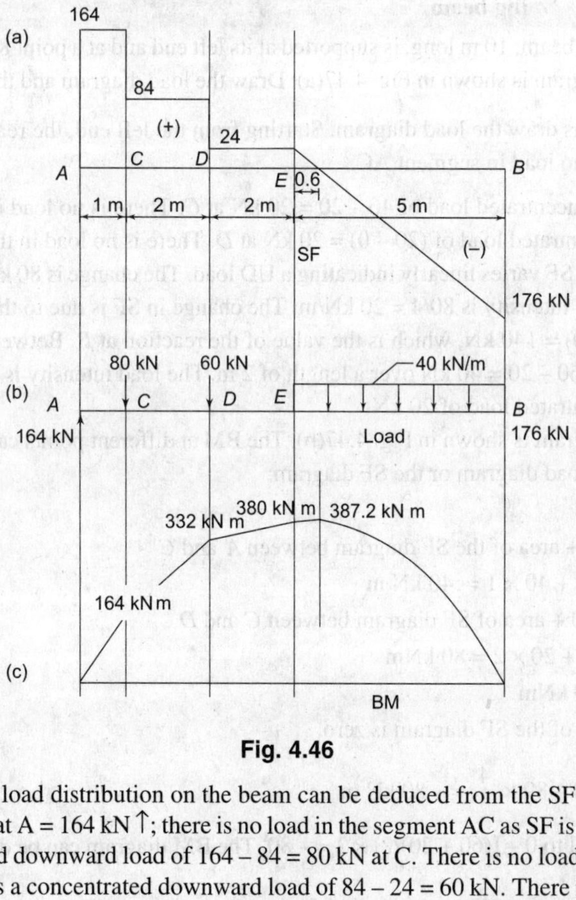

Fig. 4.46

Solution The load distribution on the beam can be deduced from the SF diagram as follows. Reaction at A = 164 kN ↑; there is no load in the segment AC as SF is constant. There is a concentrated downward load of 164 – 84 = 80 kN at C. There is no load between C and D. At D, there is a concentrated downward load of 84 – 24 = 60 kN. There is no load in the segment DE. Between E and B, the SF varies linearly indicating a UD load. The variation over a length of 5 m is 24 – (– 176) = 200 kN. The load intensity is 200/5 = 40 kN/m. This can also be found from the fact that the load is equal 24/0.6 = 40 kN/m. The load diagram is shown in Fig. 4.46(b). The reaction at B is 176 kN ↑.

The BM at different points can now be calculated.

BM at A = 0

BM at C = 164 × 1 = 164 kNm

BM at D = 164 × 3 – 80 × 2 = 332 kNm

or

164 + 84 × 2 = 332 kNm

BM at E = 164 × 5 – 80 × 4 – 60 × 2 = 380 kNm

or

332 + 24 × 2 = 380 kNm

The BM is maximum at the point of zero shear. The point of zero shear is at a distance of 24/40 from E, that is 0.6 m from E. Calculating from the right, maximum BM = 176(5 – 0.6) – 40 × $(4.4)^2/2$ = 387.2 kNm. The BM diagram is shown in Fig. 4.46(c).

Example 4.32 Load and BM diagram from shear force diagram for an over hanging beam

An overhanging beam, 10 m long, is supported at its left end and at a point 8 m from the left end. The SF diagram is shown in Fig. 4.47(a). Draw the load diagram and the BM diagram.

Solution Let us draw the load diagram. Starting from the left end, the reaction at A = 40 kN ↑. There in no load in segment AC.

There is a concentrated load of 40 – 20 = 20 kN at C. There is no load in segment CD. There is a concentrated load of (20 – 0) = 20 kN at D. There is no load in the segment DE. From E to B, the SF varies linearly indicating a UD load. The change is 80 kN over a length of 4 m. The load intensity is 80/4 = 20 kN/m. The change in SF is due to the reaction at B, and is 60 – (– 80) = 140 kN, which is the value of the reaction at B. Between B and F, the change in SF is 60 – 20 = 40 kN over a length of 2 m. The load intensity is 20 kN/m. At F, there is a concentrated load of 20 kN.

The load diagram is shown in Fig. 4.47(b). The BM at different points can be calculated either from the load diagram or the SF diagram:

BM at A = 0

BM at C = 0 + area of the SF diagram between A and C

= 0 + 40 × 1 = 40 kN m

BM at D = 40 + area of SF diagram between C and D

= 40 + 20 × 2 = 80 kNm

BM at E = 80 kNm

because the area of the SF diagram is zero.

$$\text{BM at } B = 80 - 80 \times \frac{4}{2} = -80 \text{ kN m}$$

This is also equal to 0 – [(60 + 20)/2] × 2 = – 80. The BM diagram can be drawn as in Fig. 4.47(c).

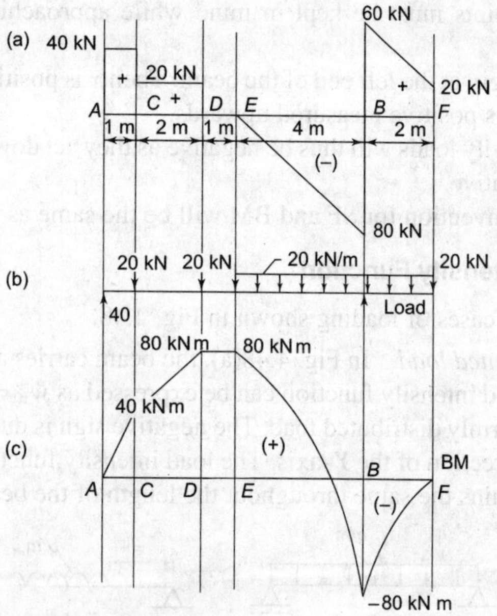

Fig. 4.47

4.11 SINGULARITY FUNCTION APPROACH FOR SF AND BM

We have seen the differential relationships between w_x (load intensity), SF (shear force F_x), and BM (bending moment M_x). These are $dM/dx = F_x$ and $dF/dx = w_x$.

One can thus intuitively understand that if the load intensity can be expressed as a function of x, then F and M can be obtained by integration of that function.

If $w_x = f(x)$, then

$$F_x = \int f(x)dx + C_1 \text{ and } M_x = \int [\int f(x)dx + C_1]dx + C_2,$$

C_1, C_2 being constants of integration.

To put it in another way,

$$dM/dx = F_x;\ dF/dx = d[dM/dx]/dx = d^2M/dx^2 = w_x,$$

where w_x is the load intensity function at x. To simplify the notation, we use M' for dM/dx and M'' for d^2M/dx^2.

Now, $\quad d^2M/dx^2 = M'' = w_x$

Integrating,

$$\int dM/dx = M' = F_x = \int w_x\, dx + C_1,$$

where F_x is the shear force at x.

Integrating again, bending moment at x,

$$M_x = \int [\int w_x\, dx + C_1]\, dx + C_2$$

The constants can be found from the boundary conditions, at supports, etc. Thus, we arrive at equations for SF and BM.

The following points must be kept in mind while approaching the problem in this way.

(i) Take the origin at the left end of the beam. Then x is positive measured to the right and y is positive measured upwards.

(ii) Normal gravity loads will thus be negative as they act downwards and w will thus be negative.

(iii) The sign convention for SF and BM will be the same as given earlier.

4.11.1 Load Intensity Function

Consider the five cases of loading shown in Fig. 4.48.

Uniformly distributed load In Fig. 4.48(a), the beam carries a uniformly distributed load. The load intensity function can be expressed as $w_x = -q$, where q is the value of the uniformly distributed load. The negative sign is due to the load acting in the negative direction of the Y-axis. The load intensity function in this case is a constant and remains the same throughout the length of the beam.

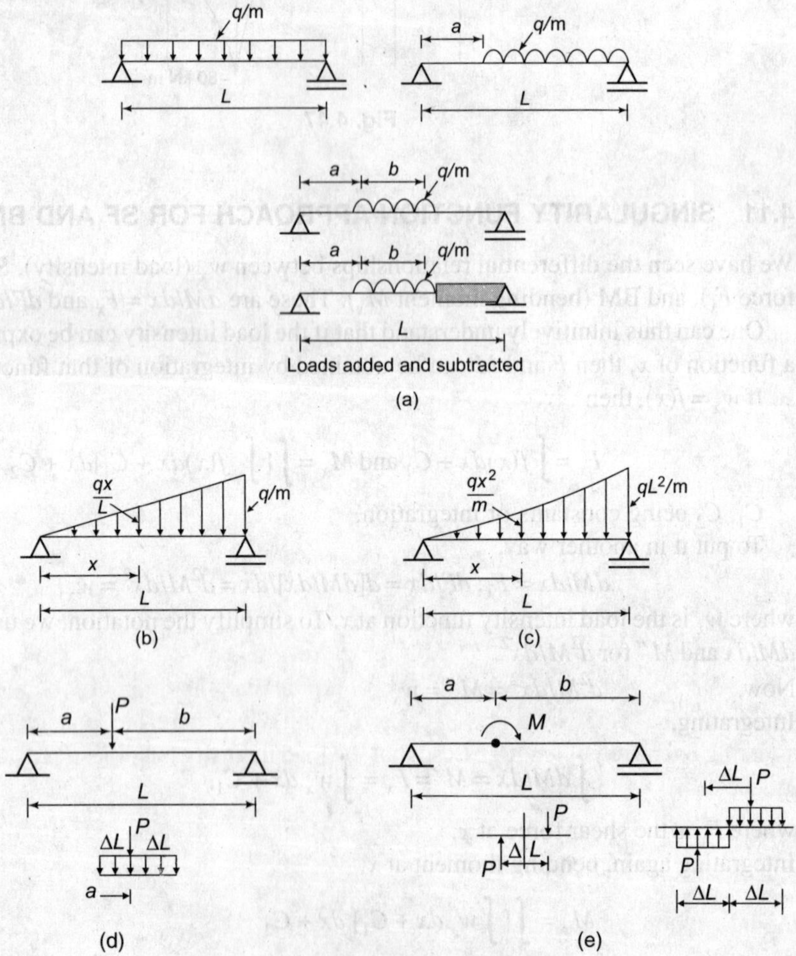

Fig. 4.48 Load intensity functions

The other cases of this load are shown in Fig. 4.48(a). If the UD load is only on a part of the beam, extending to the right to full length, the load intensity function becomes $w_x = 0$ $(0 < x < a)$ and $w_x = -q$ $(a < x < L)$. The load intensity function is thus expressed as $w_x = -q(x - a)^0$, with the condition that w becomes zero for $x < a$. Note that $(x - a)^0$ is unity and dimensional consistency is maintained. This condition is the same as saying that the bracketed term becomes zero when it is negative.

If the load is only on part of the beam not extending to the right, then a solution is found by adding equal downward and upward loads as shown in Fig. 4.48(a). Both these loads can be tackled as in the previous case and the load intensity function can be written as $w_x = -q(x - a)^0 + q(x - a - b)^0$, with the proviso that the terms become zero when the bracketed terms become negative.

To be consistent with this notation, the load intensity function for UD load acting over whole length is also written as $w_x = -q(x - 0)^0$.

Uniformly varying load Figure 4.48(b) shows a uniformly varying load varying from 0 at the left end to q at the right end. The load intensity at any section x from the origin is $w = -q(x/L)$ and remains valid for the whole beam from $x = 0$ to $x = L$. Note that the load intensity is not constant but varies with x. To be consistent with the earlier notation, this load will be written as $-q(x - 0)^1/L$.

In case the load does not start from the left end or does not extend up to the right end or both, the case can be tackled in a similar way as shown in the case of UD load.

Non-uniformly varying load Consider the case shown in Fig. 4.48(c). The load intensity function in this case is $w_x = -qx^2$. The load intensity function depends on x but the expression is valid for the whole beam. Any variation in this load for other cases can be tackled in a similar way as shown in the earlier two cases. The load intensity function is expressed as $w_x = -q(x - 0)^2$.

Concentrated load or point load For the case shown in Fig. 4.48(d), the bending moment equation can be written as

$$M_x = R_A x \ (0 < x < a) \text{ and } M_x = R_A x - P(x - a) \ (a < x < L)$$

In order to express this as a function valid for the whole beam, we have to introduce some additional concepts.

A concentrated load is a load acting at a point and theoretically the load intensity is infinite (as the load acts over zero length). We, therefore, consider the load as acting over a small length ΔL, so that the load intensity becomes $P/\Delta L$. The load P is obtained by integration as

$$\lim_{\Delta L \to 0} \int (P/\Delta L)dx = P,$$

the integration being done over $(a - \Delta L/2)$ to $(a + \Delta L/2)$.

The approach now is to write the concentrated load as a distributed load q as $q = P(x - a)^{-1}$. Note that the expression on the right has the same dimension as load per length. This is called a *singularity function* because the expression becomes infinity at $x = a$ and is zero everywhere else.

The integration of the singularity function is done by defining the integral as follows:

$$\int_0^x P\,(x - a)^{-1} dx = P(x - a)^0$$

This is done to ensure that as the function is a bounded function within the limits ΔL, it has a special meaning. One should remember to integrate the function as shown above. Some identification of the singularity function may be made to remind of the nature of the function by writing $P < x–a >$ (angular brackets) or using any other form of identification.

Couple load The fifth case in Fig. 4.48(e) is a couple load again giving a singularity function. The load in this case is a couple μ acting at a point. Again we convert the couple into two large loads (unlike parallel loads) acting at a small distance ΔL apart. This gives $P\Delta L = \mu$.

As the concentrated loads P themselves are singularity functions, we convert these two loads into distributed loads, distributed over small lengths ΔL as shown in Fig. 4.48(e). For the couple load, the load intensity function can be written as $w = \mu\,(x - a)^{-2}$. This function is also a singularity function, being μ at $x = a$ and zero elsewhere. The integration of this function is also done in the same way.

$$\int_0^x \mu(x - a)^{-2} dx = \mu(x - a)^{-1} \text{ and } \int_0^x \mu\,(x - a)^{-1} dx = \mu(x - a)^0$$

Note that the load intensity function is dimensionally correct being $\text{Nm/m}^2 = \text{N/m}$.

Table 4.3 summarizes the load intensity functions for all loads.

Table 4.3 Load intensity function

S.No.	*Load*	*Function*
1	UD Load (a) 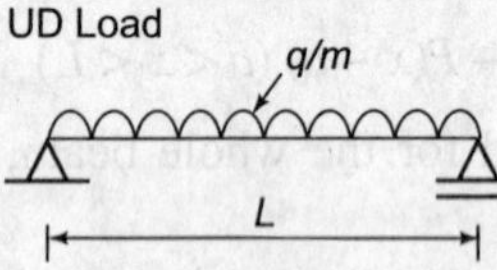 (b) 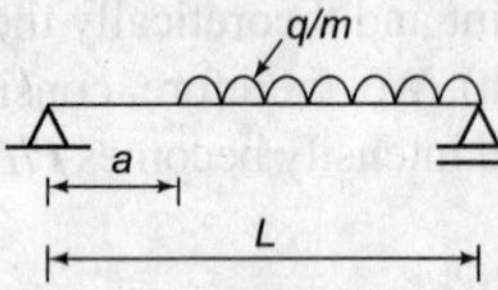 (c) 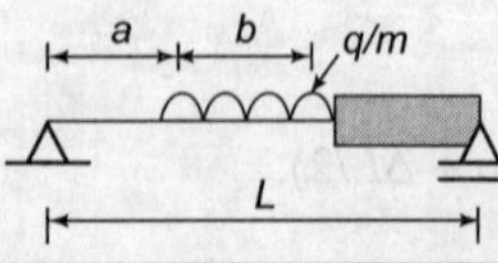	(a) $W_x = -q\ (x - 0)^0$ (b) $W_x = -q\ (x - a)^0$ (c) $W_x = -q\ (x - a)^0 + q(x - a - b)^0$ (Add and subtract the loads to the right end.)

(contd)

Table 4.3 (*contd*)

S.No.	Load	Function
2	Point load 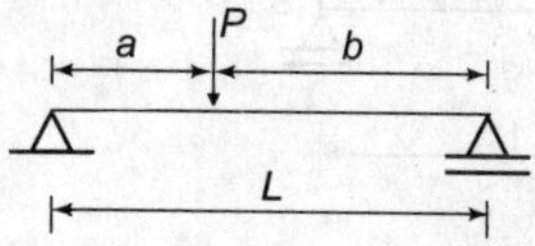	$W_x = -P\,(x-a)^{-1}$
3	Uniformly varying load (a) 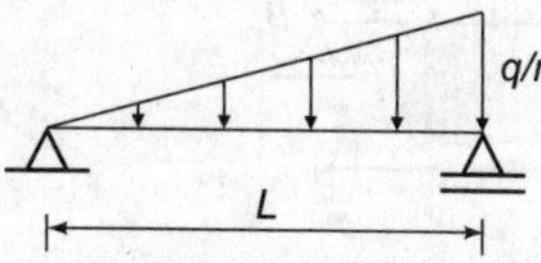 (b) 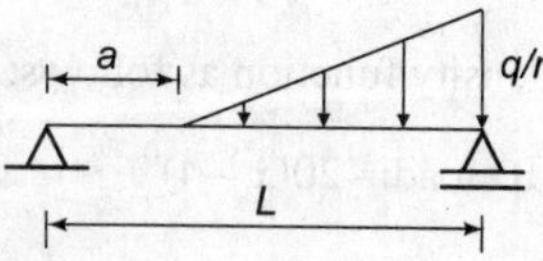 (c) 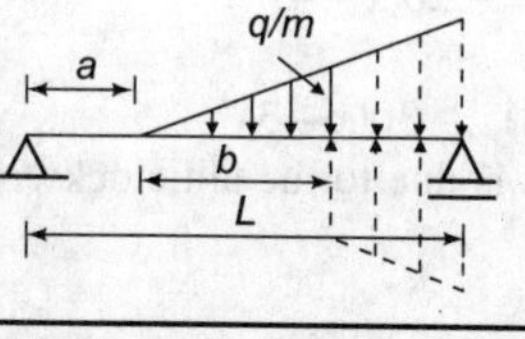	(a) $W_x = -q\,\dfrac{(x-0)^1}{L}$ (b) $W_x = -q\,\dfrac{(x-a)^1}{(L-a)}$ (c) $W_x = -\dfrac{q}{b}\,\dfrac{(L-a)(x-a)^1}{L-a} + q(x-a-b)^0 + \dfrac{q}{b}(x-a-b)^1$ (Adding and subtracting loads as shown. The added upward load is taken in two parts: a UD load and a triangular load.)
4	Couple load 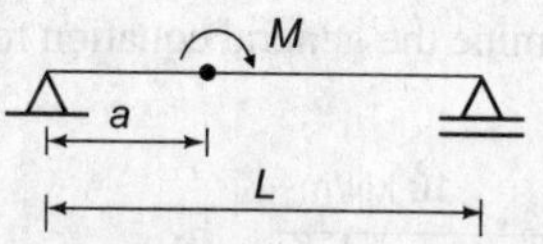	(a) $W_x = M\,(x-a)^{-2}$

The general rule for integration of singularity functions is

$$\int (x-a)^n = \frac{(x-a)^{n+1}}{n+1} \quad \text{for } n \geq 0$$

$$= (x-a)^{n+1} \quad \text{for } n < 0$$

Overhanging beams In the case of overhanging beams, shown in Fig. 4.49, the reactions are not at the ends of the beam. The reactions themselves are singularity functions and should be treated in the same way.

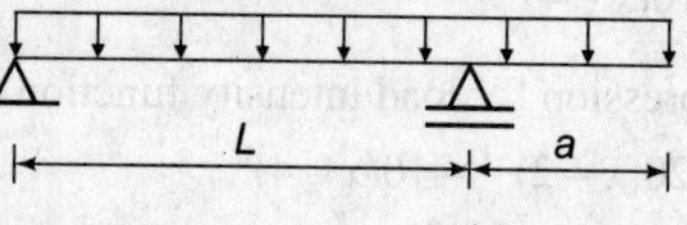

Fig. 4.49 Overhanging beams

The following examples illustrate the use of this method to find SF and BM in beams.

Example 4.33 Singularity functions: load intensity

Write the load intensity function w_x for the beams carrying loads as shown in Fig. 4.50.

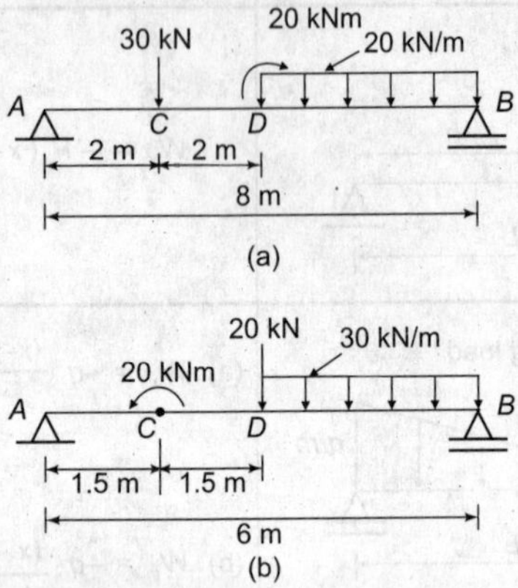

Fig. 4.50

Solution For Fig. 4.50(a), we write the load intensity function as follows:

Point load: $-30(x-2)^{-1}$; Couple: $+20(x-4)^{-2}$; UD load: $-20(x-4)^0$

Thus, w_x can be written as

$$w_x = -30(x-2)^{-1} + 20(x-4)^{-2} - 20(x-4)^0$$

Similarly, for case (b), we can write

$$w_x = -20(x-1.5)^{-2} - 20(x-3)^{-1} - 30(x-3)^0$$

Note the negative sign for the couple load. This is due to the anticlockwise nature of the couple.

Example 4.34 Singularity functions: SF and BM

For the beam loaded as shown in Fig. 4.51, determine the general equation for SF and BM. Find the SF and BM at C and D.

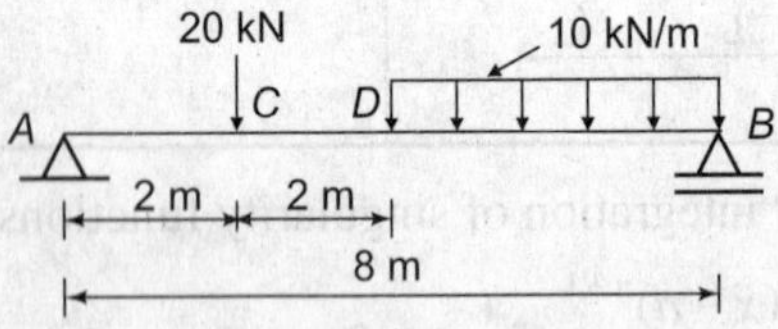

Fig. 4.51

Solution The concentrated load gives a singularity function. The load intensity corresponding to this load is $w(x) = -20(x-2)^{-1}$ (being ↓). The uniformly distributed load goes up to the right end and does not require any special technique. The load intensity function can be written as $w(x) = -10(x-4)^0$.

Having decided the expression for load intensity function for w, we can write

$$M'' = -20(x-2)^{-1} - 10(x-4)^0$$

Integrating, $$M' = F = -20(x-2)^0 - 10(x-4)^1 + C^1$$

Integrating again,

$$M = -20(x-2) - 10(x-4)^2/2 + C_1x + C_2$$

There are two conditions to determine the constants:

(i) At $x = 0, M = 0$

(ii) At $x = 8, M = 0$

Now,

(i) $x = 0, M = 0$ gives $C_2 = 0$

(ii) $x = 8, M = 0$ gives $0 = -20 \times 6 - 10(8-4)^2/2 + C_1 \times 8;\ C_1 = +25$

Note that C_1 given the reaction R_A at A.

The general equations for SF and BM can be written as

SF at $x = -20(x-2)^0 - 10(x-4)^1 + 25$

BM at $x = -20(x-2) - 10(x-4)^2/2 + 25x$

The SF and BM at any section can be found from these two equations as follows.

SF at C $(x = 2) = -20 + 0 + 25 = 5$ kN (when $x > 2$) and

$= -0 - 0 + 25 = 25$ kN (when $x < 2$)

SF at D $(x = 4) = 5$ kN

BM at C $(x = 2 \text{ m}) = -20(x-2) + 25 \times 2 = 50$ kNm

BM at D $(x = 4 \text{ m}) = -20(4-2) - 0 + 25 \times 4 = 60$ kNm

Example 4.35 Singularity functions: SF and BM

An SS beam of 6 m span carries a UD load of 18 kN/m for the left half of span and a couple of 36 kNm at 4.5 m from the left end. Derive the general equation for SF and BM using singularity functions. Determine the position and magnitude of maximum BM in the beam.

Solution The beam with the loading is shown in Fig. 4.52. As the UD load exists only on the left half, we add and subtract loads as shown in Fig. 4.52(b).

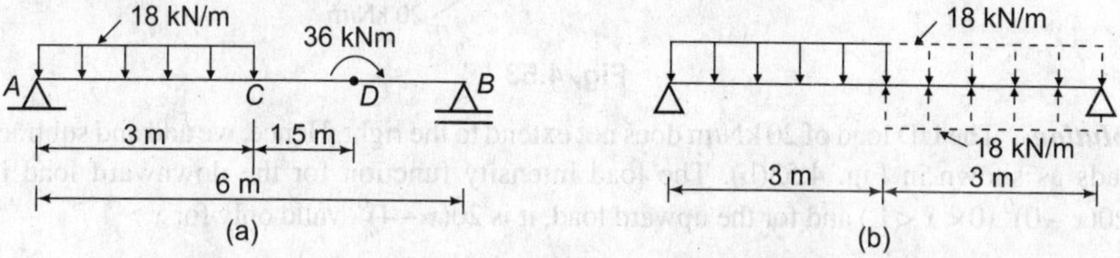

Fig. 4.52

We can now write the load intensity function as

$$w(x) = -18(x-0)^0 + 18(x-3)^0 + 36(x-4.5)^{-2}$$

Therefore,

$$M'' = -18(x-0)^0 + 18(x-3)^0 + 36(x-4.5)^{-2}$$

Integrating,

$$M' = -18(x-0)^1 + 18(x-3)^1 + 36(x-4.5)^{-1} + C_1$$
$$M = -18(x-0)^{2/2} + 18(x-3)^{2/2} + 36(x-4.5)^0 + C_1 x + C_2$$

As $M = 0$ at $x = 0$, $C_2 = 0$

Also $M = 0$ at $x = 6$ m

Therefore,

$$0 = -18(6-0)^2/2 + 18(6-3)^2/2 + 36(6-4.5)^0 + C_1 \times 6;\ C_1 = 34.5$$

The general equation for BM and SF are

SF at $x = -18(x-0)^1 + 18(x-3)^1 + 36(x-4.5)^{-1} + 34.5$

BM at $x = -18(x-0)^2/2 + 18(x-3)^2/2 + 36(x-4.5)^0 + 34.5x$

Position of maximum BM can be found by equating SF = 0.

$-18(x-0)^1 + 18(x-3)^1 - 36(x-4.5)^{-1} + 34.5 = 0; -18x + 34.5 = 0$
(assuming that SF becomes zero at $x < 3$ m.)

So, $x = 1.92$ m

Maximum BM is at 1.92 m from A.

Substituting this value of x in the BM equation,

Max BM $= -18(1.92 - 0)^2/2 + 34.5 \times 1.92 = 33$ kNm

Example 4.36 Singularity functions: SF and BM

An SS beam is loaded as shown in Fig. 4.53. Derive a general equation for SF and BM using singularity functions.

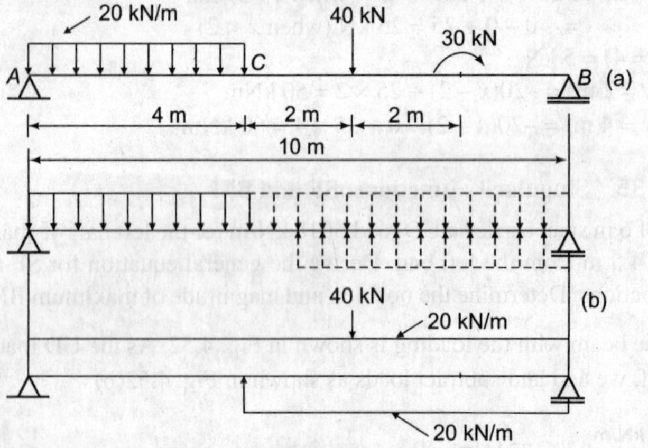

Fig. 4.53

Solution The UD load of 20 kN/m does not extend to the right. Hence, we add and subtract loads as shown in Fig. 4.53(b). The load intensity function for the downward load is $-20(x-0)^0$ $(0 < x < L)$ and for the upward load, it is $20(x-4)^0$, valid only for $x > 4$.

The concentrated load gives a singularity function $-40(x-6)^{-1}$.

The couple load gives $30(x-8)^{-2}$.

The total load intensity function is

$$w(x) = -20(x-0)^0 + 20(x-4)^0 - 40(x-6)^{-1} + 30(x-8)^{-2}$$

We can thus write

$$M'' = -20(x-0)^0 + 20(x-4)^0 - 40(x-6)^{-1} + 30(x-6)^{-2}$$

$$M' = -20(x-0)^1 + 20(x-4)^1 - 40(x-6)^0 + 30(x-8)^{-1} + C_1$$

$$M = -20(x-0)^2/2 + 20(x-4)^2/2 - 40\,(x-6)^1 + 30(x-8)^0 + C_1x + C_2$$

The constants can be found from the conditions:

(i) $x = 0, M = 0$

(ii) $x = 10, M = 0$

The first condition gives $C_2 = 0$. The second condition gives

$$0 = -20(10-0)^2/2 + 20(10-4)^2/2 - 40(10-6)^1 + 30(10-8)^0 + C_1(10) = 0$$

or $C_1 = 77$ kN, which is the reaction at A.

The general SF and BM equations are

SF at $x = M' = -20(x-0)^1 + 20(x-4)^1 - 40(x-6)^0 + 30\,(x-8)^{-1} + 77$

BM at $x = M = -20(x-0)^2/2 + 20(x-4)^2/2 - 40(x-6)^1 + 30(x-8)^0 + 77x$

Example 4.37 Singularity functions: SF and BM

An SS beam carries a uniformly varying load for the whole length of the beam. Determine the SF and BM equations and find the position and magnitude of the maximum BM.

Solution The beam with the load is shown in Fig. 4.54.

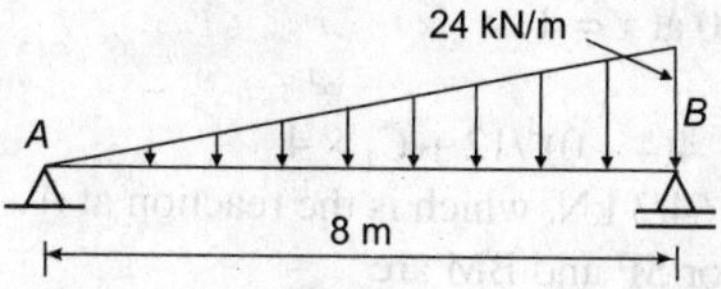

Fig. 4.54

The load intensity function is $-q(x-0)^1/L$

Therefore,

$M'' = -24(x-0)^1/8$; $M' = -24(x-0)^2/(2\times 8) + C_1$; and $M = -24(x-0)^3/(6\times 8) + C_1x + C_2$

As $M = 0$ at $x = 0$, $C_2 = 0$.

Also $M = 0$ at $x = 8$ m

Therefore,

$$0 = -24(8-0)^3/(6\times 8) + C_1 8;\ C_1 = 32$$

This is the reaction at A. The general equations for SF and BM are

SF at $x = -24(x-0)^2/16 + 32$; BM at $x = -24(x-0)^3/(48) + 32x$

Position of maximum BM is obtained by equating SF to zero.

$-24(x-0)^2/(2\times 8) + 32 = 0$; $x = \sqrt{[64/3]} = 8/\sqrt{3}$ m $= 4.62$ m

Max BM $= -24(4.62-0)^3/48 + 32\times 4.62 = 98.53$ kNm

Example 4.38 Singularity functions: SF and BM

An SS supported beam carries loads as shown in Fig. 4.55. Derive the general equations for SF and BM.

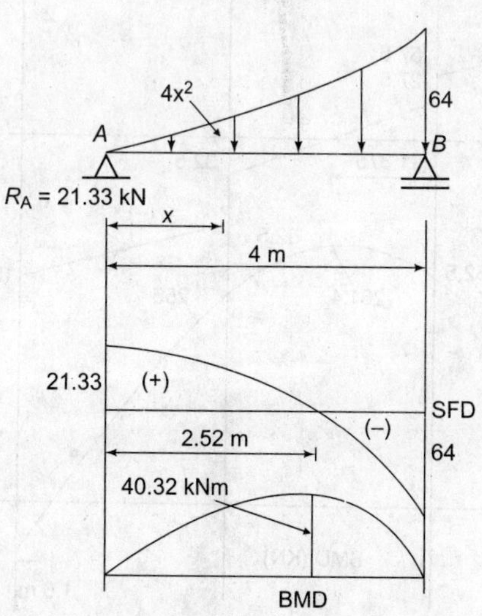

Fig. 4.55

Solution The load intensity function can be written as $4(x-0)^2$.

Therefore,

$$M'' = -4(x-0)^2$$

Integrating,

$$M' = SF = -4(x-0)^3/3 + C_1$$

$$M = BM = -4(x-0)^4/12 + C_1x + C_2$$

As $M = 0$ at $x = 0$, $C_2 = 0$

Also $M = 0$ at $x = 4$

Therefore,

$$0 = -4(4-0)^4/12 + C_1 \times 4$$

or $C_1 = 64/3$ kN, which is the reaction at A.

The general equations for SF and BM are

$$F_x = \text{SF at } x = -4(x-0)^3/3 + 64/3 \text{ and}$$

$$M_x = \text{BM at } x = -4(x-0)^4/12 + 64x/3$$

Maximum BM is at the point where SF is zero.

$$-4(x-0)^3/3 + 64/3 = 0; x = 2.52 \text{ m}$$

Putting this value of x in the BM equation, we get the maximum BM.

$$M_{max} = -4(2.52-0)^4/12 + 63 \times 2.52/3 = 40.32 \text{ kNm}$$

The SF and BM diagram can be drawn as shown in Fig. 4.55.

Example 4.39 Singularity functions: SF and BM in overhanging beam

An overhanging beam carries loads as shown in Fig. 4.56. Derive a general equation for SF and BM using singularity functions.

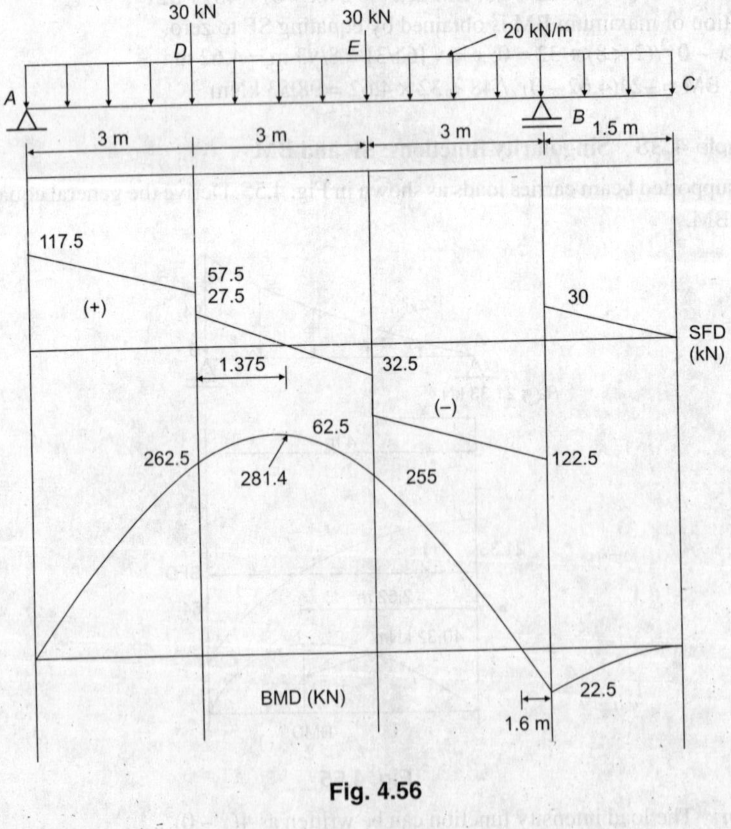

Fig. 4.56

Solution In this case, we consider the reactions as point loads and singularity functions. No other constants need to be introduced during integration. The load intensity function can be written as

$$w_x = R_A(x-0)^{-1} - 20(x-0)^0 - 30(x-3)^{-1} - 30(x-6)^{-1} + R_B(x-9)^{-1}$$

Therefore,

$$M'' = R_A(x-0)^{-1} - 20(x-0)^0 - 30(x-3)^{-1} - 30(x-6)^{-1} + R_B(x-9)^{-1}$$
$$M' = R_A(x-0)^0 - 20(x-0)^1 - 30(x-3)^0 - 30(x-6)^0 + R_B(x-9)^0$$
$$M = R_A(x-0)^1 - 20(x-0)^2/2 - 30(x-3)^0 - 30(x-6)^0 + R_B(x-9)^0$$

The two conditions we can use to find the reactions are SF = 0 at $x = 105$ m and BM = 0 at $x = 10.5$. Therefore,

$$0 = R_A(10.5-0)^0 - 20(10.5-0)^1 - 30(10.5-3)^0 - 30(10.5-6)^0 + R_B(10.5-9)^0$$

This gives $R_A + R_B = 270$

Also, from the second condition,

$$0 = R_A(10.5 - 0)^1 - 20(10.5 - 0)^2/2 - 30(10.5 - 3)^1 - 30(10.5 - 6)^1 + R_B(10.5-9)^1$$

or $\quad 10.5R_A + 1.5R_B = 1102.5 + 225 + 135 = 1462.5$

Solving these two equations,

$$R_A = 117.5 \text{ kN and } R_B = 152.5 \text{ kN}$$

General SF and BM equations at x from A are

$$F_x = 117.5(x-0)^0 - 20(x-0)^1 - 30(x-3)^0 - 30(x-6)^0 + 152.5(x-9)^0$$
$$M_x = 117.5(x - 0)^1 - 20(x - 0)^2/2 - 30(x - 3)^1 - 30(x - 6)^1 + R_B(x - 9)^1$$

The shear force and bending moment diagrams are shown in Fig. 4.56. ■

Summary

While designing beams, it is essential to determine the effects of external loads acting on them. When a section is passed through any part of a beam and the equilibrium of either part obtained by the section is considered, it is found that the internal stress resultants at the section must be an axial force (AF), a shear force (SF), and a bending moment (BM) developed as stress resultants. The effects of external loads are exactly equal and opposite to these stress resultants.

The AF at any section of a beam is the algebraic sum of the forces, parallel to the axis of the beam, acting or either side of the section.

The SF at any section of a beam is the algebraic sum of the forces, transverse to the longitudinal axis of the beam, acting on either side of the section.

The bending moment (BM) at any section of a beam is the algebraic sum of the moments of all the forces, acting on either side of the section, about the section.

It is necessary to use a consistent sign convention for these quantities. BM, SF, and AF diagrams are drawn to know the variation of the quantities along the length of the beam.

The two differential relationships $dF/dx = w$ and $dM/dx = F$ are helpful in drawing BM and SF diagrams. The maximum BM occurs at a point in the beam where the SF is zero. The point of contraflexure is a point along the length of the beam where the BM changes sign (or becomes zero).

Statically determinate beams and frames are analysed using the same principles to draw the AF, SF, and BM diagrams. One can also use graphical methods to draw SF and BM diagrams. The funicular polygon drawn for the force system acting on the beam also represents the BM diagram for the beam.

General SF and BM equations can also be derived from the load intensity functions. But a point load and a couple lead to singularity functions, which have different rules for integration. The singularity function approach uses the load intensity function, and integrating the function, the equations for SF and BM are obtained.

Exercises

Review Questions

1. Define the terms axial force, shear force, bending moment, and point of contraflexure.
2. Explain the sign convention for BM and SF. Explain why it is necessary to adopt a sign convention for BM like 'clockwise moment from the left or anticlockwise moment from the right is positive'.
3. The BM in a beam is maximum or minimum where the SF is zero. Is the converse true? Why?
4. If Fig. 4.57 shows the BM over a part of a beam, what can you say about the loading over this length?

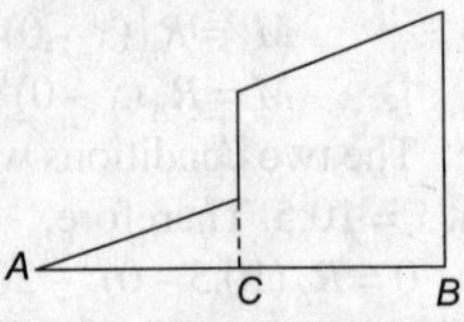

Fig. 4.57

5. The SF diagrams over a 3 m length of beam segments are shown in Figures 4.58(a) to (f).
   Determine the loading on the segment, the BM at *B* (the right end of the segment), and draw the shape of the BM diagram. The SF is given in kN in all the figures.

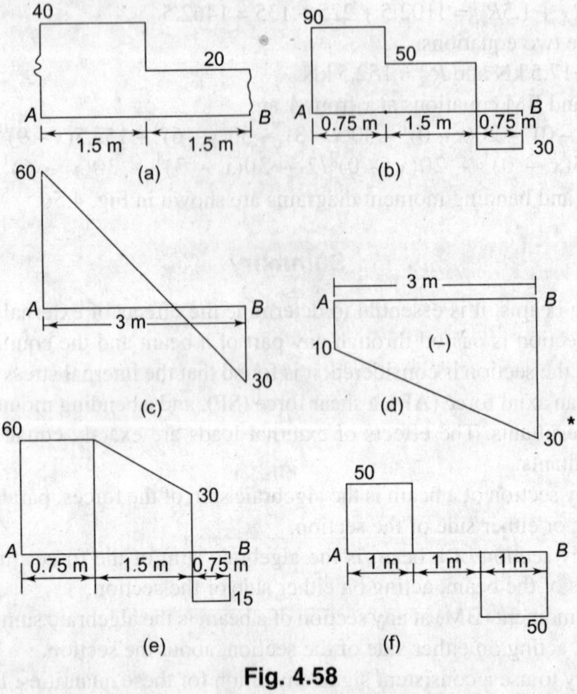

Fig. 4.58

6. Explain whether or not the beams shown in Fig. 4.59 are statically determinate.

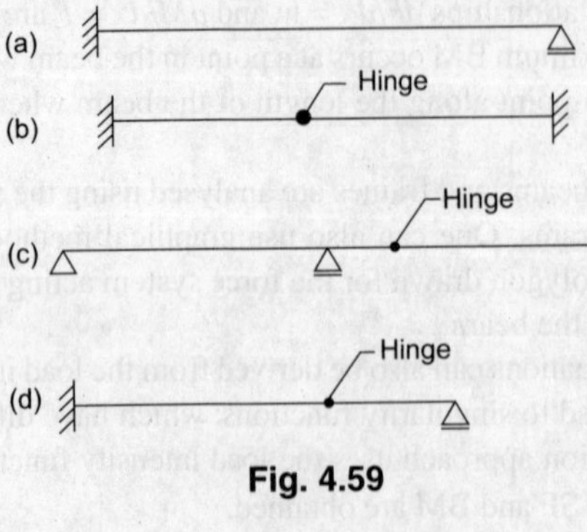

Fig. 4.59

7. What is the value of the reaction at A in the beam shown in Fig. 4.60? What will be the shape of the BM diagram for this beam?

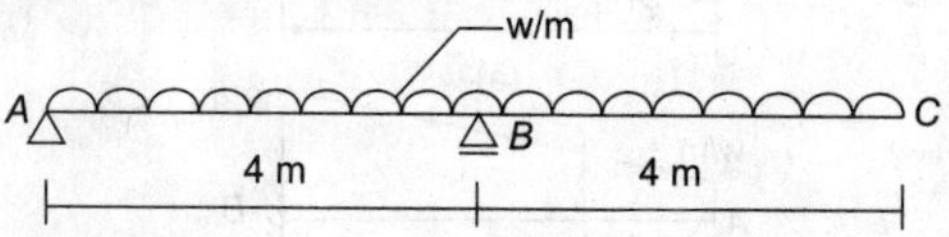

Fig. 4.60

8. For the cantilever beam loaded as shown in Fig. 4.61, draw approximately the shape of the SF and BM diagrams.

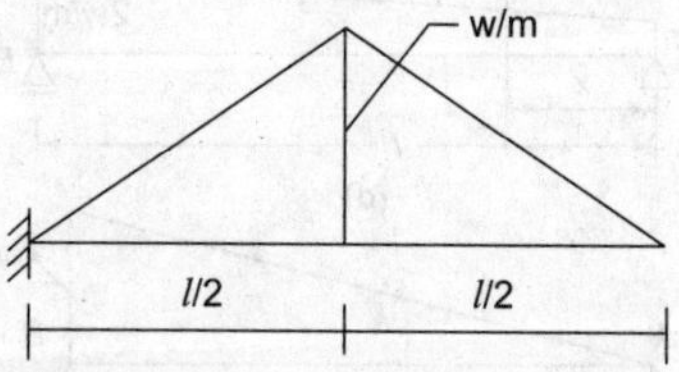

Fig. 4.61

9. Answer the following without making any calculations with reference to Fig. 4.62.
   In (a), what is the SF at the left end?
   In (b), what is the BM between C and D?
   In (c), what is the SF and BM between C and D?
   In (d), what is the value of the SF at x m from A?

In each of these cases, draw the SF and BM diagrams for the beams approximately.

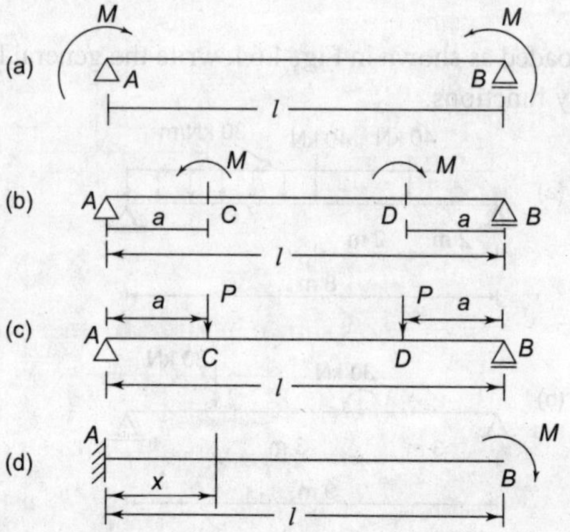

Fig. 4.62

10. In each of the cases shown in Fig. 4.63, give the general equation for BM and SF at a section x from A.

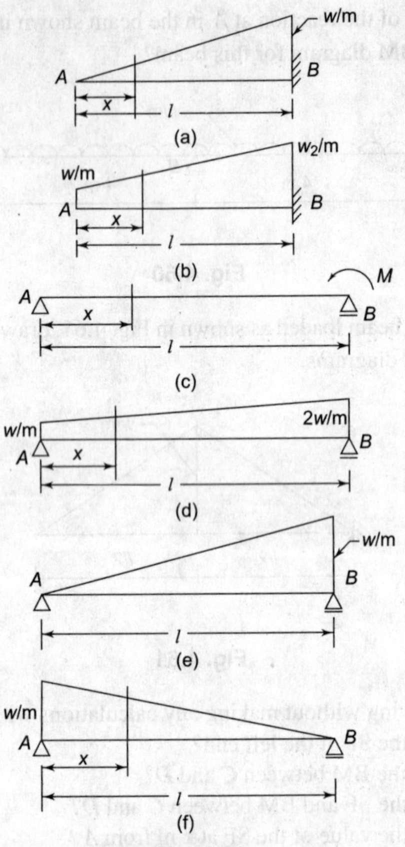

Fig. 4.63

11. Explain the concept of singularity functions with examples of point load and couple load.
12. For the beams loaded as shown in Fig. 4.64, write the general load intensity function using singularity functions.

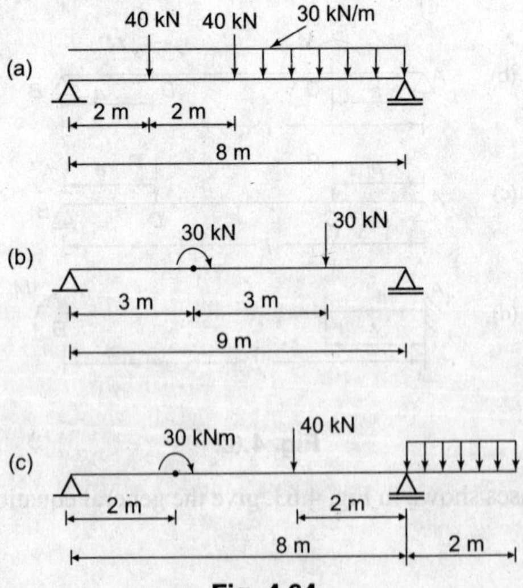

Fig. 4.64

Problems

1. A cantilever, 3 m long, carries a concentrated load of 40 kN 1.5 m from the fixed end and a concentrated load of 30 kN at the free end. Draw the SF and BM diagrams.
2. A cantilever is 3 m long and carries a UD load of 20 kN/m for a length of 2 m from the free end. Draw the SF and BM diagrams for the beam.
3. A cantilever is 4 m long and carries a UD load of 40 kN/m for a length of 2.5 m as shown in Fig. 4.65. Draw the SF and BM diagrams for the beam.
4. A cantilever, of span 5 m, carries loads as shown in Fig. 4.66. Draw the SF and BM diagrams for the beam.

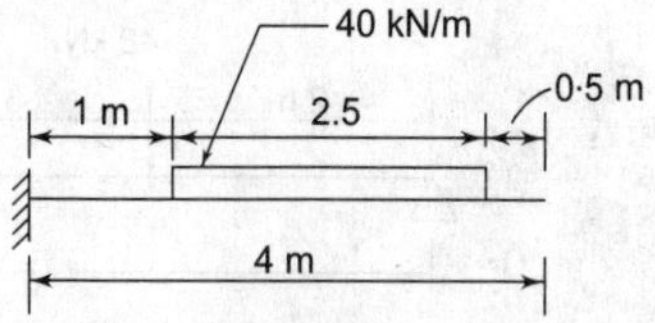

Fig. 4.65

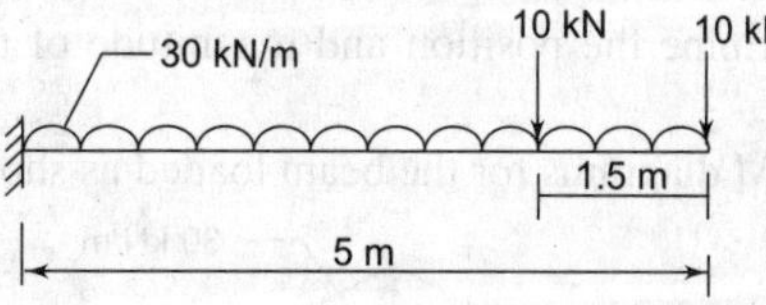

Fig. 4.66

5. Draw the SF and BM diagrams for the cantilever loaded as shown in Fig. 4.67.
6. A cantilever, of span 4 m, carries a UD load of 10 kN/m for the whole length in addition to a point load of 40 kN at the free end, and a couple of 30 kNm at mid-span. Draw the SF and BM diagrams indicating salient values.
7. A simply supported (SS) beam has a span of 10 m and carries the loadings as shown in Fig. 4.68. Draw SF and BM diagrams indicating salient values. Find the position and magnitude of the maximum BM in the beam.

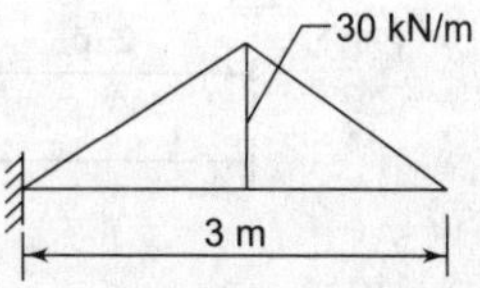

Fig. 4.67

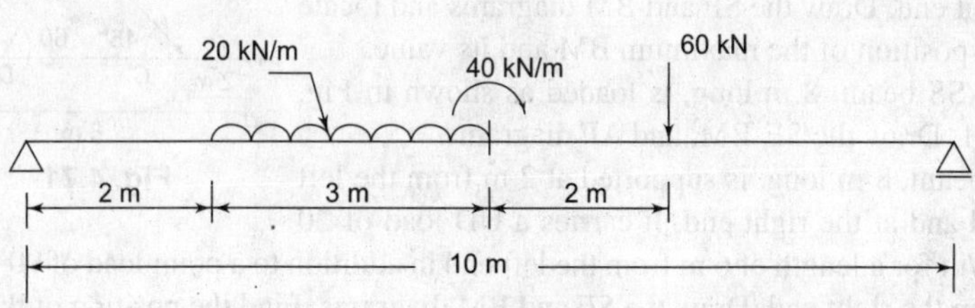

Fig. 4.68

8. A simply supported (SS) beam has a span of 5 m and carries a UD load of 20 kN/m in the left half and a UD load of 40 kN/m in the right half of its length. Draw the SF and BM diagrams and find the position and magnitude of maximum BM in the beam.
9. An SS beam, l m long, carries a clockwise couple M kNm at a m from the left end and an equal anticlockwise couple M at a m from the right end. Draw the SF and BM diagrams for the beam.
10. An SS beam is subjected to a clockwise couple load of 60 kNm at 2 m from the left end. The span of the beam is 5 m. Draw the SF and BM diagrams. What modifications in these diagrams are needed if the load is applied (a) at the left end and (b) at the right end?

11. An SS beam of 9 m span carries a load, uniformly varying from zero at the left end to 84 kN/m at the right end. Draw the SF and BM diagrams. Determine the position and magnitude of the maximum BM in the beam.
12. An SS beam, of 10 m span, carries three point loads as shown in Fig. 4.69. Draw the SF and BM diagrams indicating salient values.

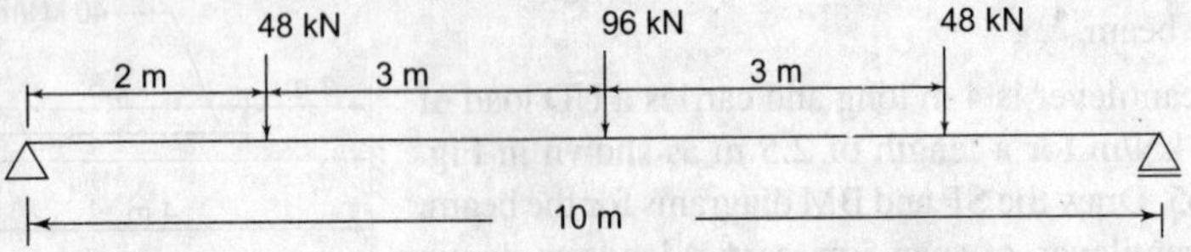

Fig. 4.69

13. An SS beam, of 8 m span, carries a UD load of 24 kN/m over the left half of its length in addition to a point load of 60 kN at 6 m from the left end. Draw the SF and BM diagrams, and determine the position and magnitude of the maximum BM in the beam.
14. Draw the SF and BM diagrams for the beam loaded as shown in Fig. 4.70.

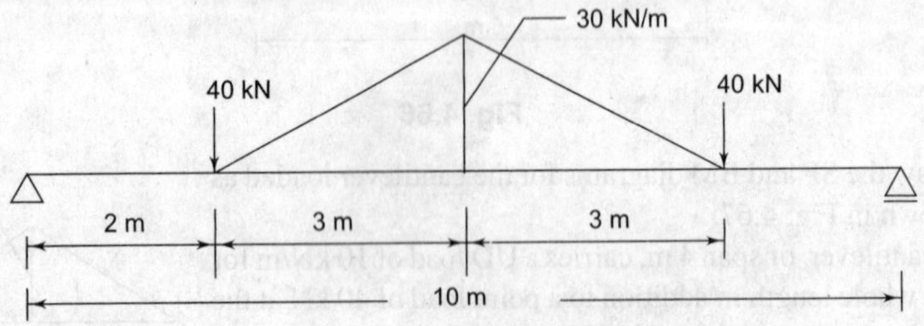

Fig. 4.70

15. An SS beam of 3 m span carries a load uniformly varying from 50 kN/m at the centre to 20 kN/m at either end. Draw the SF and BM diagrams, indicating salient values.
16. An SS beam, 6 m long, carries a load uniformly varying from 10 kN/m at the left end to 40 kN/m at the right end. Draw the SF and BM diagrams and locate the position of the maximum BM and its value.
17. An SS beam, 8 m long, is loaded as shown in Fig. 4.71. Draw the SF, BM, and AF diagrams.

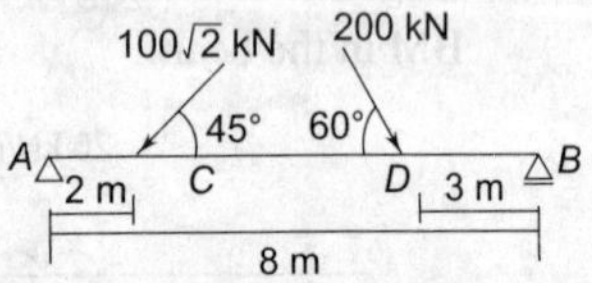

Fig. 4.71

18. A beam, 8 m long, is supported at 2 m from the left end and at the right end. It carries a UD load of 20 kN/m for a length of 6 m from the left end in addition to a point load of 90 kN at 1 m from the right end. Draw the SF and BM diagrams. Find the position of the point of contraflexure and the location and magnitude of the maximum BM.
19. A 10 m long overhanging beam is loaded as shown in Fig. 4.72. Draw the SF and BM diagrams. Find the location and magnitude of the maximum BM and locate the point of contraflexure in the BM diagram.

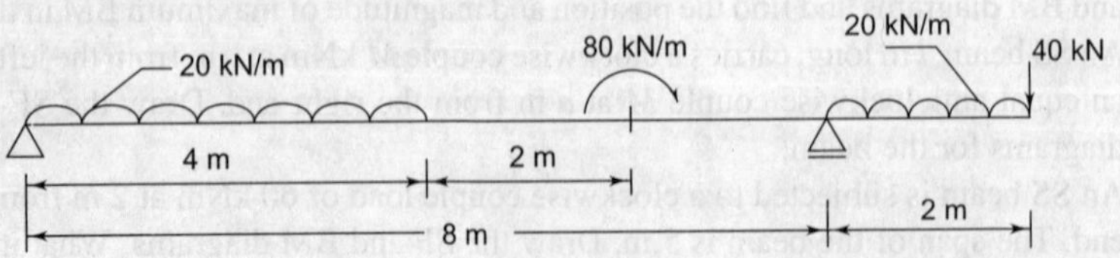

Fig. 4.72

20. A doubly overhung beam, 9 m long, is supported at 1.5 m from either end. It carries a UD load of 30 kN/m for a length of 3 m from the right end and the left end. Draw the SF and BM diagrams.
21. A beam, 14 m long, is supported at 2 m and 12 m from the left end. It carries UD loads of 20 kN/m over both overhanging lengths in addition to a clockwise couple load of 200 kN m at mid-span. Draw the SF and BM diagrams for the beam. Find the position and magnitudes of maximum BMs and the position of the point of contraflexure.
22. A beam 10 m long is supported at the left end and at 8 m from the left end. It carries a load uniformly varying from zero at the left support to 60 kN/m at the right support, and a UD load of 30 kN/m over the overhanging length. Draw the SF and BM diagrams. Find the position and magnitude of the maximum BM and the position of the point of contraflexure.
23. A pile 12 m long is being lifted in a horizontal position by cables attached to points 2 m from each end. If the pile weighs 4 kN/m of its length, draw the SF and BM diagrams. At what distance from the ends should the cables be attached so that the pile has minimum BM while being lifted?
24. A pile 12 m long weighs 4 kN/m. It is being lifted by a cable attached to a point 2 m from one end, the other end resting on rough ground. Draw the SF, BM, and AF diagrams when the pile has been lifted to a position at 30° to the horizontal, when the cable is at right angles to the length. What changes will be there if the lifting cable is vertical?
25. For the pile given in Problem 24, at what point should the cable be attached to have minimum BM while lifting?
26. A ladder 4 m long weighs 120 N/m. It is kept at an angle of 60° against a smooth vertical wall, the other end resting on rough ground. A man weighing 750 N has climbed the ladder for a distance 3 m along it. Draw the AF, SF, and BM diagrams for the ladder.
27. Draw the AF, SF, and BM diagrams for the beam loaded and supported as shown in Fig. 4.73.

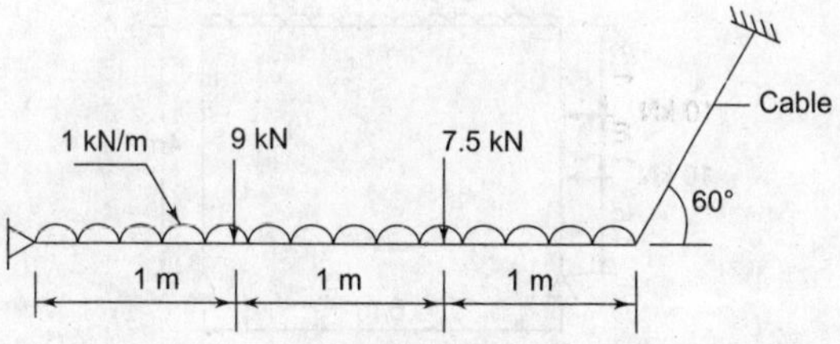

Fig. 4.73

28. Draw the SF and BM diagrams for the beam loaded as shown in Fig. 4.74.

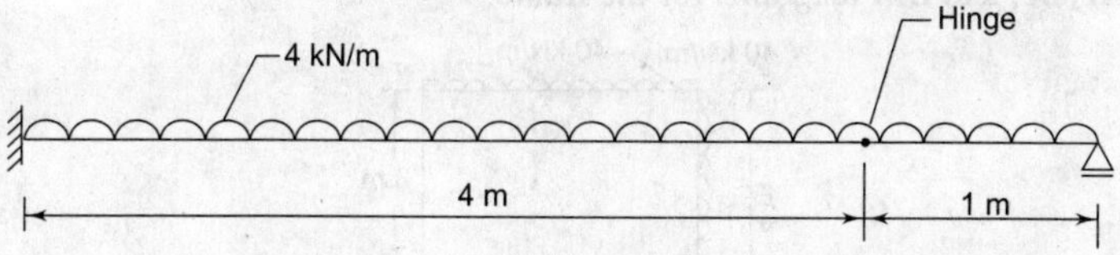

Fig. 4.74

29. For the hinged beam loaded as shown in Fig. 4.75, draw the SF and BM diagrams.

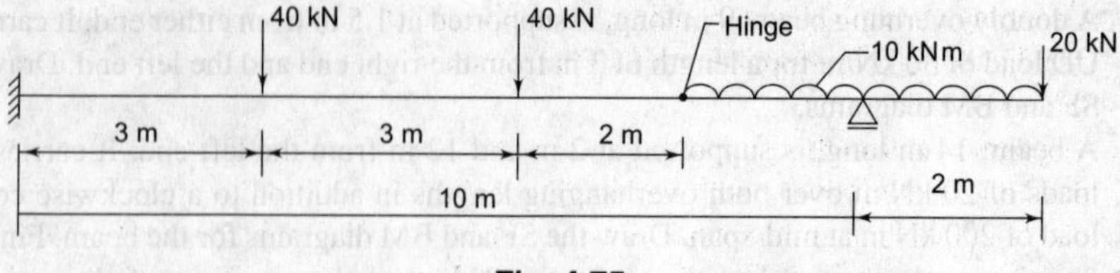

Fig. 4.75

30. Draw the SF and BM diagrams for the hinged beam shown in Fig. 4.76, indicating salient values.

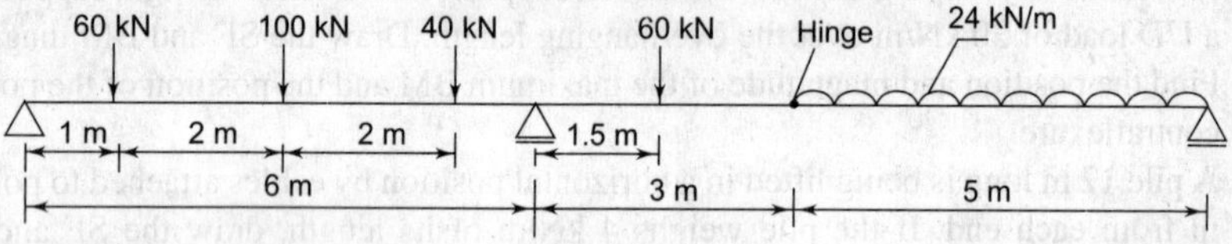

Fig. 4.76

31. Draw the AF, SF, and BM diagrams for a plank kept as shown in Fig. 4.77. Assume that the reaction at B is normal to the plank and the ground is rough at A. Compute values and draw diagrams (i) if the weight of the plank is neglected and (ii) if the plank weighs 180 N/m.

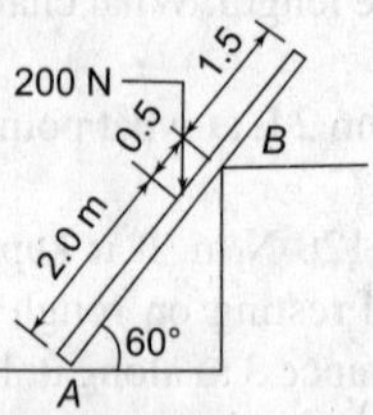

Fig. 4.77

32. Draw the AF, SF, and BM diagrams for the rigid frame loaded as shown in Fig. 4.78.

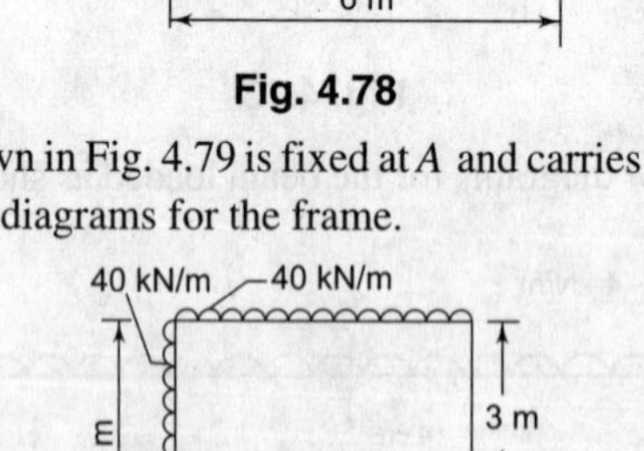

Fig. 4.78

33. The rigid frame shown in Fig. 4.79 is fixed at A and carries UD loads as shown. Draw the AF, SF, and BM diagrams for the frame.

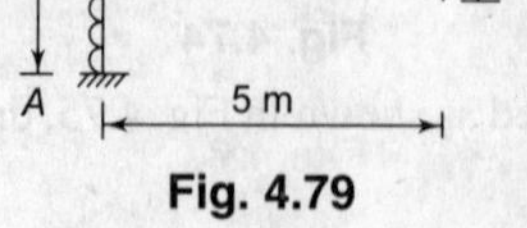

Fig. 4.79

34. Draw the AF, SF, and BM diagrams for the frame loaded as shown in Fig. 4.80.

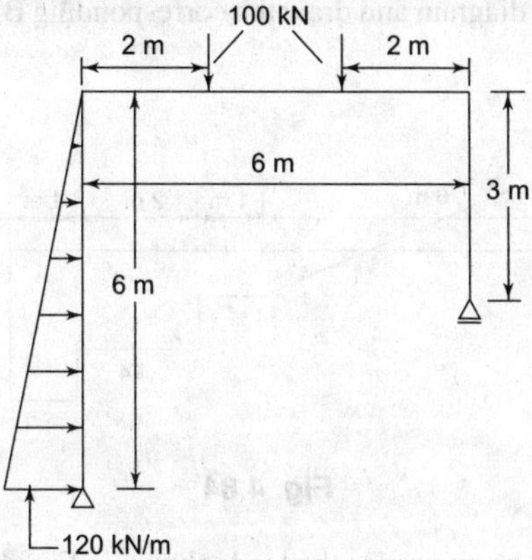

Fig. 4.80

35. For the rigid frame loaded as shown in Fig. 4.81, draw the AF, SF, and BM diagrams.

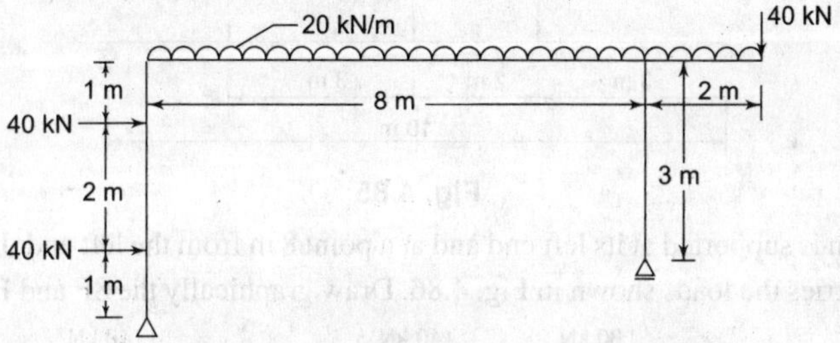

Fig. 4.81

36. The SF diagram for an SS beam is shown in Fig. 4.82. Deduce the loading on the beam and draw the corresponding BM diagrams.

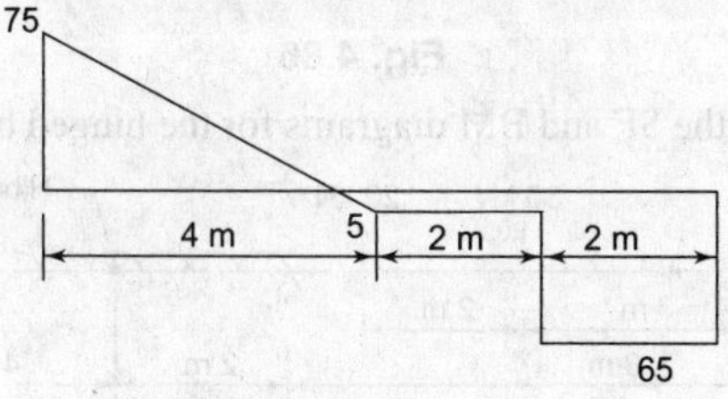

Fig. 4.82

37. The SF diagram for an overhanging beam supported at *A* and *B* is shown in Fig. 4.83. Derive the loading and BM diagrams for the beam from the given SF diagram. All values are in kN.

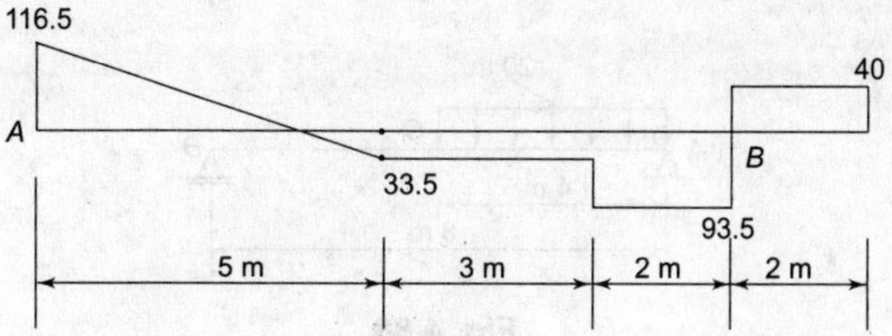

Fig. 4.83

38. The diagram shown in Fig. 4.84 is the SF diagram for a beam supported at *A* and *B*. Deduce the loading diagram and draw the corresponding BM diagram. All values are in kN.

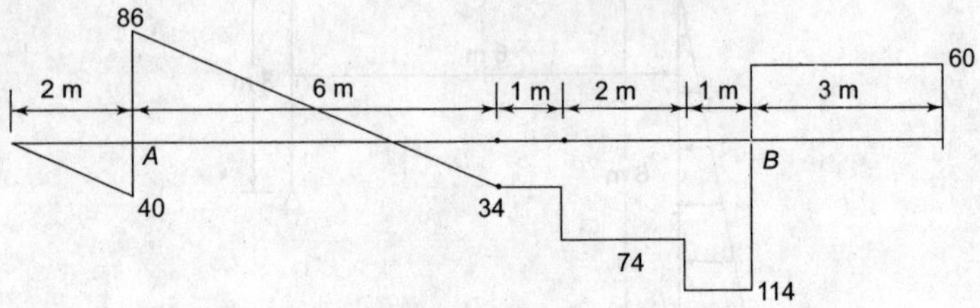

Fig. 4.84

39. A simply supported beam carries the loads shown in Fig. 4.85. Draw graphically the SF and BM diagrams.

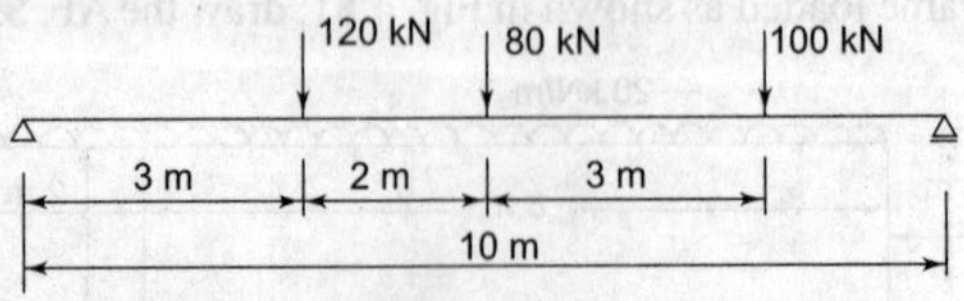

Fig. 4.85

40. A beam is supported at its left end and at a point 8 m from the left end. It is 10 m long and carries the loads shown in Fig. 4.86. Draw graphically the SF and BM diagrams.

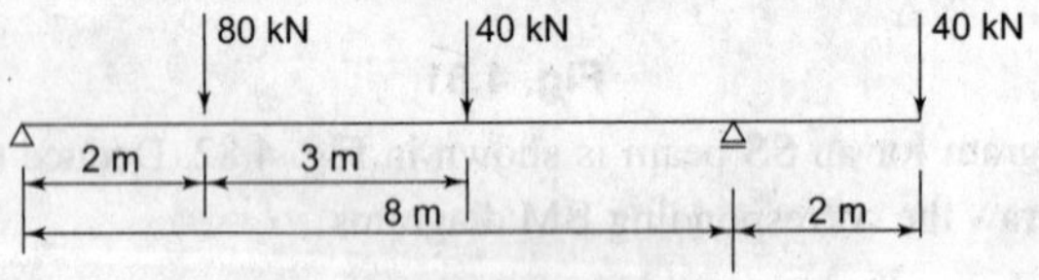

Fig. 4.86

41. Draw graphically the SF and BM diagrams for the hinged beam shown in Fig. 4.87.

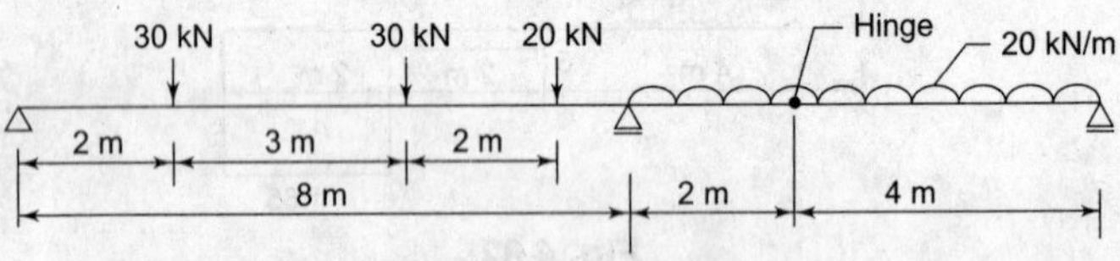

Fig. 4.87

Solve the following problems using singularity functions:

42. An SS beam carries loads as shown in Fig. 4.88. Derive the general equations for SF and BM.

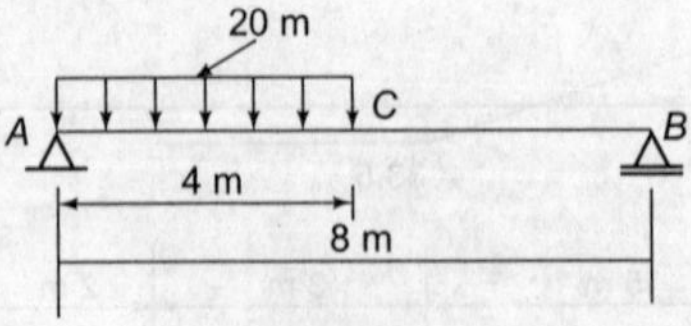

Fig. 4.88

43. An SS beam carries loads as shown in Fig. 4.89. Derive the general equations for SF and BM.

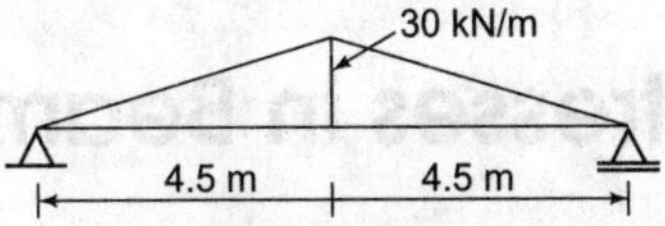

Fig. 4.89

44. An overhanging beam carries loads as shown in Fig. 4.90. Derive the general equation for SF and BM using singularity functions.

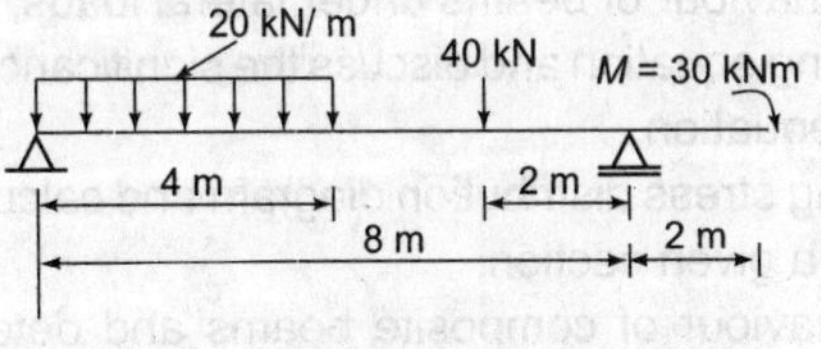

Fig. 4.90

45. An SS beam carries loads as shown in Fig. 4.91. Derive general equation for SF and BM by singularity functions.

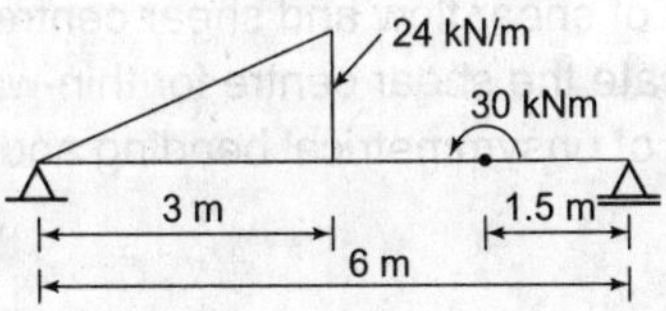

Fig. 4.91

46. A doubly overhung beam carries a UD load throughout its length as shown in Fig. 4.92. The span between the supports is 8 m while the overhang is 1 m at the left side and 1.5 m on the right. Derive general equations for SF and BM if the loading is 30 kN/m.

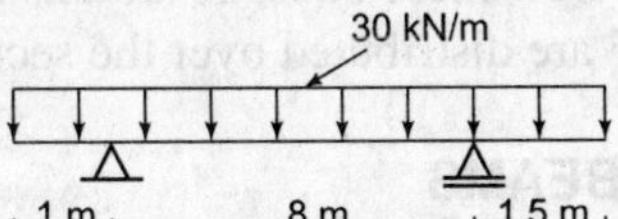

Fig. 4.92

CHAPTER 5

Stresses in Beams

Learning Objectives

After going through this chapter, the reader will be able to

- describe the behaviour of beams under lateral loads,
- derive the bending equation and discuss the significance of assumptions made in deriving the equation,
- draw the bending stress distribution diagram and calculate the maximum normal stresses in a given section,
- explain the behaviour of composite beams and determine the stresses in such beams,
- derive the equation for shear stress distribution and calculate the maximum shear stress,
- design beam sections for bending moment (BM) and shear force (SF),
- explain the concept of shear flow and shear centre, and calculate shear flow at any point and locate the shear centre for thin-walled sections, and
- explain the concept of unsymmetrical bending and determine the stresses in such cases.

5.1 INTRODUCTION

In the last chapter, we discussed methods to find the effect of external loads on beams. The predominant effect of lateral loads on beams are bending moments and shear forces. We discussed methods to calculate the BM and SF at any section of the beam and to also find the variation of these effects over the length of the beam. The beams resist these effects by developing internal stress resultants. In this chapter, we will study how these stress resultants are developed and how the stresses due to BM and SF are distributed over the section.

5.2 BEHAVIOUR OF BEAMS

We briefly discussed the behaviour of beams in the last chapter. To recapitulate, under the action of the external planar loads, a beam bends. If we take any section of a beam, as shown in Fig. 5.1(a), and consider the free body on either side, we find that there must be three internal stress resultants at the section in general to maintain the equilibrium of the free body. These internal stress resultants are equal and opposite to the external effects of loads on the section, i.e. the bending moment, shear force, and axial force [Fig. 5.1(b)].

In this chapter, we will discuss how these internal stresses are distributed and how the resisting forces and moments are developed. We will discuss only two major effects in beams—due to the BM and SF. Axial stresses are generally small. Beams essentially carry loads in the lateral direction transverse to their longitudinal axis [see Fig. 5.1(a)]. As we know, the effects of the lateral loads are a bending moment, a shear force, and an axial force, (in case there are loads inclined to the longitudinal axis). The loads and reactive forces lie in a plane which also contains an axis of symmetry of the section. In cases where there is no axis of symmetry, a different situation will arise, which is discussed later in this chapter.

Under the effect of the lateral loads, the beam bends and the longitudinal axis takes a curved shape. The beam produces an internal resisting couple to balance the BM due to the applied loads. This couple is generated by the normal compressive and tensile forces in the section, as shown in Fig. 5.1(d). This couple is equal and opposite to the external BM.

To generate a force to balance the SF due to applied loads, the section has tangential stresses whose resultant is equal and opposite to the SF [Fig. 5.1(e)].

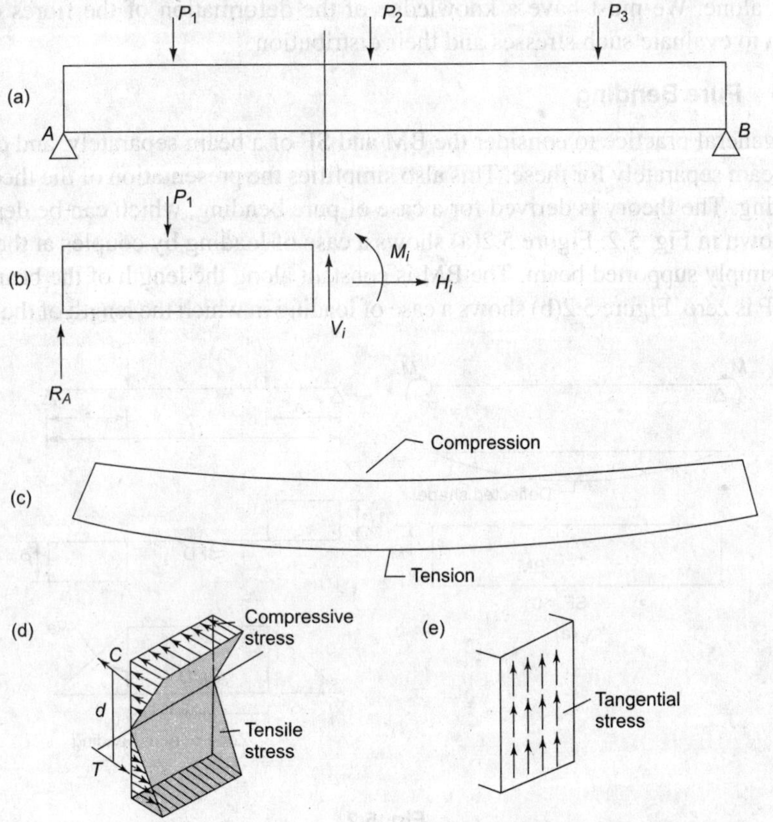

Fig. 5.1

In this chapter, we will discuss how these normal, bending stresses and tangential, shear stresses are developed and how to calculate their maximum values and their distribution. We will start with a discussion of the effects due to bending moment.

5.3 BENDING STRESSES

The beams bend due to the effect of external loads and take a curved shape. In the case of a positive or sagging BM, the beam bends concave upwards. This bending results in compression in the top fibres of the beam, which shorten in length and tension in the bottom fibres, which stretch. The effect is the opposite in the case of a negative (hogging) BM. The compressive stresses acting over the top part (area) of the beam cross section form a resultant C. Similarly, tensile stresses acting over the bottom part (area) of the cross section form a stress resultant T. Forces C and T are parallel and act at a distance apart. Neglecting axial forces, C and T must be equal for there to be an equilibrium of forces in the longitudinal direction. These equal and opposite forces form the internal resisting couple. The distance between C and T is known as the lever arm. Force C or T multiplied by the lever arm gives a couple equal and opposite to the BM at the section [Fig. 5.1(d)].

While this discussion covers the equilibrium conditions, how the internal stresses are distributed within the section cannot be known from the equilibrium conditions alone. We must have a knowledge of the deformation of the fibres of the beam to evaluate such stresses and their distribution.

5.3.1 Pure Bending

It is general practice to consider the BM and SF of a beam separately, and design the beam separately for these. This also simplifies the presentation of the theory of bending. The theory is derived for a case of pure bending, which can be depicted as shown in Fig. 5.2. Figure 5.2(a) shows a case of loading by couples at the ends of a simply supported beam. The BM is constant along the length of the beam and the SF is zero. Figure 5.2(b) shows a case of loading in which the length of the beam

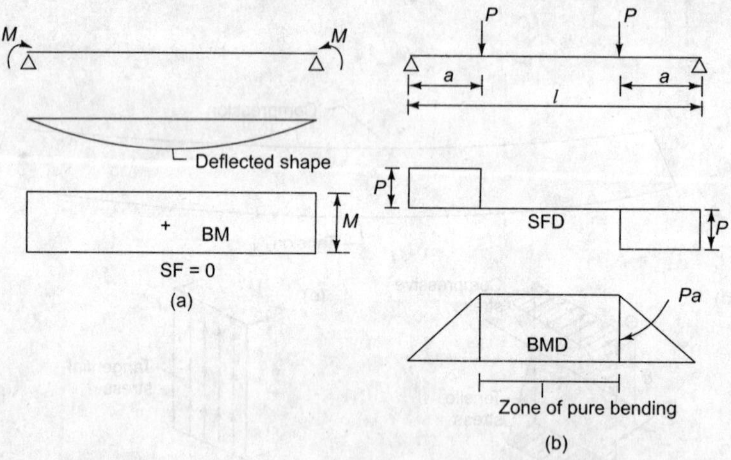

Fig. 5.2

between the loads is subjected to pure bending and can be used for experimental purposes to simulate pure bending. The BM is constant between the loads and the SF is zero. The absence of SF ensures that some of the assumptions made in the theory of pure bending are valid.

In the case of most beams met with in practice, BM and SF coexist at any section. However, the theory of bending can be applied to such cases with the accuracy required for engineering design. The case of pure bending is also valid because, in most cases, the BM is maximum where the SF is zero or changes sign.

5.3.2 Theory of Pure Bending: Bernoulli's Equation

Consider a beam bent by pure couples so that the BM is constant and the SF is zero. The beam bends in such cases in the form of an arc of constant radius. The beam is prismatic, i.e., of constant cross section and is symmetrical about a vertical plane through its centroid. In deriving the theory of bending, the following assumptions are made.

(i) Sections which are plane before bending remain plane after bending [Fig. 5.3(a)]. This is one of the most important assumptions made in the theory of bending. If we consider a beam with horizontal and vertical lines drawn through it, we will find that the bent arc will appear as shown in Fig. 5.3(b). The cross section of the beam rotates about a line (parallel to the Z-axis) but remains plane during this deformation. As we have seen earlier, the top fibres of the beam are under compression as these are shortened and the bottom fibres are under tension as these are stretched. Obviously, since the sections remain plane, there must be one layer which is neither stretched nor shortened.

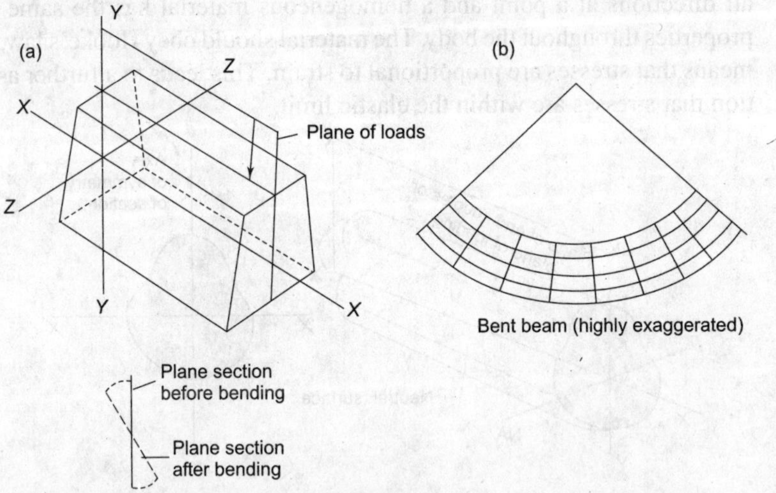

Fig. 5.3

Consider a small element of length dx of the beam [Fig. 5.4(a)], enlarged and shown in the deformed position in Fig. 5.4(c). The longitudinal fibre n-n is

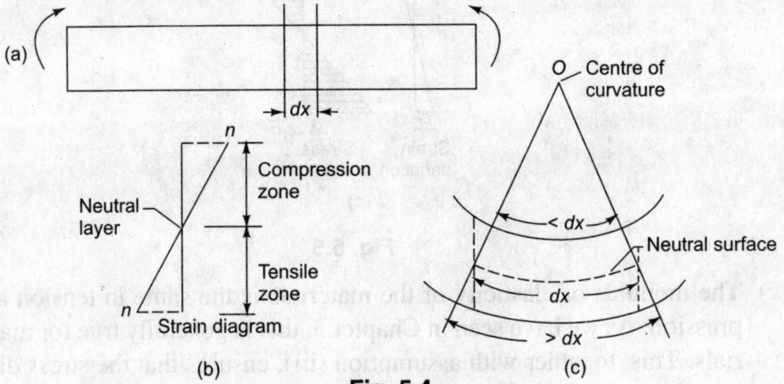

Fig. 5.4

neither stretched nor compressed and is known as the neutral layer. It is obvious from the figure that the stretching or shortening of fibres is proportional to the distance from this layer. This leads us to the important conclusion that the strains in the layers vary linearly and the strain diagram will be as shown in Fig. 5.4(b).

(ii) The plane of loading must contain a principal axis of the beam cross section. This assumption is illustrated in Fig. 5.5. Since the loading is by pure couples applied at the ends, the plane of this couple also contains the axis Y-Y, an axis of symmetry of the section. The beam may not be symmetrical about the axis X-X. But symmetry about the axis Y-Y is important; otherwise bending transverse to the plane of load may result. In case we consider the loading case in Fig. 5.5(a) or the loading cases that we come across in practice, the plane containing the axis of symmetry should also contain the loads and reactive forces.

(iii) The material is homogeneous and isotropic and obeys Hooke's law. As explained in Chapter 3, an isotropic material has the same elastic properties in all directions at a point and a homogeneous material has the same elastic properties throughout the body. The material should obey Hooke's law, which means that stresses are proportional to strain. This leads to a further assumption that stresses are within the elastic limit.

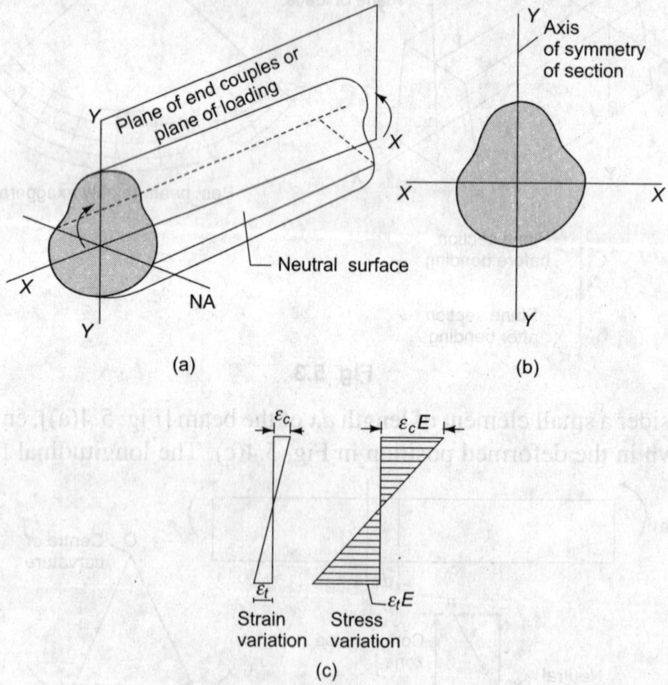

Fig. 5.5

(iv) The modulus of elasticity of the materials is the same in tension and compression. As we have seen in Chapter 3, this is generally true for many materials. This, together with assumption (iii), ensures that the stress diagram is linear and has the same shape as the strain diagram. Stress values are obtained by multiplying the strains by E; the stress diagram is thus obtained by multiplying the strain values by a constant which is the same in tension and compression. The stress diagram will be thus linear and consist of two triangles, as shown in Fig. 5.5(c).

(v) The beam is initially straight and has the same cross section. As was stated earlier, we consider a beam of constant cross section, i.e., a prismatic beam. In addition, we assume that the beam, i.e., its longitudinal axis, is straight. Otherwise, bending in a horizontal plane also results.

(vi) Each layer of the material is free to expand or contract under stress. This is necessary to ensure that we are able to calculate the stress at any layer using the elastic properties. The layers are not affected by the presence of adjoining layers, and are free to deform.

Bending equation Based upon these assumptions, the bending equation can be derived as follows.

In Fig. 5.6, the beam is loaded by a couple at each end to produce a constant BM [Figures 5.6(a) and (b)]. We consider a length dx of the beam [Fig. 5.6(c)]. Since plane sections remain plane, the length dx undergoes deformation as shown. The line n-n is the neutral layer and undergoes no deformation. The layers above n-n are compressed and those below n-n are stretched in the case shown of positive BM. If the BM is negative, the upper layers will be in tension and the bottom layers will be in compression, with the beam bending convex upwards.

R is the radius to which the neutral layer bends. Considering the deformation of a layer y above n-n, we have

$$dx = R\,d\theta$$

where $d\theta$ is the angle subtended by the length dx at the centre of curvature and m-$m = (R - y)d\theta$. The length of the layer m-m before bending is $dx = R\,d\theta$.

$$\text{Change in length} = R\,d\theta - (R - y)d\theta$$
$$= y\,d\theta$$

and $$\text{Strain } \varepsilon = \frac{\Delta L}{L} = \frac{y\,d\theta}{R\,d\theta} = \frac{y}{R}$$

Since R is a constant, strain at any layer is proportional to its distance y from the neutral layer. Since E is constant and is the same in tension and compression, stress σ at layer m-m = strain × E.

$$\sigma = \frac{Ey}{R}$$

E and R being constant, this expression shows that σ varies linearly with y. This is generally expressed as

$$\frac{\sigma}{y} = \frac{E}{R}$$

The stress diagram is thus linear following the variation in strain. This diagram is obtained by multiplying the strain diagram by the constant E. It is shown in Fig. 5.6(d).

It may be noted that we have used the deformation characteristics of the beam to arrive at this relation. We have now to relate the external BM to the stress distribution and internal stress resultant or resisting moment. We use the equilibrium equations for this purpose.

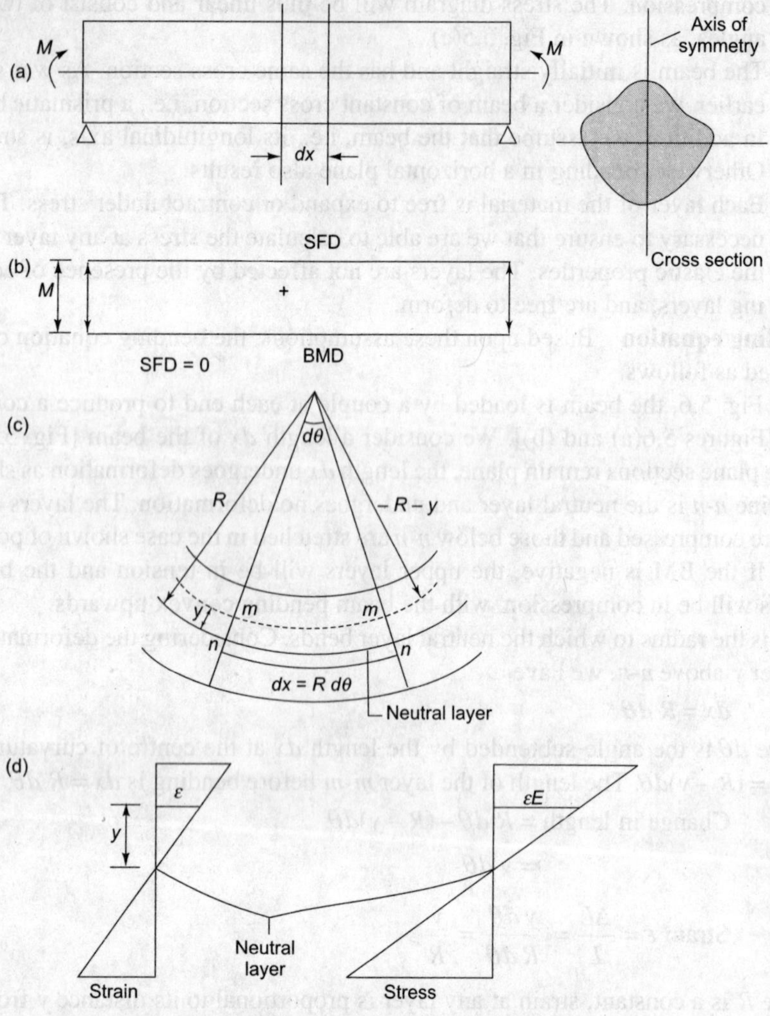

Fig. 5.6

For a given cross section, the stress value at any layer, at a distance y from the neutral axis, is constant and varies linearly for different layers. The compressive stresses act on the cross-sectional area over layers above the neutral layer.

The line along which the neutral layer intersects the cross section is called the neutral axis (NA). Obviously, the bending stress is zero at this axis.

The diagrams in Fig. 5.7(b) show the compressive and tensile stresses acting over the respective areas of cross section. These are known as the stress blocks and it is easy to see that the volume of the compressive stress block is equal to the total compressive force. Similarly, the volume of the tensile stress block is equal to the total tensile force. Two equations of equilibrium can be formulated.

(i) Since no external normal forces act on the section, the total compressive force C is equal in magnitude to the total tensile force T.

(ii) Since C and T are equal and opposite, and non-collinear, they give rise to a couple, which is the internal resisting moment. This moment must be equal to the external BM.

In Fig. 5.7(c), the stress at any layer, at a distance y from the NA is equal to $\sigma = (E/R)y$. Considering an elementary strip dy of width b of the cross section, the force acting on the elementary area is given by

$$dF = \sigma b\, dy = \frac{E}{R} y\, b\, dy$$

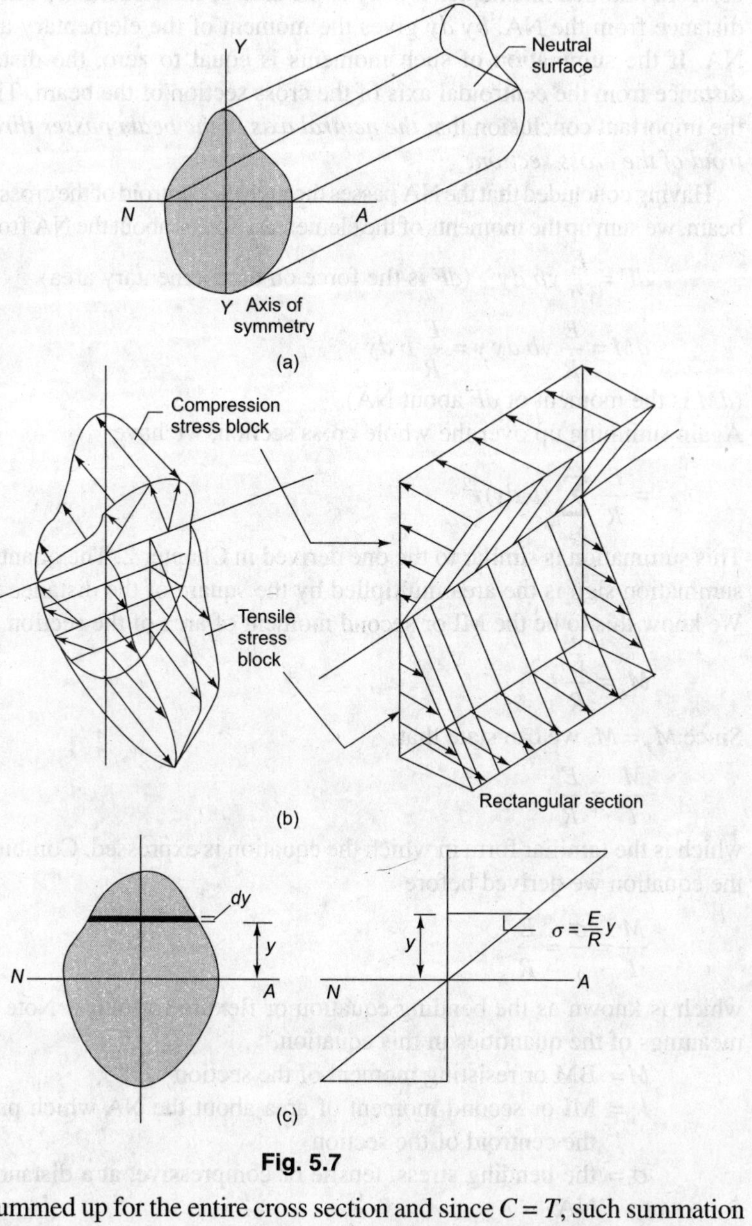

Fig. 5.7

This can be summed up for the entire cross section and since $C = T$, such summation is equal to zero. If y_1 and y_2 are the extreme fibre depths above and below the NA, then

$$\sum_{y_2}^{y_1} \frac{E}{R} by\, dy = 0, \quad \frac{E}{R}\sum_{y_2}^{y_1} b\, dy\, y = 0$$

since E and R are constant.

or
$$\sum_{y_2}^{y_1} by\, dy = 0$$

as $E \neq 0$ and $R \neq 0$. $\Sigma_{y_2}^{y_1} (b\ dy\ y)$ is a familiar expression from the concept of the centroid outlined in Chapter 1. $b\ dy$ is the area of the elementary strip and y is the distance from the NA. $by\ dy$ gives the moment of the elementary area about the NA. If the summation of such moments is equal to zero, the distance y is the distance from the centroidal axis of the cross section of the beam. This leads is to the important conclusion that *the neutral axis of the beam passes through the centroid of the cross section.*

Having concluded that the NA passes through the centroid of the cross section of the beam, we sum up the moments of the elementary forces about the NA from Fig. 5.7(c),

$$dF = \frac{E}{R} yb\ dy \quad (dF \text{ is the force on the elementary area})$$

$$dM = \frac{E}{R} yb\ dy\ y = \frac{E}{R} b\ dy\ y^2$$

(dM is the moment of dF about NA).

Again summing up over the whole cross section, we have

$$= \frac{E}{R} \sum_{y_2}^{y1} (b\ dy)y^2$$

This summation is similar to the one derived in Chapter 2. The quantity within the summation sign is the area multiplied by the square of the distance from an axis. We know this to be the MI or second moment of area of the section. Therefore,

$$M_r = \frac{E}{R} I$$

Since $M_r = M$, we can state that

$$\frac{M}{I} = \frac{E}{R}$$

which is the familiar form in which the equation is expressed. Combining this with the equation we derived before

$$\frac{M}{I} = \frac{\sigma}{y} = \frac{E}{R}$$

which is known as the bending equation or flexure formulae. Note carefully the meanings of the quantities in this equation.

M = BM or resisting moment of the section

I = MI or second moment of area about the NA which passes through the centroid of the section

σ = the bending stress, tensile or compressive, at a distance y from the NA

y = the distance of a layer from the NA, with stress σ

E = Young's modulus of elasticity of the material

$\frac{1}{R}$ = curvature of the bent beam axis; R is the radius of curvature

The bending equation expresses the following two equalities:

$$\frac{M}{I} = \frac{\sigma}{y}$$

relates the strength of the material to the BM. This is the strength equation with which we will be mainly concerned in this chapter.

$$\frac{M}{I} = \frac{E}{R}$$

relates the deformation of the beam with the applied BM and is concerned with the stiffness of the beam or its flexural rigidity. This equation will be taken up in more detail in Chapter 7 to determine deformations in beams.

We must remember the assumptions stated earlier and used in arriving at this result. In actual beams, where BM and SF coexist, the behaviour will be different. But these results are applicable within the limits of engineering accuracy required. The important conclusions arrived at are as follows.

(i) The beam bends to a circular arc of curvature $1/R$. The strain diagram is linear, with compressive strains above the NA and tensile strains below NA for a sagging moment. For a hogging moment, the reverse is true.

(ii) The stress diagram is correspondingly linear, the compressive stresses varying from zero at the NA to a maximum value at the top. The tensile stresses vary from zero at the NA to a maximum at the extreme bottom fibre [Fig. 5.8(a)].

(iii) The NA passes through the centroid of the cross section [Fig. 5.8(b)].

(iv) The total compressive force is equal to the total tensile force; $C = T$.

(v) C and T form a couple, which is equal to the internal resisting moment. The distance between C and T is known as the lever arm.

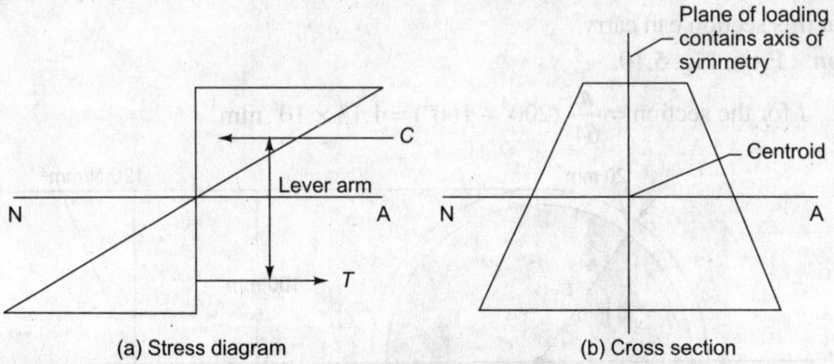

Fig. 5.8

Example 5.1 Bending of a steel bar

A steel plate of dimensions 20 mm × 2 mm is bent into a circular arc of radius 2 m by applying end couples. If E = 200 GPa, find the end couples applied and the maximum stress in the bar.

Solution From Fig. 5.9,

$$I \text{ for the section} = \frac{20 \times 2^3}{12} = \frac{40}{3}\ \text{mm}^4$$

$R = 2$ m = 2000 mm, $E = 200$ GPa = 200,000 N/mm^2

$$\frac{M}{I} = \frac{E}{R}, \quad M = \frac{EI}{R} = \frac{200{,}000}{2000} \times \frac{40}{3} = 1333.33\ \text{Nmm}$$

For maximum stress, we use the relation

$$\frac{E}{R} = \frac{\sigma}{y}$$

$y = 1$ mm. Therefore,

$$\sigma = \frac{200{,}000}{2000} \times 1 = 100\ \text{N/mm}^2 = 100\ \text{MPa}$$

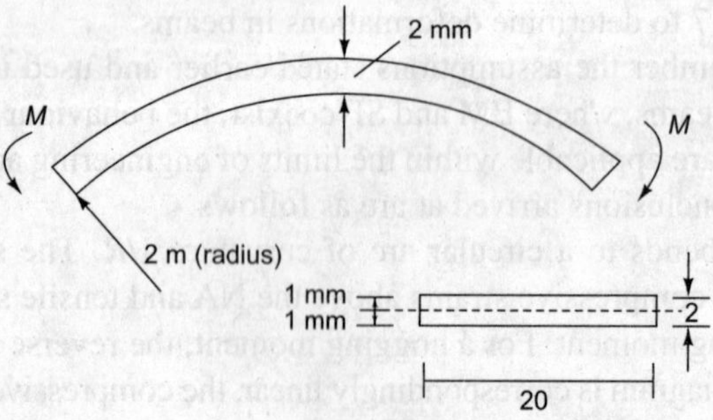

Fig. 5.9

Example 5.2 Moment capacity of a hollow circular section

A hollow circular section, of external diameter 200 mm, and thickness 20 mm, is to be used as a beam. If the maximum stress permissible in the material is 120 MPa, find the maximum BM that this section can carry.

Solution From Fig. 5.10,

$$I \text{ for the section} = \frac{\pi}{64}(200^4 - 160^4) = 1.13 \times 10^8\ \text{mm}^4$$

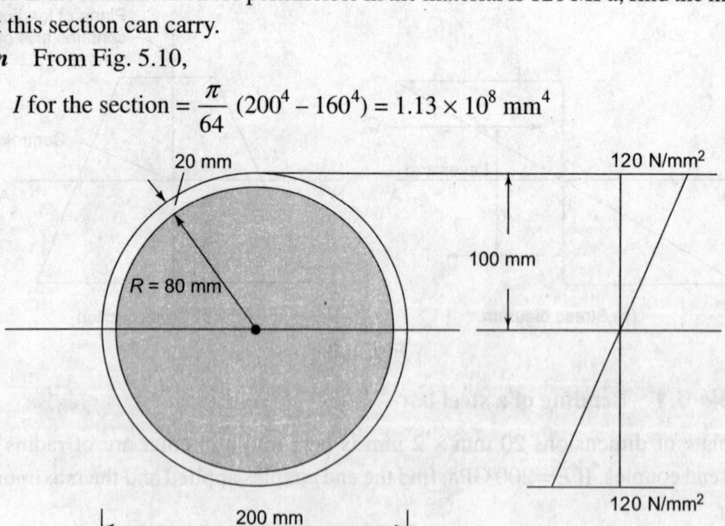

Fig. 5.10

$$\frac{M}{I} = \frac{\sigma}{y}$$

$$\Rightarrow \quad M = \frac{\sigma I}{y}$$

$$\sigma = 120 \text{ MPa}, \quad y = 100 \text{ mm}$$

$$\Rightarrow \quad M = \frac{120 \times 1.13 \times 10^8}{100} = 135.6 \times 10^6 \text{ Nmm} = 135.6 \text{ kNm}$$

Example 5.3 Load capacity of a rectangular section on an SS span

A rectangular section 200 mm wide and 400 mm deep is used on a span of 6 m for the loading shown in Fig. 5.11. Find the maximum value of W so that the permissible stress of 50 MPa is not exceeded in the material.

Solution From Fig. 5.11(a), the maximum BM in the beam is $W \times 2$ kNm.

$$I \text{ for the section} = \frac{200 \times 400^3}{12} = \frac{32}{3} \times 10^8 \text{ mm}^4$$

$$y = 200 \text{ mm}$$

$$M = \frac{\sigma I}{y} = \frac{50}{200} \times \frac{32}{3} \times 10^8 = 266.67 \text{ kNm}$$

$$W \times 2 = 266.67$$

$$\Rightarrow \quad W = 133.33 \text{ kN}$$

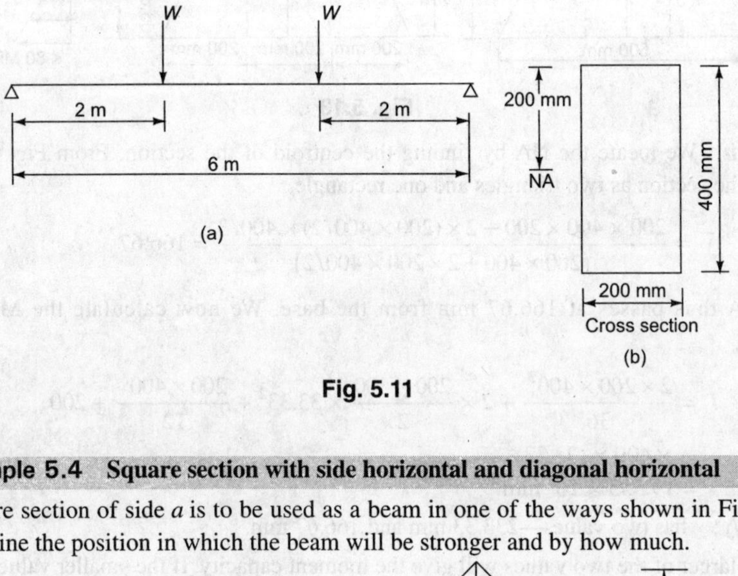

Fig. 5.11

Example 5.4 Square section with side horizontal and diagonal horizontal

A square section of side a is to be used as a beam in one of the ways shown in Fig. 5.12. Determine the position in which the beam will be stronger and by how much.

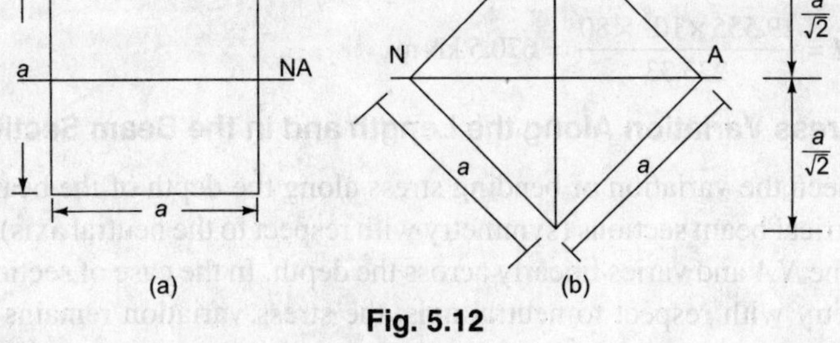

Fig. 5.12

Solution Since M = σ I/y and σ is constant, the moment-carrying capacity of the sections is proportional to their I/y values. For the section of Fig. 5.12(a),

$$I = \frac{a^4}{12}, \quad y = \frac{a}{2}, \quad \frac{I}{y} = \frac{a^4 \times 2}{12 \times a} = \frac{a^3}{6}$$

For the section shown in Fig. 5.12(b),

$$I = 2 \times \frac{\sqrt{2}a}{12} \times \left(\frac{a}{\sqrt{2}}\right)^3 = \frac{a^4}{12}, \quad y = \frac{a}{\sqrt{2}}, \quad \frac{I}{y} = \frac{a^4 \times \sqrt{2}}{12 \times a} = \frac{a^3}{6\sqrt{2}}$$

The section in Fig. 5.12(a) is thus $(2)^{1/2}$ times stronger than that in Fig. 5.12(b). ■

Example 5.5 Moment capacity of a trapezoidal section

Find the moment-carrying capacity of the trapezoidal section shown in Fig. 5.13(a) if the permissible stress in the material is 80 MPa in tension and compression.

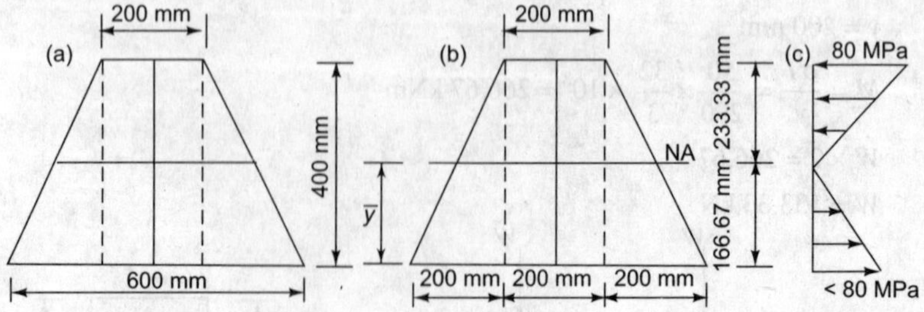

Fig. 5.13

Solution We locate the NA by finding the centroid of the section. From Fig. 5.13(b), taking the section as two triangles and one rectangle,

$$\bar{y} = \frac{200 \times 400 \times 200 + 2 \times (200 \times 400/2) \times 400/3}{(200 \times 400 + 2 \times 200 \times 400/2)} = 166.67$$

The NA thus passes at 166.67 mm from the base. We now calculate the MI about the NA.

$$I = \frac{2 \times 200 \times 400^3}{36} + 2 \times \frac{200 \times 400}{2} \times 33.33^2 + \frac{200 \times 400^3}{12} + 200$$
$$\times 400 \times (33.33)^2$$
$$= 19.555 \times 10^8 \text{ mm}^4$$

$M = \sigma I/y$; y has two values—233.33 mm and 166.67 mm.

The larger of the two values will give the moment capacity. If the smaller value is used then the stress in the top fibre will be more than the permissible value. Therefore,

$$M = \frac{19.555 \times 10^8 \times 80}{233.33} = 670.5 \text{ kN m}$$ ■

5.3.3 Stress Variation Along the Length and in the Beam Section

We have seen the variation of bending stress along the depth of the beam section. In symmetrical beam sections (symmetry with respect to the neutral axis), the stress is zero at the *NA* and varies linearly across the depth. In the case of sections having no symmetry with respect to neutral axis, the stress variation remains the same.

The neutral axis in this case is not at half the depth as in the case of symmetrical sections. Both these cases are shown in Fig. 5.14.

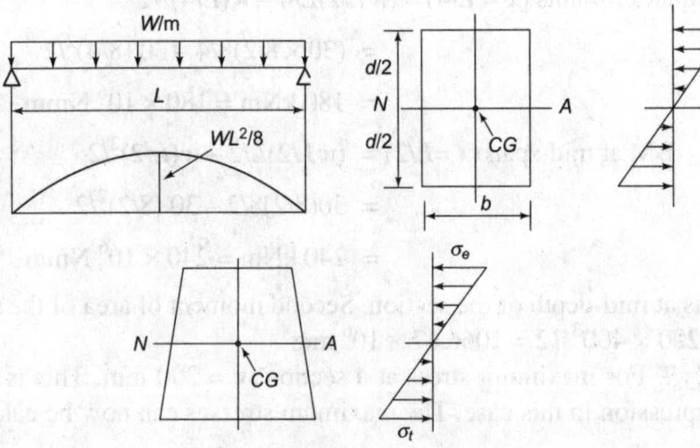

Fig. 5.14 Bending stress variation along the depth

You will notice that maximum stress is only at the top and bottom fibres. The rest of the section is stressed to lesser values of the stress than the maximum stress. The material is not used efficiently in bending. Compare this with axial stress where the stress distribution is uniform across the section. An I-section is more efficient because it has less material at depths where the stresses are less.

The bending equation was derived for a case of pure bending where the BM is constant over the length. In actual beams, this is not the case. The bending moment varies along the length. Let us illustrate with an example.

Example 5.6 Bending stress variation along length

A beam of rectangular section, 200 mm × 400 mm carries a uniformly distributed load of 30 kN/m over the whole span. Find the maximum bending stresses at mid-span and at quarter span points.

Solution The beam with the loading is shown in Fig. 5.15. BM at any point to a distance x from A can be found as

$$M_x = (wL/2)x - wx^2/2$$

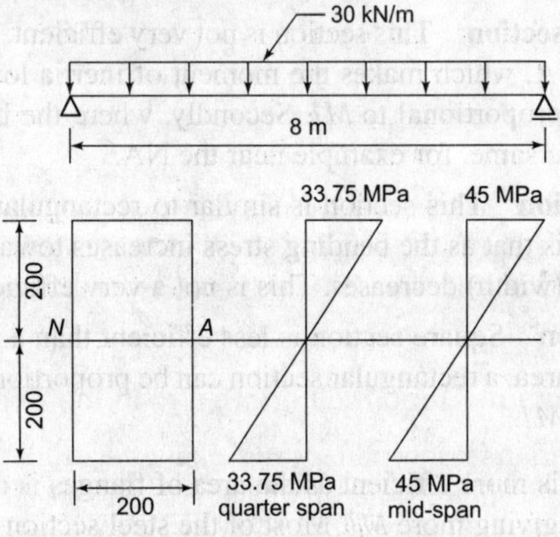

Fig. 5.15

The BM diagram is a parabola and is shown in Fig. 5.15. BM at quarter points and at mid-span are ass follows.

$$\text{BM at quarter points } (x = L/4) = (wL/2)L/4 - w(L/4)^2/2$$
$$= (30 \times 8/2)8/4 - 30\,(8/4)^2/2$$
$$= 180 \text{ kNm} = 180 \times 10^6 \text{ Nmm}$$
$$\text{BM at mid-span } (x = L/2) = (wL/2)L/2 - w(L/2)^2/2$$
$$= 30(8/2)8/2 - 30\,(8/2)^2/2$$
$$= 240 \text{ kNm} = 240 \times 10^6 \text{ Nmm}$$

Neutral axis is at mid-depth of the section. Second moment of area of the section about NA = $bd^3/12 = 200 \times 400^3/12 = 1066.67 \times 10^6$ mm^4.

Stress $\sigma = My/I$. For maximum stress at a section, y = 200 mm. This is the same for tension and compression in this case. The maximum stresses can now be calculated at the three sections.

At quarter points, $\sigma = 180 \times 10^6 \times 200/1066.67 \times 10^6 = 33.75$ N/mm^2

At mid-span, $\sigma = 240 \times 10^6 \times 200/1066.67 \times 10^6 = 45$ N/mm^2

The stress diagrams are shown in Fig. 5.15.

At support points, theoretically, BM is zero and bending stresses are zero. However, SF is high at these points and section is required to resist the same. ■

5.3.4 Effect of Shape of Beam Section on Stress Induced

The bending equation derived and used so far requires that the beam section is symmetrical about the vertical (perpendicular to the NA) axis through the centroid. If the section is not symmetrical about the vertical axis, this will cause bending about the vertical axis also.

From $\sigma = (M/I)y$, the bending stress varies linearly with y and remains the same for all sections. The shape of the section affects the distribution because it has an influence on the value of I, the second moment of area of the section. A number of sections are shown in Fig. 5.16 and the stress distribution remains the same for all sections. We consider each one of the sections.

(a) **Rectangular section** This section is not very efficient. Firstly, large area is close to the NA, which makes the moment of inertia low. Bending stresses are inversely proportional to *MI*. Secondly, where the bending stresses are low, area is the same, for example near the NA.

(b) **Circular section** This section is similar to rectangular section. An added disadvantage is that as the bending stress increases towards the top and bottom, the area (width) decreases. This is not a very efficient section.

(c) **Square section** Square section is less efficient than a rectangular section. For the same area, a rectangular section can be proportioned more efficiently to give larger *MI*.

(d) **I-section** It is more efficient as the area of flanges is distributed far away from the NA, giving more *MI*. Most of the steel sections used for structural use are of this shapes.

(e) **T-section** A T-section is an I-section without one of the flanges. It is an unsymmetrical section but is less efficient than the I-section.

(f) **Hollow circular section** A hollow circular section is more efficient than a solid circular section. Pipes carrying water and supported at some distance is an example of a hollow circular section.

The above discussion is only with respect to bending stresses. When you also consider shear stress (see Section 5.6), the requirement of sectional dimensions changes. A balance has to be made in proportioning the beam.

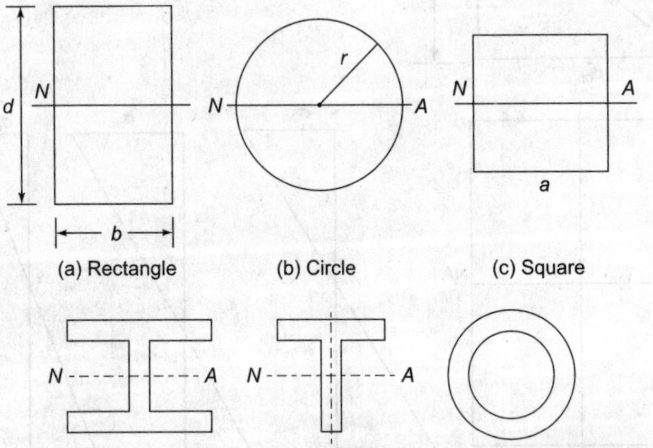

Fig. 5.16 Beam sections

5.4 DESIGN OF BEAMS FOR STRENGTH

Beams are designed in such a way that the tensile and compressive stresses do not exceed the permissible limits at the section of maximum BM. The permissible stresses in tension and compression may be the same or different. The distance to the extreme fibre in tension and compression may also be different in the case of sections unsymmetrical with reference to the NA.

Let y_c and y_t be the extreme fibre distances (from the NA) to the compressive and tensile fibres, and σ_{bc} and σ_{bt} be the compressive and tensile stresses permissible for the material [Fig. 5.17(a)].

$$M = \sigma\frac{I}{y}, \quad M_c = \sigma_{bc}\frac{I}{y_c}, \quad \text{and} \quad M = \sigma_{bt}\frac{I}{y_t}$$

If the section is symmetrical, $y_c = y_t$, and if $\sigma_{bc} = \sigma_{bt}$, any of these relationships can be used to find the moment of resistance. If $\sigma_{bc} \neq \sigma_{bt}$, the lesser of the two values calculated from the equations is the moment capacity of the section [Fig. 5.17(b)].

In the case of unsymmetrical sections, if $\sigma_{bc} = \sigma_{bt}$, then the larger of the two values y_c or y_t, should be used as this will give the lower value for M_r. If $\sigma_{bc} \neq \sigma_{bt}$, then both values of M may be calculated and the lower value taken [Fig. 5.17(c)].

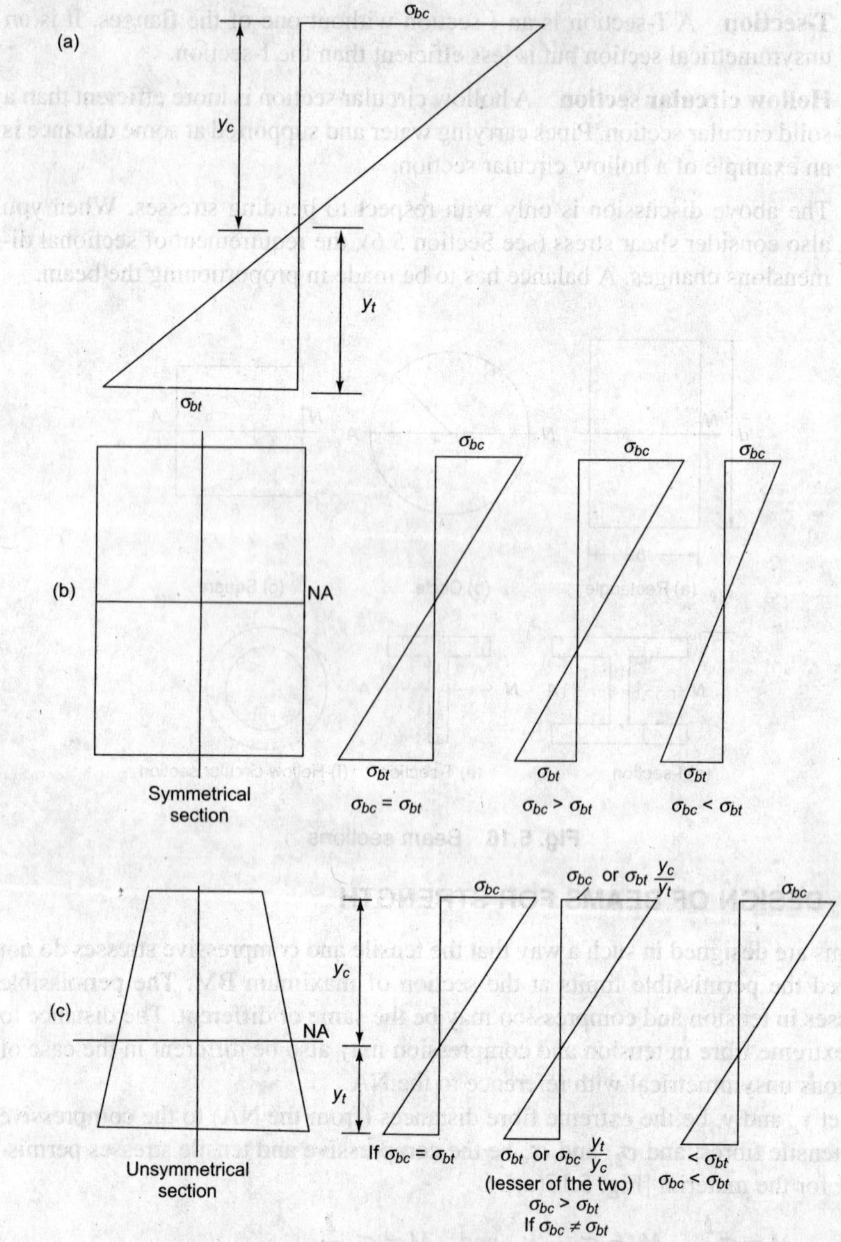

Fig. 5.17

5.4.1 Section Modulus

The calculation of section modulus has been discussed in Chapter 2. The values I/y_c and I/y_t are known as section modulus or modulus of section. Please note that extreme fibre distances should be used to calculate the section modulus. Denoting by Z_c and Z_t the section moduli in compression and tension,

$$M_c = \sigma_{bc} Z_c$$

or

$$M_t = \sigma_{bt} Z_t$$

To design a beam section, the value of the BM is known and these equations are used to calculate the required section modulus or second moment of area.

Section moduli for some common sections are given in Table 5.1

Table 5.1 Section modulus (I/y)

(Z_t is about bottom/tension edge and Z_c is about top/compression edge.)

S. No.	*Section*	*Moment of inertia*	*Section modulus Z*
1	Rectangular section of width b and height d	$bd^3/12$	$Z_t = Z_c = bd^2/6$
2	Square section of side a	$a^4/12$	$Z_t = Z_c = a^3/6$
3	Diamond section (square with diagonal horizontal) of side a	$a^4/12$	$Z_t = Z_c = \sqrt{2}a^3/12$
4	Circular section of diameter d	$\pi d^4/64$	$Z_t = Z_c = \pi d^3/32$
5	Hollow circular section of internal diameter d_i and external diameter d_e	$\pi(d_e^4 - d_i^4)/64$	$Z_t = Z_c = \dfrac{\pi(d_e^4 - d_i^4)}{32\, d_e}$
6	Triangle of base b and height h	$bh^3/36 = bh^2/12$	$Z_t = bh^2/24$; $Z_c = bh^2/12$
7	Equilateral triangle of side a	$\sqrt{3}a^4/96$	$Z_t = a^3/32$; $Z_c = a^3/16$
8	Regular hexagon of side a	$5\sqrt{3}a^4/16$	$Z_t = Z_c = 5a^3/8$
9	Semicircle of diameter d	$0.11d^4$	$Z_t = 0.52d^3$; $Z_c = 0.38d^3$
10	Quarter circle of diameter d	$0.0123d^4$	$Z_t = 0.058d^3$; $Z_c = 0.044d^3$

5.4.2 Modulus of Rupture

When a beam is loaded up to failure, the stresses in the beam section cannot be calculated by using the flexure formula. This is, however, done sometimes to compare strengths of beams of different materials. The stress calculated at rupture using the bending equation is known as the modulus of rupture. The stresses calculated are the maximum values in the section due to failure load.

Example 5.7 Design of a timber rectangular beam for BM

A timber beam has to carry a load of 2 kN/m over a span of 3 m. The permissible stresses are 12 MPa in compression and 8 MPa in tension. Design the section if the width is half of the depth.

Solution The beam with the loading is shown in Fig. 5.18.

$$\text{Maximum BM} = \frac{wl^2}{8} = \frac{2 \times 3^2}{8}\ \text{kN m} = \frac{18}{8} \times 10^6\ \text{N mm}$$

As the section is symmetrical about the NA, the lower value of $\sigma_{bt} = 8\ \text{N/mm}^2$ is to be used.

$$M = \sigma Z, \quad \sigma = 8\ \text{N/mm}^2, \quad Z = \frac{M}{\sigma} = \frac{18}{8} \times \frac{10^6}{8} = 0.2812 \times 10^6\ \text{mm}^3$$

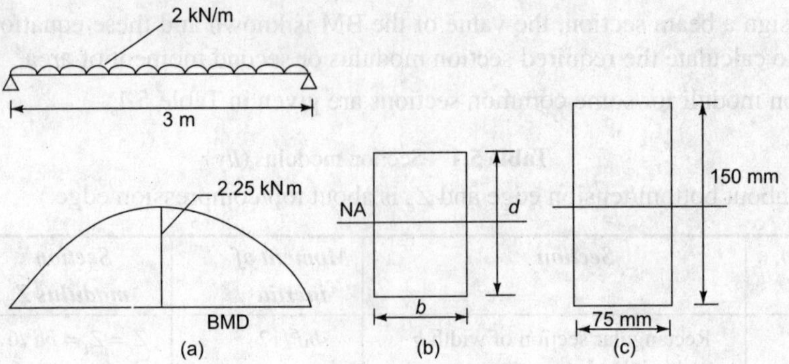

Fig. 5.18

If the cross section of the beam is $b \times d$ mm, then

$$z = \frac{I}{y} = \frac{bd^3 \times 2}{12 \times d} = \frac{bd^2}{6} = \frac{d^3}{12}$$

since $b = d/2$. Therefore,

$$\frac{d^3}{12} = 0.2812 \times 10^6$$

$$\Rightarrow \quad d = 150 \text{ mm}, \quad b = 75 \text{ mm}$$

The section is shown in Fig. 5.18(c).

Example 5.8 Design of steel I-section for BM

A steel beam has a span of 16 m and carries a load of 14 kN/m. Design a suitable I-section if the permissible stress is 150 MPa.

Solution From Fig. 5.19,

$$\text{Maximum BM} = \frac{wl^2}{8} = \frac{14 \times 16^2}{8} \times 10^6 \text{ N mm}$$

$$\text{Section modulus required} = \frac{M}{\sigma} = \frac{14 \times 16^6}{8} \times \frac{10^6}{150} = 2.986 \times 10^6 \text{ mm}^3$$

Select an I-section. ISMB 600 @ 1230 N/m has a modulus of section Z_{XX} of 3.06×10^6 mm^4.

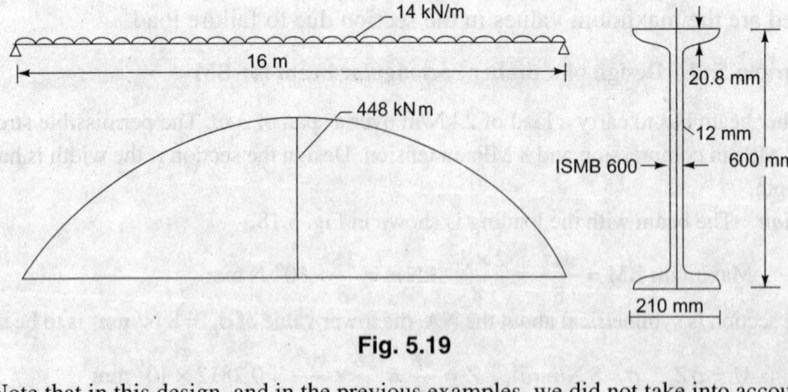

Fig. 5.19

Note that in this design, and in the previous examples, we did not take into account the self-weight of the beam. The selected beam weighs 1.23 kN/m. The beam has to withstand a load of 14 + 1.23 = 15.23 kN/m. For this load, the BM (maximum) is

$$M_{\max} = \frac{Wl^2}{8} = \frac{15.23 \times 16^2}{8} = 487.36 \text{ kN m}$$

$$\text{Section modulus } Z \text{ required} = \frac{487.36 \times 10^6}{150}$$
$$= 3.249 \times 10^6 \text{ mm}^4$$

You will notice that the selected section is not adequate as the Z_{XX} of this section is only 3.06×10^6 mm^4. We can either add extra plates to the I-section or select an I-section with a higher Z_{XX} value. ISWB 600 @ 1.312 kN/m will be found to be sufficient with a revised calculation.

The reader should appreciate from this example that design is an iterative process. Either an assumed value of self-weight can be added in the beginning to the load or the section can be checked and revised later.

Example 5.9 Load carrying capacity of a given section

The section shown in Fig. 5.20 is used as a beam over a span of 6 m. The height h is so adjusted that the NA falls at the position shown. Find the UD load that can be put on the beam without exceeding the permissible stress of 14 N/mm^2 in compression and 9 N/mm^2 in tension.

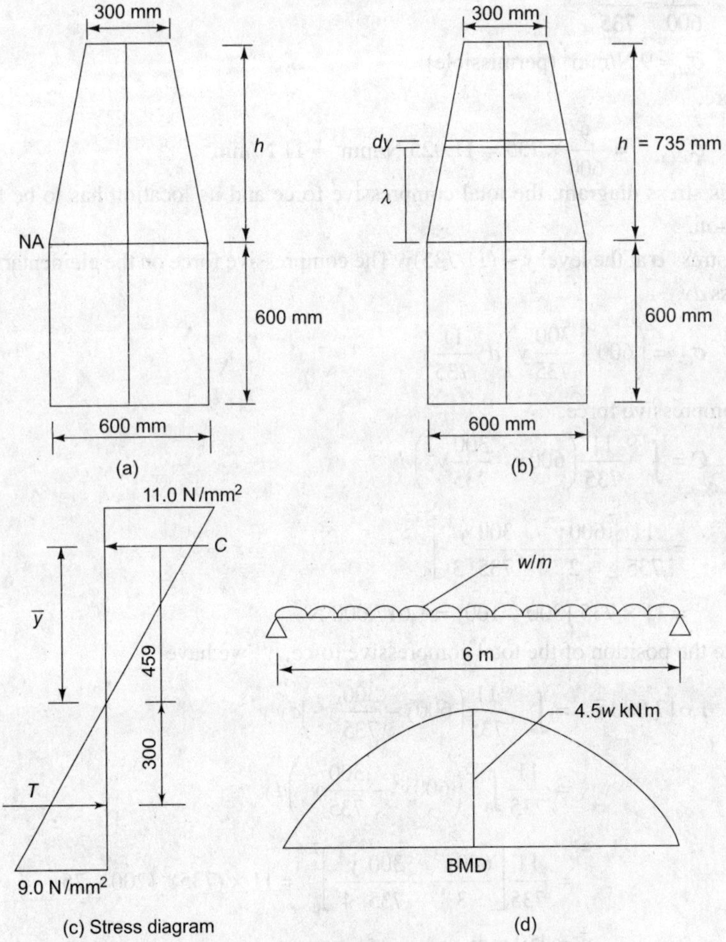

Fig. 5.20

Solution We first determine the height h by equating the moments about the NA of the portions above and below. From Fig. 5.20(b), consider an elementary strip at a distance y and of thickness dy.

$$\text{Width of the elementary strip} = 600 - \frac{(600-300)}{h}y$$

$$\text{Moment of elementary strip about NA} = \left(600 - \frac{300y}{h}\right)dy\,y$$

The moment of the part above the NA is obtained by integrating this from 0 to h.

$$\text{Moment of area about NA} = \int_0^h \left(600y - \frac{300y^2}{h}\right)dy$$

$$= \left[\frac{600y^2}{2} - \frac{300y^3}{3h}\right]_0^h = 300h^2 - 100h^2 = 200h^2$$

This is equal to the moment of the area below the NA, about the NA.

$$200h^2 = 600 \times 600 \times 300, \quad h = 735 \text{ mm}$$

The stress diagram for the beam is shown in Fig. 5.20(c). If the maximum stress in compression is 11 N/mm^2, the corresponding maximum tensile stress σ_{bt} can be obtained as follows.

$$\frac{\sigma_{bt}}{600} = \frac{11}{735}$$

$$\Rightarrow \quad \sigma_{bt} \approx 9 \text{ N/mm}^2 \text{ (permissible)}$$

Therefore,

$$\sigma_{bc(\max)} = \frac{9}{600} \times 735 = 11.025 \text{ N/mm}^2 \approx 11 \text{ N/mm}^2$$

With this stress diagram, the total compressive force and its location has to be found by integration.

The stress σ at the level $y = (11/735)y$. The compressive force on the elementary strip of thickness dy,

$$\sigma_C = \left(600 - \frac{300}{735}y\right)dy\frac{11}{735}y$$

Total compressive force,

$$C = \int_0^{735} \frac{11}{735}\left(600y - \frac{300}{735}y^2\right)dy$$

$$= \frac{11}{735}\left[\frac{600y^2}{2} - \frac{300}{735}\frac{y^3}{3}\right]_0^{73.5}$$

$$= 11 \times 735\,[300 - 100] = 1{,}617{,}000 \text{ N}$$

To locate the position of the total compressive force, $\bar{y}$, we have

$$1{,}617{,}000\,\bar{y} = \int_0^{735} \frac{11}{735}\left(600y - \frac{300}{735}y^2\right)dy\,y$$

$$= \frac{11}{735}\int_0^{735}\left(600y^2 - \frac{300}{735}y^3\right)dy$$

$$= \frac{11}{735}\left[\frac{600y^3}{3} - \frac{300}{735}\frac{y^4}{4}\right]_0^{735} = 11 \times (735)^2\,(200 - 75)$$

$$\bar{y} = 459 \text{ mm}$$

Thus, the lever arm = 300 + 459 = 759 mm.

Moment of resistance = 1,617,000 × 759 = 1227 kN m

$$\frac{wl^2}{8} = 1227, \quad w = \frac{1227 \times 8}{36} = 272.7 \text{ kN/m}$$

Permissible load = 272.7 kN/m

Example 5.10 Channel section carrying water

The channel section shown in Fig. 5.21(a) carries water to full depth. If the permissible stress in the material is 150 MPa, find the spacing at which the channel should be supported.

Solution We first locate the centroid of the channel section. From Fig. 5.21(a),

$$\bar{y} = \frac{2 \times 500 \times 10 \times 250 + 980 \times 10 \times 5}{2 \times 500 \times 10 + 980 \times 10} = 128.7 \text{ mm}$$

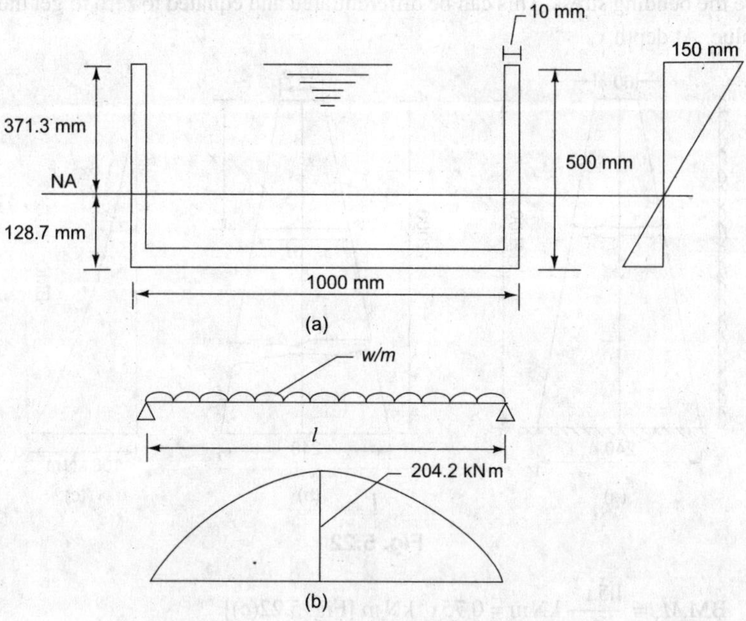

Fig. 5.21

We will now calculate the MI of the channel section about the NA.

$$I = \frac{980 \times 10^3}{12} + 980 \times 10 \times 123.7^2 + 2\left(\frac{10 \times 500^3}{12} + 10 \times 500 \times 121.3^2\right)$$

$$= 5.0555 \times 10^8 \text{ mm}^4$$

Z = section modulus is calculated from the larger value of the extreme fibre distance of 371.3 mm.

$$Z = \frac{5.0555 \times 10^8}{371.3} = 1.36145 \times 10^6 \text{ mm}^3$$

Moment capacity of the section $= \sigma Z = 150 \times 1.36145 \times 10^6$

$= 204.2$ kN m

Load/meter length due to water $= 0.98 \times .49 \times 10 = 4.8$ kN/m

From Fig 5.18(b), if l is the length between supports, assuming SS end conditions,

$$\text{BM} = \frac{4.8 \times l^2}{8} \text{ kN m}$$

$$\frac{4.8 \times l^2}{8} = 204.2$$

$$\Rightarrow \quad l = 18.4 \text{ m}$$

Example 5.11 Maximum stress in a tapering flagstaff

A tapering flagstaff is 12 m long and has a diameter of 240 mm at the base and 60 mm at the top. Find the maximum stress in the staff due to a uniform horizontal wind load of 1.5 kN/m.

Solution This situation is shown in Fig. 5.22(a). As the section increases in width towards the bottom, so does the BM. We take a section x below the top [Fig. 5.22(b)] and calculate the bending stress. This can be differentiated and equated to zero to get the maximum value. At depth x,

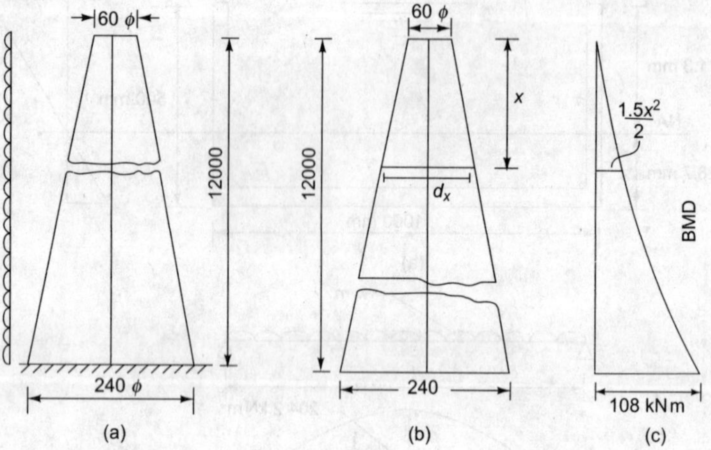

Fig. 5.22

$$\text{BM } M_x = \frac{1.5x^2}{2} \text{ kN m} = 0.75x^2 \text{ kN m [Fig. 5.22(c)]}$$

Diameter at x,

$$d_x = 60 + \frac{(240 - 60)}{12000} \times 1000x = 60 + 15x$$

MI at x,

$$I_x = \frac{\pi d^4}{64} = \frac{\pi(60+15x)^4}{64}$$

$$Z_x = \frac{I_x}{y_x} = \frac{\pi(60+15x)^4 \times 2}{64 \times (60+15x)}$$

$$= \frac{\pi(60+15x)^3}{32}$$

Maximum stress,

$$\sigma_x = \frac{M_x}{Z_x} = \frac{0.75 \times x^2 \times 10^6 \times 32}{\pi(60+15x)^3}$$

The maximum stress varies with x as both M_x and Z_x change with x. For the absolute maximum value of σ_x,

$$\frac{d\sigma_x}{dx} = 0$$

$$\frac{d\sigma_x}{dx} = \frac{0.75 \times 32 \times 10^6}{\pi} [(60 + 15x)^3 \times 2x - x^2 \times 3(60 + 15x)^2 \times 15] = 0$$

$$(60 + 15x) \times 2x - 45x^2 = 0, \quad 2(60 + 15x) - 45x = 0$$

$$120 + 30x - 45x = 0$$

$$x = 8 \text{ m}$$

$$\text{Maximum stress} = \frac{0.75 \times 8^2 \times 10^6 \times 32}{\pi(60 + 15 \times 8)^3} = 125.7 \text{ N/mm}^2$$

Example 5.12 Strength comparison of rectangular and hollow circular sections

Compare the strengths of the two sections shown in Fig. 5.23 if they are of the same material, area, and thickness.

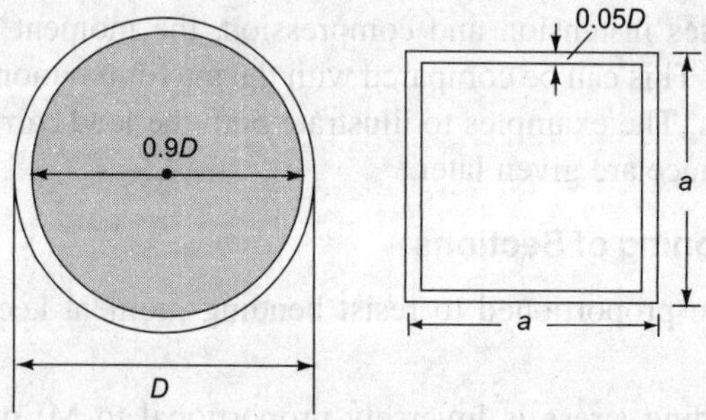

Fig. 5.23

Solution The strengths of the sections shown will be proportional to their section moduli, since the permissible stresses are the same.

$$\text{Area of hollow circular section} = \frac{\pi D^2}{4}(1 - 0.81)$$

$$= \frac{\pi D^2}{4} \times 0.19$$

$$\text{Area of square tubular section} = a^2 - (a - 0.1D)^2$$

$$= a^2 - a^2 - 0.01D^2 + 0.2aD$$

Since the areas are equal,

$$0.2aD - 0.01D^2 = \frac{\pi D^2}{4} \times 0.19$$

$$a = 0.796D$$

Section moduli For the hollow circular section, the MI is given by

$$I_c = \frac{\pi}{64} D^4 [1 - (0.9)^4] = 0.01688D^4 \text{ units}$$

For the square section, the MI is

$$I_s = \frac{a^4}{12} - \frac{(a - 0.1D)^4}{12}$$

Setting $a = 0.796D$,

$$I_s = \frac{D^4}{12}(0.796^4 - 0.696^4) = 0.0139D^4$$

$$Z_c = \frac{0.01688D^4}{D} \times 2 = 0.03376D^3$$

for the hollow circular section, and

$$Z_s = \frac{0.139D^4 \times 2}{D} = 0.0278D^3$$

for the square section.

The hollow circular section will have more strength, as it has a higher value of Z. The ratio of strengths is $0.03376D^3/0.0278D^3 = 1.214$.

5.4.3 Load Carrying Capacity

Given a beam section and the span on which it is required to support loads, we can find the load that the section can carry. This is generally done for UD load. More appropriate is the moment carrying capacity called *moment of resistance* of a section. This is independent of the span and the loads. Given a section and the maximum permissible stresses in tension and compression, the moment carrying capacity can be calculated. This can be compared with the maximum moment in a span and the load condition. The examples to illustrate both the load carrying capacity and moment of resistance are given later.

5.4.4 Proportioning of Sections

Sections are to be proportioned to resist bending moment keeping in mind the following.

(a) As the bending stress is inversely proportional to MI of the section, the sections are proportioned to give larger MI. An I-section is more efficient compared to a rectangular section having the same area because the I-section has area distribution that gives larger MI.

(b) While attaining larger MI, the section should not become too thin. For example, in the case of a rectangular section, as MI varies as the cube of the depth, section is proportioned to have larger depth. But the width to depth ratio should not be too small in which case instability may arise due to any accidental eccentricity from the plane of the beam. The width to depth ratio is thus normally kept between 1/2 and 2/3.

The above concepts are now illustrated with examples.

Example 5.13 Load capacity of beam section

A rectangular section, 200 mm × 400 mm, spans a distance of 2 m. Find what UD load can the beam section carry on this span. Permissible stresses are 7 N/mm^2 in tension and 15 N/mm^2 in compression.

Solution The situation is shown in Fig. 5.24. On a simply supported span, the maximum BM due a UD load is $wL^2/8$, where $L = 2$ m.

So, the maximum BM = $w2^2/8 = w/2$ kNm = $(w/2) \times 10^6$ Nmm. Here w is the load in kN/m.

For the section, MI = $200 \times 400^3/12 = 1066.67 \times 10^6$ mm^4

The stress in tension, 7 N/mm^2, governs the load capacity because of the symmetrical nature of the section.

$$M = I\sigma/y = (w/2) \times 10^6;\ w = 2I\sigma/y \times 10^6$$

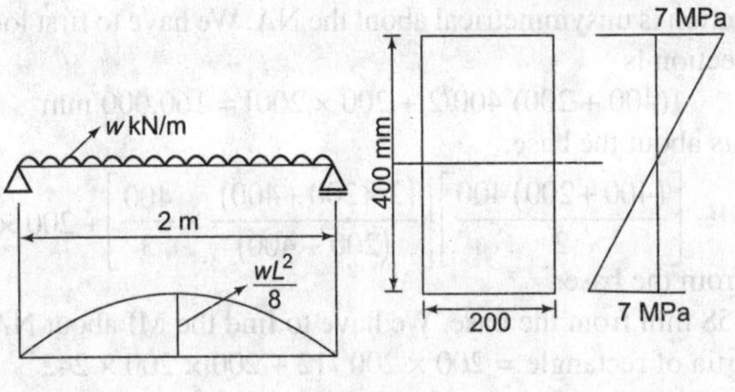

Fig. 5.24

Substituting values,

$$w = 2 \times 1066.67 \times 10^6 \times 7 /(200 \times 10^6) = 75 \text{ kN/m}$$

Thus, the load carrying capacity of the section is 75 kN/m. ■

Example 5.14 Moment of resistance of circular section

A circular log of wood is used as a beam. If the diameter of the log is 200 mm, find the moment of resistance of the section. Permissible stresses are 10 N/mm^2 in tension and 18 N/mm^2 in compression.

Solution The beam is shown in Fig. 5.25. From $M/I = \sigma/y$, we get $M = I\sigma/y$

For the given section,

$$\text{MI} = \pi d^4/64 = \pi(200)^4/64 = 78.54 \times 10^6 \text{ mm}^4$$

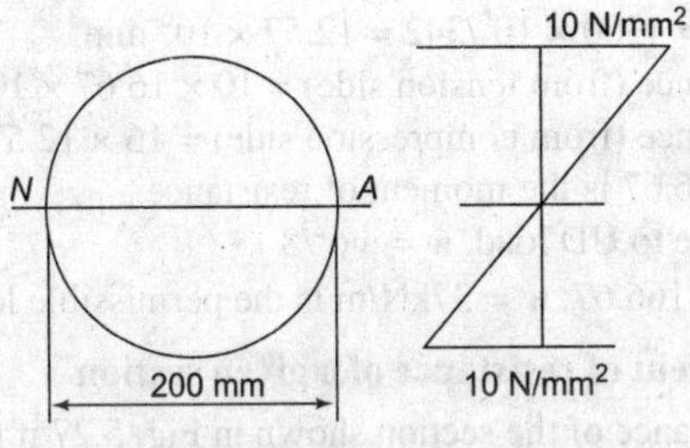

Fig. 5.25

The value of σ is the lower of the two values. If we take the maximum stress as 18 N/mm^2, due to symmetry of the section, the maximum tensile stress will also be 18 N/mm^2, which is more than the permissible value. To calculate the moment of resistance, we take the maximum value of y, which is 100 mm.

Moment of resistance $= 78.54 \times 10^6 \times 10 /100 = 7.854 \times 10^6$ Nmm

$= 7.854$ kNm ■

Example 5.15 Moment of resistance of a given section

The section shown in Fig. 5.26 is used as a beam of span 6 m. Find the moment of resistance of the section. The permissible stresses are 10 N/mm^2 in tension and 16 N/mm^2 in compression. Find the UD load it can safely carry on this span.

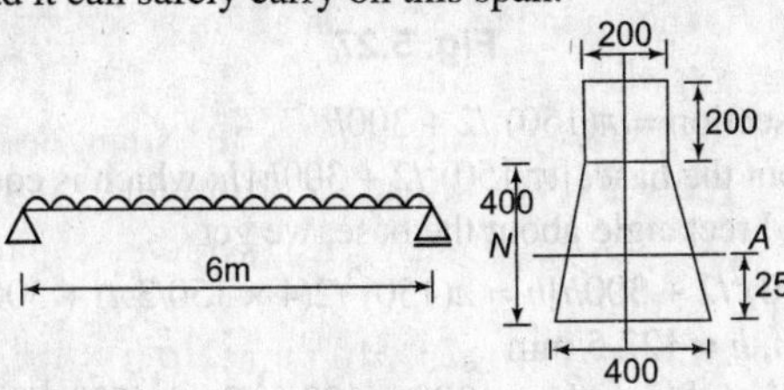

Fig. 5.26

Solution The section is unsymmetrical about the NA. We have to first locate the centroid. Total area of the section is

$$[(400 + 200)\,400/2 + 200 \times 200] = 160{,}000 \text{ mm}^2$$

Taking moments about the base,

$$160000\,\tilde{y} = \left[\frac{(400+200)\,400}{2}\right]\left[\frac{(2\times 200+400)}{(200+400)}\times\frac{400}{3}\right] + 200\times 200\times 500$$

$\tilde{y}$ = 258 mm from the base.

NA is thus at 258 mm from the base. We have to find the MI about NA.

Moment of inertia of rectangle = $200 \times 200^3/12 + 200 \times 200 \times 242^2$
$= (133.33 + 2342.56)\,10^6 = 2475.89 \times 10^6 \text{ mm}^6$

MI of trapezium can be considered as the sum of the MIs of a rectangle and two equal triangles. The rectangle has the dimensions 200 mm × 400 mm and the two equal triangles have base 100 mm and height 400 mm.

For the rectangle, MI (NA) = $200 \times 400^3/12 + 200 \times 400 \times 58^2$
$= 1336 \times 10^6 \text{ mm}^4$

For the two triangles, MI (NA) = $2[100 \times 400^3/36 + (100 \times 400/2) \times (258 - 400/3)^2]$
$= 488.6 \times 10^6 \text{ mm}^4$

Total MI = $(2475.89 + 1336 + 488.6) \times 10^6 \text{ mm}^4 = 4300 \times 10^6 \text{ mm}^4$

As the section is unsymmetrical and the stresses in tension and compression are also different, it is a better practice to calculate the moment of resistance with both values and take the least value.

Z_t (tension) = $4300 \times 10^6/258 = 16.67 \times 10^6 \text{ mm}^3$

Z_c (compression) = $4300 \times 10^6/342 = 12.57 \times 10^6 \text{ mm}^3$

Moment of resistance (from tension side) = $10 \times 16.67 \times 10^6$ = 166.7 kNm

Moment of resistance (from compression side) = $16 \times 12.57 \times 10^6$ = 201.12 kNm

The lower value of 166.67 is the moment of resistance.

Maximum moment due to UD load, $w = w6^2/8$

Therefore, $w \times 36/8 = 166.67$; w = 37kN/m is the permissible load on the beam.

Example 5.16 Moment of resistance of a given section

Find the moment of resistance of the section shown in Fig. 5.27 if the height of the rectangular portion is so adjusted that the NA is as shown. The permissible stress in tension is 120 MPa and in compression it is 160 MPa.

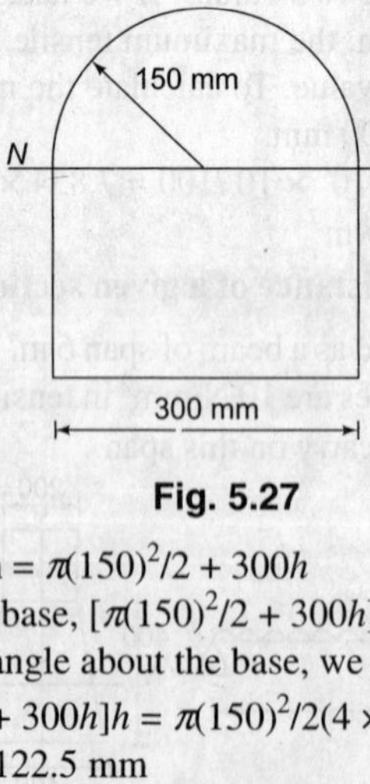

Fig. 5.27

Solution Area of the section = $\pi(150)^2/2 + 300h$

Taking moments about the base, $[\pi(150)^2/2 + 300h]\,h$, which is equal to the moment of the semicircular area and rectangle about the base, we get

$$[\pi(150)^2/2 + 300h]h = \pi(150)^2/2(4 \times 150/3\pi) + 300hh/2$$

Solving this equation, h = 122.5 mm

MI of the section = $300 \times 122.5^3/3 + \pi(150)^4/8 = 382.63 \times 10^6 \text{ mm}^4$

$Z_t = 382.63 \times 10^6 / 122.5 = 3.1235 \times 10^6$ mm^3 (tension/bottom fibre)

$Z_c = 382.63 \times 10^6 /150 = 2.55 \times 10^6$ mm^3 (compression/top fibre)

Moment of resistance is the lower value of $10 \times 3.1235 \times 10^6$ and $16 \times 2.55 \times 10^6$ Nmm.

MR = 31.235 kNm (The other value is higher at 40.8 kNm.)

Example 5.17 Design of a rectangular section for BM

Design a rectangular section to carry a load of 32 kN/m over its whole span of 3 m, including the self-weight. Permissible stresses are 8 N/mm^2 in tension and 15 N/mm^2 in compression.

Solution The beam is shown in Fig. 5.28.

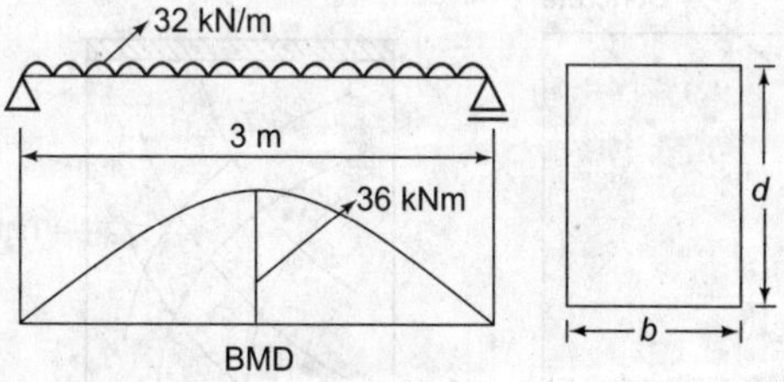

Fig. 5.28

Maximum BM in the beam = $32 \times 3^2/8 = 36$ kNm $= 36 \times 10^6$ Nmm

We select a section with height 2 × breadth ($d = 2b$).

MI = $bd^3/12$; $Z = I/(d/2) = bd^2/6 = b(2b)^2/6 = 4b^3/6$

M = $\sigma Z = 8 \times 4b^3/6$ (lower value of stress is taken as the section is symmetrical)

$32b^3/6 = 36 \times 10^6 = 6.75 \times 10^6$; $b = 190$ mm; $d = 380$ mm

A rectangular section of width 190 mm and depth 380 mm is required.

Example 5.18 Moment of resistance

Find the moment resistance of the section shown in Fig. 5.29. Permissible stresses are 100 MPa in tension and compression. What UD load can it carry over a span of 8 m?

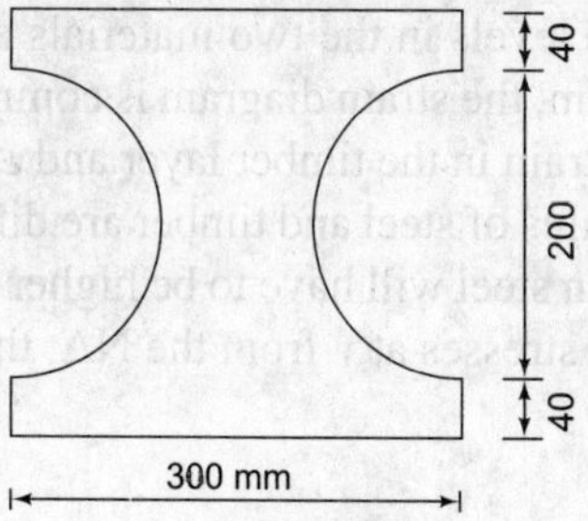

Fig. 5.29

Solution The section is symmetrical. The NA will be at 140 mm from the base or the top edge.

Moment of inertia = $300 \times 280^3/12 - 2\,[\pi(100)^4/8] = 4.715 \times 10^8$ mm^4

As the section is symmetrical, the moment of resistance can be calculated from either side.

Section modulus, $Z = 4.715 \times 10^8/140 = 3.368 \times 10^6$ mm^3

Moment of resistance = $\sigma Z = 100 \times 3.368 \times 10^6$ Nmm = 336.8 kNm

If w kN/m is the UD load on the span of 8 m, maximum BM = $w8^2/8 = 8w$ kNm

This must be equal to the moment of resistance of the section.

$8w = 336.8$

So, $w = 42$ kN/m is the UD load that the section can carry. ■

5.5 COMPOSITE SECTIONS

When beams are made of two materials which are rigidly joined so that they act as one section, they are said to form a composite section. Examples are reinforced concrete beams and flitched beams (timber sections strengthened using steel plates), shown in Fig. 5.30.

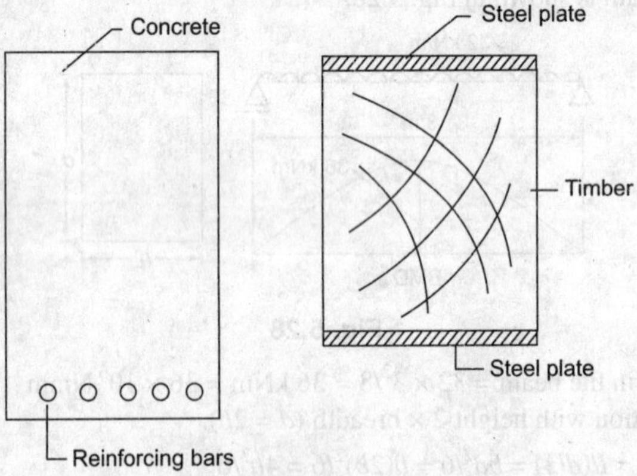

Fig. 5.30

5.5.1 Behaviour of a Composite Beam

It is assumed that the two parts of different materials are so joined that they act together. If it were not so, each part would act individually and the total strength would be the sum of the strengths of the individual parts.

In a composite beam, the behaviour is governed by the deformation condition that the strains at the same levels in the two materials are the same. As shown in Fig. 5.31, for a flitched beam, the strain diagram is common for both the materials. At any level y, if ε_t is the strain in the timber layer and ε_s the strain in steel, $\varepsilon_t = \varepsilon_s$.

Since the Young's modulus of steel and timber are different ($E_s \approx 20E_t$), it naturally means that the stress in steel will have to be higher at the same level for equal strains. If σ_s and σ_t are the stresses at y from the NA, then

$$\sigma_s = E_s \varepsilon_s$$

and $$\sigma_t = E_t \varepsilon_t$$

Since $\varepsilon_s = \varepsilon_t$,

$$\frac{\sigma_s}{E_s} = \frac{\sigma_t}{E_t}$$

or

$$\sigma_s = \frac{E_s}{E_t} \sigma_t$$

The quantity E_s/E_t is known as the modular ratio m ($\sigma_s = m\,\sigma_t$). m is approximately 20, so that stress in steel is 20 times that in timber at the same level, as shown in the stress diagram. In finding the strength of a composite beam, it must be ensured that the stress in either material is not exceeded.

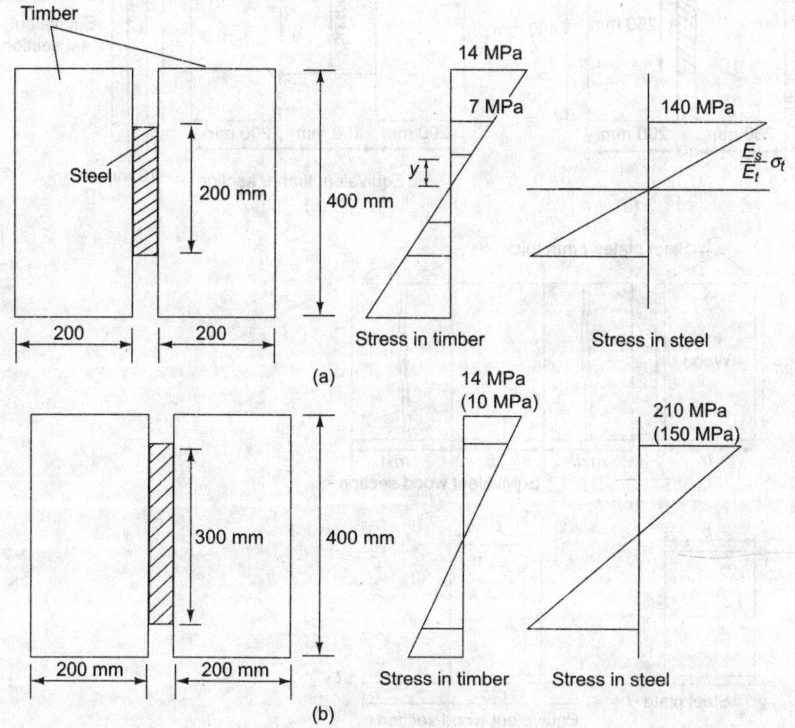

Fig. 5.31

If we consider the cases shown in Fig. 5.31, we will obtain the strain diagrams as shown in each case. In the first case, if the permissible stresses in steel and timber are 150 MPa and 14 MPa respectively, assuming the stress in timber at the extreme fibre to be 14 MPa,

$$\text{Maximum stress in steel} = \frac{14}{200} \times 100 \times 20 = 140 \text{ MPa}$$

which is within permissible limits. Steel cannot be stressed to its full value because the stress in timber will exceed, in such a case, its permissible value.

In the second case, with the same permissible stresses,

$$\text{Maximum stress in steel} = \frac{14}{200} \times 150 \times 20 = 210 \text{ N/mm}^2$$

which is more than the permissible value. Hence the stress in steel should be taken as 150 MPa and the extreme fibre stress in timber is

$$\frac{150}{150} \times \frac{200}{20} = 10 \text{ N/mm}^2$$

as against the permissible value of 14 N/mm^2. The values shown in brackets in the stress diagram should be used to calculate the moment capacity of the section.

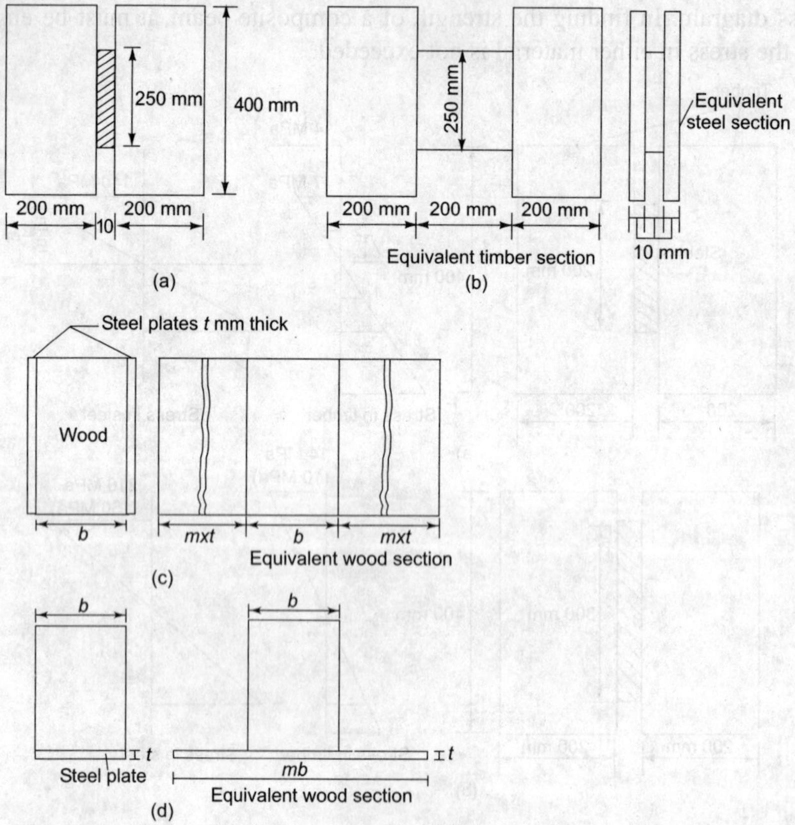

Fig. 5.32

Equivalent section As a second approach to solving composite beam problems, we can think of an equivalent section in one material only. Consider the flitched beam shown in Fig. 5.32. Steel can take 20 times the stress in timber or m (modular ratio) times the stress. Every unit area of steel is thus equivalent to 20 times this area in timber. However, this increase cannot be made in depth because that will change the strain, and correspondingly the stress, at any layer. The increase can be only in width. The equivalent section is thus constructed as shown in Fig. 5.32(b). Note that the strain and stress diagrams do not change. The equivalent section is in one material, i.e., timber. The equivalent sections in two more cases are shown in Fig. 5.32(c) and (d).

Example 5.19 Load carrying capacity of flitched beam

Determine the UD load that the flitched beam shown in Fig. 5.33 can carry over a span of 3 m. The permissible stress in steel is 150 MPa and for timber, the permissible stresses are 9 MPa in compression and 14 MPa in tension. Take $m = 20$.

Solution The strain and stress diagrams are shown in Fig. 5.33. The section being symmetrical about the NA, the stress in timber cannot exceed 9 MPa at the extreme fibres at the top and bottom.

With this stress in timber,

$$\text{Stress in steel} = \frac{9}{75} \times 50 \times 20 = 120 \text{ MPa} < 150 \text{ MPa}$$

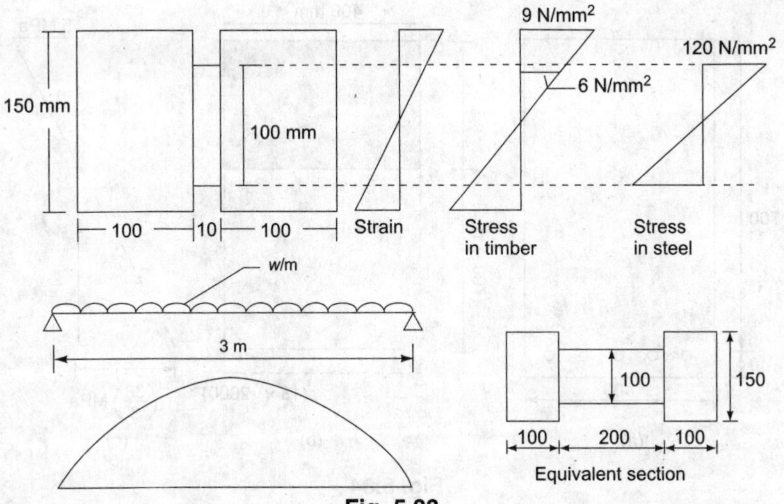

Fig. 5.33

Moment of resistance $M_r = \sigma Z$

where $Z = bd^2/6$.

$$\text{MR of steel section} = 120 \times 10 \times \frac{100^2}{6} = 2 \times 10^6 \text{ N mm}$$

$$\text{MR of timber section} = 2\left[9 \times 100 \times \frac{150^2}{6}\right] = 6.75 \times 10^6 \text{ N mm}$$

$$\text{Maximum BM on a 3 m span} = \frac{wl^2}{8} = \frac{w \times 9}{8} \text{ kN m}$$

$$\frac{w \times 9}{8} = (2 + 6.75)$$

$$\Rightarrow \quad w = 7.77 \text{ kN/m}$$

Alternately, considering the equivalent section shown in the figure, the timber section is formed by increasing the width of steel 20 times, which is the modular ratio.

$$\text{M } I = 2\left[\frac{100 \times 150^3}{12}\right] + \frac{200 \times 100^3}{12} = 72.92 \times 10^6 \text{ mm}^4$$

Taking the stress in the extreme fibre as 9 N/mm²

$$M = \frac{\sigma I}{Y} = \frac{9 \times 72.92 \times 10^6}{75} = 8.75 \times 10^6 \text{ Nmm} = 8.75 \text{ kNm}$$

Example 5.20 Moment capacity of a rectangular RCC section

The section shown in Fig. 5.34 is a reinforced concrete beam section. Find the moment capacity of the section if the permissible stress in concrete is 5 N/mm² in compression and 140 N/mm² in steel in tension. The modular ratio is 18.

Solution In a concrete section, the tension zone of the concrete is neglected as the tensile resistance of concrete is very small. The effective section is therefore, as shown in Fig. 5.34(b). There is concrete above the NA and steel below the NA. The steel area is equivalent to m times the concrete area.

To locate the NA, we find the centroid of the effective section. Equating the moments about the NA,

$$\frac{bx^2}{2} = mA_s(d - x), \quad \frac{400x^2}{2} = 18 \times 2000\,(700 - x)$$

$$x^2 + 180x - 126{,}000 = 0, \quad x = 276 \text{ mm}$$

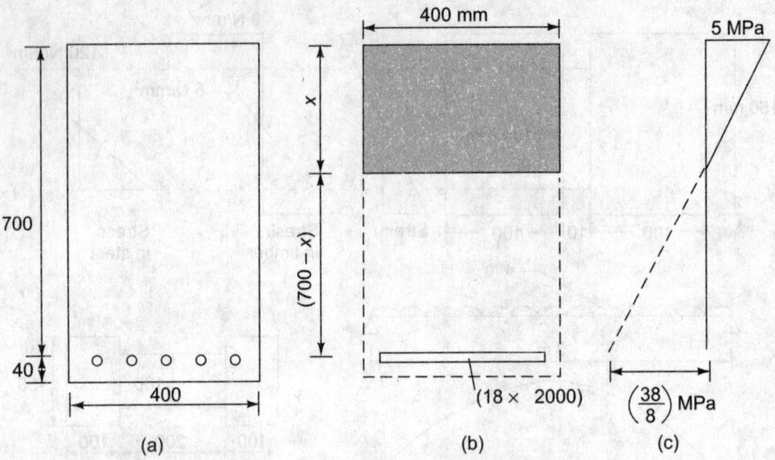

Fig. 5.34

If the extreme fibre stress is concrete is 5 N/mm², the stress in steel,

$$\sigma_{bt} = \frac{5}{276} \times (700 - 276) \times 18 = 138 \text{ N/mm}^2$$

which is less than 140 N/mm².

Total compressive force, $C = \dfrac{(5+0)}{2} \times 400 \times 276$

The moment of this force about tensile steel will give the moment of resistance.

$$\text{MR} = 2.5 \times 400 \times 276 \left(276 \times \frac{2}{3} + 424\right)$$

$$= 167.8 \times 10^6 \text{ N mm} = 167.8 \text{ kN m}$$

5.6 SHEAR STRESS IN BEAMS

There exists an SF and a BM at any section of most beams. We have seen the distribution of bending stress in beam cross sections. We can similarly work out the resistance of the beam to shear and the distribution of the shear stress on the beam section.

Consider the beam shown in Fig. 5.35. If squares are etched on the beam sides before loading, they deform as shown after loading, showing the existence of shear.

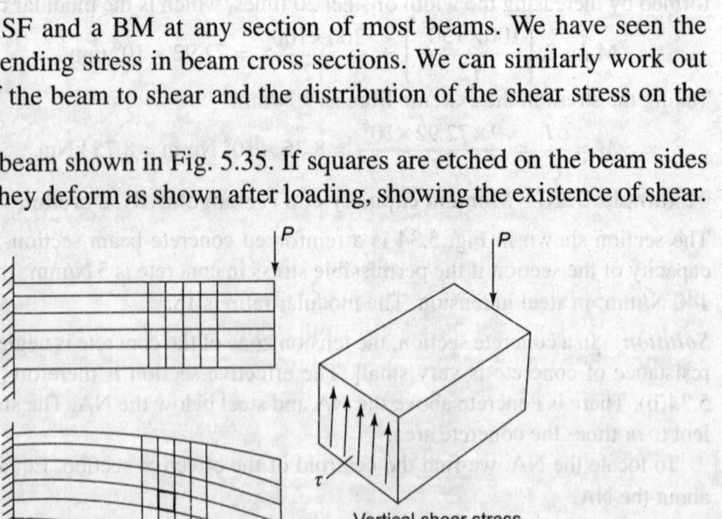

Fig. 5.35

However, determining the shear stress distribution is more complex than working out the bending stress distribution. The shear stress acts tangential to the cross section, i.e., in a vertical direction, to provible a shear stress resultant to balance the SF. Determining the vertical shear stress is difficult and is obtained indirectly, considering shear in the horizontal direction. Once the horizontal shear stress is known, the existence of vertical shear (as complementary shear stress) for equilibrium is established, as has already been discussed.

Equation for horizontal shear stress Consider the beam shown in Fig. 5.36. We isolate a differential length dx of the beam. There is a change in the BM between sections 1-1 and 2-2 on the left and right of the element, and a corresponding change in the bending stress. This is obvious for the existence of shear because if the BM is the same, there will be no shear. If $dM/dx = 0$, then $F = 0$ as we have seen in Chapter 4.

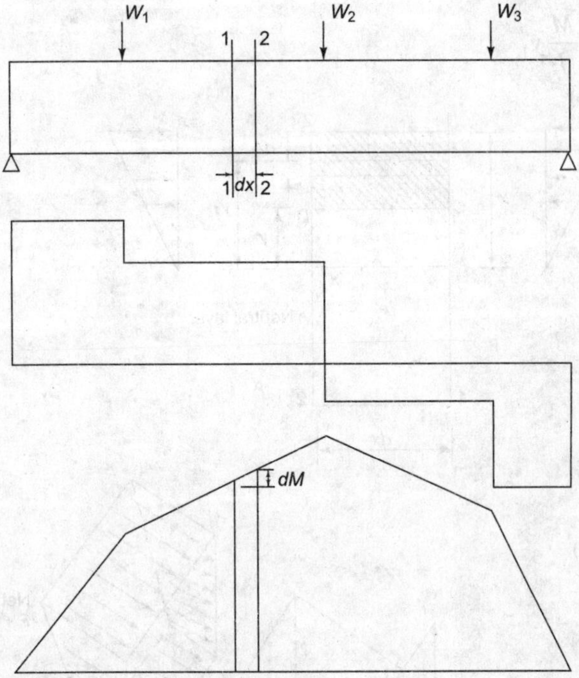

Fig. 5.36

Let M be the BM at section 1-1 and $M + dM$ that at section 2-2. Correspondingly the bending stress will also be different at the two sections as shown [Fig. 5.37(a)]. We consider the equilibrium of the top part of the length dx obtained by cutting a horizontal section at a distance y from the neutral layer. The forces on either side of this part are shown in Fig. 5.37(b).

The stresses on section 1-1 and 2-2 act on the areas of cross section of the beam, which are equal in the case of a prismatic beam. Obviously, since the stresses on the section 2-2 are different due to different BMs, the resultant horizontal force due to these stresses will also be different. Let F and $F + dF$ be the resultant forces on sections 1-1 and 2-2. The load acting on the beam is assumed transverse to the longitudinal axis and hence acts vertically.

The equilibrium of the isolated part requires that horizontal stresses be developed at the section *CD* over the length dx, giving a stress resultant of dF. One basic assumption made is that the horizontal shear stress is uniform along the width b of the beam. With this assumption, if τ is the horizontal shear stress, then

Resultant horizontal force at $CD = b\tau\, dx$

For equilibrium,

$$dF = b\tau\, dx$$

The horizontal force dF due to bending stresses can be calculated from the bending stress distribution. From the distribution shown in Fig. 5.37,

$$F = \int_y^{y_c} \sigma_1 dA \quad \text{and} \quad F + dF = \int_y^{y_c} \sigma_2 dA$$

where $y_c = x$, distance to the top fibre.

$$\sigma_1 = \frac{M}{I} y_1$$

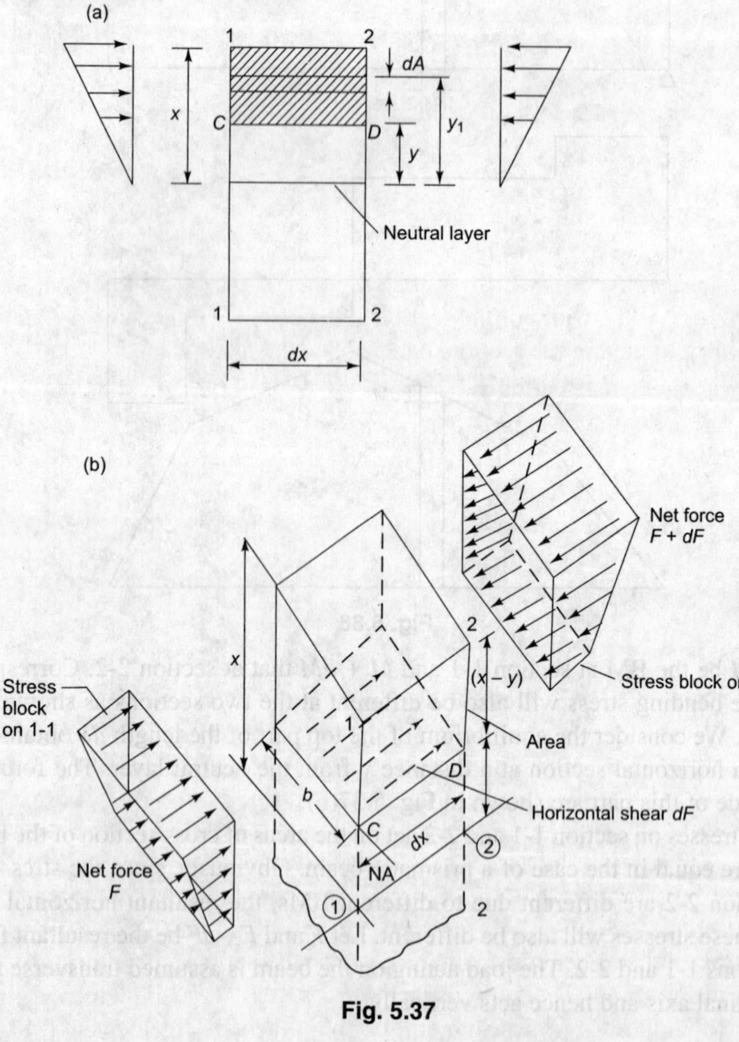

Fig. 5.37

and $\sigma_2 = \dfrac{M + dM}{I} y_1$

at the layer y_1. Therefore,

$$F = \frac{M}{I}\int_y^{y_c} y_1\,dA \text{ and } F + dF = \frac{M + dM}{I}\int_y^{y_c} y_1\,dA$$

From these equations,

$$dF = \frac{dM}{I}\int_y^{y_c} y_1\,dA$$

$\int_y^{y_c} y_1\,dA$ is the moment of the area above y about the NA. Setting this equal to $A\bar{y}$, we have

$$dF = \frac{dM}{I} A\bar{y}$$

Setting this value of dF in the equation for horizontal equilibrium, $dM/IA\bar{y} = b\,dx\,\tau$, from which,

$$\tau = \frac{dM}{dx}\frac{A\bar{y}}{Ib}$$

dM/dx is the SF at the section (as explained in Chapter 4). Using the symbol V for this, we have

$$\tau = \frac{V}{Ib} A\bar{y}$$

In this equation, as shown in Fig. 5.38, τ is the shear stress (horizontal) at a distance y from the NA, V is the SF at the section, I is the MI of the cross section about the NA, b is the width of the section at the distance y from the NA, and $A\bar{y}$ is the moment of area of the sectional area above the layer y about the NA.

The shear stress distribution is parabolic, as $A\bar{y}$ will be a second degree equation for sections with constant b. The shear stress is zero at the top and bottom layers and will vary along the depth. If the width b is constant, shear stress will be maximum at the NA as in the case of rectangular section. If the width b varies along the depth, the maximum shear stress will occur at some point along the depth, but not necessarily at the NA.

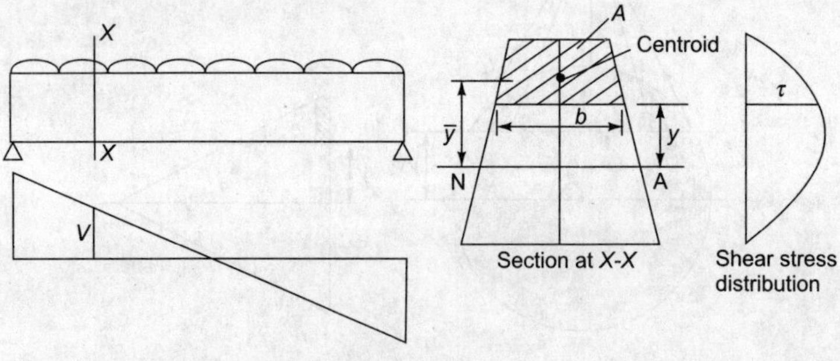

Fig. 5.38

The horizontal shear stress so obtained is also the vertical shearing stress. This is shown in Fig. 5.39. The vertical shearing stress is the complementary shear stress, and both these are equal for equilibrium of the small element shown.

Assumptions and limitations of the shear stress formula All the assumptions made in deriving the bending equation are valid in deriving the shear stress formula. In addition, it is assumed that the shear stress is uniform across the width at a layer distant *y* from the NA.

However, these assumptions are not valid at certain points along the beam, such as sections where point loads are applied, at support points, at the fixed end support of a cantilever, etc. Particularly, the assumption of a uniform shear stress across the width is obviously not valid, for example, in a circular section or in a triangular section, shown in Fig. 5.40. In the case of a circular section, the shear stress must be tangential at the extreme edges for equilibrium. In the case of a triangular section, the force calculated across the width for horizontal shear force is not uniform and, consequently, nor is the vertical shear stress.

Because of the uncertainty in shear stress distribution, most design codes stipulate an empirical method of determining average shear stress rather than the shear stress formula we have derived. The average shear stress is obtained by dividing the shear force by the area of cross section.

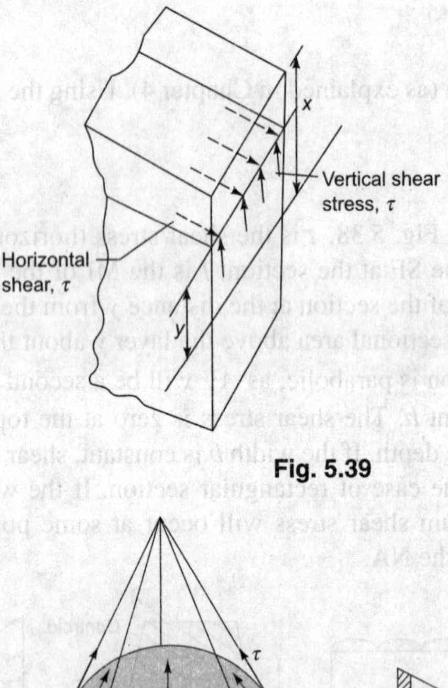

Fig. 5.39

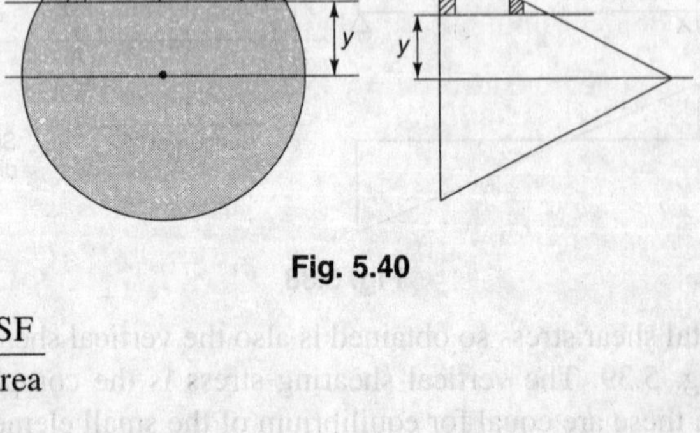

Fig. 5.40

$$\tau_{av} = \frac{\text{SF}}{\text{area}}$$

Design specifications prescribe limits to the average shear stress from experimental values rather than specifying a maximum permissible shear stress.

5.7 SHEAR STRESS DISTRIBUTION

First we will discuss the shear stress distribution in a rectangular section.

Rectangular section In the rectangular section shown in Fig. 5.41, the shear stress distribution at a layer distant y from the NA can be obtained as

$$\tau = \frac{V}{Ib}A\bar{y} = \frac{Vb(d/2 - y)\,1/2\,(d/2 + y)}{(bd^3/12)b}$$

This is a parabolic distribution with zero values at top fibre ($y = d/2$) and at the bottom fibre ($y = -d/2$). The maximum value can be seen to be at the NA, where

$$y = 0 \quad \text{and} \quad \tau_{max} = \frac{12V}{bd^3}\frac{d^2}{8} = \frac{3}{2}\frac{V}{bd}$$

This also shows that the maximum shear stress is 3/2 times the average shear stress (V/bd). The shear stress distribution is shown in Fig. 5.41.

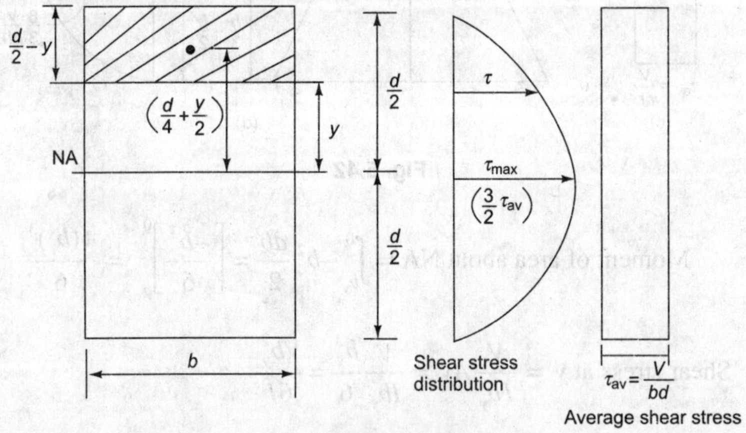

Fig. 5.41

Circular section In the circular section shown in Fig. 5.42(a), if we consider a layer at a distance y, the quantity $A\bar{y}$ is for the area shown hatched. To calculate this quantity, we have, for the elementary strip shown in the figure, $b^2 = 2(r^2 - z^2)$.

$$\text{Area of strip} = \sqrt{2(r^2 - z^2)}\,dz$$

$$\text{Moment of area about NA} = \left(\sqrt{2(r^2 - z^2)}\right) z\,dz$$

This will have to be integrated from $z = y$ to $z = r$. It will be easier to carry out the integration in terms of b:

$$b^2 = 2\,(r^2 - z^2)$$

Differentiating,

$$2b\,db = -4z\,dz, \quad z\,dz = -\frac{b\,db}{2}$$

$$\text{Moment of area about NA} = bz\,dz = -\frac{b^2\,db}{2}$$

The limits of integration for b are b_y at $Z = y$ to 0 at $Z = r$.

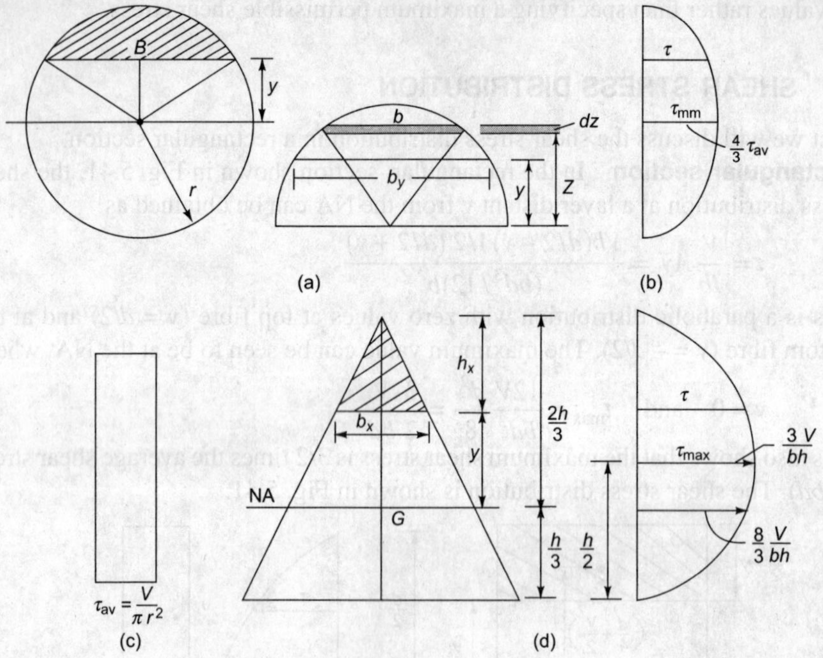

Fig. 5.42

$$\text{Moment of area about NA} = \int_{b_y}^{0} -b^2 \frac{db}{2} = \left[\frac{-b^3}{6}\right]_{b_y}^{0} = \frac{+(b_y)^3}{6}$$

$$\text{Shear stress at } y = \frac{V}{Ib_y} A\bar{y} = \frac{V}{Ib_y}\frac{b_y^3}{6} = \frac{Vb_y^2}{6I}$$

$$= \frac{V}{6\pi r^4/4} \times 2(r^2 - y^2) = \frac{4}{3}\frac{V(r^2 - y^2)}{\pi r^4}$$

setting $b_y^2 = 2(r^2 - y^2)$ and $I = \pi r^4/4$. This expression has a maximum value when $y = 0$.

$$\tau_{max} = \frac{4}{3}\frac{V}{\pi r^2} = \frac{4}{3}\tau_{av}$$

The shear stress distribution is as shown in Fig. 5.42(b).

Triangular section The NA passes through the centroid G if the section as shown in Fig. 5.42(d). For a layer at a distance y from the NA, the quantity $A\bar{y}$ is the moment of the triangular area shown hatched. If b_x and h_x are the base and height of this triangle, then

$$A\bar{y} = \frac{b_x h_x}{2}\left(\frac{2}{3}h - \frac{2}{3}h_x\right) = \frac{b_x h_x}{3}(h - h_x)$$

$$h_x = \frac{2}{3}h - y, \quad A\bar{y} = \frac{b_x}{3}\frac{(2h-3y)}{3}\left[h - \left(\frac{2}{3}h - y\right)\right]$$

$$= \frac{b_x}{27}(2h-3y)(h+3y)$$

Shear stress $\tau = \dfrac{V}{Ib}A\bar{y} = \dfrac{V}{Ib_x}\dfrac{b_x}{27}(2h-3y)(h+3y)$

$$I = \frac{bh^3}{36}, \quad \tau = \frac{36V}{bh^3}\frac{(2h-3y)(h+3y)}{27}$$

τ is zero at the top and bottom fibres, where $y = 2(h/3)$ or $h/3$. $d\tau/dy = 0$ at the layer of maximum shear stress.

$$\frac{d}{2y}[(2h-3y)(h+3y)] = 0, \quad 3h - 18y = 0$$

$$\Rightarrow \quad y = \frac{h}{6}$$

Height from base = $h/3 + h/6 = h/2$. The maximum shear stress occurs at half the height.

$$\tau_{max} = \frac{36V}{bh^3}\frac{(2h-3h/6)(h+3h/6)}{27} = \frac{36V}{bh^3}\frac{(3h/2)(3h/2)}{27}$$

$$= \frac{3V}{bh} = \frac{3}{2}\frac{V}{(bh/2)} = \frac{3}{2}\times\frac{V}{\text{area}} = \frac{3}{2}\tau_{av}$$

This means that the maximum shear stress is 1.5 times the mean shear stress τ_{av}. The shear stress at the NA, where $y = 0$,

$$\tau \text{ at NA} = \frac{36V}{bh^3}\frac{2h^2}{27} = \frac{8}{3}\frac{V}{bh}$$

The shear stress distribution is parabolic as shown in Fig. 5.42(d).

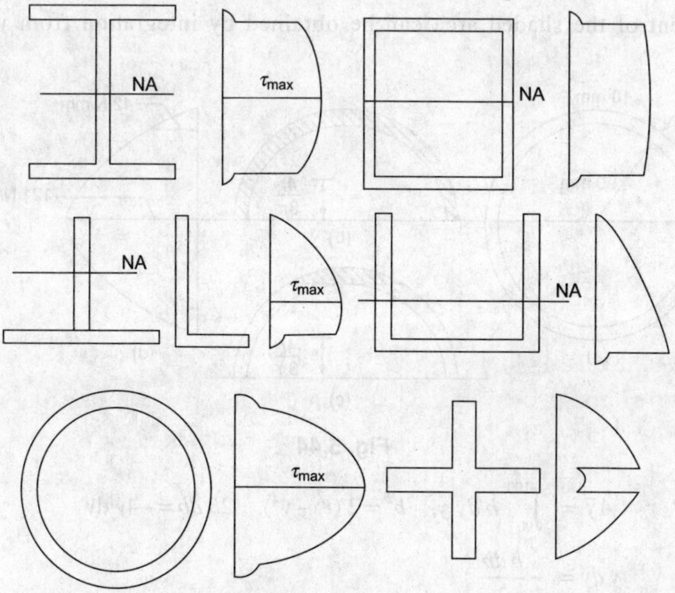

Fig. 5.43

The stress distributions for some other sections are shown in Fig. 5.43. You should note the similarity in distribution for sections such as box and I-sections, T- and channel sections, etc.

The similarities occur because one section can be obtained from the other by simply moving the vertical members. The box-section is obtained by splitting the web of an I-section and moving it to the edges of the flanges. The shear stress distribution can be easily sketched if we remember that it is proportional to $A\bar{y}$ and inversely proportional to the width.

Example 5.21 Shear stress distribution in a hollow circular section

A hollow circular section of outside diameter 200 mm and thickness 10 mm carries an SF of 25 kN. Find the maximum shear stress and the shear stress at the inner edge, and draw the shear stress distribution diagram.

Solution The section is shown in Fig. 5.44(a).

$$I_{XX} = \frac{\pi}{64}(200^4 - 180^4) = 3.7306 \times 10^6 \text{ mm}^4$$

The maximum shear stress occurs at the NA in such a section. $b = 2 \times 10 = 20$ mm and $A\bar{y}$ is due to the shaded area shown in Fig. 5.44(b). The position of the centroid of a semicircular area is $4r/3\pi$ from the base. Therefore, the moment of shaded area about the NA is given by

$$A\bar{y} = \frac{\pi \times 200^2}{4} \times \frac{4 \times 100}{3\pi} - \frac{\pi \times 180^2}{4} \times \frac{4 \times 90}{3\pi} = 0.3613 \times 10^6 \text{ mm}^4$$

$$\tau_{max} = \frac{V}{Ib}A\bar{y} = \frac{25{,}000}{3.7306 \times 10^6 \times 20} \times 0.3613 \times 10^6 = 121 \text{ N/mm}^2$$

To find the shear stress at the inner edge, the moment is due to the shaded area shown in Fig. 5.44(c). The width b of the section may be worked out from

$$b^2 = 2(100^2 - 90^2)$$

$$\Rightarrow \quad b = 61.6 \text{ mm}$$

The moment of the shaded area can be obtained by integration from $y = 90$ to $y = 100$.

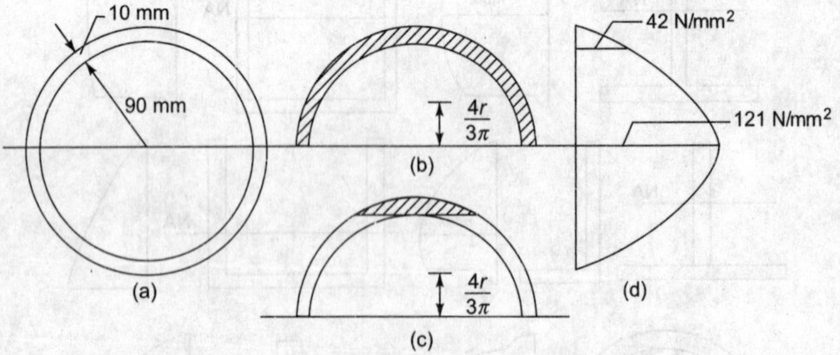

Fig. 5.44

$$A\bar{y} = \int_{90}^{100} b\,dy\,y, \quad b^2 = 2(r^2 - y^2), \quad 2b\,db = -4y\,dy$$

$$y\,dy = -\frac{b\,db}{2}$$

The limits for b are $y = 90$, $b = 61.6$ and $y = 100$, $b = 0$. Therefore,

$$A\bar{y} = \int_{61.6}^{0} -b^2 \frac{db}{2} = \left[-\frac{b^3}{6}\right]_{61.6}^{0} = \frac{61.6^3}{6} = 39{,}013 \text{ mm}^4$$

$$\text{Shear stress at the inner edge} = \frac{V}{Ib} A\bar{y}$$

$$= \frac{25{,}000 \times 39013}{3.7306 \times 10^6 \times 61.6} = 4.2 \text{ N/mm}^2$$

The shear stress distribution is shown in Fig. 5.44(d). ■

Example 5.22 Shear stress distribution in a T-section

The T-section shown in Fig. 5.45(a) is subjected to an SF of 100 kN. Draw the shear stress distribution diagram and find the maximum shear stress.

Solution We first locate the NA by finding the centroid of the section. If the centroid is $\bar{y}$ below the top edge, then

$$(20 \times 100 + 130 \times 20)\bar{y} = 20 \times 100 \times 10 + 130 \times 20 \times 85$$

$$\Rightarrow \quad \bar{y} = 52.4 \text{ mm}$$

$$I_{\text{NA}} = 100 \times \frac{20^3}{12} + 2000 \times (42.4)^2 + \frac{20 \times 130^3}{12} + 2600 \times (32.6)^2$$

$$= 10.087 \times 10^6 \text{ mm}^4$$

The shear stress is zero at the top edge of the flange. At the bottom edge of the flange,

$$\tau = \frac{100{,}000 \times 2000 \times 42.4}{10.087 \times 10^6 \times 100} = 8.4 \text{ N/mm}^2$$

At the top edge of the web, the width is reduced to 20 mm, everything else remaining the same.

$$\tau \text{(at top edge of web)} = \frac{8.4}{20} \times 100 = 42 \text{ N/mm}^2$$

$$\tau_{\text{NA}} = \frac{100{,}000 \times [2000 \times 42.4 + 32.4 \times 20 \times 16.2]}{10.087 \times 10^6 \times 20} = 47 \text{ MPa}$$

Below the NA, the stresses are tensile. The SF reduces to zero at the bottom edge. The shear stress distribution is shown in Fig. 5.45(b).

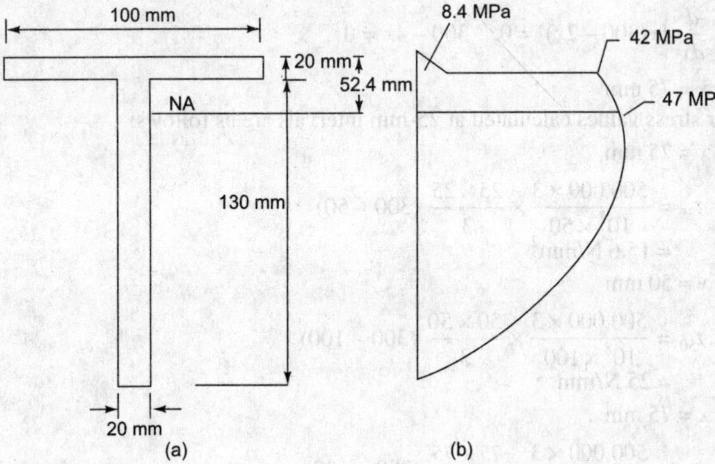

Fig. 5.45 ■

Example 5.23 Shear stress distribution in a diamond-shaped section

The diamond-shaped section shown in Fig. 5.46 carries an SF of 500 kN at a section. Draw the shear stress distribution diagram.

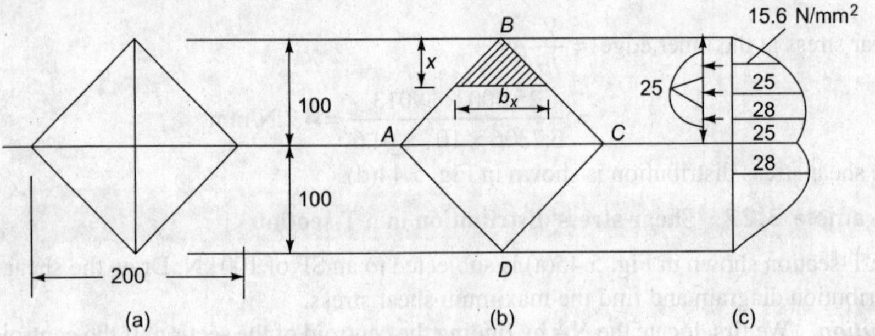

Fig. 5.46

Solution By symmetry, the NA is along the diagonal AC. The MI about the NA is given by

$$I_{\text{NA}} = \frac{2b\,(b/2)^3}{12} = \frac{2 \times 200 \times 100^3}{12} = \frac{10^8}{3}\ \text{mm}^4$$

Consider a layer at a depth x from vertex B. Width at x,

$$b_x = \frac{200}{100}\,x = 2x$$

The shaded area $= 2x\,\dfrac{x}{2} = x^2$

Moment of shaded area about NA $= x^2\left(100 - \dfrac{2x}{3}\right)$

$$= \frac{x^2}{3}\,(300 - 2x)$$

If V is the shear force and I is the MI about the NA, then the shear stress at x from B is

$$\tau_x = \frac{V}{Ib}\,A\bar{y} = \frac{V}{I \times 2x}\,\frac{x^2}{3}\,(300 - 2x)$$

To obtain the maximum value, $d\tau_x/dx = 0$

$$\frac{d}{dx}\,[x(300 - 2x)] = 0, \quad 300 - 4x = 0$$

$$\Rightarrow \qquad x = 75\ \text{mm}$$

The shear stress values calculated at 25-mm intervals are as follows:

$x = 25$ mm

$$\tau_{25} = \frac{500{,}000 \times 3}{10^8 \times 50} \times \frac{25 \times 25}{3}\,(300 - 50)$$
$$= 15.6\ \text{N/mm}^2$$

$x = 50$ mm

$$\tau_{50} = \frac{500{,}000 \times 3}{10^8 \times 100} \times \frac{50 \times 50}{3}\,(300 - 100)$$
$$= 25\ \text{N/mm}^2$$

$x = 75$ mm

$$\tau_{75} = \frac{500{,}000 \times 3}{10^8 \times 150} \times \frac{75 \times 75}{3}\,(300 - 150)$$
$$= 28\ \text{N/mm}^2$$

$x = 100$ mm

$$\tau_{100} = \frac{500{,}000 \times 3}{10^8 \times 200} \times \frac{100 \times 100}{3} (300 - 200)$$

$$= 25 \text{ N/mm}^2$$

The shear stress distribution is shown in Fig. 5.46(c).

Example 5.24 Shear stress distribution in a trapezoidal section

Draw the shear stress distribution diagram for the trapezoidal section shown in Fig. 5.47 if the section carries an SF of 600 kN.

Solution We locate the centroidal distance from the top edge BC as

$$\bar{y} = \frac{2 \times 200 + 100}{200 + 100} \times \frac{210}{3} = 116.7 \text{ mm}$$

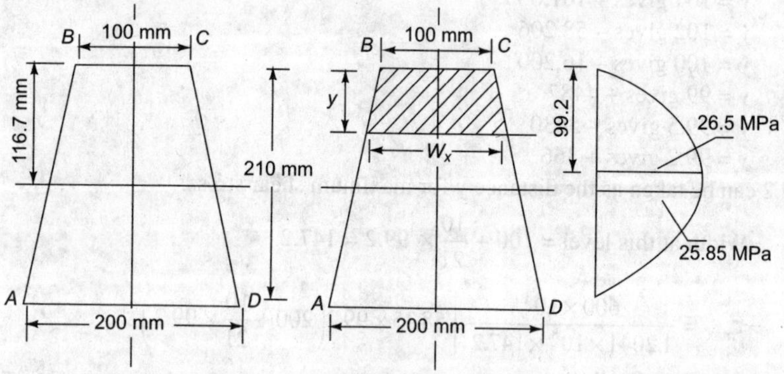

Fig. 5.47

The NA passes through the centroid.

$$I_{NA} = 2 \times \left(\frac{50 \times 210^3}{36} + 50 \times 210 \times (26.4)^2 \right) + \frac{100 \times 210^3}{12} 100 \times 210 \times (11.7)^2$$

$$= 1.2041 \times 10^8 \text{ mm}^4$$

At a depth y from the top fibre BC, the width of the section,

$$W_x = 100 + (200 - 100) \frac{y}{210} = 100 + \frac{10}{21} y = 100 + ky$$

where $k = 10/21$. The centroidal distance of the hatched area from the top,

$$\bar{y} = \frac{2(100 + ky) + 100}{100 + ky + 100} \frac{y}{3} = \frac{300 + 2ky}{200 + ky} \frac{y}{3}$$

Moment of this area about the neutral axis,

$$A\bar{y} = \frac{(100 + 100 + ky)}{2} y \left[-\frac{(300 + 2ky)}{200 + ky} \frac{y}{3} + 116.7 \right]$$

$$= \frac{116.7}{2} y (200 + ky) - (300 + 2ky) \frac{y^2}{6}$$

$$\tau_y = \frac{V}{Ib} A\bar{y} = \frac{V}{I(100 + ky)} \left[\frac{116.7}{2} y(200 + ky) - (300 + 2ky) \frac{y^2}{6} \right]$$

To find the maximum value of τ_y, we have to set $d(\tau_y)/dy = 0$.

$$\frac{d}{dy} \left[\frac{A\bar{y}}{(100 + ky)} \right] = 0 = \frac{d}{dy} \left\{ \frac{11{,}670y + 27.78y^2 - 50y^2 - 0.1587y^3}{[100 + (10y/21)]} \right\}$$

$$\left(100+\frac{10}{21}y\right)(11{,}670+55.56y-100y-0.4761y^2)$$
$$-(11{,}670y+27.78y^2-50y^2-0.1587y^3)\frac{10}{21}=0$$

This reduces to

$$1{,}167{,}000-4444y-58.16y^2-0.1572y^3=0$$

This has to be solved by trial and error. As a first approximation, neglecting the term containing y^3,

$$58.16y^2+4444y-1{,}167{,}000=0$$

This gives y = 108.5. The value of y will have to be less since the term containing ($-ky^3$) was not considered. Taking different values of y,

y = 106 gives – 144,777 on the right side,
y = 104 gives – 101,063
y = 102 gives – 58,206
y = 100 gives – 16,200
y = 99 gives + 4487
y = 99.5 gives – 5830
y = 99.2 gives + 366

y = 99.2 can be taken as the distance y for maximum shear stress.

$$\text{Width at this level} = 100+\frac{10}{21}\times 99.2 = 147.2$$

$$\tau_{max}=\frac{600\times10^3}{1.2041\times10^8\times147.2}\left[58.35\times99.2\left(200+\frac{10}{21}\times99.2\right)-\left(300+\frac{20}{21}\times99.2\right)\times\frac{99.2^2}{6}\right]$$
$$=26.5\ \text{N/mm}^2$$

$$\tau_{NA}=\frac{600\times10^3}{1.2041\times10^8\times155.57}\left[58.35\times116.7\left(200+\frac{10}{21}\times116.7\right)-\left(300+\frac{20}{21}\times116.7\right)\times\frac{116.7^2}{6}\right]$$
$$=25.85\ \text{N/mm}^2$$

The shear stress distribution diagram is shown in Fig. 5.47.

Example 5.25 Shear stress in a thin circular tube

In a thin circular tube, show that the maximum shear stress is twice the average shear stress over the cross section.

Solution Consider the thin tubular section shown in Fig. 5.48.

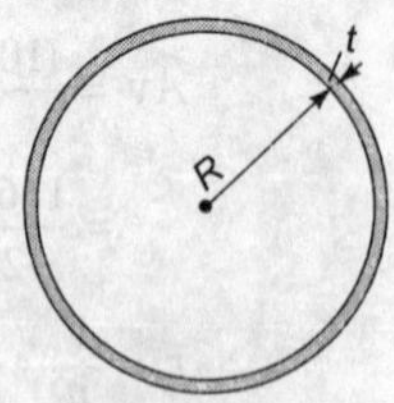

Fig. 5.48

For this section,

$$\text{Area} = \pi[(R+t)^2-R^2]=\pi(R^2+t^2+2Rt-R^2)$$
$$=2\pi rt$$

neglecting terms of higher order in t, as t is very small.

$$I_{XX}=\frac{\pi}{4}[(R+t)^4-R^4]=\frac{\pi}{4}[R^4+t^4+4R^3t+6R^2t^2+4Rt^3-R^4]$$
$$=\pi R^3t$$

neglecting higher order terms in t.

Moment of shaded area about NA,

$$A\bar{y} = \frac{\pi}{2}(R+t)^2 \frac{4(R+t)}{3\pi} - \frac{\pi}{2}R^2 \frac{4R}{3\pi}$$

$$A\bar{y} = \frac{4}{6}[(R+t)^3 - R^3] = \frac{4}{6}[R^3 + t^3 + 3R^2t + 3Rt^2 - R^3]$$

$$= 2R^2t$$

$$\text{Maximum shear stress} = \frac{V}{Ib}A\bar{y} = \frac{V \times 2R^2t}{\pi R^3 t \times 2t} = \frac{V}{\pi Rt}$$

$$\text{Average shear stress} = \frac{V}{\text{area}} = \frac{V}{2\pi Rt}$$

Thus the maximum shear stress is twice the average shear stress in the section.

5.8 ECONOMICAL SECTIONS

Generally, bending stress governs the design of a beam. Since $M = \sigma z$, the moment of resistance is directly proportional to $Z = I/y$, and the larger the moment of inertia, the greater is the moment of resistance.

If we compare the sections shown in Fig. 5.49, we will find that for the same weight, a rectangular section is more economical than a square section. Since the

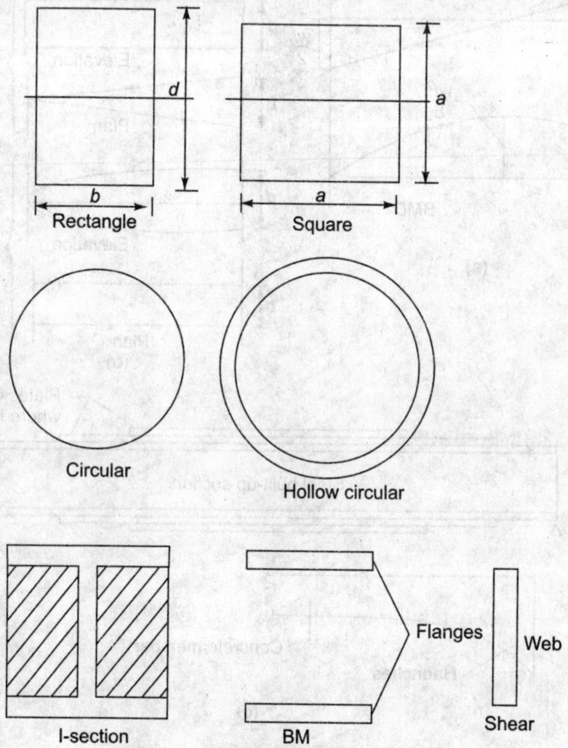

Fig. 5.49

MI of a rectangular section increases in proportion to the cube of the depth, a larger MI is obtained by increasing the depth and decreasing the width for the same area. This reduction, however, cannot be large as the section will then become too slender and unstable.

For the same reason, a hollow circular section is more economical than a solid circular section for the same area. This is so because the area is distributed at a greater distance in the hollow circular section than in the solid section from the NA. Since $I = \int y^2 \, \delta A$, the value of the MI increases if the area is away from the NA.

For the same reason, you will find that steel beams are not made in rectangular shapes but in the form of an I. In one of the sections shown in Fig. 5.49, it is clear that the shaded area of a rectangular section is removed and put in the flanges in an I-section. Consequently, the MI of an I-section is much larger than that of an equivalent rectangular section. Also, the BM is resisted mainly by the flanges (> 90%) and shear resistance is provided by the web.

5.9 BEAMS OF UNIFORM STRENGTH

If we consider a beam of span L m and carrying a uniformly distributed load w/m, the BM and SF diagrams are as shown in Fig. 5.50(a). It is clear that the BM is

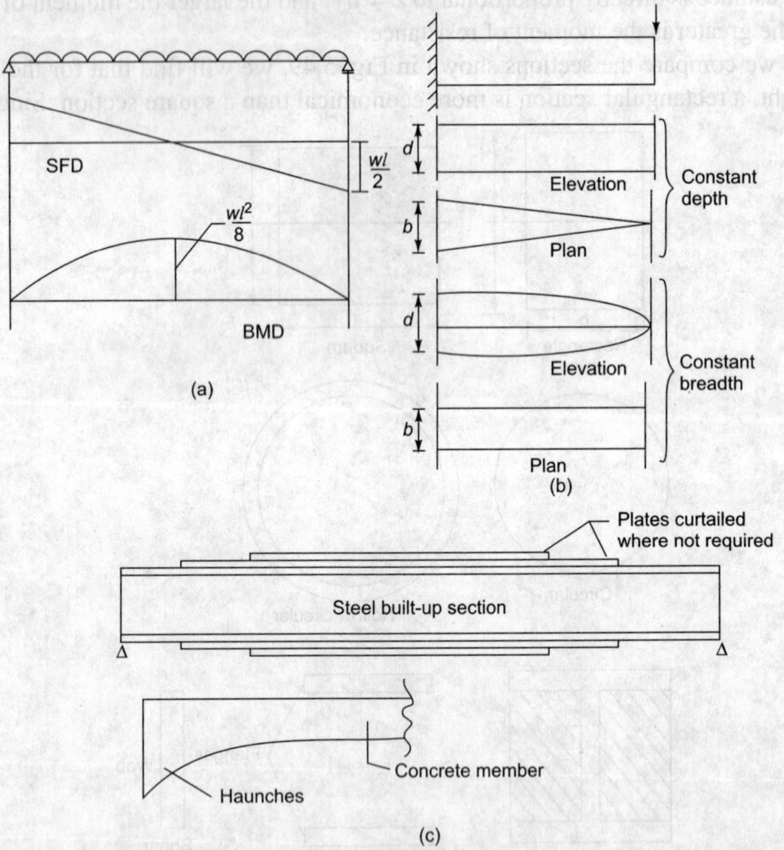

Fig. 5.50

maximum at mid-span and decreases parabolically to zero at the ends. The SF, on the other hand, is maximum at the supports, and decreases linearly to zero at mid-span. Designing a beam for maximum BM or SF and keeping a uniform section is, therefore, very uneconomical.

A beam of uniform strength, where sectional dimensions are varied so that the material is used to its full capacity at all sections, is generally impractical. The depth of the section in the case shown in Fig. 5.50(b) will have to vary in a non-linear fashion, which may be very difficult to achieve. In the case of heavy beams and beams built up of plates, it is possible to curtail the plates at points along the length so that some economy is achieved. This is particularly possible, and should be done, in the case of steel beams with added plates or in plate girders, made up of plates welded or riveted together, as shown in Fig. 5.50(c).

In the case of concrete sections, haunches are provided in continuous, rigid frame structures, providing more depth at the support points where the BM is larger [Fig. 5.50(c)].

5.10 DESIGN OF BEAMS FOR BM AND SF

A beam has to be designed for both the effects separately using the methods we have outlined. Generally, the BM dictates the design and the SF considerations are satisfied, particularly if the span is large. Only in the case of short and heavily loaded beams can the shear stress call for a change in a section designed for BM.

As an illustration, consider a rectangular section beam carrying a UD load w used to span a length L, shown in Fig. 5.51.

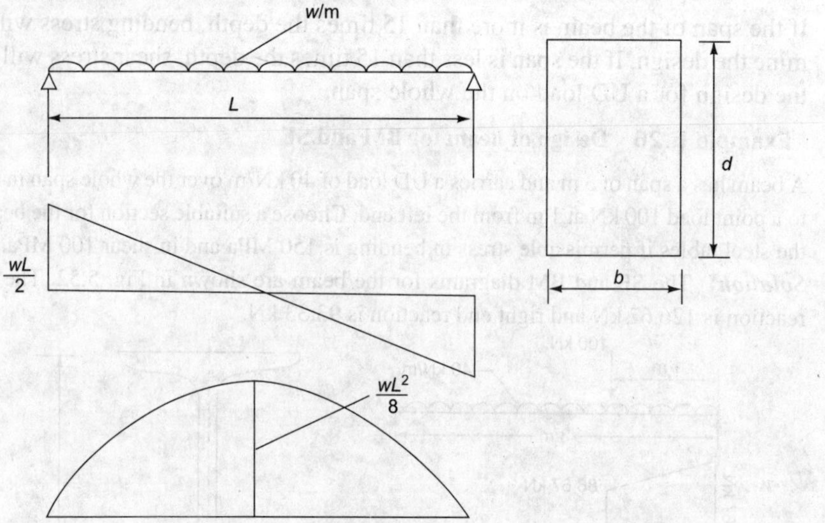

Fig. 5.51

$$\text{Maximum BM} = \frac{wL^2}{8}$$

$$\text{Maximum SF} = \frac{wL}{2}$$

If the permissible bending stress and shear stress are σ and τ respectively, then

$$\text{Moment capacity} = \sigma z = \frac{\sigma b d^2}{6}$$

and

$$\text{Shear stress } \tau = \frac{3}{2} \times \text{average shear stress}$$

$$\tau = \frac{3}{2}\frac{V}{bd} = \frac{3}{2}\frac{wL}{2bd} = \frac{3wL}{4bd}$$

Maximum load $(wL) = (4/3)\,\tau bd$, from the consideration of limiting maximum shear stress to τ. To limit the bending stress to σ,

$$\frac{wL^2}{8} = \sigma\frac{bd^2}{6}$$

Substituting for wL in terms of τ,

$$\frac{4}{3}\tau bd\frac{L}{8} = \frac{\sigma bd^2}{6}$$

$$\Rightarrow \qquad L = \frac{\sigma}{\tau}d$$

If L is larger than this value, then BM will govern the design. If L is less than this value, SF will govern the design.

If we take a timber beam of rectangular section with $\sigma = 12$ MPa and $\tau = 0.8$ MPa, then

$$L = \frac{12}{0.8}d = 15d$$

If the span of the beam is more than 15 times the depth, bending stress will determine the design. If the span is less than 15 times the depth, shear stress will govern the design for a UD load on the whole span.

Example 5.26 Design of beam for BM and SF

A beam has a span of 3 m and carries a UD load of 40 kN/m over the whole span in addition to a point load 100 kN at 1 m from the left end. Choose a suitable section for the beam from the steel tables if permissible stress in bending is 150 MPa and in shear 100 MPa.

Solution The SF and BM diagrams for the beam are shown in Fig. 5.52. The left end reaction is 126.67 kN and right end reaction is 93.33 kN.

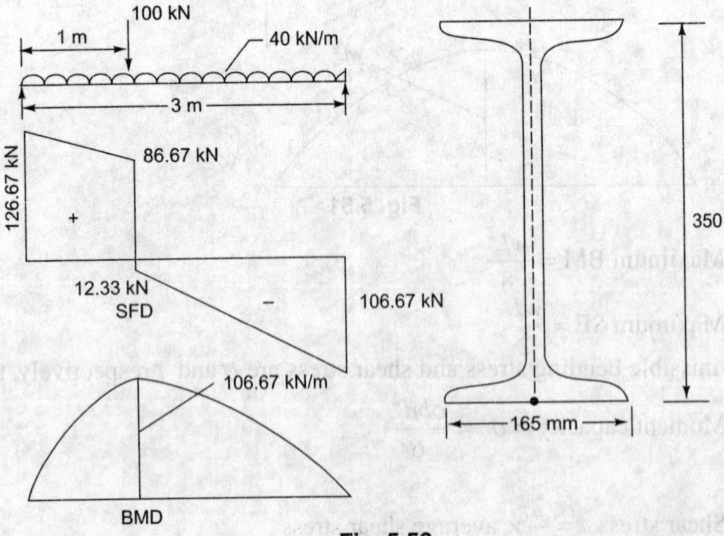

Fig. 5.52

Maximum SF = 126.67 kN

Maximum BM = $126.67 \times 1 - 40 \times 1 \times \frac{1}{2} = 106.67$ kN m

Section modulus required = $\frac{M}{\sigma} = \frac{106.67 \times 10^6}{150} = 711{,}133\ \text{mm}^3$

We select an I-section. ISLB 350 @ 495 N/m has a Z_{XX} value of 751.9 cm^3.

Area of the section = 6301 mm^2

Average shear stress = $\frac{126.67\ \text{kN}}{\text{area}} = \frac{126.67 \times 10^3}{6301} = 20\ \text{N/mm}^2$

which is quite safe.

Example 5.27 Design of a rectangular timber beam for BM and SF

A timber beam carries a UD load of 15 kN/m (including self-weight) over its entire span of 4 m. If the permissible stresses for timber are 12 MPa in compression, 10 MPa in tension, and 0.8 MPa in shear, design a suitable rectangular beam. Take the width of rectangular beam as one-third the depth.

Solution The BM and SF diagrams for the beam are shown in Fig. 5.53(a).

Maximum BM = $\frac{Wl^2}{8} = \frac{15 \times 4^2}{8} = 30$ kN m $= 30 \times 10^6$ N mm

Maximum SF = $\frac{15 \times 4}{2} = 30$ kN $= 30 \times 10^3$ N

For BM,

Section modulus $Z = \frac{M}{\sigma} = \frac{30 \times 10^6}{10} = 3 \times 10^6\ \text{mm}^3$

$$Z = \frac{bd^2}{6} = \frac{d}{3}\frac{d^2}{6} = 3 \times 10^6$$

$d = 378$ mm, $b = 126$ mm

For SF,

$$\tau_{\max} = \frac{3}{2} \times \frac{V}{\text{area}} = \frac{3}{2} \times \frac{30 \times 10^3}{bd}$$

$\tau_{\max} = 0.8$, $0.8 = \frac{3}{2} \times \frac{30 \times 10^3}{(d/3) \times d}$, $d = 411$ mm, $b = 137$ mm

The shear stress governs the design in this case. The section is shown in Fig. 5.53(b).

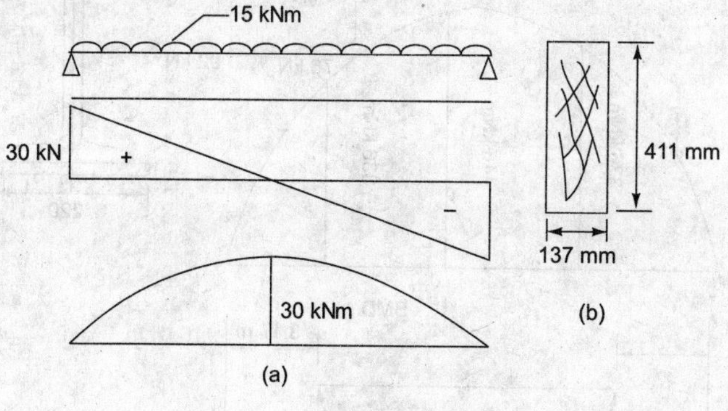

Fig. 5.53

Example 5.28 Design of a steel I-section with plates

Design a beam carrying the loads shown in Fig. 5.54(a). The steel section available is ISMB 500. The permissible stresses are 150 MPa in tension and compression, and 100 MPa in shear (average). If cover plates are used, at what point along the span can they be curtailed?

Solution The SF and BM diagrams are shown in Figures 5.54(b) and (c). The maximum BM is $M_{(max)} = 92 \times 8 - 40 \times 4 - 8 \times 4 \times 4 = 448$ kN m.

$$\text{Section modulus required} = \frac{M}{\sigma} = \frac{448 \times 10^6}{150} = 2.9867 \times 10^6 \text{ mm}^3$$

For ISMB 500, $Z = 1.81 \times 10^6$ mm^3.

We use extra plates at bottom and top to get the extra section modulus. If t is the thickness of the plates, then approximately

$$2 \times 200 \times t \times 250 = 1.1767 \times 10^6$$

$$\Rightarrow \quad t = 11.76 \text{ mm}$$

We use two cover plates of dimensions 200 × 12 mm, one each on the tension and compression side. The actual value of the section modulus provided can now be worked out by calculating the MI of the section.

$$I_{NA} = \text{MI of 1SMB 500} + \text{MI of plates}$$

$$= 452.183 \times 10^6 + 2\left[200 \times \frac{12^3}{12} + 200 \times 12 \times (256)^2\right]$$

$$= 7.66183 \times 10^8 \text{ mm}^4$$

$$Z_{XX} = \frac{7.66183 \times 10^8}{262} = 2.9267 \times 10^6 \text{ mm}^3$$

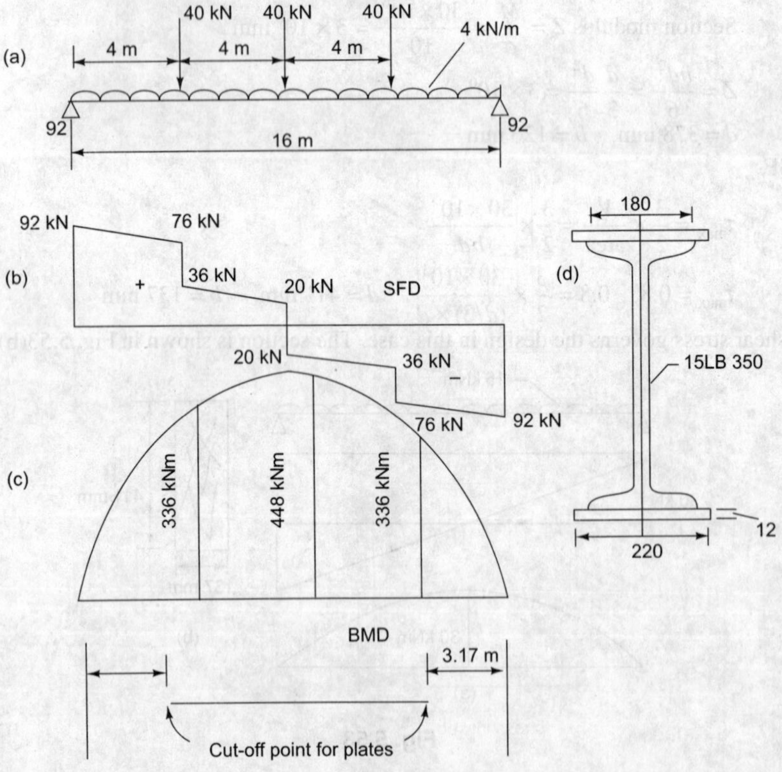

Fig. 5.54

This is slightly less than the 2.9867×10^6 mm^3 required. Let us increase the width of the plates to 220 mm. If we do so, the value of Z becomes 3.0468×10^6 mm^4. This can be considered satisfactory.

$$\text{Average shear stress} = \frac{\text{SF}}{\text{area}} = \frac{92 \times 10^3}{11100 + 2 \times 220 \times 12} = 5.6 \text{ N/mm}^2$$

which is far less than the permissible shear stress. Let us find the curtailment point for the plates. Z_X of ISMB 500 = 1.81×10^6 mm^3.

$$\begin{aligned}\text{Moment capacity of I-section} &= 1.81 \times 10^6 \times 150\\ &= 271.5 \times 10^6 \text{ Nmm}\\ &= 271.5 \text{ kNm}\end{aligned}$$

From the BM diagram, it is evident that the BM will have this value within 4 m from the support. At a distance x, $x \le 4$ from the left support,

$$M_x = 92x - 4x\,\frac{x}{2}$$

$$92x - 2x^2 = 271.5$$

$$\Rightarrow \quad x = 3.17 \text{ m}$$

The plates can be theoretically cut off at 3.17 m from the supports. ■

Example 5.29 Percentage of BM and SF carried by flange and web

An I-section has a depth of 200 mm, flange width of 120 mm, flange thickness of 15 mm and web thickness of 10 mm. Determine what percentage of the BM and SF are carried by the flange and web individually.

Solution The section is shown in Fig. 5.55. The bending stress at any layer, distant y from the NA is given by

$$\sigma = \frac{M}{I} y$$

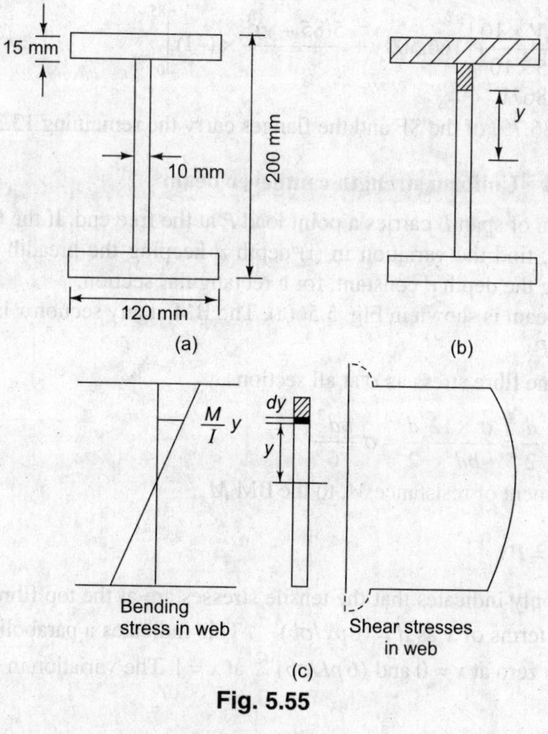

Fig. 5.55

$$I_{NA} = 2\left[120 \times \frac{15^3}{12} + 120 \times 15 \times (92.5)^2\right] + \frac{10 \times 170^3}{12} = 35 \times 10^6 \text{ mm}^4$$

The bending resistance of the web can be calculated as

$$M_w = \int_{-85}^{+85} (10\,dy)\frac{M}{I}y^2 = 2\int_0^{85} 10\frac{M}{I}y^2 dy = \frac{2 \times 10}{35 \times 10^6} M\left[\frac{y^3}{3}\right]_0^{85}$$

$$= \frac{2 \times 10}{35 \times 10^6} M \frac{(85)^3}{3} = 0.1169M$$

The bending resistance of the flanges $= M - 0.1169 = 0.8831$. Thus the flange carries 88.31% of the BM and the web carries 11.69%.

Shear resistance The shear resistance at any layer y is, as shown in Fig. 5.55(c), for the web,

$$\delta V_w = \left[\frac{V}{Ib} A\bar{y}\right] 10\,dy$$

where $A\bar{y}$ is the moment of the area shaded in the figure.

$A\bar{y}$ = moment of the area of the flanges + moment of area of the web above y

$$= 120 \times 15 \times 92.5 + 10 \times (85 - y)\frac{(85 - y)}{2}$$

$$= 166{,}500 + 5(85 - y)^2$$

The total shear resistance of the web is given by

$$V_w = 2\int_0^{85} \frac{V}{35 \times 10^6 \times 10}[166{,}500 + 5(85 - y)^2]dy \times 10$$

$$= \frac{2V \times 10}{35 \times 10^7}\int_0^{85}[166{,}500 + 5(85 - y)^2]dy$$

$$= \frac{2V \times 10}{35 \times 10^7}\left[166{,}500y + \frac{5(85 - y)^3}{3} \times (-1)\right]_0^{85}$$

$$= 0.867V$$

The web carries 86.7% of the SF and the flanges carry the remaining 13.3%.

Example 5.30 Uniform strength cantilever beam

A cantilever beam of span L carries a point load P at the free end. If the beam is to have a uniform strength, find the variation in (i) depth d keeping the breadth constant and (ii) breadth b keeping the depth d constant, for a rectangular section.

Solution The beam is shown in Fig. 5.56(a). The BM at any section x is

$$M_x = -Px$$

Taking the extreme fibre stress as σ at all sections

$$M_r = \frac{\sigma}{I}\frac{d}{2} = \frac{\sigma \times 12}{bd^3}\frac{d}{2} = \sigma\frac{bd^2}{6}$$

Equating the moment of resistance M_r to the BM M_x,

$$\sigma\frac{bd^2}{6} = Px$$

The sign of M_x only indicates that the tensile stresses are at the top fibres. This gives the variation of d in terms of x as $d = (6px/\sigma b)^{1/2}$. This indicates a parabolic variation of the depth d, which is zero at $x = 0$ and $(6pL/\sigma b)^{1/2}$ at $x = 1$. The variation in depth is shown in Fig. 5.56(b).

If the depth is kept constant and the breadth is varied, then from the same expression, $b = 6px/\sigma d^2$. This shows a linear variation of b. The breadth is zero at $x = 0$ and $6pL/\sigma d^2$ at $x = L$. The variation in breadth is shown in Fig. 5.56(c).

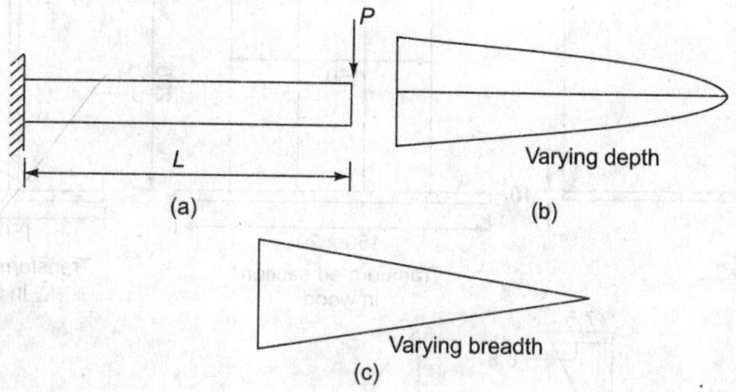

Fig. 5.56

Example 5.31 Comparison of strengths of two flitched beam sections

Compare the strengths of the two flitched beam sections shown in Fig. 5.57(a), in carrying bending moment and shear force. Draw the shear stress distribution diagram for the maximum shear stress carried by the sections. $\sigma_{(max)} = 150$ MPa for steel and 14 MPa for wood. $\tau_{max} = 1.0$ MPa for wood. (All dimensions are in mm and N/mm^2.)

Solution We use transformed sections to calculate the shear and moment capacities.

Case (i) In this case, the transformed section in terms of wood and steel are both shown. Using the transformed section in wood,

$$I_{NA} = 2\left[\frac{3000 \times 10^3}{12} + 30{,}000 \times 105^2\right] + \frac{150 \times 200^3}{12} = 762 \times 10^6 \text{ mm}^4$$

From the stress diagram shown, note that the extreme fibre stress in wood cannot exceed 7.5 N/mm^2 (150/20 N/mm^2).

$$\text{Moment capacity} = \sigma Z = \frac{7.5 \times 762 \times 10^6}{110} = 51.95 \text{ kNm}$$

$$\tau_{max} = \frac{V}{Ib} A\bar{y}$$

$$\Rightarrow \quad V = \frac{\tau_{max} Ib}{A\bar{y}}$$

$$A\bar{y} \text{ (at the NA)} = 3000 \times 10 \times 105 + 150 \times 100 \times 50 = 390 \times 10^4$$

$$\text{Shear capacity } V = \frac{1.0 \times 762 \times 10^6 \times 150}{390 \times 10^4} = 29.3 \text{ kN}$$

Case (ii) In this case, the transformed section in terms of wood and steel are shown. Considering the transformed section in terms of wood,

$$I_{NA} = \frac{2 \times 75 \times 200^3}{12} + \frac{400 \times 150^3}{12} = 212.5 \times 10^6 \text{ mm}^4$$

The extreme fibre stress in wood,

$$\sigma_{max} = \frac{150}{75} \times \frac{100}{20} = 10 \text{ N/mm}^2 < 14 \text{ MPa}$$

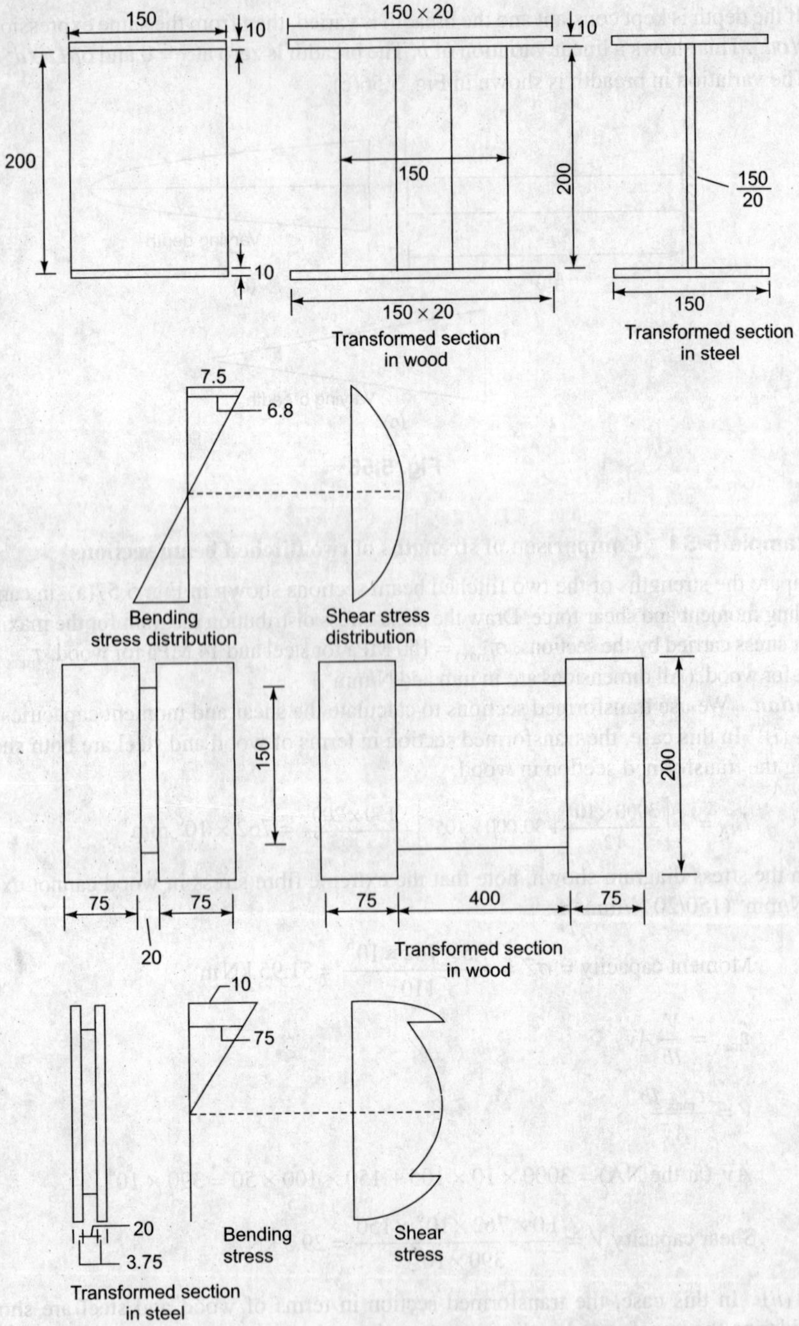

Fig. 5.57

$$\text{Moment capacity} = \sigma Z \quad = \frac{10 \times 212.5 \times 10^6}{100} = 21.25 \times 10^6 \text{ N mm}$$

$$= 21.25 \text{ kN m}$$

$$\text{Shear capacity } V = \frac{\tau_{\max} Ib}{A\bar{y}}$$

$$A\bar{y} = 100 \times 75 \times 50 \times 2 + 400 \times \frac{75 \times 75}{2} = 187.5 \times 10^4$$

$$\text{Shear capacity } V = \frac{1.0 \times 212.5 \times 10^6 \times 170}{187.5 \times 10^4} = 19.3 \text{ kN}$$

Note that the value of b in this equation is the actual width (75 + 75 + 20) = 170 mm.

Comparing the two section values, we find that in case (i), the arrangement is stronger. In terms of moment,

$$\frac{M(\text{i})}{M(\text{ii})} = \frac{51.95}{21.25} = 3.26$$

In terms of SF,

$$\frac{V(\text{i})}{V(\text{ii})} = \frac{29.3}{19.3} = 1.52$$

The relevant shear stress distribution diagrams with the maximum shear permissible are shown in Fig. 5.48. In the case of composite sections, the shear stress distribution is worked out using the formula $\tau = VA\bar{y}/Ib$, where I and $A\bar{y}$ are calculated on the basis of the transformed section, and b is taken as the actual width.

Example 5.32 Timber I-section made with planks

Five planks are joined together to form the I-section shown in Fig. 5.58. Find the load this beam can carry on a span of 4 m, distributed over its length if the permissible stress in bending is 12 MPa and in shear 0.8 MPa.

Solution We calculate the second moment of area of the section as

$$I_{\text{NA}} = 4\left[\frac{40 \times 40^3}{12} + 40 \times 40 \times 170^2\right] + \frac{40 \times 380^3}{12}$$

$$= 368.72 \times 10^6 \text{ mm}^4$$

$$\sigma(\text{allowable}) = 12 \text{ MPa} = 12 \text{ N/mm}^2$$

$$M = \sigma Z = \frac{12 \times 368.72 \times 10^6}{190} = 23.28 \text{ kN m}$$

$$\frac{wl^2}{8} = 23.28$$

$$\Rightarrow \quad w = \frac{23.28 \times 8}{16} = 11.64 \text{ kN/m}$$

The maximum shear stress,

$$\tau_{\max} = \frac{V}{Ib} A\bar{y}$$

$$\Rightarrow \quad V = \frac{\tau_{\max} bI}{A\bar{y}}$$

The maximum shear stress will be at the NA.

$$A\bar{y} = 2\,(40 \times 40 \times 170) + 190 \times 40 \times 95 = 1.266 \times 10^6 \text{ mm}^2$$

$$\text{Maximum } V = \frac{0.8 \times 40 \times 368.72 \times 10^6}{1.266 \times 10^6} = 9.32 \text{ kN}$$

$$\frac{wl}{2} = 9.32$$

$$\Rightarrow \quad w = \frac{9.32 \times 2}{4} = 4.66 \text{ kN/m}$$

The maximum load permissible on the span is 4.66 kN/m. The shear stress governs the permissible load in this case.

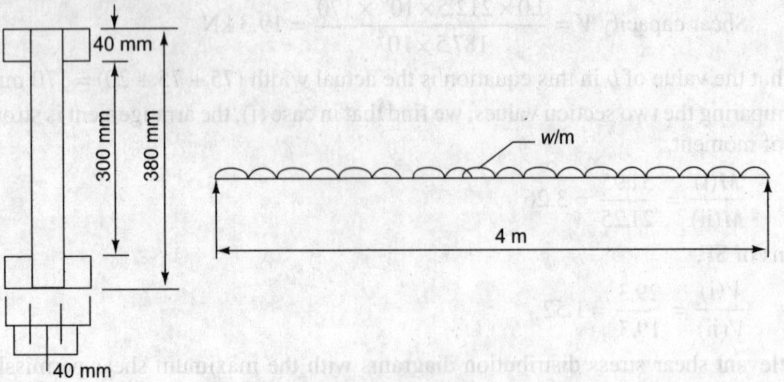

Fig. 5.58

5.11 SHEAR FLOW IN THIN-WALLED SECTIONS

The concept of shear flow was introduced in Chapter 3. In the case of a beam built up of thin plates, such as an I-section, a channel, box or T-section, shown in Fig. 5.59, shear flow can be easily worked out as follows.

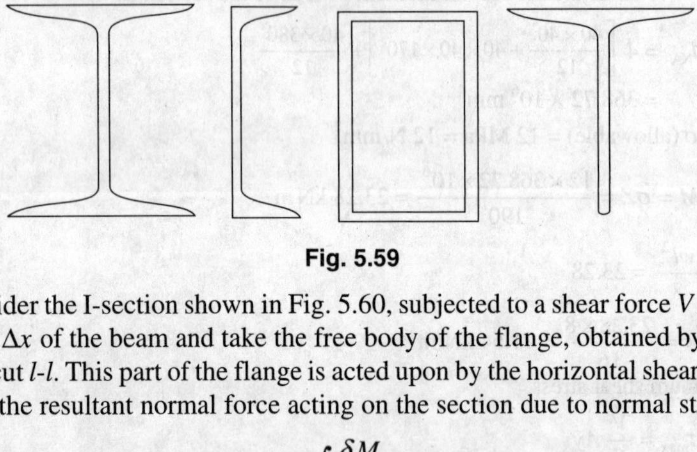

Fig. 5.59

Consider the I-section shown in Fig. 5.60, subjected to a shear force V. We take a length Δx of the beam and take the free body of the flange, obtained by a longitudinal cut *l-l*. This part of the flange is acted upon by the horizontal shearing force ΔH and the resultant normal force acting on the section due to normal stresses σ.

$$\Delta H = \tau \Delta x \times \tau \quad \text{and} \quad \Delta\sigma = \int \frac{\delta M}{I} \bar{y}\, dA$$

over the cross-sectional area of the isolated flange. Equating the two for equilibrium, we have

$$\tau \Delta x \times t = \frac{\delta M}{I} (A\bar{y})$$

$$\tau t = \frac{\delta M}{I \Delta x} (A\bar{y}) = \frac{V}{I} (A\bar{y})$$

as $dM/\delta x = V$. τt is the shear force per unit length, i.e., the shear flow, and is equal to $VA\bar{y}/I$. It should be noted that the value of τ obtained in this fashion is an average shear stress over the thickness t. The variation of shear stress over the thickness can be neglected in the case of thin-walled sections. Corresponding to this shear stress in the longitudinal direction, the shear stress on the cross section, as a complementary shear, must be directed as shown. These two being equal, the shear stresses on the flange cross section are directed as shown.

In the lower flange, where the fibres are in tension, the direction is reversed but have the same magnitude at a corresponding cut, if the section is symmetrical.

The shear flow $q = VA\bar{y}/I$ thus depends upon the area moment $A\bar{y}$, V and I being constant at a section. This makes it easier to work out the shear flow in thin-walled sections. In the case of an I-section, the shear flow is zero at the edges of the flanges and increases towards the centre. In the vertical leg, it is downwards, increasing up to the NA and decreasing from the NA downwards. In the case of a box-section, the shear flow varies from zero at the middle of the flanges, increasing to a maximum in either direction up to the NA [Figures 5.60(e) and (d)].

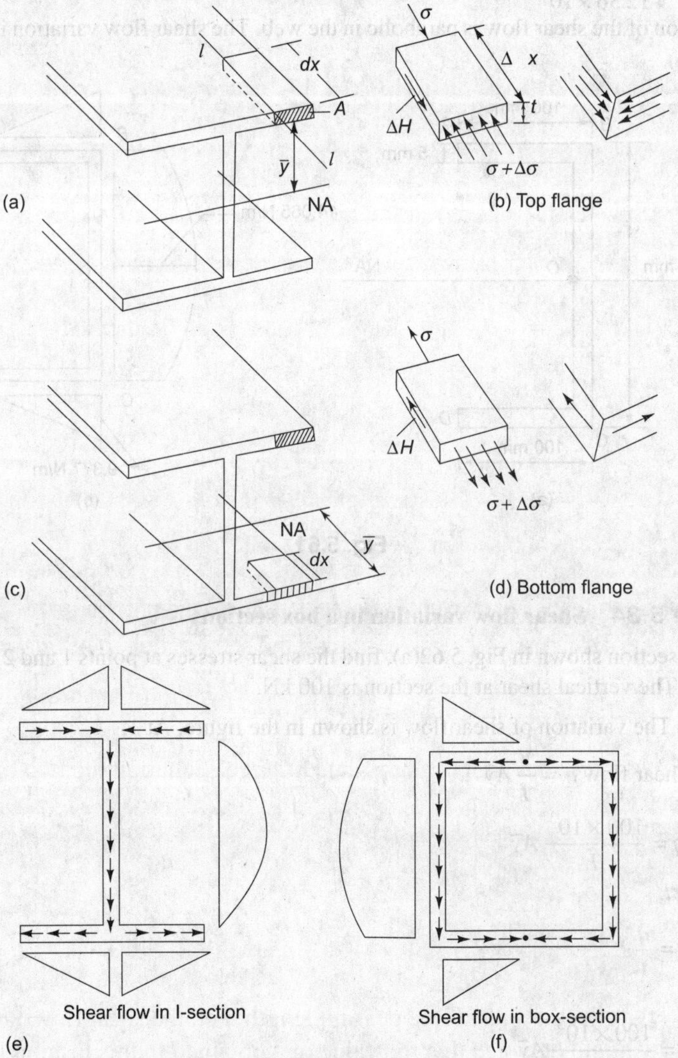

Fig. 5.60

Example 5.33 Shear flow variation in a channel section

Find the shear flow variation and sketch the same for the channel section shown in Fig. 5.61(a). The vertical shear force at the section is equal to 2500 N.

Solution $I_{XX} = \dfrac{5\times 200^3}{12} + 2\times 5\times 100\times 100^2$

$= 13.33\times 10^6\ \text{mm}^4$

$\tau_A = \tau_D = 0$

$$q_B = \frac{V}{I}A\bar{y} = \frac{2500\times 100\times 5\times 100}{13.33\times 10^6}$$

$= 9.377\ \text{N/m}$

This will be numerically the same at *C*. At a point *O* on the NA,

$$q = \frac{2500}{12.36\times 10^6}[100\times 5\times 100 + 100\times 5\times 50] = 14.065\ \text{N/m}$$

The variation of the shear flow is parabolic in the web. The shear flow variation is shown in Fig. 5.61(b).

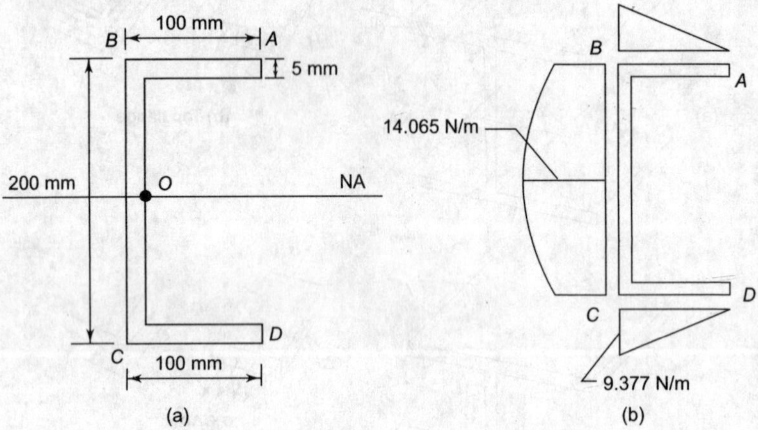

Fig. 5.61

Example 5.34 Shear flow variation in a box section

In the box-section shown in Fig. 5.62(a), find the shear stresses at points 1 and 2 marked in the figure. The vertical shear at the section is 100 kN.

Solution The variation of shear flow is shown in the figure. At 1,

$$\text{Shear flow } q = \frac{V}{I}A\bar{y}$$

$$q = \frac{100\times 10^3}{I}A\bar{y}$$

Since $q = \tau t$,

$$\tau = \frac{q}{t}$$

Therefore,

$$\tau = \frac{100\times 10^3}{It}A\bar{y}$$

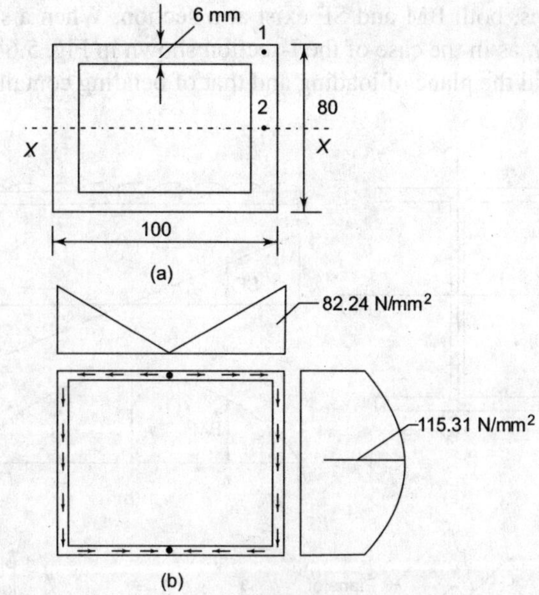

Fig. 5.62

I for the section about the NA,

$$I_{XX} = 2 \times 6 \times \frac{80^3}{12} + 2 \times 100 \times 6 \times 40^2$$
$$= 51.2 \times 104 + 192 \times 10^4$$
$$= 243.2 \times 10^4 \text{ mm}^4$$

$$\tau_1 = \frac{100 \times 10^3 \times 50 \times 6 \times 40}{243.2 \times 10^4 \times 6} = 82.24 \text{ N/mm}^2$$

$$\tau_2 = \frac{100 \times 10^3}{243.2 \times 10^4 \times 6}(50 \times 6 \times 40 + 40 \times 6 \times 20) = 115.31 \text{ N/mm}^2$$

5.12 THE CONCEPT OF SHEAR CENTRE

The concept of shear centre is important while considering the bending of members with sections having only one or no axis of symmetry, which is the NA. Typical examples of such sections are shown in Fig. 5.63.

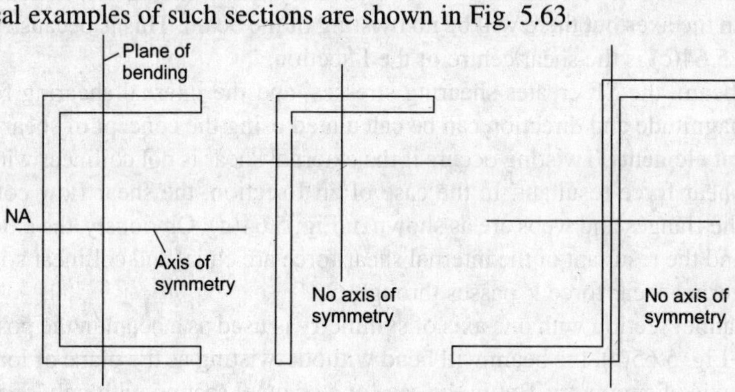

Fig. 5.63

In actual beams, both BM and SF exist at a section. When a section has two axes of symmetry, as in the case of the I-section shown in Fig. 5.64(a), one of the axes is the NA and the plane of loading and that of bending contain the other axis of symmetry.

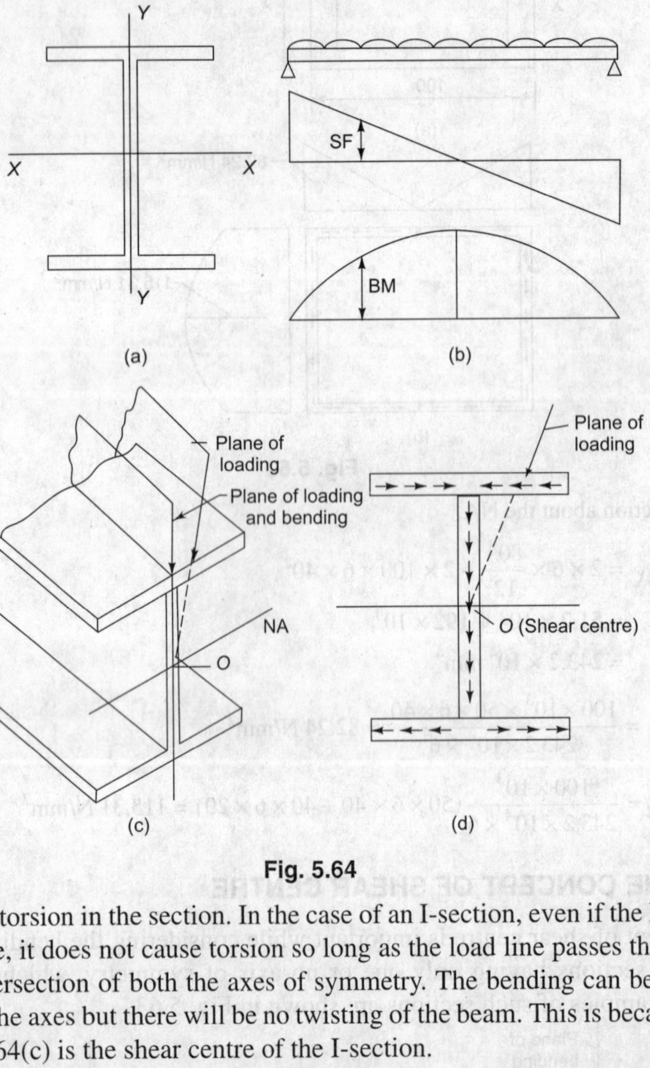

Fig. 5.64

This avoids torsion in the section. In the case of an I-section, even if the loading is off the plane, it does not cause torsion so long as the load line passes through the point of intersection of both the axes of symmetry. The bending can be resolved about both the axes but there will be no twisting of the beam. This is because point O in Fig. 5.64(c) is the shear centre of the I-section.

In 1a beam, the SF creates shearing stresses, and the internal shearing forces both in magnitude and direction can be calculated using the concept of shear flow in different elements. Twisting occurs if the external shear is not collinear with the internal shear force resultant. In the case of an I-section, the shear flow components in the flanges and webs are as shown in Fig. 5.64(d). Obviously, the external shear V and the resultant of the internal shear force are equal and collinear so long as the external shear force V passes through O.

If a channel section with one axis of symmetry is used as a beam in the position shown in Fig. 5.65(a), the beam will bend without twisting as the plane of loading is on the axis of symmetry. But in the case of a channel section with only one axis of symmetry, which becomes the NA if used as a beam in the position shown in Fig. 5.65(b), the shear flow in the flange and webs will be in the direction shown in Fig. 5.65(b).

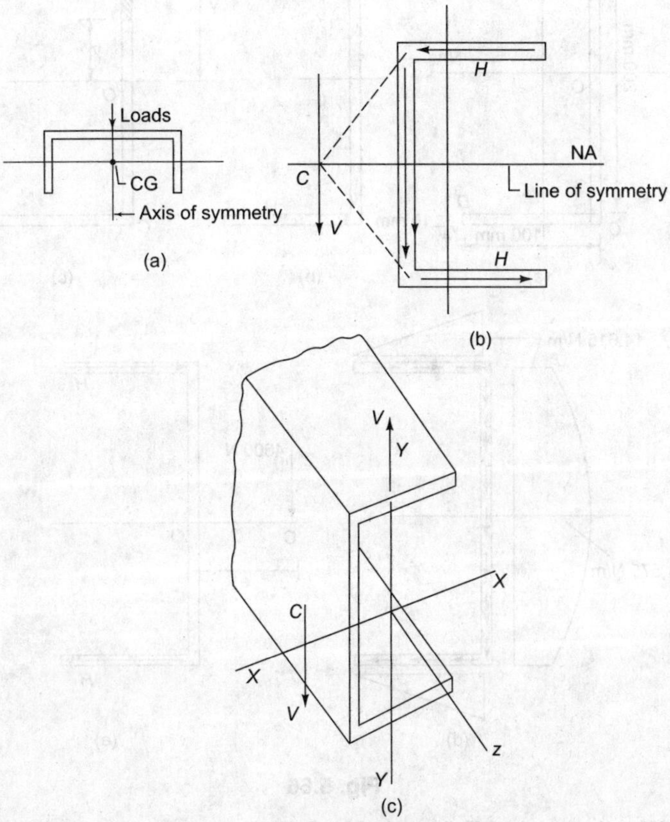

Fig. 5.65

The resultant of the forces H and V can be seen to be at a point C on the line of symmetry, and is a vertical force V. If the external shear passes through point C, there will be no twisting of the section. Otherwise, the internal shear resultant and external shear force V will form a couple, which is a moment about an axis perpendicular to the plane of the paper. Such a moment is called a twisting moment. As shown in Fig. 5.56(c), the moment is about the Z-axis and can be avoided if the load is applied through point C. Point C is known as the shear centre.

The shear centre is a point in the cross section through which the internal shear force resultant passes. When the transverse load passes through this point, there is no twisting of the beam. In the case of sections with two orthogonal axes of symmetry, the shear centre lies at their intersection. If the section has only one plane of symmetry, the shear centre lies on the axis of symmetry but its location can be found from shear flow in the elements.

Example 5.35 Shear centre of equal channel section

For the channel section shown in Fig. 5.66, calculate and show in a diagram the shear flow due to an SF of 4 kN. Also locate the shear centre for this section.

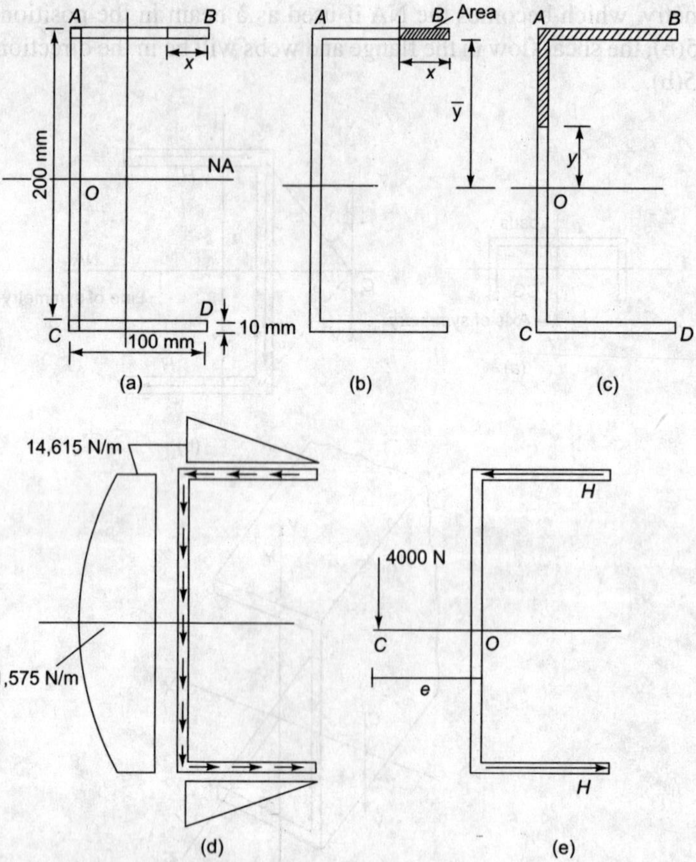

Fig. 5.66

Solution We calculate the MI about the NA first.

$$I_{\text{NA}} = 2\left(\frac{100\times10^3}{12} + 100\times10\times105^2\right) + \frac{10\times200^3}{12} = 28.74\times10^6\ \text{mm}^4$$

$$= 28.74\times10^{-6}\ \text{m}^4$$

In the flanges, at a distance x from the edge [Fig. 5.66(b)],

$$\text{Shear flow } q = \frac{V}{I}A\bar{y}$$

$$= \frac{4000}{28{\cdot}74\times10^{-6}}(0.01x\times0.105)$$

This varies linearly with x and reaches a maximum value at the junction with the web.

$$q_B = 0, \quad q_A = \frac{4000\times0.01\times0.1\times0.105}{28.74\times10^{-6}} = 14{,}615\ \text{N/m}$$

In the web at any distance y from the NA [Fig. 5.66(c)],

$$q = \frac{V}{I}(A\bar{y})$$

$A\bar{y}$ is the moment of the shaded area shown in Fig. 5.66(c).

$$q = \frac{4000}{28.74 \times 10^{-6}}\left[0.1 \times 0.01 \times 105 + (0.100 - y)(0.01)\frac{(0.1 + y)}{2}\right]$$

This varies parabolically with y, attaining a maximum value at the NA, where $y = 0$.

$$q_{NA} = \frac{4000}{28.74 \times 10^{-6}}\left[0.105 + 0.1 \times 0.01 \times \frac{0.1}{2}\right]$$

$$= 21{,}575 \text{ N/m}$$

The shear flow variation can be shown as illustrated in Fig. 5.66(d). To locate the shear centre, from Fig. 5.66(e), taking moment of the forces about O,

$$4000e = H \times 200, \quad H = \frac{q_A}{2} \times \text{length} = \frac{14{,}615}{2} \times 0.1$$

$$H = 730.75 \text{ N}$$

$$e = \frac{730.75 \times 200}{4000} \text{ mm} = 36.5 \text{ mm}$$

Example 5.36 Shear centre of a thin circular ring

Find the location of the shear centre of the thin semicircular ring shown in Fig. 5.67.

Solution X-X is an axis of symmetry and is the NA. The MI of the section about the NA can be calculated as

$$I_{NA} = \frac{\pi}{8}[(R + t)^4 - R^4] = \frac{\pi}{8} 4R^3 t$$

neglecting terms containing higher order terms in t. From Fig. 5.670(b), at a cut at an angle θ from the vertical, the shear flow can be calculated as

$$q = \frac{V}{I} A\bar{y}$$

$A\bar{y}$ being the moment of area of the shaded section.

Considering an elementary angular strip $r\,d\theta$, as shown in Fig. 5.67(b),

$$q = \frac{V}{I} \int_0^{\theta} r\,d\,\theta tr \cos\theta = \frac{V}{I} r^2 t \sin\theta$$

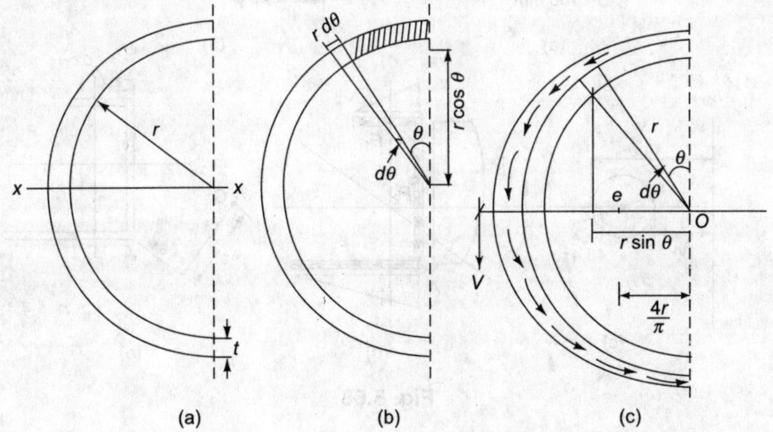

Fig. 5.67

This shows that q varies from zero at the top edge to a maximum value at the NA and then decreases to zero at the bottom edge. The shear centre is located by taking moment of the resultant of the shear flow about O and equating it to Ve, where V is the external shear and e is the eccentricity of the shear centre from the sgeometrical centre.

On the elementary strip, shear force $qr\,d\theta$ is tangential to the arc and its moment about O is $qr\,d\theta \times r$, where

$$q = \frac{Vr^2 t \sin\theta}{I}$$

The total moment can be integrated from $\theta = 0$ to $\theta = 180°$.

$$\text{Moment about } O = \int_0^{180} \frac{Vr^2 t \sin\theta}{I} r^2\, d\theta$$

$$= \frac{Vr^4 t}{I}[-\cos\theta]_0^{180} = \frac{2Vr^4 t}{I}$$

$$Ve = \frac{2Vr^4 t}{I}, \quad e = \frac{2r^4 t}{I} = \frac{2\,r^4 t}{(\pi/2)\,r^3 t} = \frac{4r}{\pi}$$

The shear centre is located as shown in Fig. 5.67(c).

Example 5.37 Shear centre of an unequal channel section

Locate the shear centre for the unsymmetrical channel section shown in Fig. 5.68(a).

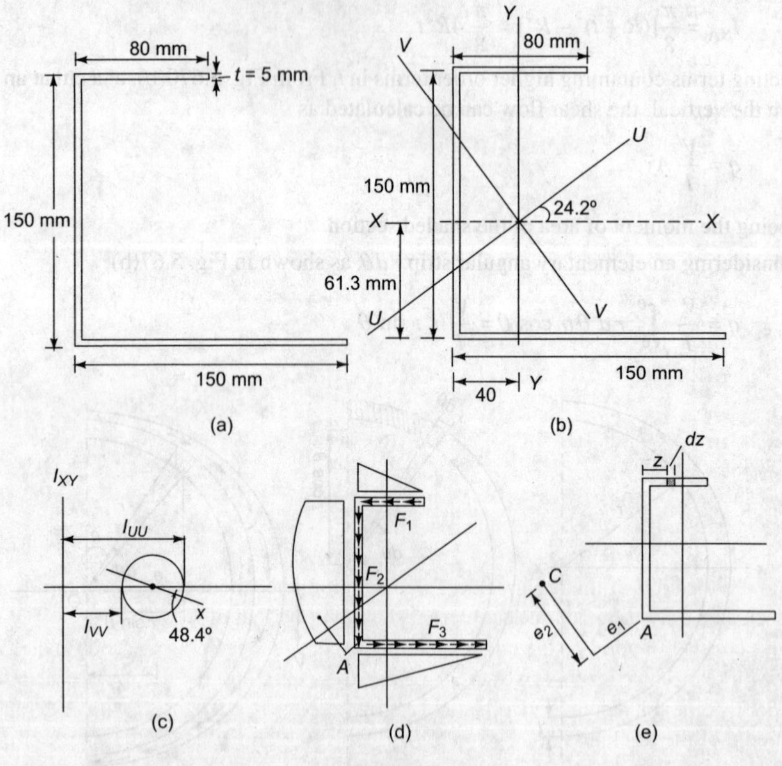

Fig. 5.68

Solution We first locate the centroid of the section. Its distance from the bottom edge is given by

$$\bar{y} = \frac{150 \times 5 \times 75 + 75 \times 5 \times 147.5 + 145 \times 5 \times 2.5}{5(150 + 75 + 145)} = 61.3 \text{ mm}$$

The distance from the leftmost edge is given by

$$\bar{x} = \frac{150 \times 5 \times 2.5 + 75 \times 5 \times 42.5 + 145 \times 5 \times 77.5}{5(150 + 75 + 145)} = 40 \text{ mm}$$

We now calculate the MI about the X- and Y-axis and the product of inertia about the X-Y axes.

$$I_{XX} = \frac{5 \times 150^3}{12} + 5 \times 150 \times 12.7^2 + \frac{75 \times 5^3}{12} + 75 \times 5 \times 86.2^2 + \frac{145 \times 5^3}{12} + 145 \times 5 \times 58.8^2$$

$$= 6.822 \times 10^6 \text{ mm}^4$$

$$I_{YY} = \frac{150 \times 5^3}{12} + 150 \times 5 \times 37.5^2 + \frac{5 \times 75^3}{12} + 5 \times 75 \times 2.5^2 + \frac{5 \times 14.5^3}{12} + 5 \times 145 \times (37.5)^2$$

$$= 3.524 \times 10^6 \text{ mm}^4$$

$$I_{XY} = 150 \times 5 \times (-37.5)(12.5) + 75 \times 5 \times (86.2) \times (2.5)$$

$$= -1.856 \times 10^6 \text{ mm}^4$$

If U-U and V-V are the principal axes through the centroid, then the angle between the U-U and X-X axes may be worked out from the equation

$$\tan 2\theta = \frac{-2I_{XY}}{(I_X - I_Y)} = \frac{-2 \times (-1.856 \times 10^6)}{10^6 (6.822 - 3.524)} = 1.1255$$

$$2\theta = 48.38°$$

$$\Rightarrow \quad \theta = 24.2° \text{ and } 114.2°$$

$$I_{UU} = \frac{I_{XX} + I_{YY}}{2} + \sqrt{\left(\frac{I_{XX} - I_{YY}}{2}\right)^2 + (I_{XY})^2}$$

$$= 10^6 \left[\frac{6.822 + 3.524}{2} + \sqrt{\left(\frac{6.822 - 3.524}{2}\right)^2 + (-1.856)^2}\right]$$

$$= 10^6 (5.173 + 2.482)$$

$$I_{UU} = 7.655 \times 10^6 \text{ mm}^4$$

and

$$I_{VV} = 2.69 \times 10^6 \text{ mm}^4$$

These values can be easily verified by drawing Mohr's circle as shown in Fig. 5.68(c).

The shear flow in the channel section is shown in Fig. 5.68(d). Let P_U and P_V be the resolved components of external shear along the U-U and V-V axes.

$$P_U = P \sin \theta = 0.4097P, \quad P_V = P \cos \theta = 0.9121P$$

To find the location of the shear centre, we take moments about point A, where F_2 and F_3 intersect. The only moment due to internal shear will be that of F_1.

If we consider a point on the cross section with coordinates (x, y), the coordinates of this point with respect to the U-V axes are $u = y \cos \theta - x \sin \theta$ and $v = x \cos \theta + y \sin \theta$. Let us consider a point at a distance z from d and a strip of length dz. Then, $x = 40 - z$, $y = 86.2$ at the middle of the thickness.

$$u = 86.2 \cos (24.19) - (40 - z) \sin (24.19)$$

$$= 62.24 + 0.4097z$$

$$v = (40 - z) \times 0.9122 + 86.2 \times 0.4097$$

$$= 71.8 - 0.9122z$$

The moment of the shaded area of length dz about the U-axis,

$$A\bar{y} = \int_0^z 5\,dz(62 \times 24 + 0.4097z)$$

$$= 311.2z + 1.024z^2$$

$$\text{Shear flow } q \text{ at } z = \frac{P_U}{I_U}(311.2z + 1.024z^2)$$

$$\text{Total force } F_1 = \int_0^{77.5} \frac{P_U}{I_U}(311.2z + 1.024z^2)\,dz$$

$$= \frac{P_U}{I_U}\left[311.2 \times \frac{77.5^2}{2} + 1.024\frac{(77.5)^3}{3}\right]$$

If e_1 is the distance of the shear centre from A parallel to the U-U axis, then

$$P_U e_1 = \frac{P_U}{I_U}\left[311.2 \times \frac{77.5^2}{2} + 1.024\frac{(77.5)^3}{3}\right] \times 145$$

$$\Rightarrow \quad e_1 = 20.7 \text{ mm}$$

Consider the shear acting parallel to the V-axis. The moment of area will be taken about the U-U axis, which will be the NA.

$$A\bar{y} \text{ of shaded area } = \int_0^z 5\,dz\,(71.8 - 0.9122z)$$

$$= 359z - 22.8z^2$$

$$\text{Force } F_1 = \int_0^{77.5} \frac{P_V}{I_V}(359z - 22.8z^2)dz$$

$$= \frac{P_V}{I_V}\left[359\frac{z^2}{2} - 22.8\frac{z^3}{3}\right]_0^{77.5}$$

$$= 724{,}353\frac{P_V}{I_V}$$

$$P_V e_2 = 724{,}353 \times \frac{P_V}{I_V} \times 145$$

$$\Rightarrow \quad e_2 = 39 \text{ mm}$$

The shear centre is located as shown in Fig. 5.68(e).

5.13 UNSYMMETRICAL BENDING

In applying the condition of bending, an important assumption was made regarding the planes of symmetry of the cross section of the beam. The section was considered to be symmetric with reference to the plane of loading, i.e., the plane of loading was assumed to be a plane of symmetry of the cross section.

Consider the two positions of the channel section shown in Figures 5.69(a) and (b). The X- and Y-axes are the centroidal axes of the cross section. In case (a), the load acts in a plane containing the X-axes, which is a plane of symmetry of the cross section. The bending equations apply to this beam without any modifications.

In case (b), the beam is rotated through 90°. The load acts in a plane containing the Y-axis, which is not a plane of symmetry. If you experiment by loading a beam, say the cantilever shown in Fig. 5.69(c), the section will tend to twist in addition to bending. Note that the section is symmetrical about the X-X axis, which would have been the NA if the section had been subjected to pure bending couples in the plane of the Y-axis.

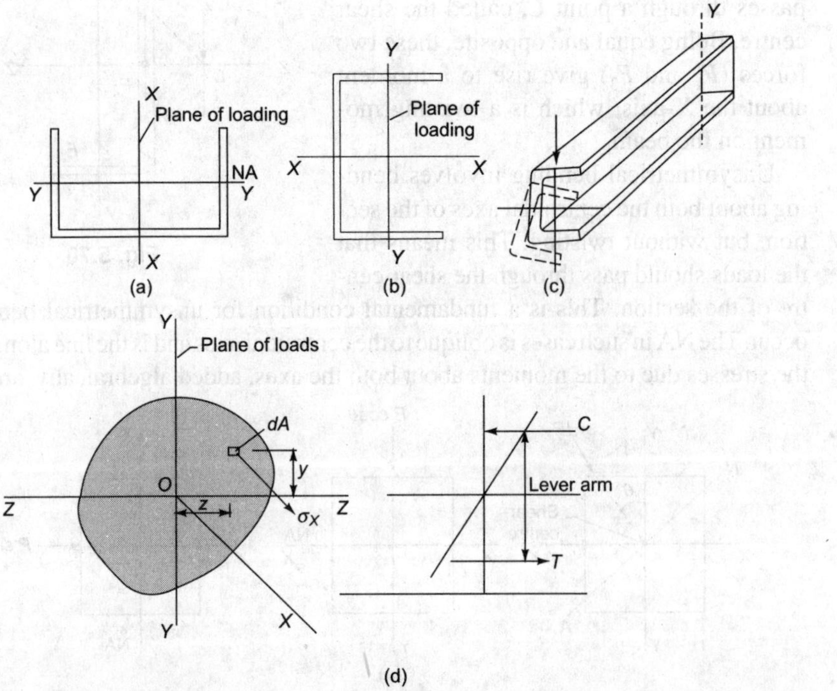

Fig. 5.69

The twisting of the beam can be easily explained. Consider the cross section shown in Fig. 5.69(d). The Z-axis is the NA of the section, or the axis of bending. If we consider an elementary area dA and the stress on this elemary area (normal stress) is σ_X, acting in the direction of X-axis, then $\Sigma\, \sigma_X\, dA = 0$, which is satisfied if the NA passes through the centroid of the cross section. If M is the external BM, then

$$\Sigma\, \sigma_X\, dA\, y = M$$

The normal stress is compressive above the NA and tensile below the NA. This gives resultant compressive stresses C and T, which are equal and C or T multiplied by the lever arm is equal to the moment of resistance M, which resists the external BM.

The difference in structural action comes with respect to the moment about the Y-axis. $\Sigma\, \sigma_X\, dA\, z$ will be zero if the Y-axis, the plane of loading containing it, is an axis of symmetry. The Y-axis being an axis of symmetry in a symmetrical section, the moment of forces to the left about the axis is balanced by that to its right. If the plane containing the Y-axis, the plane of loads, is not a plane of symmetry, then $\Sigma M \neq 0$ about the Y-axis. Also, $\Sigma\, \sigma_X\, dA\, z = \Sigma(Kyz\, dA) = 0$ means that $K\, \Sigma yz\, dA = 0$. This means that the product of inertia about the Y-Z axes is zero. From Chapter 2, we know that the product of inertia is zero about the set of principal axes of the section.

If the Y-axis is not a plane of symmetry, twisting can still be prevented if the loads pass through the shear centre of the section. Considering the channel section shown in Fig. 5.70, if the loads are in the plane containing the Y-axis, the beam twists because the internal shear stress resultant passes through a point C, called the shear centre. Being equal and opposite, these two forces (F_1 and F_3) give rise to a moment about the X-axis, which is a twisting moment on the beam.

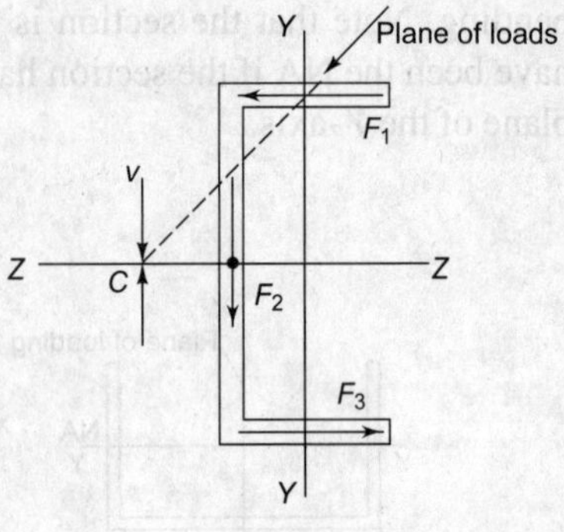

Fig. 5.70

Unsymmetrical bending involves bending about both the centroidal axes of the section, but without twisting. This means that the loads should pass through the shear centre of the section. This is a fundamental condition for unsymmetrical bending to occur. The NA in such cases is oblique to the centroidal axes and is the line along which the stresses due to the moments about both the axes, added algebraically, are zero.

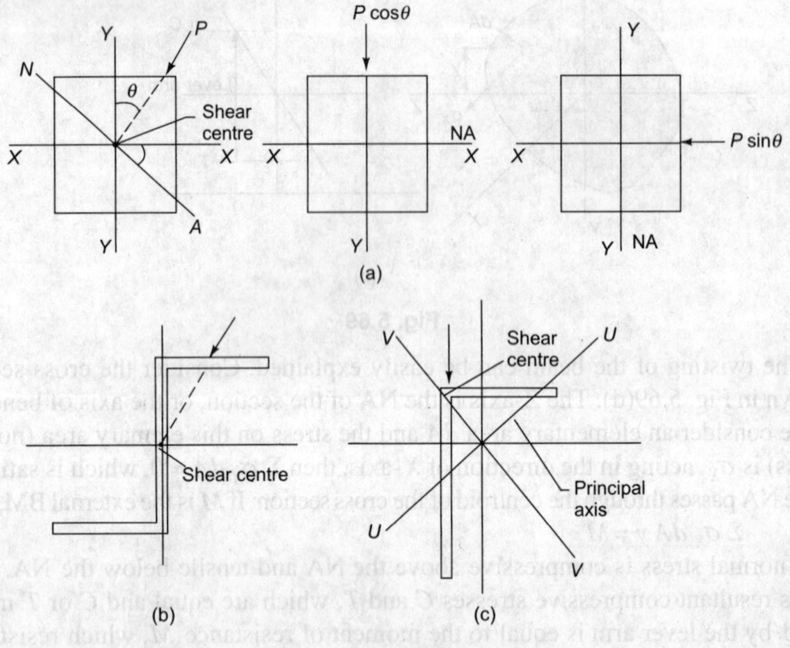

Fig. 5.71

In the case of sections with two planes of symmetry, unsymmetrical bending can occur when the plane of loads is not along the principal centroidal axes, but passes through the centroid, which is the shear centre in such cases. The load or the bending couple can be resolved along these axes. For bending in a plane containing one of the axes, the other axis will be the NA as shown in Fig. 5.71(a).

For sections with only one plane of symmetry, the plane of loading should pass through the shear centre for unsymmetric bending to occur. A Z-section, such as the one shown in Fig. 5.71(b), has no planes of symmetry but a point of symmetry, which is the shear centre.

In the case of sections with no plane of symmetry, unsymmetrical bending occurs so long as the load passes through the shear centre. Such sections have principal centroidal axes about which product of inertia is zero. It is possible to resolve the BM or loads along these axes and find the bending stress distribution [Fig. 5.71(c)].

5.13.1 General Equations for Unsymmetrical Bending

Taking the X-direction along the longitudinal axis of the beam, the Y- and Z-axis are taken as the principal centroidal axis of the section. If both axes are in planes of symmetry, then the principal centroidal axes are the same. Otherwise, the principal entroidal axes can be found using the methods outlined in Chapter 2. The loads in all cases pass through the shear centre.

5.13.2 Resolving Moments Along Principal Axes

If the plane of loading is inclined at an angle θ to the vertical, then, resolving it into components and finding the moments,

$$M_Y = M \sin\theta$$

and

$$M_Z = M \cos\theta$$

for the case shown in Fig. 5.72(b). Since Y and Z are the principal centroidal axes in cases (a) and (b) the normal stresses along x-direction are added algebraically to give

$$\sigma_X = \frac{-M_Z y}{I_Z} + \frac{M_Y z}{I_Y}$$

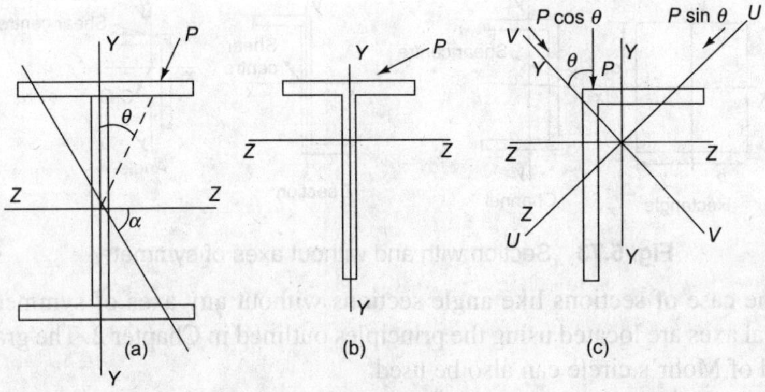

Fig. 5.72

The same formula applies to the angle section in Fig. 5.72(c), where UU and VV are the principal centroidal axes. The moment can be resolved along the U-U and V-V axes and the above formulae applied with the corresponding coordinates. The principle of superposition can be applied so long as the stresses are within the elastic limit.

The equation for σ_X shows that the stresses are distributed linearly and the position of that NA, at which the stresses are zero, will not, in general be the Y- and Z-axis. Setting $\sigma_X = 0$,

$$\frac{-M_Z y}{I_X} + \frac{M_Y z}{I_Y} = 0$$

$$-M\cos\theta\frac{y}{I_X}+M\sin\theta\frac{z}{I_Y}=0$$

We get

$$y=+z\left(\frac{I_Z}{I_Y}\right)\tan\theta$$

If α is the angle made by the NA with the Z-axis, then

$$\tan\alpha=+\left(\frac{I_Z}{I_Y}\right)\tan\theta$$

The following examples illustrate the procedure for solving problems in eases of unsymmetrical bending.

5.13.3 Centroidal Principal Axes of Section

Centroidal axes are the axes passing through the centroid of the section. In a section, there are principal axes through the centroid. As we have seen in Chapter 2 (Properties of Sections), when a section has axes of symmetry, they are the principal axes of sections. This is the case with rectangular, circular, and I-sections shown in Fig. 5.73(a).

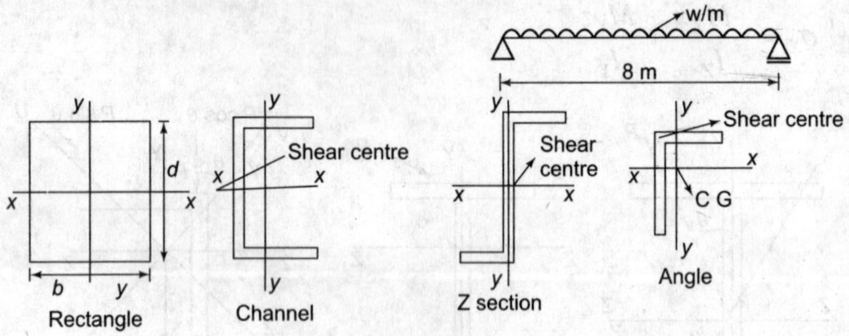

Fig. 5.73 Section with and without axes of symmetry

In the case of sections like angle sections without any axes of symmetry, the principal axes are located using the principles outlined in Chapter 2. The graphical method of Mohr's circle can also be used.

To recapitulate, if I_{xx}, I_{yy}, and P_{xy} are the MI and product of inertia about orthogonal axes through the centroid, then the principal axes and principal MIs can be found as follows.

Principal MI, $I_1, I_2=(I_{xx}+I_{yy})/2\pm\sqrt{\{[(I_{xx}+I_{yy})/2]^2+P^2{}_{xy}\}}$

The inclination of the principal axes to the X-axis is given by

$$\tan 2\theta=\frac{P_{xy}}{\left(I_{xx}-I_{yy}\right)/2}$$

In the case of sections with no axes of symmetry, we need to find the principal axes and principal moments of inertia. Of course, the product of inertia is zero about the principal axes.

5.13.4 Location of Neutral Axis

Obviously, in the case of bending about both principal axes, the neutral axis, or the line of zero, bending stress will be, in general, inclined to the principal axes. The formula for locating the neutral axis has been given in the earlier section.

In unsymmetrical bending, the neutral axis may be inclined to the three planes—the plane of loading and the planes containing the principal axes. The method to locate the neutral axis is to find the points of zero stress. Both the analytical method and graphical method (ellipse of stress method) can be used. We consider the analytical method here.

For simplicity, we consider a rectangular section. Principal axes are *U-U* and *V-V*, being axes of symmetry. Consider the principal axes *u-u* and *v-v* shown in Fig. 5.74.

If the load is inclined at an angle θ to the vertical (passing through *O*), the vertical and horizontal components are $P \cos \theta$ and $P \sin \theta$. It is clear that, for bending due to $P \cos \theta$, *U-U* is the neutral (bending) axis and for bending due to horizontal component $P \sin \theta$, *V-V* is the neutral axis. Thus, M_{uu} is due to $P \cos \theta$ and M_{vv} is due to $P \sin \theta$.

At any point having coordinates (u, v), the bending stress due to these moments is

$$\sigma_A = \frac{M_{uu} \cos \theta v}{I_{uu}} + \frac{M_{vv} \sin \theta u}{I_{vv}}$$

Due to both these moments, tension zone in the section is shown shaded. From this, note that the possibility of stress being zero is only in the top left and bottom right quarters. The NA will thus be directed as shown making an angle α with the *U-U* axis. If a point has coordinates (u, v) on this line, then the stress σ is zero at that point. Thus, at point *A* shown in the figure,

$$\sigma_A = \frac{M_{uu} \cos \theta v}{I_{uu}} + \frac{M_{vv} \sin \theta u}{I_{vv}}$$

Rearranging and simplifying, we get
$\tan \alpha = v/u = (I_{uu}/I_{vv}) \tan \theta$
The neutral axis can thus be located.

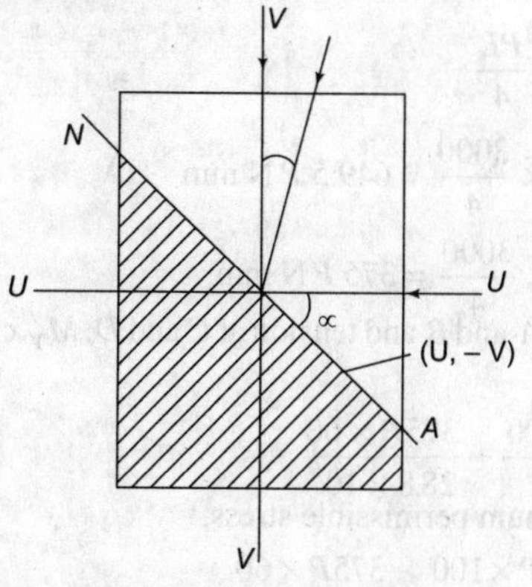

Fig. 5.74 Neutral axis

Example 5.38 Unsymmetrical bending: unequal angle section

A rectangular section of dimensions 120 × 200 mm is used as a beam on a 3 m span [Fig. 5.75(a)]. If the beam is loaded by a concentrated load at the centre at 30° to the vertical as shown in Fig. 5.75(b), find the maximum value of the load *P* if the maximum bending stress is not to exceed 12 MPa.

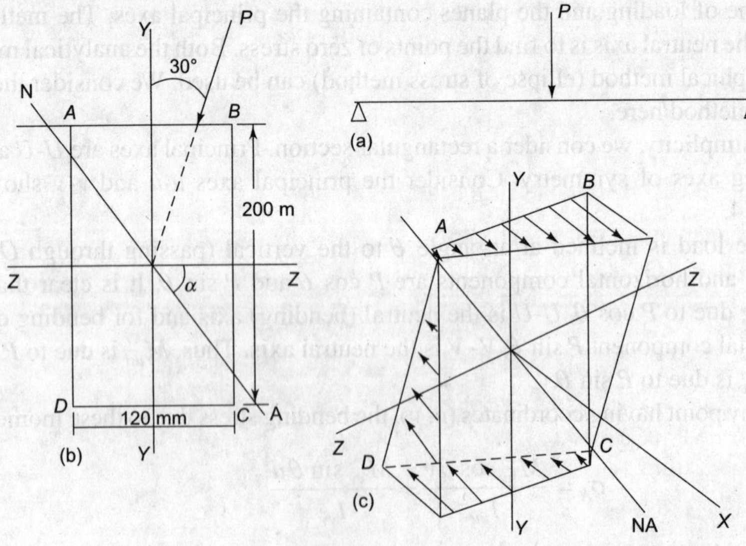

Fig. 5.75

Solution For the given section,

$$I_Z = \frac{120 \times 200^3}{12} = 80 \times 10^6 \text{ mm}^4$$

$$I_Y = \frac{200 \times 120^3}{12} = 28.8 \times 10^6 \text{ mm}^4$$

The *Y*- and *Z*-axis are the principal axes of the section through the centroid. If α is the inclination of the NA, then

$$\tan \alpha = \left(\frac{I_Z}{I_Y}\right) \tan \theta = \frac{80 \times 10^6}{28.8 \times 10^6} \times \tan 30 = 1.6037$$

$\Rightarrow$ $\alpha = 58°$

$$\text{Maximum BM} = \frac{PL}{4}$$

$$M_Z = P \cos 30° \times \frac{3000}{4} = 649.5\, P \text{ N mm}$$

$$M_Y = P \sin 30° \times \frac{3000}{4} = 375\, P \text{ N mm}$$

M_Z causes compression at *A* and *B* and tension at *C* and *D*. M_Y causes compression at *B* and *C* and tension at *A* and *B*.

$$\sigma_B = \frac{649.5P \times 100}{80 \times 10^6} + \frac{375P \times 60}{28.8 \times 10^6}$$

Equating this to the maximum permissible stress,

$$\sigma_B = 12 = \frac{649.5P \times 100}{80 \times 10^6} + \frac{375P \times 60}{28.8 \times 10^6}$$

$\Rightarrow \qquad P = 7532$ N

The stresses at the different points are

$\sigma_A = 0.22$ MPa (tensile)

$\sigma_B = 12$ MPa (compressive)

$\sigma_C = 0.22$ MPa (compressive)

$\sigma_D = 12$ MPa (tensile)

The position of the NA can be located from the equation

$$\frac{649.5\,Py}{I_Z} + \frac{375\,Pz}{I_Y} = 0$$

When $y = \pm 100$, $z = \mp 62.4$. The position of the NA and the stress distribution are shown in Fig. 5.75(c).

Example 5.39 Unsymmetrical bending: T-section

The T-section shown (Fig. 5.76) is used as a simply supported beam on a span of 6 m. Find the maximum value of w if the permissible stress in the material is 120 MPa. The plane of loading is inclined at an angle of 40° to the vertical plane, and passes through the shear centre.

Solution The location of the centroid can be obtained as, from Fig. 5.76,

$$\bar{y} = \frac{150 \times 10 \times 5 + 190 \times 10 \times 105}{1500 + 1900} = 60.88 \text{ mm}$$

The MIs about the Z- and Y-axis are

$$I_{ZZ} = \frac{150 \times 10^3}{12} + 1500 \times (55.88)^2 + 10 \times \frac{190^3}{12} + 1900 \times (44.12)^2$$

$$= 14.11 \times 10^6 \text{ mm}^4$$

$$I_{YY} = \frac{10 \times 150^3}{12} + \frac{190 \times 10^3}{12} = 2.828 \times 10^6$$

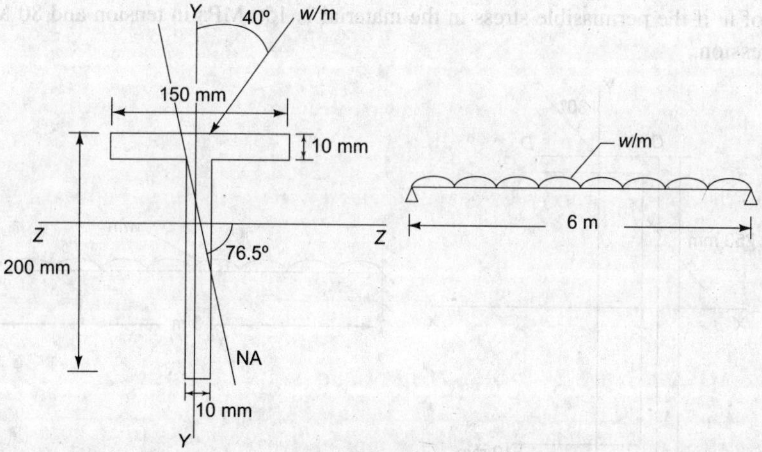

Fig. 5.76

The load can be resolved along the vertical and horizontal planes for bending in the respective planes. For bending in the Y-plane (BM about the NA) for which the Z-axis is the NA,

$$M_Z = \frac{(w \cos 40) \times 36 \times 1000}{8} \text{ N mm}$$

is the maximum BM at the centre of the span. Similarly,

$$M_Y = \frac{(w \sin 40) \times 36 \times 1000}{8} \text{ N mm}$$

is the maximum BM for bending in the perpendicular plane.

If α is the inclination of the NA to the Z-axis,

$$\tan \alpha = \frac{I_Z}{I_Y} \tan \phi = \frac{14.11 \times 10^6}{2.828 \times 10^6} \times \tan 40$$

$$\Rightarrow \quad \alpha = 76.5°$$

The stress at points A, B, C and D due to $M_Z = 3447.2w$ and $M_Y = 2892.5w$ are

$$\sigma_A = \frac{3447.2w}{14.11 \times 10^6} \times 60.88 - \frac{2892.5w}{2.828 \times 10^6} \times 75 = -0.0618w$$

$$\sigma_B = \frac{3447.2w}{14.11 \times 10^6} \times 60.88 + \frac{2892.5w}{2.828 \times 10^6} \times 75 = 0.09167w$$

$$\sigma_C = \frac{-3447.2w \times 139.12}{14.11 \times 10^6} - \frac{2892.5w \times 5}{2.828 \times 10^6} = -0.039w$$

$$\sigma_D = \frac{-3447.2w \times 139.12}{14.11 \times 10^6} + \frac{2892.5w \times 5}{2.828 \times 10^6} = -0.029w$$

Maximum stress,

$$0.0916w = 120$$

$$\Rightarrow \quad w = 1310 \text{ N/m}$$

Example 5.40 Unsymmetrical bending: channel section

The channel section shown in Fig. 5.77 is used as a cantilever beam of span 3 m. The plane of the UD load on the span is inclined at 30° to the vertical. Find the maximum allowable value of w if the permissible stress in the material is 150 MPa in tension and 80 MPa in compression.

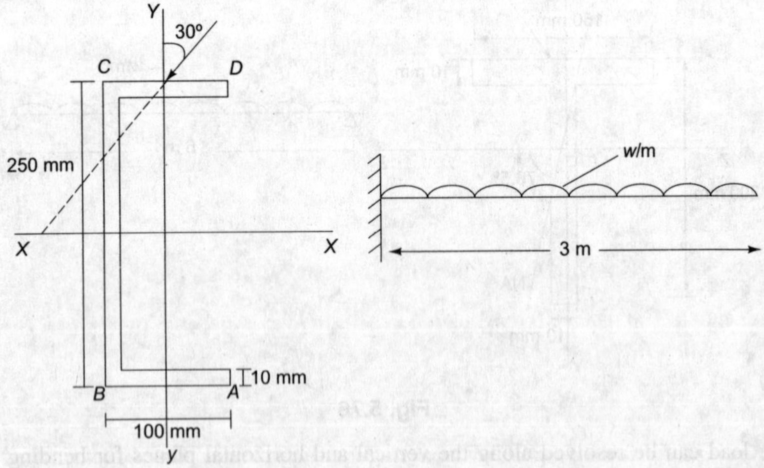

Fig. 5.77

Solution The section is symmetrical about the X-X axis. The distance of the centroid can be found as

$$\bar{x} = \frac{250 \times 10 \times 5 + 90 \times 10 \times 55 \times 2}{2500 + 2 \times 900} = 25.93 \text{ mm}$$

The MIs about the vertical and horizontal axes through the centroid are

$$I_X = 2\left(\frac{90 \times 10^3}{12} + 900 \times 120^2\right) + \frac{10 \times 250^3}{12} = 26 \times 10^6 \text{ mm}^4$$

$$I_Y = 2\left(\frac{10 \times 90^3}{12} + 900 \times 29.07^2\right) + \frac{250 \times 10^3}{12} + 2500 \times 20.93^2$$
$$= 2.484 \times 10^6 \text{ mm}^4$$

Assuming the loads pass through the shear centre, the moments for bending in the X- and Y-planes can be calculated as

$$M_X = (w \cos 30) \times 3 \times \frac{3000}{2} = 3897w \text{ N mm}$$

$$M_Y = (w \sin 30) \times 3 \times \frac{3000}{2} = 2250w \text{ N mm}$$

Of the four points A, B, C, and D, B will have tensile stress and D will have compressive stress due to M_X and M_Y. At A and C, one moment will cause tension while the other will cause compression.

$$\sigma_B = \frac{3897w}{26 \times 10^6} \times 125 + \frac{2250w}{2.484 \times 10^6} \times 25.93 = 0.0422w \text{ (tensile)}$$

$$\sigma_D = \frac{3897w}{26 \times 10^6} \times 125 + \frac{2250w}{2.484 \times 10^6} \times 74.07 = 0.0858w \text{ (compressive)}$$

Equating these stress values to the respective permissible stresses,

$0.0422w = 150, \quad w = 3555$ N/m

$0.0858w = 80, \quad w = 932$ N/m

The permissible value of the UD load = 932 N/m. ■

5.13.5 Sections with No Axis of Symmetry: Unsymmetrical Sections

In the case of sections with no axes of symmetry, the procedure is to find the principal axes of the section through the centroid. The applied load can then be resolved along the principal axes and the moment components about these axes can then be found. The neutral axis can be located as indicated earlier. It is illustrated with an example of an angle section.

Example 5.41 Unsymmetrical bending: unequal angle section

An angle section, $100 \times 80 \times 10$ mm, is used as a beam over a span of 5 m. If the permissible stress is 100 MPa, find the UD load the beam can carry. The load passes through the shear centre.

Solution The distances to the centroid can be calculated as

$$\bar{x} = \frac{80 \times 10 \times 40 + 90 \times 10 \times 5}{1700} = 21.47 \text{ mm}$$

$$\bar{y} = \frac{80 \times 10 \times 5 + 90 \times 10 \times 55}{1700} = 31.47 \text{ mm}$$

The angle section has no plane of symmetry. We can find the principal centroidal axes *U-U* and *V-V*, and consider bending about these axes.

$$I_{XX} = \frac{80 \times 10^3}{12} + 800 \times (26.47)^2 + \frac{10 \times 90^3}{12} + 900 \times (23.53)^2$$

$$= 167.3 \times 10^4 \text{ mm}^4$$

$$I_{YY} = \frac{10 \times 80^3}{12} + 800 \times (18.53)^2 + \frac{90 \times 10^3}{12} + 900 \times (15.47)^2$$

$$= 95.3 \times 10^4 \text{ mm}^4$$

Let us calculate the product of inertia I_{XY}. The coordinates of the centroid of the two rectangles about *O* are (18.53, 26.47) and (–16.47, –23.53).

$$I_{XY} = 800 \times 18.53 \times 26.47 + 900 \times (-16.47) \times (-23.53)$$

$$= 74.11 \times 10^4 \text{ mm}^4$$

The inclination of the principal axes *U-U* and *V-V* may be worked out from

$$\tan 2\theta = \frac{-2 \times 74.11 \times 10^4}{(167.3 \times 10^4 - 95.3 \times 10^4)} = -2.0586$$

$$\tan(180 - 2\theta) = 64°$$

$$\Rightarrow \quad 2\theta = 116°$$

$$\Rightarrow \quad \theta = 58° \text{ and } 148°$$

$$I_{UU}, I_{VV} = \left(\frac{167.3 + 95.3}{2} + \sqrt{\left(\frac{167.3 - 95.3}{2}\right)^2 + 74.11^2} \right) \times 10^4$$

$$= 213.7 \times 10^4 \text{ and } 48.9 \times 10^4$$

$$I_{UU} = 213.7 \times 10^4$$

and

$$I_{VV} = 48.9 \times 10^4$$

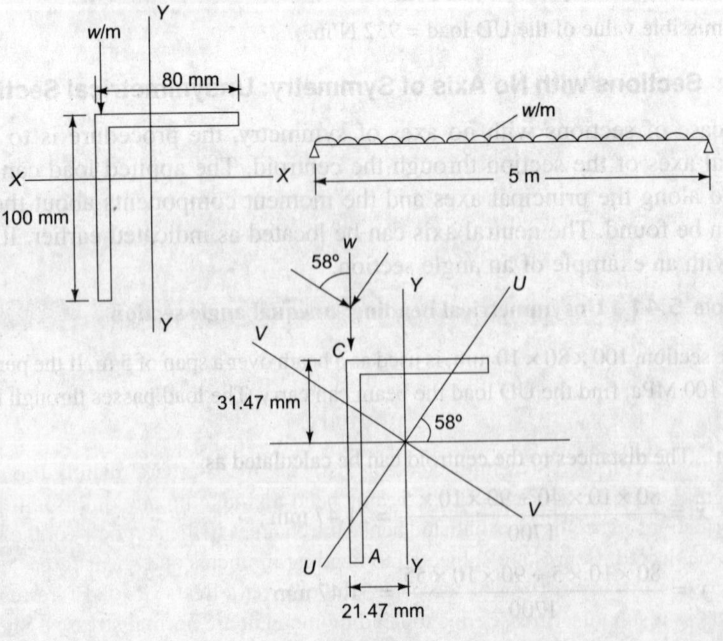

Fig. 5.78

Resolving the load W along the U-U and V-V axes,

$$w_U = w \sin 58° = 0.848w$$

$$w_V = w \cos 58° = 0.5299w$$

$$M_U = \frac{0.5299w \times 25 \times 1000}{8} = 1656w$$

$$M_V = \frac{0.848w \times 25{,}000}{8} = 2650w$$

The permissible stress in tension and compression being the same, the critical points for stress will be A and C. Point A has tensile stresses due to both BMs, and point C has compressive stresses due to both BMs. To calculate the stresses, we need the u and v coordinates of these points. From the transformation of coordinates,

$$u = x \cos \theta + y \sin \theta$$

and

$$v = y \cos \theta - x \sin \theta$$

At point A,

$$x = -11.47, \quad y = -68.53$$

$$u = -64.19, \quad v = -26.58$$

At C,

$$x = -21.47, \quad y = +31.47$$

$$u = 15.3, \quad v = 34.88$$

The stresses are

$$\sigma_A = \frac{M_U}{I_U} A_U + \frac{M_V}{I_V} A_U = \frac{1655.93w}{213.7 \times 10^4} \times 26.58 + \frac{2650w \times 64.19}{48.9 \times 10^4}$$

$$= 3684 \times 10^{-4} w$$

$$\sigma_C = \frac{1655.93w}{213.7 \times 10^4} \times 34.88 + \frac{2650w}{48.9 \times 10^4} \times 15.3 = 1099 \times 10^{-4} w$$

The larger of the two values must be equal to the permissible stress

$$3684 \times 10^{-4} w = 100$$

$$\Rightarrow \quad w = 270 \text{ N/m}$$

The permissible load is 270 N/m.

Summary

Beams are very common structural elements carrying loads predominantly transverse to their length. Under the action of loads, beams bend and take up an equilibrium position. The deformations in beams are small and hence the changes in the geometry of the loads are generally neglected. Beams are subjected to bending moments and shear forces under the action of loads. At any section of the beam, the external effects of loads are resisted by internal stress resultants. Stresses due to bending moment are normal stresses. Shear force is resisted by tangential stresses developed in the section. The two effects are generally considered separately for designing beams.

In the theory of bending, a case of pure bending couples at the ends is taken so that the SF is absent. The bending equation is derived in such cases with certain assumptions. One is that plane sections before bending remain plane after bending. If that be so, we obtain a linear strain diagram. If we take it that E is constant in tension and compression, our first assumption leads to a linear stress distribution for bending stresses. The bending equation $M/I = \sigma/y = E/R$ is derived based on such assumptions. $M/I = \sigma/y$ helps us arrive at the stress distribution and maximum bending stresses (σ_{max}) in the section either in tension or compression. The neutral layer in the beam section is one which has no strain. $M/I = \sigma/y$ relates the stresses to the applied BM while $M/I = E/R$ relates the deformation in beams to the applied BM.

Based on similar assumptions, stresses due to SF can be worked out. The shear stress τ is tangential to the section and is distributed over the section with maximum value at the neutral axis. It is assumed to be constant across the width. The shear stress distribution equation $\tau = (V/Ib)A\bar{y}$ shows a parabolic distribution of shear stress over the depth of the section.

Beams can be designed separately for BM and SF. The larger section satisfying the permissible stress values in bending and shear is selected.

Composite beams are so called because they are made of two materials, such as timber and steel or concrete and steel. Such beams can be analysed for BM and SF, based on the same principles as for single-material sections. The strains at any depth being the same in the two materials, the stresses will be in the ratio of the modulus of elasticity of the materials (known as modular ratio). Composite beams can be made into a section of one of the materials by converting the areas (but without altering the depth) by increasing the width as modular ratio × actual width.

In the case of thin-walled sections, such as I, T, [, and angles, the shear flow can be worked out assuming that the shear stress is constant across the depth which is small. Shear flow $q = \tau t$. The concept of shear centre comes from that of shear flow. Sections with two axes of symmetry, such as I-sections, have the shear centre at the intersection of the axes. In the case of a channel section, the shear centre can be located by applying the concept of shear flow, and is not at the intersection of the centroidal axes. If the applied loads do not pass through the shear centre, the section will be subjected to twisting.

Unsymmetrical bending occurs when a beam of unsymmetrical or symmetrical sections is subjected to loads which are out of the plane of symmetry but pass through the shear centre of the section. The beam bends about the principal axes of the section in two planes but without twisting. The bending stresses can be obtained by superposition. The NA may be an oblique line in the section and can be located by finding points of zero stress.

Exercises

Review Questions

1. Explain the structural behaviour of a beam subjected to transverse loads.
2. Distinguish between pure bending and ordinary bending.
3. State the assumptions made in deriving the bending equation. Explain the significance of each assumption.
4. Define the terms section modulus, modulus of rupture, bending moment, and moment of resistance.
5. Explain the basic principle involved in the analysis of a composite beam.
6. Draw equivalent sections in terms of both the materials for the composite sections shown in Fig. 5.79.

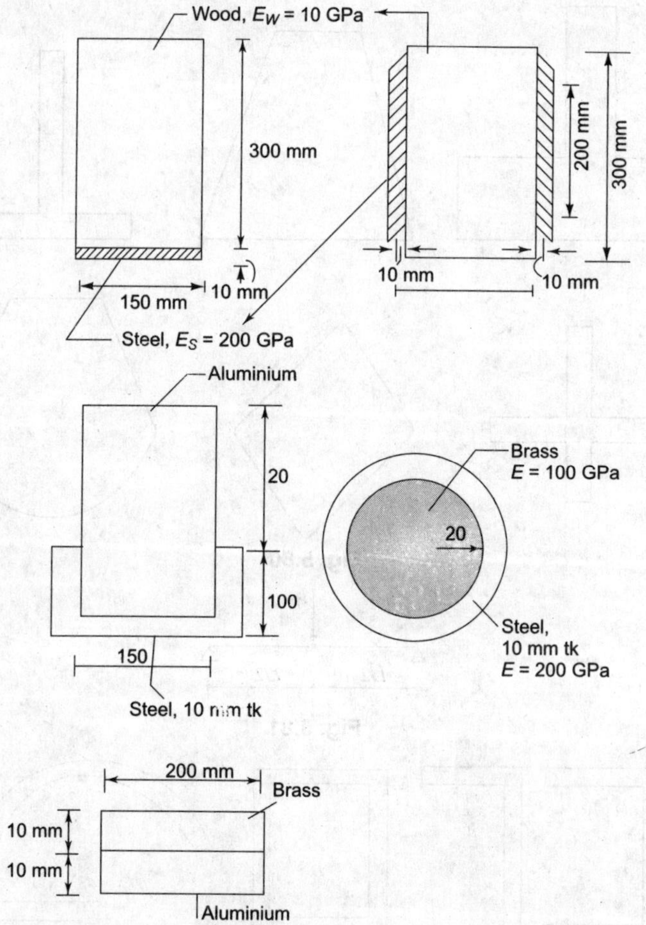

Fig. 5.79

7. State the bending equation. Give the meaning of each term and a set of consistent units for the terms.
8. The bending equation represents two equalities. Explain the meaning and use of each equality.
9. State the equation describing the shear stress distribution over the cross section of a beam.
10. State the assumptions made in deriving the equation for shear stress distribution.
11. Explain with examples the limitations of the shear stress distribution equation.
12. Make rough sketches of the shear stress distribution for the sections shown in Fig. 5.80.
13. Sketch approximately the plan and sections (longitudinal) of a beam of constant strength
    (i) if the width is constant and the depth is varied,
    (ii) if the depth is constant and the width is varied for the beam loaded as shown in Fig. 5.81.
14. What is shear flow? State and explain the general equation for shear flow.
15. Make rough sketches of the shear flow diagrams for the sections shown in Fig. 5.82.

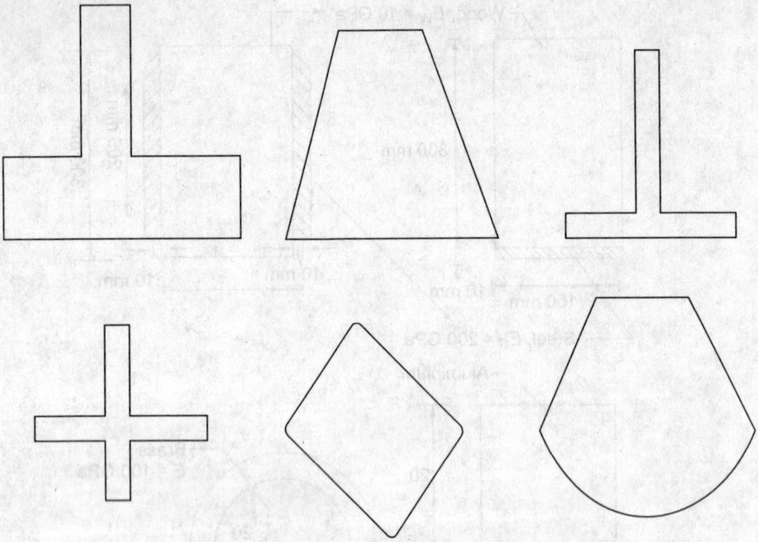

Fig. 5.80

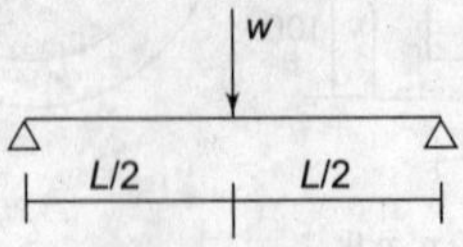

Fig. 5.81

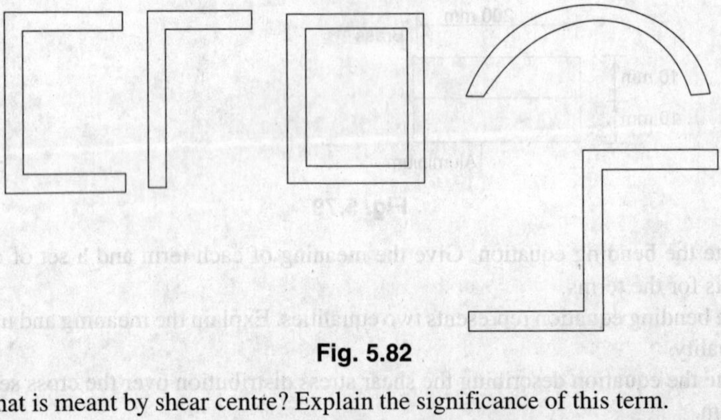

Fig. 5.82

16. What is meant by shear centre? Explain the significance of this term.
17. Show approximately the location of the shear centres in the sections shown in Fig. 5.83.

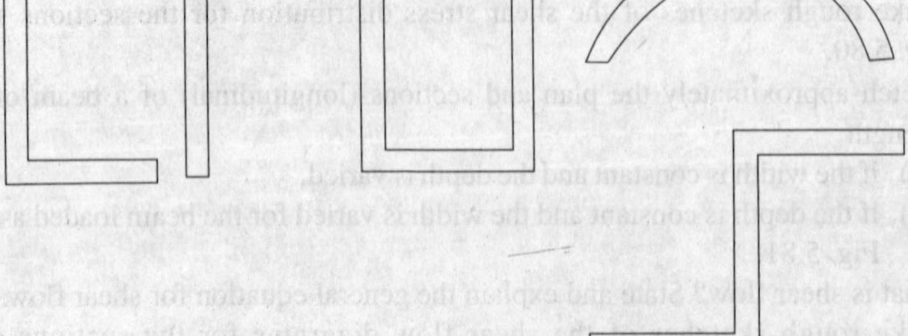

Fig. 5.83

18. Explain the concept of unsymmetrical bending. What are the conditions that should be satisfied for a beam to bend without twisting?
19. State the assumptions made in analysing a beam for unsymmetrical bending.
20. In which of the cases shown in Fig. 5.84 are the equations of unsymmetrical bending applicable? What is the behaviour of the beam in the other cases.

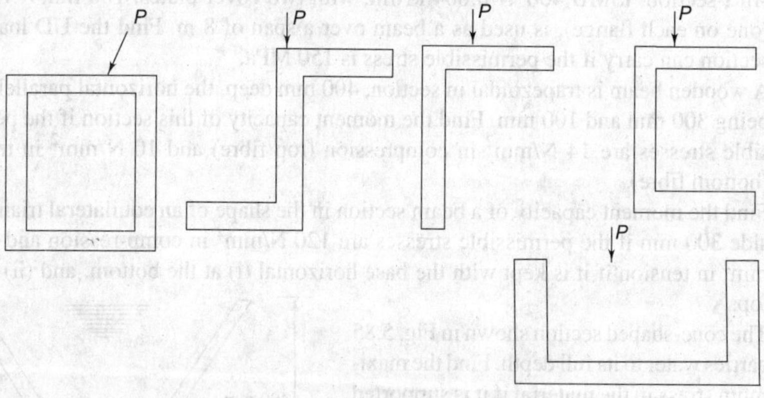

Fig. 5.84

Problems

1. An aluminium rod of square section is bent into a circular arc by applying couples at the ends. If the radius of the arc is 5 m, find the minimum value of the side of the section, so that the stress in the bar does not exceed 35 MPa. What is the magnitude of the couple applied? E for aluminium is 70 GPa.
2. A rectangular strip of a material, of width b and thickness one-tenth of the width, is bent into a circular arc of radius 1.5 m by applying end couples of 6000 N mm. The maximum stress in the material is 100 MPa. Find the value of E for the material and the value of b.
3. A circular steel bar of diameter 5 mm and length 2 m is bent into an arc of radius R m such that the chord length becomes 90% of its length. Find the maximum stress in the material if E = 200 GPa.
4. A rectangular section of wood, 400 mm × 300 mm, is used as a simply supported beam over a span of 4 m. Find the moment capacity of the section if the permissible stresses in tension and compression are 9 N/mm^2 and 14 N/mm^2. What is the maximum load that can be put on the span as a (i) UD load over the whole span and (ii) concentrated load at the centre?
5. (a) Find the dimensions of the strongest rectangular beam that can be cut out of a log of wood of 2.4 m diameter.

   (b) A T-beam having flange 150 mm × 20 mm and web 20 mm × 160 mm is simply supported over a span of 6.2 m. It carries a UDL of 5 kN/m, including the self-weight over its entire span, together with a load of 3.5 kN at mid-span. Find the maximum tensile and compressive stresses occurring in the beam sections and sketch the stresses across the section.
6. A rectangular beam, 100 mm deep, 50 mm wide, and 1.5 m long, is simply supported at its ends. Vertical loads of 5 kN each are applied at 0.5 m and 1 m from one end. Determine and plot the distribution of longitudinal stress across the section at mid-span.
7. A T-section is made up of two planks of wood, 300 × 20 mm and 200 × 20 mm, with the larger of the planks kept horizontal. If the permissible stresses in tension and

compression are 8 N/mm^2 and 12 N/mm^2, find the maximum BM it can carry as a simply supported beam with the flange (i) on top and (ii) at the bottom.

8. A simply supported beam of span 6 m carries a load uniformly varying from zero at the left end to 30 kN/m at the right end. If the permissible stress is 125 N/mm^2, find a suitable I-section to carry the maximum BM.
9. An I-section, ISMB 400 @ 0.604 kN/m, with two cover plates, 180 mm × 10 mm (one on each flange), is used as a beam over a span of 8 m. Find the UD load this section can carry if the permissible stress is 150 MPa.
10. A wooden beam is trapezoidal in section, 400 mm deep, the horizontal parallel sides being 300 mm and 100 mm. Find the moment capacity of this section if the permissible stresses are 14 N/mm^2 in compression (top fibre) and 10 N/mm^2 in tension (bottom fibre).
11. Find the moment capacity of a beam section in the shape of an equilateral triangle of side 300 mm if the permissible stresses are 120 N/mm^2 in comp-ression and 80 N/mm^2 in tension if it is kept with the base horizontal (i) at the bottom, and (ii) at the top.
12. The cone-shaped section shown in Fig. 5.85 carries water to its full depth. Find the maximum stress in the material if it is supported over a span of 3 m. Neglect the weight of the beam.
13. A hollow steel pipe, of internal diameter 100 mm and thickness 5 mm, carries water under pressure. Find the distance at which the pipe should be supported so that the bending stress does not exceed 120 MPa.
14. An unsymmetrical I-section has a top flange of dimensions 100 mm × 10 mm, bottom flange 60 × 10 mm and web 180 mm × 10 mm. If it is used as a cantilever and the permissible stress is 120 N/mm^2, find the maximum span of the cantilever if the load including the self-weight is 10 kN/m.

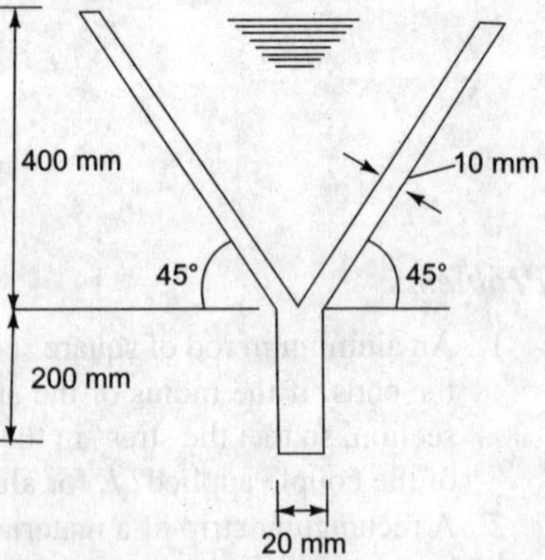

Fig. 5.85

15. An overhanging beam with 4 m between supports and 1 m overhang has to carry a UD load of 4 kN/m. Find a suitable T-section for the beam if the permissible stress is 150 MPa.
16. Compare the strengths of beam sections of the same material and weight of (i) a solid circular section, (ii) a hollow circular section of the same inside diameter as the solid section, (iii) a rectangular section of depth twice the breadth, and (iv) a square section.
17. Compare the strengths of a channel section made of three wooden planks as shown in Fig. 5.86 in two positions if the permissible stresses are 14 N/mm^2 in compression and 10 N/mm^2 in compression.

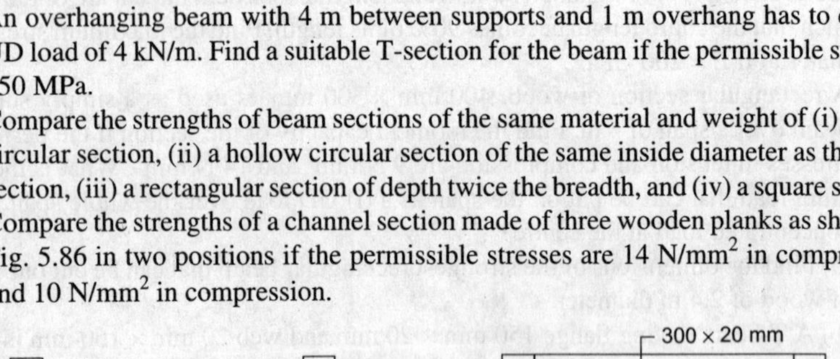

Fig. 5.86

18. Prove that the strength of a square section used as a beam to bend about a diagonal can be increased by flattening the corners in the plane of loading and that the maximum strength is obtained by removing one-ninth of the diagonal length, equally divided at the top and bottom.
19. Find the moment capacity of the sections shown in Fig. 5.87, if the permissible stress is 100 MPa.

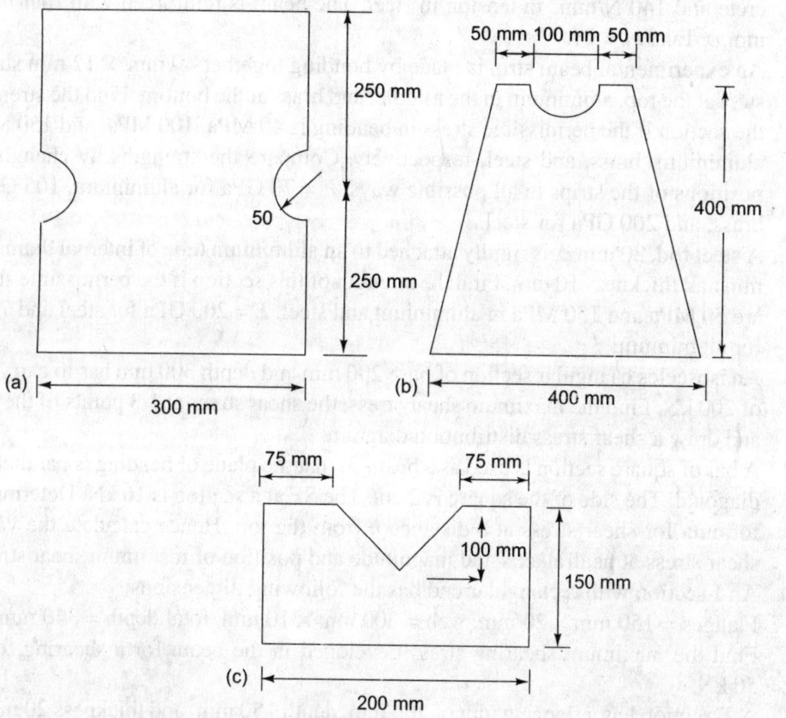

Fig. 5.87

20. Three wooden strips of dimensions 200 mm × 50 mm are available for use as a beam. Compare their strengths if they are used with the 200 mm side horizontal and (i) bonded together, and (ii) kept loose.
21. A wooden beam of size 200 mm × 150 mm is strengthened by a steel plate of dimensions 150 mm × 8 mm. The permissible stress in wood is 14 N/mm^2 in compression and 10 N/mm^2 in tension, and the permissible stress in steel is 150 MPa. Compare the strengths if the steel plate is fixed (i) at the top, and (ii) at the bottom. Take modular ratio = 20.
23. A wooden beam is strengthened by two channel sections as shown in Fig. 5.88. Find the moment capacity of the section. The permissible stresses are 14 N/mm^2 and 10 N/mm^2 in compression and tension, respectively, for wood and 150 MPa for steel in both compression and tension.

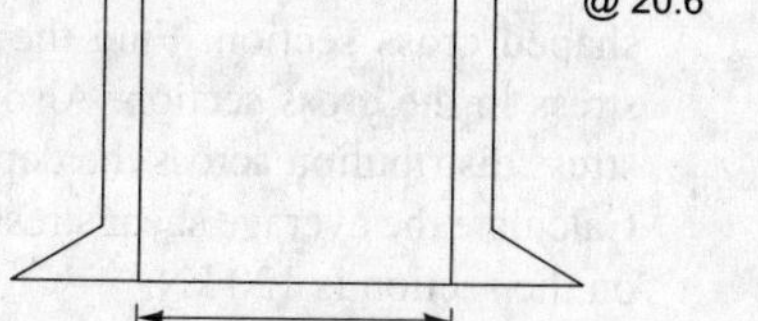

Fig. 5.88

23. A wooden beam, 200 mm square, is strengthened by four angles ISA 50 × 50 × 16 at the four corners. Find the strength of the section, if the permissible stress in timber is 12 N/mm^2 in compression, 8 N/mm^2 in tension and in steel, it is 100 MPa.

24. A brass square section, 30 × 30 mm, has two steel strips of dimensions 30 × 5 mm attached to its top and bottom. For brass, E = 100 GPa and permissible stress = 100 MPa, and for steel E = 200 GPa and permissible stress = 140 MPa. Find the moment capacity of the section.
25. Find the moment of resistance of a reinforced concrete section of size 200 × 400 mm up to the centre of steel if the permissible stress is 5 N/mm^2 in compression in concrete and 140 N/mm^2 in tension in steel. The beam is reinforced with four bars 20 mm ϕ. Take m = 18.
26. An experimental beam strip is made by bonding together 40 mm × 12 mm strips of steel at the top, aluminium in the middle, and brass at the bottom. Find the strength of the section if the permissible stress in bending is 40 MPa, 100 MPa, and 150 MPa in aluminium, brass, and steel, respectively. Compare the strengths by changing the positions of the strips in all possible ways. E = 70 GPa for aluminium, 105 GPa for brass, and 200 GPa for steel.
27. A steel rod, 20 mm ϕ, is rigidly attached to an aluminium tube of internal diameter 20 mm and thickness 10 mm. Find the strength of this section if the permissible stresses are 50 MPa and 150 MPa in aluminium and steel. E = 200 GPa for steel and 70 GPa for aluminium.
28. An isosceles triangular section of base 200 mm and depth 300 mm has to carry an SF of 200 kN. Find the maximum shear stress, the shear stress at 1/3 points of the depth, and draw a shear stress distribution diagram.
29. A bar of square section is used as a beam so that the plane of bending is parallel to the diagonal. The side of the square is 2 cm. The SF at a section is 16 kN. Determine the formula for shear stress at a distance h from the top. Hence calculate the value of shear stress at neutral axis and magnitude and position of maximum shear stress.
30. An I-section with rectangular end has the following dimensions:
    Flanges = 150 mm × 20 mm, web = 300 mm × 10 mm, total depth = 340 mm
    Find the maximum shearing stress developed in the beam for a shearing force of 50 kN.
31. A T-section has a flange width of 200 mm, depth 150 mm, and thickness 20 mm. If it is subjected to an SF of 120 kN at a section, find the maximum intensity of stress and draw the shear stress distribution diagram.
32. A beam of hexagonal cross section has a span of 2 m and is used as a cantilever to support a concentrated load at the free end. The hexagon has sides of 40 mm and is used with a diagonal horizontal. Find the maximum value of the load if the maximum shear stress is not to exceed 50 N/mm^2.
33. The section of a beam is trapezoidal, 200 mm wide at the bottom, 100 mm wide at the top, and 250 mm deep. Draw the shear stress distribution diagram and find the maximum shear stress if it is subjected to a shear force of 50 kN at a section.
34. A beam of square section of 30 mm side is used with one diagonal horizontal making a diamond-shaped cross section. Find the maximum shear stress in the cross section. Also sketch the shear stress distribution across the depth of the section. Calculate the average shear stress. The shear force on the section is 120 kN.
35. For the section shown in Fig. 5.89, determine the average shear stress at A, B, C, D for a shear force of 20 kN and find the ratio of maximum to mean shear stress.

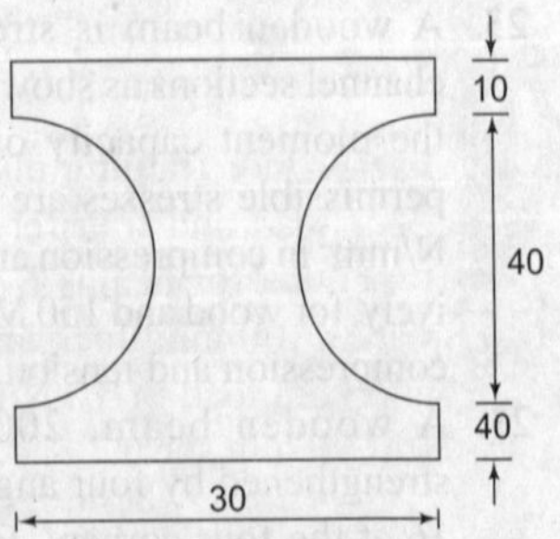

Fig. 5.89

36. Design a timber beam to span a length of 3 m. The beam carries a load varying uniformly from 1 kN/m at the centre to zero at the ends. For the available timber, permissible stresses are 12 N/mm^2, 8 N/mm^2, and 0.8 N/mm^2 in compression, tension, and shear.
37. A beam carries the loading shown in Fig. 5.90. Design a suitable I-section for the beam. The permissible stresses may be taken as 120 MPa in tension and compression, and 100 MPa in shear.

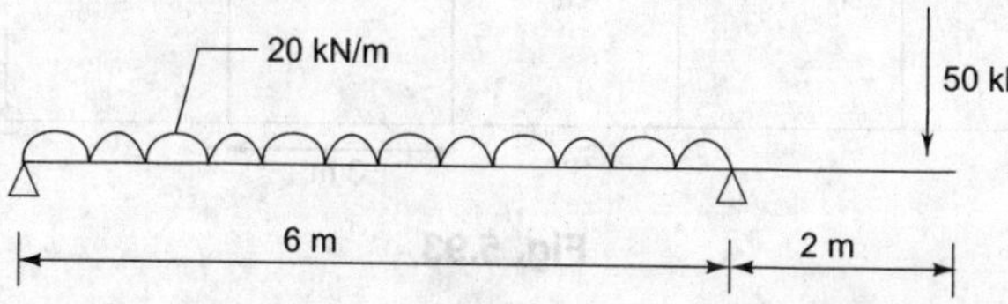

Fig. 5.90

38. A simply supported steel beam has to carry a load of 50 kN/m on a span of 10 m. The maximum section available has a depth of 500 mm. Design a beam section with cover plates. Determine the cut-off point for the cover plates. The permissible stress in steel is 150 MPa.
39. Design the beam shown in Fig. 5.91. The timber available has permissible stresses of 14 N/mm^2 in compression, 10 N/mm^2 in tension, and 1 N/mm^2 in shear.

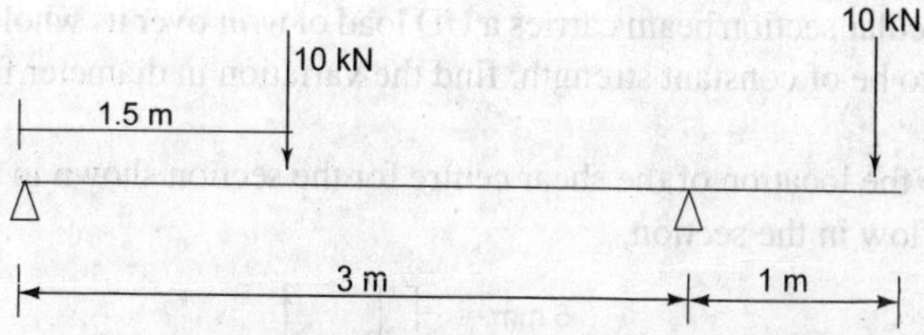

Fig. 5.91

40. Determine the maximum load w that the box section can carry when used in the beam shown in Fig. 5.92 on a span of 4 m.

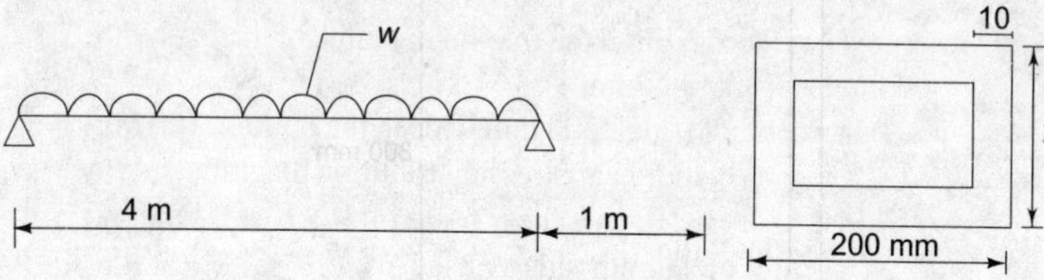

Fig. 5.92

41. In the floor framing plan shown in Fig. 5.93, the beams B_1 of 6 m span carry the floor load and transfer them as reactions to the main girders G_1. Select suitable I-sections for beams B_1 and G_1. The live load on the floor is 4 kN/m^2. The RCC slab is 100 mm thick and weighs 25 kN/m^3. The floor furnishings may be taken to impose a load of 1 kN/m^2. Take the permissible stress as 150 MPa in bending and 100 MPa in shear.

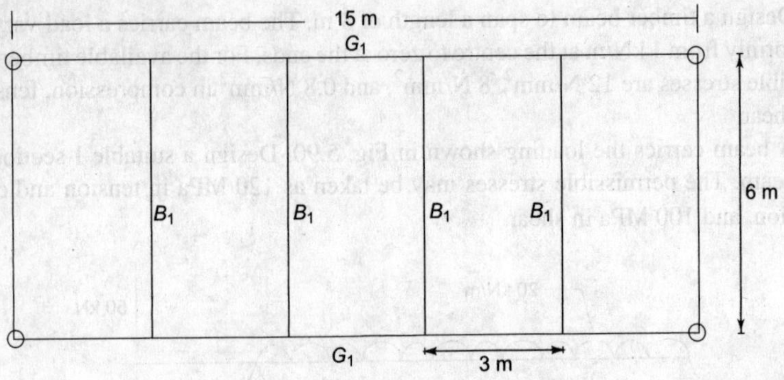

Fig. 5.93

42. A cantilever beam of span L m carries a UD load of w/m over the whole span. Design a beam of constant strength if (i) only the depth varies, and (ii) only the breadth varies. Sketch the plan and longitudinal section of the beams.
43. A simply supported beam of span L carries a UD load of w/m. Find the variation in depth, keeping the width constant, of a rectangular section for constant strength. Sketch the plan and section of the beam.
44. A cantilever beam carries a concen-trated load at its free end. Neglecting the self-weight of the beam, find the variation in diameter of a solid circular section, to have a beam of constant strength.
45. A solid circular section beam carries a UD load of w/m over its whole span L m. If the beam has to be of constant strength, find the variation in diameter from the centre to either end.
46. Determine the location of the shear centre for the section shown in Fig. 5.94. Sketch the shear flow in the section.

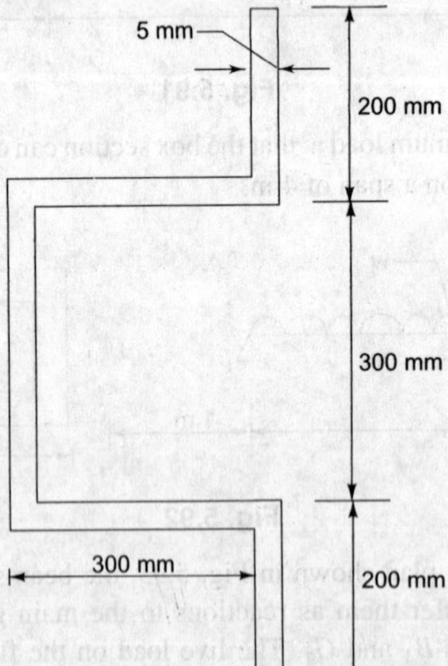

Fig. 5.94

47. For the channel section shown in Fig. 5.95, find the position of the shear centre. Show the shear flow in the section.

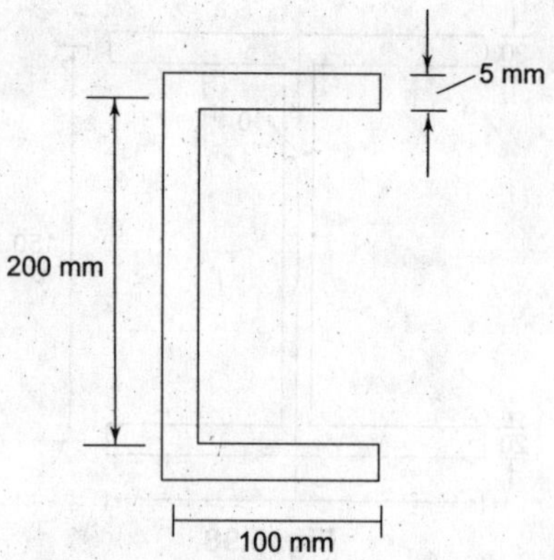

Fig. 5.95

48. The cross section shown in Fig. 5.96 is a channel with turned ends. Locate the shear centre for the section and sketch the shear flow in the section.

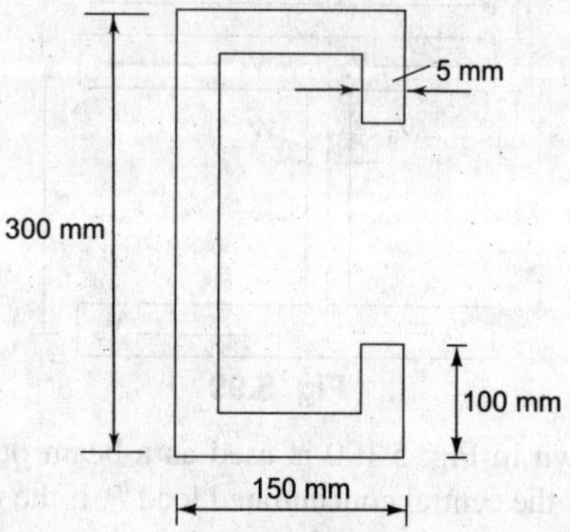

Fig. 5.96

49. For the H-section (Fig. 5.97) of different flange lengths, find the position of the shear centre.

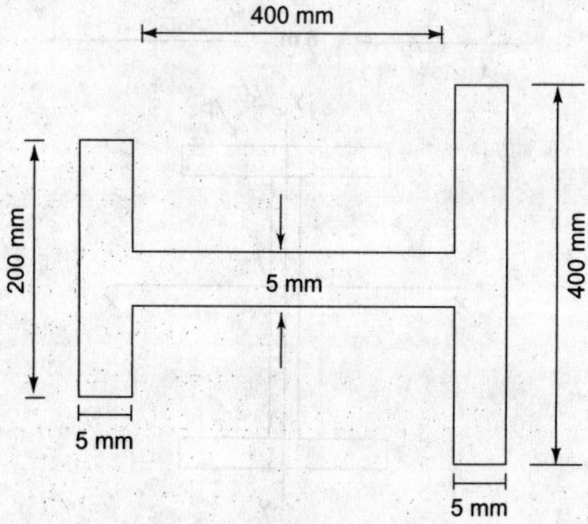

Fig. 5.97

50. Determine the shear centre for the section shown in Fig. 5.98.

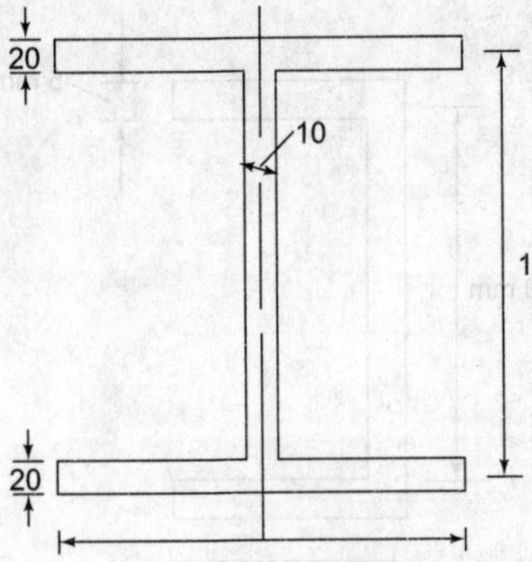

Fig. 5.98

51. For the section shown in Fig. 5.99, show that the shear centre is to the left side of the web given by

$$e = 3(b^2 - b_1^2)\,/[wh/t + 6(b + b_1)]$$

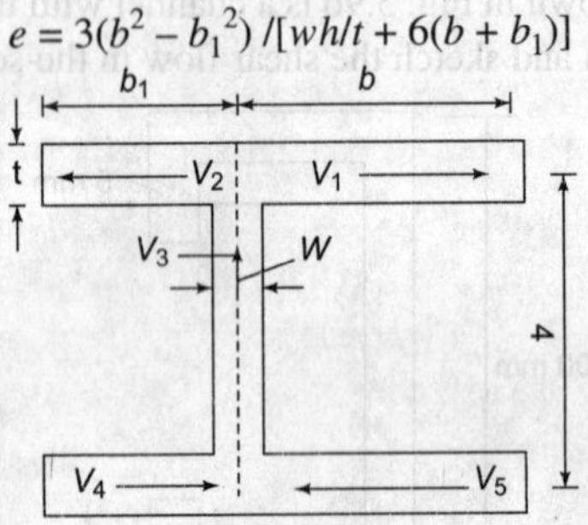

Fig. 5.99

52. The I-section shown in Fig. 5.100 is used as a beam on a span of 8 m. Find the maximum value of the central concentrated load P, if the plane of load is inclined at 40° to the Y-axis, and the permissible stress in bending is 120 MPa. Assume the load passes through the shear centre.

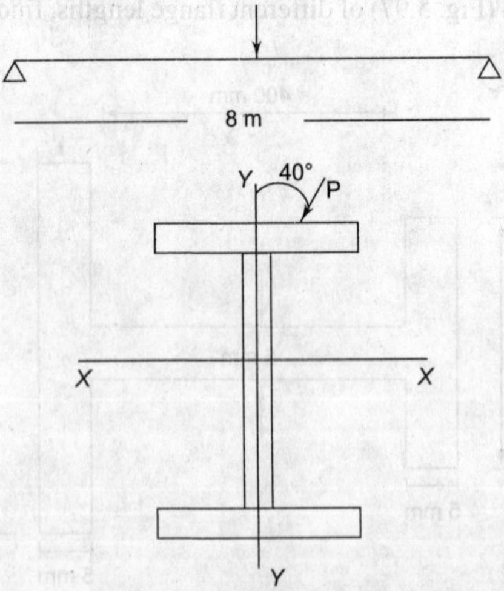

Fig. 5.100

53. The channel section shown in Fig. 5.101 is used as a cantilever of span 2 m to support a concentrated load of 20 kN at the free end. The load passes through the shear centre

but is inclined at an angle of 30° to the vertical. Find the maximum bending stress in the beam.

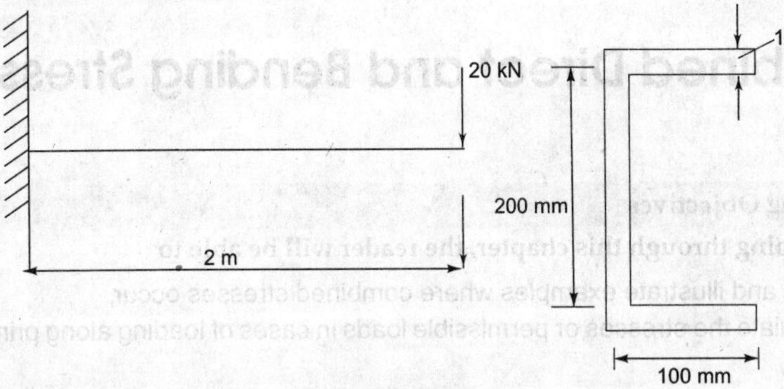

Fig. 5.101

54. The Z-section shown in Fig. 5.102 is used as a simply supported beam over a span of 4 m. Find the maximum value of the UD load w when the plane of loading is inclined at 35° to the vertical but passes through the shear centre. The permissible stress is 150 MPa.

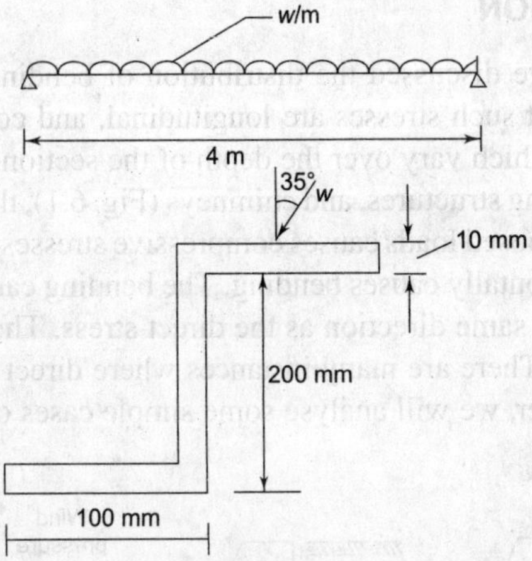

Fig. 5.102

55. The angle section shown in Fig. 5.103 is used as a simply supported beam over a span of 4 m. It carries a central concentrated load P, which is vertical and passes through the shear centre. Find the maximum value of P, if the permissible stress is 120 MPa.

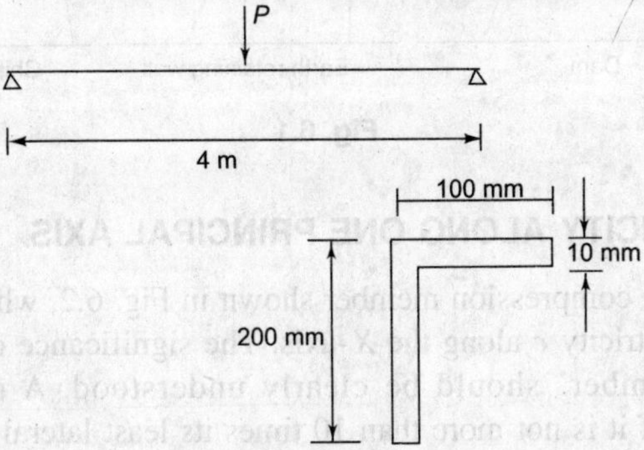

Fig. 5.103

CHAPTER 6

Combined Direct and Bending Stresses

Learning Objectives

After going through this chapter, the reader will be able to

- state and illustrate examples where combined stresses occur,
- calculate the stresses or permissible loads in cases of loading along principal axes,
- derive the middle-third rule and state its significance,
- compute the stresses and locate the neutral axes in cases of biaxial bending, and
- find the shape of the kern or core for a given section.

6.1 INTRODUCTION

In the last chapter, we discussed the distribution of bending stresses over beam sections. We saw that such stresses are longitudinal, and consist of compressive and tensile stresses which vary over the depth of the section. In many cases, such as dams, earth retaining structures, and chimneys (Fig. 6.1), the weight of the structure and any superimposed loads causes compressive stresses; water, wind, or earth pressure acting horizontally causes bending. The bending causes compressive and tensile stresses in the same direction as the direct stress. The stresses can thus be algebraically added. There are many instances where direct and bending stresses coexist. In this chapter, we will analyse some simple cases of structures in which this happens.

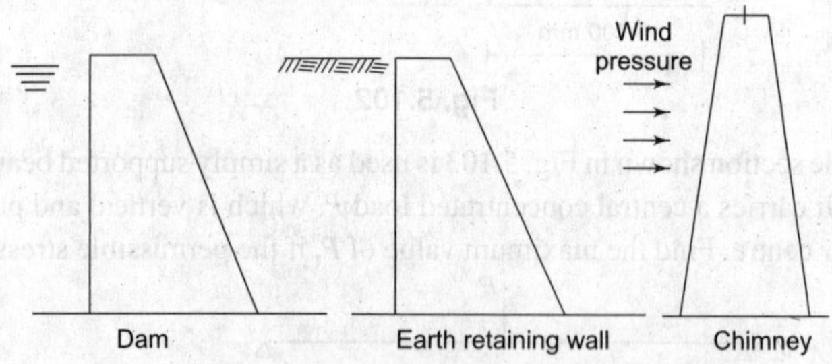

Fig. 6.1

6.2 ECCENTRICITY ALONG ONE PRINCIPAL AXIS

Consider the short compression member shown in Fig. 6.2, which carries a load acting at an eccentricity e along the X-axis. The significance of the term 'short compression member' should be clearly understood. A member may be considered short if it is not more than 10 times its least lateral dimension. Long members behave like columns, buckling being a characteristic feature in such elements. We will discuss long columns in Chapter 11.

6.2.1 Changing Eccentric Load into Axial load and Couple

In Fig. 6.2, it is possible to shift the load P to the centre of the section, as discussed in Chapter 1. We introduce two equal and opposite forces P at O as shown in Fig. 6.2(b). The given load P and the equal and opposite force at O result in a couple of magnitude Pe. The net effect of shifting the load P to the centre is to cause a couple of Pe, which acts as a bending moment about the Y-axis, as shown in Fig. 6.2(c). The member is now subjected to a compressive load P, which is centric, and a BM Pe.

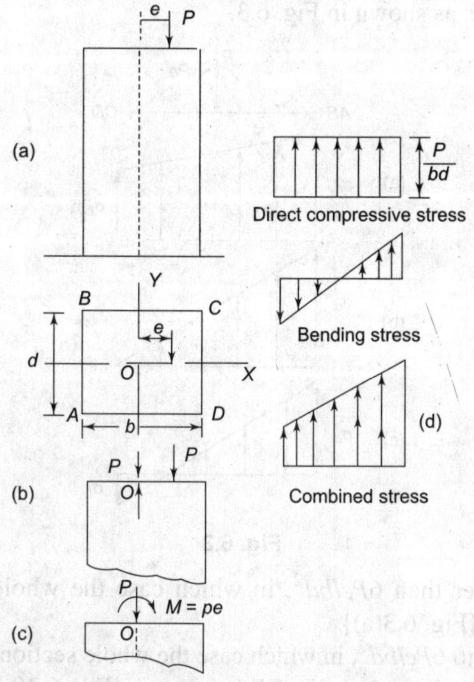

Fig. 6.2

6.2.2 Resultant Stresses in Rectangular Section

The stresses in the member due to these two effects can be easily found. Due to the central compressive load, there is a uniform compressive stress of P/bd, as shown in Fig. 6.2(d). Due to the BM about the Y-axis, the stresses $\sigma = My/I$ are distributed as shown. The maximum stresses are at the end fibres parallel to the Y-axis. As shown in the figure, for the symmetric section of a rectangle, the maximum stresses (compressive or tensile) are

$$\sigma_{\max} = \pm \frac{M}{bd^3/12}\frac{d}{2} = \pm \frac{6Pe}{bd^2}$$

The maximum compressive stress is along the edge CD and the maximum tensile stress is along the edge AB. These stresses act normal to the cross section. The net stresses along the edges are

$$\sigma \text{ along } AB = \frac{P}{bd} - \frac{6Pe}{bd^2}$$

$$\sigma \text{ along } CD = \frac{P}{bd} + \frac{6Pe}{bd^2}$$

6.2.3 Middle-third Rule: No Tension in the Section

In the case of the rectangular section we are considering, the direct compressive stress is constant and the tensile stress due to bending reduces the compressive stress at the section. Depending upon the magnitude of the BM, three stress distributions are possible, as shown in Fig. 6.3.

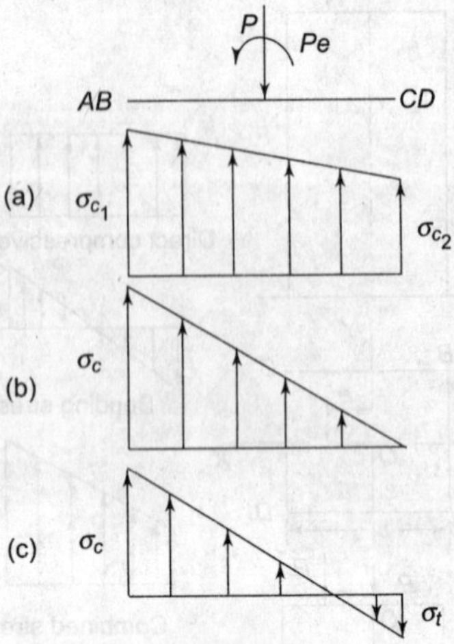

Fig. 6.3

(i) P/bd is greater than $6Pe/bd^2$, in which case the whole section is under compression [Fig. 6.3(a)].
(ii) P/bd is equal to $6Pe/bd^2$, in which case the whole section is under compression, the stress along the edge CD being zero [Fig. 6.3(b)].
(iii) P/bd is less than $6Pe/bd^2$, in which case there is tensile stress developed on a part of the section [Fig. 6.3(c)].

In many cases, as in masonary structures, or soil reaction at the base, tension is not desirable or cannot be developed. The maximum BM or the eccentricity e of the load is obtained from case (ii).

$$\frac{P}{bd} = \frac{6Pe}{bd^2}$$

$$\Rightarrow \quad e = \frac{d}{6}$$

When $e = d/6$, the stress is zero at the edge CD and compressive stress is maximum along AB. Since eccentricity can be on either side of the Y-axis, we can conclude that there will be no tensile stresses in the section if the load P lies with the middle-third of the section, as shown in Fig. 6.4(a). Note that this applies to loads lying along the X-axis, causing BM about the Y-axis, in which case $e \le d/6$ and P should lie within the middle $b/3$, the middle-third of the width for no tension in the section for a load lying along the Y-axis [Fig. 6.4(b)].

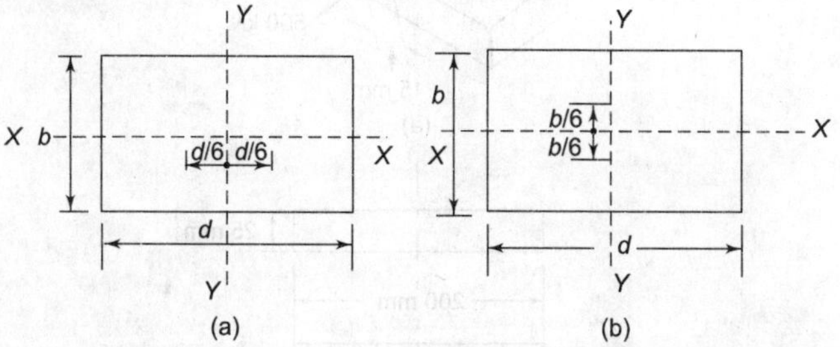

Fig. 6.4

Example 6.1 Combined stresses in a bent rod

A short rod is bent in the form of an arc such that the central deflection is about 15 mm. If the maximum permissible stress is 150 MPa, find the proportionate maximum load it can carry as a compression member compared to the same rod had it been straight. The rod is square in section of side 15 mm.

Solution From Fig. 6.5, if P is the compressive load, the bent rod is subjected to a maximum BM of $P \times 15$ N mm.

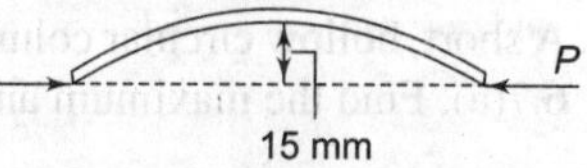

Fig. 6.5

$$\text{Direct stress} = \frac{P}{225}$$

$$\text{Bending stress} = \frac{P \times 15 \times 15}{15 \times (15^3/12) \times 2} = \frac{6P}{225}$$

$$\text{Maximum compressive stress} = \frac{P}{225} + \frac{6P}{225} = \frac{7P}{225}$$

$$\frac{P \times 7}{225} = 150$$

$$\Rightarrow \quad P = \frac{150 \times 225}{7} = 4821 \text{ N}$$

If the rod had been straight, the load carried by this column would have been

$$P = 15 \times 15 \times 150 = 33{,}750 \text{ N}$$

The bent rod carries only one-seventh of the load the straight rod can carry.

Example 6.2 Steel plate subjected to eccentric compressive load

A steel plate of dimensions 200×25 mm carries an eccentric tensile force of 500 kN, as shown in Fig. 6.6(a). Find the maximum and minimum stresses in the section.

Solution In the section shown in Fig. 6.6,

$$\text{Maximum BM} = 500 \times 15 = 7500 \text{ kN mm} = 75 \times 10^5 \text{ N mm}$$

$$\text{Bending stress (maximum)} = \pm \frac{75 \times 10^5}{25 \times 200^3} \times 12 \times \frac{200}{2}$$

$$= \pm 45 \text{ N/mm}^2$$

$$\text{Direct tensile stress} = \frac{500 \times 1000}{200 \times 25} = 100 \text{ N/mm}^2$$

Maximum tensile stress = 145 N/mm^2

Minimum tensile stress = 55 N/mm^2

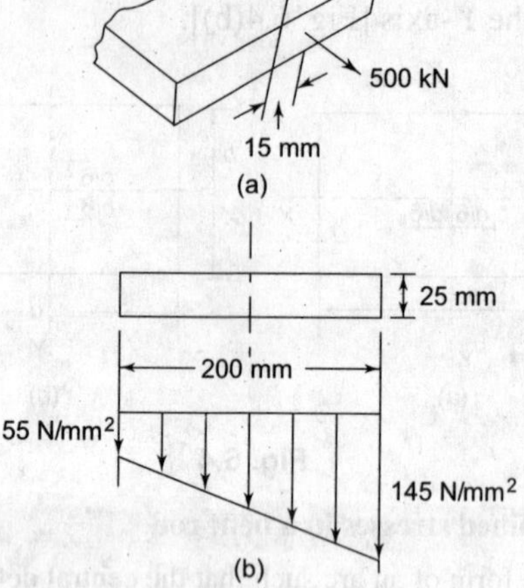

Fig. 6.6

The stress distribution is shown in Fig. 6.6(b).

Example 6.3 Eccentric load on a short, hollow circular column

A short, hollow circular column carries a load of 800 kN through a bracket as shown in Fig. 6.7(a). Find the maximum and minimum stresses due to this load.

Solution Area of the section = $\frac{\pi}{4}(400^2 - 350^2) = 2.94 \times 10^4 \text{ mm}^2$

MI of the section = $\frac{\pi}{64}(400^4 - 350^4) = 105.94 \times 10^8 \text{ mm}^4$

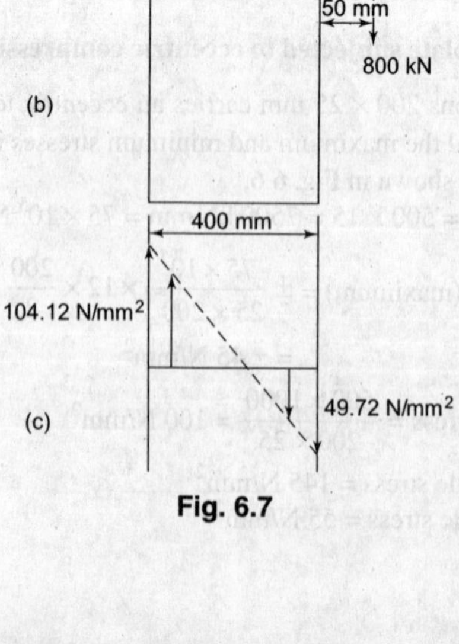

Fig. 6.7

Axial load = 800,000 N

BM = 800,000 × 250 = 200 × 10^6 N mm

$$\text{Direct compressive stress} = \frac{800{,}000}{2.94 \times 10^4} = 27.2 \text{ N/mm}^2$$

$$\text{Maximum bending stress} = \pm \frac{200 \times 10^6}{105.94 \times 10^8} \times 200 = \pm 76.92 \text{ N/mm}^2$$

Maximum compressive stress = 104.12 N/mm²

Maximum tensile stress = 49.72 N/mm²

Example 6.4 Permissible eccentricities in a rectangular column

A short column of rectangular section (Fig. 6.8) is constructed of a material with a maximum permissible compressive stress of 90 N/mm² and tensile stress of 25 N/mm². If the compressive load is 1500 kN, at what eccentricity can it be applied along the two principal axes? If the load is increased to 3000 kN, what is the permissible eccentricity along the principal axes?

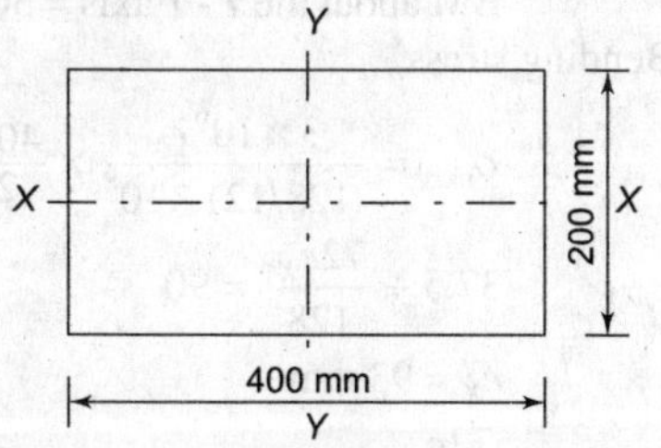

Fig. 6.8

Solution Area of the section = 400 × 200 = 8 × 10^4 mm²

$$I_{XX} = \frac{400 \times 200^3}{12} = \frac{32}{12} \times 10^8 \text{ mm}^4$$

$$I_{YY} = \frac{200 \times 400^3}{12} = \frac{128}{12} \times 10^8 \text{ mm}^4$$

Let e_X and e_Y be the maximum eccentricities along the X- and Y-axes.

$$\text{Direct stress} = \frac{1500 \times 1000}{8 \times 10^4} = 18.75 \text{ N/mm}^2$$

BM about the Y-Y axis = $1500 e_X \times 10^3$ N mm

$$\text{Maximum stresses} = \pm \frac{15 \times 10^5 e_X}{(128/12) \times 10^8} \times \frac{400}{2} = \pm \frac{36 e_X}{128}$$

$$\text{Maximum compressive stress} = 18.75 + \frac{36 e_X}{128}$$

Equating this to the permissible compressive stress,

$$18.75 + \frac{36 e_X}{128} = 90$$

⇒ $e_X = 253$ mm

In terms of permissible tensile stress,

$$\frac{36 e_X}{128} - 18.75 = 25$$

⇒ $e_X = 155$ mm

Permissible eccentricity,

$e_X = 155$ mm

Eccentricity along the Y-axis

BM about the X-X axis = $1500 \times 1000 e_Y = 15 \times 10^5 e_Y$

$$\text{Maximum bending stresses} = \pm \frac{15 \times 10^5 e_Y}{(32/12) \times 10^8} \times \frac{200}{2} = \frac{18}{32} e_Y$$

Therefore, from permissible compressive stress,

$$18.75 + \frac{18}{32} = 90;\ e_Y = 127 \text{ mm}$$

and $\frac{18}{32} e_Y - 18.75 = 25$ (from permissible tensile stress)

$\Rightarrow$ $e_Y = 78$ mm

Permissible eccentricity = 78 mm

If the axial load is raised to 3000 kN, then

$$\text{Direct compressive stress} = \frac{3000 \times 1000}{8 \times 10^4} = 37.5 \text{ N/mm}^2$$

$$\text{BM about the } Y\text{-}Y \text{ axis} = 3000 \times 1000 e_X = 3 \times 10^6 e_X$$

Bending stress,

$$\sigma_{max} = \frac{3 \times 10^6 e_X}{(128/12) \times 10^8} \times \frac{400}{2} = \frac{72 e_X}{128}$$

$$37.5 + \frac{72 e_X}{128} = 90$$

$\Rightarrow$ $e_X = 93$ mm

and $\frac{72 e_X}{128} - 37.5 = 25$

$\Rightarrow$ $e_X = 111$ mm

Permissible eccentricity,

$e_X = 93$ mm

For eccentricity e_Y,

$$\text{BM about the } X\text{-}X \text{ axis} = 3 \times 10^6 e_Y$$

$$\text{Bending stresses } \sigma_{max} = \frac{3 \times 10^6 e_Y}{(32/12) \times 10^8} \times \frac{200}{2} = \frac{36}{32} e_Y$$

Therefore,

$$37.5 + \frac{36}{32} e_Y = 90$$

$\Rightarrow$ $e_Y = 47$ mm

and $\frac{36}{32} e_Y - 37.5 = 25$

$\Rightarrow$ $e_Y = 55.5$ mm

Permissible eccentricity,

$e_Y = 47$ mm

Example 6.5 Rectangular box section

A short column is made of a rectangular box section of outside dimensions 200 mm × 250 mm and inside dimensions 180 mm × 230 mm. Determine the permissible eccentricity along either of the principal axes of a load of 500 kN, if the permissible stresses in tension and compression are 5 N/mm^2 and 50 N/mm^2.

Solution The situation is shown in Fig. 6.9. As the section is symmetrical, the principal centroidal axes are X-X and Y-Y as shown. We find MI about both axes.

$$I_{XX} = \frac{250 \times 200^3}{12} - \frac{230 \times 180^3}{12} = 54.89 \times 10^6 \text{ mm}^4$$

$$I_{YY} = \frac{200 \times 250^3}{12} - \frac{180 \times 230^3}{12} = 79.92 \times 10^6 \text{ mm}^4$$

Area of section = $250 \times 200 - 230 \times 180 = 15{,}500$ mm^2

Direct compressive stress = $500{,}000/15{,}500 = 32.6$ N/mm^2

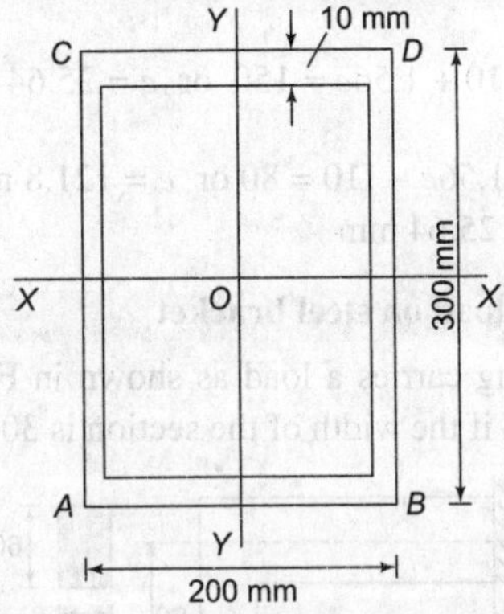

Fig. 6.9

As I_{xx} is less, we consider eccentricity along the Y-axis. If e_y be the eccentricity, then BM = $500{,}000 \times e_y$

Bending stress, $\sigma = 500{,}000 e_y \times 100/54.89 \times 10^6$, at the corners A, B, C, and D.

Assuming that e_y is positive if the load lies above the X-axis, then the stress is compressive along the edge CD and tensile along the edge AB. The maximum compressive stress is at C or D and the maximum tensile stress is at A or B. Therefore,

$$32.6 + 50e_y/54.89 < 50;\ e_y \leq 19 \text{ mm}$$

At A or B, if there is tension, $50e_y/54.89 - 32.26 < 5$; $e_y < 41$ mm

Maximum allowable eccentricity = 19 mm, along the Y-axis.

Similarly, the maximum allowable eccentricity = 22.25 mm along the X-axis.

Example 6.6 Hollow circular section

A hollow circular short column of outside diameter 300 mm and thickness 10 mm carries a load of 1000 kN. Determine at what eccentricity along a diameter can the load be placed if the permissible stresses are 150 N/mm^2 in compression and 80 N/mm^2 in tension.

Solution The column is shown in Fig. 6.10.

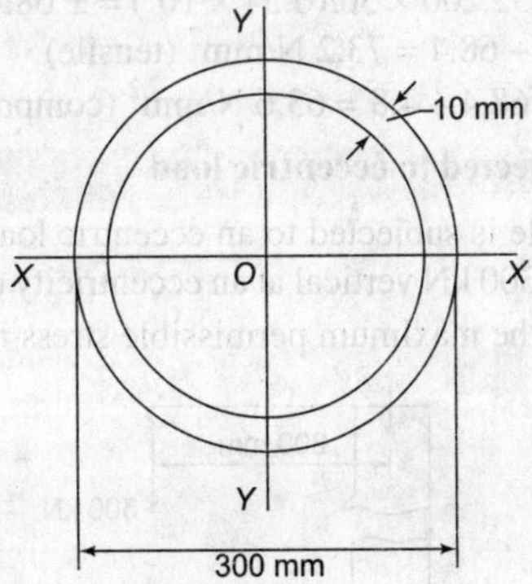

Fig. 6.10

Area of section = $\pi(300^2 - 280^2)/4 = 9111$ mm^2

Moment of Inertia = $\pi(300^4 - 280^4)/64 = 0.96 \times 10^8$ mm^4

Direct compressive stress = $1000 \times 1000/9111 = 110$ N/mm^2

If e is the eccentricity along any diameter, the bending stress

$$\sigma_b = 1{,}000{,}000 \times e \times 150/(0.96 \times 10^8) = 1.56e$$

On the compression side,

$$110 + 1.56e = 150 \text{ or } e = 25.64 \text{ mm}$$

On the tension side,

$$1.56e - 110 = 80 \text{ or } e = 121.8 \text{ mm}$$

Permissible eccentricity = 25.64 mm

Example 6.7 Eccentric load on steel bracket

A steel bracket 0.5 metre long carries a load as shown in Fig. 6.11. Find the maximum stresses at the support section if the width of the section is 30 mm and the depth is 60 mm.

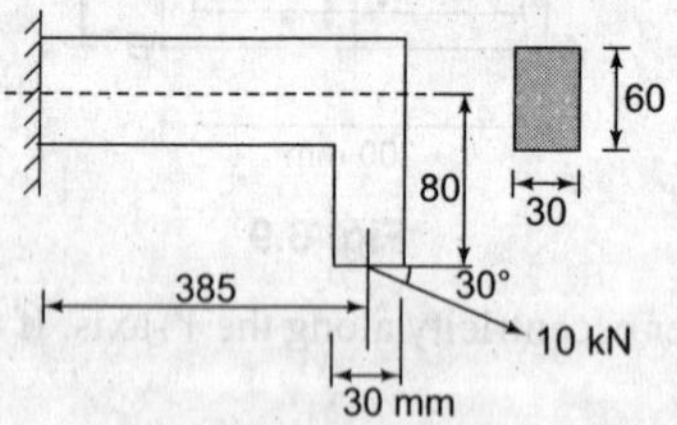

Fig. 6.11

Solution The force can be resolved into horizontal and vertical components.

Horizontal component = 10 cos 30° = 8.66 kN

Vertical component = 10 sin 30° = 5 kN

When we move the horizontal component to the centroid of the section, the accompanying moment = 10 × 1000 × cos 30° × 80 = 692,800 Nmm (anticlockwise)

The horizontal component acts as a tensile force on the section.

Direct tensile stress = 8660/(30 × 60) = 4.8 N/mm^2

The vertical component gives a clockwise moment of 5000 × 385

= 1,925,000 Nmm

Net moment at the section = 1,925,000 – 692,800 = 1,232,200 Nmm (clockwise)

This causes tension at the upper end and compression at the lower end.

MI of section = $30 \times 60^3/12 = 0.54 \times 10^6$ mm^4

Bending stress, $\sigma_b = 1{,}232{,}200 \times 30/(0.54 \times 10^6) = \pm 68.4$ N/mm^2

Fibre stress at top = 4.8 + 68.4 = 73.2 N/mm^2 (tensile)

Fibre stress at bottom = 68.4 – 4.8 = 63.6 N/mm^2 (compression)

Example 6.8 Pole subjected to eccentric load

The circular section of a pole is subjected to an eccentric load as shown in Fig. 6.12. The maximum estimated load is 300 kN vertical at an eccentricity of 200 mm. Find the diameter of the section of the pole if the maximum permissible stress is 60 N/mm^2.

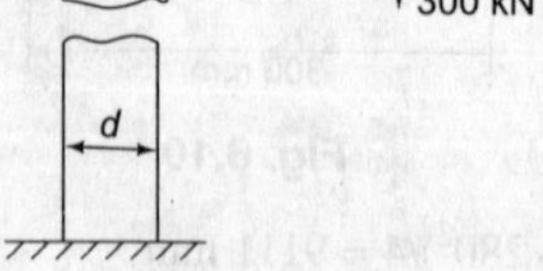

Fig. 6.12

Solution We have

Load = 300 kN = 300,000 N

When we move this load to the centre of the section, the accompanying moment

$= 300{,}000 \times 200 = 60 \times 10^6$ Nmm

Let d be the diameter of the pole. The diameter will be chosen so that the maximum stress does not exceed the permissible stress of 60 N/mm^2.

Due to the direct force and BM, the stress at extreme fibres of the section is

$$\sigma = \frac{300{,}000}{(\pi d^2/4)} + \frac{60 \times 10^6 \times 32}{(\pi d^3)} = 60$$

On simplification, this can be reduced to a cubic equation,

$$60d^3 - 0.38 \times 10^6 d = 611.15 \times 10^6$$

This is solved by trial and error to get d = 227 mm.

Minimum diameter of the pole = 230 mm

6.3 BIAXIAL BENDING: LOAD ECCENTRIC TO BOTH AXES

When the load on a short column lies on the X- (or Y-) axis, it leads to bending about one of the axes. But when the load lies as shown in Fig. 6.13(a), it causes bending about both the axes. The coordinates of the point of the load are (e_x, e_y). As there is eccentricity about both the axes, the load causes bending moments about them. This is known as *biaxial bending*.

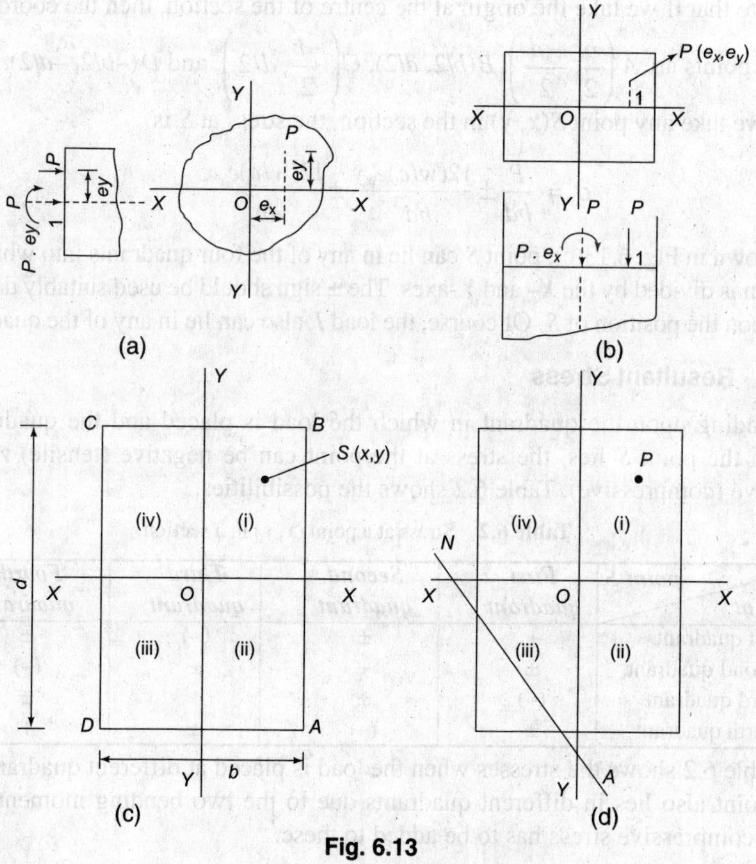

Fig. 6.13

In order to bring the load to the centre of the section, the load is first moved parallel to the Y-axis to point 1. This movement of the load causes a moment about the X-axis equal to Pe_y. We now move the load to the centre O. This introduces a moment Pe_x about the Y-axis. Thus, movement of the load to the centre of the section introduces moments about both the X- and Y-axes.

6.3.1 Rectangular Section

For the sake of simplicity, we consider a rectangular section for the column. The section has a width b (along the X-axis) and depth d (along the Y-axis). Let us see what stresses are caused by the load and the moments about different points. We consider the four corners first. Table 6.1 shows the stresses caused at these corner points.

Table 6.1 Stress at extreme points

(The compressive stress is taken positive.)

S.No.	*Stress caused by*	*At A*	*At B*	*At C*	*At D*
1	P	P/bd	P/bd	P/bd	P/bd
2	Pe_x	$+6Pe_y/bd^2$	$+6Pe_y/bd^2$	$-6Pe_y/bd^2$	$-6Pe_y/bd^2$
3	Pe_y	$-6Pe_x/b^2d$	$+6Pe_x/b^2d$	$+6Pe_x/b^2d$	$-6Pe_x/b^2d$

At any of these points, we can write the final stresses as

$$\sigma = P/bd \pm 6Pe_y/bd^2 \pm 6Pe_x/db^2$$

Note that if we take the origin at the centre of the section, then the coordinates of the points are $A\left(\frac{b}{2}, \frac{-d}{2}\right)$, $B(b/2, d/2)$, $C\left(\frac{-b}{2}, d/2\right)$, and $D(-b/2, -d/2)$.

If we take any point $S(x, y)$ in the section, the stress at S is

$$\sigma_s = \frac{P}{bd} \pm \frac{12(w/c)e_y y}{bd^3} \pm \frac{12(w/c)e_x x}{db^3}$$

As shown in Fig. 6.13(c), point S can lie in any of the four quadrants into which the section is divided by the X- and Y-axes. The $\pm$ sign should be used suitably depending upon the position of S. Of course, the load P also can lie in any of the quadrants.

6.3.2 Resultant Stress

Depending upon the quadrant in which the load is placed and the quadrant in which the point S lies, the stress at the point can be negative (tensile) zero or positive (compressive). Table 6.2 shows the possibilities.

Table 6.2 Stress at a point (x, y) in a section

Load at \ *point S*	*First quadrant*	*Second quadrant*	*Third quadrant*	*Fourth quadrant*
First quadrant	+	±	(–)	±
Second quadrant	±	+	±	(–)
Third quadrant	(–)	±	+	±
Fourth quadrant	±	(–)	±	+

Table 6.2 shows the stresses when the load is placed at different quadrants and the point also lies in different quadrants due to the two bending moments. The axial compressive stress has to be added to these.

You will note that in considering bending about any axis, the part of the section having the load has compressive stress.

6.3.3 Location of Neutral Axis

As we have seen, due to biaxial bending, the total stress at different points can be calculated using the equations derived. Neutral axis is the locus of the points having zero stress. From Table 6.2, it is clear that the points located at diagonally opposite quadrant have negative stress due to bending moments. Thus, when the load is in the first quadrant, the points in the third quadrant are likely to have tensile stress [Fig. 6.13(d)].

To locate the neutral axis, we equate the stress equation to zero. Thus,

$$\sigma = \frac{P}{bd} + \frac{\rho e_x x}{I_{yy}} + \frac{\rho e_y y}{I_{xx}} = 0$$

where I_{yy} and I_{xx} are the moments of inertia of the section about the Y-Y and X-X axes.

$$I_{yy} = A r_y^2 \text{ and } I_{xx} = A r_x^2, \text{ where } A = \text{area of section} = bd$$

So,

$$\frac{P}{bd} + \frac{\rho e_x x}{bd r_y^2} + \frac{\rho e_y y}{bd r_x^2} = 0$$

or

$$1 + \frac{e_x x}{r_y^2} + \frac{e_y y}{r_x^2} = 0$$

This is the equation of a straight line. The neutral axis can thus be located.

Example 6.9 Compression and biaxial bending

A short column of rectangular section 200 mm × 300 mm carries a compressive load of 800 kN. The load is applied at a point (–50, 100) considering the centroid of the section as the origin. Find the stresses at the four corners of the section.

Solution The column is shown in Fig. 6.14.

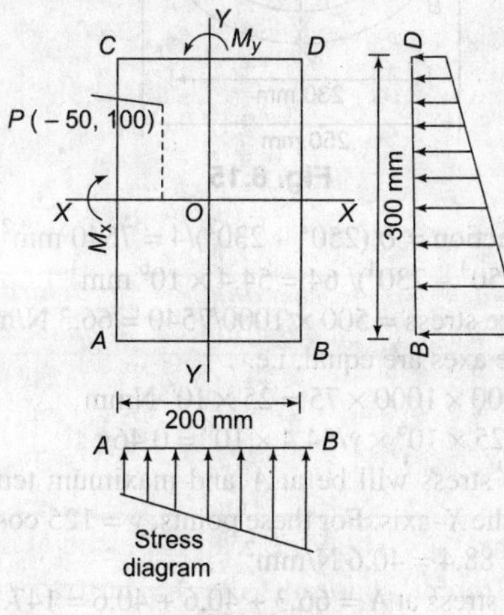

Fig. 6.14

Area of the section = $300 \times 200 = 6 \times 10^4$ mm^2

If the X- and Y-axes are as shown through the centroid,

$$I_{xx} = 200 \times 300^3/12 = 450 \times 10^6 \text{ mm}^4$$

$$I_{yy} = 300 \times 200^3/12 = 200 \times 10^6 \text{ mm}^4$$

The coordinates of the load being (–50, 100), the load is placed as shown in the figure. There is bending moment about both axes.

Moment @ the X-axis, $M_x = 800 \times 1000 \times 100 = 80 \times 10^6$ Nmm

Moment @ the Y-axis, $M_y = 800 \times 1000 \times 50 = 40 \times 10^6$ Nmm

The directions of these moments are as shown in the figure. It is clear that M_x causes tension at A and B and compression at C and D. Similarly, M_y causes tension at B and D and compression at A and C.

Direct compressive stress = $800 \times 1000/(6 \times 10^4) = 13.33$ N/mm^2

Bending stress due to $M_x = 80 \times 10^6 \times 150/(450 \times 10^6) = 26.67$ N/mm^2

Bending stress due to $M_y = 40 \times 10^6 \times 100/(200 \times 10^6) = 20$ N/mm^2

Net stress at A, $\sigma_A = 13.33 - 26.67 + 20 = +6.66$ N/mm^2

Net stress at B, $\sigma_B = 13.33 - 26.67 - 20 = +33.34$ N/mm^2

Net stress at C, $\sigma_C = 13.33 + 26.67 + 20 = +60$ N/mm^2

Net stress at D, $\sigma_D = 13.33 + 26.67 - 20 = +20$ N/mm^2

Example 6.10 Compression and biaxial bending

A compression member of hollow circular section, of 250 mm outer diameter and 10 mm thickness, is subjected to a compressive load at (75, 75) with reference to the centre of the circular section. Find the maximum stress in the section if the load is 500 kN.

Solution The column is shown in Fig. 6.15.

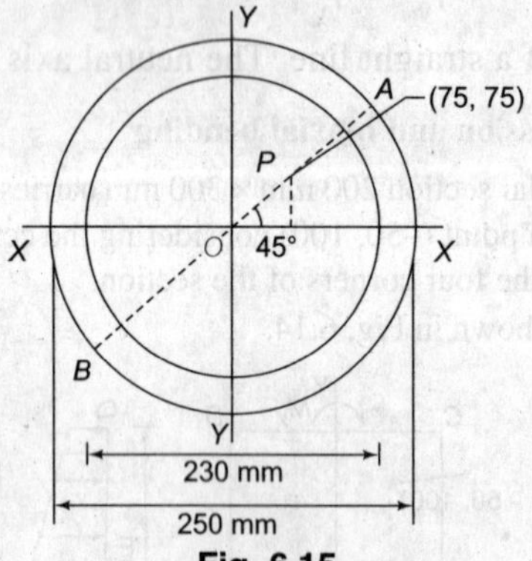

Fig. 6.15

Area of column section = $\sigma(250^2 - 230^2)/4 = 7540$ mm^2

$I_{xx} = I_{yy} = I = \pi(250^4 - 230^4)/64 = 54.4 \times 10^6$ mm^4

Direct compressive stress = $500 \times 1000/7540 = 66.3$ N/mm^2

Moments about both the axes are equal, i.e.

$M_x = M_y = M = 500 \times 1000 \times 75 = 25 \times 10^6$ Nmm

Bending stress = $25 \times 10^6 \times y/54.4 \times 10^6 = 0.46y$

Maximum compressive stress will be at A and maximum tensile stress will be at B. These points are at 45° to the X-axis. For these points, $y = 125 \cos 45° = 88.4$ mm

Bending stress = $0.46 \times 88.4 = 40.6$ N/mm^2

Maximum compressive stress at $A = 66.3 + 40.6 + 40.6 = 147.5$ N/mm^2

Maximum tensile stress at $B = 66.3 - 40.6 - 40.6 = (-)14.9$ N/mm^2

6.4 RULES FOR NO TENSION IN SECTIONS: CORE/KERNEL OF SECTIONS

We have seen that due to biaxial bending, tensile stress may develop in some parts of the section. If tensile stress has to be avoided, then the eccentricity of the load has to be limited. For different sections, the position of the load can be determined for this condition.

6.4.1 Rectangular Section: Middle-third Rule

Consider the rectangular section shown in Fig. 6.16.

The equation for the neutral axis is $1+\dfrac{e_x x}{r^2_y}+\dfrac{e_y y}{r^2_x}=0$

$$r_Y^2=\frac{bd^3}{12bd}=\frac{d^2}{12}$$

$$r_X^2=\frac{db^3}{12bd}=\frac{b^2}{12}$$

Therefore,

$$1+\frac{e_X d/2}{d^2/12}+\frac{e_Y b/2}{b^2/12}=0,\quad 1+\frac{e_X}{(d/6)}+\frac{e_Y}{(b/6)}=0$$

The situation when the NA passes through the corner A opposite the segment in which P lies is important because this is the limiting case for no tension in the section. $x=-d/2$, $y=-b/2$ for point A, and this yields

$$1+\frac{e_X(-d/2)}{d^2/12}+\frac{e_Y(-b/2)}{b^2/12}=0$$

$$\frac{e_X}{(d/6)}+\frac{e_Y}{(b/6)}=1$$

This is the equation for a straight line mn which intersects the X- and Y-axes at $d/6$ and $b/6$, respectively. If the load is positioned in the quadrants containing points B, A, and D, respectively, we get similar equations and limits. These lines intersect the principal axes at $d/6$ and $b/6$ limits and we get the diamond-shaped figure shown hatched in Fig. 6.16. If the load lies within this area, there will be no tension in the section.

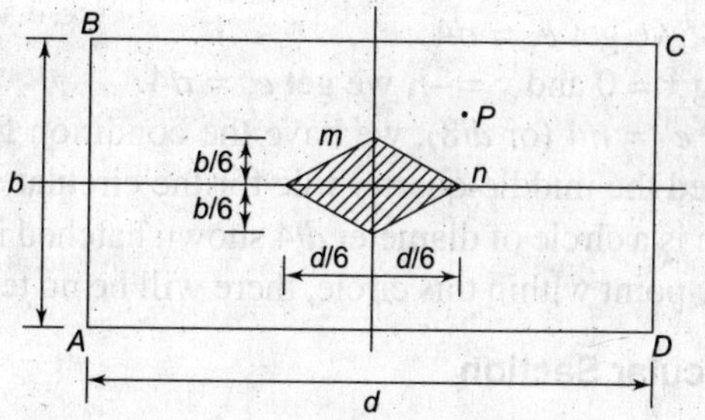

Fig. 6.16 Core of a rectangular section

The shaded area in Fig. 6.16 is known as the *kern* or *core* of the section. The kern is the area enclosed by the locus of points on limits representing the above equation.

6.4.2 Circular Section: One-fourth Diameter Rule

In the case of a circular section, $I = \pi r^4/4$, $A = \pi r^2$ and radius of gyration $r_x^2 = r_y^2$ is equal to $(\pi r^4/4)/\pi r^2 = r^2/4$.

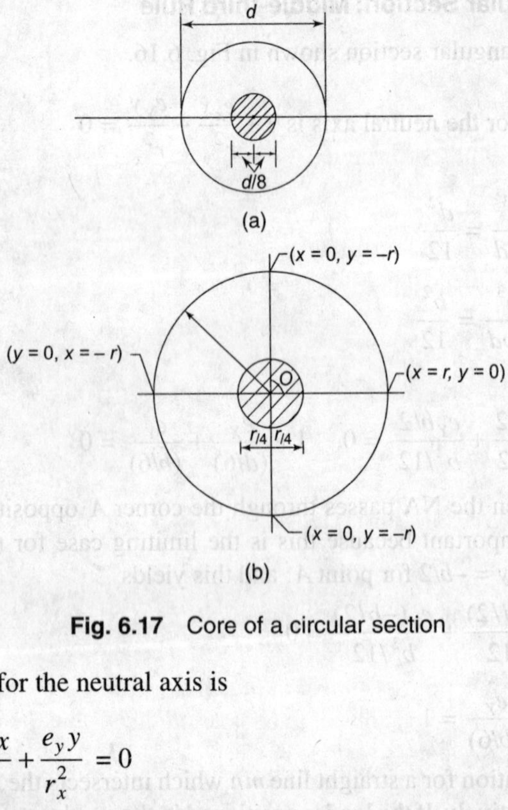

Fig. 6.17 Core of a circular section

The equation for the neutral axis is

$$1+\frac{e_x x}{r_y^2}+\frac{e_y y}{r_x^2}=0$$

When the load lies along the X-axis at eccentricity e_x, for no tension in the section, the neutral axis is tangential to the circle as a limiting case. This is shown in Fig. 6.17(b). In this position, $y = 0$ and $x = -r$. Substituting,

$$1 + e_x(-r)/r_y^2 = 0$$

Putting $r_y^2 = r^2/4$, we get $e_x = r/4$.

Similarly, putting $x = 0$ and $y = -r$, we get $e_y = r/4$.

Thus, when $e_x = e_y = r/4$ (or $d/8$), we have the condition for no tension in the section. This is called the middle-quarter rule for the circular section. The core or kernel of the section is a circle of diameter $d/4$ shown hatched in the section. When the load falls on any point within this circle, there will be no tension in the section.

6.4.3 Hollow Circular Section

In the case of a hollow circular section, of external radius r_e and internal radius r_i, a similar approach can be used to find the core of the section.

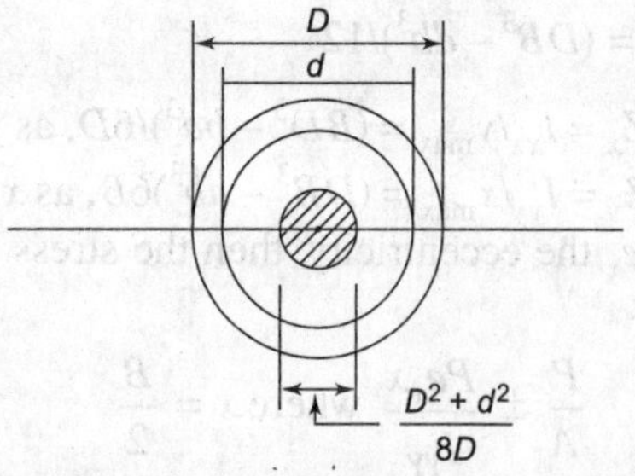

Fig. 6.18 Core of a hollow circular section

Area of the section = $\pi(r_e^2 - r_i^2)$; MI = $\pi(r_e^4 - r_i^4)/4$

Radius of gyration, $r^2 = \pi(r_e^4 - r_i^4)/[4\pi(r_e^2 - r_i^2)] = (r_e^2 + r_i^2)/4$

The equation from the neutral axis is $1 + (e_x x/r_y^2) + (e_y y/r_x^2) = 0$

When the load lies on the X-axis, $y = 0$ and the neutral axis is tangential to the section as shown in Fig. 6.17(b), $x = -r_e$ in the limiting case. Similarly, when the load lies on the Y-axis, $x = 0$ and the neutral axis is tangential to the circle in the limiting case, $y = -r_e$. In either case,

$$1 - \frac{4e_x r_e}{r_e^2 + r_i^2} = 0; \quad e_x = (r_e^2 + r_i^2)/4r_e$$

Similarly, putting $x = 0$ and $y = -r_e$, we get

$$e_y = (r_e^2 + r_i^2)/4r_e$$

These are the conditions for no tension in the section. The core or kernel of the section is a circle of radius $(r_e^2 + r_i^2)/4r_e$. In terms of diameters, core or kernel has a diameter given by $(d_e^2 + d_i^2)/4d_e$. The core is shown hatched in the section.

6.4.4 Box Section

In the case of a box section, as shown in Fig. 6.19, we use a different approach for illustration. The outer dimensions of the box are B and D and the inner dimensions are b and d.

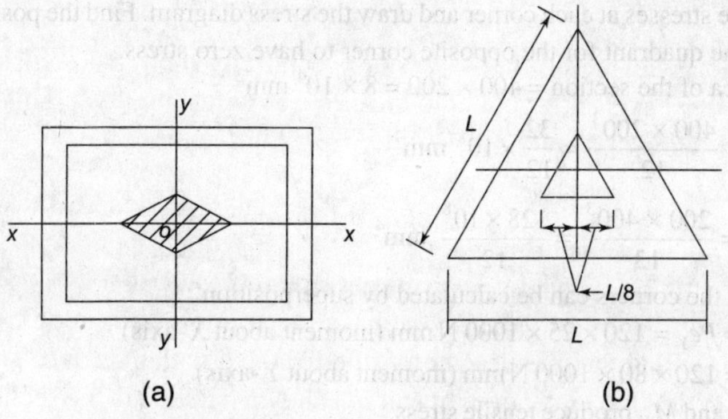

Fig. 6.19 Core of a box section

Area of box section = $(BD - bd)$

$$I_{xx} = (BD^3 - bd^3)/12$$

$$I_{yy} = (DB^3 - db^3)/12$$

Modulus of section $Z_x = I_{xx}/y_{max} = (BD^3 - bd^3)/6D$, as $y_{max} = D/2$

Modulus of section $Z_y = I_{yy}/x_{max} = (DB^3 - db^3)6B$, as $x_{max} = B/2$

If P is the load and e_x the eccentricity, then the stress in the section along the right and left edges is

$$\frac{P}{A} \pm \frac{Pe_x x}{I_{YY}} \text{ where } x = \frac{B}{2}$$

$$\sigma = \frac{P}{A} \pm \frac{Pe_x}{(I_{YY}(B/2))} = \frac{P}{A}\left[1 \pm \frac{e_x A}{Z_Y}\right]$$

Minimum stress is obtained when the sign inside is negative. When the bracketed term is zero, there is no tension in the section.

Thus, $1 - e_x A/Z_y = 0$ for no tension in the section

This gives $e_x = Z_y/A$

Similarly, due to the moment about the X-axis, we can show that $e_y = Z_x/A$ for no tension in the section.

$$e_X = \frac{Z_Y}{A} = \frac{(DB^3 - db^3)}{12 \times B} \quad \frac{2}{(BD - bd)}$$

$$= \frac{DB^3 - db^3}{6B(BD - bd)}$$

$$e_Y = \frac{BD^3 - bd^3}{6D(BD - bd)}$$

The core of the section is a diamond-shaped figure as shown in Fig. 6.19(a).

Similarly, it can be shown that for an equilateral triangular section, the core is a triangle with sides equal to $\frac{L}{4}$.

Example 6.11 Biaxial bending: rectangular column

In the rectangular section shown in Fig. 6.20, the compressive load $P = 80$ kN is applied as shown. Find the stresses at each corner and draw the stress diagram. Find the position of the load in the same quadrant for the opposite corner to have zero stress.

Solution Area of the section = 400 × 200 = 8 × 10⁴ mm²

$$I_{XX} = \frac{400 \times 200^3}{12} = \frac{32}{12} \times 10^8 \text{ mm}^4$$

$$I_{YY} = \frac{200 \times 400^3}{12} = \frac{128 \times 10^8}{12} \text{ mm}^4$$

The stresses at the corners can be calculated by superposition

$$M_X = Pe_Y = 120 \times 25 \times 1000 \text{ N mm (moment about } X\text{-axis)}$$

$$M_Y = 120 \times 80 \times 1000 \text{ N mm (moment about } Y\text{-axis)}$$

At A, both M_X and M_Y produce tensile stress.

$$\sigma_A = \frac{120{,}000}{8 \times 10^4} - \frac{12 \times 25 \times 10^4 \times 100}{(32/12) \times 10^8} - \frac{96 \times 10^5 \times 200}{(128/12) \times 10^8}$$

$$= -1.425 \text{ N/mm}^2 \text{ (tensile)}$$

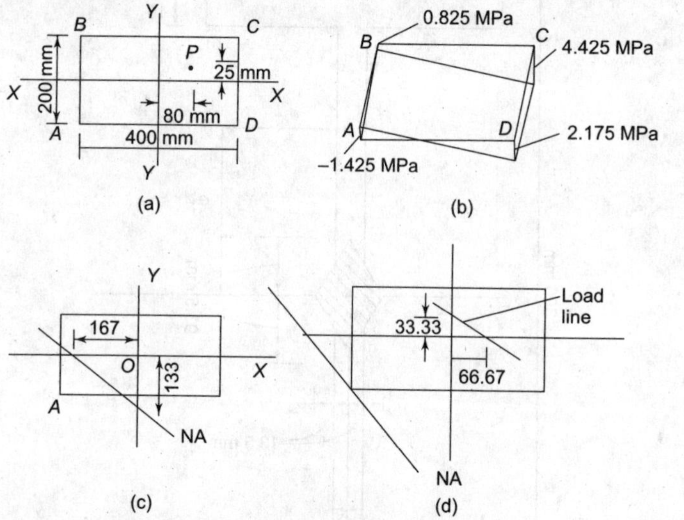

Fig. 6.20

At B, M_X produces compressive stress while M_Y produces tensile stress.

$$\sigma_B = \frac{120{,}000}{8\times 10^4} - \frac{12\times 25\times 10^4\times 100}{(32/12)\times 10^8} + \frac{96\times 10^5\times 200}{(128/12)\times 10^8}$$

$$= 0.825 \text{ N/mm}^2$$

Similarly, $\sigma_C = 4.425$ N/mm^2 (compressive) and $\sigma_D = 2.175$ N/mm^2 (compressive).

The stress diagram is shown in Fig. 6.20(b). The neutral axis is given by the equation

$$1 + \frac{e_X x}{r_Y^2} + \frac{e_Y y}{r_X^2} = 0$$

Setting $x = 0$ and $y = 0$ in this equation, we obtain points of intersection with the Y- and X-axis. These are $-r_Y^2/e_X$ and $-r_X^2/e_Y$; i.e., $-128\times 10^8/(12\times 80\times 8\times 10^4) = -166.67$ mm and $-32\times 10^8/(12\times 25\times 8\times 10^4) = -133.33$ mm. The NA is shown in Fig. 6.20(c).

For no tension in the section, with the load in the same quadrant, the NA passes through A. The maximum eccentricities are 400/6 along the X-axis and 200/6 along the Y-axis and on a line joining these. This is shown in Fig. 6.20(d). The direction of the NA changes as the load moves along this line, but it always passes through A.

Example 6.12 Kern of an I-section

Find the kern of the I-section shown in Fig. 6.21.

Solution Area = $2\times 150\times 10 + 280\times 10 = 5800$ mm^4

$$I_{XX} = 2\left(\frac{150\times 10^3}{12} + 1500\times 145^2\right) + \frac{10\times 280^3}{12} = 8139.3\times 10^4 \text{ mm}^4$$

$$I_{YY} = 2\left(\frac{10\times 150^3}{12}\right) + \frac{280\times 10^3}{12} = 564.8\times 10^4 \text{ mm}^4$$

The stress at point A on the section is given by, due to a load P at eccentricities e_X and e_Y as shown,

$$\sigma_A = \frac{P}{5800} - \frac{Pe_X\times 75}{564.8\times 10^4} - \frac{Pe_Y\times 150}{8139.3\times 10^4}$$

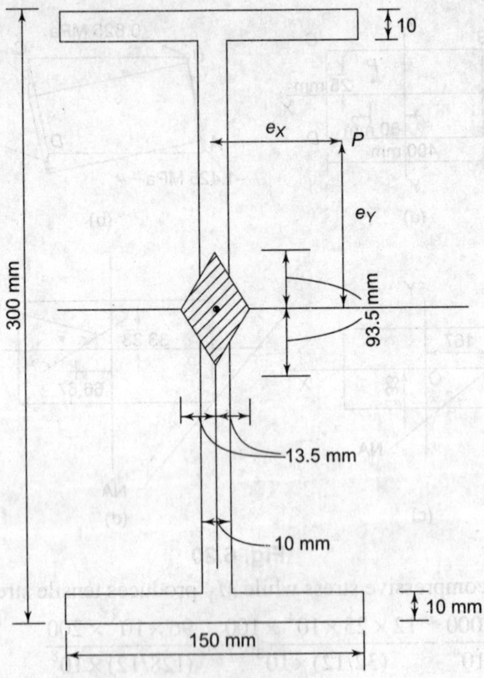

Fig. 6.21

The kern of the section is obtained when $\sigma_A = 0$.

$$0 = \frac{P}{5800} - \frac{75Pe_X}{564.8 \times 10^4} - \frac{150 \times Pe_Y}{8139.3 \times 10^4}$$

$$0 = 1 - \frac{75 \times 5800e_X}{564.8 \times 10^4} - \frac{150 \times 5800e_Y}{8139.3 \times 10^4}$$

The maximum eccentricities along the principal axes are obtained by setting $e_Y = 0$ and $e_X = 0$ separately. Thus,

$$e_X = \frac{564.8 \times 10^4}{75 \times 5800} = 13 \text{ mm}$$

and $$e_Y = \frac{819.3 \times 10^4}{150 \times 5800} = 93.5 \text{ mm}$$

Similar points can be obtained in all quadrants. The kern of the section is shown in Fig. 6.21.

Example 6.13 NA in biaxial bending

A rectangular section 200 mm × 300 mm carries a load of 250 kN at a point having coordinates (50, 100) with the centroid as the origin. Find the stresses at the four corners. Locate the neutral axis of the section.

Solution The section is shown in Fig. 6.22.

Area of section = 200 × 300 = 6 × 10^4 mm^2

Moment of inertia, I_{xx} = 200 × 300^3/12 = 450 × 10^6 mm^4

Moment of inertia, I_{yy} = 300 × 200^3/12 = 200 × 10^6 mm^4

Load = 250 kN; Direct compressive stress = 250 × 1000/6 × 10^4 = 4.17 N/mm^2

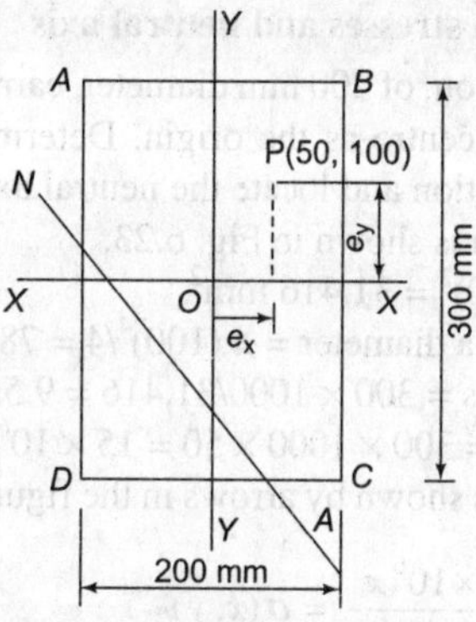

Fig. 6.22

Moment about the X-axis, $M_x = 250 \times 1000 \times 100 = 25 \times 10^6$ Nmm (It produces tensile stress at C and D and compressive stress at A and B.)

Moment about the Y-axis, $M_y = 250 \times 1000 \times 50 = 12.5 \times 10^6$

Nmm (It produces tensile stress at A and D and compressive stress at B and C.).

Table 6.3 shows the calculation of stress.

Table 6.3 Calculation of stress

Point	*P/A*	*Due to M_x*	*Due to M_y*	*Total stress*
A	+4.17	+8.33	–6.25	+6.25
B	+4.17	+8.33	+6.25	+18.75
C	+4.17	–8.33	+6.25	+2.09
D	+4.17	–8.33	–6.25	–10.41

For example, at A, the stress due to $M_x = 25 \times 10^6 \times 150/(450 \times 10^6) = 8.33$ N/mm^2

To locate the neutral axis, we use the stress equation

$$\sigma = P/A + M_x y/I_{xx} + M_y x/I_{yy} = 0$$

or
$$4.17 + \frac{25\times10^6\, y}{450\times10^6} + \frac{12.5\times10^6\, x}{200\times10^6} = 0$$

or
$$4.17 + 0.0555y + 0.0625x = 0$$

If we put $x = 0$ and $y = 0$ alternately, we get the intercepts on the Y- and X-axes made by the NA. Thus,

$$y = -4.17/0.0555 = -75 \text{ mm}; \; x = -4.17/0.0625 = -66.7 \text{ mm}$$

The neutral axis being a straight line, it can be sketched as shown.

Note We can also use the equation for NA developed earlier as follows.

$$1 + e_x x/(r_y)^2 + e_y y/(r_x)^2 = 0$$

$$(r_x)^2 = 450 \times 10^6/ 6 \times 10^4 = 7500 \text{ mm}$$

$$(r_y)^2 = 200 \times 10^6/(6 \times 10^4) = 3333.33 \text{ mm}$$

Therefore, $1 + \dfrac{50x}{3333.33} + \dfrac{100y}{7500} = 0$

Putting $x = 0$ and $y = 0$ alternately, we get

$$x = -66.7 \text{ mm and } y = -75 \text{ mm}$$

Example 6.14 Maximum stresses and neutral axis

A short column of circular section, of 200 mm diameter, carries a load of 300 kN at point (r/2, $-r$/2) with reference to the centre as the origin. Determine the maximum tensile and compressive stresses in the section and locate the neutral axis.

Solution The column section is shown in Fig. 6.23.

Area of the section = $\pi(100)^2 = 31{,}416$ mm^2

Second moment area about a diameter = $\pi(100)^4/4 = 78.54 \times 10^6$ mm^4

Load = 300 kN; direct stress = $300 \times 1000/31{,}416 = 9.55$ N/mm^2

$e_x = e_y = 50$ mm; $M_x = M_y = 300 \times 1000 \times 50 = 15 \times 10^6$ Nmm

The direction of moments is shown by arrows in the figure. The stress at a point is given

$$\text{by } +9.55 + \frac{15 \times 10^6\, y}{78.54 \times 10^6} + \frac{15 \times 10^6\, x}{78.54 \times 10^6} = \sigma(x, y)$$

where the point has coordinates (x, y).

or $+9.55 + 0.191y + 0.191x = 0$

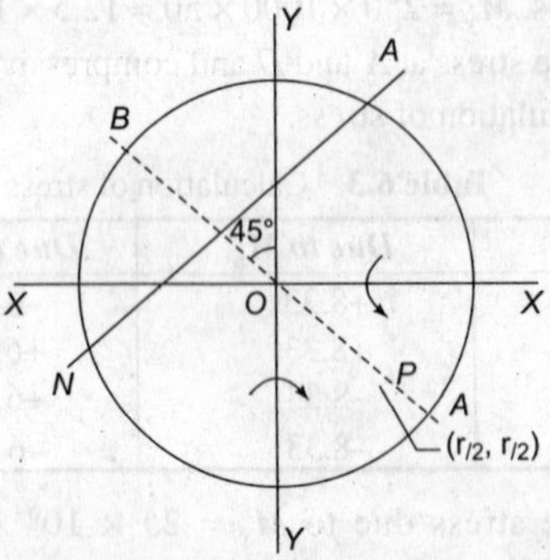

Fig. 6.23

To get the maximum value, take a point at the circumference at an angle θ.

$$x = r\cos\theta;\ y = r\sin\theta;\ r = 100 \text{ mm}$$

$$\sigma_\theta = 9.55 + 19.1 \sin\theta + 19.1 \cos\theta = 0$$

For maximum value, $d\sigma_\theta/d\theta = 0$

$$19.1 \cos\theta - 19.1 \sin\theta = 0;\ \tan\theta = 1;\ \theta = 45°$$

The points on the circumference cut by a 45° line have maximum stress. Since the load is in quadrant II, the maximum compressive stress is at A and maximum tensile stress at B.

$$x = 100 \cos 45° = 70.71 \text{ mm};\ y = 100 \sin 45° = 70.71 \text{ mm}$$

Maximum compressive stress = $9.55 + 0.191 \times 70.71 + 0.191 \times 70.71 = 36.6$ N/mm^2

Maximum tensile stress = $9.55 - 0.191 \times 70.71 - 0.191 \times 70.71 = 17.46$ N/mm^2

Example 6.15 Core of section

Find the core of the box section and sketch the core in the section.

Solution The box section is shown in Fig. 6.24.

Area of the section = $300 \times 200 - 280 \times 180 = 9600$ mm^2

$$I_{xx} = (300 \times 200^3)/12 - (280 \times 180^3)/12 = 63.9 \times 10^6 \text{ mm}^4$$

$$I_{yy} = (200 \times 300^3)/12 - (180 \times 280^3)/12 = 121 \times 10^6 \text{ mm}^4$$

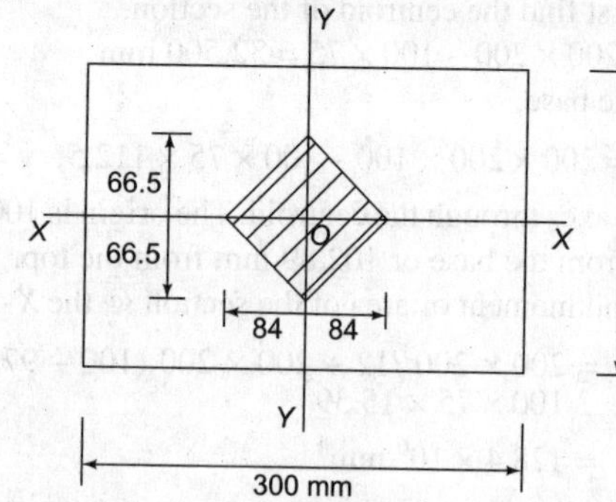

Fig. 6.24

Equation for stress at any point $S(x, y)$ due to load P placed at (e_x, e_y) is

$$\frac{P}{A} + \frac{Pe_x x}{I_{yy}} + \frac{Pe_y y}{I_{xx}} = \sigma$$

Core of the section is formed by lines joining the maximum e_x point on the X-axis and maximum e_y point on the Y-axis. We put $x = 0$ and $y = 0$ alternately in the above equation to get points on the X- and Y-axes. When $x = 0$, $y = y_{max}$ and when $y = 0$, $x = x_{max}$. Thus, when $x = 0$, $y = 100$ and when $y = 0$, $x = 150$,

$$1/A \pm e_y y/I_{xx} = 0;\ e_y = I_{yy}/(Ay) = 63.9 \times 10^6/(150 \times 9600) = \pm 66.5 \text{ mm}$$

Also $$e_x = 121 \times 10^6/(100 \times 9600) = \pm 84 \text{ mm}$$

The core of the section can be sketched as shown.

6.5 UNSYMMETRICAL SECTIONS

In the case of unsymmetrical sections, no new principles are involved. The CG of the section has to be located first. Then the moments of inertia of the section about the orthogonal axes through the centroid are determined. The formula to be used for finding the combined stress is the same as for symmetrical sections. The following examples illustrate the procedure.

Example 6.16 Square section with eccentric hole

A square section with an eccentric hole as shown in Fig. 6.25 carries a load of 300 kN. Find the stresses at the four outer corners if (i) the load is at point 1 and (ii) if the load is at point 2.

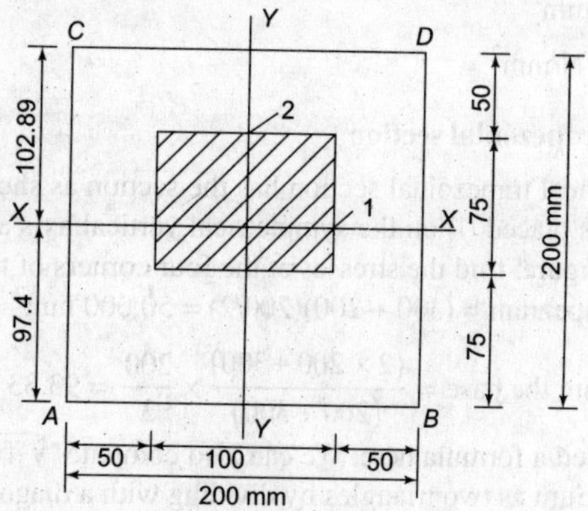

Fig. 6.25

Solution We have to first find the centroid of the section.

Area of the section = $200 \times 200 - 100 \times 75 = 32{,}500$ mm^2

Taking moments @ the base,

$$32{,}500\ \tilde{y} = 200 \times 200 \times 100 - 100 \times 75 \times 112.5;\ \tilde{y} = 97.11 \text{ mm}$$

We take the X- and Y- axes through the centroid. The origin is 100 mm to the right of the left edge and 97.11 mm from the base or 102.89 mm from the top.

We find now the second moment of area of the section @ the X- and Y- axes.

$$I_{xx} = 200 \times 200^3/12 + 200 \times 200\ (100 - 97.11)^2 - 100 \times 75^3/12 - 100 \times 75 \times 15.39^2$$

$$= 128.4 \times 10^6 \text{ mm}^4$$

(The centre of the hole is at $102.89 - 50 - 75/2 = 15.39$ mm.)

$$I_{yy} = 200 \times 200^3/12 - 75 \times 100^3/12 = 127 \times 10^6 \text{ mm}^4$$

(i) When the load is at point 1:

Direct stress = $300{,}000/32{,}500 = 9.23$ N/mm^2

For this load position, $e_x = 50$ mm, $e_y = 0$

$M_y = 300{,}000 \times 50 = 15 \times 10^6$ Nmm

The position of this load causes tension at A and C and compression at B and D. Note that $y = 100$ for all points.

$$\sigma_A = 9.23 - 15 \times 10^6 \times 100/127 \times 10^6 = -2.58 \text{ N/mm}^2 \text{ (tensile)}$$

$$\sigma_B = 9.23 + 15 \times 10^6 \times 100/127 \times 10^6 = 9.23 + 11.81 = 21 \text{ N/mm}^2$$

$$\sigma_C = 9.23 - 11.81 = -2.58 \text{ N/mm}^2$$

$$\sigma_D = \sigma_B = 21 \text{ N/mm}^2$$

(ii) When the load is at point 2:

Direct stress remains the same, i..e. 9.23 N/mm^2. The moment is now about the X-axis.

$e_x = 0$; $e_y = 102.89 - 50 = 52.89$ mm

$M_x = 300{,}000 \times 52.89 = 15.87 \times 10^6$ Nmm

In this position of the load, there is tension at A and B and compression at C and D.

$$\sigma_A = 9.23 - 15.87 \times 10^6 \times 97.11/128.4 \times 10^6 = 9.23 - 12 = -2.77 \text{ N/mm}^2$$

$$\sigma_B = 9.23 - 12 = -2.77 \text{ N/mm}^2 \text{ (for both } A \text{ and } B,\ y = 97.11 \text{ mm)}$$

$$\sigma_C = 9.23 + 15.87 \times 10^6 \times 102.89/128.4 \times 10^6 = 9.23 + 12.71 = 22 \text{ N/mm}^2$$

$$\sigma_D = 22 \text{ N/mm}^2$$

Example 6.17 Trapezoidal section

A column of symmetrical trapezoidal section has the section as shown in Fig. 6.26. If the load is 3000 kN and is placed (i) on the symmetrical vertical axis at mid-depth and (ii) at point 1 shown in the figure, find the stresses at the four corners of the section.

Solution Area of trapezium = $(300 + 200)(200/2) = 50{,}000$ mm^2

$$\text{Distance of CG from the base} = \frac{(2 \times 200 + 300)}{(200 + 300)} \times \frac{200}{3} = 93.33 \text{ mm}$$

Note We have used a formula here. We can also compute $\tilde{y}$ from the first principles considering the trapezium as two triangles by dividing with a diagonal as shown below.

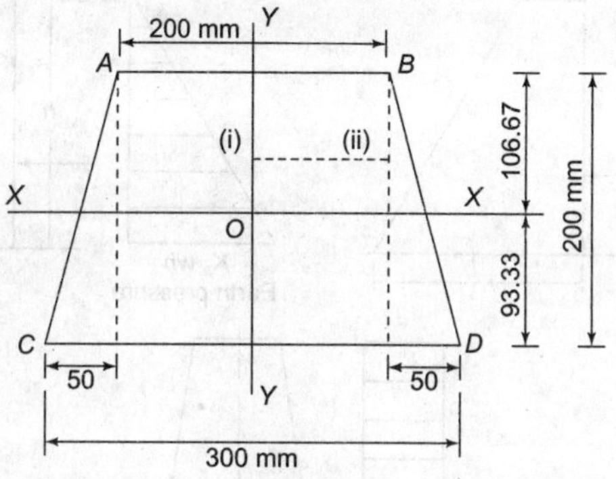

Fig. 6.26

$$50{,}000\,\tilde{y} = 300 \times (200/2)(200/3) + 200 \times (200/2)(200/3); \quad \tilde{y} = 93.33 \text{ mm}$$

Distance to the top edge = 200 – 93.33 = 106.67 mm

We find the moment of inertia about both axes. For this, we consider that the section is composed of a rectangle and two triangles as shown.

$$I_{xx} = [200 \times 200^3/12 + 200 \times 200 \times (100 - 93.33)^2] + 2\,[50 \times 200^3/36 + (50 \times 200/2)(93.33 - 200/3)^2] = 164.43 \times 10^6 \text{ mm}^4$$

$$I_{yy} = 200^4/12 + 2[200 \times 50^3/36 + (200 \times 50/2)\,(100 + 50/3)^2] = 270.83 \times 10^6 \text{ mm}^4$$

Case (i) When the load is on the Y-axis, $e_x = 0$ and $e_y = 100 - 93.33 = 6.67$ mm

$$M_x = 3000 \times 1000 \times 6.67 = 20 \times 10^6 \text{ Nmm}$$

This moment causes compression at A and B and tension at C and D.

Direct stress = 3000 × 1000/50,000 = 60 N/mm^2

$$\sigma_A = 60 + 20 \times 10^6 \times 106.67/164.43 \times 10^6 = 73 \text{ N/mm}^2 = \sigma_B$$

$$\sigma_C = 60 - 20 \times 10^6 \times 93.33/164.43 \times 10^6 = 48.65 \text{ N/mm}^2 = \sigma_D$$

Case (ii) When the load is placed as shown, $e_x = 100$ mm and $e_y = 6.67$ mm

$$M_x = 3000 \times 1000 \times 6.67 = 20 \times 10^6 \text{ Nmm}$$

$$M_y = 3000 \times 1000 \times 100 = 300 \times 10^6 \text{ Nmm}$$

$$\text{Stress at } A,\ \sigma_A = 60 + \frac{20 \times 10^6 \times 106.67}{164.43 \times 10^6} - \frac{300 \times 10^6 \times 100}{270.83 \times 10^6} = 60 + 13 - 111$$

$$= -38 \text{ N/mm}^2$$

$$\sigma_B = 60 + 13 + 111 = 184 \text{ N/mm}^2$$

$$\sigma_C = 60 - \frac{20 \times 10^6 \times 93.33}{164.43 \times 10^6} - \frac{300 \times 10^6 \times 150}{270.83 \times 10^6} = 60 - 11.35 - 166.15$$

$$= -117.5 \text{ N/mm}^2$$

$$\sigma_D = 60 - 11.35 + 166.15 = 214.8 \text{ N/mm}^2$$

6.6 STRUCTURES SUBJECTED TO LATERAL PRESSURE

The structures shown in Fig. 6.27 are all subjected to lateral pressures, due to water, earth, or wind. The stress distributions at the base sections of such structures are similar to the variations due to direct and bending stresses.

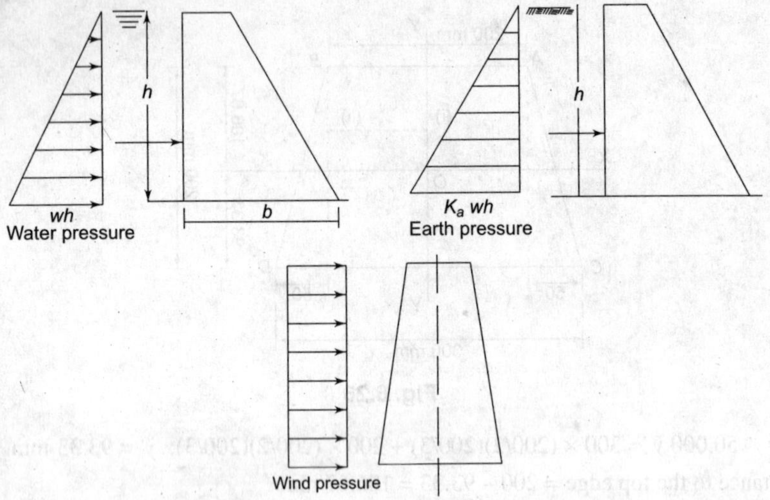

Fig. 6.27

6.6.1 Behaviour of Structures Subjected to Lateral Pressure

Water pressure varies linearly with height, and at any level the pressure intensity is wh, where w is the unit weight and h is the height of water above the point. In the case of earth pressure also the intensity varies linearly but is multiplied by the coefficient of active earth pressure K_a, which is equal to $(1 - \sin\phi)/1 + \sin\phi$, where ϕ is the angle of repose of the retained material.

In both these cases, the resultant lateral pressure acts through the centroid of the pressure intensity diagram, which is at $(h/3)$ from the base of the triangle.

In the case of walls, chimneys, and towers, the wind pressure acts as a lateral force. This pressure may be considered to be uniform or may vary with the height. It may also vary in direction so that pressures acting from different directions may have to be considered. The wind pressure is generally taken to be acting on the projected area of the structure on a vertical plane, and may also be multiplied by a factor. The total wind pressure, in the case of uniform pressure distribution, is thus $kpbh$, where bh is the projected area, p is the pressure intensity, and k is a factor which is 1.0 for a rectangular structure and 2/3 for a circular one.

The lateral pressure in all these cases causes a BM at the base section, which may have any shape depending upon the structure (cross section). The axial thrust due to the self-weight and the BM due to lateral pressure cause a non-uniform variation of base pressures [Fig. 6.28(a)].

To look at this phenomenon in another way, the lateral force and resultant self-weight [Fig. 6.28(b)] have a resultant as shown. This resultant strikes the base section at an eccentricity. The vertical component of this resultant, which is the self-weight, causes a BM about the centroidal axis of the base section. The horizontal pressure, of course, tends to translate the structure against friction and any keys provided to prevent such sliding. From this reasoning, it is evident that if the base section is rectangular, the resultant should strike it within the middle-third to avoid tension.

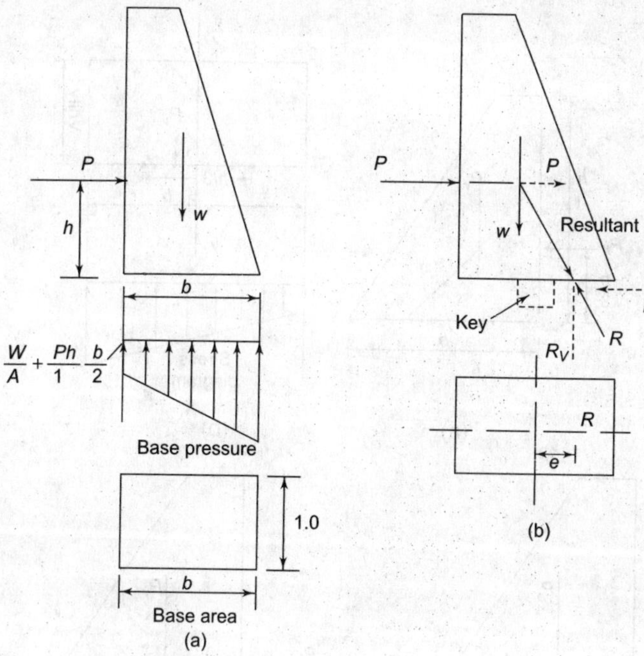

Fig. 6.28

In designing such structures, a major consideration is to avoid tension at the base. The soil pressure cannot develop tensile stresses, and if there is tension, it is equivalent to reduction in the area of the base that is transferring the load.

6.6.2 Conditions for Stability

Consider the case of a dam or a retaining wall subjected to lateral pressure as shown in Fig. 6.29. P_h is the resultant horizontal pressure due to water on the earth and P_v is the resultant (vertical) gravity load.

The structure is in equilibrium under the action of these two forces and the resultant reaction from the ground. The stability of dams and walls is dependent on the following conditions.

(a) Stress condition There are two conditions under this category. There should be no tensile stress in the base and the compressive stress must be within the permissible levels. The base of dams and walls is usually rectangular in shape. Hence, the condition of no tension requires that the resultant load of horizontal pressure and gravity load of self-weight and any superimposed load, strikes the base within the middle third of the section. This is shown in Fig. 6.29(b).

Due to the loads, the maximum compressive stress is at the toe. The maximum compressive stress is

$$\sigma_{\max} = P_v/bd + P_h(h/3)(b/2)/I_{YY}$$

This must be less than the compressive stress that is permissible to avoid crushing of the ground. The terms in this equation are given in Fig. 6.29(b).

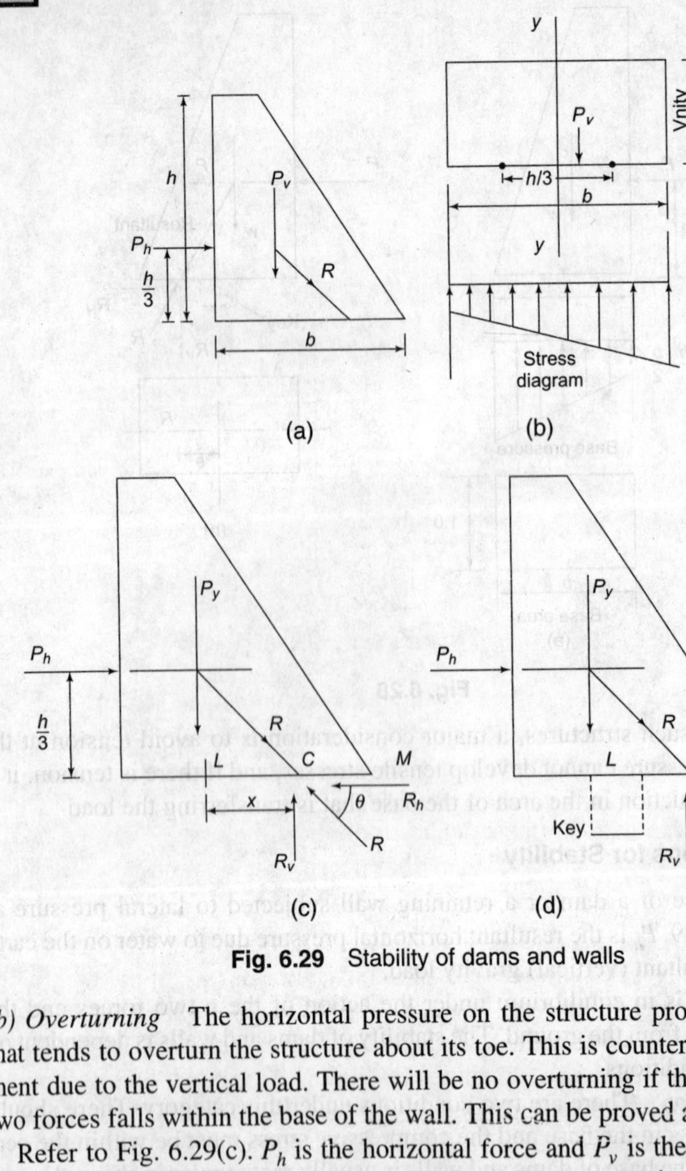

Fig. 6.29 Stability of dams and walls

(b) Overturning The horizontal pressure on the structure produces a moment that tends to overturn the structure about its toe. This is counteracted by the moment due to the vertical load. There will be no overturning if the resultant of the two forces falls within the base of the wall. This can be proved as follows.

Refer to Fig. 6.29(c). P_h is the horizontal force and P_v is the resultant gravity load. Let the resultant of the two forces fall at C. The net reaction from the ground is R to balance this resultant force. It can be shown that so long as the resultant of P_h and P_v strikes the base within the structure, there will be no overturning.

Consider the force system acting on the structure. The resultant of the forces P_h and P_v strikes the base at a distance x as shown.

Overturning moment due to P_h @ $m = P_h h/3$

Restoring moment due to P_v @ $m = P_v\, x$

When these two moments are equal, there will be no overturning of the structure.

$$P_h h/3 = P_v x$$

In the limiting case, when overturning takes place about the toe, the overturning moment is $P_h h/3$. The corresponding restoring moment is $P_v LM$. There will be no overturning if $P_h h/3 < P_v\ LM$. So, $P_v\ LM > P_v x$ (as $P_h h/3 = P_v x$)

This shows that if $x < LM$, there will be no overturning, that is, if the resultant force strikes the base within the limits of the base, overturning can be avoided.

(c) Sliding The sliding of the dam means a free body movement of the structure by sliding along the ground. Considering the free body diagram of the dam [Fig. 6.29(d)], one can see that the horizontal force due to water or earth pressure tends to slide the structure while the restoring force is the horizontal component of the reaction from the base. If $P_h < R \cos \theta$, there will be no sliding. Many times, keys are provided to add to the stability of the dam or retaining wall against sliding as shown in Fig. 6.29(d).

6.6.3 Analysis of Dams, Walls, and Chimneys

Analysis of dams, walls, and chimneys requires finding the forces acting on them. The forces in each case are horizontal forces due to water pressure, earth pressure, or wind forces. The stabilizing force is due to the self-weight of the structure, which is generally massive. There could also be loads due to other loads on the structure like moving loads.

Dams Figure 6.30(a) shows a dam that retains water pressure. The dam retains water to some height, generally less than its full height. The water pressure varies linearly and is triangular in shape. The force due to the water pressure over a unit length of the dam is given by $wh^2/2$ and acts at $h/3$ from the base. The dam normally is in the shape of a trapezium increasing in width as the depth increases. The weight of the dam can be calculated knowing the unit weight of the dam material, which is either concrete or earth. The water face is normally vertical but can also be slanting.

Retaining walls In the case of retaining wall, the horizontal pressure is due to the retained earth. The earth pressure is calculated using any of the theories like Rankine's earth pressure theory.

In the case of the earth retained, the earth has an *angle of repose*, φ, which is the angle at which the earth can remain in equilibrium by internal friction only. The pressure due to the earth is dependent on this angle. Because of this, a retaining wall can also retain the earth to its full height and at an angle (less than or equal to φ), known as surcharge.

Without going into the details of soil mechanics, here only the formulae for calculating the earth pressure are given.

Earth level As in Fig. 6.30(b), earth pressure = $(wh^2/2)\ [(1 - \sin \varphi)/(1 + \sin \varphi]$, where w is the unit weight of the earth retained, h is the height, and φ is the angle of repose. It is assumed to act at $h/3$ from the base.

Earth at a surcharge angle α In this case,

$$\text{Earth pressure} = (wh^2 \cos \alpha/2)\ \frac{[\{\cos \alpha - \sqrt{(\cos^2 \alpha - \cos^2 \varphi)}\}]}{[\cos \alpha + \sqrt{(\cos^2 \alpha - \cos^2 \varphi)}]}$$

This earth pressure is parallel to the surcharge surface and acts at $h/3$ from the base as shown in Fig. 6.30(b).

Chimneys Chimneys are tall structures designed to throw away flue gases from burning of fuel into the atmosphere at great height. The horizontal pressure in the case of chimneys is due to the wind pressure, which is due to the wind acting on the exposed area. The wind pressure is extremely variable but an average wind pressure is taken for calculation. It also depends upon what is called *form factor*, depending upon whether the exposed surface is rectangular or circular.

Wind pressure = KpA,

where K is a form factor, p is the wind pressure intensity, and A is the exposed area. The value of K is 1 for rectangular or square chimneys and 2/3 for circular chimneys.

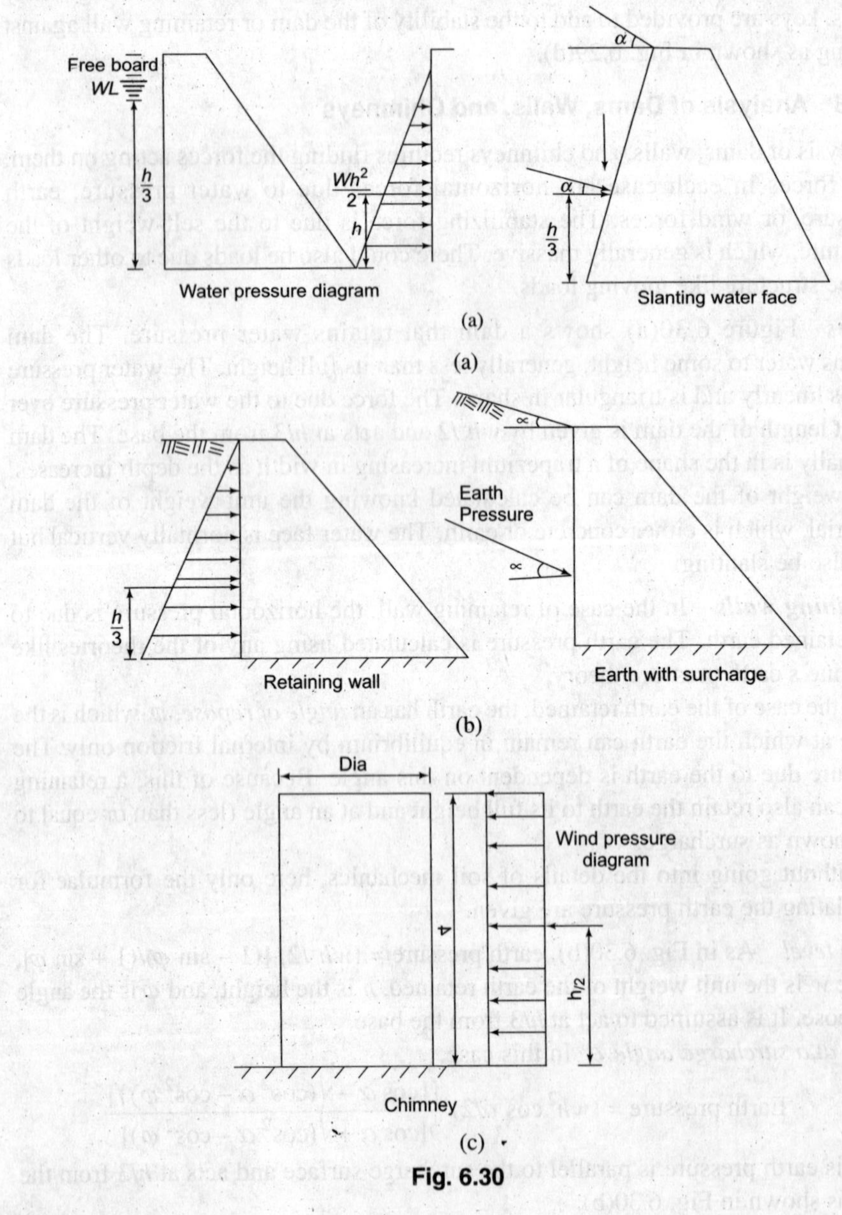

Fig. 6.30

The wind pressure is taken from the wind pressure diagrams, depending upon where the chimney is located. Wind pressure also depends on the height but is taken uniform. Thus, unlike the water and earth pressure, the force due to the wind pressure acts at mid-height of the chimney. Figure 6.30(c) shows these details.

The following examples illustrate the above principles.

Example 6.18 Chimney of hollow circular section

Show that in a hollow circular section chimney of external diameter D and internal diameter d, the resultant line of thrust due to lateral pressure and self-weight should fall within $(D^2 + d^2)/8D$ of the centre line of the chimney, for no tension at the base.

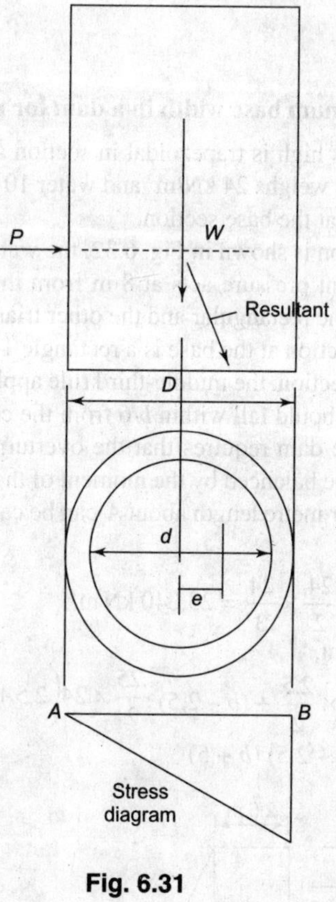

Fig. 6.31

Solution Referring to Fig. 6.31, let the resultant thrust fall at an eccentricity of e from the centre line. If W is the total self-weight of the chimney, then it is easy to see that

$$\text{Direct compressive stress} = \frac{W}{A}$$

$$\text{Maximum bending stress} = \frac{We}{I}y$$

This will be tensile at A so that the maximum value of e will be obtained when the stress at A is zero. If e exceeds this value, there will be tension in the section.

The weight W of the chimney is proportional to the cross sectional area, so that

$$W = k\frac{\pi}{4}(D^2 - d^2)$$

$$I \text{ for the section} = \frac{\pi}{64}(D^4 - d^4)$$

and $$y = \frac{D}{2}$$

Therefore,

$$\frac{k(\pi/4)(D^2 - d^2)}{(\pi/4)(D^2 - d^2)} - \frac{k(\pi/4)(D^2 - d^2)e}{(\pi/64)(D^4 - d^4)}\frac{D}{2} = \sigma_A = 0$$

$$1 - \frac{8(D^2 - d^2)}{(D^4 - d^4)} \times De = 0$$

$$\Rightarrow \qquad e = \frac{D^2 + d^2}{8D}$$

Example 6.19 Minimum base width in a dam for no tension

A concrete dam 25 metres high is trapezoidal in section and retains water to a maximum height of 24 m. If concrete weighs 24 kN/m^3 and water 10 kN/m^3, find the minimum width of the base for no tension at the base section.

Solution The dam section is shown in Fig. 6.32, the water pressure diagram is triangular as shown, and the resultant pressure acts at 8 m from the base. The dam section can be considered in two parts, one rectangular and the other triangular. Considering a length of 1 m of the dam, the cross section at the base is a rectangle 1 m wide and b m deep as shown. This being a rectangular section, the middle-third rule applies and the resultant of the water pressure and self-weight should fall within $b/6$ from the centre line of the base.

The equilibrium of the dam requires that the overturning moments about A of water pressure and self-weight be balanced by the moment of the reactive force at the base. From Fig. 6.32, the moments per metre length about A can be calculated as follows. Moment due to water pressure,

$$M_w = 10 \times 24 \times \frac{24}{2} \times \frac{24}{3} = 23{,}040 \text{ kN m}$$

Moment due to self-weight,

$$= 2.5 \times 25 \times 24 \times \frac{2.5}{2} + (b - 2.5)\frac{.25}{2} \times 24\left(2.5 + \frac{b - 2.5}{3}\right)$$

$$= 1875 + 100\,(b - 2.5)\,(b + 5)$$

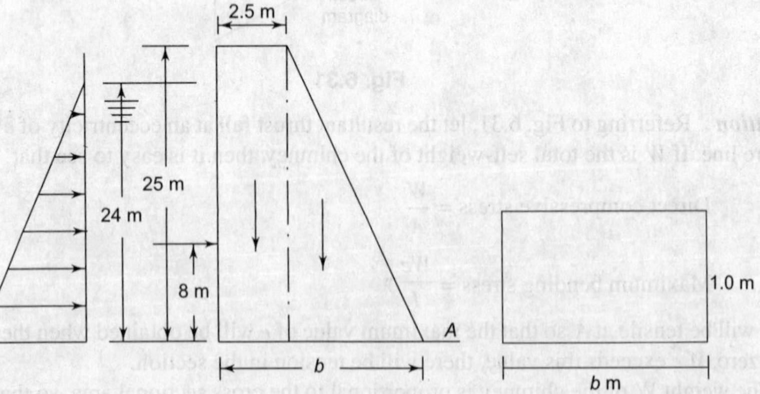

Fig. 6.32

Only the vertical component of the reaction has a moment about A, and the vertical component R_v = self-weight of the dam.

$$\text{Self-weight} = \left(\frac{b+2.5}{2}\right) \times 25 \times 24 = 300\,(b+2.5)$$

In the extreme case this acts at $(2/3)b$ from A.

$$\text{Moment due to } R_v = 300\,(b+2.5) \times \frac{2}{3}b$$
$$= 200b\,(b+2.5)$$

Therefore,

$$200b\,(b+2.5) = 23040 + 1875 + 100\,(b-2.5)\,(b+5)$$

$\Rightarrow$ $b = 14.15$ m

This is the minimum width required for no tension at the base. ■

Example 6.20 Design of a hollow square chimney

A square chimney, 30 m high, has a flue opening of size 1.5 m × 1.5 m. Find the minimum width required at the base for no tension if the masonry weighs 20 kN/m^3 and the wind pressure is 1.5 kN/m^2. The permissible stress in the masonry is 1 N/mm^2.

Solution Let b be the side of the square at the base as shown in Fig. 6.33.

Weight of the shaft = $(b^2 - 1.5^2) \times 30 \times 20$ kN

Wind pressure = $30b \times 1.5$ kN

$$\text{Direct stress} = \frac{(b^2 - 1.5^2) \times 30 \times 20}{b^2 - 1.5^2} = 600 \text{ kN/m}^2$$

BM due to wind pressure $= 30b \times 1.5 \times 15$

$= 675b$ kNm

$$I_{XX} = \left(\frac{b^4}{12} - \frac{1.5^4}{12}\right)$$

$$\text{Bending stress} = \pm \frac{675b}{(b^4 - 1.5^4)/12}\,\frac{b}{2} = \frac{4050b^2}{b^4 - 1.5^4}$$

$$\text{Maximum compressive stress } \sigma_e = 600 + \frac{4050\,b^2}{(b^4 - 1.5^4)}$$

$$\text{Tensile stress } \sigma_t = \frac{4050\,b^2}{(b^4 - 1.5^4)} - 600$$

$\sigma_c = 1$ N/mm^2 = 1000 kN/m^2 $\quad \sigma_t = 0$

$$600 + \frac{4050\,b^2}{(b^4 - 1.5^4)} = 1000$$

$$\frac{4050\,b^2}{(b^4 - 1.5^4)} - 600 = 0$$

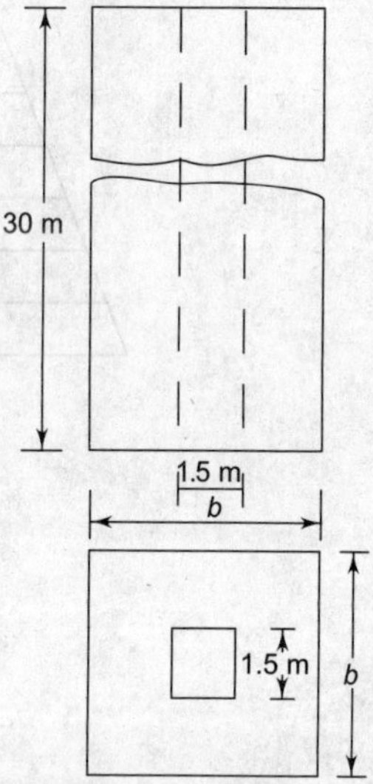

Fig. 6.33

Solving these equations

$b = 3.256$ m and 2.72 m

Minimum width required = 3.256 m ■

Example 6.21 Masonry earth retaining wall

A masonry retaining wall has the trapezoidal section shown in Fig. 6.34. It retains earth to its full height of 6 m. Find the maximum and minimum stress intensities at the base. The masonry weighs 20 kN/m^3 and the earth has a density of 18 kN/m^3. The angle of repose of the retained material is 30°.

Solution From Fig. 6.34,

$$\text{Base earth pressure} = wh\left(\frac{1-\sin\theta}{1+\sin\theta}\right) = 18\times 6\times\frac{(1-1/2)}{(1+1/2)}$$

$$= 36 \text{ kN/m}^2$$

$$\text{Total pressure} = 36\times\frac{6}{2} = 108 \text{ kN}$$

$$\text{Moment about } A = 108\times\frac{6}{3} = 216 \text{ kN m}$$

$$\text{Total weight of the wall} = 20\left(1.5\times 6 + 6\times\frac{2.5}{2}\right)$$

$$= 330 \text{ kN/m length}$$

$$\text{Moment of weight of wall about } A = 20\times 1.5\times 6\times 0.75 + 6\times\frac{2.5}{2}\times 20\times\left(1.5+\frac{2.5}{3}\right)$$

$$= 135 + 350 = 485 \text{ kN m}$$

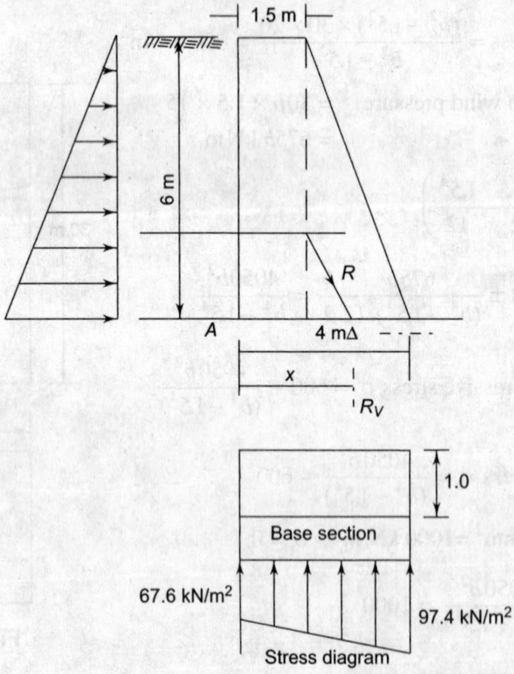

Fig. 6.34

If the resultant strikes the base at x m from A, then the moment of the reactive force about A is given by

$$\text{Reactive moment} = 330x$$

Hence, equating the moments for equilibrium,

$$330x = 216 + 485$$

$$\Rightarrow \quad x = 2.12 \text{ m}$$

Eccentricity at the base = 0.12 m

$$\text{Maximum stress at } B = \frac{330}{4 \times 1} + \frac{330 \times 0.12}{1 \times 4^3/12} \times 2$$

$$= 82.5 + 14.9 = 97.4 \text{ kN/m}^2$$

Stress at A = 82.5 – 14.9 = 67.6 kN/m^2

Example 6.22 Dam retaining water

The section of a dam retaining water is shown in Fig. 6.35. Determine the maximum and minimum stresses at the base of the dam. Take density of water as 10 kN/m^3 and density of the dam material as 24 kN/m^3.

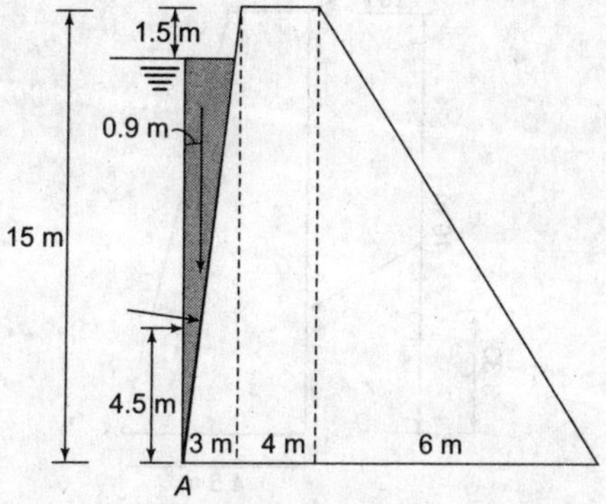

Fig. 6.35

Solution As the water face is not vertical, the water pressure is normal to the dam face. However, it is found that this pressure can be resolved into horizontal and vertical components. The horizontal component is equal to the water pressure on an assumed vertical face. The vertical component is equal to the weight of the water in the triangular area shown shaded. Consider 1 m length of the dam.

Water pressure on a vertical face = 10 × 13.5^2/2 = 911.25 kN

Weight of the water on the triangular wedge = b × 13.5/2 × 10

[b = 3 × 13.5/15 = 2.7 m]

= 2.7 × 13.5 × 10/2 = 182.25 kN

This acts at 2.7/3 = 0.9 m from the assumed vertical face.

Weight of the dam is now calculated.

Weight of the left triangular wedge = (15 × 3/2) × 1 × 24 = 540 kN

Weight of rectangle = 15 × 4 × 1 × 24 = 1440 kN

Weight of triangle on right = (15 × 6/2) × 1 × 24 = 1080 kN

To find the line of action of the total force, let it act at x m from A.

Total vertical weight = 3060 kN

Taking moments @ A,

$$3060\,x = 540 \times 2 + 1440 \times 5 + 1080 \times 9; \; x = 5.88 \text{ m}$$

If the resultant of the horizontal and vertical forces strikes the base at y m from point A, then the moment of all the forces @ F is zero. We have

$$911.25 \times 4.5 - 182.25 \times (y - 0.9) - 3060 \times 5.88 = 0$$

This gives $y = 6.86$ m

Eccentricity of the load = 6.86 – 13/2 = 0.36 m

The base section is rectangular of sides 13 m × 1 m.

So, area = 13 m^2; $I = 1 \times 13^3/12$

The maximum and minimum stresses at the base are given by

$$\sigma = \frac{3242.25}{13} + \frac{3242.25 \times 0.36 \times 6.5 \times 12}{13^3} = 291 \text{ kN/m}^2 \text{ and } 208 \text{ kN/m}^2$$

Example 6.23 Earth retaining wall

A trapezoidal retaining wall retains earth of density 20 kN/m^3 at a surcharge angle of 20° (Fig. 6.36). If the angle of repose of the soil is 30° and the density of concrete is 24 kN/m^3, find the maximum and minimum stresses at the base of the wall.

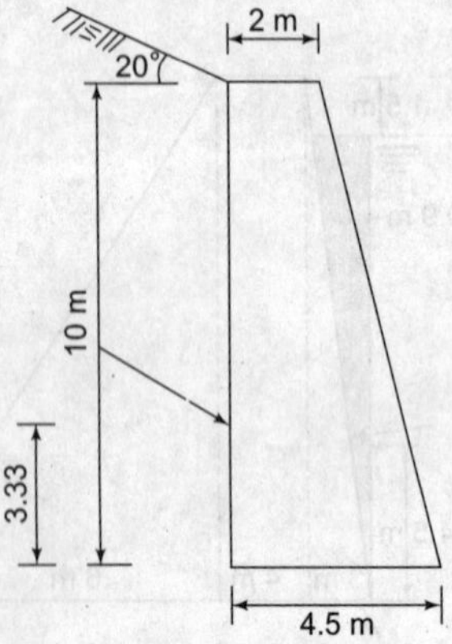

Fig. 6.36

Solution Earth pressure in the case of surcharged earth is given by

$$\text{Earth pressure} = \frac{wh^2}{2} \cos\alpha \frac{[\cos\alpha - \sqrt{(\cos^2\alpha - \cos^2\varphi)}]}{[\cos\alpha + \sqrt{(\cos^2\alpha - \cos^2\varphi)}]}$$

In this case, $w = 20$ kN/m^3; $h = 10$ m; $\varphi = 30°$; and $\alpha = 20°$

Substituting,

$$\text{Earth pressure} = \frac{20 \times 10^2}{2} \cos 20° \frac{[\cos 20° - \sqrt{(\cos^2 20° - \cos^2 30°)}]}{[\cos 20° + \sqrt{(\cos^2 20° - \cos^2 30°)}]}$$

$$= 414.3 \text{ kN}$$

This acts at 20° to the horizontal as shown.

Horizontal component, $P_h = 414.23 \times \cos 20° = 389.25$ kN

Vertical component $P_v = 414.23 \times \sin 20° = 141.67$ kN

Self-weight of the wall (per metre length)

Rectangular part = 2 × 10 × 1 × 24 = 480 kN

$$\text{Triangular part} = \frac{10 \times 2.5}{2} \times 1 \times 24 = 300 \text{ kN}$$

Total weight of the wall = 780 kN

If the total weight acts at x m from A,

$780x = 480 \times 1 + 300 \times 2.833$; $x = 1.7$ m

If the resultant of the earth pressure and self-weight falls at dm from A, then from $\Sigma M = 0$ @ A,

$$780(d-1.7)+141.7\times d-389.3\times\frac{10}{3}=0;\ d=2.846\text{ m}$$

Eccentricity = 2.846 – 2.25 = 0.596 m

Maximum and minimum stresses are:

$$\sigma=\frac{921.7}{4.5}\pm\frac{921.7\times0.596\times2.25}{7.6}$$

$$\left[\text{Area of section}=4.5\times1=4.5\text{ m}^2;\ I=\frac{1\times4.5^3}{12}=7.6\text{ m}^4\right]$$

$$\sigma=204.8\pm162.6=367.43\text{ kN/m}^2,\ 42.2\text{ kN/m}^2$$

Summary

Many structural elements such as dams, retaining walls and chimneys are subjected to axial stresses and bending moments due to self-weight as well as pressures, due to water, earth, wind, etc. In such cases, the axial stress and the stress due to bending are combined algebraically. These formulae are applicable to short compression members. Axial stress = P/bd and bending stress = $\pm My/I$.

The axial compressive stress and the compressive stress due to bending together yield a larger compressive stress. In the case of a rectangular section of dimensions $b\times d$, the maximum stress can be calculated as

$$\sigma=\frac{P}{bd}\pm\frac{6Pe}{bd^2}$$

If $P/bd > 6Pe/bd^2$, the whole section will be under compression and if they are equal, stress will be zero at one end. If $P/bd < 6Pe/bd^2$, tensile stresses will be produced.

For a rectangular section, the middle-third rule states that if the eccentricity of the load is less than or equal to $d/6$ on either side of the bending axis, no tension will be produced. The above formula for single-axis bending can be extended to biaxial bending, where the axial load has eccentricity about both the axes of the section. In such a case,

$$\sigma=\frac{P}{A}\pm\frac{Pe_Xx}{I_Y}\pm\frac{Pe_Yy}{I_Y}$$

The neutral axis in such cases is obtained by

$$1+\frac{e_Xx}{r_Y^2}\pm\frac{e_Yy}{r_X^2}=0$$

which is the equation for a line of zero stress.

The kern or core of a section gives the outer limits for eccentricity of the load. No tensile stresses are produced in the section within these limits. The kern of a rectangular section is a diamond-shaped area of diagonal lengths $d/3$ and $b/3$. In structures subjected to lateral pressure, these formulae are used—the direct stresses being produced by self-weight or other causes and the BM being produced by the lateral load.

Exercises

Review Questions

1. Why is it necessary to assume that an eccentrically loaded compression member is short to arrive at the equations derived in this chapter?
2. If the same load is put at the same eccentricity along the X- and Y-axes of a short compression member of rectangular section (sides a and b), in which case will the maximum stress be more and why?
3. Show that in the case of a short compression member of square section a, if the initial

central eccentricity is half the dimension of the side, the permissible load is only one-fourth of the load it can carry if it had been straight.

4. What is the kern of a section? Draw the kerns of square, circular, hollow circular, and triangular sections.
5. Take the case of the rectangular section shown in Fig. 6.37. Determine the position of an eccentric load *P* if the NA of the section is along (i) *AB*, (ii) *EH*, and (iii) *BH*.
6. In the rectangular section shown in Fig. 6.37, find approximately the position of the NA if the load is placed (i) at *G* and (ii) at *F*.

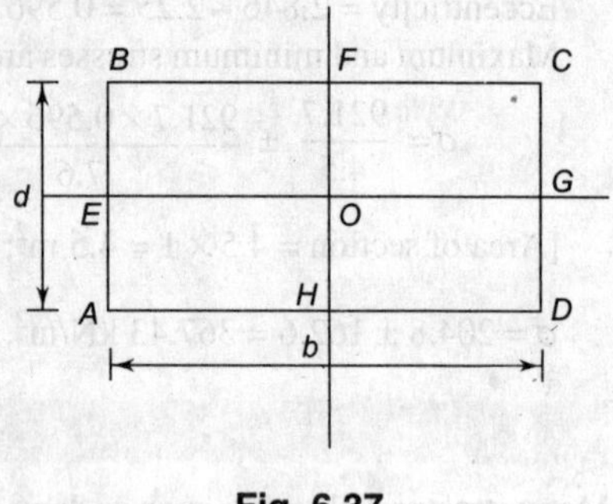

Fig. 6.37

Problems

1. A short, hollow circular column has an outside diameter of 180 mm and an inside diameter of 150 mm. If the allowable compressive stress is 140 MPa, find the maximum load that can be applied at the middle of the thickness of the pipe. What is the maximum tensile stress in this case?
2. A billboard weighing 1000 N is supported on a solid circular wooden rod. If the permissible stress is 8 MPa and 12 MPa in tension and compression, respectively, find the diameter of the rod required for the situation shown in Fig. 6.38.
3. Due to an overhanging roof structure only on one side, the effective loading on a column is as shown in Fig. 6.39. Find the diameter required of a cast-iron hollow circular column. The permissible stresses in tension and compression are 25 N/mm^2 and 100 N/mm^2, respectively. The inner diameter of the tube is 0.8 times the outer diameter.

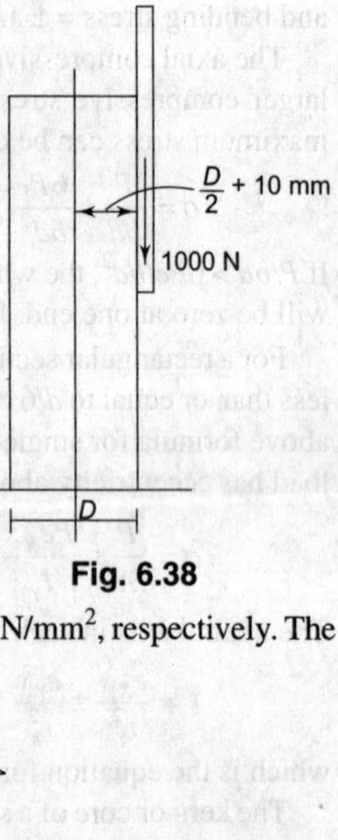

Fig. 6.38

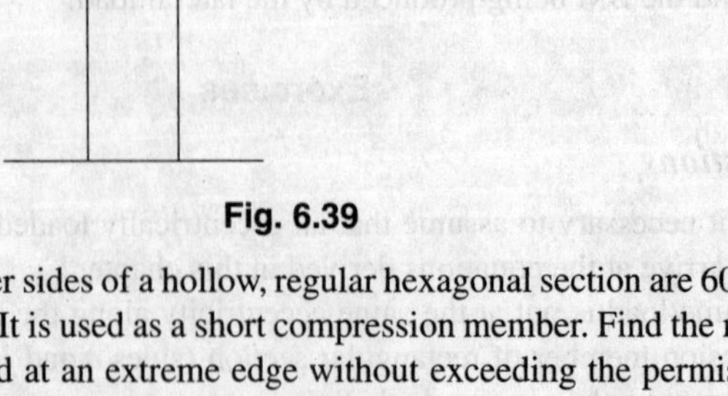

Fig. 6.39

4. The outer and inner sides of a hollow, regular hexagonal section are 600 mm and 350 mm, respectively. It is used as a short compression member. Find the maximum load that can be applied at an extreme edge without exceeding the permissible stress of 100 MPa.

5. Show that the kern of an equilateral section of side L is an equilateral triangle of side $L/4$.
6. A hollow, square section of side a and thickness 1/12 of the side is used as a short column. Find the kern of this section.
7. An I-section of area 500 mm × 200 mm and of uniform thickness 10 mm is used as a short column. Find the kern of this section and sketch it.
8. In the cantilever beam shown in Fig. 6.40 find the stresses at A and B of the section.

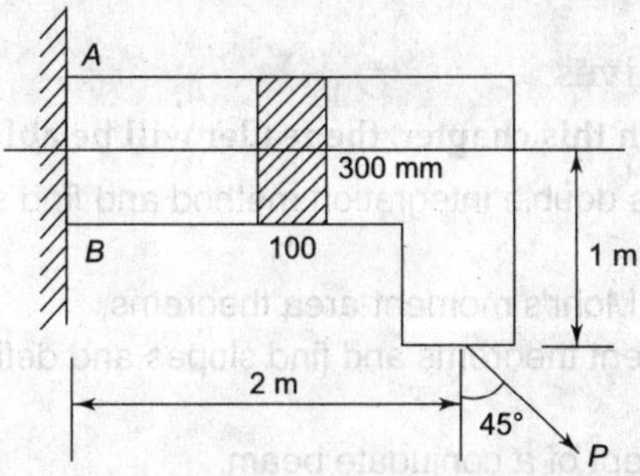

Fig. 6.40

9. In a rectangular section of dimensions 200 × 100 mm, a load of 80 kN is applied 40 mm and 20 mm off the centroid parallel to the 200 and 100 mm sides, respectively. Find the stresses at the four corners. What is the additional compressive load that can be placed at the centroid of the section to make the tensile stress zero?
10. A masonry retaining wall has a trapezoidal section 1.5 m wide at the top, 5 m wide at the base, and is 7 m high. The earth face is vertical and the angle of repose of the retained material is 30°. If the masonry weighs 21 kN/m^3 and earth has a density of 20 kN/m^3, find the stress intensities at the base.
11. Determine the minimum base width required for a masonry retaining wall 8 m high, retaining earth to the full height. The retained material has a density of 20 kN/m^3 and an angle of repose of 30°. Take the density of masonry as 22 kN/m^3. The wall is 2 m wide at the top.
12. A small concrete dam 12 m high has a top width of 2 m and a base width of 7 m, with the water face vertical. Determine the stress intensities at the base.
13. Determine the external diameter required for a hollow circular chimney, of internal diameter 2 m and height 20 m, for no tension at the base. The maximum design wind pressure is 3 kN/m^2 of the projected area. The density of masonry may be taken as 20 kN/m^3.
14. A masonry wall, 7 m high, 3 m wide, and 1.0 m thick, is subjected to design wind pressure of 2000 N/m^2. If the density of the masonry is 21 kN/m^3, find the extreme stress intensities at the base.
15. To what height can the retaining wall in Fig. 6.41 be constructed without exceeding the maximum compressive stress of 120 kN/m^2 and no tension at the other end.

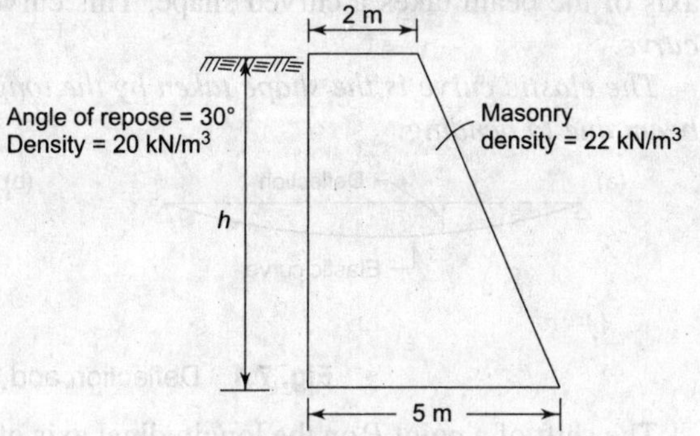

Fig. 6.41

CHAPTER 7

Deformations in Beams

Learning Objectives

After going through this chapter, the reader will be able to

- apply Macaulay's double integration method and find slopes and deflections in beams,
- state and derive Mohr's moment-area theorems,
- apply area-moment theorems and find slopes and deflections in beams with given loads,
- explain the concept of a conjugate beam,
- apply the conjugate-beam method and calculate slopes and deflections for beams with given loads,
- appreciate the importance of computing deformations in beams, and
- compute deflections in beams subjected to asymmetrical bending.

7.1 INTRODUCTION

In Chapter 4, while discussing the structural action of a beam, we mentioned that, under the action of applied loads, the vertical movement of a point on the elastic curve is called deflection. The slope at a point on the elastic curve is the angle that the tangent to this curve makes with the unbent beam axis. Under the action of the loads, a beam bends and the bent shape is known as the elastic curve. In this chapter, we will study the elastic curve in detail to enable us to locate points on it and its slope at a point.

7.1.1 Slope and Deflection

It is necessary to understand what is meant by deformations in a structure. Taking the case of a simple beam as in Fig. 7.1, the deformations considered are called vertical deflection and slope at any point. As the beam bends, the original straight axis of the beam takes a curved shape. This curved axis is known as the *elastic curve*.

The elastic curve is the shape taken by the longitudinal, centroidal axis of the beam due to bending.

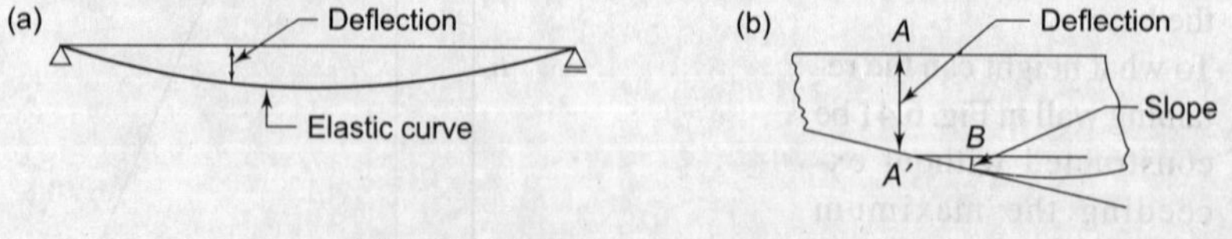

Fig. 7.1 Deflection and slope

The shift of a point P on the longitudinal axis of the beam perpendicular to that axis is called *deflection*. Here y-y' is the deflection of point P.

The deflection at a point in a beam can be defined as *the vertical distance between the point on the un-deflected axis and the corresponding point on the deflected axis.*

If we draw a tangent to the curve at point y', the deflected position of P, the angle this tangent makes with the original un-deflected axis is known as the *slope* at point P. Thus, in Fig. 7.1, angle θ is the slope at P. If the beam axis is set in the X - Y plane as shown, then this rotation (slope) is due to a rotation about the Z-axis.

The slope of beam axis at a point is defined as the angle between the tangent to the elastic curve at that point and the un-deflected beam axis.

7.1.2 Strength and Stiffness

In Chapters 5 and 6, we discussed the design of beams for strength, both shear and bending. In the bending equation $M/I = \sigma/y = E/R$, we used the part $M/I = \sigma/y$. The bending equation represents two equations $M/I = \sigma/y$ and $M/I = E/R$. While the first part deals with strength, the second concerns itself with bending or flexural stiffness. There are two considerations in the design of beams, strength and stiffness. In design for strength, the external moment is related to the permissible stress in the material. This ensures that the stresses due to bending or shear are within the permissible limits for the material of the beam. In design for stiffness, it is ensured that the deformations in beam are limited. The various design codes limit the deformations either by limiting the span to depth ratios or directly limiting the deflection as a fraction of the span. In the stiffness equation, $M/I = E/R$, the external moment is related to the curvature $1/R$ of the beam, i.e., $M = EI/R$. The term EI is known as flexural stiffness, or rigidity. We will use this equation to arrive at the equation for the elastic curve. A large part of the deflection of beams is due only to BM. Deflections due to shear are very small and are usually neglected.

7.2 EQUATION OF THE ELASTIC CURVE

Equation of the elastic curve is derived from mathematical concepts in differential calculus. Essentially, the equation relates the curvature to differential coefficient of the coordinates of a point.

7.2.1 Elastic Curve

Elastic curve, as we have defined earlier, is the bent form of the longitudinal centroidal axis of the beam. Deflection and slope are measured at points on the elastic curve relative to the original unbent beam axis. Any curve is defined by its radius at any point. In the case of elastic curve, the curvature changes from point to point on the curve due to variation of BM in the beam. The curvature can be related to the BM in the beam by the equation derived in the next section. We already know the differential relationships between load intensity, SF, and BM:

$$dM/dx = F \text{ and } dF/dx = w$$

Also we have derived the Bernoulli's equation in Chapter 5.

$$\frac{M}{I} = \frac{\sigma}{y} = \frac{E}{R}$$

From these relationships and the equation for curvature of any curve as derived in mathematics, we obtain the relationship relating BM with the curvature.

7.2.2 Differential Equation of Elastic Curve

To derive the equation for the elastic curve, we take recourse to the curvature formula for any curve as derived in mathematics. Taking any curve as shown in Fig. 7.2, take any point P on the curve. Set in the X - Y axis system, P' is a point at an incremental distance ds along the curve. Also dx and dy are the distances along the X- and Y-axes from P to P'. As the distance ds is very small, we can take

$$ds^2 = dx^2 + dy^2$$

If the radius of the curve at P is R and the angle between the radii to P and P' is $d\alpha$, then $Rd\alpha$ is equal to ds if $d\alpha$ is in radians.

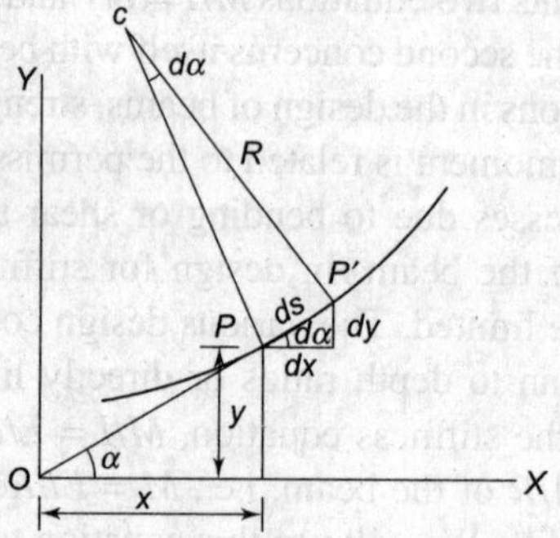

Fig. 7.2 Elastic curve

Now $\quad ds = \surd(dx^2 + dy^2) = Rd\alpha$

or $\quad \dfrac{1}{R} = \dfrac{d\alpha}{\surd(dx^2 + dy^2)}$

Also $\quad \tan\alpha = dy/dx$

Differentiating, $\sec^2\alpha\, d\alpha = \dfrac{d^2y\,dx}{dx^2}$

or $\quad (1 + \tan^2\alpha)d\alpha = \dfrac{d^2y}{dx^2}\,dx \quad$ as $\sec^2\alpha = 1 + \tan^2\alpha$

or $\quad [1 + (dy/dx)^2]d\alpha = \dfrac{d^2y}{dx^2}\,dx$

or $\quad d\alpha = \dfrac{(d^2y/dx^2)\,dx}{[1 + (dy/dx)^2]}$

or $\quad \surd(dx^2 + dy^2) = Rd\alpha = \dfrac{R(d^2y/dx^2)\,dx}{[1 + (dy/xd)^2]}$

or $$dx\sqrt{[1+(dy/dx)^2]} = \frac{R(d^2y/dx^2)\,dx}{[1+(dy/xd)^2]}$$

This gives

$$\frac{1}{R} = \frac{d^2y/dx^2}{[1+(dy/dx)^2]^{3/2}}$$

The application of this equation to the elastic curve yields a relationship between the BM and the deflection y (Fig. 7.3). Some simplification of this expression for curvature is justified in the case of the elastic curve because the deflection as well as the slope of the beams are extremely small. Thus dy/dx, which gives the slope of the elastic curve at a point, is very small and its square is still smaller. Therefore, the term $[1 + (dy/dx)^2]^{3/2}$ can be taken equal to 1, reducing the curvature expression to d^2y/dx^2. Thus,

$$\frac{1}{R} = \frac{d^2y}{dx^2}$$

$$M = \frac{EI}{R} = EI\frac{d^2y}{dx^2}$$

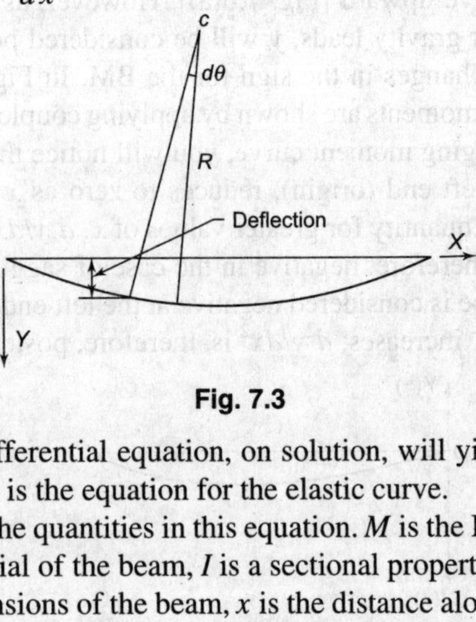

Fig. 7.3

This second-order differential equation, on solution, will yield a relationship between y and x, which is the equation for the elastic curve.

Figure 7.4 shows the quantities in this equation. M is the BM at a section, E is a property of the material of the beam, I is a sectional property depending upon the cross-sectional dimensions of the beam, x is the distance along the beam axis, and y is the deflection at x.

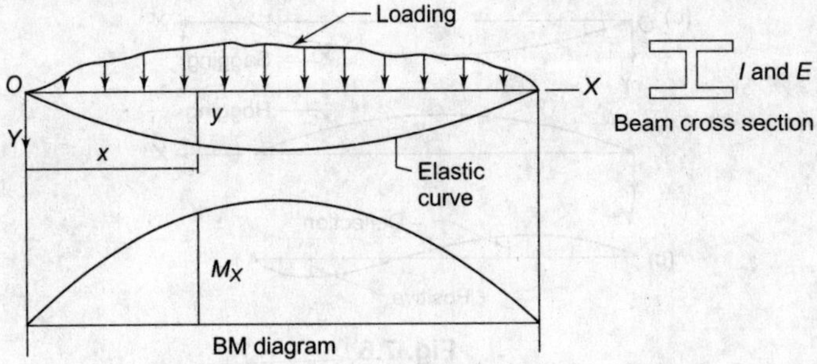

Fig. 7.4

7.3 SIGN CONVENTION

It is important to adopt uniform sign conventions for all quantities. The sign conventions already adopted for BM, M, are shown in Fig. 7.5(a). According to this, sagging BM is positive and hogging BM is negative.

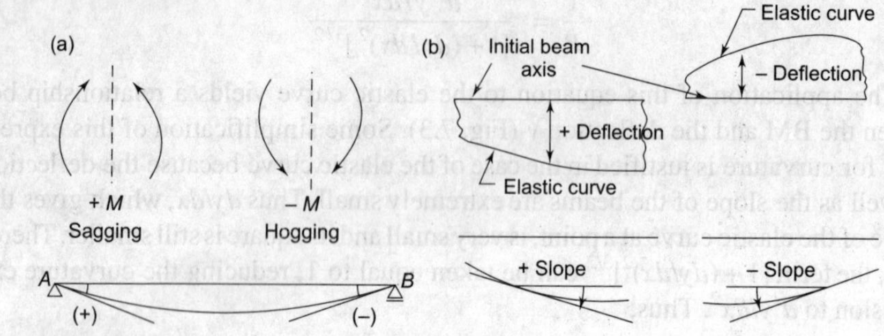

Fig. 7.5

In the expression for curvature, $1/R = d^2y/dx^2$, the sign convention for x and y is according to the conventional X-Y axes system used. x is positive to the left of the origin and y is positive upward [Fig. 7.6(a)]. However, as the deflection is normally downwards for gravity leads, y will be considered positive downwards by making appropriate changes in the sign for the BM. In Fig. 7.6(b), the effect of sagging and hogging moments are shown by applying couples at the ends of beams. In the case of the sagging moment curve, you will notice that the slope is considered positive at the left end (origin), reduces to zero as x increases and further reduces to a negative quantity for greater values of x. d^2y/dx^2, which is the rate of change of slope, is, therefore, negative in the case of sagging BM. Similarly, for hogging BM, the slope is considered negative at the left end, increases to zero and becomes positive as x increases; d^2y/dx^2 is, therefore, positive for a hogging BM.

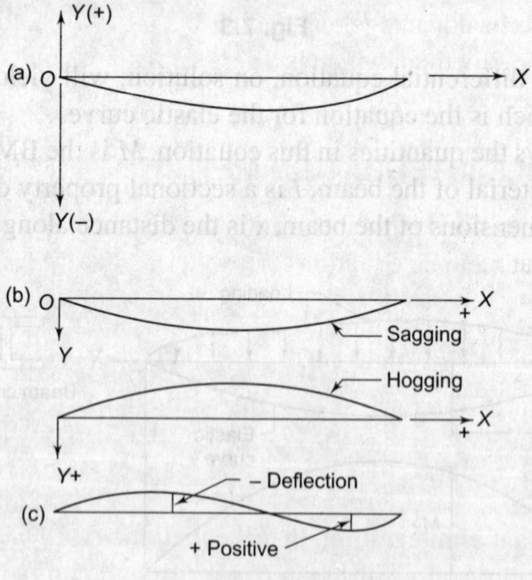

Fig. 7.6

Beams are generally subjected to gravity loads, leading to deflection downwards. To make the deflection downwards positive in the expression for $EI\,d^2y/dx^2$, a negative sign is prefixed to M to be consistent with the sign convention for BM. Thus,

$$EI\,\frac{d^2y}{dx^2} = -M$$

where M is positive if sagging and negative if hogging. The deflection in this case is positive if measured downwards [Fig. 7.6(c)].

The units used for evaluating slope and deflections must be consistent. The loading on the beam is generally measured in kilonewtons and the span of the beam in metres. The BM, therefore, has the unit kN m. To be consistent, E must be expressed in kN/m^2 and I in m^4. The slope of the beam obtained by solving the differential equation will be in *radians*. The deflection of the beam obtained will be in metres. If any of the quantities are given in units different from the above, appropriate conversions will have to be made.

7.4 METHODS FOR CALCULATING DEFLECTION

Several methods are available for calculating the deflection and slope in beams. They are all based upon the fundamental expression we have derived for curvature. We will discuss three of them in this chapter. These are

(i) the double integration method based upon the fact that d^2y/dx^2 when integrated twice yields y,

(ii) the area-moment method, wherein the slope and deflection are calculated from the area of the BM diagram and the moment of the area of the BM diagram, and

(iii) the conjugate-beam method whereby a conjugate beam derived from the actual beam is loaded with the M/EI diagram, and the slope and deflection in the actual beam are calculated as SF and BM, respectively, of the conjugate beam.

Each method has some advantage, depending upon the problem. Let us start with the double integration method, which is a very general method that yields an equation for the elastic curve.

7.5 THE DOUBLE INTEGRATION METHOD

The basic principle in the double integration method is to integrate the expression $EI\,d^2y/dx^2$ twice to obtain an equation for y in terms of x.

$$EI\,\frac{d^2y}{dx^2} = -M_x$$

where M_x is the expression for BM at x. Integrating,

$$EI\,\frac{dy}{dx} = \int (-M_x\,dx) + C_1$$

where C_1 is a constant of integration. In this equation, $dy/dx = \tan\theta$, where θ is the slope of the elastic curve at x. Since $\tan\theta$ is very small, $\tan\theta \approx \theta$ in radians (slope is, therefore, obtained in radians as mentioned earlier). Integrating once more,

$$EIy = \int\left[\int -M_x dx + C_1\right] + C_2$$

where C_2 is the second constant of integration.

Thus, once the expression for M_x, which will be in terms of x, is integrated twice, we get an expression for y in terms of x, i.e., the equation for the elastic curve.

The constants of integration, C_1 and C_2, are evaluated from the boundary conditions. More precisely, they are obtained from conditions which apply due to restraints imposed on the bending by the supports. These conditions can be in terms of either slopes or deflections. Figure 7.7 illustrates these conditions. Taking the origin at the left end of the beam, x positive to the right and y positive downwards, they are as follows.

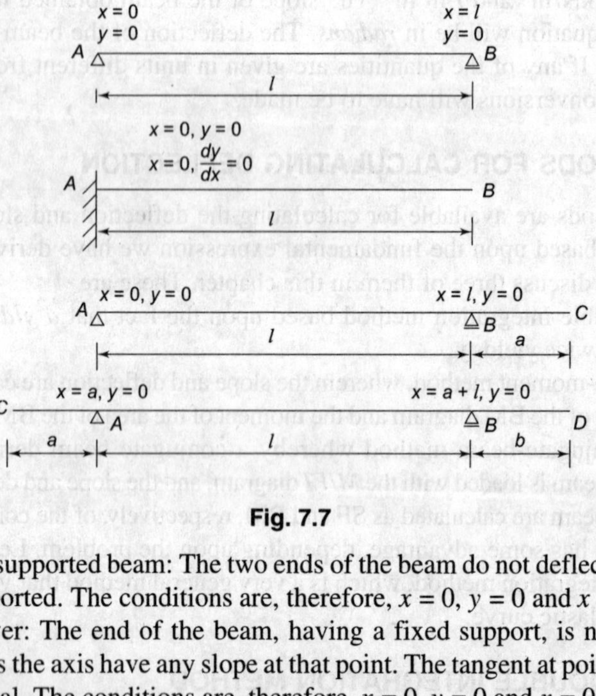

Fig. 7.7

(i) Simply supported beam: The two ends of the beam do not deflect where they are supported. The conditions are, therefore, $x = 0$, $y = 0$ and $x = l$, $y = 0$.

(ii) Cantilever: The end of the beam, having a fixed support, is not deflected. Nor does the axis have any slope at that point. The tangent at point A remains horizontal. The conditions are, therefore, $x = 0$, $y = 0$ and $x = 0$, $dy/dx = 0$.

(iii) Overhanging beam: The overhanging beam has the conditions $x = 0$, $y = 0$ and $x = l$, $y = 0$.

(iv) Doubly overhanging beam: In this case, the conditions are $x = a$, $y = 0$ and $x = l + a$, $y = 0$.

7.5.1 Bending Moment Equation

The first step in the double integration method is writing the general equation for M_x. We first illustrate the writing of BM equation and finding slopes and deflections with two examples.

We use a simplified notation for writing the equations: d^2y/dx^2 is written as y'' and dy/dx is written as y'.

Example 7.1 SS beam with UD load

A simply supported beam carries a uniformly distributed load over the whole span. Find the slopes at the supports and the maximum deflection. Sketch the elastic curve.

Solution The beam with the load is shown in Fig. 7.8. The reactions at the two ends are equal and can be easily worked out to be $wl/2 \uparrow$. The bending moment at any section x can be written as

$$M_x = \frac{wlx}{2} - \frac{wx^2}{2}$$

Please note that this equation is valid from $x = 0$ to $x = l$. The bending moment is sagging and hence positive. The bending moment equation can now be written as

$$EIy'' = -(wlx/2 - wx^2/2) = -wlx/2 + wx^2/2$$

Integrating once,

$$EIy' = -wlx^2/4 + wx^3/6 + C_1$$

Integrating again,

$$EIy = -wlx^3/12 + wx^4/24 + C_1x + C_2$$

The constants of integration can be determined from the deflection conditions which are $x = 0$, $y = 0$ and $x = l$, $y = 0$.

Now, $x = 0$ and $y = 0$ gives

$$C_2 = 0$$

and $x = l$ and $y = 0$ gives

$$-wl^4/12 + wl^4/24 + C_1 l = 0 \text{ from which } C_1 = wl^3/24.$$

The equations for slope and deflection are

$$EIy' = -wlx^2/4 + wx^3/6 + wl^3/24$$

$$EIy = -wlx^3/12 + wx^4/24 + (wl^3/24)x$$

Slope at A (when $x = 0$) = $EI\theta_A = wl^3/24$; $\theta_A = wl^3/(24EI)$. Being positive, the slope is clockwise.

Slope at B (when $x = l$) = $EI\theta_B = -wl^3/4 + wl^3/6 + wl^3/24$; $\theta_B = -wl^3/24EI$. Being negative, the slope is anticlockwise.

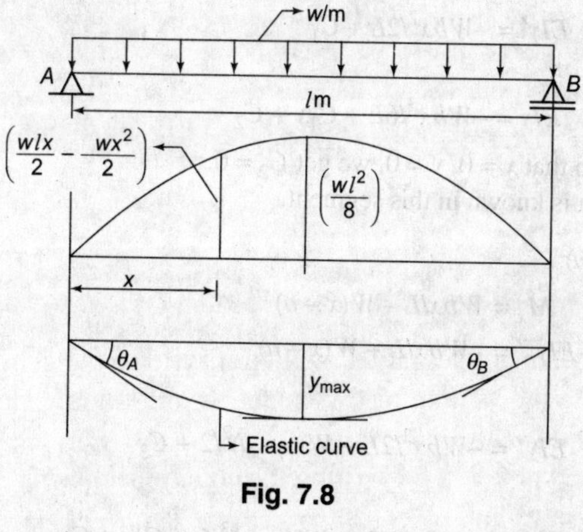

Fig. 7.8

Deflection at any section x is given by the elastic curve equation. By symmetry, the maximum deflection is at the centre when $x = l/2$. (This can also be obtained by setting slope equation to zero as y is maximum where $dy/dx = 0$.)

Maximum deflection, EIy_{max} (when $x = l/2$) $= -wl^4/96 + wl^4/384 + wl^4/48 = 5wl^4/384$

So, $y_{max} = 5wl^4/384EI$

The deflected shape is shown in Fig. 7.8.

As a more illustrative example of writing the BM equation, we take the following case.

Example 7.2 SS beam with point load

A simply supported beam AB carries a single concentrated load W at a distance a from the left support A. Find the slopes at the ends, deflections under the load, and maximum deflection (Fig. 7.9).

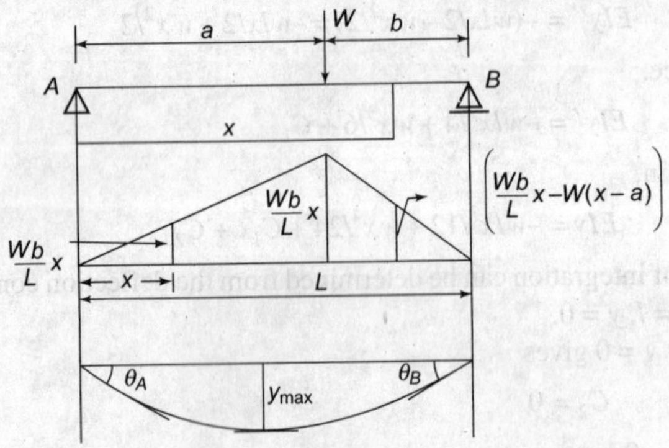

Fig. 7.9

Solution The reactions can be easily calculated as $R_A = Wb/L \uparrow$ and $R_B = Wa/L \uparrow$. You will notice that a single equation for bending moment valid for the whole beam cannot be written. Thus,

In segment AC ($x \le a$), $M_x = R_A x = Wbx/L$

In segment CB ($x \le a$), $M_x = R_A x - W(x - a) = Wbx/L - W(x - a)$

These two equations can be integrated separately.

Segment AC ($x \le a$):

We have $EIy'' = -M_x = -Wbx/L$

Integrating,

$$EIy' = -Wbx^2/2L + C_1$$

Integrating again,

$$EIy = -Wbx^3/6L + C_1x + C_2$$

From the condition that $x = 0$, $y = 0$, we get $C_2 = 0$.

No other condition is known in this segment.

Segment CB ($x \ge a$):

We have $M_x = Wbx/L - W(x - a)$

$$EIy'' = -Wbx/L + W(x - a)$$

Integrating,

$$EIy' = -Wbx^2/2L + W(x - a)^2/2 + C_3$$

Integrating again,

$$EIy = -Wbx^3/6L + W(x - a)^3/6 + C_3x + C_4$$

We have one condition in this segment also. The condition is when $x = L$, $y = 0$.

As there are four constants, we need four conditions to evaluate them. The four conditions can be found as follows:

(i) When $x = 0$, $y = 0$ in segment AC (It has been used to evaluate C_2.)

(ii) When $x = L$, $y = 0$ in segment CB.

(iii) When $x = a$, applicable to both the segments (The slope determined from both equations must be equal.)

(iv) When $x = a$, applicable to both the segments (The deflection y determined from both the equations must be equal.)

From the first condition, we have $C_2 = 0$ as determined earlier.

The second condition gives

$$-WbL^2/6 + W(L-a)^3/6 + C_3L + C_4 = 0$$

The third condition gives

$$-Wba^2/2L + C_1 = -Wba^2/2L + 0 + C_3$$

This gives $C_1 = C_3$.

The fourth condition gives

$$-Wba^3/6L + C_1a = -Wba^3/6L + 0 + C_3a + C_4$$

This equation shows that $C_4 = 0$ as $C_1 = C_3$.

From the second condition, noting that $C_1 = C_3$, we have

$$C_3L = WbL^2/6 - W(L-a)^2/6;\ C_3 = C_1 = Wb(L^2 - b^2)/6L$$

or $$C_3 = C_1 = Wab(L+b)/6L$$

We can now determine the slopes and deflections at any point in the two segments.

Segment AC ($x = a$):

We have

$$EIy' = -Wbx^2/2L + Wab(L+b)/6L$$

$$EIy = -Wbx^3/6L + Wabx(L+b)/6L$$

When $x = 0$, we get the slope at A.

$$EI\theta_A = Wab(L+b)/6L$$

or $$\theta_A = Wab(L+b)/6EIL$$

When $x = a$, we get the deflection under the load.

$$EIy_A = -Wba^3/6L + Wa^2b(L+b)/6L$$

$$= Wa^2b(-a + L + b)/6L$$

$$= Wa^2b^2/3L \text{ (knowing that } L = a + b)$$

Slope at B at $x = L$ in segment CB,

$$EI\theta_B = -\frac{WbL^2}{2L} + \frac{W(L-a)^2}{2} + \frac{Wab(L+b)L}{6L}$$

$$= Wb\,[-L + b + a(L+b)/3)]/2$$

$$= Wb[-a + a(L+b)/3]/2$$

$$= Wb(-3aL + aL + ab)/6L$$

or $$\theta_B = -Wab(2L - b)/6EIL$$

The location of the maximum deflection depends upon the values of a and b. If $a < b$, then the maximum deflection occurs in segment CB. If $a > b$, then the maximum deflection occurs in segment AC. We take the second case when $a > b$ and the maximum deflection occurring in segment AC.

$EIy' = -Wbx^2/2L + Wab(L + b)/6L = 0$ to get the value of x for the maximum deflection.

Now, $x^2 = a(L + b)/3$ as $a = L - b$, $x^2 = (L^2 - b^2)/3$; $x = \sqrt{[(L^2 - b^2)/3]}$

$$EIy_m = -Wb(L^2 - b^2)^{3/2}/18\sqrt{3}L + Wb[L^2 - b^2]^{3/2}/6\sqrt{3}L$$

$$= \frac{Wb(L^2 - b^2)^{3/2}(1 - 1/3)}{6\sqrt{3}L}$$

or $$y_m = \frac{Wb(L^2 - b^2)^{3/2}}{6\sqrt{3}EIL}$$

The above example shows the difficulty in applying the double integration method in its present form. If there are two concentrated loads, there are three segments, three sets of equations, and six constants of integration. Of course, we will have six conditions to evaluate the constants. However, if there are more concentrated loads or loads in combination, the method becomes very laborious.

This difficulty is overcome by a method known as Macaulay's method, which is the double integration method modified in the way it is applied.

7.6 MACAULAY'S METHOD

Macaulay's method allows one to write a BM equation that is valid for the whole beam. It involves using the terms selectively, depending upon the segment under consideration. In the case of the single concentrated load on the simply supported beam just considered, Macaulay proposed that the expression $M_x = (Pb/L)x - P(x - a)$ be used for the BM and be valid for the whole beam with the condition that the term $P(x - a)$ becomes zero if $0 \le x \le a$. In other words, the term $(x - a)$ becomes zero whenever this term is negative. It must be remembered that this applies to evaluating the constants and obtaining values from the equations for dy/dx and y.

In using this method, you must remember the following points.

(i) Take the origin at the leftmost end of the beam.

(ii) Express the BM equation for the rightmost segment of the beam. For example, in the beam loaded as shown in Fig. 7.10, state the BM equation for x lying in the segment FB.

(iii) The terms in brackets as in the term $P(x - a)$ in the case just discussed should be integrated keeping them in brackets throughout. It is only in numerical calculations, when the value of x has been specified, that the brackets are opened for computation.

(iv) Use the terms selectively depending upon their validity. If the bracketed term is positive, it is included, and if negative, it becomes zero, depending upon the value of x.

Even with Macaulay's method, certain cases of loading need special techniques for arriving at the BM equation valid for the whole beam. These cases are discussed below.

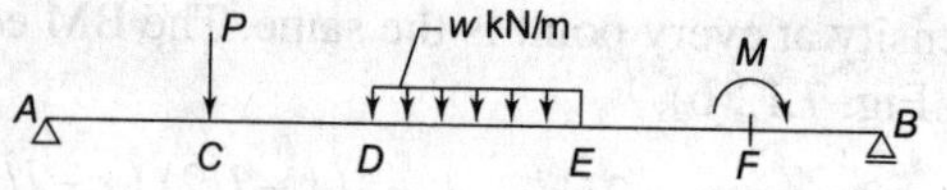

Fig. 7.10

7.6.1 Uniformly Distributed Load not Extending to the Right End

The case is shown in Fig. 7.11(a). The starting point of the load at the left does not pose any problem. In stating the BM equation, while the left end point of the load can be represented by the expression $(x - a)$ as in $w(x - a)^2/2$, the limit of the load on the right side cannot. To get over this difficulty, equal downward and upward loads are added as in Fig. 7.11(b). Note that the set of loads added is in equilibrium and does not affect the beam behaviour in any way. From Fig. 7.11(b), we get the BM as

$$M_x = R_A x - \frac{w(x-a)^2}{2} + \frac{w[x-(a+c)]^2}{2}$$

Note that the second term is valid from $x = a$ to $x = l$ while the third term is valid from $x = (a + c)$ to $x = l$.

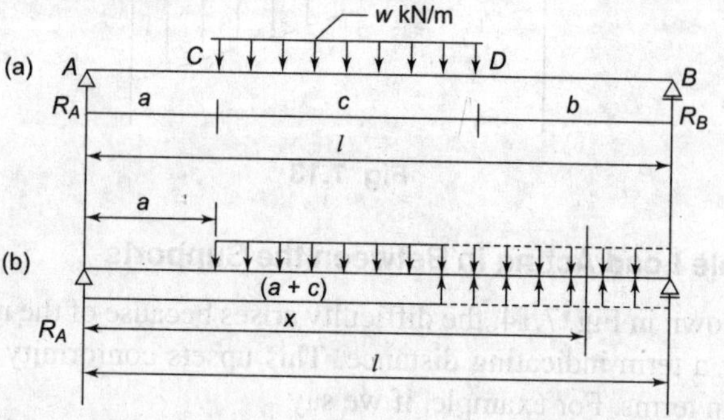

Fig. 7.11

7.6.2 Triangular Load

Such a load is shown in Fig. 7.12(a). In this case, while the load extends over the whole length of the beam, there are two different load lines, one on each half, which cannot be expressed mathematically by a single expression. Here again the procedure is to add and subtract equal loads as shown in Fig. 7.12(b). The load line on the left half is extended up to the full length of the beam. The load that is added is a triangular load, with intensity $2w$ on the right support and zero at the middle of the span. An equal load is added acting upwards. Note that the shapes of the two triangles are not simi-

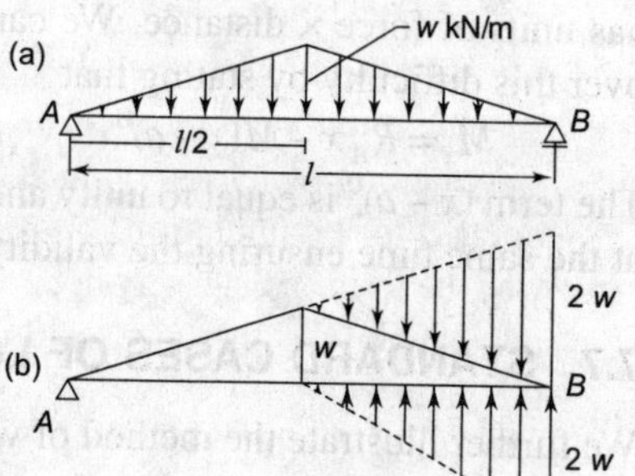

Fig. 7.12

lar but that the intensity at every point is the same. The BM equation can now be expressed as, from Fig. 7.12(b),

$$M_x = R_A x - \frac{2w}{l} x \frac{x}{2} \frac{x}{3} + \frac{2w}{l/2}(x - l/2) \frac{(x - l/2)}{2} \frac{(x - l/2)}{3}$$

$$= R_A x - \frac{2wx^3}{6l} + \frac{2w}{3l}(x - l/2)^3$$

In the same way, equal loads will have to be added and subtracted for the case shown in Fig. 7.13(a). This is done in Fig. 7.13(b).

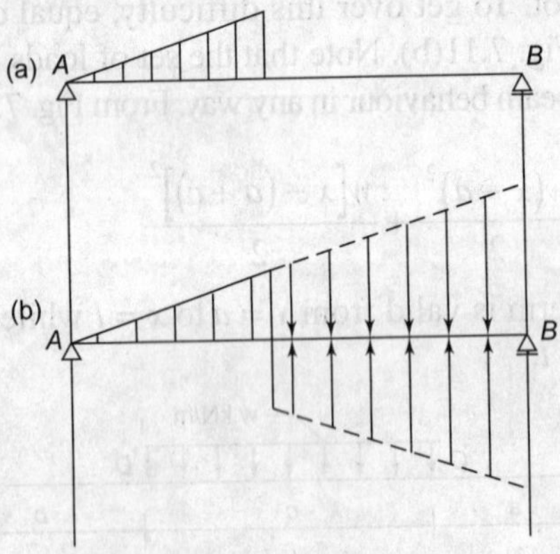

Fig. 7.13

7.6.3 Couple Load Acting in Between the Supports

In the case shown in Fig. 7.14, the difficulty arises because of the need to multiply a moment by a term indicating distance. This upsets conformity to dimensional equivalence in terms. For example, if we say

$$M_x = R_A x - M(x - a)$$

the validity of the term containing M is appropriately limited. However, in the expression for BM, all terms should have the units equivalent to force × distance. The expression containing M in this has units of force × (distance)^2 since M already has units of force × distance. We can get over this difficulty by stating that

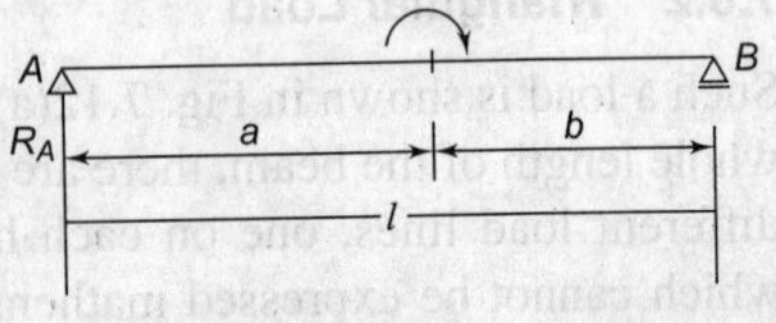

Fig. 7.14

$$M_x = R_A x - M(x - a)^0$$

The term $(x - a)^0$ is equal to unity and thus dimensional conformity is maintained, at the same time ensuring the validity of the term for $x \geq a$ only.

7.7 STANDARD CASES OF LOADING

We further illustrate the method of writing the BM equation and solving for slope and deflection with examples of standard cases of loading. These are summarized in Table 7.2 ahead.

7.7.1 Cantilevers

Example 7.3 Cantilever carrying UD load

A cantilever beam of span l carries a UD load of w kN/m. Derive the general equations for slope and deflection along the length of the beam and find the slope and deflection at the free end. EI is constant.

Solution The beam is shown in Fig. 7.15. The reactions at the fixed support are a vertically upward force of wl and an anticlockwise couple $wl^2/2$. The BM equation at a section at a distance x from A is

$$M_x = -\frac{wl^2}{2} + wlx - \frac{wx^2}{2}$$

$$EI\frac{d^2y}{dx^2} = -M_x = +\frac{wl^2}{2} - wlx + \frac{wx^2}{2}$$

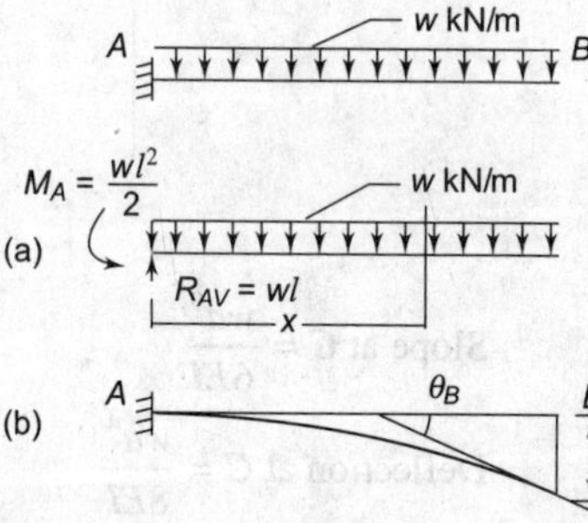

Fig. 7.15

Integrating,

$$EI\frac{dy}{dx} = \frac{wl^2}{2}x - \frac{wlx^2}{2} + \frac{wx^3}{6} + C_1$$

Integrating again,

$$EIy = \frac{wl^2}{2}\frac{x^2}{2} - \frac{wlx^3}{6} + \frac{wx^4}{24} + C_1x + C_2$$

The boundary conditions are at $x = 0$, $dy/dx = 0$ and at $x = 0$, $y = 0$. Substituting these values in the equations for dy/dx and y, $C_1 = 0$ and $C_2 = 0$.

The general expressions for slope and deflection are

$$EI\frac{dy}{dx} = \frac{wl^2x}{2} - \frac{wlx^2}{2} + \frac{wx^3}{6}$$

$$\Rightarrow \quad EIy = \frac{wl^2x^2}{4} - \frac{wlx^3}{6} + \frac{wx^4}{24}$$

To find the slope and deflection at the free end, we substitute x by l in these equations.

$$EI\frac{dy}{dx}\text{ (at } x = l) = \frac{wl^2l}{2} - \frac{wll^2}{2} + \frac{wl^3}{6}, \quad \frac{dy}{dx} = \frac{wl^3}{6EI} = \theta_B$$

This is positive, and so the tangent rotates clockwise at B.

$$EIy\text{ (at } x = l) = \frac{wl^2l}{4} - \frac{wll^3}{6} + \frac{wl^4}{24} = \frac{wl^4}{8}, \quad y = \frac{wl^4}{8EI}$$

This is positive. The deflection is, therefore, downwards at B. The deflected shape is shown in Fig. 7.15(b).

Example 7.4 Cantilever carrying UD load over part of span

If a cantilever carries a UD load of w kN/m for a length a, $a < l$, find the slope and deflection at the free end.

Solution The cantilever is shown in Fig. 7.16(a). Note that there is no BM in the beam from $x = a$ to $x = l$. The deflected shape is shown in Fig. 7.16(b). From $x = a$ to $x = l$, the beam is a straight line. The deflection, however, increases due to the slope at C.

The slope and deflection at C can be found from the expressions derived in Example 7.3 but only for a span a.

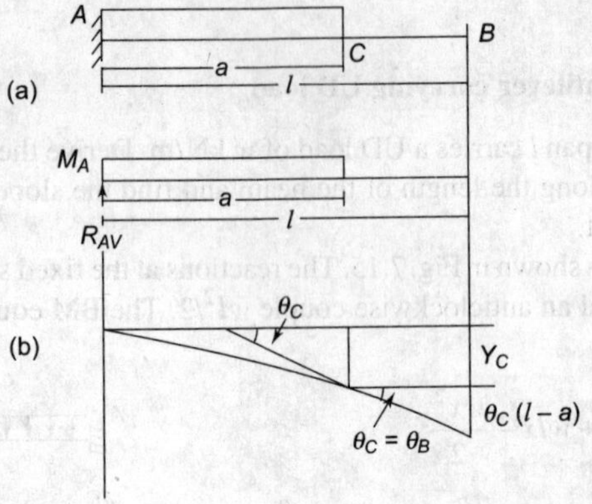

Fig. 7.16

$$\text{Slope at } C = \frac{wa^3}{6EI}$$

$$\text{Deflection at } C = \frac{wa^4}{8EI}$$

The slope at B is the same as at C as there is no BM in the length CB. The deflection at B is that at C plus the deflection due to the slope at C, as shown in Fig. 7.16(c).

$$\text{Deflection at } B = \frac{wa^4}{8EI} + \frac{wa^3}{6EI}(l - a) = \frac{wa^3}{2EI}\left(\frac{a}{4} + \frac{l - a}{3}\right)$$

$$= \frac{wa^3}{24EI}(4l - a)$$

Example 7.5 Cantilever with point load

A cantilever carries a concentrated load at the free end. Derive general expressions for slope and deflection, and find the deflection and slope at the free end.

Solution The beam is shown in Fig. 7.17. The reactions at A are an upward force of P and an anticlockwise couple of Pl. The general BM equation at x is

$$M_x = -Pl + Px$$

$$EI\frac{d^2y}{dx^2} = Pl - Px$$

Integrating,

$$EI\frac{dy}{dx} = Plx - \frac{Px^2}{2} + C_1$$

Integrating again,

$$EIy = \frac{Plx^2}{2} - \frac{Px^3}{6} + C_1x + C_2$$

The boundary conditions are at $x = 0$, $y = 0$ and at $x = 0$, $dy/dx = 0$. Substituting these in the above equations, $C_1 = 0$ and $C_2 = 0$.

The general expressions for slope and deflection are

$$EI\frac{dy}{dx} = Plx - \frac{Px^2}{2}$$

$$\Rightarrow \quad EIy = \frac{Plx^2}{2} - \frac{Px^3}{6}$$

Substituting x by l,

$$EI\frac{dy}{dx}\text{ (at } B) = Pl^2 - \frac{Pl^2}{2}$$

$$\text{Slope at } B = \frac{Pl^2}{2EI}$$

$$EIy\text{ (at } B) = \frac{Pll^2}{2} - \frac{Pl^3}{6}$$

$$\text{Deflection at } B = \frac{Pl^3}{3EI}$$

The deflected shape is shown in Fig. 7.17(b).

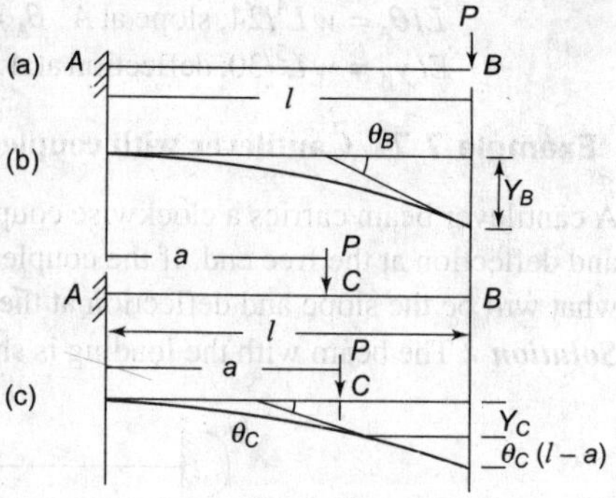

Fig. 7.17

Note If the load acts at a distance a from A as in Fig. 7.17(c), the same procedure as in Example 7.4 is adopted to find the slope and deflection at B. From Fig. 7.17(d),

$$\theta_C = \frac{Pa^2}{2EI},\quad y_C = \frac{Pa^3}{3EI}$$

$$\theta_B = \theta_C = \frac{Pa^2}{2EI},\quad y_B = y_C + \theta_C\,(l-a) = \frac{Pa^3}{3EI} + \frac{Pa^2}{2EI}\,(l-a)$$

$$y_B = \frac{Pa^2}{EI}\left(\frac{a}{3} + \frac{l-a}{2}\right) = \frac{Pa^2}{6EI}\,(3l-a)$$

Example 7.6 Cantilever with varying load

A cantilever beam carries a load varying from zero at the free end to w/m at the fixed end. Find the slope and deflection at the free end.

Solution The beam with the loading is shown in Fig. 7.18. The bending moment at section x from A can be written as

$$M_x = wx\,(x/2)\,(x/3) = wx^3/6$$

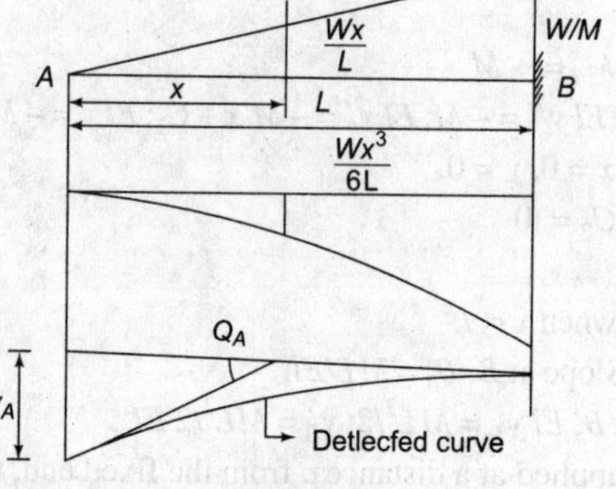

Fig. 7.18

Now, $EI\,y'' = -\,wx^3/6$

Integrating, $EI\,y' = -\,wx^4/24 + C_1$

$$EI\,y = -\,wx^5/120 + C_1x + C_2$$

$$x = L,\ y' = 0 \text{ gives, } C_1 = wL^4/24$$

$$x = L,\ y = 0 \text{ gives, } -\,wL^5/120 + wL^5/24 + C_2 = 0;\ C_2 = wL^5/30$$

Therefore,

$$EI\,y' = -\,wx^4/24 + wL^4/24;\ EI\,y = -\,wx^5/120 + wL^4x/24 + wL^5/30$$

Slope and deflection at A are obtained by putting $x = 0$ in these equations.

$EI\theta_A = wL^4/24$; slope at A, $\theta_A = wL^4/24EI$

$EI\,y_A = wL^5/30$; deflection at A, $y_A = wL^5/30EI$

Example 7.7 Cantilever with couple load

A cantilever beam carries a clockwise couple of moment M at the free end. Find the slope and deflection at the free end. If the couple were applied at a distance a from the fixed end what will be the slope and deflection at the free end.

Solution The beam with the loading is shown in Fig. 7.19.

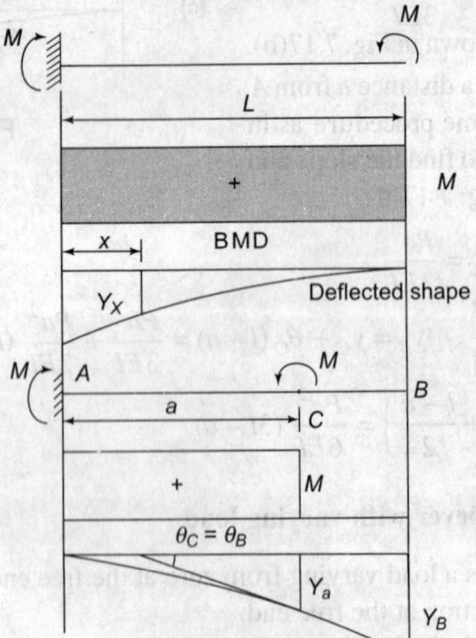

Fig. 7.19

The reaction at the fixed support is a couple M (anticlockwise). The bending moment at x is

$$M_x = + M$$

Therefore, we can write $EI\,y'' = -M$; $EI\,y' = -M\,x + C_1$; $EI\,y = -M\,x^2/2 + C_1x + C_2$

At $x = 0$, $y' = 0$ and at $x = 0$, $y = 0$.

Therefore, $C_1 = 0$ and $C_2 = 0$

$EI\,y' = Mx$

Slope at B is obtained when $x = L$

$$EI\,\theta_B = ML;\ \text{Slope at } B,\ \theta_B = ML/EI$$

Similarly, deflection at B, $EI\,y_B = ML^2/2$; $y_B = ML^2/2\,EI$

If the couple were applied at a distance a from the fixed end, then the BM diagram and deflected shape will be as shown. Note that there is no bending moment from a to L and hence the beam remains straight in this part.

Slope and deflection at the point of application of the couple can be obtained by applying the above formulae by putting a in place of L.

Slope at C, $\theta_C = Ma/\text{EI}$; deflection at C, $y_C = Ma^2/2\,EI$.

Slope at B will be the same as at C; $\theta_B = Ma/EI$.

Deflection at B, y_B = deflection at C + Slope at Cx length $(L - a)$.

$$y_B = Ma^2/2EI + Ma\,(L - a)/EI = Ma\,(2L - a)/2EI$$

7.7.2 Simply Supported Beams

Example 7.8 SS beam with eccentric point load

A simply supported beam of span l carries an eccentric concentrated load P as shown in Fig. 7.20. Derive the general expressions for slope and deflection.

Solution The reaction at the left end A is Pb/L. The general expression for BM, according to Macaulay's method, is

$$M_x = \frac{Pb}{l}x - P(x-a)$$

$$EI\frac{d^2y}{dx^2} = -M_x = -\frac{Pb}{l}x + P(x-a)$$

Integrating,

$$EI\frac{dy}{dx} = -\frac{Pb}{l}\frac{x^2}{2} + \frac{P(x-a)^2}{2} + C_1$$

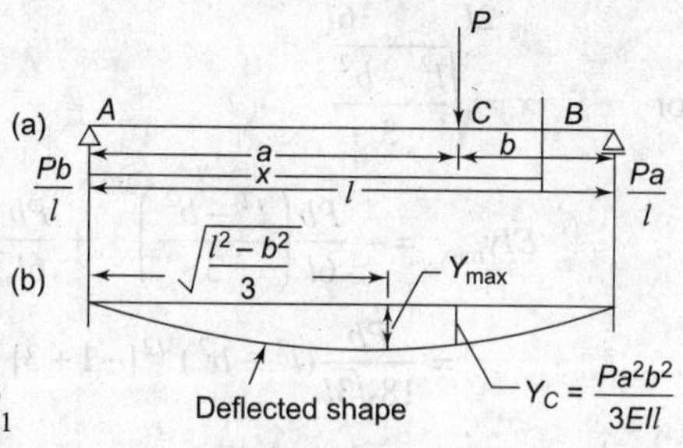

Fig. 7.20

Integrating again,

$$EIy = -\frac{Pb}{l}\frac{x^3}{6} + \frac{P(x-a)^3}{6} + C_1x + C_2$$

From the condition $x = 0$, $y = 0$, $C_2 = 0$. Note that $P(x-a)^3/6$ is zero as $(x-a)$ is negative for $x = 0$. From the condition $x = l$, $y = 0$, we get

$$0 = -\frac{Pb}{l}\frac{l^3}{6} + \frac{P(l-a)^3}{6} + C_1l$$

or $$C_1 = \frac{Pbl}{6} - \frac{P(l-a)^3}{6l} = \frac{Pb}{6l}(l^2-b^2)$$

The general expressions for slope and deflection are

$$EI\frac{dy}{dx} = -\frac{Pb}{2l}x^2 + \frac{P(x-a)^2}{2} + \frac{Pbl}{6} - \frac{P(l-a)^3}{6l} \qquad [\because b = l - a]$$

$$EIy = -\frac{Pb}{6l}x^3 + \frac{P(x-a)^3}{6} + \frac{Pbl}{6}x - \frac{P(l-a)^3}{6l}x$$

The slope at A $(x = 0)$,

$$EI\theta_A = \frac{Pbl}{6} - \frac{P(l-a)^3}{6l}$$

$$\theta_A = \frac{Pb}{6EIl}(l^2-b^2) = \frac{Pab}{6EIl}(l+b)$$

setting $l - b = a$. The slope at $B(x = l)$,

$$EI\theta_B = -\frac{Pb}{2l}l^2 + \frac{P(l-a)^2}{2} + \frac{Pb}{6l}(l^2-b^2)$$

$$\theta_B = -\frac{Pab}{6EIl}(l+a)$$

Let us obtain the deflection under the load, $x = a$.

$$EIy_C = \frac{-Pb}{6l}a^3 + \frac{Pb}{6l}(l^2-b^2)a = \frac{Pab}{6l}(-a^2+l^2-b^2)$$

or $$y_C = \frac{Pab}{6EIl} \times 2ab = \frac{Pa^2b^2}{3EIl}$$

To obtain the maximum deflection, we have to locate the value of x where $dy/dx = 0$. In the beam shown, it will be in the portion AC where $x < a$.

$$-\frac{Pb}{2l}x^2 + \frac{Pb}{6l}(l^2 - b^2) = 0$$

or $$x = \sqrt{\frac{l^2 - b^2}{3}}$$

$$EIy_{max} = -\frac{Pb}{6l}\left(\frac{l^2 - b^2}{3}\right)^{3/2} + \frac{Pb}{6l}(l^2 - b^2)\sqrt{\frac{l^2 - b^2}{3}}$$

$$= \frac{Pb}{18\sqrt{3}l}(l^2 - b^2)^{3/2}[-1+3] = \frac{Pb}{9\sqrt{3}l}(l^2 - b^2)^{3/2}$$

$$y_{max} = \frac{Pb(l^2 - b^2)^{3/2}}{9\sqrt{3}EIl}$$

If $a = b$, i.e., if the load is acting at the centre, then $a = b = l/2$ [Fig. 7.20(b)], then

$$\theta_A = \theta_B = \frac{Pl^3}{16EIl}$$

and $$y_{max} \text{ (at } x = l/2 \text{ under the load)} = \frac{Pl^3}{48EI}$$

Example 7.9 SS beam with couple load

A simply supported beam carries a couple at a distance a from the left support. Derive general expressions for slope and deflection.

Solution The beam is shown in Fig. 7.21(a). The reactions are $(M/l)\downarrow$ at A and $(M/l)\uparrow$ at B. The general equation for BM is

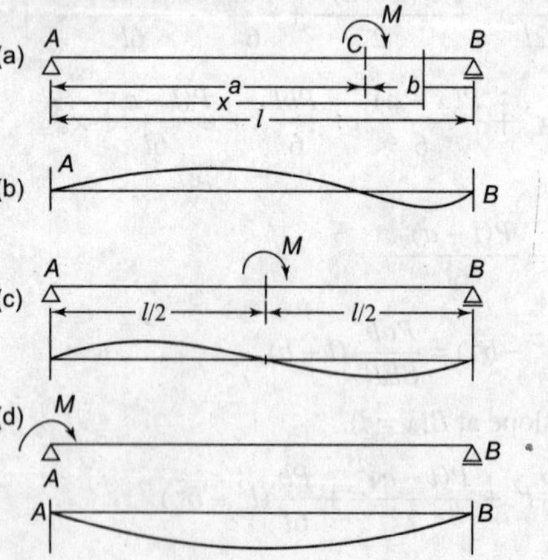

Fig. 7.21

$$M_x = -\frac{M}{l}x + M(x - a)^0$$

$$EI\frac{d^2y}{dx^2} = -M_x = \frac{M}{l}x - M(x - a)^0$$

$$EI\frac{dy}{dx} = \frac{M}{l}\frac{x^2}{2} - M(x-a) + C_1$$

$$EIy = \frac{M}{l}\frac{x^3}{6} - \frac{M(x-a)^2}{2} + C_1x + C_2$$

The condition $x = 0$, $y = 0$ yields $C_2 = 0$. The condition $x = l$, $y = 0$ gives

$$0 = \frac{M}{l}\frac{l^3}{6} - \frac{M(l-a)^2}{2} + C_1 l$$

$$\Rightarrow \qquad C_1 = \frac{M}{6l}(3b^2 - l^2)$$

The general expressions are

$$EI\frac{dy}{dx} = \frac{M}{2l}x^2 - \frac{M}{2}(x-a) + \frac{M}{6l}(3b^2 - l^2)$$

$$EIy = \frac{M}{6l}x^3 - \frac{M}{2}(x-a)^2 + \frac{M}{6l}(3b^2 - l^2)x$$

Let us obtain the slopes. At $x = 0$,

$$EI\theta_A = \frac{M}{6l}(3b^2 - l^2), \quad \theta_A = \frac{M}{6EIl}(3b^2 - l^2)$$

At $x = l$,

$$EI\theta_B = \frac{M}{2l}l^2 - M(l-a) + \frac{M}{6l}(3b^2 - l^2)$$

$$= \frac{M}{6l}(3l^2 - 6bl + 3b^2 - l^2)$$

$$\theta_B = \frac{M}{6lEI}(3b^2 - 2l^2 - 6lb) = \frac{M}{6EIl}(3a^2 - l^2)$$

The deflection at C, the point of application of the couple, may be worked out as follows.

$$EIy_C = \frac{M}{6l}a^3 + \frac{M}{6l}(3b^2 - l^2)a$$

or $\qquad y_C = \dfrac{Ma}{6EIl}(a^2 + 3b^2 - l^2)$

The deflected shape is shown in Fig. 7.21(b). If the couple were applied at the centre, $a = b = l/2$,

$$\theta_A = \frac{M}{6EIl}\left(\frac{3l^2}{4} - l^2\right) = \frac{-Ml}{24EI}$$

$$\theta_B = \frac{M}{6EIl}\left(\frac{3l^2}{4} - 2l^2 - \frac{6l^2}{2}\right) = \frac{-Ml}{24EI}$$

$$y_C = \frac{M}{6EIl}\left(\frac{l^2}{4} + \frac{3l^2}{4} - l^2\right) = 0$$

The loading and the deflected shape would then be as shown in Fig. 7.21(c).

If the simply supported beam were to be subjected to a couple at A, i.e, $a = 0$, then setting $a = 0$ and $b = l$ in these equations [Fig. 7.21(d)],

$$EI\frac{dy}{dx} = \frac{M}{2l}x^2 - Mx + \frac{Ml}{3}$$

$$EIy = \frac{M}{6l}x^3 - \frac{Mx^2}{2} + \frac{Ml}{3}x$$

Setting $x = 0$,

$$\theta_A = \frac{Ml}{3EI}$$

Setting $x = l$,

$$EI\theta_B = \frac{M}{2l}l^2 - Ml + \frac{Ml}{3} = -\frac{Ml}{6}$$

or $$\theta_B = -\frac{Ml}{6EI}$$

For maximum deflection, $dy/dx = 0$. Therefore,

$$\frac{M}{2l}x^2 - Mx + \frac{Ml}{3} = 0$$

$$x^2 - 2lx + \frac{2l^2}{3} = 0$$

from which

$$x = \left(\frac{\sqrt{3}-1}{\sqrt{3}}\right)l$$

Substituting this value in the equation for y,

$$EIy_{max} = \frac{M}{6l}\left(\frac{\sqrt{3}-1}{\sqrt{3}}\right)^3 l^3 - M\left(\frac{\sqrt{3}-1}{\sqrt{3}}\right)^2 l^2 + \frac{Ml}{3}\left(\frac{\sqrt{3}-1}{\sqrt{3}}\right)l$$

or $$y_{max} = \frac{Ml^2}{9\sqrt{3}EI}$$

Example 7.10 SS beam with varying load

A simply supported beam carries a varying load varying from w_1/m at the left end to w_2/m at the right end. Find the slopes at the supports. If $L = 9$ m, $w_1 = 20$ kN/m, and $w_2 = 40$ kN/m, find the slopes at the ends and position and magnitude of maximum deflection. $E = 200$ GN/m^2 and $I = 12{,}000$ cm^4.

Solution The beam with the loading is shown in Fig. 7.22.

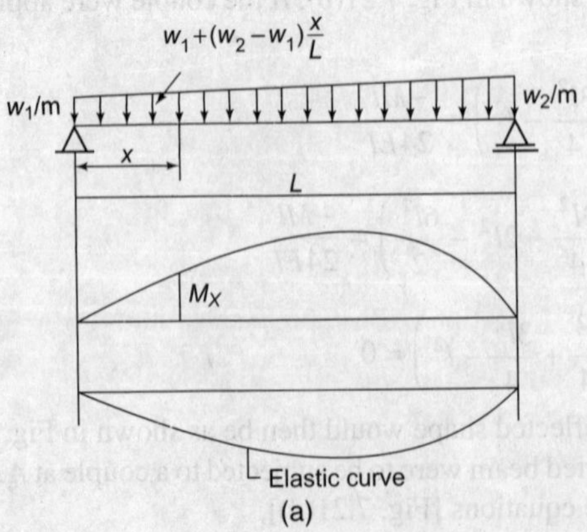

(a)

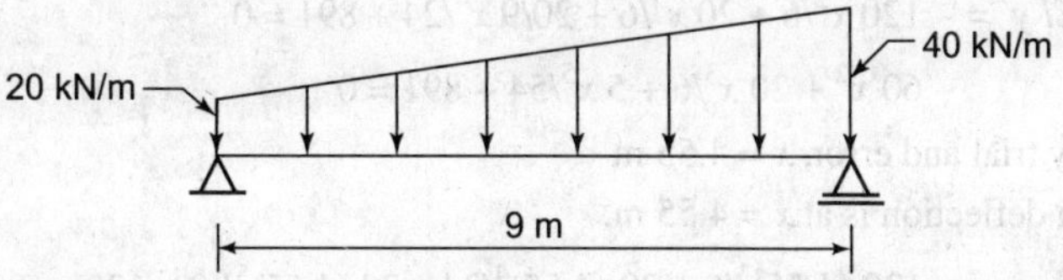

Fig. 7.22

At a distance x from A, load intensity is $w_x = w_1 + (w_2 - w_1)\, x/L = w_1 + kx$, where $k = (w_2 - w_1)/L$. The reactions can be calculated as $R_A = w_1L/2 + (w_2 - w_1)L/2 \times L/3 \times 1/L = w_1L/2 + kL^2/6$.

$$M_x = R_A\, x - w_1x^2/2 - kx^3/6$$

$$EI\, y'' = -R_Ax + w_1\, x^2/2 + kx^3/6$$

Integrating, $EI\, y' = -R_A\, x^2/2 + w_1\, x^3/6 + kx^4/24 + C_1$

$$EI\, y = -R_A\, x^3/6 + w_1\, x^4/24 + kx^5/120 + C_1x + C_2$$

As $y = 0$ when $x = 0$, $C_2 = 0$

Also $y = 0$ when $x = L$ gives $0 = -R_AL^3/6 + w_1L^4/24 + kL^5/120 + C_1L$

$$C_1 = [w_1L/2 + (w_2 - w_1)\, L/6]L^3/6 - w_1\, L^4/24 - [(w_2 - w_1)/L]\, L^5/120$$

$$= w_1L^3/45 + 7\, w_2L^3/360$$

$$EI\, y' = -R_A\, x^2/2 + w_1\, x^3/6 + kx^4/24 + L^3(w_1/45 + 7w_2/360)$$

$$EI\, y = -R_A\, x^3/6 + w_1\, x^4/24 + kx^5/120 + L^3\, x\, (w_1/45 + 7w_2/360)$$

When $x = 0$, $EI\, \theta_A = C_1 = L^3\, (8w_1 + 7w_2)/360EI$; $\theta_A = L^3(8w_1 + 7w_2)\, /360\, EI$

When $x = L$, $EI\, \theta_B = -R_AL^2/2 + w_1L^3/6 + (w_2 - w_1)L^3/24 + L^3(8w_1 + 7w_2)/360$

$$\theta_B = L^3(7w_1 + 8w_2)/360EI$$

If $w_1 = 0$, which means that the beam carries a triangular load as in Fig. 7.22(b), then

$$\theta_A = 7wL^3/360\, EI \text{ and } \theta_B = wL^3/45\, EI$$

If $w_1 = w_2$, the beam carries a UD load and $\theta_A = \theta_B = wL^3/24EI$

If w_1 =20 kN/m and w_2 = 40 kN/m and $k = (40 - 20)/9 = 20/9$

$$R_A = 20 \times 9/2 + (40 - 20)\,(9/2)\,(9/3)/9 = 90 + 30 = 120 \text{ kN}$$

Slope at A, $\theta_A = C_1/EI = [20 \times 10^3\, (9^3/45) + 7\, (40 \times 10^3)\, (9^3)/360]/EI$

$$\theta_A = 891/EI = 891 \times 10^3/(200 \times 10^9)\, (12{,}000 \times 100^{-4})$$

$$= 37 \times 10^{-3} \text{ radians}$$

$$\theta_B = (9^3)\, (7 \times 20 \times 10^3) + 8 \times 40 \times 10^3)/360$$

$$EI = 931.5/EI = 931.5 \times 10^3/\, [(200 \times 10^9)\, (12{,}000 \times 100^{-4}]$$

$$= 39 \times 10^{-3} \text{ radians}$$

The point of maximum deflection can be obtained by putting $y' = 0$

$$R_A = 120 \text{ kN};\ w_1 = 20 \text{ kN/m};\ w_2 = 40 \text{ kN/m};$$

$$C_1 = 9^3[20/45 + 7 \times 40/360] = 891$$

$$EI\,y' = -120\,x^2/6 + 20\,x^3/6 + 20/9\,x^4/24 + 891 = 0$$

or
$$-60\,x^2 + 20\,x^3/6 + 5\,x^5/54 + 891 = 0$$

Solving by trial and error, $x = 4.55$ m

Maximum deflection is at $x = 4.55$ m.

$$EI\,y_{max} = -120\,(4.55)^3/6 + 20\,(4.55)^4/24 + 20\,(4.55)^5/(9 \times 120) + 891 \times 4.55$$

Maximum deflection, $y_{max} = 2563/EI$

Example 7.11 SS beam carrying triangular load

A simply supported beam is subjected to a load varying uniformly from w at the centre to zero at the supports. Derive general expressions for slope and deflection, and find the slope at the supports and the maximum deflection. EI is constant.

Solution The beam with the loading is shown in Fig. 7.23(a). As discussed earlier, the general equation for BM can be formulated by adding and subtracting loads as shown in Fig. 7.23(b). The reaction at A is $wl/4$. The general equation for BM is

$$M_x = \frac{wl}{4}x - \frac{2w}{l}x\frac{x}{2}\frac{x}{3} + \frac{2w}{l/2}(x - l/2)\frac{(x-l/2)}{2}\frac{(x-l/2)}{3}$$

$$= \frac{wlx}{4} - \frac{wx^3}{3l} + \frac{2w}{3l}(x - l/2)^3$$

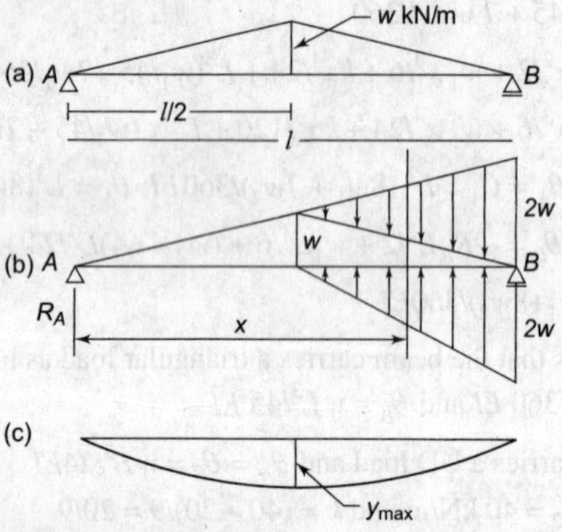

Fig. 7.23

$$EI\frac{d^2y}{dx^2} = -M_x = -\frac{wlx}{4} + \frac{wx^3}{3l} - \frac{2w}{3l}(x - l/2)^3$$

$$EI\frac{dy}{dx} = -\frac{wlx^2}{8} + \frac{wx^4}{12l} - \frac{w}{6l}(x - l/2)^4 + C_1$$

$$EIy = -\frac{wlx^3}{24} + \frac{wx^5}{60l} - \frac{w}{30l}(x - l/2)^5 + C_1 x + C_2$$

The condition $x = 0$, $y = 0$ gives $C_2 = 0$. The condition $x = l$, $y = 0$ yields

$$0 = -\frac{wl^4}{24} + \frac{wl^5}{60l} - \frac{wl^5}{960l} + C_1 l$$

$$\Rightarrow \quad C_1 = \frac{5}{192}wl^3$$

The general expressions for slope and deflection are

$$EI\frac{dy}{dx} = \frac{-wlx^2}{8} + \frac{wx^4}{12l} - \frac{w}{6l}(x - l/2)^4 + \frac{5}{192}wl^3$$

$$EIy = \frac{-wlx^3}{24} + \frac{wx^5}{60l} - \frac{w}{30l}(x - l/2)^5 + \frac{5}{192}wl^3x$$

For $x = 0$,

$$EI\theta_A = \frac{5}{192}wl^3$$

$$\theta_A = \frac{5wl^3}{192EI}$$

$$\theta_B = -\frac{5wl^3}{192EI}$$

from consideration of symmetry.

The maximum deflection, from considerations of symmetry, is at the centre, at $x = l/2$.

$$EIy_{\max} = \frac{-wll^3}{24 \times 8} + \frac{wl^5}{60l \times 32} - 0 + \frac{5wl^4}{192 \times 2} = \frac{wl^4}{120}$$

$$y_{\max} = \frac{wl^4}{120EI}$$

The deflected shape is shown in Fig. 7.23(c). ■

7.7.3 Overhanging Beams

In the case of overhanging beams, standard formulae are difficult to arrive at. The following examples illustrate the method of solution.

Example 7.12 Overhanging beam with UD load

An overhanging beam of span L between supports and overhang length of a carries a UD load w/m over the whole length. Find the slope at the supports, deflection at the free end, and maximum deflection. Take L = 8 m, a = 2 m, amd w = 40 kN/m.

Solution The beam is shown in Fig. 7.24.

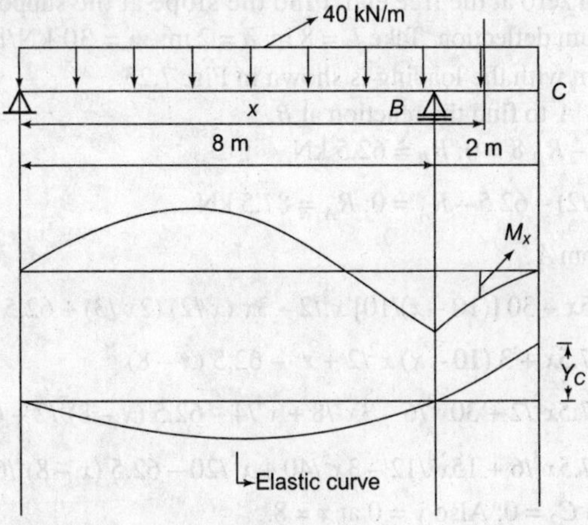

Fig. 7.24

The reactions are calculated first. Taking moments @ A,

$$40 \times 10 \times 5 - R_b \times 8 = 0;\ R_b = 250\text{ kN};\ R_a = 40 \times 10 - 250 = 150\text{ kN}$$

Moment at any distance x from A,

$$M_x = -150x + 40x^2/2 - 250(x-8)$$

Therefore, $$EIy'' = -M_x = +150x - 40x^2/2 + 250(x-8)$$

$$EIy' = 150x^2/2 - 40x^3/6 - 250(x-8)^2/2 + C_1$$

$$EIy = 150x^3/6 - 40x^4/24 - 250(x-8)^3/6 + C_1x + C_2$$

At $x = 0$, $y = 0$ gives $C_2 = 0$. At $x = 8$, $y = 0$ gives

$$150 \times 8^3/6 - 40 \times 8^4/24 - 250 \times 0 + C_1 8;\ C_1 = 747$$

The general equations for slope and deflections can be written as

$$EIy' = 75x^2 - 6.667x^3 - 125(x-8)^2 + 747$$

$$EIy = 25x^3 - 1.667x^4 - 41.667(x-8)^3 + 747x$$

When $x = 0$, we get slope at A, $EI\,\theta_A = 747$; $\theta_A = 747/EI$

When $x = 8$, we get slope at b, $EI\,\theta_B = 75(8)^2 - 6.667(8)^3 + 747$

or $\theta_B = -640/EI$

When $x = 10$, we get slope at the free end θ_C.

$$EI\theta_C = 75(10)^2 - 6.667(10)^3 - 125(10-8)^2 + 747 = -587/EI$$

Deflection at the free end is obtained when $x = 10$.

$$EIy_C = 25(10)^3 - 1.667(10)^4 - 41.667(10-8)^3 + 747 \times 10$$

or $y_C = -1200/EI\ (\uparrow)$

To find maximum deflection between supports, we set the equation for slope to zero.

$$75x^2 - 6.667x^3 = 747 = 0. \text{ (The third term is zero as } x < 8.)$$

Solving by trial and error, we get $x = 3.9$ m. [To solve by trial and error, start with the mid-span value as the maximum is likely to be closer to 4 m.)

Substituting this value,

$$y_{max} = 25(3.9)^3 - 1.667(3.9)^4 + 747(3.9) = 1815/EI$$

Example 7.13 Overhanging beam with triangular load

An overhanging beam of span L and overhanging length a carries *a* load varying from w/m at the left support to zero at the free end. Find the slope at the supports, deflection at the free end and maximum deflection. Take $L = 8$ m; $a = 2$ m; $w = 30$ kN/m.

Solution The beam with the loading is shown in Fig. 7.25.

Take moments @ A to find the reaction at B,

$30(10/2)(10/3) - R_B\,8 = 0;\ R_B = 62.5$ kN

From $\Sigma V = 0$, $30(10/2) - 62.5 - R_A = 0$; $R_A = 87.5$ kN

At any distance x from A,

$$M_x = 87.5x - 30[(10-x)/10]\,x^2/2 - 3x(x/2)(2x/3) + 62.5(x-8)$$

$$EIy'' = -87.5x + 3(10-x)x^2/2 + x^3 - 62.5(x-8)$$

$$EIy' = -87.5x^2/2 + 30x^3/6 - 3x^4/8 + x^4/4 - 62.5(x-8)^2/2 + C_1$$

$$EIy = -87.5x^3/6 + 15x^4/12 - 3x^5/40 + x^5/20 - 62.5(x-8)^3/6 + C_1x + C_2$$

At $x = 0$, $y = 0$ gives $C_2 = 0$; Also $y = 0$ at $x = 8$.

This gives, $0 = -87.5 \times 8^3/6 + 15 \times 8^4/12 - 3 \times 8^5/40 + 8^5/20 - 0 + C_1 8;\ C_1 = 396$

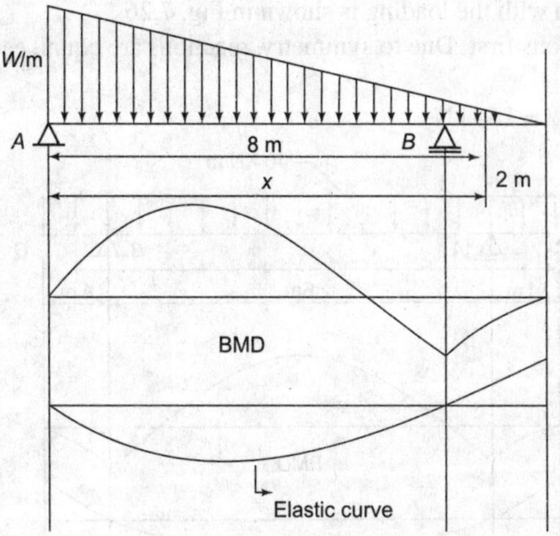

Fig. 7.25

General equations for slope and deflection can be written as

$$EIy' = -87.5x^2/2 + 30x^3/6 - 3x^4/8 + x^4/4 - 62.5\,(x-8)^2/2 + 396$$

$$EIy = -87.5x^3/6 + 15\,x^4/12 - 3x^5/40 + x^5/20 - 62.5\,(x-8)^3/6 + 396x$$

Slopes can be calculated from the first equation.

When $x = 0$, we get the slope at A; $EI\,\theta_A = 396$; $\theta_A = 396/EI$ (clockwise)

When $x = 8$, we get the slope at B

$$EI\theta_B = -87.5\,(8)^2/2 + 30\,(8)^3/6 - 3\,(8)^4/8 + 396;$$

$$\theta_B = -356/EI \text{ (anticlockwise)}$$

When $x = 10$, we get the slope at the free end C.

$$EI\,\theta_C = -87.5\,(10)^2/2 + 30\,(10)^3/6 - 3(10)^4/8 + (10)^4/4 - 62.5\,(10-8)^2/2 + 396$$

$$\theta_C = -354/EI$$

Deflections can be calculated from the second equation. Deflection at C is obtained when $x = 10$.

$$EIy_C = -87.5\,(10)^3/6 + 15\,(10)^4/12 - 3(10)^5/40 + (10)^5/20 - 62.5\,(2)^3/6 + 396\,(10)$$

$$y_C = -707/EI$$

Maximum positive deflection is between supports. To locate the point of maximum deflection, we equate the slope equation to zero.

$$0 = -87.5\,x^2/2 + 30\,x^3/6 - 3\,x^4/8 + x^4/4 - 62.5\,(x-8)^2/2 + 396$$

As $x < 8$, the fifth term is zero. This can be solved by trial and error.

We get $x = 3.85$ m.

$$EI\,y_{max} = -87.5\,(3.85)^3/6 + 15\,(3.85)^3/12 - 3\,(3.85)^5/40 + (3.85)^5/20 + 396\,(3.85)$$

Maximum deflection, $y_{max} = 946/EI$

Example 7.14 Doubly-overhanging beam with UD load

A doubly-overhanging beam of span L between supports and overhanging by length a at both ends carries UD load over the whole span. Find the slope at the supports, deflection at the free ends and maximum deflection. Take $L = 6$ m, $a = 1$ m, and $w = 40$ kN/m.

Solution The beam with the loading is shown in Fig. 7.26.

We find the reactions first. Due to symmetry, reactions are equal, each equal to half the load on the beam.

$R_A = R_B = 40 \times 8/2 = 160$ kN

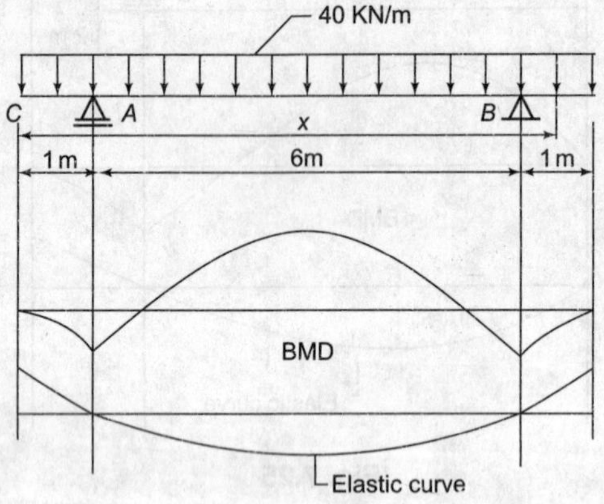

Fig. 7.26

Bending moment equation at any point x can be written as

$$M_x = -40\,x^2/2 + 160\,(x-1) + 160\,(x-7)$$

Therefore we can write

$$EIy'' = 40\,x^2/2 - 160\,(x-1) - 160\,(x-7)$$

Integrating,

$$EIy' = 40\,x^3/6 - 160\,(x-1)^2/2 - 160\,(x-7)^2/2 + C_1$$

Integrating again,

$$EIy = 40\,x^4/24 - 160\,(x-1)^3/6 - 160\,(x-7)^3/6 + C_1\,x + C_2$$

The constants C_1 and C_2 can be evaluated from the two conditions:

(i) $x = 1$ m, $y = 0$; (ii) $x = 7$ m, $y = 0$.

So, $0 = 40/24 - 0 - 0 + C_1 + C_2$.

and $0 = 40\,(7)^4/24 - 160\,(7-1)^3/6 - 160\,(7-7)^3/6 + C_1\,7 + C_2$.

$C_1 + C_2 = 1.6667$; $7C_1 + C_2 = 1758.33$

Solving, $C_1 = 293.33$; $C_2 = -295$

General equation for slope is $EI\,y' = 40\,x^3/6 - 160\,(x-1)^2/2 - 160\,(x-7)^2/2 + 293.33$

This equation yields, slope at C ($x = 0$), $\theta_C = -295/EI$

Slope at A ($x = 1$) = 0; slope at B ($x = 7$) = 0; slope at D ($x = 8$) = $-295/EI$

The general equation for deflection can be written as

$$EIy = 40\,x^4/24 - 160\,(x-1)^3/6 - 160\,(x-7)^3/6 + 293.33x - 295$$

Deflection at $C(x = 0)$, $y_C = -295/EI$ = Deflection at D ($x = 8$), y_D due to symmetry.

Due to symmetry, maximum deflection will be at the centre of the span at $x = 4$.

$$EIy_{max} = 40\,(4)^4/24 - 160\,(4-1)^3/6 - 0 + 293.33 \times 4 - 295;\ y_{max} = 1065/EI$$

7.7.4 More Examples

Example 7.15 SS beam with two point loads

A simply supported beam, 8 m long, carries two concentrated loads of 80 kN and 60 kN at distances of 3 m and 6 m from the left end, respectively. Calculate the slopes at the ends and the deflections under the loads. $E = 200$ GN/m^2 and $I = 30{,}000$ cm^4.

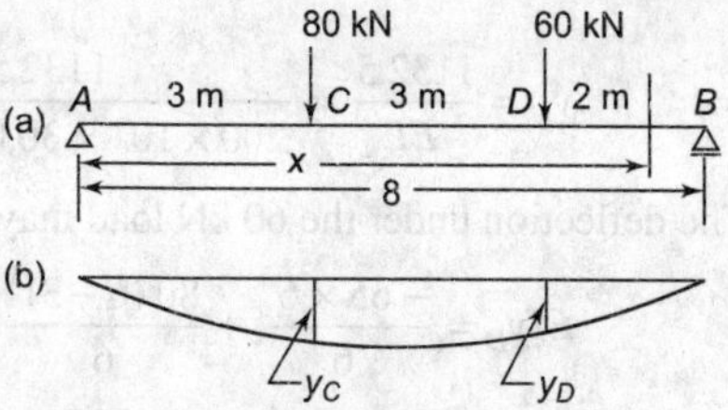

Fig. 7.27

Solution The beam with the loading is shown in Fig. 7.27(a). The reactions are

$$R_{AV} = \frac{80 \times 5 + 60 \times 2}{8} = 65 \text{ kN}$$

$$R_{BV} = 140 - 65 = 75 \text{ kN} \uparrow$$

The general BM equation is

$$M_x = 65x - 80(x - 3) - 60(x - 6)$$

$$EI\frac{d^2y}{dx^2} = -65x + 80(x-3) + 60(x-6)$$

$$EI\frac{dy}{dx} = -\frac{65x^2}{2} + \frac{80(x-3)^2}{2} + \frac{60(x-6)^2}{2} + C_1$$

$$EIy = -\frac{65x^3}{6} + \frac{80(x-3)^3}{6} + \frac{60(x-6)^3}{6} + C_1x + C_2$$

$x = 0, y = 0$. Therefore, $C_2 = 0$.

$$x = 8, y = 0$$

$$0 = -\frac{65 \times 8^3}{6} + \frac{80(x-3)^3}{6} + \frac{60(x-6)^3}{6} + 8\,C_1$$

$$\Rightarrow \quad C_1 = 475$$

The general slope and deflection equations are

$$EI\frac{dy}{dx} = -65\frac{x^2}{2} + \frac{80(x-3)^2}{2} + \frac{60(x-6)^3}{2} + 475$$

$$EIy = -65\frac{x^3}{6} + \frac{80(x-3)^2}{6} + \frac{60(x-6)^3}{6} + 475x$$

At $x = 0$,

$$\frac{dy}{dx} = \theta_A$$

$$\theta_A = \frac{475}{EI}$$

$$E = 200 \times 10^6 \text{ kN/m}^2$$

$$I = 30{,}000 \times 10^{-8} \text{ m}^4$$

Therefore,

$$\theta_A = \frac{475}{200 \times 10^6 \times 30{,}000 \times 10^{-8}} = 0.0079 \text{ rad}$$

At $x = 8$,

$$\frac{dy}{dx} = \theta_B$$

$$EI\theta_B = \frac{-65 \times 8^2}{2} + \frac{80(8-3)^2}{2} + \frac{60(8-6)^2}{2} + 475 = -485$$

$$\theta_B = \frac{-485}{200 \times 10^6 \times 30{,}000 \times 10^{-8}} = 0.0081 \text{ rad}$$

The deflection under the 80 kN load may be worked out as follows:

$$EIy_C = \frac{-65 \times 3^3}{6} + 475 \times 3 = 1132.5$$

$$\Rightarrow \quad y_C = \frac{1132.5}{EI} = \frac{1132.5}{200 \times 10^6 \times 30{,}000 \times 10^{-8}} = 18.8 \text{ mm}$$

The deflection under the 60 kN load may be calculated in the following manner:

$$EIy_D = \frac{-65 \times 6^3}{6} + \frac{80\,(6-3)^3}{6} + 475 \times 6$$

$$\Rightarrow \quad y_D = \frac{870}{EI} = \frac{870}{200 \times 10^6 \times 30{,}000 \times 10^{-8}} = 14.9 \text{ mm}$$

The deflected shape is shown in Fig. 7.27(b).

Example 7.16 SS beam with UD load and point load

A simply supported beam of span 10 m carries a UD load of 40 kN/m over a length of 5 m of its left half and a concentrated load of 80 kN at 7.5 m from the left end. Determine the slopes at the ends, and the deflections under the concentrated load and at mid-span. $E = 200 \times 10^6$ kN/m^2 and $I = 56{,}000$ cm^4.

Solution We first calculate the reactions [Fig. 7.28(a)].

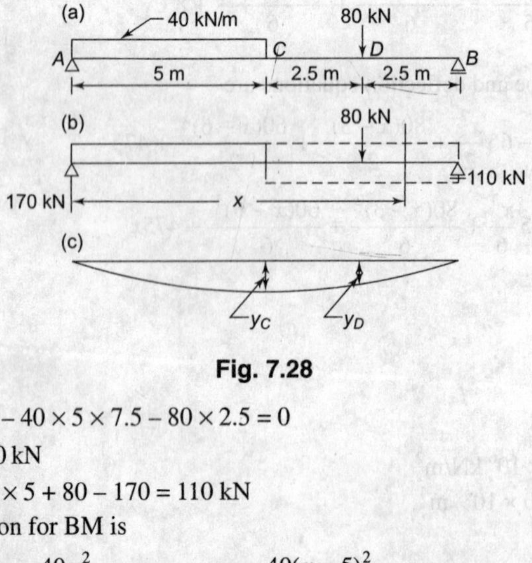

Fig. 7.28

$$R_{AV} \times 10 - 40 \times 5 \times 7.5 - 80 \times 2.5 = 0$$

$$\Rightarrow \quad R_{AV} = 170 \text{ kN}$$

$$R_{BV} = 40 \times 5 + 80 - 170 = 110 \text{ kN}$$

The general equation for BM is

$$M_x = 170x - \frac{40x^2}{2} - 80(x-7.5) + \frac{40(x-5)^2}{2}$$

from Fig. 7.28(b). Note that equal upward and downward loads are added on the right half of the span.

$$EI\frac{d^2y}{dx^2} = -M_x = -170x + \frac{40x^2}{2} + 80(x-7.5) - \frac{40(x-5)^2}{2}$$

$$EI\frac{dy}{dx} = -\frac{170x^2}{2} + \frac{40x^3}{6} + \frac{80(x-7.5)^2}{2} - \frac{40(x-5)^3}{6} + C_1$$

$$EIy = -\frac{170x^3}{6} + \frac{40x^4}{24} + \frac{80(x-7.5)^3}{6} - \frac{40(x-5)^4}{24} + C_1x + C_2$$

At $x = 0$, $y = 0$. This gives $C_2 = 0$. At $x = 10$, $y = 0$. This yields

$$0 = -\frac{170 \times 10^3}{6} + \frac{40 \times 10^4}{24} + \frac{80 \times 2.5^3}{6} - \frac{40 \times 5^4}{24} + C_1 \times 10$$

$$C_1 \times 10 = 12{,}500.34$$

$$\Rightarrow \quad C_1 = 1250$$

The general equations for slope and deflection are

$$EI\,\frac{dy}{dx} = -\,170\,\frac{x^2}{2} + \frac{40x^3}{6} + \frac{80(x-7.5)^2}{2} - \frac{40(x-5)^3}{6} + 1250$$

$$EIy = -\,170\,\frac{x^3}{6} + \frac{40x^4}{24} + \frac{80(x-7.5)^3}{6} - \frac{40(x-5)^4}{24} + 1250x$$

At $x = 0$,

$$EI\theta_A = 1250$$

$$\Rightarrow \quad \theta_A = \frac{1250}{EI} = \frac{1250}{200 \times 10^6 \times 56{,}000 \times 10^{-8}}$$

$$= 0.0111 \text{ rad}$$

At $x = 10$,

$$EI\theta_B = -\,\frac{170 \times 10^2}{2} + \frac{20 \times 10^3}{3} - \frac{80\,(10-7.5)^2}{2} + \frac{20(10-5)^3}{3} + 1250$$

$$\theta_B = \frac{-1166.67}{200 \times 10^6 \times 56{,}000 \times 10^{-8}} = -\,0.0104 \text{ rad}$$

At $x = 5$,

$$EIy_C = \frac{-170 \times 125}{6} + \frac{20 \times 5^4}{12} + 1250 = 3750$$

Deflection at mid-span,

$$y_C = \frac{3750}{200 \times 10^6 \times 56{,}000 \times 10^{-8}} = 33.5 \text{ mm}$$

Deflection under the 80 kN load,

$$y_D = \frac{[-170 \times 7.5^3/6 + 20 \times 7.5^4/12 + 1250 \times 7.5 - (20/12) \times 2.5^4]}{EI}$$

$$= \frac{2630}{200 \times 10^6 \times 56{,}000 \times 10^{-8}} = 23.5 \text{ mm}$$

Example 7.17 SS beam with uniformly varying load

A simply supported beam carries a load varying uniformly from zero at the left end to 90 kN/m at the right end, over a span of 9 m. Derive general expressions for slope and deflection, and find the slopes at the ends and the maximum deflection in the beam in terms of EI, which is constant.

Solution The beam is shown in Fig. 7.29(a).

The reactions are

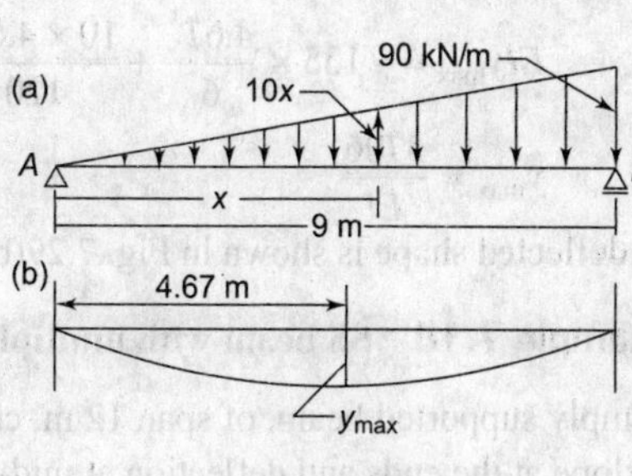

Fig. 7.29

$$R_{AV} = \frac{90 \times (9/2) \times (9/3)}{9}$$

$$= 135 \text{ kN} \uparrow$$

$$R_{BV} = 405 - 135 = 270 \text{ kN} \uparrow$$

The general equation for BM is

$$M_x = 135x - 10x\,\frac{x}{2}\,\frac{x}{3}$$

$$EI\frac{d^2y}{dx^2} = -M_x = -135x + \frac{10x^3}{6}$$

$$EI\frac{dy}{dx} = -135\frac{x^2}{2} + \frac{10x^4}{24} + C_1$$

$$EIy = -135\frac{x^3}{6} + \frac{10x^5}{120} + C_1x + C_2$$

At A, $x = 0$, $y = 0$. Therefore, $C_2 = 0$. At B, $x = 9$, $y = 0$.

$$0 = -135 \times \frac{9^3}{6} + \frac{10 \times 9^5}{120} + C_1 \times 9$$

$$\Rightarrow \quad C_1 = 1275.75$$

The general equation for slope is

$$EI\frac{dy}{dx} = \frac{-135x^2}{2} + \frac{10x^4}{24} + 1275.75$$

At $x = 0$,

$$EI\theta_A = 1275.75$$

$$\Rightarrow \quad \theta_A = \frac{1275.75}{EI}$$

At $x = 8$,

$$EI\theta_B = \frac{-135 \times 9^2}{2} + \frac{10 \times 9^4}{24} + 1275.75$$

$$\Rightarrow \quad \theta_B = -\frac{3918.375}{EI}$$

The general equation for deflection is

$$EIy = -\frac{135x^3}{6} + \frac{10x^5}{120} + 1275.75x$$

The maximum deflection will be at the point where $dy/dx = 0$.

$$-135\frac{x^2}{2} + \frac{10x^4}{24} + 1275.75 = 0$$

Setting $x^2 = z$,

$$-135\frac{z}{2} + \frac{10z^2}{24} + 1275.75 = 0, \quad z = 21.81, \quad x = \sqrt{z} = \sqrt{21.81} = 4.67$$

The maximum deflection is at a point 4.67 m from A.

$$EIy_{max} = -135 \times \frac{4.67^3}{6} + \frac{10 \times 4.67^5}{120} + 1275.75 \times 4.67$$

$$\Rightarrow \quad y_{max} = \frac{3706}{EI}$$

The deflected shape is shown in Fig. 7.29(b).

Example 7.18 SS beam with multiple loads

A simply supported beam, of span 12 m, carries loads as shown in Fig. 7.30(a). Calculate the slope at the ends and deflection at mid-span. EI is constant.

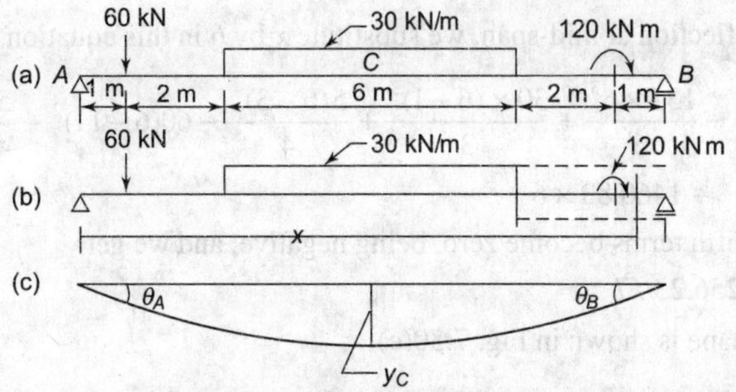

Fig. 7.30

Solution Let us calculate the reactions first.

$$R_{AV} \times 12 - 60 \times 11 - 30 \times 6 \times 6 + 120 = 0$$

$$\Rightarrow \quad R_{AV} = 135 \text{ kN} \uparrow$$

$$R_{BV} = 240 - 140 = 105 \text{ kN} \uparrow$$

The general equation for BM is stated after adding and subtracting equal loads as shown in Fig. 7.30(b). The BM equation is expressed for the last segment at the right as

$$M_x = 135x - 60(x-1) - \frac{30(x-3)^2}{2} + 120(x-11)^0 + 30\,\frac{30(x-9)^2}{2}$$

$$EI\frac{d^2y}{dx^2} = -M_x = -135x + 60(x-1) + \frac{30(x-3)^2}{2} - 120(x-11)^0 - \frac{30(x-9)^2}{2}$$

$$EI\frac{dy}{dx} = \frac{-135x^2}{2} + \frac{60(x-1)^2}{2} + \frac{30(x-3)^3}{6} - 120(x-11) - \frac{30(x-9)^3}{6} + C_1$$

$$EIy = \frac{-135x^3}{6} + \frac{60(x-1)^3}{6} + \frac{30(x-3)^4}{24} - \frac{120(x-11)^2}{2} - \frac{30(x-9)^4}{24} + C_1x + C_2$$

At $x = 0$, $y = 0$. Therefore, $C_2 = 0$. At $x = 12$, $y = 0$.

$$\text{Therefore, } 0 = -\frac{135\times 12^3}{6} + \frac{60(12-1)^3}{6} + \frac{30(12-3)^4}{24} - \frac{120(12-11)^2}{2} - \frac{30(12-9)^4}{24} + C_1 \times 12$$

$$\Rightarrow \quad C_1 = 1460.83$$

The general equation for slope is

$$EI\frac{dy}{dx} = -\frac{135x^2}{2} + 30(x-1)^2 + 5(x-3)^3 - 120(x-11) - 5(x-9)^3 + 1460.83$$

At $x = 0$, this equation gives $\theta_A = 1460.83/EI$. At $x = 12$, this equation gives θ_B.

$$EI\theta_B = \frac{135}{2} \times 12^2 + 30 \times 11^2 + 5 \times 9^3 - 120 \times 1 - 5 \times 3^3 + 1460.83$$

$$\Rightarrow \quad \theta_B = -\frac{3000 + 1578.33}{EI} = -1104.17/EI$$

The general equation for deflection is

$$EIy = -\frac{135x^3}{6} + \frac{30}{3}(x-1)^3 + \frac{5}{4}(x-3)^4 - 60(x-11)^2 - \frac{5}{4}(x-9)^4 + 1460.83x$$

To obtain the deflection at mid-span, we substitute x by 6 in this equation.

$$EIy_C = -\frac{135\times 6^3}{6} + \frac{30\times(6-1)^3}{3} + \frac{5(6-3)^4}{4} - 60(6-11)^2 - \frac{5}{4}(6-9)^4$$

$$+ 1460.83\times 6$$

The fourth and fifth terms become zero, being negative, and we get

$$y_C = 5256.23/EI$$

The deflected shape is shown in Fig. 7.30(c).

Example 7.19 Overhanging beam

An overhanging beam is loaded as shown in Fig. 7.31(a). Calculate the deflections under the loads and the slopes at A and B. EI is constant.

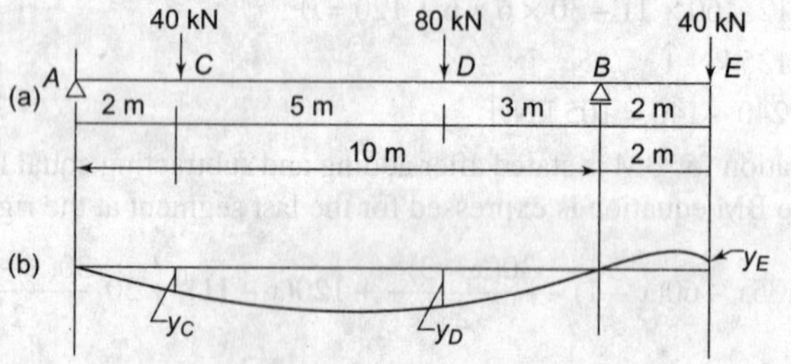

Fig. 7.31

Solution Let us calculate the reactions at A and B.

$$R_{AV} = \frac{40\times 8 + 80\times 3 - 40\times 2}{10} = 48 \text{ kN}\uparrow$$

$$R_{BV} = 160 - 48 = 112 \text{ kN}\uparrow$$

The general equation for BM for the segment BE is

$$M_x = 48x - 40(x-2) - 80(x-7) + 112(x-10)$$

$$EI\frac{d^2y}{dx^2} = -48x + 40(x-2) + 80(x-7) - 112(x-10)$$

$$EI\frac{dy}{dx} = -\frac{48x^2}{2} + \frac{40(x-2)^2}{2} + \frac{80(x-7)^2}{2} - \frac{112(x-10)^2}{2} + C_1$$

$$EIy = -\frac{48x^3}{6} + \frac{40(x-2)^3}{6} + \frac{80(x-7)^3}{6} - \frac{112(x-10)^3}{6} + C_1x + C_2$$

The boundary conditions are (i) $x = 0$, $y = 0$ and (ii) $x = 10$, $y = 0$. The first condition gives $C_2 = 0$, as the second, third, and fourth terms are invalid for $x = 0$. The second condition gives

$$0 = -\frac{48\times 10^3}{6} + \frac{40\times 8^3}{6} + \frac{80\times 3^3}{6} - 0 + C_1\times 10, \quad C_1 = 422.7$$

The general equations for slope and deflection are

$$EI\frac{dy}{dx} = -24x^2 + 20(x-2)^2 + 40(x-7)^2 - 56(x-10)^2 + 422.7$$

$$EIy = -8x^3 + \frac{20}{3}(x-2)^3 + \frac{40}{3}(x-7)^3 - \frac{56}{3}(x-10)^3 + 422.7x$$

The slope at B, θ_B, is obtained by setting $x = 10$ in the slope equation.

$$EI\theta_B = -24\times 10^2 + 20\times 8^2 + 40\times 3^2 - 56\times 0 + 422.7 = -337.3$$

$$\theta_B = -\frac{337.3}{EI},\quad \theta_A = \frac{422.7}{EI}$$

setting $x = 0$ in the slope equation.

The deflection at C ($x = 2$) may be calculated as follows:

$$EIy_C = -8\times 2^3 + 422.7\times 2 = 781.34$$

$$\Rightarrow \quad y_C = \frac{781.34}{EI}$$

The deflection at D ($x = 7$) may be worked out in the following manner:

$$EIy_D = -8\times 7^3 + \frac{20}{3}\times 5^3 + 422.7\times 7$$

$$\Rightarrow \quad y_D = \frac{1048.23}{EI}$$

The deflection at E ($x = 12$) may be calculated as follows:

$$EIy_E = -8\times 12^3 + \frac{20}{3}\times 10^3 + \frac{40}{3}\times 5^3 - \frac{56}{3}\times 2^3 + 422.7\times 12 = -567.98$$

$$\Rightarrow y_E = -\frac{568}{EI}$$

The deflected shape is shown in Fig. 7.31(b).

Example 7.20 Cantilever with uniformly varying load

A cantilever, of span 3 m, carries a load varying uniformly from zero at the free end to 60 kN/m at the fixed end. Calculate the slope and deflection at the free end, if $E = 200\times 10^6$ kN/m^2 and $I = 5000$ cm^4.

Solution The beam with the loading is shown in Fig. 7.32(a). The BM equation can be directly formulated from the left end. Note that the load intensity at x from A is $(60/3)x = 20x$ as shown in Fig. 7.32(b).

$$M_x = -20x\,\frac{x}{2}\,\frac{x}{3} = -\frac{10x^3}{3}$$

$$EI\,\frac{d^2y}{dx^2} = \frac{10x^3}{3}$$

$$EI\,\frac{dy}{dx} = \frac{10x^4}{12} + C_1$$

$$EIy = \frac{10x^5}{60} + C_1x + C_2$$

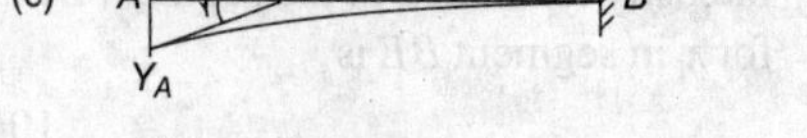

Fig. 7.32

The conditions are

(i) At $x = 3$, $dy/dx = 0$

(ii) At $x = 3$, $y = 0$

Substituting these conditions,

$$0 = \frac{10\times 3^4}{12} + C_1$$

$$\Rightarrow \quad C_1 = -67.5$$

$$0 = \frac{10\times 3^5}{60} - 67.5\times 3 + C_2$$

$$\Rightarrow \quad C_2 = 162$$

The general equations for slope and deflection are

$$EI\frac{dy}{dx} = \frac{10x^4}{12} - 67.5$$

$$\Rightarrow \quad EIy = \frac{10x^5}{60} - 67.5x + 162$$

The slope and deflection at the free end are obtained by setting $x = 0$ in these equations.

$$EI\theta_A = -67.5$$

$$\Rightarrow \quad \theta_A = \frac{-67.5}{EI} = \frac{-67.5}{200 \times 10^6 \times 5000 \times 10^{-8}} = 0.0067 \text{ rad}$$

$$EIy_A = 162$$

$$\Rightarrow \quad y_A = \frac{162}{EI} = \frac{162}{200 \times 10^6 \times 5000 \times 10^{-8}} = 16.2 \text{ mm}$$

The deflected shape is shown in Fig. 7.32(c).

Example 7.21 Doubly overhanging beam

A doubly overhanging beam is loaded as shown in Fig. 7.33(a). Compute the slope at A, and the deflection at the free end and at mid-span. EI is constant.

Solution Note that the loading is symmetrical and therefore the reactions at A and B are equal.

$$R_{AV} = R_{BV} = 80 \text{ kN} \uparrow$$

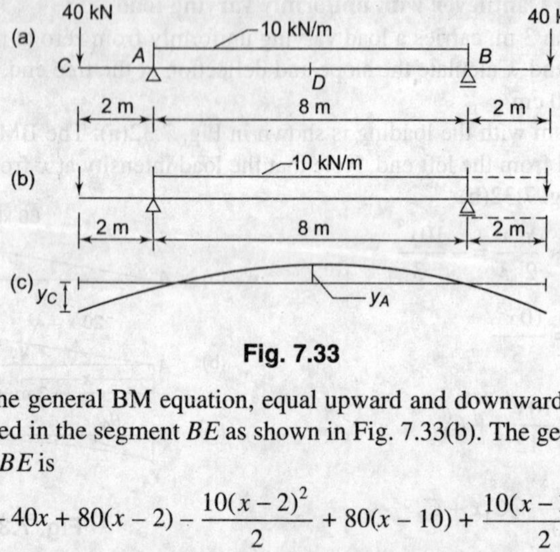

Fig. 7.33

To arrive at the general BM equation, equal upward and downward loads of 10 kN/m intensity are added in the segment BE as shown in Fig. 7.33(b). The general BM equation for x in segment BE is

$$M_x = -40x + 80(x-2) - \frac{10(x-2)^2}{2} + 80(x-10) + \frac{10(x-10)^2}{2}$$

$$EI\frac{d^2y}{dx^2} = -M_x = 40x - 80(x-2) + \frac{10(x-2)^2}{2} - 80(x-10) - \frac{10(x-2)^2}{2}$$

$$EI\frac{dy}{dx} = \frac{40x^2}{2} - \frac{80(x-2)^2}{2} + \frac{10(x-2)^3}{6} - \frac{80(x-10)^2}{2} - \frac{10(x-10)^3}{6} + C_1$$

$$EIy = \frac{40x^3}{6} - \frac{80(x-2)^3}{6} + \frac{10(x-2)^4}{24} - \frac{80(x-10)^3}{6} - \frac{10(x-10)^4}{24} + C_1x + C_2$$

The boundary conditions are (i) $x = 2$, $y = 0$ and (ii) $x = 10$, $y = 0$. The first of these yields

$$0 = \frac{40 \times 2^3}{6} + C_1 \times 2 + C_2$$

The second condition gives

$$0 = \frac{40 \times 10^3}{6} - \frac{80 \times 8^3}{6} + \frac{10 \times 8^4}{24} + C_1 \times 10 + C_2$$

$$2C_1 + C_2 = -53.33$$

$$10C_1 + C_2 = -1546.67$$

Solving, we get

$$C_1 = -186.7, \quad C_2 = 320$$

The general equations for slope and deflection are

$$EI\frac{dy}{dx} = 20x^2 - 40(x-2)^2 + \frac{5}{3}(x-2)^3 - 40(x-10)^2 - \frac{5}{3}(x-10)^3 - 186.67$$

$$EIy = \frac{20x^3}{3} - \frac{40}{3}(x-2)^3 + \frac{5}{12}(x-2)^4 - \frac{40}{3}(x-10)^3 - \frac{5}{12}(x-10)^4 - 186.67x + 320$$

The slope at C ($x = 0$) is

$$\theta_C = -\frac{186.67}{EI}$$

The slope at A ($x = 2$) is worked out as follows:

$$EI\theta_A = 20 \times 2^2 - 186.67 = -106.67$$

$$\Rightarrow \quad \theta_A = \frac{-106.67}{EI}$$

The deflection at C ($x = 0$) is

$$y_C = \frac{320}{EI}$$

The deflection at D ($x = 6$) may be worked out as follows:

$$EIy_D = \frac{20}{3} \times 6^3 - \frac{40}{3} \times 4^3 + \frac{5}{12} \times 4^4 - 186.67 \times 6 + 320$$

$$\Rightarrow \quad y_D = \frac{-106.67}{EI} \text{ (upwards)}$$

Note that the slopes at B and E are numerically equal but opposite to those at A and C, respectively. The deflection at E is equal to that at C. This is because the arrangement of the beam and loading is symmetrical. The deflected shape is shown in Fig. 7.33(c). ■

The many examples we have discussed illustrate the different features of the double integration method. With the modifications suggested by Macaulay, this is a very powerful method for finding slopes and deflections. The great advantage of the double integration method lies in the fact that it gives directly an equation for the elastic curve. The slope and deflection at any point can be found from the equation derived. The method is not useful in the case of beams of variable cross section and is sometimes very cumbersome.

We shall study two more methods for finding slopes and deflections. Instead of the mathematical approach of solution of differential equations, we will use the BM diagrams, its area and the moment of this area. The same principles are used in the two methods—the area-moment method and the conjugate-beam method. However, they involve the use of different techniques.

7.8 THE AREA-MOMENT METHOD

The area-moment method involves the use of a simple technique of finding slopes and deflections due to bending, from the BM diagram. Starting from the basic equation $M/I = E/R$, it relates the area and the moment of area of the BM diagram to the deformations in beams. While the method does not directly give the slope and deflection, it is very useful, particularly when one is interested in finding slopes and deflections at specified points only. It is also useful when the cross-sectional dimensions of the beam vary, giving different values of moment of inertia in different lengths or in a continuously varying pattern. It is also a very simple method.

7.8.1 The Basic Principle

The basic principle of this method can be explained with reference to Fig. 7.34. Figure 7.34(a) shows a beam with some arbitrary loading. The BM diagram due to this loading is shown in Fig. 7.34(b) and the deflected shape of the beam in Fig. 7.34(c).

Taking a small element ds of this curve as shown in Fig. 7.34(d), we note that

$$R\,d\theta = ds$$

$$\Rightarrow \quad d\theta = \frac{ds}{R}$$

Since the slope and deflection in a beam are very small, the length ds can be approximately taken to be equal to dx.

$$d\theta = \frac{dx}{R}$$

and since $1/R = M/EI$,

$$d\theta = \frac{M}{EI}dx$$

Note that

$$\frac{dy}{dx} = \theta, \quad \frac{d\theta}{dx} = \frac{d^2y}{dx^2} = \frac{M}{EI}$$

and hence

$$d\theta = \frac{M}{EI}\,dx$$

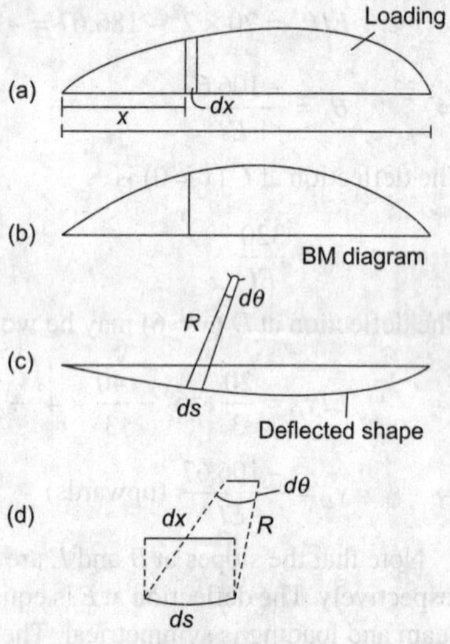

Fig. 7.34

Looking at the same elemental length ds as shown in Fig. 7.35(a), it is easy to see that the central angle $d\theta$ between the radial lines is equal to the angle between the tangents drawn at the ends of the element. If we consider any two arbitrary points of the elastic curve, say points 1 and 2, shown in Fig. 7.35(b), we can state that

$$\int_{\theta_1}^{\theta_2} d\theta = \int_1^2 \frac{M}{EI}\,dx$$

$$\theta_2 - \theta_1 = \int_1^2 \frac{M}{EI}\,dx$$

(Here $\int_1^2$ means that the limits of integration are the values at 1 and 2.)

If we consider the part BM diagram for a beam, as shown in Fig. 7.35(c), we can find the meaning of the integral on the right side. Let us consider the BM diagram divided by the flexural rigidity *EI* of the beam. Let us call this diagram the *M*/*EI* diagram. If the flexural rigidity is constant, except for the difference in scale of the ordinates, there will be no change in the BM diagram since every ordinate is divided by the same quantity. Where the value of *EI* varies along the length, the *M*/*EI* diagram will be a correspondingly modified version of the BM diagram.

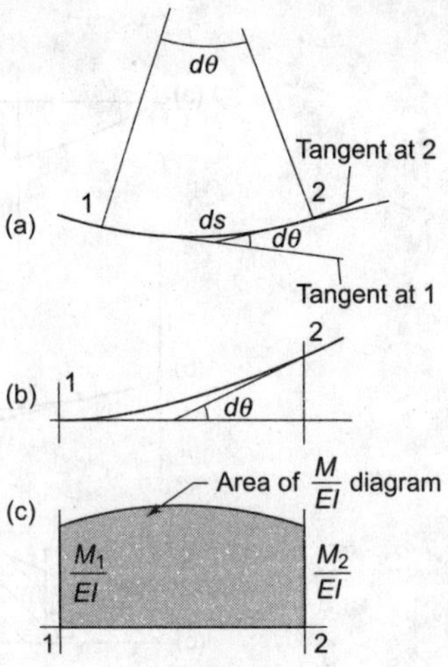

Fig. 7.35

The integral on the right-hand side of the equation we are discussing is the area of the *M*/*EI* diagram. Here $(\theta_2 - \theta_1)$ is the angle between the tangents drawn at points 1 and 2 of the elastic curve and $\int_1^2 (M/EI)\,dx$ is the area of the *M*/*EI* diagram between these points. This equation states what is called the first area-moment theorem (Mohr's theorem).

7.8.2 The First Area-Moment Theorem

This theorem states that *the angle between the tangents to the elastic curve between any two points is equal to the area of the M/EI diagram between the two points.*

As a second aspect of the same beam, consider the two points 1 and 2 of the elastic curve [Fig. 7.36(a)]. Draw the tangent at point 1 to the elastic curve and draw a vertical line through point 2. Let the two lines intersect at point 3 as in Fig. 7.36(a). Also consider an elemental length subtending an angle $d\theta$ at the centre. The tangents at the two ends of the elemental length intersect the vertical line through 2 at two points, giving a vertical intercept dv. Since the angles involved are very small, the length dv can be considered equal to $x_1 d\theta$, where x_1 is the distance of point 2 from the elemental length ds, along the beam. Therefore, from Fig. 7.36(b),

$$dv = x_1\,d\theta, \quad dv = x_1 \frac{M}{EI}\,dx$$

$(M/EI)\,dx$ being the area of the *M*/*EI* diagram over the length dx, the quantity on the right side of this equation is the moment of the area of the *M*/*EI* diagram about point 2. If we integrate both these expressions over the length 1-2,

$$\int_1^2 dv = \int_1^2 \frac{M}{EI}\,dx\,x_1$$

The left side integrated over 1-2 gives us the intercept 3-2 while the right side is the moment of the area of the *M*/*EI* diagram between 1 and 2 about 2. Figures 7.36(b) and (c) show the elemental length and integrated *M*/*EI* diagrams. This leads us to proposition 2.

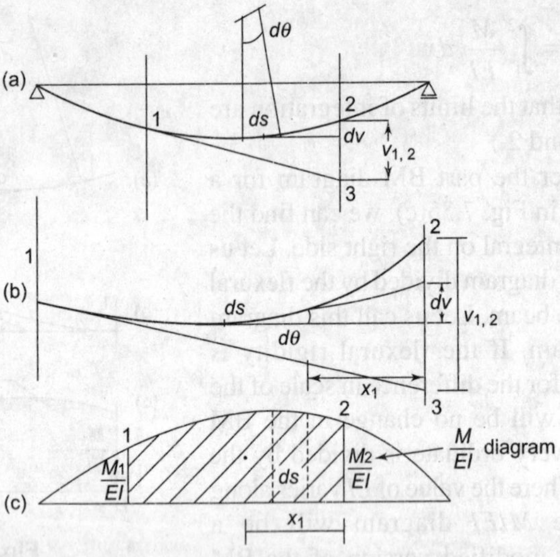

Fig. 7.36

7.8.3 The Second Area-Moment Theorem

This theorem states that *the moment of the area of the M/EI diagram between two points of a beam about one of these points is equal to the vertical intercept made by the tangent drawn at one point on a vertical line through the second point (about which moment is taken).* Note that the intercept will be made at the point about which the moment is taken. If we consider $v_{1,2}$ as the intercept made by the tangent at 1 on the vertical through 2, $v_{2,1}$ is the intercept made by the tangent at 2 on the vertical through 1 (Fig. 7.37). These two intercepts need not be equal.

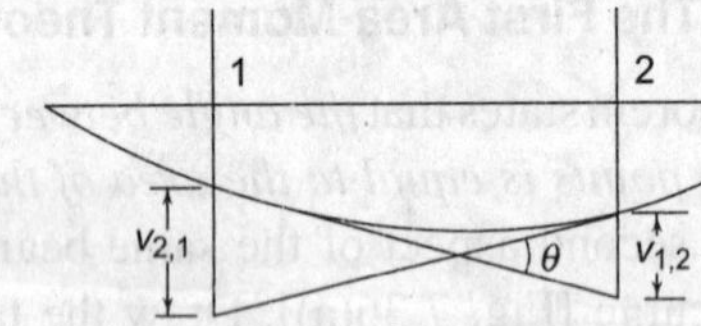

Fig. 7.37

To summarize, the first area-moment theorem states that the area of the *M/EI* diagram between any two points 1 and 2 (as shown in Fig. 7.37) is equal to the angle between the tangents drawn to the elastic curve at points 1 and 2. The second area-moment theorem states that the vertical intercept between the tangent drawn at one point and the vertical line drawn through the second point is equal to the moment of the area of the *M/EI* diagram between points 1 and 2 about the second point.

The two area-moment theorems help us find the slopes and deflections in beams. As mentioned earlier, the area-moment theorems do not directly give slopes and deflections. But they can be used to find slopes and deflections as will be illustrated by examples. Before starting with examples, however, we must decide the sign conventions.

We have already assigned sign conventions to BMs as in Fig. 7.38(a). Sagging BM is positive and hogging BM is negative. Looking at the sagging and hogging elastic curves shown in Fig. 7.38(b), we note that an area due to positive BM gives a positive moment of the area and in such a case, the intercept is above the tangent. A negative moment gives an intercept below the tangent.

In the case of angles between tangents, as we go from left to right, a positive BM area due to sagging BM causes an anticlockwise rotation of the tangent [Fig. 7.38(c)]. In the case of negative BM, as we proceed from left to right, the tangent rotates clockwise as shown in Fig. 7.38(c). We will illustrate these concepts with a number of examples. Let us start with the methods to apply the moment-area theorems to different types of beams.

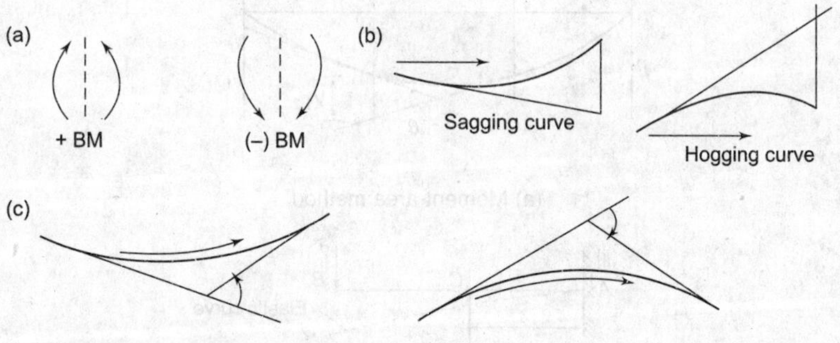

Fig. 7.38

Application of Moment-Area Method

As we have seen, the moment-area theorems do not give slope or deflection directly. The area of *M/EI* diagram is equal to the angle between the tangents and the moment of area of *M/EI* diagram is equal to the vertical intercept at the point about which the moment is taken. Refer to Fig. 7.39(a).

(i) ***Cantilever beams*** Referring to Fig. 7.39(b), we notice that the fixed end has no slope. This means that the tangent at A to the elastic curve is the same as the unbent beam axis. In the case of a cantilever beam, if you take a point at any distance x from the fixed end, the angle between the tangents gives directly the slope and the vertical intercept directly gives the deflection. The area of the shaded *M/EI* diagram between A and C is the slope at point C. Similarly, the moment of the *M/EI* diagram between A and C about C directly gives the deflection at C. The moment-area method is thus apt for application to cantilever beams.

(ii) ***SS beams*** In SS beams, as shown in Fig. 7.39(c), to find the slope at A, we proceed as follows.

(a) Find the moment of the area of the *M/EI* diagram between A and B about B. This gives the vertical intercept v_{ab} (vertical distance at B from the tangent at A).

(b) From the triangle $AB'B$, $\tan\theta = B'B/AB = v_{AB}/L$. As θ is very small, $\tan\theta =$ θ in radians. This gives the slope at A.

(c) Slope at B can be similarly found by taking moment about A.

To find the deflection at x from A, we proceed as follows [Fig. 7.39(d)].

(a) If you find the moment of the area of the *M/EI* diagram between A and C about C, this gives the length $C'C''$.

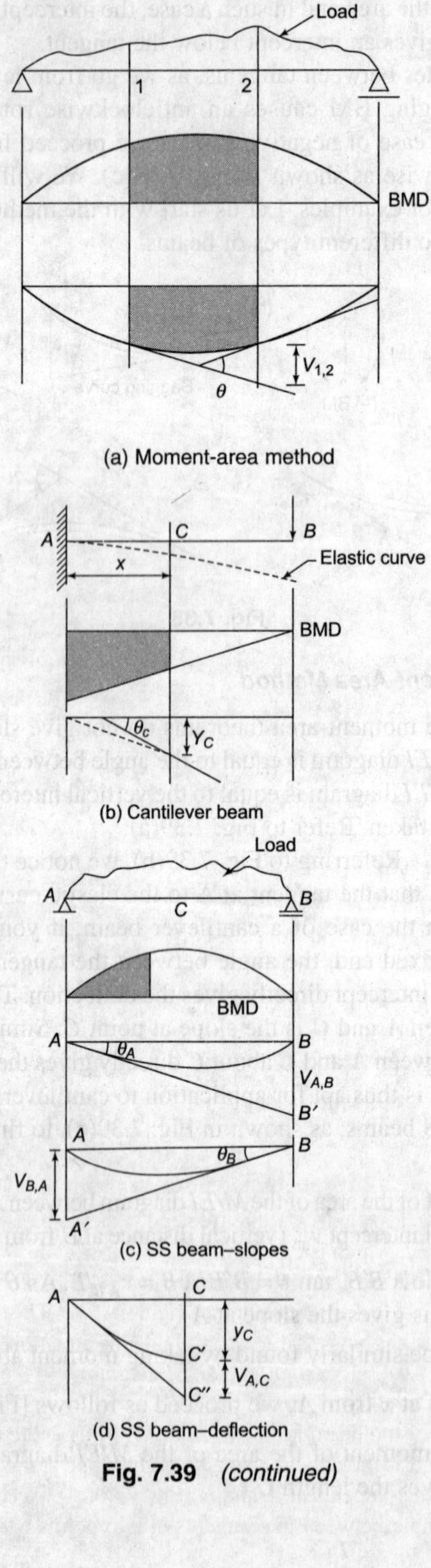

Fig. 7.39 *(continued)*

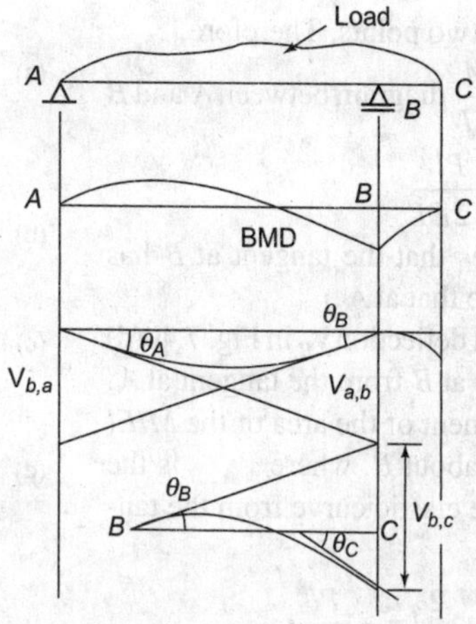

(e) Overhanging beam with loading

Fig. 7.39 Application of moment-area method

(b) From the triangle $A\,C''C$, $C''C/A\,C = \tan\theta$. Knowing the slope at A, $C''C$ can be calculated.

(c) Deflection at $C = CC' = C''C - C'C''$.

(d) Deflection at any point can be calculated from a similar procedure.

(iii) ***Overhanging beams*** The overhanging beam with loading is shown in Fig. 7.39(e).

To find the slope and deflections, we proceed as follows.

(a) To find the slope at A, use the method outlined for SS beams. Take moment of M/EI diagram about B and divide by the span. The slope at B can be similarly found by taking moment about A.

(b) To find the slope at C, The area of the M/EI diagram between B and C gives the angle between the tangents, θ, at B and C. We find that $\theta_B + \theta_C = \theta$. Knowing θ_B, slope at C can be found. Clockwise slope should be taken positive.

(c) Deflections at any point can be similarly found using the method outlined for SS beams.

The examples below illustrate the above principles.

Example 7.22 Moment-area method: cantilever with point load

A cantilever of span l carries a load P at the free end. Calculate the slope and deflection at the free end in terms of EI, which is constant.

Solution The beam with the load is shown in Fig. 7.40(a). The BM at a section at a distance x from the free end, calculating from the right, is $-Px$. This is negative because it is clockwise from the right (hogging). This is a straight-line equation and the BM diagram is shown in Fig. 7.40(b). EI being constant, the M/EI diagram will have the same shape as in Fig. 7.40(b).

The deflected shape of the beam is shown in Fig.7.40(d). Note that the tangent at A is horizontal and is the same as the unbent beam axis. The slope at the free end is θ_B, which is also the angle between the tangents at A and B [Fig. 7.40(c)]. According to the first area-moment theorem, the angle between the tangents at two points is equal to the area of the

M/EI diagram between the two points. Therefore,

$$\theta_B = \text{area of the } \frac{M}{EI} \text{ diagram between } A \text{ and } B$$

$$= -\frac{Pl}{EI}\frac{l}{2} = -\frac{Pl^2}{2EI}$$

(The negative sign indicates that the tangent at B has rotated clockwise relative to that at A.)

Also note that the vertical deflection y_B in Fig. 7.40(d) is also the vertical intercept at B from the tangent at A. Therefore $y_B = V_{A,B}$ = moment of the area of the M/EI diagram between A and B about B, where $V_{A,B}$ is the vertical intercept at B of the elastic curve from the tangent drawn at A. Therefore,

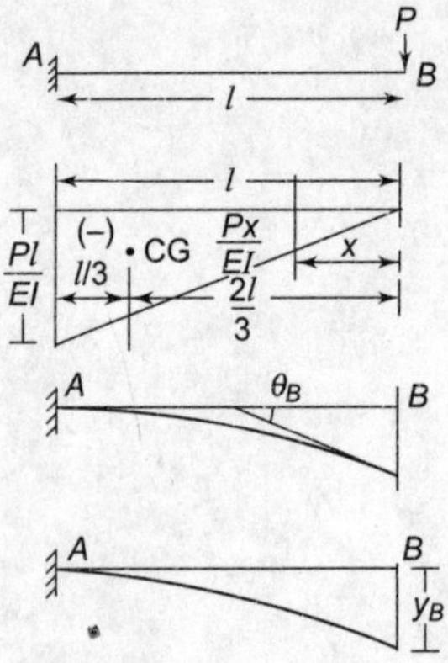

Fig. 7.40

$$y_B = V_{A,B} = -\frac{Pl}{EI}\frac{l}{2}\frac{2}{3}l = -\frac{Pl^3}{3EI}$$

The negative sign indicates that the vertical intercept is below the tangent at A, as in Fig. 7.40(d).

Example 7.23 Moment-area method: cantilever with point load

Calculate the slope and deflection for the cantilever of Example 7.22 under the load and at the free end, if the load acts at a distance a from A, $a < l$.

Solution The beam and the loading are shown in Fig. 7.41(a), and the BM diagram in Fig. 7.41(b), and is negative. As in the previous problem [Fig. 7.41(c)],

$$\theta_C = \text{area of the } \frac{M}{EI} \text{ diagram between } A \text{ and } C$$

$$= -\frac{Pa}{EI}\frac{a}{2} = -\frac{Pa^2}{2EI}$$

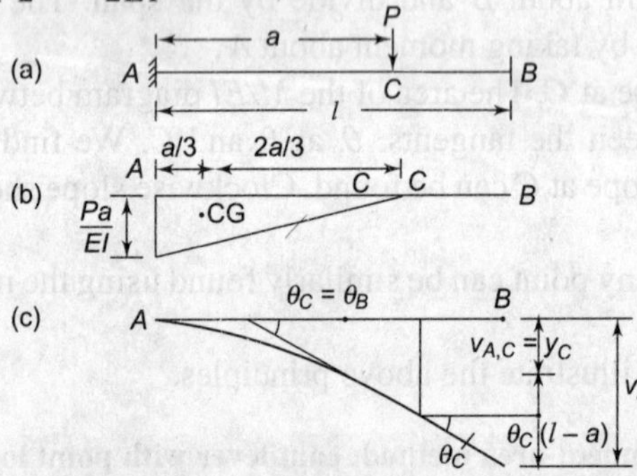

Fig. 7.41

$$y_C = V_{A,C} = \text{moment of the area of the } \frac{M}{EI} \text{ diagram between } A \text{ and } C \text{ about } C$$

$$= -\frac{Pa}{EI}\frac{a}{2}\frac{2a}{3} = -\frac{Pa^3}{3EI}$$

To calculate the slope and deflection at B, we note that $\theta_B = \theta_C$, since the areas of the M/EI diagram between A and C, and A and B, are the same. (The area of the BM diagram between C and B is zero.)

$$\theta_B = \theta_C = -\frac{Pa^2}{2EI}$$

and $\quad y_B = V_{A,B}$ = moment of area of $\dfrac{M}{EI}$ diagram between A and B about B

$$= -\frac{Pa}{EI}\frac{a}{2}\left(l - \frac{a}{3}\right) = -\frac{Pa^2}{6EI}(3l - a)$$

noting that the centroid of the triangle is at $(l - a/3)$ from B.

or $\quad y_B = y_C + \theta_C(l - a)$

$$= -\frac{Pa^3}{3EI} - \frac{Pa^2}{2EI}(l - a)$$

$$= -\frac{Pa^2}{6EI}[2a - 3(l - a)]$$

$$= -\frac{Pa^2}{6EI}(3l - a)$$

as before.

Example 7.24 Moment-area method: cantilever with UD load

A cantilever of uniform cross section carries a UD load w throughout its length. Calculate the slope and deflection at the free end.

Solution The beam with the loading is shown in Fig. 7.42(a). At a distance x from B, the BM is $-wx^2/2$. This is zero at B where $x = 0$ and $-wl^2/2$ at A where $x = l$. The BM diagram is shown in Fig. 7.42(b). Note that it is parabolic. The area in this case is one-third the area of the enclosing rectangle and its centroid is at $l/4$ from A. (Refer to Table 7.1.)

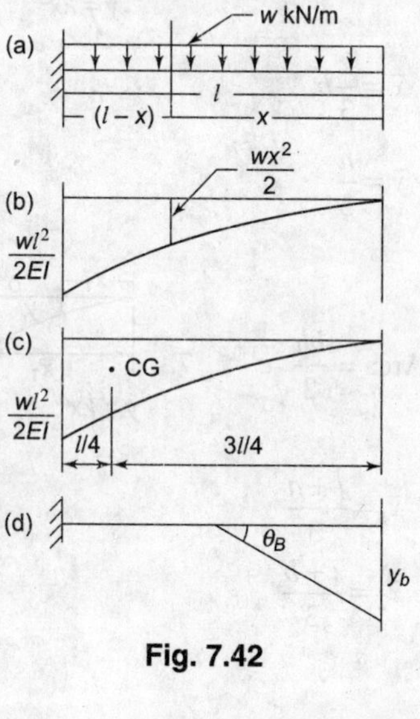

Fig. 7.42

As before,

$$\theta_B = \text{area of } \frac{M}{EI} \text{ diagram between } A \text{ and } B \text{ [Fig. 7.42(c)]}$$

$$= -\frac{1}{3}\frac{wl^2}{2EI}l = \frac{-wl^3}{6EI}$$

and $\quad y_B = V_{A,B}$ = moment of area of $\dfrac{M}{EI}$ diagram between A and B about B

$$= -\frac{1}{3}\frac{wl^2}{2EI}l\frac{3l}{4} = \frac{-wl^4}{8EI}$$

Note that the negative signs agree with the deflected shape shown in Fig. 7.42(d). ■

As in this problem, we need the areas and centroids of areas of many diagrams. Some of the common areas met with are given in Table 7.1.

Table 7.1 Centroids of areas

	Diagram	*Areas and centroids*		*Diagram*	*Areas and centroids*
1.	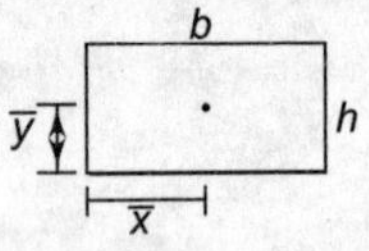	Area = bh $\bar{x}=\frac{b}{2}, \bar{y}=\frac{h}{2}$	5.	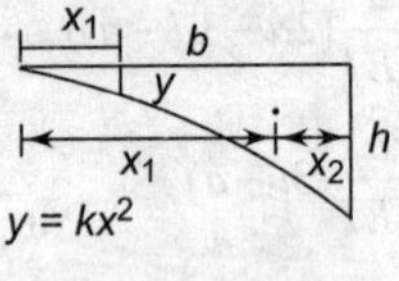	Area = $\frac{1}{3}bh$ $\bar{x}_1=\frac{3}{4}b$ $\bar{x}_2=\frac{b}{4}$
2.	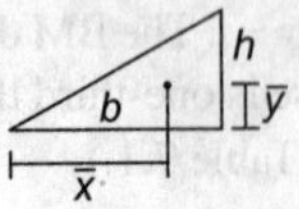	Area = $\frac{bh}{2}$ $\bar{x}=\frac{2}{3}b,$ $\bar{y}=\frac{h}{3}$	6.	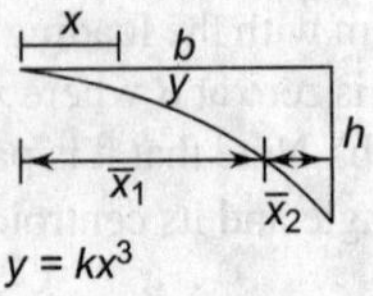	Area = $\frac{1}{4}bh$ $\bar{x}_1=\frac{4}{5}b$ $\bar{x}_2=\frac{b}{5}$
3.	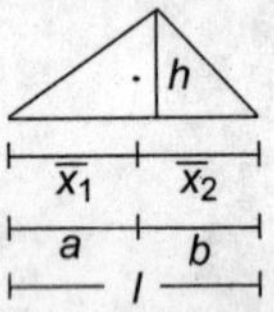	Area = $\frac{bh}{2}$ $\bar{x}_1=\frac{l+a}{3}$ $\bar{x}_2=\frac{l+b}{3}$	7.	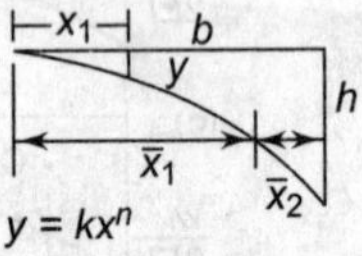	Area = $\frac{bh}{n+1}$ $\bar{x}_1=\left(\frac{n+1}{n+2}\right)b$ $\bar{x}_2=\frac{b}{n+2}$
4.	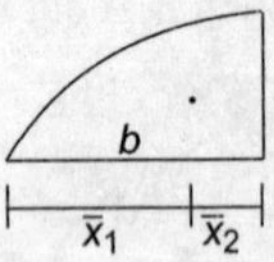	Area = $\frac{2}{3}bh$ $\bar{x}_1=\frac{5b}{8}$ $\bar{x}_2=\frac{3b}{8}$			

Example 7.25 Moment area-method: SS beam with eccentric point load

A simply supported beam carries an eccentric load P at a distance a from the left end. Compute the slopes at the ends and the deflection under the load. EI is constant.

Solution The beam is shown in Fig. 7.43(a). The reactions are $R_{AV} = (Pb/l) \uparrow$ and $R_{BV} = (Pa/l) \uparrow$. The BM diagram is positive and is a triangle with ordinate Pab/L at C as shown in Fig. 7.43(b). The M/EI diagram will have the same shape with ordinate C as Pab/EIl as in Fig. 7.43(c).

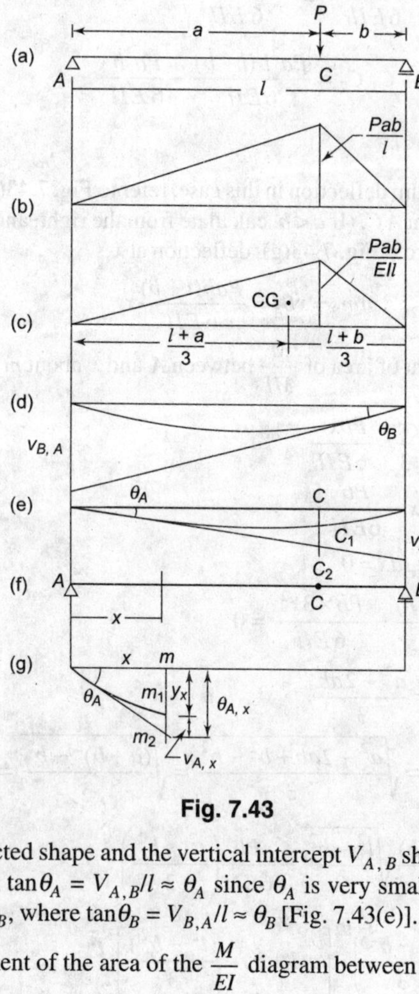

Fig. 7.43

Consider the deflected shape and the vertical intercept $V_{A,B}$ shown in Fig. 7.43(d). If θ_A is the slope at A, then $\tan\theta_A = V_{A,B}/l \approx \theta_A$ since θ_A is very small. The same procedure is adopted to calculate θ_B, where $\tan\theta_B = V_{B,A}/l \approx \theta_B$ [Fig. 7.43(e)].

$$V_{A,B} = \text{moment of the area of the } \frac{M}{EI} \text{ diagram between } A \text{ and } B \text{ about } B$$

$$= \frac{Pab}{EIl}\frac{l}{2}\frac{(l+b)}{3} = \frac{Pab\,(l+b)}{6EI}$$

$$\theta_A = \frac{V_{A,B}}{l} = \frac{Pab\,(l+b)}{6EI}\frac{1}{l} = \frac{Pab(l+b)}{6EIl}$$

Similarly,

$$\theta_B = \frac{V_{B,A}}{l} = \frac{Pab}{EIl}\frac{(l+a)}{3}\frac{l}{2} = \frac{Pab(l+a)}{6EIl}$$

To obtain the deflection under the load, consider the diagram shown in Fig. 7.43(d). Note that, in this diagram, CC_1 is the deflection at C and C_1C_2 is the vertical intercept at C of the tangent at $A = V_{A,C}$. Also, from triangle $A\,C_2\,C$, $CC_2 = a \tan\theta_A \approx a\theta_A$. Therefore,

$$C_1C_2 = V_{A,\,C} = \text{moment of the area of } \frac{M}{EI} \text{ diagram between } A \text{ and } C \text{ @ } C$$

$$= \frac{Pab}{EIl}\frac{a}{2}\frac{a}{3} = \frac{Pa^3b}{6lEI}$$

$$CC_2 = a\theta_A = a\,\frac{Pab(l+b)}{6EIl} = \frac{Pa^2b(l+b)}{6\,EIl}$$

and $$CC_1 = y_C = CC_2 - C_1\,C_2 = \frac{Pa^2b(l+b)}{6EIl} - \frac{Pa^3b}{6\,EIl}$$

$$y_C = \frac{Pa^2b^2}{3\,EIl}$$

Note To find the maximum deflection in this case, refer to Fig. 7.43(f) since the maximum deflection is in the segment $A\,C$. (If $a < b$, calculate from the right, and find the deflection at any distance x from B.) From Fig. 7.43(g), deflection at x,

$$y_x = mm_2 - m_1m_2, \quad mm_2 = x\theta_A = \frac{Pab(l+b)}{6EIl}x$$

$$m_1m_2 = \text{moment of area of } \frac{M}{MI} \text{ between } A \text{ and } x \text{ about } m$$

$$= \frac{Pbx}{EIl}\frac{x}{2}\frac{x}{3} = \frac{Pbx^3}{6EIl}$$

$$y_x = \frac{Pab(l+b)}{6\,EIl}x - \frac{Pb}{6EIl}x^3$$

For y_x to be maximum, $dy_x/dx = 0$.

$$\frac{dy_x}{dx} = \frac{Pab(l+b)}{6EIl} - \frac{Pb \times 3x^2}{6\,EIl} = 0$$

$$x^2 = \frac{a(l+b)}{3} = \frac{a^2 + 2ab}{3}$$

$$x = \sqrt{\frac{a^2+2ab}{3}} = \sqrt{\frac{a^2+2ab+b^2-b^2}{3}} = \sqrt{\frac{(a+b)^2-b^2}{3}} = \sqrt{\frac{l^2-b^2}{3}}$$

$$y_{\max} = \frac{Pab(l+b)}{6EIl}\sqrt{\frac{l^2-b^2}{3}} - \frac{Pb}{6EIl}\left(\frac{l^2-b^2}{3}\right)^{3/2}$$

$$y_{\max} = \frac{Pb}{6\,EIl}\left(\frac{l^2-b^2}{3}\right)^{1/2}\left[a(l+b) - \frac{(l^2-b^2)}{3}\right]$$

But $a(l+b) = 3x^2$. Therefore,

$$a\,(l+b) = 3\left(\frac{l^2-b^2}{3}\right) = (l^2-b^2)$$

$$y_{\max} = \frac{Pb}{6\,EIl}\left(\frac{l^2-b^2}{3}\right)^{1/2}\left[l^2-b^2-\frac{(l^2-b^2)}{3}\right]$$

$$= \frac{Pb}{9\sqrt{3}\,EIl}(l^2-b^2)^{3/2}$$

Example 7.26 Moment-area method: SS beam with UD load

A simply supported beam carries a UD load of 20 kN/m over its span of 8 m. Determine the slope at the ends and the deflection at mid-span if $E = 200$ GN/m^2 and $I = 30{,}000$ cm^4.

Solution The beam with the loading is shown in Fig. 7.44(a).

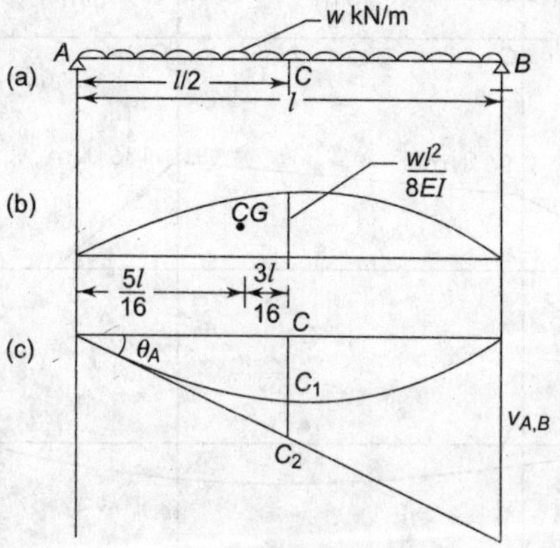

Fig. 7.44

$$\text{Maximum BM in the beam} = \frac{wl^2}{8} = \frac{20 \times 8^2}{8} = 160 \text{ kN m}$$

The *M/EI* diagram is shown in Fig. 7.44(b). From Fig. 7.44(c),

$$\theta_A = \frac{V_{A,B}}{l}$$

$$V_{A,B} = \text{moment of area of } \frac{M}{EI} \text{ diagram between } A \text{ and } B \text{ about } B$$

$$= \frac{2}{3}\frac{wl^2}{8EI}l\frac{l}{2} = \frac{wl^4}{24EI}$$

$$\theta_A = \frac{wl^4}{24EI}\frac{1}{l} = \frac{wl^3}{24EI} = \frac{20 \times 8^3}{24 \times 200 \times 10^3 \times 30{,}000 \times 10^{-8}} = 0.007 \text{ rad}$$

From considerations of symmetry, $\theta_B = -\theta_A = -0.007$ rad. The maximum deflection will be at the middle of the span. Referring to Fig. 7.44(c),

$$CC_1 = \text{deflection} = CC_2 - C_1C_2$$

$$CC_2 = \theta_A\frac{l}{2} = \frac{wl^3}{24EI}\frac{l}{2} = \frac{wl^4}{48EI}$$

$$C_1C_2 = V_{A,C} = \text{moment of area of } \frac{M}{EI} \text{ diagram between } A \text{ and } C, @\ C$$

$$C_1C_2 = \frac{2}{3} \times \frac{wl^2}{8EI}\frac{l}{2}\frac{3}{8}\frac{l}{2} = \frac{wl^4}{128EI}$$

$$CC_1 = CC_2 - C_1C_2 = \frac{wl^4}{48EI} - \frac{wl^4}{128EI} = \frac{5wl^4}{384EI}$$

$$y_{\max} = \frac{5 \times 20 \times 8^4}{384 \times 200 \times 10^6 \times 30{,}000 \times 10^{-8}} = 17.8 \text{ mm}$$

Example 7.27 Moment-area method: overhanging beam with point loads

An overhanging beam is loaded as shown in Fig. 7.45(a). Find the deflection under the loads. EI is constant. (All distances are in m and ordinates in kN m.)

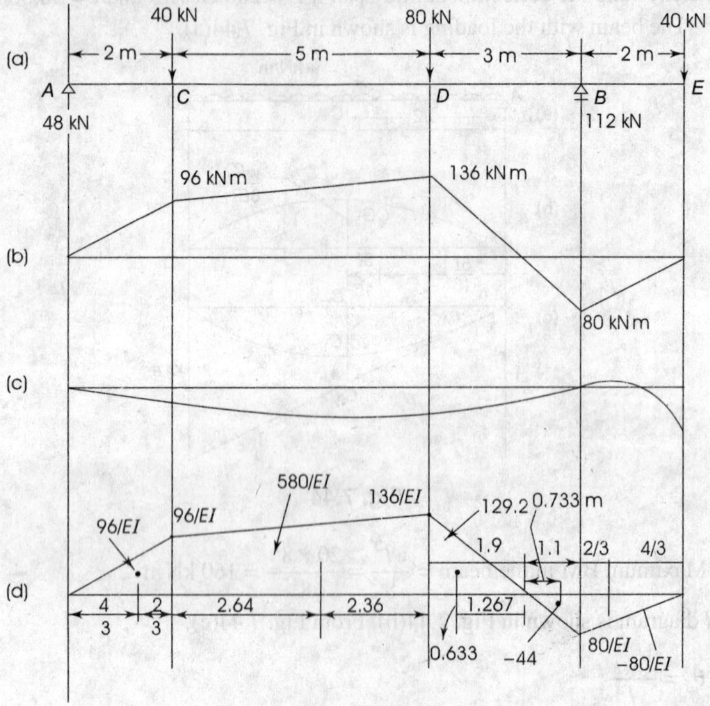

Fig. 7.45

Solution The reactions at A and B are $R_{AV} = 48$ kN ↑ and $R_{BV} = 112$ kN ↑. The BM diagram for the beam is shown in Fig. 7.45(b), the ordinates of which when divided by EI give the M/EI diagram. The deflected shape of the beam is shown in Fig. 7.45(c). From this figure, we note that $\theta_A = V_{A,B}/10$.

$$V_{A,B} = \text{moment of area of } \frac{M}{EI} \text{ diagram between } A \text{ and } B \text{ about } B$$

In Fig. 7.45(d), the areas and centroids of the different segments of the BM diagram are shown. From this diagram,

$$V_{A,B} = \frac{1}{EI}\left(96 \times \frac{26}{3} + 580 \times 5.36 + 129.2 \times 2.37 - 44 \times 0.367\right) = \frac{4227}{EI}$$

$$\theta_A = \frac{V_{A,B}}{10} = \frac{422.7}{EI} \text{ rad}$$

Similarly,

$$\theta_B = \frac{V_{B,A}}{10}$$

$$V_{B,A} = \frac{1}{EI}\left(96 \times \frac{4}{3} + 580 \times 4.64 + 129.2 \times 7.63 - 44 \times 9.633\right) = 3380$$

$$\theta_B = \frac{338}{EI}$$

The deflection at C,

$$y_C = \theta_A \times 2 - V_{A,C} = \frac{1}{EI}\left(2 \times 422.7 - 96 \times \frac{2}{3}\right)$$

$$= \frac{781.4}{EI}$$

Similarly, deflection at D,

$$y_D = 7\theta_A - V_{A,D} = \left(7 \times 422.7 - 96 \times \frac{17}{3} - 580 \times 2.36\right) \times \frac{1}{EI}$$

$$= \frac{104.6}{EI}$$

Deflection under the 40 kN load (free end),

$$y_E = \theta_B \times 2 - V_{B,E}$$

$$= \frac{338}{EI} \times 2 - \frac{80}{EI} \times \frac{4}{3}$$

$$= \frac{569.33}{EI}$$

Example 7.28 Moment-area method: SS beam with couple load

A simply supported beam is subjected to a couple of 120 kN m at a point 4 m from the left end. Calculate the slopes at the ends and the deflection at the point of application of load. *EI* is constant.

Solution The beam with the loading is shown in Fig. 7.46(a). The reactions are 120/6 = 20 kN each, downwards at A and upwards at B. The M/EI diagram is shown in Fig. 7.46(b).

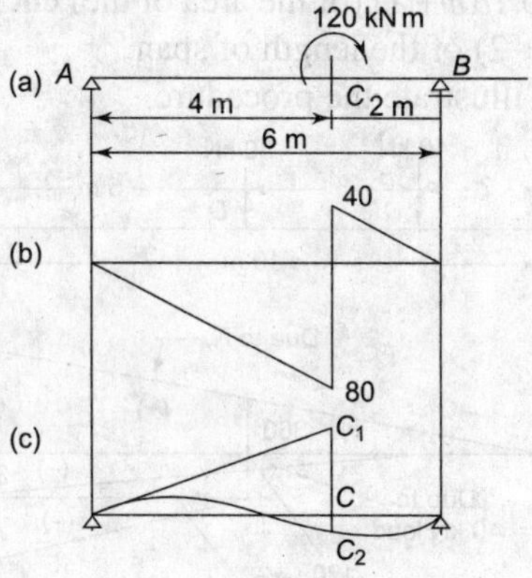

Fig. 7.46

$$\text{Slope at } A = \theta_A = \frac{V_{A,B}}{l} = \frac{[40\times(2/2)\times(4/3) - 80\times(4/2)\times(2+4/3)]}{6EI} = \frac{-80}{EI}$$

showing an anticlockwise rotation of the tangent at A.

$$\text{Slope at } B = \theta_B = \frac{V_{B,A}}{l} = [40\times(2/2)\times(14/3) - 80\times(4/2)\times(8/3)] \times \frac{1}{6EI} = \frac{-40}{EI}$$

$$\text{Deflection at } C = y_C = C_1C_2 - C_1C = -80 \times \frac{4}{2} \times \frac{4}{3} + \frac{80 \times 4}{EI} = \frac{320}{3EI}$$

The deflected shape of the beam is shown in Fig. 7.46(c).

7.8.4 Drawing Moment Diagrams by Parts

While solving problems on deflection by the area-moment method, we have to find areas and locate their centroids. While Table 7.1 can help in simple cases, they cannot be used directly in many cases due to complexity of loading. To facilitate the determination of areas and location of their centroids, a very simple technique is used, involving the drawing of what is known as a moment diagram by parts. The method is very simple in the sense that it needs drawing of a separate moment diagram, with appropriate signs, for each of the loads or reactive forces acting on the beam.

The drawing of a moment diagram by parts is based upon the following two principles.

(i) The BM at any section is calculated as the algebraic sum of the moments of all the forces to the right or left of the section. For example, considering Fig. 7.47, the BM under the 80 kN load is expressed as $72 \times 5 - 40 \times 3$. The moment due to each of the loads is calculated separately and added algebraically as $(360 - 120) = 240$ kN m. Note that the result of such addition is equal to what is calculated from the right side as $48 \times 5 = 240$ kN m. In the moment diagram by parts, each of the terms in the equation representing the moment about any section $[72\,x - 40\,(x - 2)]$ of each load is drawn separately. This is shown in Fig. 7.47(b).

(ii) Whatever be the nature of loading, the moment equation is always of the form $M = kx^n$, if the loads are considered individually. In Table 7.1, the areas of centroids of diagrams for this equation for various values of n are given. The area is equal to $1/(n + 1)$ (of the area of the) enclosing rectangle and the centroid is at $1/(n + 2)$ of the length of span.

The following examples illustrate the procedure.

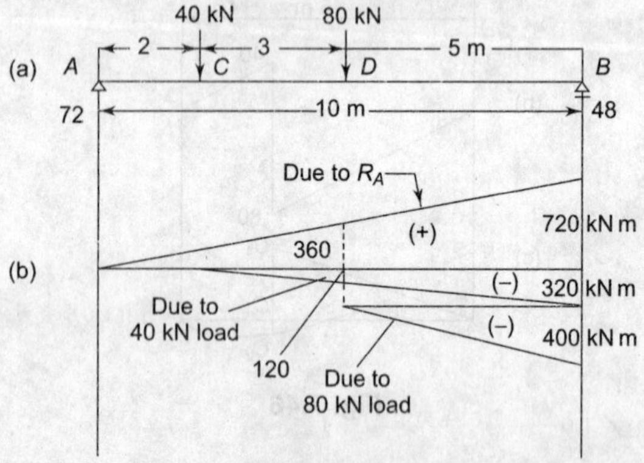

Fig. 7.47

Example 7.29 Moment-area method: moment diagram by parts

A simply supported beam of span 6 m carries a load varying uniformly from zero to 30 kN/m over the whole span. Determine the slopes at the ends and the deflection at the middle of the span. $E = 200$ GN/m^2 and $I = 8000$ cm^4.

Solution The beam with the loading is shown in Fig. 7.48(a). The reactions are

$$R_{AV} = 30 \times \frac{6}{2} \times 2 \times \frac{1}{6} = 30 \text{ kN} \uparrow$$

$$R_{BV} = 30 \times \frac{6}{2} - 30 = 60 \text{ kN} \uparrow$$

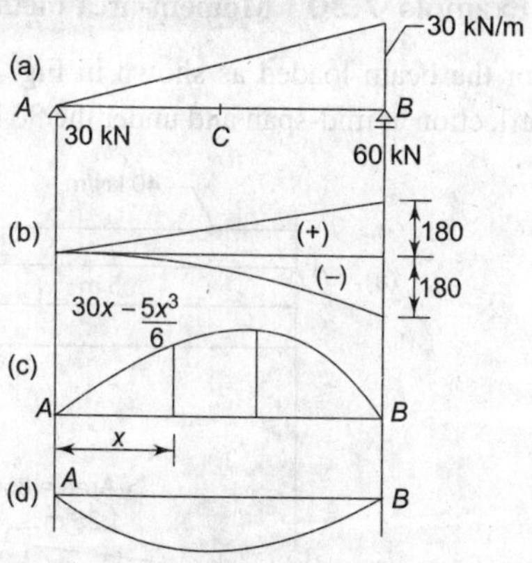

Fig. 7.48

The BM by parts is shown in Fig. 7.48(b). The triangle which is positive is due to the reaction at A of 30 kN. The BM at any section due to this load is $30x$. This ($M = 30x$) is a straight-line equation giving a value of zero at A and 180 kN m at B. The negative diagram is due to the varying load. At any section distant x from A, the BM due to this load from the left is given by

$$M_x = \frac{30}{6} x \frac{x}{2} \frac{x}{3} = \frac{5x^3}{6}$$

The curve shown in Fig. 7.48(b) represents this third-degree equation.

The conventional BM diagram is shown in Fig. 7.48(c). The ordinate of this diagram at x is given by $30x - 5x^3/6$. The area of this diagram can be obtained only by integration. *Two expressions will have to be integrated* — $\int M_x\, dx$ and $\int x\, M_x\, dx$—to obtain the area and the moment of the area about a point.

The advantages of drawing the mom-ent diagram by parts are evident in this example. The conventional BM diagram has been reduced to two geometrical shapes, both of which have standard formulae for area and centroid of area (refer to Table 7.1). The triangular and parabolic areas are shown in Fig. 7.48(b). From the deflected shape shown in Fig. 7.48(d),

$$\theta_A = \frac{V_{A,B}}{6} = \frac{1}{6EI}\left(540 \times 2 - 270 \times \frac{6}{5}\right) = \frac{126}{EI} = \frac{126}{200 \times 10^6 \times 8000 \times 10^{-8}}$$

$$= 0.0078 \text{ rad}$$

$$\theta_B = \frac{V_{B,A}}{6} = \frac{1}{6EI}\left(540 \times 4 - 270 \times \frac{24}{5}\right) = \frac{144}{EI} = \frac{144}{200 \times 10^6 \times 8000 \times 10^{-8}}$$

$$= 0.009 \text{ rad}$$

Deflection at C, mid-span, at 3 m from A,

$$y_C = 3\theta_A - \left(90 \times \frac{3}{2} \times 1 - \frac{22.5 \times 3}{2} \frac{3}{5}\right)\frac{1}{EI} = \frac{378}{EI} - \frac{124.875}{EI} = \frac{253}{EI} = 15.81 \text{ mm}$$

Note If the maximum deflection is required, we find the deflection at any distance x from A as

$$EIy_x = 126x - \left(\frac{30x^3}{6} - \frac{5x^3}{6} \frac{x}{4} \frac{x}{5}\right) = 126x - 5x^3 + \frac{x^5}{24}$$

Setting $dy/dx = 0$,

$$126 - 15x^2 + \frac{5x^4}{24} = 0$$

(By putting $x^2 = Z$, we get a quadratic equation in Z.)
Solving this, we get $x = 3.12$ m and

$$\text{Maximum deflection } y_{max} = \left[126 \times 3.12 - 5 \times 3.12^3 + \frac{3.12^5}{24}\right] = \frac{253.59}{EI}$$

$$= 15.85 \text{ mm}$$

Example 7.30 Moment-area method: moment diagram by parts

For the beam loaded as shown in Fig. 7.49(a), calculate the slopes at the ends, and the deflection at mid-span and under the 80 kN load. $E = 200 \times 10^6$ kN/m^2 and $I = 56{,}000$ cm^4.

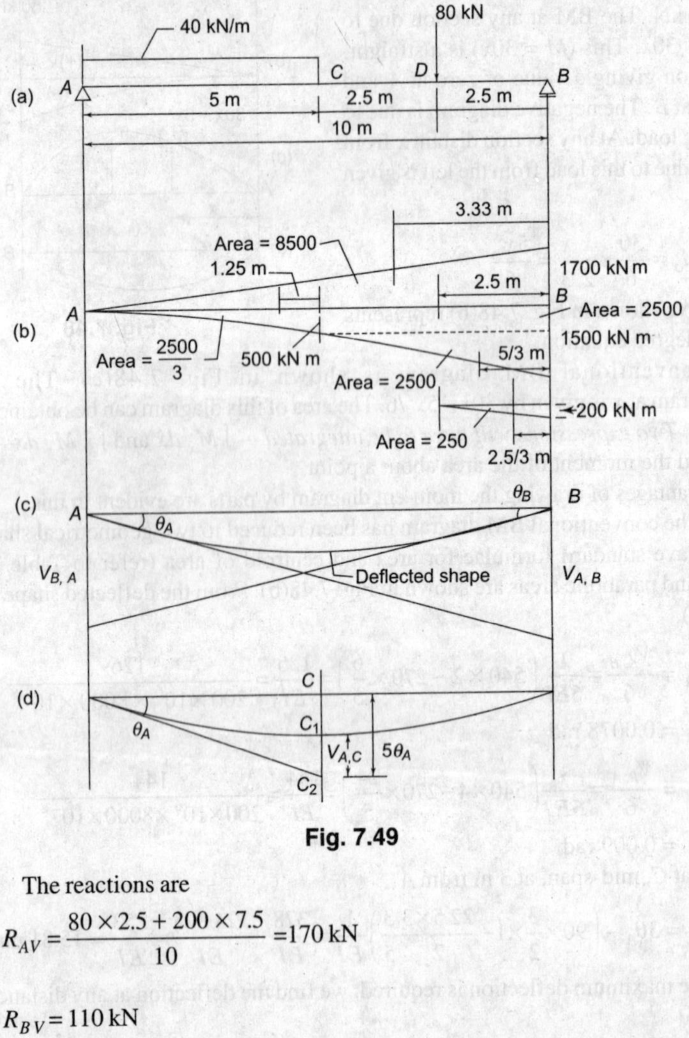

Fig. 7.49

Solution The reactions are

$$R_{AV} = \frac{80 \times 2.5 + 200 \times 7.5}{10} = 170 \text{ kN}$$

and

$$R_{BV} = 110 \text{ kN}$$

both acting upwards.

The BM diagram by parts is shown in Fig. 7.49(b). Here again, the advantage of the method is clear because a conventional BM diagram will have to be tackled by integration for finding the area and centroid, at least for the left half of the span. The BM diagram by parts consists of two triangles and a parabolic shape, which are easily tackled. Figure 7.49(b) shows the areas and centroidal distances. Figure 7.49(c) shows the calculation for θ_A and θ_B.

$$\theta_A = \frac{V_{A,B}}{l} = \frac{1}{10EI}\left[8500 \times 3.33 - \frac{2500}{3} \times 6.25 - 2500 \times 2.5 - 2500 \times \frac{5}{3} - \frac{250 \times 2.5}{3}\right]$$

$$= \frac{1247.2}{EI} = \frac{1247.2}{200 \times 10^6 \times 56{,}000 \times 10^{-8}} = 0.111 \text{ rad}$$

Similarly,

$$\theta_B = \frac{V_{B,A}}{l} = \frac{1}{10EI}\left[8500 \times 6.67 - \frac{2500}{3} \times 3.75 - 2500 \times 7.5 - \frac{2500 \times 25}{3} - 250 \times \frac{27.5}{3}\right]$$

$$= \frac{1169.5}{EI} = \frac{1169.5}{200 \times 10^6 \times 56{,}000 \times 10^{-8}} = 0.104 \text{ rad}$$

Figure 7.49(d) shows the deflected shape and calculation of y_C and y_D.
From this figure,

$$y_C = CC_2 - C_1C_2, \quad C_1C_2 = V_{A,C}, \quad \text{and} \quad CC_2 = 5\theta_A$$

Therefore,

$$y_C = \frac{5 \times 1247.2}{EI} - \frac{1}{EI}\left[2125 \times \frac{5}{3} + \frac{2500}{3} \times 1.25\right] = \frac{3736}{EI}$$

$$= \frac{3736 \times 1000}{200 \times 10^6 \times 56{,}000 \times 10^{-8}} = 33.35 \text{ mm}$$

Similarly, y_D can be calculated as

$$y_D = 7.5\theta_A - V_{A,D}$$

$$= \frac{7.5 \times 1247.2}{EI} - \frac{1}{EI}\left[4781.25 \times 2.5 + \frac{2500}{3} \times 3.75 + 1250 \times 1.25 - 625 \times \frac{2.5}{3}\right]$$

$$= \frac{2609}{EI} = \frac{2609}{200 \times 10^6 \times 56 \times 10^{-5}} = 23.29 \text{ mm}$$

Example 7.31 Moment-area method: moment diagram by parts

For the beam loaded as shown in Fig. 7.50(a), calculate the slopes at the ends and the deflection at the middle of the span. EI is constant. (All units are in m and kN.)

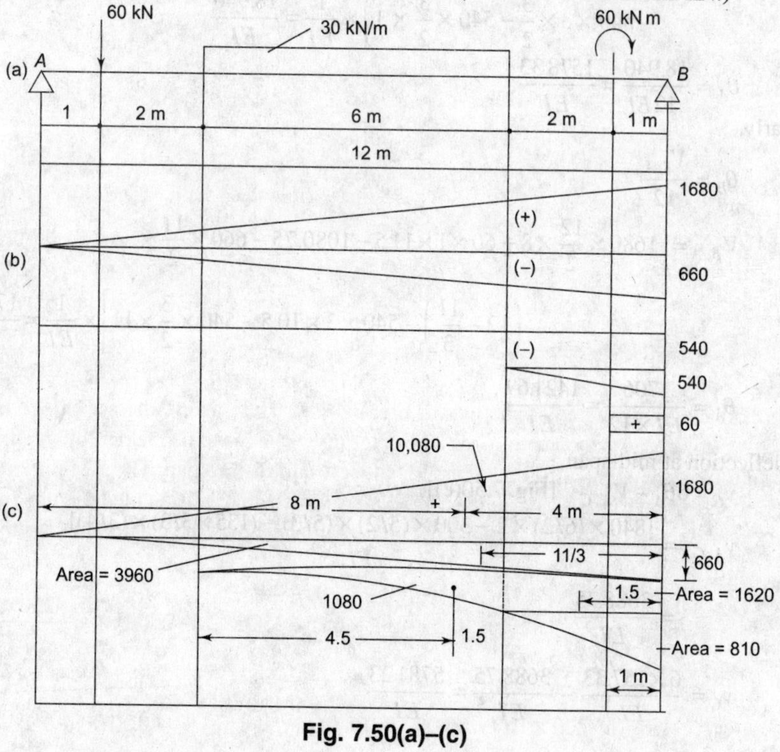

Fig. 7.50(a)–(c)

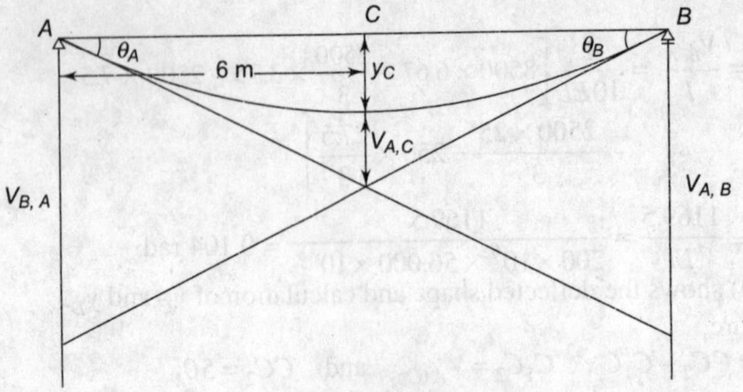

Fig. 7.50(d)

Solution The reactions are

$$R_{AV} = \frac{60 \times 11 + 30 \times 6 \times 6 - 120}{12} = 140 \text{ kN}$$

and

$$R_{BV} = 100 \text{ kN}$$

both acting upwards.

The BM diagram by parts is shown in Fig. 7.50(b). The areas and distances of centroids of these parts are shown in Fig. 7.50(c). From the deflected shape shown in Fig. 7.50(d), we can state that

$$\theta_A = \frac{V_{A,B}}{12}$$

$$V_{A,B} = \left(1680 \times \frac{12}{2} \times 4 + 60 \times 1 \times \frac{1}{2} - 660 \times \frac{11}{2} \times \frac{11}{3} - 1080 \times 4.5 - 540 \times 3 \times \frac{3}{2} - 540 \times \frac{3}{2} \times 1\right) \times \frac{1}{EI} = \frac{18{,}940}{EI}$$

$$\theta_A = \frac{18{,}940}{12EI} = \frac{1578.33}{EI}$$

Similarly,

$$\theta_B = \frac{V_{B,A}}{12}$$

$$\Rightarrow \quad V_{B,A} = \left[1680 \times \frac{12}{2} \times 8 + 60 \times 1 \times 11.5 - 1080.75 - 660 \times \frac{11}{2}\left(12 - \frac{11}{3}\right) - 540 \times 3 \times 10.5 - 540 \times \frac{3}{2} \times 11\right] \times \frac{1}{EI} = \frac{17{,}060}{EI}$$

$$\theta_A = \frac{1706}{EI \times 12} = \frac{1421.67}{EI}$$

The deflection at mid-span,

$$y_C = 6\theta_A - V_{A,C} \quad \text{[Fig. 7.50(e)]}$$

$$v_{A,C} = \frac{[840 \times (6/2) \times 2 - 300 \times (5/2) \times (5/3) - (135 \times 3/3) \times (3/4)]}{EI}$$

$$= \frac{3688.75}{EI}$$

$$y_C = \frac{6 \times 157.33}{EI} - \frac{3688.75}{EI} = \frac{5781.23}{EI}$$

7.9 THE CONJUGATE-BEAM METHOD

In the double integration method, we derived the fundamental relationship relating BM to curvature ($1/R = d^2y/dx^2$),

$$\frac{d^2y}{dx^2} = M$$

In Chapter 4, we also derived the relationships $dM/dx = F$ and $dF/dx = w$. Therefore, we have

$$\frac{d^4y}{dx^4} = \int w\text{-rate of loading}$$

$$\frac{d^3y}{dx^3} = \int w\,dx = F\text{-shear force}$$

$$EI\,\frac{d^2y}{dx^2} = \int F\,dx = M\text{-bending moment}$$

$$EI\,\frac{dy}{dx} = \int M\,dx = \text{slope}$$

$$EIy = \int\left(\int M\,dx\right) = \text{deflection}$$

These relationships show a similarity that should exist between w, F and M, and M, dy/dx (slope), and y (deflection). In fact, they also show that the deflection in certain cases can be found from the load intensity function as in simply supported beams carrying a UD load. Starting from $EI\,d^4y/dx^4 = w$, one can integrate this function four times to obtain the value of y, the deflection.

7.9.1 Basic Proposition

The conjugate-beam method is based upon the similarity of relationships between the three quantities. It is based upon the principles on which the area-moment theorems are derived but uses a different technique to find slope and deflection. Without going into a rigorous mathematical proof, let us derive the propositions of the conjugate-beam method. Considering the beam shown in Fig. 7.51(a), the M/EI diagram for the beam is shown in Fig. 7.51(b) and the deflected shape in 7.51(c). In the area-moment method, we obtain the slope at A as [from Fig. 7.51(c)]

$$\theta_A = \frac{V_{A,B}}{l} = \frac{\text{moment of the area of the } M/EI \text{ diagram between } A \text{ and } B\ @\ B}{\text{span}}$$

If we consider a beam, the load on which is the M/EI diagram of Fig. 7.51(b), we note that θ_A = reaction at the end A of such a beam. Such a beam, shown in Fig. 7.52(a), is known as the conjugate beam for the real beam shown in Fig. 7.51(a). In Fig. 7.51(c),

$$\begin{aligned}\text{Slope at } C &= \text{slope at } A - \text{the angle between the tangents drawn at } A \text{ and } C\\ &= \text{reaction at } A - \frac{M}{EI} \text{ loading between } A \text{ and } C, \text{ of the conjugate beam}\\ &= \text{SF at } C \text{ of the conjugate beam}\end{aligned}$$

Similarly, considering Fig. 7.44(c), the deflection at C is given by

$$y_C = CC_2 - C_1C_2 = \theta_A x - V_{A,C}$$

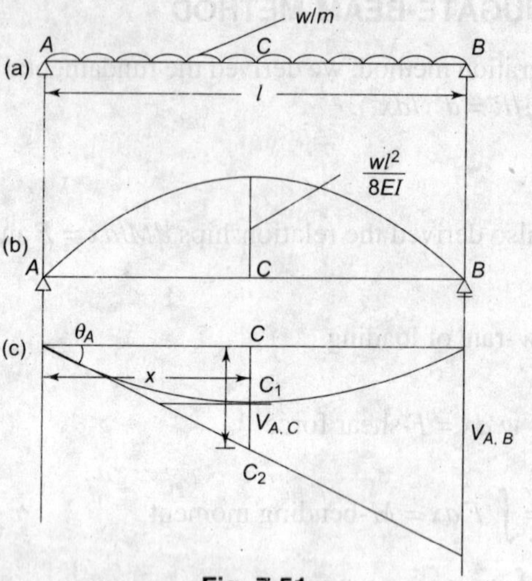

Fig. 7.51

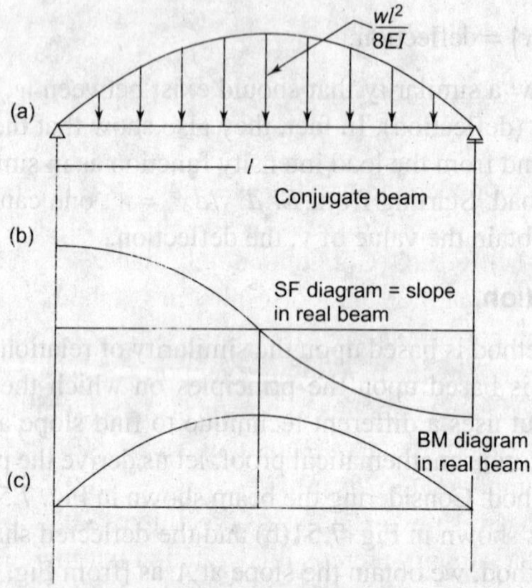

Fig. 7.52

$$= \theta_A x - \text{moment of area of } \frac{M}{EI} \text{ diagram between } A \text{ and } C \text{ about } C$$

= reaction at $A \times x$ – moment of area of M/EI diagram between A and C, @ C

y_C = BM at C of the conjugate beam

Let us look at the two propositions on which the conjugate-beam method is based, and summarize the ideas behind the concept. The conjugate beam is a beam carrying the M/EI diagram of a real beam as load.

Proposition 1. The slope at any point of a beam is equal to the SF at that point of the conjugate beam loaded with the M/EI diagram of the real beam.

Proposition 2. The deflection at any point of a beam is equal to the BM at the corresponding point of the conjugate beam loaded with the M/EI diagram of the real beam.

7.9.2 Real Beam and Conjugate Beam

Let us look at the conjugate beam in greater detail.

(i) The conjugate beam and the real beam have the same length or span because for every point of the real beam at which the slope or deflection is required to be calculated, there should be a corresponding point on the conjugate beam.

(ii) The conjugate beam should be supported in such a way as to satisfy the slope and deflection conditions of the real beam. In Fig. 7.53, the support conditions of the real beam and the corresponding conjugate beam are shown in different cases. We note the following from Fig. 7.53.

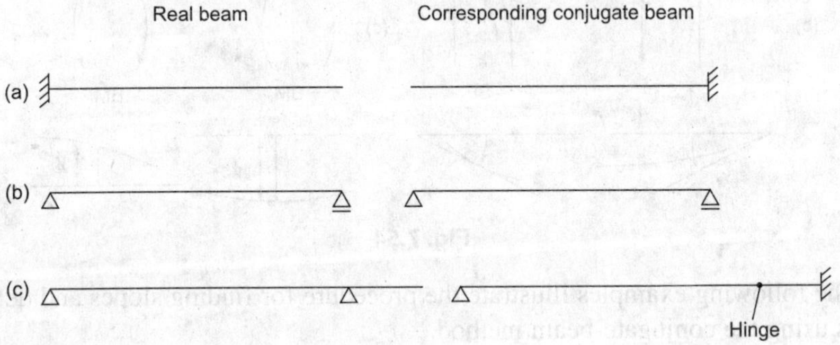

Fig. 7.53

(a) In a cantilever, the supports get interchanged. In the real beam, the deformation conditions are no slope and deflection at the fixed end but slope and deflection exist at the free end. Correspondingly, the conjugate beam should have SF as well as BM at the point corresponding to the free end but no SF and BM at the point corresponding to the fixed end of the real beam. These conditions are satisfied by exchanging the supports as shown.

(b) A simply supported beam continues to remain so. The condition that deflection is zero at both support points is satisfied, because there is no BM at the ends of a simply supported (SS) beam.

(c) The overhanging beam shown is a more revealing example of this requirement. At the interior support, there is no deflection but there is slope. So, at the corresponding point in the conjugate beam, there should be no BM. This is achieved by providing a hinge at that point, which ensures that there is no BM in the conjugate beam due to the load due to the M/EI diagram. In summary, a free end becomes a fixed end, a fixed end becomes a free end, a simple support at ends continues to remain a simple support, and a simple support in the interior becomes a hinge.

(d) The conjugate beam is loaded with the M/EI diagram of the real beam.

(e) The SF and BM at a point of the conjugate beam are equal to the slope and deflection at the corresponding point of the real beam, respectively.

Along with the sign conventions already being used for various quantities, the following sign convention will be adopted (Fig. 7.54).

(i) A positive BM in the real beam is a downward load on the conjugate beam. A negative BM becomes an upward load.

(ii) A positive SF in the conjugate beam corresponds to positive slope (clockwise rotation of the tangent). A negative SF corresponds to a negative slope.

(iii) A positive BM in the conjugate beam means downward deflection. A negative BM consequently means an upward deflection.

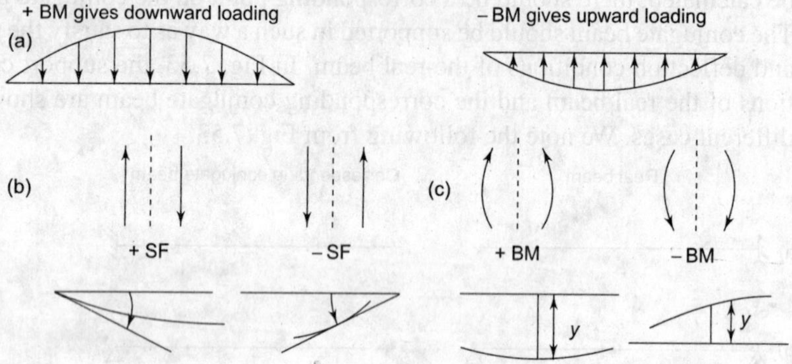

Fig. 7.54

The following examples illustrate the procedure for finding slopes and deflections using the conjugate-beam method.

Example 7.32 Conjugate-beam method: cantilever with point load

A cantilever carries a concentrated load of 60 kN at a distance of 2 m from the fixed end. The span of the cantilever is 3 m. Calculate the slope and deflection under the load and at the free end. Assume that *EI* is constant.

Solution The beam is shown in Fig. 7.55(a). The conjugate beam loaded with the *M*/*EI* diagram is shown in Fig. 7.55(b). Note that the span of the conjugate beam is the same as that of the real beam; the support conditions satisfy the deformation conditions in the real beam and the conjugate beam is loaded with the *M*/*EI* diagram of the real beam. The load of the *M*/*EI* diagram acts upwards because the BM in the real beam is negative. The reactions of the conjugate beam are shown in Fig. 7.55(b).

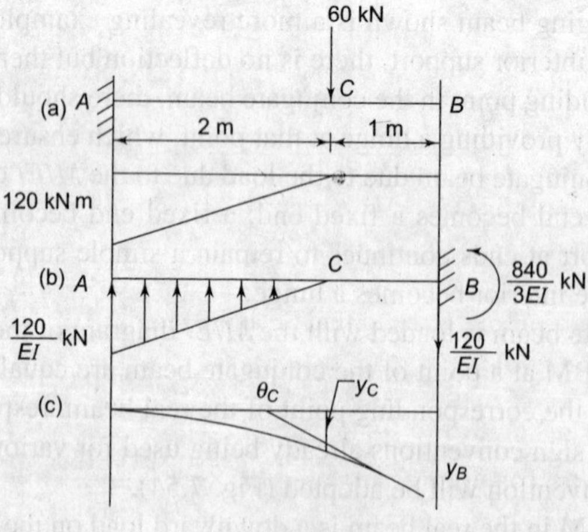

Fig. 7.55

$$\text{Total load on conjugate beam} = \frac{120}{EI}\ \text{kN} \uparrow$$

$$R_{BV} = \frac{120}{EI}\ \text{kN} \downarrow$$

$$\Sigma M = 0 \text{ @ B} \curvearrowright +, \quad \frac{120}{EI} \times \frac{2}{2} \times \frac{7}{3} - M_B = 0, \quad M = \frac{840}{EI}\ \text{kN m} \curvearrowleft$$

Slope and deflection under the load at C The slope at C, θ_C = SF in the conjugate beam at C = (120/EI) × (2/2) = 120/EI, positive, upwards from the left. See the deflected shape shown in Fig. 7.48(c). Deflection at $C = y_C$ = BM in the conjugate beam at C = (120/EI) × (4/3) = 160/EI, positive being clockwise from the left. The deflection is, therefore, downwards.

Slope and deflection at the free end B of the real beam. Slope at B, θ_B = SF at B in the conjugate beam = 120/EI. The deflection at B, y_B = BM at B in the conjugate beam = 120/EI × (2/2) × (7/3) = (840/EI), is positive and hence downwards.

Example 7.33 Conjugate-beam method: SS beam with two point loads

A simply supported beam of span 8 m carries loads of 80 kN and 60 kN at 3 m and 6 m, respectively, from the left end. Calculate the values of the slopes at the ends, and the deflection under the load if E = 200 GN/m^2 and I = 30,000 cm^4.

Solution The beam with the loading is shown in Fig. 7.56(a), and the BM diagram in Fig. 7.56(b). The BM is positive throughout and hence acts as a downward load on the conjugate beam shown in Fig. 7.56(c). The supports of the real beam and the conjugate beam are the same in this case.

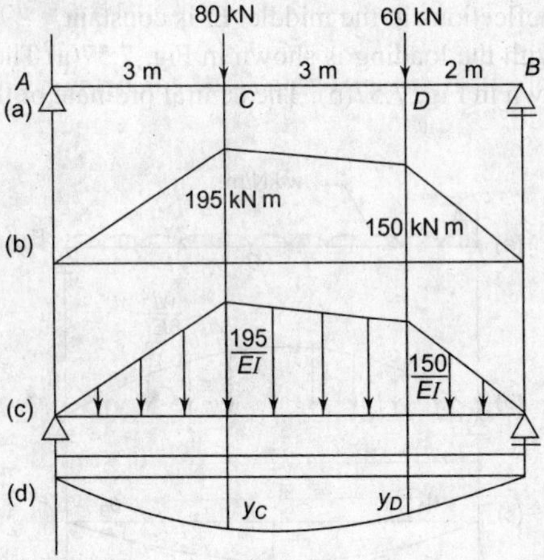

Fig. 7.56

Let us analyse the conjugate beam to find SF and BM.

$$\text{Reaction at } A = \frac{1}{8EI}\left(195 \times \frac{3}{2} \times 6 + 517.5 \times 3.57 + 150 \times \frac{4}{3}\right) = \frac{475}{EI} \uparrow$$

$$\text{Total load on the conjugate beam} = \frac{1}{EI}\left(195 \times \frac{3}{2} + \frac{195+150}{2} \times 3 + 150 \times \frac{2}{2}\right)$$

$$= \frac{960}{EI}$$

$$\text{Reaction at } B = \frac{960}{EI} - \frac{475}{EI} = \frac{485}{EI} \uparrow$$

SF at A of conjugate beam = slope at A in real beam

Slope at A,

$$\theta_A = \text{reaction at } A = \frac{475}{EI} = \frac{475}{200 \times 10^6 \times 30{,}000 \times 10^{-8}} = 0.0079 \text{ rad}$$

Slope at B,

$$\theta_B = \text{reaction at } B = \frac{-485}{EI} = \frac{-485}{200 \times 10^6 \times 30{,}000 \times 10^{-8}}$$

$$= -0.008 \text{ rad anticlockwise}$$

Deflection at C (under 80 kN load) = BM at C in conjugate beam

$$y_C = \frac{475}{EI} \times 3 - \frac{195}{EI} \times \frac{3}{2} \times 1 = \frac{1132.5}{EI} = \frac{1132.5}{200 \times 10^6 \times 30{,}000 \times 10^{-8}}$$

$$= 18.87 \text{ mm (downwards)}$$

Deflection at D = BM at D in conjugate beam. Calculating from the right,

$$y_D = \frac{485}{EI} \times 2 - \frac{150}{EI} \times \frac{2}{2} \times \frac{2}{3} = \frac{870}{EI} = \frac{870}{200 \times 10^6 \times 30{,}000 \times 10^{-8}}$$

$$= 14.5 \text{ mm (downwards)}$$

The deflected shape is shown in Fig. 7.56(d).

Example 7.34 Conjugate-beam method: SS beam with UD load

A simply supported beam carries a UD load of w/m over the whole span. Determine the slopes at the ends and deflections in the middle. EI is constant.

Solution The beam with the loading is shown in Fig. 7.57(a). The conjugate beam with the M/EI loading is shown in Fig. 7.57(b). The central ordinate of the parabolic M/EI diagram is $wl^2/8EI$.

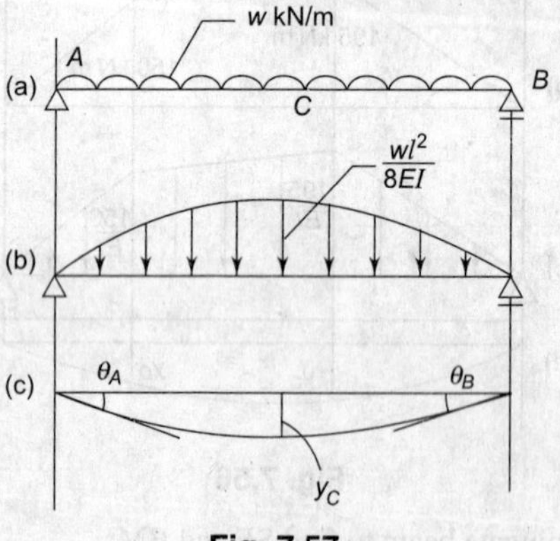

Fig. 7.57

$$\text{Total load on the conjugate beam} = \frac{2}{3}\frac{wl^2}{8EI} l = \frac{wl^3}{12EI}$$

The reactions at A and B of the conjugate beam are equal from considerations of symmetry.

Reaction at A,

$$R_{AV} = R_{BV} = \frac{wl^3}{24EI}$$

Slope at A in real beam = SF at A in conjugate beam = $\dfrac{wl^3}{24EI}$

Slope at B in real beam = SF at B in conjugate beam = $\dfrac{-wl^3}{24EI}$

negative (tangent rotated anticlockwise). Deflection at C, centre of span,

$$y_C = \text{BM at } C \text{ in conjugate beam}$$

$$= \frac{wl^3}{24EI} - \frac{2}{3}\frac{wl^2}{8EI}\frac{l}{2} \times \frac{3}{8}\frac{l}{2} \text{ (refer to Table 7.1 for areas and centroids)}$$

$$= \frac{wl^4}{48EI} - \frac{wl^4}{128EI} \times \frac{5}{384}\frac{wl^4}{EI}$$

Example 7.35 Conjugate-beam method: overhanging beam with point loads

An overhanging beam of span 12 m is supported at the left end and at 10 m from the same end. It carries loads as shown in Fig. 7.58(a). Calculate the slope at the supports and the deflections under the loads.

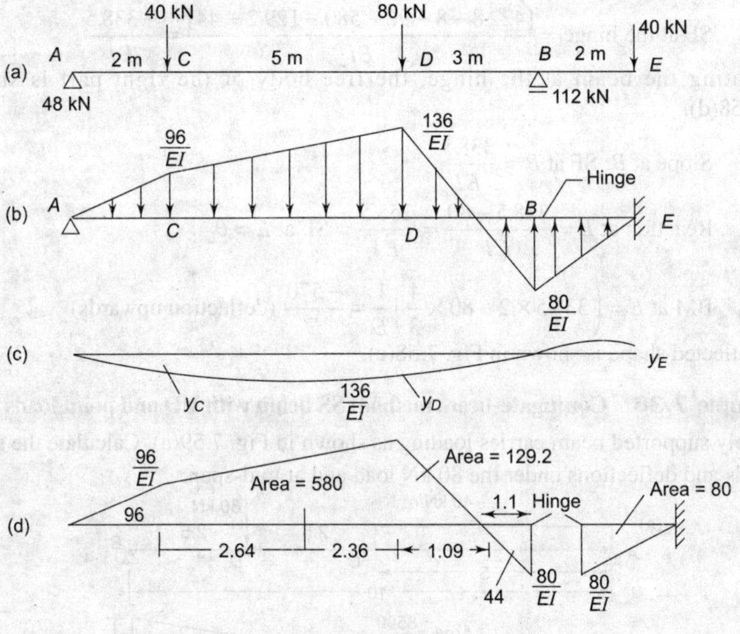

Fig. 7.58

Solution The beam and the loading is shown in Fig. 7.58(a). The conjugate beam with the M/EI loading is shown in Fig. 7.58(b). Note that in the real beam shown in Fig. 7.58(a),

$$R_{AV} = \frac{40 \times 8 + 80 \times 3 - 40 \times 2}{10} = 48 \text{ kN} \uparrow, \quad R_{BV} = 160 - 48 = 112 \text{ kN} \uparrow$$

BM at $C = 48 \times 2 = 96$ kNm

BM at $D = 48 \times 7 - 40 \times 5 = 136$ kNm

BM at $B = -40 \times 2 = -80$ kNm

The conjugate beam has a simple support at A, a fixed support at E, and a hinge at B to satisfy the deformation conditions at these points. The BM about the hinge at B in the conjugate beam is zero.

$$R_{AV} = \left(96 \times \frac{26}{3} + 580 \times 5.356 + 129.2 \times 2.37 - 44 \times \frac{1.1}{3}\right)\frac{1}{10EI} = \frac{422.8}{EI}$$

This is also the SF at A. Therefore,

$$\theta_A = \frac{422.8}{EI}$$

The deflection at C,

$$y_C = \text{BM at } C \text{ in the conjugate beam}$$

$$= \frac{422.8}{EI} \times 2 - \frac{96}{EI} \times \frac{2}{3} = \frac{781.6}{EI}$$

The deflection at D,

$$y_D = \text{BM in the conjugate beam at } D$$

$$= \frac{422.8 \times 7}{EI} - \frac{96}{EI} \times \frac{17}{3} = \frac{580 \times 2.356}{EI}$$

$$= \frac{1049}{EI}$$

$$\text{SF at the hinge} = \frac{[422.8 - 8 - 96 - 580 - 129.2 + 44]}{EI} = \frac{-338.5}{EI}$$

Separating the beam at the hinge, the free body of the right part is shown in Fig. 7.58(d).

$$\text{Slope at } B\text{: SF at } B = \frac{338.5}{EI}$$

$$\text{Reaction at } E = \frac{338.5 - 80}{EI} = \frac{258.5}{EI} = \text{SF at } E = \theta_E$$

$$\text{BM at } E = \left(338.5 \times 2 - 80 \times \frac{4}{3}\right)\frac{1}{E} = \frac{-570}{E} \text{ (deflection upwards)}$$

The deflected shape is shown in Fig. 7.58(c).

Example 7.36 Conjugate-beam method: SS beam with UD and point loads

A simply supported beam carries loading as shown in Fig. 7.59(a). Calculate the slopes at the ends and deflections under the 80 kN load and at mid-span.

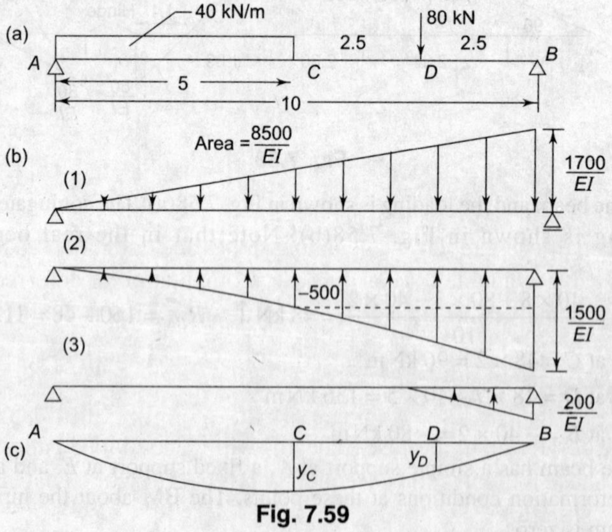

Fig. 7.59

Solution To help make calculations easy, let us draw the moment diagram by parts. The conjugate beams loaded with these parts are shown in Fig. 7.59(b). To do the calculations, all the three figures should be used. The sum of values obtained individually for the three loaded beams is the total value for the given beam. Note that the loads in (3) and (2) are upwards, due to negative BM.

$$\text{Reaction at } A \text{ in } 1 = \frac{8500}{EI} \times \frac{10}{3} \times \frac{1}{10} = 2833.33 \uparrow$$

$$\text{Reaction at } A \text{ in } 2 \quad = -\left(\frac{2500}{3} \times \frac{25}{4} + 2500 \times 2.5 + 2500 \times \frac{5}{3}\right)\frac{1}{10EI}$$

$$= -\,1562.50 \downarrow$$

$$\text{Reaction at } A \text{ in } 3 \quad = -\left(\frac{250}{EI} \times \frac{2.5}{3} \times \frac{1}{10}\right) = -\,20.83 \downarrow$$

$$\text{Total reaction at } A \quad = \frac{2833.33 - 1562.50 - 20.83}{EI} = \frac{1250}{EI} \uparrow$$

$$\text{Total load on beam} = \left[8500 - \frac{2500}{3} - 2500 - 2500 - 250\right]\frac{1}{EI} = \frac{2416.67}{EI} \uparrow$$

$$\text{Reaction at } B = \frac{2416.67 - 1250}{EI} = \frac{1166.67}{EI} \uparrow$$

$$\text{Slope at } A = \text{SF in conjugate beam at } A = \frac{1250}{EI}$$

$$\text{Slope at } B = \text{SF in conjugate beam at } B = \frac{-1166.67}{EI}$$

$$\text{Deflection at } C \quad = \text{BM in conjugate beam at } C$$

$$= \frac{1}{EI}\left[1250 \times 5 - 850 \times \frac{5}{2} \times \frac{5}{3} + \frac{2500}{3} \times \frac{5}{4}\right] = \frac{3750}{EI}$$

$$\text{Deflection at } D \quad = \text{BM in conjugate beam at } D$$

$$= \frac{1}{EI}\left(1250 \times 7.5 + \frac{2500}{3} \times \frac{15}{4} + 500 \times 2.5 \times \frac{2.5}{2}\right.$$

$$\left. + \, 500 \times \frac{2.5}{2} \times \frac{2.5}{3} - 1275 \times \frac{7.5}{2} \times \frac{7.5}{3}\right)$$

$$= \frac{2930.33}{EI}$$

■

We have seen three methods of calculating slopes and deflections. There are many other methods also available for this purpose. Computation of deformations is very important in the analysis of indeterminate structures. This is dealt with in Chapter 14.

7.10 STANDARD CASES

The slope and deflection in the case of beams with general load patterns are shown in Table 7.2. These will have to be frequently referred to while analysing statically indeterminate structures.

Table 7.2 Standard cases of deflections

	Loading	*Slope and deflection*
1.	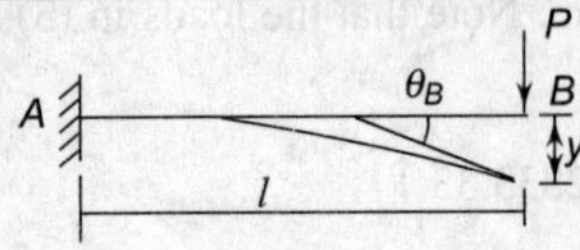	$\theta_B = \frac{Pl^2}{2EI}, \quad y_B = \frac{Pl^3}{3EI}$
2.	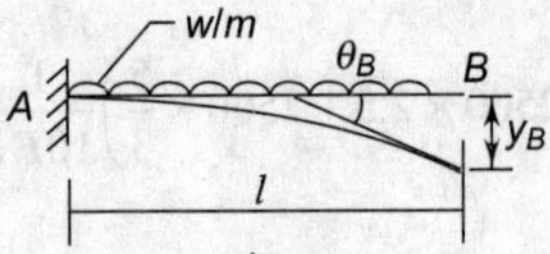	$\theta_B = \frac{wl^3}{6EI}, \quad y_B = \frac{wl^4}{8EI}$
3.	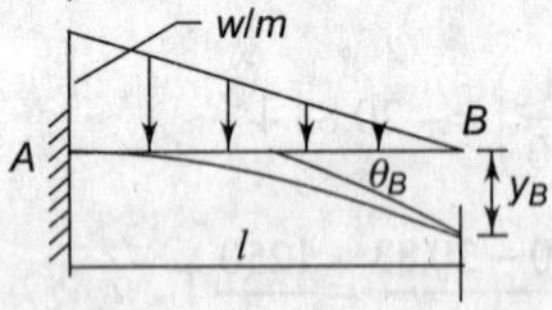	$\theta_B = \frac{wl^3}{24EI}, \quad y_B = \frac{wl^4}{30EI}$
4.	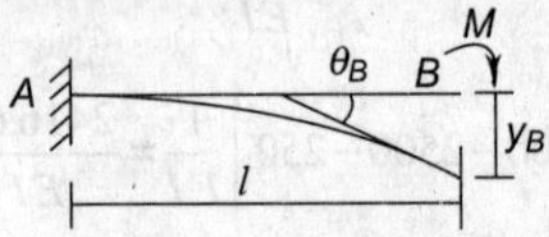	$\theta_B = \frac{Ml}{EI}, \quad y_B = \frac{Ml^2}{2EI}$
5.	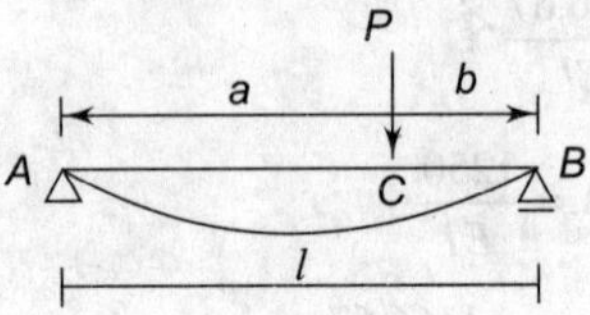	$\theta_A = \frac{Pab(l+b)}{6EIl}, \quad \theta_B = \frac{Pab(l+a)}{6EIl}$ At C (under the load), $y_C = \frac{Pa^2b^2}{3EIl}$ If $a > b$, max y at $\sqrt{\frac{l^2 - b^2}{3}}$ and $y_{max} = \frac{Pb(l^2 - b^2)^{3/2}}{9\sqrt{3}EI\,l}$ If $b > a$, max. y at $\sqrt{\frac{l^2 - a^2}{3}}$ and $y_{max} = \frac{Pa(l^2 - a^2)^{3/2}}{9\sqrt{3}EIl}$
6.	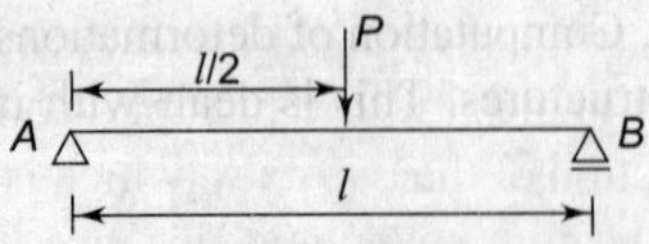	In 5, if $a = b = l/2$ $\theta_A = -\theta_B = \frac{Pl^2}{16EI}$ $y_C = y_{max} = \frac{Pl^3}{48EI}$

(contd)

Table 7.2 *(contd)*

	Loading	***Slope and deflection***
7.	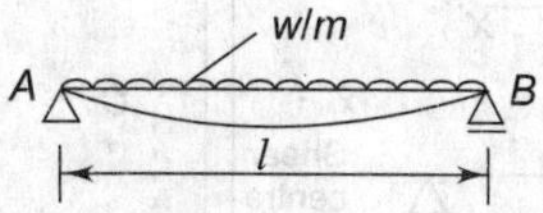	$\theta_A = -\theta_B = \frac{wl^3}{24EI}$ y_{max} at $x = l/2 = \frac{5}{384}\frac{wl^4}{EI}$
8.	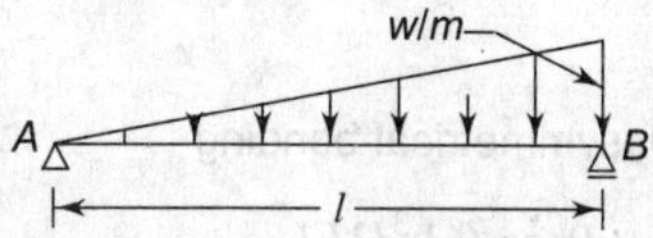	$\theta_A = \frac{7}{360}\frac{wl^3}{EI}, \quad \theta_B = -\frac{wl^3}{45EI}$ Max deflection at $0.5193l$ from A, $y_{max} = 0.00652\frac{wl^4}{EI}$
9.	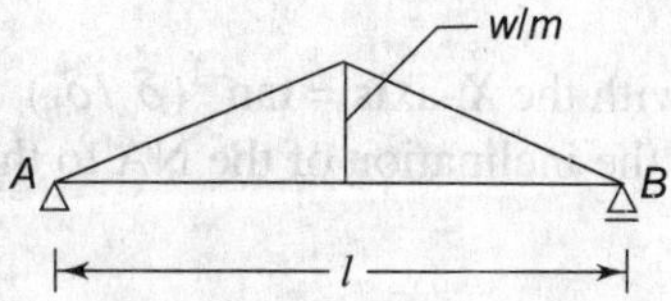	$\theta_A = \frac{5}{192}\frac{wl^3}{EI} = -\theta_B$ y_{max} at $x = l/2 = \frac{wl^4}{120EI}$
10.	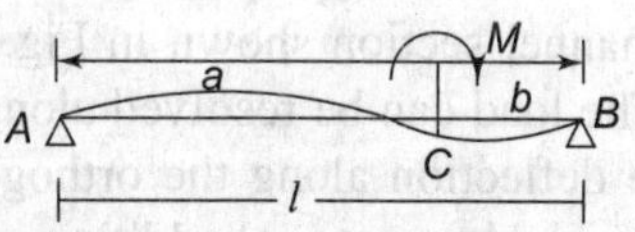	$\theta_A = \frac{-M}{6EIl}(3a^2 - l^2)$ $\theta_B = \frac{-M}{6EIl}(3a^2 - l^2)$ $y_C = \frac{Ma}{6EIl}(a^2 + 3b^2 - l^2)$ If $a = b = l/2$, $\theta_A = -\theta_B = \frac{Ml}{24EI}$ $y_C = 0$

7.11 DEFLECTIONS IN UNSYMMETRICAL BENDING

Unsymmetrical bending, as discussed in Chapter 5, is due to the plane of loads not being in the plane of one of the axes of symmetry. There is bending about both axes of symmetry. Deflections in such cases can be calculated in both directions at right angles. The net deflection will be the vector sum of the deflections in the two directions. It can be shown that the net deflection is at right angles to the neutral axis.

Considering the rectangular section with two axes of symmetry shown in Fig. 7.60, the plane of loading is inclined at an angle θ to the plane containing the Y-axis. The load passes through the intersection of the two axes, which is also the shear centre of the section. Assume that the beam is spanning a distance L and the load is a concentrated load at the centre. The load can be resolved along the two principal axes -$P \cos \theta$ parallel to the Y-axis and $P \sin \theta$ parallel to the X-axis. The deflection along the X- and Y-axes can be calculated using the principles of deflection discussed earlier.

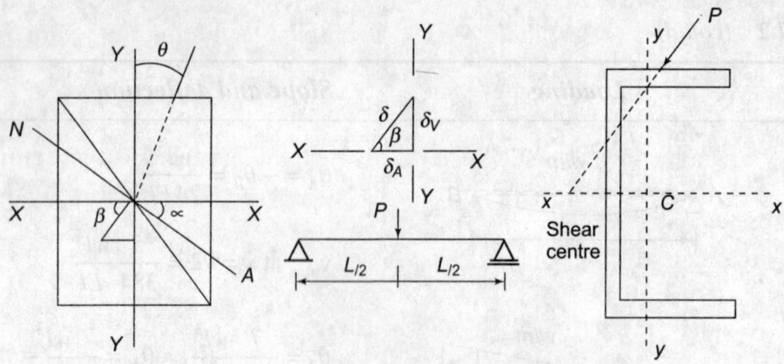

Fig. 7.60 Deflection in Unsymmetrical bending

Vertical deflection along the Y-axis, $\delta_v = P\cos\theta\, L^3/3EI_{xx}$

Horizontal deflection along the X-axis, $\delta_h = P\sin\theta\, L^3/3EI_{yy}$

Net deflection = $\sqrt{(\delta_v^2 + \delta_h^2)}$

The angle made by the net deflection with the X-axis = $\tan^{-1}(\delta_v/\delta_h)$

In this case, as discussed in Chapter 5, the inclination of the NA to the X-axis, α, is given by

$$\tan\alpha = I_{xx}\tan\theta/I_{yy}$$

It can be shown that the net deflection is in a direction perpendicular to the NA.

If the section is unsymmetrical, as a channel section shown in Fig. 7.60, the load must pass through the shear centre. The load can be resolved along the perpendicular axes through the centroid. The deflection along the orthogonal axes can be calculated and the net deflection is found by vectorial addition.

Net deflection is perpendicular to the NA It has already been discussed in Chapter 5 that the neutral axis makes an angle α with the X-axis given by

$$\tan\alpha = (I_{xx}/I_{yy})\tan\theta$$

where θ is the angle made by the plane of loading with the plane of the Y-axis.

Component of the load along the Y-axis (refer to Fig. 7.60) = $P\cos\theta$

Deflection due to this load in vertical direction, $\delta_v = P\cos\theta\, L^3/48EI_{xx}$

Component of the load in the X-direction = $P\sin\theta$

Deflection in the horizontal direction, $\delta_h = P\sin\theta\, L^3/48EI_{yy}$

Net deflection = $\sqrt{[\delta_v^2 + \delta_h^2]}$

Angle made by the net deflection with the X-axis, $\tan\beta = \delta_v/\delta_h$

$$\tan\beta = [P\cos\theta\, L^3/48EI_{xx}]/[P\sin\theta\, L^3/48EI_{yy}] = [I_{yy}/I_{xx}]\cot\theta$$

As you can see, $\tan\alpha = 1/\tan\beta = \cot\beta = \tan(90 - \beta)$, giving $\alpha + \beta = 90°$

This shows that the net deflection is normal to the neutral axis.

The following examples illustrate the principles.

Example 7.37 Deflection in unsymmetrical bending

A rectangular section 200 mm × 400 mm is subjected to a point load of 20 kN inclined to the plane of the Y-axis by 30°. The span of the beam is 6 m and the load is centrally placed on the span. If E = 200 GPa, find the maximum deflection.

Solution The beam with the loading is shown in Fig. 7.61.

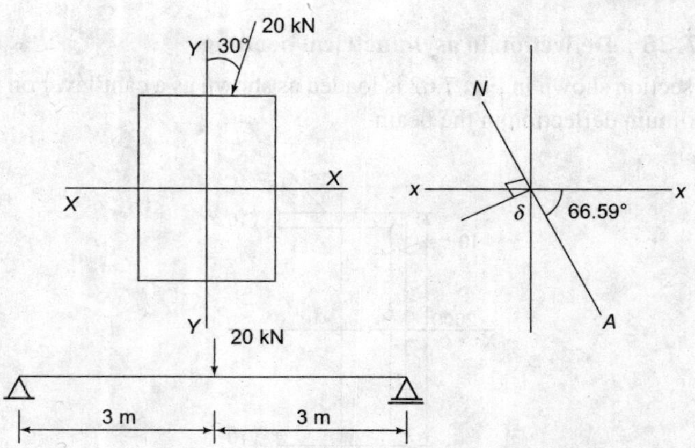

Fig. 7.61

The X-X and Y-Y axes are the principal axes of the section as both are the axes of symmetry. Point O, the intersection of these axes, is the shear centre. The load, passing through the shear centre, can be resolved into the components along the principal axes.

Component load along the Y-axis = 20 cos 30° = $10\sqrt{3}$ kN

Component load along the X-axis = 20 sin 30° = 10 kN

Second moment of area of the section can be found about the two axes as follows:

$$I_{xx} = 200 \times 400^3/12 = 1066.67 \times 10^6 \text{ mm}^4$$

$$I_{yy} = 400 \times 200^3/12 = 266.67 \times 10^6 \text{ mm}^4$$

Due to a central point load, maximum deflection = $PL^3/48EI$

Deflection along the Y-axis, vertical deflection, and deflection along the X-axis, horizontal deflection, can be found using the component loads. The net deflection is found by adding vectorially the two deflections.

Vertical deflection, $\delta_v = 10\sqrt{3}(6)^3/(48 \times 200 \times 10^6 \times 1066.67 \times 10^6 \times 10^{-12})$

$= 0.3653 \times 10^{-3}$ m = 0.3653 mm

[Note the conversions made: 200 GPa = 200×10^9 N/m^2 = 200×10^6 kN/m^2, $I = 1066.67 \times 10^6$ mm^4 = $1066.67 \times 10^6 \times 10^{-12}$ m^4. The result comes in metres.]

Horizontal deflection, $\delta_h = 10(6)^3 \times 1000/(48 \times 200 \times 10^6 \times 266.67 \times 10^6 \times 10^{-12})$

$= 0.8437$ mm

Net deflection = $\sqrt{(0.3653^2 + 0.8437^2)}$ = 0.9193 mm = 0.92 mm

Angle β made by the net deflection with the X-axis = $\tan^{-1}(0.3653/0.8437)$ = 23.41°

Angle α made by the neutral axis with the X-axis is given by the relationship

$$\tan \alpha = (I_{xx}/I_{yy})\tan \theta$$

where θ is the angle of the loading plane with the plane of the Y-axis.

So, $\tan \alpha = (1066.67/266.67)\tan 30° = 66.59°$

The directions of the net deflection and that of the neutral axis are shown in the figure. It is easy to see that the net deflection is perpendicular to the neutral axis (66.59° + 23.41° = 90°). ■

Example 7.38 Deflection in asymmetrical bending

The channel section shown in Fig. 7.62 is loaded as shown as a cantilever on a span of 2 m. Find the maximum deflection in the beam.

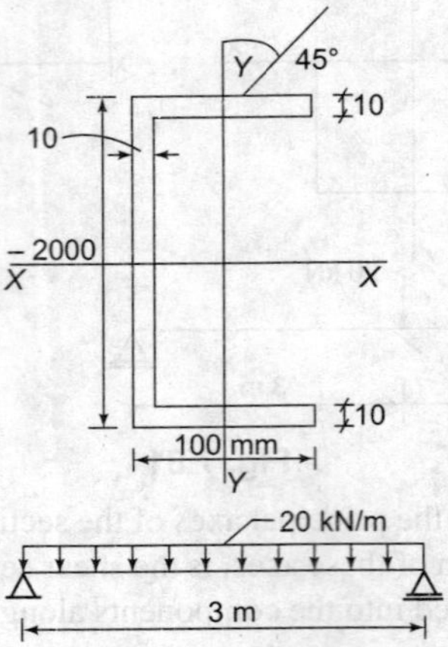

Fig. 7.62

Solution The section is symmetrical about the X-axis. The centroid is located as follows.

Area of the section = $2 \times 100 \times 10 + 180 \times 10 = 3800\ mm^2$

Now, $3800\,X = 2000 \times 50 + 1800 \times 5;\ X = 50$ mm

This is the same as the centroid of the flanges.

$I_{xx} = 2(100 \times 10^3/12 + 1000 \times 95^2) + 10 \times 180^3/12 = 28.74 \times 10^6\ mm^4$

$= 28.74 \times 10^{-6}\ m^4$

$I_{yy} = 2 \times 10 \times 100^3/12 + 10 \times 180^3/12 + 1800 \times 45^2 = 5.3267 \times 10^6\ mm^4$

$= 5.33 \times 10^{-6}\ m^4$

$w = 20$ kN/m; Vertical component, $w_v = w_h = 20\cos 45° = 14.14$ kN/m

Deflection due to UD load on a simply supported span = $5wL^4/384EI$

Vertical deflection, $\delta_v = 5 \times 14.14 \times (3)^4/\ [384 \times 200 \times 106\ (28.74 \times 10^{-6})]$

$= 2.59$ mm

Horizontal deflection, $\delta_h = 5 \times 14.14 \times (3)^4/\ [384 \times (200 \times 10^6) \times (5.33 \times 10^{-6})]$

$= 14$ mm

Net deflection = $\sqrt{[2.59^2 + 14^2]} = 14.24$ mm

Direction of the net deflection, $\beta = \tan^{-1}(2.59/14) = 10.5°$ with the X-axis

Angle made by the NA with the X-axis, α, is given by

$\tan \alpha = (I_{xx}/I_{yy}) \tan \theta$

or $\alpha = \tan^{-1}(28.74 \tan 45°/5.33) = 79.5°$ with the X-axis

Now, $\beta + \alpha = 90°$, showing that the net deflection is at right angle to the NA.

Summary

Deformations in beams, and slopes and deflections are calculated to ensure adequateness in design for stiffness. The slope is the angle that the tangent to the elastic curve makes with the initial unbent axis of the beam. The deflection is the vertical displacement of a point of the beam due to bending. The bending equation $M/I = E/R$ is used to derive the fundamental equation for calculating the deflection. Since beam deflections are small, the curvature $1/R$ can be equated to d^2y/dx^2, thus giving the equation $EI\, d^2y/dx^2 = -M$. EI is known as flexural rigidity. The BM is positive when sagging, deflection is positive downwards and the slope is positive when the tangent has rotated clockwise with respect to the original axis.

The double integration method is a mathematical procedure whereby the differential $EI\, d^2y/dx^2$ gives $EI\, dy/dx$ on first integration and EIy on second integration. Each integration yields an unknown constant of integration, which is evaluated from the restraint conditions of the beam. Using Macaulay's method greatly facilitates the double integration method—one can formulate a BM equation even in case of discontinuous loading by using the terms selectively while computing values.

The area-moment method, while based upon the same fundamental bending equation, involves the use of a different technique — that of calculating areas and moments of areas of M/EI diagrams. The area-moment theorems, also known as Mohr's theorems, give two propositions. First, that the area of the M/EI diagram between any two sections of a beam is equal to the angle between the tangents drawn at these points of the elastic curve. And secondly, that the moment of the area of the M/EI diagram between any two points is equal to the vertical intercept at the point about which the moment is taken between the elastic curve and the tangent drawn at the other point. Though these two theorems, do not directly yield slope and deflection, they can be used to evaluate them.

In the case of many types of loading, it can become difficult to evaluate areas and moments of areas of M/EI diagrams. In such cases this is done by integration. Drawing the moment diagram by parts for each load for the reactive components separately greatly facilitates computation of areas.

The conjugate-beam method is based upon the same principles as those of area-moment method but differs in technique. A conjugate beam has the same length as the actual beam, is loaded with the M/EI diagram of the actual beam, and is suitably supported, based upon the deformation conditions of the actual beam. The SF and BM in the conjugate beam are, respectively, equal to the slope and deflection in the actual beam.

When a beam is subjected to unsymmetrical bending, deflection can be calculated along the principal axes and the total deflection can be computed by vectorial addition. The total deflection is perpendicular to the neutral axis of the section.

Exercises

Review Questions

1. State the need for calculating deformations in beams.
2. Define the term stiffness. What is meant by flexural rigidity?
3. Explain the terms slope and deflection in beams with the help of a sketch.
4. In the case of the elastic curve, what is the simplification made in the expression for curvature and why?
5. Draw approximately the deflected shapes of the beams loaded as shown in Fig. 7.63.
6. State the area-moment theorems. Compare the double integration and moment-area methods and state the advantages of each.
7. Draw conjugate beams corresponding to the beams shown in Fig. 7.64. What can you say about conjugate beams for indeterminate structures.
8. Determine the boundary conditions that will be used for calculating the constants of integration obtained while using the double integration method to find slopes and deflections in the beams shown in Fig. 7.65.

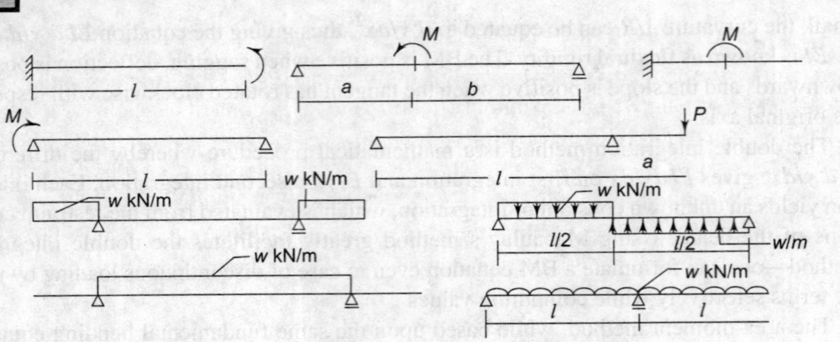

Fig. 7.63

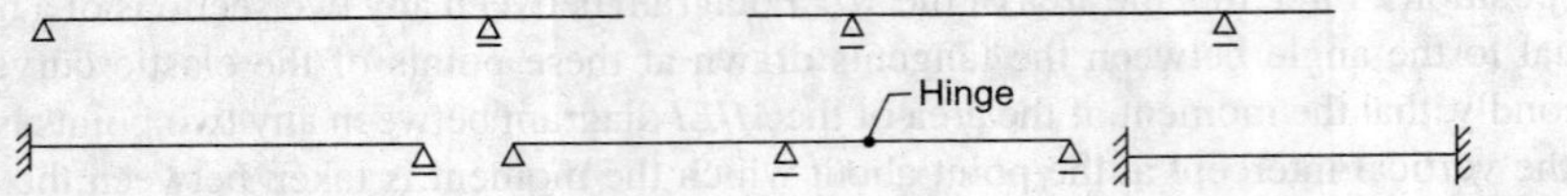

Fig. 7.64

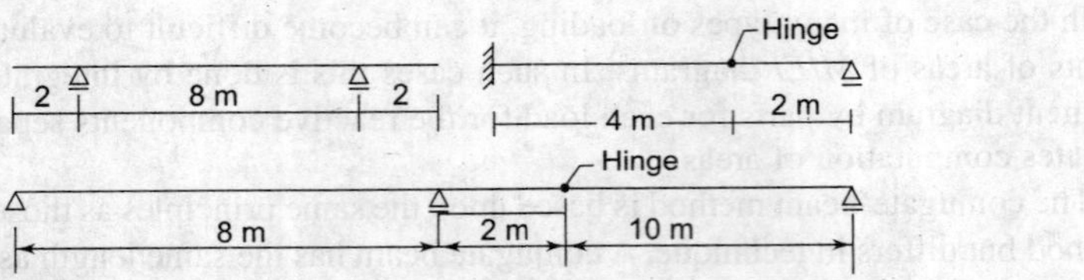

Fig. 7.65

9. Show by taking a simple example that the net deflection in unsymmetrical bending is normal to the neutral axis.

Problems

Assume *EI* is constant wherever specific values are not given, and evaluate slopes and deflections in terms of *EI*.

1. Derive the general BM equations according to Macaulay's method in the case of the beams loaded as shown in Fig. 7.66. Derive the equation of the elastic curve by the double integration method. Sketch the shape of the curve.

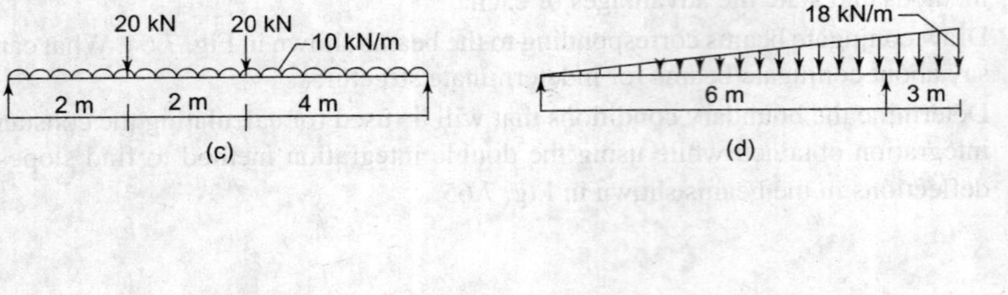

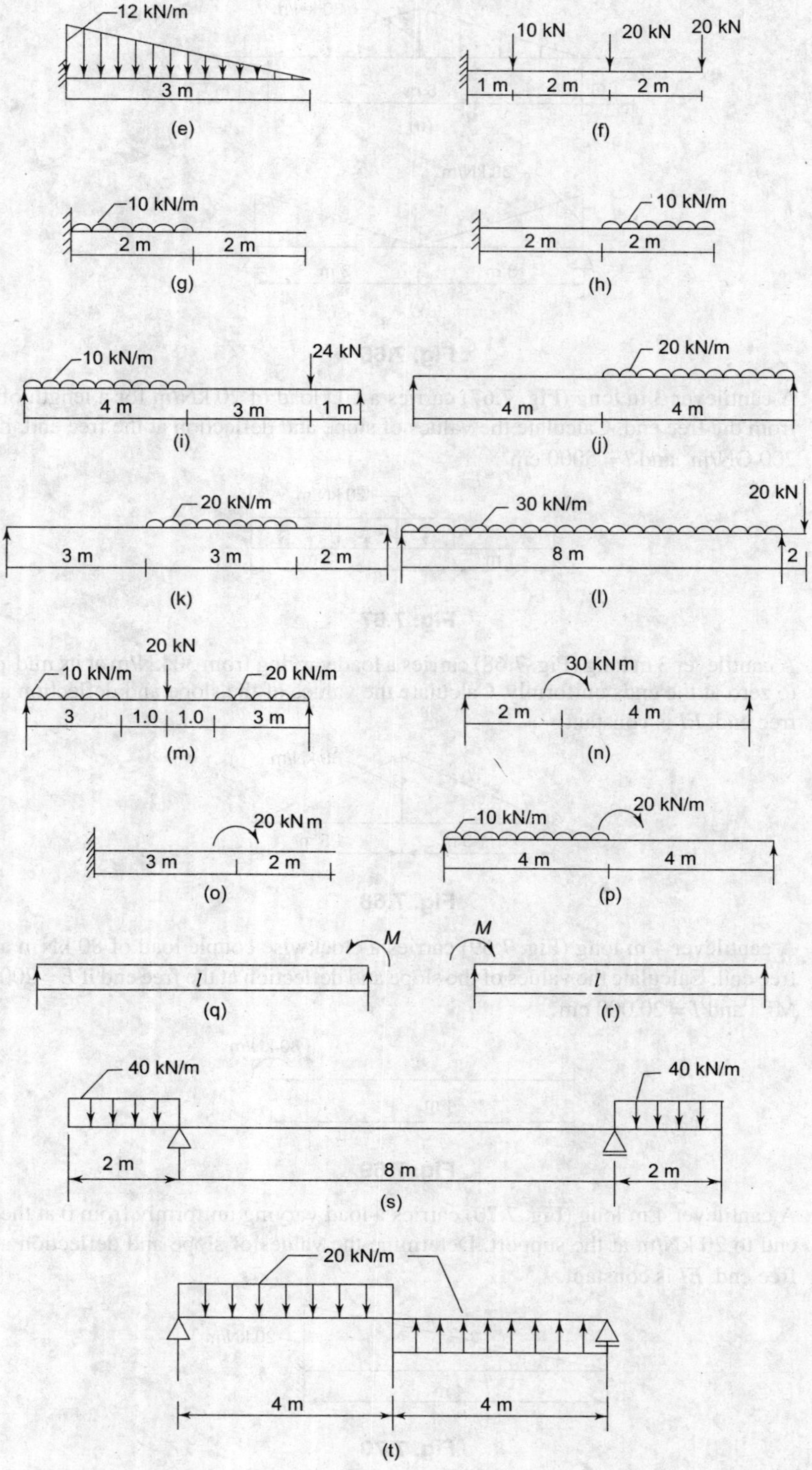
12 kN/m
3 m
(e)
10 kN
20 kN
20 kN
1 m
2 m
2 m
(f)
10 kN/m
2 m
2 m
(g)
10 kN/m
2 m
2 m
(h)
10 kN/m
24 kN
4 m
3 m
1 m
(i)
20 kN/m
4 m
4 m
(j)
20 kN/m
3 m
3 m
2 m
(k)
30 kN/m
20 kN
8 m
2
(l)
20 kN
10 kN/m
20 kN/m
3
1.0
1.0
3 m
(m)
30 kN m
2 m
4 m
(n)
20 kN m
3 m
2 m
(o)
10 kN/m
20 kN/m
4 m
4 m
(p)
M
l
(q)
M
l
(r)
40 kN/m
40 kN/m
2 m
8 m
2 m
(s)
20 kN/m
4 m
4 m
(t)

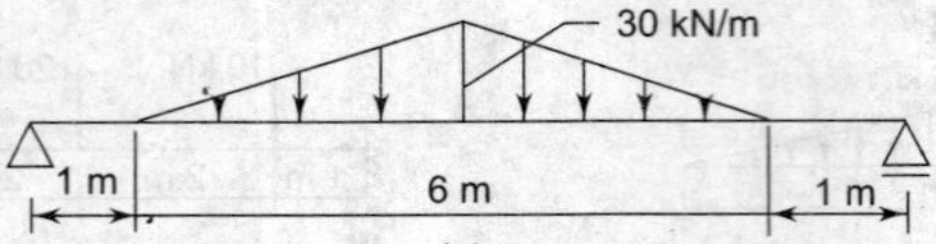

(u)

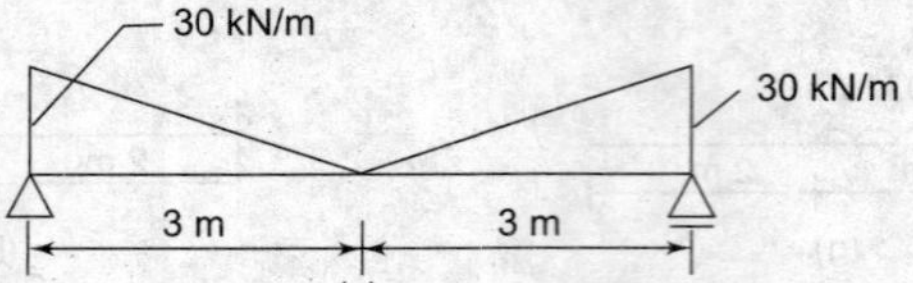

(v)

Fig. 7.66

2. A cantilever 3 m long (Fig. 7.67) carries a UD load of 20 kN/m for a length of 2 m from the free end. Calculate the values of slope and deflection at the free end, if E = 200 GN/m^2 and I = 5000 cm^4.

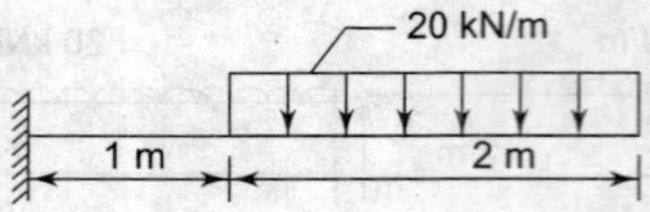

Fig. 7.67

3. A cantilever 3 m long (Fig. 7.68) carries a load varying from 30 kN/m at its mid-point to zero at the ends uniformly. Calculate the values of the slope and deflection at the free end. EI is constant.

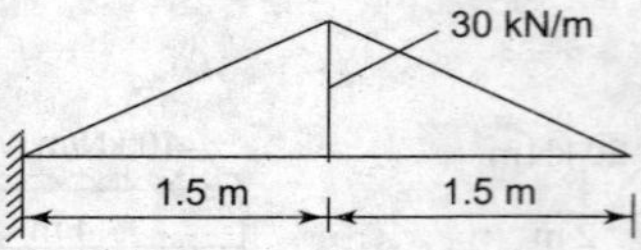

Fig. 7.68

4. A cantilever 4 m long (Fig. 7.69) carries a clockwise couple load of 80 kN m at the free end. Calculate the values of the slope and deflection at the free end if E = 200,000 MPa and I = 20,000 cm^4.

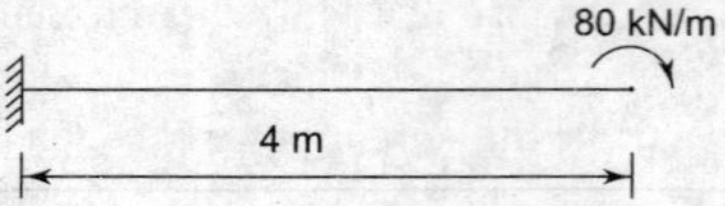

Fig. 7.69

5. A cantilever 4 m long (Fig. 7.70) carries a load varying uniformly from 0 at the free end to 20 kN/m at the support. Determine the values of slope and deflection at the free end. EI is constant.

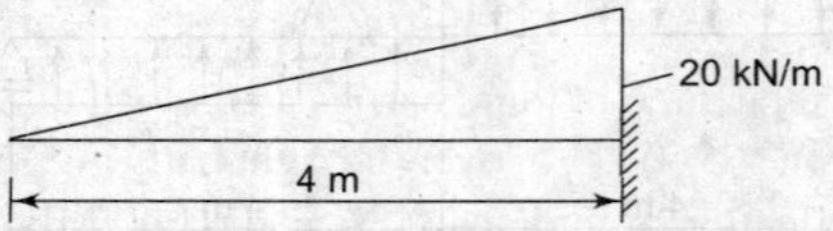

Fig. 7.70

6. A simply supported beam of span 8 m (Fig. 7.71) carries a point load of 20 kN 2.5 m from the left end and another point load of 25 kN 6 m from the same end. Calculate the slopes at the supports and the deflections under the loads. $E = 200 \times 10^6$ kN/m^2 and $I = 9000$ cm^4.

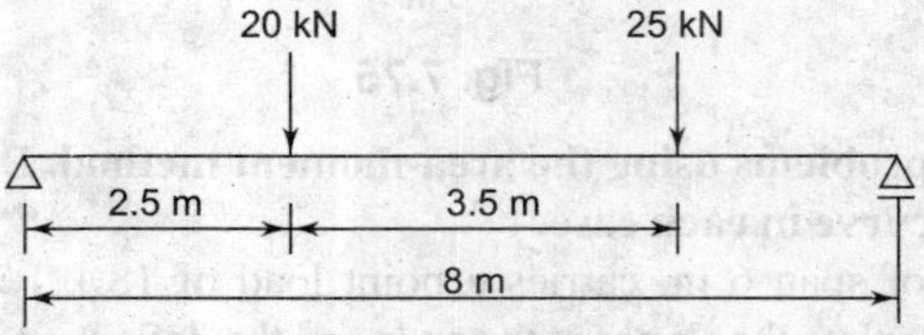

Fig. 7.71

7. A simply supported beam of span 6 m (Fig. 7.72) carries a load varying uniformly from zero at one end to 60 kN/m at the other. Calculate the values of the slopes at the supports, and find the position and magnitude of maximum deflection in terms of *EI*.

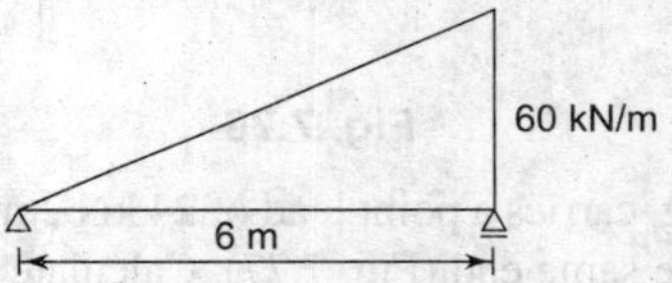

Fig. 7.72

8. A simply supported beam of span 8 m (Fig. 7.73) carries a UD load of 20 kN/m for the left half of its length and a clockwise couple of 40 kN m at a distance of 6 m from the left end. Calculate the values of slope at the supports and the deflection at mid-span. *EI* is constant.

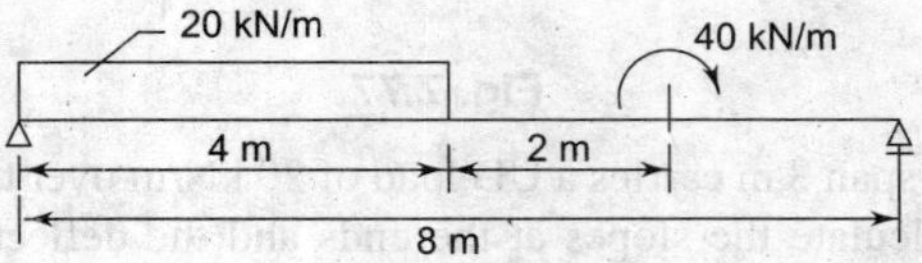

Fig. 7.73

9. A beam, 10 m long, is supported at its left end and at 8 m from this end (Fig. 7.74). The beam carries loads as follows: a UD load of 32 kN/m for a length of 4 m from the left end, a point load of 10 kN at 6 m from the left end, and a point load of 20 kN at the right end (free end). Calculate the slopes at the supports, and the deflection at the free end and at the mid-point between the supports. *EI* is constant.

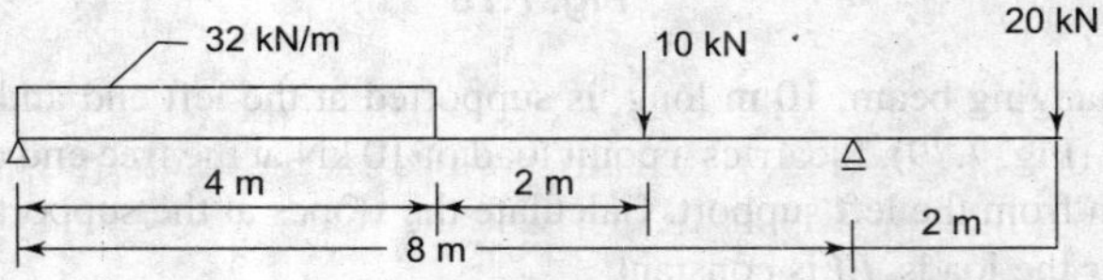

Fig. 7.74

10. A simply supported beam, 6 m long, carries a UD load of 30 kN/m over the left half of its length and a one of 60 kN/m over the right half (Fig. 7.75). Calculate the slopes at the ends and deflection at mid-span if $E = 200{,}000$ MPa and $I = 20{,}000$ cm^4.

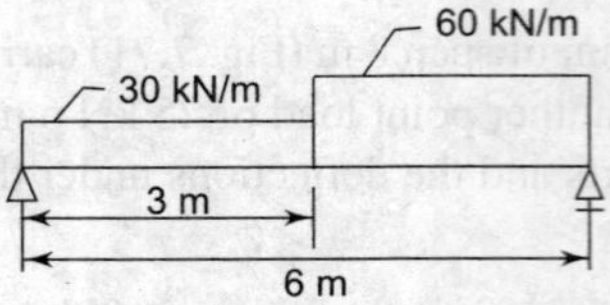

Fig. 7.75

Solve the following problems using the area-moment method. Draw the approximate shape of the elastic curve in each case.

11. An SS beam, of span 6 m, carries a point load of 18 kN 2 m from the left end (Fig. 7.76). Calculate the slopes at the ends and the deflection under the load in terms of *EI*.

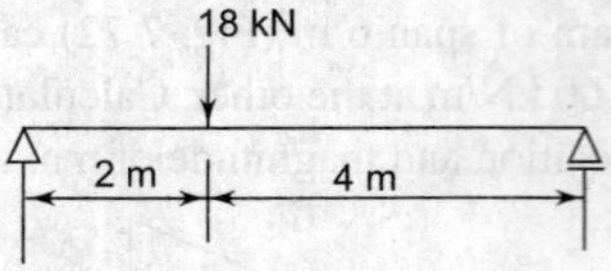

Fig. 7.76

12. An SS beam, 6 m long, carries a point load of 24 kN 2 m from the left end and another of 32 kN 5 m from the same end (Fig. 7.77). Calculate the slopes at the ends and the deflection under the loads if $E = 200$ GN/m^2 and $I = 8000$ cm^4.

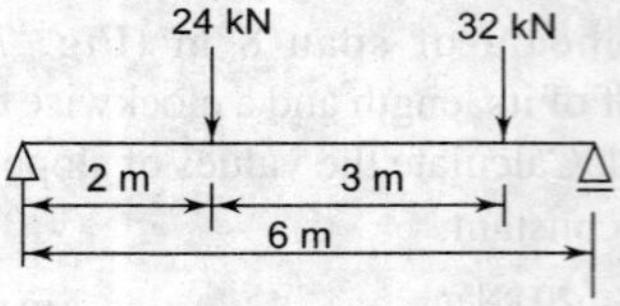

Fig. 7.77

13. An SS beam of span 8 m carries a UD load of 20 kN/m over the left half of its length (Fig. 7.78). Calculate the slopes at the ends and the deflection at mid-span. *EI* is constant.

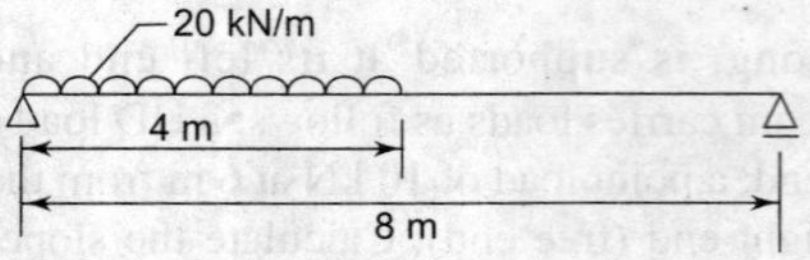

Fig. 7.78

14. An overhanging beam, 10 m long, is supported at the left end and at 8 m from the same end (Fig. 7.79). It carries a point load of 10 kN at the free end and another of 20 kN at 4 m from the left support. Calculate the slopes at the supports and the deflection under the loads. *EI* is constant.

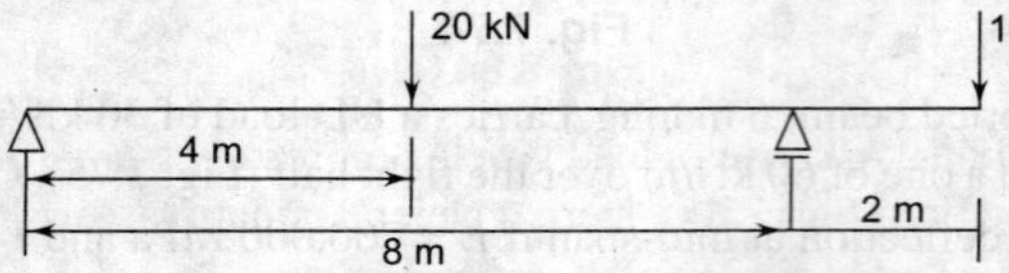

Fig. 7.79

15. A beam, 12 m long, is supported at points 2 m and 10 m from the left end (Fig. 7.80). It carries point loads of 20 kN each at the free ends and a UD load of 10 kN/m over the length between the supports. Calculate the slopes at the free ends and at the supports, and the deflections at the free ends, and at mid-span.

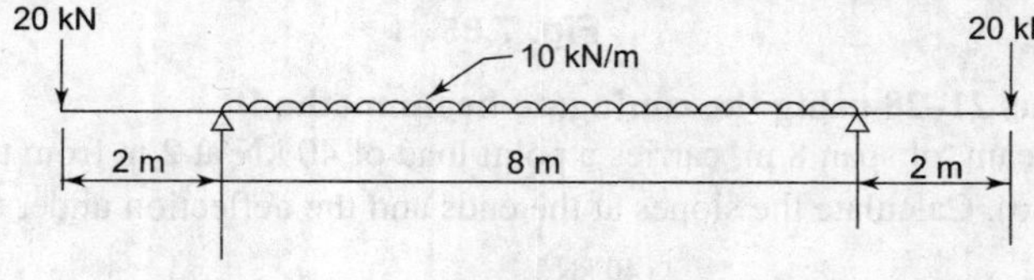

Fig. 7.80

16. An SS beam, 8 m long, is subjected to a couple load of 160 kN m at a point 6 m from the left end (Fig. 7.81). Calculate the slopes at the ends and the deflection at the point of application of the couple. At what point along the length of the beam should the couple be applied so that the slope at the right end is zero?

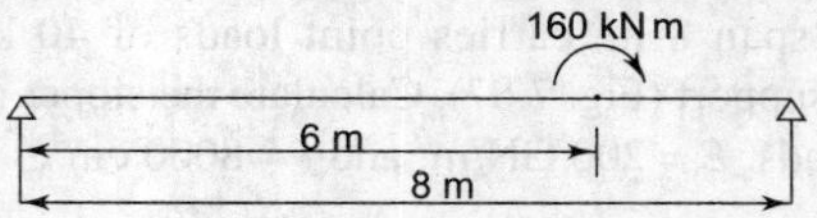

Fig. 7.81

17. A cantilever of span 3 m carries a point load of 60 kN at the free end and another of 80 kN at 1 m from the same end (Fig. 7.82). Calculate the slope and deflection at the free end.

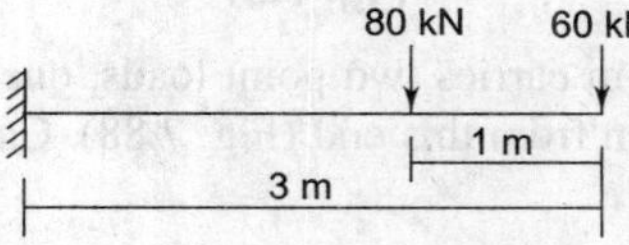

Fig. 7.82

18. A cantilever is 3 m long and is subjected to a UD load of 30 kN/m for a length of 2 m from the free end (Fig. 7.83). Calculate the slope and deflection at this end.

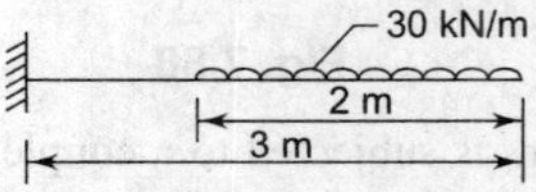

Fig. 7.83

19. A cantilever, 4 m long, carries a couple of 60 kN m at the free end (Fig. 7.84). Derive general equations for slope and deflection at any point, and calculate the values of slope and deflection at the free end.

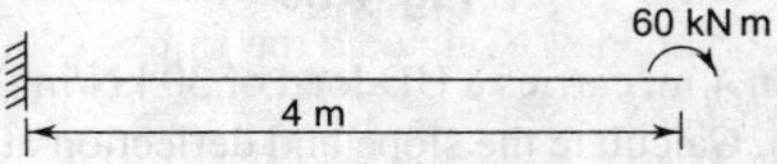

Fig. 7.84

20. A cantilever of span 3 m carries a load varying uniformly from 0 at the free end to 30 kN/m at the fixed end (Fig. 7.85). Derive general equations for slope and deflection at any point and calculate these values at the free end.

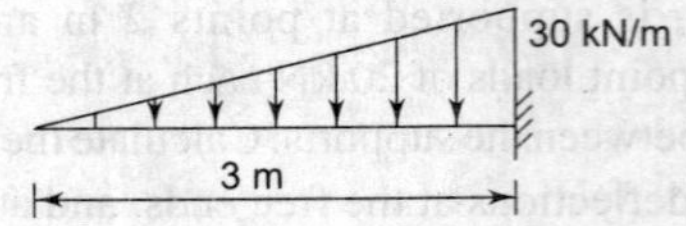

Fig. 7.85

Solve Problems 21–28 using the conjugate-beam method.

21. An SS beam, of span 8 m, carries a point load of 40 kN at 2 m from the left support (Fig. 7.86). Calculate the slopes at the ends and the deflection under the load.

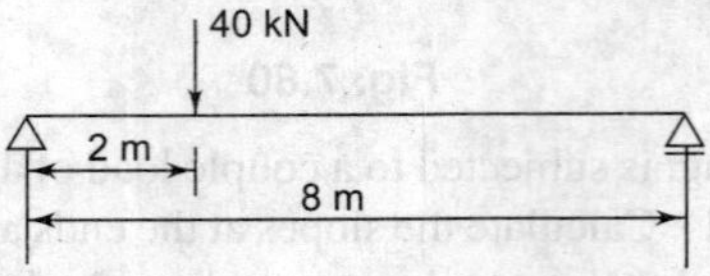

Fig. 7.86

22. An SS beam, of span 8 m, carries point loads of 40 kN at 2 m and 60 kN at 5 m from the left support (Fig. 7.87). Calculate the slopes at the ends and the deflections under the loads. $E = 200$ GN/m^2 and $I = 8000$ cm^4.

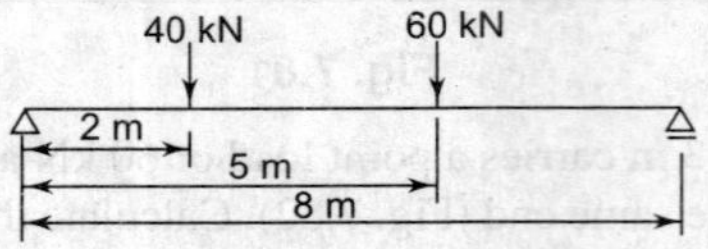

Fig. 7.87

23. A cantilever of span 4 m carries two point loads, one of 10 kN at the free end and another of 20 kN at 2 m from this end (Fig. 7.88). Calculate the slopes and deflections under the loads.

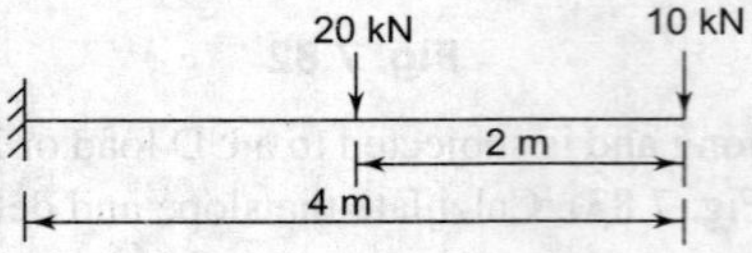

Fig. 7.88

24. A cantilever, of span 3 m, is subjected to a couple load of 60 kN m at the free end (Fig. 7.89). Calculate the slope and deflection at this end.

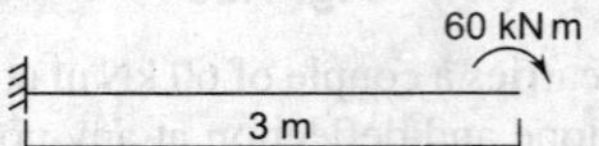

Fig. 7.89

25. A cantilever, of span 4 m, carries a UD load of 30 kN/m for a length of 2 m from the free end (Fig. 7.90). Calculate the slope and deflection at this end.

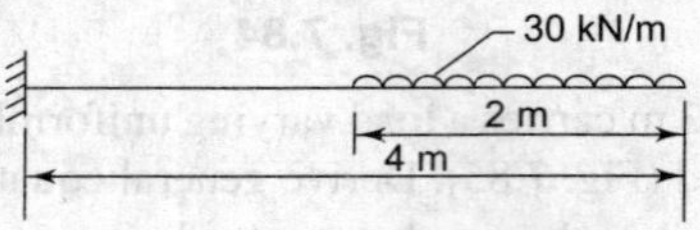

Fig. 7.90

26. A cantilever of span 3 m carries a load varying uniformly from zero at the free end to 60 kN/m at the fixed end (Fig. 7.91). Calculate the slope and deflection at the free end.

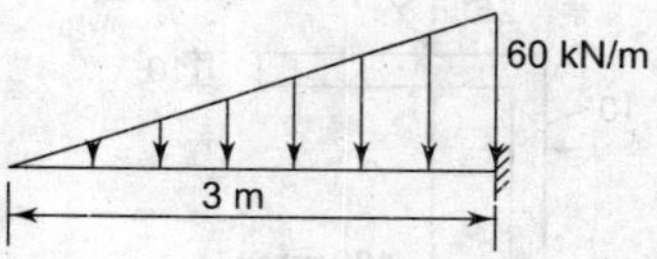

Fig. 7.91

27. A beam, 9 m long, is supported at the left end and at a point 1 m from the right end (Fig. 7.92). It carries a point load of 10 kN at the free end and another of 20 kN at the mid-point between the supports. Calculate the slopes at the supports and the deflections under the loads.

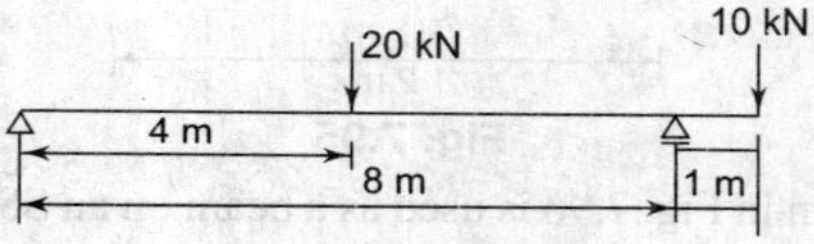

Fig. 7.92

28. A beam, 12 m long, is supported at 2 m from each end (Fig. 7.93). It carries UD loads of 20 kN/m on both the overhanging portions and a concentrated load of 60 kN at mid-span. Calculate the slopes at the supports, and the deflections at mid-span and at the free ends.

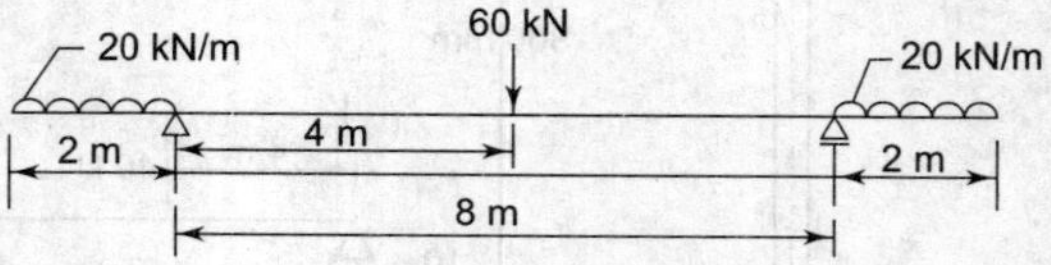

Fig. 7.93

29. The Z-section shown in Fig. 7.94 is loaded with a uniformly distributed load of 10 kN/m on a span of 4 m. Find the maximum deflection in the beam. Show that the direction of the net deflection is normal to the NA.

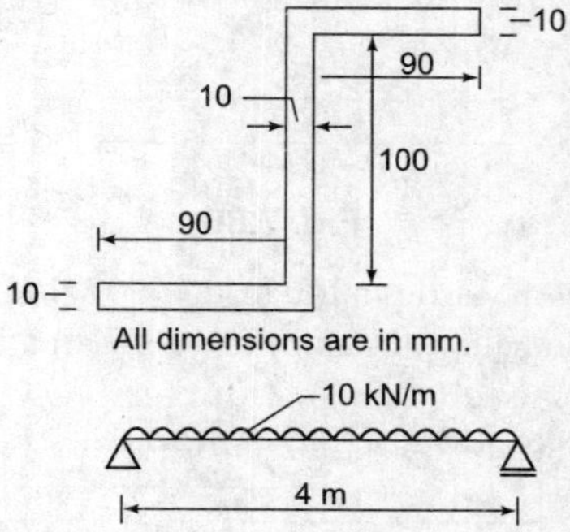

Fig. 7.94

30. The channel section shown in Fig. 7.95 is used as a cantilever beam on a span of 2 m. If the cantilever carries a point load at the free end in a plane inclined at 30° to the Y-axis, find the maximum deflection in the beam.

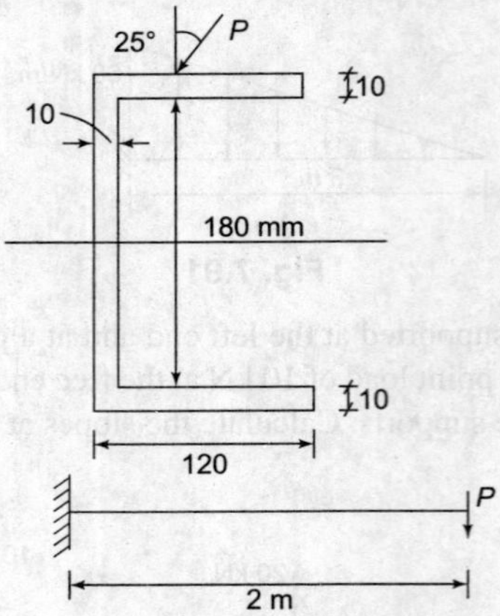

Fig. 7.95

31. The I-section shown in Fig. 7.96 is used as a beam on an SS span of 8 m. If the beam carries a point load at the centre of the span as shown, find the maximum deflection in the beam. Show that the maximum deflection is at right angles to the neutral axis.

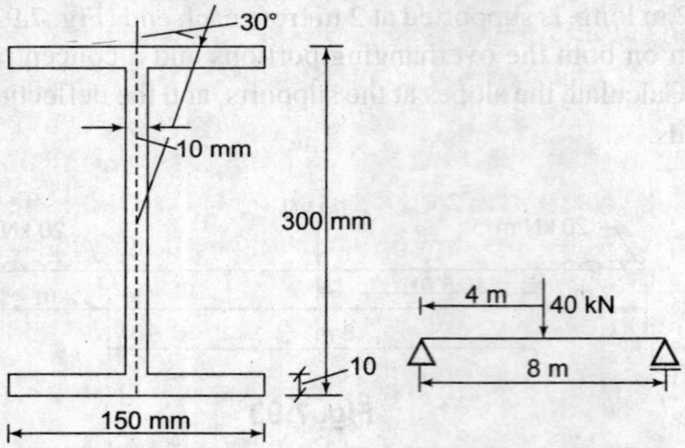

Fig. 7.96

CHAPTER 8

Torsion

Learning Objectives

After going through this chapter, the reader will be able to

- differentiate between torsion and bending,
- draw the torque diagram for a shaft,
- explain the structural behaviour of members subjected to torque,
- derive and explain the torsion equations for axisymmetric sections,
- design shafts subjected to torque, and calculate the stresses in them,
- calculate stresses in composite shafts and design such shafts, and
- design flanged couplings in shafts.

8.1 INTRODUCTION

In the last few chapters, we have studied about beams which are very common and important structural elements. We have discussed methods to calculate the BM and SF at any point, the internal stress distribution, stresses due to the combined effect of axial thrust and bending, and deformations in beams. In this chapter, we deal with another important structural element—shafts transmitting power. Shafts are commonly used to transmit power from one point to many points. While shafts also act as beams as they are supported at points along their length and carry their own weight, they predominantly act as torsion elements. In this chapter, we deal with torsion, the behaviour of elements under torsion, and the methods of designing shafts.

8.2 TORQUE AND TORSIONAL ELEMENT

Let us consider the bent bar shown in Fig. 8.1(a). The bar is fixed at A and carries a load at the free end C. Note that AB lies in the X-Y plane along the X-axis and BC lies in a plane parallel to the Y-Z plane, parallel to the Z-axis.

From the free body diagrams of AB and BC shown separately in Fig. 8.1, we find that there is a vertical upward reaction and couple at B for static equilibrium. The couple M_B lies in a plane containing the longitudinal axis of BC, and hence is the bending moment at B. The SF and BM for BC can be drawn as shown in Fig. 8.1(b).

For equilibrium of AB, there should be a vertical downward load and couple M_B at B on AB as shown, to satisfy the equal and opposite nature of internal stress resultants at B on BA and BC. The couple M_B at B on BA lies in a plane parallel to the Y-Z plane (or in a plane perpendicular to the plane containing the longitudinal axis), and obviously cannot be classified as a BM on AB.

The couple M_B acting on AB at B is called a torque or torsional moment. The difference between a torque and BM is clear in Fig. 8.1(d). While the BM acts on a plane containing the longitudinal axis and is about the NA of the section, the torque is a couple lying in a plane perpendicular to the longitudinal axis and is about the longitudinal axis.

From the free body diagram, we can draw the SF and BM diagrams for AB. We can draw a torque diagram to show the variation of the torque [Fig. 8.1(b)].

Torque diagram To draw the torque diagram of AB, we note that AB is subjected to couples M_B at B. There is no force or torque that changes the torque between A and B. The torque at A will, therefore, be the same as at B. The BM at A will be due to the load P at the end B $(= Pl_1)$. The torque diagram is a rectangle, as shown in Fig. 8.1(b).

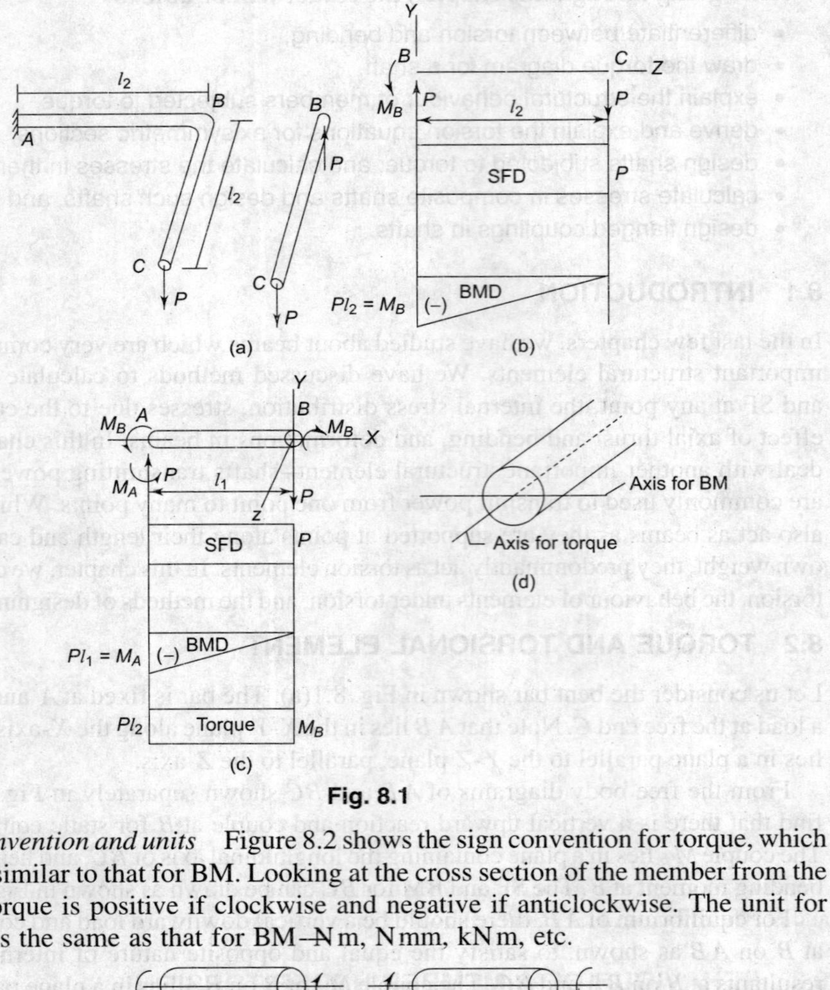

Fig. 8.1

Sign convention and units Figure 8.2 shows the sign convention for torque, which is very similar to that for BM. Looking at the cross section of the member from the right, torque is positive if clockwise and negative if anticlockwise. The unit for torque is the same as that for BM—N m, N mm, kN m, etc.

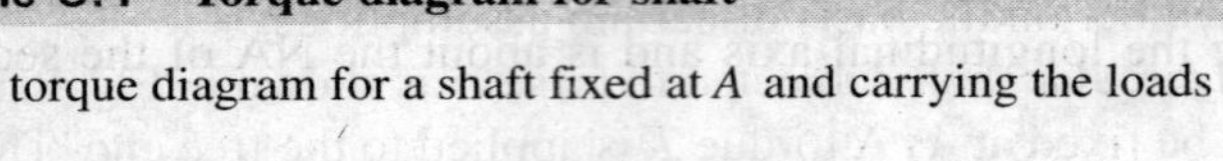

Fig. 8.2

Example 8.1 Torque diagram for shaft

Draw the torque diagram for a shaft fixed at A and carrying the loads shown in Fig. 8.3.

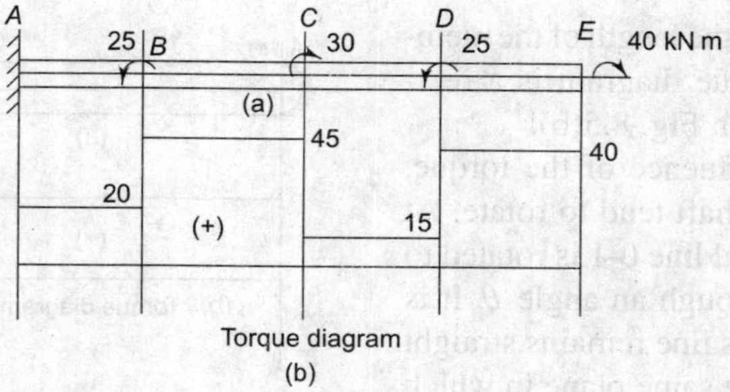

Fig. 8.3

Solution At any section of the shaft, the torque can be calculated as the net torsional moment from the right of the section. Calculating this way and according to the sign convention adopted,

Torque in $DE = +40$ kN m

Torque in $CD = +40 - 25 = 15$ kN m

Torque in $CB = 15 + 30 = 45$ kN m

Torque in $AB = 45 - 25 = 20$ kN m

The torque diagram can be drawn as shown in Fig. 8.3(b).

Example 8.2 Torque diagram for shaft

Considering the shaft system shown in Fig. 8.4, draw the torque diagram for the shaft.

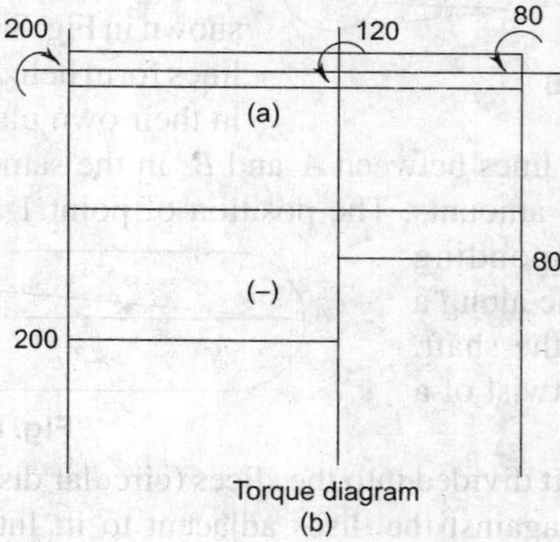

Fig. 8.4

Solution In this case, power is supplied to the shaft at the left end and drawn at points *B* and *C*. The torque diagram can be drawn as shown in Fig. 8.4(b).

8.3 BEHAVIOUR OF A MEMBER UNDER TORSION

The behaviour of a member under torsion is very complex, particularly when the section is not axisymmetrical. In the case of axisymmetrical sections, like solid circular or hollow circular sections, it is possible to simplify and explain the behaviour of the member on the basis of certain assumptions.

Consider a solid circular shaft subjected to torsion as shown in Fig. 8.5(a). The shaft is assumed to be fixed at *A*. A torque *T* is applied to the free end. The torque is constant along the length of the member and the torque diagram is a rectangle as shown in Fig. 8.5(b).

Under the influence of the torque, sections of the shaft tend to rotate. At section B, a radial line 0-1 is rotated to position 0-1′ through an angle θ. It is assumed that this line remains straight and moves in the same plane in which it lies. It may further be noted that the entire cross section, as at B, remains in a plane as the shaft deforms. This can be proved from the axisymmetric nature of the cross section.

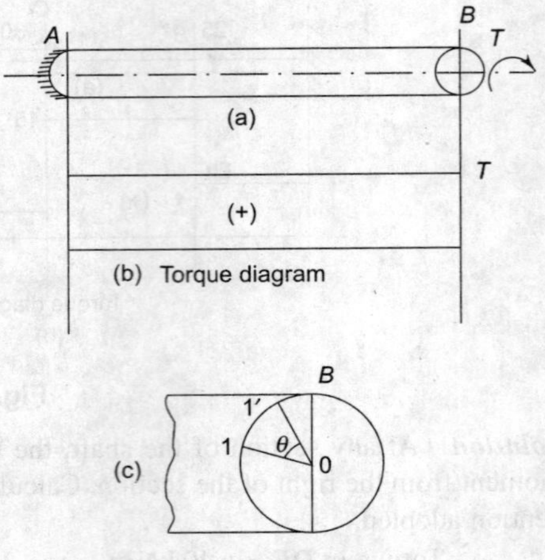

Fig. 8.5

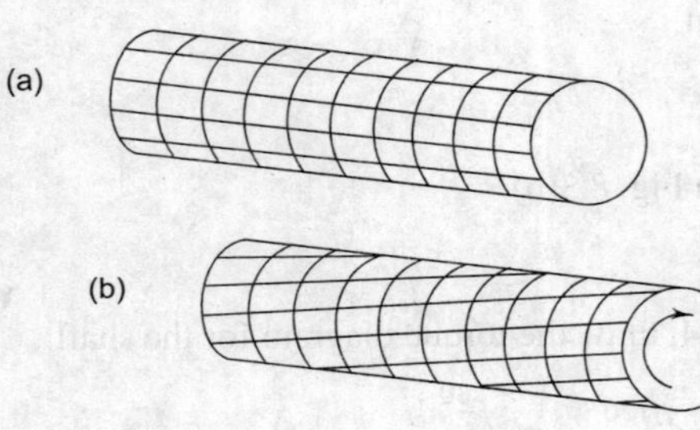

Fig. 8.6

This deformation can also be visualized from Fig. 8.6. If we draw longitudinal lines and circles perpendicular to these lines on the surface of the shaft before applying the torque, the deformed shape of these lines and circles will be as shown in Fig. 8.6(b). The longitudinal lines form helixes as the circles rotate in their own plane.

In Fig. 8.7, the radial lines between A and B, in the same plane as 0-1 before twist, rotate by different amounts. The position of point 1 at B shifts to 1′ after twisting and all corresponding points along the length lie along a helix on the surface of the shaft. Note that the maximum twist of a radial line is at B.

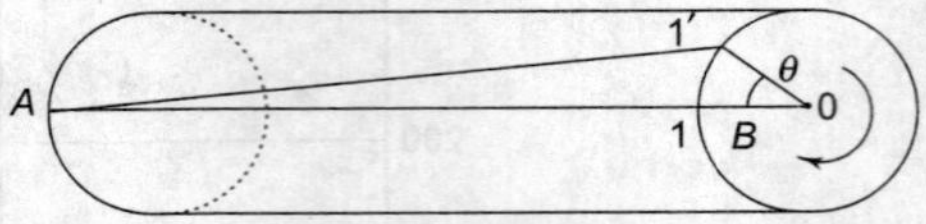

Fig. 8.7

If we consider the shaft divided into the slices (circular discs), one can visualize each such disc rotating against the discs adjacent to it. Internal stresses on the cross-sectional surface of the shaft resist such rotation. When the internal stress resultant equals the applied torque, equilibrium is established. These stresses are obviously shear stresses, being tangential to the cross section of the shaft (Fig. 8.8). The shear stresses over the section are so directed that they form a resultant couple opposite to the applied torque.

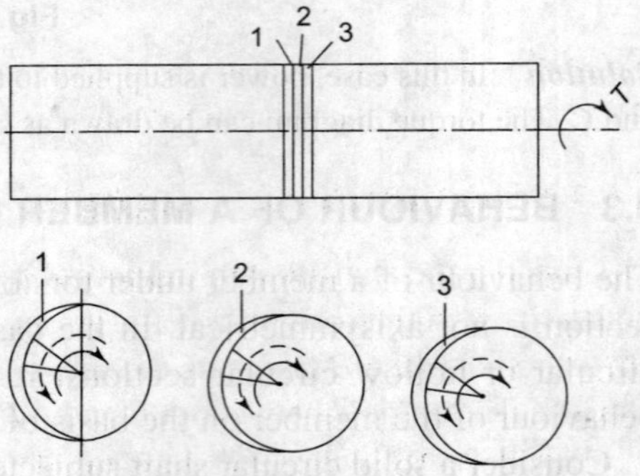

Fig. 8.8

8.4 TORSION THEORY FOR AXISYMMETRIC SECTIONS

The torsion equation we will derive will apply only to axisymmetric sections like solid circular or hollow circular sections. We will make the following assumptions while deriving this equation.

(i) Sections that are plane before twisting remain so after twisting and do not warp.

(ii) Circular sections remain circular after twisting.

(iii) Straight radial lines in a transverse section remain straight after twisting.

(iv) Stresses in the material do not exceed the proportional limit.

(v) The shaft is loaded by twisting couples perpendicular to the axis of the shaft.

(vi) Relative rotation between any two sections is proportional to the distance between them.

(vii) The material is homogeneous and isotropic.

These assumptions are illustrated in Fig. 8.9.

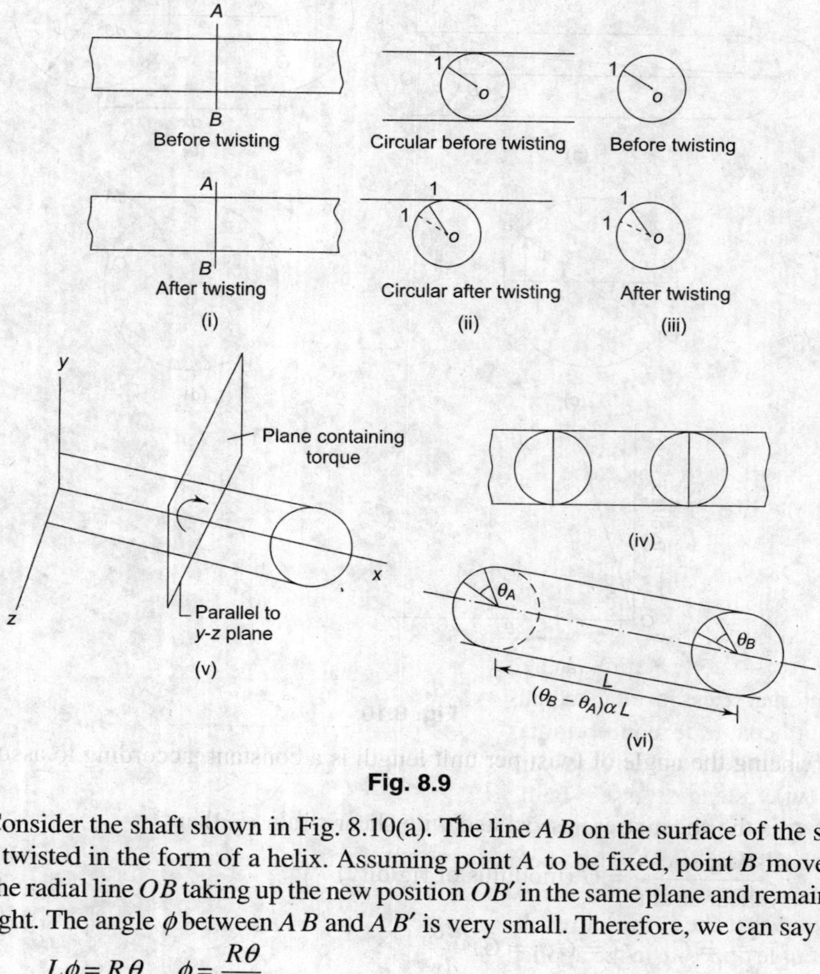

Fig. 8.9

Consider the shaft shown in Fig. 8.10(a). The line AB on the surface of the shaft gets twisted in the form of a helix. Assuming point A to be fixed, point B moves to B', the radial line OB taking up the new position OB' in the same plane and remaining straight. The angle ϕ between AB and AB' is very small. Therefore, we can say that

$$L\phi = R\theta, \quad \phi = \frac{R\theta}{L}$$

Note that ϕ and θ both should be measured in radians.

This is true for any small elementary disc of length dx shown in Fig. 8.10(b), for all longitudinal layers. Thus, in Fig. 8.10(b), a longitudinal layer 2-2, parallel to the longitudinal axis and at a radial distance r from the centre, gets shifted to the position 2-2′, due to relative twist $d\theta$. We can say that $\phi_2 = (2 - 2')/dx$ is the shear strain of the fibre 2-2.

The angle between 0-2 and 0-2′ being $d\theta$, we have

$$2\text{-}2' = r\, d\theta$$

Therefore,

$$\phi_2 = \frac{r\, d\theta}{dx}$$

Similarly, for the surface layer 1-1,

$$\phi_1 = R\frac{d\theta}{dx}$$

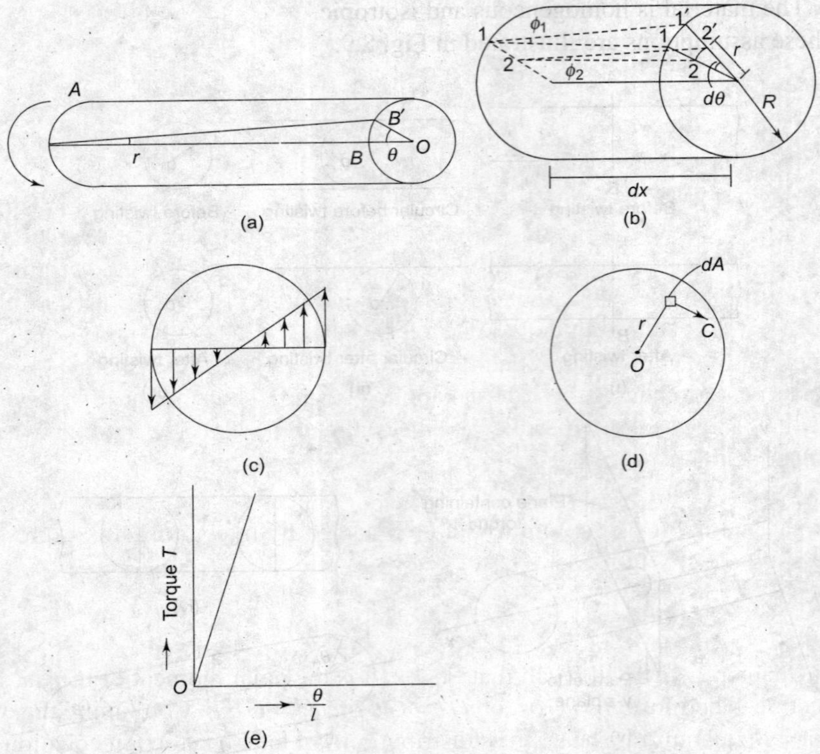

Fig. 8.10

$d\theta/dx$ being the angle of twist per unit length is a constant, according to assumption (vi).

If τ_2 is the shear stress associated with shear strain ϕ_2, then

$$\frac{\text{shear stress}}{\text{shear strain}} = G \text{ (modulus of rigidity)}$$

$$\frac{\tau_2}{\phi_2} = G, \quad \tau_2 = G\phi_2 = Gr\frac{d\theta}{dx}$$

G and $d\theta/dx$ being constant, this equation shows that the shear stress is directly proportional to the radial distance of the fibre from the centre-line of the shaft, i.e.,

$$\tau = \left(G\frac{d\theta}{dx}\right)r$$

Also, the shear stress acts in the direction of the displacement, i.e., tangential to the circle of radius r, or perpendicular to the radial line. Setting the angle of twist per unit length $d\theta/dx$ as θ/L,

$$\frac{\tau}{r} = \frac{G\theta}{L} \quad \text{and} \quad \tau = G\phi$$

Since $\phi = d\theta/dx = R\,\theta/L$, θ/L being the angle of twist per unit length,

$$\frac{\tau}{r} = \frac{G\theta}{L}$$

We have arrived at two important conclusions regarding shear stress distribution in a shaft.

(i) The shear stress τ at any point in the cross section is proportional to the distance (from the longitudinal axis) of the point along the radius.

(ii) The shear stress acts perpendicular to the radial line.

Taking any diameter of the shaft at any cross section, the shear stress distribution due to torque will be as shown in Fig. 8.10(c). It should also be noted that, as in the case of bending stress distribution, this is derived from the deformation characteristics of the shaft. The equilibrium conditions can now be used to relate the shear stress τ to the applied torque T.

Applying the equilibrium conditions, from Fig. 8.10(d), the internal twisting moment must be equal to the external applied torque. If τ is the shear stress on an elementary area dA at a distance r from the centre of the shaft, the internal twisting moment due to this force is given by

$$dT = \tau\, dA\, r$$

The total internal resisting torque can be obtained by integrating this expression over the whole cross section.

$$T = \int_A \tau\, dA\, r = \int_A (G\phi r)\, r\, dA = G\phi \int_A r^2\, dA$$

From Chapter 2, you will recall that $\int_A r^2\, dA$ is the polar moment of inertia, J, of the section. Therefore, $T = G\phi J$, or $T/J = G\phi$ or $T/J = G\theta/L$. Combining this with the earlier equation for shear stress distribution, we have the torsion equation

$$\frac{T}{J} = \frac{G\theta}{L} = \frac{\tau}{r}$$

One can immediately notice the similarity between this equation and the bending equation

$$\frac{M}{I} = \frac{E}{R} = \frac{\sigma}{y}$$

Also,

$$\frac{T}{(\theta/L)} = GJ$$

GJ is called the 'torsional stiffness', similar to EI, which is called 'bending stiffness.'

Since GJ is a constant for a given shaft, a graph between torque T and the angle of twist per unit length will be a straight line as shown in Fig. 8.10(e).

As mentioned earlier, this derivation applies to axisymmetrical sections like solid circular and hollow circular sections only, as shown in Fig. 8.11. For a solid circular section of radius R, $J = \pi R^4/2$ or $\pi D^4/32$, where D is the diameter. The maximum shear stress is on the surface of the shaft and is given by

$$\tau_{\max} = \frac{TR}{J} = \frac{TR}{(\pi D^4/32)} = \frac{16\,T}{(\pi D^3)}$$

Similarly, in the case of a hollow circular section of external diameter D and internal diameter d,

$$J = \frac{\pi}{32}(D^4 - d^4)$$

and

$$\tau_{\max} = \frac{16TD}{\pi(D^4 - d^4)}$$

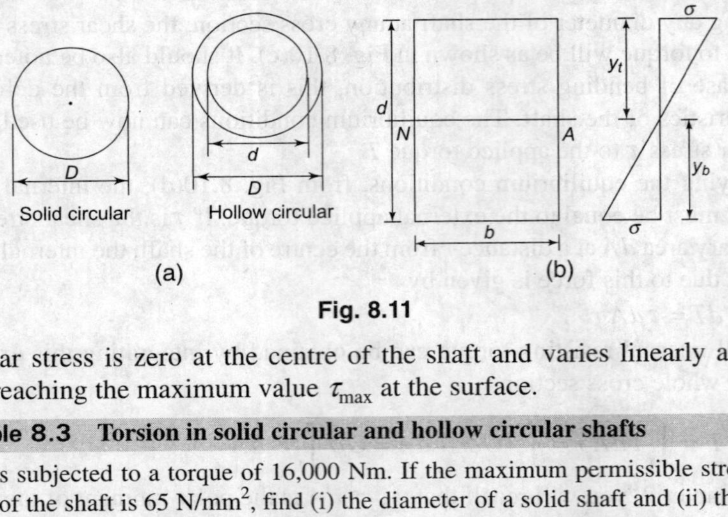

Fig. 8.11

The shear stress is zero at the centre of the shaft and varies linearly along the radius, reaching the maximum value $\tau_{\max}$ at the surface.

Example 8.3 Torsion in solid circular and hollow circular shafts

A shaft is subjected to a torque of 16,000 Nm. If the maximum permissible stress in the material of the shaft is 65 N/mm^2, find (i) the diameter of a solid shaft and (ii) the dimensions of a hollow circular shaft if the thickness is 10% of the internal diameter.

Solution Given:

Torque = 16,000 Nm = 16×10^6 Nmm

Permissible shear stress = 65 N/mm^2

We have $T/J = \tau/r$

(i) *Solid circular shaft* For this shaft, $J = \pi r^4/2$, where r is the radius of the shaft. Maximum shear stress will be on the surface, where r = radius of the shaft.

We can write $J/r = T/\tau = 16 \times 10^6/65 = 2.46 \times 10^5$ mm^3

Therefore,

$$\pi r^4/2r = 2.46 \times 10^5;\ r^3 = 2.46 \times 10^5 \times 2/\pi = 156.6 \times 10^3;\ r = 54 \text{ mm}$$

Minimum radius of the shaft = 55 mm

(ii) *Hollow shaft* Let the inner diameter be d. Outer diameter = $d + 0.2d$.

Polar moment of inertia = $\pi[(d + 0.2d)^4 - d^4]/32 = 0.1054d^4$

or $T = J\tau/[(d + 0.2d)/2]$

Maximum radial distance = $(d + 0.2d)/2$

or $16 \times 10^6 = 0.1054d^4 \times 65 \times 2/(1.2d)d^3 = 1.2 \times 16 \times 10^6/65 \times 2 \times 0.1054$

or $d = 111$ mm, say, 115 mm

Inner diameter = 115 mm; outer diameter = 140 mm

Example 8.4 Hollow circular shaft

A hollow circular shaft is of 180 mm internal diameter and thickness 10 mm. Find the maximum stress in the shaft if the torque is 12,000 Nm.

Solution Internal diameter = 180 mm; external diameter = 180 + 10 + 10 = 200 mm

From $T/J = \tau/r$, $\tau = Tr/J$

Torque = 12,000 Nm = 12×10^6 Nmm

On the surface of the shaft where the maximum shear stress is developed, $r = 100$ mm.

Polar moment of inertia for the shaft, $J = \pi(200^4 - 180^4)/32 = 54 \times 10^6$ mm^4

Maximum shear stress = $12{,}000 \times 10^3 \times 100/(54 \times 10^6) = 22.2$ N/mm^2

Example 8.5 Diameter of a solid circular shaft

A solid circular shaft is subjected to a torque of 20,000 Nm. if the maximum permissible shear stress is 50 N/mm^2, find a suitable minimum diameter for the shaft.

Solution Torque = 20,000 Nm = 20×10^6 Nmm

Maximum stress = 50 N/mm^2 on the surface, where the radius of the shaft = r

From $T/J = \tau/r$, $J/r = T/\tau = 20 \times 10^6/50 = 4 \times 10^5$ mm^3

Polar moment of inertia, $J = \pi r^4/2$; $J/r = \pi r^3/2$

Therefore, $\pi r^3/2 = 4 \times 10^5$; $r^3 = 2 \times 4 \times 10^5/\pi = 254.65 \times 10^3$; $r = 64$ mm

Minimum diameter of the shaft = 128 mm

8.4.1 Torsional Rigidity

The quantity GJ is called torsional rigidity (or stiffness). This is similar to EI, which is called flexural rigidity. From the torsion equation, $T/J = G\theta/L$, we get $GJ = T\theta/L$. When θ = unity (1 radian) and L = 1 m, we have $GJ = T$. This leads to the definition of torsional rigidity.

Torsional stiffness can be defined as *the torque required to produce unit rotation per unit length of the shaft.* Torsional stiffness is inversely proportional to the length of the shaft. When the length is more, stiffness decreases.
In the case of a solid circular section, $J = \pi r^4/2$ or $\pi d^4/32$

Torsional stiffness = $G\pi r^4/2$ or $G\pi d^4/32$

8.4.2 Polar Section Modulus

In the case of bending stress, we have used the term sectional modulus (see Chapters 2 and 5). Section modulus Z is given by moment of inertia divided by the distances to the extreme fibres from the neutral axis [Fig. 8.11(b)]. Thus, $M = \sigma(I/y_t)$ or $\sigma(I/y_b) = \sigma Z$. On similar lines, we define polar modulus of section. In the case of torsion, the torque is about the longitudinal axis and polar MI of the section is also about that axis.

Polar section modulus is defined as the polar moment of inertia divided by the distance to the extreme fibre of the section. In the case of circular and hollow circular sections, the extreme fibre distance is either the radius (r) or external radius (r_e). Thus,

Polar modulus, $Z_p = J/r$ or J/r_e

From $T/J = \tau/r$,

Torque $= J\tau/r = Z_p\tau$

For a circular section of radius r, $Z_p = \pi r^4/2r = \pi r^3/2$ or $\pi d^3/16$

For a hollow circular section, $Z_p = \pi(r_e^4 - r_i^4)/2r_e$ or $2\pi(d_e^4 - d_i^4)/32d_e$

8.4.3 Torsional Moment of Resistance

In the case of bending stress distribution, we have defined the term *moment of resistance* as the maximum moment that can be carried by the section without exceeding the maximum permissible stress in tension or compression. On similar lines, torsional moment of resistance is the maximum torque that can be carried by the section without exceeding the maximum permissible shear stress.

Torsional moment of resistance, $T_m = Z_p \times \tau$, where τ = permissible shear stress

For a solid circular section, $T_m = \pi r^3\tau/2$ or $\pi d^3\tau/16$

For a hollow circular section, $T_m = \tau\pi(r_e^4 - r_i^4)/2r_e$ or $\tau\pi(d_e^4 - d_i^4)/16d_e$

where r_e and r_i are the external and internal radii of the shaft and d_e and d_i are the external and internal diameters.

Example 8.6 Maximum shear stress in a solid circular shaft

A solid circular shaft has a diameter of 80 mm. Find the maximum shear stress and the angle of twist in a length of 2 m when the shaft is subjected to a torque of 10 kN m. G = 85 GPa.

Solution From the torsion equation $T/J = \tau/r$, τ will be maximum when r is equal to the radius of the shaft. T = 5 kN m, D = 80 mm.

$$\tau_{max} = \frac{TR}{J} = \frac{16T}{\pi D^3} = \frac{16\times10\times10^6}{\pi\times(80)^3} = 99.5 \text{ N/mm}^2$$

Also, $T/J = G\theta/L$, where θ is the angle of twist over a length L of the shaft. L = 2 m = 2000 mm and G = 85 GPa = 85,000 N/mm^2.

$$\theta = \frac{TL}{GJ} = \frac{10\times10^6\times2000\times32}{85{,}000\times\pi\times(80)^4} = 0.0585 \text{ rad or } 3.35°$$

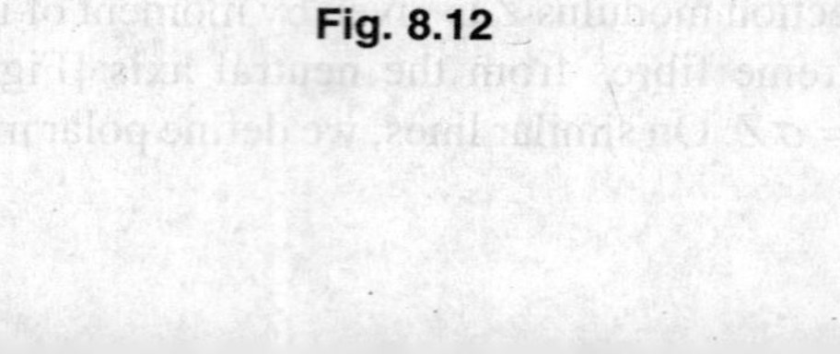

Fig. 8.12

Example 8.7 Hollow circular shaft

A hollow circular shaft has an external diameter of 120 mm and the internal diameter is three-fourths the external diameter. If the stress at a fibre inside is 36 MPa, due to a torque T applied, find this torque, the maximum shear stress, and the angle of twist per unit length. $G = 85$ MPa.

Solution External diameter = 120 mm

$$\text{Internal diameter} = \frac{3}{4} \times 120 = 90 \text{ mm}$$

$$J = \frac{\pi}{32}(120^4 - 90^4)$$

$$\tau = \frac{T}{J} r$$

$$T = \frac{\tau J}{r}$$

$$\tau = 36 \text{ N/mm}^2$$

$$r = 90 \text{ mm}$$

$$T = \frac{36\times}{90}\frac{\pi\times10^4}{32} \times 14{,}175 = 5{,}566{,}510 \text{ N mm} = 5.567 \text{ kN m}$$

The maximum shear stress τ_{max} will be at the surface fibres.

$$\tau_{max} = \frac{T}{J}R = \frac{5{,}566{,}510\times120}{(\pi/32)\times10^4\times14{,}175} = 48 \text{ N/mm}^2$$

$$= \frac{36\times120}{90} = 48 \text{ N/mm}^2$$

$$\text{Angle of twist per unit length} = \frac{T}{GJ} = \frac{5{,}566{,}510}{85{,}000\times(\pi/32)\times10^4\times14{,}175}$$

$$= 4.7 \times 10^{-6} \text{ rad/mm}$$

$$= 0.0047 \text{ rad/m} = 0.27°\text{/m}$$

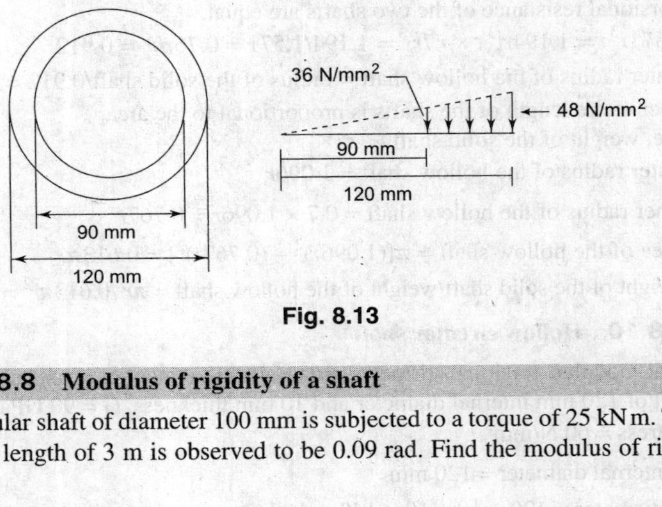

Fig. 8.13

Example 8.8 Modulus of rigidity of a shaft

A solid circular shaft of diameter 100 mm is subjected to a torque of 25 kN m. The angle of twist over a length of 3 m is observed to be 0.09 rad. Find the modulus of rigidity of the material.

Solution $T = 25 \text{ kNm} = 25 \times 10^6 \text{ N mm}$

$$J = \frac{\pi}{32} \times 100^4$$

$$\frac{T}{J} = G\frac{\theta}{L}, \quad G = \frac{TL}{J\theta} = \frac{25\times10^6\times3000}{(\pi/32)\times100^4\times0.09} = 84{,}882 \text{ N/mm}^2$$

$$= 84.88 \text{ GPa}$$

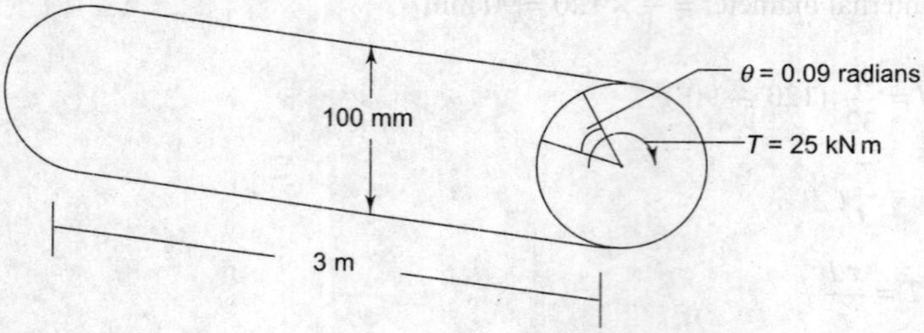

Fig. 8.14

Example 8.9 Comparison of weights of solid and hollow shafts

Compare the weight per metre length of a solid circular shaft to that of a hollow circular shaft (the inside diameter being 70% of the outside diameter) if they are made of the same material, same length, and have the same torsional resistance.

Solution For a solid circular shaft, torsional resistance = $J\tau/r$, where r is the radius of the shaft. Also $J = \tau r^4/2$ and the maximum shear stress is on the surface at r.

Torsional resistance = $\pi r^4 \tau/2r = 1.571\tau r^3$

For the hollow circular shaft, of outside diameter r_e, internal diameter $r_i = 0.7r_e$

Polar moment of inertia, $J = \pi(r_e^{\,4} - 0.7r_e^{\,4})/2 = 1.194r_e^{\,4}$

Maximum stress is on the outer surface where $r = r_e$

So, $J/r = 1.194r_e^{\,3}$

Torsional resistance = $J\tau/r = 1.194r_e^{\,3}\tau$

As the torsional resistance of the two shafts are equal,

$1.571r^3\tau = 1.194r_e^{\,3}\tau \times r^3/r_e^{\,3} = 1.194/1.571 = 0.76r/r_e = 0.912$

Outer radius of the hollow shaft = radius of the solid shaft/0.912 = 1.096r

Weight per metre length of the shafts is proportional to the area.

Therefore, weight of the solid shaft $\propto \pi r^2$

Outer radius of the hollow shaft = 1.096r

Inner radius of the hollow shaft = $0.7 \times 1.096r = 0.767r$

Area of the hollow shaft = $\pi[(1.096r)^2 - (0.767r)^2] = 0.613\pi r^2$

Weight of the solid shaft/weight of the hollow shaft = $\pi r^2/0.613\pi r^2 = 1.63$

Example 8.10 Hollow circular shaft

Find the polar modulus, torsional stiffness, and torsional moment of resistance of a hollow circular shaft of 120 mm internal diameter and 10 mm thickness. G = 90 GPa and permissible shear stress = 60 N/mm².

Solution Internal diameter = 120 mm

External diameter = 120 + 10 + 10 = 140 mm

Polar MI of the section = $\pi(140^4 - 120^4)/32 = 17.35 \times 10^6$ mm^4

Polar section modulus = $J/r = 17.35 \times 10^6/70 = 0.25 \times 10^6$ mm^3

Torsional stiffness = GJ; where G = 90 GPa = 90,000 MPa = 90,000 N/mm^2

Torsional stiffness = $90{,}000 \times 17.35 \times 10^6 = 1561.5 \times 10^9$ Nmm

$= 1.56 \times 10^6$ kNm

Torsional moment of resistance = polar section modulus × shear stress

$= 0.25 \times 10^6 \times 65$

$= 16.25 \times 10^6$ Nmm

= 16,250 Nm

8.5 STEPPED SHAFTS AND SHAFTS OF VARYING SECTIONS

In case the diameters of different lengths of shafts vary, as shown in Fig. 8.15, it is necessary to consider the shafts in different parts. Sometimes, the torque may also vary with change in section. It is only necessary to consider appropriate lengths, either due to change in diameter or change in torque, and compute the stresses at different sections and angles of twist between sections. In Fig. 8.15, the angle of twist between A and D should be calculated as the *algebraic* sum of the angles of twist between different sections. The angle of twist is considered positive when the right section rotates clockwise relative to the left section.

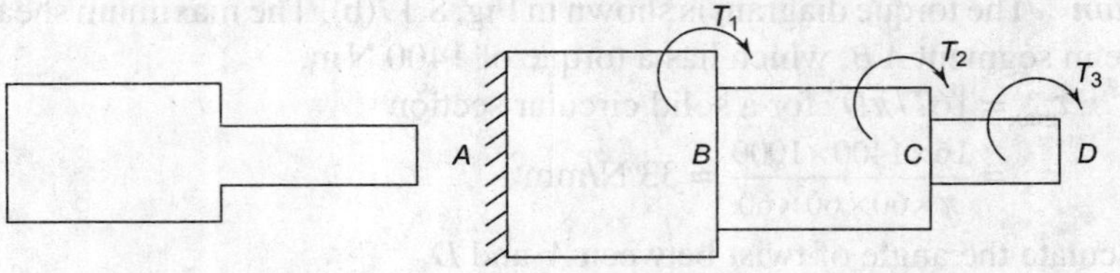

Fig. 8.15

In the case of a shaft of variable cross section, the same torsion formulae can be used so long as the polar moment of inertia can be expressed as a function of x, as shown in Fig. 8.16. We consider a length dx of the shaft. The angle of twist between the two end sections of elemental length dx is

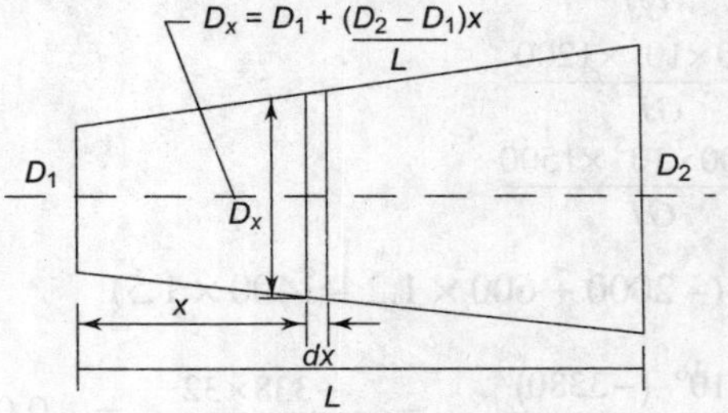

Fig. 8.16

$$d\theta = \frac{T\,dx}{JG}$$

and $$\theta = \int \frac{T\,dx}{JG}$$

where θ is the angle of twist over length L.

Example 8.11 Maximum shear stress and angle of twist

A shaft of diameter 60 mm is subjected to torques as shown in Fig. 8.17(a). Find the maximum shear stress and the angular deformation of D with respect to A. $G = 85$ GPa.

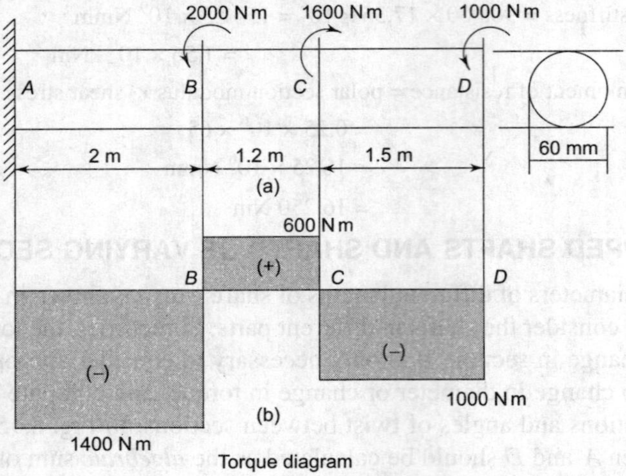

Fig. 8.17

Solution The torque diagram is shown in Fig. 8.17(b). The maximum shear stress will be in segment AB, which has a torque of 1400 N m.

$$\tau_{max} = 16T/\pi D^3 \text{ for a solid circular section}$$

$$= \frac{16 \times 1400 \times 1000}{\pi \times 60 \times 60 \times 60} = 33 \text{ N/mm}^2$$

To calculate the angle of twist between A and D,

$$\theta = \Sigma \left(\frac{TL}{GJ}\right)$$

Appropriate signs should be used between different sections. Taking clockwise rotations as positive,

$$\theta_{D/C} = \frac{-1000 \times 10^3 \times 2000}{GJ}$$

$$\theta_{C/B} = \frac{+600 \times 10^3 \times 1200}{GJ}$$

$$\theta_{B/A} = \frac{-1400 \times 10^3 \times 1500}{GJ}$$

$$\theta_{D/A} = \frac{10^6}{GJ}(-2000 + 600 \times 1.2 - 1400 \times 1.5)$$

$$= \frac{10^6\ (-3380)}{85{,}000 \times (\pi/32) \times 60^4} = \frac{-338 \times 32}{85 \times \pi \times 36 \times 36} = -0.0312 \text{ rad}$$

Example 8.12 Maximum shear stress and angle of twist for a tapering shaft

A circular shaft, 1200 mm long, varies uniformly in diameter from 40 mm to 80 mm. If it is subjected to a torque of 0.5 kN m, find the maximum stress in the shaft and the angle of twist (relative) between the end sections. $G = 85$ GPa.

Solution The shaft is shown in Fig. 8.18. The maximum stress is at the smaller end and is given by

$$\tau_{max} = \frac{16T}{\pi D^3} = \frac{16 \times 0.5 \times 10^6}{\pi \times 40^3} = 40 \text{ N/mm}^2$$

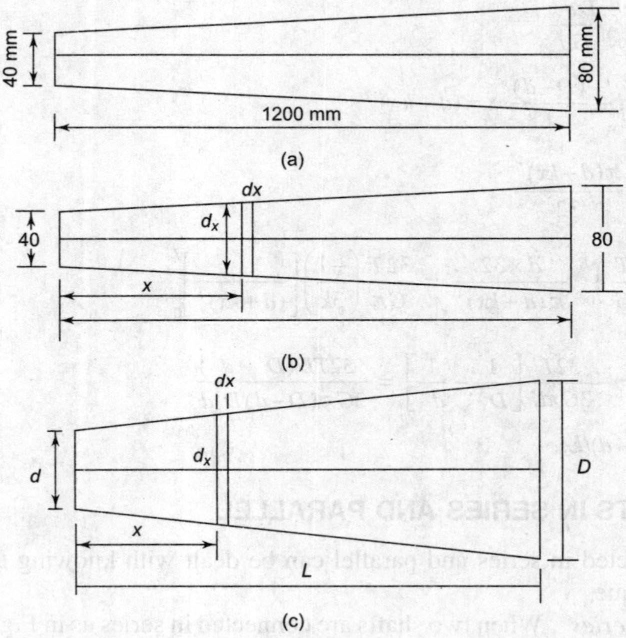

Fig. 8.18

To calculate the angle of twist, we consider a disc of elemental length dx, at a distance x from the smaller end shown. The diameter at this section is given by

$$d_x = 40 + \frac{(80-40)}{1200}x = 40 + \frac{x}{30}$$

Polar moment of inertia,

$$J_x = \frac{\pi\,(40+x/30)^4}{32}$$

$$d\theta = \frac{T\,dx}{GJ}, \quad \theta = \int_0^{1200} \frac{T\,dx}{GJ}$$

$$\theta = \frac{T}{G}\int_0^{1200} \frac{dx \times 32}{\pi(40+x/30)^4} = \frac{T \times 32}{G\pi}\left[-\frac{30}{3}\left(40+\frac{x}{30}\right)^{-3}\right]_0^{1200}$$

$$= \frac{32T \times (-10)}{G\pi}\left[\frac{1}{(40+1200/30)^3} - \frac{1}{(40)^3}\right]$$

$$= \frac{320T}{G\pi} \times \frac{7}{8 \times (40)^3} = \frac{320 \times 0.5 \times 10^6}{85{,}000 \times \pi} \times \frac{7}{8 \times 40^3} = 0.0082 \text{ rad}$$

A general expression for the angle of twist of tapering shafts can be found from Fig. 8.18(c).

$$d\theta = \frac{T\,dx}{GJ}$$

$$\theta = \int_0^L \frac{T\,dx}{GJ}$$

$$d_x = d + \frac{(D-d)}{L}\,x = (d + kx)$$

$$J_x = \frac{\pi(d+kx)^4}{32}$$

$$\theta = \frac{T}{G}\int_0^L \frac{d\times 32}{\pi\,(d+kx)^4} = \frac{32T}{G\pi}\left(\frac{-1}{3k}\right)\left[\frac{1}{(d+kx)^3}\right]_0^L$$

$$= -\frac{32T}{3G\pi K}\left[\frac{1}{D^3} - \frac{1}{d^3}\right] = \frac{32TL\,(D^3 - d^3)}{3G\pi(D-d)D^3 d^3}$$

where $K = (D - d)/L$.

8.6 SHAFTS IN SERIES AND PARALLEL

Shafts connected in series and parallel can be dealt with knowing the way they resist the torque.

(a) *Shafts in series* When two shafts are connected in series as in Fig. 8.19(a), the shafts are *subjected to the same torque.* The shafts can be of the same material but of different diameters or can be of different materials. For such a case, the condition governing their behaviour is that they are subjected to the same torque. Let the subscripts 1 and 2 relate to the two sections of the shaft. Then, we have

$$T = \tau_1 J_1/r_1 = \tau_2 J_2/r_2;\ \tau_1/\tau_2 = J_2 r_1/J_1 r_2$$

Also, the twist of the two shafts can be related by

$$T/J_1 = G_1\theta_1/L_1;\ \theta_1 = TL_1/G_1J_1$$

Similarly, for the other shaft,

$$\theta_2 = TL_2/G_2J_2$$

Therefore, we have, $\theta_1/\theta_2 = L_1 G_2 J_2/ L_2 G_1 J_1$

When both the shafts are solid circular shafts, $J_1 = \pi r_1^4/2;\ J_2 = \pi r_2^4/2$

Therefore,

$$\theta_1/\theta_2 = (\pi r_2^4/2)r_1 / (\pi r_1^4/2) = (r_2/r_1)^3$$

Similarly, the angles of twist are related by

$$\theta_1/\theta_2 = L_1G_2(\pi r_2^4/2) / [L_1G_1(\pi r_1^4/2)] = L_1G_2(r_2/r_1)^4/L_2\,G_1$$

$$\text{Total angle of twist} = \theta_1 + \theta_2 = T\left(\frac{L_1}{G_1 J_1} + \frac{L_2}{G_2 J_2}\right)$$

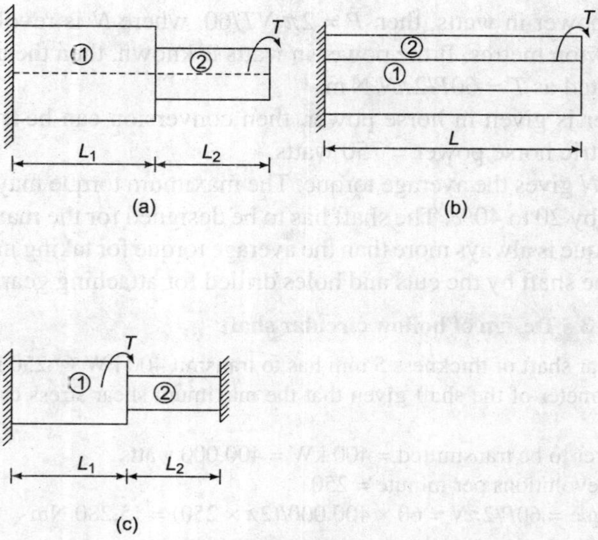

Fig. 8.19

(b) *Shafts in parallel* Two shafts are said to be in parallel configuration, when one shaft is placed surrounding the second shaft as in Fig. 8.19(b). The difference between the shafts in series and the shafts in parallel is the fact that the torque carried is the same in the first case while the applied torque is shared by the two shafts in some proportion in the parallel case. The same is true of the case shown in Fig. 8.19(c), where the two parts of the shaft share the torque applied in some proportion depending upon their properties. If T is the applied torque, then $T = T_1 + T_2$, where T_1 and T_2 are the shares of the two parts of the total torque. In this case, the angle of twist remains the same for the two shafts, i.e., $\theta_1 = \theta_2 = \theta$.

$$T_1 = G_1 J_1 \theta / L_1;\ T_2 = G_2 J_2 \theta / L_2$$

Therefore, $T_1 L_1 / G_1 J_1 = T_2 L_2 / G_2 J_2$

Generally, the two shafts will have the same length ($L_1 = L_2$). These two equations help us to evaluate T_1 and T_2.

$$T_1 / T_2 = \frac{G_1 J_1}{G_2 J_2}$$

$$T = \theta (G_1 J_1 + G_2 J_2)/L \quad \text{and} \quad \theta = TL/(G_1 J_1 + G_2 J_2)$$

8.7 POWER AND TORQUE

A shaft is subjected to torque when power is transmitted through various means to serve some useful purpose. The many attachments used for this purpose are fixed to the shaft by making cuts and holes in the shaft. This makes the shaft weak.

If a shaft is rotating at N revolutions per minute, the average torque being T N m,

Work done = (average torque) × (angle turned through in 1 s)

$$= T \times 2\pi \frac{N}{60} \text{ W}, \quad 1 \text{ W} = 1 \text{ Nm/s} = 1 \text{ J/s}$$

If P is the power in watts, then $P = 2\pi NT/60$, where N is revolutions/minute and T is in newton metres. If the power in watts is known, then the average torque can be calculated as $T = 60P/2\pi N$ N m.

If the power is given in horse power, then conversion can be made using the equation 1 metric horse power = 750 watts.

$T = 60P/2\pi N$ gives the average torque. The maximum torque may be above this average value by 20 to 40%. The shaft has to be designed for the maximum torque. The design torque is always more than the average torque for taking into account the weakness in the shaft by the cuts and holes drilled for attaching gears, pulleys, etc.

Example 8.13 Design of hollow circular shaft

A hollow circular shaft of thickness 5 mm has to transmit 400 kW @ 250 rpm. Determine the external diameter of the shaft given that the maximum shear stress cannot exceed 60 MPa.

Solution Power to be transmitted = 400 kW = 400,000 watts

Number of revolutions per minute = 250

Average torque = $60P/2\pi N = 60 \times 400{,}000/(2\pi \times 250) = 15{,}280$ Nm

We take the design torque to be 20% more than the average.

Design torque = $15280 \times 1.2 = 18{,}335$ Nm

Let d_e mm be the external diameter. So, inner diameter = $(d_e - 50)$ mm

Polar MI = $\pi[d_e^4 - (d_e - 10)^4]/32$

Polar modulus = $\pi[d_e^4 - (d_e - 50)^4]/32d_e$

We have

Torque = polar modulus × shear stress

or $18{,}335 \times 10^3 = \pi\tau[d_e^4 - (d_e - 10)^4]/16d_e$

or $[d_e^4 - (d_e - 10)^4]/d_e = 18{,}335 \times 10^3 \times 16/\pi \times 65 = 1436 \times 10^3$

or $[d_e^4 - (d_e^4 - 40d_e^3 + 600d_e^2 - 4000d_e + 10^4)] = 1436 \times 10^3\, d_e$

or $d_e^3 - 15d_e^2 - 35{,}800d_e = 250$

Solving by trial and error, d_e = 197 mm, say, 200 mm

Inner radius = 180 mm

Example 8.14 Design of hollow circular shaft

Find the required diameter of a solid circular shaft if it has to transmit 300 kW of power @ 120 rpm, given that the maximum shear stress is limited to 50 N/mm^2. The maximum torque is 30% more than the mean torque.

Solution Power in watts, $P = T(2\pi N/60)$

Therefore,

Average torque = $60P/2\pi N = 60 \times 300 \times 1000/(2\pi \times 120) = 23{,}870$ Nm

Design torque = 30% more than the average torque = $1.3 \times 23{,}870 = 31{,}035$ Nm

For a solid circular shaft, polar MI = $\pi d^4/32$

Polar section modulus = $\pi d^4/16d$ (maximum radial distance = $d/2$)

= $\pi d^3/16$

Torsional moment of resistance = $\pi d^3\tau/16$

Permissible shear stress, $\tau = 50$ N/mm^2

This is equal to the design torque.

$31{,}035 \times 10^3 = \pi d^3\tau/16$ from which d = 146 mm, say, 150 mm

Minimum diameter of the shaft = 150 mm

Example 8.15 Design of hollow shaft

A hollow circular shaft has to transmit 300 kW power at 100 rpm. It is proposed to use a hollow circular shaft of internal diameter 0.7 times the external diameter. Find the diameter and thickness of the shaft if the shear stress is not to exceed 60 MPa and the maximum torque is 40% more than the mean torque.

Solution Torque $T = 60P/2\pi N = 60 \times 300 \times 1000/(2\pi \times 100) = 28.65 \times 10^3$ Nm

Design torque = $1.4 \times 28.65 \times 10^3 = 40.11 \times 10^3$ Nm

If d_e is the external diameter of the shaft, then the internal diameter = $0.7d_e$

Polar MI = $\pi[d_e^4 - (0.7d_e)^4]/32 = 0.0746d_e^4$

Polar section modulus, $Z_p = 0.0746d_e^4/(d_e/2) = 0.1492d_e^3$

Maximum permissible shear stress = 60 MPa = 60 N/mm^2

$$T = Z_p \times \tau_{max}$$

Therefore, $0.1492d_e^3 \times 60 = 40.11 \times 10^6$; $d_e = 165$ mm

External diameter = 165 mm; internal diameter = 115 mm

8.8 DESIGN OF SHAFTS

Shafts are designed based upon the torque to be carried, permissible shear stress of the material of the shaft, and stiffness. As in the case of beams, there are two criteria for the design of shafts—strength and stiffness.

8.8.1 Strength and Stiffness

The torsion equation, $T/J = \tau/r = G\theta/L$ represents two equations—$T/J = \tau/r$ is the equation defining the strength of the shaft and $T/J = G\theta/L$ is the equation specifying the stiffness of the shaft.

Strength design In strength design, knowing the torque T to be carried and the maximum permissible stress τ_{max} for the material of the shaft, we arrive at a diameter of the shaft. The maximum shear stress occurs at the surface of the shaft. As the torsion equation is applicable only to circular and hollow circular sections, we have, for given T and τ_{max}, J and Z_p as follows.

For a solid circular section We have $T = J\tau_{max}/r$, where J and r are unknowns. For a shaft of radius r,

$$J = \pi r^4/2;\ T = \pi r^4 \tau_{max}/2r$$

As $J/r = Z_p$, the polar section modulus, $T = Z_p\tau_{max}$

Z_p for a solid circular section = $\pi r^3/2$

In terms of diameter, $Z_p = \pi d^3/16$

For a hollow circular section We have $T = Z_p\tau_{max}$. The polar section modulus of a hollow circular section is $Z_p = \pi(r_e^4 - r_i^4)/2r_e$ or $\pi(d_e^4 - d_i^4)/16d_e$. Knowing T and τ_{max}, the diameter can be found.

Stiffness criterion The stiffness criterion comes from the second part of the torsion equation, i.e., $T/J = G\theta/L$. Stiffness criterion is specified in terms of the angle of twist θ. The angle of twist θ is specified as not to exceed a certain value θ_{max} in a certain length L. We then have $J = TL/G\theta_{max}$. The required polar moment of inertia can be ascertained for the stiffness criterion. The radius or diameter of the shaft can be determined from the value of J.

When the shaft is designed from the strength and stiffness criteria, the radius (or diameter) obtained will not be the same from the two criteria. We have to select the maximum of the two values obtained from applying the criteria.

Example 8.16 Power transmitted by solid and hollow shafts

Find the maximum power transmitted by a shaft at 200 rpm without exceeding the permissible stress of 100 MPa if the shaft is (i) a solid circular shaft of diameter 60 mm and (ii) a hollow shaft of the same internal diameter and has the same weight as the solid shaft.

Solution For a solid circular shaft, the maximum internal torque that can be generated is given by

$$T = \frac{\tau J}{r}$$

$$\tau = 100 \text{ MPa} = 100 \text{ N/mm}^2, \quad r = 30 \text{ mm}, \quad J = \frac{\pi \times 60^4}{32}$$

$$T = \frac{100 \times \pi \times 60^4}{30 \times 32} = 4.24 \times 10^6 \text{ N mm} = 4.24 \text{ kN m}$$

$$\text{Power} = \frac{2\pi NT}{60} \quad \text{(where } N \text{ is in rpm and } T \text{ is in N m)}$$

$$= \frac{2\pi \times 200 \times 4240}{60} \text{ W} = 88{,}802 \text{ W} = 88.8 \text{ kW}$$

If the weights of the two shafts are the same, then their areas must be equal.

$$\frac{\pi \times 60^4}{4} = \frac{\pi}{4}(D^4 - 60^4)$$

$$\Rightarrow \quad D = 71.35 \text{ mm}$$

The hollow shaft has an external diameter of 71.35 mm and an internal diameter of 60 mm. For the hollow shaft,

$$T = \frac{100J}{r} = 100 \times \frac{\pi}{32} \frac{(71.35^4 - 60^4) \times 2}{71 \cdot 35} = 3.565 \text{ kN m}$$

$$\text{Power transmitted} = \frac{2\pi NT}{60} = \frac{2\pi \times 200 \times 3565}{60} \text{ W}$$

$$= 74.665 \text{ kW}$$

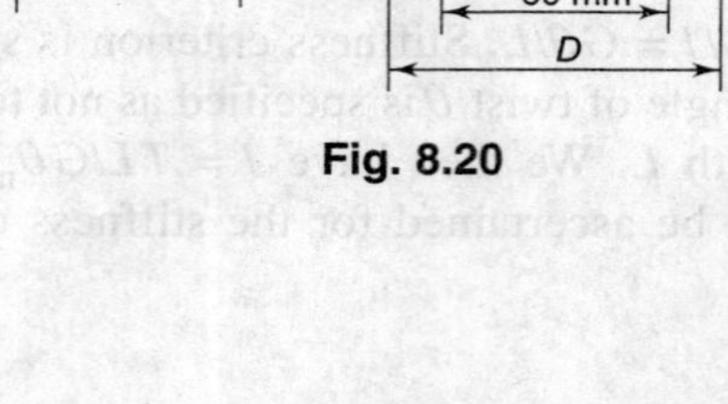

Fig. 8.20

Example 8.17 Design of solid circular shaft

A shaft has to be designed to transmit a power of 100 kW at 300 rpm. Determine the diameter required for a solid circular shaft of steel for which the permissible shear stress is 90 MPa. What is the angle of twist per unit length of this shaft? $G = 82$ GPa.

Solution Torque $= \dfrac{P\times 60}{2\pi N} = \dfrac{100\times 1000\times 60}{2\pi\times 300} = 3.183\times 10^6$ N mm

For a solid circular shaft,

$$\tau = \frac{16T}{\pi d^3}$$

$$\Rightarrow \quad T = \frac{\pi d^3}{16}\tau$$

$$3.183\times 10^6 = \frac{\pi\times d^3}{16}\times 90$$

$$\Rightarrow \quad d^3 = \frac{3.183\times 10^6\times 16}{\pi\times 90} = 0.18\times 10^6$$

$$\Rightarrow \quad d = 56.5 \text{ mm}$$

$$\text{Angle of twist per metre length} = \frac{\theta}{L} = \frac{T}{GJ} = \frac{3.183\times 10^6\times 32}{82{,}000\times\pi\times(56.5)^4}$$

$$\Rightarrow \quad \theta = \frac{1000\times 3.183\times 10^6\times 32}{82{,}000\times\pi\times(56.5)^4}$$

$$= 0.0388 \text{ rad (Fig. 8.21)}$$

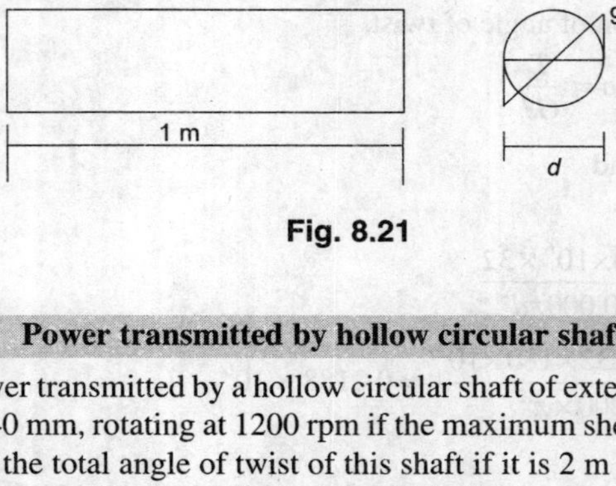

Fig. 8.21

Example 8.18 Power transmitted by hollow circular shaft

Determine the power transmitted by a hollow circular shaft of external diameter 60 mm and internal diameter 40 mm, rotating at 1200 rpm if the maximum shear stress is not to exceed 100 MPa. What is the total angle of twist of this shaft if it is 2 m long? $G = 85$ GPa.

Solution The cross section of the shaft is shown in Fig. 8.22.

Polar MI,

$$J = \frac{\pi}{32}(60^4 - 40^4) = 102.1\times 10^4 \text{ mm}^4$$

$$\text{Torque} = \frac{\tau_{max} J}{D/2} = \frac{100\times 102.1\times 10^4}{30} = 3.403 \text{ kN m}$$

$$\text{Power} = \frac{2\pi NT}{60} = \frac{2\pi\times 1200\times 3.403}{60} = 427.6 \text{ kW}$$

$$\text{Angle of twist for a length } L = \frac{TL}{GJ}$$

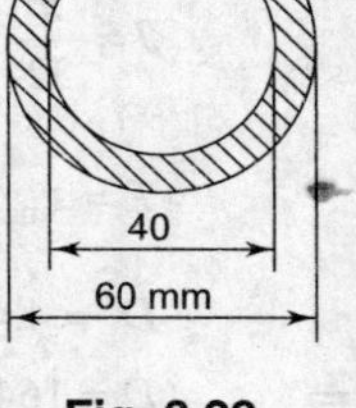

Fig. 8.22

For a 2 m length,

$$\theta = \frac{3.403\times10^6\times2000}{85,000\times102.1\times10^4} = 0.0784 \text{ rad}$$

Example 8.19 Design of circular shafts

A shaft has to transmit a torque of 30 kN m. The maximum shear stress is not to exceed 100 MPa and the angle of twist is not to exceed 1°/meter length. $G = 80$ GPa. Design the shaft according to the given specifications if it is a (i) solid circular shaft and (ii) hollow circular shaft of internal diameter 90% of the external diameter.

Solution Two conditions are stipulated for the design—the maximum diameter obtained from the two conditions will be selected.

(i) *Solid circular shaft* From the consideration of maximum stress,

$$T = \frac{\tau_{max} J}{r}$$

$$T = 30 \text{ kN m} = 30\times10^6 \text{ N mm}$$

$$\tau_{max} = 100 \text{ MPa} = 100 \text{ N/mm}^2$$

$$r = \frac{d}{2},\quad J = \frac{\pi d^4}{32}\quad \frac{J}{R} = \frac{\pi d^3}{16}$$

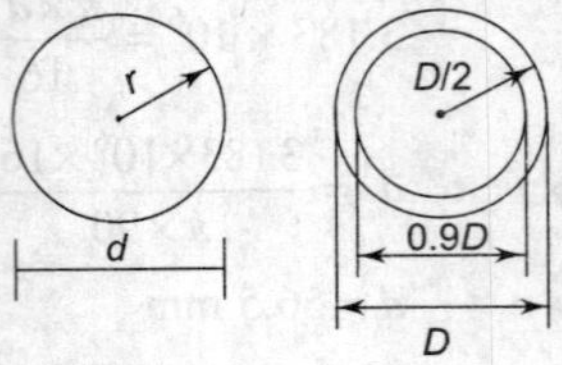

Fig. 8.23

Therefore,

$$30\times10^6 = 100\times\frac{\pi}{16}\times d^3$$

$$\Rightarrow \quad d^3 = \frac{30\times10^6\times16}{100\times\pi} = 1.528\times10^6 \text{ mm}$$

$$\Rightarrow \quad d = 152.8 \text{ mm}$$

From the consideration of angle of twist,

$$\theta \text{ per unit length} = \frac{T}{GJ}$$

$$\theta = 1° = \frac{\pi}{180} \text{ rad}$$

Therefore,

$$\frac{\pi}{10^3\times180} = \frac{30\times10^6\times32}{80,000\pi d^4}$$

$$d^4 = \frac{30\times10^6\times32\times180\times10^3}{80,000\times\pi^2} = 0.2188\times10^6$$

$$\Rightarrow \quad d = 121.6 \text{ mm}$$

Since the value obtained from stress consideration is higher, $d = 152.8$ mm.

(ii) *Hollow circular shaft* Let the external diameter of the shaft be D mm.

Internal diameter $= 0.9D$

$$J = \frac{\pi}{32}[D^4 - (0.9\,D)^4] = 0.03376\,D^4$$

$$T = \tau_{max}\frac{J}{(D/2)},\quad 30\times10^6 = 100\times\frac{0.03376\,D^4}{D}\times2$$

$$D^3 = 4.443\times10^6$$

$$\Rightarrow \quad D = 164.4 \text{ mm}$$

From the consideration of angle of twist,

$$\theta = 1°/\text{m length} = \frac{\pi}{180\times1000} \text{ in 1 mm length}$$

$$\theta = \frac{T}{GJ}, \quad \frac{\pi}{180\times1000} = \frac{30\times10^6}{80{,}000\times0.003376D^4}$$

$D = 158.8$ mm

The higher value of 164.4 mm is selected.

Internal diameter = 148 mm

The sections are shown in Fig. 8.23.

Example 8.20 Compound shaft

A compound shaft is made of two materials—on the left is 1 m length of steel, 90 mm ϕ, and on the right is 0.8 m length of aluminium, 60 mm ϕ. The shaft is fixed at the left end and a torque of 1 kN m is applied at the right end, as shown in Fig. 8.24. Determine the additional torque that can be applied at B without exceeding the permissible stress in steel of 90 MPa. Also work out the nature and magnitude of a torque T to be applied at B so that the angle of twist at C with respect to A is zero. The modulus of rigidity is 85 GPa for steel and 30 GPa for aluminium.

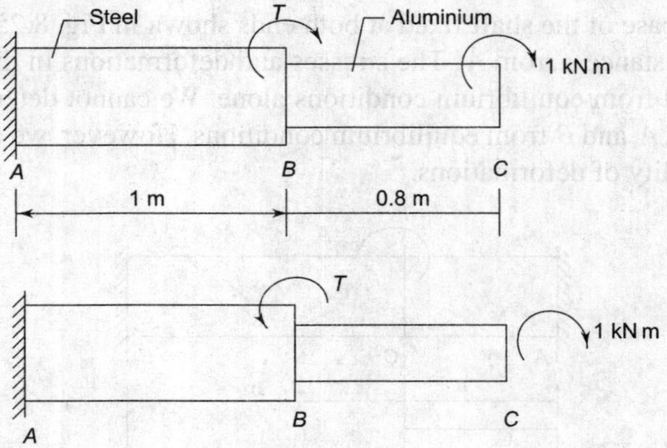

Fig. 8.24

Solution For the aluminium shaft,

$$J = \frac{\pi}{32}\times60^4 = 127.23\times10^4 \text{ mm}^4$$

$$T = 1 \text{ kN m} = 1\times10^6 \text{ N mm}$$

$$\text{Maximum stress } \tau_{max} = \frac{1\times10^6\times30}{127.23\times10^4} = 23.6 \text{ N/mm}^2$$

The angle of twist of end C with respect to B is given by

$$\theta_{C/B} = \frac{TL}{GJ} = \frac{1\times10^6\times800}{30{,}000\times127.23\times10^4} = 0.021 \text{ rad, clockwise}$$

For the steel shaft,

$$\tau_{max} = 90 \text{ MPa} = 90 \text{ N/mm}^2$$

Maximum torque,

$$T = \frac{\tau_{max} J}{r} = \frac{90 \times (\pi/32) \times (90)^4}{45} = 12.88 \times 10^6 \text{ Nm}$$

The torque T that can be applied at B is 12.88 kNm. Since point A is fixed, let $\theta_{B/A}$ be the rotation of B with respect to A. The rotation of C with respect to A is $\theta_{C/A}$ which is equal to $\theta_{B/A} + \theta_{C/B}$. From the given condition $\theta_{B/A} + \theta_{C/B} = 0$, $\theta_{B/A} + 0.021 = 0$.

$$\theta_{B/A} = -0.021 \text{ rad}$$

A torque in the anticlockwise direction has to be applied at B. If T is the magnitude of this torque, then $\theta_{B/A} = TL/GJ$.

$$-0.021 = \frac{T \times 1000}{85{,}000 \times (\pi/32)(90)^4}$$

$$\Rightarrow \quad T = \frac{-0.021 \times 85{,}000}{1000} \times \frac{\pi}{32} \times 90^4$$

$$= -11.49 \times 10^6 \text{ Nmm}$$

$$T = 11.49 \text{ kNm (anticlockwise)}$$

8.9 STATICALLY INDETERMINATE SHAFTS

Consider the case of the shaft fixed at both ends shown in Fig. 8.25. A torque T is applied at a distance a from A. The stresses and deformations in this shaft cannot be determined from equilibrium conditions alone. We cannot determine the reactive couples at A and B from equilibrium conditions. However, we can do so using the compatibility of deformations.

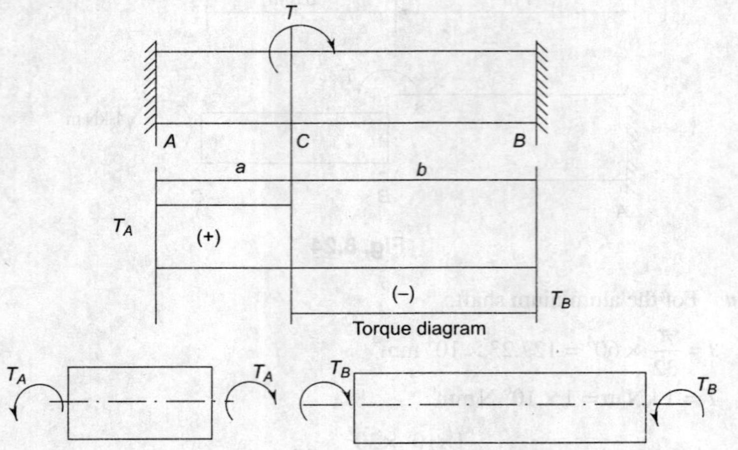

Fig. 8.25

If T_A and T_B are the reactive couples at A and B, from the free body diagrams of AC and CB,

$$T_A + T_B = T$$

A second equation can be derived from the compatibility of deformations, showing that the angle of twist at C with respect to A must be the same as that of C with respect to B.

$$\theta_{C/A} = \frac{T_A a}{GJ}, \quad \theta_{C/B} = \frac{T_B b}{GJ}$$

$$\frac{T_A a}{GJ} = \frac{T_B b}{GJ}$$

$$T_A = T_B \frac{b}{a}$$

Substituting this value in the first equation,

$$T_B \frac{b}{a} + T_B = T, \quad T_B \left(\frac{b+a}{a}\right) = T, \quad T_B = \frac{Ta}{L}$$

where $L = a + b$.

$$T_A = \frac{Tb}{L}$$

A similar approach can be used if the segments AC and CB are of different sections and different materials and in cases where there are more than two segments of different sections and materials. The following examples illustrate the procedure.

Example 8.21 Statically indeterminate compound shaft

For the shaft shown in Fig. 8.26, find the maximum value of torque T such that stress in steel is within 90 MPa and that in brass is within 50 MPa. The steel shaft is of 20 mm ϕ and the brass shaft of 30 mm ϕ. G = 85 GPa for steel and 40 GPa for brass.

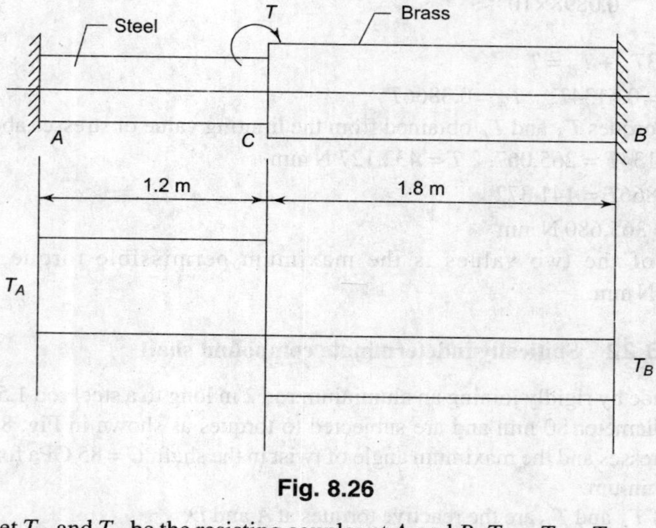

Fig. 8.26

Solution Let T_A and T_B be the resisting couples at A and B. $T_A + T_B = T$.

For the steel shaft,

$$J_s = \frac{\pi d^4}{32} = \frac{\pi \times 20^4}{32} = \frac{\pi}{2} \times 10^4 \text{ mm}^4$$

For the brass shaft,

$$J_b = \frac{\pi \times 30^4}{32} = 7.952 \times 10^4 \text{ mm}^4$$

Maximum stress in the steel shaft,

$$\tau_s(\max) = \frac{T_A r}{J}$$

$$= \frac{T_A \times 10}{(\pi/2)\times 10^4} = \frac{2T_A}{3141.6}$$

This is limited to 90 MPa or 90 N/mm^2.

$$90 = \frac{2T_A}{3141.6}, \quad T_A \le 141{,}372 \text{ N mm}$$

Similarly, $\tau_{b(\max)}$ is limited to 50 N/mm^2.

$$50 = \frac{T_B \times 15}{7.952\times 10^4}, \quad T_B \le 265{,}067 \text{ N mm}$$

From the compatibility of deformations, the angle of twist of C with respect to A must be the same as that with respect to B.

$$\theta_{C/A} = \frac{TL}{GJ} = \frac{T_A \times 1200}{85{,}000\times(\pi/2)\times 10^4} = 0.0898 \times 10^{-5} T_A$$

$$\theta_{C/B} = \frac{TL}{GJ} = \frac{T_B \times 1800}{40{,}000\times 7.952\times 10^4} = 0.5659 \times 10^{-6} T_B$$

Equating the two,

$$T_A = \frac{0.5659\times 10^{-6}\, T_B}{0.0898\times 10^{-5}} = 0.63 T_B$$

Therefore,

$$0.63T_B + T_B = T$$

$$\Rightarrow \quad T_B = 0.6134T, \quad T_A = 0.3866T$$

Equating to torques T_A and T_B obtained from the limiting value of stresses above,

$$0.6134T = 265{,}067, \quad T = 432{,}127 \text{ N mm}$$

$$0.3866T = 141{,}372$$

$$T = 365{,}680 \text{ N mm}$$

The lower of the two values is the maximum permissible torque. Therefore, $T = 365{,}680$ N mm. ■

Example 8.22 Statically indeterminate compound shaft

A shaft is made by rigidly joining an aluminium rod 2 m long to a steel rod 1.5 m long. The rods are of diameter 80 mm and are subjected to torques as shown in Fig. 8.27. Find the maximum stresses and the maximum angle of twist in the shaft. $G = 85$ GPa for steel and 30 GPa for aluminium.

Solution If T_A and T_D are the reactive torques at A and D,

$$T_A + T_D = 4\times 10^6$$

The shaft has to be divided into three sections, AB, BC, and CD, because of the changes in torque and material.

In segment AB,

$$\text{Torque} = T_A$$

and $$\theta_{B/A} = \frac{T_A L}{GJ} = \frac{T_A \times 750}{30{,}000\times J}$$

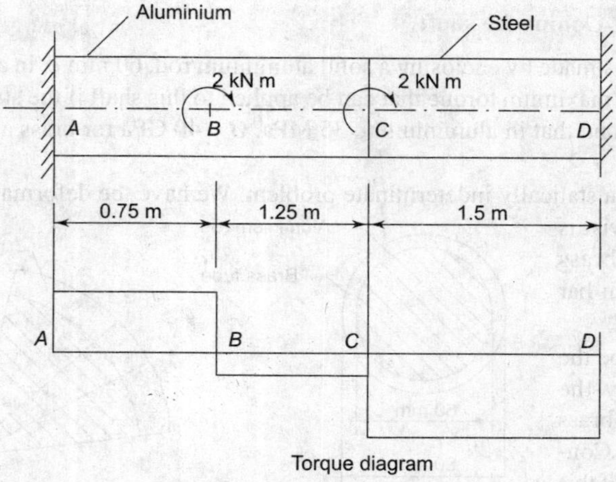

Fig. 8.27

In segment *BC*,

$$\text{Torque} = T_A - 2 \times 10^6$$

and $$\theta_{C/B} = \frac{(T_A - 2\times10^6)\times1250}{30{,}000J}$$

In segment *CD*,

$$\text{Torque} = T_D = 4 \times 10^6 - T_A$$

and $$\theta_{D/C} = -\frac{(4\times10^6 - T_A)\times1500}{85{,}000J}$$

The magnitude of *J* is the same, and equals $\pi \times 80^4/32$.

A and *D* being fixed ends, the angle of twist of *D* with respect to *A* is zero. That is,

$$\theta_{B/A} + \theta_{C/B} + \theta_{D/C} = 0$$

$$\frac{T_A \times 750}{30{,}000J} + \frac{\left(T_A - 2\times10^6\right)\times1250}{30{,}000J} - \frac{\left(4\times10^6 - T_A\right)\times1500}{85{,}000\,J} = 0$$

$$T_A = 1.825 \times 10^6\,\text{N mm}, \quad T_D = 2.175 \times 10^6\,\text{N mm}$$

The maximum stress in aluminium will be in the segment *AB*.

$$\tau_{\max}(\text{Al}) = \frac{1.825\times10^6\times40}{(\pi/32)\,(80)^4} = 18.15\ \text{N/mm}^2$$

Similarly,

$$\tau_{\max}(\text{steel}) = \frac{2.175\times10^6\times40}{(\pi/32)\,(80)^4} = 21.63\ \text{N/mm}^2$$

The angles of twist are given by

$$\theta_{B/A} = \frac{1.825\times10^6\times750}{30{,}000\times(\pi/32)\,(80)^4} = 0.01134\ \text{rad}$$

$$\theta_{C/B} = \frac{-0.175\times10^6\times1250}{30{,}000\times(\pi/32)\,(80)^4} = -\,0.00181\ \text{rad}$$

$$\theta_{D/C} = \frac{-2.175\times10^6\times1500}{85{,}000\times(\pi/32)\,(80)^4} = -\,0.00953\ \text{rad}$$

Thus, the maximum angle of twist is between *A* and *B*. ■

Example 8.23 Composite shaft

A composite shaft is made by enclosing a solid aluminium rod, 60 mm ϕ, in a brass tube 10 mm thick. Find the maximum torque that can be applied to this shaft if the stress in brass is limited to 50 MPa and that in aluminium to 35 MPa. G = 40 GPa for brass and 28 GPa for aluminium.

Solution This is a statically indeterminate problem. We have the deformation condition that the angle of twist is the same for the brass tube and aluminium bar at any section.

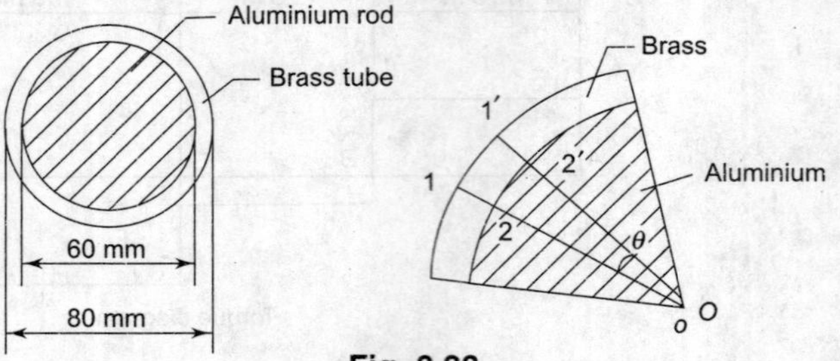

Fig. 8.28

Let T_{Al} and T_b be the torques carried by the aluminium rod and brass tube, respectively. Considering a length L of this shaft, we can state that

$$\frac{T_{Al}L}{G_{Al}J_{Al}} = \frac{T_b L}{G_b J_b}$$

where the subscripts refer to aluminium and brass respectively. Also, from $T = (\tau J)/r$,

$$T_{Al} = \frac{\tau_{Al}J_{Al}}{30}, \quad T_b = \frac{\tau_b J_b}{40}$$

in terms of the maximum stresses in the two materials. Substituting

$$\frac{\tau_{Al}J_{Al}L}{30G_{Al}J_{Al}} = \frac{\tau_b J_b L}{40G_b J_b}$$

$$\tau_{Al} = \frac{\tau_b \times 30G_{Al}}{40G_b}$$

$$\tau_{Al} = \frac{30\times 28{,}000}{40\times 40{,}000}\ \tau_b = 0.525\tau_b$$

If $\tau_b = 50$,

$$\tau_{Al} = 0.525 \times 50 = 26.25 \text{ N/mm}^2$$

The maximum stress in aluminium cannot reach the value of 35 N/mm^2 because the stress in brass will then be more than the permissible stress.

Torque carried by aluminium,

$$T_{Al} = \frac{26.25}{30}\times\frac{\pi}{32}\times(60)^4$$
$$= 1.1133 \text{ kN m}$$

Torque carried by brass,

$$T_b = \frac{50}{40}\times\frac{\pi}{32}(80^4 - 60^4)$$
$$= 3.4361 \text{ kN m}$$

Maximum torque = $T_{Al} + T_b$ = 1.1133 + 3.4361 = 4.5449 kN m ■

Example 8.24 Shafts in series

A hollow circular brass shaft of external diameter 30 mm is rigidly connected to a steel shaft of diameter 20 mm in series. The area of the two sections of shafts is the same. The length of the two shafts is the same, i.e., 500 mm. What maximum torque can be applied to the shaft without exceeding the permissible stress of 150 MPa in steel and 60 MPa in brass? Also calculate the angular rotation in the shaft.

Solution The situation is shown in Fig. 8.29.

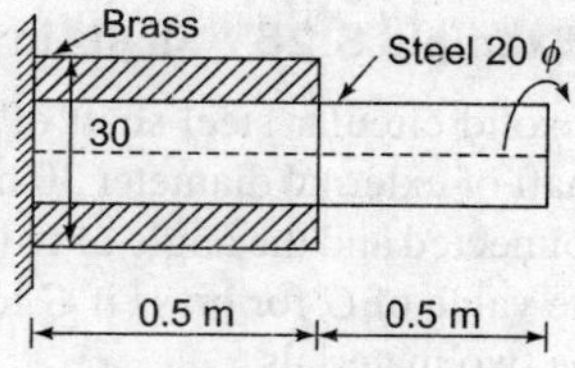

Fig. 8.29

Let d be the internal diameter of the brass shaft.

Area of the brass shaft = $\pi(30^2 - d^2)/4$

Area of the steel shaft = $\pi 20^2/4$

As the two shafts are of equal area,

$\pi(30^2 - d^2)/4 = \pi 20^2/4$; $d^2 = 900 - 400 = 500$;

$d = 22.36$ mm

Polar MI of the steel shaft = $\pi(10)^4/2 = 5000\pi$ mm^4

Polar MI of the brass shaft = $\pi(15^4 - 11.18^4)/2 = 54{,}980.5$ mm^4

From $T/J = \tau/r$, $\tau = T(r/J) = TZ_p$, where Z_p is the polar section modulus.

Now, $T = (5000\pi/10)150 = 235{,}619.4$ Nmm, considering the maximum stress in steel and

$T = 54{,}980.5 \times 60/15 = 219{,}922$ Nmm, considering the maximum stress in brass

Maximum torque = 219.922 Nm

$$\text{Angular twist, } \theta = \theta_s + \theta_b = \frac{219\cdot922\times500}{80{,}000\times5000\pi} + \frac{219\cdot922\times500}{60{,}000\times54{,}980.5}$$

$$= 0.0875 + 0.0333$$

or $\theta = 0.1208$ radians

Example 8.25 Shafts in series

A hollow, circular copper shaft of 60 mm external and 30 mm internal diameter and a steel solid shaft of 50 mm radius are rigidly connected in series and subjected to a torque of 5000 Nm as shown in Fig. 8.30. Determine the maximum stresses in the two shafts. $G = 80$ GPa for steel and 40 GPa for copper. Length of the copper shaft is 0.5 m and that of the steel shaft is 0.45 m.

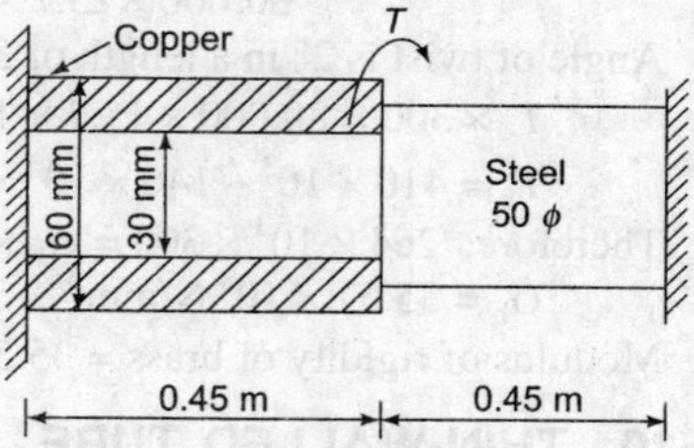

Fig. 8.30

Solution Let T_s and T_c be the torque carried by the two shafts.

$$T_s + T_c = 5000 \times 10^3 \text{ Nmm} \tag{1}$$

Polar MI of the copper shaft = $\pi(30^4 - 15^4)/2 = 1192.8 \times 10^3$ mm^4

Polar MI of the steel shaft = $\pi 25^4/2 = 613.6 \times 10^3$ mm^4

The share of the applied torque is governed by the condition that the angle of twist must be the same at the junction.

Angle of twist θ_s of the steel shaft = $T_s \times 500/(80{,}000 \times 613.6 \times 10^3)$

Angle of twist of the copper shaft = $T_b \times 450/(40{,}000 \times 1192.8 \times 10^3)$

Now, $\theta_s = \theta_c$

$$\text{or} \quad \frac{T_s\times500}{80{,}000\times613.6\times10^3} = \frac{T_b\times450}{40{,}000\times1192.8\times10^3}$$

or $T_s = 1.0288T_b$

Substituting this in the first equation,

$1.0288T_b + T_b = 5000 \times 10^3$; $T_b = 2464.5 \times 10^3$ Nmm; $T_s = 2535.5 \times 10^3$ Nmm

Knowing the torque share of the two shafts, the stresses can be found.

Maximum stress in the brass shaft = $2464.5 \times 10^3 \times 30/1192.8 \times 10^3$
$= 62$ N/mm^2

Maximum stress in the steel shaft = $2535.5 \times 10^3 \times 25/613.6 \times 10^3$
$= 103.3$ N/mm^2

Example 8.26 Shafts in parallel

A solid circular steel shaft of diameter 20 mm is enclosed within a brass hollow circular shaft of external diameter 30 mm and internal diameter 20 mm. If the two shafts are rigidly connected and the angle of twist due to a torque of 410 Nm is 2° in a length of 300 mm, find the value of G for brass if G for steel is 80 GPa. Also find the maximum shearing stress in the two materials.

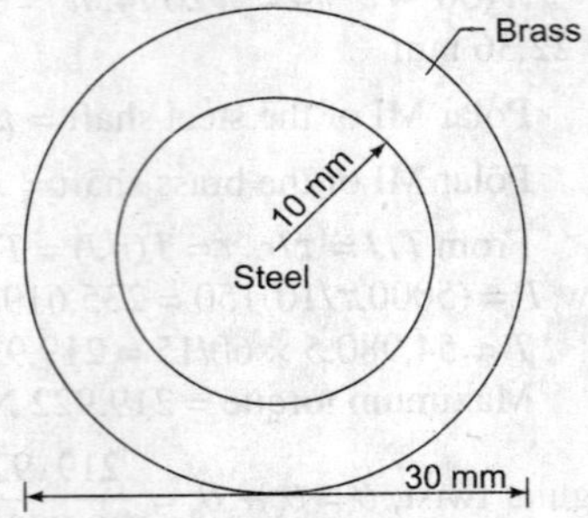

Fig. 8.31

Solution The situation is shown in Fig. 8.31.

Torque = 410 Nm = 410 × 10^3 Nmm Polar MI for the steel shaft = $\pi r^4/2$

= $\pi(10)^4/2 = 15.7 \times 10^3$ mm^4

Polar MI of the brass shaft = $\pi(r_e^4 - r_i^4)/2 = \pi(15^4 - 10^4)/2 = 63.8 \times 10^3$ mm^4

If the torque shares of the two shafts are T_s and T_b for the steel and brass shafts, then

$$T_s + T_b = 410 \times 10^3 \text{ Nmm}$$

The angular twist given by TL/GJ is the same for both the shafts.

Angular twist of the steel shaft = $T_s \times 300/(80{,}000 \times 15.7 \times 10^3)$

Angular twist of the brass shaft = $T_b \times 300/(G_b \times 63.8 \times 10^3)$

Equating the two, $$\frac{T_s \times 300}{80{,}000 \times 15.7 \times 10^3} = \frac{T_b \times 300}{G_b \times 63.8 \times 10^3}$$

Angle of twist is 2° in a length of 300 mm or 0.0349 radians in 300 mm.

$$T_s \times 300 = 80{,}000 \times 15.7 \times 10^3 \times 0.0349 = 146 \times 10^3 \text{ Nmm}$$

$$T_b = 410 \times 10^3 - 146 \times 10^3 = 264 \times 10^3 \text{ Nmm}$$

Therefore, $264 \times 10^3 \times 300 = G_b \times 63.8 \times 10^3 \times 0.0349$

$$G_b = 35.57 \times 10^3 \text{ N/mm}^2$$

Modulus of rigidity of brass = 35.57 GPa

8.10 THIN-WALLED TUBE

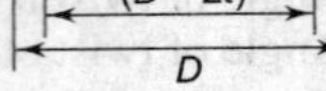

Fig. 8.32

Consider the thin-walled tube of outer radius R and thickness t shown in Fig. 8.32. The polar moment of inertia of this tube whose external diameter is $D = 2R$ and internal diameter is $D - 2t$ is

$$J = \frac{\pi}{32}[D^4 - (D - 2t)^4] = \frac{\pi}{32}[D^4 - (D^4 + 16t^4 - 8D^3 t + 24D^2t^2 - 32 Dt^3)]$$

$$= \frac{\pi}{32}(8D^3t - 24D^2t^2 + 32 Dt^3 - 16t^4)$$

Since t is very small, its higher powers can be neglected. Thus $J = \pi D^3 t/4$.

$$\text{Allowable torque} = \frac{\tau_{max} J}{R} = \tau_{max}\frac{\pi}{4}\frac{D^3 t \times 2}{D} = \frac{\pi D^2 t}{2}\tau_{max}$$

Consider an element of length dL cut out of any thin-walled tube of arbitrary cross section, and varying thickness, as shown in Fig. 8.33. The thickness of the tube is small but may vary. The shear stress on the cross-sectional surface is accompanied by an equal longitudinal shear stress as shown in the figure, from the principle of complementary shear stress. The distribution of shear stress across the thickness is unknown. We can define a quantity called shear flow, given by

$$q = \int_{-t/2}^{t/2} \tau \, dt$$

From Fig. 8.33, for longitudinal equilibrium,

$$q_1 \Delta L = q_2 \Delta L$$

and

$$q_1 = q_2$$

Thus the shear flow q must be constant around the cross section of the tube.

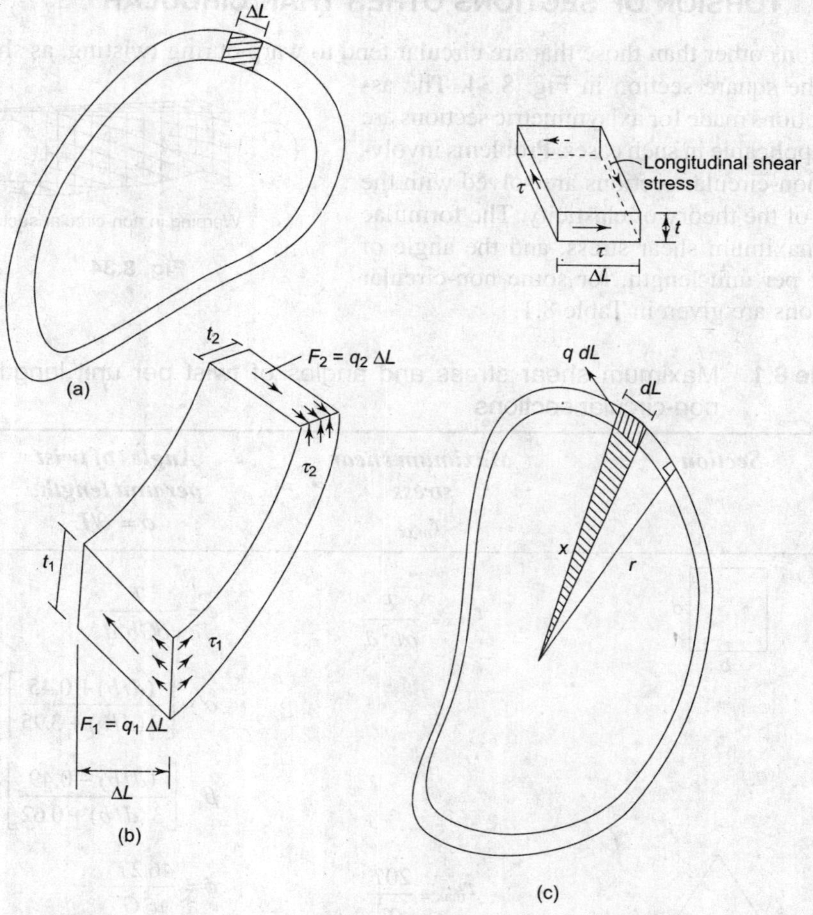

Fig. 8.33

A relationship between applied torque T and shear flow can be obtained from Fig. 8.33(c). The shear flow over a length dL provides a torsional resistance of $q\ dL\ x$ about any centre point within the tube. The total torsional resistance is equal to the applied torque.

$$T = \int q \, dL \, x$$

Since the shear flow is constant, $T = q \int dL\ x$. This integral is twice the area of the shaded triangle of base dL. Obviously, the integral will be twice the area enclosed within the centre line of the tube. Indicating this area by A,

$$T = 2Aq$$

Since $q = \tau_{av}\, t$, average shear stress,

$$\tau_{av} = \frac{q}{t} = \frac{T}{2At}$$

Note that the area is not the cross-sectional area of the tube but the area enclosed within the centre line of the tube.

8.11 TORSION OF SECTIONS OTHER THAN CIRCULAR

Sections other than those that are circular tend to warp during twisting, as shown for the square section in Fig. 8.34. The assumptions made for axisymmetric sections are not applicable in such cases. Problems involving non-circular sections are solved with the help of the theory of elasticity. The formulae for maximum shear stress, and the angle of twist per unit length, for some non-circular sections are given in Table 8.1.

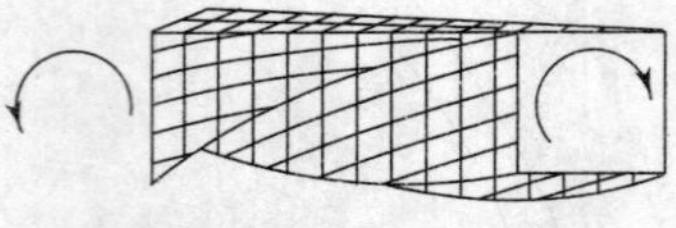

Warping in non-circular section

Fig. 8.34

Table 8.1 Maximum shear stress and angles of twist per unit length for non-circular sections

Section	*Maximum shear stress* τ_{max}	*Angles of twist per unit length,* $\phi = \theta/L$
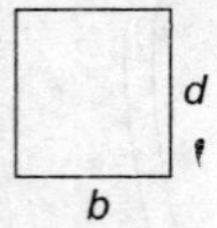	$\tau_{max} = \frac{T}{\alpha b^2 d}$	$\phi = \frac{T}{\beta G b^2 d}$ $\alpha \approx \left[\frac{(d/b) + 0.45}{3(d/b) + 3.95}\right]$ $\beta \approx \left[\frac{(d/b) - 0.49}{3(d/b) + 0.62}\right]$
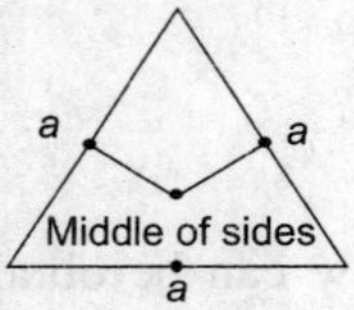	$\tau_{max} = \frac{20T}{\alpha^3}$ at middle points of sides	$\phi = \frac{46.2T}{a^4 G}$
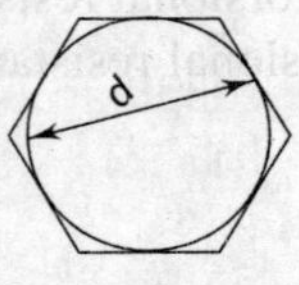 A = Area	$\tau_{max} = \frac{T}{0.217Ad}$	$\phi = \frac{T}{0.133GAd^2}$

Example 8.27 Torque capacity of a thin rectangular tube

The thin rectangular tube shown in Fig. 8.35 is made of brass. What is the maximum torque that can be applied to this section without exceeding a stress of 50 MPa?

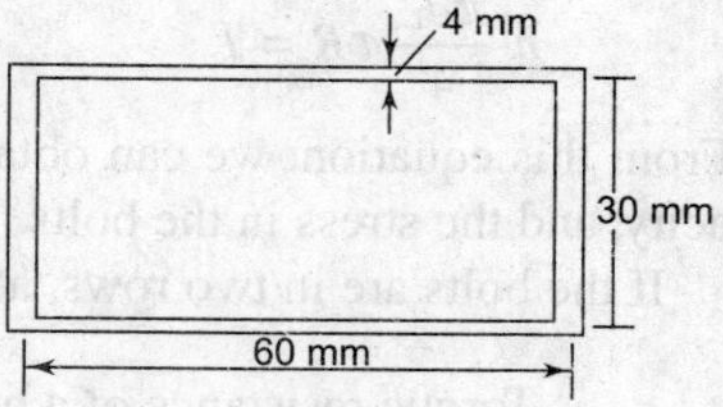

Fig. 8.35

Solution The torque is given by

$$T = 2At\tau_{av}$$

$$A = 60 \times 30 = 1800 \text{ mm}^2$$

Therefore,

$$\text{Maximum torque} = 2 \times 1800 \times 4 \times 50$$
$$= 720 \text{ Nm}$$

Example 8.28 Design of a square shaft

A square shaft is required to transmit 150 kW of power at 200 rpm. Determine the size of the shaft required if the maximum stress cannot exceed 90 MPa and the angle of twist is to be restricted to 1° in a length equal to 25 times the side of the shaft. G = 85 GPa.

Solution For a square section $d/b = 1$, from Table 8.1,

$$\alpha = \frac{1.45}{6.95} = 0.21 \text{ and } \beta = \frac{1 - 0.49}{3.62} = 0.141$$

$$\text{Torque} = \frac{60P}{2\pi n} = \frac{60 \times 150}{2\pi \times 200} \text{ kNm} = 7.162 \text{ kNm}$$

$$= 7.162 \times 10^6 \text{ N mm}$$

From Table 8.1,

$$\tau_{max} = \frac{T}{\alpha a^4}$$

where a is the side of the square and $\alpha = 0.2135$.

$$a^4 = \frac{T}{\alpha\tau_{max}} = \frac{7.162 \times 10^6}{0.21 \times 90}, \quad a = 24.8 \text{ mm}$$

From the consideration of angle of twist,

$$\theta = \frac{TL}{\beta G a^4}, \quad L = 25a, \quad \beta = \left[\frac{(30/60) - 0.49}{3 \times (30/60) + 0.62}\right]$$

$$1 \times \frac{\pi}{180} = \frac{7.126 \times 10^6 \times 25a}{0.141 \times 85{,}000 \times a^4} \quad a^3 = \frac{7.126 \times 10^6 \times 25 \times 180}{\pi \times 0.141 \times 85{,}000} = 8.512 \times 10^3$$

$$a = 20.42 \text{ mm}$$

The required size is the larger of the two values—a = 24.8 mm

8.12 FLANGED COUPLINGS FOR SHAFTS

Two shafts are often connected by using flanged couplings. The coupling consists of flanges attached rigidly to the shafts with bolt holes. The two flanges are then bolted together as shown in Fig. 8.36.

Due to the torque in the shaft, the bolts are subjected to shear stresses. The bolts are provided in a concentric circle if in a single row or in concentric circles if two or more rows of bolts are required. If the bolts are of the same diameter, they will be stressed equally. If n bolts are there in the bolt circle,

$$n\frac{\pi d^2}{4}\tau R = T$$

From this equation, we can obtain the number of bolts required, the torque capacity, and the stress in the bolts.

If the bolts are in two rows, at radii R_1 and R_2, as shown in Fig. 8.36(c),

$$\text{Torque resistance of a bolt at } R_1 = \frac{\pi d_1^2}{4}\tau_1 R_1 = T_1$$

$$\text{Torque resistance of a bolt at } R_2 = \frac{\pi d_2^2}{4}\tau R_2 = T_2$$

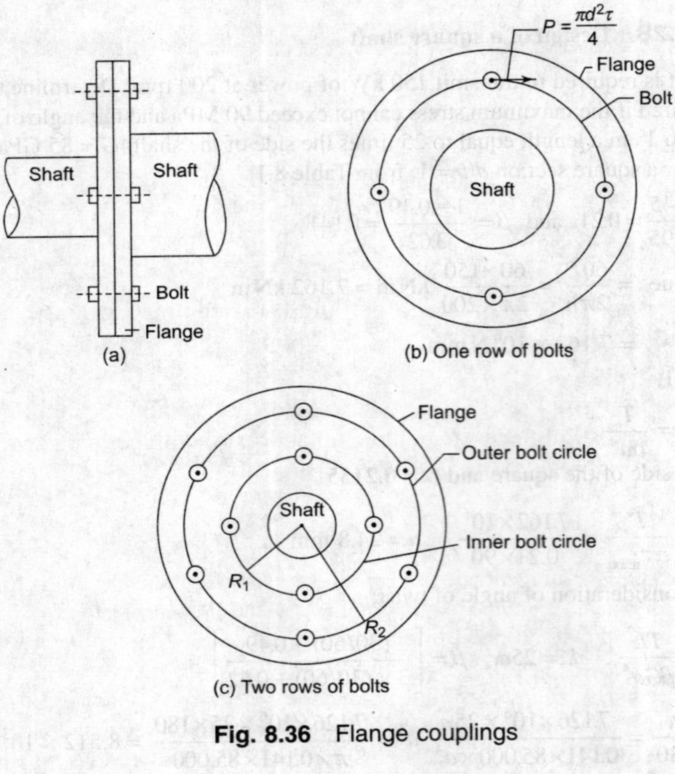

Fig. 8.36 Flange couplings

$$\text{Torque resistance of the connection} = n_1\frac{\pi d^2}{4}\tau R_1 + n_2\frac{\pi d^2}{4}\tau R_2$$

The shearing strains r_1 and r_2 in the bolts at R_1 and R_2 are related by

$$\frac{r_1}{R_1} = \frac{r_2}{R_2}$$

Since $\tau = \frac{G}{r}$

$$\frac{\tau_1}{GR_1} = \frac{\tau_2}{GR_2}$$

$$\tau_1 = \frac{T_1}{A_1}, \tau_2 = \frac{T_2}{A_2}$$

where A_1 and A_2 are the areas of the bolts at R_1 and R_2, respectively.

$$\frac{T_1}{A_1 G_1 R_1} = \frac{T_2}{A_2 G_2 R_2}$$

Example 8.29 Design of flanged bolt coupling

A flanged bolt coupling consists of ten bolts of diameter 12 mm, evenly spaced around a circle of diameter 350 mm. Determine the torque capacity of the connection if the allowable shearing stress is 48 N/mm^2.

Solution Force in each bolt $= \frac{\pi d^2}{4} \times 48 = 5429$ N

Torque capacity = force × radius × number of bolts
= 5429 × 175 × 10
= 4,750,000 N mm
= 4.75 kN m

8.13 BENDING MOMENT AND AXIAL THRUST IN SHAFTS

While determining the stresses in shafts or designing them, we considered only torsional moment. While torque is the predominant action in shafts, shafts are also subjected to BM due to self-weight and loads attached to them. They may also be subjected to axial thrusts.

While designing shafts, it is necessary to take into account the combined effect of torque, BM and axial thrust. Problems related to this topic are dealt with in the chapter on principal planes and stresses (Chapter 9).

Summary

Unlike BM, which is a moment about a centroidal axis perpendicular to the longitudinal axis, torque is a moment about the longitudinal axis of the element. The torque diagram for an element can be drawn like the BM diagram, by calculating the torque about different sections.

Torque is resisted by a resisting torque produced by shear stresses induced in the section. A simple theory of torsion can be developed for axisymmetrical sections like circular and hollow circular sections, based on certain assumptions. The basic torque equation is very similar to the bending equations, $M/I = E/R = \sigma/y$, and can be stated as $T/J = G\theta/l = \tau/r$, where τ is the torque, J is the polar MI, G is the modulus of rigidity, θ is the shear strain, l is the length, τ is the shear stress, and r = radial distance of the fibre in the section. GJ is called torsional stiffness. Shear stress increases with the radial distance and is maximum on the surface fibres.

The polar section modulus, denoted by Z_p, is the polar moment of inertia divided by the distance to the extreme fibre of the section. In the case of axisymmetric sections, $Z_p = J/r$ and hence $T = Z_p \tau_{max}$.

Power $P = 2\pi NT/60$, where N is the number of revolutions of the shaft per minute, T is the torque in newton metres, and P is in watts. Shafts to transmit power are designed based on this equation, including a factor of safety for the power for which the shaft has to be designed.

If a shaft is supported such that it is statically indeterminate, it is analysed using compatibility conditions. Composite shafts, which are made of two materials, are also analysed using compatibility conditions.

In the case of thin-walled tubes,

$$\text{Allowable torque} = \frac{\pi D^2 t}{2}\tau_{\max}$$

and

$$T = \int q\, dL x$$

where q is the shear flow.

Deriving expressions for the shear stress and angle of twist for sections which are not axisymmetrical is very difficult. So we directly use formulae to calculate them.

Flanged couplings are commonly used as connections between shafts. The torque resistance of bolted couplings can be easily calculated as

$$T = \frac{\pi d^2}{4}\tau n R$$

Exercises

Review Questions

1. Distinguish clearly between torque and BM, with examples.
2. Describe the behaviour of a solid circular section subjected to a uniform torque.
3. Bring out clearly the differences between bending stresses and torsional stresses. Show how the resisting BM and torque are developed in a section.
4. State and explain the significance of the assumptions made in deriving the torsion equation.
5. State the torsion equation, and explain what each term signifies. Provide a consistent set of units for each term.
6. The brass and steel bars shown in Fig. 8.37 are identical in length and cross-sectional area. The ratio of modulus of rigidity of steel to that of brass is approximately 2. State, without calculation, which segment will carry more torque and why.

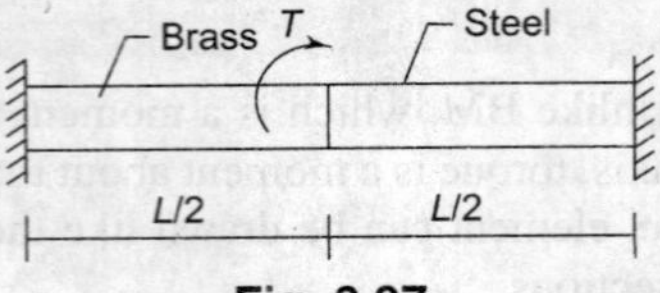

Fig. 8.37

7. For the case shown in Fig. 8.38, which segment should be longer so that they carry equal torque. Why? What should be the ratio of their lengths for this condition? G for steel is 90 GPa and for brass 40 GPa. The steel and brass rods have equal cross-sectional area.

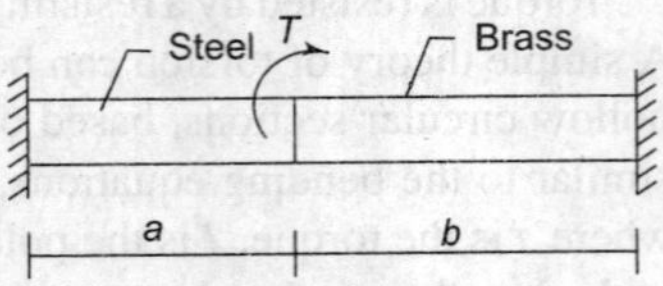

Fig. 8.38

8. Of the two sections of composite shafts shown in Fig. 8.39, which can carry more torque and why?

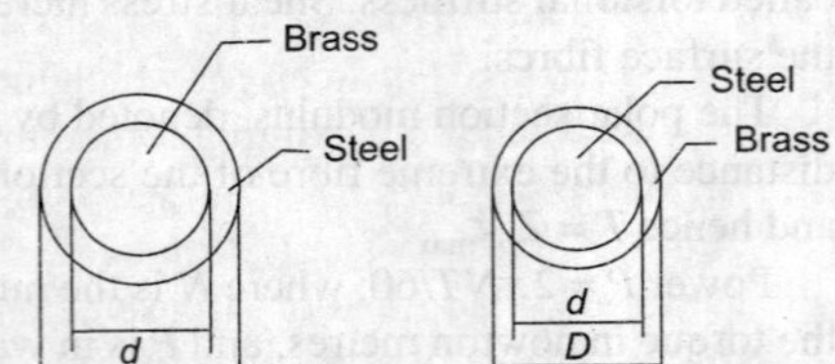

Fig. 8.39

9. If a shaft transmits 20 kW of power at 200 rpm, what is the torque generated?
10. A shaft rotating at 300 rpm is subjected to an average torque of 120 N m. What is the power transmitted?

11. If the torque in a shaft is 400 N m and the frequency is 3 Hz/s, what is the power transmitted?

Problems

1. Draw the SF, BM, and torque diagrams for the bent bar loaded as shown in Fig. 8.40.

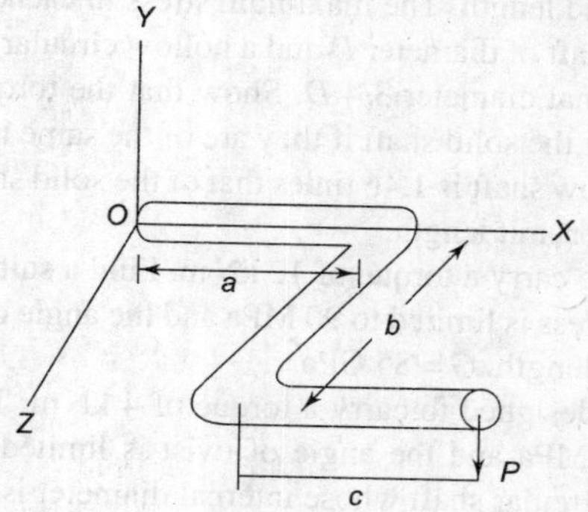

Fig. 8.40

2. Draw the torque diagram for the shaft loaded as shown in Fig. 8.41.

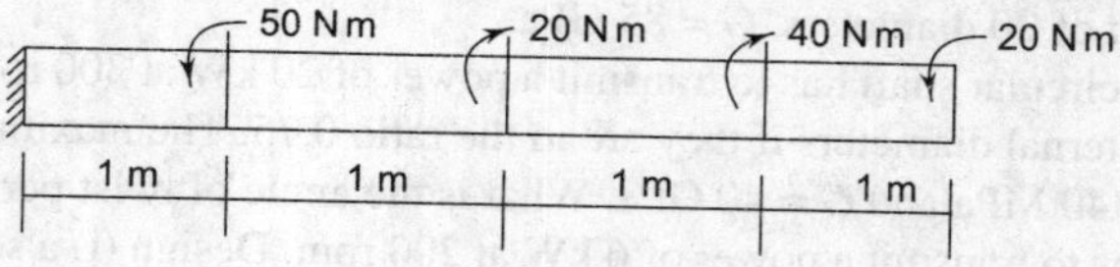

Fig. 8.41

3. In the shaft shown in Fig. 8.42, a torque of 120 kN m is supplied at *A* and power is taken off at *B, C* and *D*. Draw the torque diagram for the shaft.

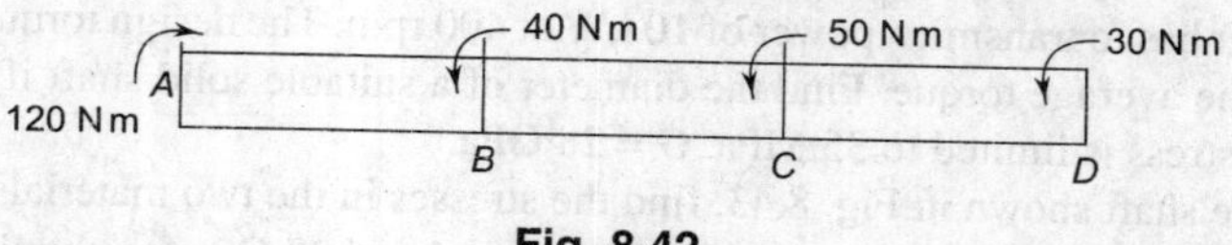

Fig. 8.42

4. A shaft, 20 mm ϕ, is made of brass and is 800 mm long. Find the maximum stress and angle of twist when it is subjected to a torque of 80 N m. G = 40GPa.
5. A steel shaft, 30 mm ϕ, was found to have an angle of twist of 2° in a length of 2 m when subjected to a torque of 12 kN m. Find the value of G for the material.
6. A shaft, of diameter 60 mm and length 1.2 m, had a maximum angle of twist of 2.5 degrees when subjected to a uniform torque. Find the maximum stress in the material. G = 85 GPa.
7. A hollow circular shaft, of outside diameter 50 mm and inside diameter 36 mm, is made of steel, for which the permissible stress in shear is 90 MPa and G = 85 GPa. Find the maximum torque that such a shaft can carry and the angle of twist per metre length.
8. A hollow circular shaft, of outside diameter 80 mm and inside diameter 60 mm, is subjected to a torque of 10 kN m, and the angle of twist is 5° over a length of 2 m. Find the value of G for the material and the maximum stress in the shaft.
9. Consider two shafts of the same material, one solid circular and the other hollow circular. The diameter of the solid circular shaft is equal to the inner diameter of the

hollow shaft. Show that for equal torque capacity the outside diameter of the hollow shaft is 1.221 times the diameter of the solid shaft.

10. Consider a solid circular shaft of diameter d and a hollow circular shaft of (external) diameter D and internal diameter pD of the same material. Show that the ratio of torque capacity of the solid shaft to the hollow shaft is $(1-p^2)^{1/2}/(1+p^2)$ if they have the same weight and length. The maximum stress in each shaft is also the same.
11. Consider a solid shaft of diameter D and a hollow circular shaft of the same external diameter and internal diameter 3/4 D. Show that the torque capacity of the hollow shaft is 68% that of the solid shaft if they are of the same material, and that the angle of twist of the hollow shaft is 1.46 times that of the solid shaft if they are subjected to the same torque, for unit length.
12. A solid shaft has to carry a torque of 12 kN m. Find a suitable diameter for the shaft if the maximum stress is limited to 90 MPa and the angle of twist should not be more than 3° per metre length. $G = 85$ GPa.
13. A shaft has to be designed to carry a torque of 4 kN m. The maximum shear stress permissible is 80 MPa and the angle of twist is limited to 2° in a length of 1 m. Design a hollow circular shaft whose internal diameter is 80% of the external diameter. $G = 85$ GPa.
14. Find the diameter for a solid shaft carrying a design torque of 4 kN m if the maximum shear stress cannot exceed 90 MPa and the angle of twist should not be more than 1° in a length of 20 diameters. $G = 85$ GPa.
15. A hollow circular shaft has to transmit a power of 20 kW at 300 rpm. Find the external and internal diameters if they are in the ratio 0.75. The maximum shear stress is limited to 40 MPa and $G = 40$ GPa. What is the angle of twist per metre length?
16. A shaft has to transmit a power of 6 kW at 200 rpm. Design (i) a solid circular shaft, and (ii) a hollow circular shaft with a 1.2 ratio of external to internal diameter. The maximum shear stress is limited to 90 MPa and the angle of twist cannot be more than 3° per metre length. Find the weight ratio of the two shafts per metre length. $G = 85$ GPa.
17. A shaft has to transmit a power of 10 kW at 400 rpm. The design torque is 25% more than the average torque. Find the diameter of a suitable solid shaft if the maximum shear stress is limited to 35 MPa. $G = 28$ GPa.
18. For the shaft shown in Fig. 8.43, find the stresses in the two materials and the angle of twist between A and B. $G = 85$ GPa for steel and 28 GPa for aluminium.

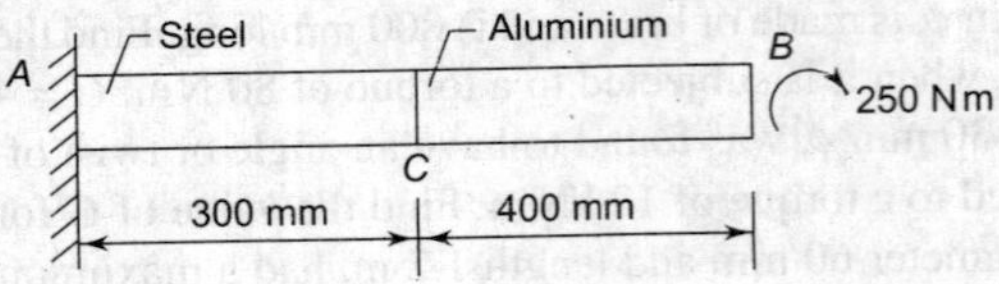

Fig. 8.43

19. Find the angle of twist between A and C and between A and B for the shaft shown in Fig. 8.44. $G = 40$ GPa for brass and 28 GPa for aluminium.

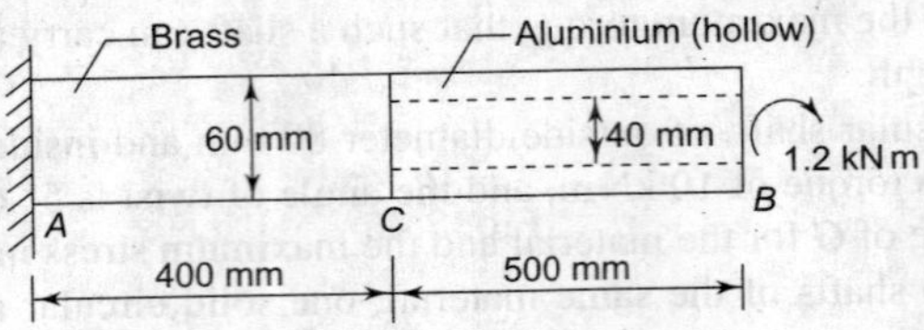

Fig. 8.44

20. A compound shaft is made of an aluminium rod, 40 mm ϕ, and 1m long, and a steel rod, 20 mm ϕ, and 1.5 m long (Fig. 8.45). If the maximum permissible shear stresses in aluminium and steel are 40 MPa and 90 MPa, and G is 28 GPa and 85 GPa, respectively, find the maximum torque that can be applied as shown. Also find the angle of twist between A and B.

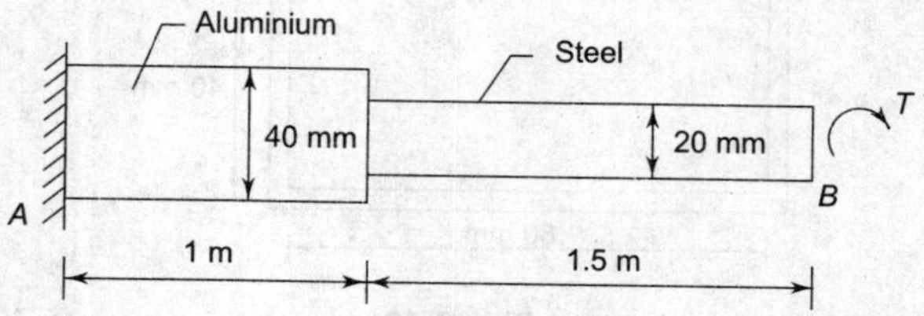

Fig. 8.45

21. Figure 8.46 shows a solid circular shaft of diameter 80 mm on the left attached to a hollow shaft of the same external diameter and an internal diameter of 64 mm, on the right. Find the ratio of lengths L_1 and L_2 so that the two segments carry equal torque.

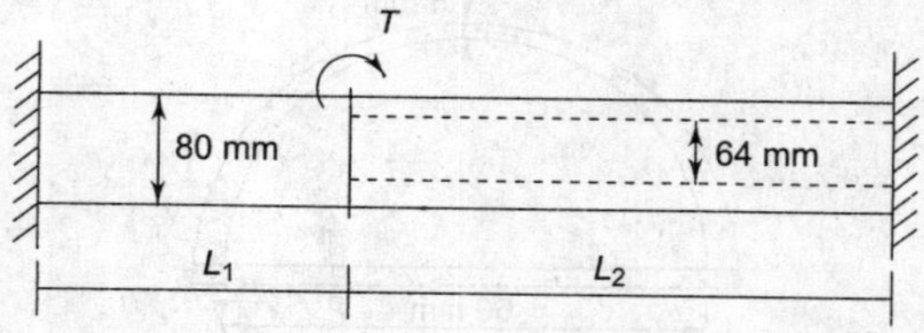

Fig. 8.46

22. The shaft shown in Fig. 8.47 is a compound shaft of brass and steel. If G for brass is half that for steel, find the ratio of the lengths L_1 and L_2 so that they carry equal torques.

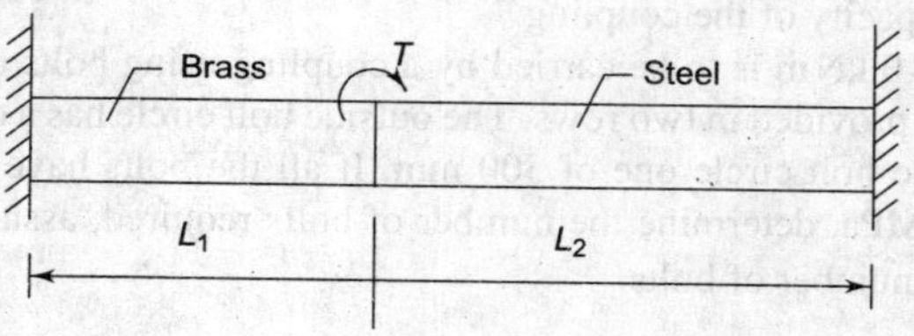

Fig. 8.47

23. In the shaft shown in Fig. 8.48, the left half is of brass and the right half is of steel. The permissible shear stress for brass is 50 MPa and that for steel is 90 MPa. Find the diameters of the two shafts so that they are subjected to their permissible limits of stress when a torque of 5 kN m is applied. $G = 85$ GPa for steel and 40 GPa for brass.

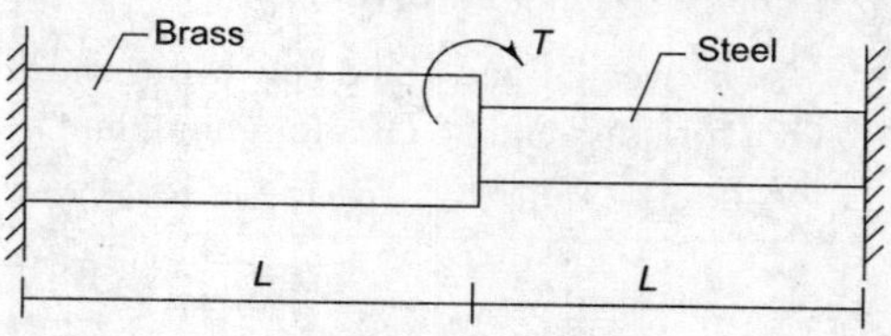

Fig. 8.48

24. The rectangular thin-walled section shown in Fig. 8.49 is subjected to a torque of 500 N m. If the shear stress is not to exceed 50 MPa, find the thickness required of the section.

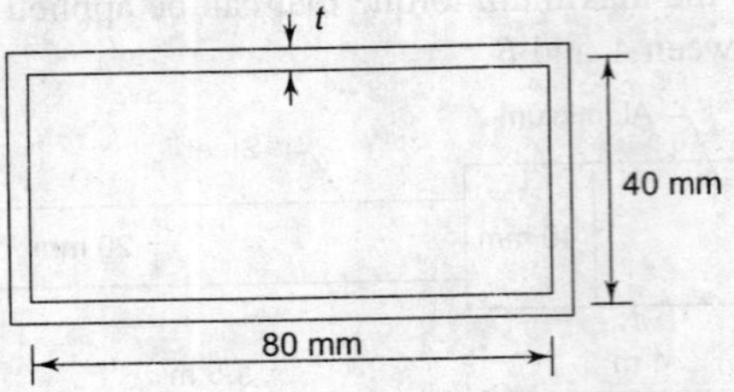

Fig. 8.49

25. The thin-walled tubular section is semicircular in shape as shown in Fig. 8.50. Find the maximum torque that can be applied to this section without exceeding the permissible stress of 90 MPa.

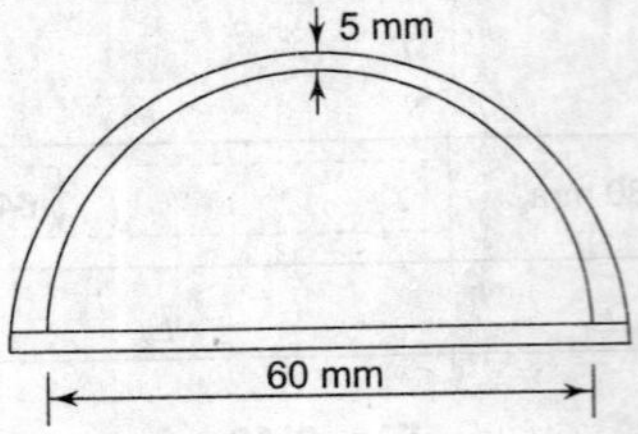

Fig. 8.50

26. A flange bolt coupling consists of ten bolts of diameter 20 mm spaced equally in a circle of diameter 300 mm. If the allowable shear stress in the bolts is 60 MPa, find the torque capacity of the coupling.
27. A torque of 10 kN m is to be carried by a coupling using bolts of diameter 10 mm. The bolts are provided in two rows. The outside bolt circle has a diameter of 400 mm and the inside bolt circle one of 300 mm. If all the bolts have an allowable shear stress of 50 MPa, determine the number of bolts required, assuming that each row has an equal number of bolts.

CHAPTER 9

Analysis of Principal Planes, Stresses, and Strains

Learning Objectives

After going through this chapter, the reader will be able to

- determine the stress on oblique plane in members subjected to uniaxial and biaxial stresses,
- derive the transformation equations for stresses in a plane stress system,
- determine the magnitude and nature of stresses on an oblique plane,
- derive the equations for principal stresses and the maximum in-plane shear stress, and calculate their magnitudes and directions,
- draw the Mohr's circle for a plane stress system and interpret this circle,
- determine the principal strains, analytically and graphically, for a given set of plain strains,
- calculate the principal strains and stresses from measurements obtained from a strain rosette,
- determine the equivalent torque and equivalent BM and find the stresses in shafts subjected to torque, BM, and axial thrust, and
- define the term stress tensor and briefly discuss the method of analysis in the case of three-dimensional stress condition.

9.1 INTRODUCTION

In the earlier chapters we studied different types of stresses—simple stresses like tensile, compressive and shear, flexural stresses, shear stresses in beams, torsional shear stress, and so on. We dealt with each of them separately. In many structures, a combination of these stresses acts at a point. It is quite possible that the combined effect of these stresses in a particular direction is more than that of the individual stresses. In this chapter, we shall examine this aspect in detail.

9.2 COMPLEX STRESSES

To recapitulate, the elementary stresses dealt with in Chapter 3 are shown in Fig. 9.1. σ_t is a tensile stress and by convention is shown by two arrows directed outwards. Note that σ_t is not a force but only the stress intensity. To obtain the force acting on the element, the stress has to be multiplied by the area on which it acts. σ_c is a compressive stress represented by two arrows directed towards each other. In shear, there will be one pair (parallel and opposite) of stresses on a pair of opposite faces, and the complementary shear stress to maintain equilibrium on the other pair of faces.

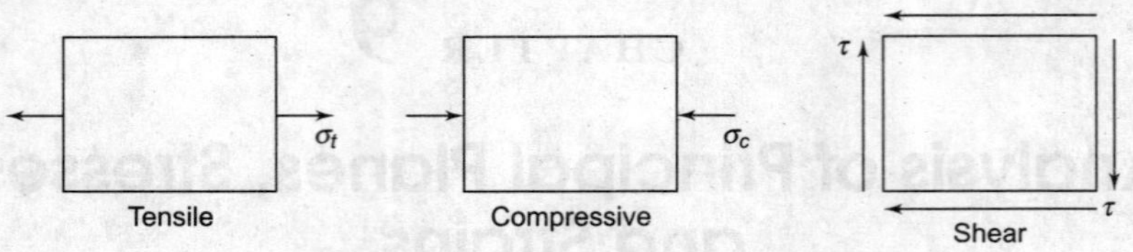

Fig. 9.1

To deal with complex stresses, it is necessary to have an appropriate symbolic representation and a sign convention. If we take a point within a body, the principal coordinate axes can be considered through this point. X-X, Y-Y, and Z-Z are the axes through O [Fig. 9.2(a)]. The plane of the paper contains the X- and Y-axes and the Z-axis is perpendicular to the plane of the paper. In a very general situation, a stressed element may have stresses as shown in Fig. 9.2(b).

Nomenclature and sign convention There must be uniformity in the naming of stresses and the signs assigned to them.

In Fig. 9.2(c), the cubical element has six surfaces. Each surface is a plane and is named by the normal to its surface. For example, the surface *DCGH* is named the X-plane because the X-axis is normal to it. By normal, we mean the line emerging from the plane (outwards from the element). Note that *ABFE* is also an X-plane, but the negative direction of the X-axis is normal to it. This way the planes are named as follows.

DCGH $\rightarrow +X$-plane; *BFGC* $\rightarrow +Y$-plane
ABFE $\rightarrow -X$-plane; *AEHD* $\rightarrow -Y$-plane
ABCD $\rightarrow +Z$-plane
EFGH $\rightarrow -Z$-plane

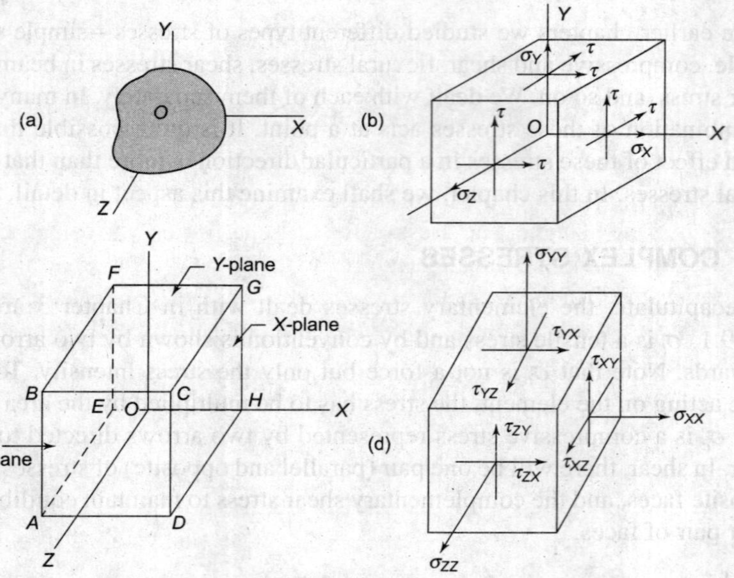

Fig. 9.2

Each surface can have one normal stress and two shear stresses. Each stress is named with two subscripts. The symbol σ is for normal stresses and τ for shear stresses. The first subscript to the symbol denotes the plane on which the stress is acting and the second subscript the direction of the stress.

Thus, σ_{XX} means the normal stress acting on the X-plane (first subscript) and in the direction of the X-axis (second subscript). τ_{ZY} means a shear stress on the Z-plane (first subscript) acting in the direction of the Y-axis (second subscript).

All the stresses in Fig. 9.2(d) are named according to this convention. The reader should verify that the stresses have been named properly.

Since for normal stresses, the two subscripts will always be the same, like σ_{XX}, σ_{YY}, and σ_{ZZ}, only one subscript is sometimes used for convenience. Thus, σ_X means σ_{XX}, σ_Y means σ_{YY}, and so on.

In case of normal stresses, tensile stress is positive and compressive stress negative (Fig. 9.3). Shear stress is positive when the stress resultant produces a clockwise couple, as shown in Fig. 9.3. Note that the shear stress resultants on one pair of opposite faces form a couple and those on the other pair form a balancing couple. Shear stress will thus be positive on one pair of opposite faces and negative on the other pair.

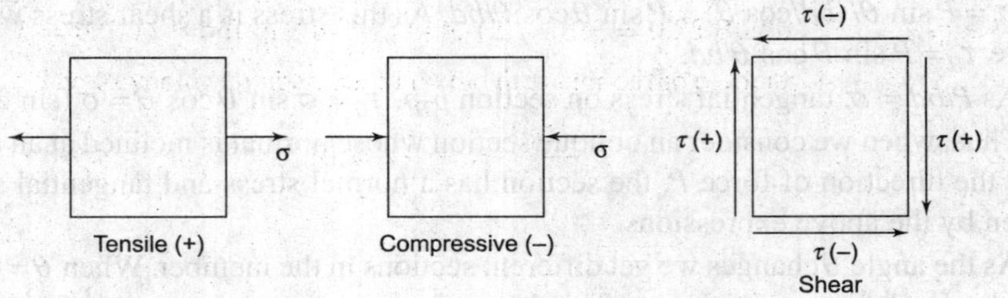

Fig. 9.3

Stress is expressed in units of N/mm^2, MPa, kN/m^2, etc.

It should be noted that the stress values given are on a unit area. While formulating equations of equilibrium, etc., the stress must be multiplied by the area.

9.3 UNIAXIAL STRESS: STRESSES ON AN OBLIQUE SECTION

Consider a bar subjected to a uniaxial stress in the X-direction (see Fig. 9.4). The X-axis is along the length of the member and a section y-y, normal to this axis, of the member is in the Y-plane. The stress on the section y-y is easy to calculate as we have seen in Chapter 3. As the force P is normal to the plane y-y, the stress in the member is a normal stress. In this case, the stress is tensile as the force is a tensile force.

Normal stress on section y-y, $\sigma = P/bd$, where b is the width of the member and d is its depth.

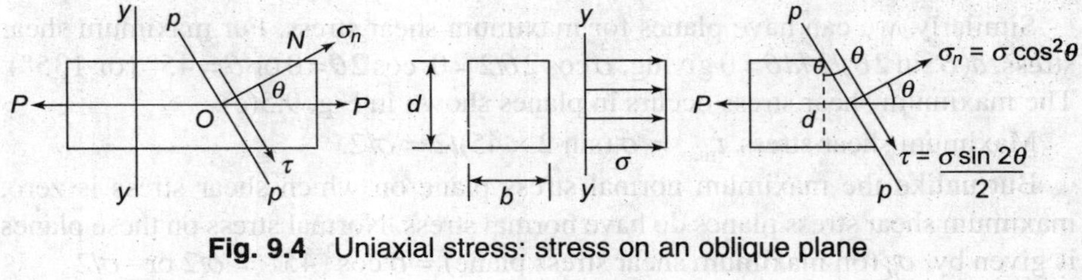

Fig. 9.4 Uniaxial stress: stress on an oblique plane

Now, we consider a section *p-p*, whose normal is inclined at an angle θ to the member axis. We identify the plane with the outward normal as we have discussed in the earlier section. This normal is ON, which makes a positive (anticlockwise) angle θ with the X-axis. We have to find the stresses on this oblique plane.

If we resolve the force P into components normal and tangential to the section *p-p*, we get, Normal component $P_n = P \cos \theta$, Tangential component $P_t = P \sin \theta$.

The plane *p-p* thus has normal stress as well as tangential stress. To find the stresses, we have to find the area on which it acts.

Now, it is clear that the width b of the section *p-p* remains the same. However, the dimension along the plane *p-p* is different from d. This is given by $d/\cos \theta$ as shown in Fig. 9.4.

Area of the section is, therefore, $bd/\cos \theta$. Knowing the forces normal and tangential to the plane and the area, we can now find the stresses.

Normal stress on section 1-1, $\sigma_\theta = P \cos \theta/(bd/\cos \theta) = P \cos^2 \theta/ bd$.

As $P/bd = \sigma$, normal stress on section *p-p*, $\sigma_\theta = \sigma \cos^2 \theta$.

Similarly, tangential stress on section *p-p*,

$\sigma_t = P \sin \theta/(bd/\cos \theta) = P \sin \theta \cos \theta/bd$. As this stress is a shear stress we can write $\tau_\theta = P \sin \theta \cos \theta/bd$.

As $P/bd = \sigma$, tangential stress on section *p-p*, $\tau_\theta = \sigma \sin \theta \cos \theta = \sigma (\sin 2\theta)/2$.

Thus, when we consider an oblique section whose normal is inclined at an angle θ to the direction of force P, the section has a normal stress and tangential stress given by the above expressions.

As the angle θ changes we get different sections in the member. When $\theta = 0$, we get the Y-plane or section *y-y*. This has a normal stress $\sigma_\theta = \sigma \cos^2 0 = \sigma$ and tangential stress $\tau_\theta = \sigma (\sin 2 \times 0)/2 = 0$. The plane *p-p* has a normal stress and tangential stress. The resultant stress can be calculated.

The resultant stress on the plane *p-p* is $\sigma_R = \sqrt{[\sigma_\theta^2 + \tau_\theta^2]}$.

The angle made by the resultant stress with the normal is $\tan^{-1}[\tau_\theta/\sigma_\theta]$.

The maximum values of these stresses are of interest.

Maximum normal stress occurs when $d(\sigma \cos^2 \theta)/d\theta = 0$. This gives

$d(\sigma \cos^2 \theta)/d\theta = \sigma 2 \cos \theta (-\sin \theta) = 0$. This happens when $\theta = 0$ or $270°$ (sin 0 or sin $270° = 0$) or when $\theta = 90°$ or $180°$ (cos $90°$ or cos $180° = 0$). As the condition is valid for maximum or minimum value, we can see that when $\theta = 0$ or $270°$, we have maximum normal stress σ (on *y-y* plane) and minimum when $\theta = 90°$ or $180°$ when $\sigma = 0$ (on *x-x* plane).

One should notice that on the planes on which the maximum normal stress occurs there is no shear stress on that plane.

Similarly, we can have planes for maximum shear stress. For maximum shear stress, $d(\sigma \sin 2\theta)/2 /d\theta = 0$ giving, $\sigma \cos 2\theta/2 = 0$. $\cos 2\theta = 0$ or $\theta = 45°$ (or $135°$). The maximum shear stress occurs in planes shown in Fig. 9.5(c).

Maximum shear stress $\tau_{max} = \sigma (\sin 2 \times 45)/2 = \sigma/2$.

But unlike the maximum normal stress plane on which shear stress is zero, maximum shear stress planes do have normal stress. Normal stress on these planes is given by, σ_n (on maximum shear stress plane) $= \sigma \cos^2 45° = \sigma/2$ or $-\sigma/2$.

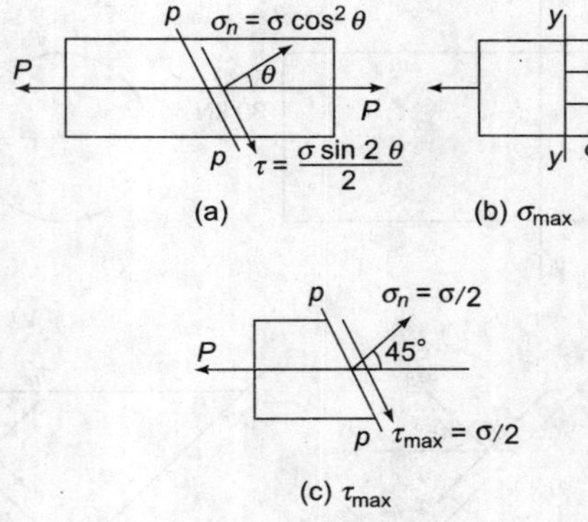

Fig. 9.5 Stresses on an oblique section

To summarize (Fig. 9.5),

1. When a member is subjected to uni-axial tensile or compressive force, the stress on a plane normal to the axis of the member is obtained by the average stress formula Force/area = $P/bd = \sigma$.
2. Any oblique plane is identified by the angle that the normal to the plane makes with the longitudinal axis (X-X axis). Positive angles are measured in the counterclockwise direction. (Fig. 9.5(b))
3. The stress on the oblique plane is a normal stress σ_n and tangential (shear) stress τ. Normal stress $\sigma_n = \sigma \cos^2\theta$ and shear stress $\tau = \sigma(\sin 2\theta)/2$.
4. As the value of the angle changes, we get different planes and different values of the stress.
5. The maximum value of the normal stress is σ, which is on a plane normal to the axis of the member ($\theta = 0$). The shear stress τ on this plane is zero.
6. Maximum shear stress is on a plane whose normal makes 45° (or 135°) with the X-axis. Maximum shear stress $\tau_{max} = \pm \sigma/2$.
7. Maximum shear stress planes have a normal stress $\sigma_n = \sigma/2$.
8. Tensile stress is taken positive. Shear stress is positive when the moment taken about a point inside the element is clockwise.

Example 9.1 Uniaxial stress: stress on an oblique plane

A bar of circular section of diameter 20 mm is subjected to axial tensile force of 30 kN. Find the maximum normal and shear stresses and the planes on which they act. What are the stresses on planes whose normal makes angles of (i) 20° and (ii) 140° with the longitudinal axis. Show the planes and stresses in a neat sketch.

Solution The situation is shown in Fig. 9.6.

Area of bar = $\pi 20^2/4 = 314.16$ mm^2

Stress on the normal section y-y, $\sigma = 30 \times 1000/314.16 = 95.5$ N/mm^2

On the plane p-p, whose normal is inclined at an angle θ, stresses are:

Normal stress $\sigma_n = \sigma \cos^2 \theta$; and shear stress $\tau = (\sigma \sin 2\theta)/2$

Maximum normal stress is on plane y-y, where $\theta = 0$

Maximum normal stress = $\sigma = 95.5$ N/mm^2

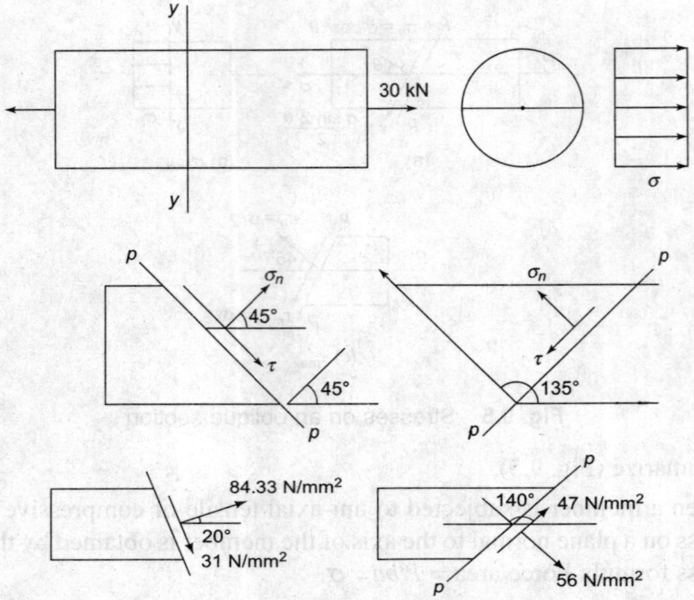

Fig. 9.6

Maximum shear stress is on a plane whose normal is inclined at $\theta = 45°$ (or 135°)

Maximum shear stress $\tau_{max} = \pm \sigma/2 = 47.75$ N/mm²

Normal stress on maximum shear stress planes = $\sigma/2 = 47.75$ N/mm²

(i) When $\theta = 20°$: Normal stress $\sigma_n = 95.5 \cos^2 20° = 84.33$ N/mm²

Shear stress $\tau = (95.5 \sin 40)/2 = 31$ N/mm²

(ii) When $\theta = 140°$: Normal stress $\sigma_n = 95.5 \cos^2 140° = 56$ N/mm²

Shear stress $\tau = (95.5 \sin 280°)/2 = -47$ N/mm²

The planes and stresses are shown in Fig. 9.6.

Example 9.2 Uniaxial stress: stress on oblique plane

A short bar of rectangular section, 25 mm × 50 mm, is subjected to a compressive force of 50 kN. Find the stresses on planes whose normal makes angles of (i) +40° and (ii) −30° with the longitudinal axis. What is the maximum shear stress in the bar and on which plane does it act? Show the planes and stresses in a neat sketch.

Solution The situation is represented in Fig. 9.7.

Area of bar = 25 × 50 = 1250 mm²

Stress $\sigma = 50 \times 1000/1250 = -40$ N/mm²

(i) Normal makes +40° with X-axis: The plane is as shown.

Normal stress = $\sigma \cos^2 \theta = -40 \times \cos^2 40 = -23.5$ N/mm²

Shear stress = $-40 \times (\sin 2\theta)/2 = (-40 \sin 80)/2 = -19.7$ N/mm²

The stresses will be as shown in Fig. 9.7(b).

(ii) Normal makes −30° with X-axis: The normal and the plane are shown in Fig. 9.7(c).

Normal stress = $-40 \cos^2 (-30) = -30$ N/mm²

Shear stress = $-40 \sin (2x - 30)/2 = -20 \sin (-60) = +17.3$ N/mm²

The stresses will be as shown in Fig. 9.7(c).

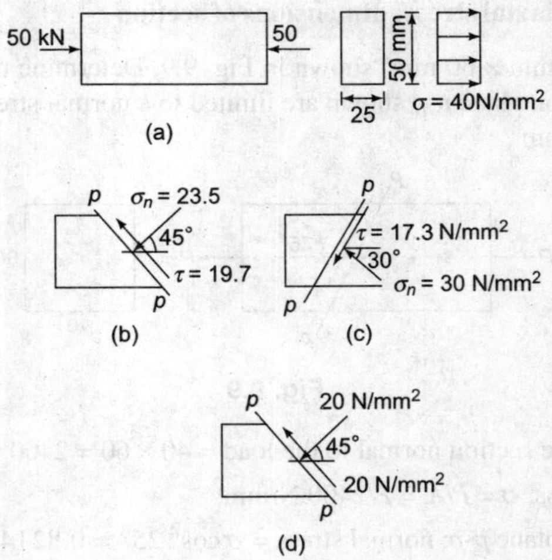

Fig. 9.7

Maximum shear stress will be on a plane making 45° (or 135°) with the X-axis. The plane is shown in Fig. 9.7(d).

Maximum shear stress $= \sigma \sin 2\theta/2 = \sigma \sin 90° \;/2 = \sigma/2 = -20$ N/mm^2

Accompanying normal stress $= \sigma \cos^2 45 = \sigma/2 = -20$ N/mm^2

When $\theta = 135°$, $\tau_{max} = (-20 \times \sin 270)/2 = 20$ N/mm^2

Accompanying normal stress $= -20 \cos^2 135 = -20$ N/mm^2

The stresses are shown in Fig. 9.7(d).

Example 9.3 Uniaxial stress: dimensions of section

If the maximum shear stress in a material is limited to 60 N/mm^2, find the dimensions of a square section subjected to an axial tensile force of 200 kN.

Solution The situation is shown in Fig. 9.8.

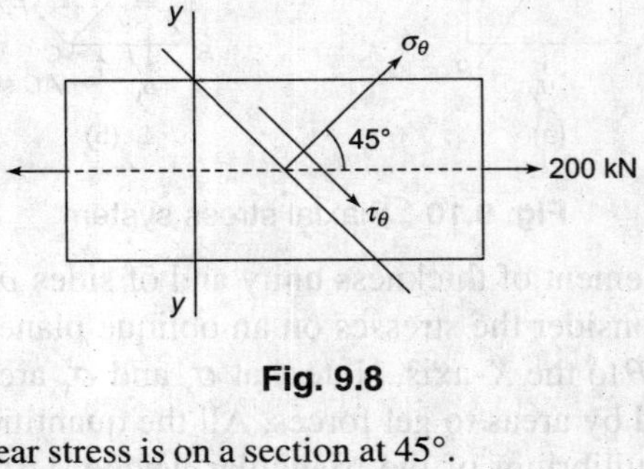

Fig. 9.8

Maximum shear stress is on a section at 45°.

If a is the side of a square section, then area a^2.

Stress on y-y plane, $\sigma = 200 \times 1000/a^2$

Maximum shear stress $= (\sigma \sin 2 \times 45)/2 = \sigma/2 = 200000/2a^2$

This should not exceed 60 N/mm^2.

$$200000/2a^2 = 60; \; a = 41 \text{ mm}$$

The side of the square section must be at least 41 mm.

Example 9.4 Uniaxial stress: dimensions of section

Consider the bar (40 mm × 60 mm) shown in Fig. 9.9. Determine the maximum value of load P if the stresses on plane p-p shown are limited to a normal stress of 30 N/mm^2 and a shear stress of 18 N/mm^2.

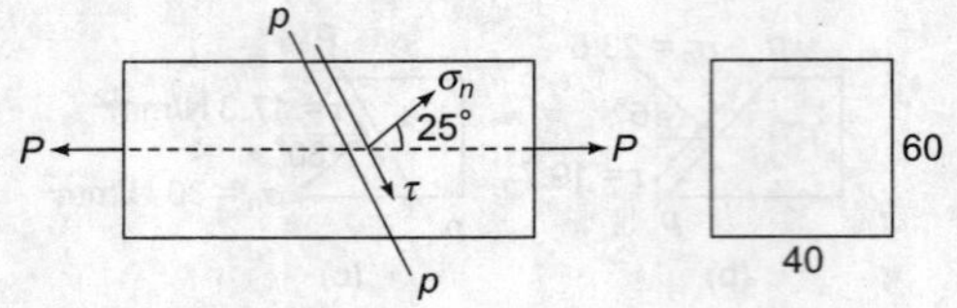

Fig. 9.9

Solution Area of the section normal to the load = 40 × 60 = 2400 mm^2

Normal stress, $\sigma = P/A = P/2400$ N/mm^2

Stresses on plane p-p: normal stress = $\sigma \cos^2 25° = 0.8214\sigma$

Shear stress = $\sigma \sin 50/2 = 0.383\sigma$

There are two conditions: (i) $0.8214\sigma = 30$ N/mm^2

This gives 0.8214 (P/2400) = 30 giving P = 87.6 kN

(ii) $0.383\sigma \leq 18$ N/mm^2, which gives 0.383 × (P/2400) = 18 or P = 112.8 kN.

The lower of the two loads is 87.6 kN, which is the maximum value of the load. ■

9.4 TWO NORMAL STESSES ON ORTHOGONAL PLANES

We now take up the case of a biaxial stress system which has two normal stresses on an element in perpendicular direction as shown in Fig. 9.10. We find out the stresses on an oblique plane and the maximum normal and shear stresses.

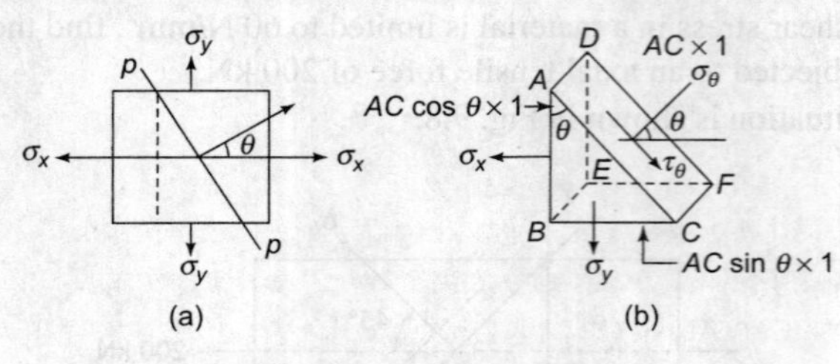

Fig. 9.10 Biaxial stress system

We consider an element of thickness unity and of sides δx and δy as shown in Fig. 9.10. We first consider the stresses on an oblique plane p-p whose normal is inclined at an angle θ to the X-axis. Note that σ_x and σ_y are stress intensities and need to be multiplied by areas to get forces. All the quantities shown in Fig. 9.10 are positive. The equilibrium of the triangular wedge ABC of unit thickness is being considered. One point to be noted is that, as the element is so small, we are actually talking about stress at a point. The sides of elementary dimensions are being considered to enable us to write the equilibrium equations. As we are essentially considering the stresses at a point, the force system on the element form a concurrent force system for which two conditions of equilibrium exist. These are $\Sigma V = 0$ and $\Sigma H = 0$.

The forces on the element and the areas on which they act are shown in Fig. 9.10(b).

$\Sigma H = 0, -\sigma_x(AC\cos\theta \times 1) + \sigma_\theta \cos\theta(AC \times 1) + \tau_\theta \sin\theta(AC \times 1) = 0$ (1)

$\Sigma V = 0, -\sigma_y(AC\sin\theta \times 1) + \sigma_\theta \sin\theta(AC \times 1) - \tau_\theta \cos\theta \times (AC \times 1) = 0$ (2)

Rearranging, $\sigma_\theta \cos\theta + \tau_\theta \sin\theta = \sigma_x \cos\theta$ (1a)

$\sigma_\theta \sin\theta - \tau_\theta \cos\theta = \sigma_y \sin\theta$ (2a)

To find σ_θ, we have to eliminate τ_θ. For this multiply (1a) by $\cos\theta$ and (2a) by $\sin\theta$.

$$\sigma_\theta \cos^2\theta + \tau_\theta \sin\theta\cos\theta = \sigma_x \cos^2\theta$$

$$\sigma_\theta \sin^2\theta - \tau_\theta \cos\theta\sin\theta = \sigma_y \sin^2\theta.$$

Adding these two equations, $\sigma_\theta(\cos^2\theta + \sin^2\theta) = \sigma_x\cos^2\theta + \sigma_y\sin^2\theta$.

We have, from trigonometry, $\sin^2\theta = (1 - \cos 2\theta)/2$ and $\cos^2\theta = (1 + \cos 2\theta)/2$.

$$\sigma_\theta = \sigma_x(1 + \cos 2\theta)/2 + \sigma_y(1 - \cos 2\theta)/2 = (\sigma_x + \sigma_y)/2 + (\sigma_x - \sigma_y)\cos 2\theta/2.$$

To find τ_θ, we eliminate σ_θ from equations 1a and 2a. For this, multiply (1a) by $\sin\theta$ and (2a) by $\cos\theta$. We get

$$\sigma_\theta \cos\theta\sin\theta + \tau_\theta \sin^2\theta = \sigma_x \cos\theta\sin\theta$$

$$\sigma_\theta \cos\theta\sin\theta + \tau_\theta \cos^2\theta = \sigma_y \sin\theta\cos\theta$$

Subtracting, $\tau_\theta(\sin^2\theta + \cos^2\theta) = \sigma_x\cos\theta\sin\theta - \sigma_y\sin\theta\cos\theta = (\sigma_x - \sigma_y)\sin 2\theta/2$.

We have thus found the stresses on an oblique plane whose normal is inclined at an angle θ to the positive direction of the X-axis. On the oblique plane, there is shear stress and normal stress given by,

$$\text{Normal stress } \sigma_\theta = \frac{(\sigma_x + \sigma_y)}{2} + \frac{(\sigma_x - \sigma_y)\cos 2\theta}{2}$$

$$\text{Shear stress } \tau_\theta = \frac{(\sigma_x - \sigma_y)\sin 2\theta}{2}$$

Maximum normal and shear stresses: To find the value of maximum normal stress, we differentiate the equation for σ_θ and set it to zero. $d(\sigma_\theta)/d\theta = 0$.

This gives, $\dfrac{(\sigma_x - \sigma_y)(-2\sin 2\theta)}{2} = 0$; $\sin 2\theta = 0$ (assuming that $\sigma_x - \sigma_y \neq 0$).

Note that this is the equation for shear stress τ_θ. This shows that maximum normal stress planes have no shear stress. We thus get $\theta = 0$ or $90°$. These are the stresses σ_x and σ_y. The maximum normal stresses are the two stresses along the X- and Y- directions.

Maximum shear stress: Maximum shear stress occurs on planes for which $\sin 2\theta$ is maximum. This happens when $\theta = 45°$ or $135°$. Maximum shear stress is, $\tau_{max} = (\sigma_x - \sigma_y)/2$. The planes on which these stresses act are shown in Fig. 9.11.

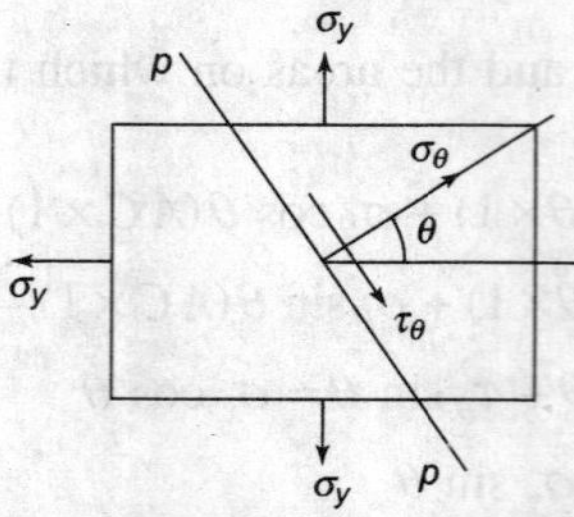

Fig. 9.11 Biaxial stress system

To summarize (Fig. 9.11):

1. We started with a biaxial stress system–two stresses along orthogonal directions (X- and Y-axes). σ_x and σ_y are normal stresses in the direction of x and y-axes.
2. We then found the stresses on an oblique plane p-p. The plane p-p is identified by the angle θ made by the normal to the plane with the X-axis. Clockwise angles are taken negative.
3. Plane p-p has in general a normal stress and a shear stress. Normal stress is given by $\sigma_\theta = (\sigma_x + \sigma_y)/2 + (\sigma_x - \sigma_y) \cos 2\theta/2$. The shear stress on plane p-p is $\tau_\theta = (\sigma_x - \sigma_y) \sin 2\theta/2$.
4. Maximum normal stresses are the stresses σ_x and σ_y. Maximum shear stress occurs on planes inclined at 45° to the X-axis. $\tau_{max} = \pm(\sigma_x - \sigma_y)/2$.
5. As the value of θ changes we get different planes and stresses on these planes can be found by the above equations by substituting the value of θ. Three cases can arise.
   (i) $\sigma_x \neq \sigma_y$ and both are of the same sign (tensile or compressive). The stresses on any plane can be found from the above equations.
   (ii) $\sigma_x \neq \sigma_y$ and are of different signs (one tensile and one compressive). This situation can be solved from the above equations by putting appropriate signs for the stresses.
   (iii) $\sigma_x = \sigma_y$. There can be two cases under this head. When they are of different nature, one tensile and the other compressive, the above equations can be used. When they are of the same nature, both tensile or both compressive, then $(\sigma_x - \sigma_y) = 0$. This leads to $\sigma_\theta = (\sigma_x + \sigma_y)/2$ and $\tau_\theta = 0$.

Please note that normal stress remains constant on all planes and shear stress is zero on all planes. This is a special case. All the cases are shown in Fig. 9.12.

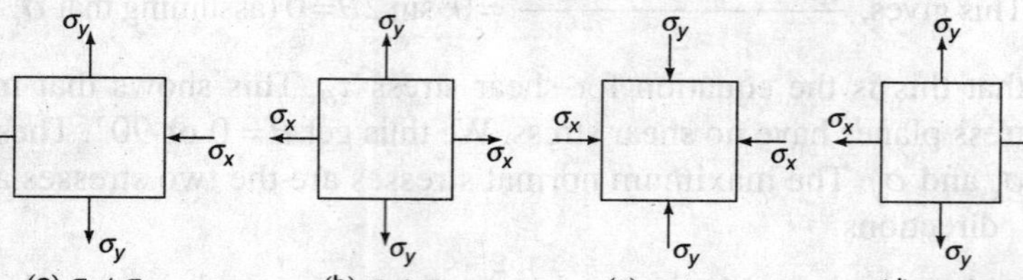

Fig. 9.12 Different cases of biaxial stresses

6. All the quantities in the equations derived are positive. If any of the quantities are negative, appropriate signs should be put. Tensile stress is positive and anticlockwise angle is positive as shown in Fig. 9.11.

Example 9.5 Biaxial stress

For the stress system shown in Fig. 9.13, find the stresses on planes whose normal makes (i) 20° and (ii) –40° with the X-axis.

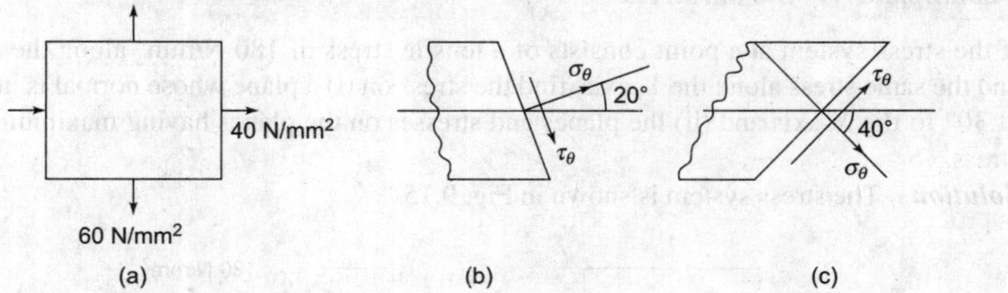

Fig. 9.13

Solution The normal stress and shear stress on a plane whose normal makes an angle θ with the X-axis are given by

Normal stress, $\sigma_\theta = (\sigma_x + \sigma_y)/2 + (\sigma_x - \sigma_y) \cos 2\theta/2$

Shear stress, $\tau_\theta = (\sigma_x - \sigma_y) \sin 2\theta/2$

In our case $\sigma_x = 60$ N/mm^2; $\sigma_y = 40$ N/mm^2

(i) $\theta = +20°$; normal stress $\sigma_\theta = (60 + 40)/2 + (60 - 40) \cos (2 \times 20)/2 = 50 + 7.66 = 57.66$ N/mm^2.

Shear stress, $\tau_\theta = (60 - 40) (\sin 40)/2 = 6.43$ N/mm^2

The stresses are shown in Fig. 9.13(b).

(ii) $\theta = -40°$. Normal stress $\sigma_\theta = 50 + (60 - 40) \cos (-80)/2 = 51.74$ N/mm^2.

Shear stress $\tau_\theta = (60 - 40) \sin (-80)/2 = -9.85$ N/mm^2.

The plane and stress are shown in Fig. 9.13(c). ■

Example 9.6 Biaxial stress

For the stress system shown in Fig. 9.14, find the planes having maximum shear stress and the magnitude of the stresses on the planes.

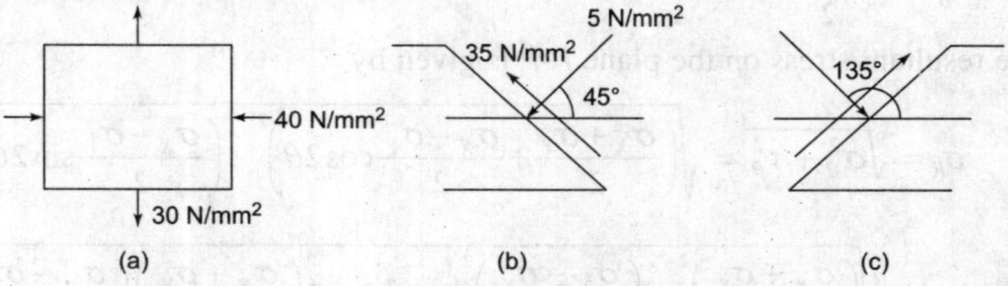

Fig. 9.14

Solution Given $\sigma_x = -40$ N/mm^2 and $\sigma_y = 30$ N/mm^2, taking tensile stress as positive. Shear stress on a plane is given by $\tau_\theta = (\sigma_x - \sigma_y) \sin 2\theta/2$. Maximum value of shear stress occurs when $\sin 2\theta$ is maximum, i.e., when $\theta = 45°$ or $135°$.

Maximum shear stress $\tau_{max} = (-40 - 30)(\sin 90)/2 = -35$ N/mm^2 or 35 N/mm^2 when $\theta =$ 135°.

Normal stress on these planes $= (\sigma_x + \sigma_y)/2 + (\sigma_x - \sigma_y)\cos 2\theta/2$

$= (-40 + 30)/2 + [(-40 - 30)\cos 90]/2$

$= -5 - 0 = -5$ N/mm^2.

The planes and stresses are shown in the figures.

Example 9.7 Biaxial stress

If the stress system at a point consists of a tensile stress of 180 N/mm^2 along the X-axis and the same stress along the Y-axis, find the stress on (i) a plane whose normal is inclined at 30° to the X-axis and (ii) the planes and stresses on the planes having maximum shear stress.

Solution The stress system is shown in Fig. 9.15.

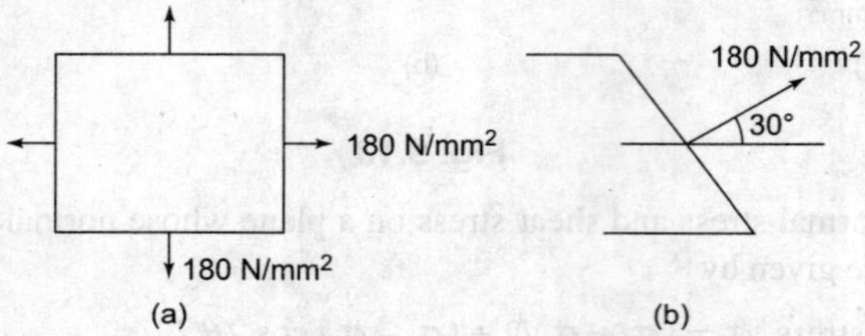

Fig. 9.15

This is a case where $\sigma_x = \sigma_y = 180$ N/mm^2. As $(\sigma_x - \sigma_y) = 0$, shear stress on all planes is zero. The normal stress remains constant on all planes equal to $(\sigma_x + \sigma_y)/2$.

Constant normal stress on all planes = (180 + 180)/2 = 180 N/mm^2.

9.4.1 Ellipse of Stress

The ellipse of stress is a graphical construction applicable to two normal stresses in orthogonal directions, as shown in Fig. 9.16(a). The stresses on any inclined plane whose normal is inclined at $+\theta$ (anticlockwise) to the X-axis can be expressed as, since $\tau_{XY} = 0$,

$$\sigma_\theta = \frac{\sigma_X + \sigma_Y}{2} + \frac{\sigma_X - \sigma_Y}{2}\cos 2\theta$$

$$\tau_\theta = \frac{\sigma_X - \sigma_Y}{2}\sin 2\theta$$

The resultant stress on the plane P-P is given by

$$\sigma_R = \sqrt{\sigma_\theta^2 + \tau_\theta^2} = \sqrt{\left(\frac{\sigma_X+\sigma_Y}{2} + \frac{\sigma_X-\sigma_Y}{2}\cos 2\theta\right)^2 + \left(\frac{\sigma_X-\sigma_Y}{2}\sin 2\theta\right)^2}$$

$$= \sqrt{\left(\frac{\sigma_X+\sigma_Y}{2}\right)^2 + \left(\frac{\sigma_X-\sigma_Y}{2}\right)^2\cos^2\theta + 2\left(\frac{\sigma_X+\sigma_Y}{2}\right)\left(\frac{\sigma_X-\sigma_Y}{2}\right)\cos 2\theta + \left(\frac{\sigma_X-\sigma_Y}{2}\right)^2\sin^2 2\theta}$$

$$= \sqrt{\left(\frac{\sigma_X + \sigma_Y}{2}\right)^2 + \left(\frac{\sigma_X - \sigma_Y}{2}\right)^2 + 2\left(\frac{\sigma_X^2 - \sigma_Y^2}{4}\right)\cos 2\theta}$$

$$= \sqrt{\frac{\sigma_X^2 + \sigma_Y^2}{2} + \frac{\sigma_X^2 - \sigma_Y^2}{2}\cos 2\theta} = \sqrt{\sigma_X^2\left(\frac{1+\cos 2\theta}{2}\right) + \frac{\sigma_Y^2(1-\cos 2\theta)}{2}}$$

$$= \sqrt{\sigma_X^2 \cos^2\theta + \sigma_Y^2 \sin^2\theta}$$

If the resultant stress makes an angle ϕ with the plane, then [Fig. 9.16(b)]

$$\tan \phi = \frac{\sigma_\theta}{\tau_\theta} = \frac{(\sigma_X + \sigma_Y)/2 + (\sigma_X - \sigma_Y \cos 2\theta)/2}{[(\sigma_X - \sigma_Y)/2]\sin 2\theta}$$

$$= \frac{\sigma_X \cos^2\theta + \sigma_Y \sin^2\theta}{(1/2)(\sigma_X - \sigma_Y)\sin 2\theta}$$

or with the normal to the plane,

$$\tan \alpha = \frac{\tau_\theta}{\sigma_\theta} = \frac{[(\sigma_X - \sigma_Y)/2]\sin 2\theta}{(\sigma_X + \sigma_Y)/2 + [(\sigma_X - \sigma_Y)/2]\cos 2\theta}$$

$$= \frac{(\sigma_X - \sigma_Y)\sin 2\theta}{2(\sigma_X \cos^2\theta + \sigma_Y \sin^2\theta)}$$

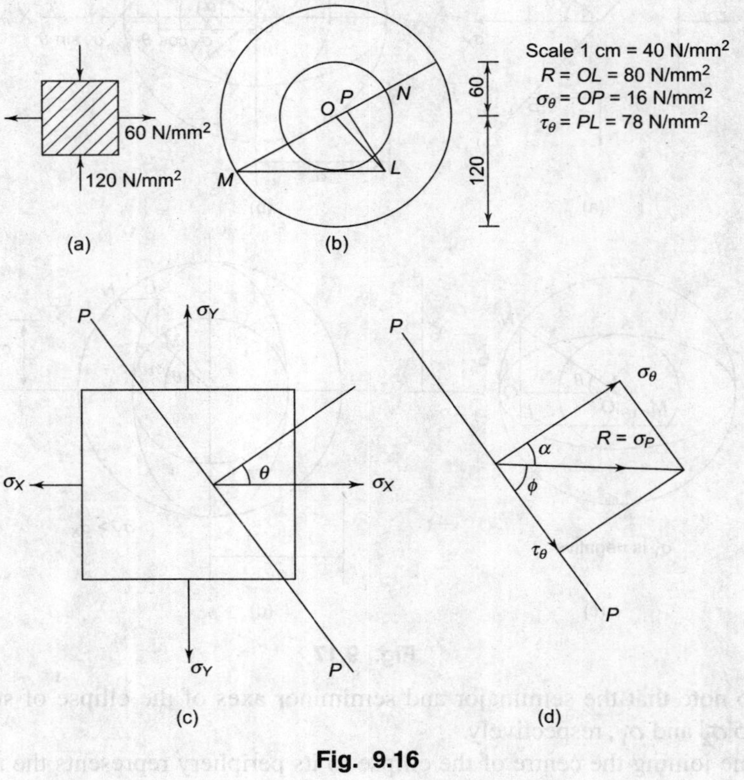

Fig. 9.16

Construction of ellipse of stress

For the stress system shown in Fig. 9.17, assuming $\sigma_X > \sigma_Y$, we draw two concentric circles with radii σ_X and σ_Y. Taking any oblique plane *P-P*, inclined at an angle θ to the *X*-axis, we draw a line parallel to this normal through *O*, intersecting the circles at *M* and *N*. Through *M* and *N*, we draw lines parallel to the *X*- and *Y*-planes, intersecting at *L*. We can easily see that the coordinates of point *L* are (σ_X cos θ, σ_Y sinθ) and $OL = \sqrt{\sigma_X^2 \cos^2\theta + \sigma_Y^2 \sin^2\theta} = \sigma_R$, the resultant stress on the plane *P-P*.

For different values of θ, the locus of point *L* is an ellipse with its centre at *O*. In Fig. 9.17,

$$x = \sigma_X \cos\theta, \quad y = \sigma_Y \sin\theta$$

$$x^2 = \sigma_X^2 \cos^2\theta, \quad y^2 = \sigma_Y{}^2 \sin^2\theta$$

$x^2/\sigma_X^2 + y^2/\sigma_Y^2 = 1$ is the equation of the ellipse of stress.

You can also easily see that $\tan\alpha = \sigma_Y \sin\theta/\sigma_X \cos\theta = (\sigma_Y/\sigma_X)\tan\theta$ is the inclination of σ_R to the *X*-axis.

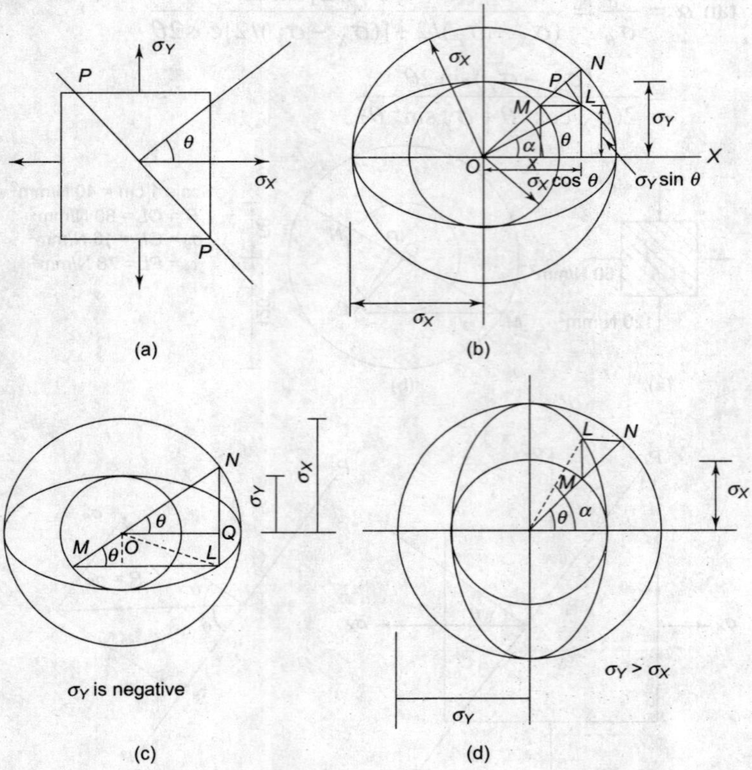

Fig. 9.17

Also note that the semimajor and semiminor axes of the ellipse of stress are equal to σ_X and σ_Y, respectively.

A line joining the centre of the ellipse to its periphery represents the resultant stress on the plane in magnitude and direction.

If one of the stresses is negative or σ_Y is greater than σ_X, the ellipse of stress is as shown in Figures 9.17(c) and (d). In Fig. 9.17(c),

$$OQ = \sigma_X \times \cos\theta$$

$$QL = -\sigma_Y \sin\theta, \quad (OL)^2 = \sigma_X^2 \cos^2\theta + \sigma_Y^2 \sin^2\theta$$

Example 9.8 Biaxial stress

An element in a stressed body is subjected to normal stresses on mutually perpendicular directions as shown in Fig. 9.18(a). Determine the normal, tangential and resultant stresses on a plane *P-P* inclined as shown, both analytically and graphically.

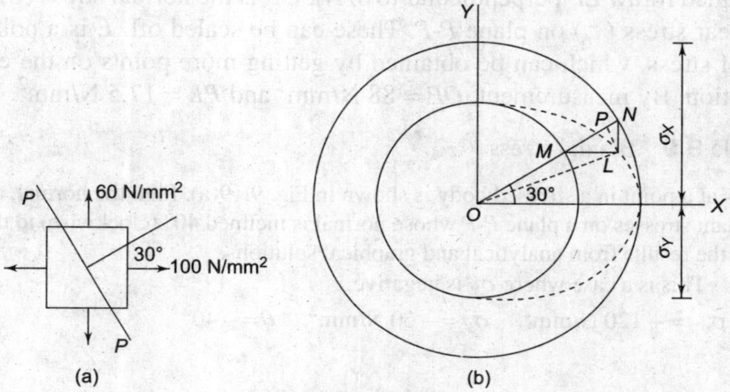

Fig. 9.18

Solution The normal to the plane is inclined at an angle 30° to the X-axis. On the plane *P-P*, normal stress,

$$\sigma_\theta = \frac{\sigma_X + \sigma_Y}{2} + \frac{\sigma_X - \sigma_Y}{2}\cos 2\theta$$

$$= \frac{100+60}{2} + \frac{100-60}{2}\cos(2\times 30)$$

$$= 90 \text{ N/mm}^2$$

Tangential stress,

$$\tau_\theta = \frac{\sigma_X - \sigma_Y}{2}\sin 2\theta$$

$$= \frac{100-60}{2}\sin 60° = 17.32 \text{ N/mm}^2$$

Resultant stress,

$$\sigma_R = \sqrt{\sigma_X^2\cos^2\theta + \sigma_Y^2\sin^2\theta}$$

$$= \sqrt{100^2\cos^2 30 + 60^2\sin^2 30}$$

$$= 91.65 \text{ N/mm}^2$$

or, $\sigma_R = \sqrt{90^2 + 17.32^2} = 91.65 \text{ N/mm}^2$

$$\tan\alpha = \frac{\tau_\theta}{\sigma_\theta} = \frac{17.32}{90} = 0.1924$$

$\Rightarrow \quad \alpha = 10.9°$

where α is the angle made by σ_R with σ_θ.

Graphical solution The graphical solution is shown in Fig. 9.18(b). We select a scale for the stresses. The scale chosen in Fig. 9.18(b) is 1 cm = 20 N/mm^2. Selecting a centre O, we draw two concentric circles, with radii equal to 100 N/mm^2 and 60 N/mm^2 to the chosen scale. X- and Y-axes are marked as shown. A line is drawn parallel to the normal to the plane through O (at 30° with the X-axis). This line intersects the circles at M and N. Though M, draw a line parallel to the normal to the Y-plane (or the X-axis) and through N, draw a line parallel to the normal to the X-plane (or the Y-axis). These lines intersect at L. OL is the resultant stress on plane P-P in magnitude and direction. If we project OL on to the line MN, a point P is obtained (draw LP perpendicular to MN). OP is the normal stress (σ_θ) and PL is the shear stress (τ_θ) on plane P-P. These can be scaled off. L is a point on the ellipse of stress, which can be obtained by getting more points on the ellipse by construction. By measurement, OP = 88 N/mm^2 and PL = 17.5 N/mm^2.

Example 9.9 Biaxial stress

The stress at a point in a stressed body is shown in Fig. 9.19(a). Find the normal, tangential, and resultant stresses on a plane P-P whose normal is inclined 40° (clockwise) to the X-axis. Compare the results from analytical and graphical solutions.

Solution This is a case where σ_Y is negative.

$$\sigma_X = +120 \text{ N/mm}^2, \quad \sigma_Y = -60 \text{ N/mm}^2, \quad \theta = -40°$$

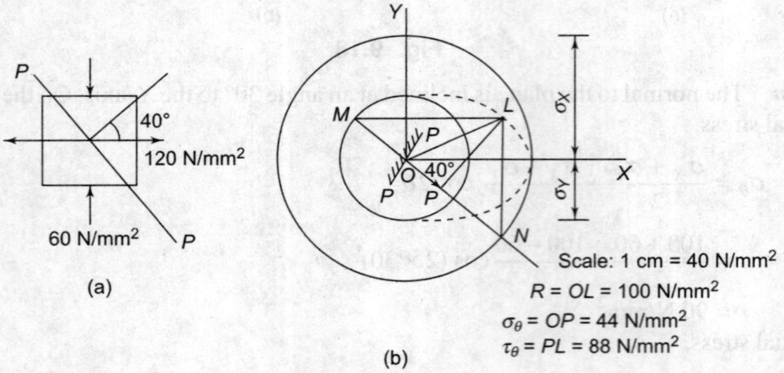

Fig. 9.19

$$\sigma_\theta = \frac{\sigma_X + \sigma_Y}{2} + \frac{\sigma_X - \sigma_Y}{2} \cos 2\theta$$

$$= \frac{120 - 60}{2} + \frac{120 + 60}{2} \cos(-80°)$$

$$= 30 + 15.63 = 45.63 \text{ N/mm}^2$$

$$\tau_\theta = \frac{\sigma_X - \sigma_Y}{2} \sin 2\theta = \frac{120 + 60}{2} \sin(-80) = -88.63 \text{ N/mm}^2$$

$$\sigma_R = \sqrt{45.63^2 + (-88.63)^2} = 99.69 \text{ N/mm}^2$$

or $$\sigma_R = \sqrt{\sigma_X^2 \cos^2\theta + \sigma_Y^2 \sin^2\theta} = \sqrt{120^2 \cos^2(-40) + (-60)^2 \sin^2(-40)}$$

$$= 99.69 \text{ N/mm}^2$$

$$\tan \alpha = \frac{\tau_\theta}{\sigma_\theta} = -\frac{88.63}{45.63} = -1.942$$

$$\Rightarrow \quad \alpha = -62.75°$$

The stresses on the plane are shown in Fig. 9.19(a). Graphically, as in Fig. 9.19(b), we draw two concentric circles, the inner circle having a radius of 60 N/mm^2 and the outer circle one of 120 N/mm^2, to scale. Since $\angle\theta = -40°$, we measure an angle of 40° from the X-axis to intersect the σ_X-circle at N. As σ_Y is negative, this normal is produced in the opposite direction to intersect the σ_Y-circle at M. A horizontal line through M and a vertical line through N intersect at L, which is a point on the ellipse of stress. OL gives the resultant stress. By measurement, OL = 100 N/mm^2.

LP is perpendicular to MN.

$$OP = \sigma_\theta = 46 \text{ N/mm}^2$$
$$LP = \tau_\theta = 88 \text{ N/mm}^2$$

Since ON is the normal to the plane, the plane is P-P as shown. On this plane, σ_θ is $\overline{OP}$ and is tensile while $\tau_\theta = \overline{PL}$ gives an anticlockwise moment at a point inside the element of which P-P is a side. Thus τ_θ is negative.

Example 9.10 Biaxial stress

The plane stress system at a point in a stressed body is shown in Fig. 9.20(a). Find the normal, shear and resultant stresses on a plane P-P whose normal is inclined at 30° (counterclockwise) to the X-axis, analytically and graphically.

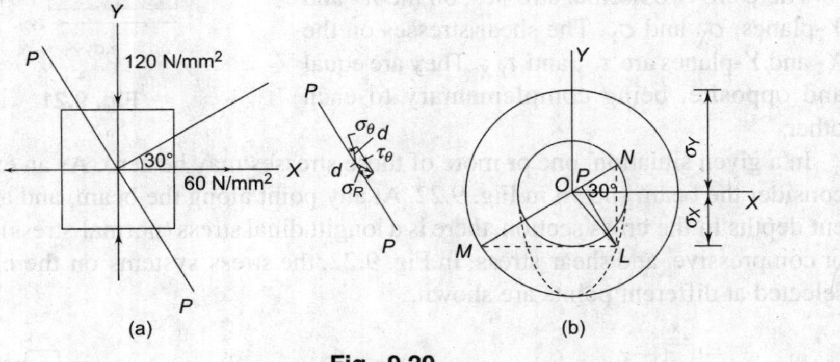

Fig. 9.20

Solution This is a case where σ_Y is negative and is also > σ_X.
From Fig. 9.20(a), σ_X = 60 N/mm^2, σ_Y = – 120 N/mm^2, and θ = 30°.

$$\sigma_\theta = \frac{\sigma_X + \sigma_Y}{2} + \frac{\sigma_X - \sigma_Y}{2}\cos 2\theta$$

$$= \frac{60 - 120}{2} + \frac{60 + 120}{2}\cos 60°$$

$$= -30 + 45 = 15 \text{ N/mm}^2$$

$$\tau_\theta = \frac{\sigma_X - \sigma_Y}{2}\sin 2\theta = \frac{60 + 120}{2}\frac{\sqrt{3}}{2} = 77.94 \text{ N/mm}^2$$

$$\sigma_R = \sqrt{15^2 + 77.94^2} = 79.37 \text{ N/mm}^2$$

or $\sigma_R = \sqrt{60^2 \cos^2 30 + 120^2 \sin^2 30} = 79.37$ N/mm^2

$$\tan \alpha = \frac{\tau_\theta}{\sigma_\theta} = \frac{77.94}{15}$$

$\Rightarrow$ $\alpha = 79.1°$

These are shown in Fig. 9.20(a).

The graphical solution by the ellipse of stress method is shown in Fig. 9.20(b). We draw two concentric circles, with centre at O and of radii 60 and 120 to scale. ON is a line parallel to the normal to the plane P-P cutting the circle of radius $\sigma_X = 60$ N/mm^2 at N. In this case, σ_Y is negative and $\sigma_Y > \sigma_X$. The normal ON is produced in the opposite direction to meet the circle of radius $\sigma_Y = 120$ N/mm^2 at M. A horizontal line through M and a vertical line through N intersect at L. OL is the resultant stress on plane p-p. By measurement,

$$\sigma_R = OL = 80 \text{ N/mm}^2, \quad OP = \sigma_\theta = 14 \text{ N/mm}^2, \quad PL = \tau_\theta = 78 \text{ N/mm}^2$$

9.5 PLANE STRESS ANALYSIS

In a plane stress situation, there is no stress in the Z-direction. This means that σ_{ZZ}, τ_{XZ} and τ_{YZ} are zero. The plane stress situation thus reduces to that shown in Fig. 9.21. This is the most general stress system in plane stress. There is no stress perpendicular to the plane of the paper.

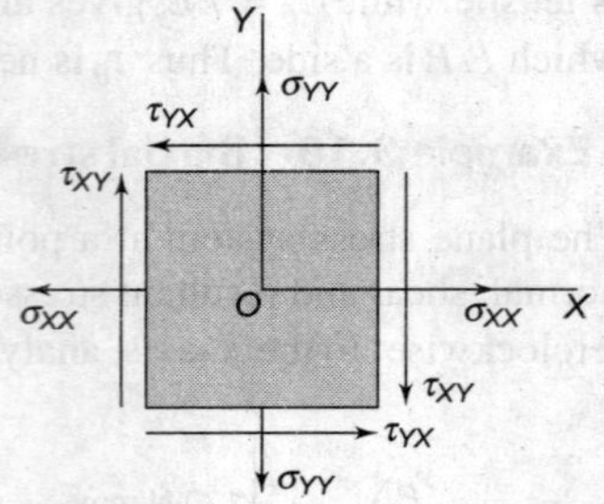

Fig. 9.21

There are two normal stresses, on the X- and Y-planes, σ_X and σ_Y. The shear stresses on the X- and Y-planes are τ_{XY} and τ_{YX}. They are equal and opposite, being complementary to each other.

In a given situation, one or more of these stresses may be zero. As an example, consider the beam shown in Fig. 9.22. At any point along the beam, and at different depths in the cross section, there is a longitudinal stress (normal stress), tensile or compressive, and shear stress. In Fig. 9.22, the stress systems on the elements selected at different points are shown.

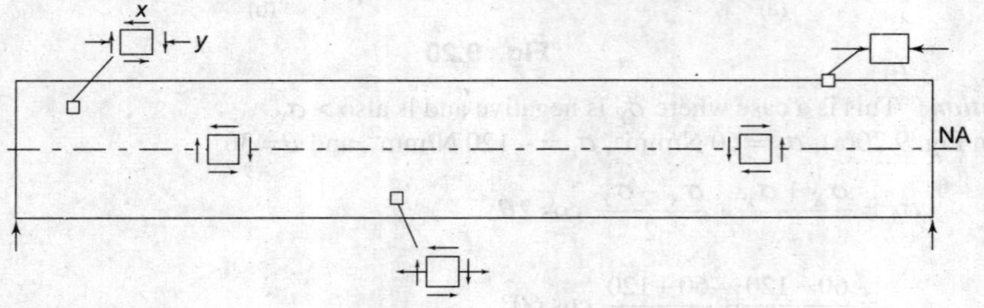

Fig. 9.22

Note that the elements shown have very small sides Δx and Δy. This has to be so to consider uniform stresses on the sides.

The objective of plane stress analysis is to find out the maximum stresses at a point and their directions. To do this, we categorize two problems for solution.

(i) Stress on an inclined plane: Considering the element shown in Fig. 9.23 subjected to a plane stress system, we find the stress on a plane *P-P* inclined as shown.

(ii) To determine the planes on which maximum stresses occur, both normal and shear, and their magnitude: Both these problems can be solved analytically and graphically. We start with the analytical solution to problem (i).

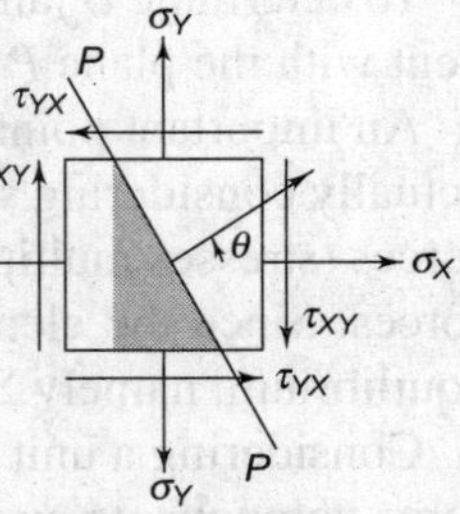

Fig. 9.23

9.6 STRESS TRANSFORMATION: STRESSES ON AN OBLIQUE PLANE

Given that an element in a structure is subjected to the general plane stress system shown, we have to find the stresses on the plane *P-P* [Fig. 9.24(a)].

The plane *P-P* is identified by the angle θ that the outward normal to the plane makes with the *X*-axis. This angle is positive if measured counterclockwise from the *X*-axis, as shown.

Note that on the plane *P-P*, there may be a normal stress σ_θ and a shear stress τ_θ. Since this is a plane stress situation, there can be no stress normal to the plane of the paper [Fig. 9.24(b)].

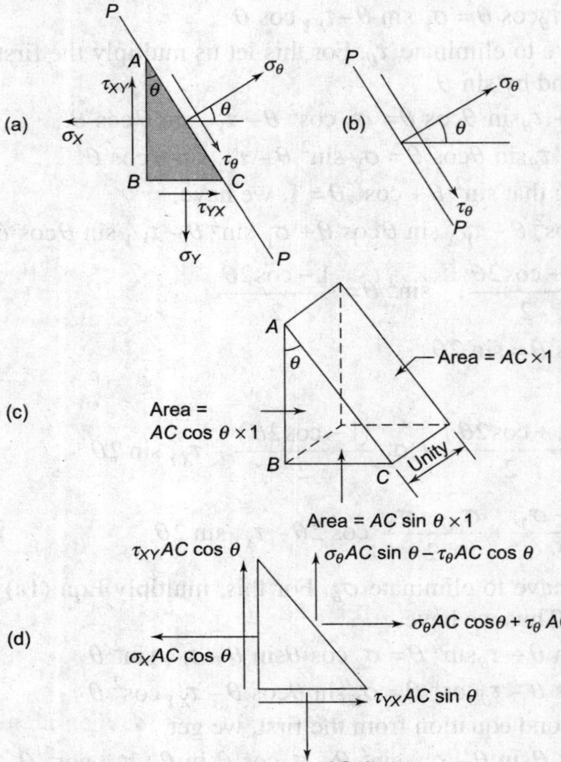

Fig. 9.24

To determine σ_θ and τ_θ, we isolate an element ABC as shown, AC being coincident with the plane P-P.

An important point to bear in mind is that since the element is so small, we are actually considering stresses at a point in different directions. This means that the forces (stresses multiplied by the areas on which they act) form a set of concurrent forces. Since the element is in equilibrium, we can apply the two conditions of equilibrium, namely $\Sigma H = 0$ and $\Sigma V = 0$, to these forces [Fig. 9.24(c)].

Considering a unit length perpendicular to the plane of the paper, the areas and force components on the faces are shown in Fig. 9.24(d).

$$\Sigma H = 0 \quad \overrightarrow{+}$$

$$\sigma_\theta AC\cos\theta + \tau_\theta AC\sin\theta + \tau_{YX}AC\sin\theta - \sigma_X AC\cos\theta = 0$$

This can be reduced to

$$\sigma_\theta\cos\theta + \tau_\theta\sin\theta + \tau_{YX}\sin\theta - \sigma_X\cos\theta = 0 \qquad (1)$$

$$\Sigma V = 0 \uparrow +$$

$$\sigma_\theta AC\sin\theta - \tau_\theta AC\cos\theta - \sigma_Y AC\sin\theta + \tau_{XY}AC\cos\theta = 0 \qquad (2)$$

This, too, can be reduced to

$$\sigma_\theta\sin\theta - \tau_\theta\cos\theta - \sigma_Y\sin\theta + \tau_{XY}\cos\theta = 0$$

These two equations can be solved to get the values of σ_θ and τ_θ. Rewriting these equations,

$$\sigma_\theta\cos\theta + \tau_\theta\sin\theta = \sigma_X\cos\theta - \tau_{YX}\sin\theta \qquad (1a)$$

$$\sigma_\theta\sin\theta - \tau_\theta\cos\theta = \sigma_Y\sin\theta - \tau_{XY}\cos\theta \qquad (2a)$$

To find σ_θ, we have to eliminate τ_θ. For this let us multiply the first equation by $\cos\theta$ and the second by $\sin\theta$.

$$\sigma_\theta\cos^2\theta + \tau_\theta\sin\theta\cos\theta = \sigma_X\cos^2\theta - \tau_{YX}\sin\theta\cos\theta$$

$$\sigma_\theta\sin^2\theta - \tau_\theta\sin\theta\cos\theta = \sigma_Y\sin^2\theta - \tau_{XY}\sin\theta\cos\theta$$

Adding, and noting that $\sin^2\theta + \cos^2\theta = 1$, we have,

$$\sigma_\theta = \sigma_X\cos^2\theta - \tau_{YX}\sin\theta\cos\theta + \sigma_Y\sin^2\theta - \tau_{XY}\sin\theta\cos\theta$$

$$\cos^2\theta = \frac{1+\cos 2\theta}{2}, \quad \sin^2\theta = \frac{1-\cos 2\theta}{2}$$

$$2\sin\theta\cos\theta = \sin 2\theta$$

$$\tau_{XY} = \tau_{YX}$$

$$\sigma_\theta = \sigma_X\frac{(1+\cos 2\theta)}{2} + \sigma_Y\frac{(1-\cos 2\theta)}{2} - \tau_{XY}\sin 2\theta$$

$$= \frac{\sigma_X+\sigma_Y}{2} + \frac{\sigma_X-\sigma_Y}{2}\cos 2\theta - \tau_{XY}\sin 2\theta$$

To find τ_θ, we have to eliminate σ_θ. For this, multiply Eqn (1a) by $\sin\theta$ and Eqn (2a) by $\cos\theta$. Thus we have

$$\sigma_\theta\cos\theta\sin\theta + \tau_\theta\sin^2\theta = \sigma_X\cos\theta\sin\theta - \tau_{YX}\sin^2\theta$$

$$\sigma_\theta\sin\theta\cos\theta - \tau_\theta\cos^2\theta = \sigma_Y\sin\theta\cos\theta - \tau_{XY}\cos^2\theta$$

Subtracting the second equation from the first, we get,

$$\tau_\theta = \sigma_X\cos\theta\sin\theta - \tau_{YX}\sin^2\theta - \sigma_Y\cos\theta\sin\theta + \tau_{XY}\cos^2\theta$$

which reduces to

$$\tau_\theta = \frac{\sigma_X - \sigma_Y}{2} \sin 2\theta + \tau_{XY} \cos 2\theta$$

(by substituting $\tau_{YX} = \tau_{XY}$, $\sin^2\theta + \cos^2\theta = 1$, $\cos^2\theta - \sin^2\theta = \cos 2\theta$, and $2 \sin\theta \cos\theta = \sin 2\theta$).

Thus, the normal stress and shear stress on a plane inclined at an angle θ (normal to the plane inclined at angle θ) to the X-axis are given by

$$\sigma_\theta = \frac{\sigma_X + \sigma_Y}{2} + \frac{\sigma_X - \sigma_Y}{2} \cos 2\theta - \tau_{XY} \sin 2\theta$$

$$\tau_\theta = \frac{\sigma_X - \sigma_Y}{2} \sin 2\theta + \tau_{XY} \cos 2\theta$$

Please note that we started with a plane stress system in which stresses and the angle θ were all positive. If any of these quantities are negative, appropriate signs must be used. These two equations in fact give the values of normal and shear stresses on planes obtained by changing the angle θ. For example, if we set $\theta = 0$, the plane *P-P* is parallel to the X-plane, and $\sigma_\theta = \sigma_X$ and $\tau_\theta = \tau_{XY}$. If we take $\theta = 90°$, which makes the plane *P-P* parallel to Y-plane, $\sigma_\theta = \sigma_Y$ and $\tau_\theta = \tau_{YX}$.

Thus, when we substitute different values of angle θ, we get different planes and stresses on these planes. This is also known as transformation of plane stress by which the coordinate axes X, Y are rotated about the Z-axis by an angle θ. The resulting stresses on planes normal to the X' and Y'-axes are given by the equations derived above [(Fig. 9.25(a)].

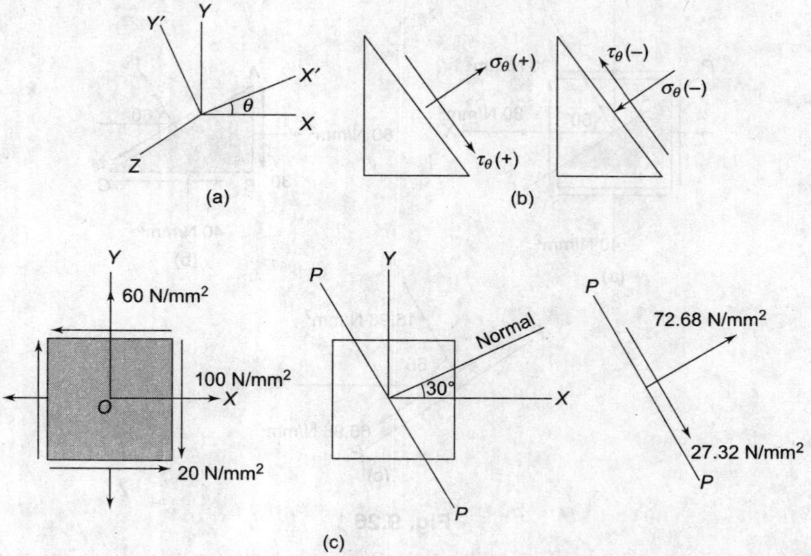

Fig. 9.25

The directions of stresses σ_θ and τ_θ on plane *P-P* will depend upon their signs and will be as shown in Fig. 9.25(b). Note that when the shear stress is positive, its moment about a point within the element is clockwise. The following examples illustrate the use of the equations we have derived.

Example 9.11 Stresses on an oblique plane

The plane stress system on an element of a stressed body is shown in Fig. 9.25(c). Determine the stresses on a plane whose normal is inclined at + 30° to the X-axis.

Solution The plane is shown in Fig. 9.25(b). Since the angle is given as + 30°, it is measured counterclockwise from the X-axis. Both the normal stresses are tensile and hence positive. The shear stress on the X-plane is positive, i.e., $\tau_{XY} = +20$ N/mm^2. Substituting the values in the equations for σ_θ and τ_θ,

$$\sigma_\theta = \frac{\sigma_X + \sigma_Y}{2} + \frac{\sigma_X - \sigma_Y}{2}\cos 2\theta - \tau_{XY}\sin 2\theta$$

$$= \frac{100+60}{2} + \frac{100-60}{2}(\cos 60° - 20 \sin 60°)$$

$$= 72.68 \text{ N/mm}^2$$

$$\tau_\theta = \frac{\sigma_X - \sigma_Y}{2}\sin 2\theta + \tau_{XY}\cos 2\theta$$

$$= \frac{100-60}{2}\sin 60° + 20\cos 60°$$

$$= 27.32 \text{ N/mm}^2$$

These stresses can be represented on the plane as shown in Fig. 9.25(c).

Example 9.12 Stresses on an oblique plane

At a point in a stressed body, the stresses are as shown in Fig. 9.26(a). Determine the stresses on the plane P-P shown.

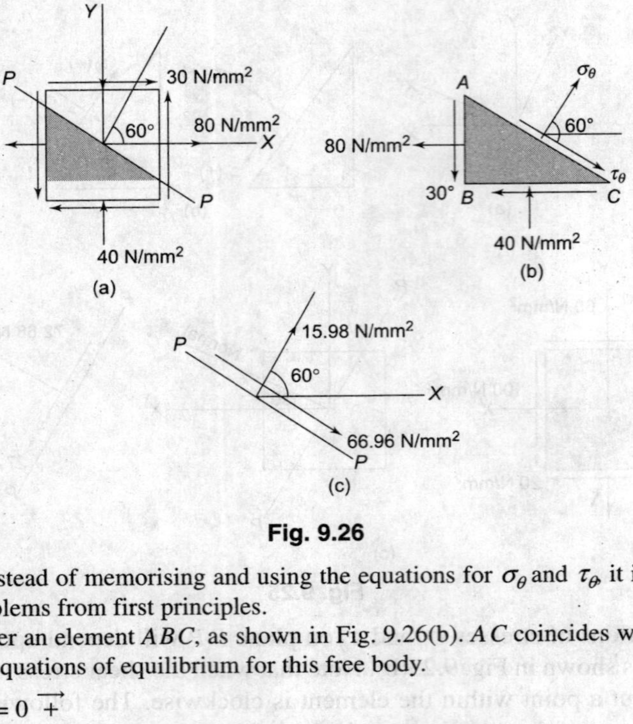

Fig. 9.26

Solution Instead of memorising and using the equations for σ_θ and τ_θ, it is desirable to solve the problems from first principles.

We consider an element ABC, as shown in Fig. 9.26(b). AC coincides with plane P-P. We state the equations of equilibrium for this free body.

$$\Sigma H = 0 \; \overrightarrow{+}$$

$$\sigma_\theta AC\cos 60 + \tau_\theta AC\sin 60 - 80AC\cos 60 - 30AC\sin 60 = 0$$

$$\sigma_\theta \frac{1}{2} + \tau_\theta \frac{\sqrt{3}}{2} = 80\frac{1}{2} + 30\frac{\sqrt{3}}{2}$$

$$\sigma_\theta + \tau_\theta\sqrt{3} = 131.96$$

$$\Sigma V = 0 \uparrow +$$

$$\sigma_\theta AC \sin 60 - \tau_\theta AC \cos 60 - 30\, AC \cos 60 + 40\, AC \sin 60 = 0$$

$$\sigma_\theta \frac{\sqrt{3}}{2} - \tau_\theta \frac{1}{2} = 30 \times \frac{1}{2} - 40\frac{\sqrt{3}}{2}$$

$$\sigma_\theta\sqrt{3} - \tau_\theta = -39.28$$

Solving the two equations, $\sigma_\theta = 15.98$ N/mm^2 and $\tau_\theta = 66.96$ N/mm^2. The stresses on the plane can be represented as shown in Fig. 9.26(c).

Example 9.13 Plane of zero shear stress

The stress system on an element of a stressed body are as shown in Fig. 9.27(a). Determine the planes on which there is no shear stress. What are the stresses acting on these planes?

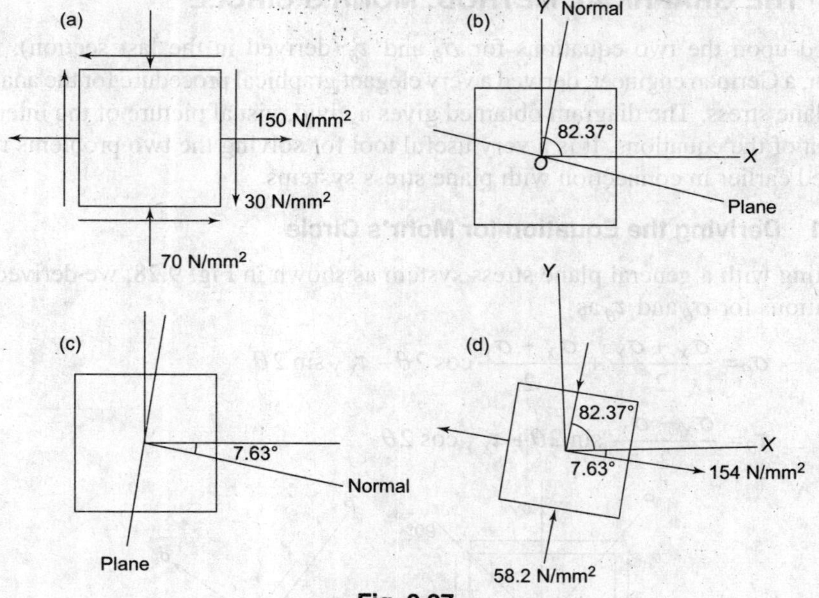

Fig. 9.27

Solution We know that the shear stress on a plane is given by

$$\tau_\theta = \frac{\sigma_X - \sigma_Y}{2} \sin 2\theta + \tau_{XY} \cos 2\theta$$

In the given case,

$$\sigma_X = +150 \text{ N/mm}^2, \quad \sigma_Y = -70 \text{ N/mm}^2, \quad \tau_{XY} = 30 \text{ N/mm}^2$$

Therefore,

$$\tau_\theta = \frac{150 - (-70)}{2} \sin 2\theta + 30 \cos 2\theta$$

This equation shows the variation of shear stresses as the angle θ, the normal to the planes, varies. We set $\tau_\theta = 0$ to determine the values of θ to identify the planes, if any, on which the shear stress is zero. Thus,

$$0 = 110 \sin 2\theta + 30 \cos 2\theta$$

$$\tan 2\theta = \frac{30}{-110} = -0.2727$$

$$2\theta = -15.26 \text{ and } 164.74°, \quad \theta = -7.63° \text{ and } 82.37°$$

These planes can be identified as shown in Figures 9.27(b) and (c). The stress on these planes will be a normal stress given by

$$\sigma_\theta = \frac{\sigma_X + \sigma_Y}{2} + \frac{\sigma_X - \sigma_Y}{2} \cos 2\theta - \tau_{XY} \sin 2\theta$$

$$\sigma_\theta = \frac{150 - 70}{2} + \frac{150 - (-70)}{2} \cos(-15.26) - 30 \sin(-15.26)$$

$$= 154 \text{ N/mm}^2$$

and $$\sigma_\theta = \frac{150 - 70}{2} + \frac{150 - (-70)}{2} \cos(164.74) - 30 \sin(164.74)$$

$$= -58.2 \text{ N/mm}^2$$

The planes and the stresses are shown in Fig. 9.27(d). ■

9.7 THE GRAPHICAL METHOD: MOHR'S CIRCLE

Based upon the two equations for σ_θ and τ_θ (derived in the last section), Otto Mohr, a German engineer, derived a very elegant graphical procedure for the analysis of plane stress. The diagram obtained gives a vivid, visual picture of the interpretation of the equations. It is a very useful tool for solving the two problems mentioned earlier in connection with plane stress systems.

9.7.1 Deriving the Equation for Mohr's Circle

Starting with a general plane stress system as shown in Fig. 9.28, we derived the equations for σ_θ and τ_θ as

$$\sigma_\theta = \frac{\sigma_X + \sigma_Y}{2} + \frac{\sigma_X - \sigma_Y}{2} \cos 2\theta - \tau_{XY} \sin 2\theta$$

$$\tau_\theta = \frac{\sigma_X - \sigma_Y}{2} \sin 2\theta + \tau_{XY} \cos 2\theta$$

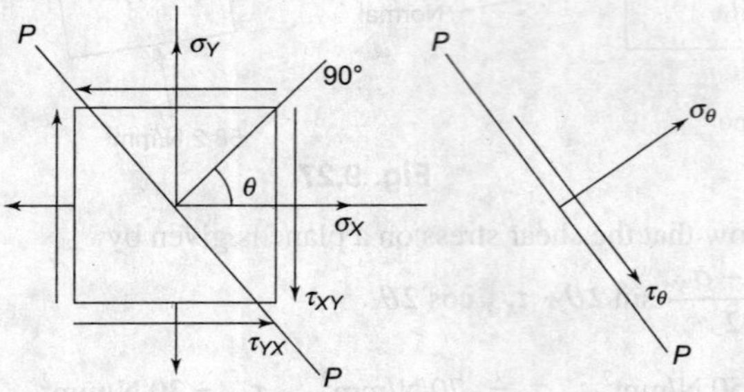

Fig. 9.28

It must be remembered that, in these equations, σ_X, σ_Y and τ_{XY} are known quantities, and σ_θ and τ_θ are the normal and shear stresses on any arbitrary plane identified by angle θ.

Rearranging and squaring these equations, we get

$$\sigma_\theta - \frac{\sigma_X + \sigma_Y}{2} = \frac{\sigma_X - \sigma_Y}{2} \cos 2\theta - \tau_{XY} \sin 2\theta$$

$$\left(\sigma_\theta - \frac{\sigma_X + \sigma_Y}{2}\right)^2 = \left(\frac{\sigma_X - \sigma_Y}{2}\right)^2 \cos^2 2\theta + \tau_{XY}^2 \sin^2 2\theta - 2\,\frac{\sigma_X - \sigma_Y}{2} \cos 2\theta\, \tau_{XY} \sin 2\theta$$

$$\tau_\theta = \frac{\sigma_X - \sigma_Y}{2} \sin 2\theta + \tau_{XY} \cos 2\theta$$

Squaring,

$$\tau_\theta^2 = \left(\frac{\sigma_X - \sigma_Y}{2}\right)^2 \sin^2 2\theta + \tau_{XY}^2 \cos^2 2\theta + 2\,\frac{\sigma_X - \sigma_Y}{2} \cos 2\theta\, \tau_{XY} \sin 2\theta$$

Adding these two equations,

$$\left(\sigma_\theta - \frac{\sigma_X + \sigma_Y}{2}\right)^2 + \tau_\theta^2 = \left(\frac{\sigma_X - \sigma_Y}{2}\right)^2 + \tau_{XY}^2$$

which is the equation for Mohr's circle.

Considering the equation for a standard circle, shown in Fig. 9.29(a), $(x - c)^2 + y^2 = R^2$, we see that (x, y) are the coordinates of a point on a circle whose radius is R and whose centre is shifted by c from the origin O along the X-axis.

The equation for Mohr's circle is identical to such an equation. Thus, the equation

$$\left(\sigma_\theta - \frac{\sigma_X + \sigma_Y}{2}\right)^2 + \tau_\theta^2 = \left(\frac{\sigma_X - \sigma_Y}{2}\right)^2 + \tau_{XY}^2$$

is the equation of a circle whose points have coordinates $(\sigma_\theta, \tau_\theta)$, whose radius is $\{[(\sigma_X - \sigma_Y)/2]^2 + \tau_{XY}^2\}^{1/2}$ and whose centre is shifted along the σ_θ axis by $(\sigma_X + \sigma_Y)/2$. This means that, given a plane stress system, the stresses on any inclined plane σ_θ and τ_θ can be represented as the coordinates of a point on a circle which is set in a σ, τ axes system [Fig. 9.29(b)].

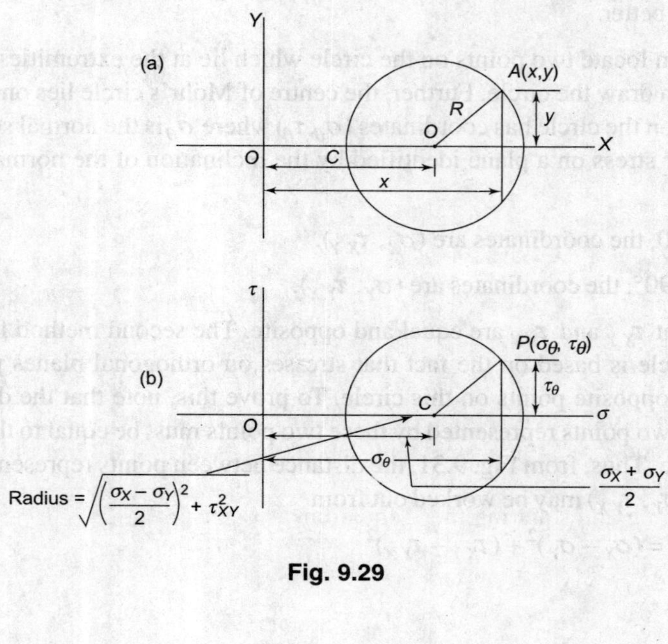

Fig. 9.29

9.7.2 Drawing Mohr's Circle

One obvious method to draw Mohr's circle is to calculate the coordinates of its centre and radius. Thus, for the plane stress system shown in Fig. 9.30,

$$\frac{\sigma_X + \sigma_Y}{2} = \frac{120 + 40}{2} = 80 \text{ units}$$

is the shift of the centre. The coordinates of the centre of Mohr's circle are (80, 0). The radius of this circle is given by

$$\sqrt{\left(\frac{\sigma_X - \sigma_Y}{2}\right)^2 + \tau_{XY}^2} = \sqrt{\left(\frac{120 - 40}{2}\right)^2 + 30^2} = 50 \text{ units}$$

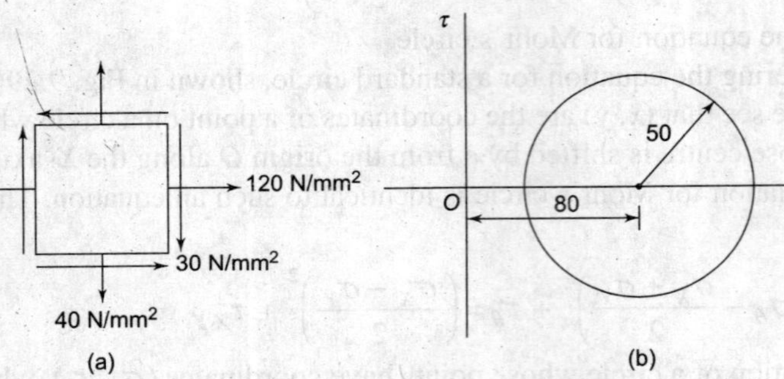

Fig. 9.30

Mohr's circle can be drawn as shown in Fig. 9.30(b). While this procedure for drawing the circle is correct, there is a more elegant way to do so, which helps to interpret it better.

If we can locate two points on the circle which lie at the extremities of a diameter, we can draw the circle. Further, the centre of Mohr's circle lies on the σ-axis. Any point on the circle has coordinates (σ_θ, τ_θ) where σ_θ is the normal stress and τ_θ is the shear stress on a plane identified by the inclination of the normal to the X-axis, θ.

For $\theta = 0$, the coordinates are (σ_X, τ_{XY}).

For $\theta = 90°$, the coordinates are (σ_Y, τ_{YX}).

Note that τ_{XY} and τ_{YX} are equal and opposite. The second method for drawing Mohr's circle is based on the fact that stresses on orthogonal planes plot as diametrically opposite points on this circle. To prove this, note that the distance between the two points represented by these two points must be equal to the diameter of the circle. Thus, from Fig. 9.31, the distance between points represented by (σ_X, τ_{XY}) and (σ_Y, τ_{YX}) may be worked out from

$$d^2 = (\sigma_X - \sigma_Y)^2 + (\tau_{XY} - \tau_{YX})^2$$

The radius of Mohr's circle as proved earlier may be obtained from

$$R^2 = \left(\frac{\sigma_X - \sigma_Y}{2}\right)^2 + \tau_{XY}^2$$

since $\tau_{YX} = -\tau_{XY}$.

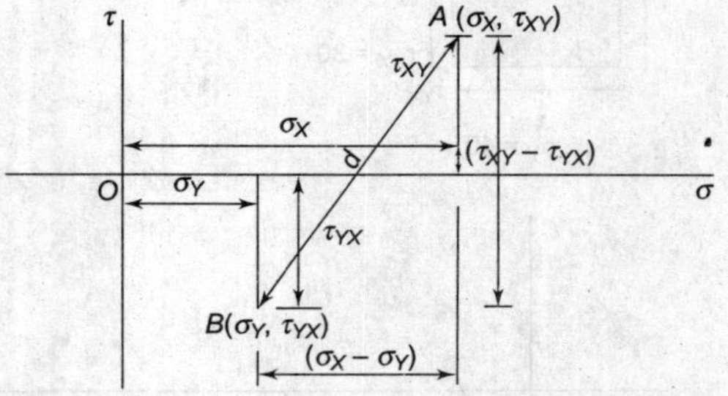

Fig. 9.31

If D is the diameter of Mohr's circle, then

$$\left(\frac{D}{2}\right)^2 = \left(\frac{\sigma_X - \sigma_Y}{2}\right)^2 + \tau_{XY}^2$$

$$D^2 = (\sigma_X - \sigma_Y)^2 + 4\tau_{XY}^2$$

Noting that $\tau_{YX} = -\tau_{XY}$, we have

$$d^2 = (\sigma_X - \sigma_Y)^2 + 4\tau_{XY}^2$$

which is the same as the square of the diameter of Mohr's circle.

Mohr's circle can thus be drawn by plotting the two points represented by (σ_X, τ_{XY}) and (σ_Y, τ_{YX}), which plot as extremities of a diameter.

We take the stress system shown in Fig. 9.30 and redraw as shown in Fig. 9.32. To draw Mohr's circle,

(i) draw the σ-τ axes conveniently,
(ii) select a scale to plot the stresses,
(iii) plot two points with coordinates A (120, 30) and B (40, – 30),
(iv) join the two points—where this line cuts the σ-axis is the centre of Mohr's circle, and
(v) with C as centre and CA or CB as radius, draw a circle, which is Mohr's circle for the stress system.

Figures 9.33(a), (b), and (c) show Mohr's circles for different stress systems. The procedure and nomenclature in all these are the same as the one used for drawing the Mohr's circle given in Fig. 9.32(d)—C is centre of Mohr's circle. A represents the stresses on the X-plane, and B those on the Y-plane. A compressive stress should be plotted to the left of the origin and a negative shear stress downward. Figure 9.33(c) shows a case of pure shear, and in it the centre of Mohr's circle C coincides with the origin, and A and B lie on the τ-axis.

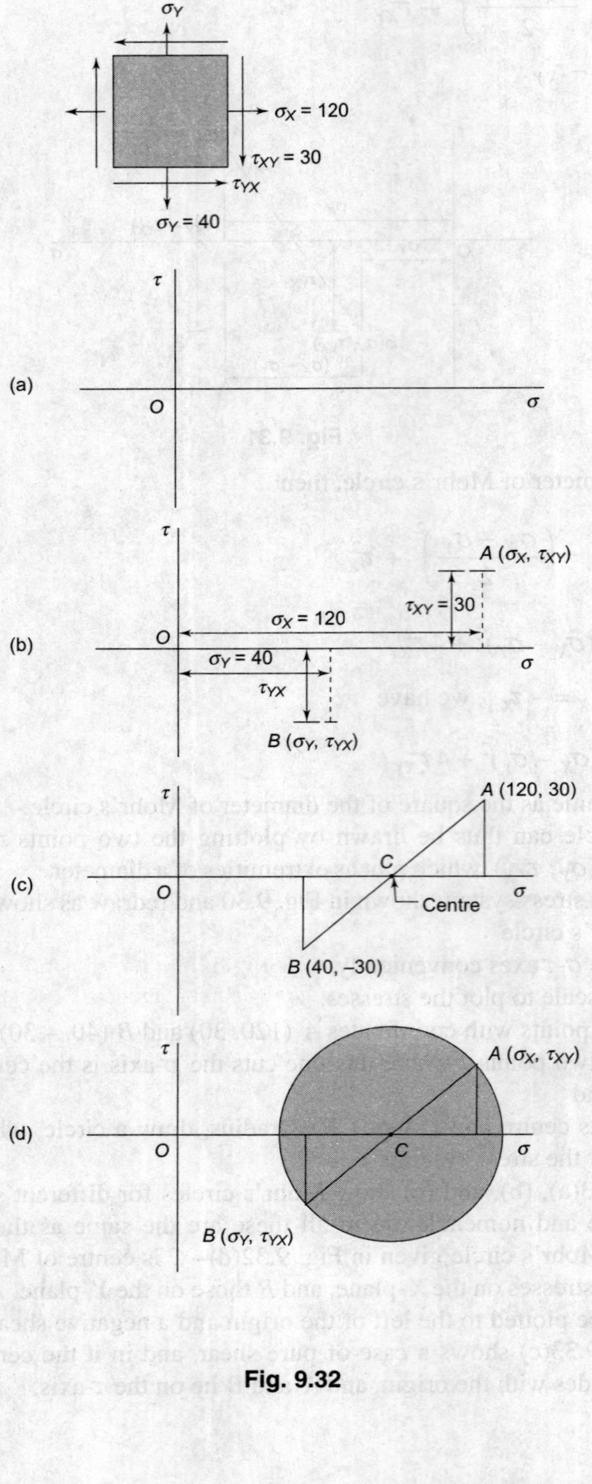

σ_Y
σ_X = 120
τ_{XY} = 30
τ_{YX}
σ_Y = 40
(a)
τ
O
σ
(b)
τ
A (σ_X, τ_{XY})
τ_{XY} = 30
σ_X = 120
O
σ_Y = 40
σ
τ_{YX}
B (σ_Y, τ_{YX})
(c)
τ
A (120, 30)
C
O
σ
Centre
B (40, –30)
(d)
τ
A (σ_X, τ_{XY})
O
C
σ
B (σ_Y, τ_{YX})

Fig. 9.32

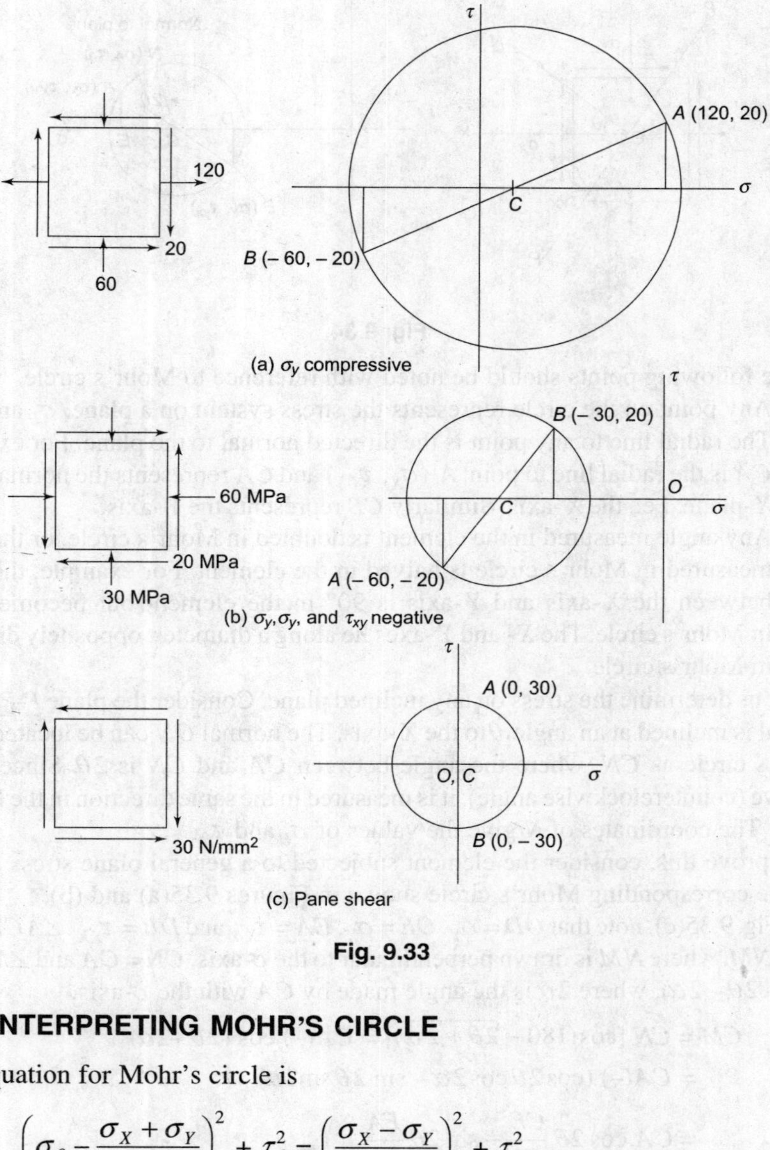

Fig. 9.33

9.8 INTERPRETING MOHR'S CIRCLE

The equation for Mohr's circle is

$$\left(\sigma_\theta - \frac{\sigma_X + \sigma_Y}{2}\right)^2 + \tau_\theta^2 = \left(\frac{\sigma_X - \sigma_Y}{2}\right)^2 + \tau_{XY}^2$$

Mohr's circle is shown in Fig. 9.34. Note that $OD = \sigma_Y$, $OE = \sigma_X$, $EA = \tau_{XY}$ and $DB = \tau_{YX}$. From this,

$$OC = OD + DC = \sigma_Y + \frac{\sigma_X - \sigma_Y}{2} = \frac{\sigma_X + \sigma_Y}{2}$$

From ΔACE, the radius of the circle may be obtained from the equation

$$CA^2 = CE^2 + EA^2 = \left(\frac{\sigma_X - \sigma_Y}{2}\right)^2 + \tau_{XY}^2$$

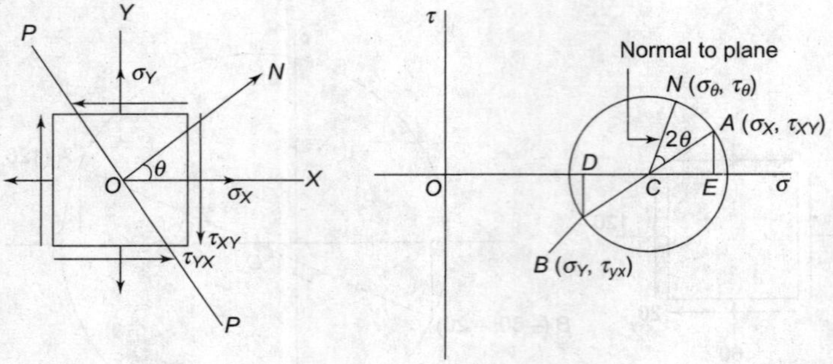

Fig. 9.34

The following points should be noted with reference to Mohr's circle.

- Any point on the circle represents the stress system on a plane, σ_θ and τ_θ.
- The radial line to any point is the directed normal to the plane. For example, CA is the radial line to point A (σ_X, τ_{XY}) and CA represents the normal to the X-plane, i.e., the X-axis. Similarly CB represents the Y-axis.
- Any angle measured in the element is doubled in Mohr's circle, or the angle measured in Mohr's circle is halved in the element. For example, the angle between the X-axis and Y-axis is 90° in the element but becomes 180° in Mohr's circle. The X- and Y-axes lie along a diameter, oppositely directed, in Mohr's circle.

Let us determine the stress on any inclined plane. Consider the plane P-P whose normal is inclined at an angle θ to the X-axis. The normal ON can be located in the Mohr's circle as CN, where the angle between CA and CN is 2θ. Since θ is a positive (counterclockwise angle), it is measured in the same direction in the Mohr's circle. The coordinates of N give the values of σ_θ and τ_θ.

To prove this, consider the element subjected to a general plane stress system and the corresponding Mohr's circle shown in Figures 9.35(a) and (b).

In Fig. 9.35(c), note that $OD = \sigma_Y$, $OE = \sigma_X$, $EA = \tau_{XY}$ and $DB = \tau_{YX}$. $\angle ACN = 2\theta$. In ΔCNM, where NM is drawn perpendicular to the σ-axis, $CN = CA$ and $\angle MCN = (180 - 2\theta - 2\alpha)$, where 2α is the angle made by CA with the σ-axis.

$$CM = CN\,[\cos(180 - \overline{2\theta + 2\alpha})] = CA(-)\cos(2\theta + 2\alpha)$$

$$= CA(-)\,(\cos 2\theta \cos 2\alpha - \sin 2\theta \sin 2\alpha)$$

$$= CA \cos 2\theta \frac{CE}{CA} - \sin 2\theta \frac{EA}{CA}$$

$$CE = OE - OC = \sigma_X - \frac{\sigma_X + \sigma_Y}{2} = \frac{\sigma_X - \sigma_Y}{2}$$

$$EA = \tau_{XY}$$

$$CM = -\frac{\sigma_X - \sigma_Y}{2}\cos 2\theta + \tau_{XY}\sin 2\theta$$

$$OM = OC - CM = \frac{\sigma_X + \sigma_Y}{2} + \frac{\sigma_X - \sigma_Y}{2}\cos 2\theta - \tau_{XY}\sin 2\theta$$

which is equal to σ_θ.

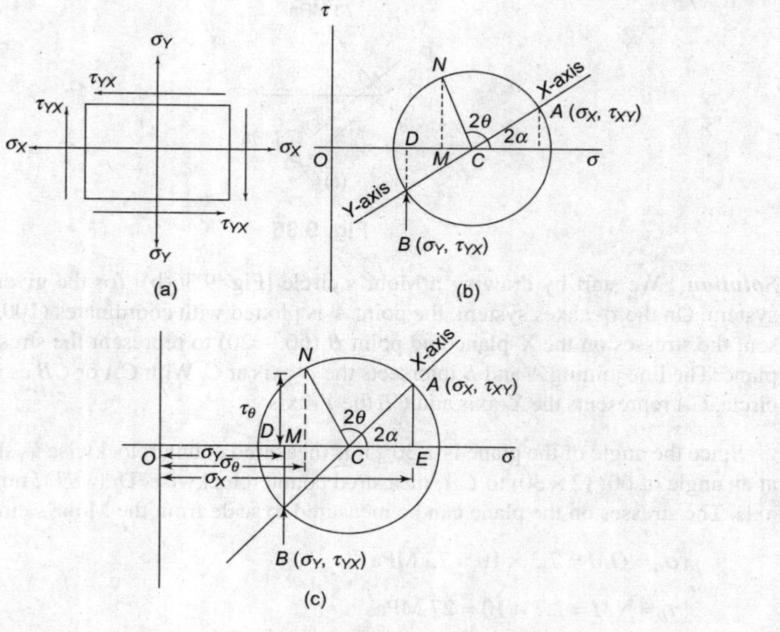

Fig. 9.35

Similarly,

$$MN = \tau_\theta = CN \sin (2\theta + 2\alpha)$$

$$MN = CA (\sin 2\theta \cos 2\alpha + \cos 2\theta \sin 2\alpha)$$

$$MN = CA \left(\sin 2\theta \frac{CE}{CA} + \cos 2\theta \frac{EA}{CA} \right)$$

$$MN = \frac{\sigma_X - \sigma_Y}{2} \sin 2\theta + \tau_{XY} \cos 2\theta$$

which is equal to τ_θ.

Thus, the coordinates of N are (σ_θ, τ_θ). The following examples illustrate the procedure to determine the stresses on an oblique plane.

Example 9.14 Mohr's circle: stresses on oblique plane

The stress system on orthogonal planes at a point in a stressed body is as shown in Fig. 9.36(a). Find the stresses on a plane inclined at + 30° to the X-axis.

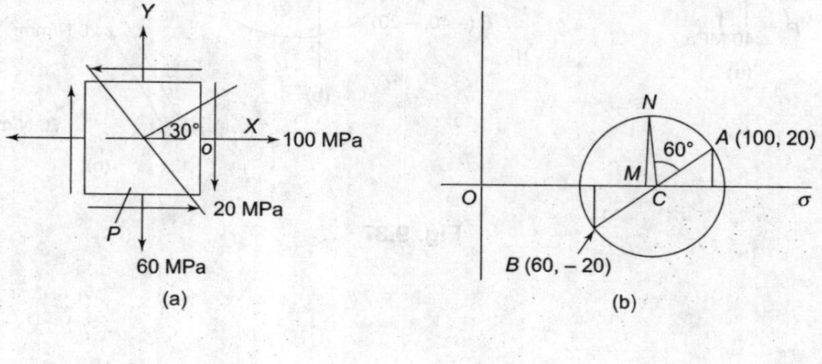

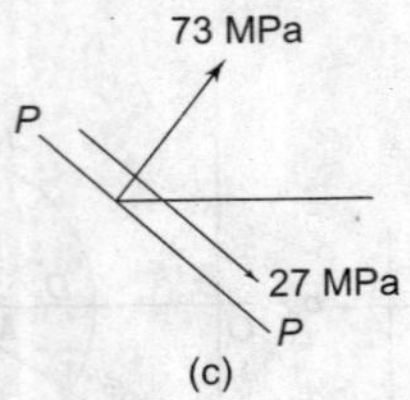

(c)

Fig. 9.36

Solution We start by drawing a Mohr's circle [Fig. 9.36(b)] for the given plane stress system. On the σ-τ axes system, the point A is plotted with coordinates (100, 20) to represent the stresses on the X-plane and point B (60, – 20) to represent the stresses on the Y-plane. The line joining A and B intersects the σ-axis at C. With CA or CB as radius, draw a circle. CA represents the X-axis and CB the Y-axis.

Since the angle of the plane is + 30°, it is measured counterclockwise as shown. Set CN at an angle of 60° (2 × 30) to CA, measured counterclockwise. Draw NM normal to the σ-axis. The stresses on the plane can be measured to scale from the Mohr's circle as

$$\sigma_\theta = OM = 7.3 \times 10 = 73 \text{ MPa}$$

$$\tau_\theta = NM = 2.7 \times 10 = 27 \text{ MPa}$$

The stresses on the plane can be represented as shown in Fig. 9.36(c).

Example 9.15 Mohr's circle: stresses on oblique plane

If the plane stresses at a point in a body are as shown in Fig. 9.37, find graphically the stresses on a plane whose normal is inclined 20° clockwise from the X-axis.

Solution Mohr's circle for the stresses is shown in Fig. 9.37(b). Note that the coordinates of A and B are (70, 20) and (– 40, – 20). CA is the X-axis and CB the Y-axis in the Mohr's circle. Draw CN at 40° measured clockwise from CA. Then,

$$\sigma_\theta = OM = 70 \text{ N/mm}^2, \quad \tau_\theta = MN = -20 \text{ N/mm}^2$$

These stresses can be represented as shown in Fig. 9.37(c).

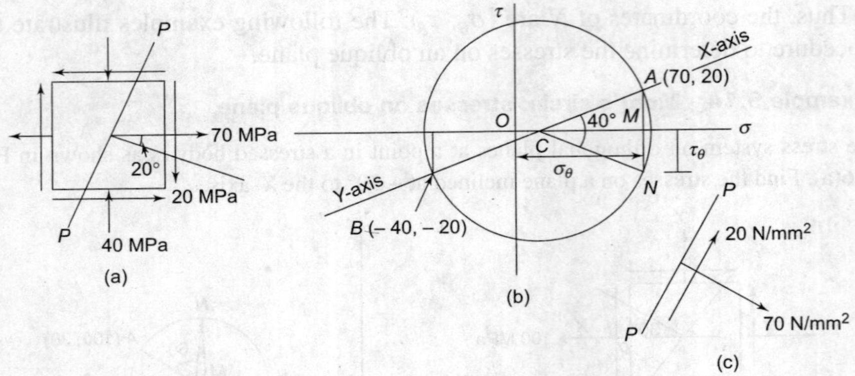

Fig. 9.37

9.8.1 Principal Planes and Stresses

We identified two situations in plane stress—(i) stresses on any oblique plane, and (ii) maximum normal stress and shear stress and the planes on which they act.

Mohr's circle helps in immediate visualisation and solution to problems relating to the second case. Consider the general plane stress system and the corresponding Mohr's circle shown in Fig. 9.38. Note that all the quantities are assumed positive in Fig. 9.38 but the principles remain the same if any of the stresses are negative.

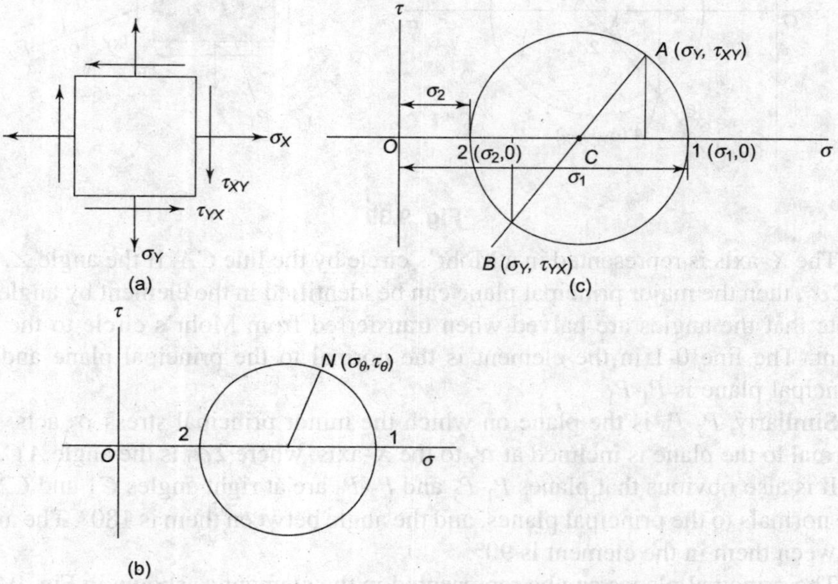

Fig. 9.38

The Mohr's circle cuts the σ-axis at two points, 1 and 2. 1 is the point at the extreme right. We know that any point on the circle has coordinates $(\sigma_\theta, \tau_\theta)$, where σ_θ is the normal stress and τ_θ is the shear stress. Studying the Mohr's circle of Fig. 9.38(b), the following two points are obvious.

(i) The two points at the ends of a horizontal diameter (on the σ-axis) have the maximum and minimum normal stress. The stress corresponding to point 1, σ_1, is the maximum normal stress and that corresponding to point 2, σ_2, is the minimum normal stress [Fig. 9.38(c)].

(ii) The planes on which the stresses σ_1 and σ_2 act have no shear stress.

Point 1 has the maximum σ coordinate and point 2 the minimum σ-coordinate. Since points 1 and 2 lie on the σ-axis, their τ-coordinates (shear stress) are zero. Thus point 1 has coordinates $(\sigma_1, 0)$ and point 2 $(\sigma_2, 0)$. These two normal stresses, σ_1 and σ_2, are called the principal stresses. σ_1, corresponding to the extreme right point, is the algebraically larger stress, and is called the major principal stress. σ_2, corresponding to the left extremity, is the algebraically smaller stress, and is called the minor principal stress.

The planes on which these stresses act are called principal planes. They can be easily located from Mohr's circle, as shown in Fig. 9.39.

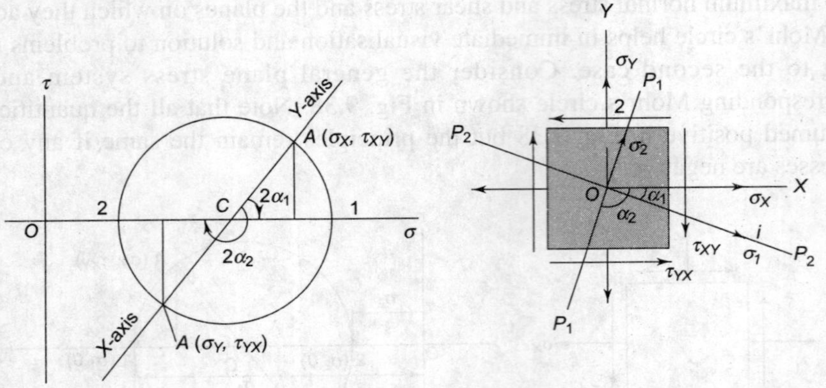

Fig. 9.39

The X-axis is represented in a Mohr's circle by the line CA. If the angle $\angle AC1$ is $2\alpha_1$, then the major principal plane can be identified in the element by angle α_1. Note that the angles are halved when transferred from Mohr's circle to the element. The line 0-1 in the element is the normal to the principal plane and the principal plane is P_1-P_1.

Similarly, P_2-P_2 is the plane on which the minor principal stress σ_2 acts. The normal to the plane is inclined at α_2 to the X-axis, where $2\alpha_2$ is the angle $AC2$.

It is also obvious that planes P_1-P_1 and P_2-P_2 are at right angles $C1$ and $C2$ are the normals to the principal planes, and the angle between them is 180°. The angle between them in the element is 90°.

The principal planes can be represented in the element as shown in Fig. 9.39.

What we have discussed may be summarized as follows.

- Considering all the planes passing through a point in a stressed material, there is one plane that has the maximum normal stress and one that has the minimum normal stress. These planes are called principal planes.
- The normals to these planes indicate the principal directions.
- The two principal planes are at right angles.
- The normal stresses acting on the principal planes are called principal stresses. The algebraically larger stress, σ_1, is called the major principal stress and σ_2, the minor principal stress.
- Shear stress is zero on the principal planes.

Example 9.16 Mohr's circle: principal stresses and plane

The stresses at a point in a stressed body are as shown in Fig. 9.40(a). Find the principal stresses and planes.

Solution Mohr's circle is drawn as shown in Fig. 9.40(b). A (100, 20) and B (60, – 20) are the stresses on the X- and Y-planes respectively. CA is the X-axis and CB the Y-axis in Mohr's circle.

$C1$ is the principal plane on which stress σ_1 is acting. $\angle AC1$ can be measured, and is found to be – 45° (clockwise).

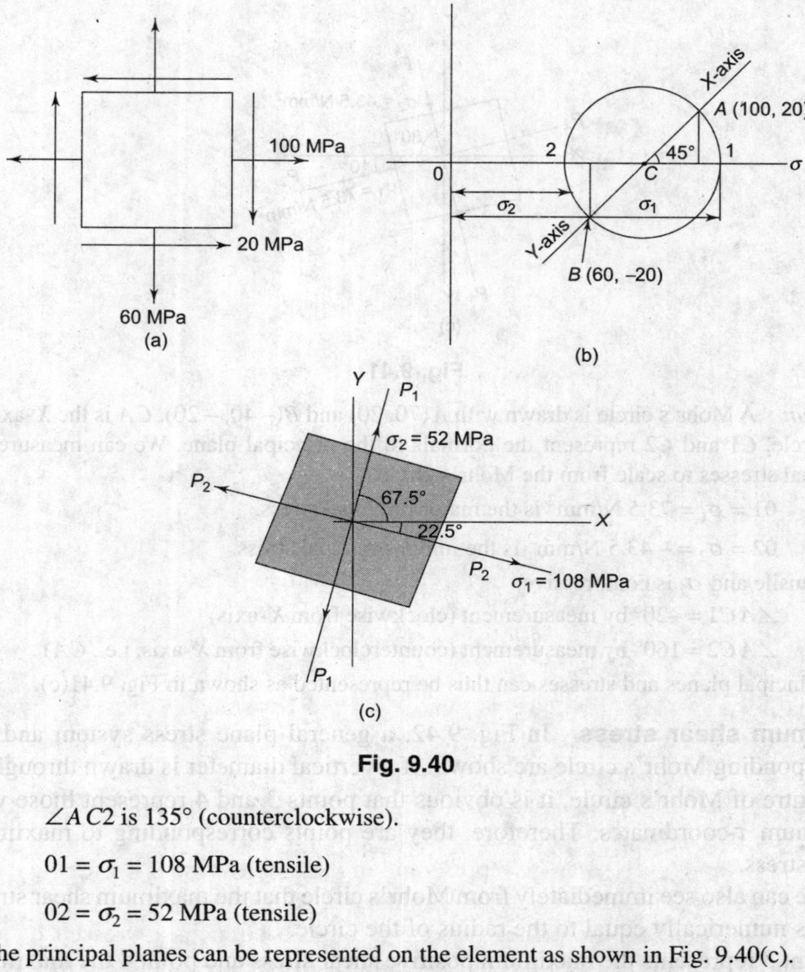

Fig. 9.40

$\angle AC2$ is 135° (counterclockwise).

$01 = \sigma_1 = 108$ MPa (tensile)

$02 = \sigma_2 = 52$ MPa (tensile)

The principal planes can be represented on the element as shown in Fig. 9.40(c). ■

Example 9.17 Mohr's circle: principal stresses and plane

Determine the principal planes and stresses if the stress system on an element is as shown in Fig. 9.41(a).

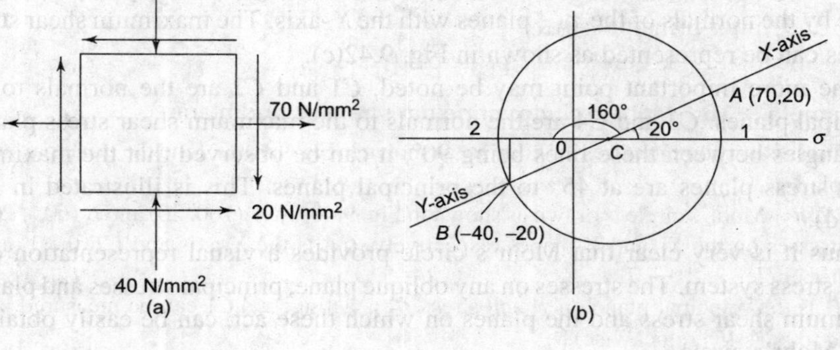

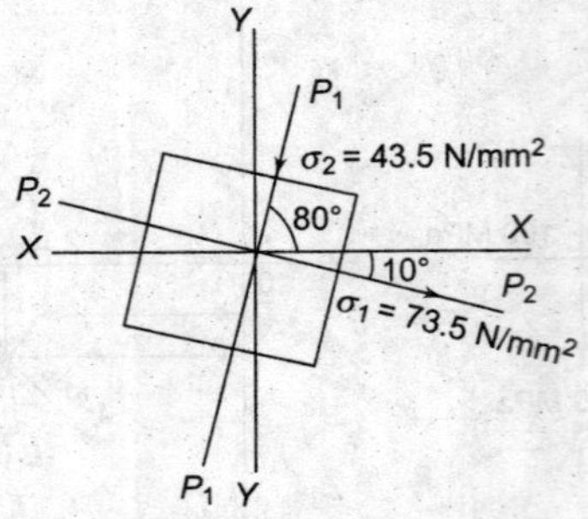

(c)

Fig. 9.41

Solution A Mohr's circle is drawn with $A(70, 20)$ and $B(-40, -20)$. CA is the X-axis in this circle. $C1$ and $C2$ represent the normals to the principal plane. We can measure the principal stresses to scale from the Mohr's circle.

$$01 = \sigma_1 = 73.5 \text{ N/mm}^2 \text{ is the major principal stress.}$$

$$02 = \sigma_2 = -43.5 \text{ N/mm}^2 \text{ is the minor principal stress.}$$

σ_1 is tensile and σ_2 is compressive.

$\angle AC1 = -20°$ by measurement (clockwise from X-axis)

$\angle AC2 = 160°$ by measurement (counterclockwise from X-axis, i.e., CA)

The principal planes and stresses can thus be represented as shown in Fig. 9.41(c). ■

Maximum shear stress In Fig. 9.42, a general plane stress system and the corresponding Mohr's circle are shown. If a vertical diameter is drawn through C, the centre of Mohr's circle, it is obvious that points 3 and 4 represent those with maximum τ-coordinates. Therefore, they are points corresponding to maximum shear stress.

One can also see immediately from Mohr's circle that the maximum shear stress, τ_{max}, is numerically equal to the radius of the circle.

Point 3 represents the maximum positive shear stress and point 4 the maximum negative shear stress. $C3$ and $C4$ represent the radial lines, which are the normals to the planes of maximum shear stress. Being collinear in a Mohr's circle, $C3$ and $C4$ are at right angles to each other in the element. This is as it should be because one shear stress is the complementary shear for the other.

The angles $AC3$ and $AC4$ are equal to $2\beta_1$ and $2\beta_2$. β_1 and β_2 are the angles made by the normals of the τ_{max} planes with the X-axis. The maximum shear stress planes can be represented as shown in Fig. 9.42(c).

One more important point may be noted. $C1$ and $C2$ are the normals to the principal planes. $C3$ and $C4$ are the normals to the maximum shear stress planes. The angles between these lines being 90°, it can be observed that the maximum shear stress planes are at 45° to the principal planes. This is illustrated in Fig. 9.42(d).

Thus it is very clear that Mohr's circle provides a visual representation of a plane stress system. The stresses on any oblique plane, principal stresses and planes, maximum shear stress and the planes on which these act, can be easily obtained from Mohr's circle.

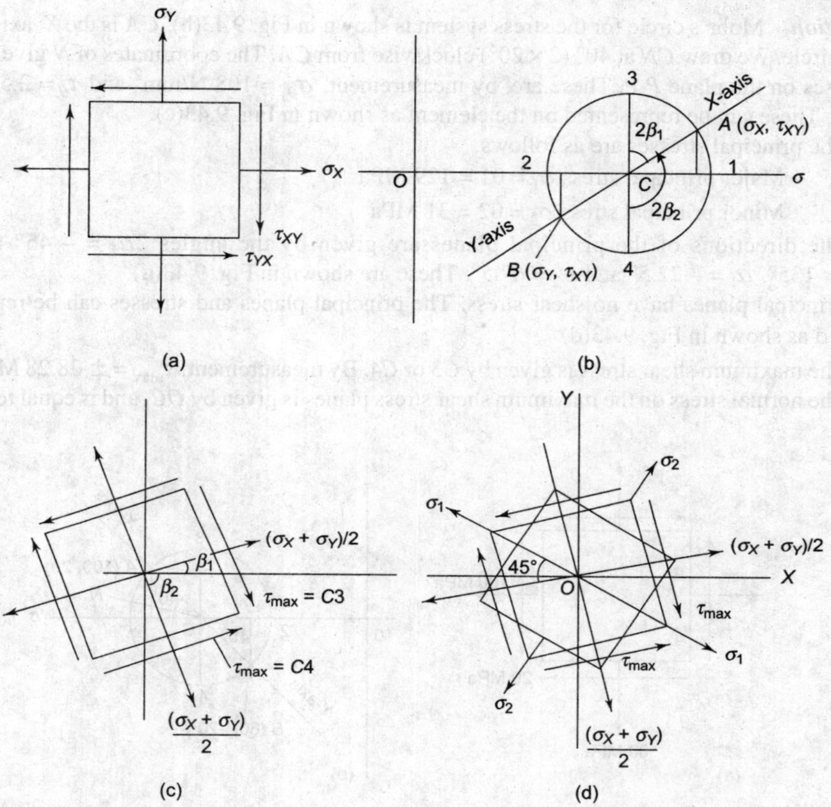

Fig. 9.42

Unlike the principal planes which have no shear stress, maximum shear stress planes are not without normal stress. From Fig. 9.42(b), it is clear that the maximum shear stress planes have a normal stress equal to OC, which as we have shown earlier is equal to $(\sigma_X + \sigma_Y)/2$. What we have discussed may be summarized as follows.

- There are two planes at right angles, one of which has the maximum positive shear stress and the other an equal negative shear stress, equal to the radius of Mohr's circle to scale.
- The two planes are at right angles.
- The normal stress on these planes is equal to the σ-coordinate of the centre of Mohr's circle.
- The maximum shear stress planes make angles of 45° with the principal planes.

Also note that the maximum shear stress is equal to $(\sigma_1 - \sigma_2)/2$ (equal to the radius), i.e., half the difference of the principal stresses.

Example 9.18 Mohr's circle: principal stresses and maximum shear stress

The stresses on orthogonal planes at a point in a stressed body are as shown in Fig. 9.43(a). Determine the stresses on a plane whose normal is inclined at –20° to the X-axis. Find the principal planes and stresses. Also find the planes of maximum shear stress and the stresses on these planes.

Solution Mohr's circle for the stress system is shown in Fig. 9.43(b). CA is the X-axis in this circle. We draw CN at 40° (2 × 20°) clockwise from CA. The coordinates of N give the stresses on the plane P-P. These are, by measurement, $\sigma_\theta = 108$ N/mm^2 and $\tau_\theta = 2.5$ N/mm^2. These can be represented on the element as shown in Fig. 9.43(c).

The principal stresses are as follows.

Major principal stress $\sigma_1 = 01 = 129$ MPa

Minor principal stress $\sigma_2 = 02 = 31$ MPa

The directions of the principal planes are given by the angles $2\alpha_1 = -45°$ and $2\alpha_2 = 135°$. $\alpha_1 = -22.5°$ and $\alpha_2 = 67.5°$. These are shown in Fig. 9.43(d).

Principal planes have no shear stress. The principal planes and stresses can be represented as shown in Fig. 9.43(d).

The maximum shear stress is given by $C3$ or $C4$. By measurement, $\tau_{max} = \pm 28.28$ MPa.

The normal stress on the maximum shear stress planes is given by OC, and is equal to 80 MPa.

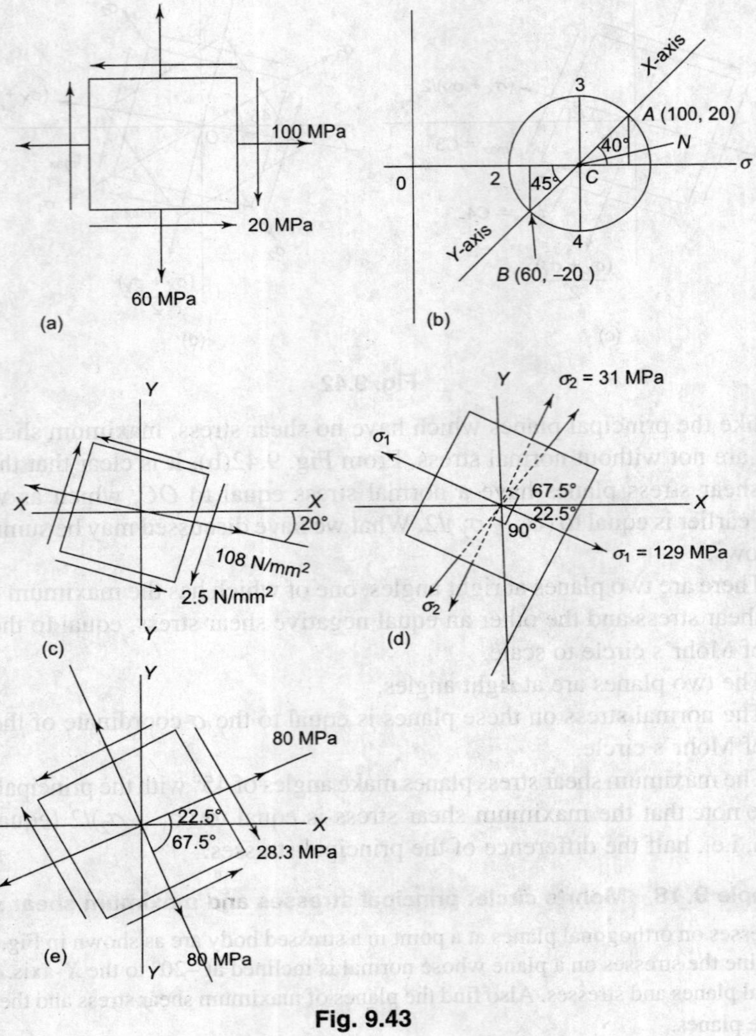

Fig. 9.43

The directions of the normals of maximum shear stress planes may be worked out from $2\beta_1 = 45°$ and $2\beta_2 = -135°$.

The maximum shear stress planes and the stresses on these planes can be represented as shown in Fig. 9.43(e).

9.8.2 More Observations from Mohr's Circle

Referring to Fig. 9.44, we can easily make the following observations.

(i) The normal stress on the maximum shear stress planes is equal to $(\sigma_1 + \sigma_2)/2$.

$$OC = \frac{\sigma_X + \sigma_Y}{2} = \sigma_2 + \frac{\sigma_1 - \sigma_2}{2} = \frac{\sigma_1 + \sigma_2}{2}$$

(ii) The resultant stress on any plane is equal to ON. ON makes an angle ϕ with the σ-axis such that $\tan\phi = \tau_\theta/\sigma_\theta$. $ON = \sqrt{\tau_\theta^2 + \sigma_\theta^2}$. The maximum and minimum values of the resultant stresses are σ_1 and σ_2.

(iii) The maximum obliquity of the resultant stress is given by $\sin(\phi_{max}) = (\sigma_1 - \sigma_2)/(\sigma_1 + \sigma_2)$. The maximum value of ϕ will be when ON is tangential to Mohr's circle. Then, $CN = (\sigma_1 - \sigma_2)/2$, $OC = (\sigma_1 + \sigma_2)/2$, and $\angle CNO = 90°$ [Fig. 9.44(b)].

(iv) The sum of the normal stresses on mutually perpendicular planes is always a constant equal to $(\sigma_X + \sigma_Y)$ or $(\sigma_1 + \sigma_2)$. Referring to the Mohr's circle shown, two mutually perpendicular planes have their normals along a diameter, like CP_1 and CP_2. P_1q_1 and P_2q_2 are perpendiculars to the σ-axis. It is obvious that $Oq_1 + Oq_2$ is a constant because Cq_1 and Cq_2 are always equals [Fig. 9.44(c)].

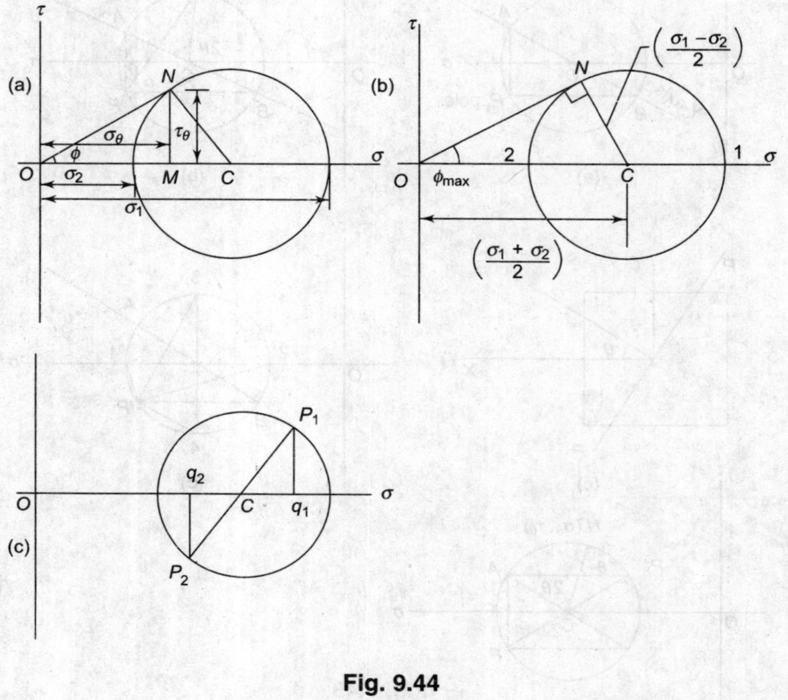

Fig. 9.44

9.8.3 Origin of Planes

Referring to the Mohr's circle shown in Fig. 9.45, *CA* represents the *X*-axis and *CB* the *Y*-axis. We draw a line parallel to the *X*-plane through *A* and another parallel to the *Y*-plane through *B*. Note that these lines are parallel to the planes on which the stresses represented by *A* and *B* act. Let these two lines intersect at *P*. Since *A B* is a diameter and the angle at *P* is 90°, *P* should lie on the circle. The point *P* is known as the pole or the origin of planes. The concept of the pole has many important advantages.

(i) If a line is drawn through the pole *P* parallel to any oblique plane on the element, it intersects the circle at a point whose coordinates are the normal and shear stresses on that plane.

This is true for *X*- and *Y*-planes, of course, because the pole *P* has been obtained by such a construction. For any oblique plane inclined at an angle θ, it is easy to see that $\angle APN = \theta$, because *PN*, the line drawn parallel to the plane, will make an angle θ with *PA* (the vertical line). From the property of a circle that the angle subtended at the centre by an arc is equal to twice the angle sustained by the same arc on the circumference, $\angle ACN = 2\theta$. We find the stresses on an oblique plane by such a construction, measuring 2θ at *C*. Therefore, by drawing a line parallel to the plane *P-P* [Fig. 9.45(c)] the line intersects the circle at *N* whose coordinates are the normal and shear stresses on the plane *P-P*.

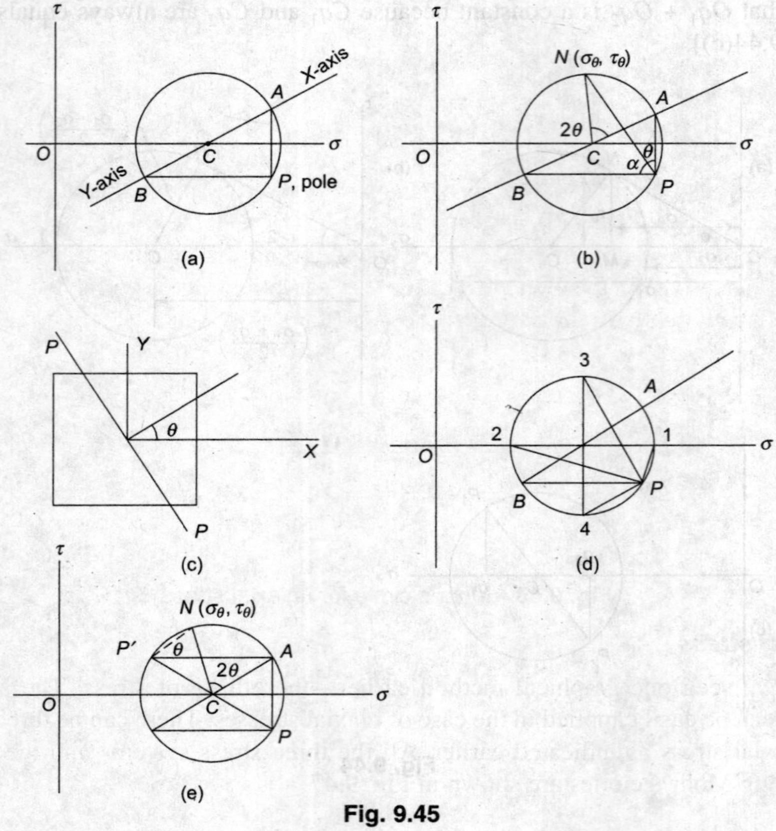

Fig. 9.45

The pole P thus provides a convenient point through which planes can be drawn from the element, or planes transferred from the circle to the element.

As shown in Fig. 9.45(d), $P1$ and $P2$ represent the principal planes, and $P3$ and $P4$ represent the maximum shear stress planes. $\angle 1P2 = 90° = \angle 3P4$. The two principal planes and the two maximum shear stress planes are at right angles.

(ii) The point P' diametrically opposite to P can also be similarly used. AP' acts as a modified X-axis.

In Fig. 9.45(e), the angle θ of an oblique plane P-P, made by its normal with the X-axis, can be directly measured. If $\angle AP'N = \theta$, then $\angle ACN = 2\theta$ from the property of a circle, and the coordinates of N give σ_θ and τ_θ. P' is the origin of the normals, which can be directly transferred through P' to give the stresses on any oblique plane.

9.8.4 Mohr's Circle for Uniaxial and Biaxial Stress Systems

We have seen the method to draw and interpret the Mohr's circle for a general plane stress system having normal stresses along orthogonal directions and a shear stress on the orthogonal planes. Uniaxial and biaxial stress systems, we discussed earlier, are specific cases of this when one or more of these stresses are zero.

Uniaxial stress

When σ_y and τ_{xy} are zero, we have only σ_x and this is a uniaxial stress system. The Mohr's circle for uniaxial stress is shown in Fig. 9.46. Point A represents the stresses on the X-plane and point O (origin) represents the stresses on the Y-plane. s-axis and X-axis coincide and Y-axis must be in the negative direction of σ-axis. Radius of the Mohr's circle is $\sigma_x/2$ and the centre of the circle is at $(\sigma_x/2, 0)$. The major principal stress σ_1 is σ_x and σ_2 is the maximum normal stress, and minor principal stress is zero. Maximum shear stress is given by the radius of the Mohr's circle and is equal to $\sigma_x/2$. The angle between OC and σ-axis being 90° in the Mohr's circle, the angle between X-axis and the normal to the maximum shear stress plane is 45° in the element. All these observations agree with what we have derived analytically in the earlier section.

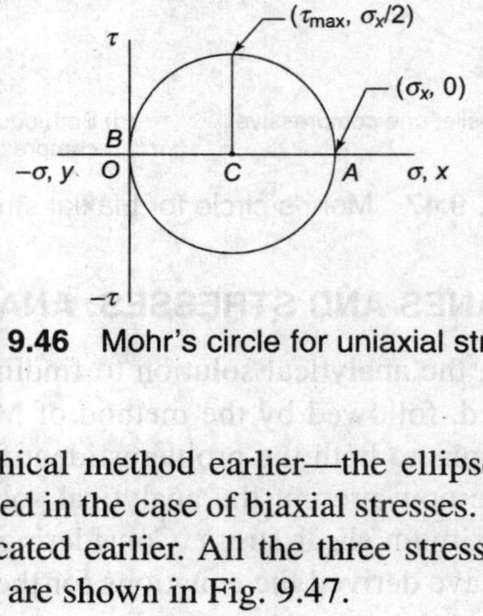

Fig. 9.46 Mohr's circle for uniaxial stress

Biaxial stress

We have seen one graphical method earlier—the ellipse of stress. The Mohr's circle can be easily applied in the case of biaxial stresses. There can be three cases of biaxial stress as indicated earlier. All the three stress systems and the corresponding Mohr's circles are shown in Fig. 9.47.

Figure 9.47(a) shows a case when the two stresses are not equal, both tensile and both compressive or one tensile and one compressive. As there is no shear stress, the X- and Y-axes coincide with the s-axis. Point A represents the stresses on the X-axis and point B represents the stresses on the Y-axis. The centre of the Mohr's circle is at the mid-point of AB and its radius $= AB/2$. σ_1 and σ_2 corresponding to points A and B represent the principal stresses. Maximum shear stress is equal to the radius of the Mohr's circle and is equal to CD or CE. The angles of the planes can be measured from the Mohr's circle. They become half in the element.

Figure 9.47(b) shows when the two stresses are equal, but one tensile and the other compressive. When the two stresses are equal and of opposite nature, the centre of the Mohr's circle coincides with the origin. Maximum shear stress is equal to the magnitude of the normal stress.

Figure 9.47(c) show the Mohr's circle for the case where the two stresses are equal and both are tensile or compressive. In this case, Mohr's circle reduces to a point. You should study these Mohr's circles and satisfy yourself that our earlier observations made analytically are true.

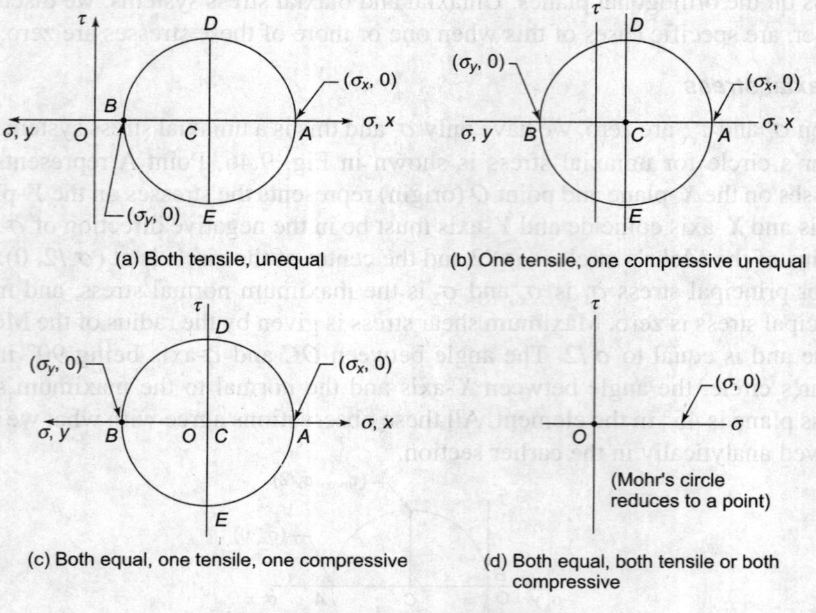

Fig. 9.47 Mohr's circle for biaxial stress

9.9 PRINCIPAL PLANES AND STRESSES: ANALYTICAL SOLUTION

In the foregoing articles, the analytical solution to finding the stresses on an oblique plane was presented, followed by the method of Mohr's circle, which was interpreted to give solutions to both the problems identified in Section 9.5 under plane stress analysis. We now present the analytical solution to Problem (ii) of principal planes and maximum shear stress. Considering the plane stress system shown in Fig. 9.48, we have derived the equations for the stresses on any oblique plane as

$$\sigma_\theta = \frac{\sigma_X + \sigma_Y}{2} + \frac{\sigma_X - \sigma_Y}{2}\cos 2\theta - \tau_{XY}\sin 2\theta$$

$$\tau_\theta = \frac{\sigma_X - \sigma_Y}{2}\sin 2\theta + \tau_{XY}\cos 2\theta$$

These two equations show the variation of normal stress and shear stress with angle θ, σ_X, σ_Y, and τ_{XY} remaining constant. To find the maximum value of σ_θ, we set $d\sigma_\theta/d\theta = 0$ to find the value of θ for which σ_θ is a maximum or minimum.

$$\frac{d\sigma_\theta}{d\theta} = -\frac{\sigma_X - \sigma_Y}{2} \times 2\sin 2\theta - \tau_{XY} \times 2\cos 2\theta = 0$$

from which

$$\tan 2\theta = \frac{-2\tau_{XY}}{(\sigma_X - \sigma_Y)}$$

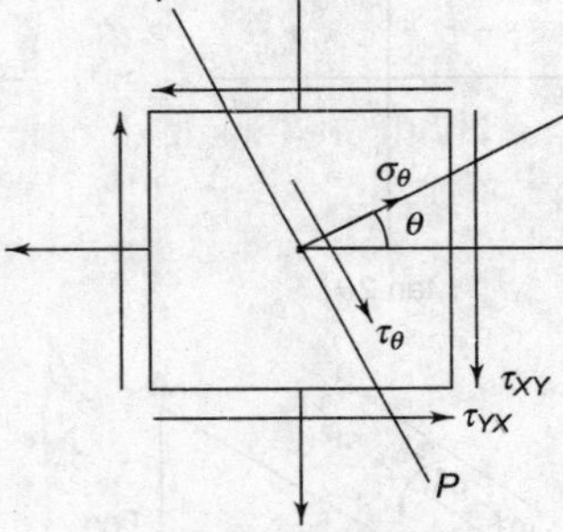

Fig. 9.48

You will notice $d\sigma_\theta/d\theta = \tau_\theta$, which means that $\tau_\theta = 0$ on the planes of maximum normal stress. Principal planes are thus planes with no shear stress.

From the equation for tan 2θ, two values of 2θ, differing by 180°, can be found. That is,

$$2\theta_1 = 2\alpha, \quad \theta_1 = \alpha$$

$$2\theta_2 = 2\alpha + \pi, \quad \theta_2 = \alpha + \frac{\pi}{2}$$

Taking the stress system shown in Fig. 9.49,

$$\tan 2\theta = \frac{-2\tau_{XY}}{(\sigma_X - \sigma_Y)} = \frac{-2 \times 20}{(100 - 40)} = -0.6667$$

$$2\theta = 146.31° \text{ or } 326.3°, \quad \theta = 73.15° \text{ or } 163.15°$$

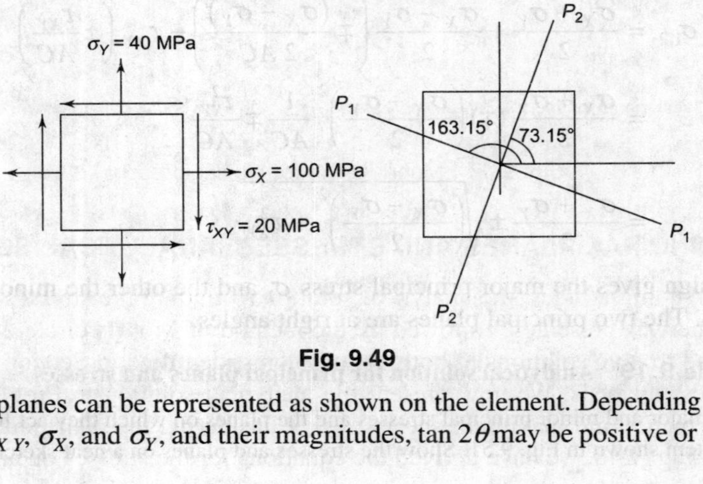

Fig. 9.49

These planes can be represented as shown on the element. Depending upon the signs of τ_{XY}, σ_X, and σ_Y, and their magnitudes, tan 2θ may be positive or negative.

Correspondingly, sin 2θ and cos 2θ may be positive or negative. This is shown in Fig. 9.50. From the right-angled triangle shown,

$$\sin 2\theta = \pm \frac{\tau_{XY}}{AC}$$

and

$$\cos 2\theta = \pm \frac{\sigma_X - \sigma_Y}{2AC}$$

where $AC = \sqrt{\left(\dfrac{\sigma_X - \sigma_Y}{2}\right)^2 + \tau_{XY}^2}$

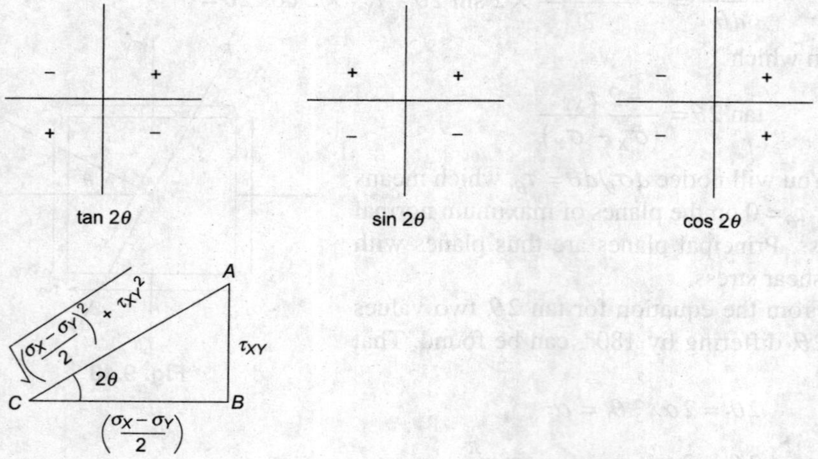

Fig. 9.50

These values can be substituted in the expression for σ_θ to get the maximum or minimum values. Assuming all the quantities— τ_{XY}, σ_X, and σ_Y —to be positive and tan 2θ to be negative, sin 2θ and cos 2θ will have opposite signs. Then, the principal stresses will be

$$\sigma_{1,2} = \frac{\sigma_X + \sigma_Y}{2} + \frac{\sigma_X - \sigma_Y}{2}\left(\mp\frac{(\sigma_X - \sigma_Y)}{2\,AC}\right) - \tau_{XY}\left(\pm\frac{\tau_{XY}}{AC}\right)$$

$$= \frac{\sigma_X + \sigma_Y}{2}(\mp)\left(\frac{\sigma_X - \sigma_Y}{2}\right)^2\frac{1}{AC} \mp \frac{\tau_{XY}^2}{AC}$$

$$= \frac{\sigma_X + \sigma_Y}{2} \pm \sqrt{\left(\frac{\sigma_X - \sigma_Y}{2}\right)^2 + \tau_{XY}^2}$$

One sign gives the major principal stress σ_1 and the other the minor principal stress σ_2. The two principal planes are at right angles.

Example 9.19 Analytical solution for principal planes and stresses

Find the major and minor principal stresses and the planes on which they act for the plane stress system shown in Fig. 9.51. Show the stresses and planes on a neat sketch.

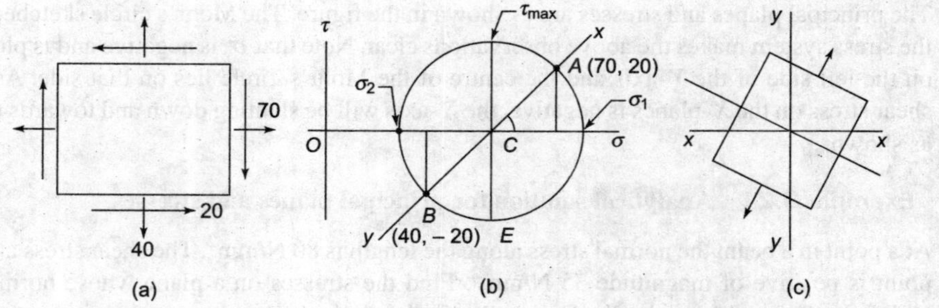

Fig. 9.51

Solution The principal stresses are given by

$$\sigma_{1,2} = (\sigma_x + \sigma_y)/2 \pm \sqrt{[(\sigma_x - \sigma_y)^2/4 + \tau_{xy}^2]}$$

In this case, $\sigma_x = +70$ N/mm^2 $\sigma_y = 40$ N/mm^2, and $\tau_{xy} = 20$ N/mm^2

$$\sigma_{1,2} = (70 + 40)/2 \pm \sqrt{[(70 - 40))^2]/4 + 20^2]} = +\ 80 \text{ N/mm}^2,\ 30 \text{ N/mm}^2$$

Major principal stress is the algebraically larger principal stress + 80 N/mm^2

Minor principal stress is the algebraically smaller principal stress 30 N/mm^2

The planes on which these stresses act is given by

$$\tan 2\alpha = -2\tau_{xy}/(\sigma_x - \sigma_y) = -\ 2 \times 20/[(70 - 40)] = -1.3333 = -53°, -233°.$$

The negative sign for the angles show that the normal to the planes are directed at clockwise angles from the X-axis. The planes and stresses are as shown in Fig. 9.51.

The Mohr's circle shown (not drawn to scale) makes these observations very clear. ■

Example 9.20 Analytical solution for principal planes and stresses

Find the maximum and minimum normal stresses for the plane stress system shown in Fig. 9.52. Show the stresses and planes on a neat sketch.

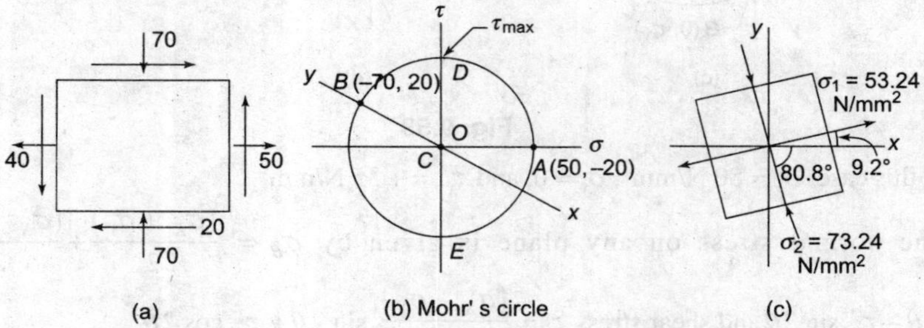

Fig. 9.52

Solution In this case, $\sigma_x = +\ 50$ N/mm^2, $\sigma_y = -70$ N/mm^2, and $\tau_{xy} = -20$ N/mm^2

The principal stresses are given by

$$\sigma_{1,2} = (\sigma_x + \sigma_y)/2 \pm \sqrt{[(\sigma_x - \sigma_y)^2/4 + \tau_{xy}^2]}$$

$$= (50 - 70)/2 \pm \sqrt{[(50 - (-70))^2/4 + (-20)^2]}$$

$$= +\ 53.24 \text{ N/mm}^2,\ -73.2 \text{ N/mm}^2$$

The planes on which these stresses act make angles with X-axis given by

$$\tan 2\alpha = -2\tau_{xy}/(\sigma_x - \sigma_y) = (-2 \times (-20))/(50 - (-70)) = 18.4°,\ 198.4°$$

The principal planes and stresses are as shown in the figure. The Mohr's circle sketched for the stress system makes the above observations clear. Note that σ_y is negative and is plotted on the left side of the Y-axis and the centre of the Mohr's circle lies on that side. As the shear stress on the X-planes is negative, the X-axis will be slanting down and towards right as shown.

Example 9.21 Analytical solution for principal planes and stresses

At a point in a beam the normal stress along the length is 80 N/mm^2. The shear stress at that point is positive of magnitude 35 N/mm^2. Find the stresses on a plane whose normal is inclined at 30° to the longitudinal axis. Also find the principal stresses and planes on which they act.

Solution The stress system is shown in Fig. 9.53.

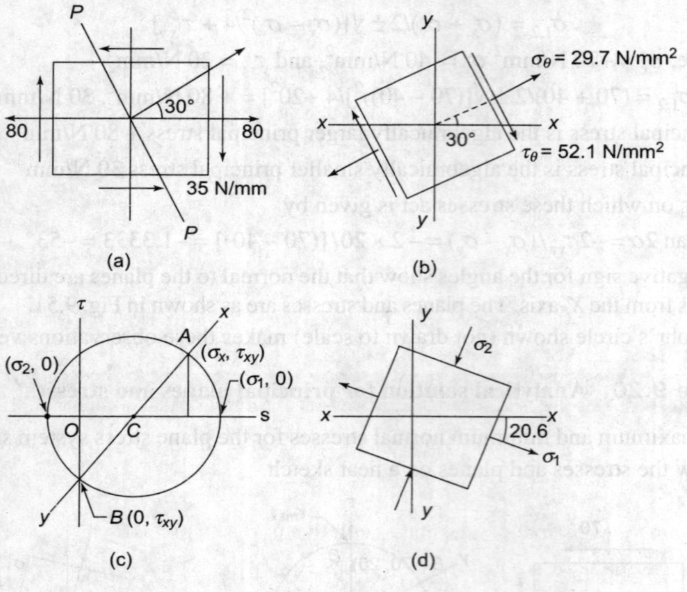

Fig. 9.53

In this case, σ_x = 80 N/mm^2, σ_y = 0, and τ_{xy} = + 35 N/mm^2.

The normal stress on any plane is given by $\sigma_\theta = \dfrac{(\sigma_X + \sigma_Y)}{2} + \dfrac{(\sigma_X - \sigma_Y)}{2}$ $\cos 2\theta - \tau_{xy} \sin 2\theta$ and shear stress $\tau_\theta = \dfrac{(\sigma_X - \sigma_Y)}{2} \sin 2\theta + \tau_{xy} \cos 2\theta$

In our case θ = 30°. Therefore,

$$\text{Normal stress } \sigma_\theta = 80/2 + (80 \cos 60)/2 - 35 \sin 60 \ [\sigma_y = 0.]$$
$$= 40 + 20 - 30.3 = 29.7 \text{ N/mm}^2$$

$$\text{Shear stress } \tau_\theta = (80 \sin 60)/2 + 35 \cos 60 = 34.6 + 17.5 = 52.1 \text{ N/mm}^2.$$

The plane and stresses are shown in the figures.

The principal stresses can be obtained by

$$\sigma_{1,2} = (\sigma_x + \sigma_y)/2 \pm \sqrt{[(\sigma_x - \sigma_y)^2/4 + \tau^2_{xy}]} = 80/2 \pm \sqrt{[(80)^2/4 + 35^2]}$$
$$= 40 \pm 53.2 = 93.2 \text{ N/mm}^2, -13.2 \text{ N/mm}^2$$

The planes on which these stresses act is given by

$$\tan 2\alpha = -2\,\tau_{xy}/(\sigma_x - \sigma_y) = -2\times 35/80 = -41.2°,\ 138.8°$$

The principal planes and stresses will be as shown in the figure. The Mohr's circle stretched makes things easier to visualize. Note that as $\sigma_y = 0$, we plot (80, 35) and (0, –35) and join them to get the centre c of the Mohr's circle on the σ-axis.

Example 9.22 Analytical solution for principal planes and stresses

The stress system at a point is given by a normal stress of 120 N/mm^2 (compressive) along the X-axis, 80 N/mm^2 (tensile) along the Y-axis and a shear stress of –40 N/mm^2 on the X-planes. Find the principal stresses and the planes on which they act. Show the stresses and planes in a neat sketch.

Solution The stress system is shown in Fig. 9.54.

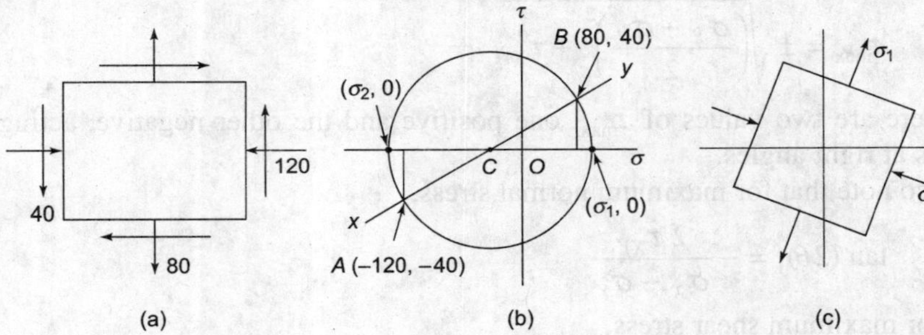

Fig. 9.54

The principal stresses are given by $\sigma_{1,2} = (\sigma_x + \sigma_y)/2 \pm \sqrt{[(\sigma_x - \sigma_y)^2/4 + \tau_{xy}^2]}$

$$\sigma_{1,2} = (-120 + 80)/2 \pm \sqrt{[(-120 - 80)^2/4 + (-40)^2]}$$
$$= +87.7 \text{ N/mm}^2,\ -107.7 \text{ N/mm}^2$$

The angle that the principal planes make with the X-axis is given by,

$$\tan 2\alpha = -2\tau_{xy}/(\sigma_x - \sigma_y) = -2(-40)/(-120 - 80)$$
$$= -0.4;\ 2\alpha = -21.8°,\ 158.2°$$

The principal planes and stresses are shown in the figure. The Mohr's circle for the stress system will be as shown in Fig. 9.54(b).

9.9.1 Maximum Shear Stress

To find the maximum shear stress, we set $d\tau_\theta/d\theta = 0$.

$$\tau_\theta = \frac{\sigma_X - \sigma_Y}{2}\cos 2\theta + \tau_{XY}\sin 2\theta$$

$$\frac{d\tau_\theta}{d\theta} = -\frac{\sigma_X - \sigma_Y}{2}\sin 2\theta + \tau_{XY}\cos 2\theta = 0$$

$$\tan 2\theta = \frac{\sigma_X - \sigma_Y}{2\,\tau_{XY}}$$

This will give two values of 2θ, which differ by 180° and two values of θ differing by 90°. From the triangle shown in Fig. 9.55, we have

$$\sin 2\theta = \pm \frac{\sigma_X - \sigma_Y}{2AC}$$

and $$\cos 2\theta = \pm \frac{\tau_{XY}}{AC}$$

where $$AC = \sqrt{\left(\frac{\sigma_X - \sigma_Y}{2}\right)^2 + \tau_{XY}^2}$$

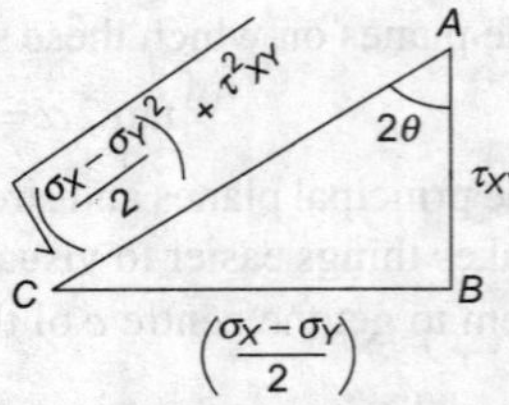

Fig. 9.55

Substituting these values in the expression for τ_θ, we get

$$\tau_\theta = \left(\frac{\sigma_X - \sigma_Y}{2}\right)\left(\frac{\sigma_X - \sigma_Y}{2\,AC}\right) + \frac{\tau_{XY}\tau_{XY}}{AC} = \frac{1}{AC}\left[\left(\frac{\sigma_X - \sigma_Y}{2}\right)^2 + \tau_{XY}^2\right]$$

$$\tau_{\max} = \pm\sqrt{\left(\frac{\sigma_X - \sigma_Y}{2}\right)^2 + \tau_{XY}^2}$$

There are two values of $\tau_{\max}$, one positive and the other negative, acting on planes at right angles.

Also note that for maximum normal stress,

$$\tan (2\theta)_1 = -\frac{2\,\tau_{XY}}{\sigma_X - \sigma_Y}$$

and for maximum shear stress,

$$\tan (2\theta)_2 = \frac{\sigma_X - \sigma_Y}{2\,\tau_{XY}}$$

$\tan (2\theta)_1 \tan (2\theta)_2 = -1$, showing that these two 2θ values differ by 90°, or are at right angles. Thus the maximum shear stress planes are at 45° to the principal planes.

Example 9.23 Principal planes and stress: analytical solution

In the plane stress situation at a point shown in Fig. 9.56, find (i) the principal planes and stresses, and (ii) the maximum shear stress planes and the maximum shear stress.

Solution In the case shown in Fig. 9.56(a), $\sigma_X = 200$ N/mm², $\sigma_Y = -80$ N/mm², and $\tau_{XY} = 60$ N/mm².

The principal stresses are given by

$$\sigma_{1,2} = \frac{\sigma_X + \sigma_Y}{2} \pm \sqrt{\left(\frac{\sigma_X - \sigma_Y}{2}\right)^2 + \tau_{XY}^2}$$

$$= \frac{200 - 80}{2} \pm \sqrt{\left[\frac{200 - (-80)}{2}\right]^2 + 60^2} = 60 \pm 10\sqrt{232}$$

The major principal stress,

$\sigma_1 = 212.3$ N/mm²

The minor principal stress,

$\sigma_2 = -92.3$ N/mm²

The inclination of the principal planes may be worked out as follows.

$$\tan 2\theta = -\frac{2 \times 60}{(200 - (-80))} = -\frac{3}{7}$$

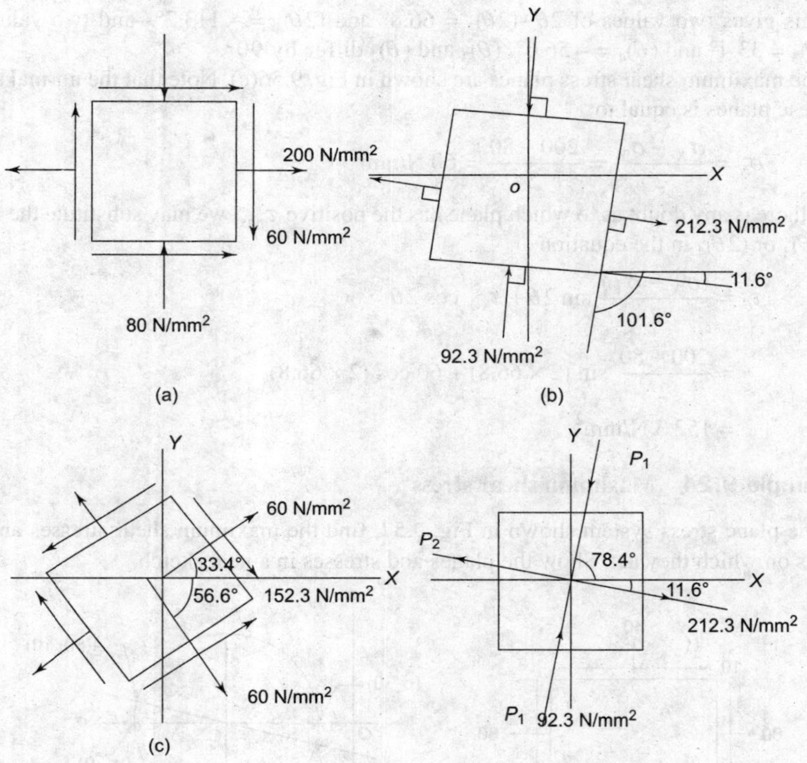

Fig. 9.56

This yields two values of 2θ—$(2\theta_1) = -23.2°$ and $(2\theta)_2 = -203.2°$—and two values of θ—$(\theta_1) = -11.6°$ and $(\theta)_2 = -101.6°$. The two values of θ differ by 90°. The principal planes can be identified by the normals to these planes inclined at these angles, as shown in Fig. 9.56(b). To identify which is the major principal plane and which is the minor principal plane, one value of θ may be substituted in the equation

$$\sigma_\theta = \frac{\sigma_X + \sigma_Y}{2} + \frac{\sigma_X - \sigma_Y}{2}\cos 2\theta - \tau_{XY}\sin 2\theta$$

$$= \frac{200 - 80}{2} + \frac{200 - (-80)}{2}\cos[2(-11.6)] - 60\sin[2(-11.6)]$$

$$= 212.3 \text{ N/mm}^2$$

The principal planes and stresses are shown in Fig. 9.56(b).

Maximum shear stress The maximum shear stress is given by

$$\tau_{max} = \pm\sqrt{\left(\frac{\sigma_X - \sigma_Y}{2}\right)^2 + \tau_{XY}^2} = \pm\sqrt{\left(\frac{200 - (-80)}{2}\right)^2 + 60^2}$$

$$= \pm\ 152.3 \text{ N/mm}^2$$

The inclination of the maximum shear stress planes is given may be obtained from

$$\tan 2\theta = \frac{\sigma_X - \sigma_Y}{2\,\tau_{XY}} = \frac{200 - (-80)}{2 \times 60} = \frac{7}{3}$$

This gives two values of 2θ—$(2\theta)_3 = 66.8°$ and $(2\theta)_4 = -113.2°$—and two values of θ—$(\theta)_3 = 33.4°$ and $(\theta)_4 = -56.6°$. $(\theta)_3$ and $(\theta)_4$ differ by 90°.

The maximum shear stress planes are shown in Fig. 9.56(c). Note that the normal stress on these planes is equal to

$$\sigma_N = \frac{\sigma_X - \sigma_Y}{2} = \frac{200 - 80}{2} = 60 \text{ N/mm}^2$$

If there is any doubt as to which plane has the positive τ_{max}, we may substitute the value of $(2\theta)_3$ or $(2\theta)_4$ in the equation

$$\tau_\theta = \frac{(\sigma_X - \sigma_Y)}{2} \sin 2\theta + \tau_{XY} \cos 2\theta$$

$$= \frac{200 + 80}{2} \sin (2 \times 66.8) + 60 \cos (2 \times 66.8)$$

$$= 152.3 \text{ N/mm}^2$$

Example 9.24 Maximum shear stress

For the plane stress system shown in Fig. 9.57, find the maximum shear stresses and the planes on which they act. Show the planes and stresses in a neat sketch.

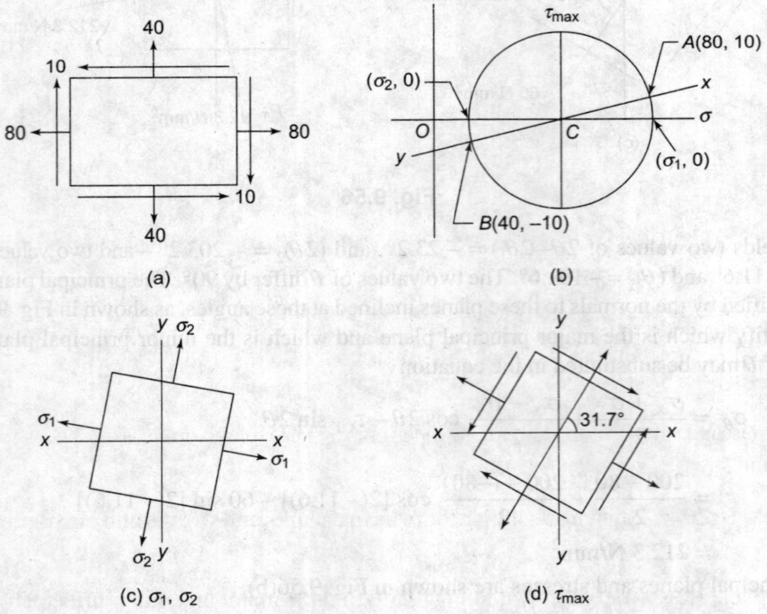

Fig. 9.57

Solution Maximum shear stress is given by $\tau_{max} = \pm \sqrt{[(\sigma_x - \sigma_y)^2/4 + \tau_{xy}^2]}$

In our case, $\sigma_x = +80$ units, $\sigma_y = +40$ units, and $\tau_{xy} = 10$ units

Therefore, Maximum shear stress $\tau_{max} = \pm \sqrt{[(80 - 40)^2/4 + 10^2} = \pm 22.4 \text{ N/mm}^2$.

The angles made by the maximum shear stress planes is given by

$$\tan 2\beta = (\sigma_x - \sigma_y)/ 2\tau_{xy} = (80 - 40)/(2 \times 10) = 2;\ 2\beta = 63.4° \text{ and } 243.4°$$

The maximum shear stress planes and the stresses are shown in Fig. 9.57. The Mohr's circle sketched clearly shows the planes and stresses.

Example 9.25 Maximum shear stress

For the plane stress system shown in Fig. 9.58, find the maximum shear stresses and the planes on which they act. Show the planes and stresses in a neat sketch.

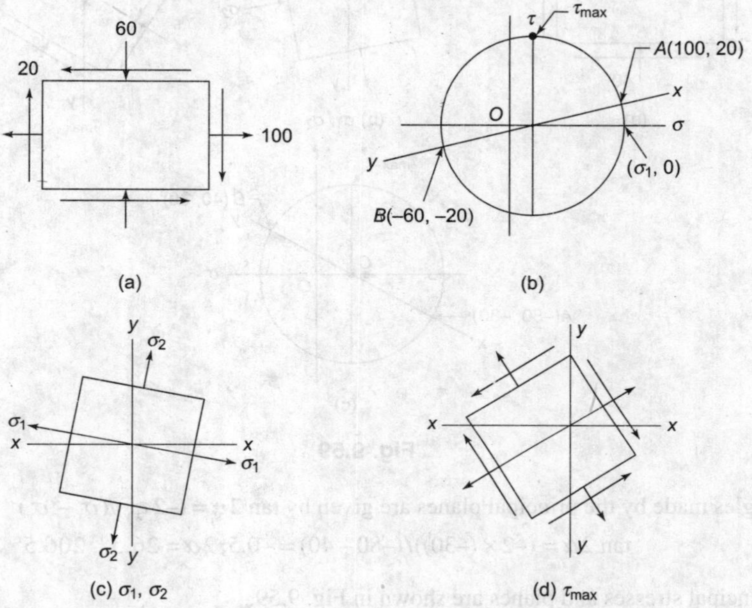

Fig. 9.58

Solution The maximum shear stress $\tau_{max} = \pm\sqrt{[(\sigma_x - \sigma_y)^2/4 + \tau_{xy}^2]}$

In our case, $\sigma_x = 100$ N/mm^2; $\sigma_y = -60$ N/mm^2 and $\tau_{xy} = +20$ N/mm^2. Therefore

Maximum shear stress $\tau_{max} = \pm\sqrt{[(100 - (-60))^2/4 + 20^2]} = \pm 82.5$ N/mm^2

The angle made by the maximum shear stress planes are given by

$$\tan 2\beta = (\sigma_x - \sigma_y)/\, 2t_{xy};\ \tan 2\beta = (100 - (-60))/(2 \times 20)$$

$$2\beta = 76° \text{ and } 256°$$

The planes and stresses are shown in the figure. The Mohr's circle sketched shows the planes and stresses very clearly. ■

Example 9.26 Analytical solution for principal planes and stresses and maximum shear stress

For the plane stress system shown in Fig. 9.59, find the major and minor principal stresses and the planes on which they act. Also find the maximum shear stresses and the planes on which they act. Show the planes and stresses in a neat sketch.

Solution For the given plane stress system, $\sigma_x = -80$ N/mm^2; $\sigma_y = +40$ N/mm^2, and $\tau_{xy} = -30$ N/mm^2.

Principal stresses: The major and minor principal stresses are given by

$$\sigma_{1,2} = (\sigma_x + \sigma_y)/2 \pm \sqrt{[(\sigma_x - \sigma_y)^2/4 + \tau_{xy}^2]}$$

$$= (-80 + 40)/2 \pm \sqrt{[(-80 - 40)^2/4 + (-30)^2]}$$

$$= 47.1 \text{ N/mm}^2, -87.1 \text{ N/mm}^2$$

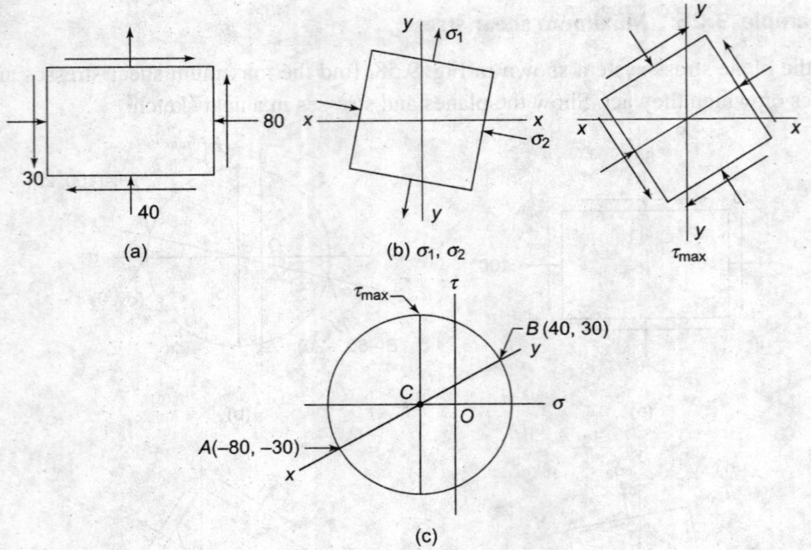

Fig. 9.59

The angles made by the principal planes are given by $\tan 2\alpha = (-2\tau_{xy})/(\sigma_x - \sigma_y)$

$$\tan 2\alpha = (-2 \times (-30))/(-80 - 40) = -0.5; 2\alpha = 26.5°, 206.5°$$

The principal stresses and planes are shown in Fig. 9.59.

$$\text{Maximum shear stress } \tau_{max} = \pm \sqrt{[(\sigma_x - \sigma_y)^2/4 + \tau_{xy}^2]}$$

$$= \pm \sqrt{[-80 - 40)^2/4 + (-30)^2]} = \pm 67.1 \text{ N/mm}^2$$

The angles made by the maximum shear stress planes are

$$\tan 2\beta = (\sigma_x - \sigma_y)/ 2\tau_{xy} = (-80 - 40)/ (2 \times (-30)) = 63.4°; 243.4°$$

The maximum shear stress planes and stresses are shown in Fig. 9.59. The Mohr's circle sketched shows the above concepts very clearly.

Example 9.27 Analytical solution for principal planes and stresses and maximum shear stress

For the plane stress system shown in Fig. 9.60, find the principal stresses and the planes on which they act. Also find the maximum shear stress and the planes on which they act. Show the stresses and planes in a neat sketch.

Solution In this case, $\sigma_x = -110$ N/mm^2; $\sigma_y = 0$ and $\tau_{xy} = +40$ N/mm^2

$$\text{Principal stresses are given by } \sigma_{1,2} = (\sigma_x + \sigma_y)/2 \pm \sqrt{[(\sigma_x - \sigma_y)^2/4 + \tau_{xy}^2]}$$

$$\sigma_{1,2} = (-110)/2 \pm \sqrt{[(-110)^2/4 + 40^2]} = 13 \text{ N/mm}^2 \text{ and } -123 \text{ N/mm}^2$$

The angles made by the principal planes are: $\tan 2\alpha = -2 \times 40 / (-110) = +0.7272$, $2\alpha = 36°, 216°$. The principal planes and stresses are shown in Fig. 9.60(b).

$$\text{Maximum shear stress, } \tau_{max} = \pm \sqrt{[(\sigma_x - \sigma_y)^2/4 + t^2_{xy}]} = \pm \sqrt{[(-110)^2/4 + 40^2]}$$

$$= \pm 68 \text{ N/mm}^2$$

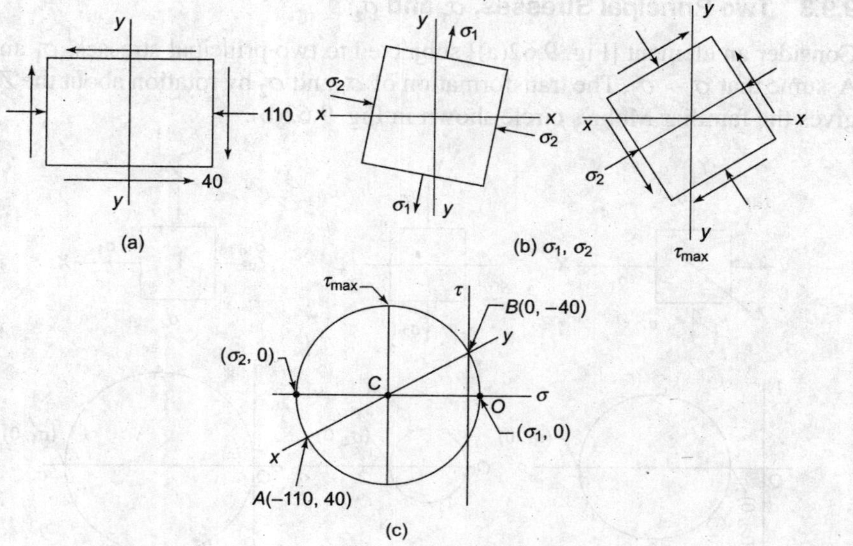

Fig. 9.60

The angles made by the maximum shear stress planes, $\tan 2\beta = -110/(2 \times 40) = 1.375$. $2b = 54°, 234°$. The maximum shear stress planes and shear stresses are shown in Fig. You can sketch the Mohr's circle for the stress system as shown. It is very easy to see the computations of stresses from such a sketch. ■

9.9.2 Maximum Shear Stress Values

Until now, we have considered a plane stress system and have discussed methods to find the stress on any oblique plane. We have also determined the planes with maximum or minimum normal and shear stress. From Fig. 9.61(a), given a plane stress system of σ_X, σ_Y, and τ_{XY}, we can find the stress on the plane *P-P*. From Mohr's circle, we have found that maximum shear stress is equal to the radius of such a circle.

This operation, as has been mentioned earlier, is called a transformation of stress components by rotation about the *Z*-axis, which is perpendicular to the plane of the paper. A similar transformation can be done by rotation about the *X*- and *Y*-axes as well.

Also note that the maximum shearing stress obtained is in the plane on which the transformation is being done. It is not necessary that the maximum shearing stress so obtained is the absolute maximum shearing stress.

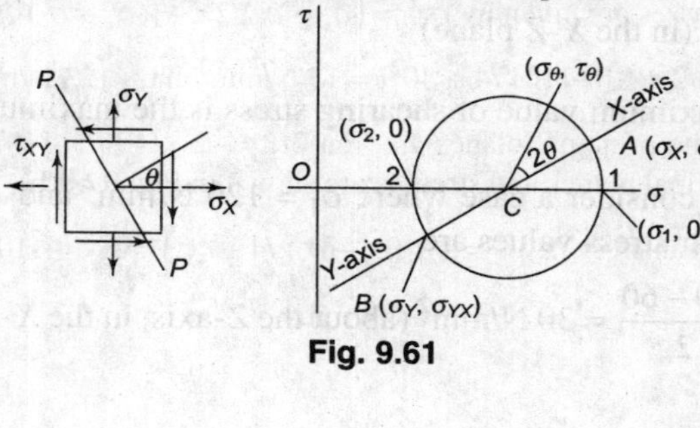

Fig. 9.61

9.9.3 Two Principal Stresses, σ_1 and σ_2

Consider an element [Fig. 9.62(a)] subjected to two principal stresses, σ_1 and σ_2. Assume that $\sigma_1 > \sigma_2$. The transformation of σ_1 and σ_2 by rotation about the Z-axis gives the familiar Mohr's circle shown in Fig. 9.62(b).

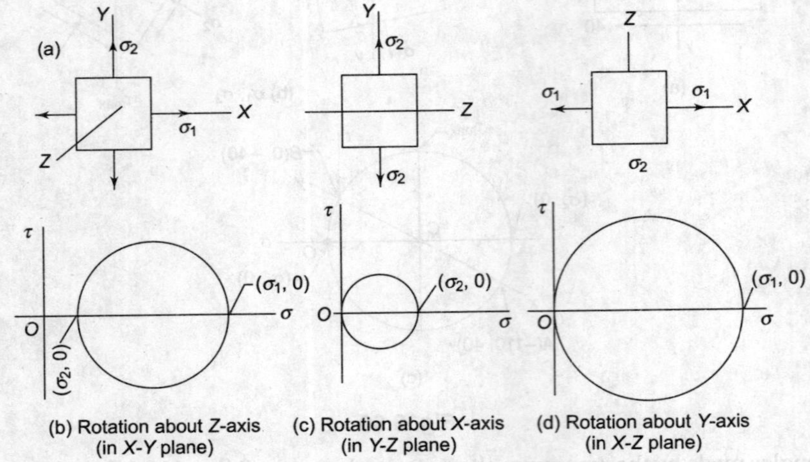

Fig. 9.62

Similar translations about the X- and Y-axes will give Mohr's circles shown in Figures 9.62(c) and (d), respectively. To facilitate understanding of these transformations, the corresponding element stress under consideration is shown by the side. For example, rotation about the X-axis will not affect σ_1. Only σ_2 need be considered for such transformation.

From the three Mohr's circles shown, the maximum shearing stress in each case is as follows.

About the Z-axis,

$$\tau_{max} = \frac{\sigma_1 - \sigma_2}{2} \text{ (in the } X\text{-}Y \text{ plane)}$$

About the X-axis,

$$\tau_{max} = \frac{\sigma_2}{2} \text{ (in the } Y\text{-}Z \text{ plane)}$$

About the Y-axis

$$\tau_{max} = \frac{\sigma_1}{2} \text{ (in the } X\text{-}Z \text{ plane)}$$

The absolute maximum value of shearing stress is the maximum of these three values.

As an example, consider a case where $\sigma_1 = 120$ N/mm^2 and $\sigma_2 = 60$ N/mm^2. The maximum shear stress values are

$$\tau_{max} = \frac{120 - 60}{2} = 30 \text{ N/mm}^2 \text{ (about the } Z\text{-axis, in the } X\text{-}Y \text{ plane)}$$

$$\tau_{max} = \frac{120}{2} = 60 \text{ N/mm}^2 \text{ (in the } X\text{-}Z \text{ plane)}$$

$$\tau_{max} = \frac{60}{2} = 30 \text{ N/mm}^2 \text{ (in the } Y\text{-}Z \text{ plane)}$$

Example 9.28 Absolute maximum shear stress

The element shown in Fig. 9.63 is subjected to two principal stresses of 100 N/mm^2 and 80 N/mm^2. Find the absolute maximum shear stress.

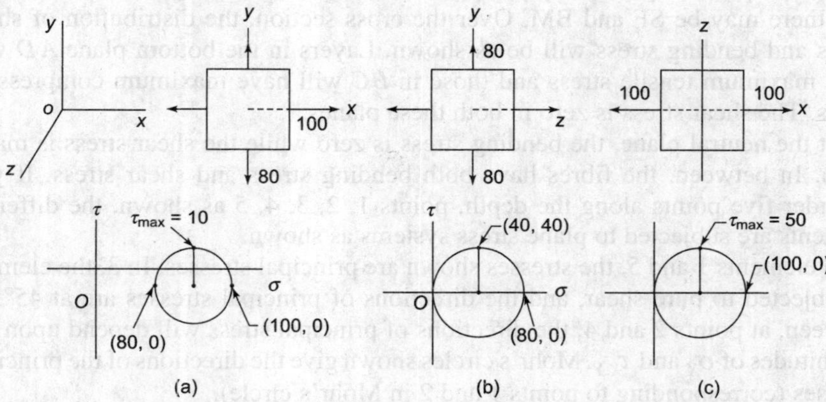

Fig. 9.63

Solution The element shown has stresses along the X- and Y-directions. The conventional rotation of the axes for transformation of stresses is done in Fig. 6.63(a).

The maximum shear stress in this case is $(\sigma_1 - \sigma_2)/2 = (100 - 80)/2 = 10$ N/mm^2. This has been obtained by transformation through a rotation about Z-axis. Similar rotation or transformation can be done by rotation about X- and Y-axes.

While transforming with a rotation about X-axis, σ_1 will not be affected. The stress system to be considered is on σ_2 as shown in Fig. 6.63(b). From the Mohr's circle shown, maximum shear stress in this case is the radius of the Mohr's circle and is equal to $80/2 = 40$ N/mm^2.

A similar transformation by rotation with respect to Y-axis will give s tress system shown in Fig. 6.63(c). σ_y is not affected by this transformation. From the Mohr's circle shown, maximum shear stress = $100/2 = 50$ N/mm^2. Thus the absolute maximum shear stress at the point is 50 N/mm^2 as against 10 N/mm^2 we get from the rotation about the Z-axis. ■

9.10 STRESS TRAJECTORIES

A stress trajectory is a line through different points of a stressed body, indicating the direction of the principal stress at any point on the body. Since, in general, there are two principal stresses at any point at right angles to each other, the stress trajectory shows the direction of one of the principal stresses. Through any point, there are two stress trajectories intersecting at right angles, as shown in Fig. 9.64. The following points are obvious from the definition of stress trajectory:

(i) Stress trajectories are lines showing the directions of principal stresses.

(ii) At different points along a stress trajectory, the principal stress magnitudes are different. The stress trajectory is *not* a line showing points of constant stress.

(iii) Two stress trajectories through a point in a body intersect at right angles along the directions of major and minor principal stress.

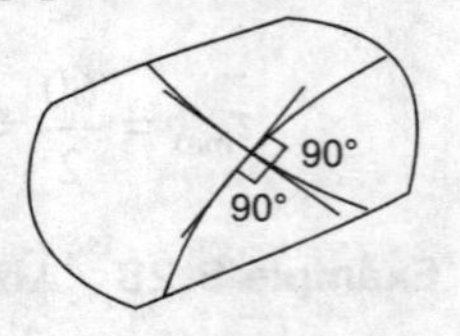

Fig. 9.64

The following examples illustrate the concept of stress trajectories.

Beam Considering the simply supported beam shown in Fig. 9.65, at a section *x*-*x*, there may be SF and BM. Over the cross section, the distribution of shear stress and bending stress will be as shown. Layers in the bottom plane *AD* will have maximum tensile stress and those in *BC* will have maximum compressive stress. The shear stress is zero in both these planes.

At the neutral plane, the bending stress is zero while the shear stress is maximum. In between, the fibres have both bending stress and shear stress. If you consider five points along the depth, points 1, 2, 3, 4, 5 as shown, the different elements are subjected to plane stress systems as shown.

In elements 1 and 5, the stresses shown are principal stresses. In 3, the element is subjected to pure shear, and the directions of principal stresses are at 45°. In between, at points 2 and 4, the directions of principal stress will depend upon the magnitudes of σ_X and τ_{XY}. Mohr's circles shown give the directions of the principal stresses (corresponding to points 1 and 2 in Mohr's circle).

Along the length of the beam, the bending stress varies in proportion with the BM. The variation of SF causes the variation of shear stress along the length.

The stress trajectory for a beam can be drawn as shown in Fig. 9.65.

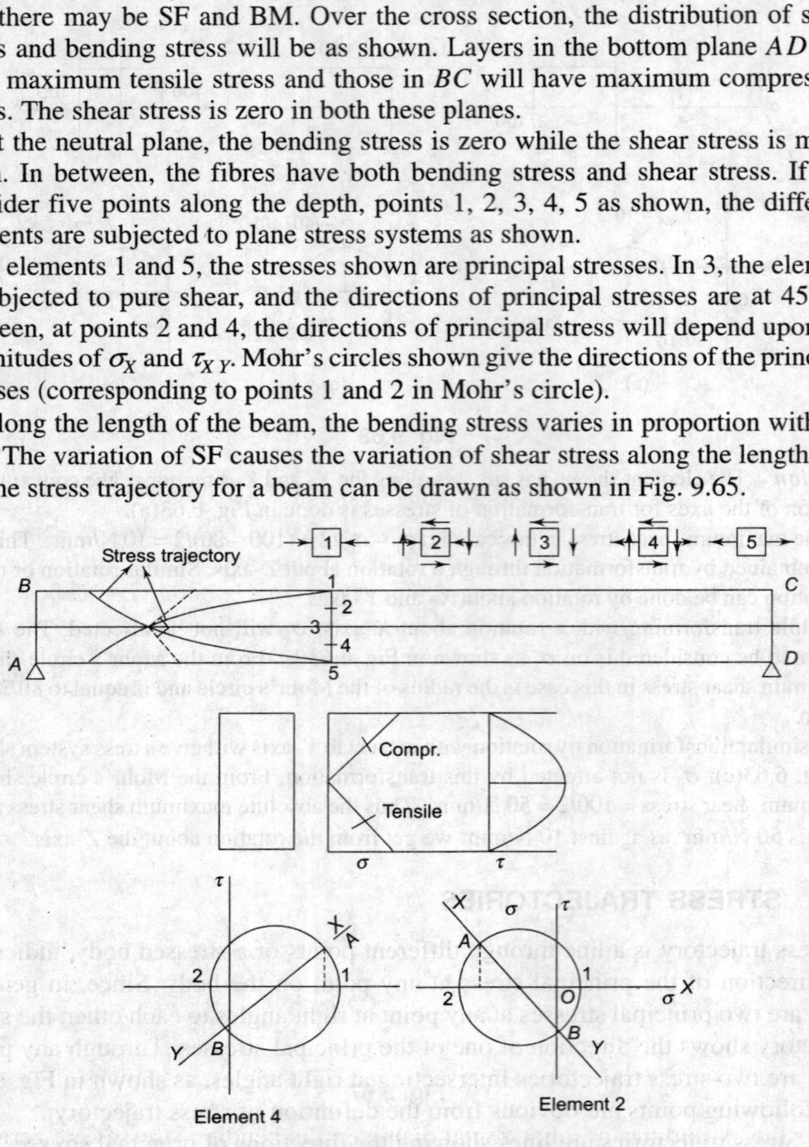

Fig. 9.65

Stress trajectories for torsion If we consider a shaft subjected to torque, the maximum shear stress is on the surface of the shaft, tangential to the cross section. This should be accompanied by longitudinal complementary shear stresses as shown in Fig. 9.66. Due to torsion, an element in a shaft is, thus, subjected to pure shear. This, as we have seen, results in tensile and compressive stresses along the diagonals. These principal stresses have the same magnitude as the torsional shear stresses. The stress trajectories on a shaft are, therefore, 45° helixes intersecting at right angles, as shown in Fig. 9.66.

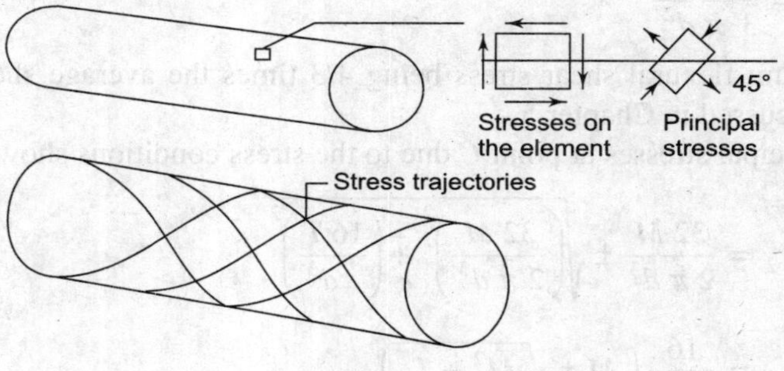

Fig. 9.66

9.11 COMBINED STRESSES DUE TO BENDING AND TORSION

A shaft in practice is not in pure torsion but is also subjected to BM. It is supported at bearings at points along its length and is subjected to BM M along its length, and also flexural shear stresses in addition to torsional shear stresses.

If we consider a shaft subjected to torque T, BM M, and SF V at any section as shown in Fig. 9.67, different elements in the shaft are subjected to different types of stress systems. Taking the cross section of the shaft, a point C on the extremity of a vertical diameter is subjected to a maximum compressive stress σ_C, normal to the cross section, maximum torsional shear stress, τ_{max}, tangential to the section. The flexural shear stress at point C is zero. The corresponding point D on the vertical diameter is similarly subjected to stresses τ_{max} and σ_t, which is tensile.

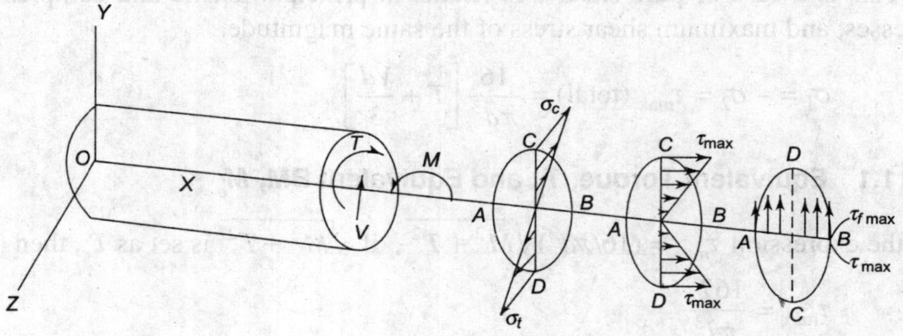

Fig. 9.67

At the extremities of a horizontal diameter, the bending stresses are zero. The flexural shear stress τ_f is maximum and so is the torsional shear stress τ_{max}.

The values of the stresses are as follows.

$$\tau_{max} = \frac{16\,T}{\pi\, d^3}$$

for the solid circular section.

$$\sigma_c \text{ or } \sigma_t = \frac{32\,M}{\pi\, d^3}$$

$$\tau_f = \frac{4\,V \times 4}{3\,\pi\, d^2}$$

the maximum flexural shear stress being 4/3 times the average shear stress, as already discussed in Chapter 5.

The principal stresses at point C due to the stress conditions shown are

$$\sigma_{1,2} = \frac{32\,M}{2\,\pi\, d^3} \pm \sqrt{\left(\frac{32\,M}{2\,\pi\, d^3}\right)^2 + \left(\frac{16\,T}{\pi d^3}\right)^2}$$

$$= \frac{16}{\pi\, d^3}\left(M \pm \sqrt{M^2 + T^2}\right)$$

$$\tau_{max\ (combined\ effect)} = \frac{\sigma_1 - \sigma_2}{2} = \frac{16}{\pi d^3}\sqrt{M^2 + T^2}$$

At point D, or the lower extremity of a vertical diameter, the magnitudes will be the same but the directions will be reversed.

At point A on the horizontal diameter, the torsional shear stress and flexural shear stress are both maximum and have the same directions.

$$\text{Total shear stress at } A = \frac{16\,T}{\pi d^3} + \frac{16\,V}{3\,\pi d^2}$$

$$= \frac{16}{\pi d^3}\left[T + \frac{Vd}{3}\right]$$

This is a case of pure shear and results in principal tensile and compressive stresses, and maximum shear stress of the same magnitude.

$$\sigma_1 = -\,\sigma_2 = \tau_{max}\ \text{(total)} = \frac{16}{\pi d^3}\left[T + \frac{Vd}{3}\right]$$

9.11.1 Equivalent Torque, T_e, and Equivalent BM, M_e

In the expression $\tau_{max} = (16/\pi d^3)\sqrt{M^2 + T^2}$, if $\sqrt{M^2 + T^2}$ is set as T_e, then

$$\tau_{max} = \frac{16 T_e}{\pi d^3}$$

T_e can be called equivalent torque due to the combined effect of BM and torque. Equivalent Torque,

$$T_e = \sqrt{M^2 + T^2}$$

Similarly, from the expression for $\sigma_{1,2}$,

$$\sigma_{1,2} = \frac{16}{\pi d^3}\left(M \pm \sqrt{M^2 + T^2}\right)$$

$$= \frac{32}{\pi d^3}\frac{1}{2}\left(M \pm \sqrt{M^2 + T^2}\right)$$

$$= \frac{32}{\pi d^3}\frac{1}{2}(M \pm T_e)$$

$(M \pm T_e)/2$ is called the equivalent moment, M_e.

Example 9.29 Combined BM and torque in a circular shaft

A solid circular shaft of diameter 100 mm is subjected to a BM of 12 kN m and a SF of 10 kN in addition to a torque of 10 kN m at a section. Find the maximum normal stress and the maximum shear stress in the section.

Solution $I = \frac{\pi}{64} \times (100)^4 = 490 \times 10^4 \text{ mm}^4$

$J = 2I = 980 \times 10^4 \text{ mm}^4$

$$\sigma_{max} = \frac{M}{I}y = \frac{12 \times 10^6 \times 50}{490 \times 10^4} = 122.47 \text{ N/mm}^2$$

The maximum normal stress will be at the top and bottom fibres. The flexural shear stress is zero at this level. Due to the torque of 10 kN m,

$$\tau_{max} = \frac{Tr}{J} = \frac{10 \times 10^6 \times 50}{980 \times 10^4} = 51 \text{ N/mm}^2$$

$$\text{Principal stress } \sigma_1 = \frac{\sigma_X + \sigma_Y}{2} \pm \sqrt{\left(\frac{\sigma_X - \sigma_Y}{2}\right)^2 + \tau_{XY}^2}$$

$\sigma_X = 122.47 \text{ N/mm}^2$

$\sigma_Y = 0$

and

$\tau_{XY} = 51 \text{ N/mm}^2$

$$\sigma_1 = \frac{122.47}{2} + \sqrt{\left(\frac{122.47}{2}\right)^2 + (51)^2} = 141 \text{ N/mm}^2$$

This will be the maximum normal stress at the top and bottom fibres.

At the neutral axis, the flexural shear stress due to the SF of 10 kN is maximum.

$$\tau_{max} \text{ (flexural)} = \frac{4}{3}t_{av} = \frac{4}{3} \times \frac{10 \times 10^3}{\pi \times (50)^2}$$

$$= 1.7 \text{ N/mm}^2$$

$$\tau_{max} \text{ at the NA} = 51 + 1.7 = 52.7 \text{ N/mm}^2$$

At the top and bottom fibres,

$$\tau_{max} = \pm\sqrt{\left(\frac{122.47}{2}\right)^2 + 51^2} = 79.83 \text{ N/mm}^2$$

This will be the maximum shear stress in the section.

Example 9.30 Design of solid circular shaft for combined BM and torque

Design a solid circular shaft subjected to a BM of 20 kNm and a torque of 12 kN m at a section. The maximum normal stress and shear stress are limited to 150 MPa and 120 MPa, respectively.

Solution From the concept of equivalent torque,

$$T_e = \sqrt{M^2 + T^2} = \sqrt{20^2 + 12^2} = 23.33 \text{ kNm}$$

If d is the diameter of the shaft,

$$J = \frac{\pi}{32} d^4$$

$$\tau_{max} = \frac{Tr}{J} = \frac{23.33 \times 10^6 \times d}{(\pi/32) \times d^4 \times 2} = \frac{16 \times 23.33 \times 10^6}{\pi d^3}$$

This is limited to 120 N/mm^2. Therefore,

$$120 = \frac{16 \times 23.33 \times 10^6}{\pi d^3}$$

$$\Rightarrow \quad d^3 = \frac{16 \times 23.33 \times 10^6}{\pi \times 120}$$

$$\Rightarrow \quad d = 99.67 \text{ mm}$$

From equivalent BM,

$$M_e = \frac{1}{2}(M + T_e) = \frac{1}{2}(20 + 23.33) = 21.665 \text{ kNm}$$

$$\sigma_{max} = \frac{M}{I} r = \frac{21.665 \times 10^6}{(\pi d^4/64)} \frac{d}{2} = \frac{32 \times 21.665 \times 10^6}{\pi d^3}$$

This is limited to 150 MPa. Therefore,

$$150 = \frac{32 \times 21.665 \times 10^6}{\pi d^3}$$

$$\Rightarrow \quad d^3 = \frac{32 \times 21.665 \times 10^6}{150 \times \pi}$$

$$d = 113.7 \text{ mm}$$

The larger of the two values should be taken. Therefore,

$$d = 114 \text{ mm}$$

9.11.2 Torque, BM, and Axial Thrust

When torque and BM are combined with axial thrust, the stress due to the axial thrust is in the same direction as the stresses due to bending (in the longitudinal direction). Considering a solid circular section subjected to axial load P, bending moment M, and torque T, the stresses in the section are

(a) Due to the axial load: $\sigma_a = P/\text{area} = P/(\pi d^2/4)$ and is uniform across the section.

(b) Due to bending: $\sigma_b = M/Z = 32M/\pi d^3$ and is maximum on the surface.

(c) Due to torque: $\tau = T/Z_p = 16T/\pi d^3$ and is maximum on the surface of the shaft.

In the solid circular section, the fibres on a vertical diameter ends (A and B) have the maximum stresses due to all the three effects. In the extremities of a horizontal diameter, the stress due to BM is zero, while due to P and T, the stresses will be maximum. The principal stresses due to these effects can be found as

Total longitudinal stress, $\sigma_L = 4P/(\pi d^2) + 32M/\pi d^3$ at A

Combining with shear stress due to torsion, we get the principal stresses as

$$\sigma_{1,2} = \sigma_L/2 \pm \sqrt{[(\sigma_L)^2/4 + \tau^2]}$$

These stresses will be at A. At B, the longitudinal stress will be less as the tensile stress due to bending will be reduced due to the axial thrust.

$$\text{At } C \text{ and } D,\ \sigma_{1,2} = \sigma_a/2 \pm \sqrt{[(\sigma_a)^2/4 + \tau^2]}$$

Stress due to BM is zero at C and D.

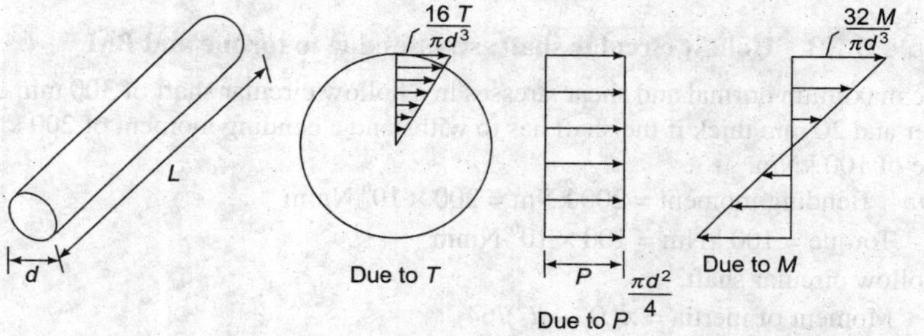

Fig. 9.68 BM, torsion, and axial thrust

Example 9.31 Torque, BM, and axial thrust on shaft

A shaft has to withstand a torque of 10,000 Nm in addition to a BM of 5 kNm. In addition, the shaft is also subjected to an axial thrust of 30 kN. Find the maximum principal stress and shear stress if the diameter of the shaft is 100 mm.

Solution For a 100 mm diameter circular section,

$$\text{Area} = \pi(100)^2/4 = 7854 \text{ mm}^2$$

Moment of inertia = $\pi(100)^4/64 = 491 \times 10^4$ mm^4

Polar moment of inertia = $\pi(100)^4/32 = 982 \times 10^4$ mm^4

Axial thrust = 30,000 N and axial stress = 30,000/7854 = 3.8 N/mm^2

Maximum bending stress = $My/I = 5000 \times 10^3 \times 50/491 \times 10^4 = 51$ N/mm^2

Maximum shear stress due to torque = $Tr/J = 10{,}000 \times 10^3 \times 50/982 \times 10^4$ = 51 N/mm^2

Total normal stress = 51 + 3.8 = 54.8 N/mm^2

Principal stress, $\pi_{1,2} = 54.8/2 \pm\sqrt{(54.8^2/4 + 51^2)} = 85.3$ N/mm^2, –30.5 N/mm^2

Maximum shear stress = ±57.9 N/mm^2

Example 9.32 Design of shaft for torque and BM

Find a suitable diameter for a shaft if it has to withstand a BM of 16 kNm and a torque of 30,000 Nm. The maximum permissible bending stress is 150 N/mm^2 and shear stress is 60 N/mm^2.

Solution Bending moment = 16 kNm = 16×10^6 Nmm and torque = 30×10^6 Nmm

(a) To find the diameter from shear stress point of view, we find the equivalent torque.

Equivalent torque, $T_e = \sqrt{(T^2 + M^2)} = \sqrt{[(16^2 + 30^2)](10^6)^2]} = 34 \times 10^6$ Nmm

Maximum shear stress due to this torque will be on the surface of the shaft and should not exceed 60 N/mm^2.

Maximum shear stress, τ_{max} = torque/polar section modulus

Polar section modulus = polar MI/$(d/2)$ = $(\pi d^4/32)/(d/2) = \pi d^3/16$

$$\tau_{max} = 60 = 34 \times 10^6/(\pi d^3/16);\ d^3 = 34 \times 10^6 \times 16/(\pi \times 60) = 142.4 \text{ mm}$$

(b) To find the diameter from the bending stress point of view, we find the equivalent BM as

$$M_e = [M + \sqrt{(M^2 + T^2)}/2 = [16 + \sqrt{(16^2 + 30^2)}] \times 10^6/2 = 25 \times 10^6 \text{ Nmm}$$

If d is the diameter required, then

$$(\pi d^3/32) \times \sigma = M_e \quad \text{or} \quad d^3 = M_e \times 32/(\sigma\pi)$$

So, $d^3 = 25 \times 10^6 \times 32/(150 \times \pi);\ d = 119.3$ mm

Minimum diameter for the shaft = 143 mm ■

Example 9.33 Hollow circular shaft: stresses due to torque and BM

Find the maximum normal and shear stresses in a hollow circular shaft of 300 mm external diameter and 20 mm thick if the shaft has to withstand a bending moment of 200 kNm and a torque of 100 kNm.

Solution Bending moment = 200 kNm = 200×10^6 Nmm

Torque = 100 kNm = 100×10^6 Nmm

For a hollow circular shaft,

Moment of inertia = $\pi(D^4 - d^4)/64$

Section modulus, $Z = \pi(D^4 - d^4)/(32D)$

For the given shaft, D = 300 mm and internal diameter = 260 mm

Section modulus, $Z = \pi(300^4 - 260^4)/32 \times 300 = 1155 \times 10^4$ mm^3

Polar section modulus, $Z_p = \pi(300^4 - 260^4)/(16 \times 300) = 2310 \times 10^4$ mm^3

Maximum longitudinal stress, $\sigma = M/Z = 40 \times 10^6/(1155 \times 10^4) = 17.3$ N/mm^2

Maximum torsional shear stress, $\tau = T/Z_p = 100 \times 10^6/2310 \times 10^4 = 4.3$ N/mm^2

Principal stresses due to these stresses, which are at right angles, are

$$\sigma_{1,2} = 17.3/2 \pm \sqrt{[(17.3)^2/4 + 4.3^2]} = 18.3 \text{ N/mm}^2, -0.95 \text{ N/mm}^2$$

Maximum shear stress = $\pm\sqrt{[(17.3)^2/4 + 4.3^2]} = \pm 9.6$ N/mm^2

Maximum normal stress = 18.3 N/mm^2 tensile

Maximum shear stress = 9.6 N/mm^2 ■

9.12 PRINCIPAL STRAINS

In the previous sections, we discussed plane stresses and derived equations to find stresses along any direction—principal stresses and maximum shear stress. In a similar fashion, plain strain can be analysed.

Consider two mutually perpendicular directions, the X- and Y-directions. Let us consider strains in the plane of X-Y axes due to applied loads, assuming that strains in the directions perpendicular to the X-Y plane are absent (Fig. 9.69).

The strains, ε_X, ε_Y and γ_{XY}, are the normal and shear strains in the plane, as shown. ε_X and ε_Y are positive if tensile and negative if compressive. γ_{XY}, the shear strain, is positive if the right angle at O is increased and negative if the angle decreases.

The problem is to find the strains along OX' and OY' which are inclined at an angle θ to the X- and Y-axes. The values of the strains ε_X' and ε_Y' and $\gamma_{X'Y'}$ can be shown to be equal to

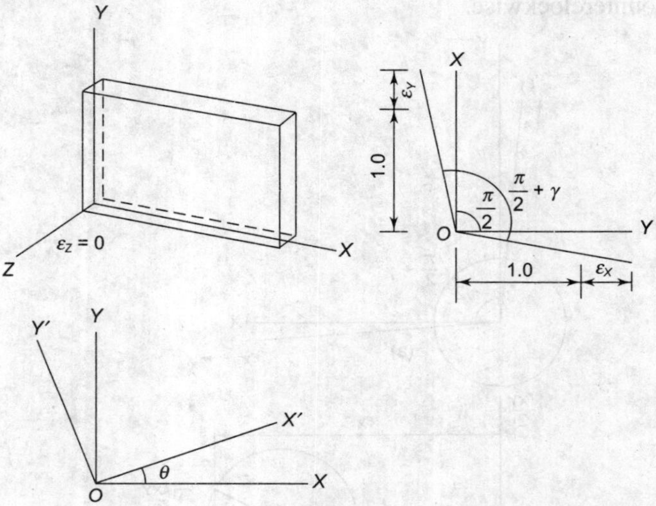

Fig. 9.69

$$\varepsilon_X' = \frac{\varepsilon_X + \varepsilon_Y}{2} + \frac{\varepsilon_X - \varepsilon_Y}{2} \cos 2\theta - \frac{\gamma_{XY}}{2} \sin 2\theta$$

$$\varepsilon_Y' = \frac{\varepsilon_X + \varepsilon_Y}{2} - \frac{\varepsilon_X - \varepsilon_Y}{2} \cos 2\theta + \frac{\gamma_{XY}}{2} \sin 2\theta$$

$$\frac{\gamma_{X'Y'}}{2} = \frac{\varepsilon_X - \varepsilon_Y}{2} \sin 2\theta + \frac{\gamma_{XY}}{2} \cos 2\theta$$

As in the case of principal stresses discussed earlier, it can be proved that the principal strains are ε_1 and ε_2 given by

$$\varepsilon_{1,2} = \frac{\varepsilon_X + \varepsilon_Y}{2} \pm \sqrt{\left(\frac{\varepsilon_X - \varepsilon_Y}{2}\right)^2 + \left(\frac{\gamma_{XY}}{2}\right)^2}$$

The directions of the principal strains may be worked out from

$$\tan 2\theta = -\frac{\gamma_{XY}}{(\varepsilon_X - \varepsilon_Y)}$$

Maximum shear strain $= \sqrt{(\varepsilon_X - \varepsilon_Y)^2 + \gamma_{XY}^2}$ and the direction of the maximum shearing strain may be obtained from $\tan 2\theta = (\varepsilon_X - \varepsilon_Y)/\gamma_{XY}$.

We have presented these results without proof, only to show that, just as in the case plane stress analysis, a Mohr's circle of strains can be drawn for the analysis of strains.

The equations for stress and strain are very similar except that only half the value of shear strain γ_{XY} is to be plotted in Mohr's circle.

A plane strain condition and the corresponding Mohr's circle of strain are shown in Figures 9.70(a) and (b). Note that all observations regarding the interpretation of stresses are applicable to Mohr's circle of strains also. In drawing such a circle and using it to find the strain along a given direction, the angle θ is positive when measured counterclockwise.

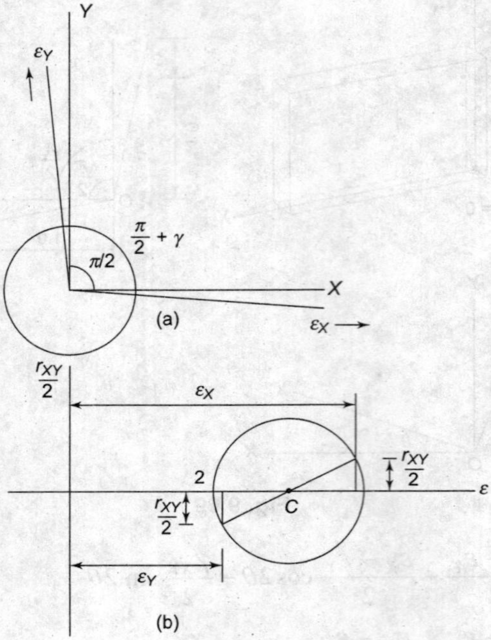

Fig. 9.70

9.13 MEASUREMENT OF STRAIN, AND STRAIN ROSETTES

While studying a complex structure where analysis is difficult, we resort to an experimental technique of finding stresses. Stress is a mathematical concept and cannot be measured. The quantity that can be measured is the strain as change in length per unit length. Many type of gauges are used to measure strain. Strain can be measured mechanically, but electrical strain gauges are more commonly used nowadays.

An electrical strain gauge is shown in Fig. 9.71. It consists of a filament of thin wire or metallic foil mounted on non-conductive paper or plastic. The gauge is fitted on to a structural component by means of an adhesive.

The strain gauge works on the principle that there is a change in electrical resistance due to a change in length of the filament. The effect of change in the cross-sectional area of the wire is very negligible. The change in the resistance of the strain gauge due to strain caused by applied loads can be measured using appropriate instruments.

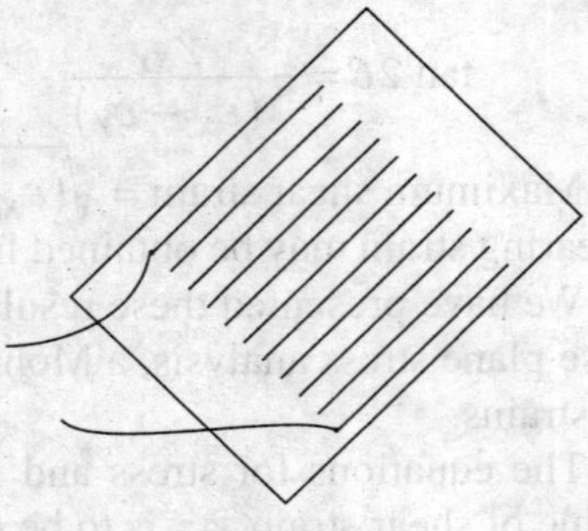

Fig. 9.71

9.13.1 Strain Rosettes

In practice, when the directions of the principal strains are unknown, the linear strains in three directions are measured. The measurement of shear strain is difficult and hence only linear strains are measured, and used to compute the principal strains. The strain rosette is a device used to compute strain.

Obtaining principal strains from three linear strains As shown in Fig. 9.72, the linear strains in three directions are measured along p, q and r. The directions q and r make angles α and β respectively with p.

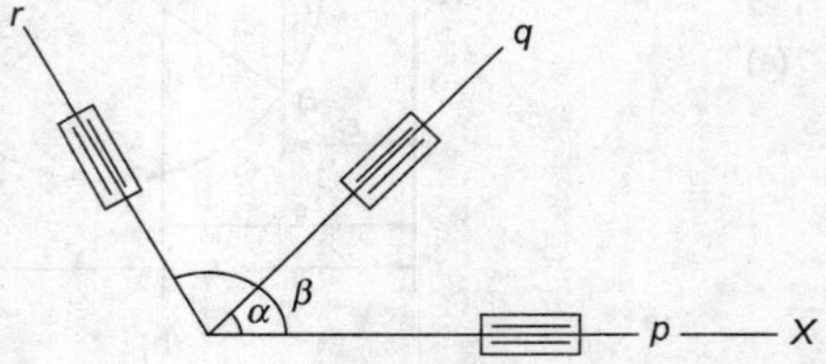

Fig. 9.72

Considering the X-direction to coincide with direction p, angle θ is zero for p, α for q, and β for r. Using the general equation for strain in any direction θ,

$$\varepsilon_\theta = \frac{\varepsilon_X + \varepsilon_Y}{2} + \frac{\varepsilon_X - \varepsilon_Y}{2}\cos 2\theta - \frac{\gamma_{XY}}{2}\sin 2\theta$$

We have

$$\varepsilon_{p(\theta=0)} = \frac{\varepsilon_X + \varepsilon_Y}{2} + \frac{\varepsilon_X - \varepsilon_Y}{2} = \varepsilon_X$$

$$\varepsilon_{q(\theta=\alpha)} = \frac{\varepsilon_X + \varepsilon_Y}{2} + \frac{\varepsilon_X - \varepsilon_Y}{2}\cos 2\alpha - \frac{\gamma_{XY}}{2}\sin 2\alpha$$

$$\varepsilon_{r(\theta=\beta)} = \frac{\varepsilon_X + \varepsilon_Y}{2} + \frac{\varepsilon_X - \varepsilon_Y}{2}\cos 2\beta - \frac{\gamma_{XY}}{2}\sin 2\beta$$

$\varepsilon_p = \varepsilon_X$ and ε_q and ε_r are measured and hence known. ε_Y and r_{XY} can be calculated from these equations. Once ε_X, ε_Y and γ_{XY} are known, the principal strains can be calculated using equations or Mohr's circle.

While many types of rosettes can be used, we consider two of the most commonly used forms. These are the 45° strain rosette and the 60° or the delta rosette, shown in Fig. 9.73. These are only special cases of the general rosette, shown in Fig. 9.72.

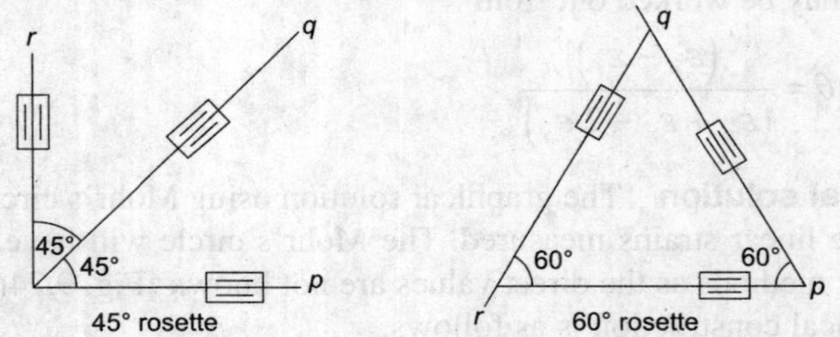

Fig. 9.73

The 45° strain rosette Considering one of the directions to coincide with the X-axis, the angles θ are zero, 45° and 90° for p, q, and r, respectively. Thus, in Fig. 9.74, $\varepsilon_p = \varepsilon_X$ for $\theta = 0$.

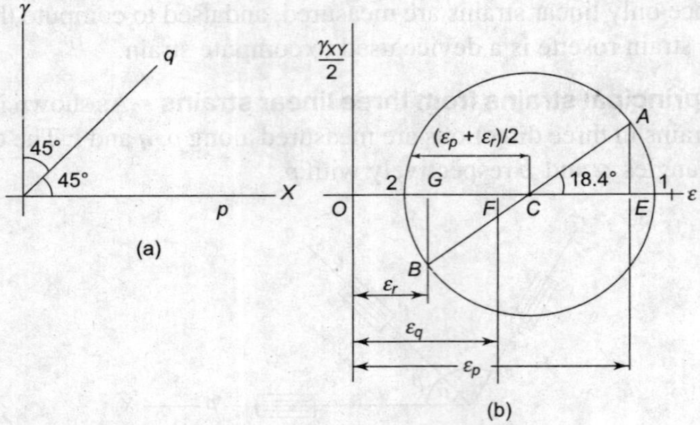

Fig. 9.74 The 45° strain rosette

$$\varepsilon_q = \frac{\varepsilon_X + \varepsilon_Y}{2} - \frac{\varepsilon_X - \varepsilon_Y}{2}\cos 90° - \frac{\gamma_{XY}}{2}\sin 90° = \frac{\varepsilon_X + \varepsilon_Y}{2} - \frac{\gamma_{XY}}{2}$$

$$\varepsilon_r = \frac{\varepsilon_X + \varepsilon_Y}{2} - \frac{\varepsilon_X - \varepsilon_Y}{2}\cos 180° - \frac{\gamma_{XY}}{2}\sin 180° = \varepsilon_Y$$

Solving these equations,

$$\gamma_{XY} = \varepsilon_p + \varepsilon_r - 2\varepsilon_q$$

On the $(\varepsilon - \gamma_{XY}/2)$ axis, the principal strain can be found as

$$\varepsilon_{\max/\min} = \frac{\varepsilon_p + \varepsilon_r}{2} \pm \sqrt{\left(\frac{\varepsilon_p - \varepsilon_r}{2}\right)^2 + \left(\frac{\varepsilon_p + \varepsilon_r - 2\varepsilon_q}{2}\right)^2}$$

The principal directions may be worked out from

$$\tan 2\theta = \frac{(\varepsilon_p + \varepsilon_r - 2\,\varepsilon_q)}{(\varepsilon_p - \varepsilon_r)}$$

Maximum shear strain,

$$\gamma_{\max} = \sqrt{(\varepsilon_p - \varepsilon_r)^2 + (\varepsilon_p + \varepsilon_r - 2\,\varepsilon_q)^2}$$

Its direction may be worked out from

$$\tan 2\theta = \frac{(\varepsilon_p - \varepsilon_r)}{(\varepsilon_p + \varepsilon_r - 2\,\varepsilon_q)}$$

The graphical solution The graphical solution using Mohr's circle is possible with the three linear strains measured. The Mohr's circle will have to be drawn using indirect methods as the direct values are not known [Fig. 9.74(b)].

The graphical construction is as follows.

(i) Lay out two mutually perpendicular lines, the horizontal line being the ε-axis and the vertical line being the $(\gamma/2)$-axis.

(ii) To a suitable scale, make ε_q, ε_q and ε_r on the ε-axis. Points E, F and G are obtained. $OE = \varepsilon_p$, $OF = \varepsilon_q$ and $OG = \varepsilon_r$.

(iii) In a 45° strain rosette, $\varepsilon_p = \varepsilon_X$ and $\varepsilon_r = \varepsilon_Y$. The centre of Mohr's circle is at C where $OC = (\varepsilon_p + \varepsilon_r)/2$.

(iv) $\gamma_{XY}/2 = (\varepsilon_p + \varepsilon_r)/2 - \varepsilon_q$ from the equation derived above.

(v) Mark $EA = GB = \gamma_{XY}/2$ on either side of the ε-axis. Join BA to obtain the diameter of Mohr's circle.

(vi) With C as centre and CA or CB as radius, draw Mohr's circle. CA is the X-axis and CB is the Y-axis in Mohr's circle.

Principal strains and their directions maximum shear strains and their directions can be obtained from Mohr's circle.

The 60° rosette In Fig. 9.75, ε_p, ε_q, and ε_r are the measured strains, and are at 120° to each other. Taking ε_p to coincide with the X-direction,

$$\varepsilon_p = \frac{\varepsilon_X + \varepsilon_Y}{2} + \frac{\varepsilon_X - \varepsilon_Y}{2}\cos 0° - \frac{\gamma_{XY}}{2}\sin 0° = \varepsilon_X$$

$$\varepsilon_q = \frac{\varepsilon_X + \varepsilon_Y}{2} + \frac{\varepsilon_X - \varepsilon_Y}{2}\cos 240° - \frac{\gamma_{XY}}{2}\sin 240°$$

$$\varepsilon_r = \frac{\varepsilon_X + \varepsilon_Y}{2} + \frac{\varepsilon_X - \varepsilon_Y}{2}\cos 480° - \frac{\gamma_{XY}}{2}\sin 480°$$

which gives

$$\varepsilon_X = \varepsilon_p$$

$$\varepsilon_Y = \frac{1}{3}(2\varepsilon_q + 2\varepsilon_r - \varepsilon_p)$$

$$\gamma_{XY} = \frac{2}{\sqrt{3}}(\varepsilon_q - \varepsilon_r)$$

Knowing ε_X, ε_Y and r_{XY}, we can calculate the principal strains and the maximum shear strains.

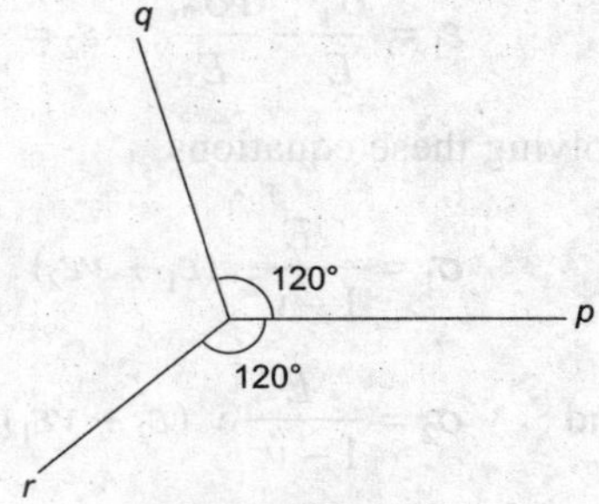

Fig. 9.75

9.13.2 Graphical Construction (Murphy's Construction)

We now describe a general graphical method applicable to all strain rosettes.

(i) Lay out a vertical line to represent the γ-axis.

(ii) Selecting a scale to represent the strains, draw three vertical lines, measuring the strains horizontally from the γ-axis. The lines 1-1, 2-2, 3-3 are obtained by plotting the strain values ε_p, ε_q and ε_r [Fig. 9.76(a)]. The graphical construction is shown in Fig. 9.76(b).

(iii) Selecting any point P_2 on the middle line, draw two lines at angles α and β to intersect the lines 1-1 and 3-3 at P_3 and P_4.

(iv) Bisect P_2-P_3 and P_2P_4. The bisectors intersect at C, which is the centre of Mohr's circle. Draw a horizontal line through C to get the ε-axis. Point O represents the origin.

(v) Draw Mohr's circle with C as centre and CP_3 (or CP_4) as the radius. The circle passes through points P_2, P_3 and P_4.

(vi) P_5 and P_6 represent principal planes. CP_4 represents the X-axis, and the directions of principal strains can be obtained.

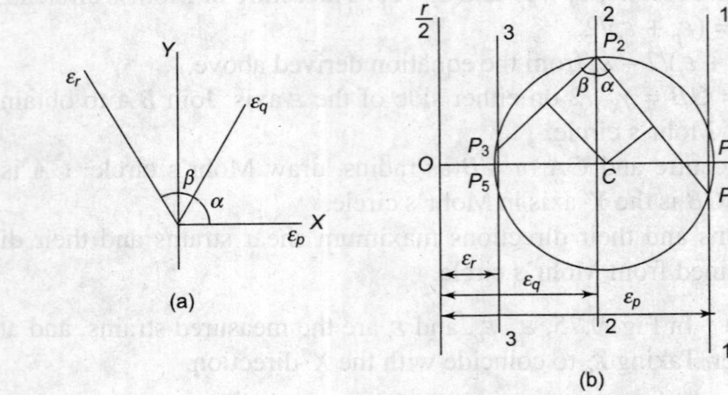

Fig. 9.76 Murphy's construction

9.13.3 Calculation of Stresses from Strains

Once the principal strains ε_1 and ε_2 are calculated, the principal stresses σ_1 and σ_2 along these directions can be calculated, knowing the values of E and ν for the material. From Fig. 9.77,

$$\varepsilon_1 = \frac{\sigma_1}{E} - \frac{\nu\sigma_2}{E}, \quad \varepsilon_2 = \frac{\sigma_2}{E} - \frac{\nu\sigma_1}{E}$$

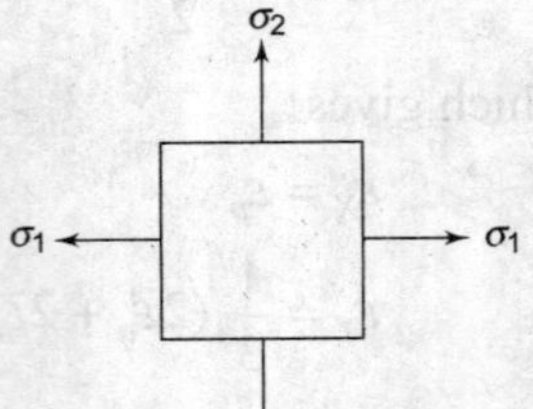

Fig. 9.77

Solving these equations,

$$\sigma_1 = \frac{E}{1-\nu^2}\,(\varepsilon_1 + \nu\varepsilon_2)$$

and
$$\sigma_2 = \frac{E}{1-\nu^2}\nu\,(\varepsilon_2 + \nu\varepsilon_1)$$

Example 9.34 Principal strains

The strains at a point in a stressed body are as follows.

$\varepsilon_X = -640\times10^{-6}$, $\varepsilon_Y = 300\times10^{-6}$ and $\gamma_{XY} = 160\times10^{-6}$. Determine the strains along the directions inclined at 25° to the X-axis. Find the principal strains and the maximum shear strain, and their directions.

Solution The strain condition is shown in Fig. 9.78.

$$\varepsilon'_X = \frac{\varepsilon_X + \varepsilon_Y}{2} + \frac{\varepsilon_X - \varepsilon_Y}{2}\cos 50 - \frac{160}{2}\sin 50$$

$$= 10^{-6}\left(-\frac{640+360}{2} + \frac{-640-360}{2}\cos 50 - 80\sin 50\right) = -523\times10^{-6}$$

$$\varepsilon'_Y = \frac{\varepsilon_X + \varepsilon_Y}{2} - \frac{\varepsilon_X - \varepsilon_Y}{2}\cos 50 + \frac{r_{XY}}{2}\sin 50$$

$$= \left(-\frac{640+360}{2} + \frac{(-640-360)}{2}\cos 50 + \frac{160}{2}\sin 50\right)\times10^{-6} = 243\times10^{-6}$$

$$r'_{Y'} = \frac{\varepsilon_X - \varepsilon_Y}{2}\sin 2\theta + \frac{r_{XY}}{2}\cos 2\theta$$

$$= \left(-\frac{640-360}{2}\sin 50 + \frac{160}{2}\cos 50\right)\times 10^{-6} = -332\times 10^{-6}$$

The principal strains are as follows.

$$\varepsilon_{1,2} = \frac{(\varepsilon_X + \varepsilon_Y)}{2} \pm \sqrt{\left(\frac{\varepsilon_X - \varepsilon_Y}{2}\right)^2 + \left(\frac{r_{XY}}{2}\right)^2}$$

$$= 10^{-6}\left[-\frac{640+360}{2} \pm \sqrt{\left(\frac{-640-360}{2}\right)^2 + \left(\frac{160}{2}\right)^2}\right]$$

$$= 10^{-6}(-140 \pm 506) = 366\times 10^{-6} \text{ and } -646\times 10^{-6}$$

The principal directions may be worked out as follows.

$$\tan 2\theta = -\frac{r_{XY}}{(\varepsilon_X - \varepsilon_Y)} = \frac{-160\times 10^{-6}}{(-640-360)\times 10^{-6}} = 0.16$$

$$2\theta = +9.1^\circ \text{ and } +189.1^\circ$$

$$\theta = +4.55^\circ \text{ and } +94.55^\circ$$

These directions are shown in Fig. 9.78(b).

$$\text{Maximum shear strain} = \pm\sqrt{(\varepsilon_X - \varepsilon_Y)^2 + (r_{XY})^2}$$

$$= \pm\left[\sqrt{(-640-360)^2 + (160)^2}\right]\times 10^{-6}$$

$$= \pm 1013\times 10^{-6}$$

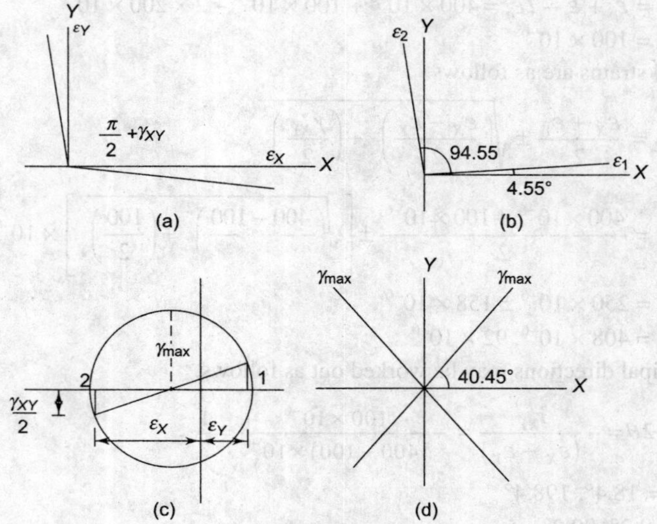

Fig. 9.78

The shear strain directions may be worked out as follows.

$$\tan 2\theta = \frac{\varepsilon_X - \varepsilon_Y}{r_{XY}} = \frac{10^{-6}}{10^{-6}}\left(\frac{-640-360}{160}\right)$$

$2\theta = 80.9, 260.9$

$\theta = 40.45°, 130.45$ [Fig. 9.78(d)]

The Mohr's circle is shown in Fig. 9.78(c).

Example 9.35 A 45° strain rosette

The strains measured using a 45° strain rosette shown in Fig. 9.79 are $\varepsilon_p = 400 \times 10^{-6}$, $\varepsilon_q = 200 \times 10^{-6}$, and $\varepsilon_r = 100 \times 10^{-6}$. Find the principal strains and maximum shear strains. Find also the principal stresses at the point. $E = 200$ GPa, $r = 0.3$.

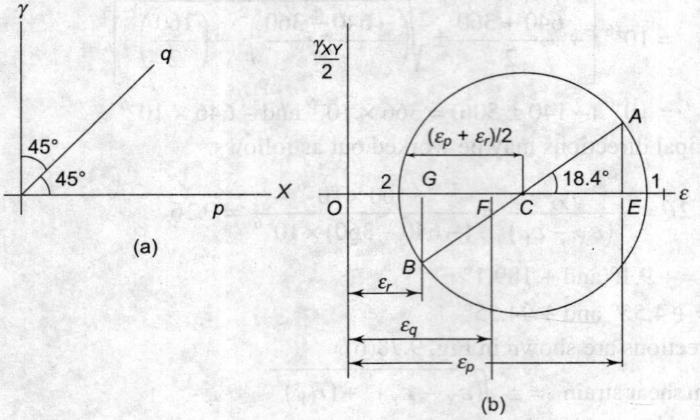

Fig. 9.79

Solution Taking the X-direction along the direction of p, we have

$$\varepsilon_X = \varepsilon_p = 400 \times 10^{-6}, \quad \varepsilon_Y = \varepsilon_r = 100 \times 10^{-6}$$

$$r_{XY} = \varepsilon_p + \varepsilon_r - 2\varepsilon_q = 400 \times 10^{-6} + 100 \times 10^{-6} - 2 \times 200 \times 10^{-6}$$

$$= 100 \times 10^{-6}$$

The principal strains are as follows.

$$\varepsilon_{1,2} = \frac{\varepsilon_X + \varepsilon_Y}{2} \pm \sqrt{\left(\frac{\varepsilon_X - \varepsilon_Y}{2}\right)^2 + \left(\frac{\gamma_{XY}}{2}\right)^2}$$

$$= \frac{400 \times 10^{-6} + 100 \times 10^{-6}}{2} \pm \left[\sqrt{\left(\frac{400-100}{}\right)^2 + \left(\frac{100}{2}\right)^2}\right] \times 10^{-6}$$

$$= 250 \times 10^{-6} \pm 158 \times 10^{-6}$$

$$= 408 \times 10^{-6}, 92 \times 10^{-6}$$

The principal directions may be worked out as follows.

$$\tan 2\theta = -\frac{r_{XY}}{(\varepsilon_X - \varepsilon_Y)} = -\frac{100 \times 10^{-6}}{(400-100) \times 10^{-6}} = \frac{1}{3}$$

$2\theta = 18.4°, 198.4°$

$\theta = 9.2°, 99.2°$

These are shown in Fig. 9.79.

The maximum shear strains are

$$\pm\sqrt{(\varepsilon_X - \varepsilon_Y)^2 + (r_{XY})^2}$$

$$= \pm 10^{-6}\sqrt{(400-100)^2+(100)^2}$$
$$= \pm 316 \times 10^{-6}$$

The corresponding directions may be worked out from

$$\tan 2\theta = \frac{\varepsilon_X - \varepsilon_Y}{r_{XY}} = \frac{400-100}{100} = 3$$

$$2\theta = 71.5°,\ 251.5°$$
$$\theta = 35.75°,\ 125.75°$$

Mohr's circle is shown in Fig. 9.55. The principal stresses are as follows.

$$\sigma_1 = \frac{E}{1-\nu^2}(\varepsilon_1 + \nu\varepsilon_2) = \frac{200{,}000}{1-(0.3)^2}(408 + 0.3 \times 92) \times 10^{-6} = 94 \text{ N/mm}^2$$

$$\sigma_2 = \frac{E}{1-\nu^2}(\varepsilon_2 + \nu\varepsilon_1) = \frac{200{,}000}{1-(0.3)^2}(92 + 0.3 \times 408) = 47 \text{ N/mm}^2$$ ■

9.14 THREE-DIMENSIONAL STRESS ANALYSIS

The analysis done so far has been for uniaxial, biaxial, and general plane stress systems. We now look at the most general stress system that we started with. Such a stress system illustrated in Fig. 9.80 has stresses in the Z-direction as well. The third dimension makes the analysis quite complicated. This topic is essentially dealt with in theory of elasticity. Here only a brief introduction to three-dimensional stress analysis is provided.

9.14.1 General Stress System

The nomenclature for stresses was given in the beginning of this chapter. The same will be followed. Here σ_x, σ_y, and σ_z are the normal stresses on the X-, Y-, and Z-planes and τ_{xy}, τ_{xz}, τ_{yx}, τ_{yz}, and τ_{zx}, τ_{zy} are the shear stresses . These are shown in Fig. 9.80. They are set in the X-, Y-, Z-coordinate system. These are the given stresses on the planes. The analysis is to find out stresses on any plane and the maximum normal and shear stresses. In order to do that, we take an elementary area dA of the triangle abc. The points a, b, and c lie on the X-, Y-, and Z-axes and the plane abc is identified by the outward directed normal onto the plane from the origin (Fig. 9.81).

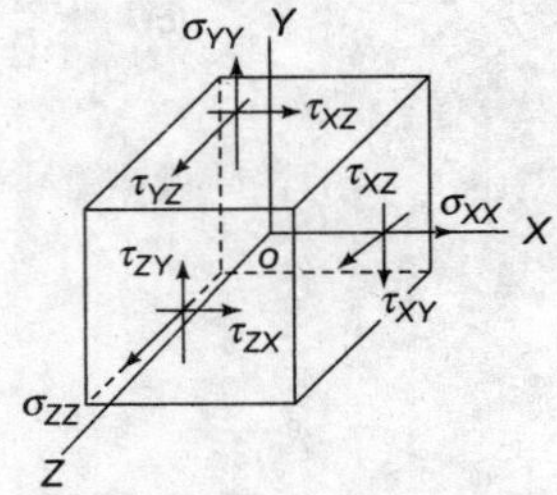

Fig. 9.80 General stress system

We recall some basic concepts from vector algebra that you would have studied earlier. Referring to Fig. 9.81, ON is the line from the origin of the coordinate axes to a point N in space. The angles made by this line with the axes OX, OY, and OZ are α, β, and γ. The line is identified by its direction cosines, which are cos α, cos β, and cos γ. The direction cosines are also represented by l(cos α), m (cos β), and n(cos γ), respectively. If the length ON is unity (unit vector), then we can see that $l^2 + m^2 + n^2 = 1$.

The projection of the area abc on the XY, YZ, and ZX planes are oab, obc, and oca. These projected areas are given by dA cos α, dA cos β, and dA cos γ. This can be easily seen from the volume calculated from the different bases *oab*, *obc*, and *oac*. The volume of the tetrahedron (a triangular pyramid) is $k \times$ area $abc \times ON$. If the volume is calculated from the base *Oab*, the volume is area $k \times Oab \times Oc = k \times$ area $Oab \times ON/\cos\gamma$. Equating the two volumes, we get area Oab = area $abc \times \cos\gamma$ = area $abc \times n$.

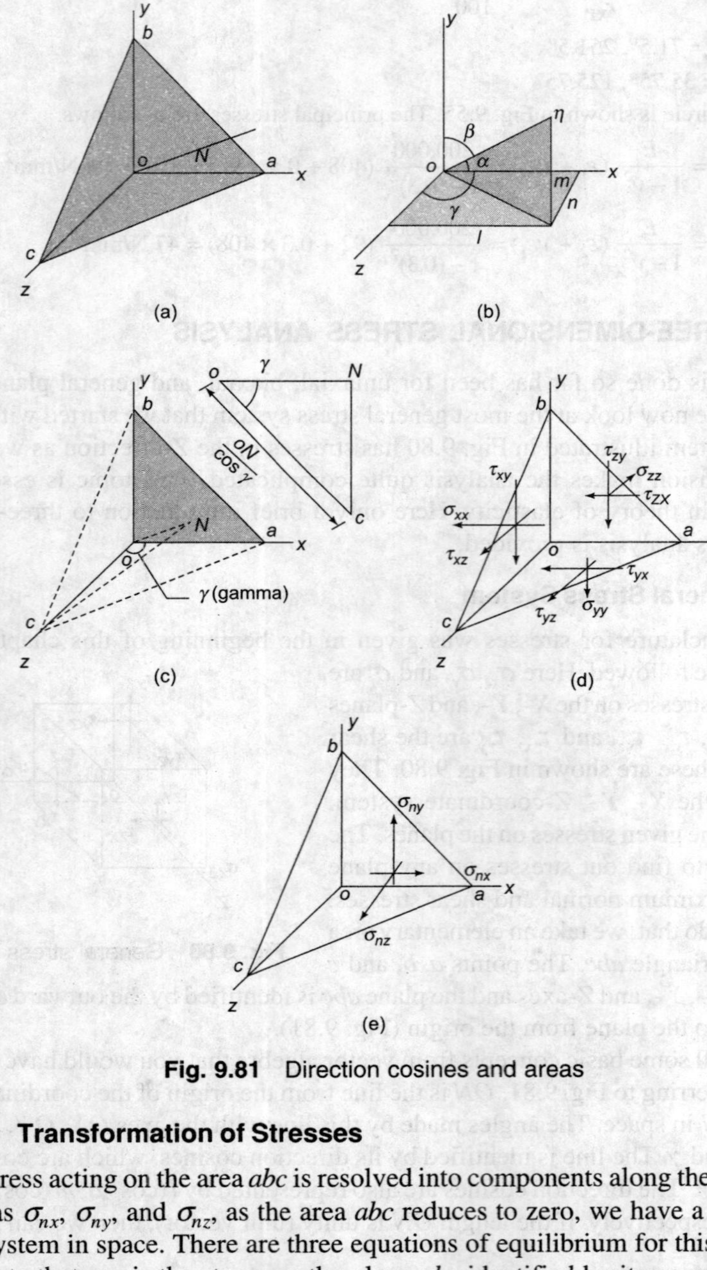

Fig. 9.81 Direction cosines and areas

9.14.2 Transformation of Stresses

If the stress acting on the area *abc* is resolved into components along the axes X, Y, and Z as σ_{nx}, σ_{ny}, and σ_{nz}, as the area *abc* reduces to zero, we have a concurrent force system in space. There are three equations of equilibrium for this force system. Note that σ_{nx} is the stress on the plane *abc* identified by its normal *ON* and acting along the X-axis. The conditions $\Sigma X = 0$, $\Sigma Y = 0$, and $\Sigma Z = 0$ can be applied as follows:

$$\Sigma X = 0;\ \sigma_{nx} dA - \sigma_{xx} dA \cos\alpha - \tau_{yx} dA \cos\beta - \tau_{zx} dA \cos\gamma = 0$$
$$\Sigma Y = 0;\ \sigma_{ny} dA - \sigma_{yy} dA \cos\beta - \tau_{xy} dA \cos\alpha - \tau_{zy} dA \cos\gamma = 0$$
$$\Sigma Z = 0;\ \sigma_{nz} dA - \sigma_{zz} dA \cos\gamma - \tau_{xz} dA \cos\gamma - \tau_{yz} dA \cos\beta = 0$$

This can be rewritten by rearranging the terms and removing dA as a common term as follows:

$$\sigma_{nx} = \sigma_{xx}\, l + \tau_{yx}\, m + \tau_{zx}\, n$$
$$\sigma_{ny} = \tau_{xy}\, l + \sigma_{yy}\, m + \tau_{zy}\, n$$
$$\sigma_{nz} = \tau_{xz}\, l + \tau_{yz}\, m + \sigma_{zz}\, n$$

This can be written in matrix notation as

$$\begin{vmatrix} \sigma_{nx} \\ \sigma_{ny} \\ \sigma_{nz} \end{vmatrix} = \begin{vmatrix} \sigma_{xx} & \tau_{yx} & \tau_{zx} \\ \tau_{yx} & \sigma_{yy} & \tau_{zy} \\ \tau_{xz} & \tau_{yz} & \sigma_{zz} \end{vmatrix} \begin{vmatrix} l \\ m \\ n \end{vmatrix}$$

Now, if we select the axes U, V, and W on the area abc with the U-axis along ON and the V- and W-axes on the plane abc (Fig. 9.82), we can find the stress components along these axes giving the normal and shear stresses on the plane abc. This can be done by finding the angles between the X-, Y-, and Z-axes and the new set of axes U, V, and W. For this, we use notations as follows.

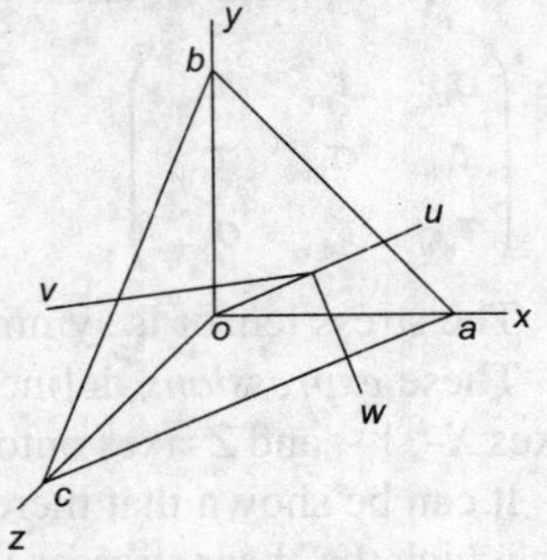

Fig. 9.82

The U-axis makes angles α_1, β_1, and γ_1 with the X-, Y-, and Z-axes. The direction cosines are l_1, m_1, and n_1. Note that as the U-axis coincides with ON, $l = l_1$, $m = m_1$, and $n = n_1$. Similarly, the direction cosines of the V- and W-axes are l_2, m_2, n_2 and l_3, m_3, n_3 with the X-, Y-, and Z-axes, respectively.

Thus, we can write, for σ_{uu}, the normal stress on abc and the shear stresses τ_{uv} and τ_{uw},

$$\sigma_{uu} = \sigma_{nx} l_1 + \sigma_{ny} m_1 + \sigma_{nz} n_1$$

Since the U-axis is normal to the plane abc, $l = l_1$, $m = m_1$, and $n = n_1$.

Putting the values of σ_{nx}, σ_{ny}, and σ_{nz}, which we have derived before in terms of the stresses on the orthogonal XY-, YZ-, and ZX-planes, we get

$$\sigma_{uu} = (\sigma_{xx} l + \tau_{yx} m + \tau_{zx} n) l_1 + (\tau_{xy} l + \sigma_{yy} m + \tau_{zy} n) m_1 + (\tau_{xz} l + \tau_{yz} m + \sigma_{zz} n) n_1$$
$$= (\sigma_{xx} l_1^2 + \sigma_{yy} m_1^2 + \sigma_{zz} n_1^2) + 2\tau_{xy} m_1 l_1 + 2\tau_{yz} n_1 m_1 + 2\tau_{zx} n_1 l_1$$

We have reduced this expression by using the relationship between the direction cosines and the fact that $\tau_{yx} = \tau_{xy}$, $\tau_{zx} = \tau_{xz}$, and $\tau_{yz} = \tau_{zy}$.

Now, τ_{uv} is the shear stress on the plane abc in the V-direction and τ_{uw} is the shear stress on the plane abc in the W-direction. The expression for these can be similarly derived.

$$\tau_{uv} = \sigma_{nx} l_2 + \sigma_{ny} m_2 + \sigma_{nz} n_2$$

$$= (\sigma_{xx}l + \tau_{yx}m + \tau_{zx}n)l_2 + (\tau_{xy}l + \sigma_{yy}m + \tau_{zy}n)m_2 + (\tau_{xz}l + \tau_{yz}m + \sigma_{zz}n)n_2$$

$$= (\sigma_{xx}l_1l_2 + \sigma_{yy}m_1m_2 + \sigma_{zz}n_1n_2) + \tau_{xy}(m_1l_2 + m_2l_1) + \tau_{yz}(n_1m_2 + n_2m_1) + \tau_{zx}(n_1l_2 + n_2l_1)$$

$$\tau_{uw} = \sigma_{nx}l_3 + \sigma_{ny}m_3 + \sigma_{nz}n_3$$

$$= (\sigma_{xx}l + \tau_{yx}m + \tau_{zx}n)l_3 + (\tau_{xy}l + \sigma_{yy}m + \tau_{zy}n)m_3 + (\tau_{xz}l + \tau_{yz}m + \sigma_{zz}n)n_3$$

$$= (\sigma_{xx}l_1l_3 + \sigma_{yy}m_1m_3 + \sigma_{zz}n_1n_3) + \tau_{xy}(m_1l_3 + m_3l_1) + \tau_{yz}(n_1m_3 + n_3m_1) + \tau_{zx}(n_1l_3 + n_3l_1)$$

We have thus found expressions for the normal stress σ_{uu} and the shear stresses τ_{uv} and τ_{uw} on the plane *abc*.

We can also consider an area perpendicular to V and another area perpendicular to W. We can then similarly find expressions for normal stress σ_{vv} and shear stresses τ_{vu} and τ_{vw} on the first plane and normal stress σ_{ww} and shear stresses τ_{wu} and τ_{wv} on the plane normal to the W-axis.

There will thus be nine equations giving stresses on the planes that are normal to the U-, V-, and W-axes. We thus get the stress tensor as

$$\begin{pmatrix} \sigma_{uu} & \tau_{uv} & \tau_{uw} \\ \tau_{vu} & \sigma_{vv} & \tau_{vw} \\ \tau_{wu} & \tau_{wv} & \sigma_{ww} \end{pmatrix}$$

The stress tensor is symmetric as $\tau_{uv} = \tau_{vu}$, etc.

These *expressions* define the transformation of the stresses on the orthogonal axes X-, Y-, and Z-axes onto the mutually perpendicular U-, V-, and W-axes.

It can be shown that there are three planes that are mutually perpendicular and on which the shear stresses vanish. Such planes are called *principal planes* and the stresses on these planes are called *principal stresses*. We denote the principal stresses on the principal planes by σ_1, σ_2, and σ_3. The stress tensor on these planes can be represented by

$$\begin{pmatrix} \sigma_1 & 0 & 0 \\ 0 & \sigma_2 & 0 \\ 0 & 0 & \sigma_3 \end{pmatrix}$$

If $\sigma_1 = \sigma_2 = \sigma_3 = \sigma$, we have

$$\begin{pmatrix} \sigma & 0 & 0 \\ 0 & \sigma & 0 \\ 0 & 0 & \sigma \end{pmatrix}$$

This is called a *spherical tensor*.

If the area *abc* is a principal area, the total stress acting on it is only σ_{uu}. Then the projections of this stress along the X-, Y-, and Z-axes are given by

$$\sigma_{nx} = \sigma_{uu}l,\ \sigma_{ny} = \sigma_{uu}m, \text{ and } \sigma_{nz} = \sigma_{uu}n$$

If we substitute these values in the equations derived earlier for σ_{nx}, σ_{ny}, and

σ_{nz}, we get

$$(\sigma_{xx} - \sigma_{uu})l + \tau_{yx}m + \tau_{zx}n = 0$$
$$\tau_{xy}l + (\sigma_{yy} - \sigma_{uu})m + \tau_{zy}n = 0$$
$$\tau_{zx}l + \tau_{yz}m + (\tau_{zz} - \tau_{uu})n = 0$$

We also know that $l^2 + m^2 + n^2 = 1$.

Designating σ_{uu} by σ_n, we can find the principal stress σ_1 and the direction cosines of the principal plane from these four equations.

As l, m, n all cannot be zero, the determinant should be equal to zero. We have

$$\begin{vmatrix} \sigma_{xx} - \sigma_n & \tau_{yx} & \tau_{zx} \\ \tau_{xy} & \sigma_{yy} - \sigma_n & \tau_{zy} \\ \tau_{xz} & \tau_{yz} & \sigma_{zz} - \sigma_n \end{vmatrix} = 0$$

If this determinant is expanded, we get a cubic equation

$$\sigma_1^{\,3} - k_1\sigma_1^{\,2} + k_2\sigma_1 - k_3 = 0$$

The coefficients k_1, k_2, and k_3 are given by

$$k_1 = \sigma_1 + \sigma_2 + \sigma_3$$
$$k_2 = \sigma_{xx}\sigma_{yy} + \sigma_{yy}\sigma_{zz} + \sigma_{zz}\sigma_{xx} - (\tau^2_{\ xy} + \tau^2_{\ yz} + \tau^2_{\ zx})$$
$$k_3 = \sigma_{xx}\sigma_{yy}\sigma_{zz} + 2\tau_{xy}\tau_{yz}\tau_{zx} - (\sigma_{xx}\tau^2_{\ yz} + \sigma_{yy}\tau^2_{\ zx} + \sigma_{zz}\tau^2_{\ xy})$$

This equation has three real roots and they give the values of the principal stresses σ_1, σ_2, and σ_3. Also, since the values of the principal stresses do not depend upon the choice of the coordinate axes X, Y, and Z, which are purely arbitrary, the constants k_1, k_2, and k_3 are invariants. This also means that the sum of the principal stresses (k_1) is a constant.

9.14.3 Spherical and Deviatric Components

The spherical tensor is obtained when the three principal stresses are equal. Thus,

$$\begin{pmatrix} \sigma & 0 & 0 \\ 0 & \sigma & 0 \\ 0 & 0 & \sigma \end{pmatrix}$$

is the spherical tensor. The values in the tensor are called *spherical components*.

In this case, the stresses over all the areas around the point are the same.

If we take the arbitrary stress tensor,

$$\begin{pmatrix} \sigma_{xx} & \tau_{yx} & \tau_{zx} \\ \tau_{xy} & \sigma_{yy} & \tau_{zy} \\ \tau_{xz} & \tau_{yz} & \sigma_{zz} \end{pmatrix} = ST$$

and add and subtract the spherical tensor from it, we have the following expression:

$$\begin{pmatrix} \sigma_{xx} & \tau_{yx} & \tau_{zx} \\ \tau_{xy} & \sigma_{yy} & \tau_{zy} \\ \tau_{xz} & \tau_{yz} & \sigma_{zz} \end{pmatrix} = \begin{pmatrix} \sigma_{xx} - \sigma & \tau_{yx} & \tau_{zx} \\ \tau_{xy} & \sigma_{yy} - \sigma & \tau_{zy} \\ \tau_{xz} & \tau_{yz} & \sigma_{zz} - \sigma \end{pmatrix} + \begin{pmatrix} \sigma & 0 & 0 \\ 0 & \sigma & 0 \\ 0 & 0 & \sigma \end{pmatrix}$$

Expressing this tensor as the sum of two stress tensors is important. The first tensor is known as a *stress deviator* and the second tensor is the *spherical tensor*.

The spherical tensor is a state of stress where there is change in volume but without distortion. The stress deviator, on the other hand, shows a state of stress where there is change in shape but no change in volume.

9.14.4 Maximum Shearing Stress

The maximum shearing stresses in a three-dimensional stress situation are given by the following relations:

$$\tau_{max} = (\sigma_1 - \sigma_2)/2 \quad \text{or} \quad (\sigma_2 - \sigma_3)/2 \quad \text{or} \quad (\sigma_3 - \sigma_1)/2$$

There are thus three maximum shearing stress values at the point. The absolute maximum shearing stress value depends upon the values of the principal stresses.

SUMMARY

In structural elements and machine components subjected to complex loading systems, normal and shear stresses coexist. The stresses at a point in a body may have components in all the orthogonal directions as normal and shear stresses. When such stresses are combined, the normal and shear stresses in some directions may be greater than those along the orthogonal directions.

When a member is subjected to uniaxial stress, tension, or compression, shear stress exists in oblique planes, maximum being on the planes inclined at 45° to the axial direction. In case of biaxial stress, the stress on oblique planes can be found and the maximum shear stress occurs in the planes different from the planes of biaxial stress.

In a plane stress system, the stresses are considered in a plane (say the X-Y plane containing the X- and Y-axis), and the stresses in the Z-direction are zero. Such a plane stress system has normal stresses in mutually perpendicular directions (say, σ_{XX} and σ_{YY}) and shear stresses (τ_{XY} and τ_{YX}). By analysis, the stresses on any plane oblique to the X-axis can be obtained as

$$\sigma_\theta = \frac{\sigma_X + \sigma_Y}{2} + \frac{\sigma_X - \sigma_Y}{2} \cos 2\theta - \tau_{XY} \sin 2\theta$$

and $$\tau_\theta = \frac{\sigma_X - \sigma_Y}{2} \sin 2\theta + \tau_{XY} \cos 2\theta$$

where θ is the angle of the normal to the oblique plane with the X-axis. These are also known as transformation equations for plane stresses (obtained by rotating the element about the Z-axis). These equations show the variation of normal and shear stress with θ.

Mohr's circle is a graphical construction to find σ_θ and τ_θ. Knowing σ_{XX}, σ_{YY} and τ_{XY}, such a circle can be drawn on a σ-τ axes system.

Principal stresses are the maximum or minimum normal stresses at a point subjected to a stress system. The plane on which they act are known as principal planes. Principal stresses σ_1 and σ_2 are given by

$$\sigma_{1,2} = \frac{\sigma_X + \sigma_Y}{2} \pm \sqrt{\left(\frac{\sigma_X - \sigma_Y}{2}\right)^2 + (\tau_{XY})^2}$$

Principal planes have no shear stress.

The inclination of the normal to the principal planes with the X-axis may be obtained from

$$\tan 2\alpha = -\frac{2\tau_{XY}}{(\sigma_X - \sigma_Y)}$$

The maximum shear stress at a point is given by

$$\sqrt{\left(\frac{\sigma_X - \sigma_Y}{2}\right)^2 + (\tau_{XY})^2}$$

and the inclination of maximum shear stress planes may be obtained from

$$\tan 2\beta = \frac{\sigma_X - \sigma_Y}{2\tau_{XY}}$$

The maximum shear stress planes also have normal stresses = $(\sigma_X + \sigma_Y)/2$.

A similar procedure is available for normal strains ε_X, ε_Y and shear strain γ_{XY}. The Mohr's circle of strains can be drawn as for stresses but by plotting only half the shear strain ($\gamma_{XY}/2$).

In the case of beams, principal stresses can be calculated knowing the normal stress σ and shear stress τ. In the case of shafts subjected to bending and torsion, principal stresses can be worked out using the equations for σ_1 and σ_2.

As stresses cannot be measured, they are calculated by measuring strain. Strain rosettes are used to measure strains in different directions.

A three-dimensional stress system has three normal stresses and two shear stresses on each of the three orthogonal planes. The stresses can be determined on an oblique plane, oblique to all the three axes by considering a tetrahedral element.

EXERCISES

Review Questions

1. For the stress system shown in Fig. 9.83, give appropriate symbolic names to the stresses.
2. In the plane system of stresses shown in Fig. 9.84, give appropriate names to the stress values. What are the stresses on planes whose normals are inclined at + 90° and –180° to the X-axis?

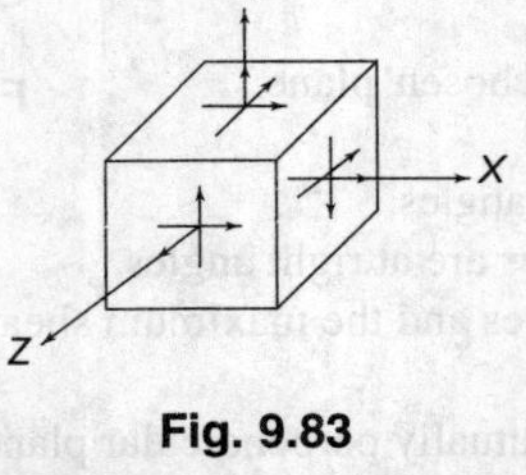

Fig. 9.83

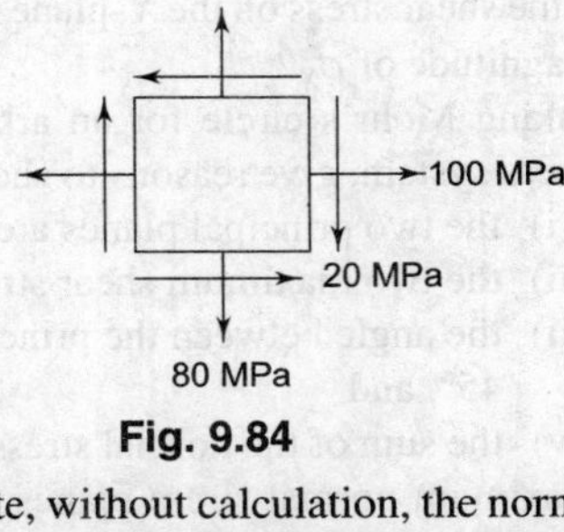

Fig. 9.84

3. For the stress system shown in Fig. 9.85, state, without calculation, the normal stress on a plane whose normal is inclined at θ to the X-axis. Will there be any shear stress on this plane?
4. If the stresses on the plane *P-P* (Fig. 9.86) are $\sigma_\theta = -80$ MPa and $\tau_\theta = 40$ MPa, represent them on the plane.

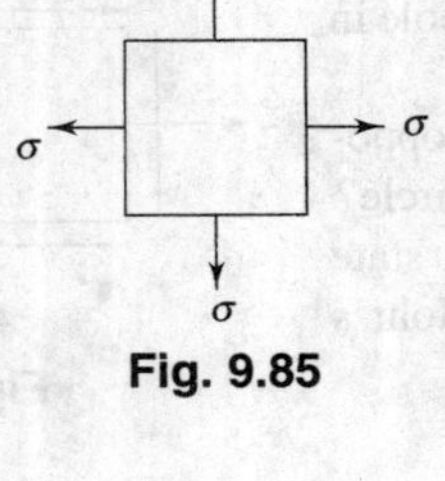

Fig. 9.85

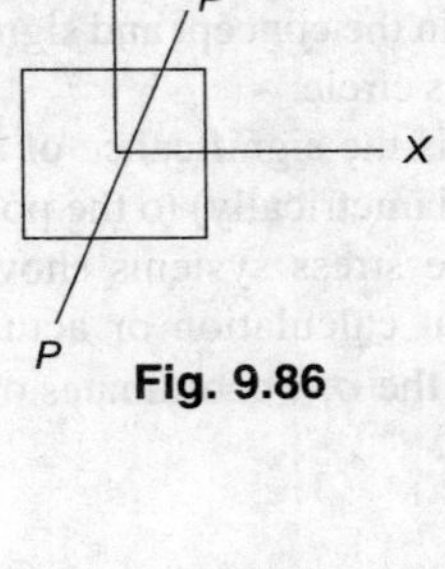

Fig. 9.86

5. In a Mohr's circle, what do the following indicate?
   (i) The coordinate axes, (ii) the coordinates of a point on the circle, and (iii) a radial line.
6. From the Mohr's circles shown in Fig. 9.87, construct the nature, without magnitude, of the stresses on a differential element in X- and Y-plane.

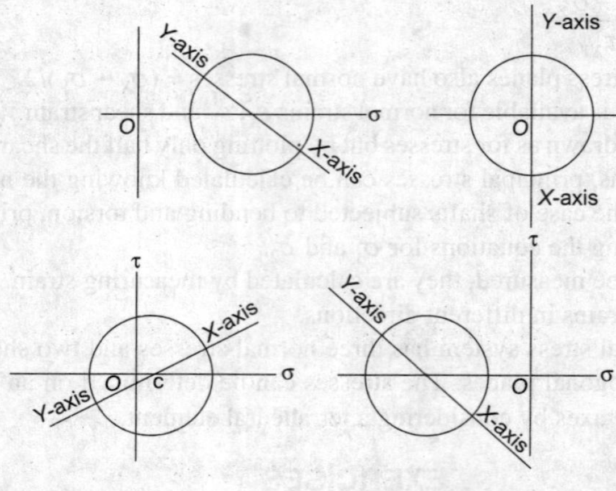

Fig. 9.87

7. Can Mohr's circle be tangential to the τ-axis? If yes, show the stress system for which this happens.
8. For the Mohr's circle shown in Fig. 9.88, (i) draw the stress system on the X-and Y-planes on a differential element, (ii) state the magnitudes of the principal stresses, (iii) find the maximum shear stresses, and (iv) if the shear stress on the X-plane is 20 MPa, what is the magnitude of σ_X?

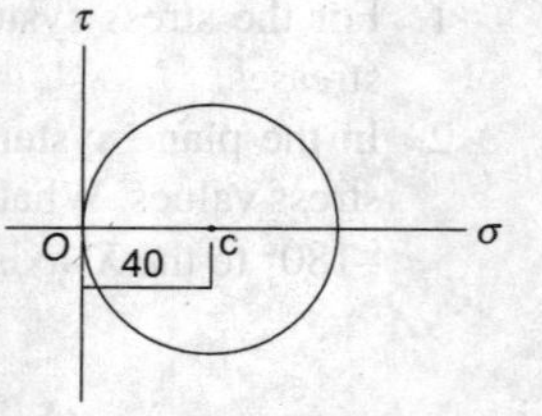

Fig. 9.88

9. Taking Mohr's circle for an arbitrarily chosen plane stress system, give reasons to show that
   (i) the two principal planes are at right angles,
   (ii) the two maximum shear stress planes are at right angles,
   (iii) the angle between the principal planes and the maximum shear stress planes is 45°, and
   (iv) the sum of the normal stresses on mutually perpendicular planes is a constant.
10. Maximum normal stress planes have no shear stress. Give an example of a stress system for which the converse is true.
11. If in a plane stress system, the principal stresses are (i) + 90 MPa, + 30 MPa and (ii) + 80 MPa, – 40 MPa, find in each case the maximum shear stress and normal stress on the maximum shear stress planes.
12. For a given plane stress system, the maximum shear stress values are ± 100 MPa and the normal stress on these planes is + 30 MPa. Find the principal stresses.
13. Explain the concept and significance of the pole in Mohr's circle.
14. What is the significance of a point directly opposite (diametrically) to the pole in a Mohr's circle?
15. For the stress systems shown in Fig. 9.89, state without calculation or actually drawing Mohr's circle, the σ, τ coordinates of the pole.

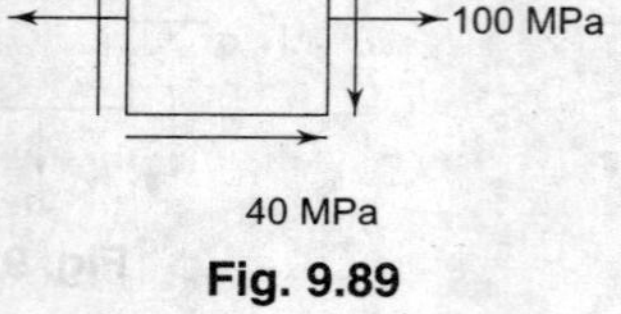

Fig. 9.89

16. Explain the concept and construction of the ellipse of stress.
17. What is a stress trajectory? Show the stress trajectories in a cantilever beam carrying a concentrated load at the free end in a freehand sketch.
18. Sketch the stress trajectories on the surface of a circular shaft and explain how you have obtained them.
19. Explain the working principle of a strain gauge.
20. State the major difference in Mohr's circles for stress and strain.
21. What is a strain rosette? Explain how principal strains and stresses are determined with a three-strain gauge general strain rosette.
22. Show that there is shear stress and normal stress on an oblique plane in the case of uniaxial stress. Also show that the maximum shear stress occurs on the planes inclined at 45° to the stress axis.
23. Explain the term biaxial stress. Derive the expressions for normal and shear stresses on planes inclined to the main axes.
24. Express the stress tensor as the sum of the deviatric and spherical components in the case of a three-dimensional stress system. State the significance of these components.

Problems

1. For the plane stress system shown in Fig. 9.90, derive from first principles the equations for σ_θ and τ_θ on the plane *P-P* shown. Find the values if $\theta = 30°$. Sketch the stresses on the plane.
2. Find the stresses on the plane *P-P* due to the plane stress system shown in Fig. 9.91
3. Find the stresses on the plane *P-P* for the plane stress system shown in Fig. 9.92. Show the stresses on the plane.

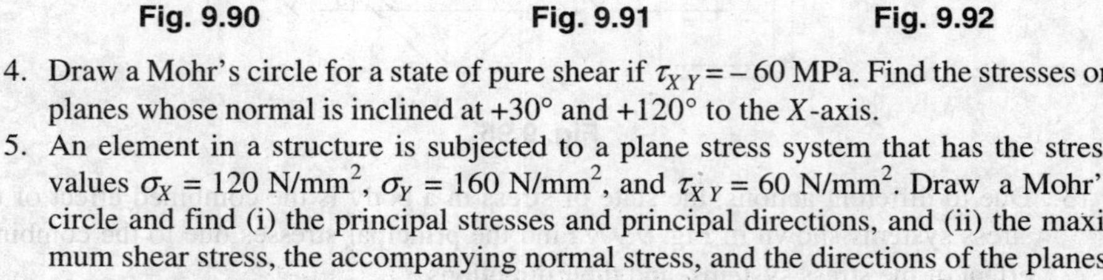

Fig. 9.90 **Fig. 9.91** **Fig. 9.92**

4. Draw a Mohr's circle for a state of pure shear if $\tau_{XY} = -60$ MPa. Find the stresses on planes whose normal is inclined at +30° and +120° to the *X*-axis.
5. An element in a structure is subjected to a plane stress system that has the stress values $\sigma_X = 120$ N/mm^2, $\sigma_Y = 160$ N/mm^2, and $\tau_{XY} = 60$ N/mm^2. Draw a Mohr's circle and find (i) the principal stresses and principal directions, and (ii) the maximum shear stress, the accompanying normal stress, and the directions of the planes.
6. An element subjected to a plane stress system has the stresses $\sigma_X = -40$ MPa, $\sigma_Y = +80$ MPa, and $\tau_{XY} = 20$ MPa. Draw a Mohr's circle and determine (i) the stresses on a plane whose normal is inclined at – 30° to the *X*-axis, and (ii) the principal stresses and directions.
7. An element in a structure is subjected to a tensile stress of 100 MPa accompanied by a shear stress of 40 MPa on the *X*-plane. Draw a Mohr's circle and find the principal stresses, the maximum shear stress and the directions of the corresponding planes.
8. In each of the cases of the plane stress systems shown in Fig. 9.93, draw a Mohr's circle and find the stresses on planes whose normal is inclined at + 20°,– 30°, and +

60° to the X-axis. What will be the stresses on the planes at right angles to the planes indicated?

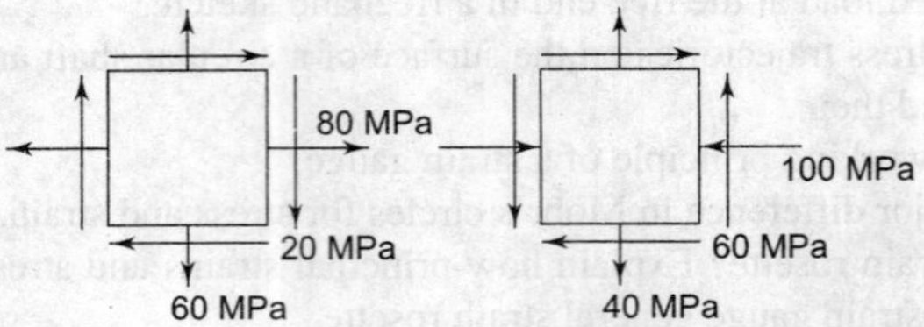

Fig. 9.93

9. An element in a stressed material has the stress system shown in Fig. 9.94. Determine analytically the principal stresses and directions, the maximum shear stresses, and the directions of the planes on which these act.
10. In a plane stress system, the normal stresses on X- and Y-planes are + 200 MPa and – 100 MPa, respectively, and the shear stress on the X-planes is + 80 MPa. Determine analytically the directions of the principal planes and the principal stresses.
11. The plane stress system shown in Fig. 9.95 is found to exist in an element of a stressed body. Find the maximum values of normal and shear stresses, and the planes on which they act.

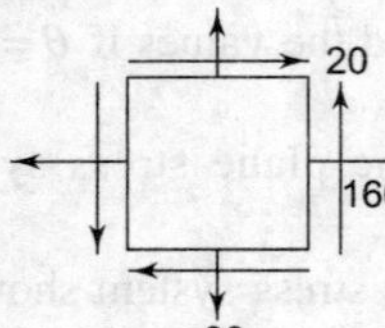

Fig. 9.94

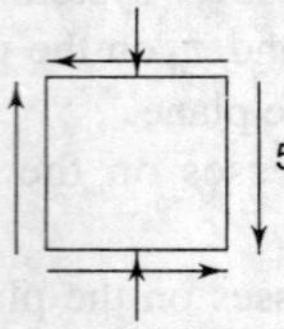

Fig. 9.95

12. If the two plane stress systems shown in Fig. 9.96 are superposed, find the principal stresses and directions.

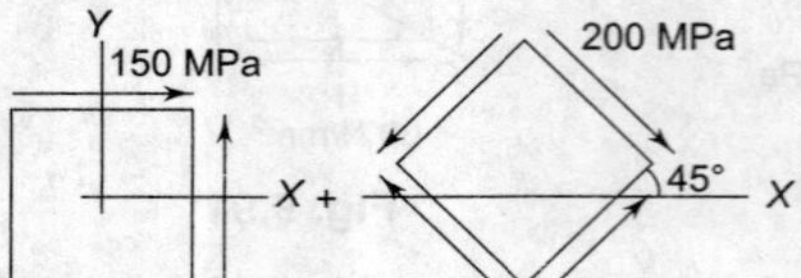

Fig. 9.96

13. Due to different actions, the state of stress in a body is the combined effect of the stress systems shown in Fig. 9.97. Find the principal stresses due to the combined effect of the stress systems, and their directions.

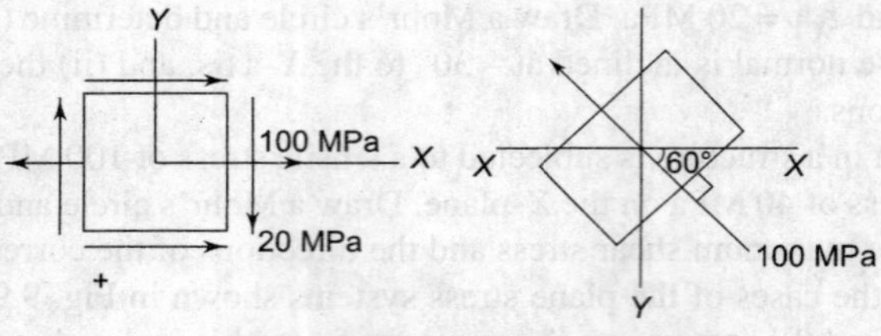

Fig. 9.97

14. A state of strain is defined by $\varepsilon_X = 400 \times 10^{-6}$, $\varepsilon_Y = -200 \times 10^{-6}$, and $r_{XY} = -400 \times 10^{-6}$. If $E = 200$ GPa and $\nu = 0.3$, determine the principal stresses and the maximum shear stress.
15. The strain values at a point in a body are $\varepsilon_X = -600\ \mu$, $\varepsilon_Y = 200\ \mu$ and $r_{XY} = 800\ \mu$, where μ means 10^{-6}. Draw a Mohr's circle and find the principal strains. Find the principal stresses and maximum shear stress, if $E = 200$ GPa and $\nu = 0.3$.
16. The strains measured in a 60° strain rosette are $\varepsilon_p = 300 \times 10^{-6}$, $\varepsilon_q = -400 \times 10^{-6}$ and $\varepsilon_r = 200 \times 10^{-6}$. Compute the principal stresses if $E = 200$ GPa and $\nu = 0.3$.
17. The strain readings in a 45° strain rosette are $\varepsilon_p = -300 \times 10^{-6}$, $\varepsilon_{45°} = 120 \times 10^{-6}$ and $\varepsilon_{90°} = 180 \times 10^{-6}$. Determine the principal stresses analytically and graphically.
18. Calculate the equivalent BM and equivalent torque in the case of a shaft of 80 mm diameter subjected to a BM of 10 kNm and a twisting moment of 15 kNm. Find also the maximum values of direct and shear stresses induced and specify the planes on which they act.
19. A tie bar of cross-sectional area 1000 mm^2 is subjected to an axial tensile load of 70 kN. Find the normal, tangential, and resultant stresses on a plane the normal of which makes an angle of 30° with the axis of the bar. Find also the maximum values of these stresses and the planes on which they act.
20. At a point in a strained material, the principal stresses are 100 N/mm^2 tensile and 40 N/mm^2 compressive. Determine the resultant stress in magnitude and direction on a plane inclined at 60° to the axis of the major principal stress. What is the maximum intensity of the shear stress in the material at the point?
21. A propeller shaft of 200 mm diameter transmits 10 MW power at 100 rpm. In addition, it is subjected to a BM of 8 kNm and an end thrust of 120 kN. Find
    (a) the principal stresses and their planes and
    (b) the maximum shear stress and its plane.
22. The stress components at a point are given by $s_x = 25$ MPa, $s_y = 15$ MPa, $s_z = 10$ MPa, $\tau_{xy} = 10$ MPa, $\tau_{xz} = 15$ MPa, and $\tau_{yz} = 5$ MPa. Also $E = 200$ GPa and Poisson's ratio $= 0.25$. Find the strain components.

CHAPTER 10

Strain Energy

Learning Objectives

After going through this chapter, the reader will be able to

- understand the concept of strain energy,
- calculate the strain energy due to normal stresses for gradually applied, suddenly applied, and impact loads,
- calculate the strain energy due to bending for bending stresses and shear stresses,
- calculate the elastic strain energy due to torsion,
- explain and illustrate the applications of strain energy,
- enunciate the energy theorems,
- prove and apply Castigliano's theorem to determine beam deflections,
- prove and apply the unit load method for determining beam deflections, and
- state and prove Maxwell–Betty theorem of reciprocal deflections.

10.1 INTRODUCTION

The three elementary straining actions detailed in Chapter 3, namely tension, compression and shear, are the basic concepts using which we can calculate stresses and deformations in structures. In this chapter, we will introduce another basic concept—that of strain energy—which will enable to calculate deformations in structures. The concept and application of strain energy is particularly useful and effective in the case of structures subjected to impact loads.

10.2 STRAIN ENERGY

A structure deforms when subjected to loads, and the points of application of the loads consequently move. The applied loads thus do work, and this work is stored as strain energy in the structure. So long as the stresses in the structure are within elastic limits so that no permanent deformation occurs, i.e., the structure can regain its original shape on unloading, the strain energy is recoverable as work.

To elaborate this concept further, consider the simple structure loaded as shown in Fig. 10.1(a). The load P is gradually applied and the structure deforms by an amount δ. We know, from concepts developed in Chapter 3, that the average stress in the member, $\sigma = P/A$ and the deformation $\delta = PL/AE$. If the load is gradually reduced to, say, half its value, then $\delta = PL/2AE$. As the load is reduced, the point of application of the load moves up, and thus work is done on the load. The work done in raising the load is obtained by computing the release of strain energy stored in the structure.

If the member is cut by a normal section, each part of the body is in equilibrium under the action of external loads and the internal stress resultant at the cut section as shown in Fig. 10.1(b). The internal stress resultant is σA, where σ is the average stress at the section and A is the area of cross section of the body. The external work done is $P\delta$, and assuming that no energy or work is lost, according to the principle of conservation of energy, the strain energy is equal to the external work done.

It should be noted that the deformation of the body always takes place against the internal stress resultant. The work done by the internal stress resultant is thus negative. It is also necessary to appreciate that all structures—simple compression members, beams, plates, shells—store energy when strained by applied loads. In this chapter, we will discuss simple cases of strain energy due to axial loading, bending, shear and torsion. Energy and work are scalar quantities, and have important applications in structural analysis.

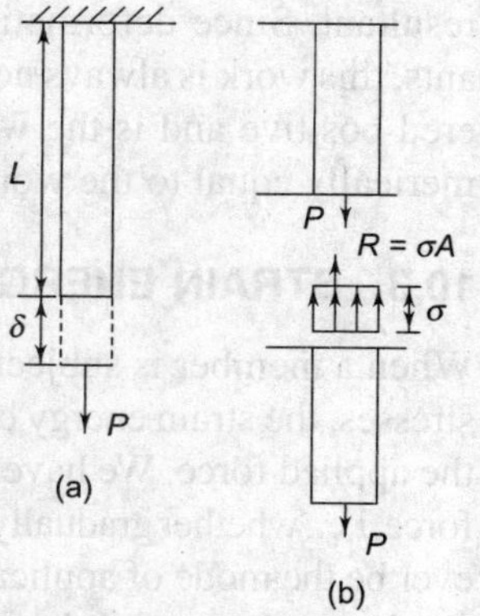

Fig. 10.1

Load deformation diagram If the load deformation diagram of the structure is non-linear, as shown in Fig. 10.2(a), then at the instant the load on the structure is P, let the deformation be equal to x. Where the load is increased by a small amount δP, the deformation increases by δx. The work W done by the load P at this instant is $P\,\delta x$, neglecting quantities of small order. It may be noted that $P\,\delta x$ is the area of the load deformation diagram hatched in Fig. 10.2(a). The elemental work $dW = P\,\delta x$.

The total work done by the applied load, as the member undergoes deformation from $x = 0$ to $x = \delta$ is, thus, $W = \int_0^\delta P\,\delta x$, which is equal to the area of the load deformation diagram from $x = 0$ to $x = \delta$. The work done by the applied loads is equal to the strain energy stored in the body as it deforms. Thus,

$$\text{Strain energy} = W = \int_0^\delta P\,\delta x$$

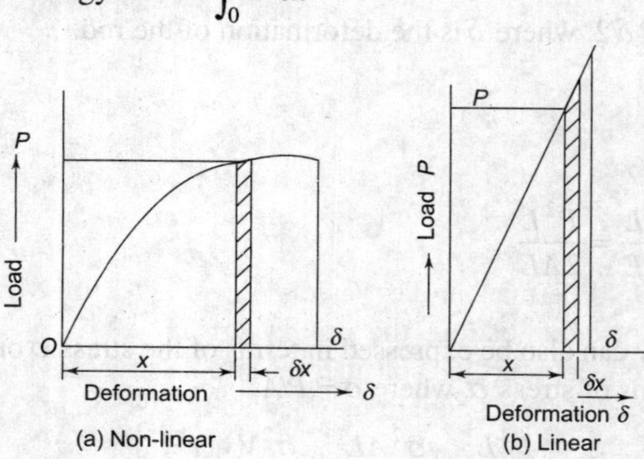

Fig. 10.2

If the deformation is linear, as in Fig. 10.2(b), the area under the load deformation curve is that of a triangle. As the body deforms on the application of load P, work done = $P\delta/2$ = strain energy. Though the strain energy is numerically equal to the work done by external forces, it is actually the work done by the internal stress resultant. Since deformation always take place against the internal stress resultants, this work is always negative. However, by convention, strain energy is considered positive and is the work done by the internal stress resultants, which is numerically equal to the work done by the external forces.

10.3 STRAIN ENERGY DUE TO NORMAL STRESSES

When a member is subjected to normal stresses, i.e., direct tensile or compressive stresses, the strain energy can be calculated by computing the external work done by the applied force. We have also to determine the mode of application of the external force, i.e., whether gradually applied, suddenly applied or applied with an impact. Whatever be the mode of application of the load, the point to note is that the internal stress resultants always develop gradually from a value of zero to the maximum value.

10.3.1 Gradually Applied Load

When the load is applied gradually, considering the member shown in Fig. 10.3(a), the internal stress resultant is R, which is numerically equal to P. The load deformation diagram is linear so long as the stresses are below the proportional limit for the material.

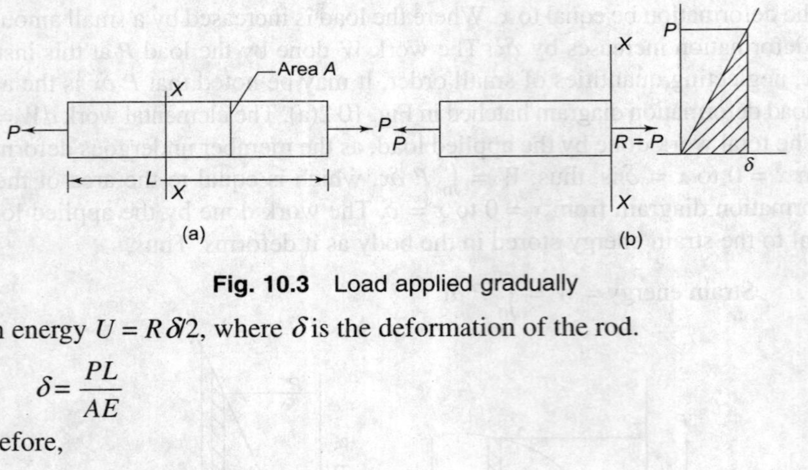

Fig. 10.3 Load applied gradually

Strain energy $U = R\delta/2$, where δ is the deformation of the rod.

$$\delta = \frac{PL}{AE}$$

Therefore,

$$U = \frac{1}{2} R \frac{PL}{AE} = \frac{P^2 L}{2AE}$$

since $R = P$.

The strain energy can also be expressed in terms of the stress σ or the deformation δ. Thus, in terms of stress σ, where $\sigma = P/A$,

$$\text{Strain energy } U = \frac{P^2 L}{2AE} = \frac{\sigma^2 AL}{2E} = \frac{\sigma^2 V}{2E}$$

where V = volume of bar = AL. In terms of deformation δ, where $\delta = PL/AE$ and $P = \delta AE/L$,

$$\text{Strain energy } U = \frac{1}{2} P\delta = \frac{1}{2} \frac{\delta AE}{L} \delta = \frac{\delta^2 AE}{2L}$$

The SI unit of work is the newton metre (N m), which is equal to one joule (J). Thus the unit of strain energy will be the newton metre, or the joule, or its multiples.

10.3.2 Suddenly Applied Load

If the external load is suddenly applied, there is an instantaneous deflection δ and instantaneous stress σ. There is no change in the development and nature of strain energy. Thus

$$\text{Strain energy } U = \frac{1}{2} R\delta = P\delta$$

from which $R = 2P$.

$$R = \sigma A, \quad \sigma = \frac{R}{A} = \frac{2P}{A} = 2\sigma_g$$

where σ_g is the stress developed if the load is applied gradually. The instantaneous stress developed in the material if the external load is applied suddenly is twice the stress developed for a gradual application of the load.

The instantaneous deformation,

$$\delta = \varepsilon L = \frac{\sigma}{E} L$$

$$\delta = 2\frac{PL}{AE} = 2\delta_g$$

where δ_g is the deformation if the load is applied gradually. The instantaneous deformation in the case of a suddenly applied load is twice that in the case of gradually applied load.

Referring to Fig. 10.4, with the suddenly applied load, the instantaneous deflection is δ. If we take a section of the bar at this instant, the internal stress resultant is $2P$.

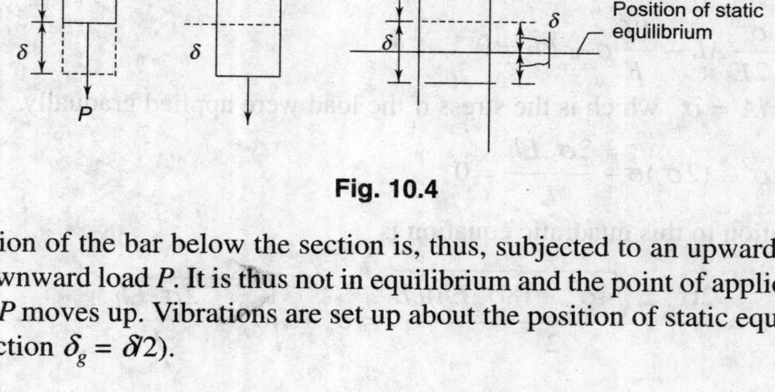

Fig. 10.4

The portion of the bar below the section is, thus, subjected to an upward load $2P$ and a downward load P. It is thus not in equilibrium and the point of application of the load P moves up. Vibrations are set up about the position of static equilibrium (at deflection $\delta_g = \delta/2$).

10.3.3 Load Applied with Impact

Consider the case shown in Fig. 10.5(a). The short bar of cross-sectional area A and length L is fixed at the lower portion and stands vertically. A load P is dropped from a height h vertically above the bar to strike it with an impact.

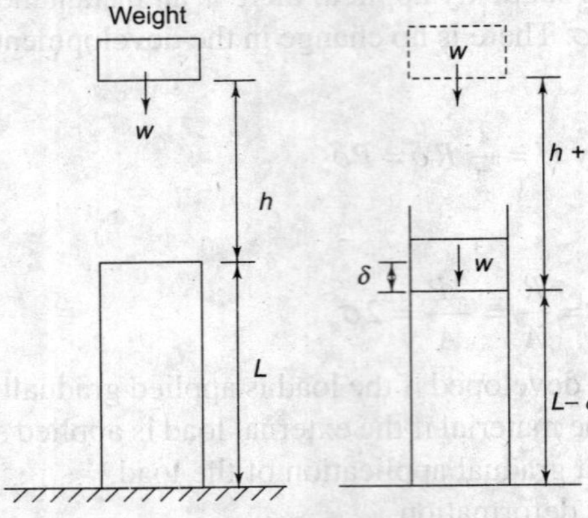

Fig. 10.5 Load applied with impact

As the load P moves down, it loses potential energy and gains kinetic energy. Let the instantaneous deflection of the bar be δ as shown. The total potential energy lost by the load = $P(h + \delta)$. This must be equal to the strain energy in the bar. If σ is the instantaneous stress, then the strain energy

$$U = \frac{\sigma^2}{2E} \times \text{volume}$$

Thus,

$$P(h + \delta) = \frac{\sigma^2}{2E} AL$$

The instantaneous deflection δ and stress σ are related by the expression $\delta = (\sigma/E)L$:

$$P\left(h + \frac{\sigma}{E}L\right) = \frac{\sigma^2}{2E} AL$$

This gives a quadratic expression by rearranging as

$$\frac{\sigma^2}{2E} AL - \frac{PL}{E}\sigma - Ph = 0$$

Setting $P/A = \sigma_g$, which is the stress if the load were applied gradually,

$$\sigma^2 - (2\sigma_g)\sigma - \frac{2\sigma_g Eh}{L} = 0$$

The solution to this quadratic equation is

$$\sigma = \frac{2\sigma_g \pm \sqrt{4\sigma_g^2 + (8\sigma_g Eh)/L}}{2} = \sigma_g + \sqrt{\sigma_g^2 + \frac{2\sigma_g Eh}{L}}$$

since the negative value is not valid. The instantaneous deformation

$$\delta = \frac{\sigma L}{E}$$

$$= \frac{\sigma_g L}{E} + \sqrt{\left(\frac{\sigma_g L}{E}\right)^2 + \frac{2\sigma_g LH}{E}}$$

The expression $\sigma_g L/E$ is the deformation δ_g if the load were applied gradually. Hence δ may be expressed as $\delta = \delta_g + [(\delta_g)^2 + 2\delta_g h]^{12}$.

In this derivation, we have assumed that all the energy of the falling weight is converted into strain energy. We have made the following two specific assumptions in this case.

(i) No energy is dissipated during the impact.
(ii) The striking body does not bounce off the structure and retain part of the energy.

In practice, the striking body does not transfer all the energy to the bar. The mass of the bar should be small compared to the inertia of the striking load for this assumption to be valid.

10.4 UNIT STRAIN ENERGY: MODULUS OF RESILIENCE OR PROOF RESILIENCE

The strain energy stored in a body is known as resilience. Resilience is, therefore, equal to

$$\text{Resilience} = \frac{\sigma^2}{2E} \times \text{volume}$$

Strain energy per unit volume or strain energy density = $\sigma^2/2E$

The maximum strain energy that can be stored in a material without permanent set is when the stress σ is equal to the elastic limit of the material. This is sometimes called the proof stress. When the stress is equal to the proof stress, the strain energy per unit volume is known as *proof resilience*.

Proof resilience = $(\sigma_{EL})^2 \times$ volume/$2E$

Proof resilience per unit volume is known as *modulus of resilience*.

Modulus of resilience = $(\sigma_{EL})^2/2E$

If we consider the stress-strain diagram for any material, the total strain energy absorbed is given by the area under the stress-strain diagram [Fig. 10.6(a)]. The strain energy that can be absorbed per unit volume till rupture is known as the modulus of toughness. The modulus of toughness is a measure of the resistance of the structure to impact loading, and is dependent on the ductility of the material.

To illustrate, consider the stress-strain diagrams of mild steel and a high-strength steel. The value of E, Young's modulus of elasticity, is the same for both types of steel. However, high-strength steels lack the ductility of mild steel. The former have a high yield strength but the maximum strain at rupture is low. The area under the stress-strain diagram for mild steel is comparatively large, showing a higher

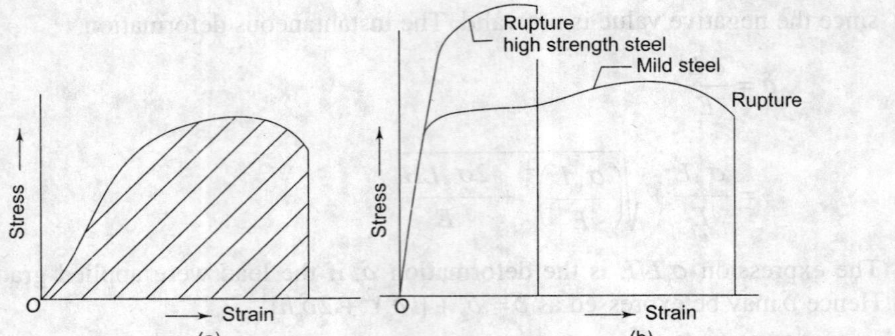

Fig. 10.6

value of toughness. Mild steel is, thus, better suited to situations requiring shock-absorbing capacity (toughness) than high-strength steel. Similarly, rubber has a much larger strain corresponding to a given stress, and hence a greater shock-absorbing capacity.

Example 10.1 Strain energy of a circular bar

A mild steel bar of diameter 30 mm and length 2.4 m is subjected to a tensile load of 90 kN. Find the strain energy stored in the bar if the load is applied gradually. What is the modulus of resilience if the proportional limit is 220 MPa? $E = 200$ GN/m^2.

Solution Strain energy $= \dfrac{P^2L}{2AE} = \dfrac{(90{,}000)^2 \times 2400}{2(\pi/4)(30)^2 \times 2\times10^5}$

$$= 68{,}755 \text{ N mm}$$

$$= 68.76 \text{ N m or J}$$

$$\text{Modulus of resilience } = \frac{(\sigma_{\text{pl}})^2}{2E} = \frac{220\times220}{2\times2\times10^5}$$

$$= 0.121 \text{ N mm}$$

Example 10.2 Comparing strain energies stored in two bars

Compare the strain energies of the two bars of the same material, shown in Fig. 10.7, when they are subjected to the same load. If $d = 10$ mm, find the strain energies of the two bars when the maximum stress in both the bars is equal to 200 MPa. Take $E = 200$ GN/m^2.

Solution If the bars are subjected to the same load, say, P, then strain energy = $P^2L/2AE$. P and E are the same for all sections of the bars. For the first bar,

$$\text{Strain energy } = \frac{P^2}{2E}\left(\frac{200}{\pi d^2/4} + \frac{400}{\pi(2d)^2/4}\right)$$

$$= \frac{P^2\times1200}{2E\pi d^2}$$

For the second bar,

$$\text{Strain energy } = \frac{P^2}{2E}\left(\frac{400}{\pi(2d)^2/4} + \frac{400}{\pi(3d)^2/4}\right)$$

$$= \frac{P^2}{2E\pi d^2}\times\frac{5200}{9}$$

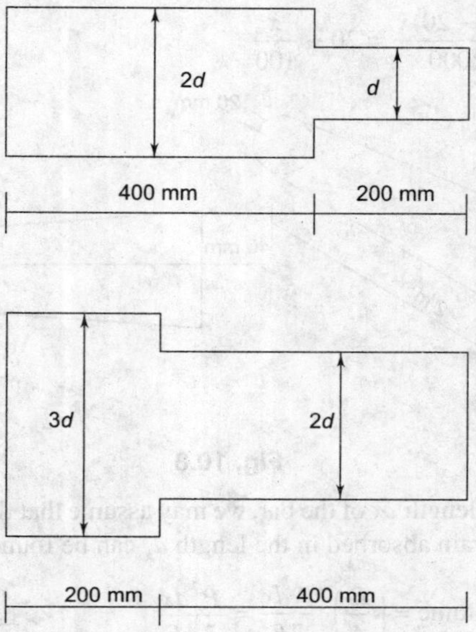

Fig. 10.7

$$\frac{\text{Strain energy of bar 1}}{\text{Strain energy of bar 2}} = \frac{P^2 \times 1200}{2E\pi d^2} \times \frac{2E\pi d^2 \times 9}{P^2 \times 5200} = \frac{27}{13}$$

When the maximum stresses in the bars are the same, they are subjected to different loads. However, the strain energy can be calculated from the stress as $(\sigma^2/2E) \times$ volume. The stress is inversely proportional to area, and area is proportional to d^2.

For bar 1 The stress is 200 MPa in the 200 mm long part, and that in the other part can be calculated as

$$\sigma = \frac{200d^2}{(2d)^2} = 50 \text{ MPa}$$

$$\text{Strain energy} = \frac{(200)^2}{2 \times 2 \times 10^5} \times 25\pi \times 200 + \frac{(50)^2}{2 \times 2 \times 10^5} \times 100\pi \times 400$$

$$= 2356.2 \text{ N mm}$$

For bar 2 The stress is 200 MPa in the part of length 400 mm, and that in the other part is

$$\sigma = \frac{200 \times (2d)^2}{(3d)^2} = \frac{800}{9} \text{ MPa}$$

$$\text{Strain energy} = \frac{(200)^2 \times 100 \times \pi \times 400}{2 \times 2 \times 10^5} + \frac{(800/9)^2 \times 225\pi \times 200}{2 \times 2 \times 10^5}$$

$$= 15{,}358.9 \text{ N mm}$$

Example 10.3 Strain energy of a tapering bar

A bar tapers from a diameter of 40 mm to 20 mm over a length of 2 m. If it is subjected to a tensile load of 50 kN, find the strain energy stored in it. $E = 200$ GN/m.^{2}

Solution From Fig. 10.8, we note that the stress in the bar varies along its length. Consider a section of the bar at a distance x from the smaller end. The diameter at section x is given by

$$d_x = 20 + \frac{(40-20)x}{2000} = 20 + \frac{x}{100}$$

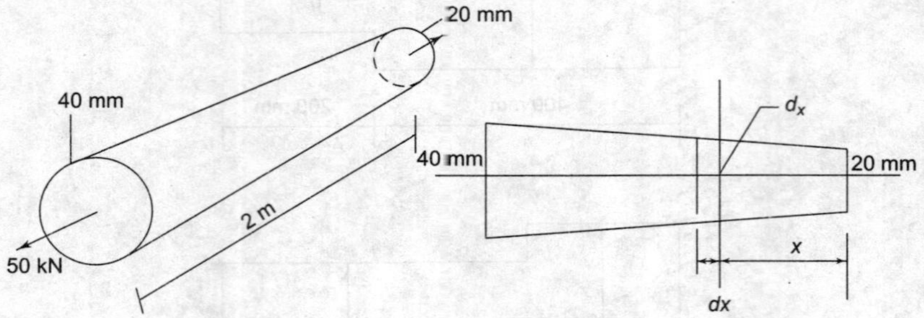

Fig. 10.8

Taking an elemental length δx of the bar, we may assume that the stress remains constant over this length. The strain absorbed in the length d_x can be found as

$$U_x = \frac{\sigma^2}{2E} \times \text{volume} = \left(\frac{P}{A}\right)^2 \frac{A\,dx}{2E} = \frac{P^2 dx}{2AE}$$

$$\text{Area at section } x = \frac{\pi (dx)^2}{4} = \frac{\pi (20 + x/100)^2}{4}$$

$$U_x = \frac{P^2}{2E} \frac{dx}{(\pi/4)(20 + x/100)^2}$$

This can be integrated from the length $x = 0$ to $x = 2000$ to obtain the strain energy in the whole bar.

$$U = \int_0^{2000} \frac{P^2}{2E} \frac{dx}{(\pi/4)(20 + x/100)^2} = \frac{2P^2}{\pi E} \int_0^{2000} \frac{dx}{(20 + kx)^2}$$

where $k = 1/100$.

$$U = \frac{2P^2}{\pi E}\left[-\frac{1}{k(20+kx)}\right]_0^{2000} = \frac{2F^2}{\pi E k}\left[-\frac{1}{40} + \frac{1}{20}\right]$$

Substituting $P = 50{,}000$ N and $E = 2 \times 10^5$ N/mm^2,

$$U = \frac{2 \times (50{,}000)^2}{\pi \times 2 \times 10^5 \times 1/100} \times \frac{1}{40} = 19{,}894 \text{ N mm}$$ ■

Example 10.4 Strain energy due to suddenly applied load

A box weighing 10 kN is attached to the end of a steel rope of diameter 25 mm. The box is lowered at a rate of 1 m/s. When the freely hanging length of the rope is 12 m, the machine gets jammed and the load is brought to a stop instantaneously. Find the maximum stress in the rope and the instantaneous deformation due to jamming. Take $E = 200$ GN/m^2.

Solution When the wire gets jammed, the load is brought to a stop. The kinetic energy of the load has to be absorbed by the wire.

$$\text{Kinetic energy of load} = \frac{1}{2} m v^2 = \frac{1}{2}\left(\frac{w}{g}\right) v^2$$

$$= \frac{1}{2}\left(\frac{10{,}000}{9.81}\right)\frac{(1000)^2}{1000}$$

$$= \frac{10^7}{19.62}\ \text{N mm}$$

If the instantaneous stress in the rope is σ, then

$$\text{Strain energy } U = \frac{\sigma^2}{2E} \times \text{volume} = \frac{\sigma^2}{2\times 2\times 10^5}\left(\frac{\pi}{4}\times 25^2 \times 12{,}000\right)$$

Therefore,

$$\frac{\sigma^2}{2\times 2\times 10^5}\left(\frac{\pi}{4}\times 25^2\times 12{,}000\right) = \frac{10^7}{19.62}$$

$\Rightarrow \quad \sigma^2 = 34{,}610$

$\Rightarrow \quad \sigma = 186\ \text{N/mm}^2$

$$\text{Instantaneous elongation} = \frac{\sigma L}{E} = \frac{186\times 12{,}000}{2\times 10^5} = 11.16\ \text{mm}$$

Example 10.5 Strain energy due to self-weight

A conical bar of base radius *r* and length *L* is hung as shown in Fig. 10.9. If the weight density of the material is γ, find the strain energy of the bar deforming under its own weight.

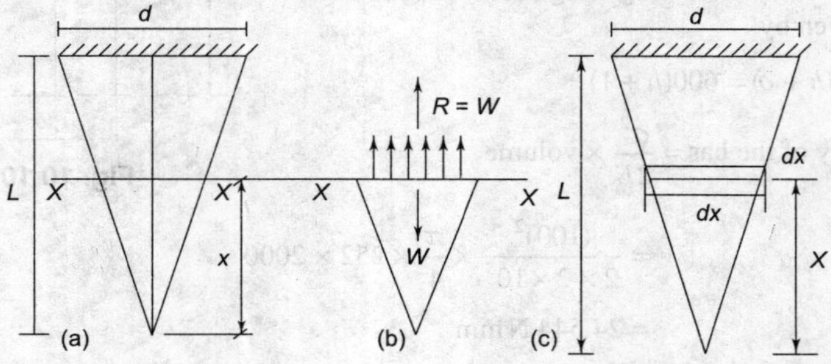

Fig. 10.9

Solution The bar is shown in Fig. 10.9(a). The free body diagram of the bar below section *x*-*x* is shown in Fig. 10.9(b). Since the stress varies along the length, we start with an elemental length *dx* as shown in Fig. 10.9(a). This elemental length deforms under the weight of the conical bar below the section *x*-*x*. The diameter at section *x*-*x*, $d_x = dx/L$ [Fig. 10.9(c)].

$$\text{Area} = \pi d^2x^2/4L^2$$

If γ is the unit weight of the material, weight of the conical portion below section *x*-*x*,

$$P = \frac{1}{3}\pi\frac{d^2x^2}{4L^2}x\gamma = \frac{\pi d^2\gamma x^3}{12L^2}$$

$$\text{Strain energy } U_x = \frac{P^2L}{2AE}$$

where $L = dx$ for the elementary strip. Therefore,

$$U_x = \left[\frac{\pi^2 d^2\gamma x^3}{12L^2}\right]^2 \frac{dx}{2\times(\pi d^2x^2/L^2)E}$$

$$= \frac{\pi^2 d^4 \gamma^2 x^6}{144L^4} \frac{d \times L^2}{2\pi d^2 x^2 E}$$

$$= \frac{\pi d^2 \gamma^2 x^4 dx}{288L^2 E}$$

The strain energy of the whole bar is obtained by integrating from $x = 0$ to $x = L$,

$$U = \int_0^L \frac{\pi d^2 \gamma^2 x^4 dx}{288L^2 E} = \frac{\pi d^2 \gamma^2}{288L^2 E}\left[\frac{x^5}{5}\right]_0^L = \frac{\pi d^2 \gamma^2 L^3}{1440E} \text{ units}$$

Example 10.6 Strain energy due to impact load

A vertically hung bar is 2 m long and has a diameter of 25 mm. A weight of 600 N is dropped from a height h on a collar attached to the lower end of the bar. Find the height of drop, if the stress in the bar is not to exceed 100 MPa. Also find the maximum weight that can be dropped from this height without causing any permanent deformation. The elastic limit is 220 N/mm^2 and $E = 200$ GN/m^2.

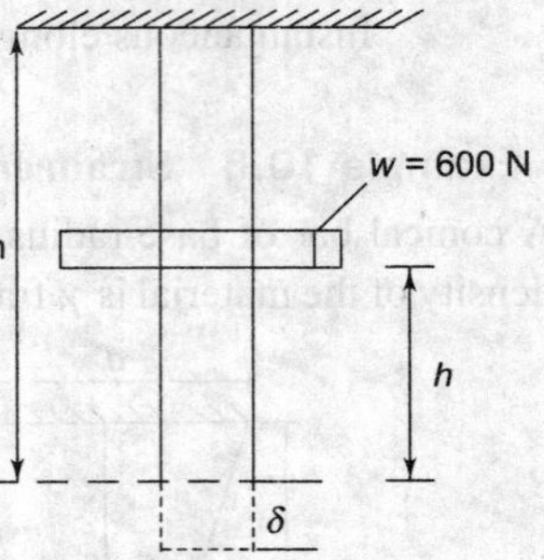

Fig. 10.10

Solution If the stress in the bar is 100 MPa, the elongation will be

$$\delta = \frac{100 \times 2000}{2 \times 10^5} = 1 \text{ mm}$$

The work done by the falling weight, referring to Fig. 10.10, is given by

$$W(h + \delta) = 600(h + 1)$$

$$\text{Strain energy of the bar} = \frac{\sigma^2}{2E} \times \text{volume}$$

$$= \frac{(100)^2}{2 \times 2 \times 10^5} \times \frac{\pi}{4} \times 252 \times 2000$$

$$= 24{,}544 \text{ N mm}$$

Equating the work done to the strain energy,

$$600(h + 1) = 24{,}544$$

$$\Rightarrow \quad h = 39.9 \text{ mm}$$

For the stress in the bar at 220 N/mm^2, the instantaneous elongation will be

$$\delta = \frac{220 \times 2000}{2 \times 10^5} = 2.2 \text{ mm}$$

If the weight is equal to W,

$$\text{Work done} = W(39.9 + 2.2) = 42.1W$$

$$\text{Strain energy of the bar} = \frac{(220)^2}{2 \times 2 \times 10^5} \times \frac{\pi}{4} \times 25^2 \times 2000$$

Equating the work done to the strain energy,

$$42.1W = \frac{(220)^2}{2 \times 2 \times 10^5} \times \frac{\pi}{4} \times 25^2 \times 2000$$

$$W = 28{,}216 \text{ N}$$

Example 10.7 Strain energy of a system of bars

In the simple bracket arrangement shown in Fig. 10.11, find the strain energy of the system due to the 50 kN load acting at B. Also find the vertical deflection of point B. Both the bars have a diameter of 25 mm. E for the material of the rods is 200 GN/m^2.

Solution From the free body diagram of point B shown in Fig. 10.11(b), the forces in the rods can be determined.

$$F_{AB} \sin 30 = 50$$

$$\Rightarrow \quad F_{AB} = 100 \text{ kN}$$

$$F_{AB} \cos 30 = F_{BC}$$

$$\Rightarrow \quad F_{BC} = 100 \cos 30 = 86.6 \text{ kN}$$

$$\text{Length of member } AB = \frac{2}{\cos 30} = 2.31 \text{ m}$$

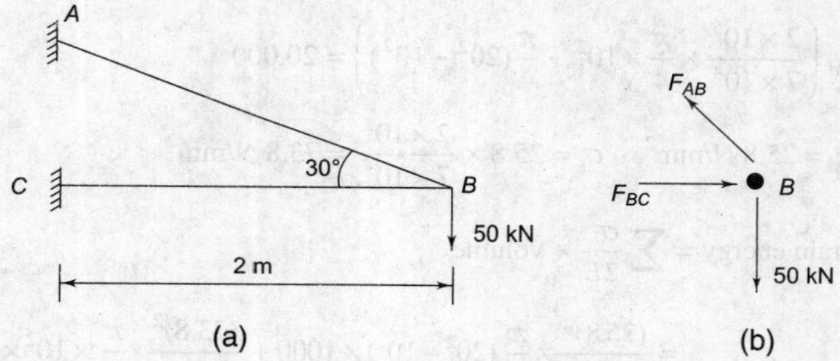

Fig. 10.11

The strain energy U of the system is the sum of the strain energies stored in the bars AB and BC. Thus,

$$U = \sum \frac{P^2 L}{2AE} = \frac{(100{,}000)^2 \times 2310}{2 \times 491 \times 200{,}000} + \frac{(86{,}600)^2 \times 2000}{2 \times 491 \times 200{,}000}$$

$$= 194 \text{ N m}$$

To find the vertical deflection of point B, we equate the external work to the strain energy.

$$\frac{1}{2} \times 50{,}000 Y_B = 194 \text{ N m}$$

$$\Rightarrow \quad Y_B = 0.00776 \text{ m}$$

$$= 7.76 \text{ mm}$$

Example 10.8 Strain energy in a composite bar

The composite section shown in Fig. 10.12 is subjected to a load of 20 kN. Find the strain energy of the composite bar under this load. E = 200 GPa for steel and 70 GPa for aluminium.

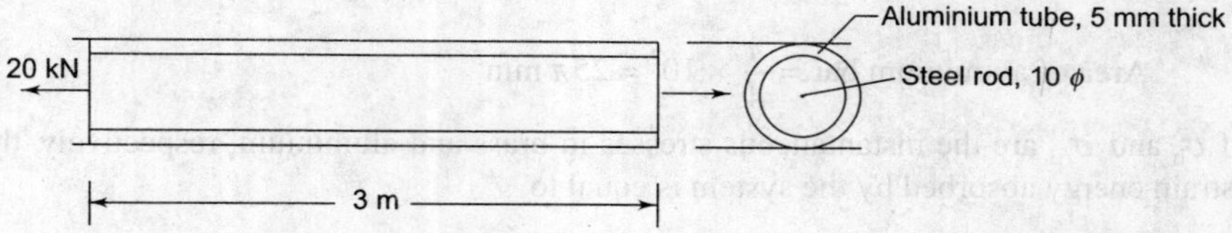

Fig. 10.12

Solution We first analyse the composite section to find the stresses in the two materials. If σ_s and σ_{Al} are the stresses in steel and aluminium respectively, then $\sigma_s A_s + \sigma_{Al} A_{Al} = 20{,}000$.
where A_s and A_{Al} are the areas of the steel rod and aluminium tube, respectively. The strain in the two materials is the same as they have the same length.

$$\frac{\sigma_s}{E_s} = \frac{\sigma_{Al}}{E_{Al}}, \quad \sigma_s = \sigma_{Al}\frac{E_s}{E_{Al}}$$

Substituting this value of σ_s in the first equation,

$$\sigma_{Al}\frac{E_s}{E_{Al}}A_s + \sigma_{Al}A_{Al} = 20{,}000$$

$$\sigma_{Al}\left(\frac{E_s}{E_{Al}}A_s + A_{Al}\right) = 20{,}000$$

$$\sigma_{Al}\left(\frac{2\times10^5}{7\times10^4}\times\frac{\pi}{4}\times10^2 + \frac{\pi}{4}(20^2 - 10^2)\right) = 20{,}000$$

$$\sigma_{Al} = 25.8 \text{ N/mm}^2, \quad \sigma_s = 25.8\times\frac{2\times10^5}{7\times10^4} = 73.8 \text{ N/mm}^2$$

$$\text{Strain energy} = \sum\frac{\sigma^2}{2E}\times\text{volume}$$

$$= \frac{(25.8)^2}{2E}\times\frac{\pi}{4}(20^2 - 10^2)\times1000 + \frac{(73.8)^2}{2E_S}\times\frac{\pi}{4}\times10^2\times1000$$

$$= 4.73 \text{ N m}$$

Example 10.9 Strain energy in a stepped bar due to impact load

A brass bar of diameter 20 mm and length 3 m, and an aluminium bar of diameter 10 mm and length 1.5 m are rigidly attached end to end and hung as shown in Fig. 10.13. A weight of 80 N is allowed to fall freely from a height of 500 mm to the collar attached to the free end. Find the instantaneous stress and elongation of the bar. For brass, $E = 1\times10^5$ and for aluminium, $E = 0.7\times10^5$ in GPa units.

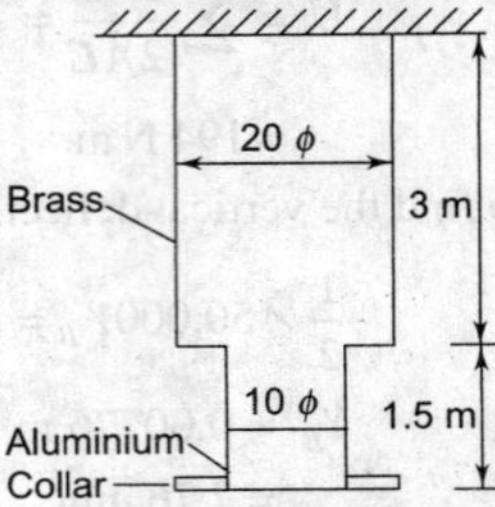

Fig. 10.13

Solution Assuming that the height of fall of 500 mm is much greater than the deflection δ, we can say that

$$\text{Potential energy lost by the falling weight} = 80\times500 = 40{,}000 \text{ Nmm}$$

$$\text{Area of brass bar} = \frac{\pi}{4}\times20^2 = 100\pi \text{ mm}^2$$

$$\text{Area of aluminium bar} = \frac{\pi}{4}\times10^2 = 25\pi \text{ mm}^2$$

If σ_b and σ_{Al} are the instantaneous stresses in brass and aluminium, respectively, then the strain energy absorbed by the system is equal to

$$U = \frac{\sigma_b^2}{2E_b}A_b L_b + \frac{\sigma_{Al}^2}{2E_{Al}}A_{Al}L_{Al}$$

Subscripts b and Al are for brass and aluminium, respectively. Since the net load acting on the bars is the same, σ_b and σ_{Al} will be inversely proportional to their areas.

$$\sigma_b = \frac{P}{100\pi}, \quad \sigma_{Al} = \frac{P}{25\pi}$$

Eliminating P,

$$\frac{\sigma_b}{\sigma_{Al}} = \frac{25\pi}{100\pi} = \frac{1}{4} \quad \sigma_{Al} = 4\sigma_b$$

$$U = \frac{\sigma_b^2 100\pi \times 3000}{2 \times 1 \times 10^5} + \frac{(4\sigma_b)^2 25\pi \times 1500}{2 \times 0.7 \times 10^5}$$

$$= 18.177\sigma_b^2$$

Equating the strain energy to the work done,

$$18.177\sigma_b^2 = 40{,}000$$

$$\sigma_b = 46.9 \text{ N/mm}^2, \quad \sigma_{Al} = 187.6 \text{ N/mm}^2$$

The instantaneous deflection,

$$\delta = \frac{\sigma_b L_b}{E_b} + \frac{\sigma_{Al} L_{Al}}{E_{Al}}$$

$$= \frac{46.9 \times 3000}{1 \times 10^5} + \frac{187.6 \times 1500}{0.7 \times 10^5}$$

$$= 5.427 \text{ mm}$$

Example 10.10 Strain energy due to impact load

A bar of length 1.2 m and diameter 25 mm hangs vertically, and has a collar attached rigidly to its lower end. When a mass of 5 kN is gradually applied to the collar, the extension is 0.05 mm. Find the maximum instantaneous deflection and stress if this load falls from a height of 2.5 mm on the collar.

Solution The value of E for the material of the bar can be calculated from the load and the corresponding extension given.

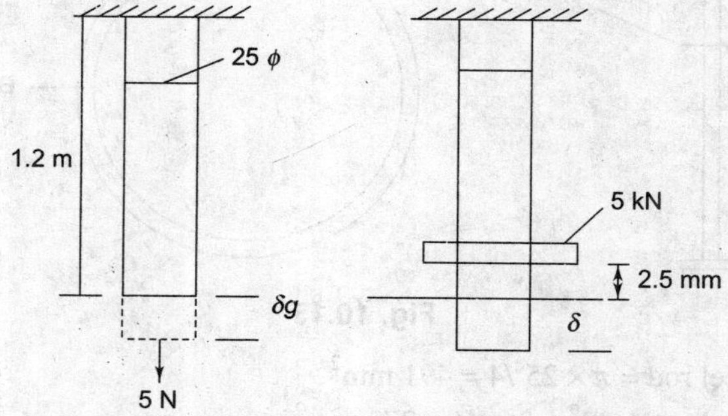

Fig. 10.14

From Fig. 10.14,

$$\delta = \frac{PL}{AE}, \quad 0.05 = \frac{5000 \times 1200}{491 \times E}$$

$$\Rightarrow \quad E = \frac{5000 \times 1200}{491 \times 0.05}$$

$$= 2.444 \times 10^5 \text{ N/mm}^2$$

If δ is the instantaneous deflection when the load falls on the collar from a height of 2.5 mm, then

Work done by the load = 5000 (2.5 + δ)

This is equal to the strain energy of the bar.

$$U = \frac{P^2 L}{2AE} = \frac{\sigma^2}{2E} AL = \frac{\delta^2 AE}{2}$$

$$\frac{\delta^2 AE}{2L} = 5000(2.5 + \delta)$$

$$\frac{\delta^2 491 \times 2.444 \times 10^5}{2 \times 1200} = 12{,}500 + 5000\,\delta$$

$$50{,}000\delta^2 - 5000\delta - 12{,}500 = 0, \quad \delta = 0.5525 \text{ mm}$$

The maximum instantaneous stress can be calculated from

$$5000(2.5 + \delta) = \frac{\sigma^2}{2E} AL$$

Setting $\delta = \sigma L/E$ and rearranging,

$$600 \times 491\sigma^2 - 5000 \times 491\sigma - 12{,}500 \times 2.444 \times 10^5 = 0$$

$$\sigma = 112.5 \text{ N/mm}^2$$

Example 10.11 Compound bar: impact load

A compound bar made of a steel rod 25 mm in diameter is enclosed in a brass tube of 26 mm internal diameter and 32 mm external diameter. The compound bar, 2400 mm long, is fixed at the upper end and is rigidly fixed to a rigid plate at the lower end. Determine the stresses in the bars when the combination is subjected to impact loading due to a weight of 10 kN falling through a height of 5 mm. *E* for steel = 200 GPa and *E* for brass is 100 GPa.

Solution The situation is illustrated in Fig. 10.15.

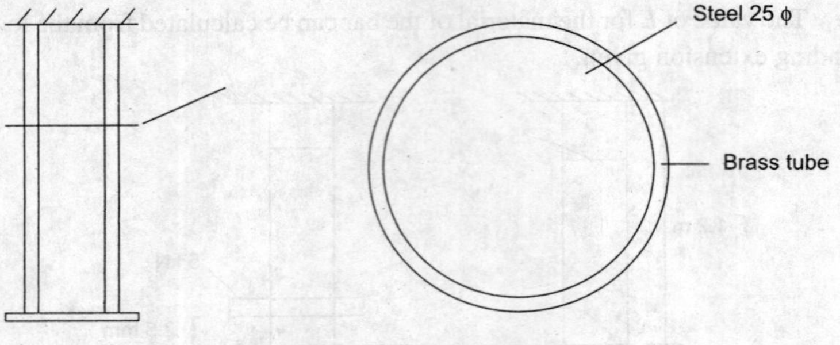

Fig. 10.15

The area of steel rod = $\pi \times 25^2/4 = 491$ mm^2

Area of brass tube = $\pi(32^2 - 26^2)/4 = 273$ mm^2

Let σ_s be the stress in the steel rod and σ_b be the stress in the brass tube.

As the bar and the tube are tied together at the ends, the strains in the two materials are equal, i.e. $\varepsilon_s = \varepsilon_b$.

or $$\frac{\sigma_s}{E_s} = \frac{\sigma_b}{E_b}$$

or $\sigma_s/(2 \times 10^5) = \sigma_b/(1 \times 10^5)$; $\sigma_s = 2\sigma_b$

($E = 200$ GPa $= 200 \times 10^9$ N/m$^2 = 2 \times 10^5$ N/mm^2)

$$\text{Strain energy} = \frac{\sigma^2 \times \text{volume}}{2E}$$

Length of the rod and tube being the same, the volume is proportional to the area.

$$\text{Strain energy of steel rod} = \frac{\sigma_s^2 \times 491L}{2\times2\times10^5} = 122.75 \times 10^{-5}\sigma_s^2 L$$

$$= 491 \times 10^{-5}\ \sigma_b^2 L \qquad (\text{as } \sigma_s = 2\sigma_b)$$

$$\text{Strain energy in the brass tube} = \frac{\sigma_b^2 \times 273L}{2\times1\times10^5} = 136.5 \times 10^{-5}\sigma_b^2 L$$

$$\text{Total strain energy} = 627.5 \times 10^{-5}\sigma_b^2 L$$

If Δ mm is the elongation of the bar due to falling weight, then the total work done is

$$\text{Work done} = 10{,}000(5 + \Delta)\ \text{Nmm}$$

$$\text{Strain in the brass tube, } \varepsilon_b = \Delta/L = \sigma_b/E_b$$

or $\varepsilon_b = 1 \times 10^{-5}\ \sigma_b L$

$$\text{Work done by falling weight} = 10{,}000(5 + 1 \times 10^{-5}\sigma_b L)$$

Equating the strain energy and the work done by the falling weight,

$$627.5 \times 10^{-5}\sigma_b^2 \times 2400 = 10{,}000(5 + 1 \times 10^{-5}\sigma_b \times 2400)$$

or $15.06\sigma_b^2 = 50{,}000 + 240\sigma_b$

This leads to a quadratic equation,

$$\sigma_b^2 - 16\sigma_b - 3320 = 0$$

Solving this equation, we get

$$\sigma_b = 66.17\ \text{N/mm}^2$$

$$\sigma_s = 2\sigma_b = 132.34\ \text{N/mm}^2$$

10.5 STRAIN ENERGY DUE TO BENDING

Consider a simply supported beam subjected to pure bending, as shown in Fig. 10.16, and a small length ds of the beam subtending an angle $d\theta$ at the centre. Recalling the bending equation $M/I = \sigma/y = E/R$ derived in Chapter 5, and if R is the radius of curvature,

$$ds = R\ d\theta, \quad d\theta = \frac{ds}{R} = \frac{M\,ds}{EI}$$

The flexural rotation is thus directly proportional to M. If the couple M is gradually applied, the rotation $d\theta$ also gradually increases linearly as shown in Fig. 10.16(c).

$$\text{External work done} = \frac{1}{2}M\,d\theta = \frac{M^2 ds}{2EI} = \frac{EI(d\theta)^2}{2ds}$$

$$\text{Strain energy } dU = \text{external work done} = \frac{M^2 ds}{2\ EI}$$

Integrating this over the whole length and noting that M is constant,

$$U = \int_0^l \frac{M^2 ds}{2\ EI} = \frac{M^2 l}{2\ EI}$$

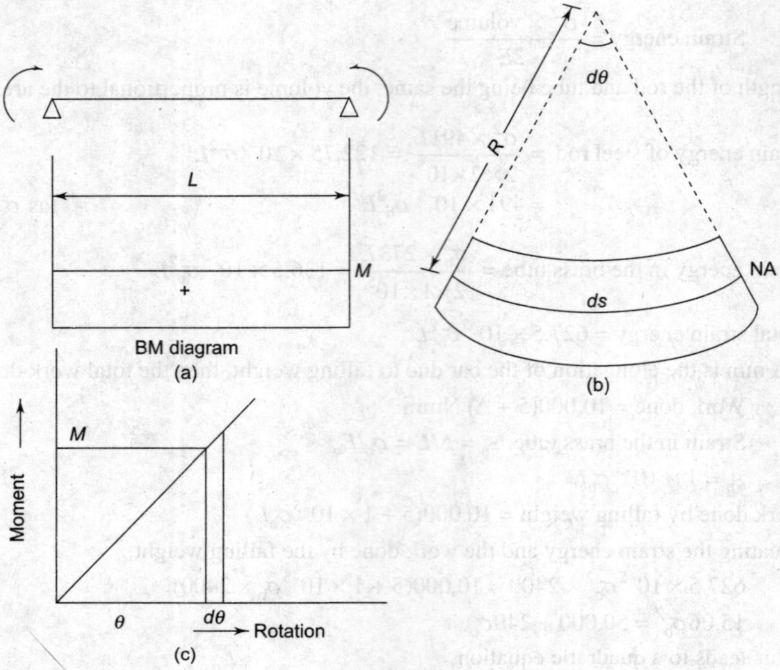

Fig. 10.16

This can also be expressed in terms of maximum stress σ_{max} in the beam as

$$\sigma_{max} = \frac{M}{I} y_{max}$$

$$M = \frac{\sigma_{max} I}{y_{max}}$$

$$U = \frac{M^2 l}{2EI} = \frac{(\sigma_{max})^2 I}{2E(y_{max})^2}$$

In the case of practical beams, shown in Fig. 10.17, the BM varies along the length. Assuming that the BM is constant over a small length dx of the beam,

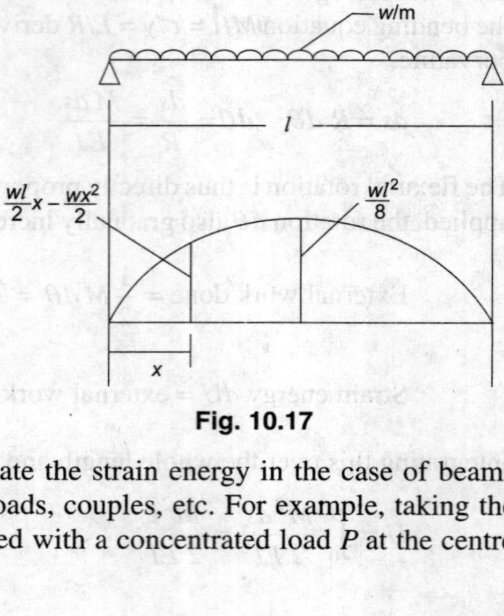

Fig. 10.17

$$dU = \frac{M^2 dx}{2E}$$

and $$U = \int_0^l \frac{M^2 dx}{2EI}$$

This expression can be used to calculate the strain energy in the case of beams loaded with concentrated loads, UD loads, couples, etc. For example, taking the case of a simply supported beam loaded with a concentrated load P at the centre shown in Fig. 10.18(a), BM = $Px/2$ at a distance x from the supports. Noting that the BM diagram is symmetrical about the midpoint, the strain energy can be calculated for half the length of the beam and then doubled.

$$U = 2\int_0^{l/2} \frac{M^2\,dx}{2EI} = 2\int_0^{l/2} \frac{P^2x^2}{8EI}\,dx = \frac{P^2l^3}{96EI}$$

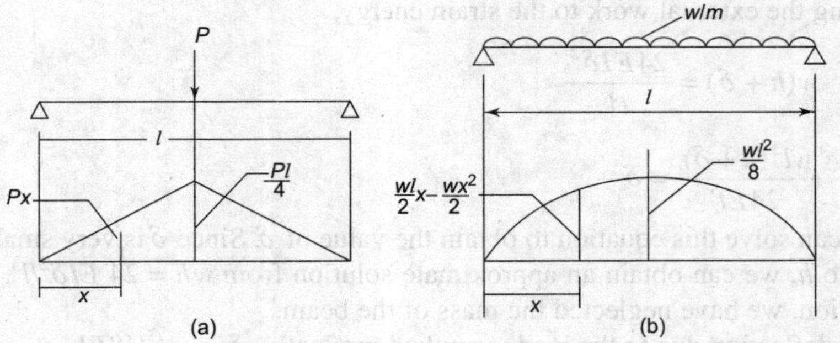

Fig. 10.18

In the case of the simply supported beam carrying a UD load w/unit length,

$$M_x = \frac{wl}{2}x - \frac{wx^2}{2}$$

and this equation is valid for the whole length of the beam. Therefore,

$$U = \int_0^l \left(\frac{wl}{2}x - \frac{wx^2}{2}\right)^2 \frac{dx}{2EI} = \int_0^l \left(\frac{w^2l^2}{4}x^2 + \frac{w^2x^4}{4} - \frac{w^2lx^3}{2}\right)\frac{dx}{EI}$$

$$= \frac{w^2l^5}{240EI}$$

10.5.1 Impact Loading on Beams

The concept of strain energy is eminently suited to tackle cases of impact loading on beams. Consider the case of the simply supported beam shown in Fig. 10.19. A load w is dropped through a height h at the midpoint of the beam. Let δ be the instantaneous central deflection due to the impact load. If the beam is subjected to a central concentrated load P, the deflection at the centre is $Pl^3/48EI$. Thus, from the instantaneous deflection δ, the equivalent static load P_e that causes this deflection can be evaluated.

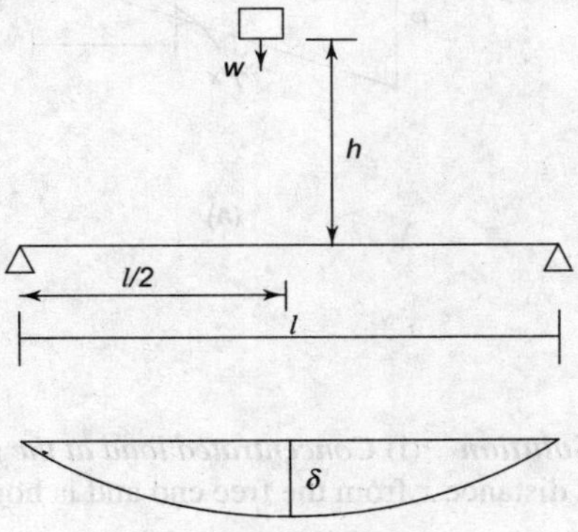

Fig. 10.19

$$\frac{P_e l^3}{48EI} = \delta$$

$$\Rightarrow \qquad P_e = \frac{48EI\delta}{l^3}$$

Work done by external load = $w\ (h + \delta)$. In terms of the equivalent static load producing the same deflection, the strain energy is equal to the work done by this equivalent static load.

$$U = \frac{1}{2} P_e \delta = \frac{24EI\delta^2}{l^3}$$

Equating the external work to the strain energy,

$$w(h+\delta) = \frac{24EI\delta^2}{l^3}$$

$$\frac{wl^3(h+\delta)}{24EI} = \delta^2$$

We can solve this equation to obtain the value of δ. Since δ is very small compared to h, we can obtain an approximate solution from $wh = 24\ EI\delta^2/l^3$. In this derivation, we have neglected the mass of the beam.

The deflection due to the load w applied gradually, $\delta_g = wl^3/48EI$.

$$\frac{wl^3}{24EI}(h+\delta) = \delta^2, \quad 2\delta_g\ (h+\delta) = \delta^2$$

From this,

$$\delta = \delta_g + \sqrt{(\delta_g)^2 + 2h(\delta_g)}$$

Note that we have taken into consideration only the bending stresses and not the effect of shear.

Example 10.12 Strain energy of a cantilever beam with point load and UD load

For the cantilever beam shown in Fig. 10.20, find the strain energy when it is subjected to (i) a point load P at the free end and (ii) a UD load over the whole length.

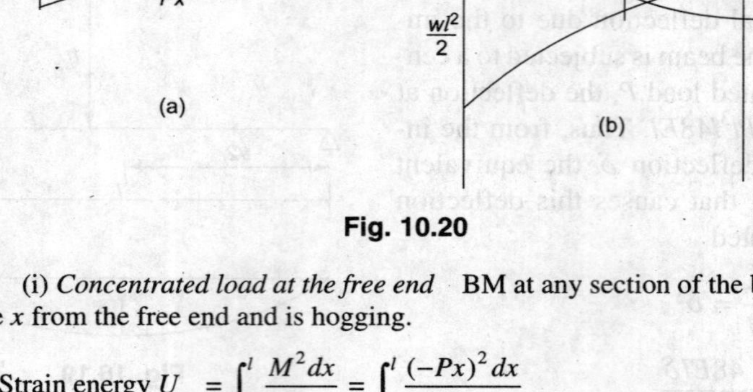

Fig. 10.20

Solution (i) *Concentrated load at the free end* BM at any section of the beam = $-Px$ at a distance x from the free end and is hogging.

$$\text{Strain energy } U = \int_0^l \frac{M^2\,dx}{2EI} = \int_0^l \frac{(-Px)^2\,dx}{2EI}$$

$$= \frac{P^2 l^3}{6EI}$$

(ii) *UD load* BM at section x-$x = -\frac{wx^2}{2}$

$$\text{Strain energy } U = \int_0^l \frac{M^2 dx}{2EI} = \int_0^l \left(\frac{-wx^2}{2}\right)^2 \frac{dx}{2EI}$$

$$= \frac{w^2 l^5}{40EI}$$

10.5.2 Finding Deformations in Beams

When a beam is subjected to a single point load or a couple, the deflection in the direction of the load or the rotation at the point of application of the couple can be evaluated by equating the strain energy to the external work done. It should be noted that it is not possible to do so by this method in the case of multiple loads acting on the beam.

Example 10.13 Deflection using work-energy principle

Find the deflection at the free end of a cantilever beam carrying a point load P at that end. Flexural rigidity = EI.

Solution Referring to Fig. 10.21,

$$\text{Work done by the applied load} = \frac{1}{2} P\delta$$

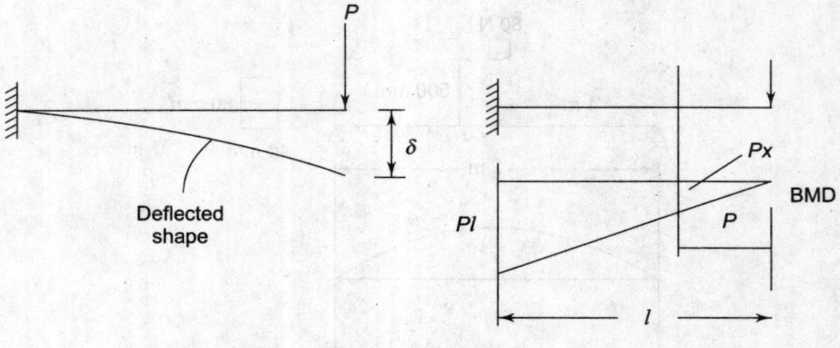

Fig. 10.21

From Example 10.12, the strain energy of a cantilever subjected to a point load at the free end is $P^2l^3/6EI$. Equating external work done to strain energy,

$$\frac{1}{2} P\delta = \frac{P^2 l^3}{6EI}$$

or $$\delta = \frac{Pl^3}{3EI}$$

Example 10.14 Impact loading on a cantilever beam

The cantilever beam shown in Fig. 10.22 has a span of 1 m and is a rod of diameter 30 mm. A weight of 20 N is dropped from a height of 50 mm at the free end of the beam. If $E = 200$ GN/m^2, find the maximum instantaneous deflection at that end.

Solution For the beam section,

$$I = \frac{\pi d^4}{64} = \frac{\pi \times 30^4}{64} = 39{,}761 \text{ mm}^4$$

If P_e is the equivalent static load,

$$\text{Strain energy} = \frac{P_e^2 l^3}{6EI} \text{ (from Example 10.12)}$$

This is equal to the work done by the external load.

$$\text{External work} = 20\,(h + \delta) = 20\,(50 + \delta)$$

$$\text{Equivalent load } P_e = \frac{3EI\delta}{l^3}$$

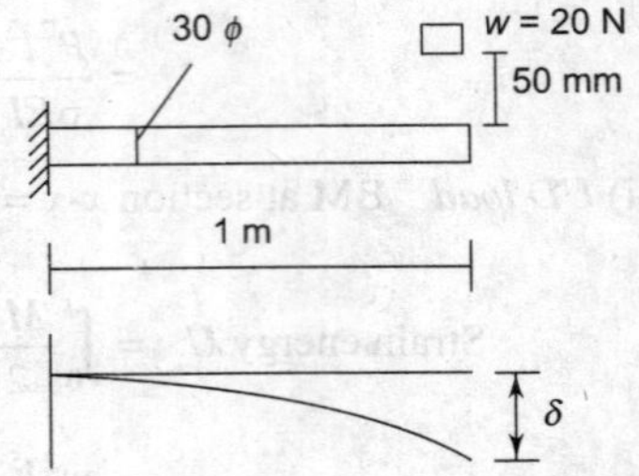

Fig. 10.22

Substituting,

$$\text{Strain energy} = \left(\frac{3EI\delta}{l^3}\right)^2 \frac{l^3}{6EI} = \frac{9EI\delta^2}{6l^3}$$

$$\frac{9EI\delta^2}{6l^3} = 1000 + 20\delta, \quad \frac{9 \times 2 \times 10^5 \times 39761}{6 \times (1000)^3}\delta^2 - 20\delta - 1000 = 0$$

$$\Rightarrow \quad \delta = 10.03 \text{ mm}$$

Example 10.15 Impact loading on an SS beam

The cross section of a simply supported beam, of span 2 m, is shown in Fig. 10.23. If a weight of 80 N is dropped from a height of 500 mm into the centre of the beam, find the maximum instantaneous deflection. Also find the maximum instantaneous stress produced in the beam. $E = 200$ GN/m^2.

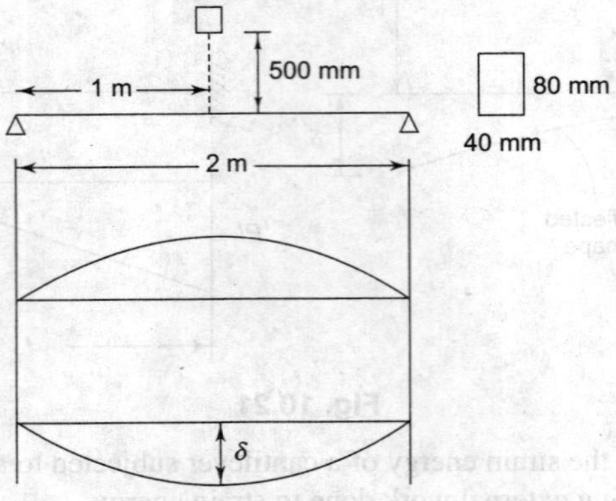

Fig. 10.23

Solution For the beam section, $I = 40 \times (80)^3/12 = 170.67 \times 10^4$ mm^4. In this case, we assume that h is much larger than the instantaneous deflection δ, and hence the external work done $= wh = 80 \times 500 = 40{,}000$ N mm.

The strain energy of a simply supported beam carrying a concentrated load at the centre was earlier shown to be $P_e^2 l^3/96EI$, where P_e is the equivalent load, applied gradually, causing the same maximum deflection. Therefore, equating the strain energy to the external work done,

$$\frac{P_e^2 l^3}{96EI} = 40{,}000$$

$$\Rightarrow \quad P_e = \sqrt{\frac{40{,}000 \times 96EI}{l^3}}$$

$$= \sqrt{\frac{40{,}000 \times 96 \times 2 \times 10^5 \times 170.67 \times 10^4}{(2000)^3}} = 12{,}800 \text{ N}$$

$$\text{Instantaneous deflection} = \frac{P_e l^3}{48EI} = \frac{12{,}800 \times (2000)^3}{48 \times 2 \times 10^5 \times 170.67 \times 10^4}$$

$$= 6.25 \text{ mm}$$

Example 10.16 Strain energy due to bending in a sheet

A steel sheet of dimensions 2000 mm × 500 mm × 0.4 mm is bent to form a cylinder 500 mm long. Find the strain energy stored in the material due to bending.

Solution Referring to Fig. 10.24,

$$\text{Radius } R \text{ of the cylinder} = \frac{2000}{2\pi} = 318.3 \text{ mm}$$

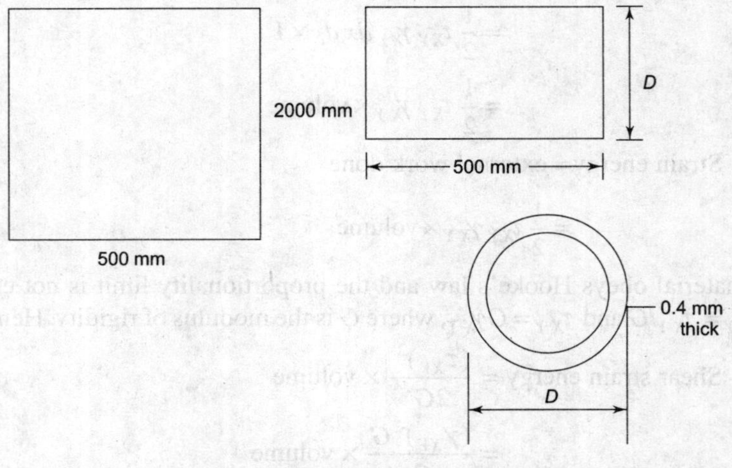

Fig. 10.24

The section of the plate has a width of 500 mm and a depth of 0.4 mm for bending. The MI of the section is, therefore, given by

$$I = \frac{1}{12} \times 500 \times (0.4)^3 = 2.67 \text{ mm}^4$$

The bending couple required is

$$M = \frac{EI}{R} = \frac{2 \times 10^5 \times 2.67}{318.3} = 1675.4 \text{ N mm}$$

The strain energy for pure bending is $M^2 l/2EI$.

$$U = \frac{(1675.4)^2 \times 2000}{2 \times 2 \times 10^5 \times 2.67} = 5256 \text{ N mm} = 5.256 \text{ N m (J)}$$

10.6 ELASTIC STRAIN ENERGY DUE TO SHEARING STRESSES

Consider an element subjected to plane shearing stresses τ_{XY}. Assuming the length perpendicular to the plane of the paper as unity, the force acting on the surface BC is

$\tau_{XY}\, dy \times 1$ (Fig. 10.25).

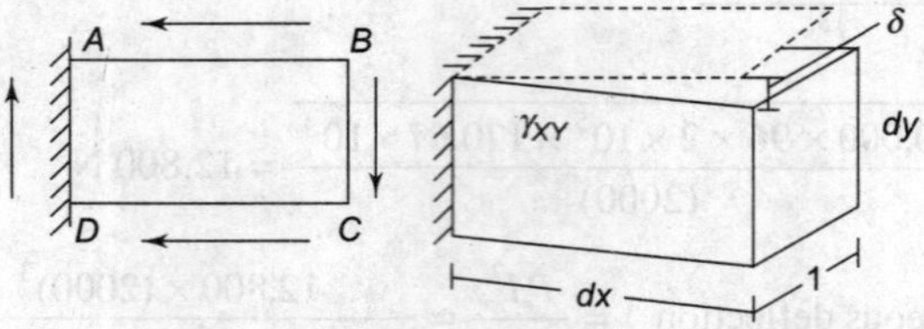

Fig. 10.25

As the shear force is gradually increased from zero to its final value, the shear stress increases from zero to τ_{XY}, and shear deformation from zero to γ_{XY}.

$$\text{Shear force } v = \tau_{XY}\, dy \times 1$$

$$\text{External work done} = \frac{1}{2} V \gamma_{XY}\, dx$$

$$= \frac{1}{2} \tau_{XY} \gamma_{XY}\, dx\, dy \times 1$$

$$= \frac{1}{2} \tau_{XY} \gamma_{XY} \times \text{volume}$$

Strain energy = external work done

$$= \frac{1}{2} \tau_{XY} \gamma_{XY} \times \text{volume}$$

If the material obeys Hooke's law and the proportionality limit is not exceeded, then $\gamma_{XY} = \tau_{XY}/G$ and $\tau_{XY} = G\gamma_{XY}$, where G is the modulus of rigidity. Hence

$$\text{Shear strain energy} = \frac{(\tau_{XY})^2}{2G} \times \text{volume}$$

$$= \frac{(\gamma_{XY})^2 G}{2} \times \text{volume}$$

In a beam subjected to lateral forces, the shear stresses vary across the section, and along the length of the beam. The shear strain energy can be calculated in such cases by integration.

Shear resilience As in the case of axial forces, shear resilience can be defined as the strain energy per unit volume.

Shear resilience = $\tau^2_{xy}/2G = \gamma^2_{xy}G/2$

Proof resilience is the strain energy within elastic limit.

Proof strain energy = $\tau^2_e/2G$, where τ_e is the shear stress within elastic limit.

10.7 ELASTIC STRAIN ENERGY DUE TO TORSION

We discussed methods of computing shearing stress and angle of twist in Chapter 8 for axisymmetric shafts. The basic torsion equation is

$$\frac{T}{J} = \frac{\tau}{r} = \frac{G\theta}{l}$$

In this expression, T is the torque, J is the polar MI, τ is the shear stress at a radial distance r, G is the modulus of rigidity, θ is the angle of twist, and l is the length of the shaft.

These quantities can be easily identified in Fig. 10.26. A torque T is applied gradually to the circular shaft at the free end. The angle of twist also increases correspondingly. So long as the elastic limits are not exceeded, the relation between the angle of twist θ and the applied torque T is linear as shown. The internal resisting torque at any instant is equal to the applied torque T.

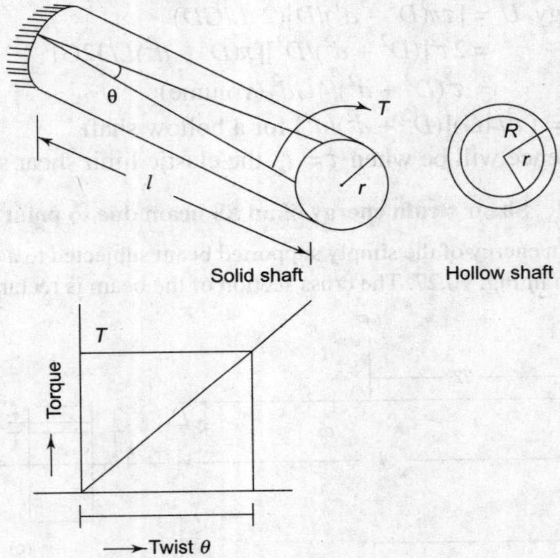

Fig. 10.26

External work done = $(T\theta)/2$ = area of triangle below the torque-twist diagram = strain energy.

$$\text{Strain energy} = \frac{1}{2}T\theta = \frac{1}{2}T\frac{TL}{GJ} = \frac{T^2L}{2GJ}$$

or in terms of the angle of twist,

$$\text{Strain energy} = \frac{1}{2}T\theta$$

$$= \frac{1}{2}\frac{G\theta J}{L}\theta = \frac{1}{2}\frac{GJ}{L}\theta^2$$

This expression is applicable only to circular shafts and hollow circular shafts.

Torsional resilience Torsional resilience (strain energy per unit volume) in the case of shafts can be calculated as follows.

(a) In the case of solid circular shaft, volume of shaft = $L\pi d^2/4$
 Strain energy, $U = T\theta/2$, for a gradually applied torque
 $U = T^2L/2GJ$, where $J = \pi d^4/32$, $T = \tau J/r$, and $\theta = \tau L/Gr$
 Substituting for T and θ,
 $U = T\theta/2 = (\tau J/r)(\tau L/Gr)/2 = (\tau^2/4G)(\pi d^2L/4) = (\tau^2/4G)$ volume

Torsional resilience = $\tau^2/4G$

Proof resilience for a solid shaft = $\tau^2/4G$, where τ_e is the elastic limit shear stress.

(b) In the case of a hollow circular shaft of external diameter D and internal diameter d,

Strain energy, $U = T\theta/2$

$T = 2\tau J/d$ and $\theta = 2\tau L/Gd$

$J = \pi(D^4 - d^4)/32$

Strain energy, $U = [\tau\pi(D^4 - d^4)/D](2\tau L/GD)$

$= 2\tau^2[(D^2 + d^2)/D^2][\pi(D^2 - d^2)L/32G]$

$= \tau^2(D^2 + d^2)/4Gd^2$ (volume)

Resilience = $(\tau^2/4G)[(D^2 + d^2)/d^2]$ for a hollow shaft

Proof resilience will be when $\tau = \tau_e$, the elastic limit shear stress.

Example 10.17 Shear strain energy of an SS beam due to point load

Find the shear strain energy of the simply supported beam subjected to a concentrated load at the centre shown in Fig. 10.27. The cross section of the beam is rectangular.

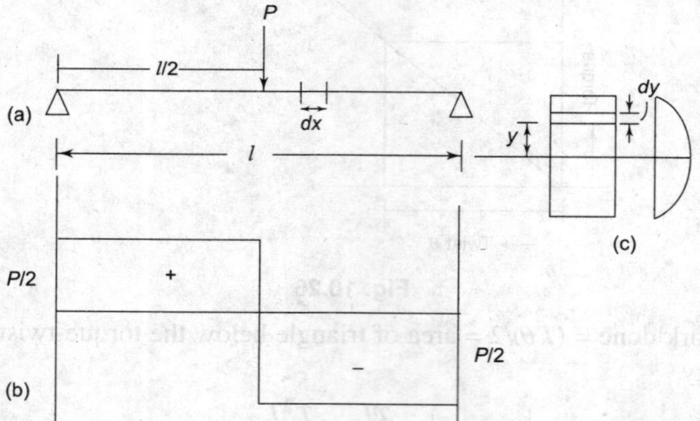

Fig. 10.27

Solution The variation of SF along the length of the beam and that of shear stress τ across the depth at any section are shown in Figures 10.27(b) and (c), respectively. We have studied in Chapter 5 that the shear stress at a section at any distance y from the neutral layer is

$$\tau = \frac{V}{Ib}A\bar{y} = \frac{V}{Ib}\frac{(d/2 - y)}{2} = \frac{V}{2I}\left(\frac{d^2}{4} - y^2\right)$$

Considering an elemental length dx and elemental depth dy as shown in Fig. 10.27, the shear strain energy of the elemental volume is

$$dU = \frac{\tau_{XY}^2}{2G} \times \text{volume} = \frac{V^2}{4I^2}\frac{(d^2/4 - y^2)^2}{2G}dx\,dy\,(b)$$

This expression can be integrated over the depth for a section and over the length for the whole beam. The expression contains the square of the SF and hence a negative sign for the SF will have no significance. Also note that the SF diagram is symmetrical about the middle of the length. Therefore, the total shear strain energy of the beam can be expressed as

$$U = 2\int_0^{l/2}\int_{-d/2}^{d/2} \frac{V^2}{8GI^2}\left(\frac{d^2}{4} - y^2\right)^2 b\,dx\,dy$$

$$= 2\int_0^{l/2} \frac{V^2 b}{8GI^2}\left(\int_{-d/2}^{+d/2}\left(\frac{d^2}{4} - y^2\right)^2 dy\right)dx$$

$$= 2\int_0^{l/2} \frac{V^2 b}{8GI^2}\frac{d^5}{30}dx = \frac{V^2 d^2 bl}{240\,GI^2}$$

For the rectangular section, $I = bd^3/12$ and $A = bd$.

$$U = \frac{3}{5}\frac{V^2 l}{AG} = \frac{3}{20} \times \frac{P^2 l}{AG}$$

where $V = \pm P/2$.

Example 10.18 Shear strain energy of a cantilever with point load

A cantilever beam of rectangular section carries a concentrated load at the free end. Find the total strain energy of the beam taking into account both bending and shear. What is the relative magnitude of bending and shear strain energies? State why shear strain energy is usually neglected.

Solution Referring to Fig. 10.28, if δ is the deflection at the free end due to the concentrated load, then

$$\text{External work done} = \frac{1}{2}P\delta = \text{strain energy } U_B$$

$$\delta = \frac{Pl^3}{3EI}$$

Therefore, $U_B = P^2l^3/6EI$ due to bending.

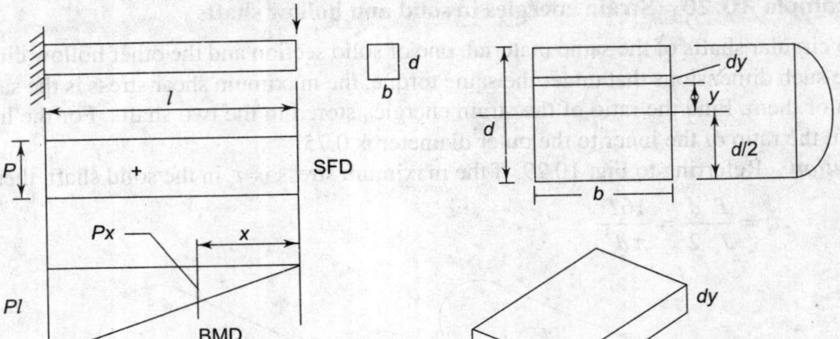

Fig. 10.28

The SF is constant along the length of the beam, and is equal to P. The variation of shear stress over the depth is parabolic as shown. We obtain the shear strain energy U_s, for a small length dx by integration from $-d/2$ to $+d/2$, and then by integration over the length l.

$$\tau = \frac{V}{Ib}A\bar{y} = \frac{Pb(d/2 - \bar{y})^2}{(1/2)\,bd^3 b \times 2} = \frac{3}{2} \times \frac{P}{bd}\left(1 - \frac{4\bar{y}^2}{d^2}\right)$$

$$dU_s = \frac{\tau \times \text{volume}}{2G} = \frac{1}{2G}\left(\frac{3}{2} \times \frac{P}{bd}\right)^2 \left(1 - \frac{4y^2}{d^2}\right)^2 b\,dy\,dx$$

where $b\,dy\,dx$ is the volume of the element as shown in Fig. 10.28. Integrating,

$$U_s = \int_0^l \int_{-d/2}^{+d/2} \frac{9P^2}{8Gbd^2}\left(1 - \frac{4y^2}{d^2}\right)^2 dy\,dx$$

$$= \frac{3P^2 l}{5Gbd} = \frac{3P^2 l}{5GA}$$

$$\text{Total strain energy} = U_B + U_S = \frac{P^2 l^3}{6EI} + \frac{3P^2 l}{5GA}$$

Noting that $I = bd^3/12$ and $A = bd$ for the rectangular section,

$$U = \frac{P^2 l^3}{GEI}\left(1 + \frac{3Ed^2}{10Gl^2}\right)$$

Taking $G \approx E/3$, the second term within parentheses is about $0.9\,(d/l)^2$. If d/l = 1/10, then the shear strain energy is only 0.9% of the bending strain energy. Thus, except in the case of short and deep beams, shear strain energy can be neglected.

Example 10.19 Torsional strain energy of a solid circular shaft

A solid circular shaft is 4 m long and has a diameter of 80 mm. Find the torsional strain energy stored in it when it is subjected to a torque of 200 Nm. G = 80 GPa.

Solution Strain energy $U = \dfrac{T^2 l}{2GJ} = \dfrac{(200{,}000)^2 \times 4000}{2 \times 0.8 \times 10^5 \times (\pi/32)(80)^4}$

$= 248.7$ Nmm

Example 10.20 Strain energies in solid and hollow shafts

Two circular shafts of the same material, one of solid section and the other hollow circular, have such dimensions that under the same torque, the maximum shear stress is the same in both of them. Find the ratio of the strain energies stored in the two shafts. For the hollow shaft, the ratio of the inner to the outer diameter is 0.75.

Solution Referring to Fig. 10.29, if the maximum stress is τ_s in the solid shaft, then

$$\tau_s = \frac{T}{J_s}\frac{d}{2} = \frac{16T}{\pi d^3}$$

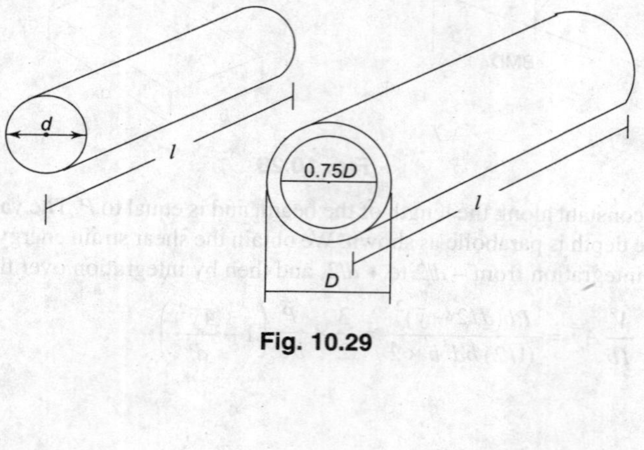

Fig. 10.29

where T is the applied torque and $J_s = \pi d^4/32$. Under the same torque, for the hollow circular shaft of outer diameter D and inside diameter $0.75D$,

$$\text{Maximum stress } \tau_h = \frac{TD/2}{(\pi/32)[D^4 - (0.75D)^4]} = \frac{16T \times 256}{\pi D^3 \times 175}$$

Since $\tau_s = \tau_h$,

$$\frac{16T}{\pi d^3} = \frac{16T}{\pi D^3} \times \frac{256}{175}$$

$$\frac{D}{d} = 3\sqrt{\frac{256}{175}} = 1.135$$

The strain energy of the solid shaft,

$$U_s = \frac{T^2 l}{2GJ_s}$$

The strain energy of the hollow shaft,

$$U_h = \frac{T^2 l}{2GJ_h}$$

$$\frac{U_s}{U_h} = \frac{J_h}{J_s} = \frac{(\pi/32)[D^4 - (0.75D^4)]}{(\pi/32)\,d^4} = \frac{175D^4}{256 \times d^4} = \frac{175}{256}\left(\frac{D}{d}\right)^4$$

$$= \frac{175}{256}(1.135)^4 = 1.1344$$

Example 10.21 Strain energy of a tapering shaft

Derive an expression for the strain energy stored in a shaft tapering from a diameter D to a diameter d over a length L, subjected to a torque T.

Solution Consider the shaft shown in Fig. 10.30. The shearing stress varies along the length of the beam due to the changing diameter. The strain energy of the shaft can be found by assuming a constant stress over a small length dx, and then integrating over the whole length.

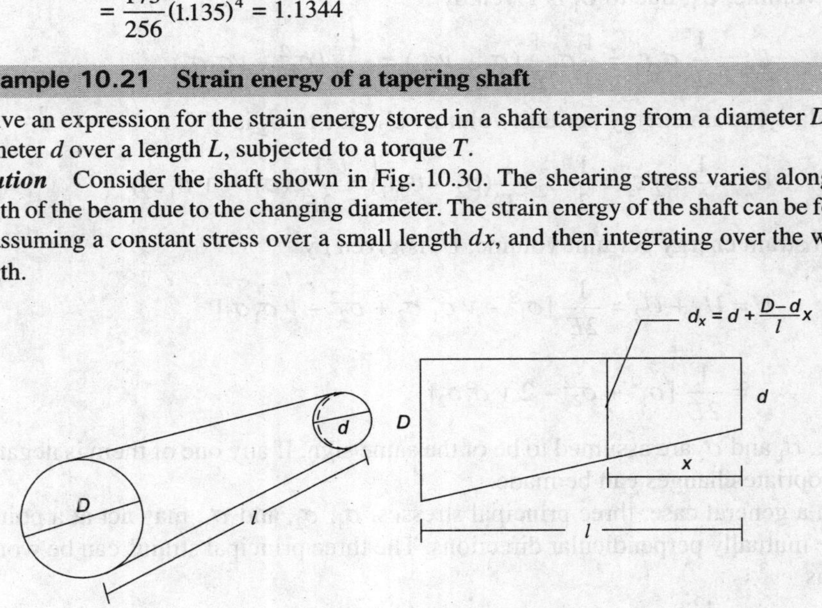

Fig. 10.30

At a distance x from the smaller end, diameter $d_x = d + kx$, where $k = (D - d)/l$. We consider a small elemental length $dxJ = \pi dx^4/32$ and assuming the shear stress is constant over dx, the torsional strain energy of the length dx of the shaft,

$$dU = \frac{T^2 dx}{2GJ} = \frac{T^2 dx}{2(\pi/32)(d + kx)^4} = \frac{16T^2 dx}{G\pi(d + kx)^4}$$

The total torsional strain energy of the shaft is given by

$$U = \int_0^l \frac{16T^2 dx}{G\pi(d+kx)^4} = \frac{16T^2}{G\pi}\int_0^l \frac{dx}{(d+kx)^4} = \frac{16T^2}{G\pi}\frac{(-1)}{3k}\left[\frac{1}{(d+kx)^3}\right]_0^l$$

$$= \frac{16T^2}{3\pi Gk}\left[\frac{-1}{D^3}+\frac{1}{d^3}\right] = \frac{16T^2 L}{3\pi G(D-d)}\left[-\frac{1}{D^3}+\frac{1}{d^3}\right]$$

10.8 STRAIN ENERGY UNDER COMPOUND STRESS

So far, we considered strain energy in terms of normal stresses, shear stress, bonding stress, torsional shear stress, etc. As discussed in Chapter 9, the combined effect of such stresses acting at a point results in principal stresses and maximum shear stress. We will briefly discuss strain energy in such cases.

(i) *Plain stress* In the plain stress situation, there are two principal stresses σ_1 and σ_2 at a point. The principal strains due to the principal stresses can be calculated as

$$\varepsilon_1 = \frac{1}{E}(\sigma_1 - \nu\sigma_2) \quad \text{and} \quad \varepsilon_2 = \frac{1}{E}(\sigma_2 - \nu\sigma_1)$$

where E is the modulus of elasticity and ν is the Poisson's ratio.

Strain energy per unit volume is given by 1/2 $\sigma\varepsilon$. Therefore, strain energy per unit volume, U_1, due to σ_1 is given by

$$U_1 = \frac{1}{2}\sigma_1\varepsilon_1 = \frac{1}{2}\sigma_1\frac{1}{E}(\sigma_1 - \nu\sigma_2) = \frac{1}{2E}(\sigma_1^2 - \nu\sigma_1\sigma_2)$$

Similarly, strain energy per unit volume, U_2, due to σ_2 is

$$U_2 = \frac{1}{2}\sigma_2\varepsilon_2 = \frac{1}{2}\sigma_2\frac{1}{E}(\sigma_2 - \nu\sigma_1) = \frac{1}{2E}(\sigma_2^2 - \nu\sigma_1\sigma_2)$$

Total strain energy per unit volume, U, is given by

$$U = U_1 + U_2 = \frac{1}{2E}[\sigma_1^2 - \nu\sigma_1\sigma_2 + \sigma_2^2 - \nu\sigma_1\sigma_2]$$

$$= \frac{1}{2E}[\sigma_1^2 + \sigma_2^2 - 2\nu\sigma_1\sigma_2]$$

Here, σ_1 and σ_2 are assumed to be of the same sign. If any one of them is negative, appropriate changes can be made.

In a general case, three principal stresses, σ_1, σ_2, and σ_3, may act at a point in three mutually perpendicular directions. The three principal strains can be worked out as

$$\varepsilon_1 = \frac{1}{E}[\sigma_1 - \nu(\sigma_2 + \sigma_3)]$$

$$\varepsilon_2 = \frac{1}{E}(\sigma_2 - \nu(\sigma_3 + \sigma_1)]$$

$$\varepsilon_3 = \frac{1}{E}[\sigma_3 - \nu(\sigma_1 + \sigma_2)]$$

The total strain energy per unit volume is

$$U = U_1 + U_2 + U_3 = \frac{1}{2}\sigma_1\varepsilon_1 + \frac{1}{2}\sigma_2\varepsilon_2 + \frac{1}{2}\sigma_3\varepsilon_3$$

Substituting for ε_1, ε_2, and ε_3, we get

$$U = \frac{\sigma_1}{2E}[\sigma_1 - \nu(\sigma_2 + \sigma_3)] + \frac{\sigma_2}{2E}[\sigma_2 - \nu(\sigma_3 + \sigma_1)] + \frac{\sigma_3}{2E}[\sigma_3 - \nu(\sigma_1 + \sigma_2)]$$

$$= \frac{1}{2E}[\sigma_1^2 + \sigma_2^2 + \sigma_3^2 - (2\,\sigma_1\,\sigma_2 + 2\,\sigma_2\,\sigma_3 + 2\,\sigma_3\,\sigma_1)\nu]$$

Here again, σ_1, σ_2, and σ_3 are assumed to be of the same sign.

Let us express σ_1, σ_2, and σ_3 as

$$\sigma_1 = \frac{1}{3}(\sigma_1 + \sigma_2 + \sigma_3) + \frac{1}{3}(\sigma_1 - \sigma_2) + \frac{1}{3}(\sigma_1 - \sigma_3)$$

$$\sigma_2 = \frac{1}{3}(\sigma_1 + \sigma_2 + \sigma_3) + \frac{1}{3}(\sigma_2 - \sigma_3) + \frac{1}{3}(\sigma_2 - \sigma_1)$$

$$\sigma_3 = \frac{1}{3}(\sigma_1 + \sigma_2 + \sigma_3) + \frac{1}{3}(\sigma_3 - \sigma_1) + \frac{1}{3}(\sigma_3 - \sigma_2)$$

Here, the first terms in these expressions are normal stresses and the second and third terms in each expression are shear stresses see discussion in Chapter 9). The normal stress $(\sigma_1 + \sigma_2 + \sigma_3)/3$ acts in each of the directions of the principal stresses σ_1, σ_2, and σ_3. The strain energy due to the normal stresses can be worked out as

$$\varepsilon_1 = \varepsilon_2 = \varepsilon_3 = \frac{\sigma_1 + \sigma_2 + \sigma_3}{3E}(1 - 2\nu)$$

as each of the normal stresses = $(\sigma_1 + \sigma_2 + \sigma_3)/3$.

$$U = \frac{1}{2}\sigma\varepsilon$$

$$U_n\text{ (normal stress)} = \frac{3}{2E}\left[\frac{\sigma_1 + \sigma_2 + \sigma_3}{3}\right]\left[\frac{(\sigma_1 + \sigma_2 + \sigma_3)}{3}(1 - 2\nu)\right]$$

This can be reduced to

$$U_n\text{ (normal stress)} = \frac{3}{2E}\left[\frac{\sigma_1 + \sigma_2 + \sigma_3}{3}\right]^2(1 - 2\nu)$$

The total strain energy due to σ_1, σ_2, and σ_3 has been worked out as

$$U_t\text{ (total)} = \frac{1}{2E}[\sigma_1^2 + \sigma_2^2 + \sigma_3^2 - 2\nu(\sigma_1\,\sigma_2 + \sigma_2\,\sigma_3 + \sigma_3\,\sigma_1)]$$

Therefore, the strain energy due to the shear stress component, U_s, (second and third terms of the expressions for σ_1, σ_2, and σ_3) is the difference between the two:

$$U_s = U_t - U_n = \frac{1}{2E}[\sigma_1^2 + \sigma_2^2 + \sigma_3^2 - 2\nu(\sigma_1\,\sigma_2 + \sigma_2\,\sigma_3 + \sigma_3\,\sigma_1)]$$

$$- \frac{3}{2E}\left[\frac{\sigma_1 + \sigma_2 + \sigma_3}{3}\right]^2(1 - 2\nu)$$

$$= \frac{(1+\nu)}{6E}\,[2(\sigma_1^2 + \sigma_2^2 + \sigma_3^2) - 2(\sigma_1\sigma_2 + \sigma_2\sigma_3 + \sigma_3\sigma_1)]$$

As $E = 2G(1 + \nu)$, this becomes

$$U_s = \frac{1}{12G}\,[(\sigma_1 - \sigma_2)^2 + (\sigma_2 - \sigma_3)^2 + (\sigma_3 - \sigma_1)^2]$$

Note that $(\sigma_1 - \sigma_2)$, $(\sigma_2 - \sigma_3)$, and $(\sigma_3 - \sigma_1)$ are the maximum shear stresses in their respective planes.

(iii) *Case of pure shear stress* In the case of pure shear stress, we have seen that the principal stresses are equal to the shear stress. In the case of pure shear, σ_1, σ_2, and σ_3 are equal to τ, $-\tau$, and zero, respectively. Substituting these values,

$$U_s = \frac{1}{12G}\,\{[\tau - (-\tau)]^2 + (-\tau - 0)^2 + (0 - \tau)^2\} = \frac{\tau^2}{2G} \text{ per unit volume.}$$

10.9 APPLICATIONS

The computation of strain energy by itself is not of much use. However, the concept of strain energy finds application in the calculation of deformations and in the analysis of indeterminate structures.

We have already shown how to find the deformation by calculating the strain energy when a single concentrated load or couple is acting on the structure. Consider, for example, the simply supported beam carrying a concentrated load at a point along its length shown in Fig. 10.31. The BM diagram for the beam is triangular with a maximum value under the load of Pab/l. The strain energy of the beam can be calculated by integration of the expression M^2dx/EI separately for the two parts AC and BC of the beam. Thus, from Fig. 10.31,

Fig. 10.31

$$U = \int_0^a \frac{M_1^2\,dx}{2EI} + \int_0^b \frac{M_2^2\,dx}{2EI}$$

$$M_1 = \frac{Pab}{l}\frac{x}{a} = \frac{Pbx}{l}$$

and $$M_2 = \frac{Pab}{l}\frac{x}{b} = \frac{Pax}{l}$$

$$U = \frac{1}{2EI}\left[\int_0^a \frac{P^2b^2}{l^2}x^2dx + \int_0^b \frac{P^2a^2}{l^2}x^2dx\right]$$

$$= \frac{1}{2EI}\left[\frac{P^2b^2}{l^2}\frac{a^3}{3} + \frac{P^2a^2}{l^2}\frac{b^3}{3}\right] = \frac{P^2a^2b^2}{6EIl^2}(a+b) = \frac{P^2a^2b^2}{6EIl}$$

noting that $a + b = l$.

If δ is the deflection under the load,

$$\text{External work done} = \frac{1}{2}P\delta = \text{strain energy}$$

$$\frac{1}{2}P\delta = \frac{P^2a^2b^2}{6EIl}, \quad \delta = \frac{Pa^2b^2}{6EIl}$$

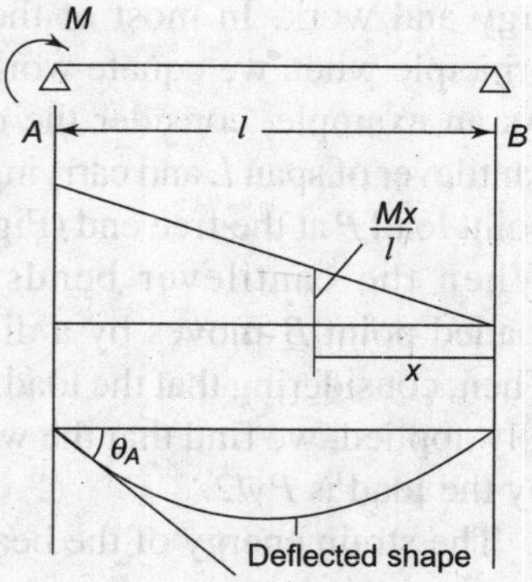

Fig. 10.32

Similarly, in the case of a simply supported beam subjected to a couple at one end, shown in Fig. 10.32, the BM diagram is as shown. The BM at any point distant x from B is $Mx/2$. The strain energy of the beam is

$$U = \int_0^l \frac{M^2 dx}{2EI} = \int_0^l \left(\frac{Mx}{l}\right)^2 \frac{dx}{2EI} = \frac{M^2 l}{6EI}$$

If θ_A is the slope of the beam at the point of application of the couple, then work done by the moment M in rotating the beam through θ_A at end A is $M\theta_A/2$. Equating the external work done and strain energy,

$$\frac{1}{2}M\theta_A = \frac{M^2 l}{6EI}$$

$$\Rightarrow \quad \theta_A = \frac{Ml}{3EI}$$

The application of strain energy can be extended to structures carrying a number of loads.

The illustrative examples of deformation calculation from strain energy discussed so far related real work and strain energy. This becomes quite tedious in case a number of loads are acting and deformations are required at a number of points on the structure. There are many elegant methods for finding deflections and analysing indeterminate structures based on strain energy.

There are a number of theorems relating strain energy of a structure with the deformations of the structure. These are discussed below with illustrative examples of their application to find the deformation of the structure. We will be looking in detail at two of the methods—Castigliano's partial derivative method and the unit load method. Application of energy principles to find deformations in pin-jointed plane frames is discussed in Chapter 13 and the analysis of indeterminate beams and trusses is discussed in Chapter 14.

10.9.1 General Energy Principles

The following general energy principles are of fundamental importance for structural applications. In all energy principles or theorems, when the terms *force* and *displacement* are used, they also mean moments and rotations. Thus, if force P and displacement $\ddot{a}$ are mentioned, they also mean moment M and corresponding rotation θ.

(a) **Law of conservation of energy** This is a fundamental law of physics and is stated in different ways in different fields. In relation to structural analysis, it can be stated as follows:

If a structure and external loads acting on it are isolated such that they neither receive nor give out energy, then the total energy of the system remains constant.

This principle is of fundamental nature and is always used when equating energy and work. In most of the applications discussed so far, we have used this principle when we equate work and strain energy.

As an example, consider the case of a cantilever of span L and carrying a single point load P at the free end (Fig. 10.33). When the cantilever bends, let the loaded point $\underline{B}$ moves by a distance y. Then, considering that the load is gradually applied, we find that the work done by the load is $Py/2$.

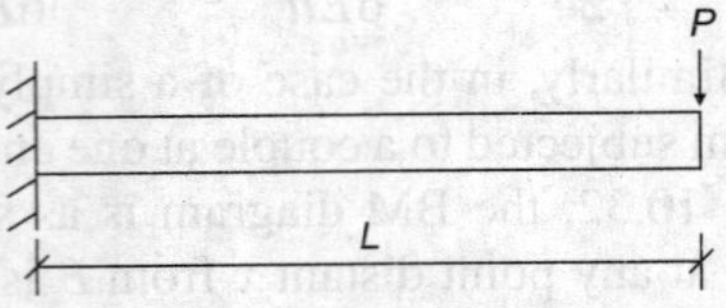

Fig. 10.33 Cantilever with point load

The strain energy of the beam due to bending moment alone is found as

Strain energy, $U = \int M^2 dx/EI$, the integration being carried out from $L = 0$ to $L = L$

We note that $M = Px$ and the integral works out to $P^2L^3/6EI$. We now use the principle of conservation of energy and equate the work done to the strain energy stored in the beam.

$$Py/2 = P^2L^3/6EI;\ y = PL^3/3EI, \text{ which is the deflection at the free end}$$

(b) **Theorem of virtual displacements and virtual work** Theorem of virtual displacement is a very powerful and versatile tool for analysing structures. This was first stated by Bernoulli and is known as Bernoulli's theorem of virtual displacements. We first define virtual displacement and virtual work.

Virtual displacement is a small displacement given to the body consistent with the boundary conditions of the body.

Virtual work is the work done during virtual (small) displacements by real forces or the work done by hypothetical forces during real small displacements.

Consider the loaded beam shown in Fig. 10.34. The beam is acted upon by a number of forces in equilibrium. Now, by some means, if we give a small displacement δy to the body as shown by the dotted lines, the forces P_1 to P_n do work. The downward displacement of load P_i is δy_i. Please note that the displacement before this displacement and after this displacement is consistent with the boundary conditions. The deflections at A and B are zero. This is called a *virtual deformation* and results in additional incremental displacements of the loads. Due to this deformation, the loads lose potential energy, which can be calculated as follows, using notations shown in the figure,

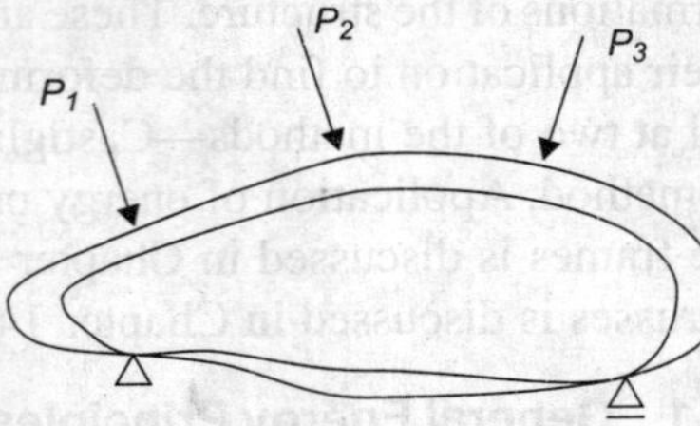

Fig. 10.34 Principle of virtual displacement and work

Decrease in potential energy = $\Sigma P_i \delta y_i$, where the summation is done for $i = 1$ to $i = n$

According to the law of conservation of energy, this decrease in potential energy must be equal to the increase in internal strain energy stored in the beam. This is stated as the theorem of virtual work (or displacement) as follows:

If a body in equilibrium under the action of a set of external loads, is given a small virtual deformation, then the work done by the external loads during this virtual deformation is equal to the increase in internal strain energy stored in the body.

Rigid body Theorem of virtual displacement for a rigid body is also known as Bernoulli's principle of virtual displacements.

In the case of a rigid body subjected to a coplanar force system, as shown in Fig. 10.35(b), there are three conditions of equilibrium. These are $\Sigma H = 0$, $\Sigma V = 0$, and $\Sigma M = 0$. If the forces $F_1, F_2, \ldots F_n$ have components $F_{1x}, F_{2x} \ldots, F_{nx}$ along the X-axis and $F_{1y}, F_{2y}, \ldots, F_{ny}$ along the Y-axis,

Component along the X-axis $= \Sigma F_{ix} = 0$

Component along the Y-axis $= \Sigma F_{iy} = 0$

Now, we assume a small virtual displacement ä s and rotation $\delta\theta$ [Fig. 10.35(b)] given to the rigid body. If the components of δs are δx and δy in the direction of coordinate axes, then the work done due to this displacement ä s is

$$\Sigma F_i \delta_x + \Sigma F_i \delta y = \delta_x \Sigma F_{ix} + \delta_y \Sigma F_{iy}$$

This is equal to zero as the sum of the components is zero in the two coordinate axes directions.

Similarly, due to rotation $\delta\theta$, the linear displacements in the X- and Y-directions are $r\delta\theta \cos\theta$ in the X-direction and $r\delta\theta \sin\theta$ in the Y-direction. As $r\cos\theta = x$ and $r\sin\theta = y$, the displacements, $\delta x = x\delta\theta$ and $\delta y = y\delta\theta$.

The work done by the components ΣF_{ix} and ΣF_{iy} and ΣM can now be written as follows:

Work done due to virtual rotation, $\delta\theta = \Sigma F_{ix} x\,\delta\theta + \Sigma F_{iy}\, y\,\delta\theta + \Sigma M\,\delta\theta$

$$= \delta\theta[\Sigma F_{ix} x + \Sigma F_{iy} y + \Sigma M]$$

$$= 0$$

because $\Sigma F_{ix} x + \Sigma F_{iy} y + \Sigma M = 0$, being the sum of the moments about a point of all the forces.

Therefore, for a rigid body in equilibrium, the work done during a small virtual displacement/rotation is zero.

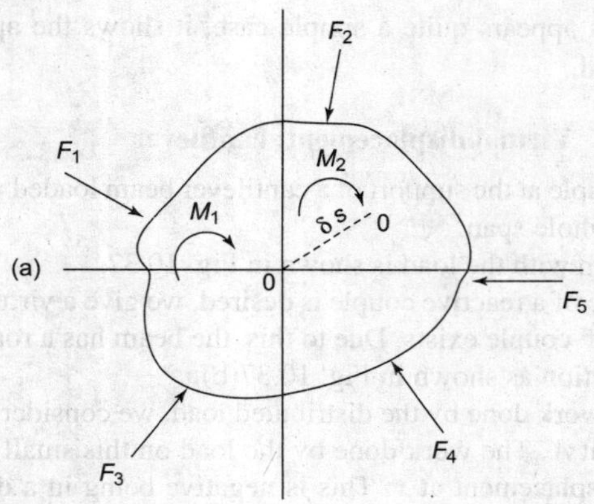

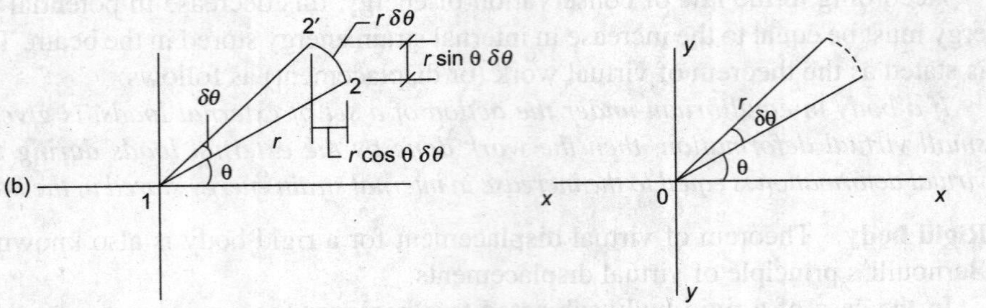

Fig. 10.35

We calculate reactions in beams considering them as rigid body. We illustrate with some examples.

Example 10.22 Virtual displacement: SS beam

In the simply supported beam loaded as shown in Fig. 10.36, find the reaction at B using the principle of virtual displacements.

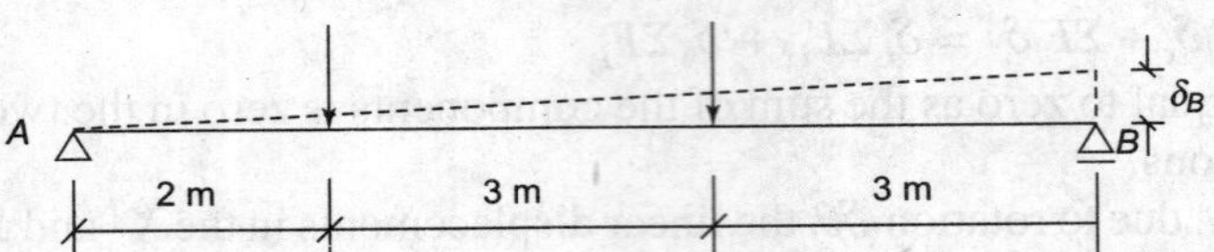

Fig. 10.36

Solution We consider a virtual displacement δB of point B upwards as shown in the figure. The displacement under the two concentrated loads can be computed from similar triangles.

Displacement under 30 kN load = $(2/8)\,\delta B = \delta B/4$

Displacement under 40 kN load = $(5/8)\,\delta B$

These displacements are opposite to the direction of the loads. Hence, the work done during this virtual displacement by real loads is negative. Assuming the direction of the reaction also as upwards; we can equate the total work done to zero.

$-30ä\ B/4 - 40 \times 5\ \delta B/8 + R_B \delta B = 0$, which gives $R_B = 32.5$ kN

Even though this appears quite a simple case, it shows the application of the virtual displacement method.

Example 10.23 Virtual displacement: cantilever

Find the reactive couple at the support of a cantilever beam loaded with a uniformly distributed load over the whole span.

Solution The beam with the load is shown in Fig. 10.37.

As the magnitude of a reactive couple is desired, we give a virtual rotation $\delta\theta$ at the end A, where the reactive couple exists. Due to this, the beam has a rotation about point A and will take up the position as shown in Fig. 10.37(b).

To calculate the work done by the distributed load, we consider a small length δx of the beam at x from point A. The work done by the load on this small segment is $(w\,\delta x)(x\,\delta\theta)$, where $x\delta\theta$ is the displacement at x. This is negative being in a direction opposite to the direction of the load.

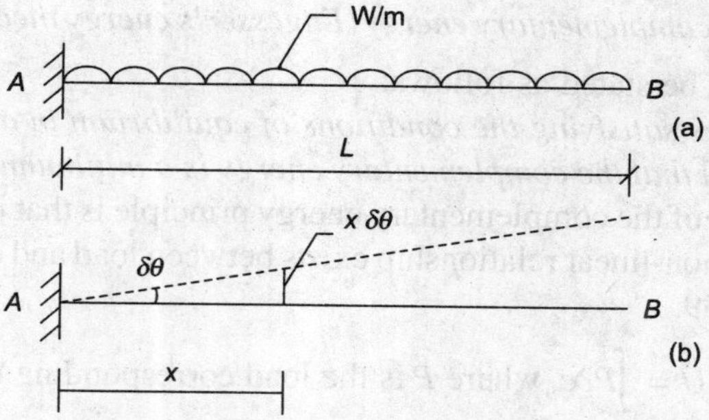

Fig. 10.37

Total work done by the UD load = $\int [w x \delta\theta dx = w L^2 \delta\theta/2$

Writing the equation of the net work done by the forces,

$M \delta\theta - w L^2 \delta\theta/2 = 0$ giving $M = w L^2/2$

Deformable body In the case of a deformable body, shown in Fig. 10.38, the body is deformed due to the application of forces F_i. The body reaches an equilibrium position when the internal stress resultants reach a value to keep the body in equilibrium position.

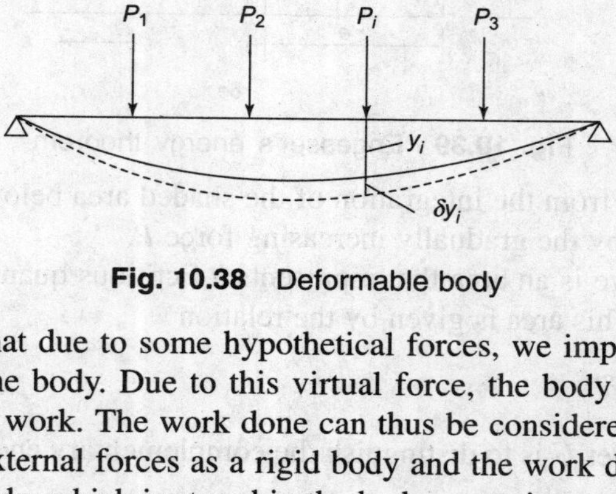

Fig. 10.38 Deformable body

Now assume that due to some hypothetical forces, we impart a small virtual displacement to the body. Due to this virtual force, the body deforms and also external forces do work. The work done can thus be considered as consisting of the work due to external forces as a rigid body and the work done due to elastic straining of the body, which is stored in the body as strain energy. We consider a small element of the body. This element has rigid body movement and strain due to elastic deformation. Therefore, the total work done on the small element is $\delta W = \delta W_1 + \delta W_2$, where δW_1 is the work done due to rigid body displacements and δW_2 is the work done due to elastic deformation. As the system is in equilibrium, $\delta W_1 = 0$ from what we have derived in the earlier section. Thus, $\delta W = \delta W_2$. Considering the total body, $W = W_2$. We can thus state:

In the case of a deformable body in equilibrium, when subjected to virtual forces causing virtual displacements, the work done by the external forces is equal to the internal virtual strain energy stored in the body.

(c) **Theorem of minimum potential energy** We state the theorem of minimum potential energy as follows:

Of all the displacements satisfying given boundary conditions, those that satisfy the equilibrium conditions make the potential energy a minimum.

(d) *Theorem of complementary energy (Engesser's energy theorem)*

This theorem can be stated as follows:

Of all the states satisfying the conditions of equilibrium in a body, the state of stress will be such that the complementary energy is a minimum.

The advantage of the complementary energy principle is that it can be applied to situations where non-linear relationship exists between load and deformation. Considering Fig. 10.39,

Strain energy, $U = \int P\delta e$, where P is the load corresponding to deformation δe.

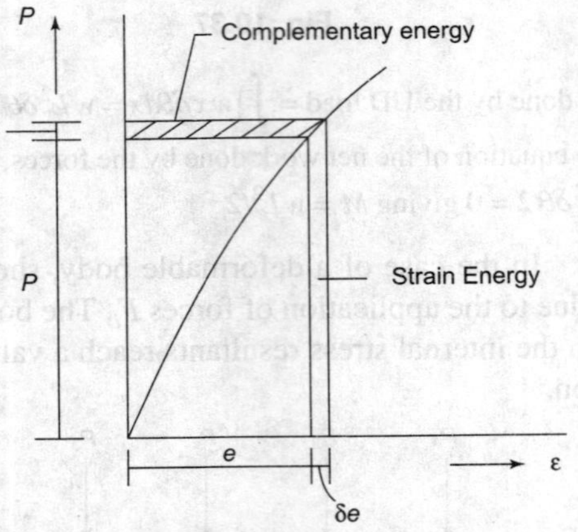

Fig. 10.39 Engesser's energy theorem

This is derived from the integration of the shaded area below the curve, which is the work done by the gradually increasing force P.

Above the curve is an area that represents a fictitious quantity called complementary energy. This area is given by the relation

$$U^* = \int e\,\delta P$$

The asterisk over U is to distinguish the complementary energy from the strain energy.

10.9.2 Castigliano's Theorems

Alberto Castigliano in 1873 described methods to find deflections and slopes in beams and trusses as his dissertation for engineering diploma, which are known as Castigliano's theorems. This is a classic example of application of energy principles to calculating deformations. The two conditions that govern his proposition are that the body is stressed within elastic limit and the deformations are linear functions of the loads.

Castigliano's first theorem is a special case of Engesser's theorem of complementary energy with the proviso that the deformations must be linear functions of the loads. The theorem can be stated as follows:

In a linearly elastic system, the partial derivative of the total strain energy stored in a structure with respect to the displacement at a point is equal to the force at that point.

Stated mathematically, $\delta U / \delta \Delta = P$

Considering a linearly elastic material, it is clear that the strain energy and complementary energy are equal. Thus, according to Engesser's theorem of complementary energy, $U^* = \int \Delta \delta P$; $U = \int P \delta \Delta$ and the two are equal in the case of linearly elastic material.

Castigliano's second theorem is again limited to linearly elastic systems. Stated simply,

In a linearly elastic system, the partial derivative of the total strain energy stored in the structure with respect to any force gives the displacement at that point in the direction of the force.

In the above theorem, displacement can mean a translation/rotation and the force can be a load or a couple.

To prove the theorem, consider the beam loaded as shown in Fig. 10.40.

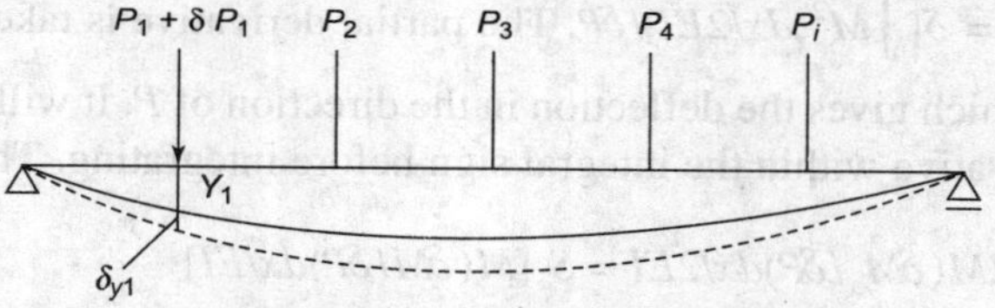

Fig. 10.40 Castigliano's theorems

$P_1, P_2, \ldots$ are the gradually applied loads acting on the beam and let $y_1, y_2, \ldots$ be the deflections under the loads. Then the total work done by the external forces is

$W_e = U = (1/2)P_1y_1 + (1/2)P_2y_2, \ldots$, where U is the strain energy stored in the beam.

Now we increase load P_1 to $P_1 + \delta P_1$ and let the additional deflections be δy_1, $\delta y_2, \ldots$, etc.

The additional work done by the loads can be expressed as

$$\delta W_e = \delta U = (1/2)\delta P_1\, \delta y_1 + P_1 \delta y_1 + P_2 \delta y_2 + \ldots$$

Total strain energy U_t stored with the two applications of the loads is

$$U_t = [(1/2)P_1y_1 + (1/2)P_2y_2 + \ldots] + [(1/2)\delta P_1 \delta y_1 + P_1 \delta y_1 + P_2 \delta y_2 + \ldots]$$

Now consider the same beam loaded simultaneously with the same loads but $P_1 + \delta P_1$ applied along with P_2, P_3 gradually. The deflection under load $P_1 + \delta P_1$ will be $y_1 + \delta y_1$, under load P_2 will be $(y_2 + \delta y_2)$, …, etc. The total external work done or strain energy stored can be expressed as

$$U_t = \frac{1}{2}(P_1 + \delta P_1)(y_1 + \delta y_1) + \frac{1}{2}P_2(y_2 + \delta y_2)$$

Since the strain energy must be the same due to the two methods of application of the load, the two strain energies obtained must be equal. Therefore,

$$[(1/2)P_1y_1 + (1/2)P_2y_2 + \ldots] + [(1/2)\delta P_1 \delta y_1 + P_1 \delta y_1 + P_2 \delta y_2 + \ldots]$$
$$= (1/2)(P_1 + \delta P_1) + (1/2)P_2(y_2 + \delta y_2) + \ldots$$

Simplifying, we get

$$(1/2)P_1 \delta y_1 + (1/2)P_2 \delta y_2 + \ldots = (1/2)\delta P_1 y_1$$

We have neglected the term $\delta P_1\,\delta y_1$ being a small quantity of second order. The left-hand side is the δU, the increase in strain energy calculated earlier.

We thus get $\delta U/\delta P = y_1$, the deflection under the load P_1.

Castigliano's theorem can thus be stated as follows:

The partial derivative of the total strain energy of any structure, which is linearly elastic, with respect to any of the applied forces is equal to the displacement of the point of application of that force in the direction of the force.

We have derived the theorem for a set of point loads. This is equally applicable for the moments applied on the structure. The displacement in this case will be the rotation of the point of application of the moment in the direction of the applied moment.

Castigliano's theorem gives deflection $y = \delta U/\delta P$. When we consider deflection due to bending, $U = \int M^2{}_x dx/2EI$, the integration being done over the segment or span length.

Therefore, $y = \delta[\int M^2{}_x dx/2EI]/\delta P$. The partial derivative is taken with respect to the load P, which gives the deflection in the direction of P. It will be convenient to take the derivative within the integral sign before integrating. Thus,

$$y = \int 2M(\delta M_x/\delta P)dx/2EI = S\ [M(\delta M/\delta P)dx/EI]$$

If the deflection is required at a point where no load is acting, a fictitious load can be assumed to be acting. This load is set to zero after differentiation but before integration.

The following examples illustrate the application of the theorem in calculating deflections and slopes.

Example 10.24 Castigliano's theorem: cantilever

A cantilever of span L carries a concentrated load W at the free end. Find the slope and deflection at the free end using Castigliano's theorem.

Solution The cantilever with the load is shown in Fig. 10.41.

Bending moment at x, $M_x = Px$

The strain energy stored in the beam due to flexure, neglecting shear strain energy, is

$$U = \int [M^2 dx/2EI] = \int [P^2x^2 dx/2EI]$$

$$dU/dP = d[\int [P^2x^2 dx/2EI]/dP = \int [Px^2 dx/EI]$$, differentiating within the integral sign

$$dU/dP = \text{deflection } y = PL^3/3EI$$, integrating from 0 to L

To find the slope at B, apply a fictitious couple μ at A as shown. The moment diagram due to this couple is uniform for the cantilever. Moment at x due to this couple $= \mu$.

Total moment at $x = Px + \mu$

Strain energy, $U = \int (Px + \mu)^2 dx/2EI$

$$dU/d\mu = d[\int (Px + \mu)^2 dx/2EI]/d\mu = \int [(Px + \mu)dx/EI]$$, differentiating within the integral

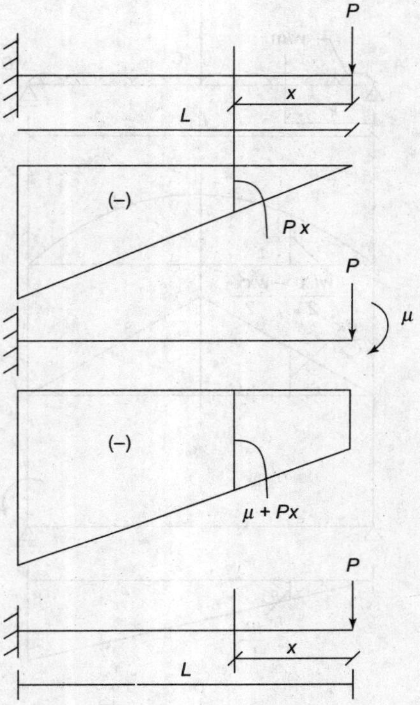

Fig. 10.41

We can now set the fictitious moment $\mu = 0$ and then integrate with respect to x.

$dU/d\mu$ = slope at B, $\theta_B = PL^2/2EI$

Example 10.25 Castigliano's theorem: SS beam with UD load

A simply supported beam is carrying a uniformly distributed load w/m throughout its span. Find the deflection at the centre and the slopes at the ends of the beam.

Solution The beam with the loading is shown in Fig. 10.42.

BM at x from $A = wLx/2 - wx^2/2$

As there is no load at the mid-point of the beam, where the maximum deflection would be, we put a fictitious load P. This load gives the BM diagram shown. We take advantage of the symmetry of the loading and integrate for strain energy from 0 to $L/2$ and double that.

$$M_x = [wLx/2 - wx^2] + Px/2$$

$$\delta M_x / \delta P = x/2$$

$$\frac{\delta U}{\delta P} = \int M\,(\delta M/\delta P)dx/EI = 2\int_0^{L/2} [wLx/2 - wx^2/2 + Px/2](x/2)dx/EI$$

After differentiating, P can be set to zero before integrating.

$$\frac{\delta U}{\delta P} = 2\int [wLx/2 - wx^2/2](x/2)dx/EI$$

Integrating from 0 to $L/2$,

$$\delta U/\delta P = \text{deflection under } P = 2[wLx^3/12 - wx^4/16]_0^{L/2}/EI$$

Deflection under $P = y_c = (2/EI)[wL^4/96 - wL^4/256] = (5/384)wL^4/EI$

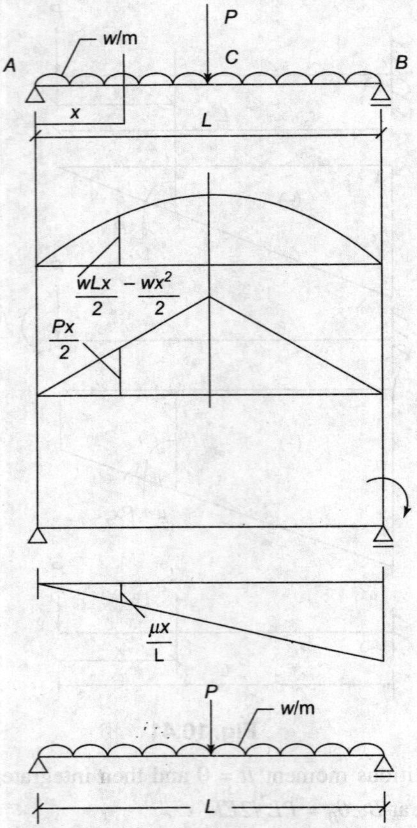

Fig. 10.42

Due to symmetry, slopes at A and B will be the same though of different signs. To find the slope at B, we apply a fictitious moment μ at B as shown. At x from A,

Moment due to $\mu = -\mu x/L$

Total moment at x, $M_x = [wL\,x/2 - wx^2/2 - \mu x/L]$

$\delta M_x / \delta\mu = -x/L$

$\delta U/\delta\mu$ = slope at B = $\int M(\delta M/\delta\mu)dx/EI$, integration being done from 0 to L

Slope at B, $\theta_B = \int [wLx/2 - wx^2/2 - \mu x/L](-x/L)dx/EI$

We can now set the fictitious couple μ to zero.

$$\theta_B = \int [wL\,x/2 - wx^2/2]\,(-x/L)dx/EI$$

$$= \frac{-Wx^3/6}{EI} + \frac{Wx^4/8L}{0}$$

$$= -wL^3/24EI$$

The sign being negative, the slope is in the opposite direction of μ, i.e. anticlockwise.

Example 10.26 Castigliano's theorem: SS beam with point load

A simply supported beam is carrying a point load P at a from the left end. Find the deflection under the load, slopes at the ends.

Solution The beam with the loading is shown in Fig. 10.43.

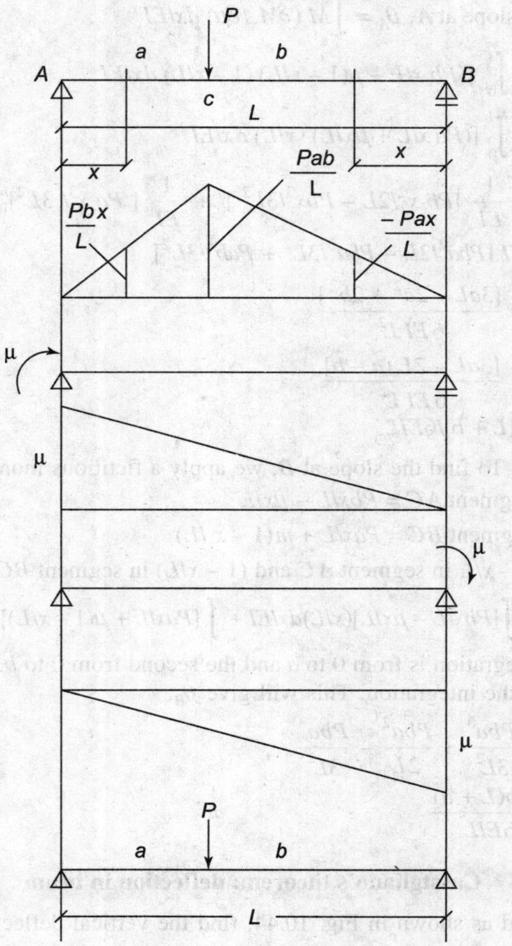

Fig. 10.43

The reaction at A is Pb/L and the reaction at B is Pa/L. As the bending equations are different in the two segments AC and CB, we have to calculate the strain energy in parts.

In segment AC, $M_x = Pbx/L\, x = 0$ to $x = a$; $\quad \delta M_x/\delta P = bx/L$

In segment BC, $M_x = Pax/L$ (calculating from right), $x = 0$ at B and $x = b$ at C; $\delta M_x/\delta P = ax/L$

(a) *Deflection under the load* $\quad U = \int M_x^2 dx/EI;\ \delta U/\delta P = \int [M(\delta M/\delta P)dx/EI]$

$$\frac{\delta U}{\delta P} = \int_0^a (Pbx/L)(bx/L)dx/EI + \int_0^b (Pax/L)(ax/L)dx/EI$$

$$y = \frac{Pb^2}{EI\,L^2}\,[x^3/3]_0 + \frac{Pa^2}{EI\,L^2}\,[x^3/3]_0 = \frac{Pa^2b^2\,(a+b)}{3EIL^2} = \frac{Pa^2b^2}{3EIL}$$

(b) *Slope at A* For finding the slope at A, apply a fictitious moment μ at A as shown.

BM at x $M_x = Pb\,x/L - \mu(1 - x/L),\ 0 < x < a$ [origin at A]

$= Pa\,x/L + \mu\,x/L,\ 0 < x < b$ [origin at B]

$\delta M_x/\delta\mu = 1 - x/L,\ 0 < x < a$ and $\delta M_x/\delta\mu = x/L;\ 0 < x < b.$

$$\delta U/\delta\mu = \text{slope at } A,\ \theta_A = \int M\,(\delta M_x/\delta\mu)\,dx/EI$$

$$= \int_0^a [(Pb\,x/L + \mu(1 - x/L)\,(1 - x/L)\,dx/EI$$

$$+ \int_0^b [(Pa\,x/L + \mu x/L)\,(x/L)\,dx/EI$$

$$= \frac{1}{EI}\,[Pb\,x^2/2L - Pbx^3/3L^2\,]_0^a + \frac{1}{EI}\,[\,Pa\,x^3/\,3L^2]_0^b$$

$$= 1/EI\,[Pba^2/2L - Pba^3/3L^2 + Pab^3/3L^2]$$

$$= Pab\,\frac{[3aL - 2a^2 + 2b^2\,]}{6\,EI\,L^2}$$

$$= Pab\,\frac{[3aL - 2L\,(a - b]}{6\,EI\,L^2}$$

$$= Pab(L + b)/6EIL.$$

(c) *Slope at B* To find the slope at B, we apply a fictitious moment μ at B.

BM in segment $AC = Pbx/L - \mu x/L$

BM in segment $BC = Pax/L + m(1 - x\,/L)$

$\delta M/\delta\mu = -x/L$ in segment AC and $(1 - x/L)$ in segment BC

$$\delta U/\delta\mu = \int [Pbx/L - \mu x/L](x/L)dx/EI + \int [Pax/L + \mu(1 - x/L)](1 - x/L)dx/EI$$

The first integration is from 0 to a and the second from 0 to b. We set $\mu = 0$ before performing the integration. This will give θ_B.

$$EI\,\theta_B = \frac{Pba^3}{3L^2} + \frac{Pba^2}{2L} - \frac{Pba^3}{3L^2}$$

$$\theta_B = \frac{Pab(L + a)}{6EIL}$$

Example 10.27 Casstgliano's theorem: deflection in beam

In the beam loaded as shown in Fig. 10.44, find the vertical deflection and slope at the free end. EI is constant.

Solution To find the vertical deflection, we apply a fictitious load P at the free end. The moment M_x can be written for the segments AB and BC separately. Due to UD load,

Reaction at $B = (30 \times 10 \times 5)/8 = 187.5$ kN ↑

Reaction at $A = (30 \times 8 \times 4 - 30 \times 2 \times 1)/8 = 112.5$ kN ↑

Due to load P at C, reaction at $A = (-P \times 2)\,/8 = -\,P/4$ ↓

Reaction at $B = P \times 10\,/8 = 1.25\,P$ ↑

M_x in $AB = 112.5x - 30x^2/2 - Px/4 = 112.5x - 15x^2 - Px/4$

$\delta M_x/\delta P = -x/4$

M_x in $BC = -Px - 30x^2/2 = -P\,x - 15x^2$

$\delta M_x/\delta P = -x$

Deflection at C, $y_c = \delta U/\delta P = \int [M(\delta M_x/\delta P)dx/EI$

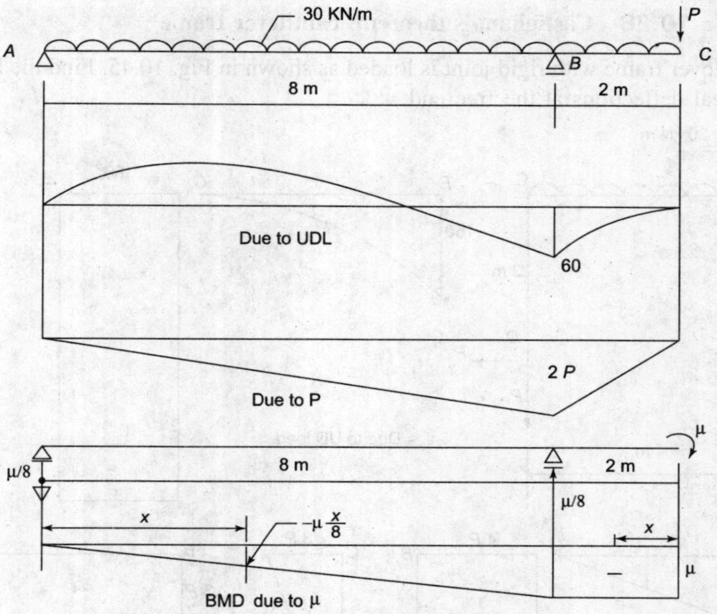

Fig. 10.44

$$y_c = \int_0^8 \left[112.5x - 15x^2 - \frac{Px}{4}\right]\left(-\frac{x}{4}\right)\frac{dx}{EI} + \int_0^2 \left[-(Px - 15x^2)\right](-x)\frac{dx}{EI}$$

Setting $P = 0$ as the load is fictitious,

$$y_c = \frac{1}{EI}\left[-\frac{112.5x^3}{24} + \frac{15x^4}{16}\right]_0^8 + \frac{1}{EI}\left[-\frac{15x^3}{3}\right]_0^2$$

$$= \frac{1}{EI}\left[-\frac{112.5(8)^3}{24} + \frac{15(8)^3}{16}\right] + \frac{1}{EI}[-5\times 8]$$

$$= \frac{1}{EI}[-2400 + 480] + \frac{-40}{EI} = \frac{-1960}{EI} \text{ (upward deflection)}$$

To find the slope at C, apply a fictitious clockwise moment μ at C.

The bending moment due to this couple at C is

$M_x = -\mu x/8$ in AB and $-\mu$ in BC (from the right)

$\delta M_x/\delta\mu = -(x/8)$ in AB and -1 in BC

Slope at C, $\theta_C = \int M(\delta M/\delta\mu)dx/EI$

$$y_c = \int_0^8 (112.5x - 15x^2 - \mu x/8)(-x/8)dx/EI + \int_0^2 (-15x^2 - \mu](-1)dx/EI$$

$$= \frac{-112.5x^3/16 + 15x^3/24]_0^8}{EI} + \frac{[+5x^3]_0^2}{EI} \text{ setting } \mu = 0 \text{ before integrating}$$

$= -2920/EI$ (anticlockwise slope)

Example 10.28 Castigliano's theorem: cantilever frame

The cantilever frame with rigid joint is loaded as shown in Fig. 10.45. Find the horizontal and vertical deflections at the free end.

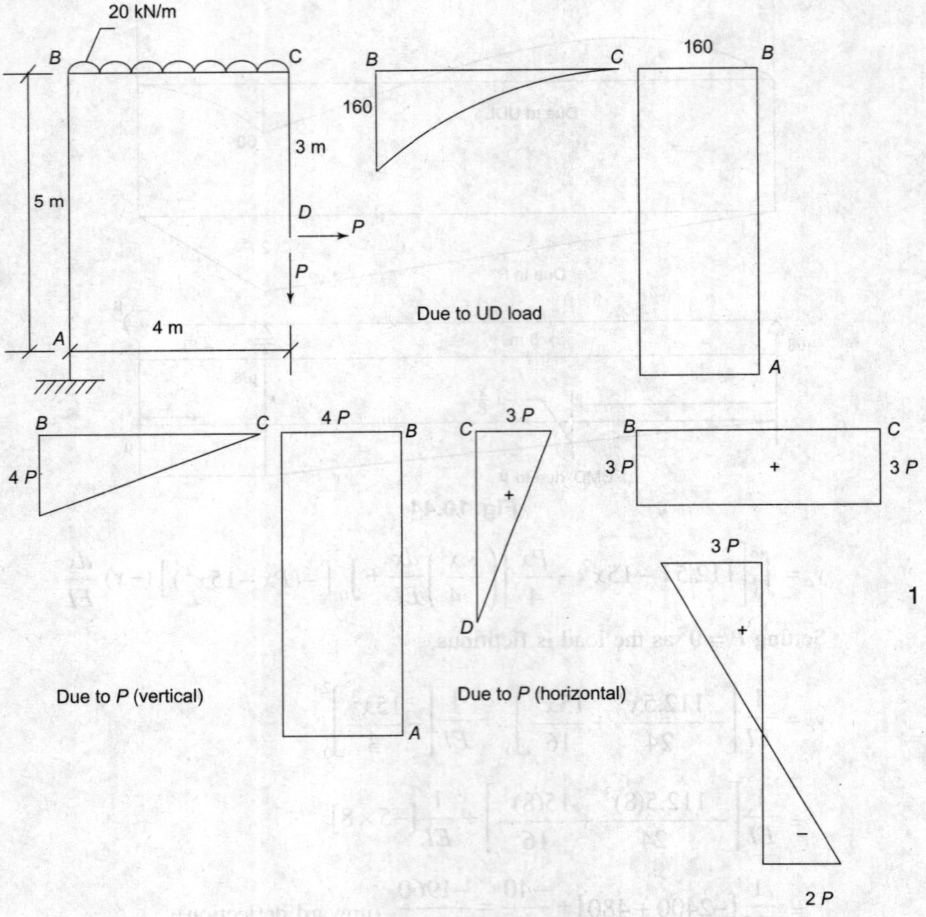

Fig. 10.45

Solution To find the vertical deflection at D, we apply a fictitious load P at the free end. In such cases, we consider the BM in each member of the frame separately. The moments due to the applied load and the BM due to P are as shown in Table 10.1.

Table 10.1

Segment	M_x	*Moment due to P*	$\Delta(M_x + M_P)\delta P$	*Origin at*	*Limits of integration*
DC	0	0	0	*D*	0 to 3
CB	$-10x^2$	$-Px$	$-x$	*C*	0 to 4
BA	–160	$-4P$	–4	B	0 to 5

Vertical deflection at $D = \delta U/\delta P = \Sigma M(\delta M/\delta P)dx/EI$

$$y_D = \int 0 + \int_0^4 (-10x^2 - Px)(-x)dx/EI + \int_0^5 (-160 - 4P)(-4)dx/EI$$

We set P to zero before integrating as P is a fictitious load.

$$= \frac{1}{EI}\,[10x^4/4]_0^4 + [640x]_0^5$$

$$= (640 + 3200)/EI = 3840/EI$$

To find the horizontal deflection at D, apply a fictitious load P horizontally at D. The moments in the segments are as shown in Table 10.2.

Table 10.2

Segment	M_x	M_P	$\Delta(M_x + M_P)/\delta P$	*Origin at*	*Limits of integration*
DC	0	$P\,x$	X	D	0 to 3
CB	$-10\,x^2$	$3P$	3	C	0 to 4
BA	-160	$3P - Px$	$3 - x$	B	0 to 5

The horizontal deflection at $D = h_D = \Sigma \int M(\delta M/\delta P)dx/EI$

$$h_D = \int_0^3 (0 + Px)(x)dx/EI + \int_0^4 (-10x^2 + 3P)(3)dx/EI$$

$$+ \int_0^5 (-160 + 3P + Px)(3 - x)dx/EI$$

$$= \left[\frac{-10x^3}{EI}\right]_0^4 + \left[\frac{-480x + 160x^2/2}{EI}\right]_0^5, \text{ setting } P \text{ to zero before integrating}$$

$$= -640 + (-2400) + 2000 = -1040/EI$$

The negative sign shows that the deflection is opposite to the direction of P assumed.

Example 10.29 Castigliano's theorem: curved bar

The beam in the shape of a quarter circle is loaded at the free end with a load of 30 kN. Find the horizontal and vertical deflections at the free end.

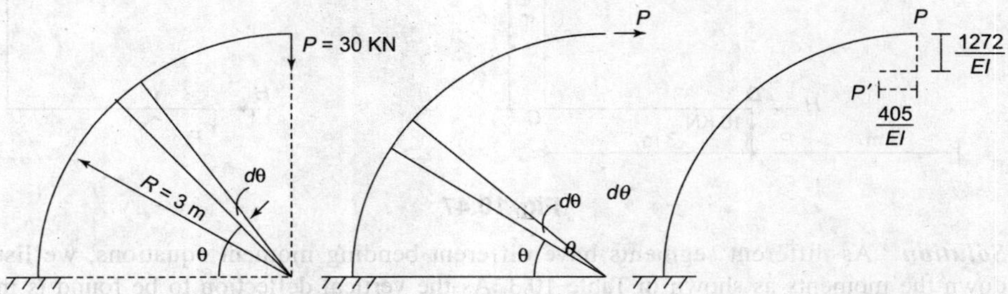

Fig. 10.46

Solution We consider an elementary strip of length ds at an angle θ from the base as shown. We have $ds = Rd\theta$, where R is the radius. The moment of the 30 kN load $= PR\cos\theta$.

As the applied load is in the direction of the deflection required, we find the strain energy as $\int M^2 ds/EI$.

$$U = \int_0^{\pi/2} [(PR\cos\theta)^2 Rd\theta/EI] = \frac{P^2R^3}{EI}\int_0^{\pi/2}\frac{(1+\cos 2\theta)}{2}$$

$$= \frac{P^2R^3}{2EI}[\theta + \sin 2\theta]_0^{\pi/2}$$

$$= P^2R^3\delta/4EI = P^2R^3\delta/4EI$$

$\delta U/\delta P$ = vertical deflection $= 2PR^3\pi/4EI = 2\times 30\times 3^3\times(\delta)/4EI = 1272/EI$

To find the horizontal deflection, we apply a fictitious horizontal load as shown.

Moment due to 30 kN load = $30R\cos\theta$

Moment due to $P = PR(1 - \sin\theta)$

$$\delta M/\delta P = R(1-\sin\theta)$$

$$\frac{\delta U}{\delta P} = \int_0^{\pi/2} [M(\delta M/\delta P)ds/EI], \text{ gives the horizontal deflection}$$

$$y_h = \int [30R\cos\theta + PR(1-\sin\theta)][R(1-\sin\theta)]Rd\theta EI$$

Setting $P = 0$ and integrating,

$$= \int_0^{\pi/2}\frac{[30R^3\cos\theta]}{EI}[1-\sin\theta]d\theta = \frac{30R^3}{EI}\int_0^{\pi/2}(\cos\theta - \cos\theta\sin\theta)d\theta$$

$$= \frac{30R^3}{EI}\left[\sin\theta + \frac{\cos 2\theta}{4}\right]_0^{\pi/2}$$

$= -30R^3/2EI = -405/EI$ (The negative sign shows that the displacement is against the direction of P assumed.) ■

Example 10.30 Castigliano's theorem: bent bar

In the bent bar shown in Fig. 10.47, find the deflection and slope at the free end.

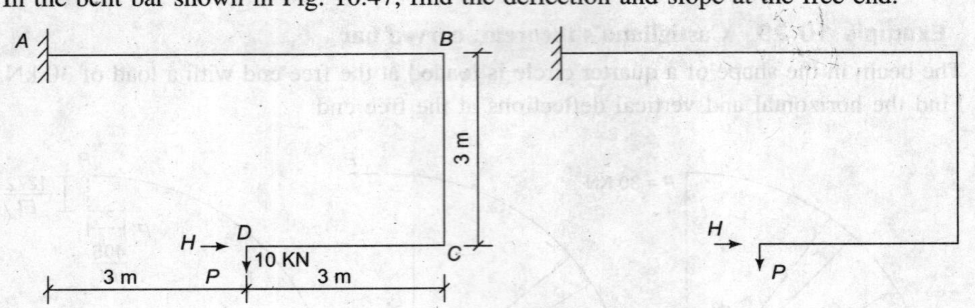

Fig. 10.47

Solution As different segments have different bending moment equations, we list down the moments as shown in Table 10.3. As the vertical deflection to be found is in the direction of the applied load, we take the load as P and substitute its value after integration.

Table 10.3

Segment	M_x	*Origin at*	*Limits of integration*
DC	Px	*D*	0 to 3 m
CB	$3P$	*C*	0 to 3 m
BA	$3P - Px$	*B*	0 to 6 m

Strain energy, $U = \int M^2 dx/2EI$

$$= \int_0^3 \frac{(Px)^2\, dx}{2EI} + \int_0^3 \frac{(3P)^2\, dx}{2EI} + \int_0^6 \frac{(3P - Px)^2\, dx}{2EI}$$

$$= \frac{[P^2 x^3]_0^3}{2EI} + \frac{[3Px]_0^3}{2EI} + \frac{[9P^2 x + P^2 x^2 - 6P^2 x^3]_0^6}{2EI}$$

$$= 54P^2/2EI$$

Here $\delta U/\delta P$ gives the vertical deflection at D.

$$\delta U/\delta P = 54P/EI = 540/EI$$

Vertical deflection of $D = 540/EI$

To find the horizontal deflection, we apply a fictitious horizontal force H at D as shown. The bending moments in the different segments are listed as shown in Table 10.4.

Table 10.4

Segment	*Moment due to loads*	*Moment due to H*	*Origin at*	*Limits of integration*
DC	$10x$	0	*D*	0 to 3 m
CB	30	Hx	*C*	0 to 3 m
BA	$30 - 10x$	$3H$	*B*	0 to 6 m

Horizontal deflection is given by $\int M(\delta M/\delta H) dx/EI$.

For the segment DC, $\delta M/\delta H = 0$. The integration has to be done for the other two segments only.

$$\text{Horizontal deflection} = \int_0^3 [(30 + Hx)x\,dx/EI] + \int_0^6 [(30 - 10x + 3H)3\,dx/EI]$$

$$= \frac{[30x^2/2]_0^3}{EI} + \frac{[90x - 30x^2/2]_0^6}{EI}$$

$$= 135/EI$$

Example 10.31 Castigliano's theorem: tapering section

A cantilever beam, 2 m long, has uniform width but its depth is varying from 200 mm at the free end to 400 mm at the fixed end. If $E = 200$ GPa and the width is 100 mm, find the deflection at the free end due to point load of 40 kN acting at the free end.

Solution The beam is shown in Fig. 10.48.

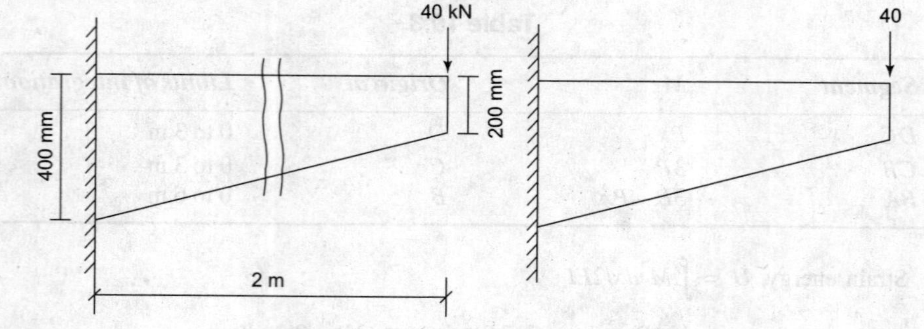

Fig. 10.48

As the depth is varying, *MI* changes with *x*. MI has to be expressed as a function of *x*.

Bending at x m from the free end = Px kNm

Depth of beam at x m = 200 + (400 – 200)x/2 = (200 + 100x) mm

MI at x, $I_x = 100(200 + 100x)^3/12 = 100^4(2 + x)^3/12\ \text{mm}^4$

$= (2 + x)^3/(12 \times 10^4)\ \text{m}^4$

Strain energy, $U = \int M^2 dx/EI$

$$= \int_0^2 \frac{P^2 x^2 (12 \times 10^4) dx}{E(2 + x)^3}$$

$$= \frac{P^2 \times 12 \times 10^4}{E} \int_0^2 \frac{x^2 dx}{(2 + x)^3}$$

To integrate this expression, we take

$y = 2 + x,\ x = y - 2,\ x^2 = y^2 - 4y + 4,\ dx = dy$

$$\int_0^4 \frac{(y^2 - 4y + 4)dy}{y^3} = \int_0^4 (y^{-1} - 4y^{-2} + 4y^{-3})dy = [\log_e y + 4/y - 4/2y^2]_2^4$$

$= [\log_e(2 + x) + 4/(2 + x) - 4/2(2 + x)^2]_0^2$

$= \log_e 4 - \log_e 2 + 1 - 2 - 32 + 1/2 = -31.8$

$U = P^2 \times 12 \times 10^4 (-31.8)/200 \times 10^6 = -0.019 \times 10^{-4} P^2$

$\delta U/\delta P = -0.038 \times 10^{-4} P\ \text{m} = 0.15\ \text{mm}$

Deflection at the free end = 0.15 mm

10.9.3 Unit Load Method

The unit load method, also known as the dummy load method, can be applied to beams and trusses. The unit load method is a versatile method and simple to apply. The application to beam deflections and slopes is illustrated here.

The unit load method can be directly derived from Castigliano's first theorem. We derive it from the first principles. Consider a body subjected to forces (and moments) as shown in Fig. 10.49. The deflection at point C is required where a unit load is applied.

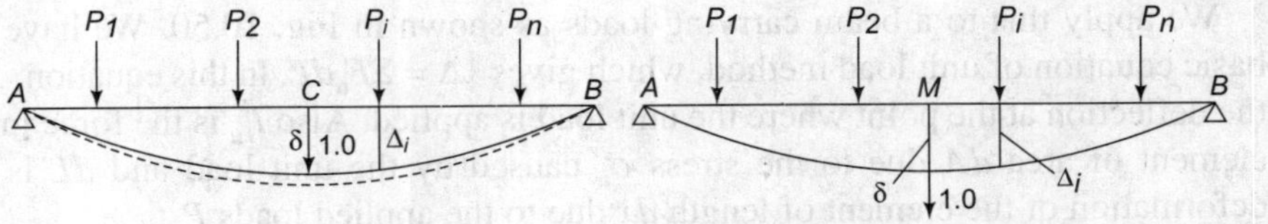

Fig. 10.49 Derivation of unit load method

The applied loads cause stresses in the body. If we consider an elemental area dA of length dx, the force in the element due to internal stresses is σdA. This force shortens or elongates the elemental length dx. Let the deformation of the element be dL. Then the elemental work done by the internal forces on the element is $(1/2)\sigma dAdL$. The total internal work, which is stored as internal strain energy, is $(1/2)\Sigma(\sigma dAdL)$, summing up the energies stored in all such elements. If the deflections due to the gradually applied loads P_1, P_2, ... are Δ_1, Δ_2, ..., then the external work done by the forces is $(1/2)P_1\Delta_1$, $(1/2)P_2\Delta_2$, and so on. The total external work is $(1/2)\Sigma P_i\Delta_i$.

By the law of conservation of energy,

$$(1/2)\Sigma p_i\Delta_i = (1/2)\Sigma(\sigma dAdL) \quad (10.1)$$

If we apply a unit load before the loads P_i are applied, then the loads P_1, P_2, etc. are gradually applied and the external work and strain energy can be calculated as

External work done due to unit load = $(1/2) \times 1 \times \delta$

Internal strain energy stored due to unit load = $(1/2)\delta_u dAdu$

We have

$$(1/2) \times 1 \times \delta = (1/2)\Sigma\sigma_u dAdu \quad (10.2)$$

If we now apply the loads P_i gradually, the deformation under M will be Δ and the deformation under P_i will be Δ_i.

External work done = $1\Delta + (1/2)\Sigma P_i\Delta_i$

You will note that both linear relationship between loads and deformations and the law of superposition are applied here.

Total strain energy stored in the beam

$$= (1/2)\Sigma\sigma_u dAdu + \Sigma\sigma_u dAdL + (1/2)\Sigma\sigma dAdL$$

By law of conservation of energy,

$$(1/2)1\ \delta + (1/2)\Sigma P_i\Delta_i + 1\ \Delta = (1/2)\sigma_u dAdu + (1/2)\Sigma(\sigma dAdL) + (1/2)\Sigma\sigma_u dAdL \quad (10.3)$$

Subtracting Equations (10.1) and (10.2) from Eq. (10.3), we get

$$1\Delta = \Sigma\sigma_u dAdL = \Sigma F_u dL$$

where F_u is the force on the element of area dA.

The force in the element can be due to bending moment, axial force, etc., depending upon the structure being analysed. The deformation sought can be deflection, slope, etc. In the case of angular deformations, we apply a unit moment in place of a unit point load.

Limiting the discussion to deformation due to bending alone, the term ΣFdL is the strain energy stored in the beam due to bending moment. This, as we have derived earlier, is given as $b\int M_x^2dx/EI$.

We apply this to a beam carrying loads as shown in Fig. 10.50. We have the basic equation of unit load method, which gives $1\Delta = \Sigma F_u dL$. In this equation, Ä is the deflection at the point where the unit load is applied. Also F_u is the force in the element of area dA due to the stress σ_u caused by the unit load and dL is the deformation of the element of length dx due to the applied loads P_i.

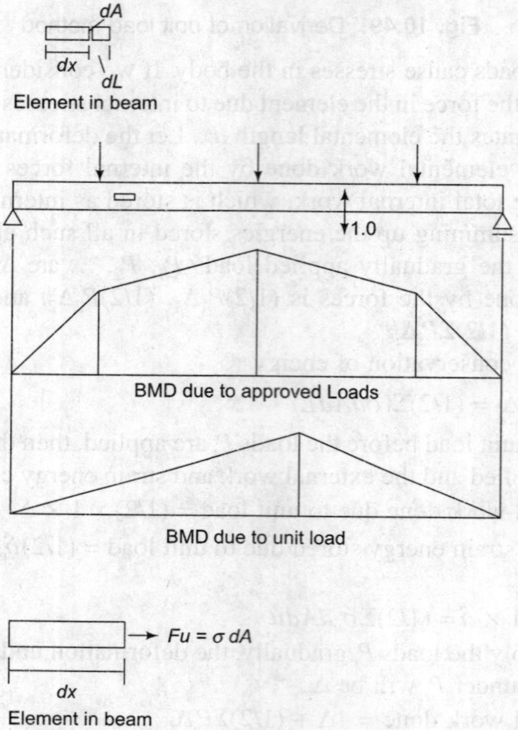

Fig. 10.50 Unit load method for a beam

If m is the moment caused by the unit load at the section of the element, then

$$F_u = (my/I)dA$$

The terms are explained in the figure.

If M is the moment caused by the loads P_i at the same section, then

$$dL = (My/I)dx/E$$

We can now write

$$1\Delta = \Sigma F_u dL = \int_0^L \int_0^A (mydA/I)(Mydx/EI)$$

$$= \int_0^L (Mmdx/EI^2) \int_0^A y^2 dA$$

$$= \int_0^L \frac{Mmdx}{EI} \text{ as } \int_0^A y^2 dA = I$$

This is the basic equation for the unit load method. The application of this equation to beam deflections and slopes is illustrated through the following examples. The application to truss deflections is given in Chapter 13.

Example 10.32 Unit load method: cantilever with uniformly varying load

A cantilever of 3 m span carries a uniformly varying load varying from 18 kN/m at the fixed end to zero at the free end. Find the deflection and slope at the free end using the unit load method. $EI = 40{,}000$ kNm^2.

Solution The beam with the load is shown in Fig. 10.51.

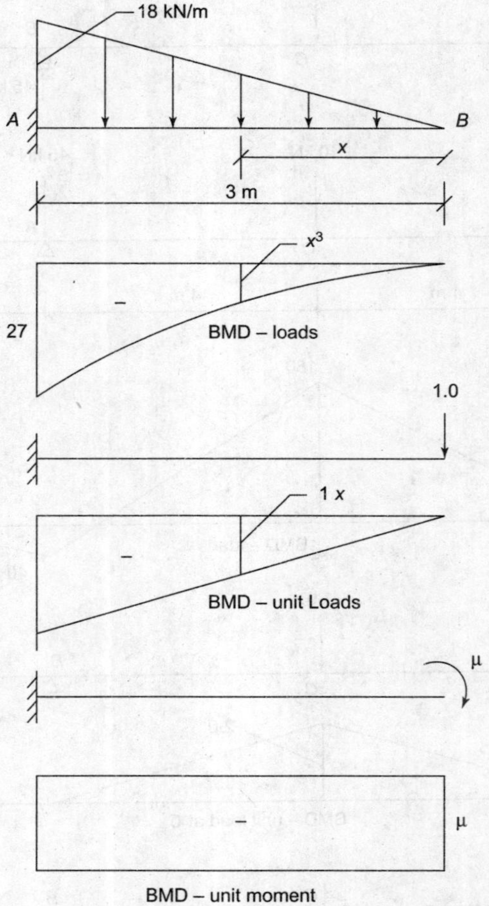

Fig. 10.51

$$m_x = -1x$$

$$y_B = \int_0^3 (-x^3)(-x)dx/EI = [x^5/5EI]_0^3 = \frac{3^5}{5\times 40{,}000}\ \text{m} = \frac{243\times 1000}{5\times 40,000} = 1.2\ \text{mm}$$

To find the slope at B, we apply a unit couple of 1 kNm at B.

Here $m_x = 1$ kNm is a constant.

$$\text{Slope at } B,\ \theta_B = \int_0^3 (-x^3)dx/EI = \frac{[-x^4]_0^3}{(4E)} = \frac{81}{4(40{,}000)} = 0.506 \times 10^{-3} \text{ radians}$$

Example 10.33 Unit load method: overhanging beam

The overhanging beam carries the loads as shown in Fig. 10.52. Find the deflection under the loads and the slope at A.

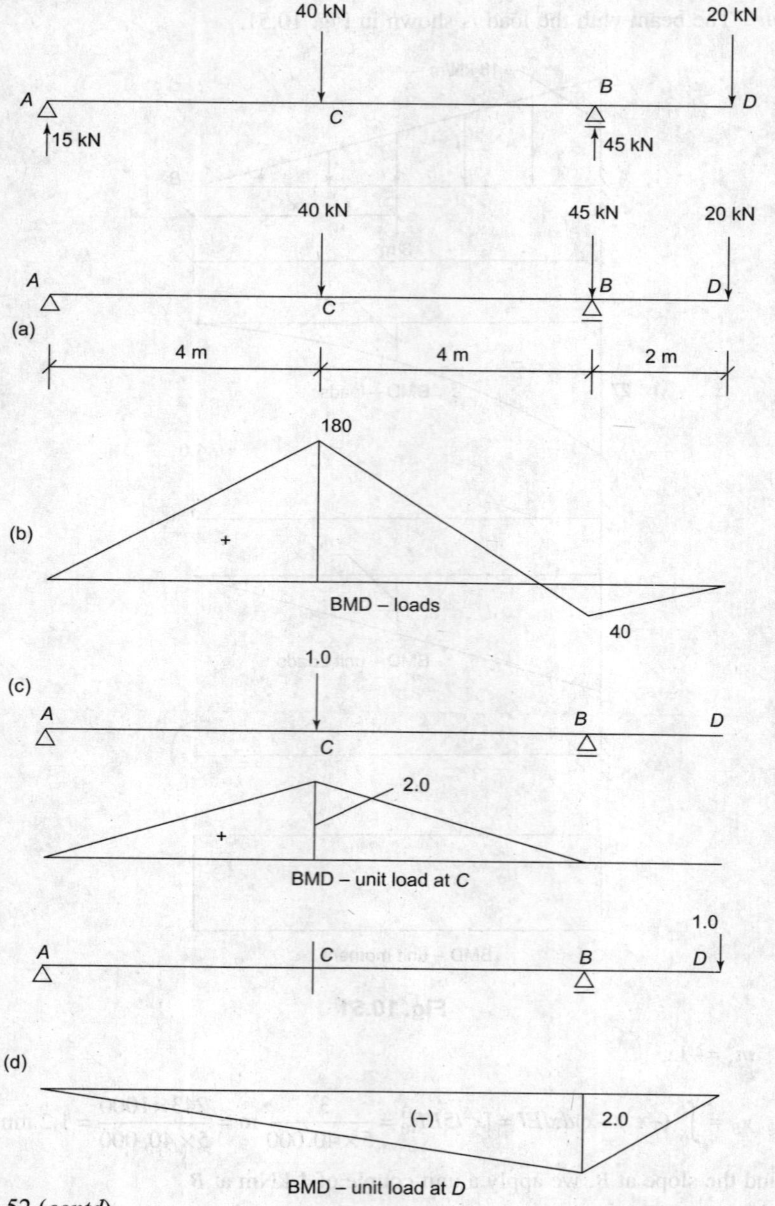

Fig. 10.52 (*contd*)

(e)

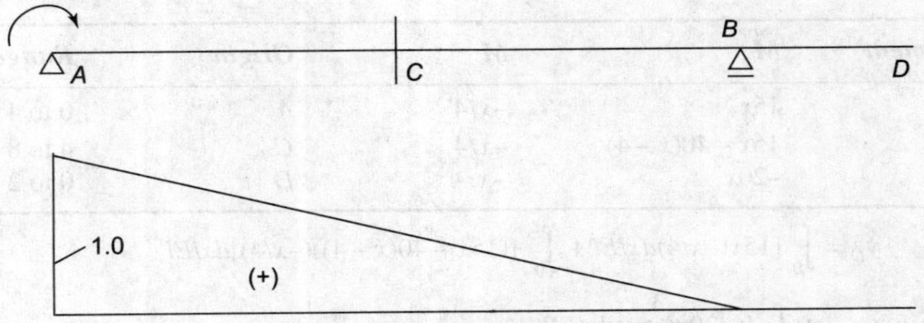

BMD – unit couple at *A*

(f)

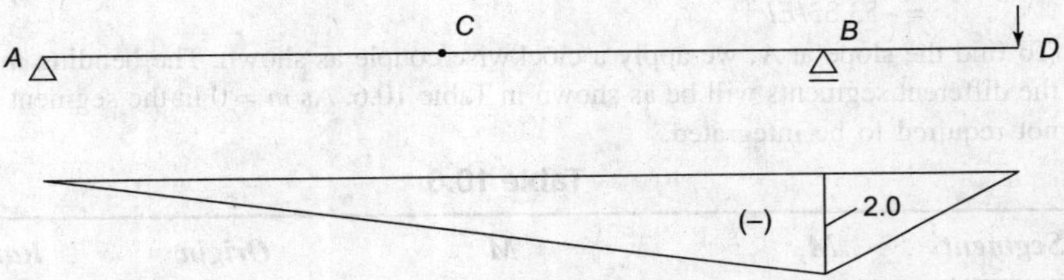

BM due to unit load at *D*

Fig. 10.52

Solution The bending moment diagram due to the applied loads is shown in Fig. 10.52(b). To find the deflection at *C*, we apply a unit load at *C*. The bending moment diagram due to the unit load is shown in Fig. 10.52(c). We have to consider the integration of the expression (*Mmdx/EI*) in three parts. Table 10.4 shows the moments.

Table 10.4

Segment	M_x	***M***	***Origin***	***Range***
AC	15*x*	*x*/2	*A*	0 to 4
CB	15x – 40(*x* – 4)	(1/2)*x* – (*x* – 4)	*B*	4 to 8
BD	–20*x*	0	*D*	0 to 2

As $m = 0$ in the segment *DC*, integration need not be done for that segment.

Vertical deflection at *C*, $y_C = \int_0^4 15x(x/2)dx/EI + \int_0^8 (15x - 40(x-4))[(1/2)x - (x-4)]dx/EI$

$$y_C = \int_0^4 \frac{15x^2 dx}{EI} + \int_4^8 \frac{[(160-25x)(40-x/2)dx}{EI}$$

$$= [15x^3/6EI]_0^4 + \frac{[640x - 90x^2 + 25x^3/6]_4^8}{EI}$$

$$= 266.67/EI \downarrow$$

To find the deflection under the 20 kN load, we apply a unit load at *D*. The bending moments at different segments are as shown in Table 10.5.

Table 10.5

Segment	M_x	M	*Origin*	*Range*
AC	$15x$	$-x/4$	A	0 to 4
CB	$15x - 40(x-4)$	$-x/4$	C	4 to 8
BD	$-20x$	$-x$	D	0 to 2

$$y_D = \int_0^4 [15x(-x/4)dx/EI + \int_0^8 [((15x - 40(x-4))(-x/4)]dx/EI$$

$$+ \int_0^2 [(-20x(-x))dx/EI]$$

$$= \frac{[-15x^3/12]_0^4}{EI} + \frac{[-20x^2 + 25x^3/12]_4^8}{EI} + \frac{[20x^3/3]_0^2}{EI}$$

$$= -53.33/EI \uparrow$$

To find the slope at A, we apply a clockwise couple as shown. The bending moment in the different segments will be as shown in Table 10.6. As $m = 0$ in the segment BD, it is not required to be integrated.

Table 10.6

Segment	M_x	M	*Origin*	*Range*
AC	$15x$	$1 - x/8$	A	0 to 4
CB	$15x - 40(x-4)$	$1 - x/8$	C	4 to 8

$$EI\theta_A = \int_0^4 [15x(1 - x/8)dx + \int_0^8 [(15x - 40(x-4))\,(1 - x/8)]dx$$

$$= [15x^2/2 - 15x^3/24]_0^4 + [160x - 45x^2/2 + 25x^3/24]_0^8$$

$$= 120 - 40 + [1280 - 1440 + 533.33 - 640 + 360 - 66.67]$$

$$= 106.67$$

$$\theta_A = 106.67/EI \text{ (clockwise as assumed)}$$

Example 10.34 Unit load method: cantilever frame

In the cantilever frame loaded as shown in Fig. 10.53, find the horizontal and vertical deflection at D.

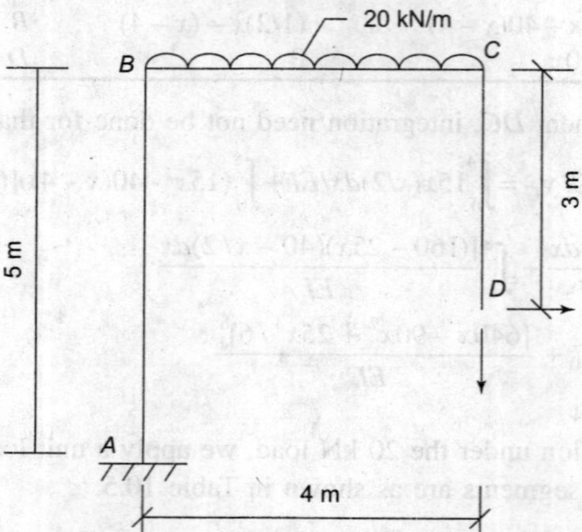

Fig. 10.53

Solution To find the vertical deflection of point D, apply a vertically downward unit load at D. The bending moments in the different segments are as shown in Table 10.7.

Table 10.7

Segment	M_x	M	*Origin*	*Limits of integration*
DC	0	0	D	0 to 3
CB	$-20x^2/2$	$-x$	C	0 to 4
BA	-160	-4	B	0 to 5

Deflection at D, $y_D = \Sigma \int M_x m dx/EI$

$$y_D = \int_0^4 \frac{(-10x^2)(-x)dx}{EI} + \int_0^5 \frac{(-160x^2)(-4)dx}{EI} = \frac{[10x^4]_0^4}{EI} + \frac{[640x]_0^5}{EI}$$

$$= 3840/EI$$

To find the horizontal deflection at D, apply a horizontal unit load as shown in the figure. The moments are as shown in Table 10.8.

Table 10.8

Segment	M_x	M	*Origin*	*Range*
DC	0	$1.x$	D	0 to 3 m
CB	$-10x^2$	$+3$	C	0 to 4 m
BA	-160	$(3 - 1.x)$	B	0 to 5 m

Horizontal deflection, $y_H = \Sigma \int Mmdx/EI$

$$EI\, y_H = \int_0^4 -30x^2 dx + \int_0^5 (-480 + 160x)dx = [-10x^3]_0^4 + [-480x + 80x^2]_0^5$$

$$= -1040$$

Horizontal deflection, $y_H = -1040/EI$ (The negative sign shows that the deflection is in a direction opposite to that of the unit load.)

Example 10.35 Unit load method: cantilever

The cantilever beam shown in Fig. 10.54 carries a UD load of 40 kN/m over the whole span. Find the deflection and slope at the free end.

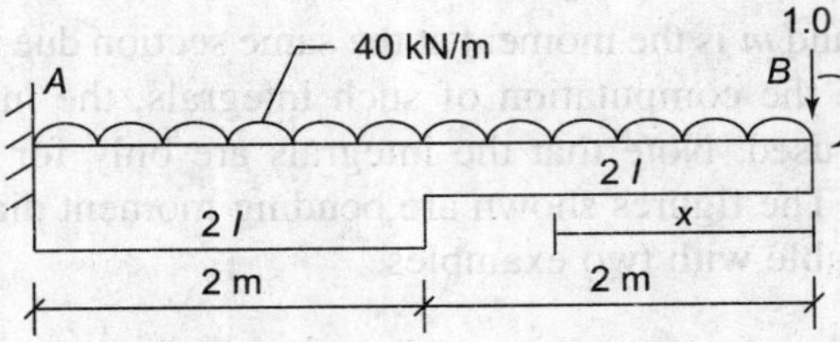

Fig. 10.54

Solution Due to change in value of MI, integration has to be done in two parts. The moments are shown in Table 10.9.

Table 10.9

Segment	M_x	*m*	*MI*	*Origin*	*Range*
BC	$-20x^2$	$-1.x$	*I*	*B*	0 to 2
CA	$-20x^2$	$-1.x$	2*I*	*C*	2 to 4

Vertical deflection at B, $y_B = \Sigma \int M_x m dx/EI$

$$y_B = \int_0^2 \frac{(-20x^2)(-x)dx}{EI} + \int_2^4 \frac{(-20x^2)(-x)dx}{EI} = \frac{[20x^4]_0^2}{EI} + \frac{[20x^4]_0^4}{EI}$$

$$= 680/EI$$

Vertical deflection of $B = 680/EI$

To find the slope at B, apply a unit clockwise couple as shown at B. The bending moments will be as shown in Table 10.10.

Table 10.10

Segment	M_x	*m*	*MI*	*Origin*	*Range*
BC	$-20x^2$	-1	*I*	*B*	0 to 2
CA	$-20x^2$	-1	2*I*	*C*	2 to 4

Slope at B, $\theta_B = \Sigma \int M_x m dx/EI]$

$$EI\,\theta_B = \int_0^2 20x^2 dx + \int_0^4 (20x^2/2)dx = [20x^3/3]_0^2 + [20x^3/6]_2^4 = 240$$

$$\theta_B = 240/EI$$

Table of product integrals In the unit load method, we find the integral $\int Mmdx/EI$. In most of the structures, material property like E and geometric property I remain constant over the length of the member. EI can then be taken out of the integral sign. We then have $(1/EI)\int Mmdx$. In this integral, M is moment at x due to the applied loads and m is the moment at the same section due to the unit load or couple. To facilitate the computation of such integrals, the integrals shown in Table 10.11 can be used. Note that the integrals are only for straight line and quadratic equations. The figures shown are bending moment diagrams. We illustrate the use of the table with two examples.

Table 10.11 Table of integrals of product Mm

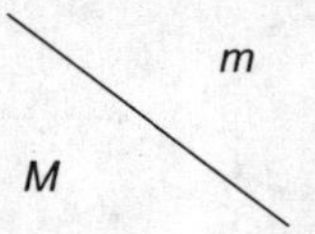 m, M	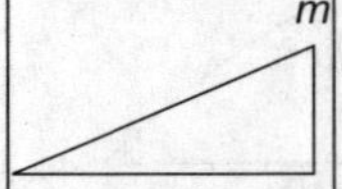 m	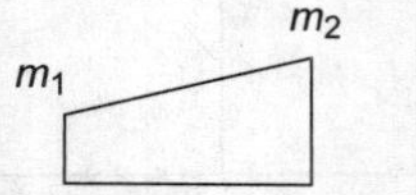 m_1, m_2	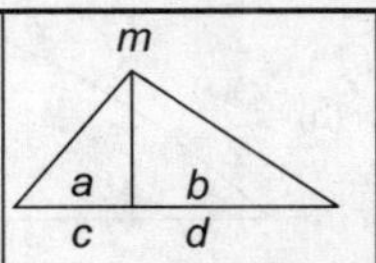 m, a, b, c, d	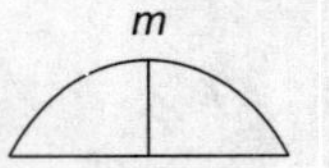 m
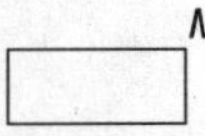 M	$\frac{L}{2}Mm$	$\frac{L}{2}(m_1+m_2)M$	$\frac{L}{2}(mM)$	$\frac{2L}{3}mM$
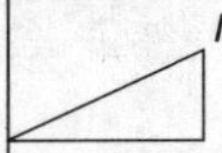 M	$\frac{L}{3}Mm$	$\frac{L}{6}(m_1+2m_2)M$	$\frac{L}{6}\left(1+\frac{a}{L}\right)mM$	$\frac{L}{3}mM$
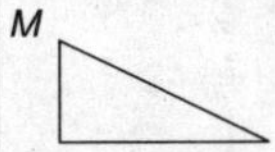 M	$\frac{L}{6}mM$	$\frac{L}{6}(2m_1+m_2)M$	$\frac{L}{6}\left(1+\frac{b}{L}\right)mM$	$\frac{L}{3}mM$
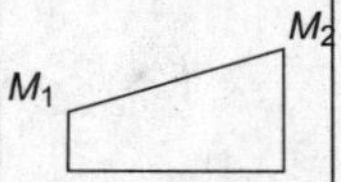 M_1, M_2	$\frac{L}{6}m(m_1+2m_2)$	$\frac{L}{6}m_1(2m_1+m_2)$	$\frac{L}{6}\left(1+\frac{b}{L}\right)mM_1$	$\frac{L}{6}m(M_1+M_2)$
		$+\frac{L}{6}m_2(M_1+2M_2)$	$+\frac{L}{6}\left(1+\frac{a}{L}\right)mM_2$	
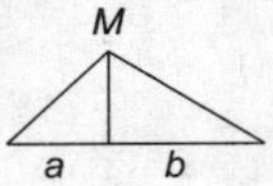 M, a, b	$\frac{L}{6}\left(1+\frac{a}{L}\right)mM$	$\frac{L}{6}\left(1+\frac{b}{L}\right)m_1M$ $+\frac{L}{6}\left(1+\frac{a}{L}\right)m_2M$	For < ≤ a $\frac{L}{3}mM -$ $\frac{L(a-c)^2}{6ad}$ mM	$\frac{L}{3}\left(1+\frac{ab}{L^2}\right)mM$
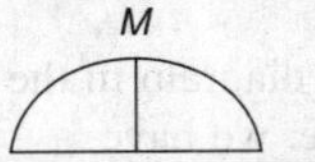 M	$\frac{L}{3}mM$	$va\frac{L}{3}(m_1+m_2)M$	$\frac{L}{3}\left(1+\frac{ab}{L^2}\right)mM$	$\frac{8L}{15}mM$
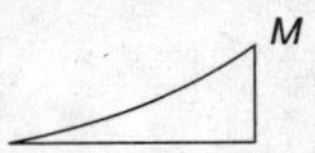 M	$\frac{1}{4}mM$	$\frac{L}{12}(m_1+3m_2)M$	$\frac{L}{12}\left(1+\frac{a}{L}+\frac{a^2}{L^2}\right)$ $\times(m.\ M)$	$\frac{L}{5}Mm$

Example 10.36 Unit load method: table of integrals

The simply supported beam of span 10 m carries a UD load 40 kN/m over the whole span. Find the deflection at the centre of the span and the slope at the left support. EI is constant and is equal to 10^6 kNm^2.

Solution The beam with the loading and the BM diagram are shown in Figures 10.55(a) and (b).

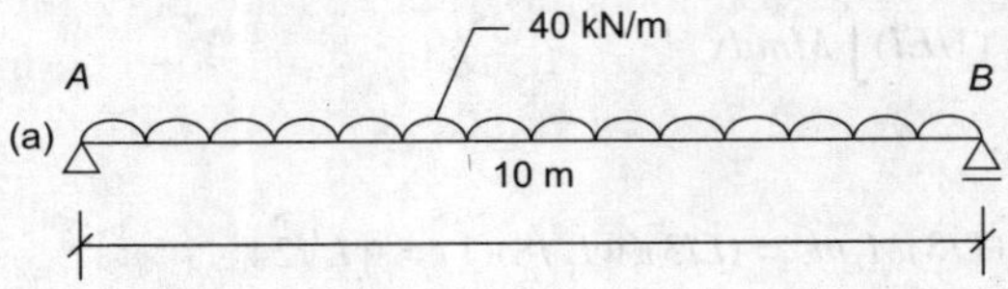

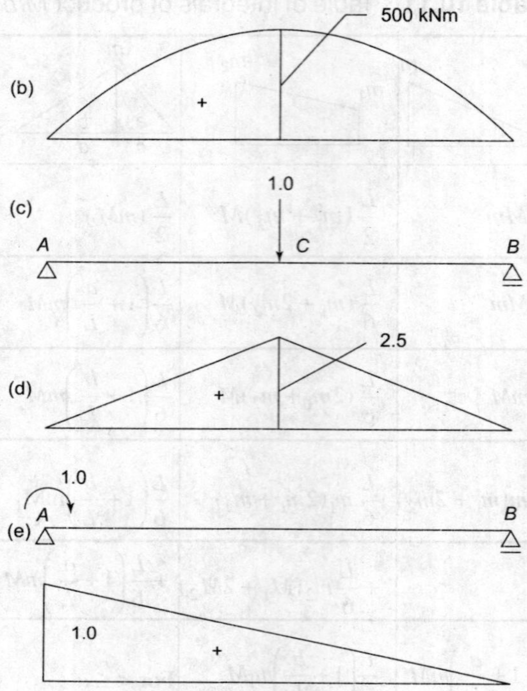

Fig. 10.55

To find the deflection at the centre of the span, we apply a unit load at C, the centre of the beam. The BM diagram due to the unit load is shown in Fig. 10.55(d). Now,

$$\text{Deflection at the centre} = \int (Mmdx/EI) = (1/EI)\int Mmdx$$

This integral can be found from Table 10.11. For the bending moment diagram in the form of a parabola and the unit load BM diagram in the form of a triangle, we have

$$\int Mmdx = L(1 + ab/L^2)m_u M_L/3$$

From the bending moment diagrams, $m_u = L/4$ and $M_L = wL^2/8$; $a = b = L/2$. Therefore,

$$\int Mmdx = L(1 + L^2/4L^2)(L/4)(wL^2/8) = 5/384wL^4$$

$$\text{Deflection at the centre} = \frac{5wL^4}{384\,EI} = \frac{5 \times 40 \times 10^4}{384 \times 10^6} = 5.2 \text{ mm}$$

To find the slope at A, we apply a unit couple at A as shown. The bending moment diagram due to this couple is shown in Fig. 10.55(e). Now,

$$\text{Slope } \theta_A = (1/EI)\int Mmdx$$

From Table 10.11,

$$\int Mmdx = (L/3)M_L m_u = (L/3)(wL^2/8)(1) = wL^3/24$$

$$\text{Slope at } A,\ \theta_A = \frac{40(10)^3}{24 \times 10^6} = 0.002 \text{ radians}$$

Example 10.37 Unit load method: table of integrals

A simply supported beam, of 8 m span, carries two point loads of 30 kN and 40 kN at 2 m and 4 m from the left end. Find the deflection under the loads and the slope at B. $EI = 0.4 \times 10^6$ kNm^2.

Solution The beam with the loads is shown in Fig. 10.56(a).

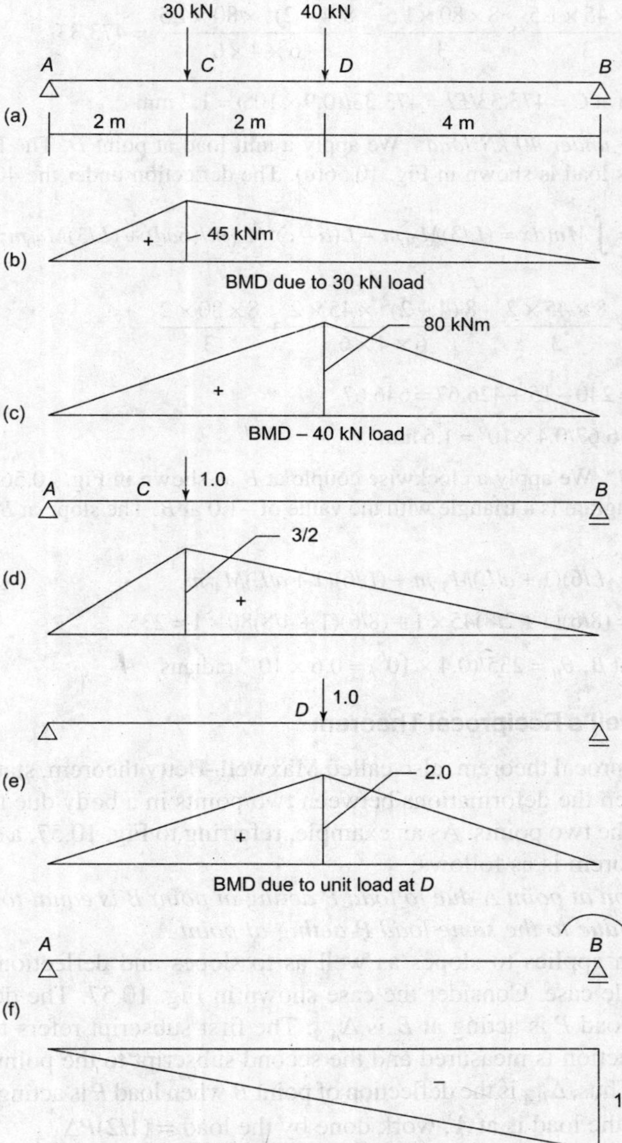

Fig. 10.56

We consider the two point loads separately. The bending moment diagrams due to the 30 kN load are shown in Figures 10.56(b) and (c).

(i) *Deflection under 30 kN load* To find the deflection under the 30 kN load, we apply a unit load at *C*. The BM diagram due to the unit load is shown in figure 10.56 (d). The integral of the product of this diagram with the two load diagrams is given by

$$EI\,y_C = \int Mm\,dx = [(L/3)(M_{30}m) + (L/3)M_{40}m - L(a-c)^2 M_{40}m/(6ad)$$

$$= \frac{8\times 45\times 1.5}{3} + \frac{8\times 80\times 1.5}{3} - \frac{8(4-2)^2\times 80\times 1.5}{6\times 4\times 6} = 473.33$$

Deflection at $C = 473.33/EI = 473.33/(0.9\times 10^6) = 1.2$ mm

(ii) *Deflection under 40 kN load* We apply a unit load at point *D*. The BM diagram due to this load is shown in Fig. 10.56(e). The deflection under the 40 kN load is

$$EI\,y_D = \int Mm\,dx = (L/3)M_{30}m - L(a-c)^2 M_{30}m/(6ad) + (L/3)M_{40}m$$

$$= \frac{8\times 45\times 2}{3} - \frac{8\,(4-2)^2\times 45\times 2}{6\times 4\times 6} + \frac{8\times 80\times 2}{3}$$

$$= 240 - 15 + 426.67 = 646.67$$

$$y_D = 646.67/0.4\times 10^6 = 1.6\text{ mm}$$

(iii) *Slope at B* We apply a clockwise couple at *B* as shown in Fig. 10.56(f). The unit load *m*-diagram is a triangle with the value of –1.0 at *B*. The slope at *B* is θ_B and is given by

$$EI\,\theta_B = (L/6)(1 + a/L)M_{30}m + (L/6)(1 + a/L)M_{40}m$$

$$= (8/6)(1 + 2/8)45\times 1 + (8/6)(1 + 4/8)80\times 1 = 235$$

Slope at B, $\theta_B = 235/(0.4\times 10^6) = 0.6\times 10^{-3}$ radians ■

10.9.4 Maxwell's Reciprocal Theorem

Maxwell's reciprocal theorem, also called Maxwell–Betty theorem, states the relationship between the deformations between two points in a body due to load systems acting at the two points. As an example, referring to Fig. 10.57, a simple way to state the theorem is as follows:

The deflection at point A due to load P acting at point B is equal to the deflection at point B due to the same load P acting at point A.

The theorem applies to slopes as well as to slopes and deflections. We first prove the simple case. Consider the case shown in Fig. 10.57. The deflection at point *A* when load *P* is acting at *B* is Δ_{AB}. The first subscript refers to the point where the deflection is measured and the second subscript to the point where the load is placed. Thus, Δ_{BA} is the deflection of point *B* when load *P* is acting at point *A*.

When the load is at A, work done by the load = $(1/2)P\Delta_{AA}$

When the load is at B, work done by the load = $(1/2)P\Delta_{BB}$

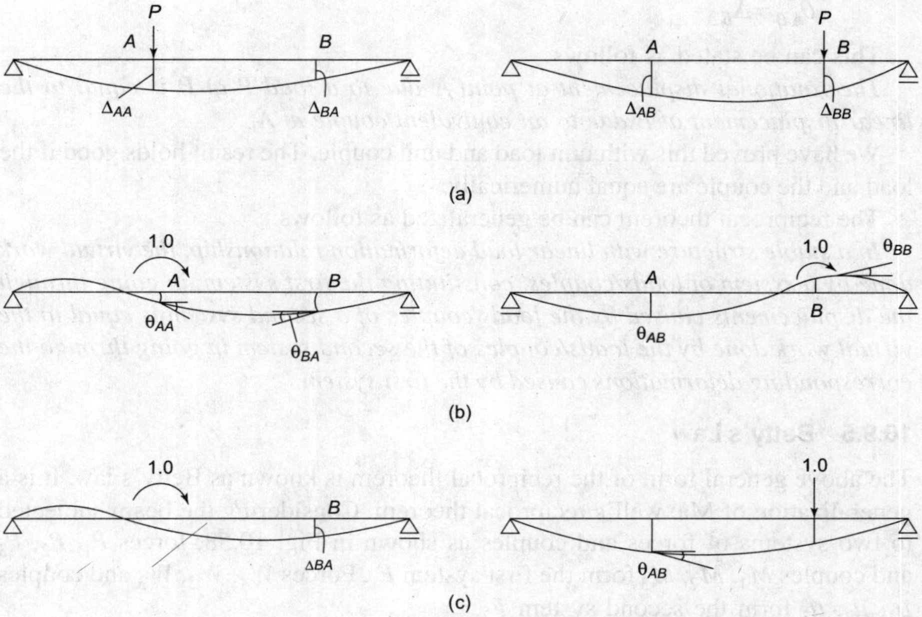

Fig. 10.57

Let load P be first gradually applied at A and then an equal load is gradually applied at B. Then the work done can be expressed as

$$W_1 = (1/2)P\Delta_{AA} + (1/2)P\Delta_{BB} + P\Delta_{AB}$$

as the load is already acting at A when the load is applied at B

Now, consider the reverse case of load P being applied first at B and then an equal load being applied at A. The work done in this case can be expressed as

$$W_2 = (1/2)P\Delta_{BB} + (1/2)P\Delta_{AA} + P\Delta_{BA}$$

As the work done must be equal irrespective of the sequence of the applying loads, $W_1 = W_2$.

$$(1/2)P\Delta_{AA} + (1/2)P\Delta_{BB} + P\Delta_{AB} = (1/2)P\Delta_{BB} + (1/2)P\Delta_{AA} + P\Delta_{BA}.$$

We get $\Delta_{AB} = \Delta_{BA}$

Now consider the case of couple M being applied in a similar fashion as shown in Fig. 10.57(b). If the rotational displacement at A when the couple is applied at B is θ_{AB} and the rotational displacement at B when the couple is placed at A is θ_{BA}, we can prove on similar lines that $\theta_{AB} = \theta_{BA}$.

Now consider the third case shown in Fig. 10.57(c). Here a unit couple is applied at point A. The rotation at A is θ_{AA} and the deflection at B is Δ_{BA}. We now apply a unit load at B. The deflection at B is Δ_{BB} and the rotation at A is θ_{AB}. The work done in this process is $W_1 = (1/2)\theta_{AA} + (1/2)\Delta_{BB} + \theta_{AB}$.

Now consider the reverse process of applying the unit load at B first and then applying the unit couple at A. Using the same notations, work done in this case is

$$W_2 = (1/2)\Delta_{BB} + (1/2)\theta_{AA} + \Delta_{BA}$$

As the total work done by these two processes must be the same, $W_1 = W_2$. This gives

$$\theta_{AB} = \Delta_{BA}$$

This can be stated as follows:

The rotational displacement at point A due to a load P at B is equal to the linear displacement at B due to an equivalent couple at A.

We have proved this with unit load and unit couple. The result holds good if the load and the couple are equal numerically.

The reciprocal theorem can be generalized as follows:

In a stable structure with linear load deformation relationship, the virtual work done by a system of loads/couples, constituting the first system, in going through the displacements caused by the loads/couples of a second system is equal to the virtual work done by the loads/couples of the second system in going through the corresponding deformations caused by the first system.

10.9.5 Betty's Law

The above general form of the reciprocal theorem is known as Betty's law. It is a generalization of Maxwell's reciprocal theorem. Considering the beam subjected to two systems of forces and couples as shown in Fig. 10.58, forces P_1, P_2, P_3 and couples M_1, M_2, M_3 form the first system F_1. Forces W_1, W_2, W_3 and couples μ_1, μ_2, μ_3 form the second system F_2.

Let W_1 be the work done by the first system of forces when acting alone.

Let W_2 be the work done by the second system of forces acting alone.

Let W_{12} be the work done by the first system of forces when they move through displacements and rotations caused by the second system of forces at their points of application.

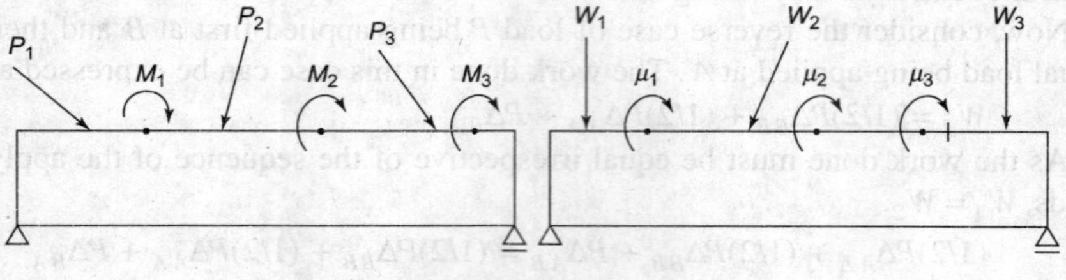

Fig. 10.58 (a)

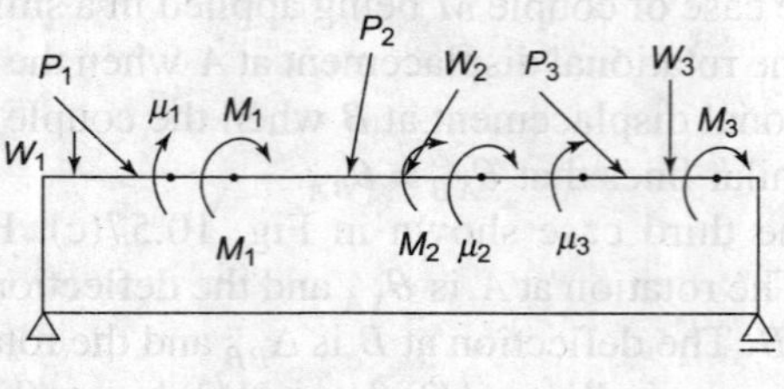

Fig. 10.58 (b)

Let W_{21} be the work done by the second system of forces when they move through displacements and rotations caused by the first system of forces at their points of application.

Refer to Fig. 10.58(b). We apply the system of forces F_1 alone first and then apply the second system of forces F_2.

Total work done in this case = $W_1 + W_2 + W_{12}$

If we apply the system of forces F_2 first and then the system of forces F_1, then

Total work done = $W_2 + W_1 + W_{21}$

As the total work done should remain the same irrespective of the order of applying the forces, we can equate the two and get $W_{12} = W_{21}$, which is the statement of Betty's law.

Summary

When a structural element is subjected to a straining action, the point of application of the external load moves due to the deformation of the structure and does work. This work is stored in the element as strain energy. Strain energy is, thus, the negative work done by the internal stress resultants, and has a direction opposite to that of deformation. It is equal to the area under the load-deformation diagram for the element.

Under an axial load, when the load is applied gradually,

$$\text{Strain energy } U = \frac{P^2L}{2AL} = \frac{\sigma^2 AL}{2E} = \frac{\sigma^2}{2E} \times \text{volume}$$

The maximum strain energy stored in a material per unit volume without permanent set (i.e., when the stress is within the elastic limit) is known as modulus of resilience. When the axial load is suddenly applied, the stress and deformation are double the values when the same load is gradually applied; the strain energy is consequently four times that in the gradually applied case. Vibrations are set up in the member due to this suddenly applied load.

When the axial load is applied with an impact, the kinetic energy of the body and the external work done due to deformation are converted into instantaneous strain energy. Vibrations are set up in the member due to large instantaneous deformation. The stresses and deformations can be calculated by equating strain energy to the energy of the striking body and the external work done.

Toughness is a characteristic of the material of its ability to resist impact loading and depends on the ductility of the material. Mild steel has more toughness than high-strength steel because it is ductile and can undergo large deformation and absorb large amounts of energy.

In the case of a member subjected to bending, the strain energy is given by $U = \int M^2 dx/2EI$, where the integral is over the whole length of the beam. The strain energy due to shear is given by $U = \tau^2/2G = r^2G/2$ for a unit volume, considering the SF to be constant. For varying SF, as in the case of a beam subjected to lateral loads, the strain energy stored is calculated by integration, taking into account the variation of shear stress over the cross section and the SF over the length. The strain energy due to torsion in the case of a circular shaft of length l subjected to torque T is given by $U = T^2l/2GJ = GJ\theta^2/2l$. Strain energy concepts are generally employed to calculate deformations. The methods are particularly useful in the case of members subjected to impact loading.

The theorems relating to strain energy are the theorem of conservation of energy, Castigliano's theorem, and Engesser's theorem. Castigliano's theorem and unit load method are useful in determining deflections and slopes in beams and frames.

Exercises

Review Questions

1. Define the terms strain energy, modulus of resilience, and toughness.
2. State from everyday experience a system wherein strain energy is stored in an element and is recovered for convenient use.
3. For a prismatic bar subjected to an axial load, state the expression for strain energy, in terms of (i) load, (ii) stress, and (iii) elongation of the bar.
4. How is the strain energy related to the load-deformation diagram of a bar?
5. Within the proportional limit, does the strain energy of bar vary linearly with the load? Explain.
6. Mild steel has more toughness than high-strength steel. Explain in terms of strain energy.
7. If the load on an axially loaded bar is doubled, by how much does the strain energy increase?
8. Two bars are made of the same material and have the same cross-sectional area. Find the ratio of their lengths, if their strain energies are equal when one is subjected to a gradually applied load and the other to the same load suddenly applied.
9. If two bars have the same sectional area and length, and are subjected to the same load, find the ratio of the strain energies stored in them, if E for one is two and a half times that of the other.
10. If two bars have the same mass, will the strain energies be the same under the same load?
11. One bar is half as long as the other, and they are made of the same material. Under the same load, if their strain energies are also in the ratio 1 : 2, what is the ratio of their cross-sectional areas.
12. Give the expressions for strain energy due to axial load, BM, shear, and torsion. Discuss the similarity in the expressions for strain energy in all cases.
13. Explain the behaviour of a bar subjected to (i) a suddenly applied load, and (ii) an impact load.
14. State the advantages of calculating the strain energy of a bar. What is its application in structural analysis?
15. State the theorem of conservation of energy.
16. State Engesser's theorem of complimentary energy. State the advantages of this theorem.
17. State and prove Castigliano's first theorem.
18. State and prove the unit load method for finding deflections.
19. State and prove Maxwell–Betty reciprocal theorem.

Problems

1. Calculate the strain energy stored in a bar 2 m long, 50 mm wide, and 40 mm thick, when it is subjected to a tensile load of 50 kN. Take E = 200 GPa.
2. A rectangular body 500 mm long, 100 mm wide, and 50 mm thick subjected to a shear stress of 80 MPa. Determine the strain energy stored in the body. G = 85 GPa.

3. Find the strain energy of a steel rod, of diameter 20 mm and length 400 mm, when it is subjected to an axial load of 40 kN. What is the elongation of the rod under this load? Determine the modulus of resilience of the rod if the elastic limit is 250 MPa. Take $E = 200$ GPa.
4. A stepped steel rod is in two parts—one has a diameter of 20 mm and a length of 1 m, and the second a length of 800 mm and a diameter of 12 mm. Find the strain energy of the bar under a load of 15 kN. Take E = 200 GPa.
5. Find the strain energy of an aluminium bar which is in three sections—one of 20 mm ϕ and length 1.2 m, another of diameter 15 mm and length 0.8 m, and the last of 12 mm ϕ and length 0.6 m. Take E = 70 GPa. The load is such that the maximum stress in the bar is 35 MPa.
6. A solid rod tapers from a diameter D to d over a length L. If it is subjected to an axial load P, show that its strain energy is

$$\frac{P^2 L}{2Ea}\frac{d}{D}$$

where a is the area of the smaller end of diameter d and E is Young's modulus of elasticity of the material.
7. A solid rod of diameter 20 mm and length 200 mm is subjected to an axial load of 20 kN, which is suddenly applied to the rod. Find the maximum stress and maximum instantaneous elongation. E = 200 GPa.
8. A plate of uniform thickness 5 mm has a tapering width varying from 50 mm to 25 mm over a length of 250 mm. It is subjected to an axial pull of 30 kN. Find the strain energy stored in the bar and the extension in length. Take E = 200 GPa.
9. A steel rod of solid circular section and diameter 20 mm is covered by a brass sleeve of internal diameter 20 mm and thickness 5 mm. The composite rod of length 200 mm is subjected to axial load such that the stress in the brass is not more than 50 MPa. Find the elongation of the bar and the strain energy stored under this load. Take E = 200 GPa for steel and 100 GPa for brass.
10. A bar 400 mm long and of diameter 500 mm is subjected to a tensile load of 200 kN. Find the stress, elongation, and strain energy produced if the load is applied gradually. What would be the instantaneous stress and elongation if the same load is applied suddenly. Take E = 200 GPa.
11. In the plane frameworks shown in Fig. 10.59, find the strain energy stored in the systems. The bars are of uniform section and have the cross-sectional areas indicated. Take E = 200 GPa.

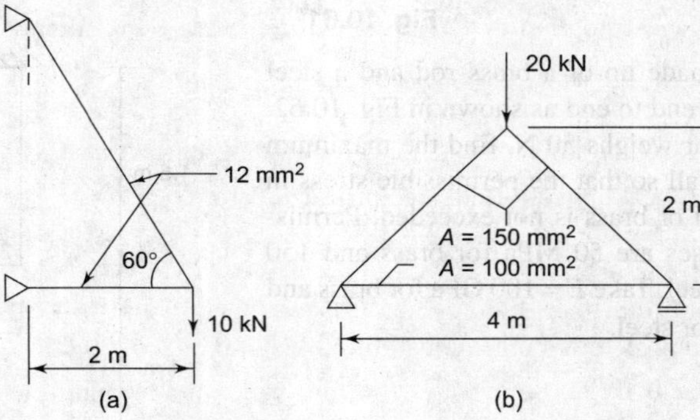

Fig. 10.59

12. The brass rod shown in Fig. 10.60 has a diameter of 20 mm. The collar D moves along the rod and strikes the metal plate attached at the lower end. If E = 100 GPa, determine the strain energy and instantaneous elongation when (i) the collar D weighs 1000 N and falls through a height of 5 mm, and (ii) the collar weighs 10 N and falls through a height of 500 mm.

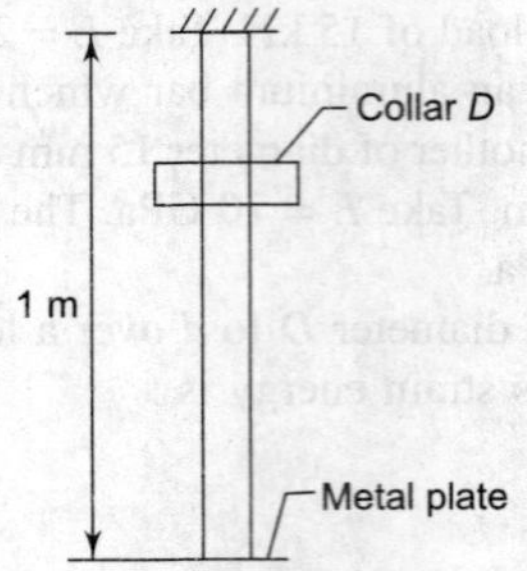

Fig. 10.60

13. A bar of length 400 mm and diameter 50 mm is subjected to a tensile load of 20 kN. Find the stress, elongation, and strain energy produced if the load is applied gradually. What would be the instantaneous stress and elongation if the same load is applied suddenly. Take E = 200 GPa.
14. In the system shown in Fig. 10.61, the collar falls through a height of 1 m. Find the maximum weight of the collar such that the stress in the material does not exceed 200 N/mm^2. Take E = 200 GPa.

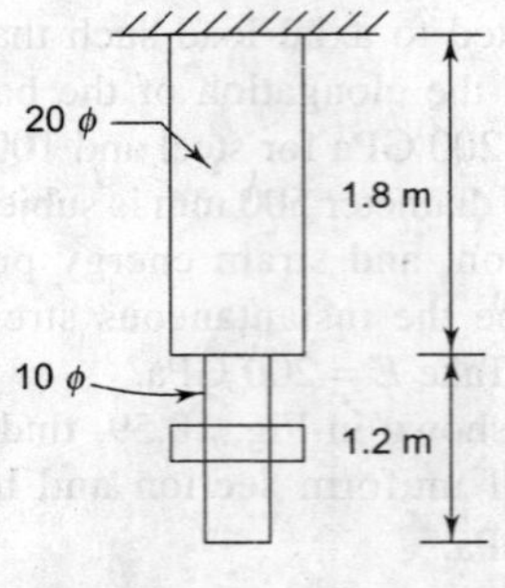

Fig. 10.61

15. A bar is made up of a brass rod and a steel rod joined end to end as shown in Fig. 10.62. If the collar weighs 50 N, find the maximum height of fall so that the permissible stress in either steel or brass is not exceeded. Permissible stresses are 50 MPa for brass and 150 MPa for steel. Take E = 100 GPa for brass and 200 GPa for steel.

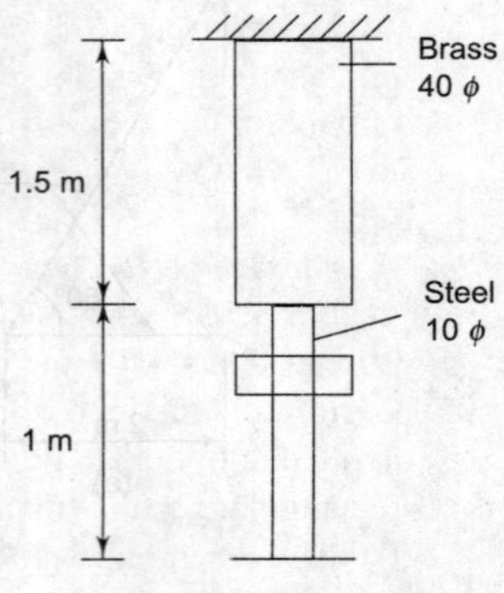

Fig. 10.62

16. In the system shown in Fig. 10.63, find the mass of the body that strikes the rod axially with a velocity of 2 cm/s. The rod is of brass with E = 100 GPa, and the maximum stress should not exceed 50 MPa.

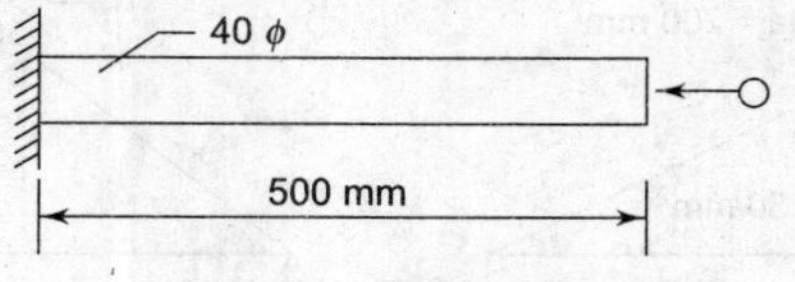

Fig. 10.63

17. A vertical tie rod rigidly fixed at the top end consists of a steel rod of 2 m length and 20 mm diameter encased throughout in a brass tube of 20 mm internal diameter and 30 mm external diameter. The rod and the casing are fixed together at the ends. The compound bar is suddenly loaded in tension by a mass of 1530 kg falling freely through 3 mm before being arrested by the tie. Compute the maximum stresses in steel and brass. Take $E_s = 2 \times 10^5$ MPa and $E_b = 1 \times 10^5$ MPa.
18. A bar of length 3 m has an area of cross section of 2400 mm^2 for a length of 1.8 m and an area of cross section of 1200 m^2 for the remaining length. Find the strain energy stored in the bar when a load of 80 kN is gradually applied. Determine also the ratio of the strain energy stored in a bar of uniform section and of the same length and volume under the same load. Take E = 200 GPa.
19. A composite bar is made up of a steel rod of diameter 20 mm, and a brass tube of external diameter 30 mm and thickness 5 mm. The bar is 1 m long and is subjected to impact due to a load of 50 N falling from a height of 200 mm. Calculate the instantaneous stresses in each material. Take E = 200 GPa for steel and 100 GPa for brass.
20. A cantilever beam of span 2 m carries a load of 1 kN at the free end. If its cross section is 80 mm wide and 120 mm deep, find the strain energy stored in the beam due to bending. What is the deflection at the end of the cantilever under this load? Take E = 200 GPa.
21. A steel pipe of external diameter 30 mm and thickness 5 mm is supported at its ends which are 2 m apart. Find the strain energy due to bending, stored in the pipe. E = 200 GPa. The pipe carries a load of 20 kN/m of its length in addition to its own weight. Assume that the weight is 78.5 kN per m^3.
22. A beam of span 3 m and section 30 mm × 60 mm carries a load varying uniformly from zero at one end to 30 kN/m at the other. Find the ratio of the strain energies stored in the beam due to bending and shear. Take E = 200 GPa and G = 80 GPa.
23. A cantilever beam has a span of 2 m and is of rectangular section—of breadth 30 mm and depth 60 mm. A load of 50 N falls at the free end from a height h. Find the maximum value of h so that the maximum stress in the beam does not exceed 200 MPa. What is the maximum instantaneous deflection in such a case. Take E = 200 GPa.
24. A beam supported over a span of 6 m has a maximum deflection of 2 mm when a load of 5 kN is applied at its mid-point. If the same load is dropped from a height of 20 mm at the mid-point of the beam, what will be the maximum deflection?
25. A simply supported beam has a span of 3 m. A concentrated load of 30 kN is applied at mid-span gradually. If E = 200 GPa and G = 80 GPa and the beam is of rectangular section (60 mm × 150 mm), find the strain energies stored in the beam (i) due to bending and (ii) due to shear. Also find the deflection under the load.

26. In the arrangement shown in Fig. 10.64, find the vertical deflection of the loaded point *P*. Take E = 200 GPa.

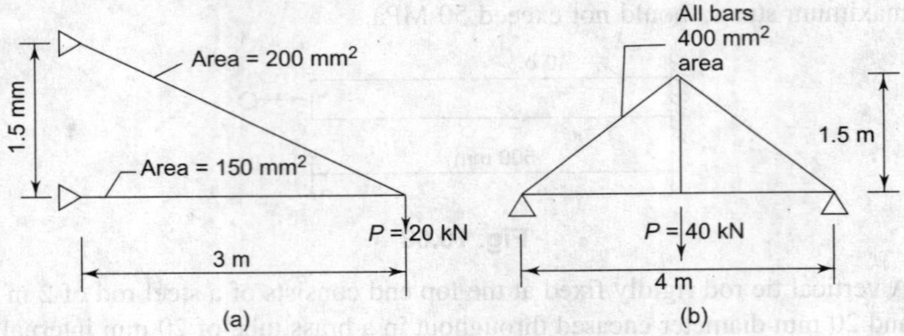

Fig. 10.64

27. A cantilever of length L carries a UD load of w/m for half its length from the fixed end. Determine the deflection at the free end. EI is constant.
28. The two bars AB and AC support a load of 20 kN as shown in Fig. 10.65. Find the vertical deflection of point A under the load.

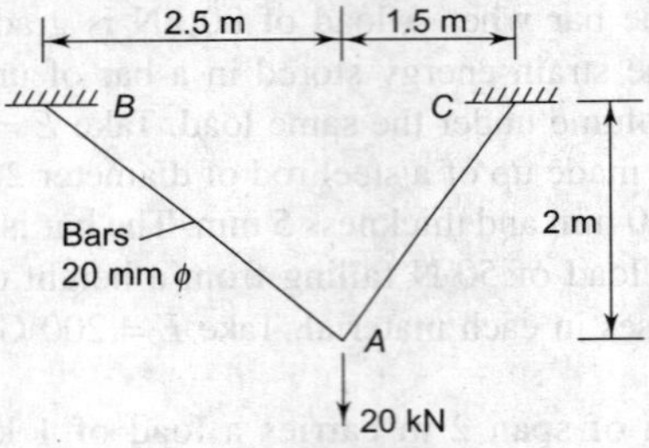

Fig. 10.65

29. In the frame shown in Fig. 10.66, find the horizontal deflection of point C under the load of 30 kN. The bars have an area of 600 mm^2. E = 200 GPa.

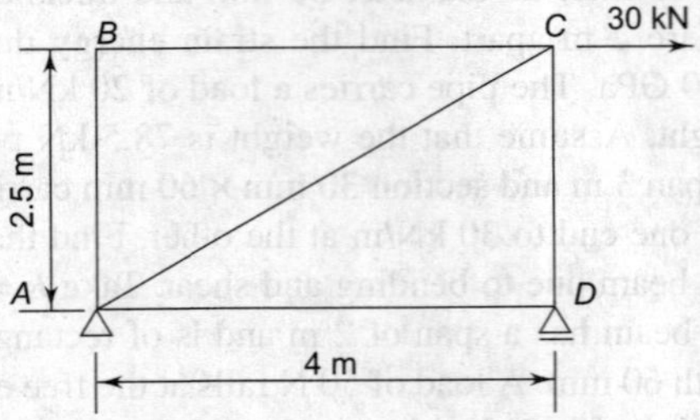

Fig. 10.66

30. A simply supported beam has a span of 6 m. It is subjected to a clockwise couple load of 48 kN m at 2 m from the left end. If E = 200 GPa and I = 100,000,000 mm^4, find the slope of the beam at the point of application of the couple.
31. Show that in the case of a simply supported beam, if a couple M is applied at one end, the slope at the point of application of the couple is $Ml/3EI$, where l is the length of the beam.
32. Using Castigliano's theorem, calculate the vertical deflection at the middle of a simply supported beam carrying a uniformly distributed load of w N/m over full span.

33. A solid circular shaft of steel, of diameter 150 mm, is 4 m long, and is subjected to a torque of 10 kN m. Find the strain energy stored in the shaft and the relative rotation between the ends. Take $G = 80$ GPa.
34. A hollow shaft of external diameter 200 mm and internal diameter 150 mm transmits 1500 kW @ 150 rev/minute. Calculate the strain energy per unit length stored in the shaft and the maximum shear stress. Take $G = 85$ GPa.
35. A shaft is in two parts, as shown in Fig. 10.67. Find the total strain energy stored in it. Take $G = 85$ GPa.

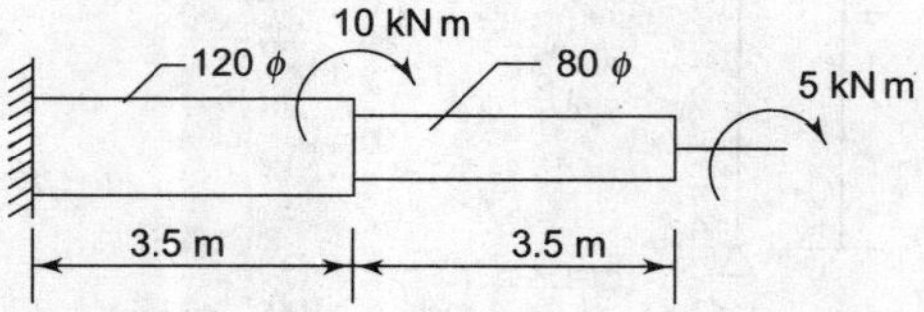

Fig. 10.67

36. The diameter of a shaft, which is 100 mm at one end, reduces to 80 mm at the other end over a length of 500 mm. If it is subjected to a torque of 20 kN m, constant over its length, find the strain energy stored in the shaft. Take $G = 80$ GPa.
37. Prove that the strain energy of a curved beam with a small initial curvature can be given by $\delta M^2 ds/2EI$.

Solve the following problems using Castigliano's method.

38. Find the deflection and slope at the free end of a cantilever of span 3 m, carrying a uniformly varying from zero at the free end to 25 kN/m at the fixed end. EI is constant.
39. Find the deflection under the loads and slopes at the supports of simply supported beam carrying loads as shown in Fig. 10.68. $EI = 0.8 \times 10^6$ Nm^2.

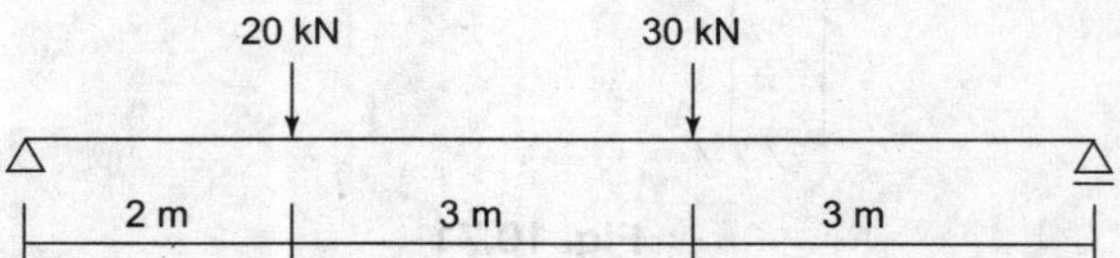

Fig. 10.68

40. Find the deflection under the loads and slopes at the supports of an overhanging beam loaded as shown in Fig. 10.69. EI is constant.

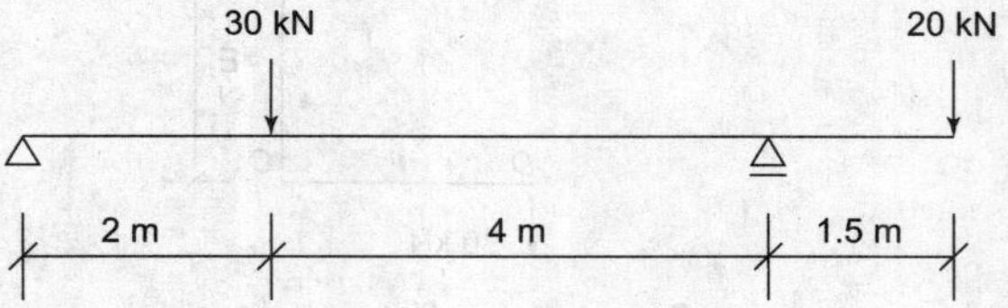

Fig. 10.69

41. Find the horizontal movement of the roller support of the frame loaded as shown in Fig. 10.70.

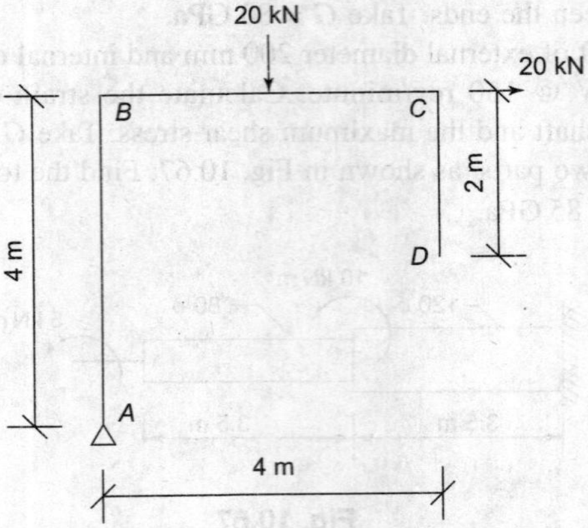

Fig. 10.70

42. Find the horizontal and vertical deflection of the free end of the frame loaded as shown in Fig. 10.71. *EI* is constant.

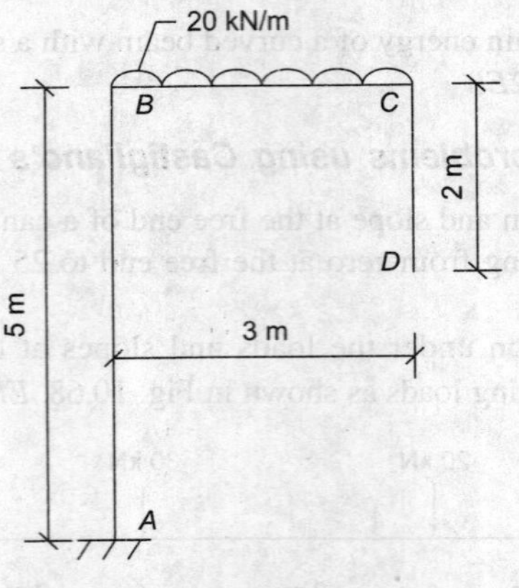

Fig. 10.71

43. The frame shown in Fig. 10.72 carries a single point load of 30 kN at the free end. Find the horizontal and vertical deflection at the free end and the slope at *B*. *EI* is constant.

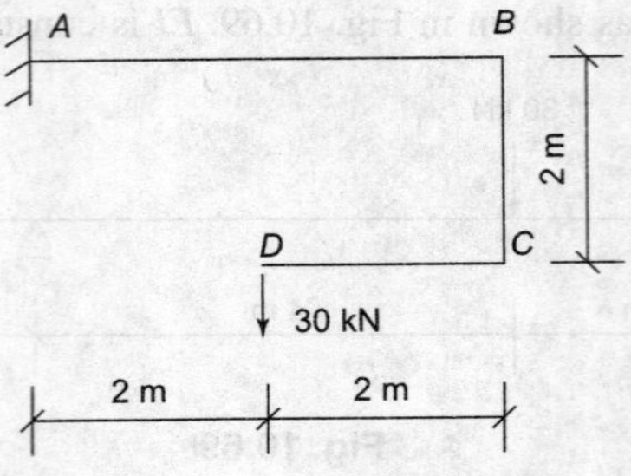

Fig. 10.72

Solve the following problems using the unit load method.

44. A simply supported beam of 9 m span carries a uniformly varying load varying from zero at the left end to 30 kN/m at the right support. Find the slopes at the ends and the deflection at the mid-span.
45. In the overhanging beam loaded as shown in Fig. 10.73, find the slope at support *A* and the deflection at the free end. *EI* is constant.

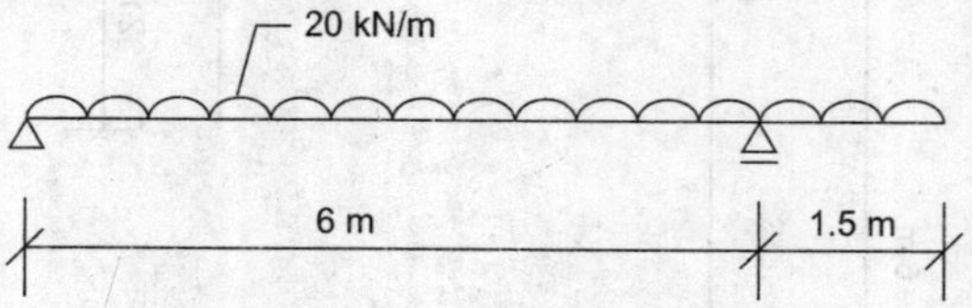

Fig. 10.73

46. Find the horizontal and vertical deflections at the free end *D* of the frame loaded as shown in Fig. 10.74. *EI* is constant.

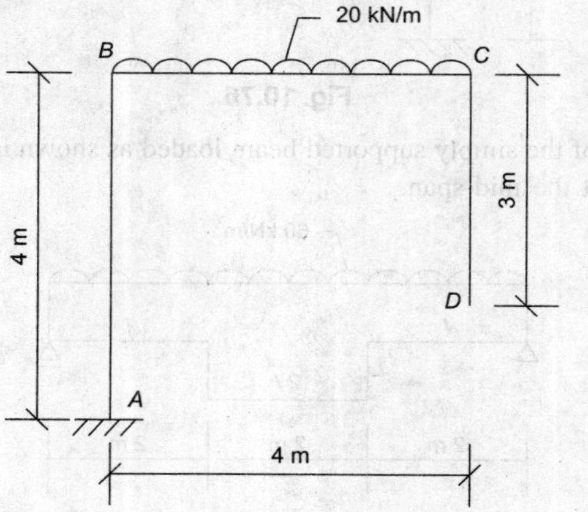

Fig. 10.74

47. Find the horizontal and vertical deflections at the free end *E* of the frame loaded as shown in Fig. 10.75. Take *EI* constant.

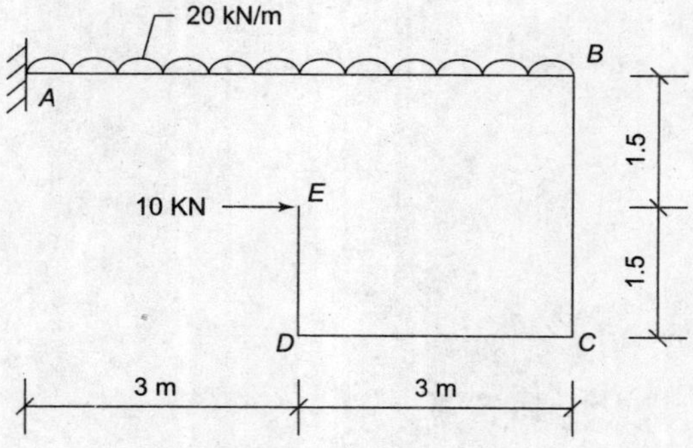

Fig. 10.75

48. In the frame loaded as shown in Fig. 10.76, find the horizontal and vertical deflections at the free end. *EI* is constant.

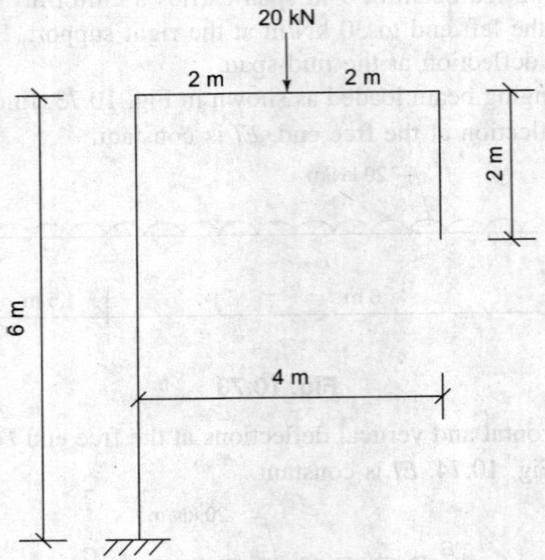

Fig. 10.76

49. In the case of the simply supported beam loaded as shown in Fig. 10.77, find the deflection at the mid-span.

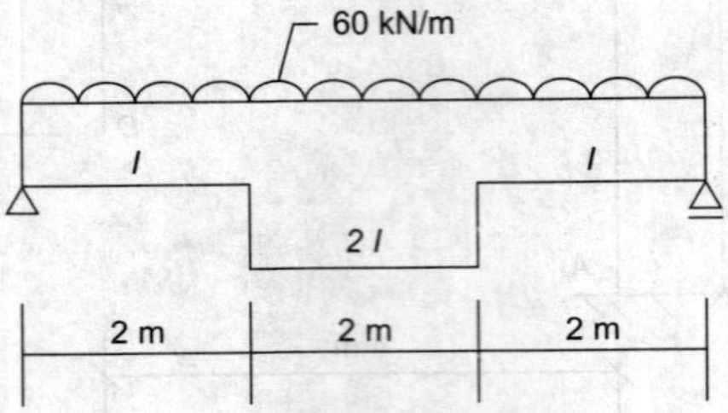

Fig. 10.77

CHAPTER 11

Columns

Learning Objectives

After going through this chapter, the reader will be able to

- understand the concept of stability of compression members,
- explain the structural behaviour of columns under load,
- derive the Euler crippling load for columns with different end conditions,
- understand the limitation and applicability of Euler's formulae,
- derive the Rankine–Gordon empirical formula for compression members,
- calculate and compare the crippling load for columns using Euler and Rankine formulae,
- derive the secant formula for columns subjected to eccentric loads, and apply it for calculating permissible loads and eccentricities,
- understand the structural action of beam columns and derive the beam-column formulae, and
- calculate the permissible load on columns with unsymmetrical sections.

11.1 INTRODUCTION

In Chapters 3 and 6, we dealt with compression members. We assumed that these members were short. This assumption was necessary, because we could not have applied the equations derived in these chapters if the members were long. Most columns, however, are not short, and their behaviour is governed more by stability than strength. In this chapter, we will discuss the behaviour of columns, and the mathematical and empirical methods for their design. Most design codes for columns are based upon these principles for the behaviour of long columns, to be derived later in this chapter, and are modified according to material properties, practical end conditions and experimental results.

Compression members differ from tension members in one respect. Any lack of straightness in a member is reduced, and the member becomes straighter if a tensile load is applied. On the other hand, a compressive load applied to such a member increases the eccentricity [Fig. 11.1(a)].

The concept of stability can be understood from a simple example. Consider the system shown in Fig. 11.1(b). AC and CB are two members joined at C by a pin-joint and a torsional spring of stiffness K. As it is, the system is stable as shown in Fig. 11.1(b). But if we give a small lateral deflection to the rods at C so that AC and CB make a small angle $\Delta\theta$ with the vertical, the system takes the form shown in

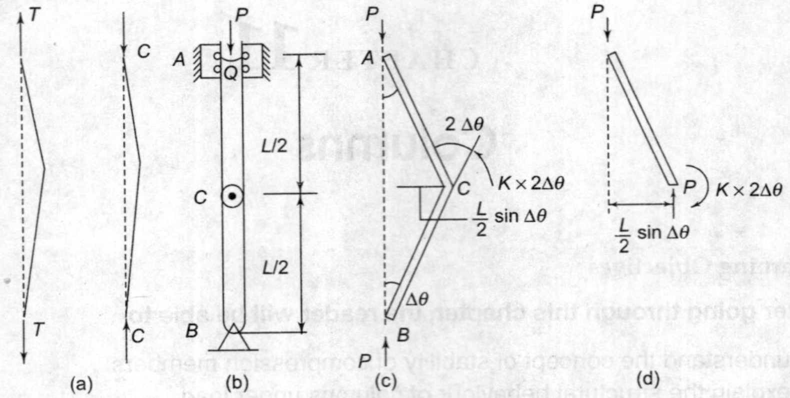

Fig. 11.1

Fig. 11.1(c). The rotation for the torsional spring is $2\,\Delta\theta$, and the restoring couple is $K(2\,\Delta\theta)$ [Fig. 11.1(d)]. The couple formed by the forces P is $P(l/2)\sin\Delta\theta = P(l/2)\,\Delta\theta$ as $\Delta\theta$ is very small [Fig. 11.1(d)]. This couple tends to increase the angle $\Delta\theta$, opposite to the couple of the torsional spring. Thus,

$$P\frac{l}{2}\sin\Delta\theta = K \times 2\,\Delta\theta$$

$$P = \frac{4K}{L}$$

So long as P is less than $4K/L$, the system is stable. $P = 4K/L$ is the critical value, and if $P > 4K/L$, the system is unstable, and will collapse if there is a small lateral deflection. The stability of an elastic column, such as the one shown in Fig. 11.2 (a), is governed by similar principles.

11.2 BEHAVIOUR AND CLASSIFICATION OF COLUMNS

Consider the case shown in Fig. 11.2(a). When the compression member carries an axial load P, its behaviour is governed by many factors such as (i) the length of the column, (ii) the size (cross-sectional), (iii) end conditions, and (iv) the material properties. It has been observed that the strength of a column is inversely proportional to the square of its length and directly proportional to the size and material properties. The end conditions, shown in Fig. 11.2(a), affect the strength because the end restraints imposed affect the stability characteristics.

When a column is short, and short is defined in all design codes based upon the ratio of length and the least lateral dimension or the radius of gyration, and the load is truly axial, the average stress can always be calculated as σ = load/area. The strength of the column or its load carrying capacity can be found as $\sigma \times$ area, where σ can have a suitable value. For example in the case of steel, σ is the yield stress of the material. Beyond the yield stress, the material will yield and the strength above this limit will have no practical use. For practical designs, the same formula can be used, σ_Y being the yield stress for ultimate strength design or yield stress divided by a factor of safety for working stress design.

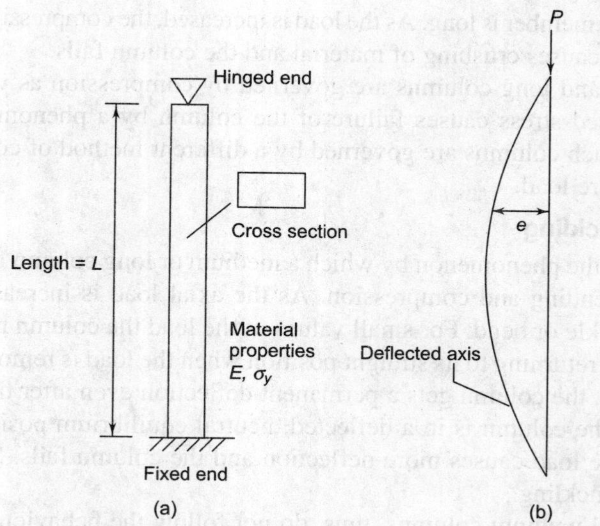

Fig. 11.2

Due to many factors, the behaviour of long columns is considerably different. The load may not be truly axial or the column itself may have a centre line which is not straight initially. These essentially lead to the phenomenon of buckling, i.e., lateral deflection of the bar. When such eccentricities are small and the column is short, the behaviour of the column is governed by the direct compressive stress we have discussed. But if the column is long, the effects due to buckling become significant, and are the principal cause of failure.

Consider the column shown in Fig. 11.2(b). Due to differences in material properties, accidental eccentricity of the load or the centroidal line being initially not straight, a typical section is subjected to a BM due to the compressive load. Under small values of load, the column is under compression. Long columns are inherently prone to buckling, and this lateral deflection causes BM. Longer the column, smaller is the value of the load required to initiate lateral deflection. The column section is thus subjected to direct stresses due to P as well as bending stresses due to the BM, Pe. The column, therefore, tends to deflect laterally. Such deflection, as we have seen in Chapter 7, varies proportionally with the cube of the length. If the length is doubled, the deflection increases eight times. This makes long columns behave much differently from short columns. In the case of short columns, lateral deflections as well as bending stresses are small, and direct stresses dominate. For long columns, however, bending stresses become significant.

11.2.1 Axially Loaded Compression Members

Axially loaded compression members can thus be classified into long, medium and short columns. Short columns generally fail by compression. The compressive stress can be calculated as load/area. Thus in Chapter 3, when we calculate the load carrying capacity of a concrete column, we specify it as a short compression member. Also the load shared by steel and concrete is in the modular ratio (or the ratio of modulii of elasticity of the materials). This does not hold good if the compression member is long. As the load is increased, the compressive stress reaches a value that causes crushing of material and the column fails.

Medium and long columns are governed by compression as well as bending. The combined stress causes failure of the column by a phenomenon known as buckling. Such columns are governed by a different method of computing allowable or failure load.

11.2.2 Buckling

Buckling is the phenomenon by which a medium or long column fails by a combination of bending and compression. As the axial load is increased, the column tends to buckle or bend. For small values of the load the column remains in stable equilibrium returning to its straight position when the load is removed. As the load is increased, the column gets a permanent deflection even after the lateral load is removed. The column is in a deflected, neutral equilibrium position. Further increase in the load causes more deflection and the column fails. Such a failure is known as buckling.

Long and medium columns, thus, do not follow the behaviour of a true compression member. Most of the design codes have taken this into account. Steel design in particular follows the behaviour of the column up to buckling by using appropriate formula for design. Even in the case of concrete columns, recent code provisions, whether for working stress method or limit state method of design, have taken this behaviour into account. This behaviour is particularly important due to accidental eccentricities which make the load being not truly axial in columns. The buckling phenomenon is also governed by the support conditions of the column.

11.2.3 Stability

Stability is an important factor in the behaviour of long columns. The three states of equilibrium—stable, neutral and unstable—are demonstrated by the ball resting on different surfaces shown in Fig. 11.3. A small displacement of the ball causes forces which bring it back to the equilibrium position in Fig. 11.3(a). A small displacement increases the displacement in Fig. 11.3(b) as the ball is in unstable equilibrium. In neutral equilibrium as in Fig. 11.3(c), the ball changes to a new equilibrium position on being given a small displacement.

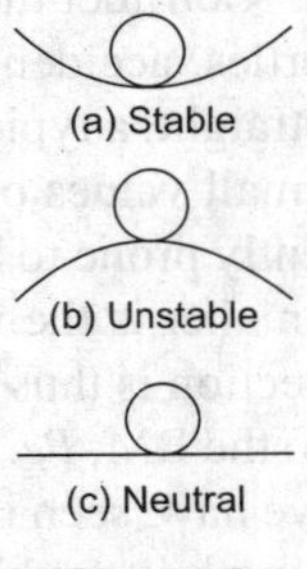

Fig. 11.3

A long column behaves in a similar fashion. Considering the column shown in Fig. 11.4, when the load P is small, the column is in stable equilibrium. This can be demonstrated by applying a lateral load and deflecting the column. The deflected shape shown by dotted lines can be maintained only so long as the lateral force exists. When the lateral force is removed, the column returns to its original straight position. The column is thus in stable equili-brium.

When the value of P is progressively increased, at a particular value of P, the column does not return to the initial position when deflected by a small lateral force even after this force is removed. The column maintains a new equilibrium position, which is the deflected position. The column is thus in neutral equilibrium. Any increase in the value of P increases the deflection and causes the failure of the column by buckling.

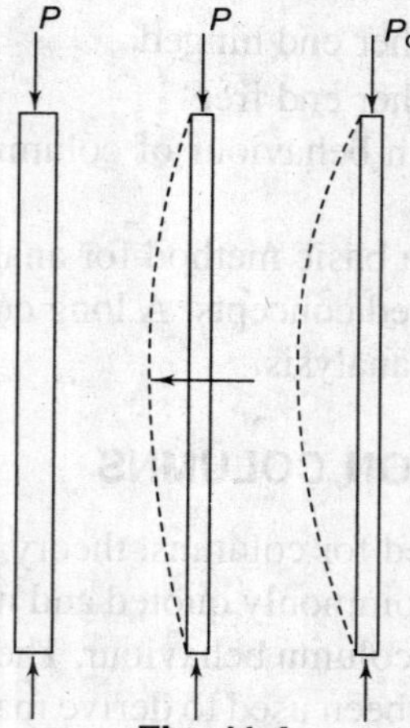

Fig. 11.4

11.2.4 Critical or Buckling Load

The maximum value of load, P_{cr}, up to which the column remains in a position of neutral equilibrium, is known as the critical load for the column. At the critical load, the column can remain in equilibrium either in a straight or a deflected position.

11.2.5 End Conditions

The end conditions of a column affect the way a column resists the load. The support conditions of a column are idealized. The column may have one end fixed, meaning it is restrained against translations and rotations. When the ends are hinged, it is restrained against translation only. A free end can both translate and rotate. Accordingly, the support conditions of a column can be of four types (Fig. 11.5):

(i) Both ends hinged
(ii) Both ends fixed

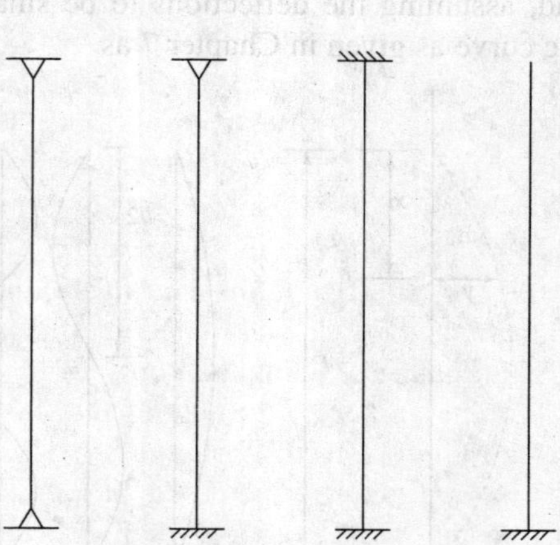

Fig. 11.5 End conditions of columns

(iii) One end fixed and the other end hinged
(iv) One end fixed and the other end free

We will see the difference in behaviour of columns due to support conditions later in the discussions.

Leonhard Euler developed a basic method for analysing long columns in 1757 based upon the above-mentioned concepts. A long column with both ends hinged is taken as a standard case for analysis.

11.3 EULER'S THEORY ON COLUMNS

Of the many theories developed for columns, theory developed by Leonard Euler (1707–1783) is the one most commonly quoted and many design codes have been based on this basic analysis of column behaviour. There is limitation of applicability of Euler's formulae but has been used to derive many empirical formulae based on this basic analysis.

11.3.1 Assumptions in Euler's Theory

Euler's theory is based on the following assumptions:

(i) The column is initially perfectly straight.
(ii) The load on the column is truly axial.
(iii) The column is prismatic, has the same cross section throughout.
(iv) The material of the column is perfectly elastic, homogeneous and isotropic.
(v) The column is long, its length being much larger compared to sectional dimensions.
(vi) The column fails by buckling.

11.3.2 Critical Load for Columns with Hinged Ends

Consider a column with hinged ends subjected to an axial load P, and remaining in a deflected position shown by dotted lines (Fig. 11.6). If y is the deflection at a distance x from the hinged end, assuming the deflections to be small, we can state the equation for the elastic curve as given in Chapter 7 as

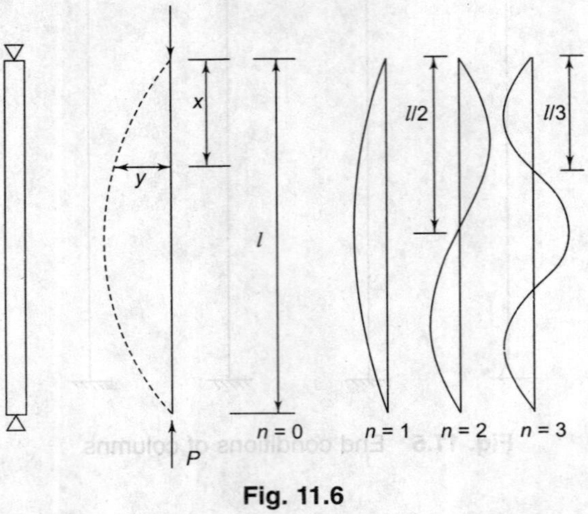

Fig. 11.6

$$EI \frac{d^2y}{dx^2} = -Py$$

$$\frac{d^2y}{dx^2} + \frac{P}{EI}y = 0$$

The solution for this equation from differential calculus is obtained as

$$y = C_1 \cos \sqrt{\frac{P}{EI}}\, x + C_2 \sin \sqrt{\frac{P}{EI}}\, x$$

The constants C_1 and C_2 can be evaluated from the end conditions of the column as (i) $y = 0$ at $x = 0$ and (ii) $y = 0$ at $x = l$. $x = 0$, $y = 0$ gives $C_1 = 0$; $x = l$, $y = 0$ gives $C_2 \sin(P/EI)^{1/2}\, l = 0$. C_2 cannot be zero because if C_1 and C_2 are both zero, the column is not in the bent configuration, and $y = 0$ everywhere. Therefore,

$$\sin \sqrt{\frac{P}{EI}}\, l = 0$$

gives $(P/EI)^{1/2}\, l = n\pi$, where $n = 0, 1, 2, 3, \ldots$.

$$P = \frac{n^2 \pi^2 EI}{l^2}$$

Thus $y = C_2(\sin n\pi/l)x$, which is the equation for the elastic curve of the column.

For different values of n, the deflected shapes are shown in Fig. 11.6. $n = 0$ means no deflection and no axial load. $n = 1$ gives the deflected shape with the elastic curve lying entirely to one side of the straight line. $n = 2$ gives the elastic shape in two halves on either side, and so on. The configurations $n = 2$, $n = 3$ and $n = 4$, etc. are possible only when the column is braced. Further, they give larger values of the load P compared to the one obtained with $n = 1$. Thus the critical load, P_{cr}, for such a case is given by

$$P_{cr} = \frac{\pi^2 EI}{l^2}$$

This critical load is for a column of length l, with both ends hinged, which we will consider to be the standard case. For the other end condition shown in Fig. 11.7, we will adopt a similar approach.

11.3.3 Column with One End Fixed and the Other End Hinged

Consider Fig. 11.7. The bent form is modified by the fixity at one end, where the slope is zero. The fixing moment M_A at A will give rise to horizontal reactions H at each end such that $H = M_A/l$. From the deflected shape shown in Fig. 11.7, we can say that

$$EI \frac{d^2y}{dx^2} = -Py + H(l - x)$$

or

$$\frac{d^2y}{dx^2} + \frac{P}{EI}y = \frac{P}{EI}\frac{H}{P}(l - x)$$

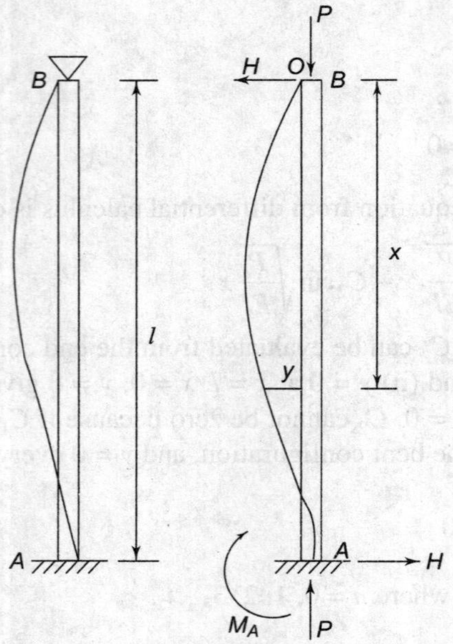

Fig. 11.7

Setting $(P/EI)^{1/2} = m$, the solution of this differential equation is

$$y = C_1 \cos mx + C_2 \sin mx + \frac{H}{P}(l - x)$$

$$\frac{dy}{dx} = -C_1 m \sin mx + C_2 m \cos mx - \frac{H}{P}$$

The end conditions in this case are $y = 0$ at $x = 0$ and $dy/dx = 0$ at $x = l$. $y = 0$ at $x = 0$ gives $C_1 = -Hl/P$. $dy/dx = 0$ at $x = l$ gives $C_2 = H/Pm$. Also at $x = l$, $y = 0$. This yields

$$C_1 \cos ml + C_2 \sin ml = 0$$

Substituting for C_1 and C_2,

$$-\frac{Hl}{P}\cos ml + \frac{H}{Pm}\sin ml = 0, \quad \tan ml = ml$$

This equation can be solved by trial and error to give $ml = 4.4934$ rad as the smallest value. $(P/EI)^{1/2}\, l = 4.4934$ and $P \approx 2\pi^2 EI/l^2$ because $(4.4934)^2 \gg 2\pi^2$. The critical load is thus

$$P_{\text{cr}} = \frac{2\pi^2 EI}{l^2}$$

11.3.4 Column with One End Fixed and the Other End Free

Figure 11.8 shows a column with the bottom end fixed and the top end free. The maximum deflection is at the free end and the BM throughout is negative. Thus,

$$EI\frac{d^2y}{dx^2} = -M = P(\delta - y)$$

where δ is the deflection at the free end. Setting $m = (P/EI)^{1/2}$,

$$\frac{d^2y}{dx^2} + m^2 y = m^2 \delta$$

The solution to this differential equation can be obtained as

$$y = C_1 \cos mx + C_2 \sin mx + \delta$$

The end conditions are (i) $x = 0, y = 0$ and (ii) $x = 0, dy/dx = 0$.
At $x = 0$, $y = 0$ gives $C_1 = -\delta$.

$$\frac{dy}{dx} = -C_1 m \sin mx + C_2 m \cos mx$$

At $x = 0$, $dy/dx = 0$ gives $C_2 = 0$. The equation for the elastic curve is, therefore,

$$y = -\delta, \quad \cos mx + \delta = \delta(1 - \cos mx)$$

At $x = l$, $y = \delta$ and this condition yields

$$\delta = \delta(1 - \cos ml), \qquad \delta \cos ml = 0$$

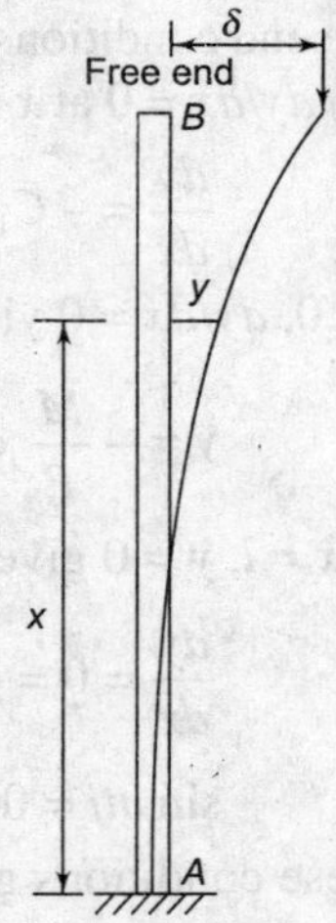

Fig. 11.8

$\cos ml = 0$ as δ is not zero. This yields $ml = n\pi/2$, where $n = 1, 3, 5, 7, \ldots$.
Setting $m = (P/EI)^{1/2}$, we get

$$P = \frac{n^2 \pi^2 EI}{4l^2}$$

The smallest value of $n = 1$ gives the smallest value of buckling load P_{cr}.

$$P_{cr} = \frac{\pi^2 EI}{4l^2}$$

11.3.5 Column with Both Ends Fixed

For the fixed ended column shown in Fig. 11.9, the deflected shape is such that the slopes at both ends are zero. There will be fixed end moments at both ends as shown, which are equal and hence do not give rise to any horizontal reaction. As before,

$$EI \frac{d^2y}{dx^2} = -Py + M$$

$$\frac{d^2y}{dx^2} + \frac{P}{EI} y = \frac{P}{EI}\frac{M}{P}$$

Setting $m^2 = P/EI$, we get

$$\frac{d^2y}{dx^2} + m^2 y = \frac{M}{P} m^2$$

The solution to this differential equation is

$$y = C_1 \cos mx + C_2 \sin mx + \frac{M}{P}$$

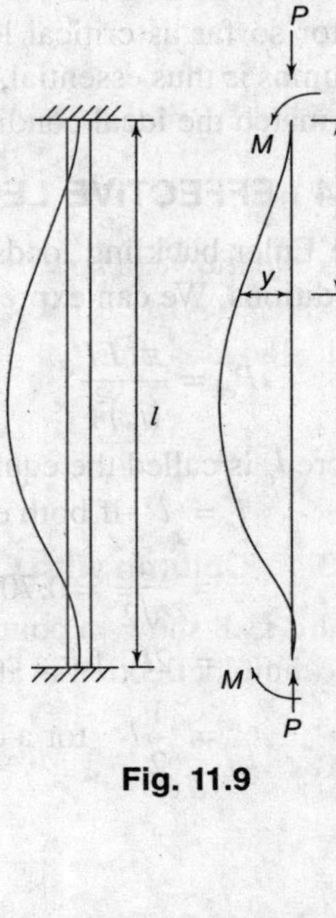

Fig. 11.9

The end conditions are (i) $y = 0$ at $x = 0$, (ii) $y = 0$ at $x = l$, (iii) $dy/dx = 0$ at $x = 0$ and (iv) $dy/dx = 0$ at $x = l$. $x = 0$, $y = 0$ gives $C_1 = -M/P$.

$$\frac{dy}{dx} = -C_1 \sin mx + C_2 m \cos mx$$

$x = 0$, $dy/dx = 0$ yields $C_2 = 0$. Substituting these values,

$$y = -\frac{M}{P}\cos mx + \frac{M}{P} = \frac{M}{P}(1 - \cos mx)$$

At $x = l$, $y = 0$ gives $\cos ml = 1$. $dy/dx = 0$ at $x = l$.

$$\frac{dy}{dx} = 0 = \frac{M}{P} m \sin ml$$

$$\Rightarrow \quad \sin ml = 0$$

These conditions give

$$ml = 0, \pi, 2\pi, 3\pi, 4\pi, \ldots$$

The minimum value satisfying both the conditions, $\cos ml = 1$ and $\sin ml = 0$, is 2π, the value zero not being possible.

$$\sqrt{\frac{P}{EI}}\, l = 2\pi, \quad P_{cr} = \frac{4\pi^2 EI}{l}$$

which is the critical load in this case.

We have considered four cases here and the Euler critical load in each case is as follows:

(a) Column with both ends hinged—$P_{cr} = \pi^2 EI/l^2$
(b) Column with one end fixed and one end hinged—$P_{cr} = 2\pi^2 EI/l^2$
(c) Column with one end fixed and the other end free—$P_{cr} = \pi^2 EI/4l^2$
(d) Column with both ends fixed—$P_{cr} = 4\pi^2 EI/l^2$.

It is clear from these values that the end conditions of columns are an important factor so far as critical loads are concerned. Determining the end conditions in columns is thus essential, especially in practical columns, as the end conditions do not match the ideal conditions assumed here.

11.4 EFFECTIVE LENGTH AND SLENDERNESS RATIO

The Euler buckling loads for columns have basically the same form for all end conditions. We can express this as

$$P_{cr} = \frac{\pi^2 EI}{(l_e)^2}$$

where l_e is called the equivalent length of the column.

$l_e = l$ if both ends are hinged

$= \frac{l}{\sqrt{2}} = 0.707l$ if one end is fixed and the other end hinged

$= 2l$ for a column with one end fixed and the other end free,

$= \frac{1}{2}l$ for a column with both ends fixed

where l is the actual length of the column. The equivalent length can be substituted in the general formula for critical load, $P_{cr} = \pi^2 EI/(l_e)^2$.

The concept of equivalent length can also be understood from the deflected shapes shown in Fig. 11.10, where the effective length is the length between the points of inflection on the deflected shape.

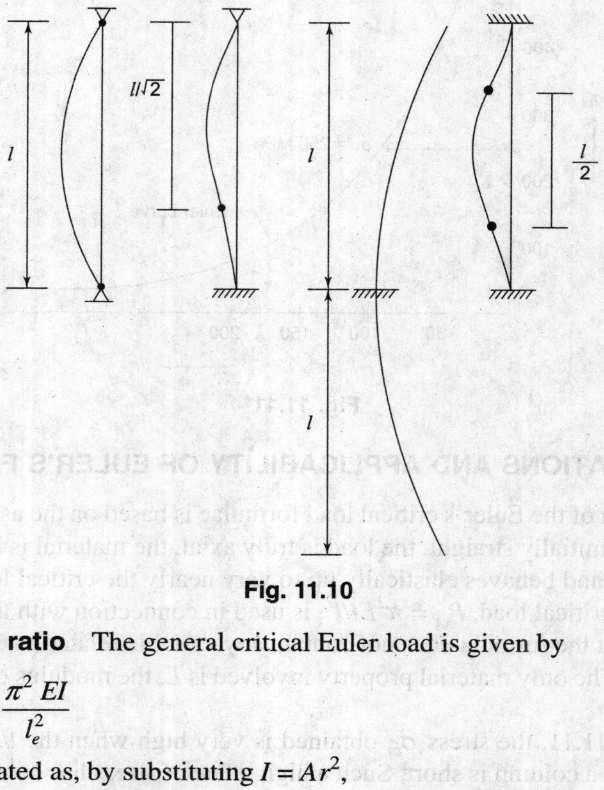

Fig. 11.10

Slenderness ratio The general critical Euler load is given by

$$P_{cr} = \frac{\pi^2 EI}{l_e^2}$$

This can be stated as, by substituting $I = Ar^2$,

$$P_{cr} = \frac{\pi^2 EAr^2}{l^2}$$

$$\frac{P_{cr}}{A} = \frac{\pi^2 E}{(l/r)^2}$$

where r is the radius of gyration and A is the area of cross section of the member. P_{cr}/A is the critical average stress σ_{cr} in the member.

$$\sigma_{cr} = \frac{\pi^2 E}{(l/r)^2}$$

l/r is known as the slenderness ratio:

$$\text{Slenderness ratio} = \frac{\text{length}}{\text{radius of gyration}}$$

The critical compressive stress is thus inversely proportional to the square of the slenderness ratio. A plot of σ_{cr} against l/r will give appear as shown in Fig.

11.11. The curve is asymptotic to both the σ_{cr} and (l/r) axes, σ_{cr} tending to infinity as l/r tends to zero and σ_{cr} tending to become zero as l/r tends to infinity.

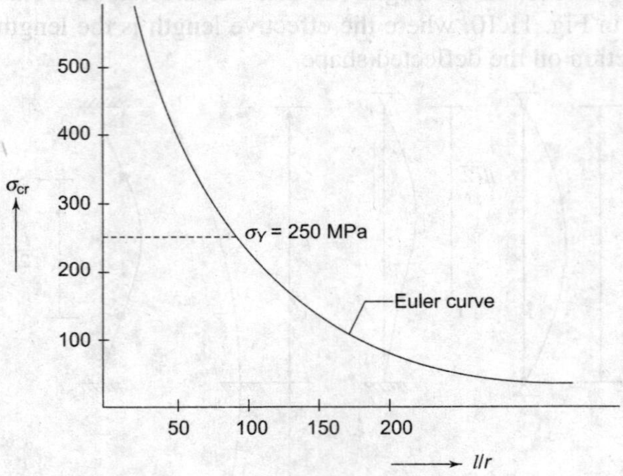

Fig. 11.11

11.5 LIMITATIONS AND APPLICABILITY OF EULER'S FORMULA

The derivation of the Euler's critical load formulae is based on the assumption that the column is initially straight, the load is truly axial, the material is homogeneous and isotropic, and behaves elastically up to very nearly the critical load.

The Euler critical load, $P_{cr} = \pi^2 EI/l^2$, is used in connection with the strength of the column but the formula does not contain any variable related to the strength of the material. The only material property involved is E, the modulus of elasticity of the material.

From Fig. 11.11, the stress σ_{cr} obtained is very high when the l/r ratio is very small or when a column is short. Such a high value of stress has no meaning since the material would have already failed by crushing or yielding of the material. The critical load, therefore, has to be such that the elastic limit is not exceeded, considering the average stress. For the method of superposition of stresses to be valid, the stresses should not exceed the proportional limit.

Taking the case of a steel column with both ends hinged and length l, if the proportional limit is equal to, say, 210 MPa, then

$$\sigma_{cr} = 210 = \frac{\pi^2 E}{(l/r)^2}, \quad \left(\frac{l}{r}\right)^2 = \frac{\pi^2 E}{210}$$

$$\left(\frac{l}{r}\right)^2 = \frac{\pi^2 \times 200{,}000}{210}, \quad \frac{l}{r} \approx 97$$

This is the lower limit of l/r ratio up to which Euler's formula can be applicable for steel. Since it is known that short columns do fail by crushing or yield, in the case of steel, the σ_{cr} v. l/r curve will be as shown in Fig. 11.12. Note that for extremely low values of slenderness ratio, the critical stress is the yield stress and in long columns the critical stress decreases according to Euler's formulae. In between is the likely zone of columns which are neither short nor long. Such columns may fail by a combination of yielding and buckling, the resulting stress very close to the yield stress. Both for short columns and intermediate columns, Euler's formulae will not give correct results.

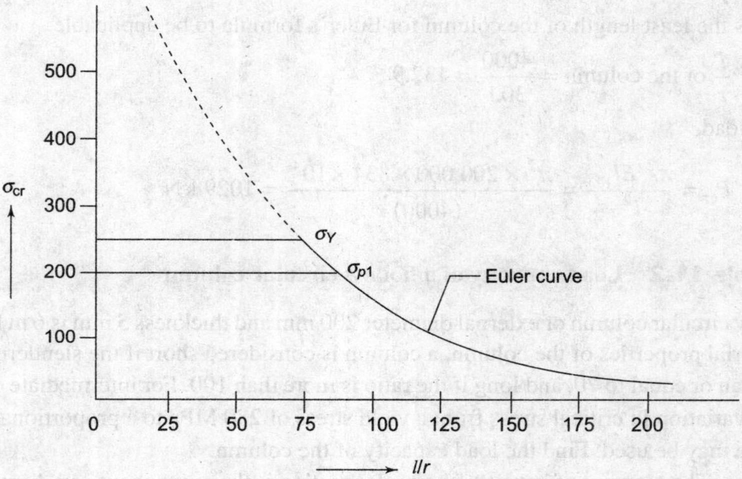

Fig. 11.12

Note that columns of the same length and cross-sectional area, one of mild steel and the other of high-strength steel, have the same load capacity according to Euler's formula because E is the same.

Example 11.1 Crippling load of a steel column

A steel section, ISMB 450 @ 72.4 kg/m, is 4 m long and is used as a column with both ends hinged. What is the minimum length of the column for Euler's formula to be applicable? Find the load carrying capacity of the column. $E = 200$ GPa. The proportional limit for the steel may be taken as 200 MPa.

Solution From steel tables, the minimum radius of gyration is about the Y-axis, shown in Fig. 11.13.

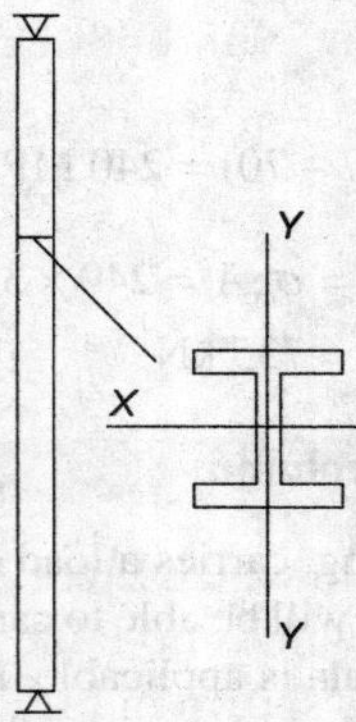

Fig. 11.13

$r_{YY} = 3.01$ cm = 30.1 mm

$$\sigma_{cr} = \frac{\pi^2 E}{(l/r)^2}, \quad \left(\frac{l}{r}\right)^2 = \frac{\pi^2 E}{\sigma_{cr}} = \frac{\pi^2 \times 200 \times 1000}{200} = 9869.6$$

$$\frac{l}{r} = 99.34, \quad l = 99.34 \times 30.1 = 2990 \text{ mm} = 2.99 \text{ m}$$

2.99 m is the least length of the column for Euler's formula to be applicable.

$$\frac{l}{r} \text{ of the column} = \frac{4000}{30.1} = 132.9$$

Critical load,

$$P_{cr} = \frac{\pi^2 EI_{YY}}{l^2} = \frac{\pi^2 \times 200{,}000 \times 834 \times 10^4}{(4000)^2} = 1029 \text{ kN}$$

Example 11.2 Load capacity of a hollow circular column

A hollow circular column of external diameter 200 mm and thickness 5 mm is 6 m long. For the material properties of the column, a column is considered short if the slenderness ratio is less than or equal to 70, and long if the ratio is more than 100. For intermediate columns, a linear variation in critical stress from a yield stress of 280 MPa to a proportional limit of 210 MPa may be used. Find the load capacity of the column.

Solution The moment of inertia of a circular section is the same about any diameter. The MI of the given section about the diameter is

$$I = \frac{\pi}{64}(200^4 - 190^4) = 14.568 \times 10^6 \text{ mm}^4$$

$$\text{Area of the section} = \frac{\pi}{4}(200^2 - 190^2) = 3.063 \times 10^3 \text{ mm}^2$$

$$r = \sqrt{\frac{I}{A}} = \sqrt{\frac{14.568 \times 10^8}{3.063 \times 10^3}} = 69$$

$$\text{Slenderness ratio of the column} = \frac{6000}{69} = 87$$

This is more than 70 but less than 100, and hence the column is an intermediate column. Euler's formula is, therefore, not directly applicable. The permissible critical stress can be calculated as

$$\sigma_{cr} = 280 - \frac{(280 - 210)}{30} \times (87 - 70) = 240 \text{ MPa}$$

$$\text{Load capacity of the column} = \sigma_{cr} A = 240 \times 3.063 \times 10^3$$
$$= 735 \text{ kN}$$

Example 11.3 Wooden circular column

A rectangular column of wood, 3 m long, carries a load of 300 kN. Determine whether or not a section of size 200 mm × 150 mm will be able to carry this load if a factor of safety of 3 is to be used, assuming Euler's formula is applicable. $E = 12.5$ GPa and the permissible stress is 12 MPa. If this section will not be able to carry this load, design a square section to do so.

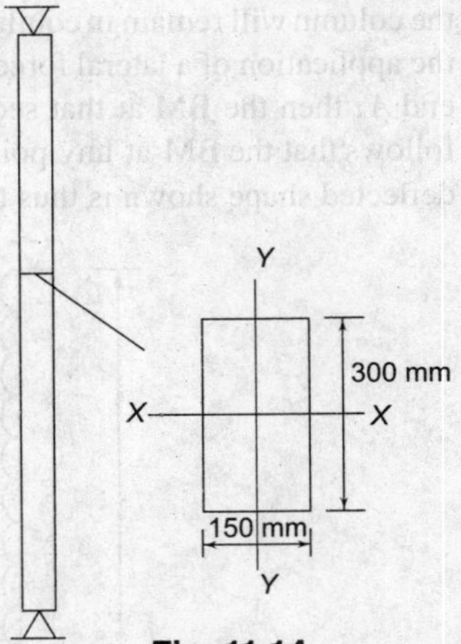

Fig. 11.14

Solution Let us find the load capacity of the column using Euler's formula. The column will bend about the axis about which the moment of inertia is minimum. We have to use I_{YY} for the section shown in Fig. 11.14.

$$I_{YY} = \frac{200 \times 150^3}{12} \text{ mm}^4$$

Assuming hinged ends, Euler critical load,

$$P_{\text{cr}} = \frac{\pi^2 EI}{l^2} = \frac{\pi^2 \times 12.5 \times 1000 \times 200 \times 150^3}{3000 \times 3000 \times 12}$$

$$= 771 \text{ kN}$$

Load to be carried in working condition = 300 kN

Note that Euler's formula does not contain any factor of safety. Therefore,

Critical load capacity required = 300 × factor of safety

= 300 × 3 = 900 kN

The given rectangular column is inadequate to carry this load.

To design a square section to carry this load, the minimum area required may be worked out as follows.

$$\sigma \times \text{area} = \text{load}$$

$$a^2 = \frac{300{,}000}{12}$$

$$\Rightarrow \quad a = 158.1 \text{ mm}$$

Let us check with the Euler critical load.

$$P_{\text{cr}} = 300 \times 3 = 900 \text{ kN} = 900{,}000 \text{ N}$$

$$= \frac{\pi^2 EI}{l^2}$$

$$\Rightarrow \quad I = \frac{P_{\text{cr}} l^2}{\pi^2 E}, \quad l = 3 \text{ m} = 3000 \text{ mm}$$

$$E = 12.5 \text{ GPa} = 12{,}500 \text{ N/mm}^2$$

$$I = \frac{900{,}000 \times 3000 \times 3000}{\pi^2 \times 12{,}500} = 0.6564 \times 10^8 \text{ mm}^4$$

$$I = \frac{a^4}{12} = 0.2188 \times 10^8, \quad a = 167.5 \text{ mm}$$

A square section of size 168 mm × 168 mm is required.

$$\text{Stress in the section} = \frac{300{,}000}{168 \times 168} = 10.6 \text{ N/mm}^2$$

Example 11.4 Euler critical load and beam with UD load

Show that an approximate value of critical load for a long column can be obtained by assuming a deflected shape due to a lateral UD load.

Solution Consider the long column with hinged ends shown in Fig. 11.15. It is initially straight, and the load is acting along this straight centroidal axis. If $P = P_{\text{cr}}$, the critical load, the column will remain in equilibrium in the deflected position. The deflection is caused by the application of a lateral force for an instant. If δ is the deflection at any point y above the end A, then the BM at that section is $P\delta$. This is so at any section. Since P is constant, it follows that the BM at any point is given by the deflected shape to a suitable scale. The deflected shape shown is thus the BM diagram for the column.

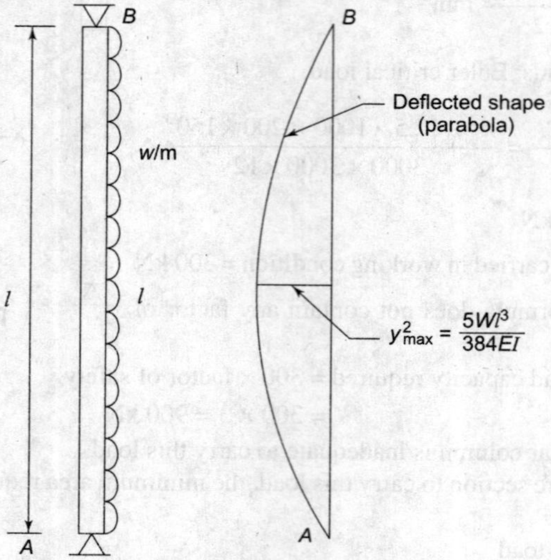

Fig. 11.15

The deflected shape cannot be a circular arc because the BM is not constant. We approximate the deflected shape to a parabolic curve, obtained due to a lateral UD load as shown. The central (maximum) deflection due to such loading is given by

$$\delta_c = \frac{5Wl^3}{384EI}$$

where $W = wl$, the total load.

$$\text{Maximum BM in the beam} = \frac{wl^2}{8} = \frac{Wl}{8} = M$$

$$W = \frac{8M}{l}$$

and $$\delta_c = \frac{5Ml^2}{48EI}$$

Maximum BM in the column,

$$M = P\delta_c$$

Substituting for M,

$$\delta_c = \frac{5P\delta_c l^2}{48EI}$$

$$P = \frac{48EI}{5l^2} = \frac{9.6EI}{l^2}$$

which is the critical load. This is approximately the same as the Euler critical load $\pi^2 EI/l^2$ $= 9.87EI/l^2$.

Example 11.5 Design of a rectangular column

A column has a rectangular section, and is supported as shown in Fig. 11.16. The end B has free movement in the Z-direction but movement along the X-direction is prevented. The end A is fixed. Find the ratio of the sides for the most efficient design. Find the dimensions of the column if it is 1 m long and $E = 100$ GPa. It has to carry a load of 30 kN and the factor of safety required is 3.

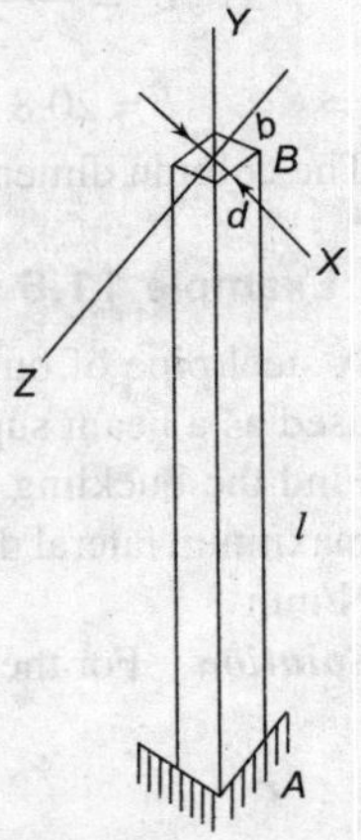

Fig. 11.16

Solution The column may bend in the X-Z or the Y-Z plane. The end conditions and section properties are different in each case. The most efficient design is obtained when the strength of the column is the same for bending in the two planes.

Bending in X-Y plane For this, the column is fixed at A and hinged at B as shown.

Area = bd,

$$I = \frac{db^3}{12}, \quad r = \sqrt{\frac{I}{A}} = \sqrt{\frac{db^3}{12bd}} = \frac{b}{\sqrt{12}}$$

If l is the length, then effective length = $0.707l$. Slenderness ratio,

$$\frac{l}{r} = \frac{0.707\,l \times \sqrt{12}}{b} = 2.45\,\frac{l}{b}$$

Critical load

$$P_{cr} = \frac{\pi^2 EI}{l^2} = \frac{\pi^2 EAr^2}{l^2} = \frac{\pi^2 EA}{(l/r)^2}$$

Bending in Y-Z plane For this, the column is fixed at A and free at B. The effective length is $2l$ in this case.

Area = bd

$$I = \frac{bd^3}{12}, \quad r = \sqrt{\frac{I}{A}} = \sqrt{\frac{bd^3}{12bd}} = \frac{d}{\sqrt{12}}$$

Effective length = $2l$

$$\text{Slenderness ratio} = \frac{2l\sqrt{12}}{d} = 6.928\,\frac{l}{d}$$

For the most efficient design, the critical loads in the two directions must be equal. That is,

$$\frac{\pi^2 EA}{(2.45l/b)^2} = \frac{\pi^2 EA}{(6.928l/d)^2}$$

which gives

$$\frac{d}{b} = 2.83$$

For the given values,

$$l = 1000 \text{ mm}, \quad E = 100 \text{ GPa} = 100{,}000 \text{ N/mm}^2, \quad A = bd$$

$$P_{cr} = 30 \times 3 = 90 \text{ kN} = 90{,}000 \text{ N}$$

Therefore,

$$90{,}000 = \frac{\pi^2 EA}{(l/r)^2} = \frac{\pi^2 \times 100{,}000\,bd}{(1000 \times 2.45/b)^2}$$

$$db^3 = 5.47 \times 10^5$$

$$\Rightarrow \quad d = 2.83b$$

$$b^4 = \frac{5.47 \times 10^5}{2.83}$$

$$\Rightarrow \quad b = 20.8 \text{ mm}, \quad d = 58.86 \text{ mm}.$$

The column dimensions are 21 mm × 60 mm.

Example 11.6 Hollow circular section

A steel pipe of outside diameter 20 mm and thickness 3 mm is deflected by 3 mm when used as a beam supported at its ends, 1 m apart, and subjected to a central load of 170 N. Find the buckling load when the pipe is used as a column with hinged ends. What is the maximum lateral deflection of this column before the material attains the yield stress of 250 N/mm^2?

Solution For the pipe section (Fig. 11.17),

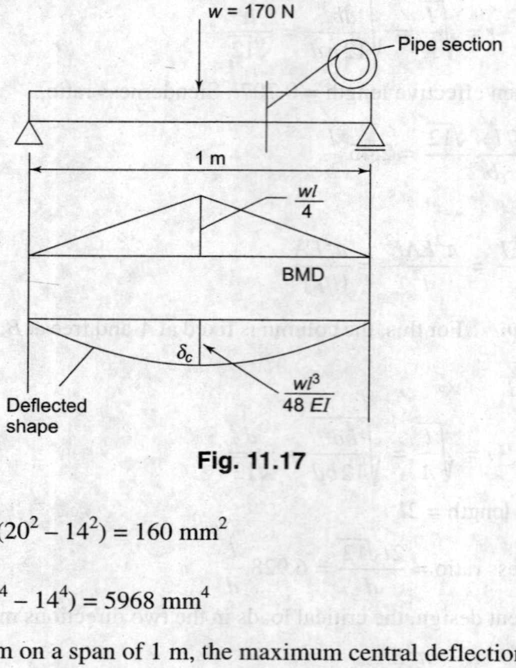

Fig. 11.17

$$\text{Area} = \frac{\pi}{4}(20^2 - 14^2) = 160 \text{ mm}^2$$

$$I = \frac{\pi}{64}(20^4 - 14^4) = 5968 \text{ mm}^4$$

When used as a beam on a span of 1 m, the maximum central deflection is 3 mm.

$$\delta = \frac{wl^3}{48EI}$$

$$EI = \frac{wl^3}{48\delta} = \frac{170 \times (1000)^3}{48 \times 3} = 1.18 \times 10^9 \text{ N mm}^2$$

$$\text{Maximum BM in the beam} = \frac{wl}{4} = \frac{170 \times 1000}{4}$$

Maximum bending stress,

$$\sigma_b = \frac{170 \times 1000 \times 10}{4 \times 5968} \times 10 = 71 \text{ N/mm}^2$$

Since this is a column with hinged ends, the critical load is given by

$$P_{cr} = \frac{\pi^2 EI}{l^2}$$

Substituting,

$$\frac{\pi^2 \times 1.18 \times 10^9}{(1000)^2} = 11{,}646 \text{ N}$$

$$\text{Average critical stress} = \frac{11{,}646}{160} = 73 \text{ N/mm}^2$$

To find the lateral deflection δ, when the stress in the material is 250 N/mm^2, let the maximum deflection at the centre be δ_c as shown.

Maximum BM at the section = $P\delta_c$

If σ_b is the bending stress at the extreme fibres and Z the modulus of section,

$$P\delta_c = \sigma_b Z$$

$$\sigma_b = \frac{P\delta_c}{Z} = \frac{11{,}646\,\delta_c}{(5968/10)} = 19.5\delta_c$$

Direct axial stress,

$$\sigma_b = \frac{P_{cr}}{A} = 73 \text{ N/mm}^2$$

$$\sigma_b + \sigma_d = 250, \quad 19.5\delta_c + 73 = 250, \quad \delta_c = 9.1 \text{ mm}$$

Example 11.7 Framework of axially loaded members

The framework shown in Fig. 11.18 is subjected to a load of 30 kN. The two members are of circular cross section and hinged at the ends. Find the minimum diameter required for the two bars. $E = 200$ GPa and the proportional limit is 210 N/mm^2. Take a factor of safety of 3.

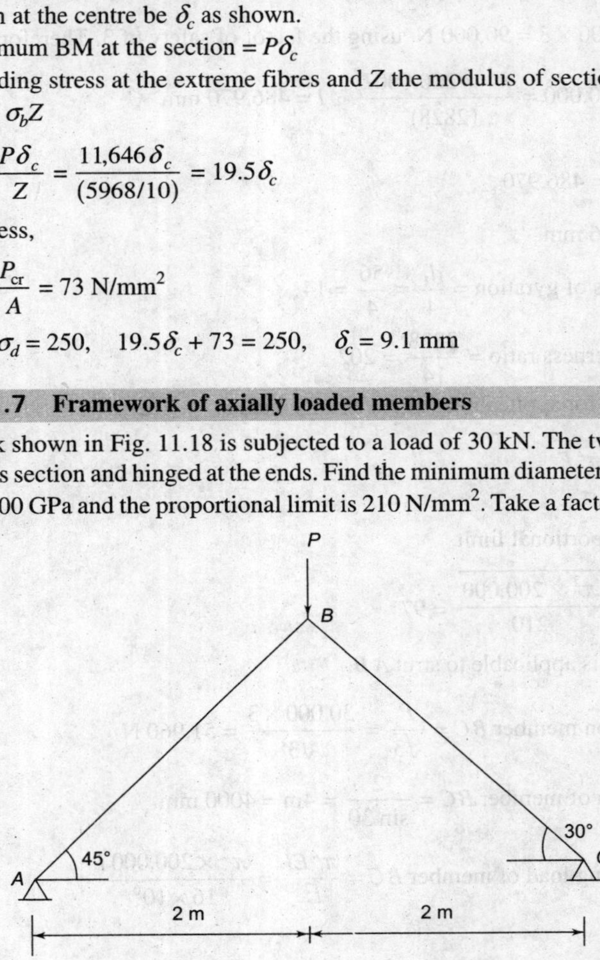

Fig. 11.18

Solution The bars BA and BC are both under compression. Resolving the forces at B,

$$P_1 \cos 45 = P_2 \cos 30$$

from $\Sigma H = 0$.

$$P_1 \sin 45 + P_2 \sin 30 = P$$

from $\Sigma V = 0$. Solving these two equations,

$$P_1 = \frac{2}{3}P \quad \text{and} \quad P_2 = \frac{P}{\sqrt{3}}$$

Let d_1 and d_2 be the diameters of the bars required.

Length of $AB = 2\sqrt{2}$ m = 2828 mm

$$\text{Buckling load for } AB = \frac{\pi^2 EI}{l^2} = \frac{\pi^2 \times 200{,}000 \times I}{(2828)^2}$$

$$\text{Load on member } AB = \frac{2}{3}P$$

where $P = 30{,}000 \times 3 = 90{,}000$ N, using the factor of safety of 3. Therefore,

$$\frac{2}{3} \times 90{,}000 = \frac{\pi^2 \times 200{,}000 I}{(2828)^2}, \quad I = 486{,}970 \text{ mm}^4$$

$$\frac{\pi d_1^4}{64} = 486{,}970$$

$$\Rightarrow \quad d_1 = 56 \text{ mm}$$

$$\text{Radius of gyration} = \frac{d_1}{4} = \frac{56}{4} = 14$$

$$\text{Slenderness ratio} = \frac{2828}{14} = 202$$

The lower limit for applicability of Euler's formula is

$$\frac{l}{r} = \sqrt{\frac{\pi^2 E}{\sigma_{cr}}}$$

where σ_{cr} = proportional limit.

$$\frac{l}{r} = \sqrt{\frac{\pi^2 \times 200{,}000}{210}} = 97$$

Euler's formula is applicable to strut AB.

$$\text{Load on member } BC = \frac{P}{\sqrt{3}} = \frac{30{,}000 \times 3}{\sqrt{3}} = 51{,}960 \text{ N}$$

$$\text{Length of member } BC = \frac{2}{\sin 30} = 4\text{m} = 4000 \text{ mm}$$

$$\text{Buckling load of member } BC = \frac{\pi^2 EI}{l^2} = \frac{\pi^2 \times 200{,}000\, I}{16 \times 10^6}$$

Therefore,

$$\frac{\pi^2 \times 200{,}000 I}{16 \times 10^6} = 51{,}960$$

$$\Rightarrow \quad I = 842{,}343 \text{ mm}^4$$

$$\frac{\pi d_2^4}{64} = 842{,}343$$

$\Rightarrow \quad d_2 = 64.4$ mm

$$\text{Radius of gyration} = \frac{d_2}{4} = \frac{64.4}{4} = 16.1$$

$$\text{Slenderness ratio for } BC = \frac{4000}{16.1} = 248$$

Euler's formula is applicable to strut BC. The diameters of bars AB and BC, respectively, should be at least 56 mm and 64.5 mm.

Example 11.8 Euler critical load

Determine the Euler critical load for the column section shown in Fig. 11.19 if its length is 3 m and (i) if its ends are hinged and (ii) if its ends are fixed. $E = 200$ GPa.

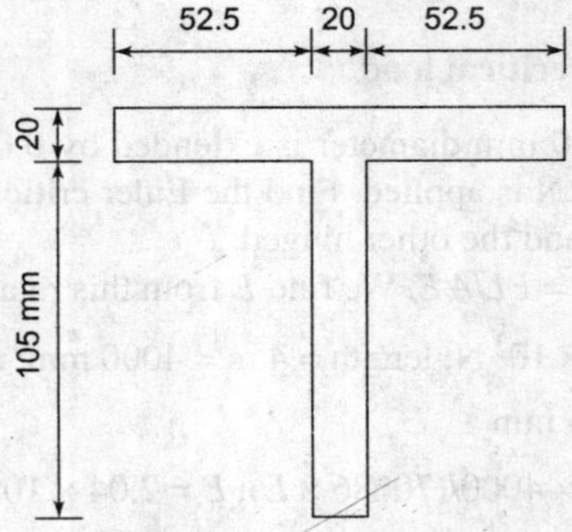

Fig. 11.19

Solution We first find the centroid of the section. Taking moments about the top, $(125 \times 20 + 105 \times 20)\, y = 125 \times 20 \times 10 + 105 \times 20 \times 72.5$; $y = 38.5$ mm.

The centroid is 38.5 mm from the top and 86.5 mm from the bottom. X-X and Y-Y are the centroidal axes. We find moments of inertia about both axes.

$$I_{xx} = 125 \times 20^3/12 + (125 \times 20)\ 28.5^2 + 20 \times 105^3/12 + (20 \times 105) \times (86.5 - 52.5)^2$$
$$= 6.5 \times 10^6 \text{ mm}^4$$
$$I_{yy} = 20 \times 125^3/12 + 105 \times 20^3/12 = 3.325 \times 10^6 \text{ mm}^4$$

(a) When both ends are hinged, effective length = actual length = 3 m = 3000 mm

$$E = 200 \text{ GPA} = 200{,}000 \text{ N/mm}^2$$
$$P_{cr} = \pi^2 EI/l^2 = \pi^2 \times 200{,}000 \times 3.325 \times 10^6 / (3000)^2 = 730 \text{ kN}$$

(b) When both ends are fixed, effective length = actual length /2 = 1500 mm.

$$P_{cr} = \pi^2 \times 200{,}000 \times 3.325 \times 10^6 / (1500)^2 = 2917 \text{ kN}$$

Example 11.9 Comparison of critical load: circular and hollow circular sections

Compare the Euler crippling loads of two columns—one of solid circular section and the second of hollow circular section of internal diameter 70% of the external diameter if they are of the same material, same length, same area, and same end conditions.

Solution Let d be the diameter of the solid column. Let D be the external diameter of hollow circular column. Interior diameter = $0.7D$. Areas of the two columns are equal.

$$\pi d^2/4 = \pi(D^2 - (0.7D)^2/4;\ D = 1.4d$$

External diameter of hollow column = $1.4d$; internal diameter = $0.7 \times 1.4d = 0.98d$.
Critical load = $\pi^2 EI/l_e^2$. All other conditions remaining the same, critical load is proportional to I.

$$\frac{\text{Critical load of solid column}}{\text{Critical load of hollow column}} = \frac{I \text{ of solid column}}{I \text{ of hollow column}}$$

I of solid column = $\pi d^4/64$, I of hollow column = $\pi(1.4d)^4 - (0.98d)^4]/64 = 2.92\,\pi d^4/64$

$$\frac{P_{cr}\,(\text{solid})}{P_{cr}\,(\text{hollow})} = \frac{\pi d^2/64}{2.92\,\pi d^4/64} = 0.34$$

The hollow circular column can carry 2.92 times the load carried by a solid column of the same area.

Example 11.10 Euler critical load

A bar of circular section 300 mm diameter is extended by 0.025 mm on a length of 4 m when a tensile load of 100 kN is applied. Find the Euler critical load of this bar used as a column with one end fixed and the other hinged.

Solution Extension of bar = PL/AE. We find E from this relationship.

$P = 100$ kN $= 100 \times 10^3$ N; length = 4 m = 4000 mm; area = $\pi 150^2 = 70{,}686$ mm^2

Elongation = 0.025 mm

$0.025 = 100 \times 10^3 \times 4000/(70686 \times E)$; $E = 2.04 \times 10^5$ N/mm^2

Moment of inertia = $\pi \times 300^4/64 = 3.976 \times 10^8$ mm^4

Effective length when one end is hinged and the other fixed is $L/\sqrt{2}$

Effecting length = $4000/\sqrt{2} = 2828$ mm

Euler critical load $= \pi^2 EI/l_e^2 = \sqrt{}^2 \times 2.04 \times 10^5 \times 3.976 \times 10^8/(2828)^2$

$= 100$ MN

11.6 EMPIRICAL FORMULAE

While Euler's formulae for columns with different end conditions are useful to understand the behaviour of columns, practical designs require a more convenient method for design of columns. Euler's formulae are basically useful for long columns and suffer from the limitations listed earlier. Many empirical formulae have been developed from theoretical considerations supported by experimentations. One of the most commonly used empirical formula was that due to Rankine. There are many other formulae developed from practical results on testing columns which are discussed below.

11.6.1 The Rankine–Gordon Formula

Euler's formulae are applicable to columns above a certain minimum value of the l/r ratio or for long columns defined by material properties and sectional dimensions, as illustrated earlier. Very short columns obviously fail by crushing. There are cases of intermediate columns which may not fail by crushing, and for which

Euler's formulae are not applicable. Such columns may fail by a combination of crushing and buckling.

For such intermediate columns, Rankine suggested an empirical relationship, taking into account the direct compressive stress and bending stress. He suggested the relationship

$$\frac{1}{P_r} = \frac{1}{P_s} + \frac{1}{P_e}$$

where P_r is the Rankine buckling load, P_s is the failure load due to direct compressive stress as a short column, and P_e is the Euler buckling load. Thus,

$$P_r = \frac{P_s P_e}{P_s + P_e} = \frac{P_s}{1 + P_s / P_e}$$

$$P_s = \sigma_Y A$$

$$P_e = \frac{\pi^2 EI}{l_e^2}$$

where σ_Y = yield stress and l_e = equivalent length. Therefore,

$$P_r = \frac{\sigma_Y A}{1 + \sigma_Y A / (\pi^2 EI / l_e^2)}, \quad \text{since } I_{\min} = A(r_{\min})^2$$

$$= \frac{\sigma_Y A}{1 + (\sigma_Y / \pi^2 E)(l_e / r_{\min})^2}$$

$\sigma_Y / \pi^2 E$ is called the Rankine constant. The Rankine constants for different materials are given in Table 11.1.

Table 11.1 The Rankine constant for some materials

Material	σ_Y	***Rankine constant***
Mild steel	330	$\frac{1}{7500}$
Medium carbon steel	500	$\frac{1}{500}$
Cast iron	550	$\frac{1}{1600}$
Wrought iron	250	$\frac{1}{9000}$
Strong timber	50	$\frac{1}{750}$

Rankine's formula can be applied to columns with different end conditions by substituting the equivalent length for a particular set of end conditions. The following examples illustrate the use of this formula.

Example 11.11 Hollow circular column: Euler's and Rankine's formula

A hollow steel strut, 2.4 m long, is pin-jointed at the ends. It has an outer diameter of 40 mm and a thickness of 5 mm. If the yield stress is 320 N/mm^2 and $E = 2 \times 10^5$ N/mm^2, compare the crippling load given by Euler's and Rankine's formulae. Also determine the minimum l/r ratio for which Euler's formula applies.

Solution $I = \dfrac{\pi}{64}(40^4 - 30^4) = 8.59 \times 10^4 \text{ mm}^4$

$$\text{Area} = \frac{\pi}{4}(40^2 - 30^2) = 550 \text{ mm}^2$$

The radius of gyration,

$$r = \sqrt{\frac{I}{A}} = \sqrt{\frac{8.59 \times 10^4}{550}} = 12.5 \text{ mm}$$

Effective length = Actual length for the end conditions (pin-jointed at both ends)

$$\frac{l}{r} = \frac{2400}{12.5} = 192$$

$$\text{Euler load} = \frac{\pi^2 EI}{l^2} = \frac{\pi^2 \times 2 \times 10^5 \times 8.59 \times 10^4}{(2400)^2}$$

$$= 29{,}437 \text{ N}$$

$$\text{Rankine load} = \frac{\sigma_Y A}{1 + 6y/\pi^2 E(l_e/r)^2}$$

$$\frac{\sigma_Y}{\pi^2 E} = \frac{1}{7500}$$

$$= \frac{320 \times 550}{1 + (1/7500)(192)^2} = 29{,}754 \text{ N}$$

$$\frac{P_e}{P_r} = \frac{29{,}437}{29{,}754} = 0.9893$$

$$\left(\frac{l}{r}\right)^2 = \frac{\pi^2 E}{320} = 6168.5$$

$$\frac{l}{r} = \sqrt{6168.5} = 78.5$$

This is the minimum l/r ratio for which Euler's formula is applicable.

Example 11.12 Comparison of Euler and Rankine loads for different l/r ratios

Compare the critical stresses using Euler's and Rankine's formulae for struts with slenderness ratios 50, 100, 150, and 200. Assume that both ends are hinged. $E = 200$ GPa, Rankine's constant = 1/7500, and $\sigma_Y = 300$ MPa.

Solution Euler's formula is

$$P_{cr} = \frac{\pi^2 EI}{l^2} = \frac{\pi^2 EAr^2}{l^2}$$

$$\frac{P_{cr}}{A} = \sigma_{cr} = \frac{\pi^2 E}{(l/r)^2} = \text{critical stress} = \sigma_e$$

For $l/r = 50$,

$$\sigma_e = \frac{\pi^2 \times 2 \times 10^5}{(50)^2} = 789.56 \text{ N/mm}^2$$

For $l/r = 100$,

$$\sigma_e = \frac{\pi^2 \times 2 \times 10^5}{(100)^2} = 197.39 \text{ N/mm}^2$$

For $l/r = 150$,

$$\sigma_e = \frac{\pi^2 \times 2 \times 10^5}{(150)^2} = 87.72 \text{ N/mm}^2$$

For $l/r = 200$,

$$\sigma_e = \frac{\pi^2 \times 2 \times 10^5}{(200)^2} = 49.34 \text{ N/mm}^2$$

Rankine's formula is

$$P_r = \frac{\sigma_Y A}{1 + (\sigma_Y/\pi^2 E)(l/r)^2}$$

$$\frac{P_r}{A} = \frac{\sigma_Y A}{1 + (\sigma_Y/\pi^2 E)(l/r)^2}, \quad \frac{\sigma_Y}{\pi^2 E} = \frac{1}{7500}$$

Therefore,

$$\sigma_r = \text{critical stress} = \frac{\sigma_Y}{1 + (l/r)^2/7500}$$

For $l/r = 50$,

$$\sigma_r = \frac{300}{1 + (2500/7500)} = 225 \text{ N/mm}^2$$

For $l/r = 100$,

$$\sigma_r = \frac{300}{1 + (10{,}000/7500)} = 128.57 \text{ N/mm}^2$$

For $l/r = 150$,

$$\sigma_r = \frac{300}{1 + [(150)^2/7500]} = 75 \text{ N/mm}^2$$

For $l/r = 200$,

$$\sigma_r = \frac{300}{1 + [(200)^2/7500]} = 47.37 \text{ N/mm}^2$$

Example 11.13 Hollow circular column

A hollow cylindrical column, with both ends hinged, is 6 m long, and has an outer diameter of 120 mm and an inner diameter of 80 mm. Compare the crippling load obtained by Euler's and Rankine's formulae. $E = 80{,}000$ N/mm^2 and $\sigma_Y = 550$ N/mm^2. The Rankine constant = 1/1600. What is the length of the column if both crippling loads are equal?

Solution For the given section,

$$\text{Area} = \frac{\pi}{4}(120^2 - 80^2) = 6283 \text{ mm}^2$$

$$I = \frac{\pi}{64}(120^4 - 80^4) = 816.8 \times 10^4 \text{ mm}^4$$

$$r = \sqrt{\frac{I}{A}} = 36 \text{ mm}$$

The Euler load,

$$P_e = \frac{\pi^2 EI}{l^2} = \frac{\pi^2 \times 80{,}000 \times 816.8 \times 10^4}{(6000)^2} = 179{,}144 \text{ N}$$

The Rankine load,

$$P_r = \frac{\sigma_Y A}{1 + (1/7600)(l/r)^2}$$

$$= \frac{530 \times 6283}{1 + (1/1600)\,(6000/36)^2}$$

$$= \frac{550 \times 6283}{1 + (1/1600) \times 27{,}778} = \frac{550 \times 6283}{18.36}$$

$$= 188{,}216 \text{ N}$$

If the Euler and Rankine loads are equal,

$$\frac{\pi^2 EI}{l^2} = \frac{A\sigma_Y}{1 + (1/1600)\,(l/r)^2}$$

$$\frac{1 + (1/1600)\,(l/r)^2}{l^2} = \frac{A\sigma_Y}{\pi^2 EI} = \frac{6283 \times 550}{\pi^2 \times 8 \times 10^4 \times 816.8 \times 10^4} = 53.58 \times 10^{-8}$$

$$1 + \frac{1}{1600}\left(\frac{l}{r}\right)^2 = 53.58 \times 10^{-9} l^2$$

$$1 + \frac{l^2}{1600 \times 36^2} = 53.58 \times 10^{-9} l^2$$

$$2{,}073{,}600 = 1.111 l^2 - l^2$$

$$l = 4.322 \text{ m}$$

11.6.2 Limitations of Rankine–Gordon Formula

While Rankine formula as outlined in the previous section is applicable to intermediate and long columns, they suffer from the limitations outlined for Euler's theory. The assumptions made are similar and may not be valid. The constant *a* in the Rankine's formula is given by $(\sigma_y/\pi^2 E)$. Rankine's formula is an empirical formula and the constant is not calculated from σ_y and E. The constants given are based upon the experimental results by testing columns.

11.6.3 Straight Line Formula

Another empirical formula that was developed was the straight line formula. Due to the inadequacy of the Euler's and Rankine's formulae and the fact that the end conditions assumed in such cases are never realized in practice, empirical formulae were evolved for ease of use in practical designs. The straight line formula can be written as

$$P_{cr} = \sigma_y A\,[1 - b(L_e/k_{min})]$$

Here, σ_y is the yield stress in compression, b is a constant, L_e is the effective length, and k_{min} is the least radius gyration of the section of the column. This equation if plotted as P against (L_e/k_{min}) gives a straight line relationship and hence is known as the straight line formula. The values of the constant b can be taken as follows

For mild steel: $\sigma_y = 320$ N/mm^2, $b = 1/189$

For cast iron: $\sigma_y = 550$ N/mm^2, $b = 1/125$

For wrought iron: $\sigma_y = 250$ N/mm^2, $b = 1/189$

11.6.4 Johnson's Parabolic Formula

Johnson's formula can be stated as

$$P_{cr} = \sigma_y A(1 - nl^2/k^2)$$

This is known as the parabolic formula as a plot P vs l/k gives a parabola. For mild steel, the constant n can be taken as 1/18,000.

11.6.5 IS Code Formula

The Indian standard code IS 800–1984 has proposed a formula for design of steel columns based on some empirical formulae. The formula is given as

$$\sigma_c = \sigma_c' = \frac{\sigma_y / m}{[1 + 0.20 \sec [L_e / k]\sqrt{(m\sigma_c / 4E)}]} \quad \text{(for } L_e/k = 160)$$

$$\sigma_c = \sigma_c'[1.2 - L_e/800k] \text{ for } L_e/k = 160$$

The IS formula is essentially meant for mild steel columns. In this equation,

σ_c is the allowable compressive stress obtained from Table 11.2,

σ_c' is the value of the stress obtained from the secant formula,

σ_y is the guaranteed yield stress for mild steel,

m is a factor of safety taken as 1.68, and

E is the modulus of elasticity.

The values of the stress σ_c have been worked out using this formula and are given in Table 11.2.

Table 11.2

L/k	σ_c *(N/mm²)*	*L/k*	σ_c *(N/mm²)*	*L/k*	σ_c *(N/mm²)*
10	150	100	80	190	30
20	148	110	71	200	28
30	145	120	64	210	24
40	139	130	57	220	21
50	132	140	51	230	20
60	122	150	45	240	19
70	112	160	41	250	18
80	101	170	37	300	17
90	90	180	33	350	8

Example 11.14 Empirical formulae

Compare the critical loads of a solid circular column and a hollow circular column (the internal diameter being 60% of the external diameter) of the same material steel using (a) the straight line formula and (b) the Johnson's parabolic formula. Both columns are 4 m long and the diameter of the solid column is 200 mm. Take the constant in the straight line formula as 1/189 and in the parabolic formula as 1/18,000. Assume the support conditions as hinged.

Solution For a solid circular column of diameter d, area $= \pi d^2/4$

If the external diameter of a hollow circular column is D, then the internal diameter $= 0.6D$.

$$\text{Area} = \pi[(D^2 - (0.6\,D)^2]/4 = 0.64\pi D^2/4$$

As the areas are equal, $\pi d^2/4 = 0.64\pi D^2/4$; $D = 1.25d$

MI of the circular section $= \pi d^4/64$; $k = \sqrt{[I/A]} = \sqrt{[(\pi d^4/64)/(\pi d^2/4)]} = d/4$

MI of the hollow section $= \pi[D^4 - (0.6D)^4]/64 = 0.0136\,\pi D^4$

$$k = \sqrt{I/A} = \sqrt{[0.0136\pi D^4/(0.16\pi D^2)]} = 0.4D = 0.5d \qquad [\text{as } D = 1.25d]$$

(a) Using the straight line formula:

The critical loads are as follows.

For the solid circular column: $P_{cr} = \sigma_y \times 0.25\pi d^2[1 - b(L_e/0.25d)]$

For the hollow column: $P_{cr} = \sigma_y \times 0.25\pi d^2[1 - b(L_e/0.5d)]$

Also the constant $b = 1/189$ for steel

$$\frac{P_{cr}\,(\text{solid})}{P_{cr}\,(\text{hellow})} = \frac{\sigma_y \times 0.25\pi d^2[1 - b(L_e/0.25d)]}{\sigma_y \times 0.25\pi d^2[1 - b(L_e/0.25d)]}$$

$$= \frac{[1 - 4000/(189 \times 0.25 \times 200)]}{[1 - 4000/(189 \times 0.5 \times 200)]}$$

$$= 0.5767/0.7883 = 0.73$$

(b) Using the parabolic formula:

According to the parabolic formula,

$P_{cr} = \sigma_y A(1 - nl^2/k^2)$, $n = 1/18{,}000$ for steel

$$\frac{P_{cr}\,(\text{solid})}{P_{cr}\,(\text{hellow})} = \frac{\sigma_y A[1 - (4000)^2/(18{,}000 \times 0.25 \times 200)^2]}{\sigma_y A[1 - (4000)/(18{,}000 \times 0.5 \times 200)^2]}$$

$$= 0.64/\,0.91 = 0.703$$

Example 11.15 Safe load capacity of column

A column has both ends hinged and is 4.2 m long. Determine the safe load-carrying capacity of its section shown in Fig. 11.18 as per the IS code formula. The column is of mild steel with the yield stress of 250 N/mm^2 and $E = 200$ GPa.

Solution First, we find the centroid of the section.

Area of section $= 125 \times 20 + 105 \times 20 = 4600$ mm^2

Taking moments about the top,

$$(125 \times 20 + 105 \times 20)y = 125 \times 20 \times 10 + 105 \times 20 \times 72.5;\ y = 38.5 \text{ mm}$$

The centroid is 38.5 mm from the top and 86.5 mm from the bottom. X–X and Y–Y are the centroidal axes. We find the moments of inertia about both axes.

$$I_{XX} = 125 \times 20^3/12 + (125 \times 20)28.5^2 + 20 \times 105^3/12 + (20 \times 105) \times (86.5 - 52.5)^2$$

$$= 6.5 \times 10^6 \text{ mm}^4$$

$I_{YY} = 20 \times 125^3/12 + 105 \times 20^3/12 = 3.325 \times 10^6$ mm^4

Least radius of gyration = $\sqrt{[I_{yy}/A]} = \sqrt{[3.325 \times 10^6/4600]} = 26.9$ mm

Slenderness ratio = 4200/26.9 = 156

Permissible stress = 43 N/mm^2

Safe load = 43 × 4600 = 198 kN

Example 11.16 Safe load capacity of I-section

Find the load capacity of an I-section column, ISMB 250 @365 N/m. The column is 3.5 m long and the support conditions are such that the lower end can be considered fixed while the upper end is hinged.

Solution From the steel tables,

Area of section = 4755 mm^2, $I_{XX} = 5131.6 \times 10^4$ mm^4, $I_{YY} = 334.5 \times 10^4$ mm^4

As the effective length = $L/\sqrt{2}$ for a column with one end fixed and the other hinged, we get

Effective length = $3500/\sqrt{2} = 2475$ mm

Also Least radius of gyration = $\sqrt{I/A} = \sqrt{[334.5 \times 10^4/4755]} = 26.5$ mm

Slenderness ratio = 2475/26.5 = 93.4

From Table 11.2,

Permissible stress = 90 – (90 – 80) × 3.4/10 = 86.6 N/mm^2

Safe load on the column = 86.6 × 4755 N = 412 kN

11.7 SECANT FORMULA FOR ECCENTRICALLY LOADED COLUMN

Consider a column loaded with an axial load acting at an eccentricity e as shown in Fig. 11.20(a). Both ends of the column are hinged. Transfer the eccentric load to the centroidal axis, as shown in Fig. 11.20(b). The column is thus subjected to an axial

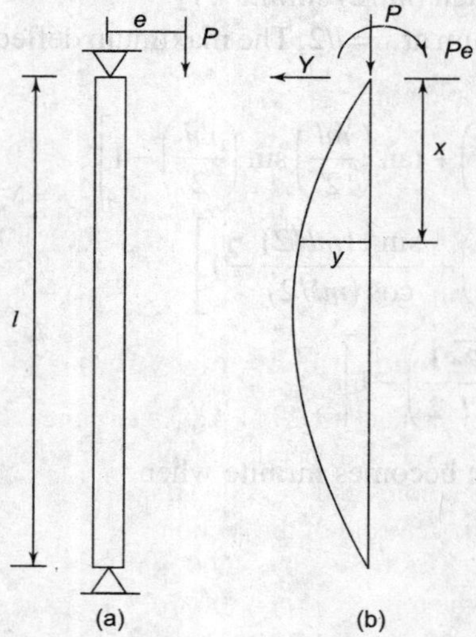

Fig. 11.20

load and a BM, Pe, as shown. Note that even when the load is small, there is an axial load and a BM, which causes the column to bend. As the load increases, the axial load as well as BM increase, and the column bends further. This is another way of looking at the column problem. Instead of analysing it from the point of view of neutral equilibrium and stability, we are basically considering the problem in terms of bending and lateral deflection.

We can state the differential equation for bending as

$$EI\,\frac{d^2y}{dx^2} = -P\,(e+y)$$

Setting $P/EI = m^2$,

$$\frac{d^2y}{dx^2} + m^2y = -m^2e$$

The solution to this differential equation is

$$y = C_1\cos mx + C_2\sin mx - e$$

where C_1 and C_2 are constants. The end conditions of deformation are (i) $x = 0$, $y = 0$ and (ii) $x = l$, $y = 0$, as both ends are hinged.

Substituting these values,

$$0 = C_1 \times 1 + 0 - e$$

$$C_1 = e \quad \text{and} \quad 0 = C_1\cos ml + C_2\sin ml - e$$

$$C_2 = \frac{e(1-\cos ml)}{\sin ml} = \frac{e \times 2\sin^2(ml/2)}{2\sin(ml/2)\cos(ml/2)} = e\tan(ml/2)$$

Substituting these values of C_1 and C_2,

$$y = e\cos mx + e\tan\left(\frac{ml}{2}\right)\sin mx - e$$

$$= e\,[\cos mx + \tan(ml/2)\sin mx - 1]$$

The deflection is maximum at $x = l/2$. The maximum deflection δ_c (at the centre) is given by

$$\delta_c = e\left[\cos\left(\frac{ml}{2}\right) + \tan\left(\frac{ml}{2}\right)\sin\left(\frac{ml}{2}\right) - 1\right]$$

$$= e\left[\cos\left(\frac{ml}{2}\right) + \frac{\sin^2(ml/2)}{\cos(ml/2)} - 1\right]$$

$$= e\left[\sec\left(\sqrt{\frac{P}{EI}}\,\frac{l}{2}\right) - 1\right]$$

Note that this expression becomes infinite when

$$\sqrt{\frac{P}{EI}}\,\frac{l}{2} = \frac{\pi}{2}$$

or $$P = \frac{\pi^2 EI}{l^2}$$

which is the Euler critical load for a column with hinged ends.

From $P_{cr} = \pi^2 EI/l^2$,

$$EI = \frac{P_{cr} l^2}{\pi^2}$$

The maximum deflection can also be put in an alternate form by substituting for *EI* as

$$\delta_c = e\left[\sec\left(\sqrt{\frac{P\pi^2}{P_{cr} l^2}}\frac{l}{2}\right) - 1\right] = e\left[\sec\left(\sqrt{\frac{P}{P_{cr}}}\frac{\pi}{2}\right) - 1\right]$$

The maximum BM and maximum stress in the column will occur at the midpoint.

$$\text{Maximum BM} = P\delta_c + P_e = P(\delta_c + e)$$

$$Z = \frac{I}{c} = \frac{Ar^2}{c}$$

where A is the area, r is the radius of gyration, and $c = d/2$. Substituting these values,

$$\sigma_{max} = \frac{P}{A} + \frac{P(\delta_c + e)c}{Ar^2} = \frac{P}{A}\left(1 + \frac{(\delta_c + e)c}{r^2}\right)$$

Substituting for δ_c, we have

$$\sigma_{max} = \frac{P}{A}\left[1 + \frac{ec}{r^2}\sec\sqrt{\frac{P}{EI}}\frac{l}{2}\right]$$

$$= \frac{P}{A}\left[1 + \frac{ec}{r^2}\sec\left(\sqrt{\frac{\pi}{2}}\sqrt{\frac{P}{P_{cr}}}\right)\right]$$

This can also be expressed as

$$\frac{P}{A} = \frac{\sigma_{max}}{1 + ec/r^2 \sec(l/2r\sqrt{P/AE})}$$

In this formula, note that P is the axial load, e is the eccentricity of the load, c is half the depth of the section perpendicular to the bending axis, r is the radius of gyration, P_{cr} is the Euler critical load, and A is the area of the section.

We have derived the formula for hinged end conditions. As is obvious, we can use the same expressions for any end conditions by substituting the appropriate value of l. Note that the argument for the secant function is in radians.

This identity is of a transcendental nature since P/A appears on both sides, and has to be solved by the trial and error method. For a given column and loading conditions, the secant formula can be solved. A plot between P/A and different values of l/r for different values of ec/r^2 will be as shown in Fig. 11.20. These curves have been drawn for mild steel with $E = 200$ GPa and $\sigma_Y = 250$ MPa. Such a graph or table can be prepared and used to find P/A for a given slenderness ratio and ec/r^2.

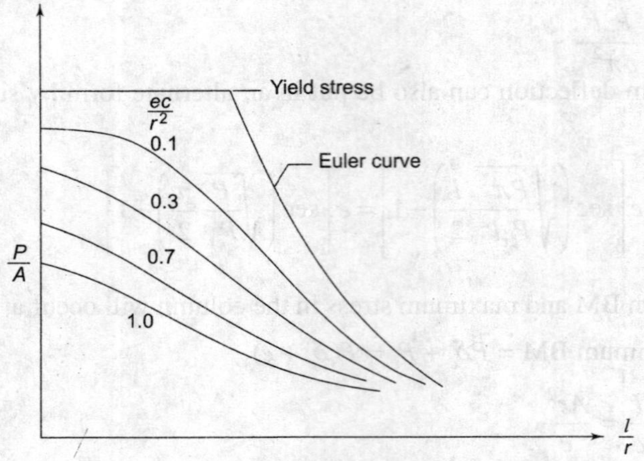

Fig. 11.21

From Fig. 11.21, we observe that the secant formula gives results very close to Euler's formula for large values of the slenderness ratio. When l/r values are small, sec $[(l/2r)\,(P/AE)^{1/2}]$ is very nearly equal to 1, and the secant formula becomes

$$\frac{P}{A} = \frac{\sigma_{\max}}{1 + ec/r^2}$$

which is what we would have obtained using the methods presented in Chapter 6.

The secant formula is mainly useful for intermediate columns. It has some limitations for practical use, such as (i) difficulty in obtaining a reasonable estimate of the eccentricity of the load, and (ii) the transcendental nature of the formula, requiring a trial and error solution.

The secant formula is useful in the calculation of stresses in columns given the values of P and e. However, it is very cumbersome to calculate the value of P using this formula. It is necessary to put the formula in a different form if P is the unknown quantity. The modification suggested is as follows.

$$\sigma_{\max} = \frac{P}{A} + \frac{Pec}{Ar^2}\sec\left[\frac{\pi}{2}\sqrt{\frac{P}{P_{\text{cr}}}}\right]$$

It is found that

$$\sec\left[\frac{\pi}{2}\sqrt{\frac{P}{P_{\text{cr}}}}\right] \approx \frac{1.2\,P_{\text{cr}}}{P_{\text{cr}} - P} \approx \frac{P_{\text{cr}}}{P_{\text{cr}} - P}$$

where P_{cr} is the Euler critical load.

$$\sigma_{\max} = \frac{P}{A} + \frac{Pec}{Ar^2}\left[\frac{1.2\,P_{\text{cr}}}{P_{\text{cr}} - P}\right]$$

Setting $\sigma_c = P/A$ and $\sigma_{\text{cr}} = P_{\text{cr}}/A$,

$$\sigma_{\max} = \sigma_c + \frac{\sigma_c ec}{r^2}\frac{1.2\,P_{\text{cr}}}{P_{\text{cr}} - P}$$

$$\frac{\sigma_{max}}{\sigma_c} - 1 = \frac{ec}{r^2} \frac{1.2\, P_{cr}}{P_{cr} - P}$$

or
$$\left(\frac{\sigma_{max}}{\sigma_c} - 1\right)\left(1 - \frac{\sigma_c}{\sigma_{cr}}\right) = \frac{1.2\, ec}{r^2}$$

This form can be used to calculate load P.

Example 11.17 Secant formula

A metal column of external diameter 300 mm and thickness 20 mm carries a load of 400 kN at an eccentricity of 50 mm. Determine the maximum and minimum stresses in the column if its length is 5 m, and both ends of the column are fixed. $E = 95$ GPa.

Solution Area of the column $= \frac{\pi}{4}(300^2 - 260^2) = 17{,}590\ \text{mm}^2$

MI of the column section $= \frac{\pi}{64}(300^4 - 260^4) = 1.732 \times 10^8\ \text{mm}^4$

Maximum BM $= Pe \sec\left(\frac{le}{2}\sqrt{\frac{P}{EI}}\right)$

$P = 400 \times 10^3$ N

$e = 50$ mm, $E = 95{,}000\ \text{N/mm}^2$

$l_e = \frac{5000}{2} = 2500$ mm

$$\begin{aligned}\text{Maximum BM} &= 400 \times 10^3 \times 50 \sec(0.195\ \text{rad}) \\ &= 400 \times 10^3 \times 50 \sec(11.17°) \\ &= 400 \times 10^3 \times 50 \times 1.019 \\ &= 20.386 \times 10^6\ \text{N mm}\end{aligned}$$

$$\begin{aligned}\sigma_{max}/\sigma_{min} &= \frac{400 \times 10^3}{17590} \pm \frac{20.386 \times 10^6 \times 150}{1.732 \times 10^8} \\ &= 22.74 \pm 17.67 \\ &= 40.41\ \text{N/mm}^2, \quad 5.07\ \text{N/mm}^2\end{aligned}$$

Example 11.18 Secant formula

A steel column of length 6 m, external diameter 200 mm and thickness 10 mm carries a load at an eccentricity of 30 mm. Find the maximum value of the load if the permissible stress is limited to 150 MPa. Both ends of the column are hinged. $E = 200$ GPa.

Solution Area $= \frac{\pi}{4}(200^2 - 180^2) = 5970\ \text{mm}^4$

$$I = \frac{\pi}{64}(200^4 - 180^4) = 2700 \times 10^4\ \text{mm}^4$$

The Euler load,

$$P_{cr} = \frac{\pi^2 EI}{l^2} = \frac{\pi^2 \times 2 \times 10^5 \times 2700 \times 10^4}{(6000)^2} = 1480\ \text{kN}$$

$$\sigma_{cr} = \frac{P_{cr}}{A} = \frac{1480 \times 10^3}{5970} = 248\ \text{N/mm}^2$$

Using the modified secant formula,

$$\left(\frac{150}{\sigma_c}-1\right)\left(1-\frac{\sigma_c}{\sigma_{cr}}\right)=\frac{1.2ec}{r^2}$$

$$e = 30 \text{ mm}, \quad c = 100 \text{ mm}, \quad r=\sqrt{\frac{I}{A}} = 67.25 \text{ mm}$$

Substituting the values,

$$\left(\frac{150}{\sigma_c}-1\right)\left(1-\frac{\sigma_c}{248}\right)=\frac{1.2\times 30\times 100}{67.25^2}$$

This gives the quadratic equation

$$\sigma_c^2 - 594\sigma_c + 37{,}200 = 0$$

$$\Rightarrow \quad \sigma_c = 71.15 \text{ N/mm}^2$$

Permissible load = 71.15 × 5970 = 424 kN ■

11.8 COLUMNS WITH INITIAL CURVATURE

A strut may have an initial curvature due to a manufacturing defect or because of self-weight if it is not vertical. Refer to Fig. 11.22. The firm line shows the initial shape of the column, and the dotted line the deflected shape when an axial thrust P is applied. We assume that the initial curvature follows a sinusoidal variation. (We can also assume a parabolic or circular variation.)

The initial deflection at x,

$$y_0 = C \sin \pi x/l$$

where C is a constant. The BM at x,

$$M_x = -P(y_0 + y)$$

$$\frac{d^2y}{dx^2} = -\frac{P}{EI}(y_0 + y)$$

$$\frac{d^2y}{dx^2} + \frac{P}{EI}y = -\frac{P}{EI}y_0$$

Setting $\lambda^2 = P/EI$,

$$\frac{d^2y}{dx^2} + \lambda^2 y = -\lambda^2 y_0 = -\lambda^2 \sin\frac{\pi x}{l}$$

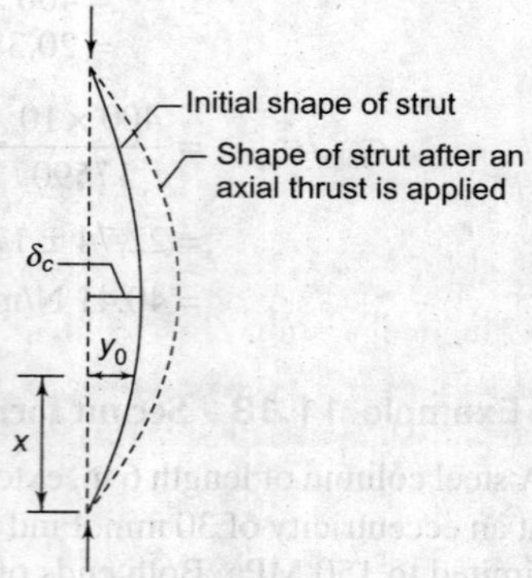

Fig. 11.22

The solution to this differential equation is

$$y = A\cos\lambda x + B\sin\lambda x + \frac{\lambda^2 C\sin(\pi x/l)}{\pi^2/l^2 - \lambda^2}$$

Substituting the boundary conditions (i) $x = 0$, $y = 0$ and (ii) $x = l$, $y = 0$, we get $A = 0$ from (i), and

$$B\sin\lambda L + \frac{\lambda^2 C\sin(\pi x/l)}{\pi^2/l^2 - \lambda^2} = 0$$

from (ii). Therefore,

$$B = 0 \text{ or } \sin(\lambda L) = 0$$

Since $\sin \lambda L \neq 0$, $B = 0$.

$$y = \frac{\lambda^2 C \sin(\pi x/l)}{\pi^2/l^2 - \lambda^2}$$

Deflection at $x = y_0 + y$

$$= C \sin\left(\frac{\pi x}{l}\right) + \frac{\lambda^2 C \sin(\pi x/l)}{\pi^2/l^2 - \lambda^2}$$

$$= C \sin\left(\frac{\pi x}{l}\right)\left(1 + \frac{\lambda^2}{\pi^2/l^2 - \lambda^2}\right)$$

Setting $\lambda^2 = P/EI$,

$$\text{Deflection at } x = C \sin\frac{\pi x}{l}\left(1 + \frac{P/EI}{\pi^2/l^2 - \lambda^2}\right)$$

$$= C \sin\frac{\pi x}{l}\left(\frac{P_e}{P_e - P}\right)$$

where $P_e = \pi^2 EI/e^2 l^2$.

The maximum deflection is at centre, giving the maximum BM:

$$M_{\max} \text{ at } x = \frac{l}{2} = PC \sin\left(\frac{\pi l}{2l}\right)\left(\frac{P_e}{P_e - P}\right) = \frac{CPP_e}{P_e - P}$$

$$\sigma = \frac{P}{A} + \frac{M}{I} y$$

can be used to calculate the stresses.

Example 11.19 Column with initial curvature

A round bar, 5 mm ϕ, is 2.5 m long, and is used as a strut with hinged ends. It has an initial curvature with a maximum deflection of 2.5 mm at the centre. Find the maximum stress in the bar when it carries a load of 60 kN. $E = 200$ GPa.

Solution Assuming sinusoidal variation,

$$y_0 = 2.5 \sin\left(\frac{\pi x}{l}\right)$$

$$\text{Maximum BM} = PC \frac{P_e}{(P_e - P)}$$

$P = 60$ kN $= 60 \times 10^3$ N

$C = 2.5$ mm

$$P_e = \frac{\pi^2 EI}{l^2} = \frac{\pi^2 \times 2 \times 10^5 \times (\pi/64) \times 50^4}{(2500)^2} = 96.8 \text{ kN}$$

$$\text{Maximum BM} = \frac{60 \times 10^3 \times 2.5 \times 96.8 \times 10^3}{(96.8 - 60) \times 10^3}$$

$$= 394.5 \times 10^3 \text{ N mm}$$

$$\sigma_{max} = \frac{P}{A} + \frac{M}{I}y = \frac{60{,}000}{(\pi/4)\times 50^2} + \frac{394.5\times 10^3}{(\pi/64)\times 50^4}\times 25 = 30.55 + 32.15$$
$$= 62.7 \text{ N/mm}^2$$

11.9 BEAM COLUMNS

A member subjected to direct thrust as well as lateral loads is called a beam column. All horizontal members subjected to axial thrust act as beam columns because the self-weight of the member acts as a lateral load. Shafts having axial thrust and subject to transverse loads are common. Let us consider a few cases.

11.9.1 Beam Column with Transverse Uniformly Distributed Load

Consider a member subjected to an axial thrust P and a transverse load uniformly distributed all along its length (Fig. 11.23).

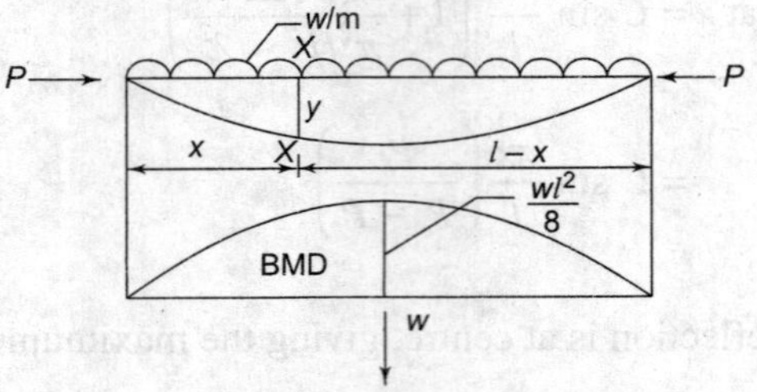

Fig. 11.23

BM at section x-x,

$$M_x = -Py - \frac{wl}{2}x + \frac{wx^2}{2}$$

$M = EI\, d^2y/dx^2$ gives

$$\frac{d^2y}{dx^2} = -\frac{P}{EI}y - \frac{wl}{2EI}x + \frac{wx^2}{2EI}$$

Setting $P/EI = \lambda^2$ and rearranging, we get

$$\frac{d^2y}{dx^2} + \lambda^2 y = -\frac{wl}{2EI}x + \frac{wx^2}{2EI}$$

This differential equation can be solved to obtain

$$y = A\sin\lambda x + B\cos\lambda x - \frac{wl}{2P}x + \frac{wx^2}{2P} - \frac{wEI}{P^2}$$

The boundary conditions are (i) $x = 0$, $y = 0$ gives $B = wEI/P^2$ and (ii) $x = l$, $y = 0$ gives

$$y = A\sin\lambda l + \frac{wEI}{P^2}\cos\lambda l - \frac{wl^2}{2P} + \frac{wl^2}{2P} - \frac{wEI}{P^2} = 0$$

$$\frac{wEI}{P^2} = A\sin\lambda l + \frac{wEI}{P^2}\cos\lambda l$$

$$\frac{wEI}{P^2}(1 - \cos\lambda l) = A\sin\lambda l$$

$$\frac{wEI}{P^2} = A\frac{2\sin(\lambda l/2)\cos(\lambda l/2)}{2\sin^2(\lambda l/2)}$$

$$A = \frac{wEI}{P^2}\tan\left(\frac{\lambda l}{2}\right)$$

The maximum deflection occurs at $x = l/2$ and the maximum BM also occurs at this section.

$$y_{\max} = \frac{wEI}{P^2}\tan\frac{\lambda l}{2}\sin\frac{\lambda l}{2} + \frac{wEI}{P^2}\cos\frac{\lambda l}{2} - \frac{wl^2}{4P} + \frac{wl^2}{8P} - \frac{wEI}{P^2}$$

$$= \frac{wEI}{P^2}\left(\sec\frac{\lambda l}{2}\right) - \frac{wl^2}{8P} - \frac{wEI}{P^2}$$

$$M_{\max} = Py_{\max} + \frac{wl}{2}\frac{l}{2} - \frac{wl}{2}\frac{l}{4}$$

$$= P\frac{wEI}{P^2}\sec\left(\frac{\lambda l}{2}\right) - P\frac{wl^2}{8P} - P\frac{wEI}{P^2} + \frac{wl^2}{8}$$

Knowing the maximum BM, we can evaluate the maximum stress.

11.9.2 Beam Column with Transverse Central Point Load

Considering Fig. 11.24, we proceed as in the previous case.

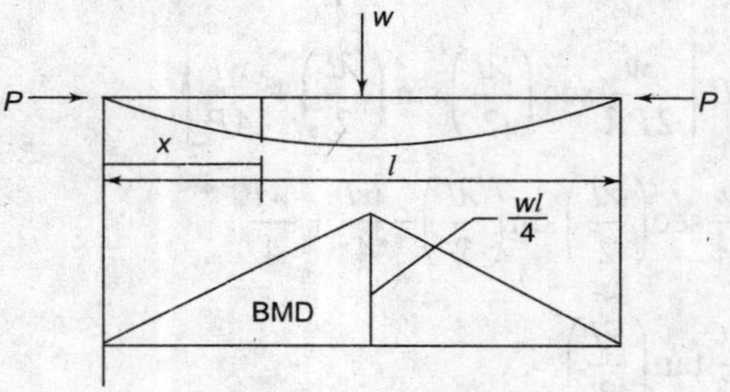

Fig. 11.24

The BM at x,

$$M_x = -Py + \frac{w}{2}x$$

Setting $M/EI = d^2y/dx^2$,

$$\frac{d^2y}{dx^2} = -\frac{Py}{EI} + \frac{w}{2EI}x$$

for $x \le l/2$.

$$\frac{d^2y}{dx^2} + \frac{P}{EI}y = \frac{w}{2EI}x$$

$$\frac{d^2y}{dx^2} + \lambda^2 y = \frac{w}{2EI}x$$

Setting $\lambda^2 = P/EI$, the solution to this differential equation is

$$y = A\cos\lambda x + B\sin\lambda x + \frac{wx}{2P}$$

Setting $x = 0$ and $y = 0$, we get $A = 0$. As the BM equation is valid only for $x \le l/2$, we use a second condition that at $x = l/2$, $dy/dx = 0$.

$$\frac{dy}{dx} = +B\lambda \cos \lambda x + \frac{w}{2P}$$

$$0 = B\lambda \cos \lambda \frac{l}{2}$$

which gives

$$B = -\frac{w}{2P\lambda} \sec\left(\frac{\lambda l}{2}\right)$$

Therefore,

$$y = -\frac{wP}{2P\lambda} \sec\left(\frac{\lambda l}{2}\right) x \sin \lambda x + \frac{wx}{2P}$$

$$= -\frac{wP}{2P\lambda} \sec\left(\frac{\lambda l}{2}\right) \sin \lambda x + \frac{wx}{2P}$$

The maximum BM is at $x = l/2$.

$$M_{\max} = -Py + \frac{wx}{2P}$$

$$= -Py_{\max} + \frac{wl}{4}$$

$$= -P\left[\frac{w}{2P\lambda} \sec\left(\frac{\lambda l}{2}\right) \sin\left(\frac{\lambda l}{2}\right) + \frac{wl}{4P}\right]$$

$$= \frac{w}{2\lambda} \sec\left(\frac{\lambda l}{2}\right) \sin\left(\frac{\lambda l}{2}\right) - \frac{wl}{4} + \frac{wl}{4}$$

$$= \frac{w}{2\lambda} \tan\left(\frac{\lambda l}{2}\right)$$

$$= \frac{w}{2\lambda} \tan\left(\frac{l}{2}\sqrt{\frac{P}{EI}}\right)$$

substituting for λ. Knowing the maximum BM, we can calculate the stress.

Example 11.20 Beam column with UD load

A strut, 30 mm ϕ and 2.2 m long, is hinged at both ends. It carries a UD load of 60 N/m in addition to an axial thrust of 8000 N. Calculate the maximum stress. $E = 200$ GPa.

Solution From Fig. 11.25, the maximum BM is at the centre due to symmetry of loading.

$$\text{Area of section} = \frac{\pi}{4} \times 30^2 = 225\,\pi\,\text{mm}^2$$

$$I \text{ of the section} = \frac{\pi}{64} \times 30^4 = 3.976 \times 10^4\ \text{mm}^4$$

$$M_{\max} = \frac{WEI}{P}\left[\sec\left(\frac{\lambda L}{2}\right) - 1\right]$$

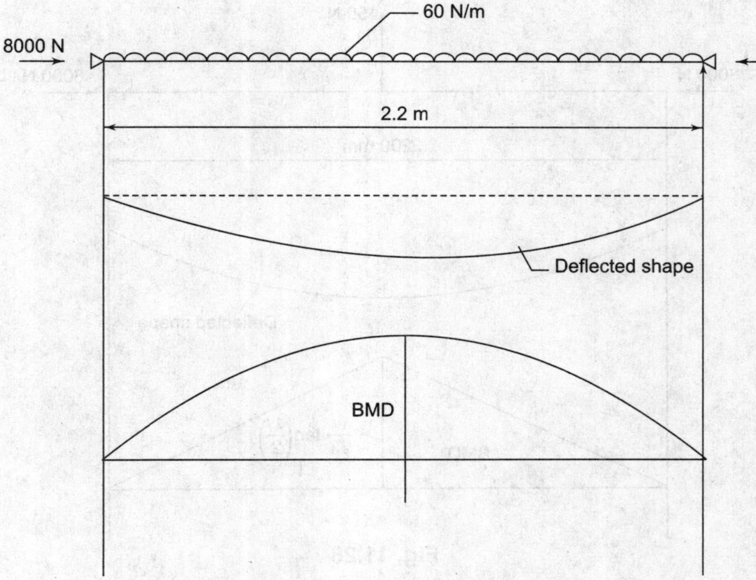

Fig. 11.25

$$\lambda = \sqrt{\frac{P}{EI}} = \sqrt{\frac{8000}{2\times10^5 \times 3.976\times10^4}} = 1.003\times10^{-3}$$

$$M_{\max} = \frac{0.06\times2\times10^5\times3.976\times10^4}{8000}\left[\sec\left(\frac{1.00\times10^{-3}\times2200}{2}\right)-1\right]$$

$$= 72{,}580 \text{ N mm}$$

$$\sigma_{\max} = \frac{8000}{225\pi} + \frac{72{,}589\times15}{3.976\times10^4} = 11.32 + 27.38 = 38.7 \text{ N/mm}^2$$

Example 11.21 Beam column with point load

If the strut in Example 11.20 carries a central point load of 150 N instead of the UD load, find the maximum stress.

Solution Figure 11.26 shows the strut with the loads. The maximum BM for a beam column with a central point load is given by

$$M_{\max} = \frac{w}{2\lambda}\tan\left(\frac{\lambda l}{2}\right)$$

$$\lambda = \sqrt{\frac{P}{EI}} = \sqrt{\frac{8000}{2\times10^5\times3.976\times10^4}} = 1.003\times10^{-3}$$

$$M_{\max} = \frac{150}{2\times1.003\times10^{-3}}\tan\left(\frac{1.003\times2200}{2}\right) = 148 \text{ N m}$$

$$\sigma_{\max} = \frac{P}{A} + \frac{My}{I} = \frac{8000}{225\pi} + \frac{148\times10^3\times15}{3.976\times10^4}$$

$$= 11.32 + 55.83 = 67.15 \text{ N/mm}^2$$

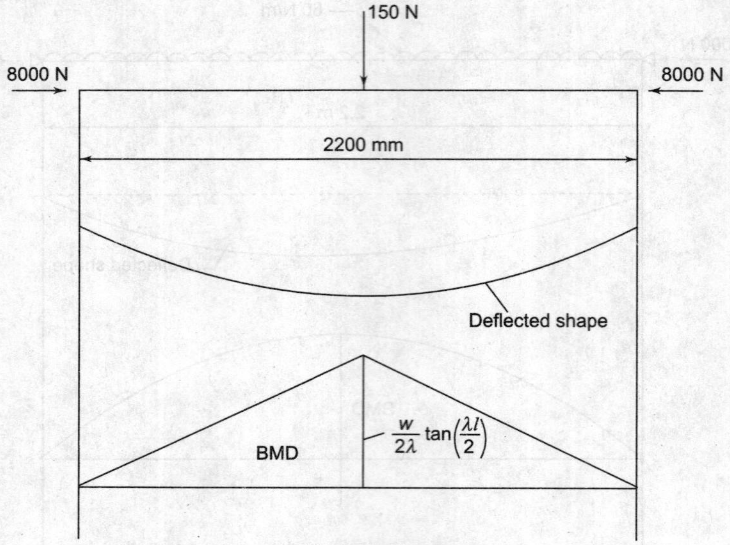

Fig. 11.26

11.10 STRUCTURAL SECTIONS AS STRUTS

Structural steel sections are commonly used as struts and columns. I-sections, T-sections and channels are symmetrical sections with at least one axis of symmetry. Angle sections when used as struts have to be looked at carefully. If double angles are used back to back, then they have one axis of symmetry. If a single angle section is used, the least moment of inertia of the section is about a principal axis, which is not an axis parallel to any side, as discussed in Chapter 2. Let us illustrate this point with an example.

Example 11.22 Angle section as struts

An equal angle section of dimensions 100 mm × 100 mm × 8 mm is used as a strut with a length of 3.76 m. The strut may be considered as hinged at both ends. Calculate the critical load for the column using Euler's formula. E = 200 GPa.

Solution The section is shown in Fig. 11.27. We can see that the X-X as well as Y-Y axes are not axes of symmetry. We have to find the principal axes U-U and V-V to locate the axis of minimum MI. Let us locate the centroid.

$$\bar{x} = \frac{100 \times 8 \times 4 + 90 \times 8 \times 53}{1520} = 27.21 \text{ mm}$$

$$\bar{y} = \frac{100 \times 8 \times 50 + 90 \times 8 \times 4}{1520} = 28.21 \text{ mm}$$

$$I_{XX} = \frac{8 \times 100^3}{12} + 800 \times (21.71)^2 + \frac{90 \times 8^3}{12} + 90 \times 8 \times 24.21^2 = 147.2 \times 10^4 \text{ mm}^4$$

$$I_{YY} = \frac{100 \times 8^3}{12} + 800 \times (23.21)^2 + \frac{8 \times 90^3}{12} + 720 \times 25.74^2 = 140 \times 10^4 \text{ mm}^4$$

P_{XY} is obtained using the transfer formula. As the P_{XY} of each rectangle about its own centroidal axes is zero, this quantity is obtained as $A\,\bar{x}\,\bar{y}$, where $\bar{x}$ and $\bar{y}$ are the coordinates of the centroid of the whole section with respect to the centroidal axes of the rectangles. Thus,

$$P_{XY} = 100 \times 8 \times 23.21 \times (-21.79) + 90 \times 8\,(-25.79)\,(24.21)$$
$$= -85.41 \times 10^4$$

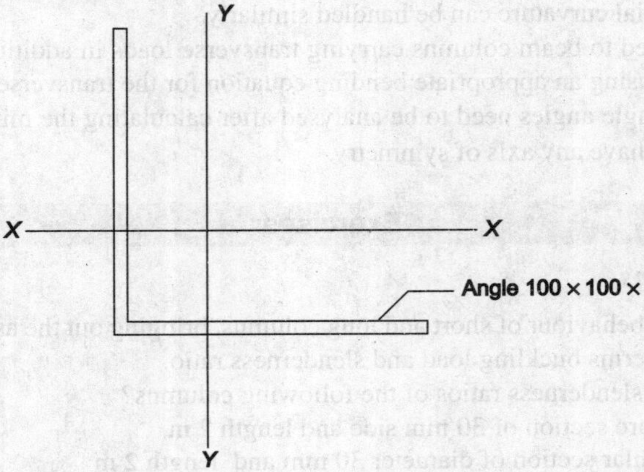

Fig. 11.27

The principal MIs I_1 and I_2 can now be calculated.

$$I_{1,2} = \frac{I_{XX} + I_{YY}}{2} \pm \sqrt{\left(\frac{I_{XX} - I_{YY}}{2}\right)^2 + \pi x^2 y}$$

$$= \left(\frac{147.2 + 140}{2}\right) 10^4 \pm 10^4 \sqrt{\left(\frac{7.2}{2}\right)^2 + (-85.41)^2}$$

$$= 143.6 \times 10^4 \pm 85.48 \times 10^4$$

Minimum MI $= 58.12 \times 10^4$ mm^4

$$\text{Euler load} = \frac{\pi^2 EI}{l^2} = \frac{\pi^2 \times 2 \times 10^5 \times 58.12 \times 10^4}{(3600)^2} = 88.52 \text{ kN}$$

Summary

The behaviour of compression members is dictated by their length and size. Long compression members fail by buckling, which is due to bending, rather than by crushing due to excessive compressive stresses.

The most common method of solving problems related to long columns involves the use of Euler's formula $P_{cr} = \pi^2 EI/(l_e)^2$, where P_{cr} is the Euler load and l_e the effective length. The effective length depends upon the manner of support provided to the column at its ends.

Euler's formula can be applied only to columns with a minimum slenderness ratio (l/r). Euler's formula may not be applicable to several columns which are neither long nor short. For such intermediate columns, Rankine proposed the empirical relationship

$$\frac{1}{P_r} = \frac{1}{P_s} + \frac{1}{P_e}$$

where P_r is the Rankine load, P_s the direct crushing load, and P_e the Euler load.

In the case of columns subjected to eccentric loading, the secant formula can be used.

Many codes use a modified form of the secant formula to take care of the accidental eccentricity of the load. The secant formula is

$$\sigma_{max} = \frac{P}{A}\left[1 + \frac{ec}{r^2}\sec\left(\frac{1}{2}\sqrt{\frac{P}{EI}}\right)\right]$$

Columns with initial curvature can be handled similarly.

Problems related to beam columns carrying transverse loads in addition to axial thrust can be solved by using an appropriate bending equation for the transverse loads.

Struts using single angles need to be analysed after calculating the minimum principal MI as they do not have any axis of symmetry.

Exercises

Review Questions

1. Discuss the behaviour of short and long columns, bringing out the aspect of stability.
2. Define the terms buckling load and slenderness ratio.
3. What is the slenderness ratios of the following columns?
   (i) A square section of 30 mm side and length 2 m
   (ii) A circular section of diameter 30 mm and length 2 m
   (iii) A hollow circular section of outer diameter 30 mm, thickness 5 mm, and length 2 m
4. What is the minimum length up to which Euler's formula is applicable for a column made of a material of proportional limit 300 MPa, given that the diameter of the column is 20 mm?
5. The sections of two columns, one square and the other solid circular, have equal area. Which column will be more flexible, if they have the same length.
6. Why is it necessary to use the minimum radius of gyration of a section to calculate the crippling load?
7. Two columns are of the same cross-sectional area and length. One has both ends hinged while the other has one end hinged and the other fixed. Which column will have the greater load bearing capacity?
8. State the secant formula and explain each of the terms in it.
9. What is a beam column? Give an example.
10. In the case of an angle used as a strut, about which axis will the strut bend and why?
11. Find the critical load for the system shown in Fig. 11.28.

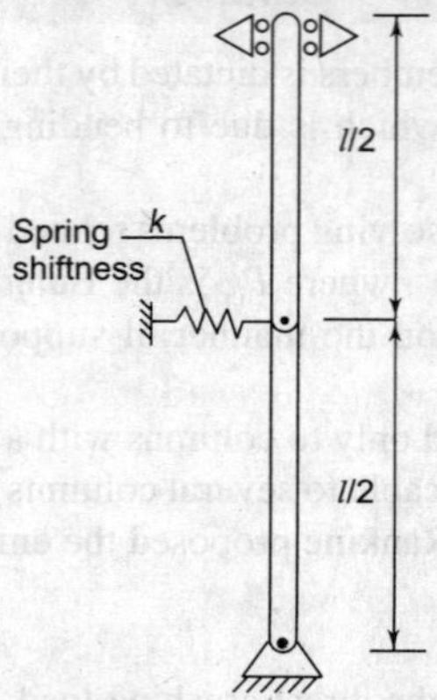

Fig. 11.28

12. For the sections shown in Fig. 11.29, about which axis will the columns bend and why?

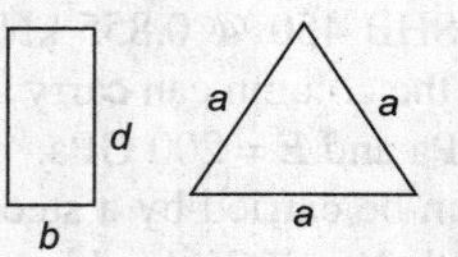

Fig. 11.29

13. State and explain the straight line and parabolic formulae for columns.
14. State and explain the IS code formulae for columns.

Problems

1. Determine the l/r ratio for the following columns.
   (a) A square column, of side 300 mm and length 4 m, with both ends hinged
   (b) A hollow circular column, of outer diameter 300 mm, thickness 5 cm, and length 6 m, with one end fixed and the other hinged
2. Determine the safe load that can be carried by a timber column, 300 mm φ and 3 m long, if both its ends are hinged. Use a factor of safety of 2.5. If the proportional limit is 35 MPa, determine the minimum length up to which Euler's formula can apply. E = 12 GPa.
3. A hollow circular column of steel, of outer diameter 200 mm and thickness 5 mm, has a length of 4 m, with both ends fixed. Find the Euler critical load if E = 200 GPa. If the yield stress is 300 MPa, determine the length below which Euler's formula can not be applied.
4. The channel section shown in Fig. 11.30 is used as a column, 3 m long, with both ends hinged. Compare the load carrying capacities obtained using Euler's and Rankine's formulae. E = 200 GPa and σ_y = 300 MPa.

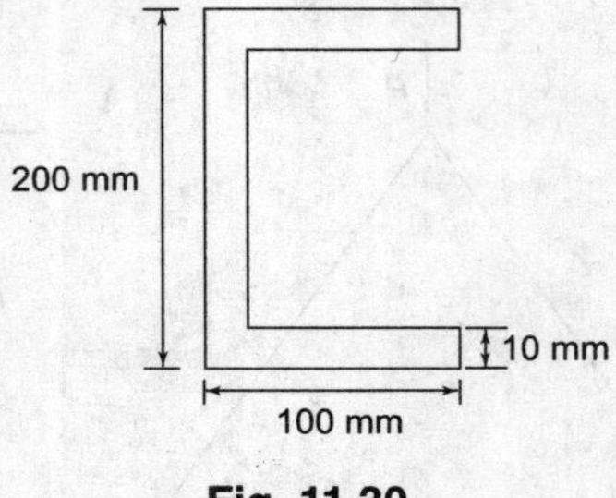

Fig. 11.30

5. Compare the critical stresses of columns with slenderness ratios of 40, 80, 120, 160, and 200 using Euler's and Rankine's formulae. E = 200 GPa and σ_y = 320 MPa.
6. Determine the load carrying capacity of an Indian standard I-section ISHB 400 @ 0.806 kN/m if it is 3.2 m long with one and fixed and the other hinged, using Euler's and Rankine's formulae. E = 200 GPa and σ_y = 320 MPa. What comments can you make on the results?
7. An Indian standard T-section, ISHT 150 @ 288 N/m, is used as a column 3.5 m long with both ends hinged. Determine the load carrying capacity using Euler's and Rankine's formulae. E = 200 GPa and σ_y = 320 N/mm^2.
8. A built-up column 5 m long is made of ISHB 450 @ 0.855 kN/m with two plates of size 250 × 10 mm, one each attached to the flanges. Calculate the safe load the

column can carry using Rankine's formula if $\sigma_y = 320$ N/mm^2. $E = 200$ GPa. Assume that both ends are hinged.

9. A column, 4 m long, is made of ISHB 450 @ 0.855 kN/m with both ends hinged. Determine the maximum load that the column can carry at an eccentricity of 50 mm along a principal axis. $\sigma_y = 300$ MPa and $E = 200$ GPa.
10. What is the maximum load that can be carried by a steel pipe of external diameter 200 mm, thickness 5 mm, and length 3 m, if the load is applied along the outer edge of the pipe? Assume the ends are hinged. $\sigma_y = 320$ MPa and $E = 200$ GPa.
11. A bar, of 30 mm square section and 4 m long, has an initial curvature with a maximum deflection at the centre of 20 mm. It carries an axial load of 30 kN. Assuming sinusoidal variation of deflection, determine the stresses in the bar. Assume hinged ends. $E = 200$ GPa.
12. A beam column, 30 mm wide, 90 mm deep, and 3 m long, is pin-jointed at the ends. The member carries an axial load of 80 kN along with a transverse UD load of 4 kN/m. Determine the maximum stress in the bar. $E = 200$ GPa.
13. If the beam in Problem 12 carries a central point load 2 kN instead of a UD load, find the maximum stresses in the section. Find the ratio of the maximum stresses if the (i) axial thrust is neglected and (ii) the transverse load is neglected.
14. An angle section, 100 mm × 100 mm × 10 mm is used as a strut 2 m long. Find the load capacity of the strut using Euler's and Rankine's formulae. $E = 200$ GPa and $\sigma_y = 320$ N/mm^2.
15. A rod, 30 mm ϕ and 2.5 m long, has a deflection of 5 mm as a simply supported beam with a central load of 100 N. If it is used as a strut with both ends hinged, find the safe load on the rod, if a factor of safety of 3 is required. What is the critical stress under this load?
16. A framework is loaded as shown in Fig. 11.31. Find the maximum value of P to avoid buckling if a factor of safety of 3 is desired. $E = 200$ GPa. All the members are 20 mm ϕ rods.

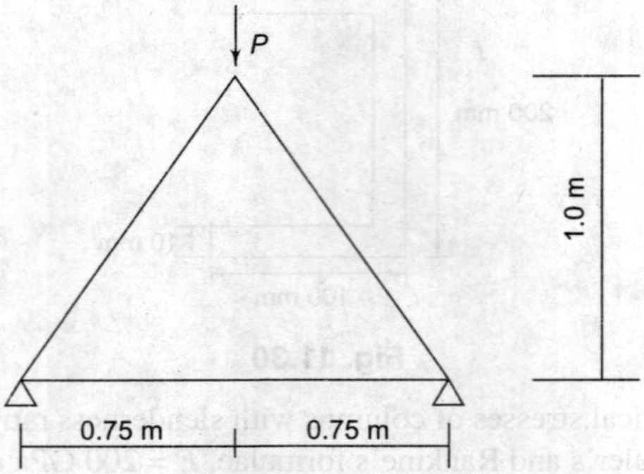

Fig. 11.31

17. A 2 m long pin-ended column of square section is to be made of wood. Assuming $E = 12$ GPa and the allowable stress being limited to 12 MPa, determine the size of the column to support the following loads safely: (i) 95 kN and (ii) 200 kN. Use the factor safety of 3 and also calculate the Euler's crippling load for buckling.
18. A hollow cast iron column, of outside diameter of 200 mm and thickness of 20 mm, is 4.5 m long and is fixed at both ends. Calculate the safe load by Rankine's formula using a factor of safety of 2.5. Find the ratio of Euler's to Rankine's loads. $E = 1 \times 10^5$ N/mm^2. Rankine's constant is 1/1600 and $\sigma_c = 550$ N/mm^2.

19. An RS T-section, 150 mm wide, 75 mm deep, thickness of flange 9 mm, and thickness of web 8.4 mm, is used as a strut of 3 m length with the ends hinged. Calculate the safe axial load by the Rankine's formula using a factor of safety of 3. Rankine's constants are $\sigma_c = 315$ N/mm^2 and $a = 1/7500$.
20. (a) What do you understand by beam columns?
    (b) A horizontal strut of length L, having hinged ends, carries an axial compressive load P and a central vertical load W. Derive the expression for the maximum values of deflection, BM, and stress.
21. A steel column of 50 mm diameter is pin-jointed at both ends. If $E = 200$ GPa and yield stress $\sigma_y = 300$ N/mm^2 for steel, calculate the length of the column for which the failure by yielding and buckling by Euler's formula is equally likely, when the column is subjected to axial compression.

CHAPTER 12

Special Topics

Learning Objectives

After going through this chapter, the reader will be able to

- analyse carriage springs and determine stresses, deformation and strain energy of the springs,
- analyse close- and open-coiled helical springs, subjected to axial load or axial twist, and compute stresses, deformations, and strain energy,
- analyse thin-walled cylindrical and spherical vessels and find stresses, strains, and change in volume,
- analyse thick cylindrical vessels using Lamé's theory, both analytically and graphically,
- derive the expression for stresses in a thick spherical shell,
- analyse wire-wound or compound cylinders to find stresses and strains,
- derive the bending equation for bending of curved bars and find maximum bending stress, and
- state and explain different theories of elastic failure and their limitations.

12.1 INTRODUCTION

In the last chapter, we discussed the analysis and design of compression members, and the important aspect of buckling. In this chapter, we will deal with some special structural elements which have important applications, such as springs and pressure vessels.

12.2 CARRIAGE OR LEAF SPRINGS

As the term signifies, a carriage spring finds wide application in all forms of transport vehicles. Such a spring is shown in Fig. 12.1. It is also called a semi-elliptic spring because of its shape.

The spring shown in Fig. 12.1(a) consists of a number of plates placed one above the other, and tied to form a single bending element. It can undergo sufficient deformation and can absorb a considerable amount of energy without permanent distortion. The energy can be recovered when required.

Let the spring consist of n plates, each of the same width b and thickness t. The plates are so arranged that each overlaps its neighbour by $L/2n$ at each end [Fig. 12.1(b)]. Neglecting friction between them, the plates are free to slide over each other.

If we consider the section at the end of any plate, and there are x plates at the section, the section is at a distance of $xL/2n$ from the end of the spring.

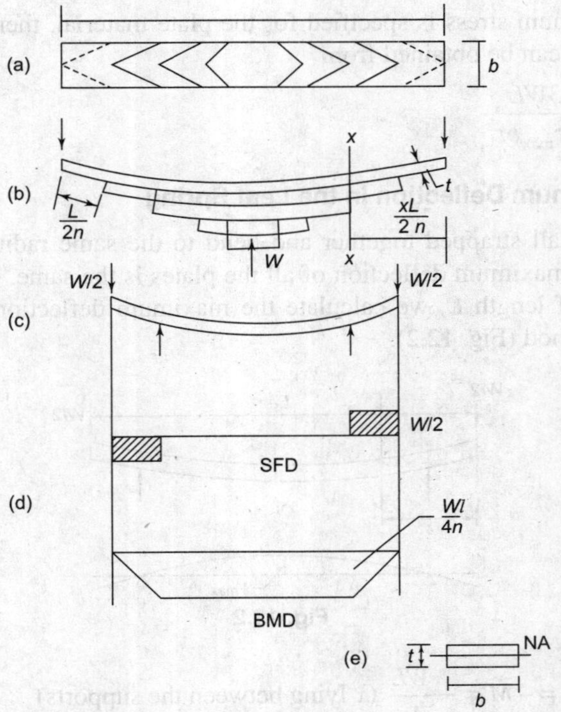

Fig. 12.1

The load on each plate is as shown in Fig. 12.1(c). If W is the central load carried by the spring, each plate receives a load of $W/2$ from the plate above and is supported by the ends of the plate below it, which are at $L/2n$ from its ends. Each plate acts like a doubly overhanging beam with equal overhang at the ends subjected to symmetrical loads and reactions. The BM will thus be uniform between the supports. The BM at xx is, therefore,

$$\frac{W}{2}\frac{xL}{2n} = \frac{WLx}{4n}$$

The BM and SF diagrams are shown in Fig. 12.1(d).

Since there are x plates and assuming the plates are of equal section and bend to the same radius of curvature, this BM is shared equally by all the plates.

$$\text{BM in one plate} = \frac{WLx}{4n}\frac{1}{x} = \frac{WL}{4n}$$

This is the maximum BM in any plate, and being independent of x, is the same in all the plates. For a plate, $I = bt^3/12$, and the maximum distance from the NA is $t/2$ [Fig. 12.1(e)].

The maximum stress,

$$\sigma_{\max} = \frac{M}{I} y_{\max} = \frac{WLt/2}{4nbt^3/12}$$

$$= \frac{3WL}{2nbt^2}$$

If the maximum stress is specified for the plate material, then the number of plates required can be obtained from

$$n = \frac{3WL}{2\sigma_{\max} bt^2}$$

12.2.1 Maximum Deflection in the Leaf Spring

The plates are all strapped together and bend to the same radius of curvature. Therefore, the maximum deflection of all the plates is the same. Considering the longest plate of length L, we calculate the maximum deflection by the double integration method (Fig. 12.2).

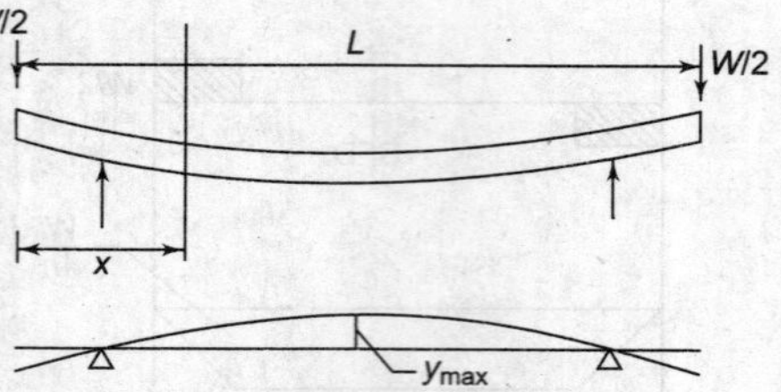

Fig. 12.2

$$EI\frac{d^2y}{dx^2} = -M = +\frac{WL}{4n} \quad (x \text{ lying between the supports})$$

$$EI\frac{dy}{dx} = \frac{WL}{4n}x + C_1$$

$$EIy = \frac{WL}{4n}\frac{x^2}{2} + C_1 x + C_2$$

$x = 0$, $y = 0$ gives $C_2 = 0$. At $x = L/2$, $dy/dx = 0$ gives

$$0 = \frac{WL^2}{8n} + C_1$$

$$C_1 = -\frac{WL^2}{8n}$$

The maximum deflection is at $x = L/2$.

$$EI_{y\,\max} = \frac{WL}{4n}\frac{(L/2)^2}{2} + \left(-\frac{WL^2}{8n}\right)\frac{L}{2}$$

$$y_{\max} = -\frac{3WL^3}{32nEI}$$

The negative sign indicates that the deflection is upwards according to the sign convention. Setting $I = bt^3/12$,

$$y_{\max} = \frac{3WL^3}{8nEbt^3}$$

The radius to which the plates must be bent initially is determined by the proof load or proof stress.

The proof load is the load required to be applied to straighten the spring. The proof stress is the maximum bending stress under the proof load. If σ_p is the proof stress, then

$$\frac{\sigma_p}{y} = \frac{E}{R}$$

where $y = t/2$ for maximum stress, and

$$R = \frac{Ey}{\sigma_p} = \frac{Et}{2\sigma_p}$$

where t is the thickness of the plate.

12.2.2 Strain Energy of Leaf Spring

The BM is constant in the leaf spring between the supports. To make for uniform bending stress, the ends of the leaves (plates) are generally tapered to account for variation of BM at the ends.

Strain energy,

$$U = \frac{M^2}{2EI} \times \text{length of the beam}$$

since the BM, M is uniform over the length of all the plates.

$$U = \left(\frac{WL}{4n}\right)^2 \frac{1}{2EI} \times \text{total length of plates}$$

For a rectangular section, the modulus of the section,

$$Z = \frac{1}{6}bt^2$$

and

$$M = \sigma_{\max} Z = \sigma_{\max} \times \frac{1}{6}bt^2$$

In terms of maximum stress, the strain energy can be expressed as

$$U = \left(\frac{\sigma_{\max} bt^2}{6}\right)^2 \times \frac{1}{2EI} \times (\text{length of plates})$$

Substituting $I = bt^3/12$,

$$U = \frac{(\sigma_{\max})^2 b^2 t^4}{36} \frac{1}{2Ebt^3/12} \times (\text{length of plates})$$

$$= \frac{(\sigma_{\max})^2}{6E} (bt \times \text{total length of plates})$$

$$= \frac{(\sigma_{\max})^2}{6E} \times \text{volume of the spring}$$

The total length of the plates can be obtained as follows.

If L is the length of the longest plate and n the number of plates, each lower plate is shorter by a length L/n (the offset is $L/2n$ on either side). Therefore,

$$\text{Total length of plates} = L + \left(L - \frac{L}{n}\right) + \left(L - \frac{2L}{n}\right) + \cdots + \left(L - \frac{(n-1)L}{n}\right)$$

$$= nL - \frac{L}{n}(1 + 2 + 3 + \cdots + n - 1)$$

$$= nL - \frac{L}{n}\frac{n(n-1)}{2}$$

$$= \frac{L}{2}(n+1)$$

where L is the length of the longest plate and n is the number of plates.

Example 12.1 Semi-elliptic leaf spring

A leaf spring is made of plates 50 mm wide and 8 mm thick. The spring has a span of 700 mm. Determine the number of plates required to carry a central load of 4.5 kN. The maximum allowable stress in the plates is 200 MPa. What is the maximum deflection under this load? Determine the strain energy stored in the spring under this load.

Solution Referring to Fig. 12.3,

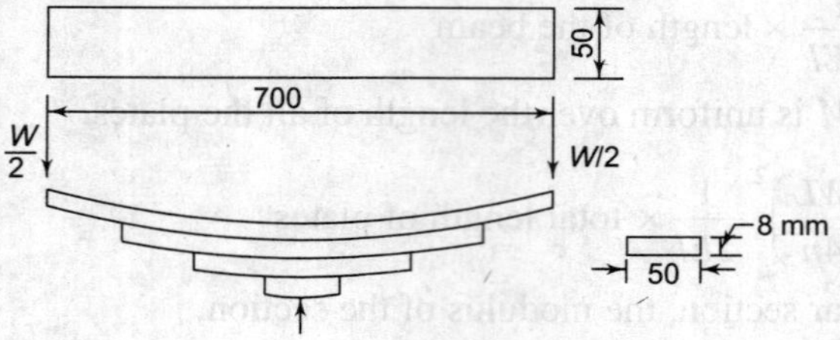

Fig. 12.3

$$\text{Maximum BM in any plate} = \frac{WL}{4n}$$

where n is the number of plates and L is the span. Maximum bending stress,

$$\sigma_{max} = \frac{M}{I}y = \frac{WL}{4nI}\frac{t}{2}$$

$$= \frac{WL}{4n(bt^3/12)}\frac{t}{2} = \frac{3WL}{2nbt^2}$$

$$\sigma_{max} = 200 \text{ N/mm}^2,$$

$$n = \frac{3WL}{2\sigma_{max}bt^2} = \frac{3 \times 4500 \times 700}{2 \times 200 \times 50 \times 64}$$

= 7.4, say 8 plates

Maximum central deflection,

$$\delta = \frac{(3WL^3)}{8n\,Ebt^3}$$

$$= \frac{3 \times 4500 \times (700)^3}{8 \times 8 \times 2 \times 10^5 \times 50 \times (8)^3}$$

= 14.13 mm

Strain energy stored in the spring = $\dfrac{(\sigma_{max})^2}{6E} \times$ volume

$$\sigma_{max} = \frac{3WL}{2nbt^2} = \frac{3 \times 4500 \times 700}{2 \times 8 \times 50 \times 64} = 184.6 \text{ N/mm}^2$$

Strain energy,

$$U = \frac{(184.6)^2}{6 \times 2 \times 10^5} \times \text{volume}$$
$$= 0.0284 \times \text{volume}$$

Length of the spring = $\dfrac{L}{2}(n + 1)$

$L = 700$ mm, $n = 8$

Length of the spring $= \dfrac{700}{2}(8 + 1)$

$= 3150$ mm

Volume of spring $= 3150 \times 50 \times 8$

$= 1{,}260{,}000$ mm^3

$U = 0.0284 \times 1{,}260{,}000$

$= 35.784$ N m

Example 12.2 Carriage spring

A carriage spring has to carry a central load of 9000 N. It has a length of 900 mm, and consists of plates 76 mm wide and 8 mm thick. If the maximum stress is limited to 220 MPa and the maximum deflection to 32 mm, find the number of plates required. Determine the actual values of maximum stress and deflection using the number of plates selected. Also find the radius to which the plates must be bent so that the spring straightens under the load. $E = 200$ GPa.

Solution We can find the number of plates required from the two conditions given—those of maximum stress and maximum deflection.

σ_{max}, the maximum stress, is limited to 220 N/mm^2. The maximum BM in any plate $= WL/4N$ (Fig. 12.4)

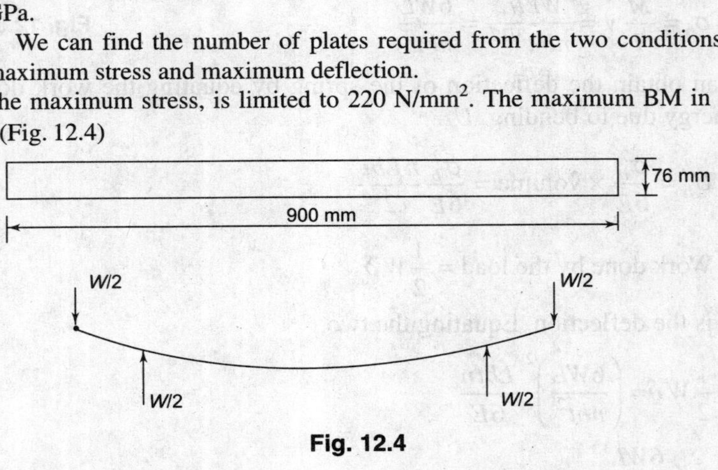

Fig. 12.4

$$\sigma_{max} = MZ = \frac{M}{I}y = \frac{WL}{4n}\frac{t/2}{bt^3/12} = \frac{3WL}{2nbt^2}$$

$$n = \frac{3WL}{2\sigma_{max}bt^2} = \frac{3 \times 9000 \times 900}{2 \times 220 \times 76 \times 8^2} = 11.35, \text{ say } 12$$

The maximum central deflection is limited to 32 mm.

$$\delta = \frac{3WL^3}{8nEbt^3}, \quad n = \frac{3WL^3}{8Ebt^3\delta} = \frac{3 \times 9000 \times (900)^3}{8 \times 2 \times 10^5 \times 76 \times 8^3 \times 32} = 9.3$$

The number of plates provided should be 12. With 12 plates,

$$\sigma_{\max} = \frac{3WL}{2nbt^2} = \frac{3 \times 9000 \times 900}{2 \times 12 \times 76 \times 64}$$
$$= 208 \text{ MPa}$$

$$\text{Actual central deflection} = \frac{3WL^3}{8nbt^3n} = \frac{3 \times 9000 \times (900)^3}{8 \times 2 \times 10^5 \times 76 \times 8^3 \times 12}$$
$$= 26.34 \text{ mm}$$

Using the flexure formula $\sigma/y = E/R$,

$$\sigma_{\max} = 208 \text{ N/mm}^2, \quad y = 4 \text{ mm}, \quad E = 2 \times 10^5 \text{ N/mm}^2.$$

$$R = \frac{Ey}{\sigma_{\max}} = \frac{2 \times 10^5 \times 4}{208} = 3.84 \text{ m}$$

12.2.3 Quarter Elliptic Springs

The quarter elliptic spring, shown in Fig. 12.5(a), makes up only half the arch functioning as a cantilever fixed at the end with multiple plates. The plan of the spring in Fig. 12.6(b) shows that the plates are gradually tapered off (as in the case of carriage springs).

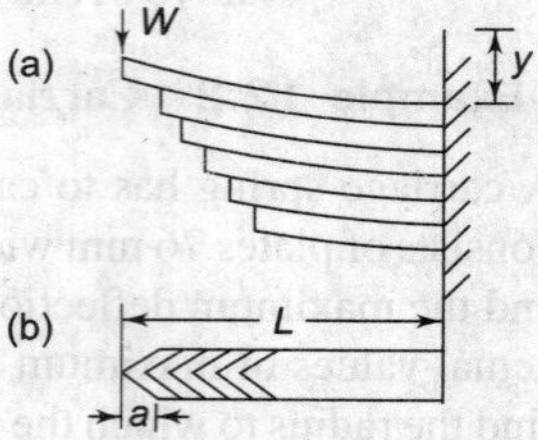

Fig. 12.5

For a load applied at the free end, the maximum moment at the fixed end is WL, where W is the load and L is the length of the spring. The bending stress,

$$\sigma_b = \frac{M}{I} y = \frac{WLt/2}{nb\,t^{3/12}} = \frac{6WL}{nbt^2}$$

We can obtain the deflection of the spring by equating the work done to the strain energy due to bending, U_b.

$$U_b = \frac{\sigma_b^2}{6E} \times \text{volume} = \frac{\sigma_b^2}{6E}\frac{nLbt}{2}$$

$$\text{Work done by the load} = \frac{1}{2} W\delta$$

where δ is the deflection. Equating the two,

$$\frac{1}{2} W\delta = \left(\frac{6WL}{nbt^2}\right)^2 \frac{Lbtn}{6E}$$

$$\delta = \frac{6WL^3}{nbt^3E}$$

$$\text{Strain energy} = \frac{1}{2} W\delta = \frac{3W^2L^3}{nbt^3E}$$

As in the case of carriage springs, overlap at the end of plates will $L/2n$ for each plate.

From Fig. 12.6(a), if y is the depth of the spring, then from the property of a circle,

$$y(2R - y) = LL$$

$$R = \frac{L^2}{2y}$$

where R is the radius of curvature of the spring, neglecting higher powers of y.

Example 12.3 Quarter elliptic spring

A quarter elliptic spring is to be 900 mm long and has to carry a load of 20 kN. The bending stress is limited to 300 MPa and the maximum deflection to 75 mm. Find the suitable size (width) and the number of plates required. Assume the width is eight times the thickness. E = 200 GPa.

Solution Maximum bending stress,

$$\sigma_{max} = \frac{6WL}{nbt^2}$$

and maximum deflection,

$$\delta_{max} = \frac{6WL^3}{nbt^3E}$$

Dividing one by the other,

$$\frac{\sigma_{max}}{\delta_{max}} = \frac{Et}{L^2}$$

$$t = \frac{\sigma_{max}}{\delta_{max}} \frac{L^2}{E}$$

Therefore, thickness

$$t = \frac{300}{75} \times \frac{900^2}{2 \times 10^5} = 16.2 \text{ mm, say 17 mm}$$

Width $b = 8 \times 17 = 136$ mm

The number of plates can be calculated from stress as well as deflection conditions, and the higher value adopted. From the stress condition that $\sigma_{max} \leq 300$ MPa,

$$n = \frac{6WL}{\sigma_{max}\, bt^2} = \frac{6 \times 20000 \times 900}{300 \times 136 \times (17)^2} = 9.15, \text{ say 10 plates}$$

From the deflection condition that $\delta_{max} \leq 75$ mm,

$$n = \frac{6WL^3}{\delta bt^3 E} = \frac{6 \times 20000 \times (900)^3}{75 \times 136 \times (17)^3 \times 2 \times 10^5} = 8.72$$

Therefore,

Number of plates = 10

Thickness of plates = 17 mm

Width of plates = 136 mm

12.3 HELICAL SPRINGS

Helical springs are commonly used to absorb shocks. For example, they are used at the ends of railway tracks to stop railway wagons. Such springs are of two types— close-coiled and open-coiled. The basic difference between the two is in the nature of stresses developed. A close-coiled spring is essentially a torsion spring while an open-coiled spring carries forces by bending and torsion.

12.3.1 Close-Coiled Springs

Consider the close-coiled spring shown in Fig. 12.6. The wires forming such a spring are closely wound so that one turn of the spring virtually lies in a plane. The ends of the spring are bent to form or are connected to short radial arms with hooks along the longitudinal axis so that loads can be applied axially, and the reactive force at the top end is also axial. The spring shown is subjected to an axial load *P*. Cutting the spring with a vertical plane through the centre and considering the equilibrium of half a turn at the top as a free body, the situation will be as shown in Fig. 12.6(b).

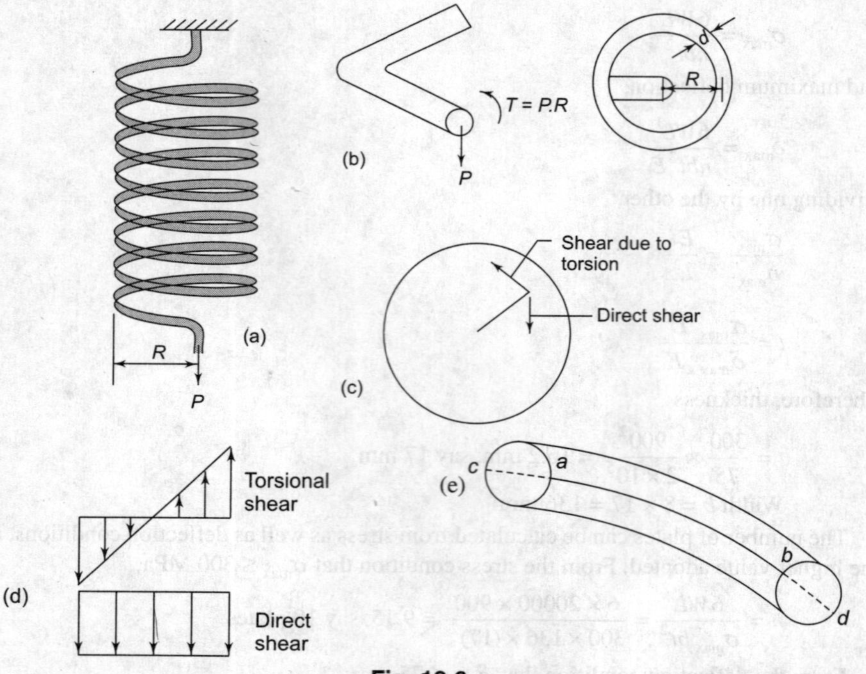

Fig. 12.6

The following symbols are used.

(i) R is the radius of the spring,

(ii) d is the diameter of the wire forming the spring, and

(iii) P is the applied load (axial).

The circular section of the wire is subjected to a vertical shear due to a load P and a torque $T = PR$ in the direction shown, the vertical shear stress,

$$\tau_1 = \frac{P}{A} = \frac{4P}{\pi d^2}$$

and a torsional shear stress, τ_{max}, which, for a solid circular section is $16\,T/\pi d^3 = 16\,PR/\pi d^3$.

These stress distributions are shown in Fig. 12.6(d). From this figure, it is clear that the average vertical shear stress and the torsional shear stress are in the same direction at the inner point of the wire. The maximum shear stress is, therefore,

$$\tau_{max} = \frac{4P}{\pi d^2} + \frac{16PR}{\pi d^3} = \frac{16PR}{\pi d^3}\left(1 + \frac{d}{4R}\right)$$

Deflection of the spring The axial load tends to elongate the spring. If the point where the load is attached to the radial arm moves down by an amount δ, then δ can be approximated as $R\theta$, where θ is the rotation of the radial arm. The radial arm actually rotates in a radial arc. However, since the angle θ is generally very small, δ can be taken to be the vertical deflection. This axial deflection is due to the cumulative rotation of each section of wire about its centre due to torsion.

To calculate the value of θ, we consider the total length of the wire subjected to a uniform torque $T = PR$ to be a shaft.

If the spring has n turns of mean radius R, then the total length of the spring = $2\pi Rn$.

The angle of twist θ is obtained from the torsion formula,

$$\frac{T}{J} = \frac{G\theta}{l}$$

$$\Rightarrow \qquad \theta = \frac{Tl}{GJ}$$

$$\theta = \frac{PR \times 2\pi Rn}{G\pi d^4/32} = \frac{64nPR^2}{Gd^4}$$

$$\delta = R\theta = \frac{64nPR^3}{Gd^4}$$

The stiffness K of a spring is defined as the force required for unit deflection. Setting $\delta = 1$ in the preceding expression, stiffness

$$K = \frac{Gd^4}{64nR^3}$$

Strain energy of close-coiled helical spring The strain energy stored in the spring has two components—one due to the direct shear P and the other due to the torsional moment $T = PR$.

$$\text{Strain energy due to torsion} = \frac{T^2 l}{2GJ} = \frac{(PR)^2 2\pi Rn}{2G\pi d^4/32}$$

$$U_1 = \frac{32P^2R^3n}{Gd^4}$$

The vertical shear is equal to P, and is also constant over the length of the wire.

The strain energy due to the constant shear P can be obtained as

$$U_2 = \frac{5P^2 l}{9AG}$$

$$= \frac{5P^2 2\pi Rn}{9P^2 2\pi Rn/4} = \frac{40}{9}\frac{P^2 Rn}{Gd^2}$$

where A is the area of cross section. The total strain energy U is given by

$$U = U_1 + U_2 = \frac{32PR^3 n}{Gd^4} + \frac{40}{9}\frac{P^2 Rn}{Gd^2}$$

$$= \frac{32PR^3 n}{Gd^4}\left(1 + \frac{d^2}{R^2}\frac{5}{36}\right)$$

The second term in brackets is the ratio of strain energy due to transverse shear to that due to torsion. It is easy to see that the second term is quite neglible compared to the first.

The deflection of the spring can be easily obtained by equating the strain energy to the external work done.

$$\text{External work done} = \frac{1}{2}P\delta = \text{strain energy}$$

$$\frac{32PR^3 n}{Gd^4} = \frac{1}{2}P\delta$$

$$\delta = \frac{64PR^3 n}{Gd^4}$$

as obtained before.

Close-coiled spring under axial twist Consider a close-coiled spring subjected to an axial twist as shown in Fig. 12.7. It is easy to see that the axial twist acts as a BM at sections of the spring tending to wind or unwind the spring, depending upon the direction of the axial twist. The twist thus results in a constant BM at sections of the spring, and is numerically equal to the axial twist.

Before the axial twist is applied, let n be the number of coils, R be the radius of curvature, and l be the length of the spring.

After the axial twist is applied, let n' be the number of coils, and R' be the radius of curvature.

The total length of the wire remains the same before and after twisting. Therefore,

$$l = 2\pi nR = 2\pi n'R'$$

or $$\frac{1}{R} = \frac{2\pi n}{l}$$

and $$\frac{1}{R'} = \frac{2\pi n'}{l}$$

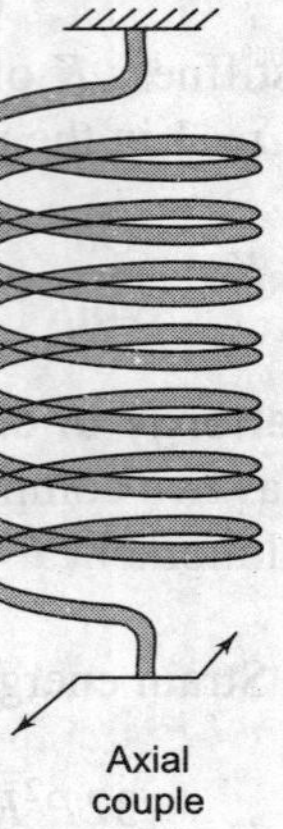

Fig. 12.7 Close-coiled spring subjected to axial couple

In the case of a member with an initial curvature, the change in curvature and BM can be related by the bending equation $M/I = E/R$ or $1/R = M/EI$.

Change in curvature $= M/EI = 1/R' - 1/R$, where $M = T$, $1/R$ is the initial curvature, and $1/R'$ is the final curvature. Therefore,

$$\frac{2\pi}{l}n' - \frac{2\pi}{l}n = \frac{M}{EI}$$

$$M = \frac{2\pi EI}{l}(n' - n) = \frac{EI(n' - n)}{nR}$$

The number of coils increases or decreases to n' from the initial value of n.

$$n' = \frac{MnR}{EI} + n$$

For a circular section of wire forming the spring,

$$I = \frac{\pi d^4}{64} \quad \text{and} \quad n' = \frac{64MnR}{E\pi d^4} + n$$

The total angular rotation or angular twist is given by the change in angular distance between the coil ends. This will be equal to

$$\phi = 2\pi n' = 2\pi n = 2\pi(n' - n)$$

As derived before,

$$\frac{2\pi EI}{l}(n' - n) = M$$

From this,

$$\phi = \frac{Ml}{EI}\frac{Tl}{EI}$$

Setting $l = 2\pi nR$ and $I = \pi d^4/64$ for a circular section, we have

$$\phi = \frac{T \times 2\pi nR}{E\pi d^4/64} = \frac{128\,TnR}{Ed^4}$$

For a circular section of diameter d, the maximum stress in the spring due to constant BM, M is given by

$$\sigma_{\max} = \frac{M}{\pi d^4/64}\frac{d}{2} = \frac{32M}{\pi d^3}$$

Strain energy due to axial twist The sections of the wire of a spring subjected to an axial twist are subjected to a constant BM. From Chapter 10, we know that the strain energy in the case of pure bending is $U = M^2l/2EI$. We treat a close-coiled spring as a straight member of length $l = 2\pi nR$ and $M = T$, the applied axial twist. Therefore,

$$U = \frac{T^2 \times 2\pi nR}{2EI}$$

For a wire of circular section, $I = \pi d^4/64$.

$$U = \frac{T^2 \times 2\pi nR}{2E\pi d^4/64} = \frac{64\,T^2Rn}{Ed^4}$$

Equating the work done to the strain energy, we can easily find the total angular twist. If T is the torque applied and ϕ is the total angular twist, then

$$\text{External work done} = \frac{1}{2}T\phi$$

Equating this to the strain energy,

$$\frac{1}{2}T\phi = \frac{64T^2Rn}{Ed^4}$$

$$\Rightarrow \qquad \phi = \frac{128TRn}{Ed^4}$$

Example 12.4 Close-coiled helical spring

A close-coiled helical spring has 20 turns wire of diameter 25 mm. The mean radius of the coils is 100 mm. Find the maximum stress and elongation of the spring under an axial load of 2 kN. $G = 85$ GPa.

Solution The section of the spring is subjected to a transverse shear equal to the axial load and a torque of PR, where P is the axial load and R is the mean radius of the coils. The maximum shear stress is the sum of these two effects.

$$\tau_{max} = \frac{P}{\text{area}} + \frac{16PR}{\pi d^3} = \frac{4 \times 2000}{\pi \times (25)^2} + \frac{16 \times 2000 \times 100}{\pi \times (25)^2}$$

$$= \frac{8000}{1963.5} + \frac{32 \times 10^5}{49087.3} = 4.074 + 65.189$$

$$= 69.26 \text{ N/mm}^2$$

The maximum deflection is calculated only due to the angular rotation of the coils due to the torque as the deflection due to transverse shear is very small.

$$\delta_{max} = \frac{64PR^3n}{Gd^4} = \frac{64 \times 2000 \times (100)^3 \times 20}{85000 \times (25)^4}$$

$$= 7.71 \text{ mm}$$

Example 12.5 Close-coiled helical spring

A close-coiled helical spring is made of a wire of diameter 20 mm. The mean radius of the coils is 100 mm. Find the number of turns required and the maximum axial load permissible if the shear stress is not to exceed 100 MPa and the maximum elongation is limited to 40 mm. $G = 83$ GPa.

Solution The maximum shear stress in a close-coiled helical spring is given by

$$\sigma_{max} = \frac{4P}{\pi d^2} + \frac{16PR}{\pi d^3} = \frac{4P}{\pi d^2}\left(1 + \frac{4R}{d}\right)$$

$$\sigma_{max} = 100 \text{ N/mm}^2$$

$$R = 100 \text{ mm}$$

$$\delta = 20 \text{ mm}$$

The maximum permissible load to limit the shear stress to 100 N/mm^2 is

$$P = \frac{\sigma_{max}\pi d^2}{4(1 + 4R/d)} = \frac{100\pi(20)^2}{4(1 + 4 \times 100/20)} = 1496 \text{ N}$$

Also,

$$\delta_{max} = \frac{64PR^3n}{Gd^4} = 40 \text{ mm}, \quad G = 83{,}000 \text{ N/mm}^2$$

$$40 = \frac{64 \times 1496 \times (100)^3 \times n}{83000 \times (20)^4}$$

$$\Rightarrow \qquad n = \frac{40 \times 83000 \times 20^4}{64 \times 1496 \times (100)^3} = 5.5, \text{ say 6 turns}$$

Example 12.6 Close-coiled helical spring

A wagon weighing 30 kN is moving at a speed of 1 m/s. It has to be stopped by absorbing the energy of motion using close-coiled helical springs. Determine the number of springs required if each spring has a mean radius of 250 mm and 20 turns of wire of diameter 30 mm. The maximum elongation of the spring is limited to 250 mm. For the material of the spring, $G = 83$ GPa.

Solution The kinetic energy of the wagon is to be absorbed by the springs to bring the wagon to a stop.

$$\text{Kinetic energy} = \frac{mv^2}{2} = \frac{30000 \times (1000)^2}{9.81 \times 1000 \times 2} = 15.29 \times 10^5$$

$$\delta = \frac{64PR^3n}{Gd^4} = \frac{64 \times P(250)^3 \times 20}{G \times (30)^4}, \quad \delta_{max} = 250 \text{ mm}$$

$$P = \frac{250 \times 83000 \times (30)^4}{64 \times (250)^3 \times 20} = 840 \text{ N}$$

$$\text{The work done by this load} = \frac{1}{2}P\delta = \frac{1}{2} \times 840 \times 250$$

$$= 105{,}000 \text{ N mm}$$

This is equal to the energy absorbed by one spring. The number of springs required is

$$\frac{15.29 \times 10^5}{1.05 \times 10^5} = 15$$

Example 12.7 Close-coiled helical spring under axial twist

A close-coiled helical spring has a mean diameter of 105 mm and 18 coils of wire of diameter 10 mm. Find the maximum stress in the wire, the increase in the number of turns, and the total rotation when the coil is subjected to an axial twist of 1.2 Nm. $E = 200$ GPa.

Solution For a wire of diameter 10 mm,

$$I = \frac{\pi d^4}{64} = \frac{\pi \times (10)^4}{64} = 491 \text{ mm}^4$$

The wire is subjected to a constant BM equal to the axial torque of 1.2 N m or 1200 N mm.

Number of turns = 18

If n' is the number of turns after the twist is applied, then

$$l = 2\pi nR = 2\pi n'R'$$

and

$$\frac{1}{R'} = \frac{2\pi n'}{l}, \quad \frac{1}{R} = \frac{2\pi n}{l}$$

where R and R' are the radii of curvature before and after the twist is applied.

$$\text{Change of curvature} = \frac{1}{R'} - \frac{1}{R} = \frac{M}{EI}$$

$$\frac{2\pi(n' - n)}{l} = \frac{M}{EI}, \quad n' = \frac{Ml}{2\pi EI} + n = \frac{MnR}{EI} + n$$

$$n' = \frac{1200 \times 18 \times 105}{2 \times 2 \times 10^5 \times 491} + n = 18.011$$

Increase in number of turns = 0.011

$$\text{Total angular shift} = 2\pi(n' - n) = 2\pi \times 0.011$$
$$= 0.07225 \text{ rad}$$
$$= 2.0698 \text{ degrees}$$

Maximum stress,

$$\sigma_{max} = \frac{32M}{\pi d^3} = \frac{32 \times 1200}{\pi \times (10)^3} = 12.2 \text{ N/mm}^2$$

Example 12.8 Axial and torsional stiffness of spring

Find the axial and torsional stiffnesses of a spring made of a wire diameter 6 mm with 20 turns of mean diameter 50 mm. Determine the maximum stress in the wire when subjected to an axial twist of 2 N m. G = 80 GPa and E = 200 GPa.

Solution The axial stiffness is given by

$$\frac{P}{\delta} = \frac{P}{64PnR^3/Gd^4} = \frac{Gd^4}{64nR^3}$$
$$= \frac{80{,}000 \times (6)^4}{64 \times 20 \times (25)^3} = 5.18 \text{ N/mm}$$

$$\text{Torsional stiffness} = \frac{Ed^4}{128NR} = \frac{200{,}000 \times (6)^4}{128 \times 20 \times 25}$$
$$= 4050 \text{ N/rad}$$

The maximum stress in the wire,

$$\sigma_{max} = \frac{32M}{\pi d^3}$$
$$= \frac{32 \times 2000}{\pi \times (6)^3} = 94.31 \text{ N/mm}^2$$

Example 12.9 Design of close-coiled helical spring

A close-coiled spring is required to have an axial stiffness of 5 N/mm and torsional stiffness of 100 N mm/degree angle of twist. The maximum bending stress should not exceed 120 MPa when subjected to an axial twist of 2.5 N m. If E = 200 GPa and G = 85 GPa for the material of the spring, find the diameter of wire, the mean radius of the coil and the number of turns.

Solution When a spring is subjected to an axial twist, every section of the spring is subjected to a constant BM of an equal magnitude. Therefore,

$$\sigma_{max} = \frac{32M}{\pi d^3}$$
$$120 = \frac{32 \times 2500}{\pi \times (d^3)}$$
$$d = 5.96 \text{ mm, say 6 mm}$$

Axial stiffness

$$K_a = \frac{P}{\delta} = \frac{Gd^4}{64\,nR^3} = 5 \text{ N/mm}$$

$$5 = \frac{G \times (6)^4}{64nR^3}, \quad \frac{G}{nR^3} = \frac{64 \times 5}{(6)^4} = 0.2469$$

Torsional stiffness,

$$K_t = \frac{T}{\phi} = \frac{Ed^4}{128nR}$$

$$= 100 \text{ N mm/degree} = \frac{100 \times 180}{\pi} \text{ per rad}$$

Therefore,

$$\frac{18{,}000}{\pi} = \frac{E \times (6)^4}{128nR}$$

$$\frac{E}{nR} = \frac{18{,}000}{\pi} \times \frac{128}{(6)^4} = 565.88$$

$$\frac{E}{nR} \div \frac{G}{nR^3} = \frac{E}{G} R^2 = \frac{565.88}{0.2469}$$

$$R^2 = \frac{565.88 \times 85{,}000}{200{,}000 \times 0.2469} = 974$$

$$\Rightarrow \quad R = 31.2 \text{ mm}$$

Mean radius of coil = 31.2 mm

From $E/nR = 565.88$,

$$n = \frac{E}{R565.88} = \frac{200{,}000}{31.2 \times 565.88}$$

$n = 11.32$, say 12 turns

12.3.2 Wahl Correction

The formula for close-coiled springs shows that the ratio of d/R (diameter of wire forming the spring to the radius of the spring) is a factor to decide whether to include the effect of direct shear stress. If the radius of the wire forming the spring is large compared to the radius of the spring, then this needs to be taken into account. This is true with many heavy springs used in railway carriages.

Another important aspect in the spring formula is that we have applied the torsion formula, $T/J = \tau/r = G\theta/L$, derived for a straight prismatic circular shaft to the curved spring. The difference comes from the effect shown in Fig. 12.6(e). In a straight bar, the shearing deformation is the same on the inside fibre *a–b* and the outside fibre *c–d* of the bar. But in the case of circular springs, the shearing deformation remaining the same on fibres *a–b* and *c–d*, the shear strains are different due to the different lengths along the curve. The inside fibre is shorter in length initially because of the curvature. The shear strain is consequently greater on the inside fibre than on the outside fibre. This difference in shear strain causes the inside fibres to have more stress than outside fibres.

This difference in stresses depends upon the curvature of the spring – how great is the difference in length between the inside and outside fibres. d/R ratio determines this difference.

A.M. Wahl developed a spring formula that takes into account this curvature effect on the stresses. His formula is

$$\text{Maximum shear stress, } \tau_{max} = \frac{16\,PR}{\pi d^3}\left[\frac{(4m-1)}{(4m-4)} + \frac{0.615}{m}\right]$$

where $m = 2R/d$ = ratio of the diameter of the spring to the diameter of the wire forming the spring.

In the case of sharply curved springs, the Wahl formula gives the correct value of stress. When m is very large (D/d ratio is large), the first term becomes unity and the formula becomes

$$\frac{16\,PR}{\pi d^3}\left(1 + \frac{0.615}{m}\right)$$

This formula is the same as the previous formula which can be written as

$$\frac{16\,PR}{\pi d^3}\left(1 + \frac{0.5}{m}\right)$$

The difference in the two formulae is because the shear stress distribution in a circular section is not uniform.

12.3.2 Open-coiled Springs

In the close-coiled springs we have considered, the pitch is very small and each turn more or less lies in one plane. The construction of the open-coiled spring does not permit the assumption of a turn lying in one plane as the pitch of the coils is larger. Short radial arms as shown in Fig. 12.8 help to ensure that the loads and twists are axial.

Open-coiled springs behave differently from close-coiled springs mainly because, in the former, the perpendicular section of the wire is inclined to the horizontal, and, therefore, is not subjected to pure twist.

Open-coiled spring subjected to axial load Consider an open-coiled spring subjected to an axial load as shown in Fig. 12.8(a). p is the pitch of the coil and is the vertical distance traversed as the wire goes around the helical path in one turn. Let d be the diameter of the wire and R the mean radius of the coils.

If α is the angle made by the helical centre line to the horizontal, then, from the triangle [Fig. 12.8(b)] showing the length of the wire and the pitch, $\tan\alpha = p/2\pi R$, where p is the pitch and R is the mean radius of the coil.

If we consider a diametrical plane cutting the coil, the axial load produces a moment $M = PR$, as in the case of a close-coiled spring. However, this moment is at an angle α to the helical central line of the coil. The vertical load P is also inclined to the plane of the section at an angle α. This leads to the following four effects on the section of the coil.

(i) Bending moment, $M_b = M \sin\alpha = PR \sin\alpha$

(ii) Torsional moment, $M_t = M \cos\alpha = PR \cos\alpha$

(iii) Axial tension, $T = P \sin\alpha$

(iv) Transverse shear, $V = P \cos\alpha$

M_b and M_t are components of M acting at an angle α to the helical centre line. The axial tension and transverse shear are components of the vertical load P acting obliquely to the normal section of the wire.

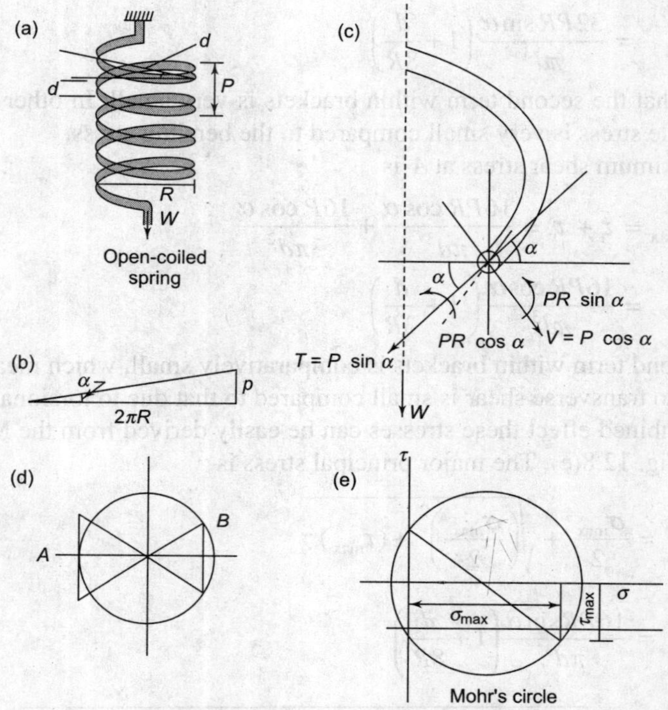

Fig. 12.8

The stresses due to these effects can be determined as follows.

(i) Maximum bending stress,

$$\sigma_b = \frac{32M_b}{\pi d^3} = \frac{32\,PR\sin\alpha}{\pi d^3}$$

which is tensile at A and compressive at B, as shown in Fig. 12.8(d).

(ii) Maximum shear stress,

$$\tau_t = \frac{16m_t}{\pi d^3} = \frac{16PR\cos\alpha}{\pi d^3}$$

(iii) Uniform tensile stress due to axial tension T,

$$\sigma_t = \frac{4P\sin\alpha}{\pi d^2}$$

(iv) The shear stress due to transverse shear V is maximum along AB, having a value of $\tau_v = (4/3) \times$ average shear stress for a circular section

$$\tau_v = \frac{4}{3}\frac{P\cos\alpha}{\pi d^2/4} = \frac{16P\cos\alpha}{\pi d^2}$$

The tensile stresses and shear stresses, acting normal and parallel to the sections, add up at A. Thus, the maximum tensile stress σ at A is given by

$$\sigma_{\max} = \sigma_b + \sigma_t = \frac{32PR\sin\alpha}{\pi d^3} + \frac{4P\sin\alpha}{\pi d^2}$$

$$= \frac{32PR\sin\alpha}{\pi d^3}\left(1+\frac{d}{8R}\right)$$

Notice that the second term within brackets is very small. In other words, the direct tensile stress is very small compared to the bending stress.

The maximum shear stress at A is

$$\tau_{max} = \tau_t + \tau_v = \frac{16PR\cos\alpha}{\pi d^3} + \frac{16P\cos\alpha}{3\pi d^2}$$

$$= \frac{16PR\cos\alpha}{\pi d^3}\left(1+\frac{d}{3R}\right)$$

The second term within brackets is comparatively small, which means that the stress due to transverse shear is small compared to that due to torsional moment.

The combined effect these stresses can be easily derived from the Mohr circle shown in Fig. 12.8(e). The major principal stress is

$$\sigma_1 = \frac{\sigma_{max}}{2} + \sqrt{\left(\frac{\sigma_{max}}{2}\right)^2 + (\tau_{max})^2}$$

$$= \frac{16PR\sin\alpha}{\pi d^3}\left(1+\frac{d}{8R}\right)$$

$$+ \sqrt{\left[\frac{16PR\sin\alpha}{\pi d^3}\left(1+\frac{d}{8R}\right)\right]^2 + \left[\frac{16PR\cos\alpha}{\pi d^3}\left(1+\frac{d}{3R}\right)\right]^2}$$

Neglecting the effects of axial tension and transverse shear,

$$\sigma_1 = \frac{16PR\sin\alpha}{\pi d^3} + \sqrt{\left(\frac{16PR\sin\alpha}{\pi d^3}\right)^2 + \left(\frac{16PR\cos\alpha}{\pi d^3}\right)^2}$$

$$= \frac{16PR}{\pi d^3}\left[\sin\alpha + \sqrt{\sin^2\alpha + \cos^2\alpha}\right]$$

$$= \frac{16PR}{\pi d^3}(\sin\alpha + 1)$$

Similarly, the maximum shear stress is the radius of the Mohr circle. Thus,

$$(\tau_{max})_{max} = \pm\sqrt{\left(\frac{\sigma_{max}}{2}\right)^2 + (\tau_{max})^2}$$

$$= \pm\sqrt{\left[\frac{16PR\sin\alpha}{\pi d^3}\left(1+\frac{d}{8R}\right)\right]^2 + \left[\frac{16PR\cos\alpha}{\pi d^3}\left(1+\frac{d}{3R}\right)\right]^2}$$

Neglecting the effects of transverse shear and axial tension,

$$(\tau_{max})_{max} = \pm\frac{16PR}{\pi d^3}$$

Deformation of the spring Under the action of the axial load, the spring undergoes deformation, which can be measured as a vertical deflection of the spring or as a total angular rotation of the lower point with respect to the fixed upper point. It can also be measured as the sum of the effects due to the four effects we have mentioned earlier, namely, the axial twist M_t, the bending moment M_b, the axial tension T, and the transverse shear V. To simplify, we neglect the effects of axial tension and transverse shear, as these are small.

Due to axial twist M_t The axial twist M_t is equal to $PR\cos\alpha$. We consider an elementary length ds of the spring as shown in Fig. 12.9. Assuming that the upper and lower portions, adjacent to length ds, are rigid, the relative rotation between sections 1 and 2, $d\phi$, from the torsion formula $M/J = G\theta/l$ is

$$\theta = \frac{Ml}{GJ}, \quad d\phi = \frac{M_t ds}{GJ} = \frac{PR\cos\alpha\, ds}{GJ}$$

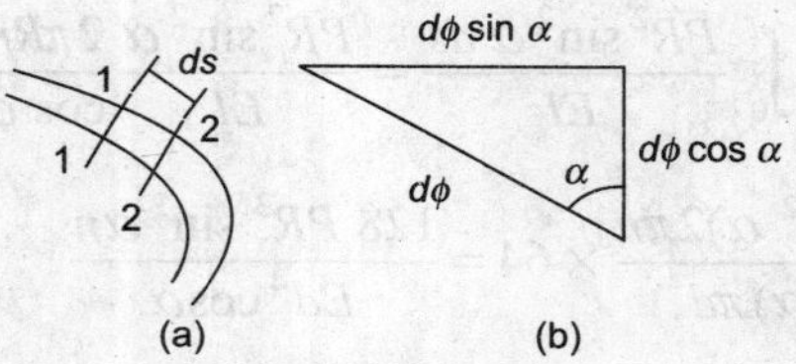

Fig. 12.9

This rotation $d\phi$ is considered as a vector and resolved into components $d\phi \cos\alpha$ and $d\phi \sin\alpha$ as rotations about horizontal and vertical axes. Rotation about the vertical axis does not cause any deflection, and it is the rotation about the horizontal axis that contributes to the vertical deflection of the coil. The vertical deflection due to this rotation component is $Rd\phi\cos\alpha$ of the lower point of the element with respect to the upper point. It is the same for the lowermost point of the spring as the portion below is considered rigid. Therefore,

$$d\Delta_1 = R\, d\theta \cos\theta$$

$$= \frac{PR^2 \cos^2 \alpha\, ds}{GJ}$$

The torque M_t is over the whole length of the spring. The whole length of the spring is given by

$$L = \frac{2\pi Rn}{\cos\alpha}$$

The axial deflection due to twist can be obtained by integrating the expression for $d\Delta$ over the whole length of the spring.

$$\Delta_1 = \int_0^L d\Delta_1 = \int_0^L \frac{PR^2 \cos^2\alpha}{Gs} dS = \frac{PR^2\cos^2\alpha}{Gs}\frac{2\pi Rn}{\cos\alpha}$$

$$= \frac{PR^2(\cos\alpha)\times 2\pi RN}{G\pi d^4} \times 32 = \frac{64PR^3 n\cos\alpha}{Gd^4}$$

Deflection due to BM, M_b Any elemental length ds is subjected to a constant BM, $M_b = PR\sin\alpha$. The relative rotation between the ends of the elementary length ds is given by

$$d\phi = \frac{M\,ds}{EI} = \frac{PR\sin\alpha\,ds}{EI}$$

This is the relative rotation between the ends about the normal to the helical axis. This again can be resolved into horizontal and vertical components [Fig. 12.9(b)] as $d\phi\sin\alpha$ and $d\phi\cos\alpha$. Only the horizontal component $d\phi\sin\alpha$ leads to any vertical deflection. Thus, the vertical deflection $d\Delta_2$ due to a constant BM. M_b acting on an element of length ds is

$$d\delta_2 = R\,d\phi\sin\alpha = \frac{PR^2\sin\alpha\,ds}{EI}$$

Integrating over the whole length and noting that $I = \pi d^4/64$ for a circular section,

$$\Delta_2 = \int_0^l d\Delta_2 = \int_0^l \frac{PR^2\sin^2\alpha\,ds}{EI} = \frac{PR^2\sin^2\alpha}{EI}\frac{2\pi Rn}{\cos\alpha}$$

$$\Delta_2 = \frac{PR^3(\sin^2\alpha)2\pi n}{E(\cos\alpha)\pi d^4}\times 64 = \frac{128\,PR^3\sin^2\alpha n}{Ed^4\cos\alpha}$$

Total deflection,

$$\Delta = \Delta_1 + \Delta_2$$

$$= \frac{64PR^3n\cos\alpha}{Gd^4} + \frac{128PR^3n\sin^2\alpha}{Ed^4\cos\alpha}$$

$$= \frac{64PR^3n}{d^4\cos\alpha}\left(\frac{\cos^2\alpha}{G} + \frac{2\sin^2\alpha}{E}\right)$$

Angular rotation The angular rotation of the loaded end of the spring with respect to the fixed top end can similarly be found. We consider the effect of only the BM and torque.

In Fig. 12.10, we consider the deformation of an elementary length ds of the spring, assuming the remaining parts to be rigid. The relative rotation of one end with respect to the other can be resolved into horizontal and vertical components as before. The component of rotation about the vertical axis results in relative rotation between the ends of the spring.

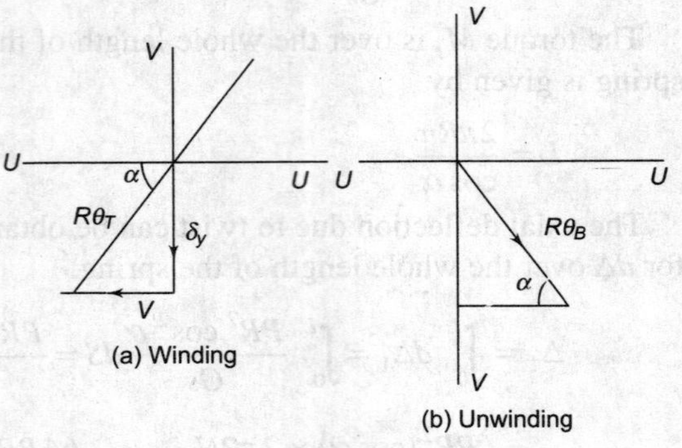

Fig. 12.10

Due to torque M_t If $d\theta_1$ is the relative rotation between the ends of the spring about the vertical axis, then $d\theta_1 = d\phi \sin\alpha$, where

$$d\phi = \frac{PR\cos\alpha\, ds}{GJ}$$

$$d\theta_1 = \frac{PR\cos\alpha\, ds\sin\alpha}{GJ}$$

To obtain the total rotation of the bottom end with respect to the top end, we integrate over the whole length of the spring.

$$\text{Total length} = \frac{2\pi Rn}{\cos\alpha}$$

$$\theta_1 = \int_0^l d\theta_1 = \int_0^l \frac{PR\cos\alpha\sin\alpha\, ds}{GJ}$$

$$= \frac{PR\cos\alpha\sin\alpha}{GJ}\frac{2\pi Rn}{\cos\alpha}$$

Substituting $J = \pi d^4/32$, we get

$$\theta_1 = \frac{64PR^2 n\sin\alpha}{Gd^4}$$

Note that, for a downward axial load, this rotation is clockwise while looking from the top, leading to a winding up of the coils of the spring (Fig. 12.10).

Due to BM M_b From Fig. 12.10(b), the rotation of the elementary length ds due to the BM M_b can be obtained as

$$d\theta_2 = d\phi\cos\alpha = \frac{PR\sin\alpha}{EI} ds\cos\alpha$$

The total rotation can be obtained by integration.

$$\theta_2 = \int_0^l d\theta_2 = \int_0^l \frac{PR\sin\alpha\cos\alpha}{EI} ds = \frac{PR\sin\alpha\cos\alpha\times 2\pi Rn}{EI\cos\alpha}$$

Substituting $I = \pi d^4/64$, where d is the diameter of the wire,

$$\theta_2 = \frac{128PR^2 n\sin\alpha}{Ed^4}$$

This rotation is anticlockwise as seen from the top and is opposite to that of θ_1. The total rotation of the bottom end of the spring with respect to the fixed top end is

$$\theta = \theta_1 + \theta_2 = \frac{64PR^2 n\sin\alpha}{Gd^4} - \frac{128PR^2 n\sin\alpha}{Ed^4}$$

$$= \frac{64PR^2 n}{d^4}\sin\alpha\left(\frac{1}{G} - \frac{2}{E}\right)$$

θ is clockwise when looked at from the top, meaning that the coil is wound up.

Due to this rotation, which is in the same direction in which the coils are wound, the number of coils increases. If n' is the number of coils after rotation, then

$$\theta = 2\pi n' - 2\pi n$$

$$\Rightarrow \qquad n' = \frac{\theta}{2\pi} + n$$

Open-coiled spring under axial torsion Consider the open-coiled helical spring subjected to an axial torsion, as shown in Fig. 12.11(a). The spring has n coils of diameter $2R$ made up of wire of diameter d. It is fixed at the top through a short radial arm and the torsion is applied axially through a short radial arm at the free bottom end. If p is the pitch of the coils, then the inclination of the helical axis may be worked out from

$$\tan\alpha = \frac{p}{2\pi R}$$

Due to the applied torque T, each section of the wire is subjected to a moment T, which is in a horizontal plane passing through the centroid of the section. The moment T acting at any section can be resolved into normal and tangential components as

$$M_b = T\cos\alpha \quad \text{and} \quad M_t = T\sin\alpha$$

where M_b is the bending moment and M_t is the torsional moment. The stresses due to the BM and torque can be determined as

$$\sigma = \frac{M}{I}y = \frac{T\cos\alpha}{\pi d^4/64}\frac{d}{2} = \frac{32T\cos\alpha}{\pi d^3}$$

$$\tau = \frac{M_t}{J}r = \frac{T(\cos\alpha)d}{\pi d^4/32\times 2} = \frac{16T\sin\alpha}{\pi d^3}$$

Due to the normal stress σ and shear stress τ, the principal stress σ_1 and maximum shear stress τ_{max} can be evaluated as

$$\sigma_1 = \frac{\sigma}{2} + \sqrt{\left(\frac{\sigma}{2}\right)^2 + \tau^2}$$

$$= \frac{32T\cos\alpha}{2\pi d^3} + \sqrt{\left(\frac{32T\cos\alpha}{2\pi d^3}\right)^2 + \left(\frac{16T\sin\alpha}{\pi d^3}\right)^2}$$

$$= \frac{16T}{\pi d^3}(1+\cos\alpha)$$

This is the maximum normal stress, tensile at the outer edge A and compressive at the inner edge B of the section shown in Fig. 12.11(b). The maximum shear stress is

$$\tau_{max} = \pm\sqrt{\left(\frac{\sigma}{2}\right)^2 + \tau^2}$$

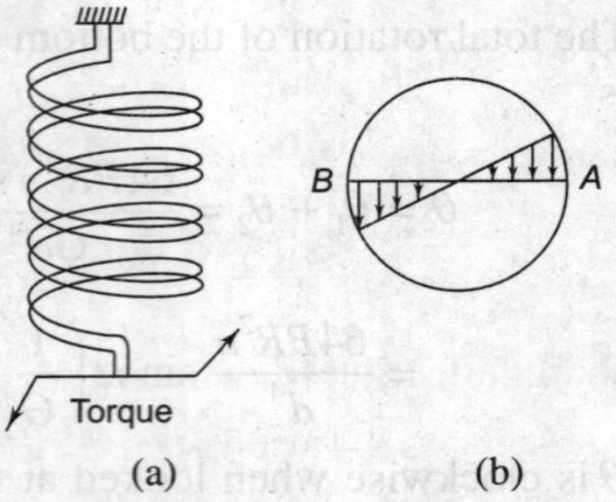

Fig. 12.11

$$= \pm \sqrt{\left(\frac{32T\cos\alpha}{2\pi d^3}\right) + \left(\frac{16T\sin\alpha}{\pi d^3}\right)^2}$$

$$= \pm \frac{16T}{\pi d^3}$$

Axial deflection of the spring The axial deflection is caused by the torque M_t as well as the BM M_b.

Due to torque M_t We consider an elementary length ds of the spring as shown in Fig. 12.11. Due to the torque M_t, the differential rotation can be calculated as

$$d\phi = \frac{M_t\,ds}{GJ} = \frac{T\sin\alpha\,ds}{GJ}$$

As explained in the last section, only the component of this rotation about the horizontal axis causes vertical deflection. The horizontal component is $d\phi\cos\alpha$, and the vertical deflection is

$$d\Delta_1 = R\,d\phi\cos\alpha = \frac{RT\sin\alpha\cos\alpha}{GJ}ds$$

Integrating over the length of the spring,

$$\Delta_1 = \int_0^l d\Delta_1 = \int_0^l \frac{RT\sin\alpha\cos\alpha}{GJ}ds = \frac{RT\sin\alpha\cos\alpha}{GJ}\frac{2\pi Rn}{\cos\alpha}$$

Setting $J = \pi d^4/32$,

$$\Delta_1 = \frac{64R^2Tn\sin\alpha}{Gd^4}$$

Due to BM M_b The relative rotation between the ends of the elementary length ds is given by

$$d\phi = \frac{M_b\,ds}{EI} = \frac{T\cos\alpha\,ds}{EI}$$

Horizontal component of $d\phi = d\phi\sin\alpha$

The vertical deflection due to this horizontal component of the rotation is

$$d\Delta_2 = R\,d\phi\sin\alpha = \frac{RT\cos\alpha\,ds}{EI}\sin\alpha$$

Note that this deflection is downward and opposite to that due to M_t.

$$\Delta_2 = \int_0^l d\Delta_2 \int_0^l \frac{RT\cos\alpha\sin\alpha}{EI}ds = \frac{RT\cos\alpha\sin\alpha}{EI}\frac{2\pi Rn}{\cos\alpha}$$

Setting $I = \pi d^4/64$,

$$\Delta_2 = \frac{128R^2Tn\sin\alpha}{Ed^4}$$

Total deflection (downward) $\Delta = \Delta_1 - \Delta_2$

$$\Delta = \frac{64Rtn\sin\alpha}{Gd^4} - \frac{128Rtn\sin\alpha}{Ed^4}$$

$$= \frac{64R^2 Tn \sin\alpha}{d^4}\left(\frac{1}{G}-\frac{2}{E}\right)$$

Rotation The vertical components of the rotation vectors due to the effects of M_t and M_b cause an angular rotation between the fixed point at the top and the lowest point where the twist is applied. This rotation can be evaluated, as before, by considering an elementary length ds of the spring.

Due to torque M_t The component causing rotation is $d\phi \sin\alpha$, where $d\phi$ is the relative rotation between the ends of the elementary length ds. The rotation between the ends of the spring,

$$\theta_1 = \int_0^l d\phi \sin\alpha = \int_0^l \frac{T \sin\alpha \sin\alpha}{GJ} ds$$

$$= \frac{T \sin^2\alpha}{GJ}\frac{2\pi Rn}{\cos\alpha} = \frac{T \sin^2\alpha \times 2\pi Rn}{G(\pi d^4/32)\cos\alpha}$$

$$= \frac{64TRn \sin^2\alpha}{Gd^4 \cos\alpha}$$

Due to BM M_b The vertical component of rotation is

$$d\theta_2 = d\phi \cos\alpha = \frac{T \cos\alpha \cos\alpha}{EI} ds$$

$$\theta_2 = \int_0^l \frac{T\cos^2\alpha}{EI} ds = \frac{T\cos^2\alpha}{EI}\frac{2\pi Rn}{\cos\alpha}$$

Setting $I = \pi d^4/64$

$$\theta_2 = \frac{T\cos^2\alpha}{E\pi d^4/64}\frac{2\pi Rn}{\cos\alpha} = \frac{128TRn\cos^2\alpha}{Ed^4\cos\alpha}$$

The total rotation,

$$\theta = \theta_1 + \theta_2$$

$$= \frac{64TRn\sin^2\alpha}{Gd^4\cos\alpha} + \frac{128TRn\cos^2\alpha}{Ed^4\cos\alpha}$$

$$= \frac{64TRn}{d^4\cos\alpha}\left(\frac{\sin^2\alpha}{G} + \frac{2\cos^2\alpha}{E}\right)$$

If n' be the new number of turns due to this rotation, then $n' = \theta/2\pi + n$, where n is the number of turns in the unloaded spring.

Strain energy of open-coiled springs We can calculate the strain energy stored in a spring on similar lines. We take into account only the effect of torque and BM.

Due to axial load If the axial load is P, then torque $M_t = PR\cos\alpha$ and $M_b = PR \sin\alpha$. The strain energy stored in the spring, over an elementary length ds, is

$$dU = \frac{M_b^2\, ds}{2EI} + \frac{M_t^2\, ds}{2GJ}$$

We integrate this expression over the whole length of the spring, $2\pi Rn/\cos\alpha$.

$$U = \int_0^l \frac{M_b^2\,ds}{2EI} + \int_0^l \frac{M_t^2\,ds}{2GJ}$$

$$= \frac{P^2R^2\sin^2\alpha}{2EI}\frac{2\pi Rn}{\cos\alpha} + \frac{P^2R^2\cos^2\alpha}{2GJ}\frac{2\pi Rn}{\cos\alpha}$$

$I = \pi d^4/64$ and $J = \pi d^4/32$. Therefore,

$$U = \frac{P^2R^2\sin^2\alpha}{2E\pi d^4/64}\frac{2\pi Rn}{\cos\alpha} + \frac{P^2R^2\cos^2\alpha}{2G\pi d^4/32}\frac{2\pi Rn}{\cos\alpha}$$

$$= \frac{32P^2R^3n}{d^4\cos\alpha}\left(\frac{2\sin^2\alpha}{E} + \frac{\cos^2\alpha}{G}\right)$$

Due to axial torque We have seen that the sections of the spring are subjected to a constant torque, $M_t = T\sin\alpha$, and a BM, $M_b = T\cos\alpha$. Over an elemental length ds, the strain energy is

$$dU = \frac{M_t^2\,ds}{2GJ} + \frac{M_b^2\,ds}{2EI}$$

$$\text{Length of the spring} = \frac{2\pi Rn}{\cos\alpha}$$

$$U = \int_0^l \left[\frac{M_t^2\,ds}{2GJ} + \frac{M_b^2\,ds}{2EI}\right] = \frac{32T^2Rn}{d^4\cos\alpha}\left(\frac{\sin^2\alpha}{G} + \frac{2\cos^2\alpha}{E}\right)$$

When an open-coiled helical spring is subjected to axial load P, if δ is the deflection,

External work done = $P\delta/2$

Equating this to the strain energy worked out above,

$$P\delta/2 = \frac{32P^2R^3n}{d^4\cos\alpha}\left[\frac{2\sin^2\alpha}{E} + \frac{\cos^2\alpha}{G}\right]$$

or $$\delta = \frac{64PR^3n}{d^4\cos\alpha}\left[\frac{2\sin^2\alpha}{E} + \frac{\cos^2\alpha}{G}\right]$$

Relative rotation between the ends of the spring,

$$\theta = \frac{64PR^3n\sin\alpha}{d^4}\left[\frac{1}{G} - \frac{2}{E}\right]$$

as already derived for deformation of springs

When the spring is subjected to axial torque, if θ is the relative rotation between the ends of the spring, then the external work done = $T\theta/2$. Equating this to the strain energy derived above, we get

$$T\theta/2 = \frac{32T^2Rn}{d^4\cos\alpha}\left[\frac{\sin^2\alpha}{G} + \frac{2\cos^2\alpha}{E}\right]$$

or $$\theta = \frac{64TRn}{d^4 \cos\alpha}\left[\frac{\sin^2\alpha}{G} + \frac{2\cos^2\alpha}{E}\right]$$

The axial deflection δ is given by

$$\delta = \frac{64TR^3 n \sin\alpha}{d^4}\left[\frac{1}{G} - \frac{2}{E}\right]$$

as already derived for open-coiled springs under torsion.

Example 12.10 Open-coiled helical springs

An open-coiled helical spring has a mean radius of 150 mm and 20 coils of wire of diameter 12 mm at a pitch of 100 mm. Find the maximum normal and shear stresses in the wire when the spring is subjected to an axial load of 100 N. Also find the vertical deflection and increase in the number of coils. E = 200 GPa and G = 80 GPa.

Solution At a section of the wire, the axial force causes a torque, BM, tensile force, and transverse shear. We calculate the effect due to all of them.

Torque $M_t = PR\cos\alpha = 100 \times 150 \cos\alpha$

α is the inclination of the helical axis and may be worked out as follows:

$$\tan\alpha = \frac{p}{2\pi R}$$

$$\Rightarrow \quad \alpha = \tan^{-1}\left(\frac{100}{2\pi \times 150}\right) = 6.056°$$

$M_t = 100 \times 150 \cos(6.056) = 14{,}916$ Nm

$M_b = PR\sin\alpha = 100 \times 150 \times \sin(6.056) = 1582.7$ Nm

Direct tensile force,

$T = P\sin\alpha = 100 \times \sin(6.056) = 10.55$ N

Transverse shear,

$V = P\cos\alpha = 100 \times \cos(6.056) = 99.44$ N

Bending stress,

$$\sigma_b = \frac{32M_b}{\pi d^3} = \frac{32 \times 1582.7}{\pi \times (12)^3} = 9.33 \text{ N/mm}^2$$

Shear stress (due to M_t),

$$\tau_t = \frac{16M_t}{\pi d^3} = \frac{16 \times 14{,}916}{\pi \times (12)^3} = 43.96 \text{ N/mm}^2$$

Direct tensile stress due to T,

$$\sigma_t = \frac{T}{\text{area}} = \frac{10.55}{(\pi/4) \times (12)^2} = 0.09 \text{ N/mm}^2$$

Shear stress due to V,

$$\tau_s = \frac{V}{\pi d^2/4} \times \frac{4}{3} = \frac{99.44}{\pi \times (12)^2/4} \times \frac{4}{3} = 1.17 \text{ N/mm}^2$$

We can easily see that the stresses due to the tensile force T and shear V are negligible. The total normal stress,

$$\sigma = \sigma_b + \sigma_t = 9.33 + 0.09 = 9.42 \text{ N/mm}^2$$

The total shear stress,

$$\tau = \tau_t + \tau_s = 43.96 + 1.17 = 45.13 \text{ N/mm}^2$$

$$\text{Principal stress } \sigma_1 = \frac{\sigma}{2} + \sqrt{\left(\frac{\sigma}{2}\right)^2 + \tau^2}$$

$$= \frac{9.42}{2} + \sqrt{\left(\frac{9.42}{2}\right)^2 + (45.13)^2}$$

$$= 4.71 + 45.38 = 50.1 \text{ N/mm}^2$$

The maximum shear stress,

$$\tau_{max} = 45.38 \text{ N/mm}^2$$

In calculating the vertical deflection and rotation, we neglect the effects of the tensile force and transverse shear. Thus, we assume that the vertical deflection δ is caused only by the torque and BM.

$$\delta = \frac{64PR^3n}{d^4\cos\alpha}\left(\frac{\cos^2\alpha}{G} + \frac{2\sin^2\alpha}{E}\right)$$

$$= \frac{64 \times 100 \times (150)^3 \times 20}{(12)^4 \times \cos(6.056)}\left[\frac{(\cos 6.056)^2}{80{,}000} + \frac{2(\sin 6.056)^2}{200{,}000}\right]$$

$$= 261 \text{ mm}$$

The rotation between the ends of the spring, θ, due only to the effect of torque and BM is given by

$$\theta = \frac{64PR^2n}{d^4}\left(\frac{1}{G} - \frac{2}{E}\right)\sin\alpha$$

$$= \frac{64 \times 100 \times (150)^2 \times 20}{(12)^4}\left(\frac{1}{80{,}000} - \frac{2}{200{,}000}\right)\sin(6.056)$$

$$= 0.3297 \text{ rad}$$

$$\text{Increase in number of turns of coils} = \frac{\theta}{2\pi} = \frac{0.3297}{2\pi}$$

$$= 0.05$$

Example 12.11 Design of open-coiled helical spring

An open-coiled helical spring is made of a wire of diameter 10 mm, and carries an axial load of 150 N. If the permissible normal and shear stresses are 100 MPa and 70 MPa, find the mean radius of the coil, inclination α of the helical axis, and the pitch. If G = 80 GPa and E = 200 GPa and an axial stiffness of 4 N/mm is required, find the number of coils required.

Solution Taking only the effect of torque M_t and BM M_b into account

$$\tau_{max} = \frac{16PR}{\pi d^3}, \quad \frac{16 \times 150 \times R}{\pi (10)^3} = 70, \quad R = 91.6 \text{ mm}$$

$$\sigma_{max} = \frac{16PR}{\pi d^3}(1 + \sin\alpha), \quad 100 = 70(1 + \sin\alpha)$$

$$\sin\alpha = \frac{30}{70} = 0.4285, \quad \alpha = \frac{25.37°}{91.6}$$

Pitch $= 2\pi R$

$\tan\alpha = 2\pi \times 91.6 \times \tan(25.3)$

$= 268.5$ mm

Axial stiffness $= \dfrac{P}{\delta} = 4$

$\Rightarrow \quad P = 4\delta$ N/mm

$$\delta = \frac{64PR^3n}{d^4\cos\alpha}\left(\frac{\cos^2\alpha}{G} + \frac{2\sin^2\alpha}{E}\right)$$

$$\frac{P}{4} = \frac{64PR^3n}{d^4\cos\alpha}\left(\frac{\cos^2\alpha}{G} + \frac{2\sin^2\alpha}{E}\right)$$

$\cos\alpha = 0.9036, \quad \sin\alpha = 0.4285$

$$n = \frac{d^4\cos\alpha}{4 \times 64 \times (91.6)^3} \times \frac{1}{(\cos^2\alpha/G + 2\sin^2\alpha/E)}$$

$= 3.8$ turns

Example 12.12 Open-coiled helical springs: deformations

An open-coiled helical spring has 20 coils of wire of diameter 10 mm at a pitch of 80 mm, the coils having a mean radius of 120 mm. If the spring is subjected to an axial twist of 5 N m, find the maximum normal and shear stress in the section of the wire. If $E = 200$ GPa and $G = 80$ GPa, determine the axial extension of the spring and the relative rotation between the ends. Find the strain energy stored in the spring.

Solution Inclination of helical axis,

$$\alpha = \tan^{-1}\left(\frac{p}{2\pi R}\right)$$

$$= \tan^{-1}\left(\frac{80}{2\pi \times 120}\right) = 6.06°$$

$\cos\alpha = 0.9944, \quad \sin\alpha = 0.1055$

Twisting moment,

$$M_t = T\sin\alpha = 5000 \times \sin 6.06 = 527.5 \text{ N mm}$$

Bending moment,

$$M_b = T\cos\alpha = 5000 \times \cos 6.06 = 4972 \text{ N mm}$$

Shear stress due to M_t,

$$\tau = \frac{16M_t}{\pi d^3} = \frac{16 \times 527.5}{\pi(10)^3} = 2.68 \text{ N/mm}^2$$

Normal stress due to M_b,

$$\sigma = \frac{32M_b}{\pi d^3} = \frac{32 \times 4972}{\pi(10)^3} = 50.64 \text{ N/mm}^2$$

Major principal stress,

$$\sigma_1 = \frac{\sigma}{2} + \sqrt{\left(\frac{\sigma}{2}\right)^2 + \tau^2}$$

$$= \frac{50.64}{2} + \sqrt{\left(\frac{50.64}{2}\right)^2 + (2.68)^2}$$

$= 25.3 + 25.44 = 50.74$ N/mm^2

Maximum shear stress,

$$\tau_{max} = \sqrt{\left(\frac{\sigma}{2}\right)^2 + \tau^2}$$

$$= 25.44 \text{ N/mm}^2$$

$$\text{Axial deflection, } \delta = \frac{64TR^2 n \sin\alpha}{d^4}\left(\frac{1}{G} - \frac{2}{E}\right)$$

$$= \frac{64 \times 5000 \times (120)^2 \times 20 \times \sin(6.06)}{(10)^4}\left(\frac{1}{80{,}000} - \frac{2}{200{,}000}\right)$$

$$= 2.189 \text{ mm}$$

$$\text{Relative rotation, } \theta = \frac{64TRn}{d^4 - \cos\alpha}\left(\frac{\sin^2\alpha}{G} + \frac{2\cos^2\alpha}{E}\right)$$

$$= \frac{64 \times 5000 \times 120 \times 20}{(10)^4 \times \cos(6.06)}\left(\frac{(\sin 6.06)^2}{80{,}000} + \frac{2(\cos 6.06)^2}{200{,}000}\right)$$

$$= 0.7744 \text{ rad}$$

$$\text{Strain energy, } U = \frac{32T^2 Rn}{d^4 \cos\alpha}\left(\frac{\sin^2\alpha}{G} + \frac{2\cos^2\alpha}{E}\right)$$

$$U = \frac{1}{2}T\theta = \frac{1}{2} \times 5000 \times 0.7744 = 1936 \text{ N mm}$$

12.3.4 Spring in Series and Parallel Configurations

Two springs of the same diameter and number of turns or of different diameters and number of turns can be combined to form a single spring.

Springs in series When two springs are connected in series [Fig. 12.12(a)], it is clear that the load on each spring is the same as the applied load. The deflection of each spring will, however, be different. In Fig. 12.12,

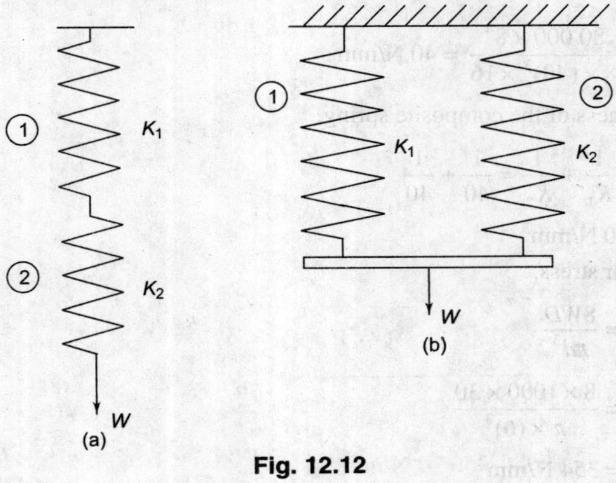

Fig. 12.12

Total deflection of the spring $\delta = \delta_1 + \delta_2$,
where δ_1 and δ_2 are the deflections of springs 1 and 2. Since $\delta = W/K$,

$$\frac{W}{K} = \frac{W}{K_1} + \frac{W}{K_2}$$

where K is the stiffness of the composite spring. Therefore,

$$\frac{1}{K} = \frac{1}{K_1} + \frac{1}{K_2}$$

Springs in parallel When the springs are laid parallel [Fig. 12.12(b)], it is assumed that the deflection of each spring is the same. The springs will share the load W in some proportion, depending upon their properties. If W_1 and W_2 are the loads taken by springs 1 and 2, then,

$$W = W_1 + W_2$$

$$\delta K = \delta K_1 + \delta K_2$$

or $$K = K_1 + K_2$$

where K is the stiffness of the composite spring.

The following examples illustrate the use of these formulae.

Example 12.13 Springs in series

Two springs connected in series carry a load of 2 kN. One spring has 12 coils of 6 mm wire wound in a diameter of 30 mm, and the second spring has 16 coils of wire of diameter 8 mm wound in a diameter of 40 mm. Find the spring constant for the composite system, and the maximum stress produced in the wire. $G = 80$ GPa.

Solution The stiffness of the spring is given by

$$K = \frac{Gd^4}{8D^3 n}$$

For the first spring, $d = 6$ mm and $D = 30$ mm. So, $n = 12$. Therefore,

$$K_1 = \frac{80{,}000 \times (6)^4}{8 \times (30)^3 \times 12} = 40 \text{ N/mm}$$

For the second spring, $d = 8$ mm and $D = 40$ mm. So, $n = 16$. Therefore,

$$K_2 = \frac{80{,}000 \times 8^4}{8 \times (40)^3 \times 16} = 40 \text{ N/mm}$$

If K is the stiffness of the composite spring,

$$\frac{1}{K} = \frac{1}{K_1} + \frac{1}{K_2} = \frac{1}{40} + \frac{1}{40}$$

$$K = 20 \text{ N/mm}$$

Maximum shear stress,

$$\tau_{\max} = \frac{8WD}{\pi d^3}$$

$$= \frac{8 \times 1000 \times 30}{\pi \times (6)^3}$$

$$= 354 \text{ N/mm}^2$$

for spring 1, and

$$\tau_{max} = \frac{8 \times 1000 \times 40}{\pi \times (8)^3}$$
$$= 199 \text{ N/mm}^2$$

for spring 2.

Example 12.14 Springs in parallel

Two springs are connected in parallel. One has 18 coils of 8 mm ϕ wire with an outside diameter of 36 mm and the second has 20 coils of 4 mm ϕ wire with an outside diameter of 40 mm. Find the maximum load that the system can carry without exceeding the shear stress of 360 MPa. $G = 80$ GPa.

Solution The stiffness of each spring can be calculated as follows:

$$\text{For spring 1, } K_1 = \frac{80{,}000 \times (8)^4}{8 \times (36)^3 \times 18} = 48.77 \text{ N/mm}$$

$$\text{For spring 2, } K_2 = \frac{80{,}000 \times (6)^4}{8 \times (40)^3 \times 20} = 10.12 \text{ N/mm}$$

Let W be the load carried by the spring without exceeding the shear stress of 360 MPa. The springs will share the load so that their deflections are the same.

$$W = W_1 + W_2$$

$$W_1 = \frac{360 \times \pi \times (8)^3}{8 \times 36} = 2010 \text{ N}$$

$$W_2 = \frac{360 \times \pi \times (6)^3}{8 \times 40} = 763 \text{ N}$$

Total load = 2773 N

$$\text{Deflection of the spring} = \frac{8 \times 2010 \times (40)^3 \times 18}{80{,}000 \times (8)^4} = 41.2 \text{ mm}$$

or
$$\delta = \frac{8 \times 763 \times (40)^3 \times 20}{80{,}000 \times (6)^4} = 75 \text{ mm}$$

The two coils will not carry maximum stress. The deflection criteria will be applied.

$$\frac{8W_1 \times (36)^3 \times 18}{80{,}000 \times (8)^4} = \frac{8W_2 \times (40)^3 \times 20}{80{,}000 \times (6)^4}$$

$$W_1 = \frac{W_2 \times (40)^3 \times 20 \times (8)^4}{(36)^3 \times (18) \times (6)^4} = 4.817 W_2$$

If $W_2 = 763$ N, $W_1 = 4.817 \times 763 = 3675$ N, which is more than 2010. Therefore, $W_1 = 2010$ and $W_2 = 417.3$ N.

$$\text{Stress in coil 2} = \frac{417.3 \times 8 \times 40}{\pi \times (6)^3} = 196.7 \text{ N/mm}^2$$

12.3.5 Flat Spiral Springs

Flat spiral springs find application in toys and clock mechanisms. Such a spring is shown in Fig. 12.13. It is in one plane and wound by applying a torque at O. It is hinged at the outer end A. The section of the spring is rectangular, of thickness t and depth d.

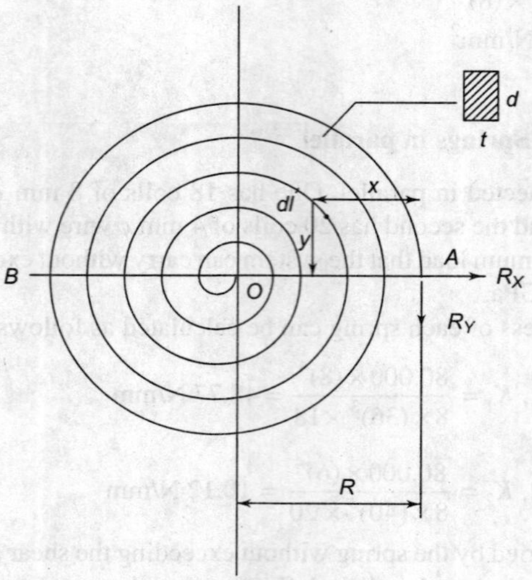

Fig. 12.13

The applied torque T will produce reactions at the pin A. Let R_X and R_Y be the horizontal and vertical reactions at A. Consider a small length dl of the spring with coordinates (x, y) as shown. The BM on element dl is

$$M = R_Y x - R_X y$$

This moment will tend to increase the number of turns of the spring. The rotation caused in the elemental length dl, relative to its ends, is

$$d\phi = \frac{M}{EI} dl = \frac{(R_Y x - R_X y)}{EI} dl$$

This can be integrated over the length to get the total rotation ϕ:

$$\phi = \int_0^l \frac{(R_Y x - R_X y)}{EI} dl$$

$\int y \ dl = 0$ if the centroid of the spring is assumed to be at the centre O due to symmetry about the horizontal axis. $\int x \ dl = Rl$, where R is the horizontal distance between O and A.

$$\phi = \frac{R_Y}{EI} Rl$$

Also,

$$T = R_Y R$$

$$\phi = \frac{Tl}{EI}$$

The maximum BM will be at B, at the other end of the horizontal axis at a distance $2R$ from O:

$$M_{\max} = R_Y \times 2R = \frac{T}{R} \times 2R = 2T$$

The maximum bending stress is given by

$$\sigma_{\max} = \frac{M}{T}y = \frac{2T \times 12}{td^3}\frac{d}{2} = \frac{12T}{td^2}$$

and $$T = \frac{\sigma_{\max} td^2}{12}$$

The strain energy stored in the spring is given by

$$U = \frac{1}{2}T\phi = \frac{1}{2}T\frac{Tl}{EI} = \frac{1}{2}\frac{T^2 A}{EI} = \frac{1}{2}\frac{T^2 l \times 12}{E + d^3}$$

$$= \frac{6T^2 l}{E\,t\,d^3}$$

Example 12.15 Flat spiral springs

A flat spiral spring 4 m long is made of a steel strip 12 mm wide and 0.5 mm thick. If E = 200 GPa, find the maximum torque that can be applied without exceeding the stress of 500 N/mm^2. Determine the number of turns given to the spindle and the energy stored.

Solution The torque is given by

$$T = \frac{\sigma_{\max} bt^2}{12} = \frac{500 \times 12 \times (0.5)^2}{12}$$

$$= 125 \text{ N mm}$$

The number of turns is given by

$$n = \frac{\phi}{2\pi}$$

$$\phi = \frac{Tl}{EI} = \frac{125 \times 4000 \times 12}{200{,}000 \times 12 \times (0.5)^3} = 20 \text{ rad}$$

$$n = \frac{20}{2\pi} = 3.18$$

Energy stored in the spring,

$$U = \frac{1}{2}T\phi$$

$$= \frac{1}{2} \times 125 \times 20 = 1.25 \text{ N m}$$

12.4 THIN-WALLED PRESSURE VESSELS

Vessels containing liquid or gas under pressure are common. They may have many shapes—cylindrical, spherical, conical, and so on. We will confine ourselves to cylindrical and spherical pressure vessels. Both these types have surfaces which can be generated as surfaces of revolution. A cylindrical surface is obtained by a line rotating about a parallel axis, a spherical surface by a semicircular arc about its diameter, and a conical surface by an inclind line rotating about a vertical axis.

If the thickness of such a vessel is less than 1/20 of the diameter, we can call it a thin-walled vessel. One important aspect in the analysis of a thin-walled vessel is that the stress distribution over the thickness can be considered to be uniform.

12.4.1 Thin-Walled Cylindrical Pressure Vessels

Consider the thin-walled pressure vessel in the form of a cylinder of diameter d, thickness t and length L shown in Fig. 12.14. It contains a fluid under pressure p. We consider the free body of a half-cylinder obtained by a longitudinal diametrical section as shown in Fig. 12.14(b). The bursting force due to the internal pressure tending to burst the cylinder along the diametrical section can be found as follows.

Consider an element at an angle θ from the horizontal diameter, subtending an angle $d\theta$ as shown. The pressure p_i is uniform along the length. The force will act normal to the elemental surface, that is in a radial direction. This force is given by

$$dF = p_i L \frac{D}{2} d\theta$$

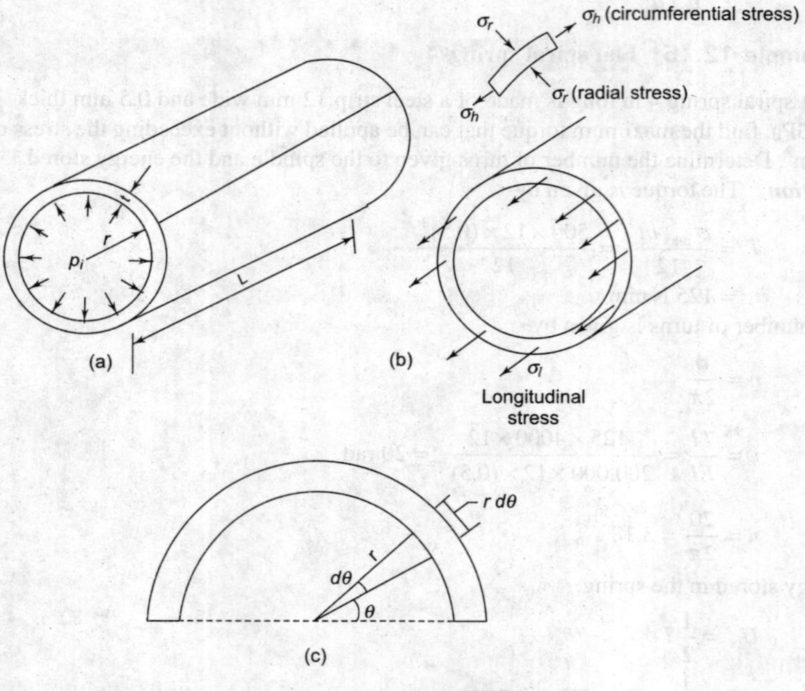

Fig. 12.14

Such a force can be resolved into horizontal and vertical components. Due to the symmetry of the surface about a vertical radial line, the horizontal forces tend to cancel, and the only remaining force will be a vertical force. The vertical component of dF is equal to

$$dF_v = p_i L \frac{D}{2} d\theta \sin\theta$$

The total vertical force is obtained by integration as θ changes from zero to π.

$$F_v = \int_0^{\pi} p_i L \sin\theta\, d\theta = p_i L \frac{D}{2} (-\cos\theta)_0^{\pi}$$

$$= p_i DL$$

which is the same as the force on the projected area LD. This force $p_i DL$ has to be resisted by the stress resultants along the two edges, each having an area Lt. If σ_h is the stress developed in the material, then

$$2\sigma_h Lt = p_i DL$$

$$\sigma_h = \frac{p_i D}{2t}$$

σ_h is the hoop stress or circumferential stress. If the cylinder has flat ends, then taking a section perpendicular to the longitudinal axis, the longitudinal force is given by, [Fig. 12.14(b)]

$$F = p_i \frac{\pi d^2}{4}$$

The resisting force is developed all along the circumference of area πDt. If σ_l is the stress developed in the longitudinal direction,

$$\pi D t \sigma_l = \frac{p_i \pi D^2}{4}$$

$$\Rightarrow \qquad \sigma_l = \frac{p_i D}{4t}$$

Thus, the longitudinal stress is half the hoop stress.

Joints in thin cylindrical vessels Joints may have to be provided in a vessel, in the longitudinal as well as circumferential directions. The longitudinal joint obviously carries the larger stress and has to be stronger. In general, if η is the efficiency of the joints, then

$$\eta = \frac{\text{strength of the joint}}{\text{strength of the plate}}$$

$$\sigma_h = \frac{pD}{2t\eta}$$

and
$$\sigma_l = \frac{pD}{4t\eta}$$

Volumetric change The cylindrical pressure vessel has stresses in two perpendicular directions, σ_h and σ_l, which are the major and minor principal stresses, respectively. If ν be Poisson's ratio, then hoop strain,

$$\varepsilon_h = \frac{\sigma_h}{E} - \nu \frac{\sigma_l}{E} = \frac{p_i D}{2tE} - \nu \frac{p_i D}{4tE}$$

$$= \frac{p_i D}{4tE}(2 - \nu)$$

which is the same as the strain along any diameter. The longitudinal strain,

$$\varepsilon_R = \frac{\sigma_l}{E} - \nu \frac{\sigma_h}{E} = \frac{p_i D}{2tE} - \nu \frac{p_i D}{4tE}$$

$$= \frac{p_i D}{4tE}(1 - 2\nu)$$

For a cylinder of length L and diameter D, the volume is

$$V = \frac{\pi}{4} D^2 L$$

If the increased diameter and length are D' and L', respectively, then

New diameter $D' = D(1 + \varepsilon_h)$

New length $L' = L(1 + \varepsilon_l)$

New volume $V' = \frac{\pi}{4}(D')^2(L')$

Volumetric strain,

$$\frac{\delta V}{V} = \frac{V' - V}{V}$$

The change in volume also can be calculated as

$$V = \frac{\pi}{4} D^2 L$$

$$\delta V = \frac{\pi}{4}(L \times 2D\,\delta D + D^2 \delta L)$$

Volumetric strain,

$$\frac{\delta V}{V} = \frac{(\pi/4)(L \times 2D\,\delta D + D^2 \delta L)}{(\pi/4) D^2 L}$$

$$= 2\frac{\delta D}{D} + \frac{\delta L}{L}$$

$$= 2\varepsilon_h + \varepsilon_L$$

Thus, the volumetric strain is the sum of twice the hoop strain and the longitudinal strain.

$$\frac{\delta V}{V} = \varepsilon_v = 2\frac{p_i D}{4tE}(2 - \nu) + \frac{p_i D}{4tE}(1 - 2\nu)$$

$$= \frac{p_i D}{4tE}(5 - 4\nu)$$

Wire-wound thin-walled cylindrical vessels Thin-walled cylindrical vessels subjected to internal pressure develop hoop tensile stress. In order to increase its capacity to carry internal pressure, the cylinder can be wound with wire under tension, which induces an initial compressive stress in the walls. This is equivalent to applying an external radial pressure on the cylinder. When the cylinder is now subjected to an internal pressure, the compressive stress initially introduced has to become zero before tensile stress is developed. This increases the capacity of the vessel to resist internal pressure.

Consider a cylinder wound with wire of diameter d (Fig. 12.15). The cylinder has a diameter D and thickness t. Let the number of turns per unit length of the cylinder be n. The compressive force on the cylinder wall is equal to the tensile force on the wires. Area of wire,

$$A_w = \frac{\pi d^2}{4}$$

Area of wires per unit length = $2nA_w$

If σ_w is the tension in the wire and σ_{c_1} is the compressive stress in the cylinder wall, then

$$2nA_w\sigma_{w_1} = 2t\sigma_{c_1}$$

$$\sigma_{c_1} = \frac{nA_w\sigma_{w_1}}{t}$$

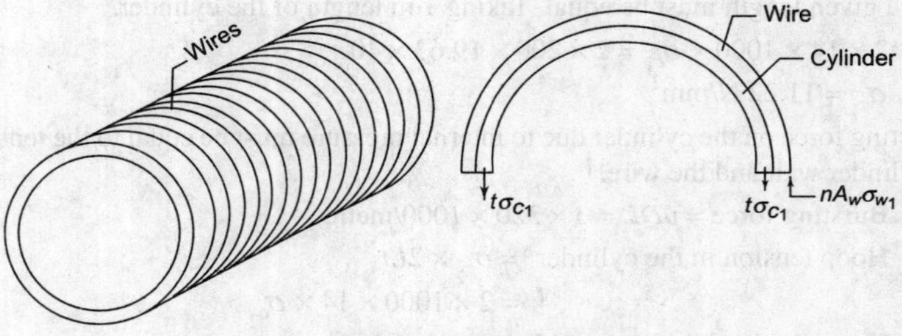

Fig. 12.15

If the cylinder is now subjected to an internal pressure p_i, the cylinder as well as the wire are subjected to tensile stresses. Let σ_{w_2} and σ_{c_2} be the stresses due to pressure p_i. Then, for equilibrium,

$$p_iD = 2\sigma_{c_2}^2 t + 2nA_w\sigma_{w_2}$$

Net stress in cylinder wall = $\sigma_{c_2} - \sigma_{c_1}$

Net stress in the wire = $\sigma_{w_1} + \sigma_{w_2}$

In order to determine the stresses σ_{c_2} and σ_{w_2}, we require the condition of deformation. Longitudinal stress in the cylinder,

$$\sigma_L = \frac{p_iD}{4t}$$

Hoop strain in the cylinder,

$$\varepsilon_c = \frac{\sigma_{c_2}}{E_c} - \frac{\nu\sigma_L}{E_c}$$

Hoop strain in the wire,

$$\varepsilon_w = \frac{\sigma_{w_2}}{E_w}$$

Assuming the cylinder and wire are in contact, these two strains must be equal.

$$\varepsilon_c = \varepsilon_w$$

$$\frac{1}{E_c}(\sigma_{c_2} - \nu\sigma_L) = \frac{\sigma_{w_2}}{E_w}$$

From the two equations relating σ_{c_2} and σ_{w_2}, these two stresses can be calculated.

Example 12.16 Stress in wire-wound thin cylinder

A steel cylinder has a diameter of 320 mm and a thickness of 14 mm. It is initially wound with steel wires of diameter 5 mm and tensioned to a stress of 40 MPa. It is then subjected to an internal pressure of 3 MPa. Determine the stresses in the cylinder wall and wire when it is subjected to this pressure. $n = 0.28$ and $E = 200$ GPa for the cylinder and wire.

Solution Area of wire,

$$A_w = \frac{\pi \times 5^2}{4} = \frac{25}{4}\pi = 19.63 \text{ mm}^2$$

There are 1000/5 = 200 turns/meter length. If σ_{c_1} is the initial compressive stress in the cylinder, then, for equilibrium, the compressive force in the cylinder and tensile force in wire for a given length must be equal. Taking 1 m length of the cylinder,

$$2 \times 14 \times 1000 \times \sigma_{c_1} = 2 \times 200 \times 19.63 \times 40$$

$$\sigma_{c_1} = 11.22 \text{ N/mm}^2$$

The bursting force on the cylinder due to internal pressure must be equal to the tensile force in the cylinder wall and the wire.

Bursting force = $pDL = 3 \times 320 \times 1000$/metre

Hoop tension in the cylinder $= \sigma_{c_2} \times 2Lt$

$= 2 \times 1000 \times 14 \times \sigma_{c_2}$

Hoop tension in the wire $= 2n\,\sigma_{w_2} A_{w_2}$

$= 2 \times 200 \times 19.63 \times \sigma_{w_2}$

where σ_{c_2} and σ_{w_2} are the tensile stresses in the cylinder walls and the wire. Therefore,

$$28{,}000\sigma_{c_2} + 7852\sigma_{w_2} = 960{,}000$$

Longitudinal stress,

$$\sigma_L = \frac{pD}{4t} = \frac{3 \times 320}{4 \times 14} = 17.14 \text{ N/mm}^2$$

$$\text{Hoop strain in the cylinder} = \frac{\sigma_{c_2}}{E} - \nu\frac{\sigma_L}{E}$$

$$= \frac{1}{E}(\sigma_{c_2} - 0.3 \times 17.14)$$

$$\text{Hoop strain in the wire} = \frac{\sigma_{w_2}}{E}$$

These two strains must be equal:

$$\frac{1}{E}(\sigma_{c_2} - 5.14) = \frac{\sigma_{w_2}}{E}, \quad \sigma_{c_2} = \sigma_{w_2} + 5.14$$

Substituting this value of σ_{c_2} in the first equation,

$$28{,}000(\sigma_{w_2} + 5.14) + 7852\sigma_{w_2} = 960{,}000$$

$$\sigma_{w_2} = 30.8 \text{ N/mm}^2$$

$$\sigma_{c_2} = 35.94 \text{ N/mm}^2$$

Net stress in cylinder wall = 35.94 – 11.22

= 24.72 N/mm^2

Net stress in the wire = 30.8 + 40 = 70.8 N/mm^2

Both these stresses are tensile. ■

Example 12.17 Thin cylinder

A cylindrical pressure vessel, of diameter 1 m and length 2 m, is subjected to an internal pressure of 2 MPa. If the hoop stress is limited to 42 MPa and the longitudinal stress to 28 MPa, find the minimum thickness required. What will be the change in volume of the cylinder under this pressure? $E = 200$ Pa and $\nu = 0.3$.

Solution The hoop stress is given by

$$\sigma_h = \frac{pD}{4t}$$

Therefore,

$$\frac{2 \times 1000}{4t} = 42, \quad t = 23.81 \text{ mm}$$

$$\sigma_l = \frac{pD}{4t}, \quad \frac{2 \times 1000}{4t} = 28$$

$$\Rightarrow \quad t = 17.86 \text{ mm}$$

Minimum thickness required = 23.81 mm

Volumetric strain,

$$\varepsilon_V = 2\varepsilon_h + \varepsilon_l$$

$$= \frac{pD}{4tE}(5 - 4\nu)$$

$$\text{Volume of cylinder} = \pi\frac{D^2}{4}L = \frac{\pi \times (1000)^2}{4} \times 2000$$

$$V = 15.708 \times 10^8 \text{ mm}^3$$

$$\varepsilon_V = \frac{2 \times 1000 \; (5 - 4 \times 0.3)}{4 \times 23.81 \times 2 \times 10^5} = 0.39899 \times 10^{-3}$$

$$\delta_V = \varepsilon_V V = 0.39899 \times 10^{-3} \times 15.708 \times 10^8 = 6.267 \times 10^5 \text{ mm}^3$$

Example 12.18 Thin cylindrical vessel

A cylinder has an internal diameter of 1.2 m and a length of 2.5 m. The internal pressure in the cylinder is 1.5 MPa. The longitudinal joint in the cylinder has an efficiency of 80% and the circumferential joint one of 50%. Find the minimum thickness required if the stresses are not to exceed 48 MPa in the circumferential direction and 32 MPa in the longitudinal direction.

Solution Longitudinal stress,

$$\sigma_L = \frac{pD}{4tM}$$

$$32 = \frac{1.5 \times 1200}{4 \times t \times 0.5}$$

$$\Rightarrow \quad t = 28.12 \text{ mm}$$

Circumferential stress,

$$\sigma_h = \frac{pD}{2t\eta}$$

$$48 = \frac{1.5 \times 1200}{2 \times t \times 0.8}$$

$$\Rightarrow \quad t = 23.43$$

The minimum thickness required is 28.12 mm.

Example 12.19 Change in volume of thin cylindrical vessel

A thin cylindrical pressure vessel has an internal diameter of 150 mm and a wall thickness of 5 mm. It is subjected to an internal pressure of 7 N/mm^2. If the cylinder is 900 mm long and E = 200 GPa, find the value of Poisson's ratio for the material if the change in volume under this pressure is 15,500 mm^3.

Solution Change in Volume = $\delta V \times V$, where δV is the volumetric strain and V = volume of the cylinder.

$$\delta V = \frac{pD}{4tE}(5-4\nu)$$

and $$V = \frac{\pi D^2}{4} l$$

Therefore,

$$\text{Change in volume} = \frac{pD}{4tE}(5-4\nu)\frac{\pi D^2}{4} l$$

$$= \frac{\pi p D^3 l}{16tE}(5-4\nu)$$

$$\frac{\pi p D^3 l}{16tE}(5-4\nu) = 15{,}500$$

$$900 \times \frac{\pi \times 7 \times (150)^3}{16 \times 5 \times 200{,}000}(5-4\nu) = 15{,}500$$

$$\nu = 0.322$$

12.4.2 Thin-Walled Spherical Vessels

In the case of the spherical vessel shown in Fig. 12.16, due to symmetry along all diametrical planes, the stress developed is the same in two mutually perpendicular directions. The bursting force across any diametrical plane is

$$F = p_i \frac{\pi D^2}{4}$$

This is resisted by the stress s acting along the circumference of thickness t.

$$\sigma \pi D t = p_i \frac{\pi D^2}{4}$$

and $$\sigma = \frac{p_i D}{4t}$$

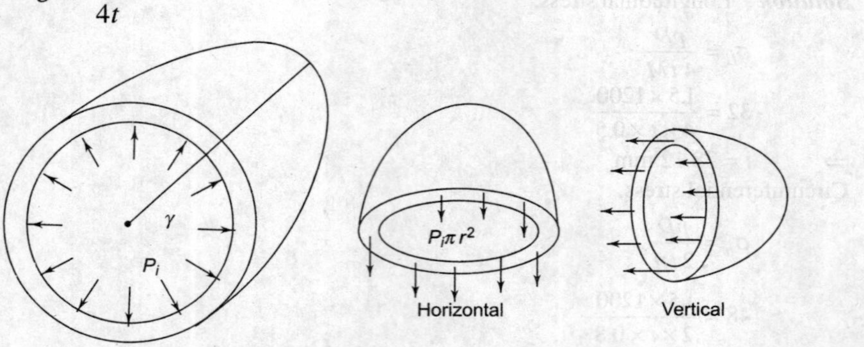

(a) Thin spherical vessel (b) Section

Fig. 12.16

Note that the maximum stress in a spherical vessel is half that in a cylindrical vessel. The stresses in two mutually perpendicular directions are the same, and so are the strains.

$$\varepsilon = \frac{\sigma}{E}(1-\nu) = \frac{p_i D}{4tE}(1-\nu)$$

Volumetric strain Volume of the spherical vessel,

$$V = \frac{4}{3}\pi r^3 = \frac{\pi D^3}{6}$$

New diameter $D' = D(1+\varepsilon)$

New volume $V' = \dfrac{\pi (D')^3}{6}$

Volumetric strain $\varepsilon_V = \dfrac{V'-V}{V}$

Alternately,

$$V = \frac{\pi D^3}{6}, \quad \delta V = \frac{\pi \times 3D^2 \delta D}{6}$$

$$\frac{\delta V}{V} = \frac{\pi \times 3D^2 \delta D}{6\pi D^3/6} = 3\frac{\delta D}{D} = 3\varepsilon, \quad \varepsilon_V = 3\varepsilon$$

Thus, the volumetric strain is three times the diametral strain.

Example 12.20 Thin spherical shell

A thin spherical shell has a diameter of 1.2 m, and is subjected to an internal pressure of 2.5 MPa. Determine the minimum thickness required if the stress is not exceed 40 MPa. Find the increase in the diameter of the sphere, and the change in volume. E = 200 GPa and ν = 0.3.

Solution Due to symmetry about its centre, the stress in the sphere in all directions is the same. The stress is equal to

$$\sigma_h = \frac{pd}{4t} = \frac{2.5 \times 1200}{4 \times t}$$

$$\sigma_h \le 40 \text{ N/mm}^2$$

Therefore,

$$t = \frac{2.5 \times 1200}{4 \times 40} = 18.75 \text{ mm}$$

Change in diameter = diameter × circumferential strain

$$\text{Circumferential strain} = \frac{\sigma_h}{E} - \frac{\sigma_h \nu}{E} = \frac{\sigma_h}{E}(1-\nu)$$

$$= \frac{40}{200{,}000}(1-0.3) = 1.4 \times 10^{-4}$$

Change in diameter = $1200 \times 1.4 \times 10^{-4} = 0.168$ mm

Change in volume = $V \times$ volumetric strain

$$\text{Volumetric strain} = \frac{\text{final volume} - \text{original volume}}{\text{original volume}}$$

$$= \frac{(\pi/6)(d+\delta d)^3 - \pi d^3/16}{\pi d^3/6}$$

$$= \frac{d^3 + 3d^2\delta d + 3d(\delta d)^2 + \delta d^3 - d^3}{d^3}$$

$$= \frac{3\,\delta d}{d}$$

$$= 3 \times \text{diametrical strain}$$

(We have neglected the terms of second order and above.)

$$\text{Change in volume} = \frac{\pi \times (1200)^3}{6} \times 3 \times 1.4 \times 10^{-4}$$

$$= 380{,}000 \text{ mm}^3$$

Thin cylindrical container with hemispherical ends A commonly used thin shell is one with hemispherical ends as shown in Fig. 12.16. The straight part acts as a thin cylindrical vessel and the thick part acts as a spherical shell. As the ends are closed, there will also be longitudinal stress in the cylindrical part. Let r be the diameter of the cylinder (also of spherical part) and t_c and t_s be the thicknesses of the cylinder and sphere.

Cylindrical part From the equilibrium of a unit length of the cylinder, we get

$$P \times 2r = \sigma_h t_c \times 2\sigma_h = \frac{Pr}{t_c} = \frac{Pd}{2t_c}$$

as the hoop stress in the cylinder

$$\text{The longitudinal stress, } \sigma_L = \frac{P \times \pi r^2}{2\pi r t_c} = \frac{Pd}{4t_c}$$

Hemispherical part In the hemispherical part, there will be hoop stresses in two mutually perpendicular directions. From the equilibrium of the free body in the diametrical plane,

$$\sigma_{\text{hs}} \times 2\pi r t_s = P \times \sigma r^2;\ \sigma_{\text{hs}} = \frac{Pr}{2t_s} = \frac{Pd}{4t_s}$$

Circumferential strain at the junction plane of the two parts must be equal.

$$\text{Circumferential strain, } \varepsilon_h = \frac{\sigma_h}{E} - \frac{\nu\sigma_L}{E} = \frac{Pd}{2t_c E} - \frac{\nu Pd}{4t_c E} = \frac{Pd(2-\nu)}{4t_c E} \quad \text{(cylinder)}$$

$$\text{Circumferential strain, } \varepsilon_h = \frac{\sigma_{hs}}{E} - \frac{\nu\sigma_{hs}}{E} = \frac{Pd(2-\nu)}{4t_c E}$$

Equating the two,

$$\frac{Pd(2-\nu)}{4t_c E} = \frac{Pd(1-\nu)}{4t_c E} \frac{t_c}{t_s} = \frac{(2-\nu)}{1-\nu}$$

This must be the ratio of the thicknesses if there has to be no distortion due to differential strains.

Example 12.21 Thin cylindrical shell with hemispherical ends

A thin cylindrical shell, of 1500 mm diameter, with hemispherical ends is subjected to an internal pressure of 2.5 N/mm^2. Find the thickness required if the maximum stress is limited to 150 MPa. What should be the thicknesses to avoid distortion at junction or the maximum stress? Poisson's ratio = 0.3.

Solution Hoop stress in the cylindrical part is the maximum.

$\sigma_h = Pd/2t_c = 2.5 \times 1500/2t_c = 150$; $t_c = 2.5 \times 1500/(2 \times 150) = 12.5$ mm

For no distortion at the junction, $t_c/t_s = (2 - 0.3)/(1 - 0.3) = 5.2$ mm

$$\text{Hoop stress in the hemispherical part} = \frac{2.5 \times 1500}{4 \times 5.2} = 180 \text{ MPa}$$

As this stress is more than permissible, the thickness has to be increased.

$$t_s = \frac{2.5 \times 1500}{4 \times 150} = 6.3 \text{ mm}$$

Example 12.22 Thin cylindrical shell with hemispherical ends

A thin cylindrical shell, of 1.8 m diameter, with hemispherical ends is subjected to an internal pressure of 2 N/mm^2. Determine the thickness required if the stress is not to exceed 120 MPa in the cylindrical and spherical parts. Determine the thickness of the two parts if the distortion at the junction is to be avoided. Find the stresses if the thickness to satisfy this condition is provided. Poisson's ratio = 0.25.

Solution Hoop stress in the cylindrical part = $Pd/2t_c$

Therefore,

$$\frac{2 \times 1800}{2t_c} = 120; \; t_c = 15 \text{ mm}$$

In the spherical part, the stress is half this value. Hence, the thickness required = 7.5 mm.

The ratio $t_c/t_s = (2 - \nu)/(1 - \nu) = 2 - 0.25/(1 - 0.25) = 2.3333$

$t_s = 15/2.3333 = 6.4$ mm

As this will increase the stress above the permissible limits, $t_s = 7.5$ mm.

Thickness of cylindrical part, $t_c = 7.5 \times 2.333 = 17.5$ mm

Stress in the cylindrical part = $2 \times 1800/(2 \times 17.5) = 102.8$ N/mm^2

12.5 THICK CYLINDRICAL VESSELS

At any point in a cylindrical pressure vessel, there are three stresses in three perpendicular directions, namely,

(a) longitudinal stress, σ_L

(b) hoop stress, σ_h, and

(c) radial stress, σ_R.

In deriving the equations for thin cylindrical shells, we made three assumptions.

(i) σ_R is negligible compared to σ_L and σ_h,

(ii) σ_h is uniform along the thickness, and

(iii) the longitudinal strain is uniform so that plane sections (transverse) remain plane before and after the application of internal pressure.

In the case of thick cylindrical vessels, assumptions (i) and (ii) are not valid, but assumption (iii) is assumed to be true.

A cylindrical shell can be classified as thick when its thickness is more than 1/20 th its diameter. Such shells are analysed using Lamé's theory.

12.5.1 Lamé's Theory

Consider a thick cylindrical shell of internal radius r_i and external radius r_e. Let the pressures at these radial distances be P_i and P_e as shown in Fig. 12.17. We consider a ring at radius r and of thickness δr. If we consider a unit length of the cylinder and the ring cut by a diametrical section, the free body diagram of the cylinder is as shown in Fig. 12.17(b). For this thin ring, the conditions are the same as that of a thin cylindrical shell. For the equilibrium of the thin ring shown in Fig. 12.17(b),

Total upward force = total downward force

Upward force = $2r\sigma_r$

Downward force = $2(r + \delta r)(\sigma_r + \delta\sigma_r) + 2\delta r\sigma_h$

Equating the two and neglecting quantities of second order, we have

$$2\delta r\sigma_r + 2r\delta\sigma_r + 2\delta r\sigma_h = 0$$

Dividing by δr and as $\delta r \to 0$,

$$r\frac{d\sigma_r}{dr} + \sigma_r + \sigma_h = 0$$

If σ_L is the longitudinal stress, then the longitudinal strain is given by

$$\varepsilon_L = \frac{\sigma_L}{E} - \frac{\nu(\sigma_h - \sigma_r)}{E}$$

In this equation, if the plane sections are to remain plane before and after the deformation, then σ_L and $(\sigma_h - \sigma_r)$ must be constant. Setting $(\sigma_h - \sigma_r)$ = constant $= 2A$, we have

$$r\frac{d\sigma_r}{dr} + \sigma_r + (2A + \sigma_r) = 0$$

$$r\frac{d\sigma_r}{dr} = -2(A + \sigma_r)$$

$$\frac{2dr}{r} = \frac{d\sigma_y}{-(A+\sigma_r)}$$

Integrating both sides,

$$2\log_e(r) + \text{constant} = -\log_e(A + \sigma_r)$$

Taking the constant as B, we get

$$\log_e[r^2(A + \sigma_r)] = B$$

$$\sigma_r = -A + B/r^2$$

From $\sigma_h = \sigma_r + 2A$, we get

$$\sigma_h = -A + B/r^2 + 2A = A + B/r^2$$

We thus have

radial stress $\sigma_r = -A + B/r^2$ and hoop stress $\sigma_h = A + B/r^2$

The constants A and B can be found from the boundary conditions in particular cases.

The equations for radial stress σ_r and hoop stress σ_h are known as Lamé's equations and the constants A and B are called Lamé's constants.

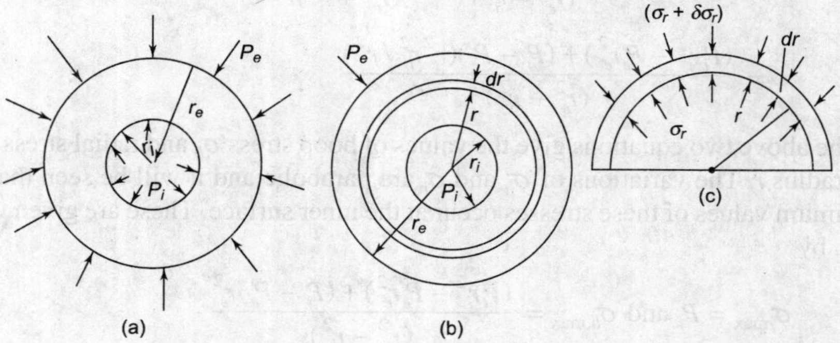

Fig. 12.17

Stresses in a thick cylindrical shell The stresses in a thick cylindrical shell depend upon the nature of the loading which can result in (i) only an internal pressure P_i, (ii) only an external pressure P_e, and (iii) both P_i and P_e.

(i) Only internal pressure, P_i
  (a) The hoop stress will be tensile.
  (b) The longitudinal stress will be tensile.
  (c) The radial stress will be compressive.

(ii) Only external pressure, P_e
  (a) The hoop stress will be compressive.
  (b) The longitudinal stress will be compressive.
  (c) The radial stress will be compressive.

(iii) Both P_i and P_e applied
  (a) The hoop stress will be tensile if $P_i > P_e$.
  (b) The longitudinal stress will be tensile for $P_i > P_e$.
  (c) The radial stress will be compressive.

Considering a case where both internal and external pressures exist, we have $\sigma_r = -P_i$ at $r = r_i$ and $\sigma_r = -P_e$ at $r = r_e$.

Substituting these values for σ_r, we get

$$P_i = -A + B/r_i^2 \text{ and } P_e = -A + B/r_e^2$$

Solving for A and B,

$$B/r_i^2 - B/r_e^2 = P_i - P_e;\ B(r_e^2 - r_i^2) = r_e^2 r_i^2 (P_i - P_e) B = \frac{(P_i - P_e) r_e^2 r_1^2}{r_e^2 - r_i^2}$$

$$A = B/r_i^2 - P_i = \frac{(P_i - P_e) r_e^2}{(r_e^2 - r_i^2)} - P_i = \frac{P_i r_i^2 - P_e r_e^2}{(r_e^2 - r_i^2)}$$

With these values of A and B, the values of σ_r and σ_h at radius r are obtained as

$$\sigma_r = \frac{-(P_i r_i^2 - P_e r_e^2)}{(r_e^2 - r_i^2)} + \frac{(P_i - P_e) r_e^2 r_i^2}{(r_e^2 - r_i^2) r^2} = \frac{-(P_i r_i^2 - P_e r_e^2) + (P_i - P_e)(r_e^2 r_i^2 / r^2)}{(r_e^2 - r_i^2)}$$

$$\sigma_h = A + B/r^2 = \frac{(P_i r_i^2 - P_e r_e^2)}{(r_e^2 - r_i^2)} + \frac{(P_i - P_e) r_e^2 r_i^2}{(r_e^2 - r_i^2) r^2}$$

$$= \frac{(P_i r_i^2 - P_e r_e^2) + (P_i - P_e)(r_e^2 r_i^2 / r^2)}{(r_e^2 - r_i^2)}$$

The above two equations give the values of hoop stress σ_h and radial stress σ_r at any radius r. The variations of σ_h and σ_r are parabolic and it will be seen that the maximum values of these stresses occur at the inner surface. These are given when $r = r_i$ by

$$\sigma_{r\max} = P_i \text{ and } \sigma_{h,\max} = \frac{(P_i r_i^2 - P_e r_e^2) + (P_i - P_e) r_e^2}{(r_e^2 - r_i^2)}$$

$$= \frac{[P_i (r_i^2 + r_e^2) - 2 P_e r_e^2]}{(r_e^2 - r_i^2)}$$

When $P_i = 0$ (only external pressure is acting on the pipe or $P_e = 0$ (only internal pressure exists), the corresponding relationships can be obtained by putting the respective pressures to zero in the above relationships.

(i) $P_e = 0$ (Only internal pressure exists.)

$$\text{At } r,\ \sigma_r = \frac{-P_i r_i^2 + P_i (r_e^2 r_i^2 / r^2)}{(r_e^2 - r_i^2)} = \frac{P_i r_i^2}{(r_e^2 - r_i^2)} \left(\frac{r_e^2}{r^2} - 1 \right)$$

$$\sigma_h = \frac{P_i r_i^2 + P_i (r_e^2 r_i^2 / r^2)}{(r_e^2 - r_i^2)} = \frac{P_i r_i^2}{(r_e^2 - r_i^2)} \left(\frac{r_e^2}{r^2} + 1 \right)$$

$$\text{At } r = r_i,\ \sigma_r = P_i;\ \sigma_h = \frac{P_i (r_e^2 + r_i^2)}{(r_e^2 - r_i^2)}$$

$$\text{At } r = r_e,\ \sigma_r = 0;\ \sigma_h = \frac{2 P_i r_i^2}{(r_e^2 - r_i^2)}$$

(ii) $P_i = 0$ (Only external pressure exists.)

$$\sigma_r = \frac{P_e r_e^2 - P_e (r_e^2 r_i^2 / r^2)}{(r_e^2 - r_i^2)} = \frac{P_e r_e^2 [1 - (r_i^2 / r^2)]}{(r_e^2 - r_i^2)}$$

$$\sigma_h = \frac{-P_e r_e^2 - P_e (r_e^2 r_i^2 / r^2)}{(r_e^2 - r_i^2)} = \frac{-P_e r_e^2 [1 + (r_i^2 / r^2)]}{(r_e^2 - r_i^2)}$$

$$\text{At } r = r_i,\ \sigma_r = 0;\ \sigma_h = \frac{-2 P_e r_e^2}{(r_e^2 - r_i^2)}$$

$$\text{At } r = r_e,\ \sigma_r = P_e;\ \sigma_h = \frac{-P_e (r_e^2 + r_i^2)}{(r_e^2 - r_i^2)}$$

The stress distributions in all cases are shown in Fig. 12.18.

Figure 12.18(a) shows the variation of σ_r and σ_h when both internal and external pressure are applied. The maximum stresses are in the interior fibres.

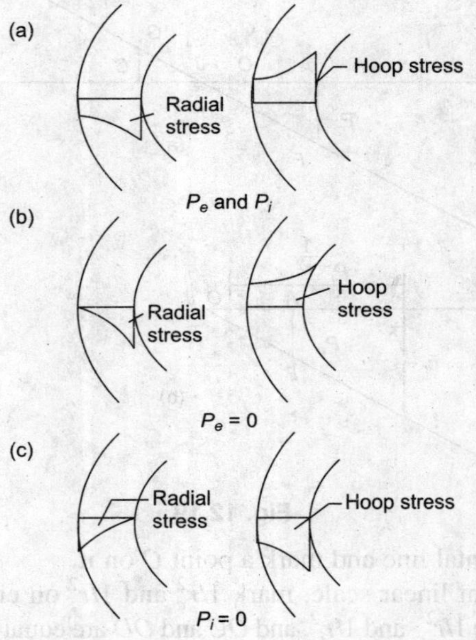

Fig. 12.18

Figure 12.18(b) illustrates the case when only internal pressure is applied, i.e., $P_e = 0$. Again, the maximum stresses occur at the interior fibres at $r = r_i$. Figure 12.18(c) shows the variation of these stresses when $P_i = 0$, i.e., when only external pressure P_e exists. In this case, the maximum radial stress occurs at $r = r_e$, while the maximum hoop stress occurs at $r = r_i$. Both the stresses are compressive in nature here.

Longitudinal stresses are developed in thick-walled cylinders with closed ends. They are normally considered to be uniformly distributed. Using the same notations as before, the longitudinal stress may be derived as follows.

$$P_i \pi r_i^2 - P_e \pi r_e^2 = \sigma_l \pi (r_e^2 - r_i^2)$$

$$\sigma_l = \frac{P_i r_i^2 - P_e r_e^2}{(r_e^2 - r_i^2)}$$

12.5.2 Graphical Method for Determining σ_h and σ_r

We can use a simple graphical procedure to calculate σ_r and σ_h based on the relevant equations. We can do so by plotting what are called Lamè's lines.

Let us consider the general case of a thick cylinder subjected to internal as well as external pressures, P_i and P_e. Let r_e and r_i be the external and internal radii of the cylinder. The procedure is as follows. (Fig. 12.19.)

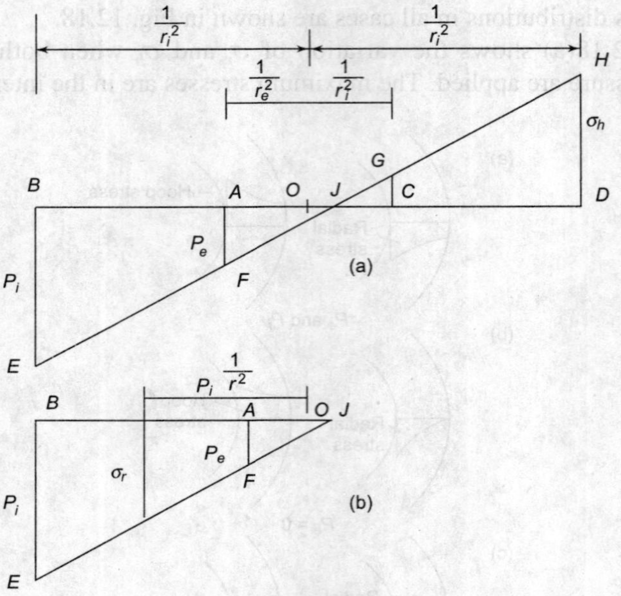

Fig. 12.19

(i) Draw a horizontal line and mark a point O on it.

(ii) To a convenient linear scale, mark $1/r_e^2$ and $1/r_i^2$ on either side of O. Thus, OA and OB are $1/r^2{}_e$ and $1/r_i^2$, and OC and OD are equal to OA and OB respectively.

(iii) To a convenient stress scale, mark $BE = P_i$ and $AF = P_e$.

(iv) Draw EF and continue the line to intersect perpendiculars drawn at C and D on the other side of the line at G and H. This line intersects the horizontal line at J.

To obtain σ_r at any radius r, between r_e and r_i, we can obtain the point by plotting $1/r^2$ from O to the left. The ordinate at that point on the line EH gives the value of σ_r at r.

We can obtain σ_h at any radius r, between r_e and r_i, by plotting $1/r^2$ from O to the right side. The ordinate at that point gives the value of hoop stress σ_h at r. We can be prove this as follows.

Let the distance $OJ = \dfrac{1}{x^2}$

The distance $JA = JO + OA = \dfrac{1}{x^2} + \dfrac{1}{r_e^2}$

The distance $JC = OC - OJ = \dfrac{1}{r_e^2} - \dfrac{1}{x^2}$

The distance $JB = JO + OB = \dfrac{1}{x^2} + \dfrac{1}{r_e^2}$

The distance $JD = OD - OJ = \dfrac{1}{r_i^2} - \dfrac{1}{x^2}$

Triangles *JBE* and *JAF* are similar. Therefore,

$$\frac{BE}{AF} = \frac{JB}{JA} = \frac{1/x^2 + 1/r_i^2}{1/x^2 + 1/r_e^2} = \frac{P_i}{P_e}$$

$$\frac{P_i}{P_e} = \frac{x^2 + r_i^2}{x^2 + r_e^2} \frac{x^2 r_e^2}{x^2 r_i^2}$$

$$x^2 = \frac{(P_e - P_i)\, r_i^2 r_e^2}{P_i r_i^2 - P_e r_e^2}$$

Triangles *JBE* and *JDH* are similar. Therefore,

$$\frac{DF}{JD} = \frac{BE}{JA}$$

$$DF = \frac{AE \times JD}{JA} = \frac{p_i(1/r_i^2 - 1/x^2)}{1/x^2 + 1/r_i^2}$$

$$DF = \frac{x^2 - r_i^2}{x^2 + r_i^2} P_i$$

Substituting the value of x^2, we get

$$CF = \sigma_h \text{ at } r_i = \frac{P_i(r_i^2 + r_e^2) - 2P_e r_e^2}{(r_e^2 - r_i^2)}$$

which is the $\sigma_{h\max}$ at the inner surface.
Similarly, at any radius *r* [Fig. 12.19(b)],

$$\sigma_r = P_e + \frac{(P_i - P_e)(1/r^2 - 1/r_e^2)}{(1/r_i^2 - 1/r_e^2)}$$

$$= \frac{P_e r^2 (r_e^2 - r_i^2) + (P_i - P_e) r_i^2 (r_e^2 - r^2)}{(r_e^2 - r_i^2) r^2}$$

$$= \frac{P_e r_e^2 r^2 - P_e r^2 r_i^2 + P_i r_i^2 r_e^2 + P_i r_i^2 r_e^2 - P_e r_i^2 r_e^2 - P_e r_i^2 r^2}{(r_e^2 - r_i^2) r^2}$$

$$= \frac{r^2 (P_e r_e^2 - P_i r_i^2) + (P_i - P_e)\, r_i^2 r_e^2}{(r_e^2 - r_i^2) r^2}$$

$$= \frac{P_i r_i^2 - P_e r_e^2 - (P_i - P_e)\, r_i^2\, r_e^2 / r^2}{(r_e^2 - r_i^2)}$$

as before.

Similarly, the value of σ_h at any radial distance *r* can be obtained.

When a thick shell is subjected to internal pressure alone, Lamè's line diagram will be as shown in Fig. 12.20(a). When a thick shell is subjected to external pressure alone, such a diagram will be as shown in Fig. 12.20(b).

In Fig. 12.20(a), we select a point *O* on a horizontal line. Mark (to same scale) $OA = OC = 1/r^2{}_e$ and $OB = OD = 1/r^2{}_i$. Mark *BE* (to some stress scale) = P_i. P_e

being zero, join E to A and continue the line to meet the vertical line at D at F. $DF = \sigma_{n(\max)}$.

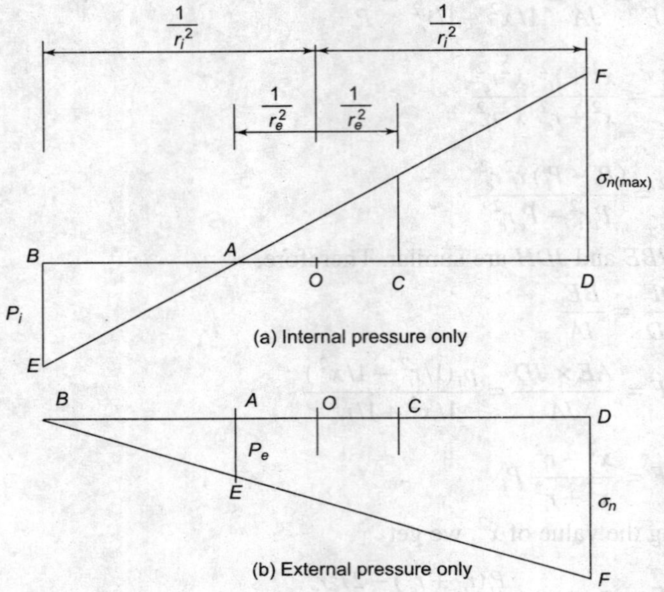

Fig. 12.20

In Fig. 12.20(b), $P_i = 0$; so the ordinate at $B = 0$. $AE = P_e$. Draw BE and produce the line to meet the vertical at D in F. $DF = \sigma_{h\,\max}$.

The following examples illustrate the graphical method.

Example 12.23 Thick cylinder: graphical solution

A thick cylinder has an internal diameter of 120 mm and an external diameter of 240 mm. It is subjected to an internal pressure of 80 N/mm^2 and an external pressure of 20 N/mm^2. Find the maximum value of the hoop stress, and the radial and hoop stresses at a radial distance of 150 mm graphically.

Solution $\dfrac{1}{(r_i)^2} = \dfrac{1}{(120)^2} = 6.94 \times 10^{-5}$, $\dfrac{1}{(r_e)^2} = \dfrac{1}{(240)^2} = 1.736 \times 10^{-5}$

and $\dfrac{1}{(r)^2} = \dfrac{1}{(150)^2} = 4.444 \times 10^{-5}$

Linear scale: 1 cm = 1×10^{-5} mm^{-2} and stress scale: 1 cm = 20 N/mm^2.

Take the following steps.

1. Draw a horizontal line and take a reference point O (Fig. 12.21).
2. Mark $OA = 1/(240)^2$ and $OB = 1/(120)^2$.
3. Mark $OK = 1/(150)^2 = OL$.
4. Mark BE = 80 N/mm^2 and AF = 20 N/mm^2. Draw EF and produce it to meet the vertical line through D at H.

DH gives the maximum hoop stress

$$DH = 4.0 \text{ cm} = 4.0 \times 20 = 80 \text{ N/mm}^2$$

KM gives the radial stress at a radial distance of 150 mm.

$$KM = 2.5 \text{ cm} = 2.5 \times 20 = 50 \text{ N/mm}^2$$

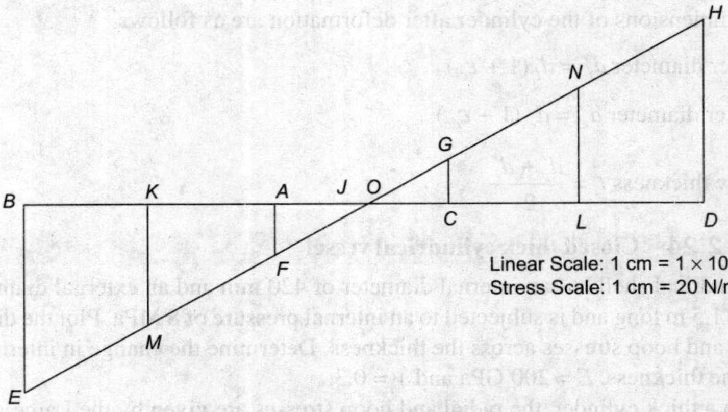

Fig. 12.21

LM gives the hoop stress σ_h at 150 mm.

$$LM = 2.5 \text{ cm} = 2.5 \times 20 = 50 \text{ N/mm}^2$$

By calculation,

$$\sigma_{h\,\max} = \frac{P_i(r_i^2 + r_e^2) - 2P_e r_e^2}{r_e^2 - r_i^2}$$

$$= \frac{80(120^2 + 240^2) - 2 \times 20 \times (240)^2}{240^2 - 120^2}$$

$$= \frac{80 \times 7.2 \times 10^4 - 230.4 \times 10^4}{4.32 \times 10^4} = 80.2 \text{ N/mm}^2$$

Longitudinal stress If the ends of the cylinder are closed, then the longitudinal stress is given by

$$\sigma_l = \frac{P_i \times \pi r_i^2}{\pi r_e^2 - \pi r_i^2} = \frac{P_i r_i^2}{r_e^2 - r_i^2}$$

Volumetric Strain Volume of the cylinder $= \frac{\pi d_i^2}{4} \times l$

where d_i is the internal diameter and l is the length.

$$V = \frac{\pi d_i^2 l}{4}$$

$$\delta V = \frac{\pi}{4}\,(2d_i\,\delta d_i + d_i^2\,\delta l)$$

$$\frac{\delta V}{4} = \text{volumetric strain} = \frac{2\delta d_i}{d_i} + \frac{\delta l}{l}$$

$$\varepsilon_V = 2\varepsilon_c + \varepsilon_l$$

where ε_V is the volumetric strain, ε_c is the circumferential strain, and ε_l is the longitudinal strain.

The circumferential strains at the inner and outer surfaces can be obtained as

$$\varepsilon_{ci} = \frac{1}{E}\,(\sigma_{hi} - \nu\sigma_{ri} - \nu\sigma_l)$$

and $$\varepsilon_{ce} = \frac{1}{E}\,(\sigma_{he} - \nu\sigma_{re} - \nu\sigma_l)$$

where the subscript *i* refers to the inner edge and *e* to the outer edge.

The new dimensions of the cylinder after deformation are as follows.

Inner diameter $d_i' = d_i\,(1 + \varepsilon_{ci})$

Outer diameter $d_e' = d_e\,(1 + \varepsilon_{ce})$

New thickness $t_i = \dfrac{d_e' + d_i'}{2}$

Example 12.24 Closed thick cylindrical vessel

A closed cylindrical shell has an internal diameter of 420 mm and an external diameter of 520 mm. It is 1.5 m long and is subjected to an internal pressure of 8 MPa. Plot the distribution of radial and hoop stresses across the thickness. Determine the change in internal volume and in the thickness. E = 200 GPa and $\nu = 0.3$.

Solution In a thick cylinder, the radial and hoop stresses are given by the Lamé's equations $\sigma_r = -A + B/r^2$ and $\sigma_h = A + B/r^2$ at any radius r. To determine the constants for this case, we use the boundary conditions

(i) At r = 260 mm, $\sigma_r = 0$
(ii) At r = 210 mm, $\sigma_r = 8$ N/mm²

We get

$$0 = -A + B/260^2 \text{ and } 8 = -A + B/210^2$$

Solving for A and B,

$$B = 8 \times 210^2 \times 260^2/(260^2 - 210^2) = 1{,}014{,}863$$

$$A = B/260^2 = 1{,}014{,}863/260^2 = 15$$

The variation of radial stress σ_r and hoop stress σ_h can be calculated by substituting different values of r. The values are tabulated as follows.

Radius, r	***Radial stress, σ_r***	***Hoop stress, σ_h***
210	8	38
220	5.97	35.97
230	4.18	34.18
240	2.62	32.62
260	zero	30.00

When the values are plotted, the distribution will appear as shown in Fig. 12.22.

The longitudinal stress remains constant over the cross section and is given by

$$\sigma_l = \frac{P\pi R_i^2}{\pi(R_e^2 - R_i^2)} = \frac{8 \times 210^2}{260^2 - 210^2} = 15 \text{ N/mm}^2$$

To find change in volume, we use the equation

$$\varepsilon_V = 2\varepsilon_{ci} + \varepsilon_l$$

ε_{ci} is the hoop strains at the inner edge and ε_l is the longitudinal strain.

$$\varepsilon_{ci} = \frac{1}{E}(\sigma_h - \nu\sigma_l - \nu\sigma_r)$$

$$= \frac{1}{200{,}000}(38 - 0.3 \times 15 + 0.3 \times 8)$$

$$= 17.95 \times 10^{-5}$$

$$\varepsilon_l = \frac{1}{E}[\sigma_l - \nu\sigma_h - \nu\sigma_r]$$

$$= \frac{1}{200{,}000}(15 - 0.3 \times 38 + 0.3 \times 8)$$
$$= 3 \times 10^{-5}$$

$$\frac{\delta V}{V} = 2 \times 17.95 \times 10^{-5} + 3 \times 10^{-5} = 38.9 \times 10^{-5}$$

Change in volume = $\varepsilon_V \times$ volume

Volume of cylinder = $\pi(210)^2 \times 1500 = 2078 \times 10^5$ mm^3

$\delta V = 38.9 \times 10^{-5} \times 2078 \times 10^5 = 80{,}834$ mm^3

To determine the change in thickness, we need to know the diametral strains at the inner and outer edges. Diameteral strain at the inner edge = circumferential strain at the inner edge

$$\varepsilon_{di} = \varepsilon_{ci} = 17.95 \times 10^{-5}$$

Strain at the external surface,

$$\varepsilon_{ce} = \frac{1}{E}[\sigma_h - \nu\sigma_l - \nu\sigma_r]$$

$$\sigma_r = 0$$

at the outer surface

$$\varepsilon_{ce} = \frac{1}{200{,}000}(30 - 0.3 \times 15) = 12.75 \times 10^{-5}$$

Internal diameter after deformation = $d_i(1 + \varepsilon_{ci})$

$= 420(1 + 17.95 \times 10^{-5})$

$= 420.075$ mm

External diameter after deformation = $520(1 + \varepsilon_{ce})$

$= 520(1 + 12.75 \times 10^{-5})$

$= 520.066$

$$\text{Thickness after deformation} = \frac{520.066 - 420.075}{2}$$

$= 49.9955$ mm

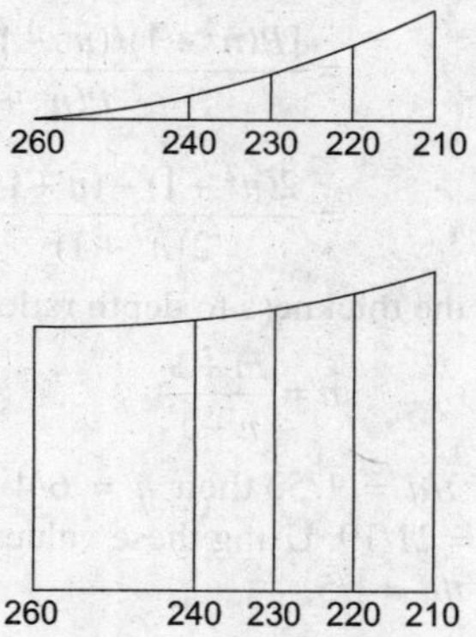

Fig. 12.22

Example 12.25 Comparison of thin and thick cylindrical formulae

Consider a cylinder of external diameter d_2 and internal diameter d_1, subjected to an internal pressure P_i. Find the percentage error in calculating the maximum hoop stress by the thin-cylinder theory compared to the one obtained by the thick-cylinder theory, if the thickness to the mean diameter ratio is 1/5, 1/10, and 1/20.

Solution External diameter = d_2

Internal diameter = d_1

Let $\frac{d_2}{d_1} = n$

$$\text{Thickness } t = \frac{d_2 - d_1}{2} = \frac{d_1(n-1)}{2}$$

$$\text{Mean diameter} = \frac{d_2 + d_1}{2} = \frac{d_1(n+1)}{2}$$

$$\text{Thickness to mean diameter ratio} = \frac{d_2 + d_1}{d_2 + d_1} = \frac{n-1}{n+1}$$

Considering the vessel as a thick cylinder, the hoop stress at any radius $r = d/2$ is

$$\sigma_h = \frac{P d_1^2 (d_2^2 + d^2)}{d^2 (d_2^2 - d_1^2)}$$

The maximum value of hoop stress is at the inner surface, i.e. at $d = d_1$.

$$\sigma_{h\,\max} = \frac{P(d_2^2 + d_1^2)}{(d_2^2 - d_1^2)} = \frac{P(n^2 + d)}{n^2 - 1}$$

Considering the vessel as a thin cylinder of diameter equal to the mean diameter,

$$\sigma_h = \frac{Pd}{2t} = \frac{Pd_1(n+1)/2}{2d_1(n-1)/2} = \frac{P(n+1)}{2(n-1)}$$

The error percentage is calculated as

$$\frac{\sigma_h \max - \sigma_h}{\sigma_h \max} \times 100$$

$$= \frac{[P(n^2+1)/(n^2-1) - P(n+1)/2(n-1)]}{P(n^2+1)/(n^2-1)} \times 100$$

$$= \frac{2(n^2+1) - (n+1)^2}{2(n^2+1)} = 1 - \frac{(n+1)^2}{2(n^2+1)}$$

If the thickness to depth ratio = $1/m$, then $1/m = (n - 1)/(n + 1)$, from which

$$n = \frac{m+1}{n-1}$$

If t/d = 1/5, then n = 6/4 = 3/2; if t/d = 1/10, then n = 11/9; if t/d = 1/20, then n = 21/19. Using these values, the following error percentages can be calculated.

If t/d = 1/5,

$$\text{Error percentage} = \frac{1-(3/2+1)^2}{2[(3/2)^2+1]} \times 100 = 4$$

If t/d = 1/10,

$$\text{Error percentage} = \frac{1-[11/9+1]^2}{2((11/9)^2+1)} \times 100 = 1$$

If t/d = 1/20,

$$\text{Error percentage} = \frac{1-[21/19+1]^2}{2[(21/19)^2+1]} \times 100 = 0.25$$

Example 12.26 Thick cylindrical shell

A thick cylindrical shell, 180 mm in diameter, is subjected to an internal pressure of 10 N/mm^2. Find the thickness required for the shell if the maximum permissible tensile stress is 30 N/mm^2.

Solution We know that the radial stress at the interior surface is compressive and given by $\sigma_r = A - B/r^2$. The hoop tensile stress is given by $\sigma_h = A + B/r^2$. The constants A and B can be determined from the boundary conditions.

(i) At $x = r_i$ = 90 mm, $\sigma_r = -10$ N/mm^2

(ii) At $x = r_i$ = 90 mm, σ_h = 30 N/mm^2, the maximum permissible tensile stress Thus, we have

$$-10 = A - B/90^2;\ 30 = A + B/90^2$$

Solving these two equations, we get

$$A = 10 \text{ and } B = 20 \times 90^2$$

To find the thickness, we know that the radial pressure is zero at the outer radius. If r_e is the external radius, we have

$$0 = 10 - 20 \times 90^2/r_e^2$$

or $r_e^2 = 90^2 \times 20/10 = 127.3$ mm

Thickness of the shell = 127.3 – 90 = 37.3 mm

At the exterior surface, hoop stress = $10 + 20 \times 90^2 / 127.3^2 = 10.08$ N/mm^2

12.5.3 Thick Spherical Shells

Thick spherical shells can be similarly analysed for stresses. In a thick spherical shell, the stress cannot be assumed to be uniform over the thickness. That is the main difference between a thin spherical shell and a thick one.

Consider the thick spherical shell shown in Fig. 12.23.

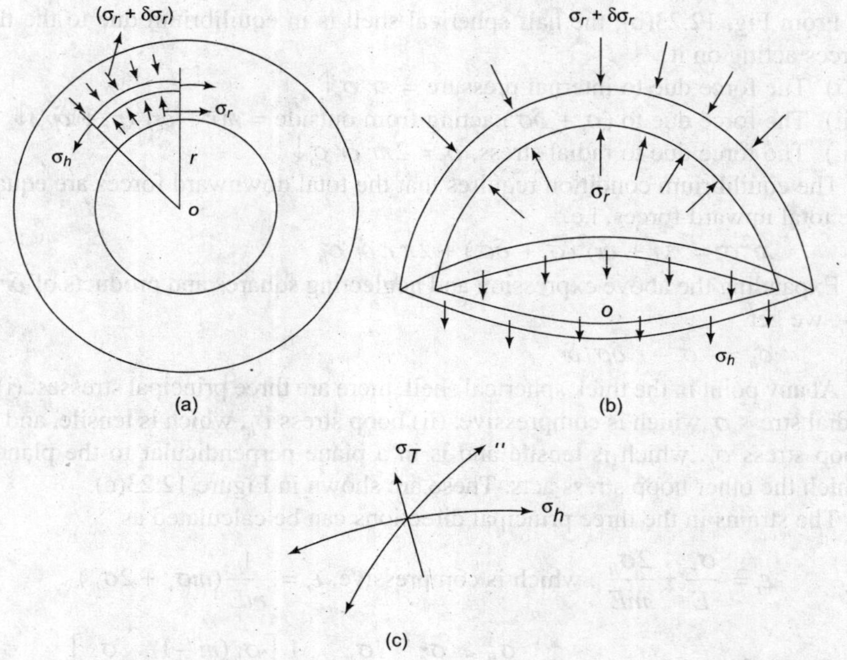

Fig. 12.23 Thick spherical shell

The following symbols are used in the discussion to follow:

r_i: Internal radius of the shell
r_e: External radius of the shell
p: Internal pressure in the shell
p_r: Pressure at radius
m: Poisson's ratio
ε_r: Radial strain
ε_h: Hoop or circumferential strain
σ_r: Radial stress
σ_h: Hoop stress

In Fig. 12.23(a), we have a diametrical section of the shell. Consider an element at radius r and of thickness δr, subtending an angle $\delta\theta$ at the centre. The pressure at r is p_r and the increase in pressure is δp_r as shown. Due to the pressure in the shell, let the radius increase to $(r + u)$. The thickness of the element will also increase from δr to $\delta r + \delta u$.

Radial strain, $\varepsilon_r = \dfrac{(\delta r + \delta u) - \delta r}{\delta r} = \dfrac{\delta u}{\delta r}$

Circumferential strain, δ_h = Change in circumference/Original circumference

$$\varepsilon_h = \frac{2\pi(r+u) - 2\pi r}{2\pi r} = \frac{u}{r}$$

$$u = r\varepsilon_h; \frac{\delta u}{\delta r} = \varepsilon_h + r\frac{\delta \varepsilon_h}{\delta r}$$

From Fig. 12.23(b), the half spherical shell is in equilibrium due to the three forces acting on it.

(i) The force due to internal pressure = $\pi r^2 \sigma_r \uparrow$
(ii) The force due to $(\sigma_r + \delta\sigma_r)$ acting from outside = $\pi(r + \delta r)^2(p_r + \delta p_r)\downarrow$
(iii) The force due to radial stress, $\sigma_r = 2\pi r\ \delta r\ \sigma_r \downarrow$

The equilibrium condition requires that the total downward forces are equal to the total upward forces, i.e.

$$\pi r^2 \sigma_r = \pi(r + \delta r)^2(\sigma_r + \delta\sigma_r) + 2\pi r\ \delta r\ \sigma_h$$

Expanding the above expression and neglecting squares and products of δr and δp_r, we get

$$\sigma_h = -\sigma_r - r\ \delta\sigma_r/\delta r$$

At any point in the thick spherical shell, there are three principal stresses: (i) the radial stress σ_r, which is compressive, (ii) hoop stress σ_h, which is tensile, and (iii) hoop stress σ_h, which is tensile and is in a plane perpendicular to the plane on which the other hoop stress acts. These are shown in Figure 12.23(c).

The strains in the three principal directions can be calculated as

$$\varepsilon_r = \frac{\sigma_r}{E} + \frac{2\sigma_h}{mE}, \text{ which is compressive. } \varepsilon_r = \frac{1}{mE}(m\sigma_r + 2\sigma_h)$$

Circumferential strain, $\varepsilon_h = \dfrac{\sigma_h}{E} + \dfrac{\sigma_r}{mE} - \dfrac{\sigma_h}{mE} = \dfrac{1}{E}\left[\dfrac{\sigma_h(m-1)}{m} + \dfrac{\sigma_r}{m}\right]$

We substitute this in the strain equation we derived earlier

$$-\varepsilon_r = \varepsilon_h + \frac{r\delta\varepsilon_h}{\delta r}$$

$$-\left[\frac{\sigma_r}{E} + \frac{2\sigma_h}{mE}\right] = \left[\frac{(m-1)\sigma_h}{mE} + \frac{\sigma_r}{mE}\right] + \frac{rd}{dr}\left[\frac{[\sigma_h(m-1)}{mE} + \frac{\sigma_r}{mE}\right]$$

This equation reduces to

$$(m+1)(\sigma_h + \sigma_r) + (m-1)r\frac{d\sigma_h}{dr} + r\frac{d\sigma_r}{dr} = 0$$

We have from the equilibrium of half shell,

$$\sigma_h = -\sigma_r - (r/2)\delta\sigma_r/\delta r$$

$$\frac{d(\sigma_h)}{dr} = \frac{-d\sigma_r}{dr} - \frac{rd^2\sigma_r}{dr^2}$$

Substituting this value in the equation above, we have

$$(m+1)(\sigma_h+\sigma_r)+(m-1)\,r\left[\frac{-d\sigma_r}{dr}-\frac{rd^2\sigma_r}{dr^2}\right]+r\,\frac{d\sigma_r}{dr}=0$$

On opening brackets and after multiplication of terms, this reduces to

$$4\,\frac{d\sigma_r}{dr}+r\,\frac{d^2\sigma_r}{dr^2}=0$$

To solve this equation, let $d\sigma_r/dr = Y$.

$$4Y + r\,dY/dr = 0;\quad \frac{dY}{Y}=\frac{-4dr}{r}$$

Integrating, $\log_e Y = -4\log_e r + \log_e C_1$

where $\log_e C_1$ is the constant of integration.

$\log_e Y = \log_e r^{-4} + \log_e C_1 = \log_e[C_1/r^4]$

or $Y = C_1/r^4$; $d\sigma_r/dr = C_1/r^4$

Integrating again,

$$\sigma_r = -C_1/3r^3 + C_2$$

where C_2 is the constant of integration.

We have

$$\sigma_h = -\sigma_r - (r/2)d\sigma_r/dr = C_1/3r^3 - C_2 - (r/2)C_1/r^4$$

$$\text{or } \sigma_h = -\frac{C_1}{6r^3} - C_2$$

This equation is generally written as $\sigma_h = A/r^3 + B$, putting $C_1 = -6A$, $C_2 = -B$.

Hence, $\sigma_r = A/r^3 + B$

The constants A and B are obtained from the boundary conditions.

In the case of thick spherical shell under consideration, we have

(i) When $r = r_i$, $\sigma_r = p$, the internal pressure

(ii) When $r = r_e$, $\sigma_r = 0$

Putting these values, we get the constants as

$$A=\frac{pr_i^3r_e^3}{(r_e^3-r_i^3)}\quad\text{and}\quad B=\frac{pr_e^3}{(r_e^3-r_i^3)}$$

The following example illustrates the application of these formulae.

Example 12.27 Thick spherical shell

A thick spherical shell, of 250 mm internal diameter, is subjected to an internal pressure of 8 N/mm^2. If the maximum permissible tensile stress is 10 MPa, find the minimum thickness required. Find the stresses in the interior and exterior surfaces of the shell.

Solution We have the radial compressive stress and hoop tensile stress given by

$$\sigma_r = (2A/r^3) - B \text{ and } \sigma_h = (A/r^3) - B$$

The constants A and B can be found from the following boundary conditions:

(i) When r = internal radius (250/2) = 125 mm, σ_r = 8 N/mm^2

(ii) When r = 125 mm, σ_h = the permissible tensile stress, 10 N/mm^2

The hoop stress will be maximum at the internal surface of the shell.

Now, we get

(i) $8 = (2A/125^3) - B$ and (ii) $10 = (A/125^3) + B$

Adding the two equations,

$18 = 3A/125^3$ or $A = 6 \times 125^3$

$B = (2A/125^3) - 8 = 12 - 8 = 4$

Knowing A and B, we have the radial stress σ_r as zero at the external surface. If r_e is the outer radius, we have

$0 = (2 \times 6 \times 125^3/r_e^3) - 4$; $4r_e^3 = 2 \times 6 \times 125^3$; $r_e = 125 \times \sqrt{3} = 180.5$ mm

Thickness of the shell = 180.5 – 125 = 55.5 mm

Hoop (tensile) stress at the outer surface = $2 \times 125^3/180.5^3 + 4 = 4.7$ N/mm^2 ■

12.6 COMPOUND CYLINDERS

In order to make pressure vessels economical in terms of thickness and ensure more uniform pressure distribution, cylinders are generally prestressed. We have discussed one method of doing so—using a wire-wound, thin cylindrical vessel. Another is to use a jacket over a cylinder. The jacket is shrunk on to the cylinder, causing prestressing. When the internal pressure is then applied, the stresses in the cylinder are much less than they would have been otherwise. The difference between the internal radius of the jacket and the outer radius of the cylinder are adjusted to achieve the required level of prestressing.

Figure 12.24 shows a compound cylinder. The shrinking of the jacket causes tensile stresses on the outer cylinder and compressive stresses on the inner cylinder.

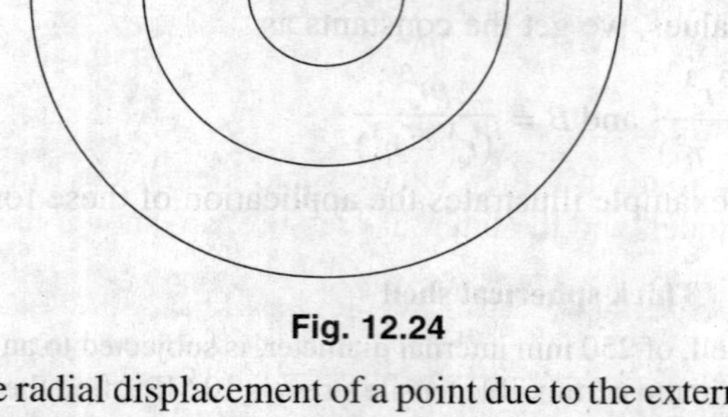

Fig. 12.24

We can find the radial displacement of a point due to the external jacket pressure and the internal pressure from the increase (or decrease) in the length of the circumference. If σ_h and σ_r are the hoop and radial stresses at any point A of the cylinder, the hoop strain is given by

$$\varepsilon_h = \frac{1}{E}(\sigma_h - \nu\sigma_r)$$

$$= \frac{\text{change in circumference}}{\text{original length of circumference}}$$

$$= \frac{2\pi\delta r}{2\pi r} = \frac{\delta r}{r}$$

$$\frac{\delta r}{r} = \frac{1}{E}\,(\sigma_h - \nu\sigma_r)$$

We have now two cases—one due to the external pressure alone because of the shrinking of the jacket and the second because of the internal pressure in the cylinder. Owing to external pressure p_e, the radius will decrease and due to internal pressure the radius will increase.

On account of external pressure p_e alone, the decrease in length of the external radius is

$$\delta_e = -\frac{p_e r_e}{E}\left(\frac{r_e^2 + r_i^2}{r_e^2 - r_i^2} - \nu\right)$$

Due to internal pressure p_i, the increase in length of the external radius is given by

$$\delta_i = \frac{p_i r_i}{E}\left(\frac{r_e^2 + r_i^2}{r_e^2 - r_i^2} + \nu\right)$$

The initial radial difference is given by

$$\delta = \delta_i + \delta_e$$

If r_3 = external radius, r_e = radius at the junction, and r_i = internal radius of the cylinder.

Shrinking allowance to be provided = $2\delta = 2\,(\delta_i + \delta_e)$

$$\text{Shrinking allowance} = \frac{2p_s r_e}{E}\left(\frac{r_3^2 + r_e^2}{r_3^2 - r_e^2} + \frac{r_e^2 + r_i^2}{r_e^2 - r_i^2}\right)$$

where p_s is the shrinkage pressure.

The stress distribution in the cylinders and jacket due to the shrinking of the jacket on the cylinder, and the internal pressure, are shown in Fig. 12.25.

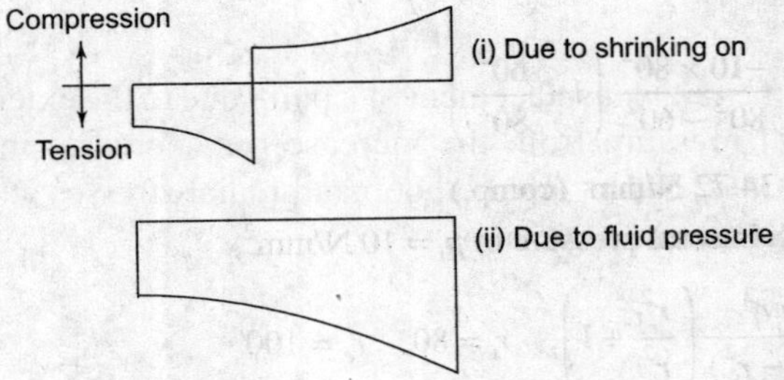

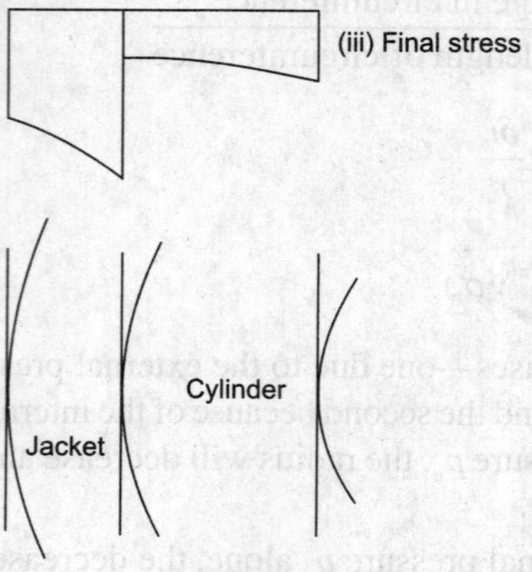

Fig. 12.25

The following examples illustrate the use of these formulae.

Example 12.28 Compound cylinder

A compound cylinder is made by shrinking a jacket on to a cylinder. For the compound cylinder, the outer and inner radii are 100 mm and 60 mm, and the radius at the junction is 80 mm. Before the fluid pressure of 40 N/mm^2 is applied, the radial pressure at the junction is 10 N/mm^2. Determine the final stresses in the cylinder. Also calculate the difference in the diameters of tubes before the jacket is shrunk on to the cylinder and the temperature at which this can be done. Take $E = 200$ GPa and $\alpha = 12 \times 10^{-6}/°\text{C}$.

Solution When the jacket is shrunk on to the cylinder, the cylinder is under compression, and the jacket under tension. For the cylinder, the radial pressure is an external pressure, while for the jacket, it acts as an internal pressure. We have to determine the hoop stresses.

For the inner cylinder, with only the external pressure $p_e = 10$ N/mm^2 acting,

$\sigma_{h\,\max}$ at $r = 60$ mm is given by

$$\sigma_{h\,\max} = \frac{-p_e r_e^2}{(r_e^2 - r_i^2)}\left(1 + \frac{r_i^2}{r^2}\right), \quad \text{(here } r = r_i\text{)}$$

$$= \frac{-2 p_e r_e^2}{r_e^2 - r_i^2} = \frac{-2 \times 10 \times 80^2}{(80^2 - 60^2)} = 45.71 \text{ N/mm}^2 \text{ (comp.)}$$

$$\sigma_h \text{ at } r = 80 = \frac{-p_e r_e^2}{r_e^2 - r_i^2} = \left(1 + \frac{r_i^2}{r_e^2}\right)$$

$$= \frac{-10 \times 80^2}{80^2 - 60^2}\left(1 + \frac{60^2}{80^2}\right)$$

$$= 34.72 \text{ N/mm}^2 \text{ (comp.)}$$

For the jacket, with internal pressure of $p_i = 10$ N/mm^2,

$$\sigma_h = \frac{p_i r_i^2}{(r_e^2 - r_i^2)}\left(\frac{r_e^2}{r^2} + 1\right), \quad r_i = 80, \quad r_e = 100$$

σ_h is maximum at $r = r_i$.

$$\sigma_{h\max} = \frac{10 \times 80^2}{(100^2 - 80^2)}\left(\frac{100^2}{80^2} + 1\right) = 45.55/\text{mm}^2 \text{ (tensile)}$$

$$\sigma_h \text{ at } r = 100 = \frac{10 \times 80^2 \times 2}{(100^2 - 80^2)} = 35.55 \text{ N/mm}^2 \text{ (tensile)}$$

With the fluid pressure of 40 N/mm^2, the inner cylinder has an internal pressure of 40 N/mm^2.

$\sigma_{h\max}$ will be at $r = 60$.

$$\sigma_{h\max} = \frac{p_i r_i^2}{(r_e^2 - r_i^2)}\left(\frac{r_e^2}{r^2} + 1\right)$$

Here $r_e = 100$ and $r_i = 60$.

$$\sigma_{h\max} = \frac{40 \times 60^2}{(100^2 - 60^2)}\left(\frac{100^2}{60^2} + 1\right) = 84.98 \text{ N/mm}^2 \text{ (tensile)}$$

At the junction, $r = 80$,

$$\sigma_h = \frac{40 \times 60^2}{100^2 - 60^2}\left(\frac{100^2}{80^2} + 1\right)$$

$$= 57.65 \text{ N/mm}^2 \text{ (tensile)}$$

At the outer end, $r = 100$.

$$\sigma_h = \frac{40 \times 60^2}{100^2 - 60^2}\left(\frac{100^2}{100^2} + 1\right)$$

$$= 45 \text{ N/mm}^2$$

The stresses can be tabulated as follows.

Condition	***Hoop stresses (N/mm^2)***			
	Inner tube		***Jacket***	
	$r = 60$	$r = 80$	$r = 80$	$r = 100$
Shrink-on pressure 10 N/mm^2	– 45.71	– 34.72	+ 45.55	+ 35.55
Fluid pressure 40 N/mm^2	+ 84.98	+ 57.65	+ 57.65	+ 45
Total	+ 39.27	+ 22.93	103.2	+ 80.55

The difference between the outer diameter of the inner tube and the inner diameter of the jacket is the shrinkage allowance.

$$\text{Shrinkage allowance} = \frac{2 p_s r_2}{E}\left(\frac{r_3^2 + r_e^2}{r_3^2 - r_e^2} + \frac{r_e^2 + r_i^2}{r_e^2 - r_i^2}\right)$$

Here,

$$p_s = 10 \text{ N/mm}^2, \quad r_3 = 100, \quad r_2 = 80, \quad r_i = 60 \text{ mm}$$

$$\text{Shrinkage allowance} = \frac{2\times 10\times 80}{2\times 10^5}\left(\frac{100^2+80^2}{100^2-80^2}+\frac{80^2+60^2}{80^2-60^2}\right)$$

$$= 0.065 \text{ mm}$$

To find the temperature difference,

$$\frac{0.065}{\pi d'\alpha T} \quad \pi\delta d'$$

$$T = \frac{\delta d'}{\alpha d'}$$

$$= \frac{0.065}{12\times 10^{-6}\times 160} = 33.8°\text{C}$$

12.7 BENDING OF CURVED BARS

We have discussed the bending of structural elements of only straight members in earlier chapters. The bending equation, $M/I = \sigma/y = E/R$, is of relevance to straight members, where $1/R$, the curvature is small or the radius is large compared to the dimensions of the members. If the member itself has a curved central line with large radius of curvature, this formula is applicable. If the ratio $d/R \approx 0.2$ or less, the straight beam formulae cannot be applied. Please note that the bars considered are curved in the vertical plane which is the plane of loads. Bars, having curvature in plan, behave quite differently, and are subjected to twisting moments in addition to BMs under transverse loads.

To understand the implications of curvature of the centre line, consider the case of a beam with a small initial curvature $1/R_i$ at the neutral surface (Fig. 12.26). The beam $ABCD$, subjected to bending couples at the ends, bends to the position $A'B'C'D'$. As the beam bends, the radius of curvature decreases and the central angle θ increases. Let the new curvature be $1/R$ and the central angle be $(\theta + \delta\theta)$. R_i, the initial radius and R the new radius are measured from the neutral layer EF.

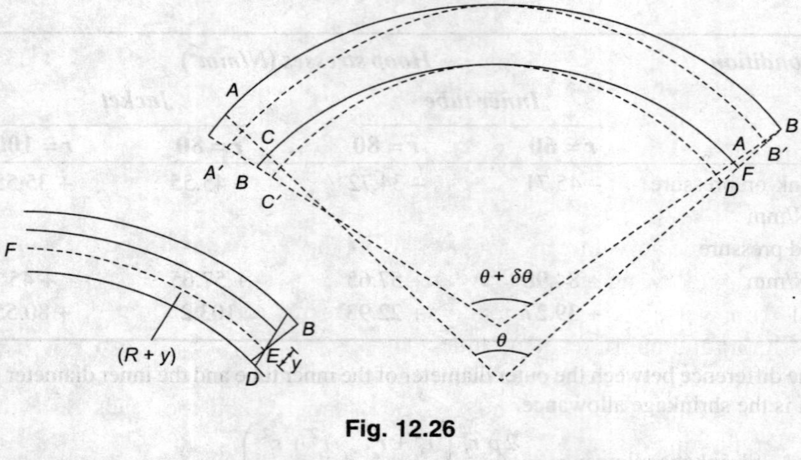

Fig. 12.26

Proceeding on the same assumptions as for a straight beam, EF is the neutral axis and has no strain. Therefore,

$$R_i\theta = R(\theta + \delta\theta)$$

We consider a fibre at a distance y from the NA. Original length of this fibre = $(R_i + y)\theta$

Increased length of the fibre = $(R + y)(\theta + \delta\theta)$

Strain in the fibre,

$$\varepsilon = \frac{(R+y)(\theta+\delta\theta) - (R_i + y)\theta}{(R_i + y)\theta}$$

$$\varepsilon = \frac{R(\theta+\delta\theta) + y\,\delta\theta - R_i\theta}{(R_i + y)\theta}$$

From $R_i\,\theta = R(\theta + \delta\theta)$, we get

$$\frac{R_i - R}{R} = \frac{\delta\theta}{\theta}$$

From these, we have

$$\varepsilon = \frac{y}{R_i + y}\frac{(R_i - R)}{R}$$

In case we consider y to be small compared to R_i, we can say that

$$\varepsilon = \frac{y}{R_i}\left(\frac{R_i - R}{R}\right) = y\left(\frac{1}{R} - \frac{1}{R_i}\right)$$

Knowing the value of strain, the stress

$$\sigma = E\varepsilon = Ey\left(\frac{1}{R} - \frac{1}{R_i}\right)$$

Then, from the simple theory of bending from which $\sigma/y = M/I$, we can state that

$$\frac{\sigma}{y} = \frac{M}{I} = E\left(\frac{1}{R} - \frac{1}{R_i}\right)$$

In this derivation we have assumed that y is small compared to R_i and the theory of simple bending is considered applicable.

In case R_i is small, as in the case of hooks and rings, we cannot make these assumptions.

12.7.1 Bars with Large Curvature

Bars with large curvature are analysed using the theory developed by Winkler and Bach. The following assumptions are made.

(i) Transverse sections remain plane before and after bending.

(ii) Longitudinal fibres bend independently.

(iii) Stresses in the beam are within the proportional limit.

(iv) The beam section has an axis of symmetry and the loads lie in a plane containing this axis.

Winkler–Bach Formula We consider the curved beam shown in Fig. 12.27(a). It is subjected to pure bending by couples applied at the ends. The transverse section is symmetric with respect to the Y-axis. O is the centre of curvature initially before bending. AD and BC are transverse sections subtending angle θ at O.

The applied couple is considered positive if it decreases the radius of curvature of the beam.

Here R_i is the initial radius of curvature of the bar and is the radius of the neutral layer. Neutral layer is the layer with no elongation or shortening. Also R is the curvature of the neutral layer after bending and R_y is the radius of a layer at y from the NA.

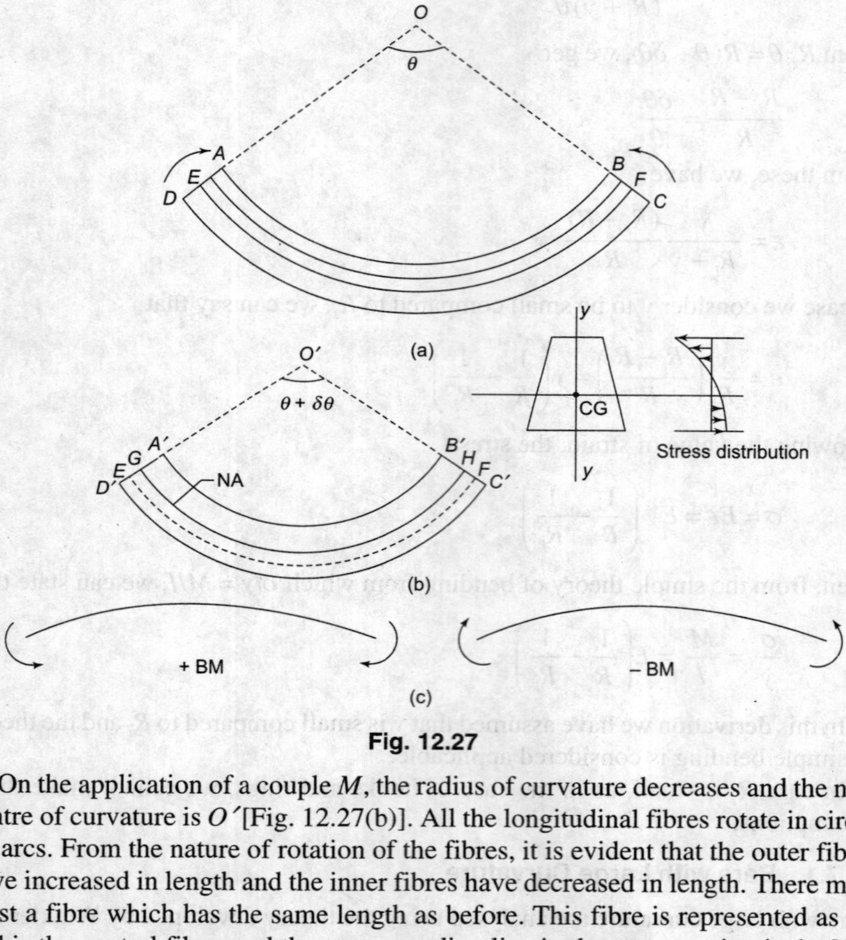

Fig. 12.27

On the application of a couple M, the radius of curvature decreases and the new centre of curvature is O'[Fig. 12.27(b)]. All the longitudinal fibres rotate in circular arcs. From the nature of rotation of the fibres, it is evident that the outer fibres have increased in length and the inner fibres have decreased in length. There must exist a fibre which has the same length as before. This fibre is represented as EF and is the neutral fibre, and the corresponding line in the cross section is the NA. The initial radius R_i and the new radius of curvature R are referred to this line. We have, therefore, since the length of EF remains the same, $R_i\theta = R(\theta + \delta\theta)$.

We now consider a fibre GH located y above the neutral surface. Length GH decreases to $G'H'$ after bending. We have, for the deformation of GH,

$$G'H' - GH = (R - y)(\theta + \delta\theta) - (R_i - y)\theta$$
$$= -y\,\delta\theta$$

This relation is an indication that the strains do not vary linearly with y. From Hooke's law,

$$\sigma = E\varepsilon = E\frac{\delta\theta}{\theta}\frac{y}{(R_i - y)}$$

The stresses σ, like ε, do not vary linearly with y. The stress variation can be seen to be hyperbolic by plotting σ against y. We have now two conditions in the case of pure bending.

(i) The sum of the forces given by $\Sigma\sigma \times \delta A = 0$, and

(ii) the sum of the moment of the forces about the NA (equal to the external couple applied), $\Sigma\sigma \times \delta A\ y$, must be equal to the moment M. That is,

$$-\int_A \sigma dA = 0$$

and $$\int_A (-y\sigma\, dA) = M$$

Substituting for σ, the first of these equations gives

$$-\int E\frac{\delta\theta}{\theta}\frac{y}{(R_i - y)}\, dA = 0$$

or $$\int \frac{y}{R_i - y}\, dA = 0$$

If R_y is the radius of the fibre GH, then

$$R_y = R - y$$

and $$\int \frac{R - R_y}{R_y}\, dA = 0$$

or $$R\int \frac{dA}{R_y} - \int dA = 0$$

This yields

$$R = \frac{\int dA}{\int dA/R_y}$$

This relation, being different from $R = (1/A)\int R_y\, dA$, which would have been the distance to the centroid, shows that the neutral surface does not contain the centroid of the section. Expressions for R for the neutral surface depend upon the cross-sectional shape, and can be derived from the above relationship.

Substituting for σ in the second condition,

$$\int \sigma dA\ y = M$$

$$\int E\frac{\delta\theta}{\theta}\frac{y}{(R_i - y)}\, y\, dA = M$$

and in terms of the radius R_y of the fibre GH,

$$\int E\frac{\delta\theta}{\theta}\frac{(R - R_y)}{R_y}(R - R_y)\, dA = M$$

$$\frac{E\,\delta\theta}{\theta}\left[\int\left(\frac{R^2+R_y^2-2RR_y}{R_y}\right)dA\right]=M$$

$$\frac{E\,\delta\theta}{\theta}\left(R^2\int\frac{dA}{R_y}+\int R_y\,dA-2R\,dA\right)=M$$

We have seen earlier that

$$R=\frac{A}{\int dA/R_y}$$

$$\int\frac{dA}{R_y}=\frac{A}{R}$$

Therefore,

$$\int R^2\frac{dA}{R_y}=RA$$

and $$\int R_y\;dA=\overline{R}A$$

where $\overline{R}$ is the radius of the centroid of the section. We thus have

$$E\frac{\delta\theta}{\theta}\,(RA-2RA-\overline{R}A)=M$$

and $$E\frac{\delta\theta}{\theta}=\frac{M}{A(\overline{R}-R)}$$

For $M>0$, $\delta\theta>0$ and $\overline{R}-R>0$ or $R<\overline{R}$. This shows that the neutral surface lies above the centroidal line and towards the centre of curvature. Setting $\overline{R}-R=S$, we get

$$E\frac{\delta\theta}{\theta}=\frac{M}{AS}$$

where S is the shift of the neutral surface from the centroidal plane.

The bending stress at y can now be expressed as

$$\sigma=\frac{My}{AS(R_i-y)}=\frac{M(R-R_y)}{ASR_y}$$

This is the Winkler–Bach formula for stress distribution in a curved bar.

In this formula, σ is the stress at a distance y from the neutral surface, A is the area of cross section, S is the shift of neutral surface from the centroidal layer, R_i is the initial radius, and R_y is the radius of the layer at which stress is being found. Also R is the radius of the neutral surface. The term y is measured from the neutral layer.

12.7.2 Sign Convention

Bending moment is considered positive when it tends to reduce the radius of curvature. Positive and negative bending moments will be as shown in Fig. 12.27(c).

Plotting σ against y, we obtain an arc of a hyperbola, which is the stress distribution (Fig. 12.27).

Let us find the curvature R of the bent beam.

$$R_i\,\theta = R(\theta + \delta\theta)$$

$$\frac{1}{R} = \frac{1}{R_i}\left(1+\frac{\delta\theta}{\theta}\right)$$

$$= \frac{1}{R_i}\left(1+\frac{M}{EAS}\right)$$

$$\frac{1}{R} - \frac{1}{R_i} = \frac{M}{EASR_i}$$

is the change in curvature.

12.7.3 Location of the Neutral Surface

R is the radial distance of the neutral surface from the centre of curvature, and can be derived for different cross-sectional shapes.

$$R = \frac{A}{\int dA/r}$$

For different cross-sectional shapes, the expressions for value of R are given in Table 12.1. For example, in the case of a rectangular section of width b and depth h,

Area $A = bh$

From Fig. 12.28(a),

$$\int_{r_1}^{r_2} \frac{dA}{r} = \int_{r_1}^{r_2} \frac{b\,dr}{r} = b[\log_e^r]_{r_1}^{r_2} = b\log_e\left(\frac{r_2}{r_1}\right)$$

$$R = \frac{A}{\int dA/r} = \frac{bh}{b\log_e(r_2/r_1)} = \frac{h}{\log_e(r_2/r_1)}$$

Similarly, in the case of a triangle of base b and depth h, from Fig. 12.28(b),

Area $A = \dfrac{bh}{2}$

$$\int_{r_1}^{r_2} \frac{dA}{r} = \int_{r_1}^{r_2} \frac{(b/h)(r_2 - r)dr}{r} = \frac{b}{h}\int_{r_1}^{r_2}\left(\frac{r_2}{r} - 1\right)dr$$

$$= \frac{b}{h}[r_2\log_e(r_2/r_1) - (r_2 - r_1)]$$

$$= \frac{b}{h}[r_2\log_e(r_2/r_1) - h]$$

$$R = \frac{A}{\int dA/r} = \frac{bh}{2b[(r_2/h)\log_e(r_2/r_1) - 1]} = \frac{h}{2[(r_2/h)\log_e(r_2/r_1) - 1]}$$

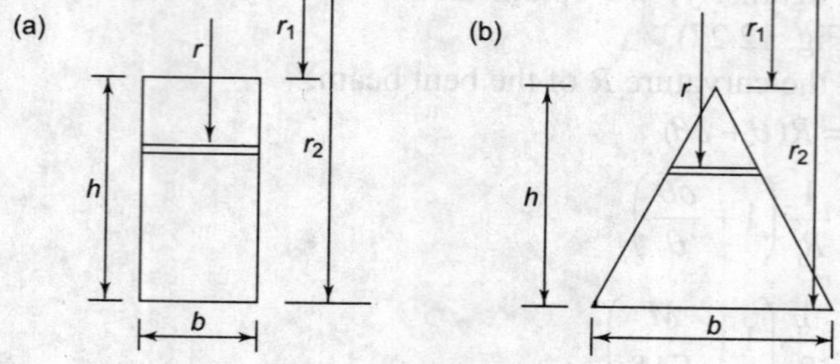

Fig. 12.28

Table 12.1 Values of R for various cross-sectional shaper

Section	***R***	***Remarks***
1. Rectangle 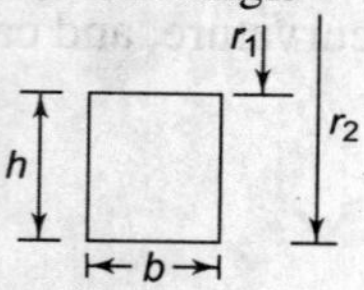	$\frac{h}{\log_e(r_2/r_1)}$	r_2 and r_1 are the radii to the bottom and top face of the beam section.
2. Triangle 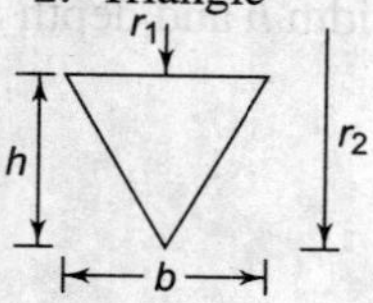	$\frac{h}{2[(r_2/h)\log_e(r_2/r_1) - 1]}$	r_1 and r_2 are the radii to the apex and base of the section.
3. Trapezium 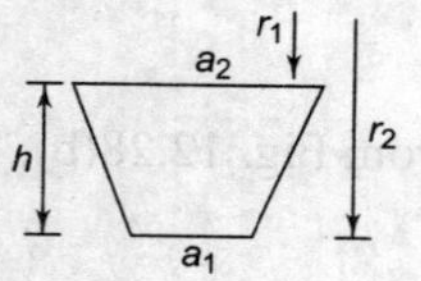	$\frac{h^2(a_1 + a_2)/2}{(a_1r_2 - a_2r_1)\log_e(r_2/r_1) - h(a_2 - a_1)}$	r_1 and r_2 are the radii to the top and bottom face of the section.
4. Circle of radius a 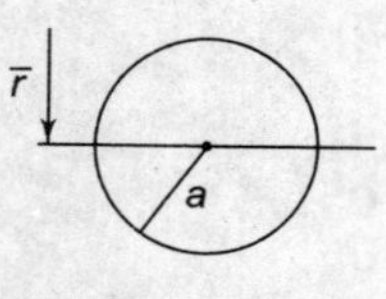	$\frac{1}{2}\left(\bar{r} + \sqrt{(\bar{r})^2 - a^2}\right)$	$\bar{r}$ is the radius to the centre of the circle.

12.7.4 Stress Distribution

The stress distribution as given by the Winkler–Bach formula is hyperbolic as seen by a plot σ across the depth of the beam. For the positive bending moment, which we have discussed and which decreases the radius of curvature, the fibres on the convex surface are in tension and the fibres on the concave surface are in compression. The following examples illustrate the procedure adopted for solving problems involving curved bars.

Example 12.29 Curved bar of a rectangular section: maximum stress

A curved bar of rectangular section 60 mm × 30 mm is subjected to a BM of 600 N m. Find the maximum stresses in the section. The mean radius of curvature, $\bar{R} = 100$ mm.

Solution We locate the neutral surface first.

$$R = \frac{h}{\log_l (r_2/r_1)} = \frac{30}{\log_l (100+15)/(100-15)} = \frac{30}{0.3023} = 99.245 \text{ mm}$$

The difference between the centroidal line and the neutral surface is only 0.755 mm. To find the maximum stress, we use the equation

$$\sigma_y = \frac{My}{Ae(R-y)}$$

or
$$\sigma = \frac{M(r-R)}{Aer}$$

where $e = 0.755$ mm.

$r = 100 + 15 = 115$ mm corresponds to the maximum tensile stress. $r = 100 - 15 = 85$ mm corresponds to the maximum compressive stress.

Thus,

$$\sigma_{max} = \frac{600(115-99.245)\times 10^3}{60\times 30\times 0.755\times 115} = 60.5 \text{ N/mm}^2 \text{ (tensile)}$$

$$\sigma_{max} = \frac{600(85-99.245)\times 10^3}{60\times 30\times 0.755\times 85} = 74 \text{ N/mm}^2 \text{ (compressive)}$$

Let us find the change from the straight bar equations.

$$\sigma_{max} = \frac{M}{I}y = \frac{600\times 10^3}{60\times 30^3/12}\times 15 = 66.66 \text{ N/mm}^2$$

Example 12.30 Curved beam of a circular section

A curved beam of circular section has an initial radius of 80 mm. The diameter of the bar is 30 mm. Find the maximum tensile and compressive stresses due to a bending moment of 400 Nm.

Solution We first find the neutral surface by finding radius R. From Table 12.1, the value of R for the section is given by

$$R = (1/2)[80 + \sqrt{(80^2 - 20^2)}] = 78.72 \text{ mm}$$

Shift = 80 – 78.72 = 1.28 mm, which is the distance between the centroidal axis and the neutral axis (NA). The neutral axis is nearer to the centre of curvature.

Bending moment = 400 Nm = 400 × 10^3 Nmm

The stress will be compressive on the concave side and tensile on the convex side.

Maximum stress is given by

$$\sigma = My/(ASR_y)$$

where y is the distance to the extreme fibre from the NA, A is the area of section, S is the shift, and R_y is the radius to the fibre.

Area of section = $\pi \times 40^2/4$ = 1256.64 mm², S = 1.28 mm

y = 20 – 1.28 = 18.72 mm for the compression fibre and 20 +1.28 = 21.28 mm to the tension fibre

R_y = 80 – 20 = 60 mm to the compression fibre and 80 + 20 = 100 mm for the tension fibre

Maximum compressive stress = $400 \times 10^3 \times 18.72/(1256.64 \times 1.28 \times 60)$
$= 77.6$ N/mm²

Maximum tensile stress = $400 \times 10^3 \times 21.28/(1256.64 \times 1.28 \times 100) = 52.9$ N/mm² ■

Example 12.31 Curved beam: T-section

A section of a curved beam is a T-section shown in Fig. 12.29. If the initial radius of curvature is 200 mm, find the maximum stresses due to a bending moment of 600 Nm.

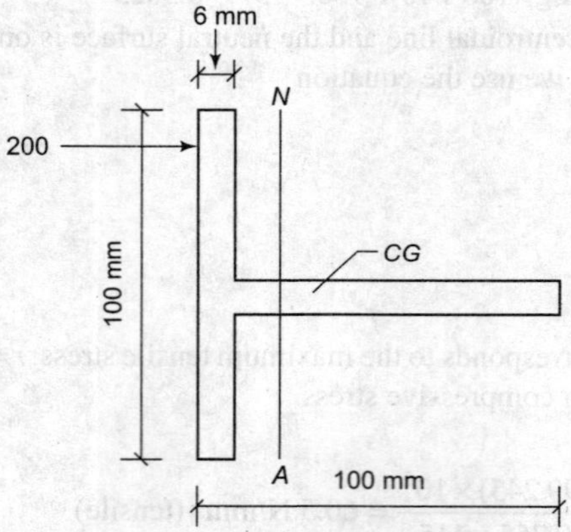

Fig. 12.29

Solution Area of the section = $100 \times 6 + 94 \times 6 = 1164$ mm²

We locate the centroid by

$\tilde{y} = (100 \times 6 \times 3 + 94 \times 6 \times 53)/1164 = 27.22$ mm from the inner surface

Radius of the neutral surface is given by

$R = A/\int (dA/R_y)$, where R_y is the radius to the elemental fibre dA.

So,

$$\int_A (dA/R_y) = \int_6^{100} \frac{6dy}{(200+y)} + \int_0^6 \frac{100dy}{(200+y)}$$

$$= 6[\log_e(200+y)]_6^{100} + 100[\log_e(200+y)]_0^6$$

$$= 2.255 + 2.956 = 5.211$$

R = 1164/5.211 = 223.4 mm

Shift = 227.22 – 223.4 = 3.8 mm

The edge of the flange will have maximum compressive stress and the edge of the web will have maximum tensile stress given by

$\sigma = M \times y/(\text{area} \times \text{shift} \times \text{radius})$

For maximum compressive stress,

$M = 600 \times 10^3$ Nmm, $y = 23.4$ mm, $A = 1164$ mm^2, shift = 3.8 mm, and radius of the edge = 200 mm

σ_{max}(compressive) = $600 \times 10^3 \times 23.4/[1164 \times 3.8 \times 200] = 15.9$ N/mm^2

Maximum tensile stress is at $y = 76.6$ mm.

Also radius = 300 mm

σ_{max}(tensile) = $600 \times 10^3 \times 76.6/[1164 \times 3.8 \times 300] = 34.6$ N/mm^2 ■

Example 12.32 Curved beam: I-section

The I-section shown is the section of the hook shown subjected to a load passing through its centre. If the permissible stress is 100 N/mm^2 in tension/compression, find the maximum load P that can be put on the hook.

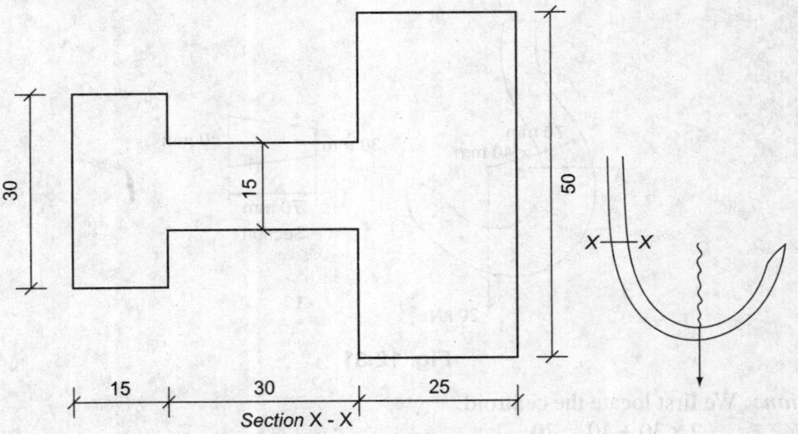

Fig. 12.30

Solution Area of the section = $30 \times 15 + 30 \times 15 + 50 \times 25 = 2150$ mm^2

The centroid of the section is at

$30 \times 15 \times 7.5 + 30 \times 15 \times 30 + 50 \times 25 \times 72.5)/2150 = 50$ mm from the edge of the smaller flange

The radius of NA,

$$R = A / \int (dA/R_y) \int (dA/R_y)$$

has to be calculated separately for the three rectangles forming the I-section.

$$\int_0^{15} \frac{30dy}{(105+y)} + \int_0^{30} \frac{15dy}{(75+y)} + \int_0^{25} \frac{50dy}{(50+y)}$$

or $\int (dA/R_y) = 30[\log_e(120/105)] + 15[\log_e(105/75)] + 50[\log_e(75/50)]$

$= 29.32$

or $R = 2150/29.32 = 73.32$ mm

Shift, $S = 85 - 73.32 = 11.68$ mm

The stress in the section will be a direct stress (tensile) with the bending stress.

Direct stress = 20,000/2150 = 9.3 N/mm^2

Bending stress will be tensile at the inner edge and compressive at the outer edge.

Bending moment = $P \times 73.32$ Nmm

$$\text{Bending stress (tensile)} = \frac{P \times 73.32 \times 28.7}{2150 \times 5.38 \times 50}$$

$$\text{Therefore, } \frac{P}{2150} + \frac{P \times 73.32 \times 28.7}{2150 \times 5.38 \times 50} = 100;\ P = 24 \text{ kN}$$

$$\text{Bending stress (compressive)} = \frac{P \times 73.32 \times 41.3}{2150 \times 5.38 \times 120}$$

$$\text{Therefore, } \frac{P}{2150} + \frac{P \times 73.32 \times 41.3}{2150 \times 5.38 \times 120} = 100;\ P = 37 \text{ kN}$$

The permissible load is the lower of the two values, i.e., $P = 24$ kN.

Example 12.33 Crane hook

A crane hook has a trapezoidal section, as shown in Fig. 12.31. For the loading shown, find the maximum stresses at section *A A*.

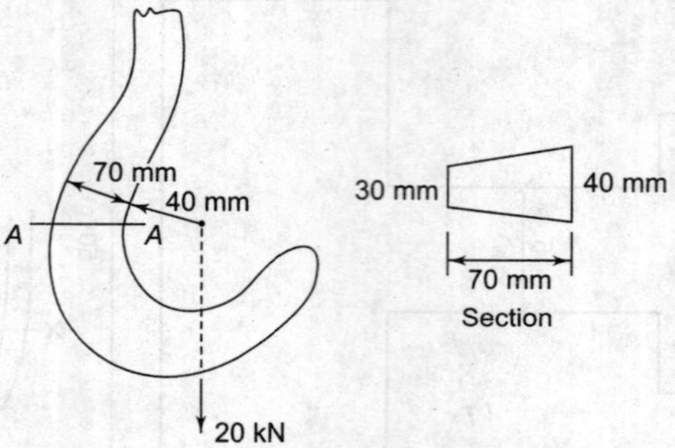

Fig. 12.31

Solution We first locate the centroid.

$$\bar{y} = \frac{2 \times 30 + 40}{30 + 40} \times \frac{70}{3} = 33.33 \text{ mm (from the 40 mm edge)}$$

The radius of the centroidal axis,

$$\bar{R} = 40 + 33.33 = 73.33 \text{ mm}$$

$$\text{Bending moment at section} = 20 \times 10^3 \times 73.33 = 1466.6 \times 10^3 \text{ N mm}$$

To locate R, the radius of the neutral surface, we use the formula for the trapezoidal section

$$R = \frac{h^2 (a_1 + a_2)/2}{(a_1 r_1 - a_2 r_1) \log (r_2 / r_1) - h(a_1 - a_2)}$$

In our case, $h = 70$ mm, $a_1 = 40$ mm, $a_2 = 30$ mm, $r_2 = 110$ mm, and $r_1 = 40$ mm. Substituting the values,

$$R = \frac{70^2 (40 + 30)/2}{(40 \times 110 - 30 \times 40) \log (110/40) - 70(40 - 30)} = 67.594$$

Therefore, the shift of the NA from the centroid = 73.333 – 67.594 = 5.739 mm.

$$\sigma_{max} (\text{at } R_y = 110) = \frac{M(R_y - R)}{ASR_y} = \frac{1466 \times 10^3 (110 - 67.594)}{2450 \times 5.739 \times 110} = 40.2 \text{ N/mm}^2$$

$$\sigma_{max} (\text{at } R_y = 40) = \frac{1466 \times 10^3 (40 - 67.594)}{2450 \times 5.739 \times 40} = 71.9 \text{ N/mm}^2 \text{ (compressive)}$$

12.7.5 Stresses in a Closed Ring

Consider a closed ring shown in Fig. 12.32(a) subjected to two loads P as shown. The ring and the loading are symmetrical. There will be no axial force in the ring at points A and B. We consider half ring shown in Fig. 12.32(b) and fix the ends A and B. At any section θ from the vertical, the section will be subjected to three internal stress resultants—a bending moment, a shear force, and an axial thrust as shown. The determination of the moment M_0 at A is statically indeterminate. An approximate solution for the determination of M_0 is made by the neglecting effect of the axial thrust. We use the theorem of minimum strain energy: $\delta U/\delta M_0 = 0$.

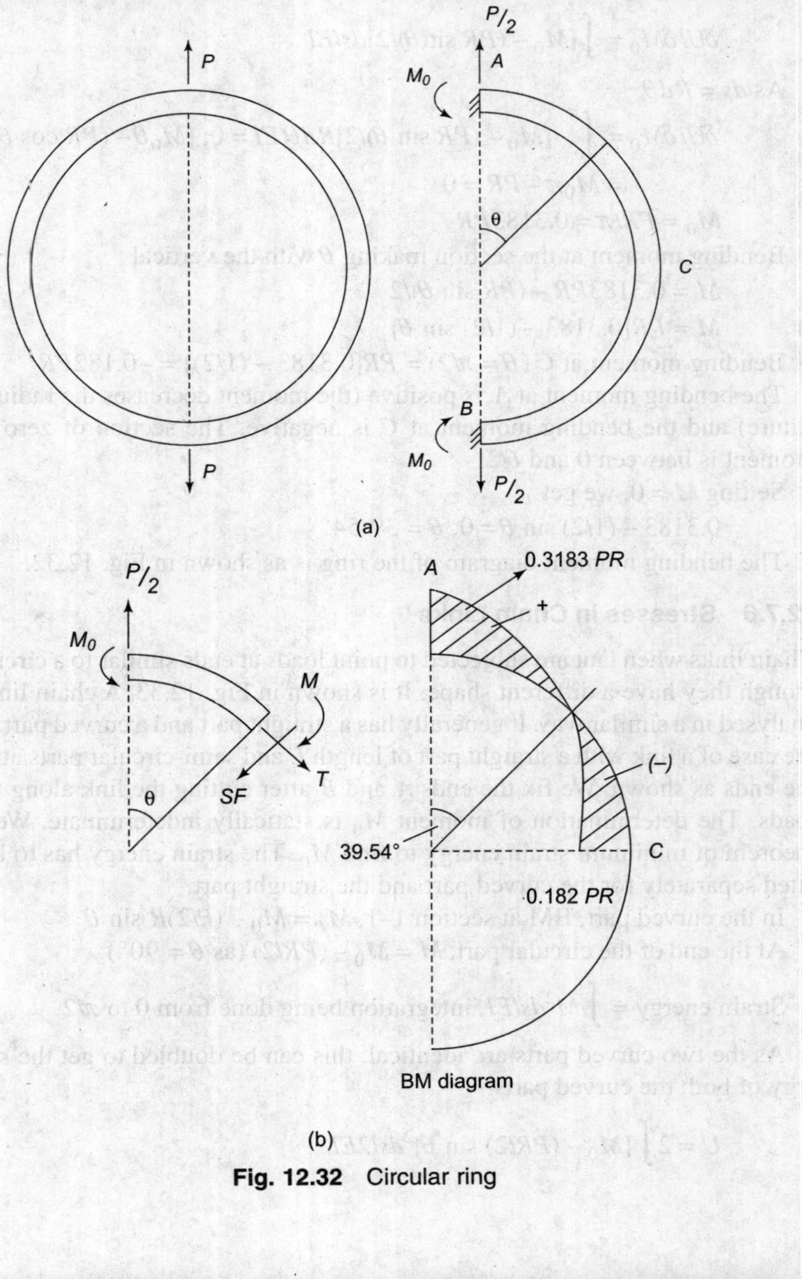

Fig. 12.32 Circular ring

The strain energy due to BM in the elemental length ds, $\delta U = M^2 ds/2EI$

Strain energy of the whole (half) ring, $U = \int M^2 ds/2EI$

$$\frac{\delta U}{\delta M_0} = \frac{d\left[\int M^2 ds/EI\right]}{dM_0} = \int M(\delta M/\delta M_0) ds/EI$$

Moment at section 1–1 (θ from vertical), $M = M_0 - (PR \sin \theta)/2$

$\delta M/\delta M_0 = 1$

$$\delta U/\delta M_0 = \int [M_0 - (PR \sin \theta)/2] ds/EI$$

As $ds = R d\theta$,

$$\delta U/\delta M_0 = \int_0^{\pi} [M_0 - (PR \sin \theta)/2] R d\theta/EI = 0; \ [M_0\theta - (PR \cos \theta)/2]_0^{\pi}$$

$$= M_0 \pi - PR = 0$$

$$M_0 = PR/\pi = 0.3183PR$$

Bending moment at the section making θ with the vertical

$$M = 0.3183PR - (PR \sin \theta)/2$$

or $\quad M = PR[0.3183 - (1/2) \sin \theta]$

Bending moment at C ($\theta = \pi/2$) $= PR[0.3183 - (1/2)] = -0.182PR$

The bending moment at A is positive (the moment decreases the radius of curvature) and the bending moment at C is negative. The section of zero bending moment is between 0 and $\theta/2$.

Setting $M = 0$, we get

$$0.3183 - (1/2) \sin \theta = 0; \ \theta = 39.54°$$

The bending moment diagram of the ring is as shown in Fig. 12.32.

12.7.6 Stresses in Chain Links

Chain links when taut are subjected to point loads at ends similar to a circular ring, though they have a different shape. It is shown in Fig. 12.33. A chain link can be analysed in a similar way. It generally has a straight part and a curved part. We take the case of a link with a straight part of length L and semi-circular parts attached at the ends as shown. We fix the ends A and B after cutting the link along the point loads. The determination of moment M_0 is statically indeterminate. We use the theorem of minimum strain energy to find M_0. The strain energy has to be calculated separately for the curved part and the straight part.

In the curved part, BM at section 1–1, $M_\theta = M_0 - (P/2)R \sin \theta$

At the end of the circular part, $M = M_0 - (PR/2)$ (as $\theta = 90°$)

Strain energy $= \int M^2 ds/EI$, integration being done from 0 to $\pi/2$

As the two curved parts are identical, this can be doubled to get the strain energy of both the curved parts.

$$U = 2\int [M_0 - (PR/2) \sin \theta]^2 ds/2EI$$

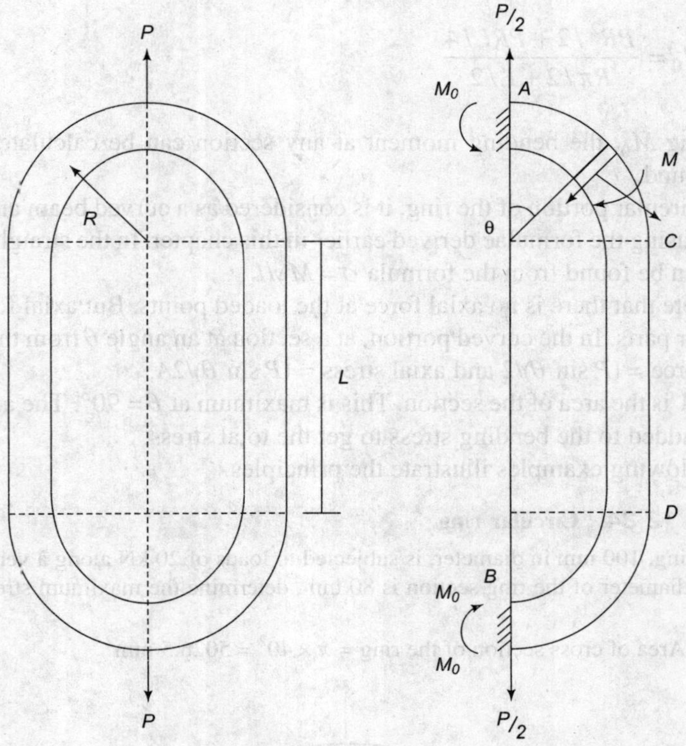

Fig. 12.33 Chain link

Now, $\delta U/\delta M_0 = 0$

Performing the differentiation within the integral sign, we have

$$\delta U/\delta M_0 = \int [M_0 - (PR/2)\sin\theta]\, ds/EI \quad (\text{as } \delta U/\delta M_0 = 1)$$

In the straight part, BM = $M_0 - (PR/2)$ and is a constant.

$$U = \int [M_0 - (PR/2)]^2 dx/2EI;\ \delta U/\delta M_0 = \int [M_0 - (PR/2)]dx/EI$$

$$(\text{as } \delta U/\delta M_0 = 1)$$

In the straight part, the integration is done from 0 to $L/2$ as the bottom half is identical to the top half and is doubled to get the strain energy of the straight part. We thus have from the theorem of minimum strain energy,

$$\frac{\delta U}{\delta M_0} = 2\int_0^{\pi/2} [M_0 - (PR/2)\sin\theta] R\,d\theta/EI + 2\int_0^{L/2} [M_0 - (PR/2)]dx/EI = 0$$

We note that EI is constant and $ds = Rd\theta$

$$[M_0 R\theta + (PR^2/2)\cos\theta]_0^{\pi/2} + [M_0 x - PRx/2]_0^{L/2} = 0$$

or $$(M_0 R\pi/2 - PR^2/2) + (M_0 L/2 - PRL/4) = 0$$

So, $$M_0 = \frac{PR^2/2 + PRL/4}{R\pi/2 + L/2}$$

Knowing M_0, the bending moment at any section can be calculated and the stresses found.

In the circular portion of the ring, it is considered as a curved beam and stresses are found using the formulae derived earlier in this chapter. In the straight part, the stresses can be found from the formula $\sigma = My/I$.

Also note that there is no axial force at the loaded points. But axial forces exist in the other parts. In the curved portion, at a section at an angle θ from the vertical,

axial force = $(P \sin \theta)/2$ and axial stress = $(P \sin \theta)/2A$

where A is the area of the section. This is maximum at $\theta = 90°$. The axial stress has to be added to the bending stress to get the total stress.

The following examples illustrate the principles.

Example 12.34 Circular ring

A circular ring, 100 mm in diameter, is subjected to loads of 20 kN along a vertical diameter. If the diameter of the ring section is 80 mm, determine the maximum stresses in the ring.

Solution Area of cross section of the ring = $\pi \times 40^2 = 5026.5\ \text{mm}^2$

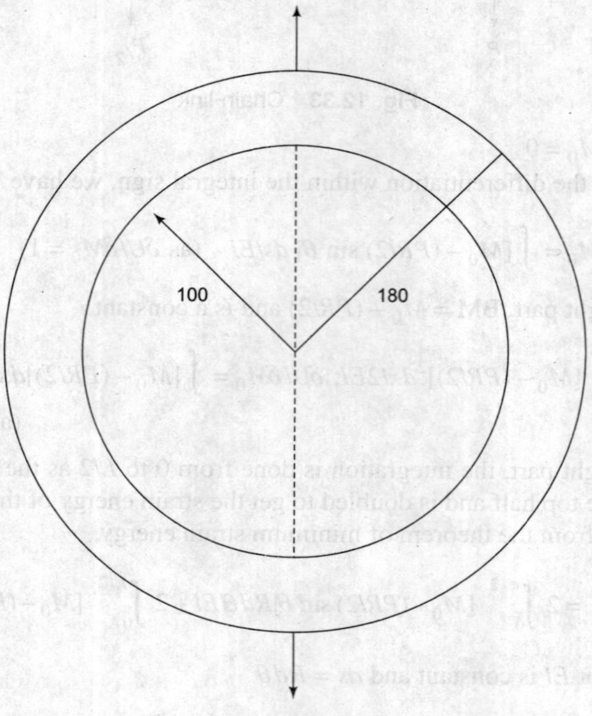

Fig. 12.34

We cut the ring into two along the load line and fix the ends as shown in Fig. 12.34. If M_0 is the fixing couple at A, the BM at any section making an angle è with the vertical is

$M = M_0 - (PR \sin \theta)/2$

This is the maximum positive moment (reducing the radius of curvature) at A and the maximum negative at the horizontal diameter section. The moment changes from positive to negative in between at 39.54° from A.

$M_0 = 0.3183PR$; $M = 0.3183PR - (PR \sin \theta)/2$

Radius of the neutral surface for a circular section,

$R = (1/2)[r^2 + \sqrt{(r^2 + a^2)}]$

where r is the radius of the centroidal line and a is the radius of the ring section.

$R = (1/2)[120 + \sqrt{(120^2 - 40^2)}] = 116.56$ mm

Shift of the NA = 120 – 116.56 = 3.44 mm

$M_0 = 0.3183 \times 20 \times 120 = 764$ kNmm

is the moment at A. Direct stress is zero at A.

The stress in the outer side of the ring will be tensile and the stress on the interior side will be compressive.

$$\text{Stress on the outer fibre} = \frac{764 \times 10^3 \times 43.44}{5026.5 \times 3.44 \times 160} = 12 \text{ N/mm}^2 \quad \text{(tensile)}$$

$$\text{Stress on the inner fibre} = \frac{764 \times 10^3 \times 36.56}{5026.5 \times 3.44 \times 80} = 20.19 \text{ N/mm}^2 \quad \text{(compressive)}$$

The bending moment at section at $C = M_0 - (PR/2) = 764 - 20 \times 120/2 = -436$ kNmm

This bending moment is negative (decreasing the radius of curvature). This will cause compression on the outer fibre and tension on the inner fibre. The section also carries direct stress, which is tensile.

Direct stress is

$P/A = 10{,}000/5026.5 = 2$ N/mm^2

$$\text{Stress at outer fibre} = \frac{436 \times 10^3 \times 43.44}{5026.5 \times 3.44 \times 160} = 6.84 \text{ N/mm}^2 \quad \text{(tensile)}$$

$$\text{Stress at the inner fibre} = \frac{436 \times 10^3 \times 36.56}{5026.5 \times 3.44 \times 80} = 4.85 \text{ N/mm}^2 \quad \text{(compressive)}$$

Total stress at the outer fibre = 6.84 - 2 = 4.84 N/mm^2

Total stress at the inner fibre = 4.85 + 2 = 6.85 N/mm^2

Example 12.35 Chain link

A chain link is of the form shown in Fig. 12.35 and is subjected to loads as shown. Find the distribution of the bending moment along the link and find the maximum stresses in the link.

Solution Area of cross section of the link = $\pi \times 30^2 = 2827.4$ mm^2

$R = (1/2)[60 + \sqrt{(60^2 - 30^2)}] = 56$ mm

Shift of NA from CG = 4 mm

Considering half part of the link and considering the ends as fixed, we have the fixing end moment as

$$M_0 = \frac{(PR^2/2) + (PRL/4)}{R(\pi/2) + (L/2)}$$

In this case, $P = 40$ kN, $R = 56$ mm, and $L = 50$ mm

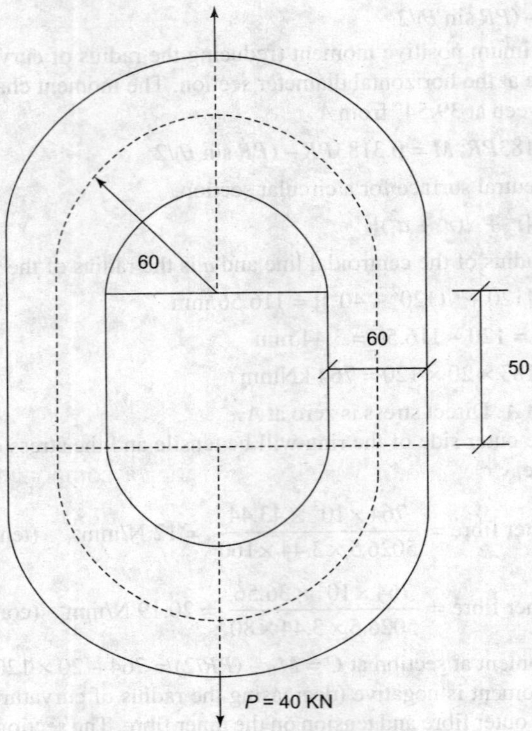

Fig. 12.35

We thus get

$$M_0 = \frac{40 \times (60^2/2) + 40 \times 60 \times (50/4)}{60 \times (\pi/2) + (50/2)} = 803 \text{ kNmm}$$

This is the moment at A or B. There is no axial force at this section. This moment is positive and causes tension on the outside and compression on the inside.

Stress, $\sigma = My/(ASR_y)$

At A, Y for top fibre = 34 mm, y for bottom fibre = 26 mm, S = 4 mm, and R_y = 60 + 30 = 90 mm

y for the bottom fibre = 26 mm, R_y = 60 – 30 = 30 mm

Maximum tensile stress at $A = \dfrac{803 \times 10^3 \times 34}{2827.4 \times 4 \times 90} = 26.83 \text{ N/mm}^2$ (tensile)

Maximum compressive stress at $A = \dfrac{803 \times 10^3 \times 26}{2827.4 \times 4 \times 30} = 61.5 \text{ N/mm}^2$

At C, the BM is

$$M = M_0 - (PR/2) = 803 - 40 \times 56/2 = -317 \text{ kNmm}$$

The bending stresses in the curved part of C is a direct stress, which is equal to 20,000/ 2827.4 = 7.07 N/mm^2

The bending stress will be tensile on the inside and compressive on the outside.

For the inside fibre, y = 26, R_y = 30, and S = 4 mm

Tensile stress = $\dfrac{317 \times 10^3 \times 26}{2827.4 \times 4 \times 30} = 24.29$

Total stress = 24.29 + 7.1 = 31.4 N/mm^2 (tensile)

Compressive stress = $\dfrac{317 \times 10^3 \times 34}{2827.4 \times 4 \times 30} = 10.6$

Total stress = 10.6 – 7.07 = 3.53 N/mm^2 (compressive)

The same negative bending moment acts in the straight portion as well. In the straight part, the bending stress is calculated as My/I.

I for the section = $\pi r^4/4 = \pi \times 30^4/4 = 636{,}172$ mm^4

Maximum compressive (tensile) stress = $317 \times 10^3 \times 30/ 636{,}172$
= 15 N/mm^2 at the outside (inside)

Total stress = 17 – 7.1 = 9.9 N/mm^2 (compressive)

Maximum tensile stress = 15 + 7.1 = 22.1 N/mm^2 (tensile) ■

12.8 THEORIES OF ELASTIC FAILURE

A structural or a machine component may fail either by actual rupture (resulting in physical separation of particles) or by yielding (or slipping of particles resulting in excessive plastic deformation). When we consider a component to have failed from the engineering point of view, rupture need not have taken place. Excessive yielding, beyond a specified limit, can be considered a condition of failure.

Generally, a component subjected to a complex loading system has principal stresses σ_1, σ_2, and σ_3 in the principal directions (σ_1 and σ_2 in the case of a plane stress system). It is common practice to compare the largest of these values to the elastic limit stress σ_y obtained in a simple tension test. However, it has been observed that σ_1 is not necessarily equal to σ_y when failure occurs. The other accompanying stresses also have a bearing on failure. No rigorous method is available to relate the value of σ_y obtained in a simple axial load test to the stress causing failure. Also, a single theory cannot satisfy the failure conditions in rupture (brittle failure) or in yielding (ductile failure).

Many hypotheses have been proposed to explain the conditions governing the elastic failure of materials. We will consider the following theories advanced by different researchers.

1. The maximum principal stress theory
2. The maximum principal strain theory
3. The maximum shear stress theory
4. The maximum total strain energy theory
5. The maximum shear strain energy theory

These theories cannot cover the conditions of failure for all the materials, and also they cover only cases other than local buckling or instability.

12.8.1 The Maximum Principal Stress Theory

This theory, also called Rankine's theory, is the oldest and the simplest of the theories of failure. According to it, failure can be established when the principal stress σ_1 (maximum tensile stress) reaches the value of the yield point stress σ_{yt} or the minimum principal stress σ_3 (compressive stress) reaches the yield point stress of σ_{yc} in a simple compression test. If $\sigma_{yc} = \sigma_{yt}$, then the condition can be expressed as

$$\sigma_1 = \sigma_y$$
$$|\sigma_3| = \sigma_y$$

If the yield stress in compression, σ_{yc}, is equal to the yield stress in tension, then in a plane stress system, $\sigma_1 = \sigma_y$ or $|\sigma_2| = \sigma_y$. This can be represented by the square shown in Fig. 12.36(a). The horizontal line represents the σ_2-axis and the vertical line represents the σ_2-axis. This representation is for a plane stress system with principal stresses σ_1 and σ_2, which are both limited by σ_y. This theory is found to be suitable only for brittle materials. Generally, in the case of such materials, $\sigma_{yc} > \sigma_{yt}$. The failure occurs when one of the principal stresses reaches the value of σ_{yc} or σ_{yt}. The material fails by fracture as there is no yielding of the material before rupture.

Generally, brittle materials are strong in compression and weak in tension. The representation in such cases will be as shown in Fig. 12.36(b), where $OA = OB = \sigma_{yt}$ and $OA' = OB' = \sigma_{yc}$ for a plane stress condition.

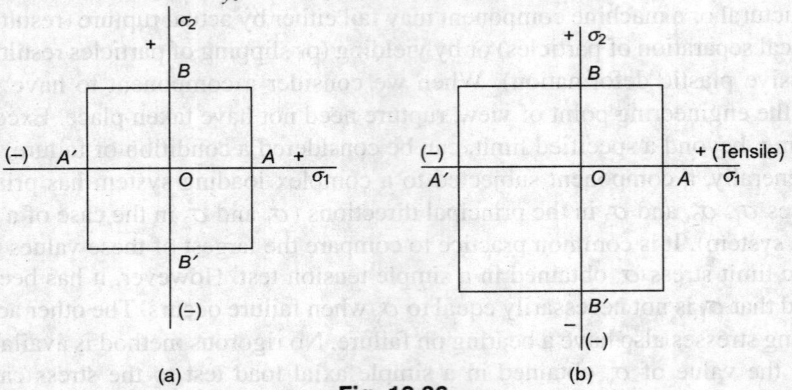

Fig. 12.36

12.8.2 The Maximum Principal Strain Theory

This theory was proposed by Saint Venant. It states that in the case of ductile material, failure can be established when the principal tensile strain reaches the value of yield strain in a simple tension test or when the principal compressive strain reaches the value of the yield compressive strain in a simple compression test.

For a compound three-dimensional stress system, the strains can be arrived at as

$$\varepsilon_1 = \frac{\sigma_1}{E} - \frac{\nu}{E}(\sigma_2 + \sigma_3)$$

$$\varepsilon_2 = \frac{\sigma_2}{E} - \frac{\nu}{E}(\sigma_1 + \sigma_3)$$

and $$\varepsilon_3 = \frac{\sigma_3}{E} - \frac{\nu}{E}(\sigma_1 + \sigma_2)$$

If ε_{yt} is the yield point strain in tension, $\varepsilon_{yt} = \sigma_{yt}/E$ and ε_{yc} that in compression, then $\varepsilon_{yc} = \sigma_{yc}/E$, where σ_{yc} and σ_{yt} are the yield stresses in compression and tension.

Thus,

$$\varepsilon_1 = \frac{\sigma_1}{E} - \frac{\nu}{E}(\sigma_2 + \sigma_3) = \varepsilon_{yt} = \frac{\sigma_{yt}}{E}$$

or $\quad \sigma_1 - \nu(\sigma_2 + \sigma_3) = \sigma_{yt}$

and $\quad \varepsilon_3 = \dfrac{\sigma_3}{E} - \dfrac{\nu}{E}(\sigma_1 + \sigma_2) = \varepsilon_{yc} = \dfrac{\sigma_{yc}}{E}$

or $\quad \sigma_3 - \nu(\sigma_1 + \sigma_2) = \sigma_{yc}$

In a two-dimensional case, $\sigma_3 = 0$ and these equations reduce to

$$\sigma_1 - \nu\sigma_2 = \sigma_{yt}$$

and $\quad \sigma_2 - \nu\sigma_1 = \sigma_{yc}$

This theory can be graphically represented as shown in Fig. 12.37. When the stresses fall within the boundary lines, the material is safe. This theory is also found to be applicable to some brittle materials.

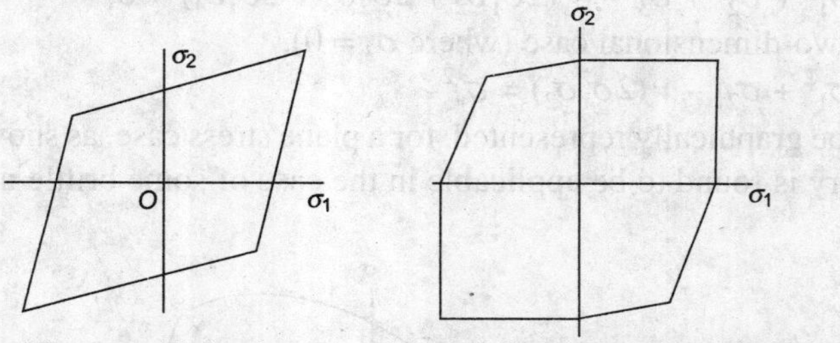

(a) σ and E are equal in tension and compression

(b) σ and E are unequal in tension and compression

Fig. 12.37

This theory, in effect, states that the stress which acting alone will produce the strain ε_1 is $(\sigma_1 - \nu\sigma_2)$.

12.8.3 The Maximum Shear Stress Theory

This theory is known as Guest's theory or Coulomb's theory, and proposes that yielding begins when the maximum shearing stress equals the maximum shearing stress in a simple tension test. In such a test, the shear stress is equal to the difference between the maximum and minimum principal stresses (σ_1 and σ_3). Since $\sigma_3 = 0$ (or in a plane stress case, $\sigma_2 = 0$), we get

$$\tau_{max} = \frac{1}{2}\sigma_1 = \frac{\sigma_y}{2}$$

Hence the condition of yielding can be expressed as $\sigma_1 - \sigma_3 = \sigma_y$ for a three-dimensional stress system and $\sigma_1 - \sigma_2 = \sigma_y$ for a two-dimensional stress system.

This theory agrees with practical results only in the case of some ductile materials, and can be represented graphically, for a plane stress case, as shown in Fig. 12.38.

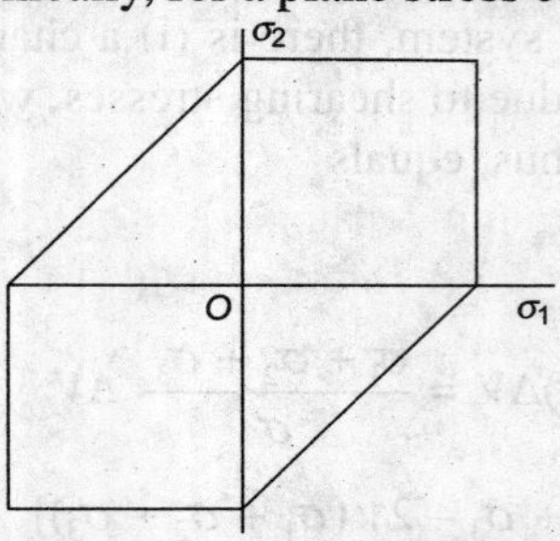

Fig. 12.38 Representation of the maximum shear stress theory

12.8.4 The Maximum Total Energy Theory

This theory, originally credited to Beltrmi, is also known as the Haigh's theory. It proposes that failure occurs when the materials under complex stresses attain strain energy more than that in a simple tension or compression test.

In a three-dimensional stress system, the strain energy per unit volume can be expressed as

$$U = \frac{1}{2E}[\sigma_1^2 + \sigma_2^2 + \sigma_3^2 - \nu(2\sigma_1\sigma_2 + 2\sigma_2\sigma_3 + 2\sigma_3\sigma_1)]$$

The yield criterion in such cases can be represented by

$$\sigma_1^2 + \sigma_2^2 + \sigma_3^2 - \nu(2\sigma_1\sigma_2 + 2\sigma_2\sigma_3 + 2\sigma_3\sigma_1) = \sigma_y^2$$

For a two-dimensional case (where $\sigma_3 = 0$),

$$\sigma_1^2 + \sigma_2^2 - \nu(2\sigma_1\sigma_2) = \sigma_y^2$$

This can be graphically represented, for a plane stress case, as shown in Fig. 12.39. This theory is found to be applicable in the case of some brittle materials.

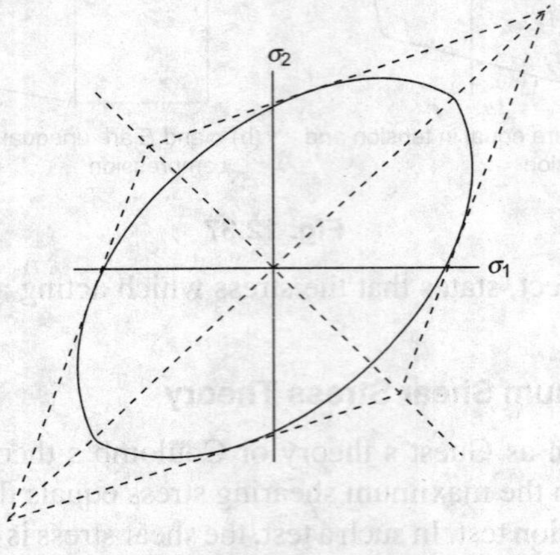

Fig. 12.39 Illustrating the total strain energy theory

12.8.5 The Maximum Shear Strain Energy Theory

This theory is also known as the Mises–Hencky theory or the distortion energy theory. It proposes that yielding begins only when the strain energy of distortion per unit volume absorbed at a point due to a complex stress system equals the distortion strain energy absorbed per unit volume in a bar stressed to the elastic limit in a simple tension (or compression) test.

Under a complex stress system, there is (i) a change in volume due to direct stresses and (ii) distortion due to shearing stresses, which does not cause volume change. The total energy, thus, equals

$$U = U_v + U_s$$

The distortion energy,

$$U_s = \frac{1}{2}(\text{average } \sigma)\Delta V = \frac{\sigma_1 + \sigma_2 + \sigma_3}{\sigma}\Delta V$$

and $$\Delta V = \frac{1}{E}[\sigma_1 + \sigma_2 + \sigma_3 - 2\nu(\sigma_1 + \sigma_2 + \sigma_3)]$$

$$= \frac{(\sigma_1 + \sigma_2 + \sigma_3)}{E} (1 - 2\nu)$$

$$U_v = \frac{(1-2\nu)}{6E} (\sigma_1 + \sigma_2 + \sigma_3)^2$$

$$= \frac{(1-2\nu)}{6E} [\sigma_1^2 + \sigma_2^2 + \sigma_3^2 + 2(\sigma_1\sigma_2 + \sigma_2\sigma_3 + \sigma_3\sigma_1)]$$

Also,

$$U = \frac{1}{2E} [\sigma_1^2 + \sigma_2^2 + \sigma_3^2 - 2\nu(\sigma_1\sigma_2 + \sigma_2\sigma_3 + \sigma_3\sigma_1)]$$

Hence the distortion energy, $U_s = U - U_v$. Subtracting, we get

$$U_s = \frac{1+\nu}{6E} [(\sigma_1 - \sigma_2)^2 + (\sigma_2 - \sigma_3)^2 + (\sigma_3 - \sigma_1)^2]$$

$$= \frac{1}{12G} [(\sigma_1 - \sigma_2)^2 + (\sigma_2 - \sigma_3)^2 + (\sigma_3 - \sigma_1)^2]$$

since $E = 2G(1 + \nu)$.

The strain energy of distortion in a simple tension test can be expressed as

$$U_s = \frac{1}{6G} \sigma_y^2$$

The failure criterion, thus, becomes,

$$\frac{1}{12G} [(\sigma_1 - \sigma_2)^2 + (\sigma_2 - \sigma_3)^2 + (\sigma_3 - \sigma_1)^2] = \frac{1}{6G} \sigma_y^2$$

In a two-dimensional case, this can be reduced to (as $\sigma_3 = 0$),

$$\sigma_1^2 + \sigma_2^2 - \sigma_1\sigma_2 = \sigma_y^2$$

This theory can be represented in a two-dimensional case as shown in Fig. 12.40.

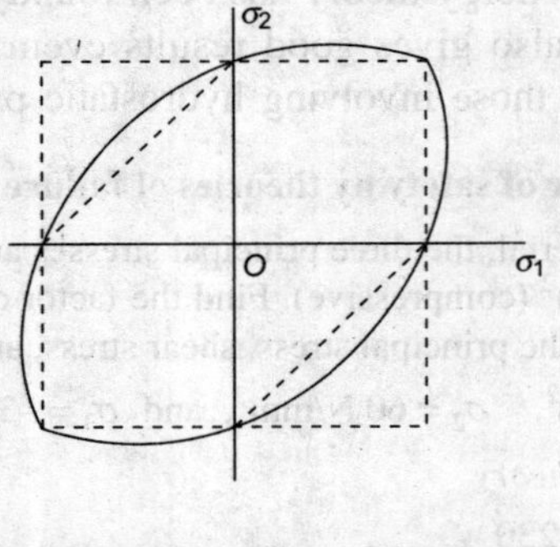

Fig. 12.40 Illustrating the maximum shear strain energy theory

A graphical representation of all the theories together is shown in Fig. 12.41. In this figure, the minimum inside area common to all the figures represents the safe region under all kinds of loading.

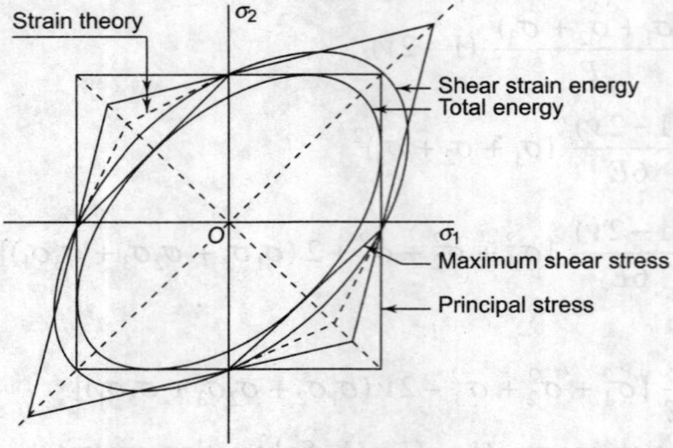

Fig. 12.41 Graphical representation of all theories of failure

12.8.6 Limitations of the Theories of Failure

The maximum principal stress theory is generally found to be applicable to brittle materials. If you compare the cases of a single tensile stress and a case of pure shear, the theory gives the same strength in both cases while the shear stress is double in the case of pure shear.

The maximum principal strain theory is applicable to some brittle materials. However, it does not cover all cases. For example, if we consider two equal tensile stresses in perpendicular directions, according to this theory, $\sigma_1(1 - \nu) = \sigma_y$, which gives $\sigma_1 > \sigma_y$, the elastic limit stress, which has been found to be unsupported by experimental results.

The maximum shear stress theory is found to be in close agreement with practical results, especially for ductile materials. However, it does not give accurate results in the case of pure shear or for nearly equal tensile stresses in mutually perpendicular directions, reducing the shear stress to a very small value.

The maximum total energy theory is applicable only to a few cases of brittle materials. The distortion energy theory has been found to be the most acceptable for ductile materials. It also gives good results even in cases of triaxial stress conditions—for instance, those involving hydrostatic pressure.

Example 12.36 Factor of safety by theories of failure

At a point in a stressed material, the three principal stresses are 90 N/mm^2 (tensile), 60 N/mm^2 (tensile) and 30 N/mm^2 (compressive). Find the factor of safety if the yield stress is 230 N/mm^2 and $\nu = 0.3$ by the principal stress, shear stress, and principal strain theories.

Solution $\sigma_1 = +90$ N/mm^2, $\sigma_2 = 60$ N/mm^2, and $\sigma_3 = -30$ N/mm^2.

Maximum principal stress theory

$$\text{Factor of safety} = \frac{230}{90} = 2.55$$

Maximum shear stress theory

$$\frac{\sigma_y}{2} = \text{maximum of } \frac{\sigma_1 - \sigma_2}{2}, \frac{\sigma_2 - \sigma_3}{2} \text{ and } \frac{\sigma_3 - \sigma_1}{2}$$

or $\frac{\sigma_y}{2}$ = maximum of $\frac{90-60}{2}$, $\frac{60-(-30)}{2}$ and $\frac{-30-90}{2}$

= 120 N/mm^2

Factor of safety = $\frac{230}{120}$ = 1.92

Principal strain theory In this case,

$$\frac{\sigma}{E} = \text{maximum of } \frac{(\sigma_1 - \nu\sigma_2 - \nu\sigma_3)}{E}, \frac{(\sigma_2 - \nu\sigma_3 - \nu\sigma_1)}{E} \text{ and } \frac{(\sigma_3 - \nu\sigma_1 - \nu\sigma_2)}{E}$$

$$\frac{\sigma}{E} = \text{maximum of } \frac{(90 - 0.3 \times 60 + 0.3 \times 30)}{E}, \frac{(60 + 0.3 \times 30 + 0.3 \times 90)}{E}$$

and $\frac{-30 + 90 \times 0.3 - 60 \times 0.3}{E}$

$$= \frac{81}{E}$$

$\sigma = 81$

and Factor of safety = $\frac{230}{81}$ = 2.84

Example 12.37 Design of steel bolt theories of failure

Design a steel bolt subjected to a tensile load of 20 kN along with a shear force of 10 kN. The yield point for the material is 300 MPa and $\nu = 0.3$. Calculate the safe diameter by all the theories of failure if a factor of safety of 2.5 is desired.

Solution The bolt is subjected to a plane stress system. If A is the cross-sectional area of the bolt, then

$$\sigma_X = \frac{20{,}000}{A}, \quad \sigma_Y = 0, \quad \tau_{XY} = \frac{10{,}000}{A}$$

From these values, the principal stresses are

$$\sigma_{1,2} = \frac{20{,}000}{2A} \pm \sqrt{\left(\frac{20{,}000}{2A}\right)^2 + \left(\frac{10{,}000}{A}\right)^2}$$

$$= \frac{10{,}000}{A}(1 \pm \sqrt{2})$$

$$\sigma_1 = +\frac{24{,}142}{A} \text{ N/mm}^2, \quad \sigma_2 = \frac{-4142}{A} \text{ N/mm}^2$$

Maximum principal stress theory

Yield stress,

$\sigma_Y = 300$

Factor of safety = 2.5

$$\sigma_{\text{allowable}} = \frac{300}{2.5} = 120 \text{ N/mm}^2$$

Therefore,

$$\frac{24{,}142}{A} = 120, \quad A = 201 \text{ mm}^2$$

$$\frac{\pi d^2}{4} = 201$$

$$\Rightarrow \qquad d = \sqrt{\frac{201 \times 4}{\pi}} = 16 \text{ mm}$$

Maximum principal strain theory

$$\sigma_1 - \nu\sigma_2 = \sigma_{\text{allowable}} = 120$$

$$\frac{24{,}142}{A} - \frac{0.3(-4142)}{A} = 120$$

$$A = \frac{24{,}142 + 0.3 \times 4142}{120} = 211.54 \text{ mm}^2$$

$$d = \sqrt{\frac{211.54 \times 4}{\pi}} = 16.41 \text{ mm}$$

Maximum shear stress theory

$$\frac{\sigma_1 - \sigma_2}{2} = \frac{\sigma_{\text{allowable}}}{2}$$

Therefore,

$$\frac{24{,}142}{A} - \frac{(-4142)}{A} = 120$$

$$A = \frac{28284}{120} = 235.7$$

$$d = \sqrt{\frac{235.7 \times 4}{\pi}} = 17.32 \text{ mm}$$

Maximum strain energy theory

$$\sigma_1^2 + \sigma_2^2 - 2\nu\,\sigma_1\sigma_2 = (\sigma_{\text{allowable}})^2$$

$$\left(\frac{24{,}142}{A}\right)^2 + \left(\frac{-4142}{A}\right)^2 + 2 \times 0.3 \times \frac{24{,}142}{A} \times \frac{4142}{A} = (120)^2$$

$$A^2 = \frac{24{,}142^2 + (4142)^2 + 0.6 \times 24{,}142 \times 4142}{(120)^2}$$

$$A = 214 \text{ mm}^2, \quad d = 16.5 \text{ mm}$$

Distortion energy theory

$$\sigma_1{}^2 + \sigma_2{}^2 - \sigma_1\sigma_2 = (\sigma_{\text{allowable}})^2$$

$$\left(\frac{24{,}142}{A}\right)^2 + \left(\frac{-4142}{A}\right)^2 + \frac{24{,}142}{A} \times \frac{4142}{A} = (120)^2$$

$$A^2 = \frac{(24{,}142)^2 + (4142)^2 + 24{,}142 \times 4142}{(120)^2}$$

$$= 48{,}610$$

$$A = 220.5$$

$$\Rightarrow \quad d = \sqrt{\frac{220.5 \times 4}{\pi}} = 16.75 \text{ mm}$$

The diameter will be the largest of the values calculated, meaning that $d = 17.32$ mm. ■

Example 12.38 Theories of failure: thin cylindrical shell

A thin cylindrical shell, of 1 m diameter, is subjected to an internal pressure of 1.8 N/mm^2. Determine the thickness of the shell plate required by the maximum principal stress, maximum shear stress, and maximum shear energy theories. Elastic limit from a simple tension test is 220 N/mm^2. Use a factor of safety of 2.5.

Solution Elastic limit = 220 N/mm^2 and factor of safety = 2.25

Permissible stress simple tension = 220/2.5 = 88 N/mm^2

At a point in the thin cylindrical shell, there are two stresses in orthogonal directions—hoop stress and longitudinal stress. Both these are principal stresses.

If t is the thickness of the shell,

Longitudinal stress, $\sigma_l = pd/4t = 1.8 \times (1/4t) = 0.45/t$ N/m^2

Hoop stress, $\sigma_h = pd/2t = 1.8 \times (1/2t) = 0.9/t$ N/m^2

Maximum principal stress theory According to the maximum principal stress theory, the larger of the two stresses is the maximum principal stress. This must be equal to the safe stress in simple tension.

So, $0.9/t = 88$; $t = 0.9/88$ m = 10.2 mm

Maximum shear stress theory The principal stresses at a point in the shell are σ_1, σ_2, and 0. The maximum shear stress is, therefore, $(\sigma_1 - 0)/2$ or $(\sigma_2 - 0)/2$

Maximum shear stress = $0.9/t$

Maximum shear stress in simple tension = 88 N/mm^2

Therefore, $0.9/t = 88$; $t = 10.22$ mm

Distortion energy theory According to this theory, if σ_1 and σ_2 are the principal stresses,

$$\sigma_1^2 + \sigma_2^2 - \sigma_1\sigma_2 = \sigma^2, \ \sigma = 88 \text{ N/mm}^2$$

$$(0.9/t)^2 + (0.45/t)^2 - (0.9 \times 0.45t^2) = 88^2 \times t = 12.7 \text{ mm}$$

Thickness of the shell required = 13 mm ■

Example 12.39 Theories of failure: circular shaft

A circular shaft, 100 mm in diameter, is subjected to a twisting moment of 8500 Nm in addition to a BM of 6000 Nm. Check whether the shaft is safe by the principal stress, principal strain, and maximum shear stress theories if the elastic limit in a simple tension test is 230 N/mm^2. Find the factor safety in each case. Poisson's ratio, $m = 0.3$.

Solution Twisting moment = 8500 Nm and diameter of the shaft = 100 mm

Maximum shearing stress = $8500 \times 10^3 \times 16/[\pi(100)^3] = 43.3$ N/mm^2

Bending moment = 6000 Nm

Maximum bending stress = $6000 \times 10^3 \times 32/[\pi \times 100^3] = 61.1$ N/mm^2

An element on the surface of the shaft is subjected to a normal stress of 61.1 N/mm^2 and a shear stress of 43.3 N/mm^2.

Principal stress theory The principal stresses on the surface are

$$\sigma_{1,2} = 61.1 \pm \sqrt{[(61.1/2)^2 + 43.3^2]} = 83.55 \text{ N/mm}^2 \text{ or } -22.4 \text{ N/mm}^2$$

Maximum principal stress = 83.55 N/mm^2

Factor of safety = 230/83.55 = 2.75

Principal strain theory Maximum principal strain, $\varepsilon = [83.55 + 22.4 \times 0.3]/E = 90.27/E$

In simple tension test, $\varepsilon = 230/(\text{factor of safety}) \times E$
Factor of safety = 230/90.27 = 2.55

Maximum shear stress theory The principal stresses are 83.55, –22.4, and 0. The maximum shear stress in this case is (1/2)(83.55 – 0) = 41.78 N/mm^2

In a simple tension test, maximum shear stress = (1/2) × 230 = 115 N/mm^2
Factor of safety = 115/41.78 = 2.75

Example 12.40 Theories of failure: case of plain stress

At a point in a stresses material, the principal stresses are 180 N/mm^2 (tensile), 90 N/mm^2 (tensile), and 40 N/mm^2 (compressive). Determine whether failure of the material will take place according to all the elastic failure theories. Take Poisson's ratio m = 0.3. Find the factor of safety in case the failure does not occur if the elastic limit stress in simple tension is 250 MPa.

Solution *Maximum principal stress theory* According to this theory, failure will not occur if the maximum principal stress in the material does not exceed the elastic limit in simple tension. As the maximum stress is 180 N/mm^2 < 250 N/mm^2, failure will not occur according to this theory.

Factor of safety = 250/180 = 1.38

Maximum principal strain theory According to this theory, failure will not occur if the maximum principal strain is less than the elastic limit strain in simple tension.

Maximum principal strain = [180 – 0.3(90 – 40)]/E = 165/E
Strain at elastic limit = 250/E and hence failure will not occur.
Factor of safety = 250/165 = 1.5

Maximum shear stress theory With three principal stresses, the maximum shear stress is half the difference between any two of the principal stresses.

Maximum shear stress = greatest of (180 – 90)/2, (180 + 40)/2, and (90 – 40)/2 = 110 N/mm^2
Maximum shear stress in simple tension = 250/2 = 125 N/mm^2
As 110 < 125, failure will not occur.
Factor of safety = 125/110 = 1.14

Maximum strain energy theory According to this theory,

$$\sigma_1^2 + \sigma_2^2 + \sigma_3^2 - m(2\sigma_1\sigma_2 + 2\sigma_2\sigma_3 + 2\sigma_3\sigma_1) = \sigma_y^2$$

or $$180^2 + 90^2 + 40^2 - 0.3(2 \times 180 \times 90 - 2 \times 90 \times 40 - 2 \times 180 \times 40) = 250^2$$

Left hand side = 40,480 and right hand side = 62,500
Hence, failure will not occur.
Factor of safety = √[62,500/40,480] = 1.24

Maximum shear stress energy theory According to this theory,

Maximum shear stress energy = $[(\sigma_1 - \sigma_2)^2 + (\sigma_2 - \sigma_3)^2 + (\sigma_3 - \sigma_1)^2]/12G$
Strain energy in simple tension = $(\sigma_y)^2/6G$
$[(180 - 90)^2 + (90 + 40)^2 + (180 + 40)^2]/2 = (250/\text{factor of safety})^2$

So, Factor of safety = √[62,500/37,700] = 1.29

Summary

This chapter dealt with a number of components where analysis and design were based on principles enunciated in earlier chapters.

Semi-elliptic springs find wide applications in transport vehicles. Such springs consist of a number of plates tied together. Each plate is shorter than the one above by a constant length. The plates are subjected to uniform BMs between the supports. The ends of the plates are tapered to achieve uniform BM. Based on the maximum bending stress, it is possible to design the number of plates, their size and thickness for a given load. Proof load is the load required to straighten the spring. Quarter elliptic springs also find application and have only half the length of a semi-elliptic spring, and are fixed at one end.

Helical springs, close-coiled and open-coiled, are used to absorb energy—for instance, to stop a moving railway wagons. Close-coiled springs have a very small pitch—each helical turn lies in one plane. They are, thus, subjected to only a torque, and the BMs are very small and are neglected. Open-coiled helical springs, on the other hand, are subjected to bending and twisting moments. Springs are fabricated in such a way that axial loads and axial twists are applied by external action. They can be analysed and designed by the same principles for bending, shear and torsional shear stresses.

Flat spiral springs are similarly used in toys and clock mechanisms. They are designed for the maximum bending stress produced due to the applied load.

Springs are also used in series and parallel combinations. In series, they carry equal loads while, in parallel configuration, they are assumed to have the same deflections.

Cylindrical and spherical vessels to withstand the pressure of liquids and gases find wide application. Thin vessels, with thickness not more than 5% of the diameter, are analysed assuming the variation of stresses to be uniform across the thickness. They are subjected to hoop or circumferential stresses and longitudinal stresses. The pressure capacity of thin vessels can be increased by applying external pressure by winding wires.

Thick cylindrical vessels have radial, hoop and longitudinal stresses. The variation of these stresses is parabolic across the thickness and cannot be assumed to be uniform. Lamé's theory is used to analyse and design such vessels. Graphical methods are also available to analyse thick cylindrical vessels. The capacity of such a vessel can be increased by shrinking an external jacket on to it, which makes for compression on the vessel. With the application of internal pressure, the stresses are much less in the cylinder.

Curved beams, whose longitudinal axis is not a straight line, are analysed using the Winkler-Bach theory. Such beams with an axis of symmetry which is also the plane of loading, have the neutral axis not coinciding with the centroidal axis, and also a nonlinear distribution of bending stress. Crane hooks which have large initial curvature are analysed using the Winkler–Bach theory.

Theories of failure relate the yield stress obtained in a simple tension or compression test to the condition obtaining at failure. The maximum principal stress theory, maximum principal strain theory, maximum shear stress theory, maximum total energy theory, and maximum distortion energy theory are the important theories of failure. They are not uniformly applicable and find agreement in the case of some brittle and ductile failures. No single theory has been found to apply to all cases of failure.

Exercises

Review Questions

1. The plates of a carriage spring are subjected to a constant moment at any section. Explain how this is achieved.
2. What is a quarter elliptic spring? How is it supported?
3. Explain the difference in the behaviour of close-coiled and open-coiled springs.
4. What is meant by the stiffness of a spring?
5. What are the advantages of springs? Enumerate their applications.
6. How do we ensure that the loads and torques are applied axially to a spring, close- or open-coiled?

7. What are the different forces and moments to which the section of an open-coiled spring is subjected?
8. When two springs are connected in series, what is condition assumed to evaluate their composite stiffness? What is the condition in case the springs are in parallel?
9. State the assumptions made in the analysis of thin and thick cylindrical shells.
10. Explain the difference in the behaviour of thin and thick cylindrical shells.
11. Which shell is more efficient in resisting pressure—the cylindrical or the spherical shell? Why?
12. In the case of cylindrical shells, what is done to enhance their pressure-bearing capacity? Explain how the shell behaves when such steps are taken.
13. State the difference in the structural action of curved and straight members. What is the difference in the bending stress distributions of the members?
14. Explain why no single theory can satisfy the failure condition for all materials.

Problems

1. A steel carriage spring has to carry a load of 8 kN. It is 1200 mm long and consists of plates 10 mm wide and 8 mm thick. Determine the number of plates required if the maximum stress is limited to 300 MPa. Find the deflection under the load. $E = 200$ GPa.
2. A laminated spring with a span of 1 m has to carry a central load of 10 kN. The bending stress has to be limited to 300 MPa, and the central deflection to 60 mm. The plates are provided with thickness in multiples of 1 mm and width in multiples of 4 mm. Assume the width to be eight times the thickness. $E = 200$ GPa. Calculate the number of plates, maximum stress and deflection, and the radius to which the plates are to be bent.
3. A carriage spring is 750 mm long, and consists of 12 plates of width 50 mm and thickness 6 mm. The bending stress is limited to 200 MPa and the deflection should not be more than 12 mm. Find the maximum load that the spring can carry at its centre. Also determine the strain energy under this load. $E = 200$ GPa.
4. A laminated spring, 1 m long, is made up of plates, each 50 mm wide and 10 mm thick. If the bending stress in the plates is limited to 105 MPa, how many plates will be required to enable the spring to carry a central point load of 2 kN? If $E = 2 \times 10^5$ MPa, what is the deflection under the given load?
5. A carriage spring is 1.3 m long and has leaves of 100 mm width and 12 mm thickness. The maximum bending stress is 105 MPa and the spring must absorb 120 Nm of energy when straightened. Calculate the number of leaves and their initial radius of curvature. Assume $E = 200$ GPa.
6. A flat spring of rectangular section is 25 mm broad, 0.5 mm thick, and 6 m long. Determine the twisting moment it may exert on a central spindle if the bending stress is not to exceed 700 MPa. How many complete turns may be given to the spring when it is run down and how much work is then stored in it? Take $E = 200$ GPa.
7. A leaf spring of semi-elliptic type consists of 8 plates, each of 40 mm width and 6 mm thickness. The length of the longest plate is 500 mm and $E = 2 \times 10^5$ MPa. The central deflection is limited to 12 mm. Determine the load that can be applied.
8. A quarter elliptic spring is 500 mm long. The plates are 100 mm wide and 8 mm thick. The spring has a static deflection of 50 mm under a load of 2 kN. Find the number of plates required and the maximum bending stress. $E = 200$ GPa.
9. A close-coiled spring has 20 turns of wire of diameter 12 mm. The mean radius of the coil is 75 mm. If the coil is subjected to an axial couple, find the strain energy stored

in the spring when one end of the spring is rotated by 90° relative to the other. $E =$ 200 GPa.

10. A close-coiled spring made of a 16 mm ϕ rod has 18 complete turns and a mean diameter of 140 mm. It is subjected to an axial pull of 500 N. Calculate the maximum stress in the spring if $G = 84$ GN/m^2. Find the deflection and the strain energy stored in the spring.
11. A wagon weighing 50 kN moving at a speed of 8 kmph has to be brought to rest. Springs made of 25 mm ϕ rods with a mean diameter of 250 mm and with 24 turns are available. Find the number of springs required in the buffer to stop the wagon at a compression of 180 mm. $G = 84$ GN/m^2.
12. A close-coiled spring is to be subjected to a maximum load of 50 N, and the maximum shearing stress is limited to 120 N/mm^2. The spring is to have a stiffness of 1 kN/m. The solid length of the spring is 45 mm. Find the diameter and mean radius of the coils if $G = 80$ GN/m^2. Also find the deflection, angle of rotation and strain energy. (Solid length is the length when the coils are touching each other.)
13. An open-coiled helical spring is made of a wire of diameter 10 mm. The coil has a mean radius of 50 mm. The spring has 12 turns and the angle of the helix is 15°. Find the deflection, and the maximum bending and shear stresses under an axial load of 300 N. $G = 85$ GN/m^2 and $E = 200$ GN/m^2.
14. If the coil of Problem 13 is subjected to an axial twist of 10 N m, find the maximum stresses, deflection, and angle of rotation.
15. An open-coiled helical spring has 12 turns. Assume the mean diameter of coils to be eight times the diameter of the wire. When subjected to an axial load, the maximum bending and shear stresses are 140 N/mm^2 and 150 N/mm^2, respectively. Find the wire diameter and the maximum axial load for a maximum extension of 18 mm. Take $E = 200$ GPa and $G = 80$ GPa.
16. A flat spiral spring is made of a 12 mm $\times$ 0.5 mm thick wire, and has a length of 6 m. If the maximum stress is not to exceed 500 MPa, find the maximum torque that can be applied and the strain energy stored. $E = 200$ GPa.
17. A composite spring made of two close-coiled springs in series has a mean coil diameter eight times the wire diameter. One spring has 18 coils of 3 mm ϕ wire. If the composite spring has a stiffneses of 1.5 N/mm, find the wire diameter of the second spring if it has 10 coils. Determine the maximum axial load that can be applied and the corresponding extension for a maximum shearing stress of 300 MPa. $G = 80$ GPa.
18. A close-coiled helical spring, whose free length when not compressed is 15 cm, is required to absorb strain energy of 500 kgcm when fully compressed with coils in contact. The maximum shear stress is 1400 kg/cm^2. Assume mean coil diameter of 10 cm. Find diameter of the steel wire required and the number of coils. $G = 8.5 \times 10^5$ kg/cm^2.
19. An open-coiled helical spring is made of wires of 10 mm diameter. It has 15 coils of mean diameter 6 cm. What is the greatest axial compressive load that can be applied to the ends of the spring if the principal stress is not to exceed 120 MPa and the maximum shear stress is not to exceed 80 MPa. Calculate for this load the axial and angular deflections of the spring. $E = 2 \times 10^5$ MPa and $N = 0.8 \times 10^5$ MPa.
20. A 2 m long cylindrical vessel with closed ends has a diameter of 250 mm and a wall thickness of 5 mm. The internal pressure is 4 N/mm^2. Find the maximum hoop stress and the longitudinal stress in the cylinder. Also determine the change in diameter, length, and volume if $E = 200$ GPa and $\nu = 0.3$.
21. A boiler shell is made of 12 mm thick plates with a maximum pressure of 1.8 MPa. If the maximum tensile stress is limited to 120 N/mm^2, find the maximum diameter of the cylinder.

22. A spherical shell, of diameter 1.2 m and wall thickness 6 mm, is filled with fluid under pressure. The volume change is 200×10^3 mm^3. Calculate the fluid pressure if $E = 200$ GPa and $v = 0.3$.
23. A tetecylindrical thin drum, 80 cm in diameter and 3 m long, has shell thickness of 1 cm. If the drum is subjected to an internal pressure of 2.5 N/mm^2, determine the change in diameter, length, and volume.
24. A pipe of internal diameter 200 mm and thickness 15 mm is closely wound with wire of diameter 5 mm under a tensile stress of 50 N/mm^2. Water under a pressure of 5 N/mm^2 is then admitted into the pipe. Find the stresses in the pipe and wire before and after admitting water under pressure.
25. A thick cylinder, of internal radius 60 mm and outside diameter 120 mm, is subjected to an external pressure of 20 N/mm^2 and internal pressure of 60 N/mm^2. Find the maximum values of hoop and radial stresses analytically and graphically.
26. A thick walled cylinder, of internal radius 120 mm, is subjected to an internal pressure of 50 N/mm^2. Find the thickness of the cylinder required if the hoop stress in the material is limited to 120 N/mm^2.
27. A compound cylinder is made by shrinking a jacket on to a cylinder. The internal diameter is 100 mm, the outside diameter 200 mm, and the diameter at the junction 80 mm. The cylinder is subjected to an internal pressure of 50 N/mm^2. If the pressure at the common junction is 12 N/mm^2 before the fluid pressure is applied, find the final stresses at the different sections.
28. The pressure within the cylinder of a hydraulic press is 9 MPa. The inside diameter of the cylinder is 25 mm. Determine the thickness of the cylinder wall if the permissible stress is 18 MPa.
29. The maximum stress permitted in a thick cylinder of internal and external radii 200 MM and 300 mm, respectively, is 16 MPa. Find the internal pressure that can be applied. Plot the curves showing variation of hoop and radial stresses through the material. What will be the change in thickness of the cylinder? $E = 2 \times 10^5$ MPa and $1/m = 0.3$.
30. A thick pipe of 300 mm outer diameter and 200 mm internal diameter is subjected to an internal pressure of 12 MPa. Determine the minimum external pressure that can be applied so that the tensile stress in the metal shall not exceed 16 MPa.
31. A compound cylinder is formed by shrinking one tube onto another, the final dimensions being internal diameter 120 mm, external diameter 240 mm, and diameter at the junction 180 mm. If after shrinking on, the radial pressure at the common surface is 8 N/mm^2, calculate the initial hoop stresses across the sections of the inner and outer tubes. If a fluid under a pressure of 60 N/mm^2 is admitted inside the compound cylinder, calculate the final stresses set up in the sections of the pipes.
32. A semicircular bar, of radius 80 mm, is of circular section of diamater 30 mm and is subjected to couples at the ends that open the ring. Find the maximum value of the couple if the bending stress is limited to 100 MPa.
33. If the semicircular ring of Problem 21 is of trapezoidal section of sides 50 mm and 25 mm and depth 40 mm, find the maximum value of the moments applied at the ends.
34. A crane hook is of trapezoidal section of sides 40 mm and 20 mm, and depth 40 mm. The radius of the hook is 50 mm and the load along the radius through the centre of curvature applied is 20 kN. Find the maximum tensile stress along a radial section (horizontal) that passes through the centre.
35. A steel bolt is subjected to a load of 20 kN along with a transverse shear of 8 kN. Calculate the safe diameter of the bolt using all the theories of failure. The yield stress is 300 MPa, $v = 0.3$, and the factor of safety required is 2.

36. A curved beam has a T-section shown in Fig. 12.42. The inner radius is 300 mm. What is the eccentricity of the section?

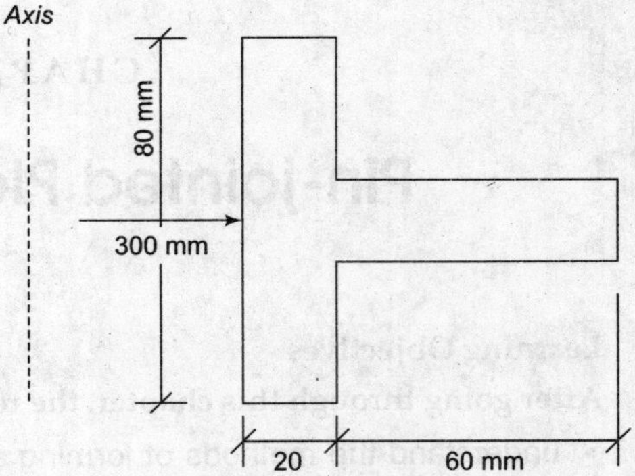

Fig. 12.42

37. A steel ring of 20 cm mean diameter has a rectangular cross section of 5 cm in the radial direction and 3 cm perpendicular to the radial direction. If the maximum tensile stress is limited to 120 MPa, determine the tensile load that the ring can carry.
38. A chain link, of the size and shape shown in Fig. 12.43, is subjected to loads P as shown in the figure. Prove that the maximum bending moment at the point of application of the load is

$PR(1 + 2R)/[1 + \pi R]$

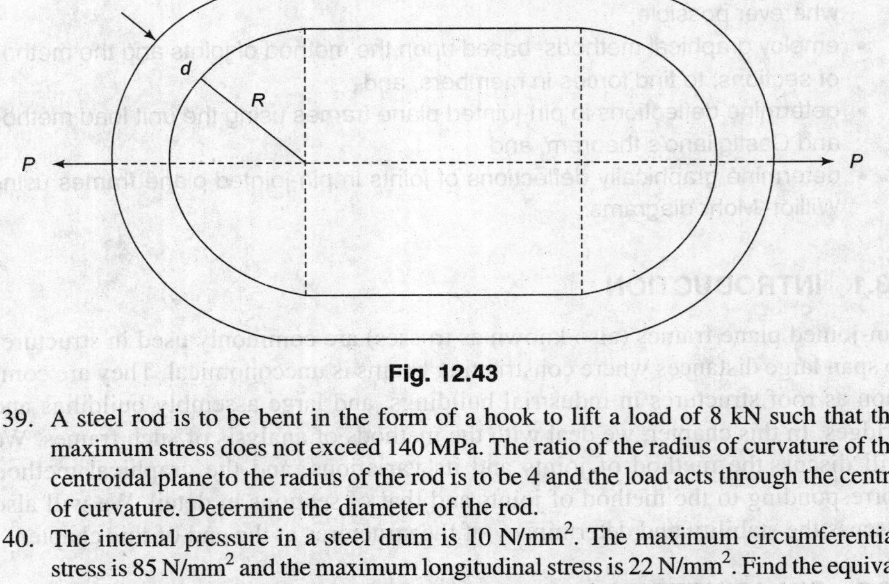

Fig. 12.43

39. A steel rod is to be bent in the form of a hook to lift a load of 8 kN such that the maximum stress does not exceed 140 MPa. The ratio of the radius of curvature of the centroidal plane to the radius of the rod is to be 4 and the load acts through the centre of curvature. Determine the diameter of the rod.
40. The internal pressure in a steel drum is 10 N/mm^2. The maximum circumferential stress is 85 N/mm^2 and the maximum longitudinal stress is 22 N/mm^2. Find the equivalent tensile stress in a simple tension test according to the maximum shear stress theory.
41. A hollow shaft, of 30 mm internal diameter and 50 mm external diameter, is subjected to a twisting moment of 800 Nm and an axial compressive force of 40 kN. Determine the factor of safety according to the strain energy theory, if the yield strength of material is 280 N/mm^2 and Poisson's ratio is 0.3.
42. The load on a bolt consists of an axial thrust of 8 kN together with a transverse shear force of 4 kN. Calculate the diameter of the bolt according to the (i) maximum principal stress theory and (ii) maximum shear stress theory. Take factor of safety (FOS) = 3, σ_y = 285 N/mm^2, and $1/m$ = 0.3.

CHAPTER 13

Pin-jointed Plane Frames

Learning Objectives

After going through this chapter, the reader will be able tos

- understand the methods of forming a pin-jointed plane frame and its structural action,
- explain the assumptions made in the analysis of a pin-jointed plane frame,
- analyse a given pin-jointed plane frame and find the forces in the members using different methods such as the method of joints, the method of tension coefficients, and Ritter's method of sections,
- use simple principles and find the forces in members by visual inspection, wherever possible,
- employ graphical methods, based upon the method of joints and the method of sections, to find forces in members, and
- determine deflections in pin-jointed plane frames using the unit load method and Castigliano's theorem, and
- determine graphically deflections of joints in pin-jointed plane frames using Williot–Mohr diagrams.

13.1 INTRODUCTION

Pin-joined plane frames (also known as trusses) are commonly used in structures to span large distances where constructing beams is uneconomical. They are common as roof structures in industrial buildings, and large assembly buildings and bridges. In this chapter, we deal with the methods of analysis of such frames. We will discuss the method of joints and its variations, and the graphical method corresponding to the method of joints and that of sections in detail. We will also discuss the stability and determinacy of the structures at the end of this chapter.

13.2 PIN-JOINTED PLANE FRAMES

Such frames, commonly made of steel and timber, are also called trusses when used in the building of roofs and bridges. They consist of a number of members connected at their ends to form a framework as in Fig. 13.1(a). The members may be built up of steel sections or wooden members as shown. The frame is represented by the centre lines of the members.The frames are called plane frames when the centre line of the members and the loading lies in one plane. Space frames [Fig. 13.1(b)] on the other hand have members which do not lie in one plane—all the loads do not lie in one plane either. An example is the transmission tower used for transmission of electric power.

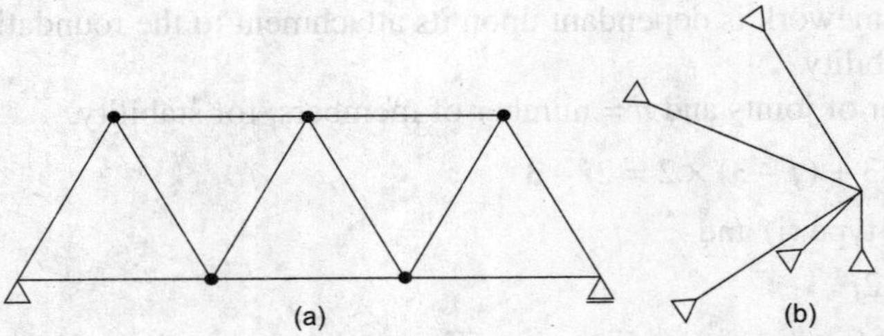

Fig. 13.1

13.2.1 Stability of Frames

For stability, such frames are generally triangulated. Considering Fig. 13.2(a), the framework of four members forming a rectangle is inherently unstable because the frame will collapse (will have large deformations) on the application of a load P if the joints are pin-jointed. The addition of a diagonal AC makes the frame stable [Fig. 13.2(b)].

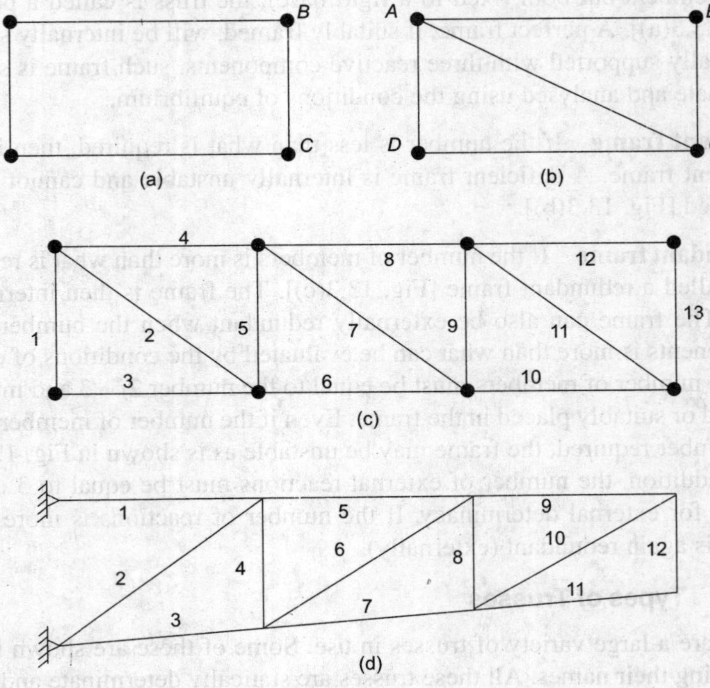

Fig. 13.2

Frames can basically be made in the following two ways.

(i) Starting with a triangle of three members and three joints [Fig. 13.2(c)], the frame can be built up to any extent by adding two members for every additional joint. This gives an internally stable framework, which can be supported suitably for external stability.

(ii) Starting from a firm foundation, two members can be made to form a joint [Fig. 13.2(d)]. The frame can be built up further as described in (i). Note that the framework is dependant upon its attachment to the foundation for internal stability.

If j = number of joints and n = number of members, for stability,

$$n = 3 + (j - 3) \times 2 = 2j - 3$$

in frames of type (i) and

$$n = 2j$$

in the case of frames of type (ii), where j is equal to the number of joints, excluding those attaching the frame to the foundation.

13.2.2 Classification of Frames: Perfect, Deficient, and Redundant Frames

Frames can be classified as perfect, deficient, and redundant based on the above discussion.

Perfect frame If the number of members is equal to the number required, $2j - 3$ (when the frame is started with three members) or $2j$ (when we start the frame with two members but both fixed to a rigid base), the truss is called a perfect frame [Fig. 13.3(a)]. A perfect frame, if suitably framed, will be internally stable. When externally supported with three reactive components, such frame is statically determinate and analysed using the conditions of equilibrium.

Deficient frame If the number is less than what is required, then it is called a deficient frame. A deficient frame is internally unstable and cannot be uniquely analysed [Fig. 13.3(b)].

Redundant frame If the number of members is more than what is required, then it is called a redundant frame [Fig. 13.3(c)]. The frame is then internally redundant. The frame can also be externally redundant when the number of reactive components is more than what can be evaluated by the conditions of equilibrium.

The number of members must be equal to the number $2j - 3$ and must be triangulated or suitably placed in the frame. Even if the number of members is equal to the number required, the frame may be unstable as is shown in Fig. 13.3(d).

In addition, the number of external reactions must be equal to 3 and suitably placed for external determinacy. If the number of reactions is more than 3, the frame is again redundant (externally).

13.2.3 Types of Trusses

There are a large variety of trusses in use. Some of these are shown in Fig. 13.4 indicating their names. All these trusses are statically determinate and find extensive use in building roofs and bridges, industrial sheds, etc.

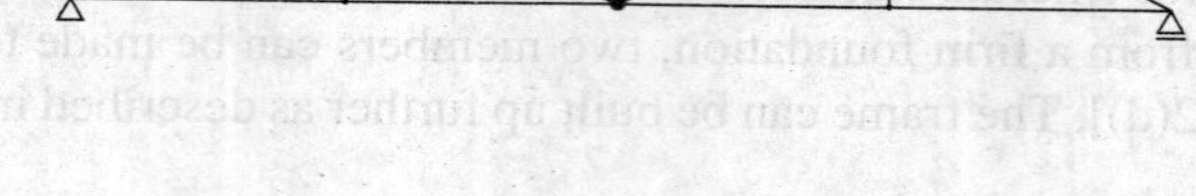

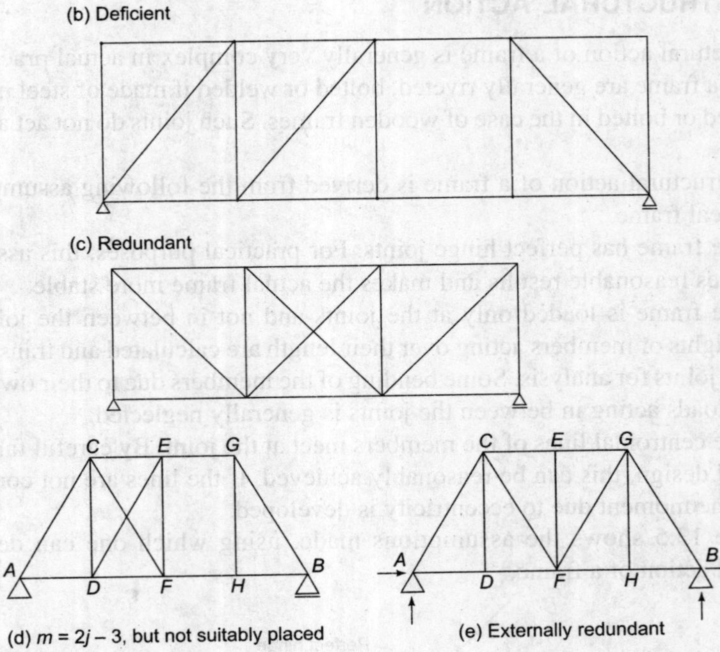

Fig. 13.3

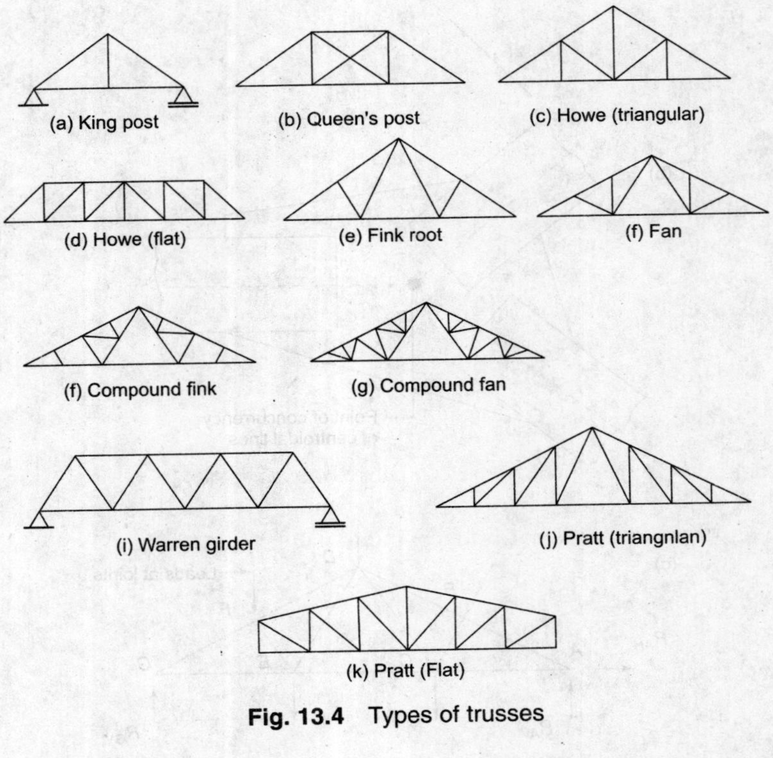

Fig. 13.4 Types of trusses

13.3 STRUCTURAL ACTION

The structural action of a frame is generally very complex in actual practice. The joints of a frame are generally riveted, bolted or welded if made of steel members, and nailed or bolted in the case of wooden frames. Such joints do not act as perfect hinges.

The structural action of a frame is derived from the following assumptions to get an ideal frame.

(i) The frame has perfect hinge joints. For practical purposes, this assumption gives reasonable results and makes the actual frame more stable.

(ii) The frame is loaded only at the joints and not in between the joints. The weights of members acting over their length are calculated and transferred to the joints for analysis. Some bending of the members due to their own weight or loads acting in between the joints is generally neglected.

(iii) The centroidal lines of the members meet at the joint. By careful fabrication and design, this can be reasonably achieved. If the lines are not concurrent, some moment due to eccentricity is developed.

Figure 13.5 shows the assumptions made, using which one can derive the structural action of a frame.

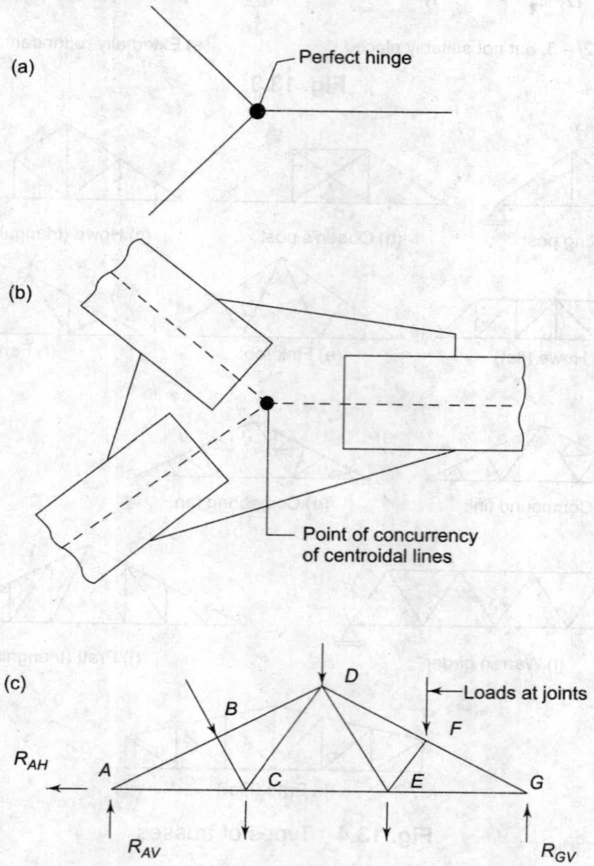

Fig. 13.5

See Fig. 13.6. Considering any two joints A and B of a frame, all the members meeting at A and B are concurrent. If we consider the free body of member AB, which is considered weightless according to our assumption, the net effect of the forces transferred at A and B by members meeting at the respective joints must be equal and opposite forces collinear with AB. If it were not so, moments would have been introduced, which is not possible because A and B are assumed to be frictionless hinges. Therefore, the force in any member of a frame has to be axial, either tensile as in Fig. 13.6(b) or compressive as in Fig. 13.6(c).

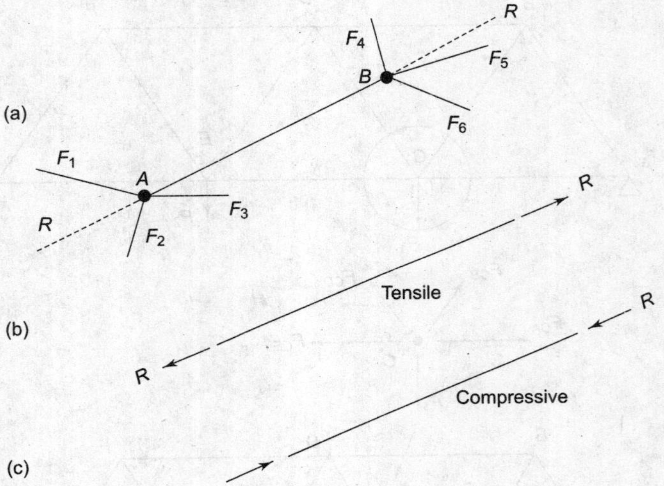

Fig. 13.6

13.3.1 Sign Convention

We will consider tensile force in members to be positive and compressive force to be negative.

By convention, the forces in a truss are represented by internal stress resultants rather than by external forces acting on the member. This is shown in Fig 13.7(a)—the member is subjected to a tensile force but is generally represented as shown in Fig. 13.7(b) on the member by arrows, which show the direction of internal stress resultants (which are opposite to the external force) in the member. The same is true for the compression member represented in Figures 13.7(c) and (d).

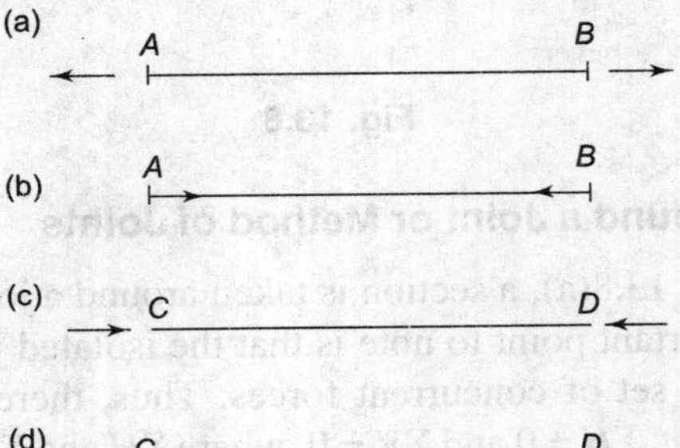

Fig. 13.7

13.4 METHODS OF ANALYSIS

The basic approach to the analysis of a frame is the section method. We take a section cutting a number of bars, and consider the equilibrium of either of the two parts so obtained. On solution, the equilibrium equations so formulated can give us values of unknown bar forces and reactions. Depending upon the method of taking a section, there are basically two methods of analysing a frame, as shown in Figures 13.8(a) and (b).

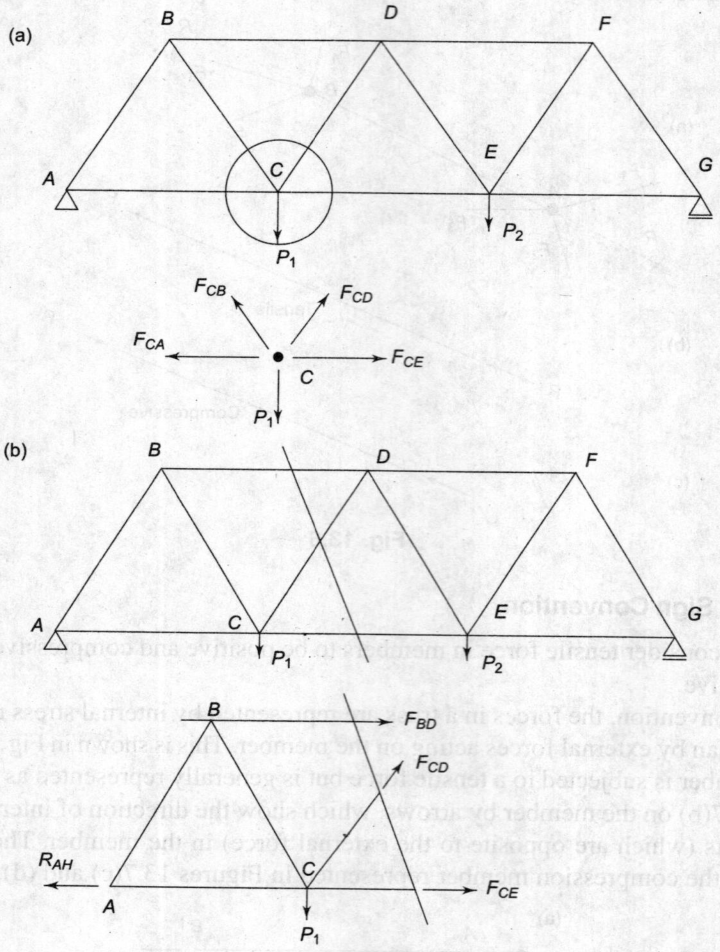

Fig. 13.8

13.4.1 Section Around a Joint or Method of Joints

In this case, as in Fig. 13.8(a), a section is taken around a joint, isolating the joint completely. The important point to note is that the isolated joint is in equilibrium under the action of a set of concurrent forces. Thus, there are two equilibrium equations for each joint, $\Sigma H = 0$ and $\Sigma V = 0$, where ΣH and ΣV are the summations of components along two mutually perpendicular directions.

In a stable, determinate frame, there are $2j$ equations available and the number of members is only $(2j - 3)$. The three extra equations available can be used to calculate the three unknown reactions or for checking.

13.4.2 Ritter's Method of Sections

Here, the section is taken as shown in Fig. 13.8(b). The truss is separated into two parts by such a section and each part is in equilibrium under the action of a general coplanar force system. There are three equilibrium equations, $\Sigma H = 0$, $\Sigma V = 0$, and $\Sigma M = 0$, available for such a force system and three unknown forces can, therefore, be determined.

If the reactions are calculated from the conditions of equilibrium for the frame as a whole, then the advantage of Ritter's method of sections is that it enables us to determine the force in any member by taking a section cutting that member. In the method of joints, it is necessary to go from joint to joint in some order, starting from the support joints or from free ends as in a cantilever truss.

Before we discuss these methods in detail, we need to look at some simple procedures to enable us to find forces in some members through visual inspection or to check the results.

13.4.3 Member force and its Components

Let us consider any member of a truss inclined to the horizontal. The orientation of such a member can be specified either by the angle that the member makes with the horizontal, θ, or by the projected distances on the horizontal and vertical axes. Figures 13.9(a) and (b) show the two cases.

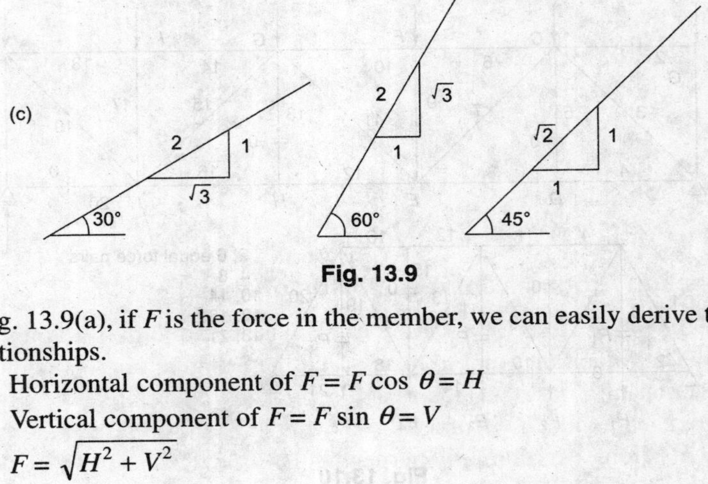

Fig. 13.9

In Fig. 13.9(a), if F is the force in the member, we can easily derive the following relationships.

Horizontal component of $F = F \cos \theta = H$

Vertical component of $F = F \sin \theta = V$

$F = \sqrt{H^2 + V^2}$

In Fig. 13.9(b), if h and v are the horizontal and vertical projections of member of length l,

$$l = \sqrt{h^2 + v^2}, \quad \sin\theta = \frac{v}{l}, \quad \cos\theta = \frac{h}{l}$$

$$\text{Horizontal component} = F\frac{h}{l}$$

$$\text{Vertical component} = F\frac{v}{l}$$

i.e., $H = Fh/l$ and $V = Fv/l$.

In Fig. 13.9(c), the common cases of inclination of a member—30°, 45°, and 60°—are shown. Note that for 30°, $h = \sqrt{3}$, $v = 1$, and $l = 2$, for 45°, $h = v = 1$, and $l = \sqrt{2}$, and for 60°, $h = 1$, $v = \sqrt{3}$, and $l = 2$. The member forces and their components can always be derived from these ratios of the sides, knowing any one of the three—F, H, or V. For example, if $\theta = 30°$ and $H = 30$ kN, then,

$$V = \frac{H}{h}v = \frac{30}{\sqrt{3}} \times 1 \quad \text{and} \quad F = \frac{H}{h}l = \frac{30}{\sqrt{3}} \times 2$$

Visual inspection In the case of many frames with particular loadings, it is possible to determine the forces in a number of members by inspection only from the conditions $\Sigma H = 0$ and $\Sigma V = 0$. Some cases are shown in Figures 13.10(a), (b), and (c).

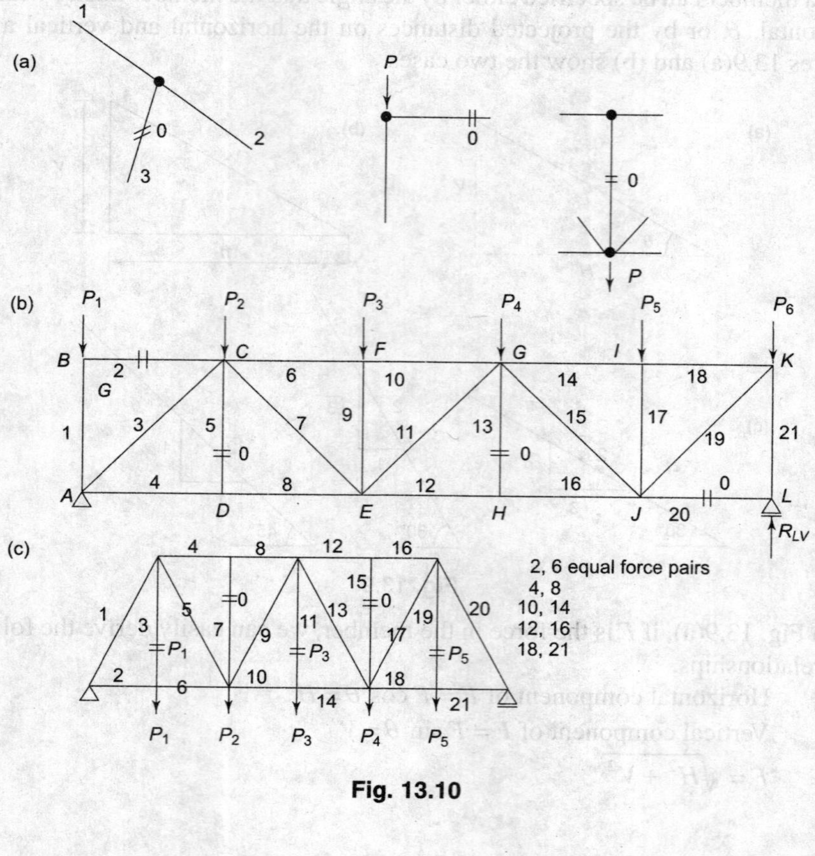

Fig. 13.10

In Fig. 13.10(a), if two of the members are collinear, say 1 and 2, and the joint is unloaded, then the force in member 3 is zero. This is so because if one of the two mutually perpendicular axes is taken along 1-2, then only 3 will have a component in the other perpendicular direction and for $\Sigma F = 0$ along this, $F_3 = 0$. This is true for any inclination of the member. Please note that this is not true if the joint is loaded or more than three members meet at the joint.

Considering Fig. 13.10(b), we can easily verify the following.

(1) For the equilibrium of joint B, the force in member 2 is zero.

(2) The force in the vertical members 5 and 13 is zero considering joints D and H.

(3) The force in member 20 is zero considering joint L and that in member 21 is equal to R_{LV}.

(4) The forces in members 9 and 17 are equal to P_3 and P_5, respectively, from joints E and I. The force in member 1 = P_1.

(5) The forces in members 4 and 8 are equal; the same is the case with forces in members 12 and 16, 6 and 10, and 14 and 18. The forces in these pairs must be equal considering the respective joints where they meet. Similar deductions for member forces can be made as shown on the frame in Fig. 13.10(c) by visual inspection.

13.5 THE METHOD OF JOINTS

In the method of joints, a section is taken around the joint to isolate it. Since the joint is subjected to a concurrent, coplanar force system, two conditions of equilibrium, $\Sigma H = 0$ and $\Sigma V = 0$, are available for the joint from which two unknown forces can be determined. The commonly used graphical method is the graphic equivalent of the method of joints. We will discuss a number of methods which use different techniques but the basic principle is the joint equilibrium.

As mentioned earlier, there are $2j$ equations for a frame with j joints, but only $(2j - 3)$ members. Three equations can be used for finding reactions or for checking. The reactions can also be determined from the equilibrium conditions for the frame as a whole, i.e., $\Sigma H = 0$, $\Sigma V = 0$, and $\Sigma M = 0$ for the loads and reactive forces. The following examples illustrate the conventional method of joints.

Example 13.1 Method of joints

Analyse the frame shown in Fig. 13.11(a), and find the forces in all the members.

Solution There are three member forces and three reactive components—R_{AH}, R_{AV}, and R_{CV}. We formulate two equations for each of the joints A, B, and C, and determine the six unknowns.

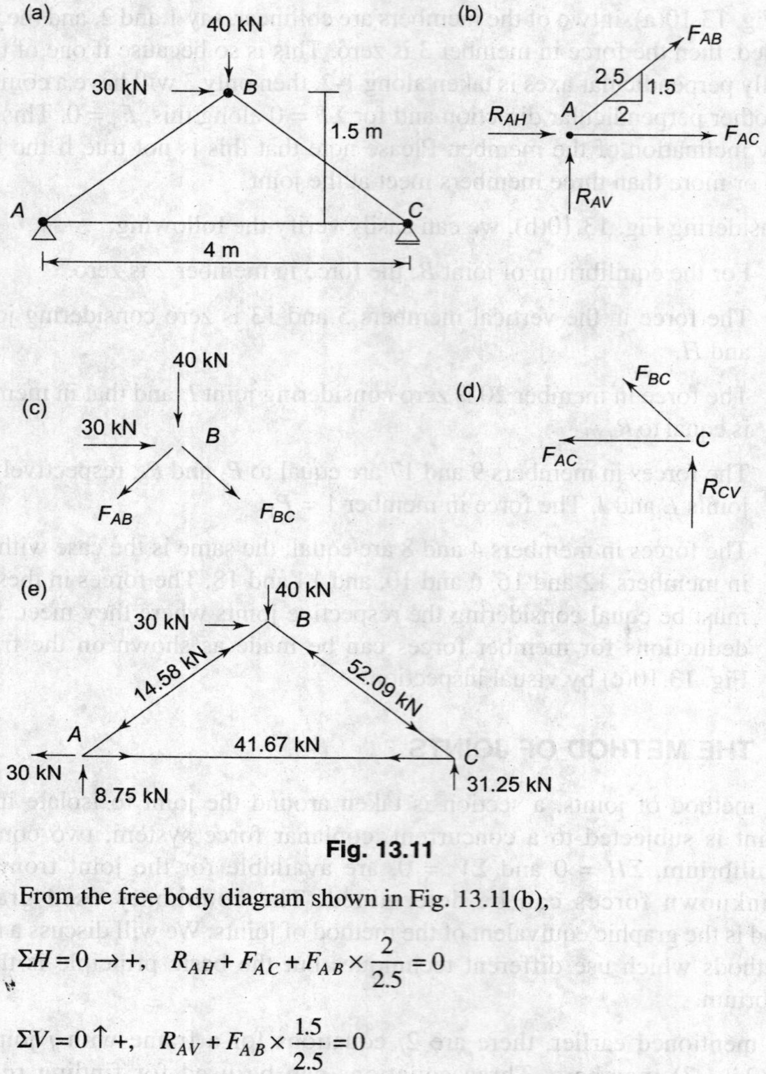

Fig. 13.11

Joint A From the free body diagram shown in Fig. 13.11(b),

$$\Sigma H = 0 \rightarrow +, \quad R_{AH} + F_{AC} + F_{AB} \times \frac{2}{2.5} = 0 \qquad \text{(i)}$$

$$\Sigma V = 0 \uparrow +, \quad R_{AV} + F_{AB} \times \frac{1.5}{2.5} = 0 \qquad \text{(ii)}$$

Note that members AB and AC are assumed to be in tension. The ratio of the length, the vertical, and horizontal projections for member AB are 2.5, 1.5, and 2.0.

Joint B From the free body diagram in Fig. 13.11(c),

$$\Sigma H = 0 \rightarrow +, \quad 80 - F_{AB} \times \frac{2}{2.5} + F_{BC} \times \frac{2}{2.5} = 0 \qquad \text{(iii)}$$

$$\Sigma V = 0 \uparrow +, \quad -40 - F_{AB} \times \frac{1.5}{2.5} - F_{BC} \times \frac{1.5}{2.5} = 0 \qquad \text{(iv)}$$

Joint C From the free body diagram in Fig. 13.11(d),

$$\Sigma H = 0 \rightarrow +, \quad -F_{AC} - F_{BC} \times \frac{2}{2.5} = 0 \qquad \text{(v)}$$

$$\Sigma V = 0 \uparrow +, \quad R_{CV} + F_{BC} \times \frac{1.5}{2.5} = 0 \qquad \text{(vi)}$$

These six equations can be solved to evaluate the six unknowns.

$$F_{AB} = -14.58 \text{ kN}$$

(The member is under compression and not tension as assumed.)

$$F_{BC} = -52.09 \text{ kN}$$

(The member is in compression.)

$$F_{AC} = 41.67 \text{ kN}$$

(The member is in tension as assumed.)

$$R_{CV} = 31.25 \text{ kN} \uparrow, \quad R_{AV} = 8.75 \text{ kN} \uparrow$$

and $R_{AH} = -30$ kN

The direction of R_{AH} is opposite to that assumed. All the member forces are shown in their correct nature in Fig. 13.11(e).

It may be easier in many cases to find the reactions from the equilibrium condi-tions for the whole frame. The member forces can then be analysed joint by joint.

Example 13.2 Method of joints

Analyse the frame loaded as shown in Fig. 13.12(a) and determine the forces in all the members.

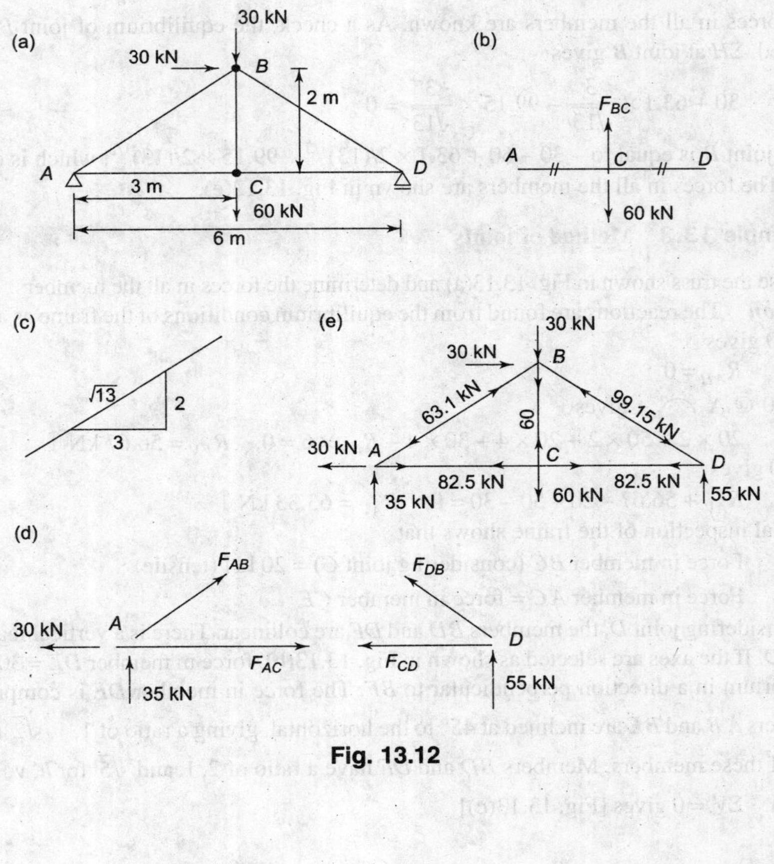

Fig. 13.12

Solution The reactions are determined from the equilibrium conditions of the frame as a whole. Let R_{AH} and R_{AV} be the horizontal and vertical reactions at A, and R_{BV} be the vertical reaction at B.

$$\Sigma H = 0 \rightarrow +, \quad R_{AH} + 30 = 0, \quad R_{AH} = -30 \text{ kN}$$

acting towards the left.

$$\Sigma M = 0 \text{ @ A } \curvearrowright +, \; 30\times2 + 30\times3 + 60\times3 - R_{BV}\times6 = 0, \; R_{BV} = 55 \text{ kN} \uparrow$$

$$\Sigma V = 0, \quad R_{AV} + 55 - 30 - 60 = 0, \quad R_{AV} = 35 \text{ kN} \uparrow$$

From visual inspection, considering joint C shown in Fig. 13.12(b), $F_{BC} = 60$ kN (tensile) and $F_{AC} = F_{CD}$, both tensile or both compressive. As shown in Fig. 13.12(c), the ratio of the length, horizontal and vertical projections are $\sqrt{13}$, 3 and 2 for AB and BD.

From the free bodies of the joints shown in Fig. 13.12(d), we observe the following.

Joint A

$$\Sigma V = 0, \; R_{AV} + F_{AB} \times \frac{2}{\sqrt{13}} = 0, \; F_{AB} = -35 \times \frac{\sqrt{13}}{2} = -63.1 \text{ kN (compressive)}$$

$$\Sigma H = 0, \quad -30 - 63.1 \times \frac{3}{\sqrt{13}} + F_{AC} = 0, \quad F_{AC} = 82.5 \text{ kN (tensile)}$$

From joint C,

$$F_{CD} = 82.5 \text{ kN (tensile)}$$

Joint D

$$\Sigma V = 0, \quad 55 + F_{DB} \times \frac{2}{\sqrt{13}} = 0, \quad F_{DB} = -99.15 \text{ kN (compressive)}$$

The forces in all the members are known. As a check, the equilibrium of joint B can be verified. ΣH at joint B gives

$$30 + 63.1 \times \frac{3}{\sqrt{13}} - 99.15 \times \frac{3}{\sqrt{13}} = 0$$

ΣV at joint B is equal to $-30 - 60 + 63.1 \times 2/(13)^{1/2} + 99.15 \times 2/(13)^{1/2}$, which is equal to zero. The forces in all the members are shown in Fig. 13.12(e). ■

Example 13.3 Method of joints

Analyse the truss shown in Fig. 13.13(a) and determine the forces in all the members.

Solution The reactions are found from the equilibrium conditions of the frame as a whole.

$\Sigma H = 0$ gives

$$R_{AH} = 0$$

$\Sigma M = 0$ @ A $\curvearrowright$ + gives

$$20 \times 2 + 50 \times 2 + 20 \times 4 + 30 \times 4 - R_{FV} \times 6 = 0, \quad R_{FV} = 56.67 \text{ kN} \uparrow$$

$\Sigma V = 0$ gives

$$R_{AV} + 56.67 - 20 - 50 - 30 = 0, \quad R_{AV} = 63.33 \text{ kN} \uparrow$$

A visual inspection of the frame shows that

Force in member BC (considering joint C) = 20 kN (tensile)

Force in member AC = force in member CE

Considering joint D, the members BD and DF are collinear. There is a vertical load of 30 kN at D. If the axes are selected as shown in Fig. 13.13(b), force in member DE = 30 kN for equilibrium in a direction perpendicular to BF. The force in member DE is compressive. Members AB and BE are inclined at 45° to the horizontal, giving a ratio of 1, 1, $\sqrt{2}$ for h, v, and l of these members. Members BD and DF have a ratio of 2, 1, and $\sqrt{5}$ for h, v, and l.

Joint A $\Sigma V = 0$ gives [Fig. 13.13(c)]

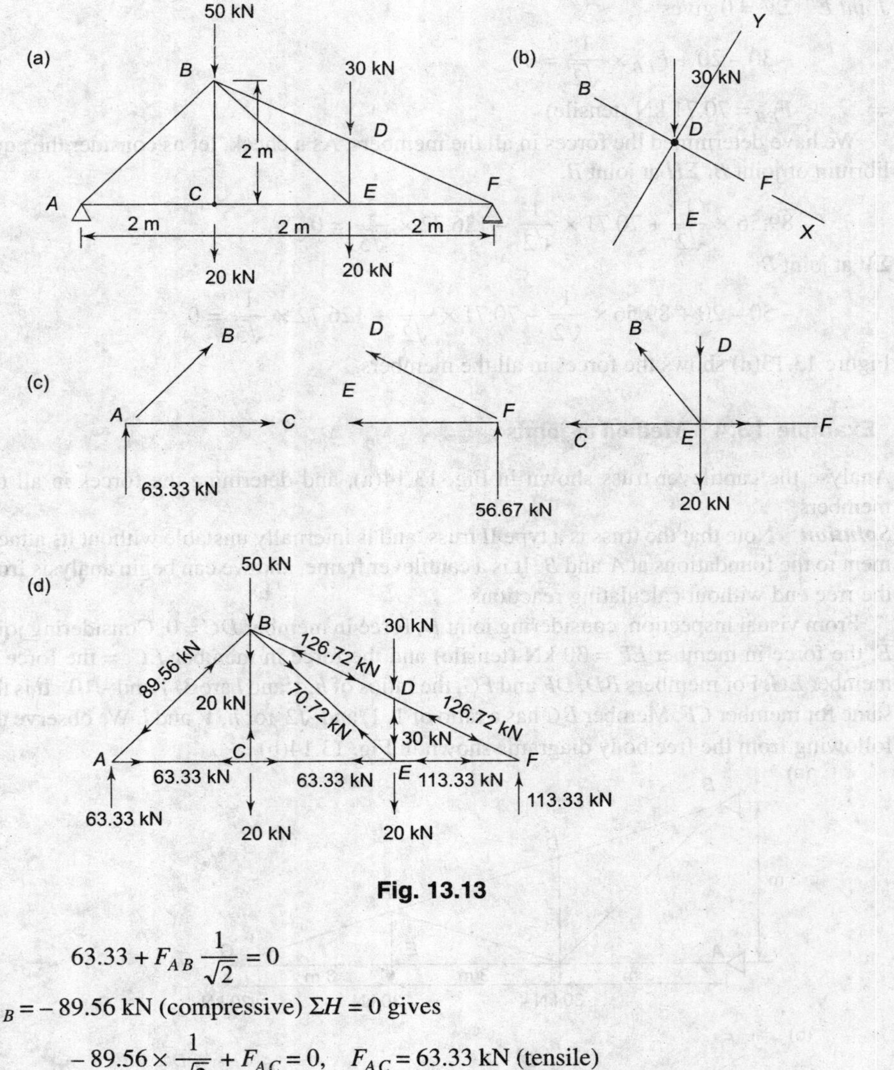

Fig. 13.13

$$63.33 + F_{AB}\frac{1}{\sqrt{2}} = 0$$

$F_{AB} = -89.56$ kN (compressive) $\Sigma H = 0$ gives

$$-89.56 \times \frac{1}{\sqrt{2}} + F_{AC} = 0, \quad F_{AC} = 63.33 \text{ kN (tensile)}$$

$$F_{CE} = 63.33 \text{ kN (tensile)}$$

from joint *C*.

Joint F From the free body shown in Fig. 13.13(c), $\Sigma V = 0$ gives

$$56.67 + F_{DF} \times \frac{1}{\sqrt{5}} = 0$$

$\Rightarrow$ $F_{DF} = -126.72$ kN (compressive)

$\Sigma H = 0$ gives

$$126.72 \times \frac{2}{\sqrt{5}} - F_{FE} = 0$$

$\Rightarrow$ $F_{FE} = 113.34$ kN (tensile)

From joint *D*,

$$F_{BD} = 126.72 \text{ kN (compressive)}$$

Joint E $\Sigma V = 0$ gives

$$-30 - 20 + F_{EB} \times \frac{1}{\sqrt{2}} = 0$$

$$\Rightarrow \quad F_{EB} = 70.71 \text{ kN (tensile)}$$

We have determined the forces in all the members. As a check, let us consider the equilibrium of joint B. ΣH at joint B,

$$89.56 \times \frac{1}{\sqrt{2}} + 70.71 \times \frac{1}{\sqrt{2}} - 126.72 \times \frac{2}{\sqrt{5}} = 0$$

ΣV at joint B,

$$-50 - 20 + 89.56 \times \frac{1}{\sqrt{2}} - 70.71 \times \frac{1}{\sqrt{2}} + 126.72 \times \frac{1}{\sqrt{5}} = 0$$

Figure 13.13(d) shows the forces in all the members. ■

Example 13.4 Method of joints

Analyse the cantilever truss shown in Fig. 13.14(a), and determine the forces in all the members.

Solution Note that the truss is a type II truss, and is internally unstable without its attachment to the foundations at A and B. It is a cantilever frame, and we can begin analysis from the free end without calculating reactions.

From visual inspection, considering joint D, force in member $DC = 0$. Considering joint E, the force in member EF = 30 kN (tensile) and the force in member EC = the force in member EG. For members BD, DF and FG, the ratios of h, v and l are 3, 1 and $\sqrt{10}$. It is the same for member CF. Member BC has a ratio of 1, 1, and $\sqrt{2}$ for h, v, and l. We observe the following from the free body diagrams shown in Fig. 13.14(b).

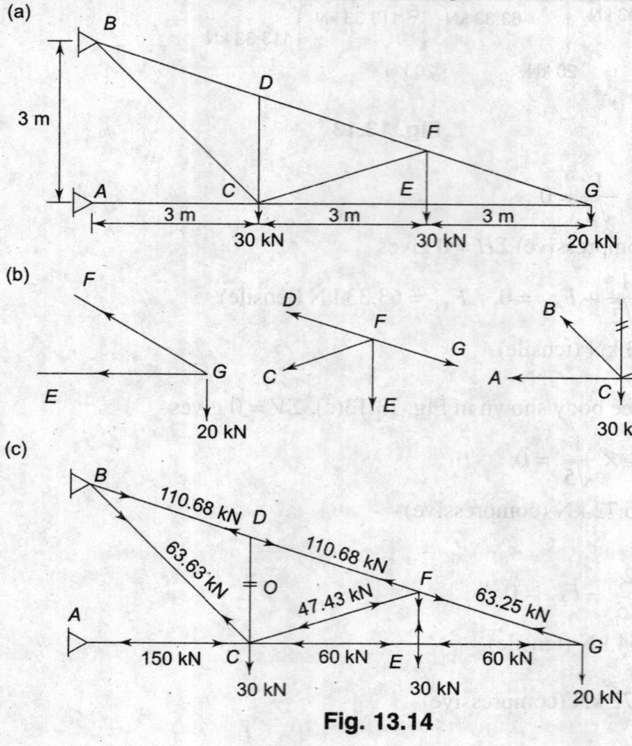

Fig. 13.14

Joint G $\Sigma V = 0$ gives

$$-20 + F_{GF} \times \frac{1}{\sqrt{10}} = 0$$

$$F_{GF} = 63.25 \text{ kN (tensile)}$$

$\Sigma H = 0$ gives

$$F_{GE} - 63.25 \times \frac{3}{\sqrt{10}} = 0$$

$\Rightarrow$ $F_{GE} = 60$ kN (compressive)

$F_{CE} = 60$ kN (compressive)

from joint E.

Joint F $\Sigma H = 0$ gives

$$63.25 \times \frac{3}{\sqrt{10}} - F_{FD} \times \frac{3}{\sqrt{10}} + F_{FC} \times \frac{3}{\sqrt{10}} = 0$$

$\Sigma V = 0$ gives

$$-30 + F_{FC} \times \frac{1}{\sqrt{10}} + F_{FD} \times \frac{1}{\sqrt{10}} - 63.25 \times \frac{1}{\sqrt{10}} = 0$$

Solving,

$F_{FC} = 47.43$ kN (compressive)

$F_{FD} = 110.68$ kN (tensile)

Joint D

$F_{BD} = 110.68$ kN (tensile)

Joint C $\Sigma V = 0$ gives

$$F_{BC} \times \frac{1}{\sqrt{2}} - F_{CF} \times \frac{1}{\sqrt{10}} - 30 = 0$$

$\Rightarrow$ $F_{BC} = 63.64$ kN (tensile)

$\Sigma H = 0$ gives

$$F_{CA} - 60 - F_{CB} \times \frac{1}{\sqrt{2}} - F_{CF} \times \frac{3}{\sqrt{10}} = 0$$

$\Rightarrow$ $F_{CA} = 150$ kN (compressive)

Reaction at A,

$R_{AH} = 150 \text{ kN} \rightarrow, \quad R_{AV} = 0$

At B,

$$R_{BV} = 80 \text{ kN} \uparrow, \quad R_{BH} = 110.68 \times \frac{3}{\sqrt{10}} + 63.64 \times \frac{1}{\sqrt{2}} = 150 \text{ kN} \leftarrow$$

The forces in all the members are shown in Fig. 13.14(c). ■

Example 13.5 Warren truss: method of joints

Analyse the Warren truss shown in Fig. 13.15(a), and find the forces in all the members.

Solution The reactions are determined as

$R_{AH} = 30 \text{ kN} \leftarrow$

$\Sigma M = 0$ @ A $\curvearrowright$ +, $30 \times 2.598 + 60 \times 3 + 60 \times 6 + 30 \times 12 - R_{BV} \times 9 = 0$

$R_{BV} = 108.66 \text{ kN}, \quad R_{AV} = 41.34 \text{ kN}$

The forces in the diagonal members meeting at any of the joints B, D, F or H must be equal and opposite to give $\Sigma V = 0$ as shown in Fig. 13.15(b). The diagonal members are inclined at 60° to the horizontal giving an h, v, l ratio of 1, $\sqrt{3}$, 2. From the free body diagrams in Fig. 13.15(c), we can observe the following.

Joint J $\Sigma V = 0$ gives

$$F_{HJ} \times \frac{\sqrt{3}}{2} - 30 = 0$$

$$\Rightarrow \quad F_{HJ} = 34.64 \text{ kN (tensile)}$$

$\Sigma H = 0$ gives

$$F_{JG} = 17.32 \text{ kN (compressive)}$$
$$F_{HG} = 34.64 \text{ kN (compressive)}$$

$\Sigma H = 0$ at joint H gives

$$F_{HF} = 34.64 \text{ kN (tensile)}$$

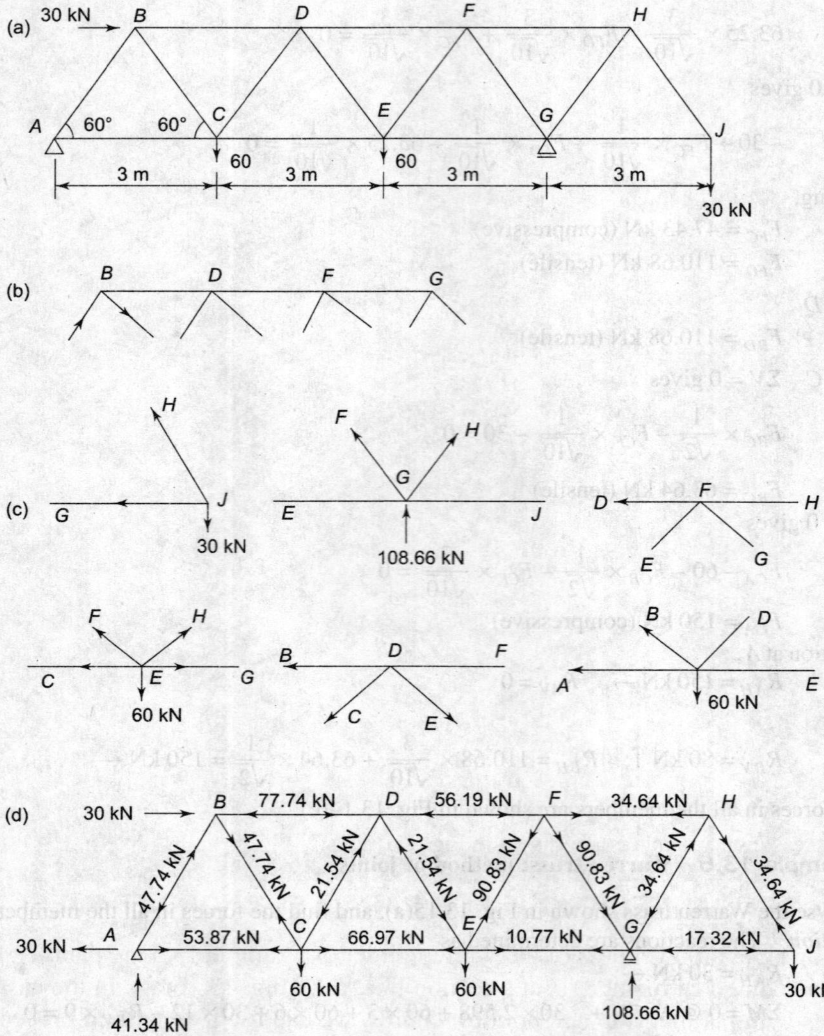

Fig. 13.15

Joint G $\Sigma V = 0$ gives

$$108.66 - F_{GH} \times \frac{\sqrt{3}}{2} - F_{GF} \times \frac{\sqrt{3}}{2} = 0$$

$$108.66 - 34.64 \times \frac{\sqrt{3}}{2} - F_{GF} \times \frac{\sqrt{3}}{2} = 0$$

$\Rightarrow$ $F_{GF} = 90.83$ kN (compressive)

$F_{FE} = 90.83$ kN (tensile)

from joint *F*.

$\Sigma H = 0$ gives

$$F_{GE} = \frac{90.83}{2} - \frac{34.64}{2} - 17.32 = 10.77 \text{ kN (tensile)}$$

Joint F $\Sigma H = 0$ gives

$$\frac{90.83}{2} + \frac{90.83}{2} - 34.64 = 56.19 \text{ kN (compressive)}$$

Joint E $\Sigma V = 0$ gives

$$-F_{ED} \frac{\sqrt{3}}{2} + 90.83 \frac{\sqrt{3}}{2} - 60 = 0$$

$F_{ED} = 21.54$ kN (compressive)

$\Sigma H = 0$ gives

$$-F_{EC} + \frac{21.54}{2} + \frac{90.83}{2} + 10.77 = 0, \quad F_{EC} = 66.97 \text{ kN (tensile)}$$

Joint D $\Sigma V = 0$ gives

$F_{DC} = 21.54$ kN (tensile)

$\Sigma H = 0$ gives

$$F_{DB} = \frac{21.54}{2} + 56.19 + \frac{21.54}{2}$$

$= 77.73$ kN (compressive)

Joint C $\Sigma V = 0$ gives

$$F_{CB} \frac{\sqrt{3}}{2} = 21.54 \frac{\sqrt{3}}{2} - 60 = 0$$

$\Rightarrow$ $F_{CB} = 47.74$ kN (tensile)

$\Sigma H = 0$

$$-F_{CB} - 47.74 \times \frac{1}{2} + 21.54 \times \frac{1}{2} + 66.97 = 0$$

$\Rightarrow$ $F_{CB} = 53.87$ kN (tensile)

$F_{AB} = 47.74$ kN (compressive)

The forces in the members are shown in Fig. 13.15(d).

13.6 SPECIAL TECHNIQUE FOR PARALLEL CHORD FRAMES

The top and bottom chords of the Warren girder of Example 13.5 are parallel, and only one diagonal member is at a section like *X*-*X* [Fig. 13.16(a)]. In the case of frames with parallel (top and bottom) chords, we can use a simplified procedure for analysis.

The free body diagram on the part to the left of X-X is shown in Fig. 13.16(b). The chord members being horizontal, they do not have a component in the vertical direction. The only vertical force at the section is F_V, the vertical component of F, the stress resultant in the diagonal member. $\Sigma V = 0$ for this free body diagram, meaning that F_V must be equal and opposite to the algebraic sum of the vertical forces acting on the free body to the left. This means that F_V must be equal to the SF in the panel in which it lies.

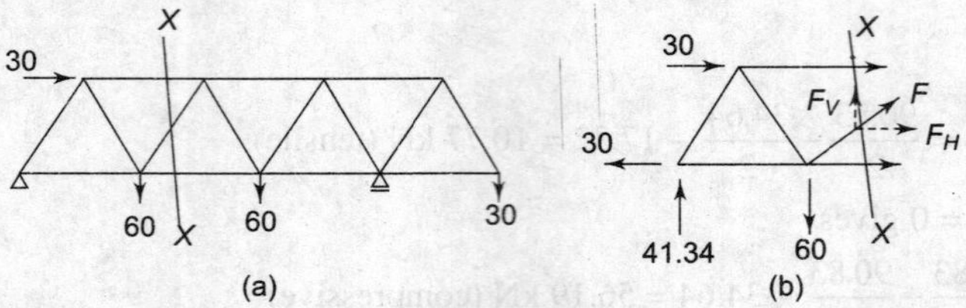

Fig. 13.16

13.6.1 Shear Force Diagram

The SF diagram for a frame can be drawn by considering all vertical forces acting on the frame, on the top and bottom chords. Figure 13.17(a) shows the SF diagram for the truss shown in Fig. 13.16(a). Figure 13.17(b) shows the SF diagram for a truss which has loads on the top and bottom chords. Figure 13.17(c) shows the SF diagram for a cantilever frame. Note that the SF diagrams are drawn as if for a beam considering only the vertical loads.

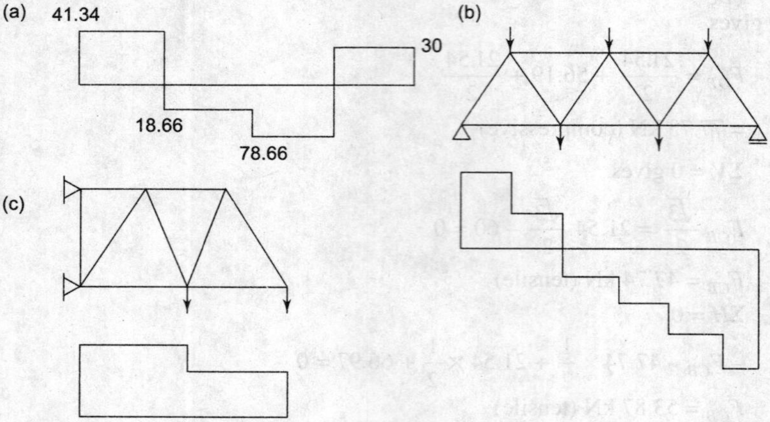

Fig. 13.17

13.6.2 Nature of Forces in a Diagonal Member

The sign convention for SF is the same as that adopted in Chapter 4, and is illustrated in Fig. 13.18(a). Figure 13.18(b) shows us how to find the nature of force in a diagonal member. If the SF in a panel is positive, then the diagonal will have a tensile force if it lies along 1-2, and a compressive force if it lies along 3-4. If the panel shear is negative, then a member lying along 1-2 will have compression and one lying along 3-4 will have tension. These diagrams we have already discussed while studying compound stresses in Chapter 9.

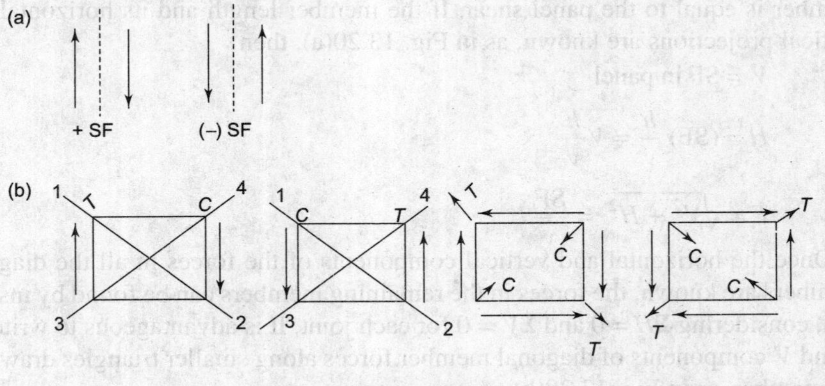

Fig. 13.18

Figures 13.19(a), (b), and (c) illustrate how to find the nature of force in a diagonal member, knowing the SF in the panel. For example, in Fig. 13.19(a), the force in member *A B* is compressive and that in *BC* is tensile as per the sign convention shown in Fig. 13.18 for positive SF. Members *DE* and *EF* are in the negative SF zone. So the force in *DE* is compressive and that in *EF* is tensile. A similar logic is used to find the nature of forces in the members of the frame shown in Fig. 13.19.

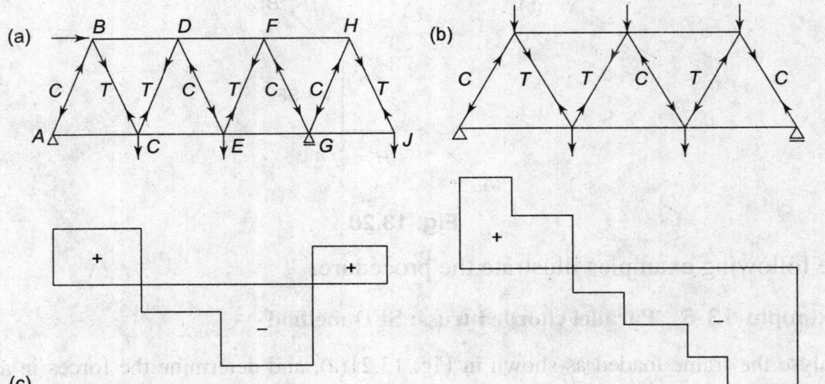

Fig. 13.19

13.6.3 Magnitude of Diagonal Member Forces

As we have already seen, the vertical component, V, of the force in the diagonal member is equal to the panel shear. If the member length and its horizontal and vertical projections are known, as in Fig. 13.20(a), then

$$V = \text{SF in panel}$$

$$H = (\text{SF})\,\frac{h}{v} = V\frac{h}{v}$$

$$F = \sqrt{V^2 + H^2} = \frac{\text{SF}}{v}\,l$$

Once the horizontal and vertical components of the forces in all the diagonal members are known, the forces in the remaining members can be found by inspection, considering $\Sigma H = 0$ and $\Sigma V = 0$ for each joint. It is advantageous to write the H and V components of diagonal member forces along smaller triangles drawn on the member, as in Fig. 13.20(b).

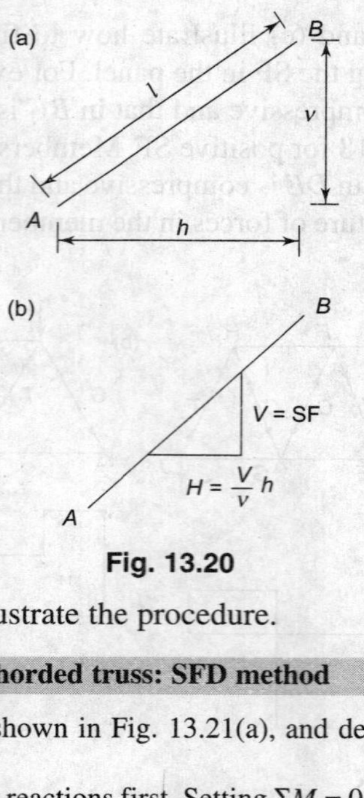

Fig. 13.20

The following examples illustrate the procedure.

Example 13.6 Parallel chorded truss: SFD method

Analyse the frame loaded as shown in Fig. 13.21(a), and determine the forces in all the members.

Solution Let us calculate the reactions first. Setting $\Sigma M = 0$ @ A,

$$R_{JV} = (40 \times 1.5 + 40 \times 3 + 60 \times 6 + 80 \times 9)/12 = 105 \text{ kN}, \quad R_{AV} = 75 \text{ kN}$$

$$R_{AH} = 40 \text{ kN} \leftarrow$$

The SF diagram is shown in Fig. 13.21(b). For all diagonal members, $v{:}h$ is 1.5:3, i.e., the horizontal component is twice the vertical component. The vertical component in a diagonal is equal to the SF in the panel, and the horizontal component is equal to twice the vertical component. For example, in the diagonal member BC, the vertical component = 75 kN and horizontal component = 150 kN. These components are written on the sides of a smaller triangle drawn on the diagonal members as in Fig. 13.21(a). The nature of force in a diagonal member is decided from the sign of the SF. All the diagonal members are in tension, as is clear from the diagrams shown in Fig. 13.21(c).

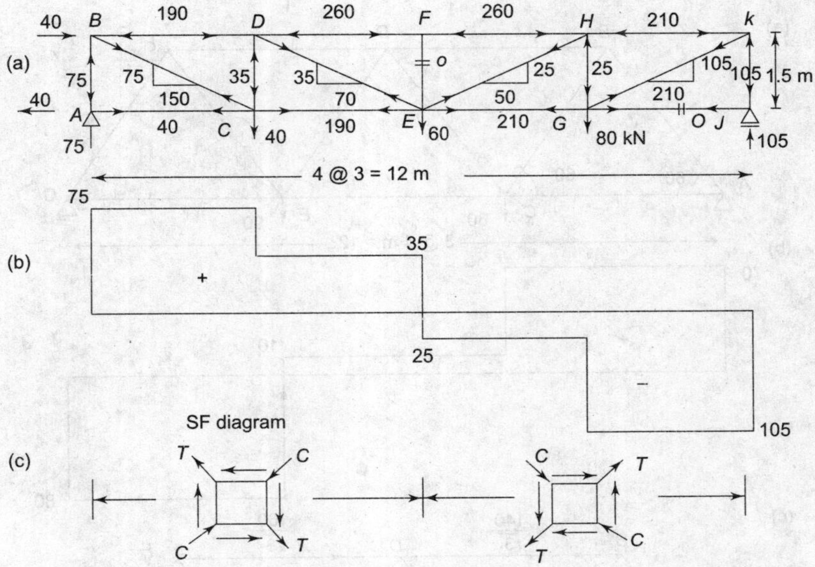

Fig. 13.21

Once we know forces in the diagonal members in terms of their horizontal and vertical components, we can easily analyse each joint to find the forces in other members. Let us consider, for example, the following joints.

Joint A For equilibrium, $F_{AB} = 75$ kN (compressive) $F_{AC} = 40$ kN (tensile)

Joint B $\Sigma V = 0$ is satisfied. $\Sigma H = 0$ requires a force of 190 kN towards B at B. Therefore, $F_{BD} = 190$ kN (compressive) = horizontal component of BC + 40 kN.

Joint C $\Sigma V = 0$ gives

$$F_{CD} = 75 - 40 = 35 \text{ kN (compressive)}$$

$\Sigma H = 0$ gives

$$F_{CE} = 150 + 40 = 190 \text{ kN (tensile)}$$

All the joints can be analysed easily from the horizontal and vertical components known. The result is shown in Fig. 13.21(a). The reader should carefully study this figure to ensure that all the joints are balanced in the horizontal and vertical directions. ■

Example 13.7 Warren girder, parallel chorded truss: SFD method

Analyse the Warren girder loaded as shown in Fig. 13.22(a) and determine the forces in all the members.

Solution Let us calculate the reactions first.

$$R_{GV} = (60 \times 4 + 90 \times 8)/12 = 80 \text{ kN}$$

$$R_{AV} = 70 \text{ kN}, \quad R_{AH} = 0$$

The SF diagram is drawn as shown in Fig. 13.22(b). The nature of force in the diagonal members is determined from the sign of the SF diagram. AB and CD are compressive while BC and CD are tensile in the zone where the SF is positive. EF is tensile and FG is compressive in the zone where the SF is negative. All this is indicated in Fig. 13.22(c). The vertical components of the forces in the diagonal members are then written equal to the panel shear. Note that while the panel length is 4 m. h:v for diagonal members is $2:2\sqrt{3}$ or $1:\sqrt{3}$. Therefore, for any diagonal member,

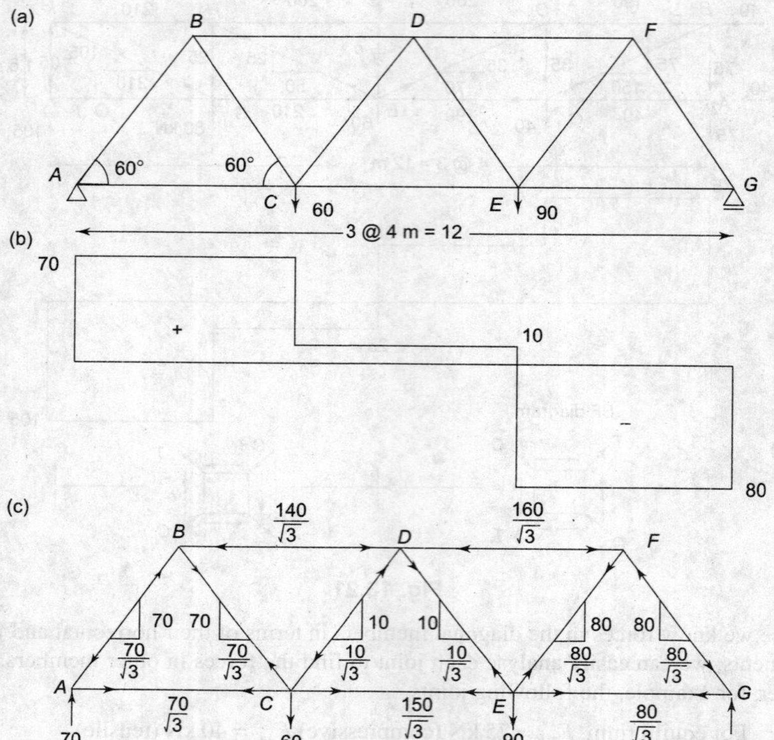

Fig. 13.22

$$\text{Horizontal component of force} = \frac{\text{vertical component}}{\sqrt{3}}$$

The horizontal components of the force in the diagonal members are expressed as shown.

A joint-by-joint inspection can now determine the forces in all the members as shown. Let us consider the forces in the following points.

Joint A $\Sigma H = 0$ at joint A requires that the force in member AC be a force of $70/(3)^{1/2}$ kN acting away from the joint at A (tensile).

Joint B The force in member BD can be determined from $\Sigma H = 0$ as

$$\frac{70}{\sqrt{3}} + \frac{70}{\sqrt{3}} = \frac{140}{\sqrt{3}} \text{ kN}$$

acting towards joint B (compressive).

Joint C

$$\text{Force in } CE = \frac{70}{\sqrt{3}} + \frac{70}{\sqrt{3}} + \frac{10}{\sqrt{3}} = \frac{150}{\sqrt{3}}$$

acting away from C. The forces in all the members can be worked out this way as shown in Fig. 13.22(c).

The forces in the diagonal members can now be obtained as $F = (x^2 + y^2)^{1/2}$ from the horizontal and vertical components.

Example 13.8 Overhanging truss, parallel chorded truss: SFD methods

Analyse the truss shown in Fig. 13.23(a) and find the forces in all the members.

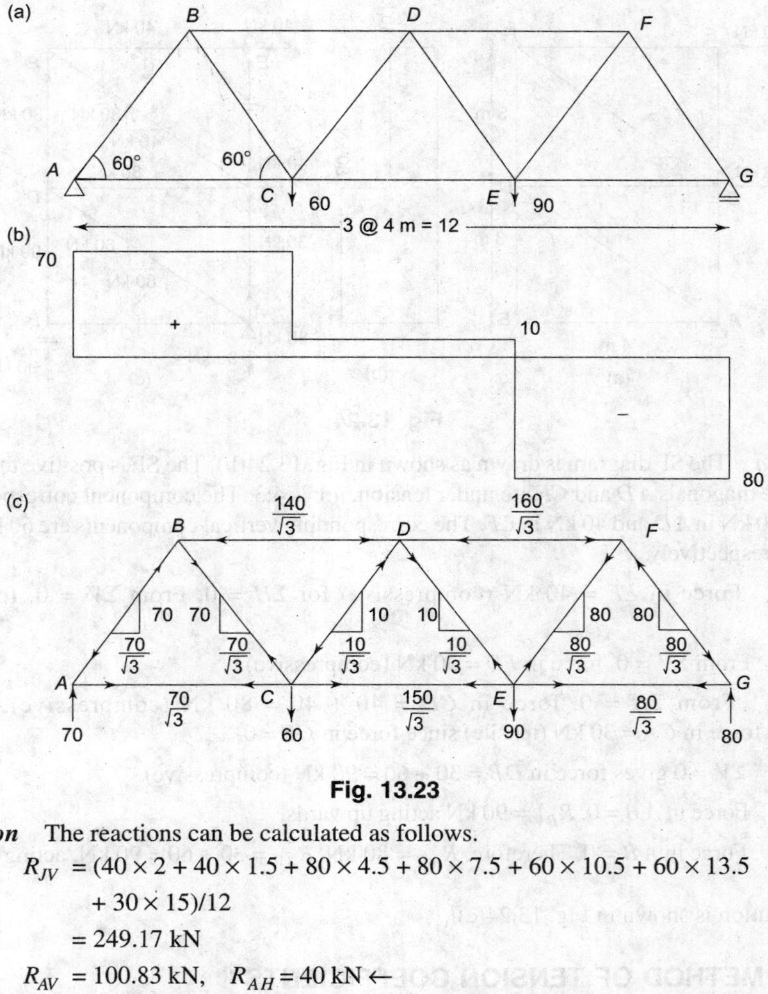

Fig. 13.23

Solution The reactions can be calculated as follows.

$$R_{JV} = (40 \times 2 + 40 \times 1.5 + 80 \times 4.5 + 80 \times 7.5 + 60 \times 10.5 + 60 \times 13.5 + 30 \times 15)/12$$

$$= 249.17 \text{ kN}$$

$$R_{AV} = 100.83 \text{ kN}, \quad R_{AH} = 40 \text{ kN} \leftarrow$$

The SF diagram is shown in Fig. 13.23(b). The nature of force (compressive or tensile) is determined from the sign of the SF and the orientation of the member. The vertical component of the force in any diagonal member is equal to the SF in the panel. Note that h:v for all the diagonal members is 1.5:2. Therefore, the horizontal component = $V \times 1.5/2$. The horizontal components are expressed as shown in Fig. 13.23(c). Once the components of the bar forces in the diagonal members are determined, the forces in the other members can be determined from the conditions $\Sigma H = 0$ and $\Sigma V = 0$ for each joint. Let us find the forces in the following joints.

Joint A From $\Sigma H = 0$, bar force in $AC = 40 + 75.62 = 115.62$ (tensile).

Joint B From $\Sigma H = 0$, force in $BD = 40 + 75.62 + 45.62 = 161.24$ kN (compressive).

Joint C From $\Sigma H = 0$, force in $AC = 115.62 + 45.62 + 45.62 = 206.86$ kN (tensile).

All the bar forces are shown in Fig. 13.23(c). ■

Example 13.9 Parallel chorded truss: SFD method

Analyse the frame loaded as shown in Fig. 13.24(a) and find the forces in all the members.

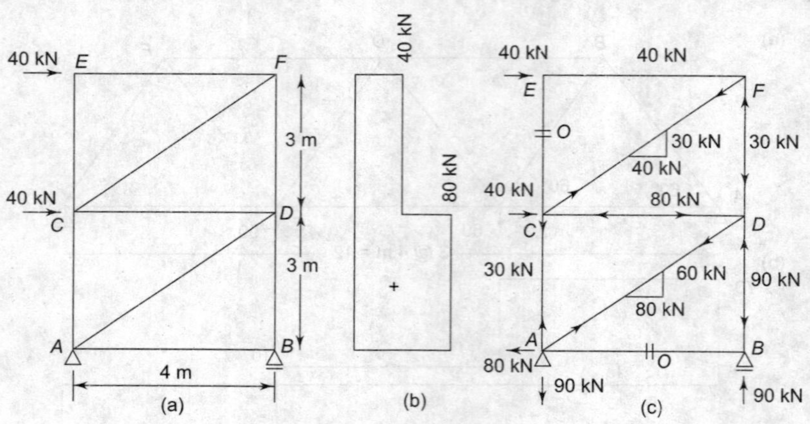

Fig. 13.24

Solution The SF diagram is drawn as shown in Fig. 13.24(b). The SF is positive throughout. The diagonals AD and CF are under tension. h:v is 4:3. The component corresponding to h is 80 kN in AD and 40 kN in CF. The corresponding vertical components are 60 kN and 30 kN, respectively.

Joint E Force in EF = 40 kN (compressive) for $\Sigma H = 0$. From $\Sigma V = 0$, force in $EC = 0$.

Joint F From $\Sigma V = 0$, force in FD = 30 kN (compressive).

Joint C From $\Sigma H = 0$, force in CD = 40 + 40 = 80 kN (compressive). From $\Sigma V = 0$, force in CA = 30 kN (tensile) since force in $CE = 0$.

Joint D $\Sigma V = 0$ gives force in DB = 30 + 60 = 90 kN (compressive).

Joint B Force in $AB = 0$. R_{BV} = 90 kN acting upwards.

Joint A Force in $AB = 0$. Therefore, R_{AH} = 80 kN, R_{AV} = 30 + 60 = 90 kN, acting downwards.

The solution is shown in Fig. 13.24(c).

13.7 METHOD OF TENSION COEFFICIENTS

The method of tension coefficients is a variation of the conventional method of joints. The principle remains the same, i.e., that of the equilibrium of a joint under the action of a set of concurrent coplanar forces. The method of tension coefficients can be easily extended to space frames.

The equations $\Sigma H = 0$ and $\Sigma V = 0$ are expressed in the method of tension coefficients using joint coordinates and tension coefficients of members.

13.7.1 Joint Coordinates

Considering the frame shown in Fig. 13.25(a), the X- and Y-coordinate axes can be chosen anywhere and the coordinates determined with respect to the origin. However, it is preferable that the origin be so selected that the entire frame is in the first quadrant. This is only for ease of computation as all the coordinates will be positive.

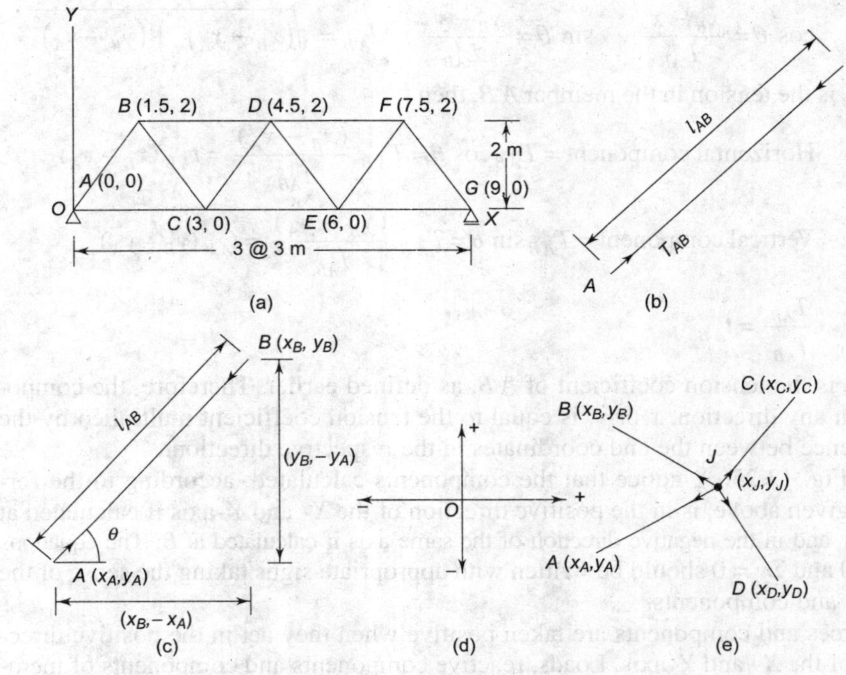

Fig. 13.25

13.7.2 Tension Coefficient

The tension coefficient of a member [Fig. 13.25(b)] is defined as the tension in the member divided by its length. Thus if T_{AB} is the tension in the member AB and l_{AB} is its length, then the tension coefficient = T_{AB}/l_{AB}. The tension coefficient is expressed by the small letter t as t_{AB}. That is,

$$t_{AB} = \frac{T_{AB}}{l_{AB}} = \frac{\text{tension in the member}}{\text{length of the member}}$$

What happens if a member is under compression? First, it is assumed that all the members are in tension. If a member is in compression, it is then the same as negative tension; its tension coefficient is negative. Therefore, the analysis is started on the assumption that every member is in tension. If the calculated value of t for any member is negative, that member is in compression.

13.7.3 Joint Equilibrium Equations

Joint equilibrium equations can be expressed in terms of tension coefficients and joint coordinates in the X- and Y-direction.

Components of forces Consider the member AB shown in Fig. 13.25(c). (x_A, y_A) are the coordinates of A and (x_B, y_B) those of B with respect to the origin O. $(x_B - x_A)$ and $(y_B - y_A)$ are the horizontal and vertical projections respectively of the length l_{AB} of the member. It is clear that if θ is the inclination of the member to the horizontal, then

$$\cos\theta = \frac{x_B - x_A}{L_{AB}}, \quad \sin\theta = \frac{y_B - y_A}{L_{AB}}, \quad l_{AB} = \sqrt{(x_B - x_A)^2 + (y_B - y_A)^2}$$

If T_{AB} is the tension in the member AB, then

$$\text{Horizontal component} = T_{AB}\cos\theta = T_{AB}\frac{(x_B - x_A)}{l_{AB}} = t_{AB}(x_B - x_A)$$

$$\text{Vertical component} = T_{AB}\sin\theta = T_{AB}\frac{(y_B - y_A)}{l_{AB}} = t_{AB}(y_B - y_A)$$

$$\frac{T_{AB}}{l_{AB}} = t_{AB}$$

which is the tension coefficient of AB, as defined earlier. Therefore, the component in any direction, x or y, is equal to the tension coefficient multiplied by the difference between the end coordinates in the respective direction.

In Fig. 13.25(c), notice that the components calculated, according to the formula given above, is in the positive direction of the X- and Y-axis if calculated at joint A and in the negative direction of the same axis if calculated at B. The equations $\Sigma x = 0$ and $\Sigma y = 0$ should be written with appropriate signs taking the sense of the forces and components.

Forces and components are taken positive when they act in the positive directions of the X- and Y-axis. Loads, reactive components and components of member forces are positive when they act towards the right or upwards. Those acting downwards or towards the left are negative [Fig. 13.25(d)].

In the case of loads and reactive forces, the signs can be assigned knowing their sense. For components of member forces, to avoid the anomaly (reversal of component signs when considered from the two end joints of a member) mentioned earlier, a simple technique is employed to get the correct sign for the components.

Expressing the equilibrium equation Consider a joint J, shown in Fig. 13.25(e). The four members meeting at the joint are JA, JB, JC and JD. All the members are in tension so that their sense at joint J is away from this joint. It will be so at all joints (as assumed initially).

We state the equilibrium equations $\Sigma x = 0$ and $\Sigma y = 0$ for joint J. The components of the tension T_{JA} of member JA are equal to $t_{JA}(x_A - x_J)$ and $t_{JA}(y_A - y_J)$ in the X-direction and the Y-direction respectively. In expressing the components, a simple rule is followed: x-component = tension coefficient (x-coordinate of the far joint – x-coordinate of the joint for which the equation is stated). The same rule is applied to the Y-direction also. This simple rule of the *coordinate of the far joint minus the coordinate of the joint for which the equation is stated* takes care of the signs of the components. Thus, $\Sigma x = 0$ gives ($\rightarrow +$) for joint J

$$t_{JA}(x_A - x_J) + t_{JB}(x_B - x_J) + t_{JC}(x_C - x_J) + t_{JD}(x_D - x_J) + P_x = 0$$

and

$$t_{JA}(y_A - y_J) + t_{JB}(y_B - y_J) + t_{JC}(y_C - y_J) + t_{JD}(y_D - y_J) - P_y = 0$$

is the equation in the Y-direction for $\Sigma y = 0$. While formulating the equation for joint A, the components of the member force in AJ are expressed as $T_{AJ}(x_J - x_A)$ and $T_{AJ}(y_J - y_A)$. Of course, $T_{JA} = T_{AJ}$ and $t_{AJ} = t_{JA}$.

The procedure for the application of the method of tension coefficients for frame analysis is as follows.

(i) Select the origin of the x-y coordinate axes system such that the entire frame is in the first quadrant.
(ii) Find the x, y coordinates of each joint.
(iii) Determine the reactions using the equilibrium conditions for the whole frame.
(iv) Make a joint-by-joint analysis, selecting a joint with not more than two unknown member forces. State the equations using tension coefficients and joint coordinates.
(v) While formulating the equations, to find the component of a member force along the X or Y direction, always use the difference between the coordinates of the far joint and the coordinates of the joint under consideration to take care of the signs.
(vi) Forces, reactions and component forces are positive in the positive directions of the respective axes.
(vii) A negative tension coefficient means that the member is in compression.
(viii) Find the member forces by multiplying the tension coefficient with the member length.

The method of tension coefficients is very simple to apply. It can be very easily modified to make a computer program. A third dimension z has to be added for space frames. The following examples illustrate the application to plane frames.

Example 13.10 Tension coefficient methods

Analyse the pin-jointed plane frame shown in Fig. 13.26(a) and find the forces in all the members using the method of tension coefficients.

Solution Taking the origin at A, the X-axis along AC and the Y-axis perpendicular to it, the coordinates of joint are $A(0, 0)$, $B(2, 1.5)$, and $C(4, 0)$. While the reactions can be calculated separately, for the purpose of illustrations we will state six equations and solve for three member forces and three reactive components. The equations $\Sigma H = 0$ and $\Sigma V = 0$ for the three joints are as follows.

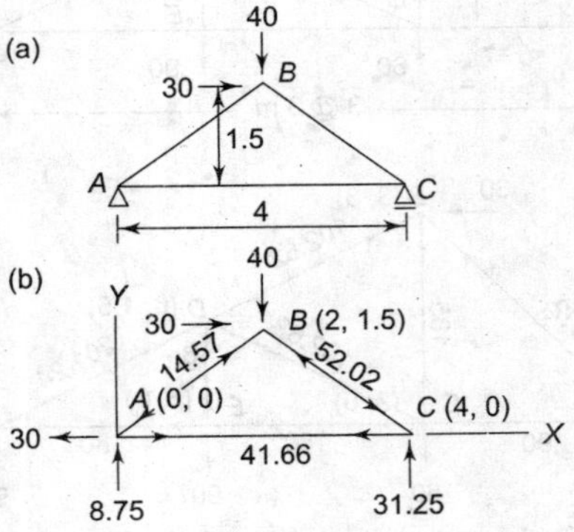

Fig. 13.26

Joint A

$$\Sigma H = 0, \quad R_{AH} + t_{AB}(x_B - x_A) + t_{AC}(x_C - x_A) = 0$$

$$R_{AH} + 2t_{AB} + 4t_{AC} = 0 \quad (1)$$

$$\Sigma V = 0, \quad R_{AV} + t_{AB}(y_B - y_A) + t_{AC}(y_C - y_A) = 0, \quad R_{AV} + 1.5t_{AB} = 0 \quad (2)$$

Joint B

$$\Sigma H = 0,\quad 30 + t_{BA}(x_A - x_B) + t_{BC}(x_C - x_B) = 0,\quad 30 - 2t_{BA} + 2t_{BC} = 0 \tag{3}$$

$$\Sigma V = 0,\quad -40 + t_{BA}(y_A - y_B) + t_{BC}(y_C - y_B) = 0,\quad 40 + 1.5t_{BA} + 1.5t_{BC} = 0 \tag{4}$$

Joint C

$$\Sigma H = 0,\quad t_{BC}(x_B - x_C) + t_{AC}(x_A - x_C) = 0,\quad 2t_{BC} + 4t_{CA} = 0 \tag{5}$$

$$\Sigma V = 0,\quad R_{CV} + t_{BC}(y_B - y_C) + t_{AC}(y_A - y_C) = 0,\quad 1.5t_{BC} + R_{CV} = 0 \tag{6}$$

From (3) and (4),

$$t_{BC} = -20.83,\quad t_{BA} = -5.83,\quad t_{AC} = 10.415$$

$$R_{AV} = 8.75,\quad R_{CV} = 31.25,\quad R_{AH} = -30$$

The forces in the members, multiplying the tension coefficients by their length, are as follows:

Force in AB = 14.57 kN (compressive)
BC = 52.07 kN (compressive)
AC = 41.66 kN (tensile)

Example 13.11 Tension coefficient method

Analyse the pin-jointed plane frame shown in Fig. 13.27(a) and determine the forces in all the members.

Solution Let us calculate the reactions first.

$$R_{FV} = (30 \times 3 + 60 \times 3 + 90 \times 6)/9 = 90 \text{ kN},\quad R_{AV} = 60 \text{ kN}$$

$$R_{AH} = 30 \text{ kN} \leftarrow$$

Taking the origin at A, the coordinates of the joint are $A(0, 0)$, $B(3, 3)$, $C(3, 0)$, $D(6, 1.5)$, $E(6, 0)$, $F(9, 0)$.

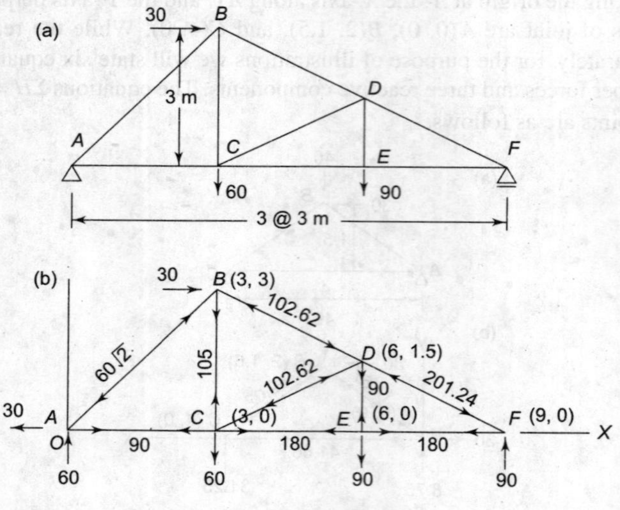

Fig. 13.27

Joint A

$$\Sigma H = 0,\quad -30 + t_{AB}(3 - 0) + t_{AC}(3 - 0) = 0$$

$$\Sigma V = 0,\quad 60 + t_{AB}(3 - 0) + t_{AC}(0 - 0) = 0,\quad t_{AB} = -20,\quad t_{AC} = 30$$

Joint B

$$\Sigma H = 0,\quad 30 + t_{AB}(0 - 3) + t_{BD}(6 - 3) + t_{BC}(3 - 3) = 0$$

$$\Sigma V = 0, \quad t_{AB}(0-3) + t_{BC}(0-3) + t_{BD}(1.5-3) = 0, \quad t_{BD} = -30, \ t_{BC} = 35$$

Joint C

$$\Sigma H = 0, \quad t_{AC}(0-3) + t_{CD}(6-3) + t_{CE}(6-3) + t_{CB}(3-3) = 0$$
$$\Sigma V = 0, \quad t_{CB}(3=0) + t_{CD}(1.5=0) + t_{CA}(0-0) + t_{CE}(0-0) - 60 = 0$$
$$t_{CD} = -30, \ t_{CE} = 60$$

Joint D

$$\Sigma H = 0, \quad t_{BD}(3-6) + t_{DF}(9-6) + t_{CD}(3-6) + t_{DE}(6-6) = 0$$
$$\Sigma V = 0, \quad t_{BD}(3-1.5) + t_{CD}(0-1.5) + t_{DE}(0-1.5) = 0, \quad t_{DF} = -60$$

Joint E

$$\Sigma H = 0, \quad t_{EC}(3-6) + t_{DE}(6-6) + t_{EF}(9-6) = 0, \quad t_{EC} = t_{EF} = 60$$

We have found all the member forces. We can use one equation of joint E and two of joint F for checking. The results are shown in Fig. 13.27(b).

Example 13.12 Tension coefficient method

Analyse the frame shown in Fig. 13.28(a) and determine the forces in all the members.

Solution We can begin the solution from joint C without calculating the reactions. The joint coordinates are A (0, 0), B(0, 2), (0, 4), D(2, 2), and E(4, 0).

Joint C

$$\Sigma H = 0, \quad 40 + t_{CD}(2-0) + t_{CB}(0-0) = 0, \quad t_{CD} = -20$$
$$\Sigma V = 0, \quad t_{CB}(2-4) + t_{CD}(2-4) = 0, \quad t_{CB} = -t_{CD} = 20$$

Joint B

$$\Sigma H = 0, \quad 40 + t_{BD}(2=0) + t_{BC}(0-0) + t_{BA}(0-0) = 0, \quad t_{BD} = -20$$
$$\Sigma V = 0, \quad t_{BC}(4-2) + t_{BA}(0-2) + t_{BD}(2-2) = 0, \quad t_{BA} = 20$$

Joint D

$$\Sigma H = 0, \quad t_{DC}(0-2) + t_{BD}(0-2) + t_{AD}(0-2) + t_{DE}(4-2) = 0$$
$$\Sigma V = 0, \quad t_{DC}(4-2) + t_{BD}(2-2) + t_{AD}(0-2) + t_{DE}(0-2) = 0$$
$$t_{AD} = 10, \quad t_{ED} = -30$$

Joint E

$$\Sigma H = 0, \quad t_{ED}(2-4) + t_{AE}(0-4) = 0, \quad t_{AE} = 15$$
$$\Sigma V = 0, \quad R_{EV} + t_{ED}(2-0) + t_{AE}(0-0) = 0, \quad R_{EV} = 60 \text{ kN}$$

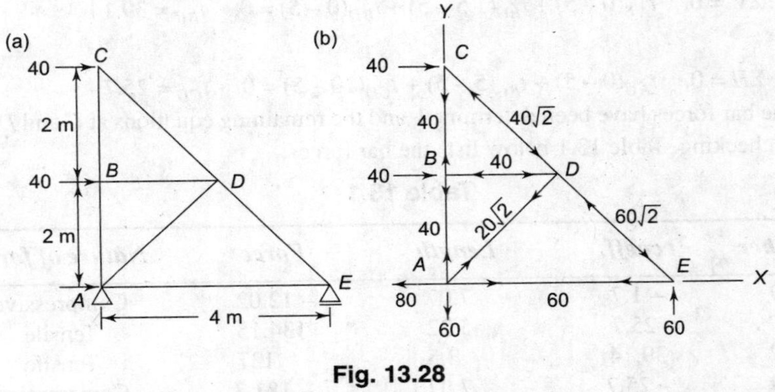

Fig. 13.28

For equilibrium of the frame as a whole, $R_{AH} = -80$ kN and $R_{AV} = -60$ kN. The results are as shown in Fig. 13.28(b).

Example 13.13 Tension coefficient method

Analyse the pin-jointed plane frame shown Fig. 13.29(a) and determine the forces in all the members.

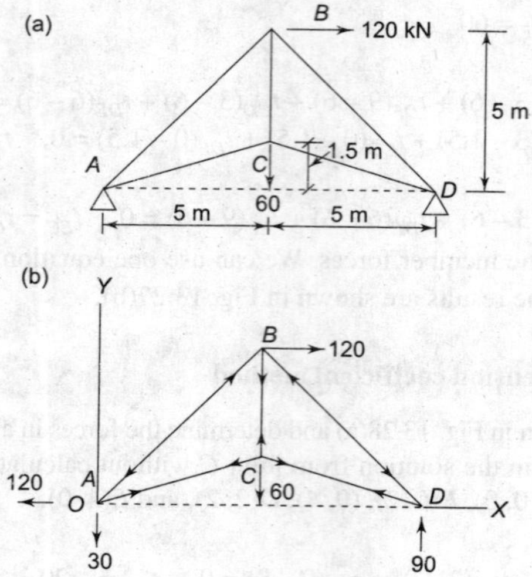

Fig. 13.29

Solution The reactions are

$$R_{BV} = (120 \times 5 + 60 \times 5)/10 = 90 \text{ kN} \uparrow$$

$$R_{AV} = 30 \text{ kN} \downarrow, \quad R_{AH} = 120 \text{ kN} \leftarrow$$

Selecting the origin at A, the coordinates of the points are $A(0, 0)$, $B(5, 5)$, $C(5, 1.5)$, and $D(10, 0)$.

Joint A

$$\Sigma H = 0, \quad -120 + t_{AB}(5-0) + t_{AC}(5-0) = 0$$

$$\Sigma V = 0, \quad -30 + t_{AB}(5-0) + t_{AC}(1.5-0) = 0, \quad t_{AC} = 25.7, \quad t_{AB} = -1.7$$

Joint B

$$\Sigma H = 0, \quad 120 + t_{AB}(0-5) + t_{BC}(5-5) + t_{BD}(10-5) = 0, \quad t_{BD} = -25.7$$

$$\Sigma V = 0, \quad t_{AB}(0-5) + t_{BC}(1.5-5) + t_{BD}(0-5) = 0, \quad t_{BC} = 39.14$$

Joint C

$$\Sigma H = 0, \quad t_{AC}(0-5) + t_{BC}(5-5) + t_{CD}(10-5) = 0, \quad t_{CD} = 25.7$$

All the bar forces have been determined, and the remaining equations at C and D can be used for checking. Table 13.1 below lists the bar forces.

Table 13.1

Member	*t coeff.*	*Length*	*Force*	*Nature of force*
AB	– 1.7	7.07	– 12.02	Compressive
AC	25.7	5.22	134.15	Tensile
BC	39.14	3.5	137	Tensile
BD	– 25.7	7.07	– 181.7	Compressive
CD	25.7	5.22	+ 134.15	Tensile

13.8 RITTER'S METHOD OF SECTIONS

As mentioned earlier, Ritter's method of sections, generally called the method of sections or that of moments, is an alternative method of taking sections, dividing the truss into two parts. We can consider the equilibrium of either part, and each part is in equilibrium, in general, under the action of a non-concurrent coplanar force system.

Considering the frame shown in Fig. 13.30(a), the section X-X divides the frame into two parts. The free body of the left part is shown in Fig. 13.30(b). The sense of the stress resultants released by sectioning the members can be assumed either way. Note that if the arrows act away from the free body (away from the joint) as shown for members *DE* and *CE*, the members are assumed to be in tension. If the sense of the force is towards the free body, as shown for member *DF*, the member is assumed to be in compression. If the calculated value comes out to be negative, then the nature of the force in the member is opposite to that assumed.

The free body, as in Fig. 13.30(b), acted upon by a general coplanar force system has three conditions of equilibrium—$\Sigma H = 0$, $\Sigma V = 0$, and $\Sigma M = 0$. From these three equations, three unknown bar forces can be determined. While the section may cut more than three members, only three unknown bar forces can be determined.

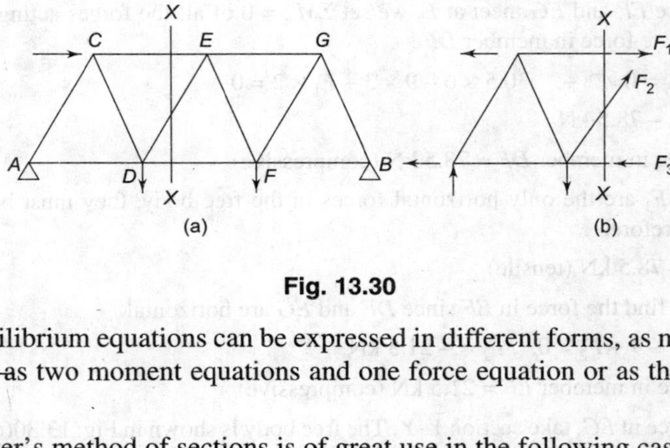

Fig. 13.30

The equilibrium equations can be expressed in different forms, as mentioned in Chapter 1—as two moment equations and one force equation or as three moment equations.

The Ritter's method of sections is of great use in the following cases.

(i) When one needs to find forces in any intermediate members,

(ii) When one needs to check values calculated by other methods, and

(iii) In special cases, as for compound frames discussed later in this chapter, to find the force in some member so that analysis by other methods is facilitated.

The following examples illustrate the procedure.

Example 13.14 Ritter's method of section

Determine the forces in the members marked 1, 2, 3, and 4, in the frame loaded as shown in Fig. 13.31(a).

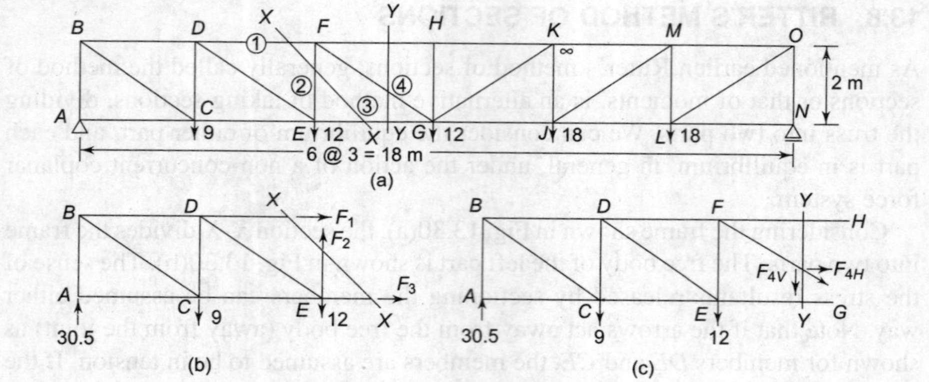

Fig. 13.31

Solution The reactions are

$$R_{NV} = (9\times 3 + 12\times 6 + 12\times 9 + 18\times 12 + 18\times 15) \div 18$$

$$= 38.5 \text{ kN}$$

$$R_{AV} = 30.5 \text{ kN}, \quad R_{AH} = 0$$

We consider the section *X-X* as shown in Fig. 13.31(a). The free body diagram to the left of the section is shown in Fig. 13.31(b). The forces in all the members are assumed outwards (tensile). Since *EF* and *EG* meet at *E*, we set $\Sigma M_E = 0$ of all the forces acting on the free body to evaluate force in member *DF*.

$$\Sigma M_E = 0 \curvearrowleft +, \quad 30.5\times 6 - 9\times 3 + F_1\times 2 = 0$$

$$\Rightarrow \quad F_1 = -78.5 \text{ kN}$$

Force in member *DF* = 78.5 kN (compressive)

Since F_1 and F_3 are the only horizontal forces in the free body, they must be equal and opposite. Therefore,

$$F_3 = 78.5 \text{ kN (tensile)}$$

Set $\Sigma V = 0$ to find the force in *EF* since *DF* and *EG* are horizontal.

$$30.5 - 9 + F_2 = 0, \quad F_2 = -21.5 \text{ kN}$$

Force in member *EF* = 21.5 kN (compressive)

To find the force in *FG*, take section *Y-Y*. The free body is shown in Fig. 13.30(c). Set $\Sigma V =$ 0 and let F_{4V} be the vertical component of force in *FG*.

$$30.5 - 9 - 12 - F_{4V} = 9.5 \text{ kN}$$

h, *v*, and *l* for member *FG* are 3, 2, and $\sqrt{13}$. Therefore,

$$F_3 = \frac{9.5}{2}\times\sqrt{13} = 17.12 \text{ kN}$$

Force in member *FG* = 17.12 kN (tensile) ■

Example 13.15 Ritter's method of section

Find the forces in the members marked 1, 2, 3, and 4 in the frame loaded as shown in Fig. 13.32(a).

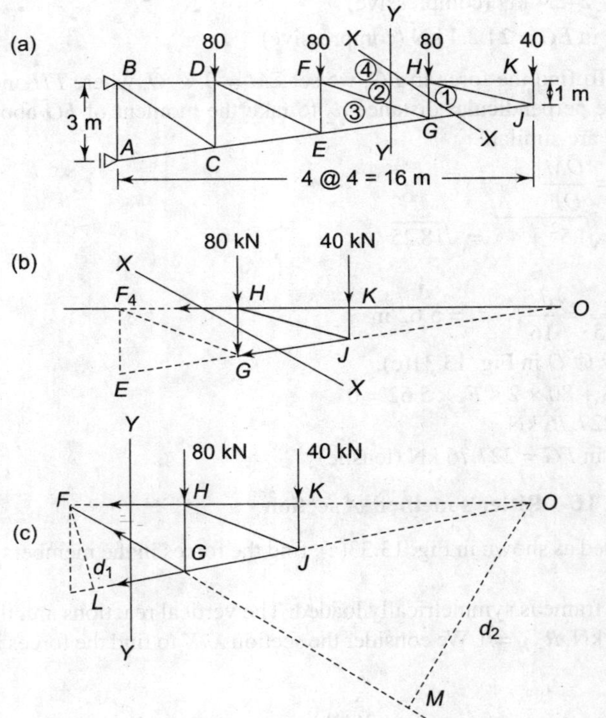

Fig. 13.32

Solution Since it is a cantilever frame, we can use the free body to the right, where reactions do not appear. Taking section *X-X*, the free body is shown in Fig. 13.32(b). The forces in the members are assumed tensile. To find the force in *HG*, set $\Sigma M = 0$ about *O*, where *FH* and *GJ* intersect. (Note that we cannot find the force in *HG* from $\Sigma V = 0$ since *GJ* also has a vertical component.) The distance of *O* from *K* = 8 m.

$\Sigma M = 0$ @ '*O*' ↺+

$80 \times 12 + 40 \times 8 + F_1 \times 12 = 0$

$\Rightarrow \quad F_1 = -106.67$

Force in *HG* is 106.67 kN (compressive)

Force in FH We set $\Sigma M = 0$ @ *G*, where *HG* and *JG* intersect.

$40 \times 4 - F_4 \times 1.5 = 0$

$\Rightarrow \quad F_4 = 106.67$

Force in *FH* = 106.67 kN (tensile)

Next we consider section *Y-Y*. The free body is shown in Fig. 13.32(c).

Force in EG To find the force in *EG*, we set $\Sigma M = 0$ @ *F*, where *GF* and *HF* meet. To find the perpendicular distance d_1, we note that triangles *FLO* and *JKO* are similar. That is,

$$\frac{JK}{JO} = \frac{FL}{FO}, \quad JO = \sqrt{1^2 + 8^2} = \sqrt{65}$$

Therefore,

$$\frac{1}{\sqrt{65}} = \frac{d_1}{16}, \quad d_1 = 1.98 \text{ m}$$

$$\Sigma M = 0 \text{ @ } F, \quad 80 \times 4 + 40 \times 4 + F_3 \times 1.98 = 0$$

⇒ $F_3 = -242.4$ kN (compressive)

Force in EG = 24.2.4 kN (compressive)

Force in FG To find the force in FG, we set $\Sigma M = 0$ @ O, where FH and EG meet. We have to find the perpendicular distance d_2 to take the moment of FG about O. Triangles FHG and FMO are similar.

$$\frac{HG}{FG} = \frac{OM}{OF}$$

$$FG = \sqrt{1.5^2 + 4^2} = \sqrt{18.25}$$

Therefore,

$$\frac{1.5}{\sqrt{18.25}} = \frac{d_2}{16}, \quad d_2 = 5.62 \text{ m}$$

Setting $\Sigma M = O$ @ O in Fig. 13.31(c),

$$40 \times 8 + 80 \times 2 - F_2 \times 5.62 = 0$$

⇒ $F_2 = 227.76$ kN

Force in FG = 227.76 kN (tensile)

Example 13.16 Ritter's method of section

In the truss loaded as shown in Fig. 13.33(a), find the forces in the members marked 1, 2, 3, and 4.

Solution The frame is symmetrically loaded. The vertical reactions are, therefore, equal. $R_{AV} = R_{KV} = 36$ kN. $R_{AH} = 0$. We consider the section X-X to find the forces in the members 1, 2, and 4.

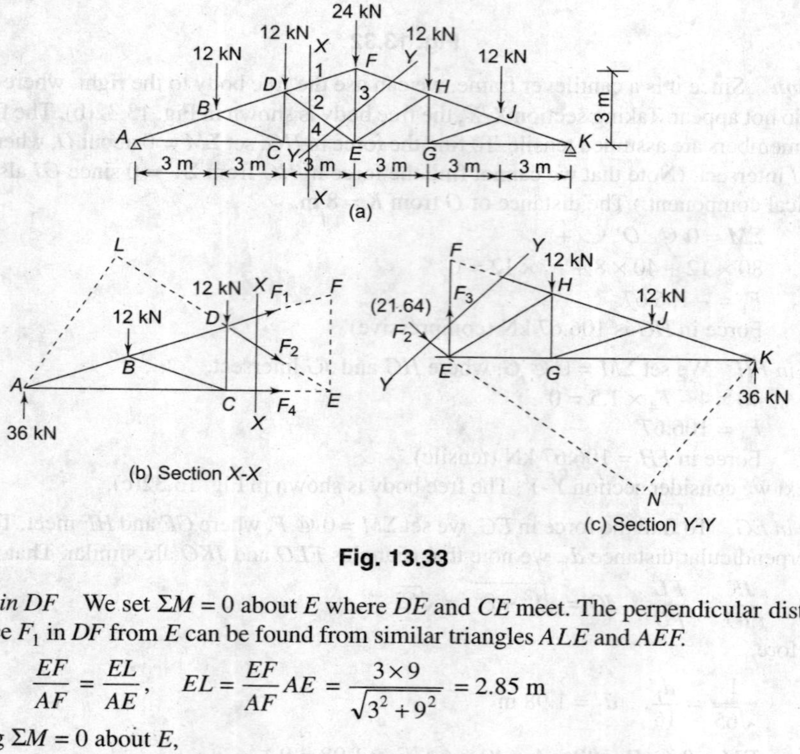

Fig. 13.33

Force in DF We set $\Sigma M = 0$ about E where DE and CE meet. The perpendicular distance of force F_1 in DF from E can be found from similar triangles ALE and AEF.

$$\frac{EF}{AF} = \frac{EL}{AE}, \quad EL = \frac{EF}{AF}AE = \frac{3 \times 9}{\sqrt{3^2 + 9^2}} = 2.85 \text{ m}$$

Setting $\Sigma M = 0$ about E,

$$F_1 \times 2.85 + 36 \times 9 - 12 \times 6 - 12 \times 3 = 0$$

$F_1 = -75.8$ N (compressive)

Force in member $DF = 75.8$ N (compressive)

Force in DE (F_2) This can be found by taking moment about A, where DF and EC meet. We need the perpendicular distance AM of DE from A. We can find this from similar triangles ECD and EMA.

$$\frac{DC}{ED} = \frac{AM}{AE}, \quad AM = \frac{DC}{ED} AE = \frac{2 \times 9}{\sqrt{2^2 + 3^2}} = 4.99 \text{ m}$$

Setting $\Sigma M = 0$ about A,

$$F_2 \times 4.99 + 12 \times 3 + 12 \times 6 = 0$$

$$\Rightarrow \quad F_2 = -21.64 \text{ kN}$$

Force in $DE = 21.64$ kN (compressive)

Force in CE To find the force in CE, set $\Sigma M = 0$ about D, where FD and ED meet.

$$F_4 \times 2 - 36 \times 6 + 12 \times 3 = 0, \quad F_4 = 90 \text{ kN (tensile)}$$

Force in $CE = 90$ kN (tensile)

To find the force in EF, we consider the section YY [Fig. 13.33(c)] and consider the right part of the free body. The section cuts four members, and the forces in ED and EC are known. We take the moment about K. FH and EC pass through K and have no moment about K. The perpendicular distance of ED from K is $KN = AM$ as triangles AME and KNE are congruent. Therefore,

$$KN = 2.85 \text{ m}$$

Setting $\Sigma M = 0$ about K,

$$12 \times 6 + 12 \times 3 + 21.64 \times 2.85 - F_3 \times 9 = 0$$

$$F_3 = +18.85$$

Force in member $EF = 18.85$ kN (tensile) ■

Example 13.17 Ritter's method of section

In the frame loaded as shown in Fig. 13.34(a), determine the forces in the members marked 1, 2, and 3.

Solution Let us calculate the reactions first

$$R_{HV} = (40 \times 3 + 60 \times 6 + 80 \times 9)/12 = 100 \text{ kN}, \quad R_{AV} = 80 \text{ kN}$$

We consider section X-X and the free body to the left as shown in Fig. 13.34(b).

Force in CE To find the force in CE, we set $\Sigma M = 0$ about D, where CD and BD meet. To find the perpendicular distance DJ from CE to D, we note that as $CC'E$ and DJE are similar [Fig. 13.34(a)].

$$\frac{CC'}{CE} = \frac{DJ}{DE}, \quad DJ = \frac{CC'}{CE} \times DE = \frac{3 \times 4}{\sqrt{10}} = 3.794 \text{ m}$$

Setting $\Sigma M = 0$ @ D,

$$80 \times 6 + F_1 \times 3.794 - 40 \times 3 = 0$$

$$\Rightarrow \quad F_1 = -94.89 \text{ kN}$$

Force in $CE = 94.89$ kN (compressive)

Force in CD To find the force in CD, we set $\Sigma M = 0$ about O, the point where EC and DB meet. Note that triangles ECC' and EOD are similar. From Fig. 13.34(b),

$$\frac{1}{4} = \frac{3}{x+6}, \quad x = 6 \text{ m}$$

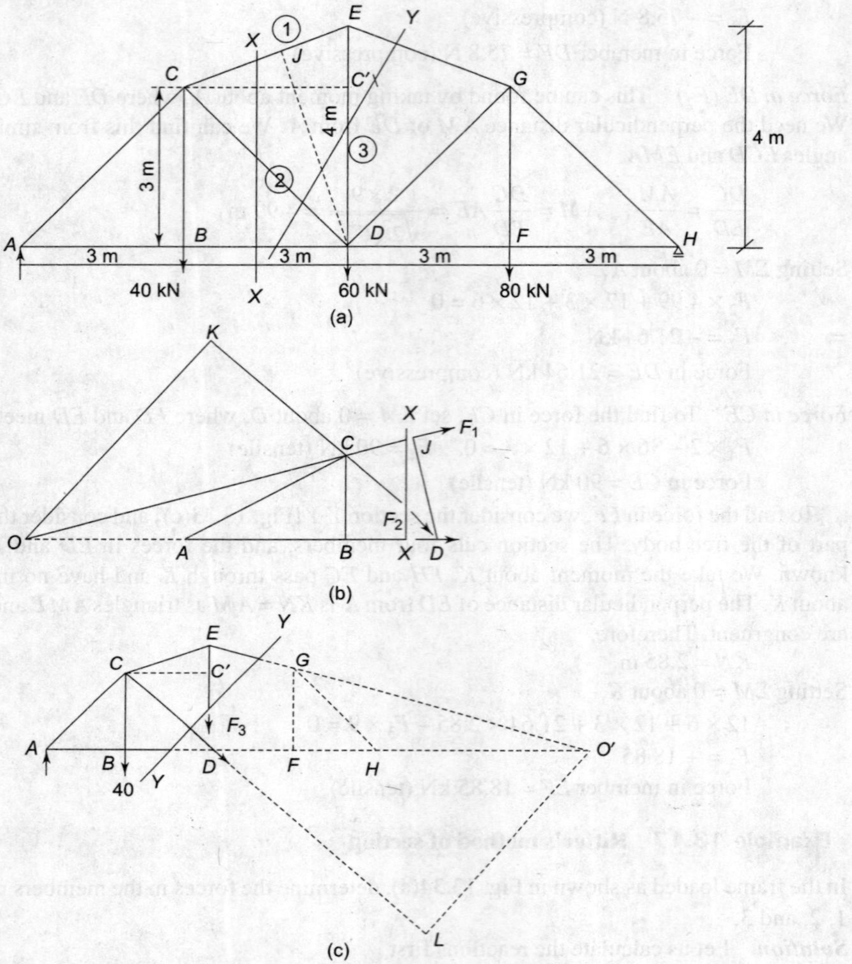

Fig. 13.34

The inclined distance,

$$EO = \sqrt{12^2 + 4^2} = 12.65 \text{ m}$$

The perpendicular distance of *CD* from *O*, *OK*, can be found from the similarity of triangles *OKD* and *CBD*.

$$\frac{CB}{CD} = \frac{OK}{OD}, \quad OK = \frac{CB}{CD} \times OD = \frac{3}{\sqrt{18}} \times 12 = 8.49 \text{ m}$$

$$F_2 \times 8.49 + 40 \times 9 - 80 \times 6$$

$\Rightarrow$ $F_2 = 14.14$ kN

Force in *CD* = 14.14 kN (tensile)

Force in DE We consider the section *Y*-*Y* and the free body to the left. We take the moment about *O*′. *O*′*L* = *OK* = 5.98 due to the symmetry of triangles *DLO*′ and *DKO*. We know the bar force in *CD*. Setting $\Sigma M = 0$ about *O*′ [Fig. 13.34(c)],

$$80 \times 18 - 40 \times 15 - 14.13 \times 8.49 - F_3 \times 12 = 0$$

$\Rightarrow$ $F_3 = 60$ kN

Force in *DE* = 60 kN (tensile)

13.9 GRAPHICAL METHODS

There are methods of solution using a graphical procedure equivalent to an analytical method to determine the member forces in bars. We have discussed two distinct analytical procedures. Each has a graphical equivalent.

Method of joints Here the section is taken around a joint. The resulting force system is a coplanar, concurrent force system, and has two conditions of equilibrium—$\Sigma X = 0$ and $\Sigma Y = 0$. The graphical equivalent is that the force polygon drawn for the force system acting at the joint closes.

Ritter's method of sections Here any general section is taken cutting not more than three bar forces (unknown). The three conditions of equilibrium for the force system are $\Sigma H = 0$, $\Sigma V = 0$, and $\Sigma M = 0$. The graphical equivalent is that the force polygon should close ($\Sigma X = 0$ and $\Sigma Y = 0$), and so should the funicular polygon ($\Sigma M = 0$).

The graphical method more commonly employed is the one corresponding to the method of joints. For the purpose of illustration, we start with examples corresponding to the method of sections.

13.9.1 Culmann's Method

Professor Culmann's method is a variation of the general graphical approach to the method of sections. Here the general force system is divided into two groups and the condition of collinearity of the resultants of the groups is used to determine the unknown forces. The following examples illustrate the procedure.

Example 13.18 Cullmann's graphical method

Determine the forces in the three members marked 1, 2, 3 graphically in the truss shown in Fig. 13.35(a).

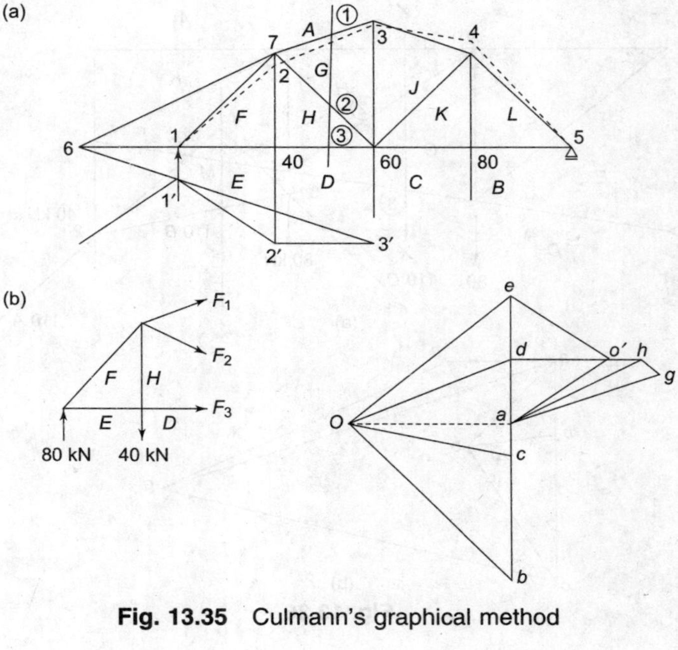

Fig. 13.35 Culmann's graphical method

Solution The truss is drawn to a scale of 1 cm = 1 m, as in Fig. 13.35(a). To find the forces in the three members, we consider section *X-X* and the free body to the left.

The reactions are found graphically by drawing the force polygon *e-d-c-b* [Fig. 13.35(b)] and the funicular polygon 1-2-3-4-5 [Fig. 13.35(a)]. Point *a* is located in the polygon by drawing a line parallel to 1-5 through *O*. The reactions are (by measurement)

$\overline{ba} = 100$ kN, $\overline{ae} = 80$ kN

The forces acting on the body are as follows.

(i) Reaction at left support $\overline{ae} = 80$ kN,

(ii) downward load of 40 kN, and

(iii) forces in the cut members 1, 2, and 3.

These are grouped into two— as forces of 80 kN, 40 kN, and the force in ③ as one group and the forces in the members ① and ② as another group.

We find the resultant of the reaction 80 kN and the load of 40 kN by the smaller funicular polygon using the pole *o′*. The direction of this resultant cuts the line of action of force in ③ at 6. 6 is a point in the line of action of the resultant of forces in the first group. The resultant of forces ① and ② must pass through point 7. The resultant of the two groups of forces must be along 6-7, as these two resultants keep the body in equilibrium and hence must be equal, opposite and collinear. A line parallel to 6-7 through *a* in the force polygon intersects a horizontal line through *d* in *h*. *dh* is the force in member ③. *ah* can be resolved into components $\overline{ag}$ and $\overline{gh}$ by drawing lines parallel to the respective members through *a* and *h*. From Fig. 13.35, by measurement,

Force in *DH* = 80 kN
Force in *AG* = 96 kN
Force in *HG* = 14 kN

Example 13.19 Cullmann's graphical method

Analyse the truss shown graphically and find the forces in the three members marked 1, 2, and 3 (Fig. 13.36).

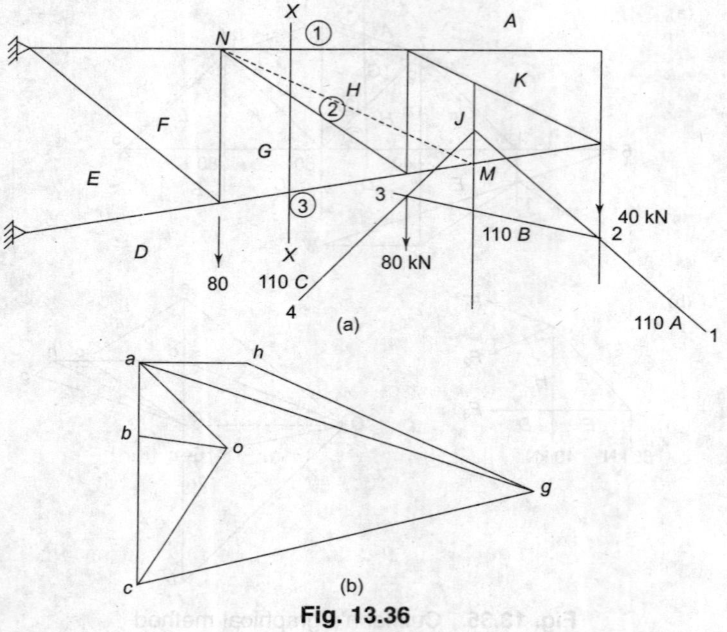

Fig. 13.36

Solution The truss is drawn to space scale [Fig. 13.36(a)]. We consider the section *X-X* and the free body to its right. The forces acting on the free body are as follows.

(i) Forces F_1 (F_{AH}), F_2 (F_{GH}), and F_3 (F_{GC}) in the members cut by the section, and
(ii) loads 80 kN and 40 kN

Let us divide the forces into two groups—F_1 and F_2 in one group, and F_3, 80 kN, and 40 kN in the other. We find the resultant of the forces 80 kN and 40 kN graphically.

Lay out the force diagram *a-b-c* to scale. *ab* = 40 kN, *bc* = 80 kN. Select a pole *o*, and draw *oa*, *ob*, and *oc*. Draw the funicular polygon, 1-2 parallel to *oa*, 2-3 parallel to *ob* and 3-4 parallel to *oc*. 4-3 and 1-2 intersect at *J*. A vertical line through *J* (parallel to *ac*) is the line of action of the resultant of 80 kN and 40 kN forces. This line intersects the line of action of *GC* in *M*. The resultant of the three forces F_3, 80 kN, and 40 kN must pass through *M*. F_1 and F_2 intersect at *N* and their resultant hence must pass through *N*.

The free body is in equilibrium under the action of two forces—one passing through *M* and the other through *N*. For equilibrium, the two forces must be collinear and opposite. The collinearity line is *NM*. *NM* is the line of action of the resultant of *AH* and *HG*.

We draw a line parallel to *NM* through *a* and a line parallel to *GC* through *c*, intersecting in *g*. $\overline{cg}$ is the force in *GC*. $\overline{ag}$ can be resolved into components $\overline{ah}$ and $\overline{hg}$ by drawing lines parallel to the members (1) and (2). $\overline{ac}$, $\overline{cg}$, and $\overline{ga}$ are in equilibrium. The force in *CG* is compressive. $\overline{gh}$ and $\overline{ha}$ both are tensile.

By measurement from the force diagram,

$$F_1 = 56 \text{ kN}, \quad F_2 = 118 \text{ kN}, \quad \text{and} \quad F_3 = 158 \text{ kN}$$

13.9.2 Graphical Method of Joints

In the graphical procedure corresponding to the method of joints, the reactions are determined using a force polygon and funicular polygon for the whole truss. In the case of a cantilever truss, the load line can be laid out and the analysis started from the free end. The graphical method corresponding to the method of joints is also known as the Maxwell stress diagram.

We consider the truss shown in Fig. 13.37(a) and the joint outlined by a circle. Each joint is in equilibrium under the action of a set of concurrent forces. A force

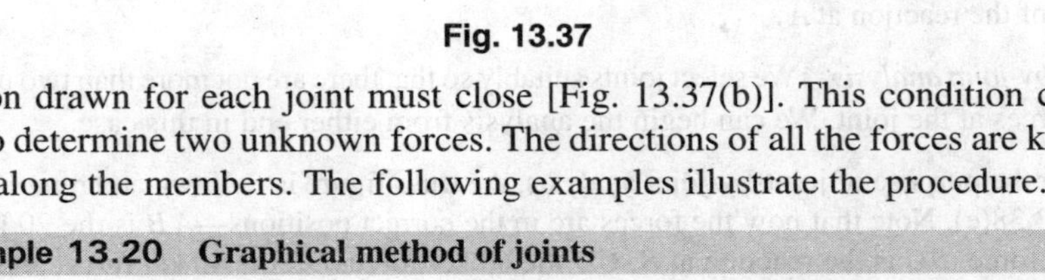

Fig. 13.37

polygon drawn for each joint must close [Fig. 13.37(b)]. This condition can be used to determine two unknown forces. The directions of all the forces are known, being along the members. The following examples illustrate the procedure.

Example 13.20 Graphical method of joints

Determine graphically the forces in the members of the Warren girder shown in Fig. 13.38(a).

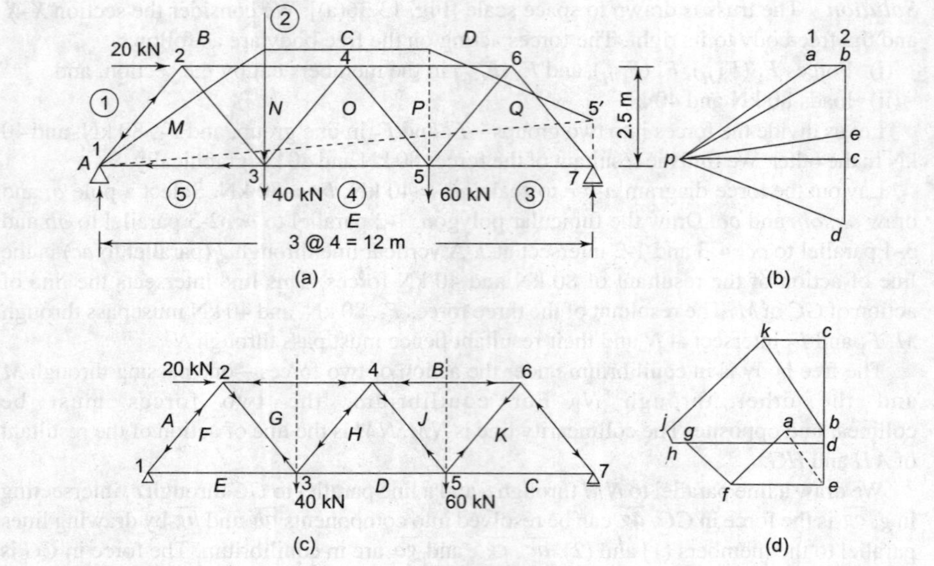

Fig. 13.38

Solution The Warren girder is drawn to a scale of 1 cm = 0.5 m in Fig. 13.38(b). This diagram is known as the space diagram, and is drawn to a linear scale. The load lines are drawn in their correct orientation but are not drawn to scale to represent the loads.

Bow's notation The loads, reactions and member forces are represented by two letters on either side by using Bow's notation. Bow's notation is shown in Fig. 13.38(a). AB represents the 20 kN horizontal force, BC, the 40 kN vertical load and CD, the 60 kN load. DE represents the vertical reactions at the right roller support and EA represents the reaction at the hinge support A. Each bar force is represented by the two letters on either side of the member.

Reactions To determine the reactions graphically, the loads are represented to scale as shown in Fig. 13.38(b). $\overline{ab}$, $\overline{bc}$ and $\overline{cd}$ represent the three loads of 20 kN, 60 kN, and 40 kN respectively.

Select a pole p. Draw pa, pb, pc and pd. The funicular polygon is started from the hinge support at 1. Draw lines parallel to the rays pa, pb, etc. in the respective spaces A, B, etc. Terminate the last ray parallel to pd on a vertical line from the hinge support, which is the direction of the reaction there. Join 1 and 5′. Draw a line parallel to 1-5′ through p. This intersects the vertical line through d at e. $\overline{de}$ is the vertical reaction at the roller and $\overline{ea}$ is the reaction at the hinge support. Obviously, $\overline{eb}$ and $\overline{ba}$ are the vertical and horizontal components of the reaction at A.

Joint-by-joint analysis We select joints suitably so that there are not more than two unknown bar forces at the joint. We can begin the analysis from either end in this case.

For the purpose of joint-by-joint analysis, the truss figure is redrawn and renumbered in Fig. 13.38(c). Note that now the forces are in the correct positions—AB is the 20 kN horizontal force, BC is the reaction at B, CD the 60 kN force, DE the 40 kN force, and EA the reaction at A. The joints are indicated by numbers. The force diagram is also redrawn to get the forces in order in Fig. 13.38(d). The analysis consists of drawing force polygons for each joint. From the known direction of any force or load, we can arrive at the nature of the forces in the members [Fig. 13.38(d)].

Joint 1 $\overline{ea} - \overline{af} - \overline{fe}$ is the force polygon. $\overline{af}$ is towards joint 1 (compressive) and $\overline{fe}$ is away from joint 1 (tensile).

Joint 2 $\overline{ab} - \overline{bg} - \overline{gf} - \overline{fa}$ is the force polygon. $\overline{bg}$ is compressive and $\overline{gf}$ is tensile.

Joint 3 $\overline{de} - \overline{ef} - \overline{fg} - \overline{gh} - \overline{he}$ is the force polygon. $\overline{gh}$ is compressive and $\overline{he}$ is tensile.

Joint 4 $\overline{hg} - \overline{gb} - \overline{bj} - \overline{jh}$ is the force polygon.

Joint 5 $\overline{cd} - \overline{dh} - \overline{hj} - \overline{jk} - \overline{kc}$ is the force polygon. $\overline{jk}$ is tensile and $\overline{kc}$ is compressive.

Joint 6 $\overline{jb} - \overline{bk} - \overline{kj}$ is the force polygon. $\overline{bk}$ is compressive.

The force can be sealed off from the force polygons. These are shown in Table 13.2.

Table 13.2 Member forces (T—tensile; C—compressive)

Member	*Force (kN)*	*Member*	*Force (kN)*
AF	54 (C)	*FE*	50 (T)
FG	54 (T)	*HD*	86 (T)
GH	3 (C)	*KC*	46 (T)
HJ	3 (T)	*BJ*	86 (C)
JK	73 (T)	*BG*	90 (C)
KB	73 (C)		

Note that there was no need to analyse the joint numbered 7. This is so because, after finding the reactions using the funicular polygon method, we need to determine only 11 member forces. The analysis of six joints enables us to find these 11 unknown member forces.

Example 13.21 Graphical method of joints

Analyse the cantilever truss shown in Fig. 13.39(a) and find the forces in all the members.

Solution The truss is drawn to scale in Fig. 13.39(b). Since this is a cantilever truss, the analysis can be started from the free-end joint 1. Using Bow's notation shown in Fig. 13.39(b), $\overline{ab}$ represents the 20 kN force in the force diagram. $\overline{bk}$ is horizontal and $\overline{ka}$ is vertical. To satisfy this condition, *k* and *b* should coincide in the polygon. The force in *ka* is away from the joint at 1 and hence tensile. $\overline{ab}$, $\overline{bk}$ and $\overline{ka}$ are the sides of the force polygon at joint 1.

At joint 2, *kj* and *ja* are the two members with unknown member forces. A line parallel to *kj* from *k* and a line parallel to *j* from *a* meet at *j*. Since the direction of force in *ak* is downward (away from joint 2) at 2, the direction of the forces are found from the closed polygon of forces at joint 2 as $\overline{ak}$, $\overline{kj}$, and $\overline{ja}$. $\overline{kj}$ is compressive and $\overline{aj}$ is tensile.

The procedure is repeated for joints 3, 4, 5, and 6 as shown in Fig. 13.39(c). The member forces can be scaled off from the force polygon. All the member forces are shown in the Table 13.3.

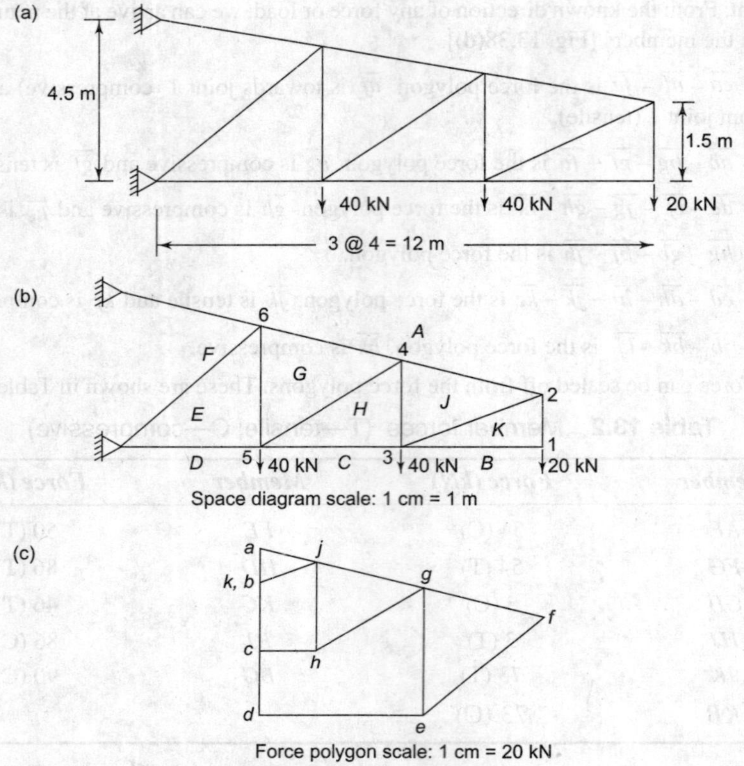

Fig. 13.39

Table 13.3

Member	*Force (kN)*	*Nature*	*Member*	*Force*	*Nature*
AK	20	Tensile	*AG*	86	Tensile
			DE	92	Compressive
BK	0	–	*EF*	92	Compressive
			EG	78	Tensile
KJ	34	Compressive	*AF*	166	Tensile
AJ	34	Tensile			
CH	34	Compressive			
HJ	52	Tensile			
GH	72	Compressive			

Example 13.22 Graphical method of joints

Analyse the truss shown in Fig. 13.40(a) for the loads shown.

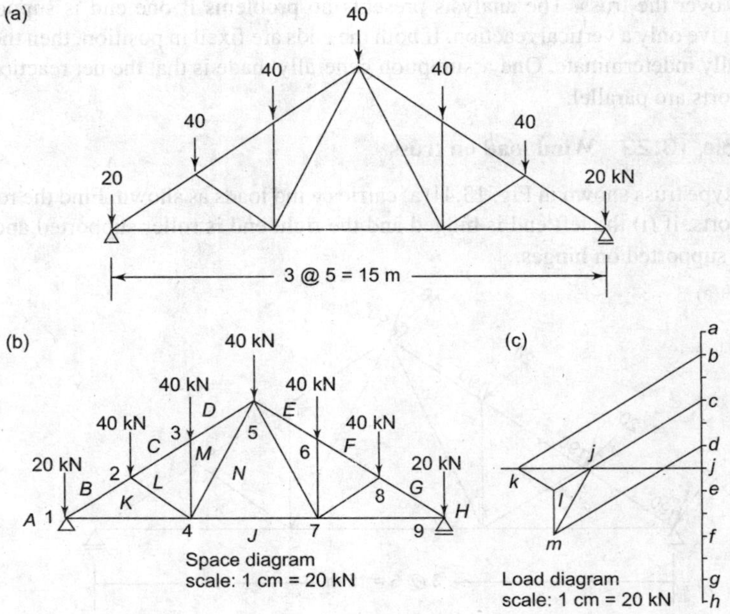

Fig. 13.40

Table 13.4 Member forces

Member	*Force (kN)*	*Nature*
BK	187	Compressive
JK	156	Tensile
KL	38	Compressive
LC	126	Tensile
DM	150	Compressive
MN	68	Tensile
NJ	94	Tensile
ML	40	Compressive

Solution The truss and the loading are symmetrical, and hence only half the truss is analysed. The truss is drawn to scale in Fig. 13.40(b). The load diagram is laid out to a load scale as shown in Fig. 13.40(c). The reactions can be determined by the funicular polygon method. This is not needed in this case as the two end reactions are equal and vertically upwards.

Starting with joint 1, we do the analysis in the order shown. For the left half of the truss, we need to analyse only four joints. The order of analysis is by joints numbered 1, 2, 3, and 4.

From joint 1, we can work out the forces in the members *BK* and *KJ*. The directions of force are $\overline{BK}$ (compressive) and $\overline{KJ}$ (tensile).

For joint 2, the force polygon is *bclkb*. The directions of forces are $\overline{ce}$ (compressive) and $\overline{ck}$ (compressive). Similarly for joints 3 and 4. The member forces can be scaled off from the force diagram. The member forces are shown in Table 13.4.

Analysis for wind forces Wind forces are assumed to act normal to the surface of the sheeting over the truss. The analysis presents no problems if one end is supported on a roller to give only a vertical reaction. If both the ends are fixed in position, then the problem is statically indeterminate. One assumption generally made is that the net reactions at both the supports are parallel.

Example 13.23 Wind load on truss

The fan-type truss shown in Fig. 13.41(a) carries wind loads as shown. Find the reactions at the supports, if (i) the left end is hinged and the right end is roller supported and (ii) both ends are supported on hinges.

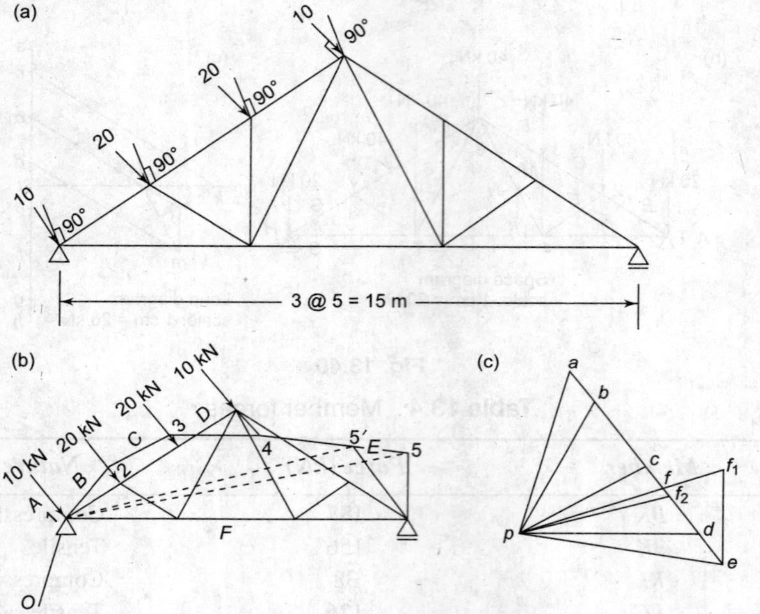

Fig. 13.41

Solution The solution is shown in Figures 13.41(b) and (c). Figure 13.41(b) shows the truss drawn to a linear scale of 1 cm = 2 m.

(i) When one end is on rollers, the reaction at that end will be vertical (perpendicular to the base on which the rollers move). *abcde* is the load line drawn to scale. The funicular polygon is drawn starting from the hinged support. The last line parallel to *oe* (line 4-5) intersects a vertical line through roller at 5. The points 1 and 5 are joined. A line parallel to 1-5 through *p* contains the point f_1, which is located by drawing a vertical line through *e*. $\overline{ef_1}$ is the reaction at the roller and $\overline{f_1a}$ that at the hinged support.

(ii) When both ends are hinged, the reactions at both supports will be inclined. It is assumed that the reactions at both the supports are parallel. In Fig. 13.41(b), in the same funicular polygon, the last line parallel to *pe* from 4 is made to intersect a line parallel to the load line from the roller at 5′. 1-5′ is the closing line of the funicular polygon. A line parallel to 1-5 through *p* locates point f_2 on the load line. f_2 will be the point on the load diagram representing *F* and $\overline{ef_2}$ and $\overline{f_2a}$ will be the reactions at the two supports. These can be scaled off from the diagram.

13.10 STABILITY AND DETERMINACY OF FRAMES

A frame or truss is basically a framework of members which are pin-jointed together in a convenient geometric pattern. A frame can be generated, as mentioned earlier, in two ways.

(i) Starting from three members and three joints forming a triangle, the framework is extended by adding two members for every additional joint. The frames shown in Fig. 13.42 are developed on this principle. In such a frame, the number of members = 2 × number of joints − 3. Such a framework is internally stable without external supports. For external stability, the frame is suitably supported on rollers and hinge supports.

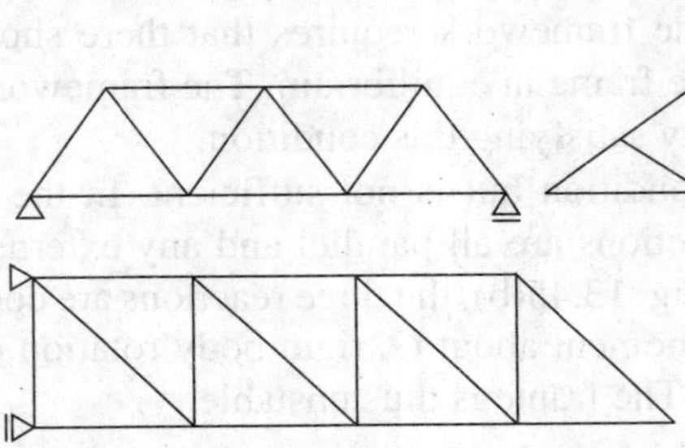

Fig. 13.42

(ii) Starting with a firm foundation, two members are attached to the foundation with hinged joints and joined together at the other end. The framework is extended by forming additional joints with two members for each joint. Such a framework requires $2j$ members, where j is the number of joints excluding those at the foundation. Such a framework is not stable by itself, and requires the foundation for its internal and external stability. The frames shown in Fig. 13.43 are developed on this principle.

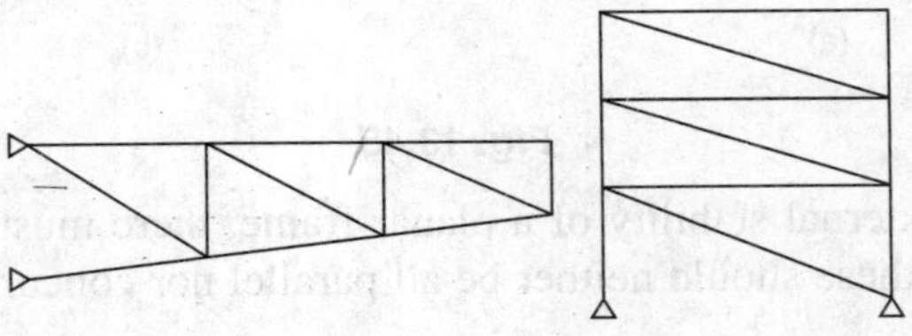

Fig. 13.43

13.10.1 Internal Stability

The internal stability of a simple framework requires that the number of joints and that of members satisfy the conditions mentioned above. However, this is only a necessary condition but not sufficient. The positioning of the members is also important, as shown in Fig. 13.44. While these frames satisfy the condition with regard to the number of members and joints, they are not stable due to the positioning of members. No shear can be transferred across the panel x shown in Fig. 13.44. There will be large and unacceptable deformations in the panel. In addition to satisfying the conditions with respect to the number of members and that of joints, the members should be suitably positioned for internal stability.

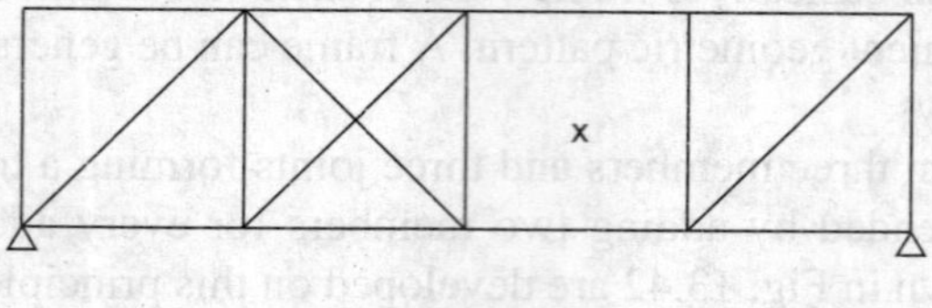

Fig. 13.44

13.10.2 External Stability

The external stability of a plane framework requires that there should be three reaction components to keep the frame in equilibrium. The frameworks shown in Fig. 13.45 have external stability satisfying this condition.

This is again a necessary condition but is not sufficient. In the frameworks shown in Fig. 13.45(a), the reactions are all parallel and any external horizontal force can lead to instability. In Fig. 13.45(b), the three reactions are concurrent and if the external loads have any moment about O, rigid body rotation of the frame about this point can take place. The frame is thus unstable.

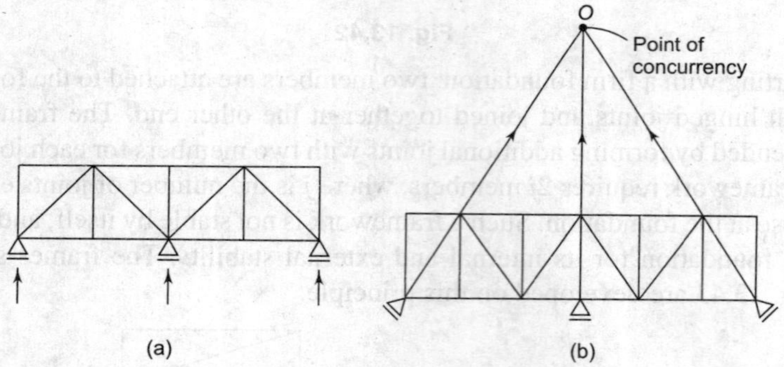

Fig. 13.45

Thus for the external stability of a planar frame, there must be three reaction components, and these should neither be all parallel nor concurrent.

13.10.3 Static Determinacy

The determinacy of a pin-jointed plane frame should also be considered in terms of internal and external determinacy.

Internal determinacy Consider the frame shown in Fig. 13.46(a). The frame has 7 joints and 11 members satisfying the condition $m = 2j - 3$. For each joint, two equations can be formulated as the joint is in equilibrium under the action of a concurrent force system. The 11 unknown member forces and the three external reactions can be determined from these 14 equations. The arrangement of bars is adequate to give stability. The frame is thus stable and statically determinate.

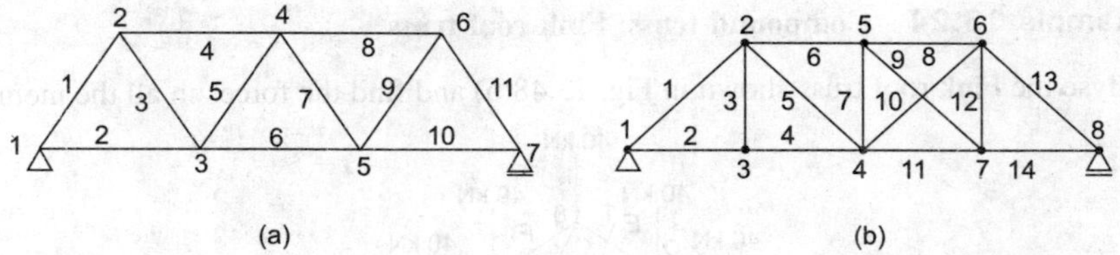

Fig. 13.46

The frame shown in Fig. 13.46(b) has more members than required for stability, i.e., $m > 2j - 3$. The frame is thus statically indeterminate. The member forces can, however, be found out using deformation conditions, etc.

External determinacy External determinacy refers to the determination of reactions using statical conditions of equilibrium which, for a plane frame, are three—$\Sigma H = 0$, $\Sigma V = 0$ and $\Sigma M = 0$. Since there are three equations of equilibrium, only three reaction components can be determined. If there are more than three reactions components, say four as in Fig. 13.47(a) or five as in Fig. 13.47(b), the frame is externally indeterminate. Here, some assumptions will have to be made or some other conditions developed to determine the reactions.

Thus, for a simple plane frame, the following conditions apply for stability and determinacy.

(i) The number of members (m) and the number of joints (j) satisfy the relations, $m = 2j - 3$ or $m = 2j$, depending upon the way the frame is developed.

(ii) The members are so positioned that there is no geometrical instability.

(iii) There are three reactions, and they are neither parallel nor concurrent.

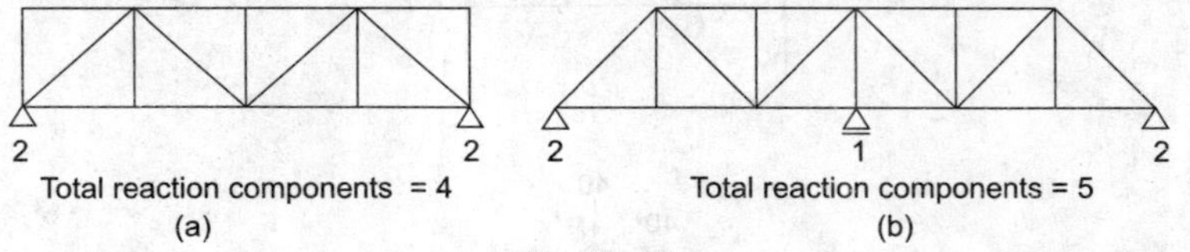

Fig. 13.47

13.10.4 Compound Frames: Fink Roof Truss

Consider the commonly used Fink roof truss shown in Fig. 13.48 (a). The number of joints is equal to 15 and the number of members is 27, satisfying the equation $m = 2j - 3$. The frame, however, cannot be directly analysed by the method of joints. After analysing joints 1, 2 and 3, joints 4 and 5 have each three unknown bar forces.

The Fink roof truss is an example of a compound frame obtained by joining two simple frames by a hinge joint at 8 and a member 5 – 11. Two frames can be joined by methods giving three reaction components between the frames. There are two reaction components at joint 8, and one force is provided by member 5 –11. The three forces provided by the connections should neither be all parallel nor concurrent. The two simple frames forming the Fink roof truss and the connections between them are shown in Fig. 13.48(b).

The following example illustrates the method of analysing a compound frame such as the Fink roof truss.

Example 13.24 Compound truss: Fink roof truss

Analyse the Fink roof truss shown in Fig. 13.48(a) and find the forces in all the members.

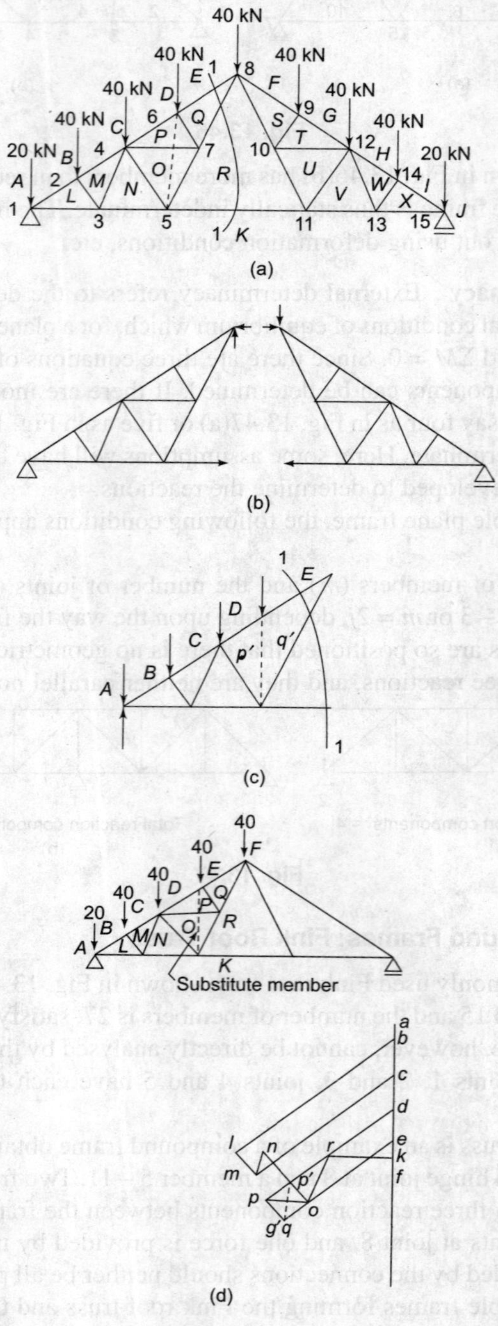

Fig. 13.48

Table 13.5 Member forces

Member	*Force (kN)*
BL	266 (C)
LK	26 (T)
CM	244 (C)
MN	28 (T)
ML	28 (C)
NK	196 (T)
DP	226 (C)
PO	32 (T)
ON	68 (C)
EQ	206 (C)
PQ	36 (C)
OR	64 (T)
QR	96 (T)
RK	130 (T)

Solution We will use the graphical method of analysis. After analysing joints 1, 2, and 3, we cannot proceed further as there are more than two unknown bar forces in each of the joints. A closed force polygon for any other joint cannot be drawn with three unknown sides.

The graphical procedure shown in Fig. 13.48(b) is known as the method of substitution. The method involves substituting a member *p′q′*, shown by dotted lines in place of members *OP* and *PQ*. With this replacement, it is possible to determine the force in the member *EQ*, which will be the same irrespective of the substitution made. This can be verified from Fig. 13.48(c), where the force in member *EQ* is determined by the method of section by isolating the left portion of section 1-1. It is clear that the force in member *EQ* will be the same in Fig. 13.48(c) or (a).

After analysing joints 1, 2, and 3, joint 4 is analysed with the substitute member. *cdp′nmc* is the force polygon for joint 4 and *deq′p′d* for joint 6. Force in member *EQ* is given by *eq′*. *q′* and *q* thus coincide.

The force in members meeting at joint 6 can now be worked back using the original member configuration with members *OP* and *PQ*. The force polygon for joint 4 is again drawn to find the force in member *NO*. The analysis can now be completed by a joint-by-joint analysis.

Only half the truss needs to be analysed due to the symmetry of the frame and the loading. The complete solution for half the frame is shown in Fig. 13.48(b), and the member forces are listed in Table 13.5.

Method of section As mentioned in the previous section, it is possible to find the force in the member *EQ* by the method of sections. Having determined this force analytically, joint 6 is analysed after joints 1, 2, and 3. Then it is possible to analyse joints 4, 5, etc. The full stress diagram is shown in Fig. 13.48(d).

Also note that if members *DP* and *EQ* are in a straight line and members *RQ* and *RO* are also in a straight line, it is possible to analyse the truss by the method of joints. After analysing joints 1, 2, and 3, we tackle joint 6. The force in member *PQ* can be found by setting $\Sigma F = 0$ in a direction perpendicular to DP and EQ. Knowing PQ, the force in OP can be similarly found from joint 7 by setting $\Sigma F = 0$ in a direction perpendicular to QR and OR. Joints 4 and 5 can now be analysed as there will be only two unknown bar forces at these joints.

13.11 DEFLECTIONS IN TRUSSES

When a truss is loaded, the joints of the truss translate and there is also rotation of members. To find the movement of truss joints, we can use energy methods outlined in Chapter 10. The unit load method or Castigliano's theorem can be used for this purpose. There is also a graphical method known as Williot–Mohr diagram for finding deflections of trusses. The following sections deals with such methods.

13.11.1 Unit Load Method

The unit load method can be used to determine the deflection of a point on the truss in the direction in which the unit load is applied. As the final position of the truss joint is not known, it will be convenient to find the horizontal deflection by applying a horizontal unit load and the vertical deflection by applying a unit vertical load. These can be vectorially added to get the final deflection and its direction.

The basic principle of the method is as follows. Referring to the truss shown in Fig. 13.49,

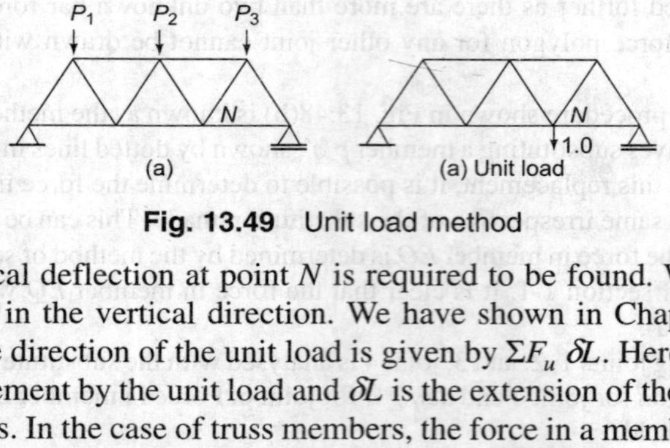

Fig. 13.49 Unit load method

Let the vertical deflection at point N is required to be found. We apply a unit load at point N in the vertical direction. We have shown in Chapter 10 that the deflection in the direction of the unit load is given by $\Sigma F_u\ \delta L$. Here F_u is the force caused in the element by the unit load and δL is the extension of the element under the applied loads. In the case of truss members, the force in a member is axial. If F is the force in the member, then $\delta L = FL/AE$. Let the force caused by the unit load at N in the member be u ($=F_u$). Then $\Sigma F_u\ \delta L = \Sigma FuL/AE$ is the deflection of the point of application of the unit load in the direction of the load.

$\Delta = \Sigma FuL/AE$, the summation being done over all the members

In the above formula, F is the force in any member due to the applied loads, u is the force in any member due to the unit load, L is the length of the member, A is the area of cross section of the member, E is the Young's modulus of elasticity, and Δ is the deflection of the joint where the unit load is applied and is in the direction of the unit load.

The following examples illustrate the procedure.

Example 13.25 Unit load method: Deflection of truss

Analyse the truss shown in Fig. 13.50 and find the horizontal and vertical deflections of joint B. Area of cross section of all the members is 1500 mm^2. $E = 200$ GPa.

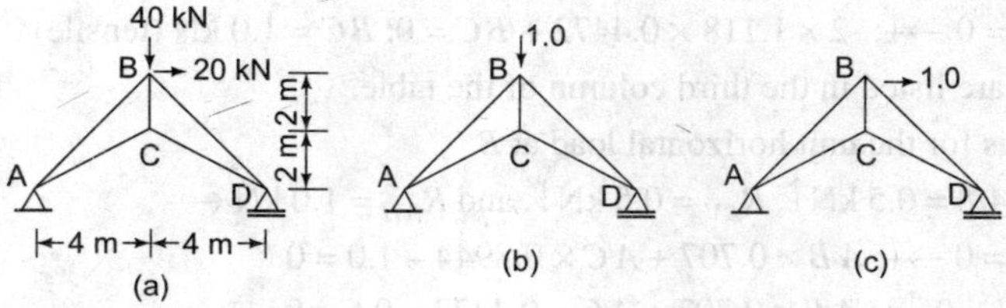

Fig. 13.50

Solution The steps involved are as follows:

(i) Analyse the truss and find the member forces F due to the applied loads.

(ii) Apply a unit vertical load at B and find the forces u_v in all the members.

(iii) Apply a unit horizontal load at B and find the member forces u_h in all the members.

The forces are tabulated in Table 13.6.

Table 13.6

Member	*Force F*	u_v	*Length*	$F u_v L$	u_h	$F u_h L$
AB	–56.57	–1.414	5.66	452.74	0	0
AC	67.1	1.118	4.47	335.28	1.118	335.32
BC	59.87	1.0	2	119.74	1.0	119.74
BD	–84.87	–1.414	5.66	679.23	–1.414	679.23
CD	67.1	1.118	4.47	335.32	1.118	335.32

(i) Analysis for the applied loads

The reactions are:

$R_{BV} = (40 \times 4 + 20 \times 4)/8 = 30$ kN↑, $R_{AV} = 40 - 30 = 10$ kN ↑, and $R_{AH} = 20$ kN ←

Inclination of members AB and BD to the horizontal is 45°. Inclination of members AC and CD to the horizontal is $\tan^{-1}(2/4) = 26.56°$ ($\cos 26.56° = 0.8944$ and $\sin 26.56° = 0.4472$).

Joint A $\Sigma H = 0$; Assuming both members to be in tension and forces to the right positive,

$AB \cos 45° + AC \cos 26.56° - 20 = 0; 0.707\,AB + 0.8944\,AC = 20$

$\Sigma V = 0 \uparrow+, 0.707\,AB + 0.4472\,AC = -10$

Solving these two equations, we get

Force in $AB = -56.57$ (compressive) and force in $AC = 67.1$ (tensile)

Joint D $\Sigma H = 0 \rightarrow+, 0.707\,BD + 0.8944\,CD = 0; BD = 1.2648\,CD$

$\Sigma V = 0 \uparrow+, 30 + 0.4472\,CD + 0.707\,BD = 0$

Solving these two equations,

Force in $CD = 67.1$ kN (tensile) and force in $BD = -84.87$ kN (compressive)

Joint C $\Sigma V = 0$?+, $-67.1 \times 0.4472 - 84.87 \times 0.4472 + CD = 0$; $CD = 67.1$ kN (tensile).

These values are tabulated in the second column.

(ii) Analysis for the unit vertical load at B

The reactions at A and B are equal to 0.5 kN↑.

Joint A $\Sigma H = 0 \rightarrow+; AB \times 0.707 + AC \times 0.8944 = 0; AB = -1.2648\,AC$

$\Sigma V = 0$?+; $AB \times 0.707 + AC \times 0.4472 + 0.5 = 0$

$AC\,(-1.2648 \times 0.707 + 0.4472) = -0.5; AC = 1.118$ (tensile)

$AB = -1.2648 \times 1.118 = -1.414$ kN (compressive)

By symmetry, $BD = -1.414$ kN (compressive) and $CD = 1.118$ kN (tensile)

Joint C $\Sigma V = 0 \rightarrow +$; $-2 \times 1.118 \times 0.4472 + BC = 0$; $BC = 1.0$ kN (tensile)

These values are listed in the third column of the table.

(iii) Analysis for the unit horizontal load at *B*

$R_{BV} = 1 \times 4/8 = 0.5$ kN$\uparrow$, $R_{AV} = 0.5$ kN$\downarrow$, and $R_{AH} = 1.0$ kN $\leftarrow$

Joint A $\Sigma H = 0 \rightarrow +$; $AB \times 0.707 + AC \times 0.8944 - 1.0 = 0$

$\Sigma V = 0 \uparrow +$; $AB \times 0.707 + AC \times 0.4472 - 0.5 = 0$

Solving these two equations, $AB = 0$ and $AC = 1.118$ kN

Joint D $\Sigma H = 0 \rightarrow +$; $-CD \times 0.8944 - BD \times 0.707 = 0$

$\Sigma V = 0 \uparrow +$; $CD \times 0.4472 + BD \times 0.707 = 0$

Solving these two equations, we get

$CD = 1.118$ kN (tensile) and $BD = 1.414$ kN (compressive)

Joint C $\Sigma V = 0 \uparrow +$; $-AC \times 0.4472 - CD \times 0.4472 + BC = 0$; $BC = 1.0$ kN (tensile)

These values are listed in the sixth column of the table.

$AE = 1500 \times 10^{-6} \times 200 \times 10^{6} = 300{,}000$ kN

We now find $F u_v L$ and $F u_h L$. These are given in the table. The summation comes out to be as follows:

$\Sigma F u_v L = 1922.31$ and the vertical deflection is $1922.31/300{,}000 = 0.0064$ mm

$\Sigma F u_h L = 1469.61$ and the horizontal deflection is $1469.61/300{,}000 = 0.005$ m

As the values have come out positive, the deflections are in the same direction as the applied unit load.

Vertical deflection of $B = 6.4$ mm $\downarrow$ and horizontal deflection of $B = 5$ mm $\rightarrow$

Example 13.26 Unit load method: deflection of truss

Analyse the cantilever truss shown in Fig. 13.51 and find the total deflection of point *E*.

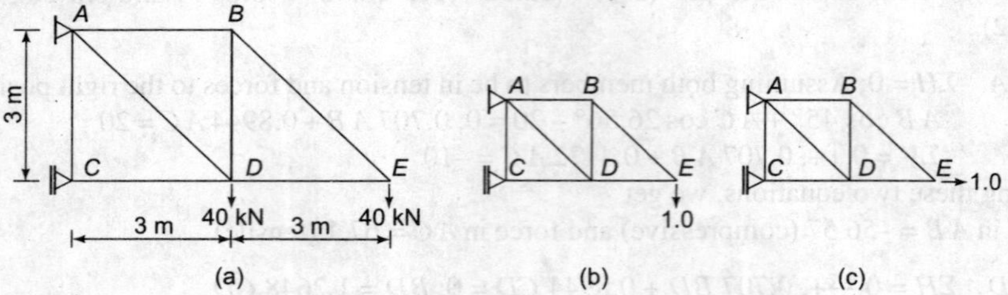

Fig. 13.51

Solution The truss can be analysed starting from the free end *E*.
The member forces due to loads are calculated below.

Joint E $BE = 40/\sin 45° = 56.57$ (tensile)

$ED = -56.57 \times \cos 45° = -40$ kN (C).

Joint B $BD = 56.57 \sin 45° = -40$ kN (compressive) and $AB = 56.57 \times \cos 45° = 40$ kN (tensile)

Joint D $AD = 80/\sin 45° = 113.14$ kN (tensile)

$CD = 113.14 \times \cos 45° + 40 = 120$ kN (compressive)

From joint *C*, $AC = 0$. These are listed in column 2 of Table 13.7.

Table 13.7

Member	*Force F*	*Force due to the unit vertical load at E*	*Force due to the unit horizontal load at E*	*Length*	Fu_vL	Fu_hL
AB	+ 40	1.0	−1.0	3	120	−40
AC	0	0	0	3	0	0
AD	113.14	1.414	0	4.243	678.8	0
BD	−40	−1.0	0	3	120	0
BE	56.57	1.414	0	4.243	339.4	0
CD	−120	−2.0	0	3	720	0
DE	−40	−1.0	0	3	120	0

$\Sigma Fu_vL = 2098.2$ and $\Sigma Fu_hL = -40$

Vertical deflection of $E = 0.0075$ m = 7.5 mm ↓

Horizontal deflection of $E = -40/280{,}000 = -0.00014$ m = −0.14 mm ←

Total deflection of $E = \sqrt{(7.5^2 + 0.14^2)} = 7.501$ mm

The direction of the total deflection = $\tan^{-1}(7.5/0.014) = 89.89°$ ■

Example 13.27 Unit load method: deflection of truss

Analyse the truss shown in Fig. 13.52 and find the horizontal and vertical deflections of point E due to the loads applied as shown on the frame. $AE = 250{,}000$ kN.

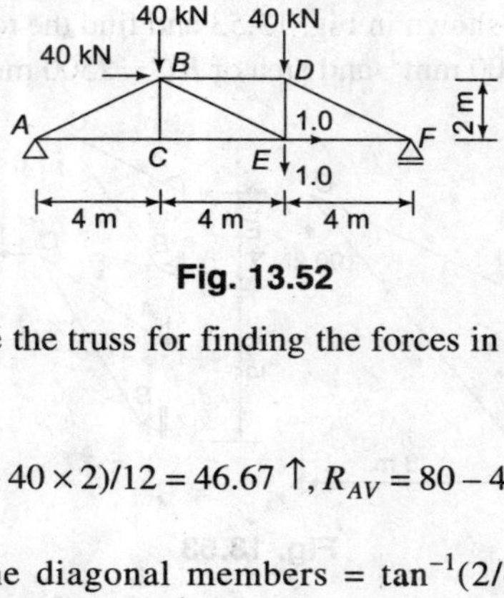

Fig. 13.52

Solution We first analyse the truss for finding the forces in the members due to the applied loads.

The reactions are

$R_{BV} = (40 \times 4 + 40 \times 8 + 40 \times 2)/12 = 46.67$ ↑, $R_{AV} = 80 - 46.67 = 33.33$ kN ↑, and R_{AH} = 40 kN ←

The angle made by the diagonal members = $\tan^{-1}(2/4) = 26.56°$ (cos 26.56° = 0.8944 and sin 26.56° = 0.4472)

Joint A $\Sigma V = 0$; $AB \times 0.4472 = 33.33$; $AB = -74.54$ kN (compressive)
$\Sigma H = 0$; $-40 - 74.54 \times 0.8944 = AC$; $AC = 106.67$ kN (tensile)

Joint C $CB = 0$ as there is no vertical force or component in the vertical direction and $CE = 106.67$ kN (tensile).

Joint B $\Sigma V = 0$; $-40 + 74.54 \times 0.4472 + BE \times 0.4472 = 0$; $BE = -14.9$ kN (compressive)
$\Sigma H = 0$; $40 + 74.54 \times 0.8944 - 14.9 \times 0.8944 = BD = -93.34$ kN (compressive)

Joint F $\Sigma V = 0$ gives $DF = 46.67/0.4472 = 104.36$ kN (compressive)
$\Sigma H = 0$ gives $EF = 104.36 \times 0.8944 = 93.34$ kN (tensile)

Joint D $\Sigma V = 0$ gives $-40 + 104.36 \times 0.4472 = DE = 6.67$ kN (tensile)
$\Sigma H = 0$ gives $BD = 104.36 \times 0.8944 = 93.34$ kN (compressive)

All the member forces have now been found. These are listed in Table 13.8.

Table 13.8

Member	*F (kN)*	U_v	u_h	*Length L*	$F u_v L$	$F u_h L$
AB	−74.54	−0.745	0	4.472	248.34	0
AC	106.67	+ 0.667	1.0	4	284.59	426.68
BC	0	0	0	2	0	0
BD	−93.34	−1.333	0	4	497.69	0
BE	−14.9	+0.745	0	4.472	−49.64	0
CE	106.67	0.667	1.0	4	284.59	426.68
DE	6.67	0.667	0	2	8.9	0
DF	−104.36	−1.49	0	4.472	695.38	0
EF	93.34	1.334	0	4	498.06	0

Vertical deflection = 2467.91/ 250,000 = 0.0099 = 9.9 mm
Horizontal deflection = $\Sigma F u_h L/AE$ = 853.36 /250,000 = 0.001 m = 1 mm

Example 13.28 Unit load method: deflection of truss

Analyse the cantilever truss shown in Fig. 13.53 and find the total deflection of point *C*. *E* = 200 GPa, area of *AC* = 3000 mm², and area of *BC* = 2500 mm².

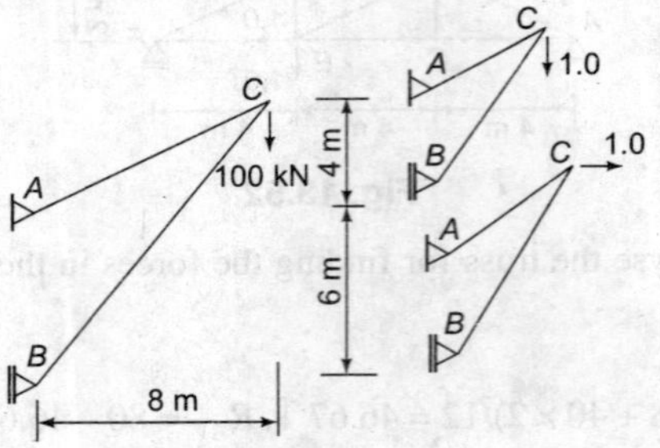

Fig. 13.53

Solution The forces in the members *AC* and *BC* can be found by analysing joint *C*. The results are given in Table 13.9.

Table 13.9

Member	*F (kN)*	U_v	u_h	*L and A*	Fu_vL/A	Fu_hL/A
AC	213.4	2.134	1.863	$L = 8.944$ m $A = 3000$ mm²	1.357×10^6	1.185×10^6
BC	−149.04	−1.49	−1.067	$L = 12.806$ m $A = 2500$ mm²	1.138×10^6	0.814×10^6
					2.737×10^6	1.999×10^6

Vertical deflection = $[\Sigma F u_v L]/E = 2.737 \times 10^6 / 200 \times 10^6 = .0137$ m = 13.7 mm

Horizontal deflection = $[\Sigma F u_h L]/E = 1.999 \times 10^6 / 200 \times 10^6 = 0.01$ m = 10 mm ■

13.11.2 Castigliano's Theorem

We have already seen the application of Castigliano's theorem to beams. The basic statement of Castigliano's theorem is $\delta U/\delta P = \Delta$. Here U is the strain energy stored in the structure due to the applied loads, P is one of the loads, and Δ is the deflection in the direction of P.

In the case of a truss, the members carry only axial forces. Assuming the loads to be gradually applied, the strain energy of a member = $F^2L/2AE$, using standard notations.

Now, $\delta U/\delta P = \delta[F^2L/2AE]/\delta P$

This has to be summed up for all the members of the truss.

$$\Delta = \delta[\Sigma F^2L/2AE]/\delta P$$

It is more convenient to do the differentiation within the summation sign, giving

$$\Delta = \Sigma[F(\delta F/\delta P)L/AE]$$

If there is no load acting at the point where the deflection is required, we can apply a fictitious load at the point and after differentiation make it zero. The following examples illustrate the procedure.

Example 13.29 Castigliano's theorem: truss deflection

Find the deflection of joint E of the truss shown in Fig. 13.54 subjected to the loads shown. Use Castigliano's theorem.

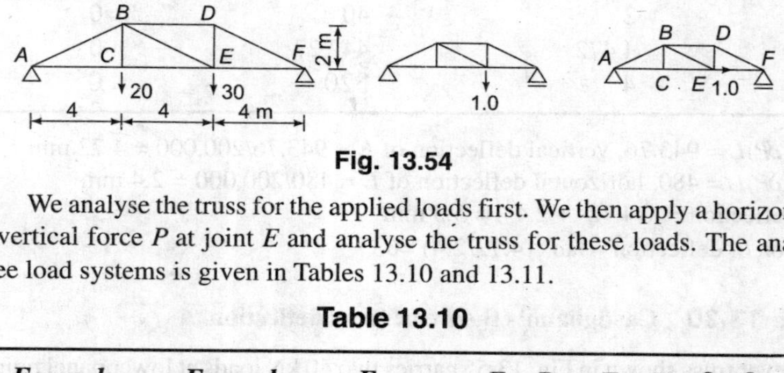

Fig. 13.54

Solution We analyse the truss for the applied loads first. We then apply a horizontal force H and a vertical force P at joint E and analyse the truss for these loads. The analysis for these three load systems is given in Tables 13.10 and 13.11.

Table 13.10

Member	*Force due to the applied loads*	*Force due to the vertical load P*	*Force due to H*	*F + P*	*F + H*	*δF/δP*	*δF/δH*
AB	−44.72	−0.1491P	0	−44.72 − 0.1491P	−44.72	−0.1491	0
AC	60	0.1333P	H	60 + 0.1333P	60 + H	0.1333	1
BC	20	0	0	20	20	0	0
BD	−40	0.2667P	0	−40 + 0.2667P	−40	0.2667	0
BE	0	0.1491P	0	0.1491P	0	0.1491	0

(Contd.)

Table 13.10 (Contd.)

Table 13.10

Member	Force due to the applied loads	Force due to the vertical load P	Force due to H	F + P	F + H	δF/δP	δF/δH
CE	60	0.1333P	H	60 + 0.1333P	60 + H	0.1333	1
DE	30	0.6667P	0	30 + 0.6667P	30	0.6667	0
DF	−67.08	−1.491P	0	−67.08 − 1.491P	−67.08	−1.491	0
EF	60	1.3333P	0	60 + 1.3333P	60 + 0	1.3333	0

Table 13.11

Member	Length	F(δF/δP)L	F(δF/δH)L
AB	4.472	29.82	0
AC	4	32	240
BC	2	0	0
BD	4	42.67	0
BE	4.472	0	0
CE	4	32	240
DE	2	40	0
DF	4.472	447.27	0
EF	4	320	0

$\Sigma F(\delta F/\delta P)L = 943.76$, vertical deflection of $E = 943.76/200{,}000 = 4.72$ mm

$\Sigma F(\delta F/\delta P)L = 480$, horizontal deflection of $E = 480/200{,}000 = 2.4$ mm

Total deflection = $\sqrt{(4.72^2 + 2.4^2)} = 5.3$ mm

Direction of deflection = $\tan^{-1}(4.72/2.4) = 63°$ ■

Example 13.30 Castigliano's theorem: truss deflection

The cantilever truss shown in Fig. 13.55 carries two 40 kN loads at lower panel points. Find the vertical deflection of point *E* and horizontal deflection of point *B*. *AE* = 200,00 kN for all members.

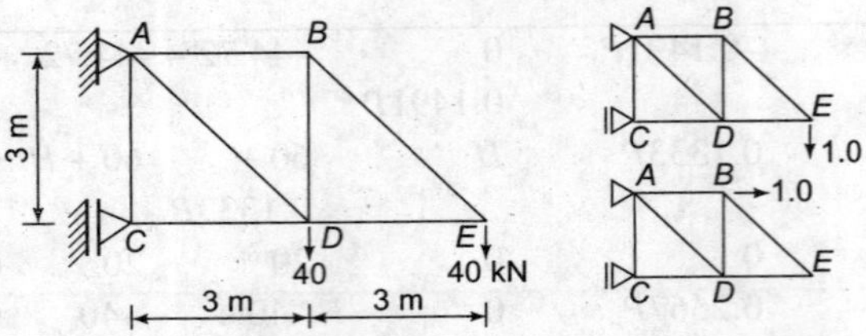

Fig. 13.55

Solution As there is a 40 kN vertical load at E, we call that as load P. Analyse the truss for the 40 kN load at D and the load P at E. The truss can be analysed by inspection if we analyse the truss for the two loads separately. The results are shown in Table 13.12.

Table 13.12

Member	*Due to 40 kN at D*	*Due to P at E*	*total*	$\delta F/\delta P$	*Length*	$F(\delta F/\delta P)L$
AB	0	P	$+P$	1	3	120
AC	0	0	0	0	3	0
AD	$40\sqrt{2}$	$P\sqrt{2}$	$40\sqrt{2} + P\sqrt{2}$	$\sqrt{2}$	$3\sqrt{2}$	678.72
BD	0	$-P$	$-P$	-1	3	120
BE	0	$P\sqrt{2}$	$P\sqrt{2}$	$\sqrt{2}$	$3\sqrt{2}$	339.41
CD	-40	$-2P$	$-40 - 2P$	-2	3	720
DE	0	$-P$	$-P$	-1	3	120

Due to 40 kN at D

As there is no load at E, the forces in BE and DE are zero. If the force in BE is zero, the force in BD must be zero for $\Sigma V = 0$ at B. From joint B, the force in AB also is zero for $\Sigma H = 0$.

Force in $AD = 40\sqrt{2}$ kN (tensile) from $\Sigma V = 0$ at D

Force in CD = horizontal component of the force in AD = 40 kN (compressive)

These forces are listed in column 2 of Table 13.12.

Due to load P at E

For joint E $\Sigma V = 0$ gives $BE = P\sqrt{2}$ (tensile)
$\Sigma H = 0$ gives $DE = P$ (compressive)

For joint B Force in $BD = P$ (compressive), from $\Sigma V = 0$
Force in $AB = P$ (tensile) from $\Sigma H = 0$

For joint D $\Sigma V = 0$ gives $AD = P\sqrt{2}$ (tensile) from $\Sigma V = 0$
$\Sigma H = 0$ gives $CD = P + P\sqrt{2}/\sqrt{2} = 2P$ (compressive)

These forces are listed in column 3 of the table.

To find the horizontal deflection of B, apply a fictitious horizontal load H at B. The member forces due to the applied loads and the load H are shown in Table 13.13.

Table 13.13

Member	*Due to loads*	*Due to H at B*	*Total*	$\delta F/\delta H$	*Length*	$F(\delta F/\delta H)L$
AB	40	H	$40 + H$	1	3	120
AC	0	0	0	0	3	0
AD	113.13	0	113.13	0	$3\sqrt{2}$	0
BD	-40	0	-40	0	3	0
BE	56.57	0	56.57	0	$3\sqrt{2}$	0
CD	-120	0	-120	0	3	0
DE	-40	0	-40	0	3	0

Due to the horizontal force H at B, only member AB has a tensile force of H. All the other member forces are zero.

$\Sigma F(\delta F/\delta P)L = 2098$ and vertical deflection of $E = 2098/280 = 7.5$ mm ↓

$\Sigma F(\delta F/\delta H)L = 120$ and horizontal deflection of $B = 120/280 = 0.43$ mm →

Example 13.31 Castigliano's theorem: truss deflection

Analyse the truss loaded as shown in Fig. 13.56 and find the vertical deflection at B and the horizontal translation of the roller support at D. AE for all the members are equal and has a value of 300,000 kN.

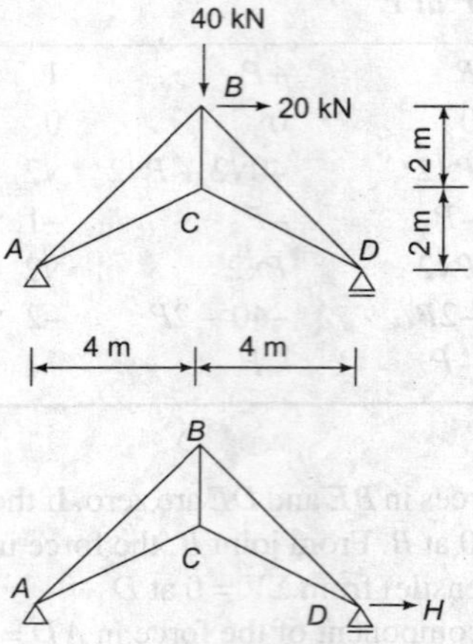

Fig. 13.56

Solution There is a vertical load at B. We call this P (= 40 kN) and its value can be substituted after differentiation. The member forces due to the 20 kN horizontal load at B and the vertical load P at B are found separately and listed in Table 13.14.

Table 13.14

Member	*Due to 20 kN at B*	*Due to P at B*	*Total*	*(δF/δP)*	*Length*	*F(δF/δP)L (P = 40)*
AB	0	$-\sqrt{2}P$	$-\sqrt{2}P$	$-\sqrt{2}$	5.656	452.34
AC	22.36	$1.118P$	$22.36 + 1.118P$	1.118	4.472	335.38
BC	20	P	$20 + P$	1	2.0	120
BD	–28.28	$-\sqrt{2}P$	$-28.28 - \sqrt{2}P$	$-\sqrt{2}$	5.656	678.51
CD	22.36	$1.118P$	$22.36 + 1.118P$	1.118	4.472	335.38

Length of AB = length of $BD = \sqrt{(4^2 + 4^2)} = 5.656$ m
Length of AC = length of $CD = \sqrt{(4^2 + 2^2)} = 4.472$ m

Angle made by AB or BD with the horizontal $= \tan^{-1}(4/4)$
$= 45°$ ($\cos 45° = \sin 45° = 1/\sqrt{2}$)
Angle made by AC or $CD = \tan^{-1}(2/4) = 26.56°$
($\cos 26.56° = 0.8944$ and $\sin 26.56° = 0.4472$)

Due to the horizontal load of 20 kN at B

The reactions are 10 kN ↑ at D and 10 kN ↓ at A.
There is also a horizontal reaction of 20 kN ← at A.

Joint A $\Sigma V = 0;\ AB/\sqrt{2} + 0.4472AC - 10 = 0$

$\Sigma H = 0; AB/\sqrt{2} + 0.8944AC - 20 = 0$

The solution of these equations gives

$AB = 0$ and $AC = 22.36$ kN (tensile)

Joint D $\Sigma V = 0; 0.4472CD + \sqrt{2}BD + 10 = 0$

$\Sigma H = 0; -0.8944CD - BD/\sqrt{2} = 0; CD = 22.36$ (tensile); $BD = -28.28$ (compressive)

From joint C $BC = 2 \times 22.36 \times 0.4472 = 20$ kN (tensile)

These are listed in column 2 of Table 13.14.

Due to the vertical load P at B

The reactions are $P/2 \uparrow$ at A and D. There is symmetry of truss and loading.

Joint A $\Sigma H = 0; AB/\sqrt{2} + 0.8944AC = 0; \Sigma V = 0; AB/\sqrt{2} + 0.4472AC + P/2 = 0$

$AB = -\sqrt{2}P; AC = 1.118P$

Due to symmetry, BD = -v2P and $CD = 1.118P$

From joint C $BC = 2 \times 1.118P \times 0.4472 = P$ (tensile)

These are listed in column 3 of Table 13.14.

In column 4, the sum of the values from columns 2 and 3 are given. $\delta F/\delta P$ is found from the values in column 4 and listed in column 5. $F(\delta F/\delta P)L$ for each member is listed in column 7 of the table.

$\Sigma F(\delta F/\delta P)L = 1921.61$ by putting $P = 40$ kN

Vertical deflection of $B = 1921.61/300 = 6.4$ mm

To find the translation of the roller joint, apply a fictitious load H at D as shown. The analysis of the truss for this load gives the member forces as shown in Table 13.15.

Table 13.15

Member	*F due to applied loads*	*Due to H*	*Total*	$\delta F/\delta H$	*Length*	$F(\delta F/\delta H)L$
AB	56.56	$-\sqrt{2}H$	$56.56 + \sqrt{2}H$	$-\sqrt{2}$	5.656	–452.34
AC	67.08	$2.236H$	$67.08 + 2.236H$	2.236	4.472	670.76
BC	60	$2H$	$60 + 2H$	2	2	240
BD	84.84	$-\sqrt{2}H$	$84.84 - \sqrt{2}H$	$\sqrt{2}$	5.656	–678.51
CD	67.08	$2.236H$	$67.08 + 2.236H$	2.236	4.472	670.76
						450.67

Analysis for the member forces has already been done in Table 13.14. In column 4 of that table, we substitute $P = 40$ kN and get the forces given in column 2 of Table 13.15.

$\Sigma F(\delta F/\delta H)L = +450.67$

Horizontal translation of the roller = 450.67/300 = 1.5 mm ■

Example 13.32 Castigliano's theorem: truss deflection

Find the horizontal and vertical deflections of joint C of the truss loaded as shown in Fig.

13.57. AE for all the members = 280,000 N.

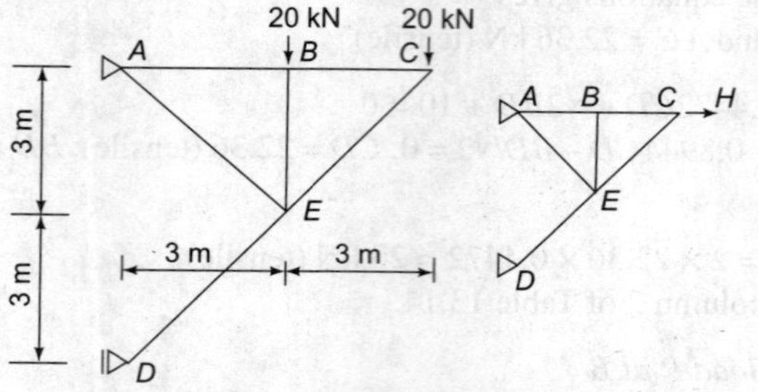

Fig. 13.57

Solution As there is a vertical load at joint C, we take that as P till we differentiate $\delta F/\delta P$. We analyse the truss for both the loads. The analysis can be done by inspection if we take the loads separately. The result is shown in Table 13.16.

Table 13.16

Member	*Force due to 20 kN at B*	*Force due to P at C*	*Total*	$\delta F/\delta P$	L	$F(\delta F/\delta P)L$	*Force due to H at C*	$F(\delta F/\delta P)L$
AB	0	P	P	1	3	60	H	60
AE	$20/\sqrt{2}$	0	$20\sqrt{2}$	0	$3\sqrt{2}$	0	0	
BC	0	P	P	1	3	60	H	60
BE	−20	0	−20	0	3	0	0	
CE	0	$-P\sqrt{2}$	$-P\sqrt{2}$	$-\sqrt{2}$	$3\sqrt{2}$	$120\sqrt{2}$	0	
DE	$-20/\sqrt{2}$	$-P\sqrt{2}$ $-P\sqrt{2}$	$-20/\sqrt{2}$	$-\sqrt{2}$	$3\sqrt{2}$	254.56	0	

$\Sigma F(\delta F/\delta P)L = 544.26$, vertical deflection of $C = 544.26/280 = 1.94$ mm

$\Sigma F(\delta F/\delta H)L = 120$, horizontal deflection of $C = 120/280 = 0.43$ mm

13.11.3 Graphical Method: Williot–Mohr Diagram

Williot–Mohr diagram is a graphical method to find the deflections at joints of a pin-jointed frame. When we know the forces in the members of a frame, we can find the elongation of every member by $\Delta L = PL/AE$. The position of a joint changes due to such changes in length of the members, which also results in rotation of the members. The location of the new position is purely based on geometry and can be handled by graphical methods. We should remember that the elongation and rotation of the members is very small. Thus, to locate the position of point C from the two points A and B (Fig. 13.58), we have to know the elongation of the members AC and BC. With A as the centre, draw an arc of length $AC + \Delta AC$ and with B as the centre, draw an arc equal to $BC + \Delta BC$. Where these two arcs intersect gives the deflected position of point C. Note that this will be the position if A and B were fixed. Also in terms of convenience of scale for drawing, we have to work with the changes in length and not with the lengths themselves. Also since the changes in length and rotations are very small, we can use perpendiculars in place of arcs to find the new position.

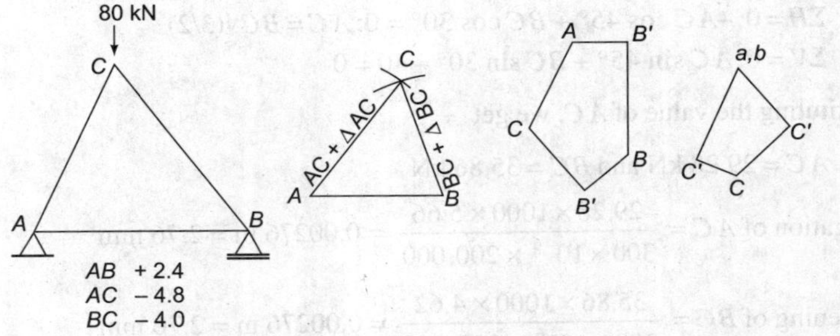

Fig. 13.58 Williot–Mohr diagram

Therefore, we plot point to represent A and B, assuming that AB does not change. We draw the two changes in lengths to a suitable scale parallel to the members. We draw perpendiculars to these lines to locate the new position of point C as shown.

We have to select a reference point with reference to which such geometrical construction can be done. For example, selecting A as the reference point, if all the deflected positions are determined, then we can check with respect to a reference line whether the deflected positions are correct. We find that AB, which should remain horizontal, is not so after this construction. This needs to be corrected by a rotation of line AB, which can be done.

The graphical construction to find the deflected position of joints is known as Williot diagram. The corrections done to make this confirm with the deflection of the truss is known as Mohr's correction diagram.

We illustrate the method with examples.

Example 13.33 Williot–Mohr diagram

Determine graphically the deflection of point C of the truss loaded as shown in Fig. 13.59.

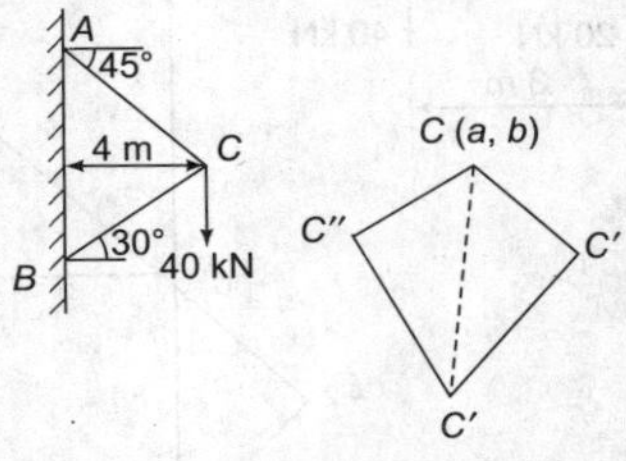

Fig. 13.59 Williot diagram

Solution We analyse the truss to find the force in the members and the change in lengths of the members. These are given in Table 13.17.

Table 13.17

Member	*Force*	*Length*	*Area*	$\Delta L = PL/AE$
AC	35.86	5.66	300	2.76
BC	29.28	4.62	300	2.76

Member forces We write $\Sigma H = 0$ and $\Sigma V = 0$ for joint C.

AC will be in tension and BC will be in compression.

$$\Sigma H = 0; -AC\cos 45° + BC\cos 30° = 0; AC = BC\sqrt{(3/2)}$$

$$\Sigma V = 0; AC\sin 45° + BC\sin 30° - 40 = 0$$

Substituting the value of AC, we get

$$AC = 29.28 \text{ kN and } BC = 35.86 \text{ kN}$$

$$\text{Elongation of } AC = \frac{29.28 \times 1000 \times 5.66}{300 \times 10^{-6} \times 200{,}000} = 0.00276 \text{ m} = 2.76 \text{ mm}$$

$$\text{Shortening of } BC = \frac{35.86 \times 1000 \times 4.62}{300 \times 10^{-6} \times 200{,}000} = 0.00276 \text{ m} = 2.76 \text{ mm}$$

In this case, we have a reference point and a reference line, which are fixed. A or B can be a reference point and AB the reference line.

To draw the Williot diagram, from point C, we lay out $CC' = 2.76$ mm to a suitable scale. This is drawn outwards as the member is in tension. Similarly, we lay out $CC'' = 2.76$ mm along CB towards B as the member is in compression. We draw perpendiculars at C' and C'' to intersect at C_1. C_1 is the deflected position of C. Net deflection of $C = CC_1$, which can be measured to the same scale to which the diagram is drawn. So,

Net deflection of $C = 3.6$ mm ■

Example 13.34 Williot–Mohr diagram

Find the deflection of joints of the truss loaded as shown in Fig. 13.60. Area of all the members = 250 mm^2 and $E = 200$ GPa.

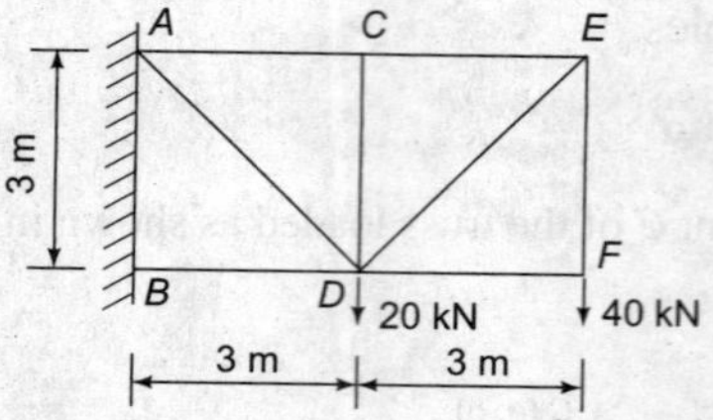

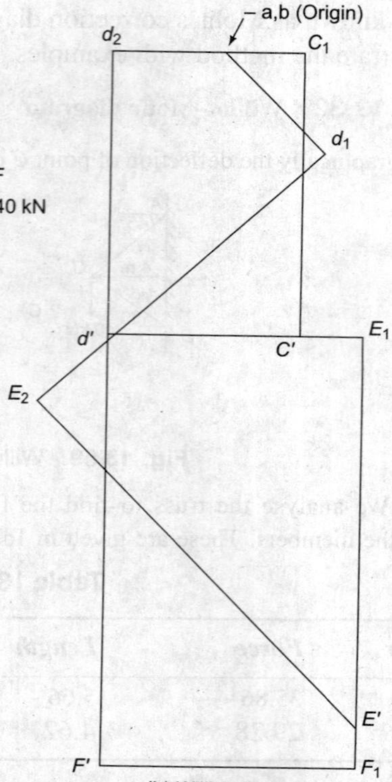

(b) Williot diagram

Fig. 13.60

Solution We find the forces in the members of the truss. The truss has to be drawn to a suitable scale. Table 13.18 shows the result of the analysis.

Table 13.18

Member	*Force (kN)*	*Length (m)*	*Area* (mm^2)	*FL/AE*
AC	40	3	250	2.4
AD	84.85	$3\sqrt{2}$	250	7.2
BD	–100	3	250	–6
CD	0	3	250	0
CE	40	3	250	2.4
DE	$-40\sqrt{2}$	$3\sqrt{2}$	250	–4.8
DF	0	3	250	0
EF	40s	3	250	2.4

The inclination of the inclined members to the horizontal is 45°. The truss can be analysed by inspection.

From joint F $\Sigma V = 0$ gives $EF = 40$ kN (tensile) and $\Sigma H = 0$ gives $DF = 0$

From joint E $\Sigma V = 0$ gives $DE = 40\sqrt{2}$ (compressive) and $\Sigma H = 0$ gives $CE = 40$ kN (tensile)

From joint C $\Sigma V = 0$ gives $CD = 0$ and $\Sigma H = 0$ gives $AC = 40$ kN (tensile)

From joint D $\Sigma V = 0$ gives $AD/\sqrt{2} = 40 + 20$, $AD = 84.85$ kN (tensile)

$\Sigma H = 0$ gives $BD = 40 + 60 = 100$ kN (compressive)

As an example, for AC,

$$\frac{FL}{AE} = \frac{40 \times 1000 \times 3000}{250 \times 200{,}000} = 2.4 \text{ mm}$$

For other members, extension or shortening can be similarly calculated.

We have a fixed line AB. Figure 13.60 shows the Williot diagram. When there is a fixed reference line and point, Mohr's correction diagram will not be required. The Williot diagram is drawn using a suitable scale for changes in length. We start with an origin, which is (*a*, *b*). To locate *d*, we draw $\Delta AD = 2.4$ units parallel to AD as shown (the member is in tension). From the same point (*a*, *b*), we draw $\Delta BD = 6$ units. This is drawn towards *B* as it is in compression. Similarly, all other points can be found. The final deflection of any point can be measured from the origin. ■

Example 13.35 Williot–Mohr diagram

Find the deflection of the joints of the truss loaded as shown in Fig. 13.61. Area of members = 280 mm^2 and $E = 200$ GPa.

Solution Due to the 80 kN load, the reactions are

$$R_{AV} = 80 \times 3/5 = 48 \text{ kN and } R_{BV} = 32 \text{ kN}$$

Member forces $AC = (48 \times \sqrt{20})/4 = 53.66$ kN (compressive) and $AB = 53.66 \times 2/\sqrt{20} = 24$ kN (tensile)

Similarly, $BC = 32 \times 5/4 = 40$ kN (compressive)

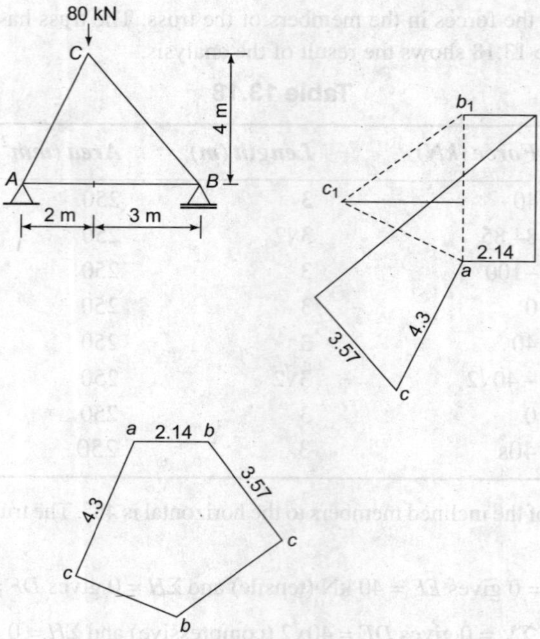

Fig. 13.61

The changes in lengths of the members are listed in Table 13.19.

Table 13.19

Member	*Force (kN)*	*Length (m)*	*Area* (mm^2)	*FL/AE (mm)*
AC	–53.66	4.472	280	4.3
AB	24	5	280	2.14
BC	– 40	5	280	3.57

The difference in this truss, compared to the previous two examples, is that there is no fixed line. There is only a fixed point in this truss, which is A. We assume AC as a fixed line.

To draw the Williot diagram, draw ac = 4.3 mm to a suitable scale. Draw ab = 2.14 mm (tensile) and cb = 3.57 mm to the same scale. Draw perpendiculars to intersect at b, where b is the deflected position of B.

This, however, does not appear correct as a and b should be at the same level. There should be only horizontal movement of the roller. The truss should be given a rigid body rotation to correct this. Each point rotates with a radius equal to its distance from the origin, which is point A. As the angles and deflections are small, such rotations can be represented by a movement perpendicular to the radial line. Mohr found that such movements result in a truss exactly equal to the original truss but at right angles to it. This is achieved by drawing a correction truss in which each member of the truss is at right angles to the original truss, keeping the fixed joint in place. This is shown by dotted lines in the Mohr's correction diagram.

Mohr's correction diagram is drawn as shown. Draw a vertical line from a and a horizontal line from b to intersect at b_1. With ab_1 as base, draw the truss as shown. Here ac_1b_1 is the truss. The deflection of any point can be measured from the corresponding point of the truss to the point of the Williot diagram. Thus, b_1b is the horizontal movement of the roller. The deflection of c is c_1c. Joint A remains fixed. ■

Example 13.36 Williot–Mohr diagram

Find the deflection of the joints of the truss shown in Fig. 13.62 using Williot–Mohr diagram. The members AB, BE, BD, and DF have area of cross section = 300 mm^2. Other members have area of section 250 mm^2. E = 200 GPa.

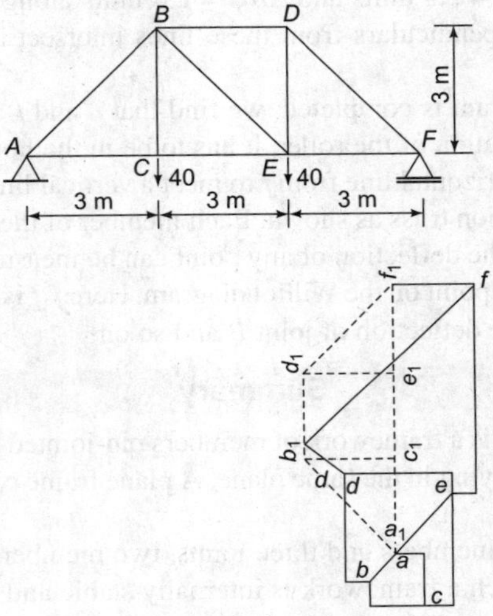

Fig. 13.62

Solution The truss is analysed and the changes in length of all the members are calculated. Table 13.20 shows the data.

Table 13.20

Member	*Force (kN)*	*Length (m)*	*Area (mm^2)*	*FL/AE*
AB	$-40\sqrt{2}$	$3\sqrt{2}$	300	–4
AC	40	3	250	2.4
BC	40	3	250	2.4
BD	–40	3	300	–2
BE	0	$3\sqrt{2}$	300	0
CE	40	3	250	2.4
DE	40	3	250	2.4
DF	$-40\sqrt{2}$	$3\sqrt{2}$	300	–4
EF	40	3	250	2.4

The analysis of the truss can be easily done. The reactions at the supports are equal to 40 kN due to symmetry.

From joint A $\Sigma V = 0$ gives $AB = 40\sqrt{2}$ kN (compressive) and $\Sigma H = 0$ gives AC = 40 kN (tensile)

From joint C CB = 40 kN (tensile) from $\Sigma V = 0$

$\Sigma H = 0$ gives CE = 40 kN (tensile)

From joint B $\Sigma V = 0$ gives $BE = 0$
$\Sigma H = 0$ gives $BD = 40$ kN (compresive)
From joint E $\Sigma V = 0$ gives $DE = 40$ kN (tensile) as $BE = 0$
$\Sigma H = 0$ gives $EF = 40$ kN (tensile)

From joint F $\Sigma V = 0$ gives $DF = 40\sqrt{2}$ kN (compressive)

The truss has a fixed point A. We assume AB to be the reference line. The Williot diagram is now drawn as shown. We set out $ab = 4$ units on a line parallel to AB. Point c can be obtained by setting $\Delta AC = 2.4$ units and $\Delta BC = 2.4$ units along parallel directions to the respective members. Perpendiculars from these lines intersect at c. Other points can be similarly located.

Once the Williot diagram is completed, we find that a and f are not at the same level. While there can be translation of the roller, it has to be in the horizontal direction. This is achieved by drawing a horizontal line from f to meet a vertical line from a at f_1. With af_1 as the base, draw the correction truss as shown. Each member of the truss is now perpendicular to the original truss. The deflection of any point can be measured from the point on the truss to the corresponding point on the Williot diagram. Here f_1f is the horizontal translation of the roller and b_1b is the deflection of joint B and so on. ■

Summary

A pin-jointed plane frame is a framework of members pin-jointed at their ends, the member axes, loads and reactions lying in the same plane. A plane frame can be made in the following two ways.

(i) Starting with three members and three joints, two members can be added for every additional joint. Such a framework is internally stable and is suitably supported for external stability.

(ii) Starting with a firm foundation, two bars are used for the first joint, the other ends of the members finned to the foundation. The rest of the framework is made up as in the first case. Such a framework is not stable without the foundation to which it is attached. In the first case $m = 2j - 3$ and in the second case $m = 2j$, where m = number of members and j = number of joints (in the second case, the attachments to the foundation are not considered as joints).

The three assumptions made in the analysis of a pin-jointed plane frame are as follows.

(i) The centroidal lines of members meet at a point at all joints,
(ii) the ends of the members are pin-jointed, and
(iii) the loads and reactive forces act only at the joints.

There are basically two methods for analysing a frame:

(i) by taking a section around the joint, resulting in a concurrent force system; the equations of equilibrium, $\Sigma H = 0$, $\Sigma V = 0$ can be used to solve for two unknowns, and
(ii) cutting a section through the truss, resulting in a general coplanar force system; the three conditions of equilibrium, $\Sigma H = 0$, $\Sigma V = 0$, and $\Sigma M = 0$ can be used to find three unknowns.

The first method is known as the method of joints and the second as Ritter's method of sections.

The tension coefficient of a member is the tension in the member divided by its length. In the method of tension coefficients, the equations $\Sigma H = 0$ and $\Sigma V = 0$ of the method of joints are expressed using joint coordinates and tension coefficients.

Corresponding to the two basic analytical methods, there are also two graphical methods—Maxwell's stress diagram for the method of joints and Culmann's graphical method corresponding to Ritter's method of sections.

A frame is supposed to be determinate and can be stable if

(i) $m = 2j - 3$ or $m = 2j$, as the case may be,

(ii) $m < 2j - 3$ (or $2j$), in which case the frame is deficient, and

(iii) $m > 2j - 3$ (or $2j$), in which case the frame is redundant.

These are necessary conditions but not sufficient. The members should be suitably placed for stability.

The external determinacy and stability requires that the truss be supported in such a way that there are three reaction components, all of which are neither parallel nor concurrent. If the number of external reaction components are more than three, the frame is statically indeterminate. A compound frame is made by joining two simple trusses suitably. In such cases, the method of joints may not be applicable directly. Problems related to compound frames can be solved by a combination of methods, or special techniques such as those involving substitute members.

Deflections in pin-jointed frames are calculated using Castigliano's theorem of partial derivative or the unit load method. Both these methods can be used to find deflections at joints in a truss. Williot diagram and Williot–Mohr diagrams are graphical procedures for determining deflections at joints of a truss.

Exercises

Review Questions

1. Define the term pin-jointed plane frame.
2. Explain the two methods of building a framework, and the relationship between the number of members and that of joints in each case, giving examples.
3. State the assumptions made in the analysis of a pin-jointed plane frame. In each case, state also the behaviour of the frame if the assumption is not valid.
4. In the cases of the frames shown in Fig. 13.63, find the forces in the members (marked by a cross) by inspection.

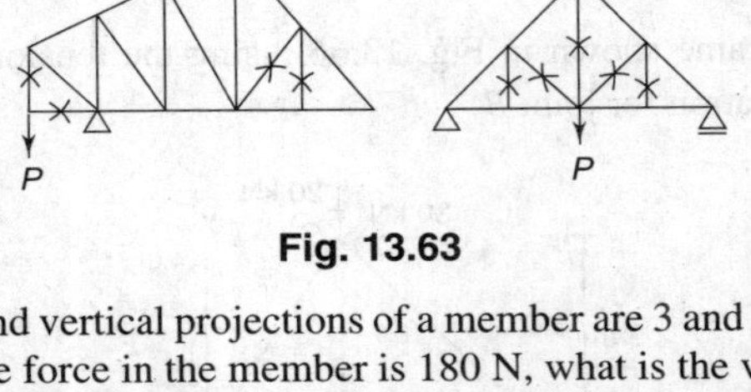

Fig. 13.63

5. The horizontal and vertical projections of a member are 3 and 2.5 m. If the horizontal component of the force in the member is 180 N, what is the vertical component?
6. The length of a member of a frame is 2.5 m and its horizontal projection is 1.5 m. If the vertical component of the force in the member is 90 N, find the force in the member.
7. State the two methods employed for the analysis of a pin-jointed frames. What is the principle involved in each case?

8. Only two unknown member forces can be evaluated at a joint by the method of joints. Explain.
9. In the diagrams shown (Fig. 13.64), the frames is cut by section X-X. The member forces cut are assumed in the directions shown. State the nature (tensile or compressive) assumed in each member.
10. Can the method of section be applied if the section cuts four bars? Explain your answer.

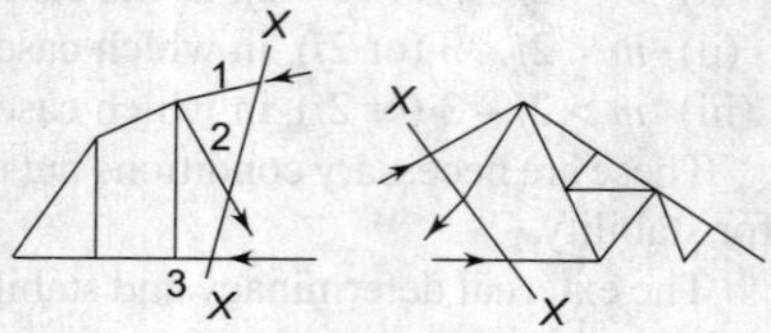

Fig. 13.64

11. From the SF diagrams shown in Fig. 13.65, find the nature and magnitude of the forces in the diagonal members.

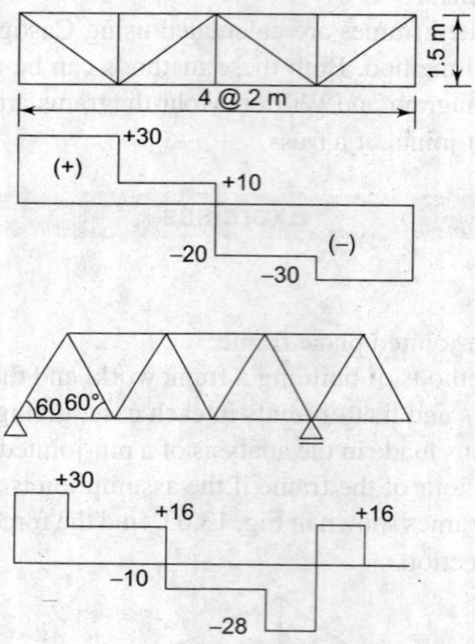

Fig. 13.65

12. Define the term tension coefficient. If the tension coefficient is prefixed by a negative sign, what does it imply?
13. If the tension coefficient and tension in a member are equal numerically, what does it mean?
14. For the plane frame shown in Fig. 13.66, using the tension coefficients, state the equilibrium equation for joint B.

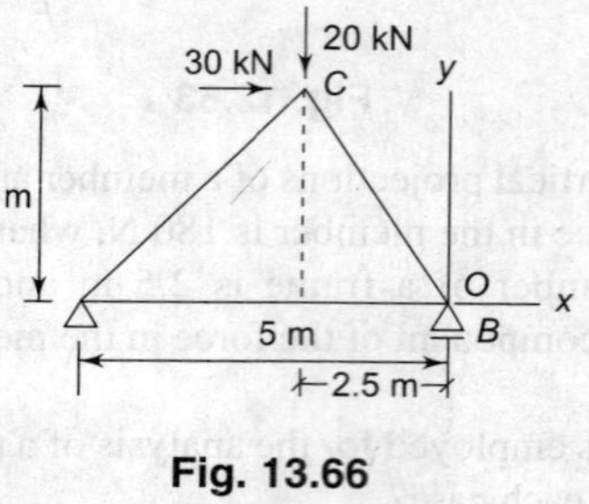

Fig. 13.66

15. Explain what a compound truss is, giving an example.
16. State whether each of the frames shown in Fig. 13.67 are stable and determinate internally and externally. Explain your answer.

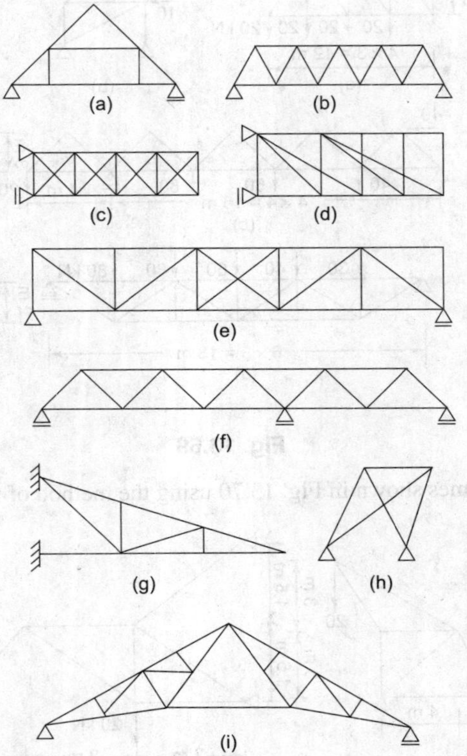

Fig. 13.67

Problems

1. Analyse the frames loaded as shown in Fig. 13.68 and find the forces in all the members.

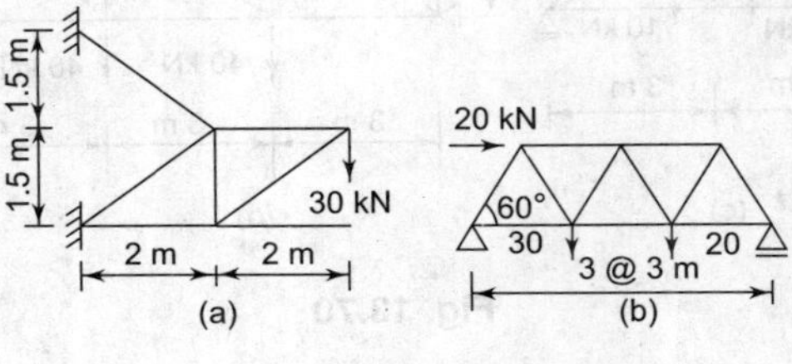

Fig. 13.68

2. Analyse the frames shown in Fig. 13.69 using the SF diagram, and find the forces in all the members.

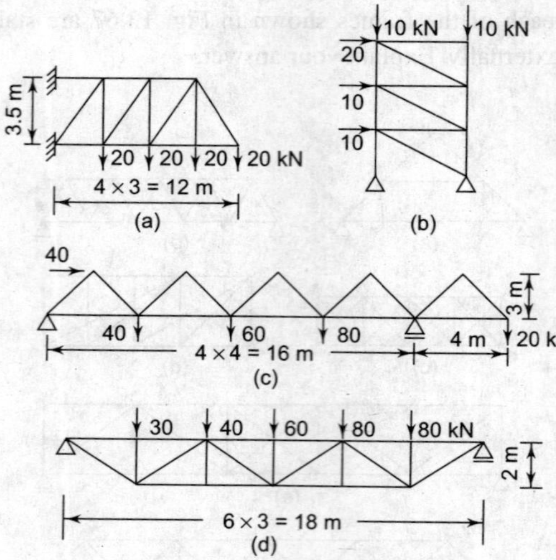

Fig. 13.69

3. Analyse the frames shown in Fig. 13.70 using the method of tension coefficients.

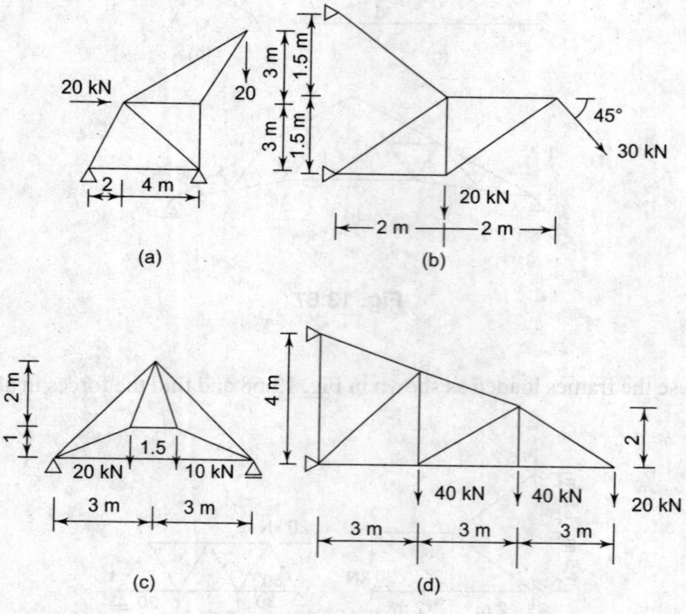

Fig. 13.70

4. In each of the frames shown in Fig. 13.71, determine the forces in the numbered members shown using the method of section.

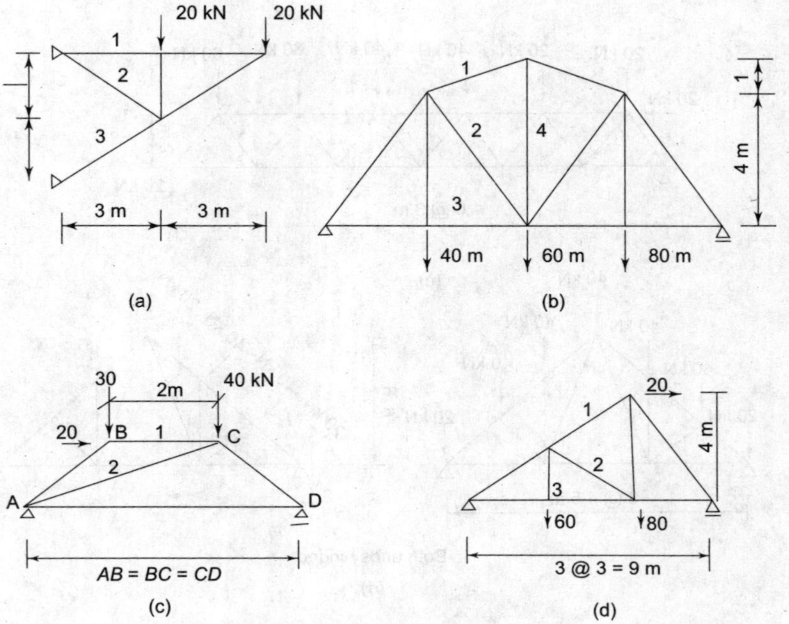

Fig. 13.71

5. Using a graphical procedure, analyse the frames shown in Fig. 13.72 to find the forces marked in members 1, 2, and 3.

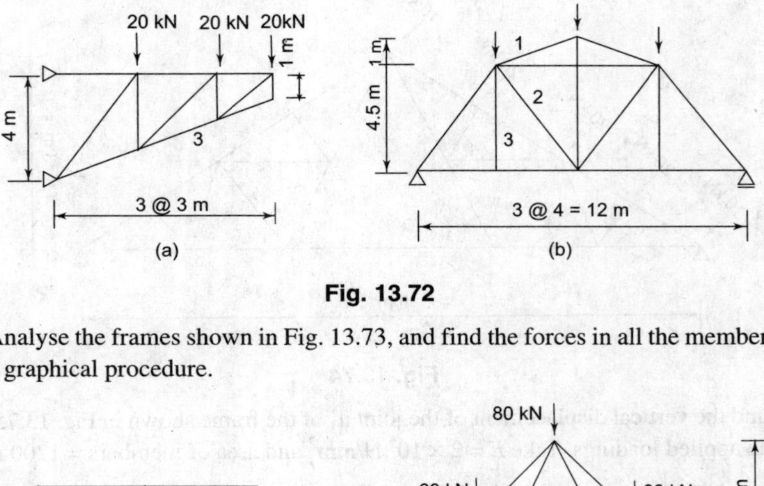

Fig. 13.72

6. Analyse the frames shown in Fig. 13.73, and find the forces in all the members using a graphical procedure.

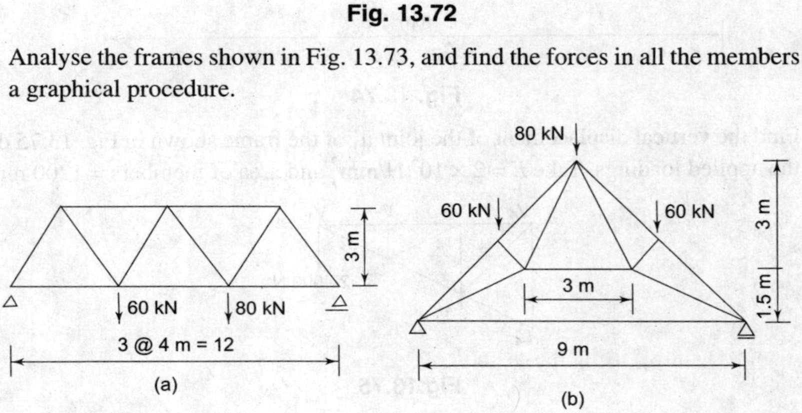

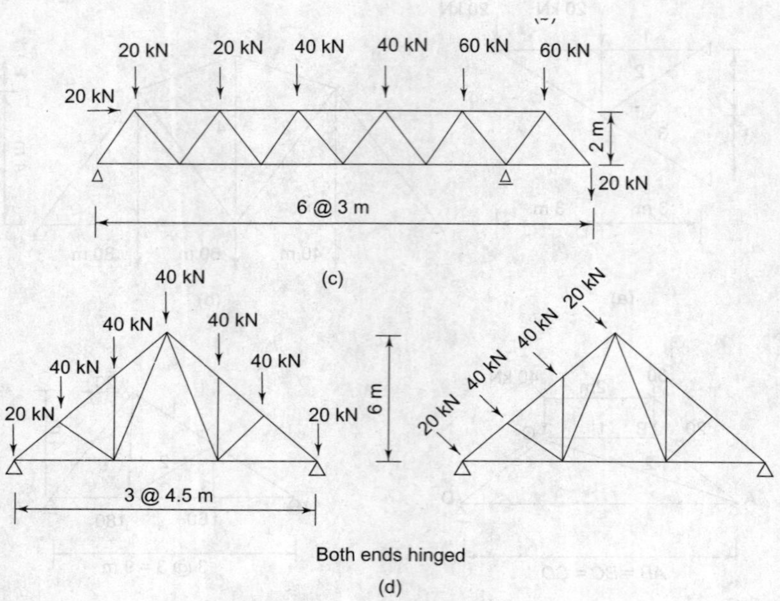

Fig. 13.73

7. Find the forces in all the members using a graphical procedure in the compound Fink roof truss shown in Fig. 13.74.

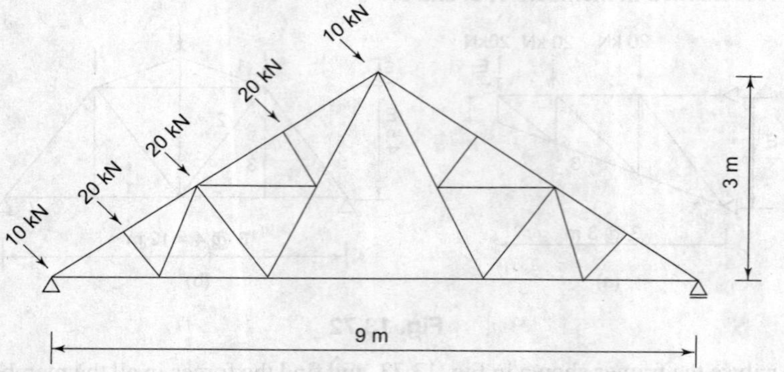

Fig. 13.74

8. Find the vertical displacement of the joint u_1 of the frame shown in Fig. 13.75 due to the applied loadings. Take $E = 2 \times 10^5$ N/mm^2 and area of members = 1200 mm^2.

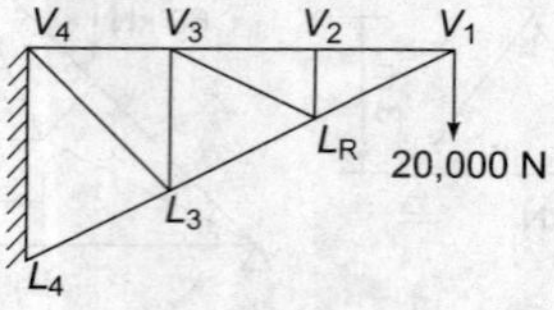

Fig.13.75

9. Analyse the frames loaded as shown in Fig. 13.76 and find the deflections of points as specified. Take $E = 200$ GPa. Areas of X-section are mentioned against the members. Use the unit load Method.

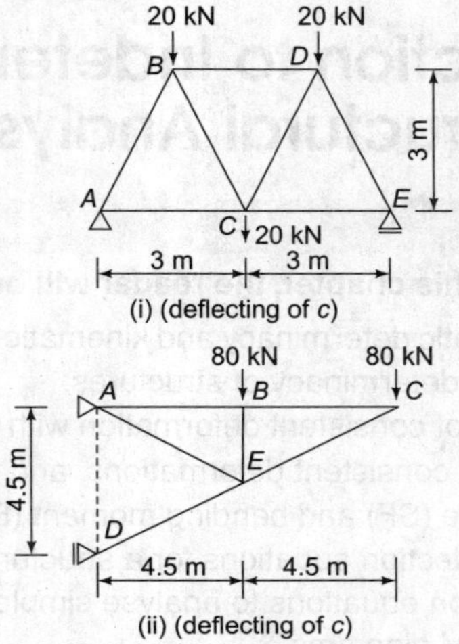

Fig. 13.76

10. Using Castigliano's theorem, find the deflections as required in the frames loaded as shown in Fig. 13.77. Take $E = 200$ GPa.

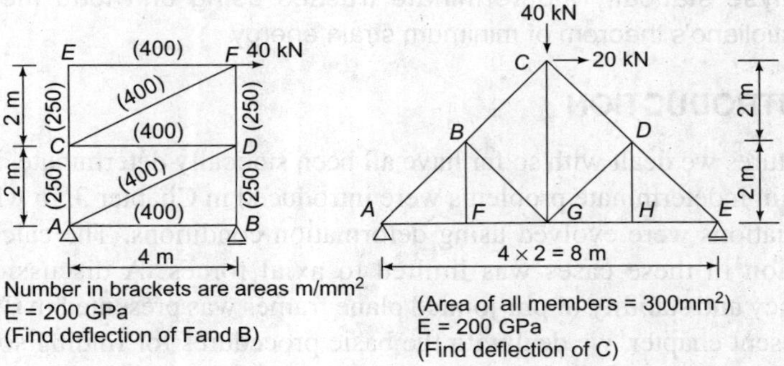

Fig. 13.77

11. Using Williot–Mohr diagram, find the deflections of the joints of the truss carrying loads as shown in Fig. 13.78. Areas of cross sections are given against the members.

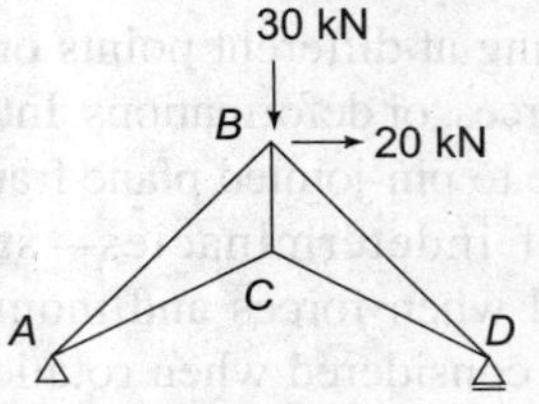

Fig. 13.78

CHAPTER 14

Introduction to Indeterminate Structural Analysis

Learning Objectives

After going through this chapter, the reader will be able to

- explain the terms static determinacy and kinematic indeterminacy, and calculate the degree of indeterminacy of structures,
- explain the method of consistent deformation with examples,
- apply the method of consistent deformations, and analyse simple structures and draw shear force (SF) and bending moment (BM) diagrams,
- derive the slope-deflection equations for a structural element,
- apply slope-deflection equations to analyse simple indeterminate structures and draw SF and BM diagrams,
- describe and distinguish between the flexibility and stiffness methods of analysis,
- define flexibility and stiffness coefficients and explain the procedure for analysing indeterminate structures using these coefficients, and
- analyse statically indeterminate trusses using unit load method and Castigliano's theorem of minimum strain energy.

14.1 INTRODUCTION

The structures we dealt with so far have all been statically determinate. Nevertheless, certain indeterminate problems were introduced in Chapter 3, in which additional equations were evolved using deformation conditions. The calculation of deformation in these cases was limited to axial forces. A discussion on the determinacy and stability of pin-jointed plane frames was presented in Chapter 13. In the present chapter, we deal with the basic procedures for finding solutions of indeterminate structures. Specific methods are dealt with in Chapters 15.

14.2 INDETERMINACY

In analysing a structure, certain unknowns have to be evaluated in order to determine the forces and moments acting at different points on the structure. These may be external reactions, internal forces, or deformations. Internal and external determinacy was explained with reference to pin-jointed plane frames in Chapter 13.

There are two types of indeterminacies—static and kinematic. Static indeterminacy is considered when forces and moments are taken as unknowns. Kinematic indeterminacy is considered when rotations and translations are taken as unknowns. There are two ways of analysing indeterminate structures based on the above-mentioned classes of indeterminacy. We will consider these two types of indeterminacies and the procedures for analysis based on them.

14.3 STATIC INDETERMINACY

In static indeterminacy, the forces and moments are the unknowns. Both external and internal indeterminacies are considered.

We will consider the case of planar structures only. Planar structures are those in which centroidal lines, forces, and moments—both actions and reactions—lie in one plane. The supports provided to planar structures, known also as body constraints, were discussed in Chapter 1 and are shown in Fig. 14.1. Fixed, hinge, and roller supports are commonly considered.

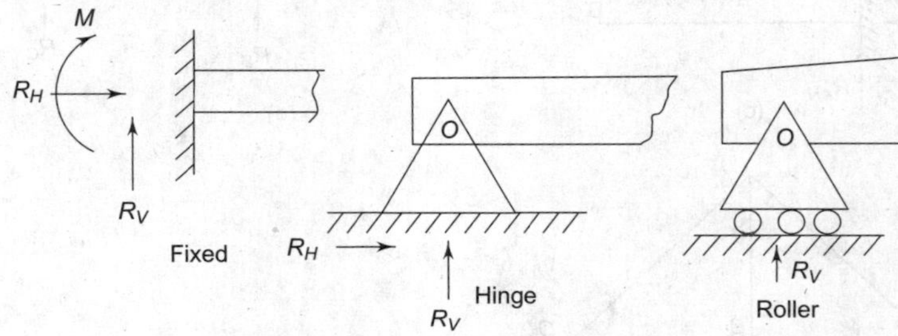

Fig. 14.1

14.3.1 Beams

Considering the simply supported beam shown in Fig. 14.2(a), we note that the hinge support at A provides two unknown components of external reactions. These may be considered as R_A and an unknown direction θ, or more conveniently the horizontal and vertical components of R_A, i.e., $R_{AH} (= R_A \cos\theta)$ and $R_{AV} (= R_A \sin\theta)$. There is one reaction component of known direction at the roller support at B. This reaction is perpendicular to the plane of the roller, in this case vertical. The three reaction components can be determined from the three equations of static equilibrium $\Sigma H = 0$, $\Sigma V = 0$, and $\Sigma M = 0$. The forces and moments (BM and SF) at any point along the beam can be determined using the conditions of static equilibrium at any point of the beam, once the external reactions are determined. The beam shown in Fig. 14.2(a) is thus statically determinate both externally and internally.

The cases shown in Figures 14.2(b) and (c) are statically determinate based on the same reasoning. In general, we can say that if a planar structure has body constraints giving only three reaction components, the structure is externally determinate. If the structure has less than three reaction components, it is unstable. The provision of three reaction components is a necessary but not sufficient condition, as can be seen from Figures 14.2(d) and (e). In (d), there are three reaction components, all parallel. The structure is unstable because any force acting perpendicular to these reaction components will cause rigid body movement in the horizontal direction, as there is no resisting force in that direction. In (e), the three reaction components, though not parallel, meet at a point. Any force causing a moment about that point will cause the structure to rotate.

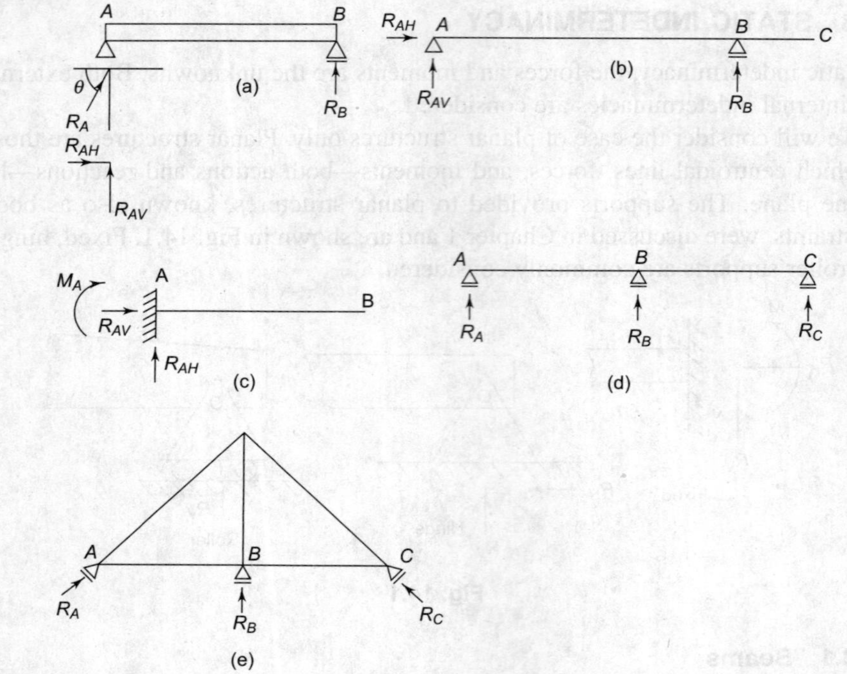

Fig. 14.2

We, therefore, conclude that a planar structure should at least have three reaction components and these should be neither parallel nor concurrent. When there are more than three external reaction components, the three conditions of static equilibrium are not adequate to evaluate all the unknown reactions. The structure is then said to be statically indeterminate. The structures shown in Fig. 14.3 fall under this category. Figures 14.3(a)–(c) show that the structures have more than three reaction components. This is more than the number of equilibrium conditions available. Such a structure is thus said to be indeterminate. The structures shown in Figs 14.3(a)–(c) have four reaction components and are said to be indeterminate to the first degree. The structure shown in Fig. 14.3(d) has six reaction components and is said to be indeterminate to the third degree. The *degree of indeterminacy* is thus the total number of unknown forces and moments (including the external and internal) minus the number of equilibrium conditions available. Based on this criterion, the structure shown in Fig. 14.3(d) is indeterminate to the third degree.

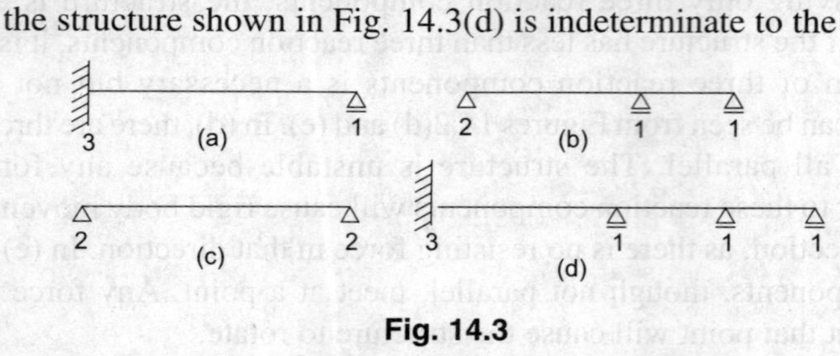

Fig. 14.3

In the case of a structure like that shown in Fig. 14.4, after counting the number of reaction components, we have to consider the effect of the hinge. The hinge, as has been described in Chapter 4, can transfer forces but no moments across it. This means that the BM calculated from either the left or the right of the hinge is zero about the hinge. (This is different from the equation $\Sigma M = 0$ about any point on the beam.) This provides an additional condition, so that the structure in Fig. 14.4(a) is statically determinate. There are four external reactions, but there are also four conditions—three conditions of static equilibrium and one stating that the BM about the hinge is zero. The structure shown in Fig. 14.4(b) is indeterminate to the first degree based on the above reasoning.

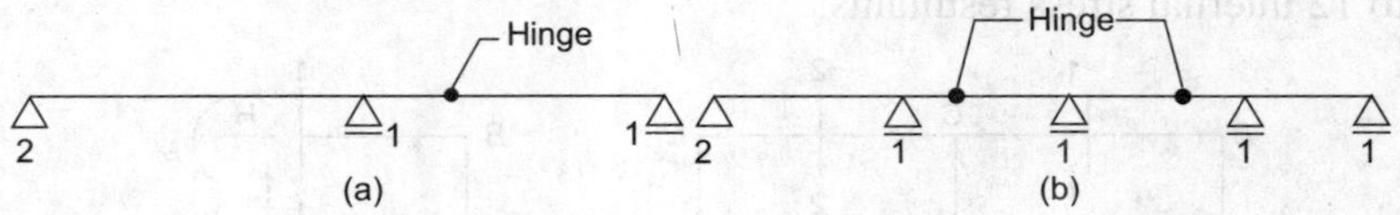

Fig. 14.4

14.3.2 Rigid Frames

Rigid frames, as explained in Chapter 1, are a framework of members with rigid joints. When the rigid frame deforms under load, the angle between the members at a joint remains the same. This is illustrated in the deformed shape of the frame shown in Fig. 14.5.

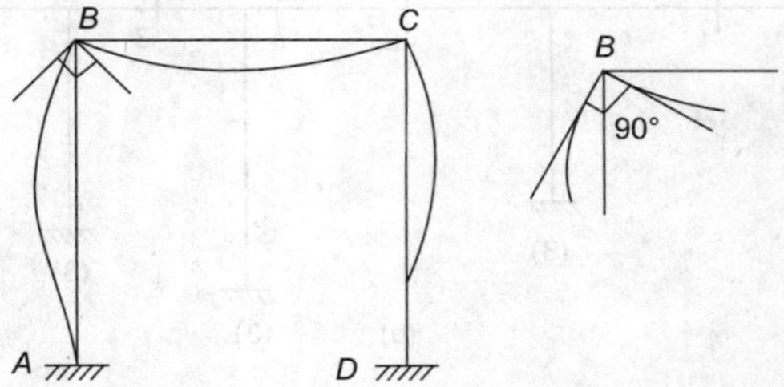

Fig. 14.5

In the case of the single-bay rigid frames shown in Fig. 14.6, the frames (a) and (b) are statically determinate, the number of external reactions being three and there being no internal indeterminacy. The frame shown in Fig. 14.6(c) is statically indeterminate to the first degree and that in (d) is so to the second degree. These frames are single-storey, single-bay portal frames.

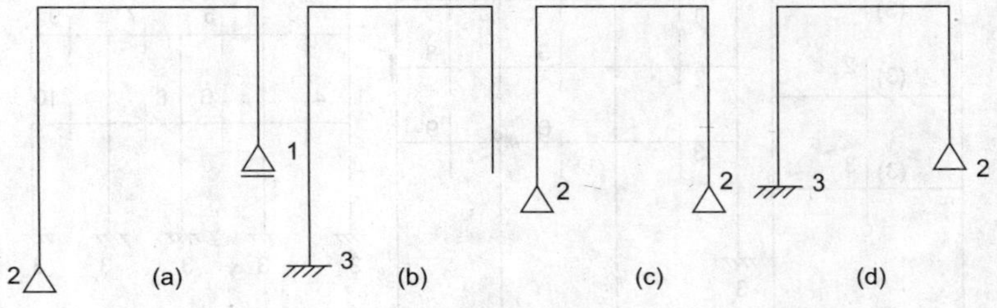

Fig. 14.6

Figure 14.7 shows a number of frames that have single storeys but multiple bays. Such frames need to be considered from both external reactions or internal redundants for finding the indeterminacy. There are more external reactions in the case of the frame shown in Fig. 14.7(a), so the frame is externally indeterminate to the sixth degree. In such a case, it is more convenient to cut the frame as shown by sections 1-1 and 2-2. These cuts give rise to the three segments into which the frame is divided, each statically determinate. Each cut, however, introduces three unknown internal stress resultants. The structure is thus indeterminate to the sixth degree. Similarly, the frame shown in Fig. 14.7(b) is indeterminate to the 12th degree. Fifteen external reactions means 12 unknown reactions, or four cuts give rise to 12 internal stress resultants.

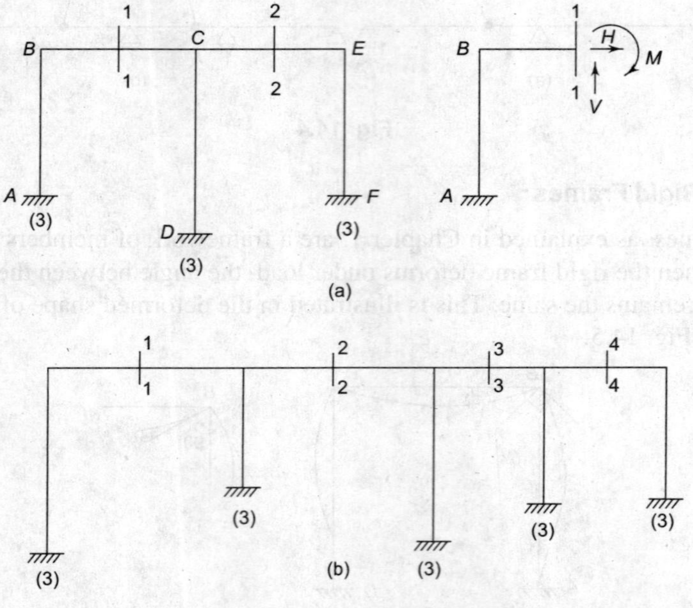

Fig. 14.7

Figure 14.8 shows multi-storey frames. The frame shown in Fig. 14.8(a) is externally indeterminate to the third degree. After making three cuts as shown, the total indeterminacy of the frame becomes 9. The frame is, thus, externally

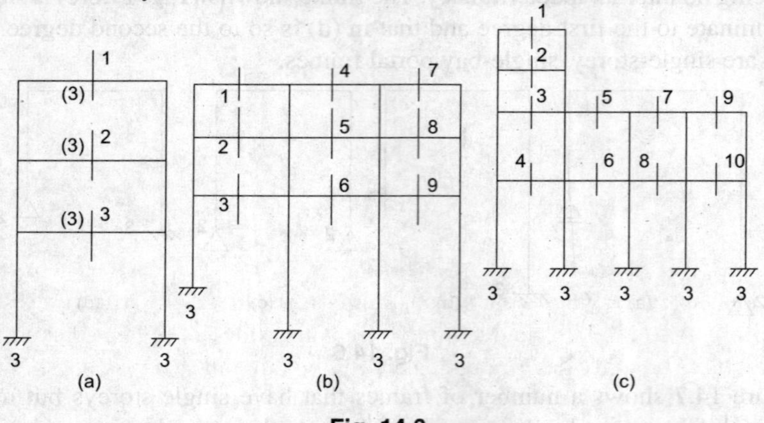

Fig. 14.8

indeterminate to the third degree and internally indeterminate to the sixth degree. In Fig. 14.8(b), the multi-storey, multi-bay frame is externally indeterminate to the ninth (12 – 3) degree. The nine cuts required to give statically determinate cantilever frames give rise to 27 internal stress resultants, which are redundant. The frame is thus internally indeterminate to the 18th degree. Similarly, the frame shown in Fig. 14.8(c) is externally indeterminate to the 12th degree and internally indeterminate to the 18th degree.

14.4 METHOD OF ANALYSIS

Based on static indeterminacy, a structure is analysed using a general method known as the *method of consistent deformation*. In a more generalized form, this procedure leads to the *flexibility method* or the *force method of analysis.*

14.4.1 Method of Consistent Deformations

The general procedure is illustrated by taking the simple case of a propped cantilever shown in Fig. 14.9(a). The first step in the solution is to make it statically

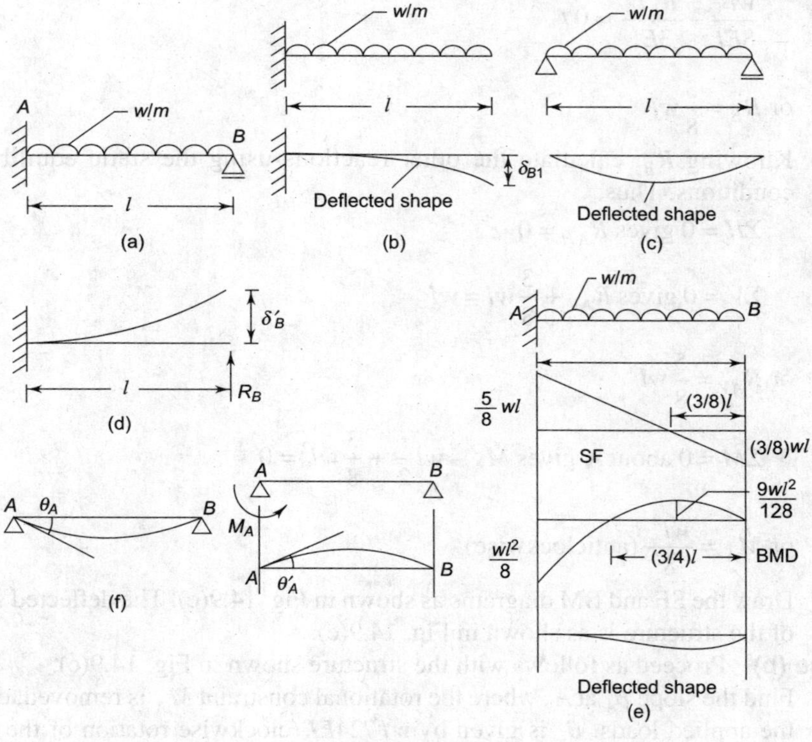

Fig. 14.9

determinate by removing as many constraints as required. The structure is statically indeterminate to the first degree. Only one constraint has to be removed to make the structure statically determinate. This can be done in two ways.

(a) By removing the roller support at B, the structure becomes a cantilever. The constraint removed is the vertical reaction at B [Fig. 14.9(b)].

(b) By removing the rotational constraint at A, the structure becomes a simply supported beam. The constraint removed is the support moment at A [Fig. 14.9(c)].

Case (a) For a solution to the given structure, the procedure is as follows.

(1) Compute the vertical deflection at B due to the applied loads. This deflection is downwards and can be calculated using any of the methods outlined in Chapter 7. The deflection δ_{B1} due to applied loads is equal to $wl^4/8EI$.

(2) Next, consider the cantilever acted upon by the unknown reaction R_B at B without the applied loads, as shown in Fig. 14.9(d). The vertical deflection at B due to R_B acting upwards, δ_B', is given by $-R_B l^3/3EI$; the minus sign is due to the deflection being upwards.

(3) In the given structure, a propped cantilever, there is no deflection at B. Therefore, write the following compatibility condition:
Net vertical deflection at $B = 0$

$$\delta_B + \delta_B' = 0$$

$$\frac{wl^4}{8EI} - \frac{R_B l^3}{3EI} = 0$$

or $R_B = \dfrac{3}{8}wl$

(4) Knowing R_B, calculate the other reactions using the static equilibrium conditions. Thus,

$\Sigma H = 0$ gives $R_{AH} = 0$

$\Sigma V = 0$ gives $R_{AV} + \dfrac{3}{8}wl = wl$

or $R_{AV} = \dfrac{5}{8}wl$

$\Sigma M = 0$ about A gives $M_A - wl\dfrac{l}{2} + \dfrac{3}{8}wl^2 = 0$

or $M_A = \dfrac{wl^2}{8}$ (anticlockwise)

(5) Draw the SF and BM diagrams as shown in Fig. 14.9(e). The deflected shape of the structure is as shown in Fig. 14.9(e).

Case (b) Proceed as follows with the structure shown in Fig. 14.9(c).

(1) Find the slope θ_A at A, where the rotational constraint M_A is removed, due to the applied loads. θ_A is given by $wl^3/24EI$ (clockwise rotation of the axis) [see Fig. 14.9(f)].

(2) Find the slope θ_A' at A due to the unknown moment M_A applied at A. Assuming the direction of M_A to be as shown [Fig. 14.9(g)],

$$\theta_A' = \frac{M_A l}{3EI} \text{ (anticlockwise)}$$

(3) Apply the compatibility equation that the net slope at A is zero because A is a fixed support. Therefore,

$$\theta_A + \theta_A' = 0$$

$$\frac{wl^3}{24EI} - \frac{M_A l}{3EI} = 0$$

$$M_A = \frac{wl^2}{8}$$

(4) Knowing M_A, calculate all the other reactions.

$$\Sigma H = 0, \quad R_{AH} = 0$$

$$\Sigma V = 0, \quad R_{AV} + R_{BV} = wl$$

$$\Sigma M = 0 \text{ @ } B, \quad -M_A + R_{AV}l - \frac{wl^2}{2} = 0$$

$$-\frac{wl^2}{8} + R_{AV}l - \frac{wl^2}{2} = 0, \quad R_{AV} = \frac{5}{8}wl, \quad R_{BV} = \frac{3}{8}wl$$

(5) Draw the SF and BM diagrams as shown in Fig. 14.9(e).

Many important points need to be noted here. These are discussed in the generalized procedure outlined below.

(i) Remove as many constraints in a given structure as required to get a statically determinate structure. The statically determinate structure so obtained is known as a *primary* or *released structure*. As shown in the simple example above, there is no unique primary structure for a given structure. With a single degree of indeterminacy, we obtained two primary structures. The number of constraints to be removed is equal to the degree of static indeterminacy. Two examples are given in Figs 14.10(a) and (b).

(ii) Compute the deformations, slopes, and deflections of the primary structure (a) due to the applied loads and (b) due to each of the constraints (removed) applied separately. Note that if a force has been removed, a linear deflection

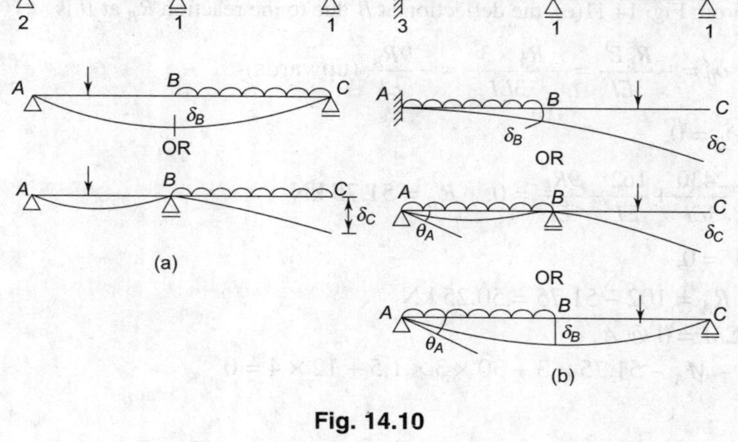

Fig. 14.10

along the direction of the force is calculated. If a restraining couple is removed, the slope corresponding to that couple at that point is calculated. It should also be noted that when a number of constraints are removed, one constraint applied may cause a slope or deflection in the direction and at the point of application of the other constraints. Therefore, with the applied loads and with each con-straint applied separately, a number of slopes and deflections need to be calculated. Figure 14.10 illustrates this.

(iii) Set up the compatibility conditions for the deformations. The number of equations available is equal to the number of constraints removed. A set of simultaneous equations with unknown constraining forces will then be obtained, which, on solution, give the values of these forces.

(iv) Now calculate the remaining constraining forces from the conditions of static equilibrium.

(v) Draw the BM, SF, and axial force (AF) diagrams. The following examples illustrate the procedure.

Example 14.1 Propped cantilever: method of consistent deformations

Analyse the beam shown in Fig. 14.11(a) and draw the SF and BM diagrams.

Solution The primary determinate structure is obtained by removing the roller support at B. Then, the vertical deflections at point B of the cantilever due to the applied loads are calculated. From Figs 14.11(c) and (d), due to the uniformly distributed (UD) load,

$$\delta_{B1} = \frac{30 \times 3^4}{8EI}$$

Due to the point load,

$$\delta_{B2} = \frac{Px^2}{6EI}(3L - x)$$

$$= \frac{12 \times 3^2}{6EI}(3 \times 4 - 3)$$

$$= \frac{162}{EI}$$

where $x = 3$ m and $L = 4$ m. The total deflection at B due to the applied loads = $2430/8EI + 162/EI$. From Fig. 14.11(e), the deflection at B due to the reaction R_B at B is

$$\delta'_B = -\frac{R_B L^3}{3EI} = -\frac{R_B \times 3^3}{3EI} = -\frac{9R_B}{EI} \text{ (upwards)}$$

Setting $\delta_B = 0$,

$$\frac{2430}{8EI} + \frac{162}{EI} - \frac{9R_B}{EI} = 0, \quad R_B = 51.75 \text{ kN}$$

From $\Sigma V = 0$,

$$R_A = 102 - 51.75 = 50.25 \text{ kN}$$

Setting $\Sigma M = 0$ @ A,

$$-M_A - 51.75 \times 3 + 30 \times 3 \times 1.5 + 12 \times 4 = 0$$

or

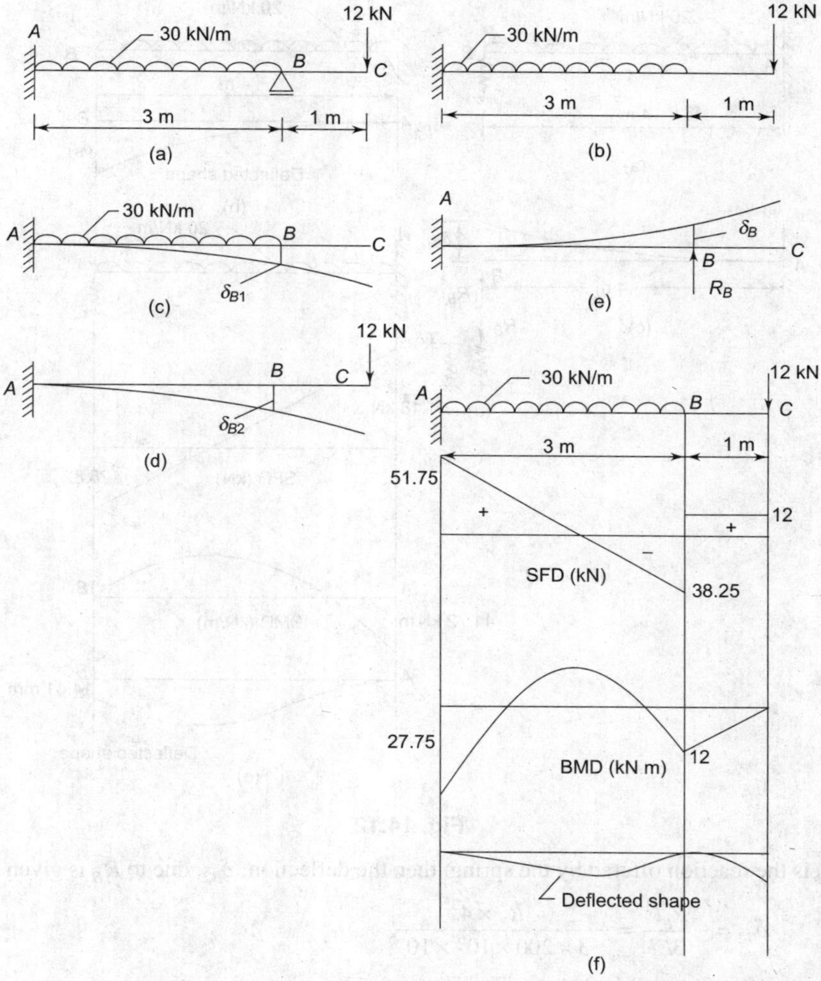

Fig. 14.11

$$M_A = 27.75 \text{ kN m}$$

The SF and BM diagrams can be drawn as shown in Fig. 14.11(f).

Example 14.2 Cantilever with spring prop

A cantilever is propped on its free end by a spring of stiffness 2 kN/mm as shown in Fig. 14.12(a). If $E = 200$ GPa and I for the cross section is 10×10^6 mm^4, find the reaction at the spring and draw the SF and BM diagrams for the beam.

Solution We remove the spring to get the statically determinate structure shown in Fig. 14.12(b). The deflection at B due to applied loads, δ_{B1}, given by

$$\delta_{B1} = \frac{20 \times 4^4}{EI} = \frac{20 \times 256}{200 \times 10^6 \times 10^{-5}}$$

where

$$E = 200 \text{ GPa} = 200 \times 10^6 \text{ kN/m}^2, \quad I = 10^7 \text{ mm}^4 = \frac{10^7}{10^{12}} = 10^{-5} \text{ m}^4$$

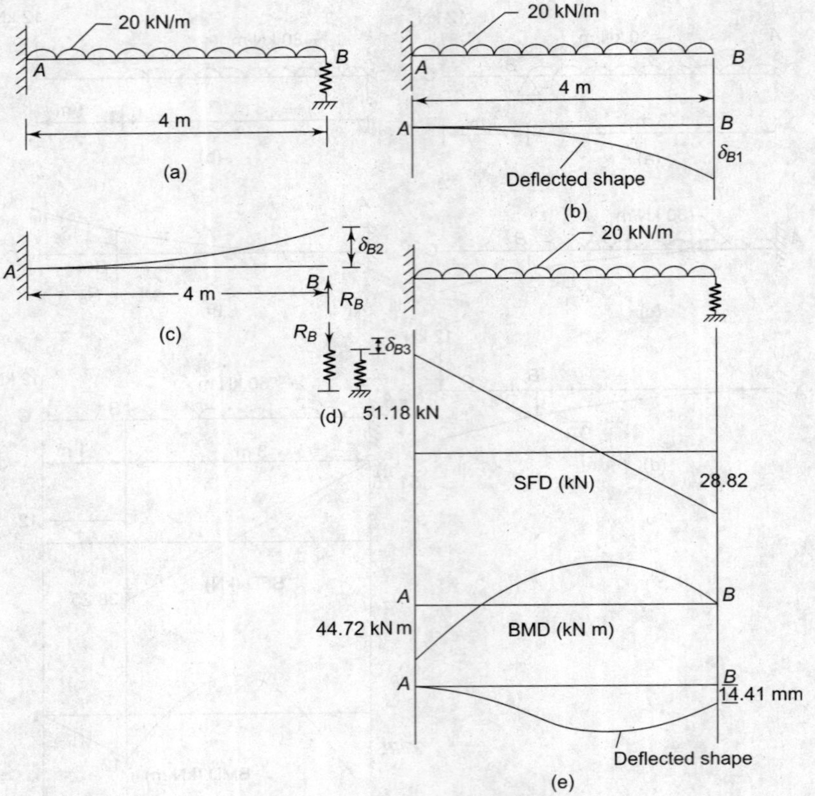

Fig. 14.12

If R_B is the reaction offered by the spring, then the deflection, δ_{B2}, due to R_B is given by

$$\delta_{B2} = -\frac{R_B l^3}{3EI} = -\frac{R_B \times 4^3}{3 \times 200 \times 10^6 \times 10^{-5}}$$

The net deflection at B is the compression of the spring. If δ_{B3} is the compression of the spring, then

$$\delta_{B3} = \frac{RB}{K} = \frac{R_B}{2000} \text{ m}$$

where $K = 2$ kN/mm or 2000 kN/m. Therefore,

$$\delta_{B1} - \delta_{B2} = \delta_{B3}$$

That is,

$$\frac{20 \times 256}{2000} - \frac{R_B \times 64}{3 \times 2000} = \frac{R_B}{2000}$$

from which $R_B = 28.82$ kN. $\Sigma V = 0$ gives

$$R_{AV} = 80 - 28.82 = 51.18 \text{ kN}, \quad R_{AH} = 0$$

$\Sigma M = 0$ about A gives

$$-M_A + 20 \times 4 \times 2 - 28.82 \times 4 = 0$$

$$M_A = 44.72 \text{ kN m}$$

The SF and BM diagrams are shown in Fig. 14.12(e).

Example 14.3 Two-span continuous beam

Analyse the two-span continuous beam shown in Fig. 14.13(a) and draw the SF and BM diagrams.

Solution We obtain a primary statically determinate structure by removing the roller support at B. This is shown in Fig. 14.13(b). The deflection at B in the simply supported beam due to applied loads is calculated. Referring to the table of deflections in Chapter 7 (Table 7.2), we note that the deflection at any distance x from A is given by

$$EI\delta_x = \frac{\omega x}{24}(L^3 - 2Lx^2 + x^3)$$

Setting $x = 4$ in the above equation, we get

$$\delta_{B1} = \frac{1}{EI}\left[\frac{12 \times 4}{24}(10^3 - 2 \times 10 \times 4^2 + 4^3)\right] = \frac{1488}{EI}$$

Next we calculate the deflection δ_{B2} due to R_B in Fig. 14.13(c). From Table 7.2, we have [Fig. 14.13(d)]

$$\delta_x = \frac{Pbx}{6LEI}(L^2 - x^2 - b^2), \quad 0 < x \leq a$$

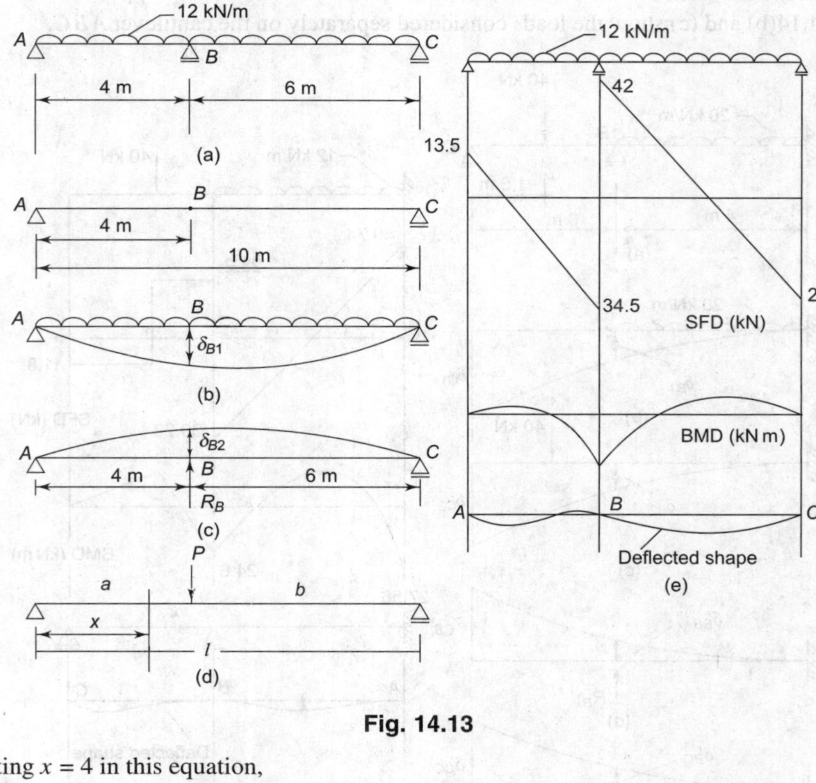

Fig. 14.13

Putting $x = 4$ in this equation,

$$\delta_{B2} = -\frac{R_B(6 \times 4)}{6 \times 10 \times EI}(10^2 - 4^2 - 6^2) = -\frac{19.2R_B}{EI}$$

The net deflection at $B = 0$. Therefore,

$$\delta_{B1} + \delta_{B2} = 0$$

$$\frac{1488}{EI} - \frac{19.2R_B}{EI} = 0, \quad R_B = 77.5 \text{ kN}$$

R_A and R_C can be calculated from the conditions of static equilibrium. Putting $\Sigma M = 0$ @ A, we have

$$12 \times 10 \times 5 - 77.5 \times 4 - R_C \times 10 = 0, \quad R_C = 29 \text{ kN}$$

$\Sigma V = 0$ gives

$$120 \times 10 - 77.5 - 29 - R_A = 0, \quad R_A = 13.5 \text{ kN}$$

The SF and BM diagrams can be drawn as shown in Fig. 14.13(e).

Example 14.4 Two-span continuous beam with fixed support

A beam ABC has a fixed support at A and roller supports at B and C. $AB = 4$ m and $BC = 3$ m. There is a uniformly distributed load of 20 kN/m on AB and a point load of 40 kN at the centre of BC. Draw the SF and BM diagrams for the beam.

Solution The beam is shown in Fig. 14.14(a). A statically determinate structure is obtained by removing the supports at B and C. The two constraints removed are R_B and R_C. This will provide the two deformation conditions that the net deflections at B and C are zero. First, we determine the vertical deflections at B and C due to the applied loads. Figures 14.14(b) and (c) show the loads considered separately on the cantilever ABC.

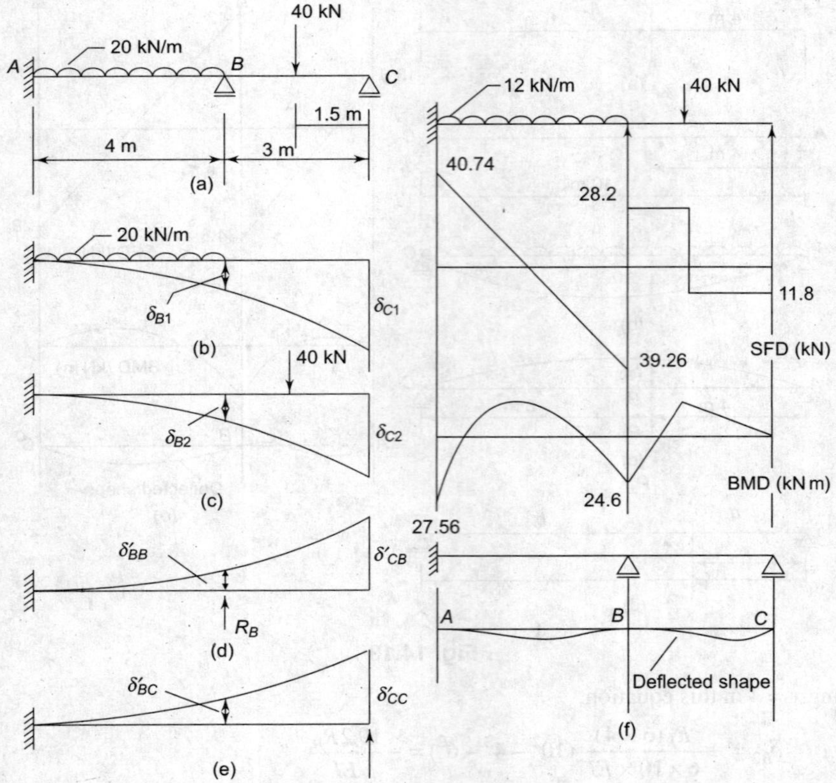

Fig. 14.14

In Fig. 14.14(b), due to the UDL on AB,

$$\delta_{B1} = \frac{wl^4}{8EI} = \frac{20 \times 4^4}{8EI} = \frac{640}{EI}$$

$$\delta_{C1} = \frac{wl^4}{8EI} + \frac{wl^3}{6EI}3 = \frac{640}{EI} + \frac{20 \times 4^3 \times 3}{6EI} = \frac{1280}{EI}$$

In Fig. 14.14(c), due to concentrated load,

$$\delta_{B2} = \frac{Px^2}{6EI}(3a - x)$$

$$= \frac{40 \times 16}{6EI}(3 \times 5.5 - 4) = \frac{1333.33}{EI}$$

where $x = 4$ m and $a = 5.5$ m.

$$\delta_{C2} = \frac{Pa^2}{6EI}(3L - a) = \frac{40 \times 5.5^2}{6EI}(3 \times 7 - 5.5) = \frac{3125.83}{EI}$$

Now each of the constraints removed is made to act independently on the cantilever as shown in Figures 14.14(d) and (e), and the deformations at B and C calculated:

$$\delta'_{BB} = -\frac{R_B a^3}{3EI} = -\frac{R_B \times 4^3}{3EI} = -\frac{64}{3}\frac{R_B}{EI}$$

and

$$\delta'_{CB} = -\frac{R_B a^2}{6EI}(3L - a) = -\frac{R_B \times 4^2}{6EI}(3 \times 7 - 4) = -\frac{272 R_B}{6EI}$$

from Fig. 14.14(d), due to R_B. From Fig. 14.14(e), due to R_C,

$$\delta'_{BC} = -\frac{R_C x^2}{6EI}(3a - x) = -\frac{R_C \times 4^2}{6\,EI}(3 \times 7 - 4) = -\frac{272 R_C}{6EI}$$

$$\delta'_{CC} = -\frac{R_C L^3}{3EI} = -\frac{R_C \times 7^3}{3EI} = -\frac{343}{3EI}R_C$$

The deformation conditions are that the net deflections at B and C are zero:

$$\delta_B = \delta_{B1} + \delta_{B2} + \delta'_{BB} + \delta'_{BC} = 0$$

$$\frac{640}{EI} + \frac{1333.33}{EI} - \frac{64}{3}\frac{R_B}{EI} - \frac{272 R_C}{6EI} = 0$$

$$\frac{64}{3}R_B + \frac{272}{6}R_C = 1973.33$$

$$\delta_C = \delta_{C1} + \delta_{C2} + \delta'_{CB} + \delta'_{CC} = 0$$

$$\frac{1280}{EI} + \frac{3125.83}{EI} - \frac{272}{6EI}R_B - \frac{343}{3EI}R_C = 0$$

$$\frac{272}{6}R_B + \frac{343}{3}R_C = 4405.83$$

Solving, we get $R_C = 11.8$ kN and $R_B = 67.46$ kN. From $\Sigma V = 0$,

$$R_A = 40.74 \text{ kN}$$

$\Sigma M = 0$ about A gives

$$-M_A - 67.46 \times 4 - 11.8 \times 7 + 20 \times 4 \times 2 + 40 \times 5.5 = 0$$

$M_A = 27.56$ kN

$M_B = 24.6$ kN m

The SF and BM diagrams are drawn as shown in Fig. 14.14(f).

14.4.2 Force (Flexibility) Method of Analysis

Even though it is not the most favoured method, it is still important to understand the basics of the flexibility method of analysis. Let us take the case of the propped cantilever that we have solved earlier by the consistent deformation method (Fig. 14.15).

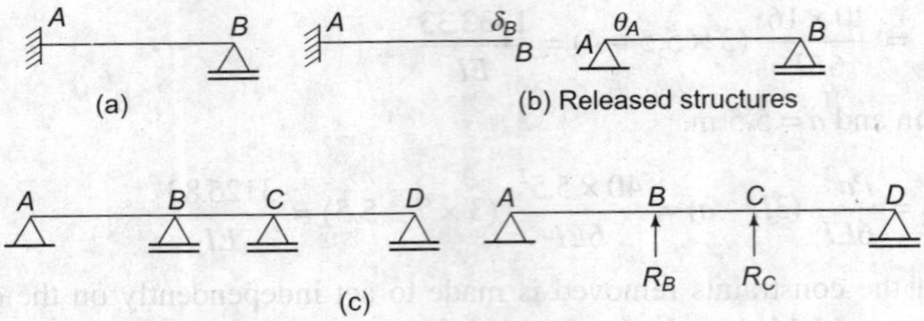

Fig. 14.15 Propped cantilever

To recapitulate, the basic method involves the following steps.

(i) Select and remove the redundants to get a primary or determinate structure. We can remove the support at B to get a cantilever or the rotational constraint at A to get a simply supported beam.

(ii) Determine the displacements and/or rotations in the direction of the redundants removed. We determine the vertical deflection at B if R_B is the redundant and slope at A if M_A is selected as the redundant.

(iii) We now apply the redundants one by one and find the deformations in the direction of the redundants removed due to the application of each redundant. If R_B is the redundant, we find the deflection at B due to the application of R_B, an unknown force. The direction of R_B can be arbitrarily assumed. If M_A is the redundant, then we find the slope at A due to the application of the unknown moment M_A.

(iv) Note that if more than one redundant is removed, then the application of one redundant may cause deformation in the direction of all the other redundants. As an example, consider the case of the beam shown in Fig. 14.15(c). We can remove the supports at B and C to get the primary, determinate structure. Let R_B and R_C are the reactions at B and C. When we apply redundant R_B on the beam, it will cause deflections at B and C. Similarly, when we apply R_C, it will cause deflections at C and B. All the four deflections need to be calculated to solve the problem.

(v) We now apply the deformation conditions, for example, the deflection at B is zero or the slope at A is zero. The condition need not always be zero deformation.

(vi) Solve the equations so obtained to evaluate the unknowns. Once the unknowns are evaluated, the other forces, bending moments, shear forces, etc., can be easily calculated.

Let us now look at this procedure in a slightly different way. Take the example of the propped cantilever shown in Fig. 14.16(a).

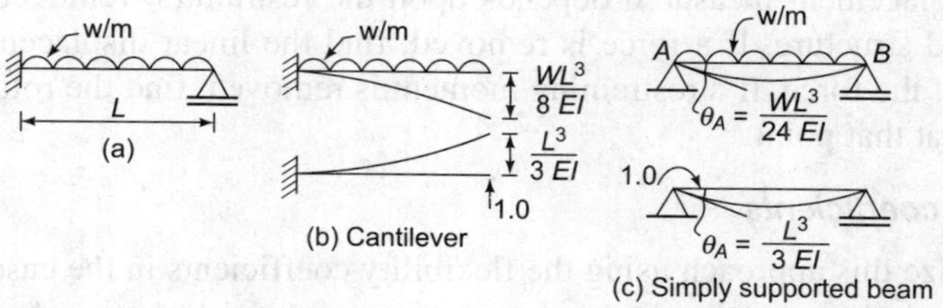

Fig. 14.16 Flexibility coefficients

Case A When R_B is the redundant

We get the cantilever as the primary structure shown in Fig. 14.16(b). We find the deflection of point B by the application of the unit load at B in the upward direction. The deflection is $L^3/3EI\uparrow$.

Deflection due to the applied load $= -wL^3/8EI\downarrow$

Deflection due to $R_B = R_B(L^3/3EI)\uparrow$

(Note that the deflections are in the direction of the unknown redundant.)

If R_B is the redundant, the deformation condition can be written as

$$-wL^3/8EI + (L^3/3EI)R_B = 0$$

This equation is the same as we have obtained earlier. The term $(L^3/3EI)$ is known as the *flexibility coefficient* in flexure (bending). In general, the flexibility coefficient can be defined as the deformation caused due to a unit force (or a unit couple).

Case B When M_A is the redundant

We find the slope at A due to a unit couple applied at A. The slope can be found to be $L/3EI$ (clockwise). The slope due to the applied load is $wL^3/24EI$ (clockwise) at A. The deformation condition can be written as

$$wL^3/24EI + (L/3EI)M_A = 0$$

which is the same equation as we have derived earlier.

The terms $(L^3/3EI)$ and $(L/3EI)$ are known as flexibility coefficients.

A flexibility coefficient is the deformation (translation or rotation) at a point due to the application of a unit force (or a unit couple) at that point or some other point in the structure.

Flexibility coefficients are generally represented by the symbol f. Subscripts are used to define the coefficient completely. The following arrangement can be used conveniently.

The flexibility coefficient $L^3/3EI$ mentioned above will be named f_{bb}, as the unit load is applied at B and the deformation is also measured at B. The term $L/3EI$ will be named f_{aa} as the deformation is measured at the same point where the couple is applied.

Referring to Fig. 14.16(c), if a unit load is applied at point 1 and the displacement at point 2 is measured, then the displacement is known as flexibility coefficient f_{21}. The first subscript is the point where the displacement is measured and the second subscript is the point where the load is applied. Similarly, f_{12} is the flexibility coefficient giving the displacement at point 1 when a unit load is applied at 2.

The displacement measured depends upon the restraint(s) removed to obtain the released structure. If a force is removed, find the linear displacement in the direction of the force. If a restraining moment is removed, find the rotational displacement at that point.

Flexibility coefficients

We formalize this approach using the flexibility coefficients in the case of different structural elements. In all structures, it is necessary to have reference axes, which are oriented to define the structure called global axes and member-oriented axes known as local axes. These are shown in Fig. 14.17(a). The flexibility coefficients for members are worked out from the member-oriented axes. One of the axes is made to coincide with the axis of the member. The X-axis is generally made to coincide with the member axis.

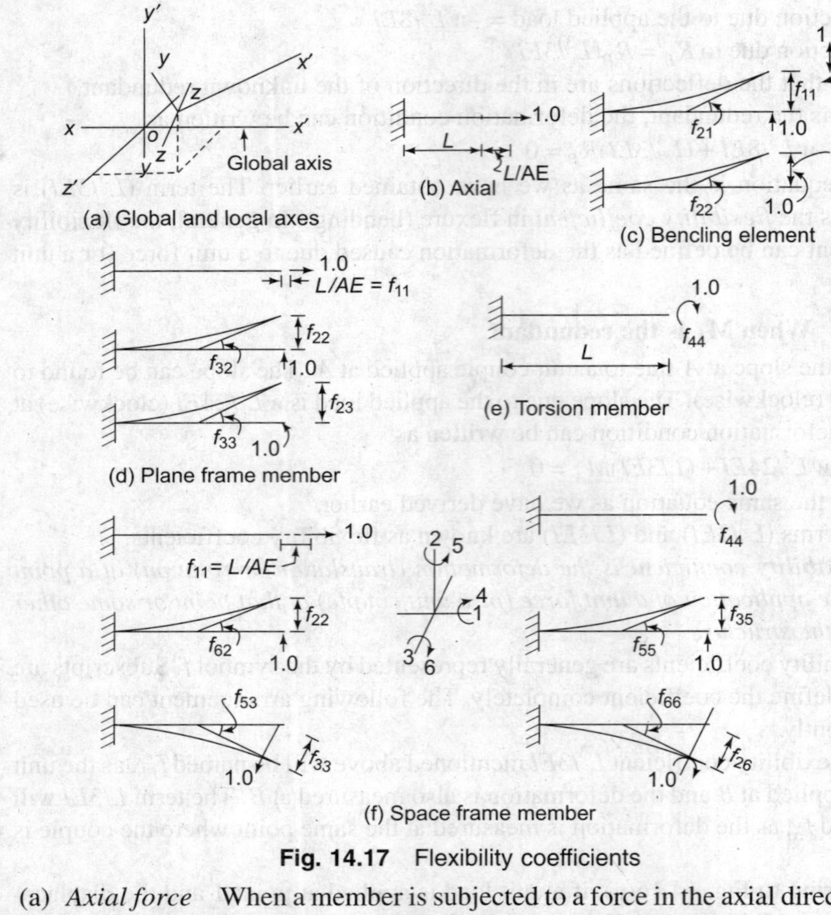

Fig. 14.17 Flexibility coefficients

(a) *Axial force* When a member is subjected to a force in the axial direction, as in compression or tension, the flexibility coefficient is by definition the shortening or extension caused due to a unit load. We have studied in Chapter 3 that the axial deformation of a bar is given by

$\Delta L = PL/AE$

When $P = 1$, $\Delta L = L/AE$ is the flexibility coefficient.

This is illustrated in Fig. 14.17(b). This is typical of the members of a pin-jointed plane frame.

(b) *Planar beam* A beam lying in one plane is subjected to bending moment and shear force. The axial force may also exist, which we neglect for the time being. There are two actions—a vertical force and a couple. The actions and displacements corresponding to these will be numbered 1 and 2. As in Fig. 14.17(c), we consider the far end A as fixed and apply a unit force first at point B. The deflection caused by the unit load at point j is $1L^3/3EI$. This will be called f_{11}. This load also will cause a slope at point B. The slope is $L^2/2EI$, which will be called f_{21}. The linear displacement and the force in that direction are given the number 1. The rotational deformation and the couple will be given the number 2.

When a unit couple is applied at B, the deflection and slope at B are given by

Deflection, $f_{21} = L^2/2EI$; $f_{22} = L/EI$

As an example, consider the beam with both ends fixed shown in Fig. 14.7(d). We remove the constraints at B to get a cantilever. Now, apply a unit force (corresponding to R_B) at B.

Deflection at $B, f_{11} = L^3/3EI$

Rotation at $B, f_{21} = L^2/2EI$

Now, we apply a unit couple (corresponding to M_B) at B.

Deflection at $B, f_{21} = L^2/2EI$

Rotation at $B = f_{22} = L/EI$

(c) *Plane frame member* The plane frame member shown in Fig. 14.17(e) has three displacements at one end—translations along the axes in the plane and a rotation. Naming them 1, 2, and 3, we can write the flexibility coefficients as follows.

(i) Axial translation: Applying unit load along the X-axis, $f_{11} = L/AE$; f_{21} and f_{31} are zero as a unit axial force does not cause any displacement in these directions.

(ii) Shear translation: Applying unit load along the Y-axis, $f_{22} = L^3/3EI$; $f_{12} = 0$ and $f_{32} = L^2/2EI$.

(iii) Rotation about the Z-axis: Applying unit couple about the Z-axis, bending in the X-Y plane is caused. $f_{13} = 0, f_{23} = L^2/2EI$, and $f_{33} = L/EI$.

(d) *Space frame member* In the case of a space frame member, there are three translations and three rotations about the three axes at one end. The flexibility coefficients can be calculated as follows [refer to Fig. 14.17(f)].

(i) Axial (translation along the X-axis): Apply unit load along the X-axis. $f_{11} = L/AE$ and f_{21} to f_{61} are zero.

(ii) Translation along the Y-axis: Apply unit load along the Y-axis. $f_{12} = 0$, $f_{22} = L^3/3EI_z, f_{32} = 0, f_{42} = 0, f_{52} = 0$, and $f_{62} = L^2/2EI_z$. (Note that we use I_{zz} here.)

(iii) Translation along the Z-axis: Apply unit load along the Z-axis. $f_{13} = 0$, $f_{23} = 0, f_{33} = L^3/3EI_y, f_{43} = 0, f_{53} = -L^2/2EI_y$, and $f_{63} = 0$. (Note that MI used is about the YY axis. Also f_{53} is negative because the moment vector is in the negative direction of the axis.)

(iv) Rotation @ the X-axis: Apply unit couple about the X-axis. This causes torsion in the member. f_{14} to $f_{34} = 0, f_{44} = L/GJ$, and $f_{54} = f_{64} = 0$.

(v) Rotation about the Y-axis: Apply unit couple about the Y-axis. This causes bending in the X-Z plane and the neutral axis is parallel to the Y-axis. $f_{15} = f_{25} = f_{35} = -L^2/2EI_y, f_{45} = 0, f_{55} = L/EI_y$, and $f_{65} = 0$.

(vi) Rotation about the Z-axis: Apply a unit couple about the Z-axis. This causes bending in the X-Y plane. The neutral axis is parallel to the Z-axis. $f_{16} = 0, f_{26} = L^2/2EI_z, f_{36} = f_{46} = f_{56} = 0$, and $f_{66} = L/EI_z$.

The flexibility coefficients can be put in a matrix as follows.

	1	2	3	4	5	6
1	L/AE	0	0	0	0	0
2	0	$L^3/3EI_z$	0	0	0	$L^2/2EI_z$
3	0	0	$L^3/3EI_y$	0	$-L^2/2EI_y$	0
4	0	0	0	L/GJ	0	0
5	0	0	$-L^2/2EI_y$	0	L/EI_y	0
6	0	$L^2/2EI_z$	0	0	0	L/EI_z

Along the diagonal are values like f_{11}, f_{22}, f_{33}, etc. The flexibility coefficients written in this form are symmetrical about the diagonal. That this is so is evident from Maxwell's reciprocal theorem whereby $f_{ij} = f_{ji}$.

The following two examples illustrate the use of flexibility coefficients in the conventional consistent deformation method.

Example 14.5 Flexibility coefficient: propped cantilever

Analyse the propped cantilever beam loaded as shown in Fig. 14.18, find the reactions and draw the shear force and bending moment diagrams.

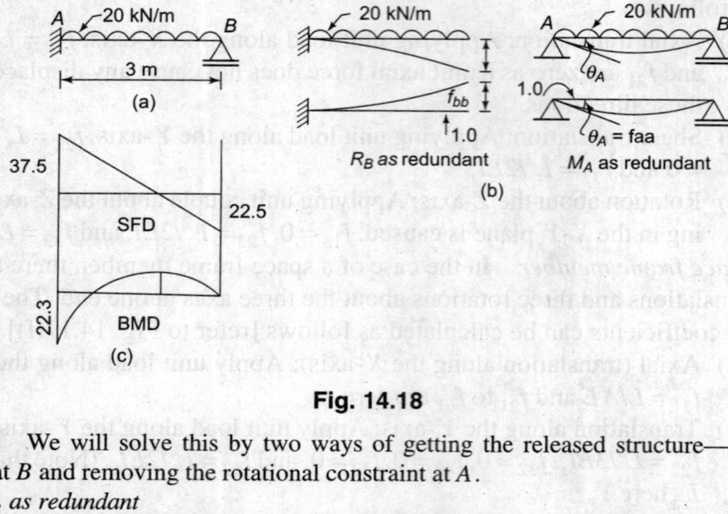

Fig. 14.18

Solution We will solve this by two ways of getting the released structure—removing support at B and removing the rotational constraint at A.

(a) *R_B as redundant*

We apply a unit upward load at B. The deflection at B (in the direction of R_B) is $L^3/3EI$.

Thus, the flexibility coefficient, $f_{bb} = 3^3/3EI = 9/EI$ (↑)
The deflection at B due to the applied load = $20 \times 3^4/8EI = -202.5/EI$ (↓)
The compatibility condition is that the deflection at $B = 0$.
$\Delta_L + f_{bb}R_B = 0$ is the condition.

$-202.5/EI + (9/EI)R_B = 0$, $R_B = 202.5/9 = 22.5$ kN

$R_A = 20 \times 3 - 22.5 = 37.5$ kN

$M_A = 20 \times 3 \times 1.5 - 22.5 \times 3 = 22.5$ kNm

The SF and BM diagrams are shown in Fig. 14.18(c).

(b) *M_A as the redundant*

When the rotational constraint at A is the redundant, the released structure is an SS beam. The deformation corresponding to the constraint removed is the slope at A. We apply a unit couple (clockwise) at A as shown. Due to this unit couple,

Slope at $A = \theta_A = f_{aa} = L/3EI = 1/EI$

In the simply supported beam carrying a UD load,

Slope at $A = 20 \times 3^3/24EI = 22.5/EI$ due to the applied load

The compatibility condition is $\Delta_L + f_{aa}M_A = 0$

$22.5/EI + (1/EI)M_A = 0$; $M_A = -22.5$ kNm

The negative sign shows that the moment is anticlockwise.
The reaction can now be calculated.

Reactions at A and B due to the UD load = $20 \times 3/2 = 30$ kN ↑ at A and B
Reaction at A due to the couple = $M_A/3 = 22.5/3 = 7.5$ kN ↑ at A and 7.5 kN ↓ at B
Net reaction at $A = 37.5$ kN
Net reaction at $B = 30 - 7.5 = 22.5$ kN at B

SF and BM diagrams are as shown in Fig. 14.18(c).

Example 14.6 Fiexibility coefficients: propped cantilever

Analyse the propped cantilever beam shown in Fig. 14.19, find the reactions, and draw the shear force and bending moment diagrams.

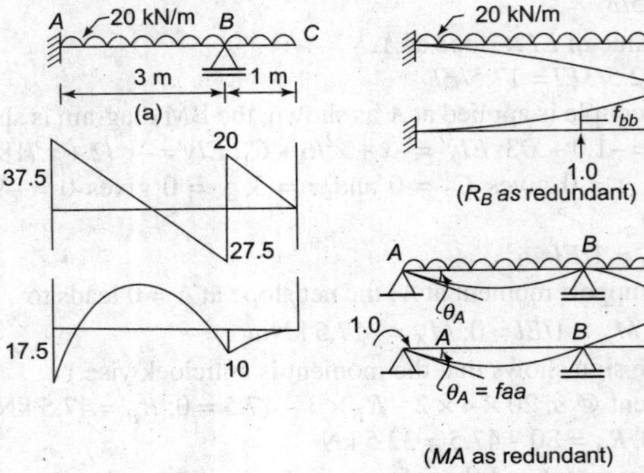

Fig. 14.19

Solution The primary released structure can be obtained in two ways—removing the support at B, we get a cantilever and removing the rotational constraint at A, we get an overhanging beam.

(a) *R_B as redundant*

In this case, we have to find vertical deflection at B due to the applied loads and due to the unit load applied at B.

From the free body of the cantilever shown,

$M_x = -160 + 80x - 20x^2/2$

$EIy'' = 160 - 80x + 10x^2$; $EIy = 160x - 40x^2 + 10x^3/3 + C_1$

$EIy = 80\,x^2 - 40\,x^3/3 + 10\,x^4/12 + C_1 x + C_2$

$EIy = 0$ when $x = 0$ and $y' = 0$ when $x = 0$. These give $C_1 = C_2 = 0$.

$EIy = 80x^2 - 40x^3/3 + 10x^4/12$; When $x = 3$, $y = 427.5/EI \downarrow$

When a unit load is applied at B, $f_{bb} = 3^3/3EI = 9/EI \uparrow$ is the deflection at B due to a unit load applied at B. The compatibility condition can be written as

$\Delta_L + f_{bb}R_B = 0$; $-427.5/EI + R_B \times 9/EI = 0$; $R_B = 47.5$ kN $\uparrow$

From $\Sigma V = 0$, $R_A = 20 \times 4 - 47.5 = 32.5$ kN $\uparrow$

Taking moment @ A,

$-M_A - 47.5 \times 3 + 20 \times 4 \times 2 = 0$; $M_A = 17.5$ kNm, anticlockwise

The shear force and bending moment diagram can be drawn as shown.

(b) *M_A as redundant*

When M_A is taken as the redundant, the primary structure is an overhanging beam shown. We have to evaluate the slope at A due to the applied loads and a unit couple applied at A.

With the applied loads,

$R_{AV} = 26.67$ kN; $R_{BV} = 53.33$ kN

We use the double integration method to find the slope.

$M_x = 26.67x - 20x^2/2 + 53.33(x - 3)$

$EIy'' = -26.67x + 10x^2 - 53.33(x - 3)$

$EIy' = -26.67x^2/2 + 10x^3/3 - 53.33(x - 3)^2/2 + C_1$

$EIy = -26.67x^3/6 + 10x^4/12 - 53.33(x - 3)^3/6 + C_1 x + C_2$

Now, $x = 0$, $y = 0$ gives $C_2 = 0$.

Also $x = 3$, $y = 0$ gives $0 = -26.67 \times 27/6 + 20 \times 81/24 + C_1 \times 3$

So, $C_1 = 52.5/3$

This is the value of $EI \times$ slope at A.

Thus, $\theta_A = 52.5/3EI = 17.5/EI$

When a unit couple is applied at A as shown, the BM diagram is shown in the figure.

Again, $EIy'' = -1.0 + x/3$; $EIy' = -x + x^2/6 + C_1$; $EIy = -x^2/2 + x^3/18 + C_1 x + C_2$

Now, $x = 0$, $y = 0$ gives $C_2 = 0$ and $x = 3$, $y = 0$ gives $0 = -9/2 + 27/18 + C_1 \times 3$; $C_1 = 1.0$

This gives $\theta_A = 1/EI$

If M_A is the support moment at A, the net slope at $A = 0$ leads to

$17.5/EI + M_A \times 1/EI = 0$; $M_A = -17.5$ kNm

(The negative sign shows that the moment is anticlockwise.)

Taking moment @ a, $20 \times 4 \times 2 - R_B \times 3 - 17.5 = 0$; $R_B = 47.5$ kN

From $\Sigma V = 0$, $R_A = 80 - 47.5 = 32.5$ kN

The bending moment and shear force diagrams can now be drawn. ■

Application of the flexibility method to continuous beams is illustrated in Chapter 15.

14.5 KINEMATIC INDETERMINACY

In the second procedure for analysis of indeterminate structures, the deformations at the joints of a structure are considered as unknowns. A joint is said to exist in a structure whenever two or more members meet at a point, at the point of supports, or at the free ends. Kinematic indeterminacy is actually a measure of the degrees of freedom of the joints in a structure.

At any joint in a structure (planar), there are three degrees of freedom—two translational movements in perpendicular directions and one rotation. These may be constrained at the points of support. The three types of supports provided in a structure give different degrees of freedom. As shown in Fig. 14.20, the fixed support provides no freedom, as it prevents both rotation and translation. The hinge support provides one degree of freedom, that of rotation. The roller support gives two degrees of freedom: (a) rotation and (b) translation parallel to the plane of the roller.

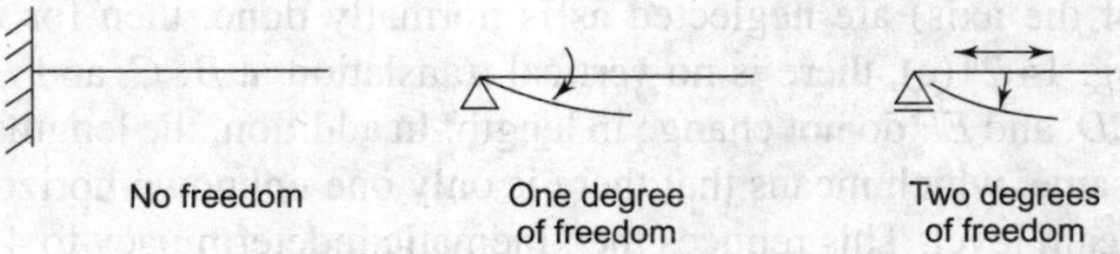

Fig. 14.20

Now, we consider the structures shown in Fig. 14.21.

(a) The simply supported beam has three degrees of freedom: a rotation at A, a rotation at B, and a horizontal translation at B. The structure is kinematically indeterminate to the third degree.

(b) The fixed beam has no freedom at the joints and is hence kinematically determinate.

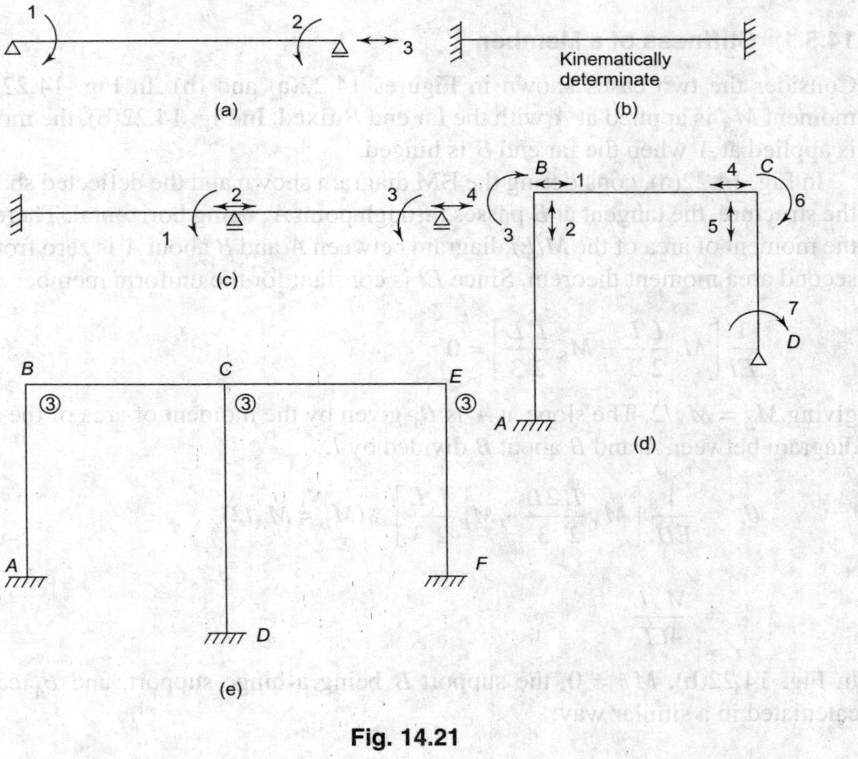

Fig. 14.21

(c) The continuous beam has two degrees of freedom at B and C and is hence kinematically indeterminate to the fourth degree.

(d) The single-bay portal frame has three degrees of freedom each at the joints B and C and one degree of freedom at D. It is, therefore, kinematically indeterminate to the seventh degree.

It may be noted that from the point of view of static indeterminacy, (a) is determinate, (b) is indeterminate to the third degree, (c) is indeterminate to the second degree, and (d) is indeterminate to the second degree.

The structure shown in Fig. 14.21(e) is kinematically indeterminate to the ninth degree, as the joints B, C, and E each have three degrees of freedom. The structure is statically indeterminate to the sixth degree. If axial deformations (shortening or stretching of the axis) are neglected as is normally done, then for the structure shown in Fig. 14.21(e), there is no vertical translation at B, C, and E, as the columns AB, CD, and EF do not change in length. In addition, the lengths BC and CE remain the same, which means that there is only one unknown horizontal translation at the beam level. This reduces the kinematic indeterminacy to 4: rotations at B, C, and E and horizontal translation at either B, C, or E. Neglecting axial deformation reduces the kinematic indeterminacy to 2 in Figs 14.21(a) and (c) and to 4 in Fig. 14.21(d).

The procedure for the analysis of structures based on kinematic indeterminacy is known as the *stiffness method*. The stiffness method is easier to put in a more generalized form for computer solutions. Only some preliminary concepts will be presented here.

14.5.1 Stiffness of a Member

Consider the two cases shown in Figures 14.22(a) and (b). In Fig. 14.22(a), a moment M_A is applied at A with the far end B fixed. In Fig. 14.22(b), the moment is applied at A when the far end B is hinged.

In Fig. 14.22(a), considering the BM diagram shown and the deflected shape of the structure, the tangent at B passes through point A, being horizontal. Therefore, the moment of area of the M/EI diagram between A and B about A is zero from the second area moment theorem. Since EI is constant for the uniform member AB,

$$\frac{1}{EI}\left[M_A\frac{L}{2}\frac{L}{3} - M_B\frac{L}{2}\frac{L}{3}\right] = 0$$

giving $M_B = M_A/2$. The slope at A is θ_A given by the moment of area of the M/EI diagram between A and B about B divided by L.

$$\theta_A = \frac{1}{EIL}\left[M_A\frac{L}{2}\frac{2L}{3} - M_B\frac{L}{2}\frac{L}{3}\right] \quad (M_B = M_A/2)$$

$$= \frac{M_A L}{4EI}$$

In Fig. 14.22(b), $M_B = 0$, the support B being a hinge support, and θ_A can be calculated in a similar way:

$$\theta_A = \frac{1}{EIL}\left(M_A \frac{L}{2}\frac{2L}{3}\right) = \frac{M_A L}{3EI}$$

When θ_A is unity, M_A is equal to $4EI/L$ when the far end is fixed and $3EI/L$ when the far end is hinged.

The moment required to be applied to the end of a member to cause unit slope at that end is known as the *stiffness of the member*. Thus, if the far end is fixed the stiffness is $4EI/L$, and if the far end is hinged the stiffness is $3EI/L$. These are known as the *flexural stiffness* of a member.

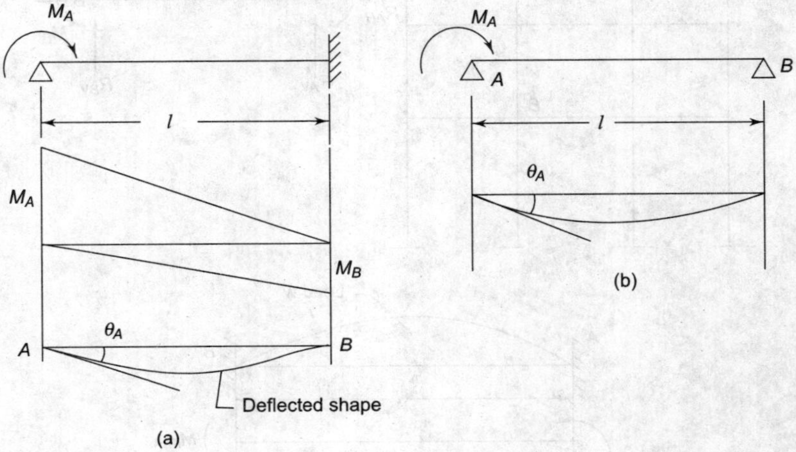

Fig. 14.22

As in the case of flexibility method, members have axial, flexural, and torsional stiffness. As the determination of these coefficients requires the analysis of a fixed beam, we detail out the stiffness method in Chapter 15.

14.5.2 Methods of Analysis

There are a number of methods of analysis based upon kinematic indeterminacy of a structure. Such methods form what are known as stiffness methods of analysis. The methods under this category are based on a basic method known as slope-deflection method. Moment distribution method and Kani's method of rotation contribution are based on this and are iterative techniques for solution of slope deflection equations.

14.6 SLOPE-DEFLECTION EQUATIONS

The slope-deflection equations give the relationship between the moment at the ends of a member and the deformations of a member.

Consider a member AB isolated from a structure and treated as a free body. Let M_A and M_B be the final moments at the ends A and B. M_A and M_B can be expressed in terms of moments due to applied loads, slopes at the ends A and B, and any relative translation between A and B, transverse to the longitudinal axis.

14.6.1 Development of Slope-Deflection Equations

In Fig. 14.23, the deformation of AB to its final position is considered in stages. We use the sign convention that clockwise moments and slopes are positive, and end translation resulting in the clockwise rotation of an axis is positive.

1. Initially, assume the member ends A and B to be fixed due to applied loads. There will be fixed end moments at A and B; let these be M_{FAB} and M_{FBA}. These have been assumed positive in Fig. 14.23.

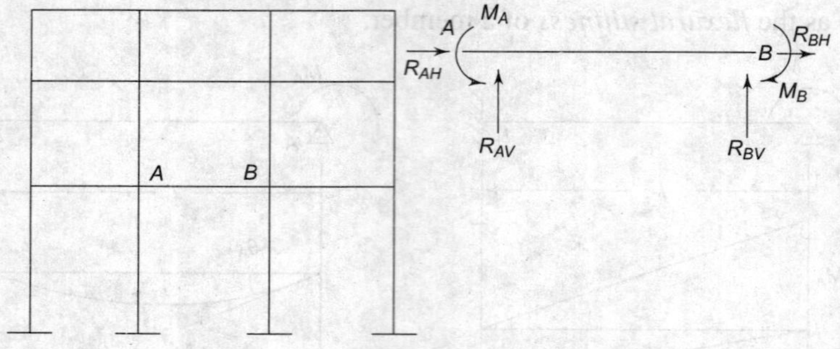

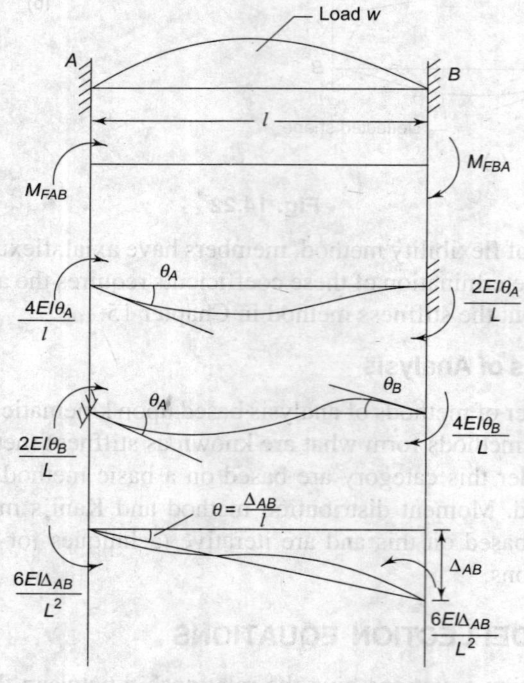

Fig. 14.23

2. Allow the end A to rotate to its final position of clockwise rotation θ_A at A. To rotate the end A by θ_A, the moment to be applied is $(4EI/L)\theta_A$ at A. As we have seen earlier, one-half of this moment, $(2EI/L)\theta_A$, is the moment at B.
3. Fix the end A at this position and free the end B to rotate clockwise through θ_B. The moment applied at B is $4EI\theta_B/L$ and the moment developed at A is $2EI\theta_B/L$.

4. Finally, any relative translation between A and B will cause equal moments at A and B. If Δ_{AB} is the relative translation, then it is positive if it results in a clockwise rotation of the beam axis. It will be shown in the next chapter, on fixed and continuous beams, that the moments developed at A and B are of the same sign to make the end slopes zero. With a positive rotation of the beam axis, the moments are negative and equal to $6EI\Delta/L^2$. If we sum up all the moments listed above at the ends A and B, we have

$$M_{AB} = M_{FAB} + \frac{4EI}{L}\theta_A + \frac{2EI}{L}\theta_B - \frac{6EI\Delta}{L^2}$$

$$= M_{FAB} + \frac{2EI}{L}\left(2\theta_A + \theta_B - 3\frac{\Delta}{L}\right)$$

$$M_{BA} = M_{FBA} + \frac{4EI}{L}\theta_B + \frac{2EI}{L}\theta_A - \frac{6EI\Delta}{L^2}$$

$$= M_{FBA} + \frac{2EI}{L}\left(2\theta_B + \theta_A - \frac{3\Delta}{L}\right)$$

These are known as slope-deflection equations, which are used in the fundamental stiffness method of analysis known as the slope-deflection method. In this method, θ_A, θ_B, etc. and Δ are the unknown deformations, which is the characteristic of kinematic indeterminacy.

14.6.2 Application of Slope-Deflection Equations

To understand the procedure for solution using the stiffness method, consider the beam shown in Fig. 14.24. The structure has only two unknown deformations, θ_B and θ_C, neglecting the axial deformations.

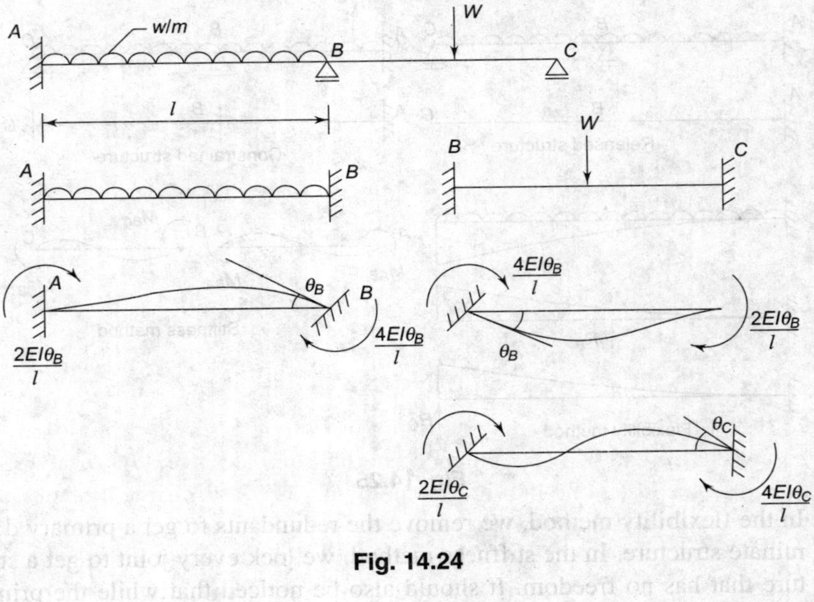

Fig. 14.24

1. Initially, consider the ends of each member AB and BC as fixed. Find the fixed end moments due to applied loads.
2. Write the slope-deflection equations for the member end moments; note that $\Delta = 0$ in both cases:

$$M_{AB} = M_{FAB} + \frac{2EI}{L}(2\theta_A + \theta_B) = M_{FAB} + \frac{2EI}{L}\theta_B \quad \text{(because } \theta_A = 0\text{)}$$

$$M_{BA} = M_{FBA} + \frac{4EI}{L}\theta_B$$

$$M_{BC} = M_{FBC} + \frac{2EI}{L}(2\theta_B + \theta_C)$$

$$M_{CB} = M_{FCB} + \frac{2EI}{L}(\theta_B + 2\theta_C)$$

3. Express the end moments in terms of the slopes at the ends. Then write the equilibrium conditions for the joint moments as

   $M_{CB} = 0, \quad M_{BA} + M_{BC} = 0$

   C being a simply supported end, the moment at it is zero. Equilibrium conditions also require that the joint moment at B in BA and BC be equal and opposite.
4. When the above equations are written, we get two equations with θ_B and θ_C as unknowns, which can be evaluated.
5. Substitute these values back in the slope-deflection equations to get the values of the end moments.
6. Draw the SF and BM diagrams using appropriate signs for the end moments.

The differences between the flexibility method and the stiffness method are summarized below. These are illustrated in Fig. 14.25.

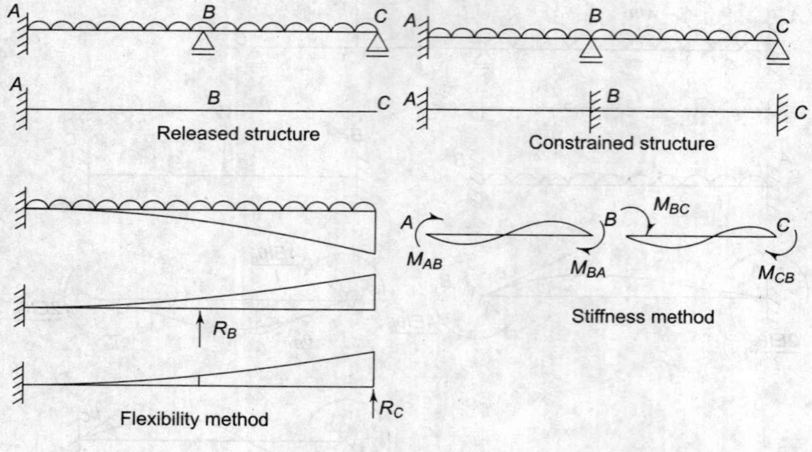

Fig. 14.25

- In the flexibility method, we remove the redundants to get a primary determinate structure. In the stiffness method, we lock every joint to get a structure that has no freedom. It should also be noticed that while the primary structure of the flexibility method is not unique, as it can be obtained in many ways, the restrained structure of the stiffness method is unique in the sense that all the joints are initially locked.

- In the flexibility method, the moments and forces are the unknowns; in the stiffness method, the deformations are the unknowns.
- In the flexibility method, the compatibility of deformations is used as the condition to obtain equations in terms of moments and forces. In the stiffness method, the equilibrium conditions of the forces and moments at the joints are used to derive equations in terms of deformations.

The following examples illustrate the stiffness method further. Information from Chapter 15 on fixed beams is used in the analysis. The table of fixed end moments in Chapter 15 (Table 15.1) may be referred to.

Example 14.7 Slope-deflection equations: propped cantilever

Analyse the propped cantilever shown in Fig. 14.26(a) and draw the SF and BM diagrams.

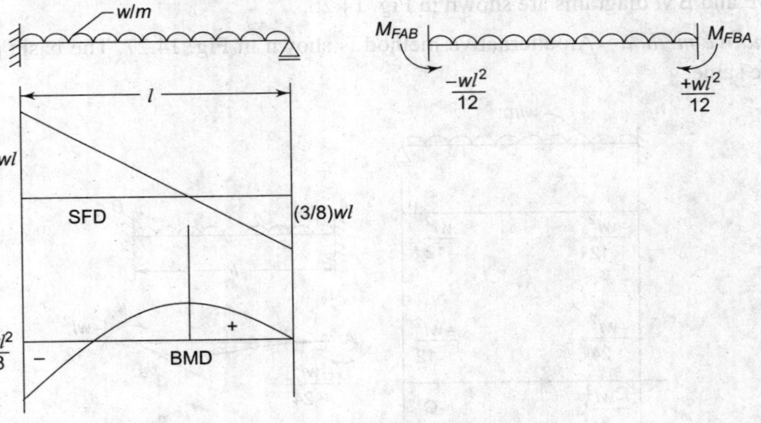

Fig. 14.26

Solution We follow the procedure outlined above. Note that the clockwise moments, slopes, and rotations of the axes are positive. The ends of the beams are locked initially; the fixed end moments due to the applied loads are $-wl^2/12$ at A and $+wl^2/12$ at B. The slope-deflection equations are

$$M_{AB} = M_{FAB} + \frac{2EI}{l}(2\theta_A + \theta_B)$$

$$= \frac{wl^2}{12} + \frac{2EI}{l}\theta_B \quad \text{(because } \theta_A = 0\text{)}$$

$$M_{BC} = M_{FBC} + \frac{2EI}{l}(\theta_A + 2\theta_B)$$

$$= \frac{wl^2}{12} + \frac{4EI}{l}\theta_B$$

The equilibrium condition is that $M_{BC} = 0$ because B is an end roller support.

$$\frac{wl^2}{12} + \frac{4EI}{l}\theta_B = 0$$

$$\theta_B = -\frac{wl^3}{48EI}$$

θ_B is negative because the slope at B is anticlockwise. Substituting this value of θ_B in the equation for M_{AB},

$$M_{AB} = -\frac{wl^2}{12} + \frac{2EI}{l}\left(-\frac{wl^3}{48EI}\right) = -\frac{wl^2}{8}$$

The moment M_{AB} is anticlockwise. The reactions at A and B can now be calculated. $\Sigma M = 0$ about B gives

$$-\frac{wl^2}{8} - \frac{wl^2}{2} + R_A = 0, \quad R_A = \frac{5}{8}wl$$

$\Sigma V = 0$ gives

$$R_B = wl - \frac{5}{8}wl = \frac{3}{8}wl$$

The SF and BM diagrams are shown in Fig. 14.26.

Alternative method An alternative method is shown in Fig. 14.27. The basic principles are the same.

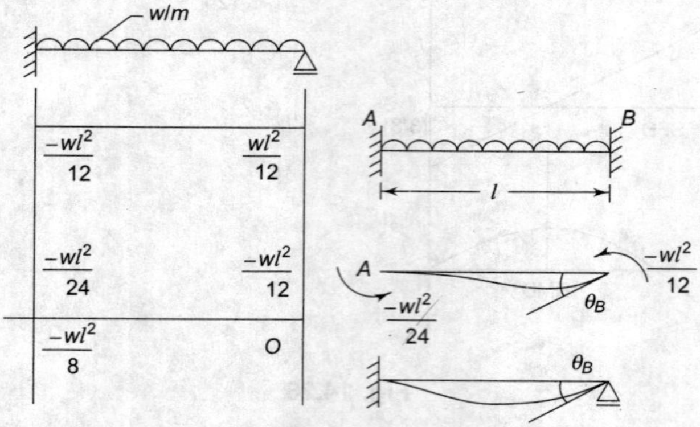

Fig. 14.27

The joints are locked initially; the fixed end moments at A and B are $-wl^2/12$ and $+wl^2/12$, respectively. These are written below the respective joints in Fig. 14.27. Joint B is not fixed and is unlocked to rotate. From the definition of flexural stiffness, if joint B is rotated by applying a moment, the moment required is $(4EI/L)\theta_B$ since the far end is fixed. Joint B was artificially locked resulting in a fixed end moment at B of $+wl^2/12$. We apply an equal moment in the opposite direction ($-wl^2/12$) to make the moment at B equal to 0. The rotation θ_B is such that

$$\frac{4EI}{l}\theta_B = -\frac{wl^2}{12}$$

where $\theta_B = -wl^3/48EI$ as before.

We have seen that when a moment is applied at the hinged end of a member, half of that moment is generated at the other end, if it is fixed. The moment added at A is

$$\frac{1}{2}\left(-\frac{wl^2}{12}\right) = -\frac{wl^2}{24}$$

If we add up the moments at A and B now,

$$M_{AB} = -\frac{wl^2}{12} - \frac{wl^2}{24} = -\frac{wl^2}{8}$$

$$M_{BA} = \frac{wl^2}{12} - \frac{wl^2}{12} = 0$$

Note that the equilibrium conditions are satisfied. Other reactions can now be calculated and the SF and BM diagrams drawn. This alternative approach is the basis of the moment distribution method, described later in Chapter 15. ■

Example 14.8 Slope-deflection equations: cantilever with spring props

The cantilever beam in Fig. 14.28 is supported by a spring of stiffness 2 kN/mm. Find the reactions and draw the SF and BM diagrams for the beam.

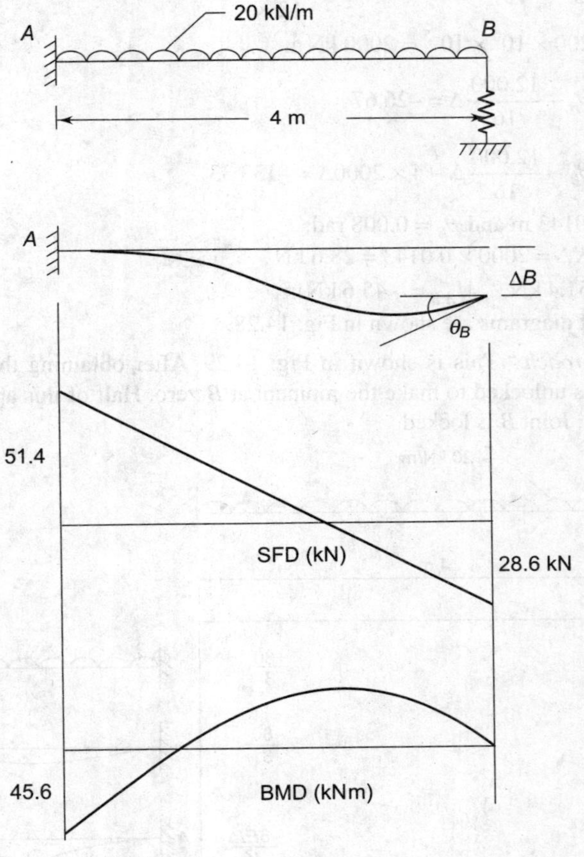

Fig. 14.28

Solution The beam has rotational freedom at A and will have a deflection equal to R_B/K, where R_B is the reaction at B and K is the stiffness of the spring. $\theta_A = 0$. Writing the slope-deflection equations,

$$M_{AB} = -\frac{wl^2}{12} + \frac{2EI}{L}\theta_B - \frac{6EI\Delta}{l^2}$$

$$= -\frac{20 \times 16}{12} + \frac{2EI}{4}\theta_B - \frac{6EI\Delta}{l^2}$$

$$M_{BA} = +\frac{20 \times 16}{12} + \frac{4EI}{4}\theta_B - \frac{6EI\Delta}{l^2}$$

We have one equilibrium condition: $M_{BA} = 0$. This gives

$$\frac{4EI}{l}\theta_B - \frac{6EI\Delta}{16} = -\frac{320}{12}$$

For a deflection Δ at B, the reaction R_B is $K\Delta = 2000\Delta$ kN. We can write an equation for M_{AB} as

$$\Sigma M = 0 \text{ @ } A, \quad -\frac{320}{12} + \frac{2EI}{4}\theta_B - \frac{6EI\Delta}{L^2} + 20 \times 4 \times 2 - K\Delta \times 4 = 0$$

$$\frac{2EI}{4}\theta_B - \frac{6EI\Delta}{L^2} - 4K\Delta = -160 + \frac{320}{12}$$

$$EI = 200 \times 10^6 \times 10^{-5} = 2000 \text{ kN m}^2$$

$$2000\theta_B - \frac{12{,}000}{16}\Delta = -26.67$$

$$1000\theta_B - \frac{12{,}000}{16}\Delta - 4 \times 2000\Delta = -133.33$$

Solving, $\Delta = 0.0143$ m and $\theta_B = 0.008$ rad;

$$R_B = K\Delta = 2000 \times 0.0143 = 28.6 \text{ kN}$$

$$R_A = 51.4 \text{ kN}, \quad M_{AB} = -45.6 \text{ kN m}$$

The SF and BM diagrams are shown in Fig. 14.28.

Alternative approach This is shown in Fig. 14.29. After obtaining the fixed end moments, joint B is unlocked to make the moment at B zero. Half of this applied moment is transferred to A. Joint B is locked.

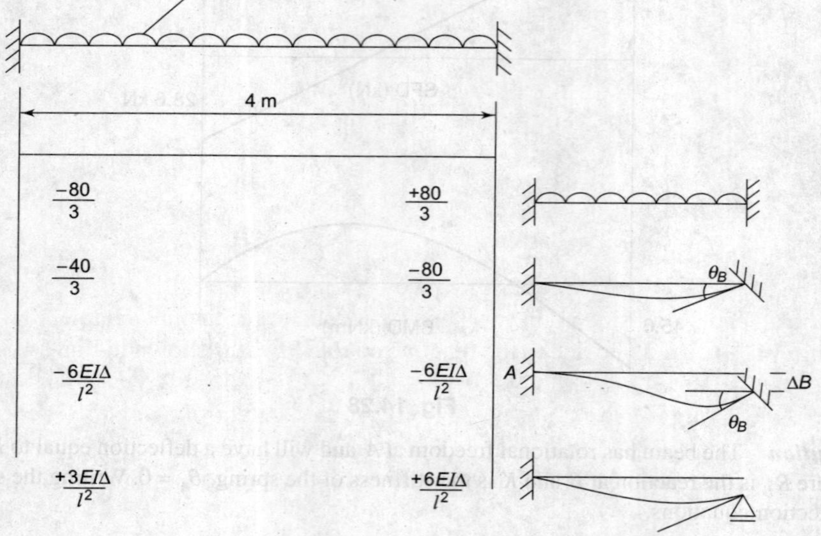

Fig. 14.29

Next, we allow end B to come down by an amount Δ. The fixed end moments are $-6EI\Delta/l^2$ at each end. Since end B has no moment finally, we unlock joint B. Unlocking is equivalent to applying a moment of $+6EI\Delta/l^2$ at B. Half of this moment is transferred to A. The final moment at A is

$$M_{AB} = -40 - \frac{6000\Delta}{16}$$

If R_B is the reaction at B, $R_B = K\Delta$. Therefore,

$$M_{AB} = -160 + 4K\Delta = -40 - \frac{6000\Delta}{16}$$

from which $\Delta = 0.0143$ m; $R_B = 2000\Delta = 28.6$ kN. Knowing R_B, other values can be calculated.

14.6.3 Iterative Techniques

We briefly outline one of the iterative techniques—the moment distribution method described by Hardy Cross. Some examples of application are given in Chapter 15 on this technique.

Preliminary concepts

Consider two cases of beams as shown in Fig. 14.30(a). Case (i) is a beam with a simple support at one end, where a couple M is applied; the far end in this case is fixed. Case (ii) is also a beam with a simple support at one end, where a couple M is applied; the far end in this case is hinged. The critical difference between the two cases is that the far end in case (i) has no rotational freedom, while the far end in case (ii) is free to rotate. The deflected shapes in the two cases are shown in Fig. 16.1(b). We analyse both these cases.

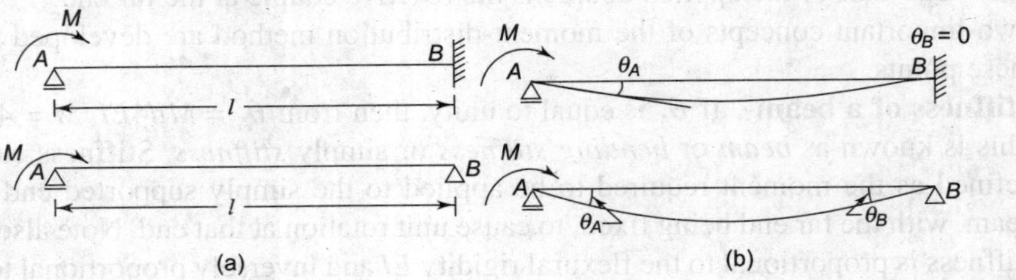

Fig. 14.30

Case (i): Far end fixed Case (i) is redrawn in Fig. 14.31(a). For the beam shown in this figure, the free body diagram is shown in Fig. 14.31(b). The BM diagram of this beam can be considered in two parts—one due to the applied couple M shown in Fig. 14.31(c) and one due to the reactive couple M_B at B shown in Fig. 14.31(d). From the deflected shape shown in Fig. 14.31, it is clear that $V_{B,A}$, the vertical distance of point A on the elastic curve from the tangent drawn at B, is zero because the tangent at B is horizontal. Therefore,

$$V_{B,A} = \text{moment of area of the } M/EI \text{ diagram between } A \text{ and } B \text{ @ } A$$

$$= \left[M\frac{l}{2}(l/3) - M_B \frac{l}{2}\frac{2l}{3} \right] \frac{1}{EI}$$

$$= 0$$

This gives $M_B = M/2$. We now calculate θ_A: $\theta_A = V_{A,B}/l$, where $V_{A,B}$ is the vertical distance of point B from the tangent drawn at A.

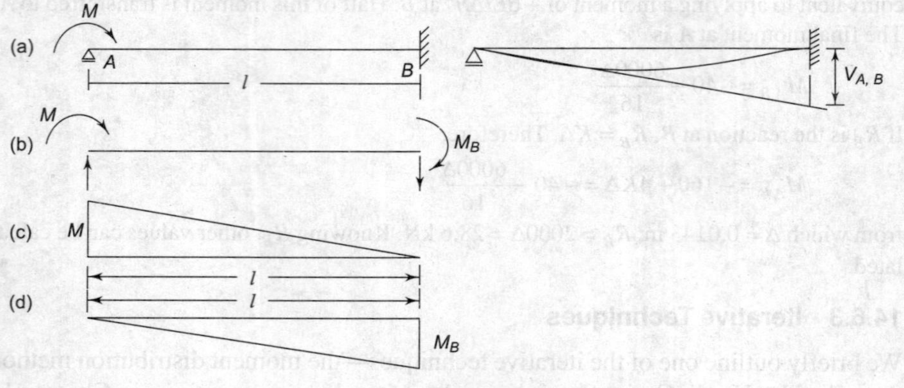

Fig. 14.31

$$V_{A,B} = \text{moment of area of the } M/EI \text{ diagram between } A \text{ and } B \text{ @ } B.$$

Therefore,

$$\theta_A = \frac{\left[M\dfrac{l}{2}\dfrac{2l}{3} - M_B\dfrac{l}{2}(l/3)\right]\dfrac{1}{EI}}{l} = \frac{Ml}{4EI}$$

Two important points should be noted here.

(i) If a couple M is applied at the freely supported end of a beam, the far end being fixed, the rotation at the end is $Ml/4EI$.

(ii) One-half of the applied couple is the reactive couple at the far end.

Two important concepts of the moment-distribution method are developed from these points.

Stiffness of a beam If θ_A is equal to unity, then from $\theta_A = Ml/4EI$, $M = 4EI/l$. This is known as *beam* or *bending stiffness* or simply *stiffness*. Stiffness can be defined as the moment required to be applied to the simply supported end of a beam, with the far end being fixed, to cause unit rotation at that end. Note also that stiffness is proportional to the flexural rigidity EI and inversely proportional to the length.

Carry-over moment From the analysis given above, if a couple M is applied at A, the reactive moment at the far end, which is fixed, is $M/2$. This is called the carry-over moment. 1/2 is known as the *carry-over factor*.

Case (ii): Far end hinged When the far end is hinged as shown in Fig. 14.32(a), the BM diagram is as shown in Fig. 14.32(b). As before, from the deflected shape shown in Fig. 16.3(c),

$$\theta_A = \frac{V_{A,B}}{l}\ \frac{(Ml/2)(2l/3)}{EIl} = \frac{Ml}{3EI}$$

The stiffness of a beam is the moment required to be applied at A for a unit rotation at A. $M = 3EI/l$ is the stiffness when the far end is hinged. Also, $3EI/l$

= (3/4)(4EI/l). Therefore, the stiffness when the far end is hinged is three-fourth the stiffness when the far end is fixed.

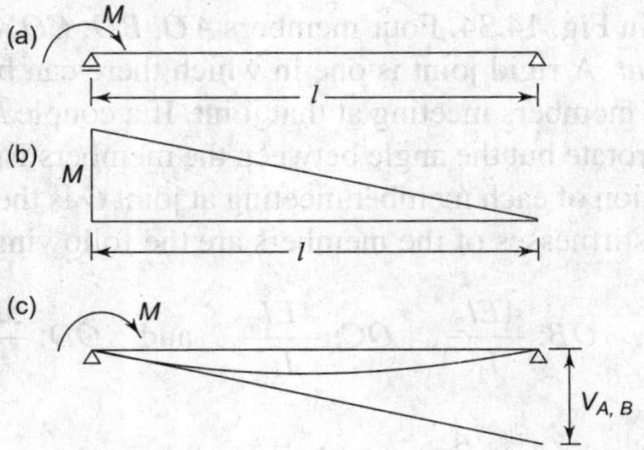

Fig. 14.32

We sum up as follows, as illustrated in Fig. 14.33.

- The bending or flexural stiffness of a member is defined as the moment to be applied to the hinged end of a beam to produce unit rotation at that end.
- Flexural stiffness is equal to 4EI/l if the far end is fixed and 3EI/l if the far end is hinged.
- The stiffness, if the far end is hinged, is three-fourth of that if the far end is fixed.
- If the far end is fixed, the carry-over moment at the fixed end is half the moment applied.
- The applied and carry-over moments have the same sign.
- The carry-over factor is +1/2. The applied moment multiplied by the carry-over factor gives the moment carried over to the far end fixed support.

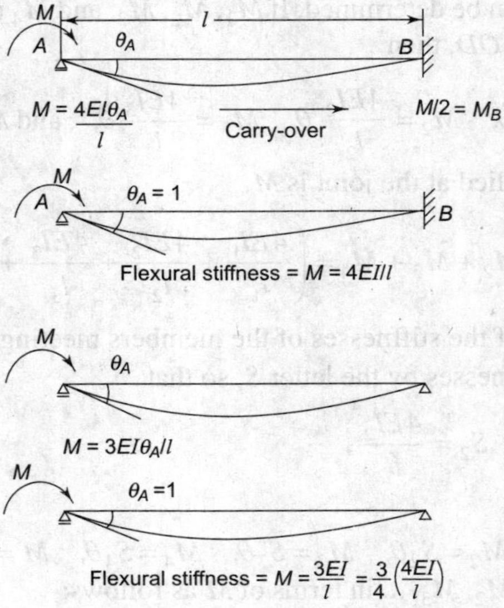

Fig. 14.33

Basic concepts of moment distribution

The basic concepts of moment distribution can be explained with reference to the framework shown in Fig. 14.34. Four members AO, BO, CO, and DO meet at O, which is a *rigid joint*. A rigid joint is one in which there can be no change in the angles between the members meeting at that joint. If a couple M is applied to this joint, the joint will rotate but the angle between the members will not change. This means that the rotation of each member meeting at joint O is the same, as shown in Fig. 14.34(b). The stiffnesses of the members are the following:

$$OA: \frac{4EI_1}{l_1}, \quad OB: \frac{4EI_2}{l_2}, \quad OC: \frac{4EI_3}{l_3}, \quad \text{and} \quad OD: \frac{3EI_4}{l_4} = \frac{3}{4}\left(\frac{4EI_4}{l_4}\right)$$

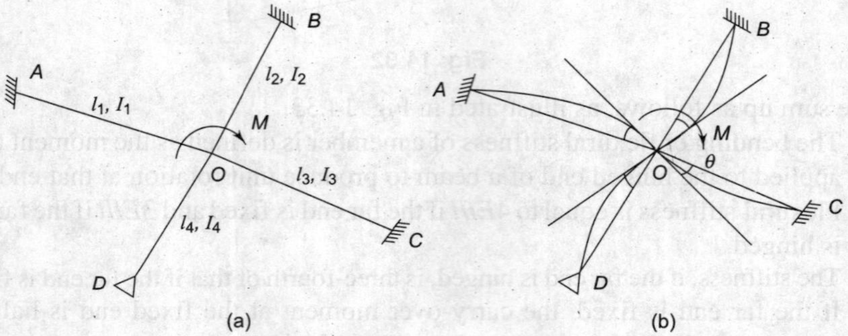

Fig. 14.34

Let us assume that the joint rotates by θ due to the couple M. Knowing the stiffness and rotation, the moment required to be applied to each member at O to cause this rotation can be determined. If M_1, M_2, M_3, and M_4 are moments applied to OA, OB, OC, and OD, then

$$M_1 = \frac{4EI_1}{l_1}\theta, \quad M_2 = \frac{4EI_2}{l_2}\theta, \quad M_3 = \frac{4EI_3}{l_3}\theta, \quad \text{and } M_4 = \frac{3EI_4}{l_4}\theta$$

Since the couple applied at the joint is M,

$$M = M_1 + M_2 + M_3 + M_4 = \left(\frac{4EI_1}{l_1} + \frac{4EI_2}{l_2} + \frac{4EI_3}{l_3} + \frac{3EI_4}{l_4}\right)\theta$$

$$= (\text{sum of the stiffnesses of the members meeting at } O) \times \theta$$

If we denote the stiffnesses by the letter S, so that

$$S_1 = \frac{4EI_1}{l_1}, \quad S_2 = \frac{4EI_2}{l_2}, \cdots$$

then

$$M_1 = S_1\theta, \quad M_2 = S_2\theta, \quad M_3 = S_3\theta, \quad M_4 = S_4\theta, \quad M = (S_1 + S_2 + S_3 + S_4)\theta$$

We can express M_1, M_2, M_3, ... in terms of M as follows:

$$\theta = \frac{M}{S_1 + S_2 + S_3 + S_4}$$

$$M_1 = \frac{S_1 M}{S_1 + S_2 + S_3 + S_4}, \quad M_2 = \frac{S_2 M}{S_1 + S_2 + S_3 + S_4}$$

$$M_3 = \frac{S_3 M}{S_1 + S_2 + S_3 + S_4}, \quad M_4 = \frac{S_4 M}{S_1 + S_2 + S_3 + S_4}$$

Distribution factor In the equation $M_1 = S_1 M/(S_1 + S_2 + S_3 + S_4)$, the term $S_1/(S_1 + S_2 + S_3 + S_4)$ is known as the distribution factor. The distribution factor is the factor by which the applied moment is multiplied to get the moment in the member. In the case discussed above, the total moment M is distributed amongst the four members in proportion to their stiffnesses.

Relative stiffness Since the distribution factor (DF) is a ratio of stiffnesses, there is no need to calculate the absolute stiffnesses of members. For example, in the case just described,

$$\text{DF for } OA = \frac{4EI_1/l_1}{4EI_1/l_1 + 4EI_2/l_2 + 4EI_3/l_3 + (3/4)(4EI_4/l_4)}$$

Any common factor in all the terms can be eliminated. For example, if all the members have the same material property E, the term $4E$ can be eliminated as a common factor and the above equation becomes

$$\text{DF for } OA = \frac{I_1/l_1}{I_1/l_1 + I_2/l_2 + I_3/l_3 + 3I_4/4l_4}$$

The stiffness for the purpose of calculating the DF can be taken as I/l. The term I/l is known as relative stiffness. The value of I need not be taken again as the absolute value, but can be a relative value. We sum up as follows.

- At a rigid joint, any moment applied is distributed among the members meeting at that joint in proportion to their stiffnesses.
- The absolute values of stiffnesses are not required except in certain cases. The stiffness calculated as I/l for each member is known as relative stiffness.
- The distribution factor (DF) for a member at any joint is the relative stiffness of the member divided by the sum of the relative stiffnesses of all the members meeting at that joint.
- The moment distributed to any member is DF $\times M$.
- The sum of the DFs for members meeting at a joint is equal to 1.

Example 14.9 Stiffness of members and carry-over factors at a joint

A moment of 200 kNm is applied to joint O shown in Fig. 14.35. Determine the proportion of this moment distributed to each member. What will be the moments at the far ends A, B, C, and D? If $I = 1 \times 10^{-4}$ m^4 and $E = 2 \times 10^{11}$ N/m^2, what is the rotation of the joint. Draw the deflected shape of the joint.

Solution The relative stiffnesses of the members are calculated as (MI of the member)/length. Note that the values of the moment of inertia itself are given in relative terms only. The relative stiffnesses are the following:

$$OA: S_1 = \frac{3I}{6} = \frac{I}{2}, \quad OB: S_2 = \frac{I}{4}, \quad OC: S_3 = \frac{3}{4}\left(\frac{2I}{5}\right)$$

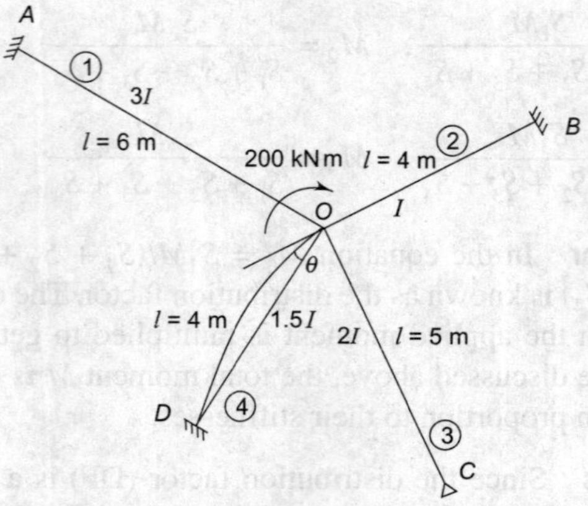

Fig. 14.35

The far end C being hinged,

$$OD: S_4 = \frac{1.5I}{4}$$

$$\text{Sum of stiffnesses} = \frac{I}{2} + \frac{I}{4} + \frac{6I}{20} + \frac{1.5I}{4} = \frac{85.5}{60} I$$

The member end moments are worked out as shown in Table 14.1.

Table 14.1

Member	***Relative stiffness***	***DF***	***Moment (kN m)***	***Moment at far end***
OA	$\frac{I}{2}$	$\frac{I}{2} \div \frac{85.5I}{60} = 0.3509$	70.18	35.09 at *A*
OB	$\frac{I}{4}$	0.1754	35.08	17.54 at *B*
OC	$\frac{6I}{20}$	0.2105	42.10	0 at *C*
OD	$\frac{1.5I}{4}$	0.2632	52.64	26.32 at *D*
Total	1.0000		200.00	

The rotation with the far end fixed,

$$\theta = \frac{Ml}{4EI}$$

$$= \frac{70.18 \times 6}{4 \times 2 \times 10^8 \times 3 \times 1 \times 10^{-4}} = 1.754 \times 10^{-3} \text{ rad}$$

We illustrate the method with examples in Chapter 15 where we discuss the analysis of continuous beams using the moment distribution method.

14.6.4 Matrix Methods of Analysis

Matrix methods of structural analysis use the matrix formulation of equations of compatibility for flexibility method and equations of equilibrium for stiffness method. Matrices are short-hand methods of writing equations.

For example, taking the case shown in Fig. 14.36, the equations as per the flexibility method can be written as

$$f_{11}R_B + f_{12}M_A + D_{L1} = 0, \quad f_{21}R_B + f_{22}M_A + D_{L2} = 0$$

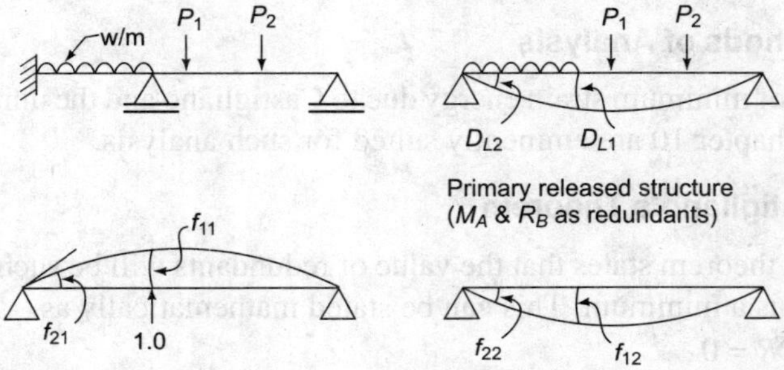

Fig. 14.36 Two-span continuous beam

Here D_{L1} is the total deflection at B due to the applied loads and D_{L2} is the total slope at A due to the applied loads. This can be written in matrix notation as

$$[F]\,[R] = [-D]$$

where

$$[F] = \begin{vmatrix} f_{11} & f_{12} \\ f_{21} & f_{22} \end{vmatrix}, \; [R] = \begin{vmatrix} R_B \\ M_A \end{vmatrix}, \text{ and } [D] = \begin{vmatrix} -D_{L1} \\ -D_{L2} \end{vmatrix}$$

This can be solved by writing

$$[R] = [F]^{-1}\,[-D]$$

where $[F]^{-1}$ is the inverse of the flexibility matrix.

Such methods are suitable for computer-based analysis as the inversion of [*F*] matrix is a very tedious job to be done by hand. Inverting a $[2 \times 2]$ matrix can be done easily. A matrix of order $[3 \times 3]$ is very tedious to invert by hand methods and is error prone. Higher-order matrices are very difficult, if not impossible, to invert by hand calculation.

Examples to illustrate the procedure for the flexibility and stiffness methods of analysis using matrices are given in Chapter 15 for continuous beams.

14.7 ANALYSIS OF INDETERMINATE TRUSSES

We have studied the analysis of determinate trusses in Chapter 13. When a truss has more number of members than required for internal stability, we have an internally indeterminate truss. The truss may also be externally indeterminate if the number of reaction components is more than the number of equations of static equilibrium. We will use the strain energy methods to analyse such trusses. Both the flexibility and stiffness methods are equally suitable for the analysis of indeterminate trusses. Figure 14.37 shows both internally and externally indeterminate trusses. The tensile force in a member will be taken positive.

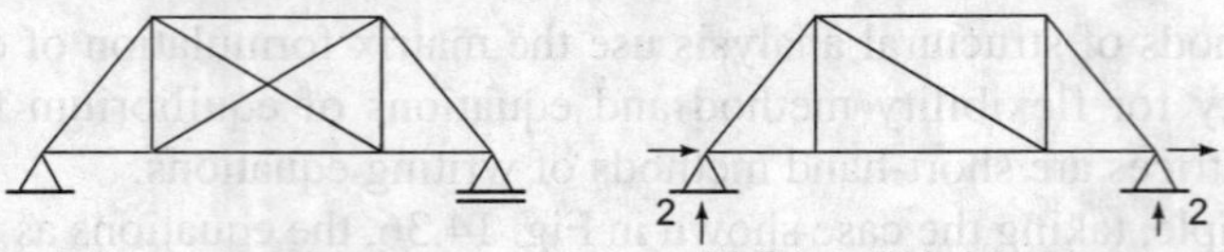

(a) Internally indeterminate (b) Externally indeterminate

Fig. 14.37 Indeterminate trusses

14.7.1 Methods of Analysis

The theorem of minimum strain energy due to Castigliano and the unit load method outlined in Chapter 10 are eminently suited for such analysis.

14.7.2 Castigliano's Theorem

Castigliano's theorem states that the value of redundants will be such that the total strain energy is a minimum. This can be stated mathematically as

$$\delta U / \delta R = 0$$

where R is a redundant. The procedure involves the following steps. Refer to the truss shown in Fig. 14.38.

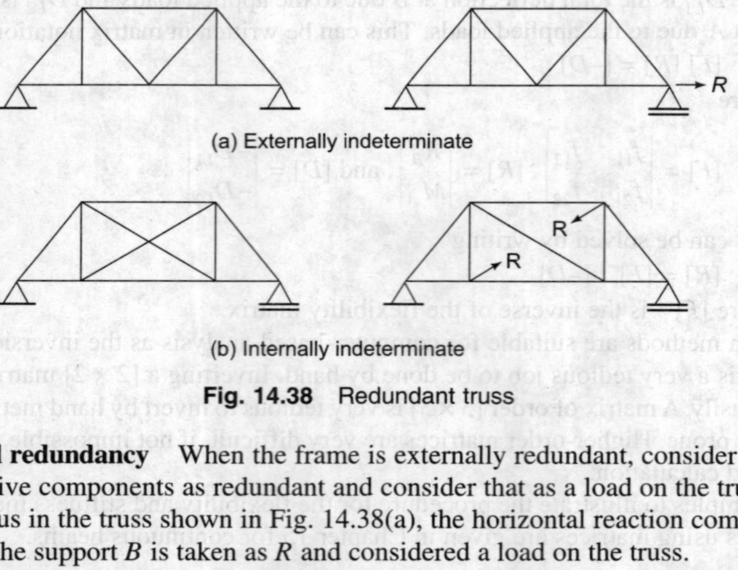

(a) Externally indeterminate

(b) Internally indeterminate

Fig. 14.38 Redundant truss

External redundancy When the frame is externally redundant, consider one of the reactive components as redundant and consider that as a load on the truss.

(i) Thus in the truss shown in Fig. 14.38(a), the horizontal reaction component at the support B is taken as R and considered a load on the truss.

(ii) Analyse the truss and determine the forces in the members. These will be in terms of the applied loads and the redundant R.

(iii) Each member of the truss is in tension or compression. The strain energy of a bar subjected to axial load is $P^2L/2AE$, where P is the force in the bar, L is its length, A is the area of cross section, and E is the Young's modulus. The total strain energy stored in the bar is $U = \Sigma P^2L/2AE$.

(iv) The expression for the strain energy contains R as an unknown. Now, as per the theorem of minimum strain energy (also sometimes called the theorem of least work) $\delta U / \delta R = 0$.

(v) If we differentiate within the integral sign, we have $\delta U/\delta R = \Sigma P(\delta P/\delta R)/AE$. Please note that P is a force in a member due to the applied loads and R.
(vi) The equation $\delta U/\delta R = 0$ gives an equation from which R can be evaluated. Once R is known, all other reaction components and forces in the members can be determined.

Internal redundancy In case the truss is internally redundant, the number of members will be more than $2j - 3$, where j is the number of joints. We can select one or more members as redundants. Remove such members and replace them with the force in such members.

(i) Select suitable members as redundants and replace them with forces R_1, R_2, etc. The forces R_1, R_2, etc. will act at both joints connected by the member and will be both towards the joints (compressive) or both away from the joints (tension).
(ii) Analyse the truss to find the member forces due to the applied loads and the redundant loads.
(iii) Determine the partial derivative of the member forces ($\delta F/\delta R$) with respect to the redundants. If there are two redundants, there will be two sets of derivatives.
(iv) We have to find the sum $\Sigma[F(\delta F/\delta R)L/AE]$. If A and E are the same for all the members, they can be taken out of the summation sign.
(v) Equate this sum to zero and find the unknown redundant.
(vi) Substituting the value of the redundant, all the member forces can be found. The following examples illustrate the procedure

Example 14.10 Castigliano's second theorem: indeterminate truss

The truss loaded shown in Fig. 14.39 is supported on two hinge supports and loaded as shown. Determine the reactions.

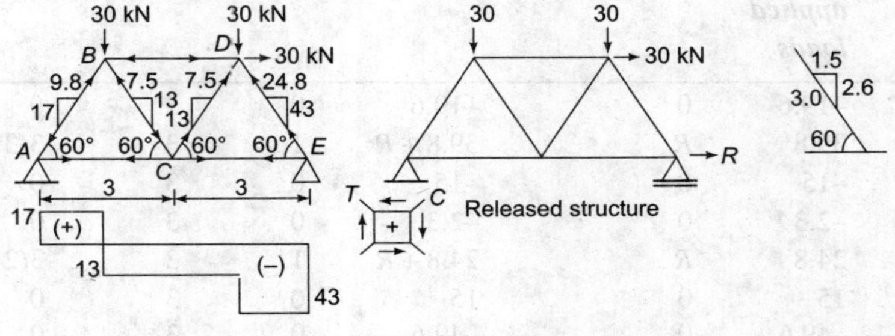

Fig. 14.39

Solution The truss is internally determinate but externally indeterminate. There are four reaction components but we have only three equations of equilibrium. We remove the horizontal reaction at E by making the hinge support at E as a roller support. The redundant R is horizontal reaction at E, R_{EH}.

The truss is analysed now for the applied loads to find the forces F in all the members. The truss is also analysed for the forces in the members due to the redundant R. The results are given in Table 14.2.

As the truss has parallel top and bottom chords, we use the SF diagram for finding the nature and magnitude of the member forces. The reactions can be found as follows.

From $\Sigma M = 0$ @ A, $R_{VE} = (30 \times 4.5 + 30 \times 1.5 + 30 \times 2.6)/6 = 43$ kN ↑

(The height of the truss is $1.5 \times \tan 60° = 1.5\sqrt{3} = 2.6$ metres.)

From $\Sigma V = 0$, $R_{AV} = 60 - 43 = 17$ kN ↑

From $\Sigma H = 0$, $R_{AH} = 30$ kN ←

The shear force diagram can now be drawn as shown. The nature of the forces in the diagonal members is found as the tension or compression from the sign of shear force in the panel in which the member lies. For positive SF in the panel, the tension and compression will be oriented as shown. For negative SF, it will be the reverse. The nature of forces in the members will be as shown in the figure.

The vertical component of the forces in the members is equal to the SF in the panel. The horizontal component can be found from the member projections of its length in the horizontal and vertical directions. As the diagonal members are all inclined at 60° to the horizontal, $v = 2.6$, $h = 1.5$, and $l = 3$ m. For example, in member AB, the vertical component is 17 kN. Also

Horizontal component = $17 \times 1.5/2.6 = 9.8$

Member force = $9.8 \times 3/1.5$ or $17 \times 3/2.6 = 19.6$ kN

Once the nature and magnitude of the components in the diagonal members are known, the remaining member forces can be found by inspection.

Joint A Force in $AC = 30 + 9.8 = 39.8$ kN (tensile)

Joint B Force in $BD = 9.8 - 7.5 = 2.3$ kN (compressive)

Joint C Force in $CE = 39.8 - 7.5 - 7.5 = 24.8$ kN (tensile)

All the member force components are shown in the figure and the member forces are listed in Table 14.2.

Table 14.2

Member	*Force, F due to applied loads*	*Force due to R*	*Total*	$\delta F/\delta R$	*Length*	$F(\delta F/\delta R)L$
AB	–19.6	0	–19.6	0	3	0
AC	39.8	*R*	39.8 + *R*	1	3	3(39.8 + *R*)
BC	–15	0	–15	0	3	0
BD	–2.3	0	–2.3	0	3	0
CE	24.8	*R*	24.8 + *R*	1	3	3(24.8 + *R*)
CD	15	0	15	0	3	0
DE	–49.6	0	–49.6	0	3	0

The sum of the last column in the table = $3(64.64 + R)$

The value of R is such that $[F(\delta F/\delta R)L/AE] = 0$.

$3(64.64 + 2R) = 0$; $R = -32.3$ kN (←)

(The negative sign means that R is in a direction opposite to that assumed.)

Knowing the redundant reaction, $R_{AH} = 2.3$ kN →

Only two members are affected by the values of R, AC, and CE.

Force in $AC = 39.8 - 32.3 = +7.5$ kN (tensile)

Force in $CE = 24.8 - 32.3 = -7.5$ kN (compressive) ■

Example 14.11 Castigliano's second theorem: indeterminate truss

Analyse the frame shown in Fig. 14.40 and find the forces in all the members. All the members have the same area of cross section and made of the same material.

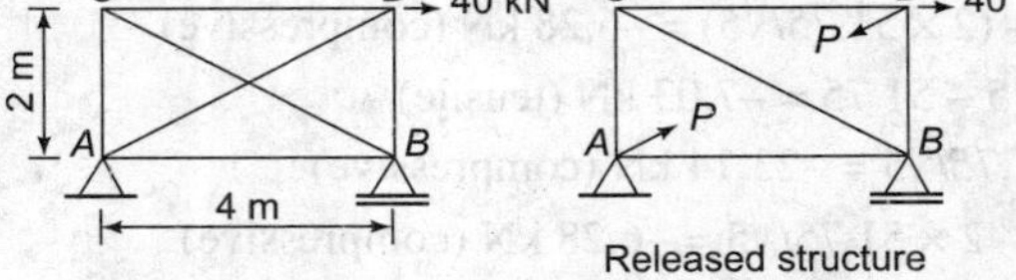

Fig. 14.40

Solution The truss is externally determinate and the number of members = 6, which is one more than 2 × 4 – 3 = 5. The degree of internal redundancy is one.

We remove member AD. Let the force in the member $AD = P$. This force is applied at the joints A and D as shown and considered as load at the joints.

The truss can now be analysed and the forces in the members found as it is a perfect frame. The calculations can be tabulated as shown in Table 14.3.

Table 14.3

Member	*Force F*	$\delta F/\delta P$	*Length (m)*	$F(\delta F/\delta P)L$
AC	$20 + \dfrac{P}{\sqrt{5}}$	$1/\sqrt{5}$	2	$40 + \dfrac{2P}{5}$
AB	$40 + \dfrac{2P}{\sqrt{5}}$	$2/\sqrt{5}$	4	$\dfrac{320}{\sqrt{5}} + \dfrac{16P}{5}$
BC	$20\sqrt{5} + P$	1	$\sqrt{20}$	$200 + P\sqrt{20}$
BD	$P/\sqrt{5}$	$1/\sqrt{5}$	2	$2P/5$
CD	$40 + \dfrac{2P}{\sqrt{5}}$	$2/\sqrt{5}$	4	$\dfrac{320}{\sqrt{5}} + \dfrac{16P}{5}$

Reactions:

Taking moments about A, $\Sigma M = 0$ gives $R_{BV} \times 4 = 40 \times 2$; $R_{BV} = 20$ kN ↑

From $\Sigma V = 0$, $R_{AV} = -20$ kN ↓

From $\Sigma H = 0$, $R_{AH} = 40$ kN ←

Member forces:

Joint A $\Sigma V = 0$ gives $F_{AC} - P/\sqrt{5} - 20 = 0$; $F_{AC} = 20 + P/\sqrt{5}$

$\Sigma H = 0$ gives $F_{AB} - 40 - P(2/\sqrt{5}) = 0$; $F_{AB} = 40 + 2P/\sqrt{5}$

Joint B $\Sigma H = 0$; $F_{BC}(2/\sqrt{5}) - 40 - 2P/\sqrt{5} = 0$; $F_{BC} = 20\sqrt{5} + P$

$\Sigma V = 0$ gives $(20\sqrt{5} + P)/\sqrt{5} - 20 - F_{BD} = 0$; $F_{BD} = P/\sqrt{5}$

Joint C $\Sigma H = 0$ gives, $-(20\sqrt{5} + P)2/\sqrt{5} + F_{CD} = 0$; $F_{CD} = 40 + 2P/\sqrt{5}$

These member forces are listed in column 2 of Table 14.3. The partial derivatives of the forces with respect to P are listed in column 3. The lengths of the members are given in column 4. The last column in the table lists the value $F(\delta F/\delta P)L$.

$\Sigma F(\delta F/\delta P)L$ = Sum of the last column = $604 + 11.67P = 0$; $P = -51.75$

The force in the member AD comes out negative. This means that the member is in tension and not in compression as assumed. The member forces can be calculated with the known value of P.

$$F_{AC} = 20 + (-51.75/\sqrt{5}) = -3.14 \text{ kN (compressive)}$$
$$F_{AB} = 40 - (2 \times 51.75/\sqrt{5}) = -6.28 \text{ kN (compressive)}$$
$$F_{BC} = 20\sqrt{5} - 51.75 = -7.03 \text{ kN (tensile)}$$
$$F_{BD} = -51.75/\sqrt{5} = -23.14 \text{ kN (compressive)}$$
$$F_{CD} = 40 - 2 \times 51.75/\sqrt{5} = -6.28 \text{ kN (compressive)}$$

Example 14.12 Castigliano's second theorem: indeterminate truss

Analyse the frame shown in Fig. 14.41 and find the forces in all the members.

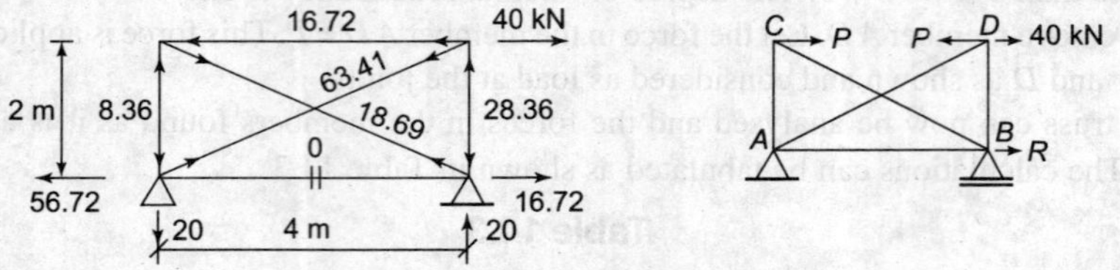

Fig. 14.41

Solution The frame is both externally and internally indeterminate. As the number of reaction components is four, the truss is externally indeterminate to the first degree. As the number of members is 6, one more than $(2 \times 4 - 3 = 5)$, the truss is internally indeterminate to the first degree. We consider the horizontal reaction component at B $(= R)$ and the member force in CD $(= P)$ as unknown loads. The truss can now be analysed to find the member forces in terms of P and R. These are tabulated in Table 14.4.

Table 14.4

Member	*Force*	$\delta F/\delta P$	$\delta F/\delta R$	*Length*	$F(\delta F/\delta P)L$	$F(\delta F/\delta R)L$
AC	$P/2$	1/2	0	2	$P/2$	0
AB	$P + R$	1	1	4	$4(P + R)$	$4(P + R)$
AD	$20\sqrt{5} - P\sqrt{5}/2$	$-\sqrt{5}/2$	0	$\sqrt{20}$	$-100\sqrt{5} + 5P\sqrt{5}/2$	0
BC	$P\sqrt{5}/2$	$\sqrt{5}/2$	0	$\sqrt{20}$	$P5\sqrt{5}/2$	0
BD	$20 - P/2$	$-1/2$	0	2	$-20 + P/2$	0

We have $\Sigma F(\delta F/\delta R)L = 0$; $4(P + R) = 0$; $P = -R$

From $\Sigma F(\delta F/\delta P)L = 203.61 + 12.18P = 0$; $P = -16.72$ kN; $R = -P = 16.72$ kN

From the signs, the assumed direction of P has to be reversed. R is in the same direction as assumed.

The member forces can now be worked out as

AC: $-16.72/2 = -8.36$ kN (compressive); $AB = 0$ and $AD = 20\sqrt{5} + 16.72\sqrt{5}/2 = 63.41$ kN (tensile)

BC: -18.69 (tensile); $BD = 20 + 8.36 = 28.36$ kN (compressive)

The final member forces are shown in the figure.

Example 14.13 Castigliano's second theorem: indeterminate truss

Analyse the frame shown in Fig. 14.42 and find the forces in all the members.

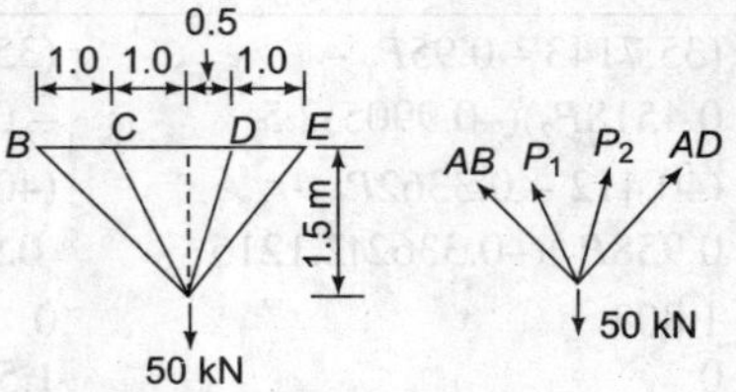

Fig. 14.42

Solution All the forces intersect at a point. There are four unknown bar forces and they cannot be determined from the two equations of equilibrium ($\Sigma H = 0$ and $\Sigma V = 0$). We take the forces in the bars AC (= P_1) and AD (= P_2) as unknowns and consider them as loads. The joint is now acted upon by the forces shown. The members and their inclinations to the horizontal are shown in Table 14.5.

Table 14.5

Member	*tan θ*	*Θ*	*sin θ*	*cos θ*
AB	1.5/2	36.87°	0.6	0.8
AC	1.5/1	56.31°	0.832	0.5547
AD	1.5/0.5	71.56°	0.9487	0.3162
AE	1.5/1.5	45°	0.707	0.707

We now write the two equations of equilibrium. We use the symbols AB for the member force in AB, AC for the member force in AC, and so on.

$$\Sigma H = 0 \rightarrow +; -0.8AB - 0.5547P_1 + 0.3162P_2 + 0.707AE = 0$$

This can be reduced to

$$0.8AB - 0.707AE = -0.5547P_1 + 0.3162P_2 \quad \text{(i)}$$

$$\Sigma V = 0 \uparrow +; -50 + 0.6AB + 0.832P_1 + 0.9487P_2 + 0.707AE = 0$$

$$0.6AB + 0.707AE = 50 - 0.832P_1 - 0.9487P_2 \quad \text{(ii)}$$

Solving Equations (i) and (ii), we get

$$AB = 35.7143 - 0.9905P_1 - 0.4518P_2$$

$$AE = 40.412 - 0.3362P_1 - 0.958P_2$$

$$\text{Length of } AB = \sqrt{(2^2 + 1.5^2)} = 2.5 \text{ m}$$

$$\text{Length of } AE = \sqrt{(1.5^2 + 1.5^2)} = 2.12 \text{ m}$$

We now tabulate these values in Tables 14.6 and 14.7.

Table 14.6

Member	*Force*	$\delta F/\delta P_1$	$\delta F/\delta P_2$
AB	$35.7143 - 0.9905P_1 - 0.4518P_2$	−0.9905	−0.4518
AE	$40.412 - 0.3362P_1 - 0.958P_2$	−0.3362	−0.958
AC	P_1	1	0
AD	P_2	0	1

Table 14.7

Member	*Length*	$F(\delta F/\delta P_1)L$	$F(\delta F/\delta P_2)L$
AB	2.5	$(35.7143 - 0.95P_1 - 0.4518P_2)(-0.9905)2.5$	$(35.7143 - 0.9905P_1 - 0.4518P_2)(-0.4518)2.5$
AE	2.1215	$(40.412 - 0.3362P_1 - 0.958P_2)(-0.3362)2.1215$	$(40.412 - 0.3362P_1 - 0.958P_2)(-0.958)2.1215$
AC	1.803	1.803	0
AD	1.581	0	1.581

The two equations are obtained by $\Sigma[F\delta F/\delta P_1 L] = 0$ and $\Sigma[F\delta F/\delta P_2 L] = 0$

$$4.4955P_1 + 1.8023P_2 = 117.2613 \quad \text{(i)}$$

$$1.7723P_1 + 4.04P_2 = 122.51 \quad \text{(ii)}$$

Solving these two equations, we get

$P_1 = 16.91$ and $P_2 = 22.91$

Substituting these values, we get

Force in the member $AB = 8.61$ kN

Force in the member $AE = 12.78$ kN

Force in the member $AC = 16.91$ kN

Force in the member $AD = 22.91$ kN ■

14.7.3 Unit Load Method

The unit load method, also called the dummy load method, is a very powerful tool to find deflections in structures as was illustrated in Chapter 10. It is derived from the strain energy concepts. The method can be used for the analysis of indeterminate trusses as follows.

Externally indeterminate

When the truss is externally indeterminate as shown in Fig. 14.43(a), we consider one of the redundant reactive components as a load on the structure. In the specific case shown in Fig. 14.43, the reactive component R_{BH} is considered the redundant. We adopt the following procedure.

(i) As the truss becomes statically determinate with the removal of the redundant reactive component, we analyse the truss with the applied loads and find the forces in the members. Let F_i be the forces in the members m_i.

(ii) We apply a unit load in the direction of the redundant reaction to the truss. Find the member forces due to the application of this load. Let u_i be the forces in the members.

(iii) As we have seen in Chapter 13, the deflection in the direction of the redundant R, due to the applied loads is given by

$$\Delta_L = \Sigma[F_i u_i L/AE]$$

(iv) Due to the application of the redundant reaction R, the forces in the members will be Ru_i. The deflection Δ_R in the direction of R due to the application of the redundant R is given by

$$\Delta_R = \Sigma(Ru_i)u_i L/AE$$

(v) As the final deflection Δ in the direction of R is zero, we have $\Delta_L + \Delta_R = 0$.
or $-\Sigma F_i u_i L/AE = \Sigma R(u_i)^2 L/AE = R\Sigma(u_i)^2 L/AE$
From the above equation, we get

$$R = \frac{-\Sigma(F_i u_i)L/AE}{\Sigma(u_i)^2 L/AE} = \frac{-\Sigma FuL/AE}{\Sigma u^2 L/AE} \quad \text{(leaving the subscripts)}$$

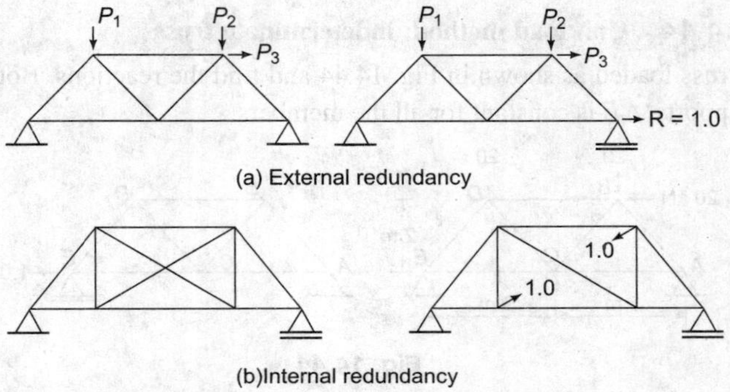

Fig. 14.43 Indeterminate trusses

Internal redundancy

Internal redundancy is dealt with in the same way. When the number of members is more than that required for stability, we have member redundancy. The following steps are involved.

(i) Remove the redundant force. The frame becomes statically determinate perfect frame. Find the forces F_i in the members of the frame due to the applied loads.

(ii) Apply unit loads in the direction of the redundant member as shown in Fig. 14.43(b). Let the forces in the members be u_i.

(iii) The displacement in the direction of the redundant member $\Delta_L = \Sigma F_i u_i L/AE$.

(iv) If R is the force in the redundant member, then the forces in the members due to the forces R as shown will be Ru_i. Again the displacement due to the forces R, Δ_R, in their direction is $\Sigma(Ru_i)u_i L/AE = R\Sigma(u_i)^2 L/AE$.

(v) We can write

Increase in distance Δ_L – decrease in distance $\Delta_R = RL_R/A_R E$

From this, we get

$$R = \frac{\Delta_L}{\Delta_R + (L_R / A_R E)}$$

(vi) Another way to visualize this concept is to assume that we cut the redundant member R into two parts. Assuming the member to be in tension, this will cause a displacement of the ends of the member by $\Sigma FuL/AE$, where F are the forces in the members due to the applied loads (without the redundant member), u are the forces in the members due to the unit loads applied as shown along the cut. If R is the force in the redundant member, then the displacement along R due to $R = \Sigma(Ru)uL/AE = R\Sigma u^2 L/AE$.

The magnitude of R must be such that this deflection is equal to the separation due to the forces F. This gives

$$\Sigma FuL/AE = R\Sigma u^2L/AE;\ R = \Sigma[FuL/AE]/\Sigma[u^2L/AE]$$

It must be kept in mind that in both the cases, $\Sigma FuL/AE$ is computed without the redundant member and the term $\Sigma u^2L/AE$ includes the force 1.0 in the member.

The following examples illustrate the procedure.

Example 14.14 Unit load method: indeterminate truss

Analyse the truss loaded as shown in Fig. 14.44 and find the reactions. Both the supports are hinged supports. AE is constant for all the members.

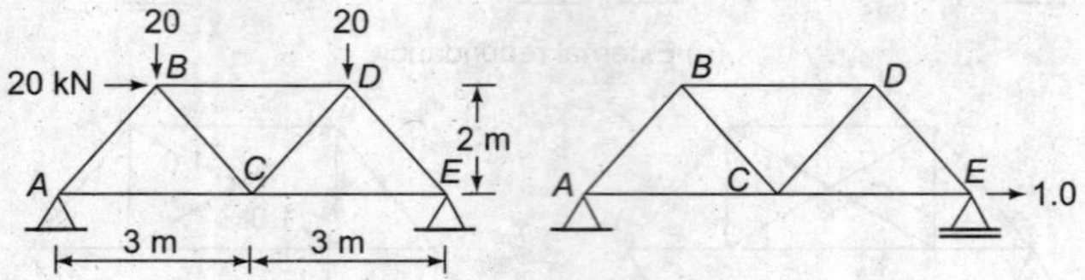

Fig. 14.44

Solution The horizontal reaction at E is taken as the redundant reactive component. We remove this component to make the truss statically determinate. We analyse the truss with the applied loads and with a unit load at E in the positive horizontal direction. The results are given in Table 14.8.

Table 14.8

Member	*Force due to the applied loads (F)*	*Force due to the unit load (u)*	*Length*	*FuL*	u^2L	*Final member forces, kN*
AB	–21.21	0	2.83	0	0	–21.21
AC	+35	1.0	4	140	4	+5
BC	–7.07	0	2.83	0	0	–7.07
BD	–30	0	4	0	0	–30
CD	7.07	0	2.83	0	0	+7.07
CE	25	1.0	4	100	4	–5
DE	–35.35	0	2.83	0	0	–35.35

(The positive sign shows tensile forces.)

Removing the redundant horizontal reaction at E, the reactions are as follows.

$$R_{BV} = (20 \times 2 + 20 \times 6 + 20 \times 2)/8 = 25 \text{ kN} \uparrow$$

$\Sigma V = 0$ gives $R_{AV} = 15$ kN $\uparrow$

$\Sigma H = 0$ gives $R_{AH} = 20$ kN $\leftarrow$

We now analyse the truss by the method of joints. The inclined members make angles of 45° with the horizontal.

Joint A $AB \sin 45° = 15;\ AB = 15/\sin 45° = -21.21$ kN (compressive)

$AC = 20 + 21.21 \cos 45° = 35$ kN (tensile)

Joint B Equilibrium in the vertical direction gives $20 - AB \sin 45° = BC \sin 45° = -7.07$ kN

$\Sigma H = 0$ gives $BD = 20 + 21.21 \cos 45° - 7.07 \cos 45° = -30$ kN

Joint C $\Sigma V = 0$ gives $CD = 7.07$ kN (tensile)

$\Sigma H = 0$ gives $35 - 7.07 \cos 45° - 7.07 \cos 45° = CE$; $CE = 25$ kN (tensile)

Joint D $\Sigma V = 0$ gives $DE \sin 45° = 20 + 7.07 \sin 45° = 35.36$ kN (compressive)

It can be verified that joints *D* and *E* are balanced.

$\Sigma FuL = 240$; $\Sigma u^2L = 8$

The redundant reaction is given by

$R = -\Sigma[FuL]/\Sigma u^2L = -200/8 = 25$ kN

The final member forces are $F + uR$ and listed in the last column.

Example 14.15 Unit load method: indeterminate truss

Analyse the truss shown in Fig. 14.45 and find the forces in all the members. *EA* is constant.

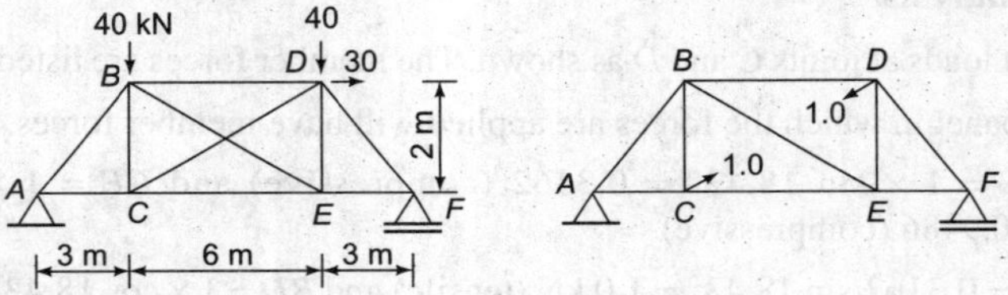

Fig. 14.45

Solution We remove the member *CD* to make the truss statically determinate. We analyse the truss for the applied loads and the application of unit loads at *C* and *D* as shown. The results are given in Table 14.9.

Table 14.9

Member	*F due to the applied loads*	*u due to the unit loads at C and D*	*Length*	*FuL*	*u^2L*	*Final member forces (kN)*
AB	–63.11	0	3.6	0	0	–63.11
AC	82.5	0	3	0	0	82.5
BC	0	–0.3162	2	0	0.2	–4.76
BD	–37.51	–0.9486	6	213.5	5.4	–51.8
BE	–15.81	1.0	6.32	–99.92	6.32	–0.75
CE	82.5	–0.9486	6	–469.56	5.4	68.2
DE	5.0	–0.3162	2	–3.16	0.2	0.24
DF	81.14	0	3.6	0	0	81.14
EF	67.5	0	3	0	0	67.5
CD	–	1.0	6.32	0	6.32	15.06

The angles made by the inclined members *AB* and *DF* are $\tan^{-1}(2/3) = 33.69°$ with the horizontal ($\cos 33.69° = 0.832$ and $\sin 33.69° = 0.5546$). The angle made by the diagonal members *BE* and *CD* is $\tan^{-1}(2/6) = 18.43°$ ($\cos 18.43° = 0.9486$ and $\sin 18.43° = 0.3161$).

The reactions are

$R_{FV} = (40 \times 3 + 40 \times 9 + 30 \times 2)/12 = 45$ kN ↑

From $\Sigma V = 0$, $R_{AV} = 80 - 45 = 35$ kN ↑

$\Sigma H = 0$ gives $R_{AH} = 30$ kN ←

Joint A $\Sigma V = 0$ gives $AB = 35/0.5546 = -63.11$ kN

From $\Sigma H = 0$, $AC = 30 + 63.11 \times 0.832 = 82.5$ kN

Joint C As the member *CD* has been removed, $CB = 0$ from $\Sigma V = 0$ and $CE = 82.5$ kN from $\Sigma H = 0$

Joint B $\Sigma V = 0$ gives $-40 + 63.11 \times 0.5546 + BE \sin 18.43° = 0$; $BE = -15.82$ kN

$\Sigma H = 0$ gives $AB \cos 33.69° - BE \cos 18.43° + BD = 0$; $BD = -37.5$ kN

Joint E $\Sigma H = 0$ gives $EF = 82.5 - 15.82 \times \cos 18.43° = 67.5$ kN

$\Sigma V = 0$ gives $DE = BE \sin 18.43° = 5$ kN

Joint F $\Sigma V = 0$ gives $DF = 45/\sin 33.69° = 81.14$ kN

Unit load analysis

We apply unit loads at joints *C* and *D* as shown. The member forces are listed in Table 14.9.

Only the panel in which the forces are applied will have member forces.

Joint *C* $BC = 1 \times \sin 18.43° = 0.3162$ (compressive) and $CE = 1 \times \cos 18.43° = 0.9486$ (compressive)

Joint B $BE = 0.3162/\sin 18.43° = 1.0$ kN (tensile) and $BD = 1 \times \cos 18.43° = 0.9486$ kN (compresive)

Joint D $DE = 1 \times \sin 18.43° = 0.3162$ kN (compressive)

All the joints are now balanced. The remaining member forces are zero.

We now calculate ΣFuL and Σu^2L.

$\Sigma FuL = -359.14$ and $\Sigma u^2L = 23.84$

Now, $R = -(-359.14)/23.84 = 15.06$ kN

The final member forces are given by $F + Ru$. These are listed in the last column of the table. ■

Example 14.16 Unit load method: indeterminate truss

Analyse the truss shown in Fig. 14.46 and find the forces in all the members. Areas of members are written against the members in mm^2. *E* is constant.

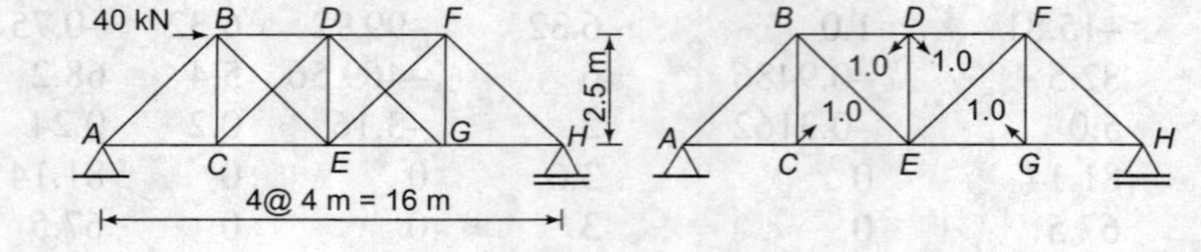

Fig. 14.46

Solution There are two internal redundants in this case. We make the forces in the diagonals *CD* and *DG* redundants. We remove the two members to get a statically determinate truss. The truss will be analysed for (i) the member forces due to the applied loads, (ii) the member forces due to the unit loads applied at *C* and *D*, and (iii) the member forces due to the unit loads applied at *D* and *G*. The results of the analysis are given in Tables 14.10 and 14.11.

Table 14.10

Member	*F due to the applied loads*	*u_1 due to the unit loads applied at C and D*	*u_2 due to the unit loads applied at D and G*	*Length*	*Fu_1L*
AB	–158.04	0	0	4.717	0
AC	174.02	0	0	4	0
BC	60	–0.53	0	2.5	–79.5
BD	–212	–0.848	0	4	+719.1
BE	44.82	1.0	0	4.717	211.42
CE	174.02	–0.848	0	4	–590.27
DE	0	–0.53	–0.53	2.5	0
DF	–212	0	–0.848	4	
EF	68.4	0	1.0	4.717	
EG	154.02	0	–0.848	4	
FG	60	0	–0.53	2.5	
FH	181.63	0	0	4.717	
GH	154.02	0	0	4	
CD (Red.)	—	1.0	—	4.717	
DG (Red.)	—	—	1.0	4.717	

Table 14.11

Member	*Fu_2L*	*u^2_1L*	*u^2_2L*	*$SF + R_1u_1 + R_2u_2$*	*Final forces*
AB	0	0	0	$-158.02 + R_1u_1 + R_2u_2$	–158.02
AC	0	0	0	174.02	174.02
BC	0	0.702	0	60 + (–21.96)(–0.53)	66.17
BD	0	2.876	0	–212 – (–0.848)(–21.96)	–193.4
BE	0	4.717	0	44.82 – 21.96	22.86
CE	0	2.876	0	174.02 + 0.848 × 21.96	192.64
DE	0	0.702	0.702	0 + (–0.53)(–21.96) + (–0.53)(–37.04)	31.27
DF	719.1		2.876	–212 + .848 × 37.04	180.6
EF	322.64		4.717	68.4 – 37.04	31.36
EG	–522.4		2.876	154.02 + 0.848 × 37.04	185.43
FG	–79.5		0.702	60 + 0.53 × 37.04	79.63
FH	0			—	181.63
GH	0			154.02 + 0	154.02
CD					–21.96
EF					–37.04

14.7.4 Forces due to Lack of Fit and Temperature Change

Lack of fit case arises when a member, too long or short, is fitted into place by force in an indeterminate truss. This causes additional forces in all the members. Temperature rise also gives rise to stresses in the members.

Consider the truss shown in Fig. 14.47. The member AB is the last member to be fitted in and is shorter/longer than its true length by δ. If this member is fitted in forcefully, the two ends A and B will subject the truss to forces. If the member is too short, the member will be in tension and will exert forces on the joints as shown. If the member is too long, it will be under compression and will exert forces on the joints as shown.

With no other force acting on the truss, we can write the following condition:

Displacement of joints of the structure + change in length of the member = δ

Displacements of the joints of the structure = $R\Sigma u^2L/AE = \Sigma u^2RL/AE$

where R is the force in the member that is short or long.

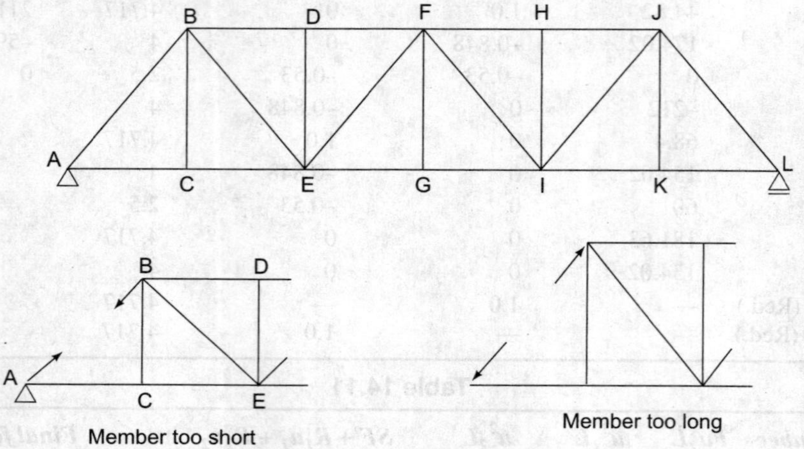

Fig. 14.47

If R is the force in the member short or long, then change in length of that member,

$$\Delta L_R = \frac{RL_R}{A_RE}$$

where L_R and A_R are the length and area of the member too short or too long. This can also be written as u^2RL_R/A_RE.

$$\text{We have } \frac{\Sigma u^2RL}{AE} + \frac{u^2RL_R}{A_RE} = \delta$$

$$\text{We have } R = \frac{\delta}{\Sigma[u^2L/AE]}$$

Here the Σ sign covers all the members, including the short or long member.

Note that when the member is short, δ is positive applying a pull at the joints and negative when long.

In the case of temperature rise/fall in a member, the elongation/shortening of the member is $L\alpha T$, where α is the coefficient of linear expansion and T is the rise or fall in temperature. Also δ in the previous expression derived for lack of fit is replaced by $L\alpha T$ and the solution remains the same.

If loads in addition to lack of fit or temperature rise act on the frame, then

$$R = \frac{\delta(\text{or } L\alpha T) - \Sigma[FuL/AE]}{\Sigma[u^2L/AE]}$$

Here again, the denominator is over the whole frame, including the redundant, while the summation in the numerator is for the members of the determinate frame only.

The following example illustrates the method.

Example 14.17 Lack of fit: indeterminate truss

In the indeterminate truss shown in Fig. 14.48, as the members are being assembled, the last member AC was found to be 2 mm too short. Find the forces in the members of the truss when this member is forcefully fitted in. Areas of the members are mentioned against the members in 1000 mm^2. E is 200 GPa.

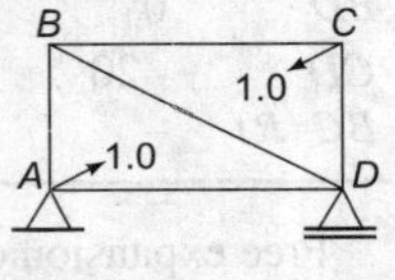

Fig. 14.48

Solution As the member is too long, it will exert forces (acting towards) at the joints. The member will be under tension. The analysis of the member due to the unit forces is done as shown in Table 14.12.

Note that $AE = A\ (\text{mm}^2) \times 10^{-6} \times 200 \times 10^6\ \text{kN/m}^2 = 200A$ kN

Table 14.12

Member	*u*	*L (m), A (mm²)*	*u²L/AE*	*Member forces*
AB	−0.4472	2, 1600	0.0125×10^{-4}	−25.2
AD	−0.8944	4, 2400	0.0667×10^{-4}	−50.4
BC	−0.8944	4, 2400	0.0667×10^{-4}	−50.4
BD	1.0	4.721, 2400	0.0983×10^{-4}	56.34
CD	−0.4472	2, 1600	0.0125×10^{-4}	−25.2
AC	1.0	4.721, 2400	0.0983×10^{-4}	56.34

$$R = \delta/(\Sigma u^2L/AE) = 0.002/[0.355 \times 10^{-4}] = 56.34 \text{ kN}$$

The member forces can be worked out as Ru.

Example 14.18 Temperature stress: indeterminate truss

In the indeterminate truss loaded as shown in Fig. 14.49, if the temperature of the bar BC rises by 40°C, find the forces in all the members. The members have the same cross-sectional area and are made of the same material.

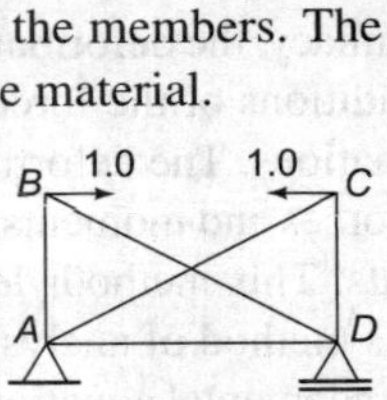

Fig. 14.49

Solution We solve this problem by removing the bar BC and making the truss statically determinate. We first analyse the truss and find the member forces due to the 40 kN load. The second analysis is done for the unit forces acting at B and C as shown. These are compressive because of the temperature rise, which will make the length of the bar to increase. As the expansion is prevented, the bar will be under compression.

The results of the analysis are given in Table 14.13.

Table 14.13

Member	*Member forces due to the applied loads (F)*	*Member forces due to the unit loads (u)*	*Length (L)*	*FuL*	u^2L	*Final member forces*
AB	0	−0.5	2	0	0.5	3.03
AC	44.72	1.118	4.472	224	5.589	37.94
AD	40	−1.0	4	−160	4.0	46.07
BD	0	1.118	4.472	0	5.589	−6.78
CD	−20	−0.5	2	20	0.5	−16.96
BC (R)	–	−1.0	4	–	4	6.07

Free expansion due to the temperature rise $= L\alpha T = 4 \times 12 \times 10^{-6} \times 40$

$$= 0.00192 \text{ m}$$

$$\text{Redundant } R = \frac{-0.00192 - \Sigma[FuLAE]}{\Sigma[u^2LAE]} = \frac{-0.00192 \times AE - \Sigma FuL}{\Sigma u^2 L}$$

$$= \frac{(-38.84 - 84)}{20.178} = -6.07 \text{ kN}$$

The final member forces are worked out as $F + Ru$ and are listed in Table 14.13. Note that the redundant is under tension because of the load acting at that joint. ■

Summary

Indeterminate structures are structures that cannot be analysed using the conditions of equilibrium alone. There are two kinds of indeterminacies: static and kinematic. The degree of indeterminacy is the number of unknown forces and moments minus the number of conditions available.

A basic method for the analysis of indeterminate structures is the method of consistent deformations. In this method, a given indeterminate structure is converted into a determinate primary structure by removing as many constraints as may be required. Compatibility of deformations is then used as conditions to evaluate the unknown forces and moments. This leads to a general approach called the flexibility method of analysis. The primary structure in the flexibility method is not a unique structure, as it can be obtained in different ways.

In the case of kinematic indeterminacy, the deformations at the joints of the structure are the unknowns. The equilibrium conditions of the forces and moments at the joints can be written to evaluate unknown deformations. The deformations can then be put back into the equations to evaluate the unknown forces and moments. Each member is initially prevented from deforming by locking the joints. This methods leads to a general approach to structural analysis, known as the stiffness method of analysis.

Slope-deflection equations are fundamental equations for the stiffness method of analysis. These can be solved using several methods, or iterative techniques such as the moment-distribution method.

Moment distribution method is an iterative technique based upon the kinematic indeterminacy of a structure. The method can be applied to beams and frames.

Indeterminate pin-jointed plane frames can be analysed using Castigliano's second theorem known as the theorem of minimum strain energy. Unit load method based on the energy principles is equally versatile in analysing indeterminate frames. Both internal and external indeterminacy as also the stresses due to the lack of fit or the temperature change can be tackled by both the methods.

All the methods of analysis are used to calculate the maximum BM and SF in a member for design purposes.

Exercises

Review Questions

1. Explain the terms statical indeterminacy, kinematic indeterminacy, and degree of indeterminacy.
2. Give two examples each of statically indeterminate and kinematically indeterminate structures. Calculate the degree of indeterminacy in each of the cases.
3. Define the term stiffness of a member. On what does the stiffness of a member depend?
4. What does the stiffness of a member indicate? Two rods, one of steel and the other of aluminium, have the same length and cross section. Will their stiffness be the same?
5. Explain, with an example, the general procedure for the analysis of an indeterminate structure by the force method.
6. Explain, with an example, the general procedure for the analysis of a structure by the stiffness method.
7. Taking the case of a two-span beam, bring out clearly the differences between the procedures of analysis using the stiffness and the flexibility method.
8. For the structures shown in Fig. 14.50, state the degree of indeterminacy and show all the possible released structures for analysis by the force method. In each case, state the deformation condition(s).

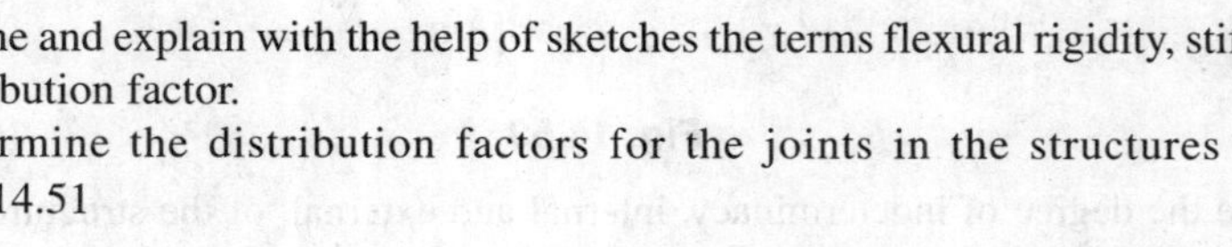

Fig. 14.50

9. Define and explain with the help of sketches the terms flexural rigidity, stiffness, and distribution factor.
10. Determine the distribution factors for the joints in the structures shown in Fig. 14.51

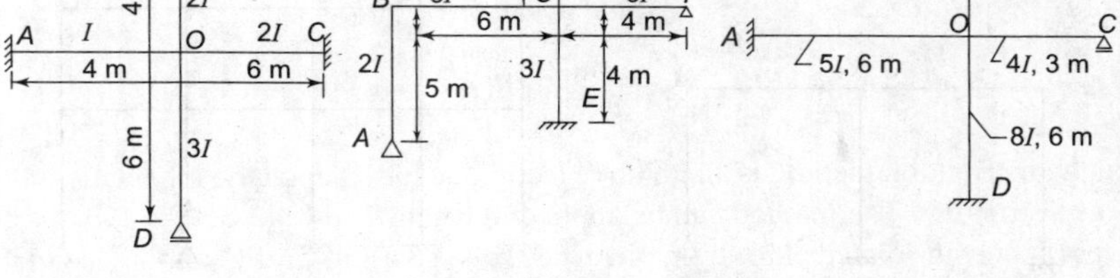

Fig. 14.51

11. Calculate the distribution factors and determine the moments distributed amongst the members in the cases shown in Fig. 14.52. What will the moments at the far ends in each case be? What is the rotation of the joint if $E = 200 \times 106$ kN/m^2 and $I = 1 \times 10^{-4}$ m^4. Draw the deflected shapes of the members.

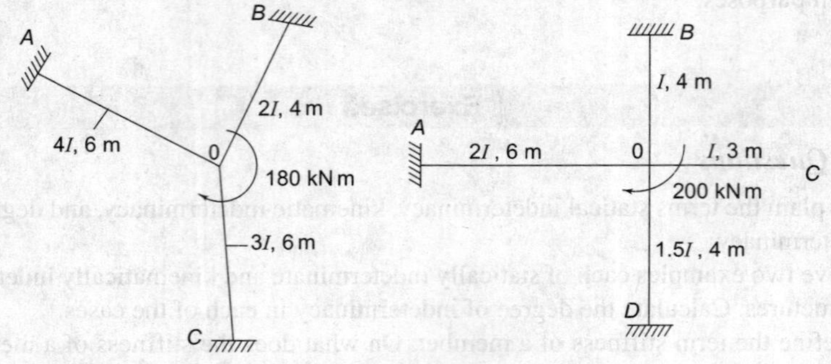

Fig. 14.52

Problems

1. State whether the structures shown in Fig. 14.53 are statically determinate. If not, state the degree of indeterminacy.

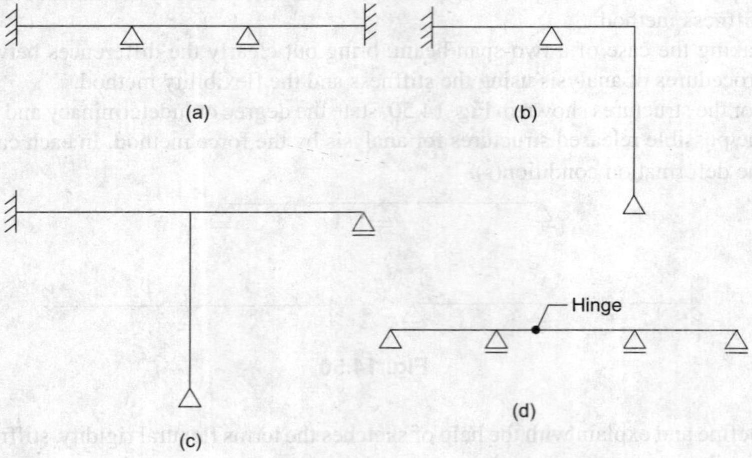

Fig. 14.53

2. State the degree of indeterminacy, internal and external, of the structures shown in Fig. 14.54.

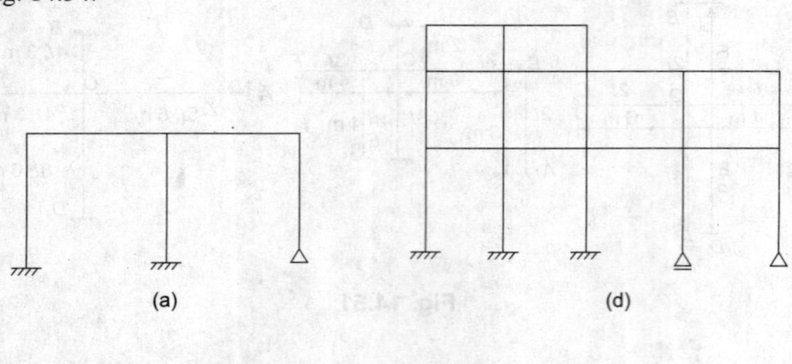

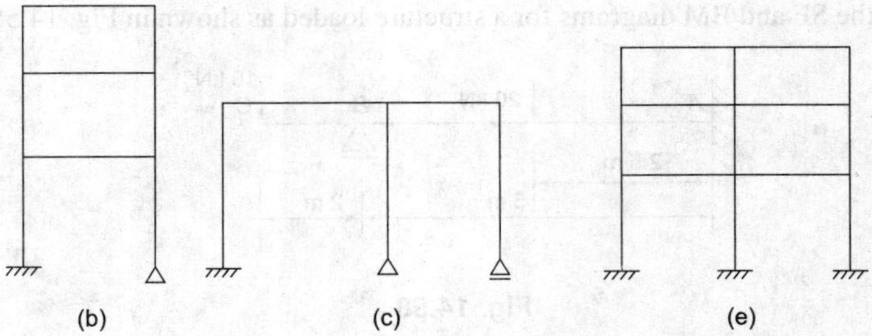

Fig. 14.54

3. State the degree of kinematic indeterminacy of the structures shown in Fig. 14.55.

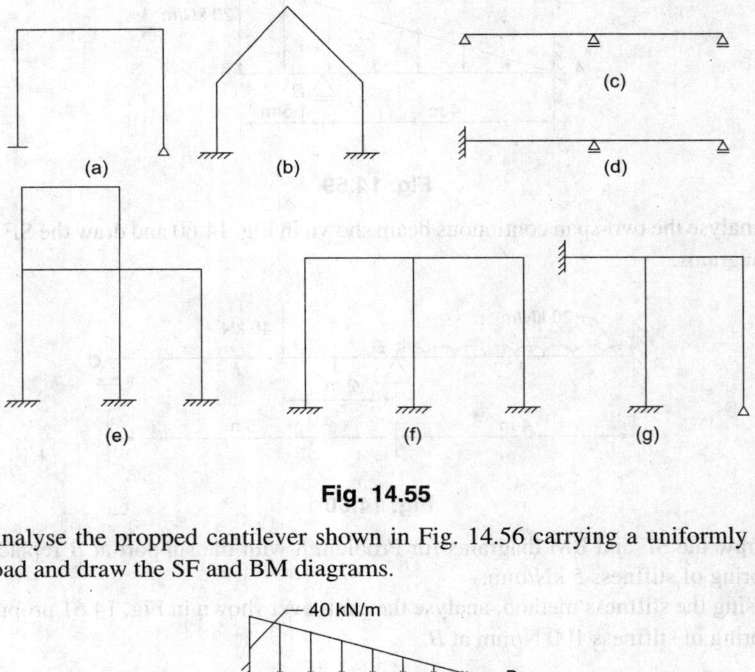

Fig. 14.55

4. Analyse the propped cantilever shown in Fig. 14.56 carrying a uniformly varying load and draw the SF and BM diagrams.

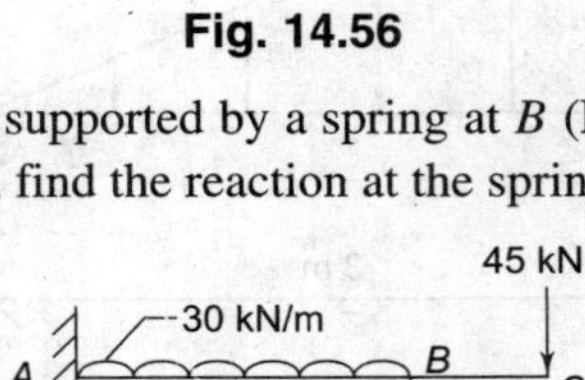

Fig. 14.56

5. A propped cantilever is supported by a spring at B (Fig. 14.57). If the spring has a stiffness of 3000 N/mm, find the reaction at the spring. EI = 60,000 kN m^2.

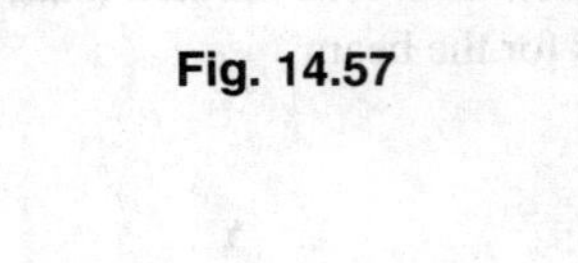

Fig. 14.57

6. Draw the SF and BM diagrams for a structure loaded as shown in Fig. 14.58.

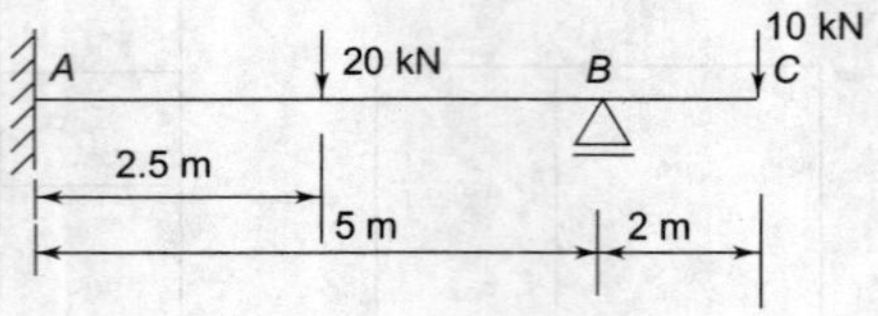

Fig. 14.58

7. Analyse the cantilever shown in Fig. 14.59 propped at *B* and carrying the loads shown. Draw the SF and BM diagrams.

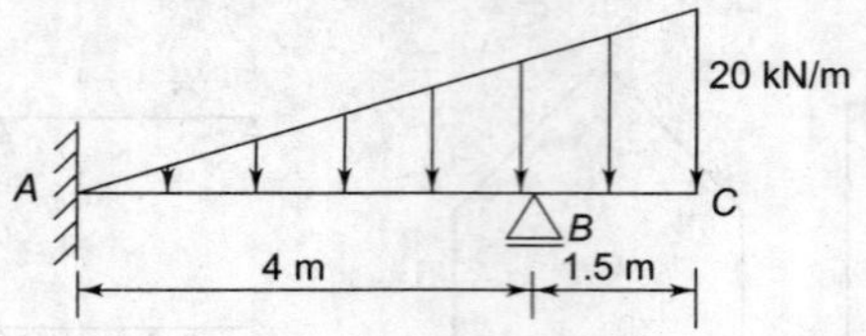

Fig. 14.59

8. Analyse the two-span continuous beam shown in Fig. 14.60 and draw the SF and BM diagrams.

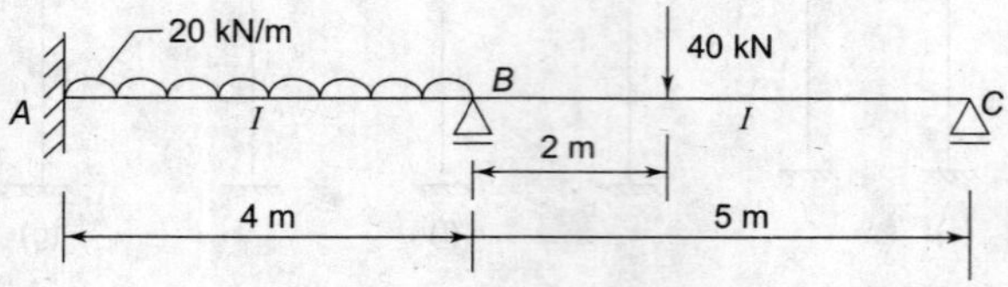

Fig. 14.60

9. Draw the SF and BM diagrams for Problem 8 with the support at *B* replaced by a spring of stiffness 5 kN/mm.
10. Using the stiffness method, analyse the cantilever shown in Fig. 14.61 propped by a spring of stiffness 100 N/mm at *B*.

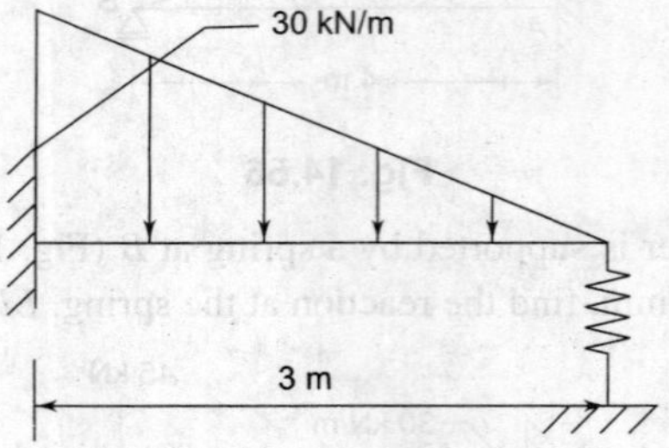

Fig. 14.61

11. A propped cantilever carries two concentrated loads as shown in Fig. 14.62. Draw the SF and BM diagrams for the beam.

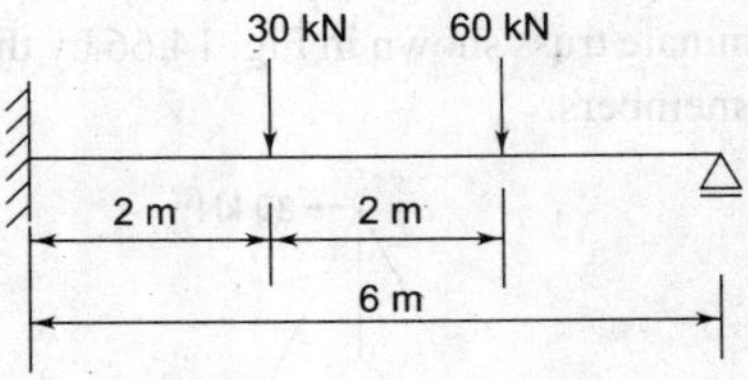

Fig. 14.62

12. Analyse the indeterminate truss shown in Fig. 14.63 by the Castigliano's method and find the forces in all the members.

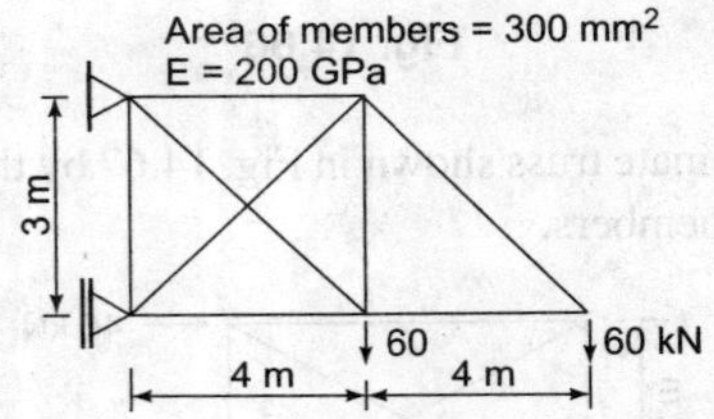

Fig. 14.63

13. Analyse the indeterminate truss shown in Fig. 14.64 by the Castigliano's method and find the forces in all the members.

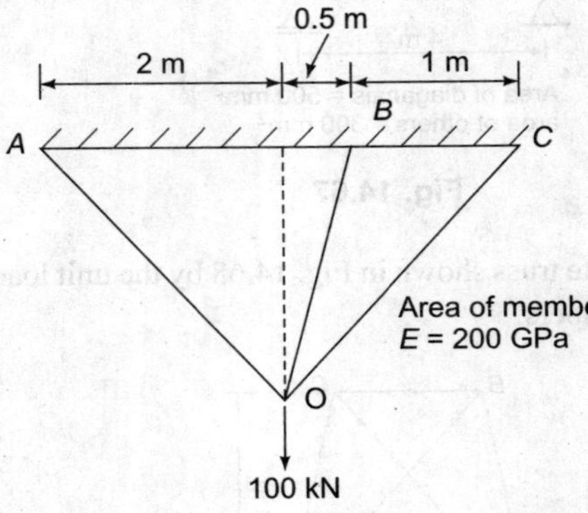

Fig. 14.64

14. Analyse the indeterminate truss shown in Fig. 14.65 by the Castigliano's method and find the forces in all the members.

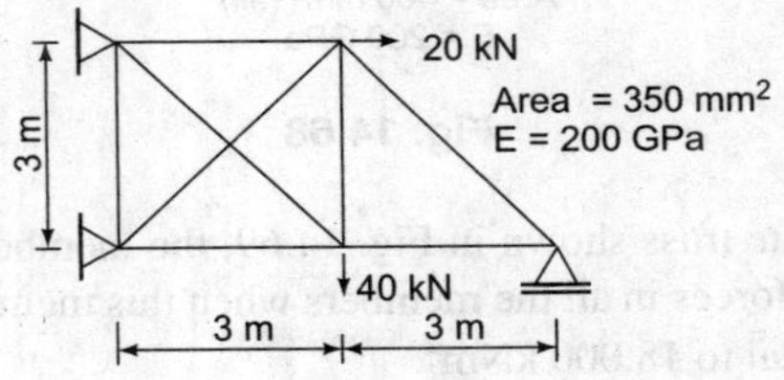

Fig. 14.65

15. Analyse the indeterminate truss shown in Fig. 14.66 by the unit load method and find the forces in all the members.

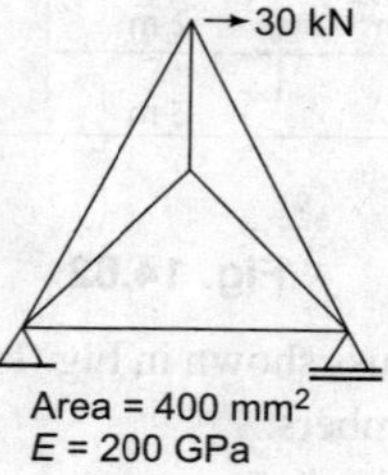

Fig. 14.66

16. Analyse the indeterminate truss shown in Fig. 14.67 by the unit load method and find the forces in all the members.

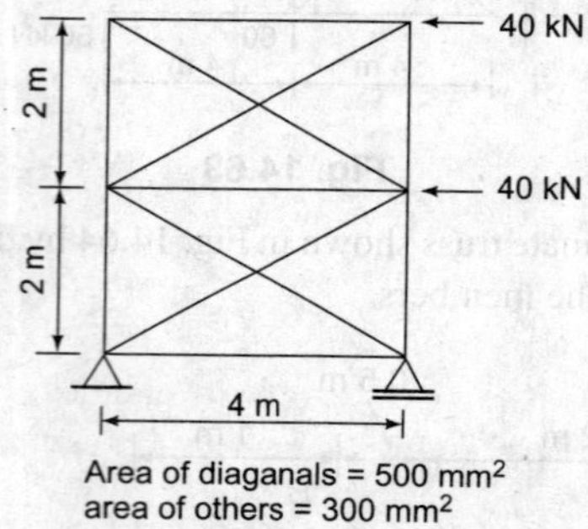

Fig. 14.67

17. Analyse the indeterminate truss shown in Fig. 14.68 by the unit load method and find the forces in all the members.

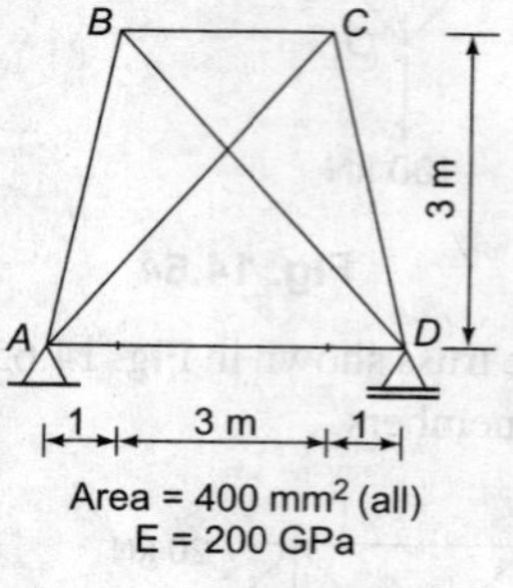

Fig. 14.68

18. In the indeterminate truss shown in Fig. 14.69, the member BC is fabricated 2 mm too long. Find the forces in all the members when this member is forcefully fitted in. EA is constant equal to 18,000 kNm^2.

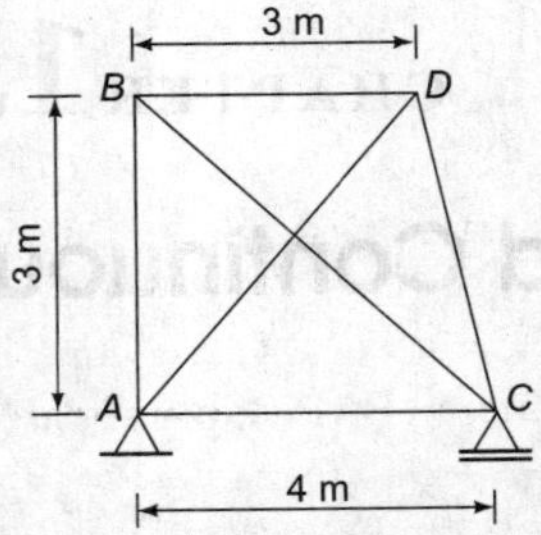

Fig. 14.69

19. In the pin-jointed plane frame shown in Fig. 14.70, find the forces in all the members when the temperature of the bar AC rises by 40°C. Also $\alpha = 12 \times 10^{-6}$/°C and $E = 200$ GPa. Area of all the members is 1600 mm^2.

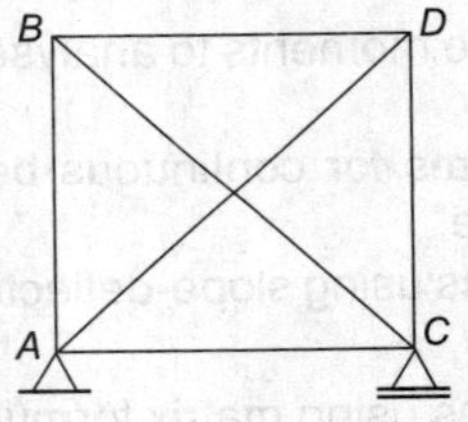

Fig. 14.70

20. The truss loaded as shown in Fig. 14.71 has one of its members, AC, 1.5 mm too long. Find the forces in the members when this member is fitted in forcefully. E = 200 GPa. Area of all the members is 1000 mm^2.

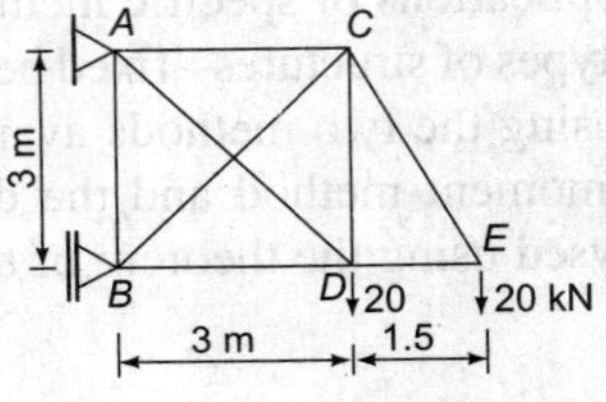

Fig. 14.71

CHAPTER 15

Fixed and Continuous Beams

Learning Objectives

After going through this chapter, the reader will be able to

- explain the structural action of fixed and continuous beams,
- use the method of consistent deformation to analyse fixed and continuous beams,
- apply the area-moment theorem and the double integration method to analyse fixed beams,
- apply the theorem of three moments to analyse continuous beams with different end conditions,
- draw SF and BM diagrams for continuous beams with given loadings and show the deflected shape,
- analyse continuous beams using slope-deflection equations and moment distribution methods, and
- analyse continuous beams using matrix formulation of flexibility and stiffness methods.

15.1 INTRODUCTION

In Chapter 14, the basic principles of indeterminate structural analysis were outlined. In this chapter, the applications of specific methods are discussed in detail. This chapter deals with two types of structures—fixed beams and continuous beams. Fixed beams are analysed using the two methods available for computing deformations, namely, the area-moment method and the double integration method. Continuous beams are analysed using the theorem of three moments.

15.2 FIXED BEAMS

Fixed beams, also called 'encastre' or 'built-in' beams, have fixed supports at both ends (Fig. 15.1). The most important characteristic of this beam is that the slope and deflection are zero at both ends. The beam is kinematically determinate but statically

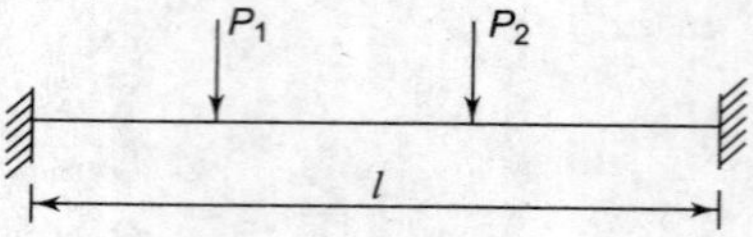

Fig. 15.1

indeterminate to the third degree. While fixed beams are not often encountered in practice, this concept is important in structural analysis, as a number of methods for analysis of indeterminate structures are based on this concept. The stiffness methods outlined in Chapter 14 are examples of such methods.

15.2.1 Structural Action of a Fixed Beam

When subjected to loads, a fixed beam (Fig. 15.2) bends, developing resisting moments and shears. The deflected shape of the beam is shown in Fig. 15.2(b). Note that the slope and deflection at the ends A and B are zero. The beam 'sags' for a part of its length in the middle and 'hogs' at the ends. This necessarily produces two points of inflection (contraflexure) corresponding to the change in the bending moment (BM). Going from A to B, the BM from A to C_1 is hogging, C_1 to C_2 is sagging, and C_2 to B is again hogging.

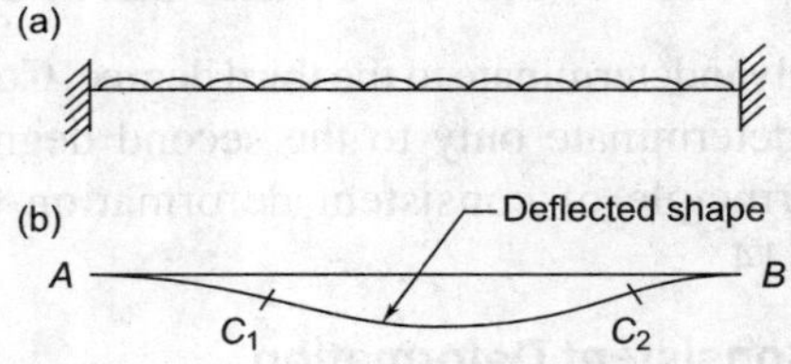

Fig. 15.2

The structural action of a fixed beam can also be visualized from the simply supported beam shown in Fig. 15.3 (a). Under the influence of the loads, the beam

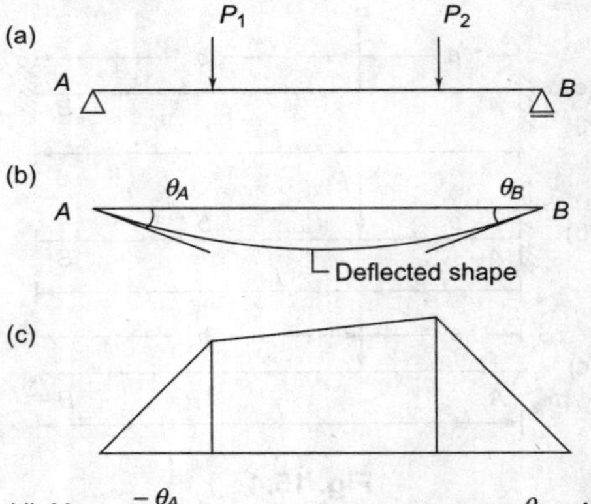

Fig. 15.3

bends as shown in Fig. 15.3(b). The BM diagram for the beam is shown in Fig. 15.3(c). The deflection at A and B is zero and θ_A and θ_B are the slopes at A and B. The deflected shape in Fig. 15.3(b) does not satisfy the conditions of a fixed beam due to the slopes at A and B.

The fixed end moments M_A and M_B are applied at A and B such that θ_A and θ_B become zero [Fig. 15.3(d)]. (Note that applying a moment at A affects θ_B also and vice versa.) The simply supported beam with the applied loads and the moments M_A and M_B at the supports A and B behaves like a fixed beam. The support moments (also called fixed end moments) M_A and M_B have values such that θ_A and θ_B are zero. Note also that M_A and M_B are both hogging (this need not always be so). The hogging effects at the ends of a fixed beam (Fig. 15.2) are due to these moments. The combination of the simply supported BM diagram and M_A and M_B will give the BM diagram shown in Fig. 15.3 (e).

15.3 METHODS OF ANALYSIS OF FIXED BEAMS

A fixed beam is statically indeterminate to the third degree. Considering only vertical loads, the beam is indeterminate only to the second degree. The beam can be analysed using the principle of consistent deformation outlined for propped cantilevers in Chapter 14.

15.3.1 Method of Consistent Deformation

This method outlines the flexibility method of analysis. Considering the fixed beam shown in Fig. 15.4(a), constraints have to be removed in such a way that the structure becomes statically determinate. There are two ways of doing this.

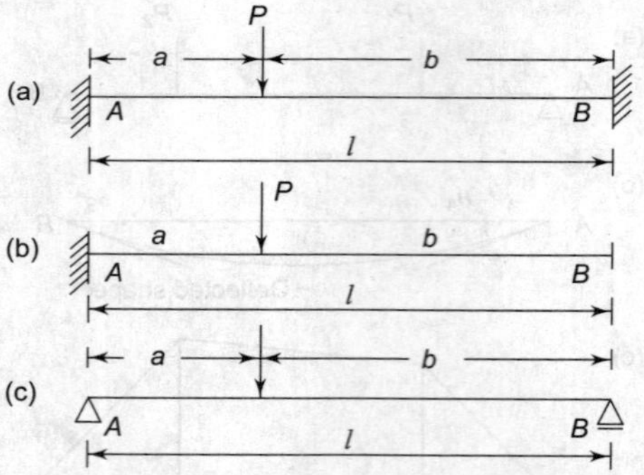

Fig. 15.4

(a) Remove the supports from one end, say, B, completely as shown in Fig. 15.4(b), so that the beam is now a statically determinate cantilever. The constraints removed are the reactions R_{BV} and the support moment M_B.

(b) Remove fixity from both ends, as shown in Fig. 15.4(c), so that the beam becomes a statically determinate simply supported beam. The constraints removed are the support moments at A and B.

We now illustrate solutions taking both these cases.

Case (a) The basic approach is outlined in three steps.

1. Find the deformations (corresponding to the constraints removed) at end B due to the applied loads.
2. Find the deformations at B by applying the removed constraints R_{BV} and M_B one by one.
3. Formulate equations using the deformation conditions for end B ($\theta_B = 0$, $y_B = 0$). These equations will have the unknowns R_{BV} and M_B, which can be evaluated.

Considering Fig. 15.5, the deformations at B due to the applied loads are θ_{B1} and y_{B1}. The beam bends only between A and C due to the concentrated load at C and is a straight line between C and B.

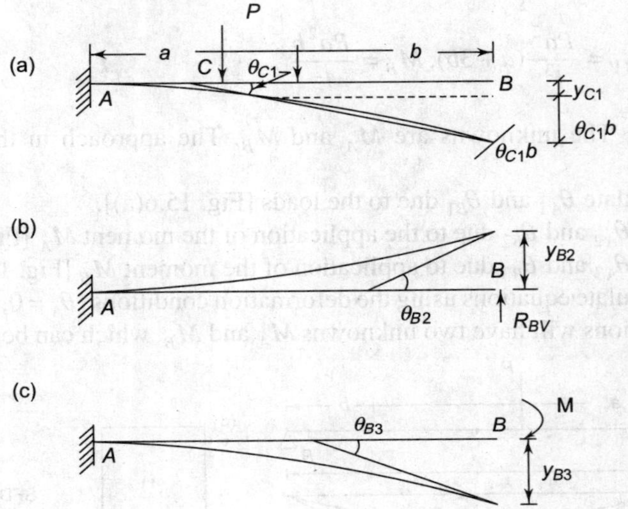

Fig. 15.5

$$\theta_{B1} = \theta_{C1} = \frac{Pa^2}{2EI}$$

$$y_{B1} = y_{C1} + \theta_{C1}b = \frac{Pa^3}{3EI} + \frac{Pa^2}{2EI}b$$

Note that θ_B is clockwise and y_B is downwards. The deformations due to R_{BV}, considering Fig. 15.5(b), are

$$\theta_{B2} = -\frac{R_{BV}l^2}{2EI} \quad \text{and} \quad y_{B2} = -\frac{R_{BV}l^3}{3EI}$$

where l is the length of the beam. Note that θ_{B2} is anticlockwise and y_{B2} is upwards. From Fig. 15.5(c), the deformations due to M_B (the direction of M_B is assumed) are

$$\theta_{B3} = +\frac{Ml}{EI} \quad \text{and} \quad y_{B3} = +\frac{Ml^2}{2EI}$$

θ_{B3} is anticlockwise and y_{B3} is downwards. The deformation conditions for the fixed beam are

$$\theta_B = 0, \quad \theta_{B1} + \theta_{B2} + \theta_{B3} = 0, \quad \frac{Pa^2}{2EI} - \frac{R_{BV}l^2}{2EI} + \frac{Ml}{EI} = 0$$

$$y_B = 0, \quad y_{B1} + y_{B2} + y_{B3} = 0, \quad \frac{Pa^3}{3EI} + \frac{Pa^2}{2EI}b - \frac{R_{BV}l^3}{3EI} + \frac{Ml^2}{2EI} = 0$$

$$\frac{Ml}{EI} - \frac{R_{BV}l^2}{2EI} = -\frac{Pa^2}{2EI}$$

$$\frac{Ml^2}{2EI}-\frac{R_{BV}l^3}{3EI}=-\frac{Pa^3}{3EI}-\frac{Pa^2b}{2EI}$$

Solving,

$$R_{BV}=\frac{Pa^2}{l^3}(a+3b),\ M_B=\frac{Pa^2b}{l^2}$$

Case (b) The unknowns are M_A and M_B. The approach in this case is the following.

1. Calculate θ_{A1} and θ_{B1} due to the loads [Fig. 15.6(a)].
2. Find θ_{A2} and θ_{B2} due to the application of the moment M_A [Fig. 15.6(b)].
3. Find θ_{A3} and θ_{B3} due to application of the moment M_B [Fig. 15.6(c)].
4. Formulate equations using the deformation conditions ($\theta_A=0$, $\theta_B=0$). These equations will have two unknowns M_A and M_B, which can be evaluated.

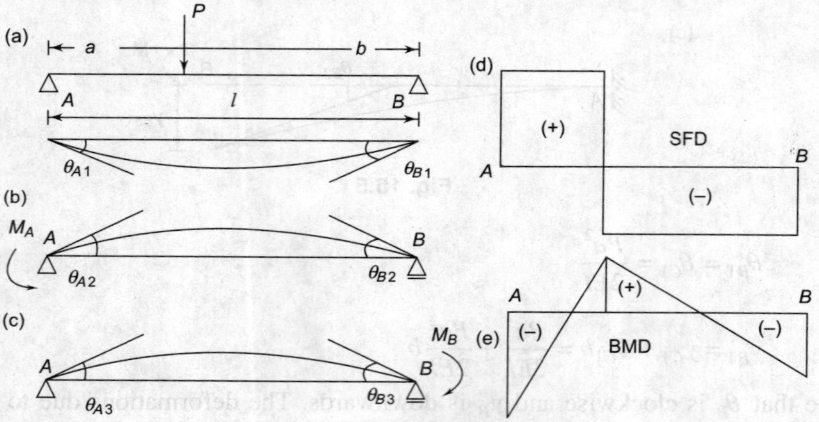

Fig. 15.6

From Fig. 15.6 (a), the deformations due to the applied loads are

$$\theta_{A1}=\frac{Pab(l+b)}{6lEI}\quad\text{and}\quad\theta_{B1}=-\frac{Pab(l+a)}{6lEI}$$

The deformations due to M_A, from Fig. 15.6(b), are

$$\theta_{A2}=-\frac{M_Al}{3EI}\quad\text{and}\quad\theta_{B2}=+\frac{M_Al}{6EI}$$

From Fig. 15.6(c), the deformations due to M_B are

$$\theta_{A3}=-\frac{M_Bl}{6EI}\quad\text{and}\quad\theta_{B3}=\frac{M_Bl}{3EI}$$

Note that the directions of M_A and M_B are assumed, as shown in the figure. The clockwise slope is positive and the anticlockwise slope is negative.

The deformation conditions are

$$\theta_A=0,\ \theta_{A1}+\theta_{A2}+\theta_{A3}=0,\quad \frac{Pab(l+b)}{6lEI}-\frac{M_Al}{3EI}-\frac{M_Bl}{6EI}=0$$

$$\theta_B = 0,\ \theta_{B1} + \theta_{B2} + \theta_{B3} = 0, \quad -\frac{Pab(l+a)}{6lEI} + \frac{M_A l}{6EI} + \frac{M_B l}{3EI} = 0$$

$$\frac{M_A l}{3} + \frac{M_B l}{6} = \frac{Pab(l+b)}{6l}, \quad 2M_A + M_B = \frac{Pab(l+b)}{l^2}$$

$$\frac{M_A l}{6} + \frac{M_B l}{3} = \frac{Pab(l+a)}{6l}, \quad M_A + 2M_B = \frac{Pab(l+a)}{l^2}$$

Solving,

$$M_A = \frac{Pab^2}{l^2}, \quad M_B = \frac{Pa^2b}{l^2}$$

Knowing M_A and M_B, the reactions at the supports can be calculated as follows:

$$R_A = \frac{Pb}{l} + \frac{(M_A - M_B)}{l} = \frac{Pb}{l} + \frac{Pab(b-a)}{l} = \frac{Pa^2}{l^3}(3a+b)$$

$$R_B = \frac{Pb}{l} + \frac{(M_A - M_B)}{l} = \frac{Pb}{l} + \frac{Pab(b-a)}{l} = \frac{Pa^2}{l^3}(a+3b)$$

The bending moment at A is M_A (hogging). The bending moment under the load is

$$\frac{Pa^2}{l^3}(3a+b)a - \frac{Pab^2}{l^2}$$

The bending moment at B is M_B (hogging).

Figures 15.6(d) and (e) show the SF and BM diagrams, respectively. While this method is very simple, there are more elegant methods. We consider two methods next—the area-moment method and the double integration method.

15.3.2 Area-moment Method

As discussed in Section 15.2.1, a fixed beam can be considered as a simply supported (SS) beam carrying applied loads and end moments. Consider the fixed beam shown in Fig. 15.7(a), which can be treated as a summation of the two cases shown in Fig. 15.7(b). From the deflected shapes of the beam, assuming that the supports are at the same level, we can deduce two important features relevant to the application of the area-moment theorem. First, since the slopes at the two ends are zero, the angle between the tangents drawn at A and B to the elastic curve is zero. [Note that in Fig. 15.7(c), the tangent lines coincide with the straight axis.] Second, the deflections at A and B being zero, the vertical intercept at B with respect to the tangent at A (or vice versa) is zero.

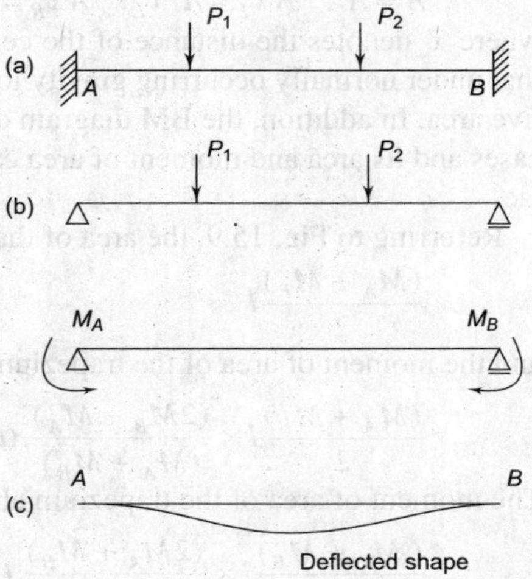

Fig. 15.7

From the area-moment theorem and the two features stated above, we observe that in the case of a fixed beam with supports at the same level,

(a) the area of the M/EI diagram between A and B is zero and

(b) the moment of the area of the M/EI diagram between A and B about A (or about B) is zero.

We have also seen that the BM diagram of a fixed beam is the sum of the BM on a simply supported beam due to the applied loads and that due to the end moments (Fig. 15.8). Figure 15.8(b) shows the BM diagram due to loads. A is the area of this diagram with its centroid at C. Figure 15.8(c) shows the BM diagram due to end moments only. A' is the area of this diagram and its centroid is located at C_1. The application of the area-moment theorem as discussed above gives

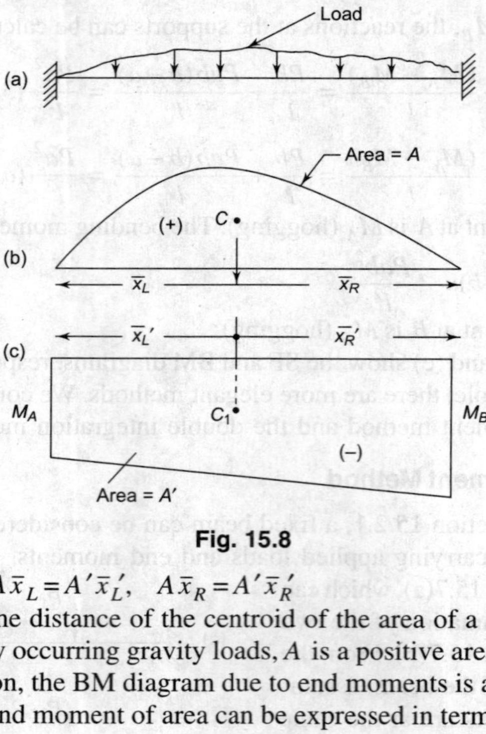

Fig. 15.8

$$A = A', \quad A\bar{x}_L = A'\bar{x}_L', \quad A\bar{x}_R = A'\bar{x}_R'$$

where $\bar{x}$ denotes the distance of the centroid of the area of a BM diagram. Note that under normally occurring gravity loads, A is a positive area, and A' is a negative area. In addition, the BM diagram due to end moments is a trapezium in such cases and its area and moment of area can be expressed in terms of simple formulae.

Referring to Fig. 15.9, the area of the trapezium is given by

$$\frac{(M_A + M_B)}{2} l$$

and the moment of area of the trapezium about A is

$$\frac{(M_A + M_B)}{2} l \times \frac{(2M_B + M_A)}{(M_A + M_B)} (l/3) = (2M_B + M_A)\frac{l^2}{6}$$

The moment of area of the trapezium about B is

$$\frac{(M_A + M_B)}{2} l \times \frac{(2M_A + M_B)}{(M_A + M_B)} (l/3) = (2M_A + M_B)\frac{l^2}{6}$$

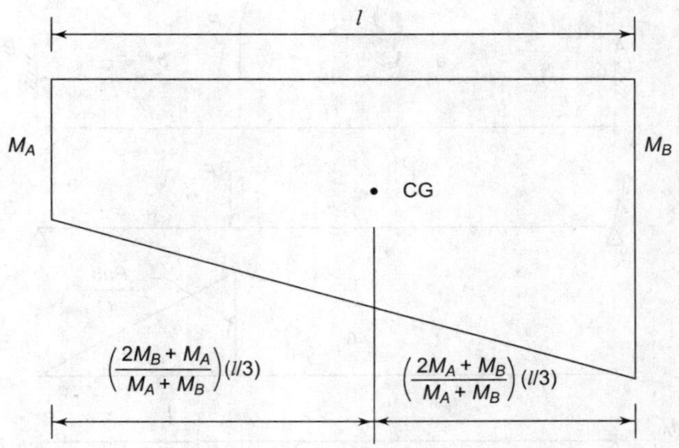

Fig. 15.9

The area and moment of area of the BM diagram due to applied loads (on an SS beam) can be determined from formulae (refer to Table 7.1 of Chapter 7) or by integration. We illustrate the procedure with the following examples.

Example 15.1 Moment area method, Fixed beam: single point load

A fixed beam is subjected to a single concentrated load P as shown in Fig. 15.10 (a). Draw BM and SF diagrams for the beam.

Solution The BM diagram due to applied loads on an SS beam is shown in Fig. 15.10(b). The BM diagram due to end moments is shown in Fig. 15.10(c).

$$\text{Area } A = \frac{Pab}{l}\frac{l}{2} = \frac{Pab}{2}$$

$$\text{Area } A' = \frac{(M_A + M_B)}{2} l 1$$

Equating the two,

$$\frac{(M_A + M_B)}{2} l = \frac{Pab}{2}, \quad M_A + M_B = \frac{Pab}{l}$$

Taking moments about B,

$$\text{Moment of } A = \frac{Pab}{l}\frac{l}{2}\frac{(l+b)}{3} = \frac{Pab}{6}(l+b)$$

$$\text{Moment of } A' = (2M_A + M_B)\frac{l^2}{6}$$

Equating the two moments of area,

$$(2M_A + M_B)\frac{l^2}{6} = \frac{Pab}{6}(l+b), \quad 2M_A + M_B = \frac{Pab}{l^2}(l+b)$$

Solving the two equations,

$$M_A = \frac{Pab^2}{l^2} \quad \text{and} \quad M_B = \frac{Pa^2b}{l^2}$$

The fixed beam is shown in Fig. 15.10(d). The reactions are

$$R_{AV} = \frac{Pb}{l} + \frac{(M_A - M_B)}{l} \quad \text{and} \quad R_{BV} = \frac{Pa}{l} - \frac{(M_A - M_B)}{l}$$

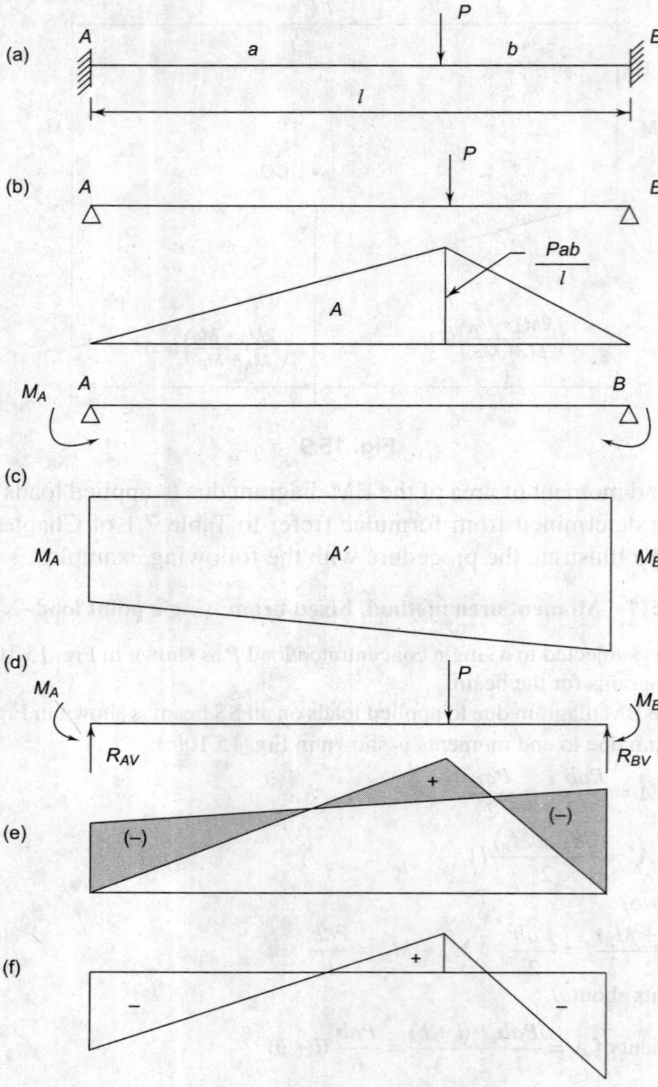

Fig. 15.10

The SF diagram can be drawn as in Fig. 15.10(e). The BM diagram can be drawn by superposing the two BM diagrams, as shown in Fig. 15.10(e). Note that the diagram due to fixed end moments is negative. The BM diagram can also be drawn conventionally as shown in Fig. 15.10(f).

Example 15.2 Moment area method, fixed beam: UD load

A fixed beam carries a uniformly distributed load over the whole span as shown in Fig. 15.11(a). Draw the SF and BM diagrams for the beam. (w denotes the load intensity of the UD load.)

Solution The BM diagrams due to applied loads and end moments M_A and M_B are shown in Figures 15.11(b) and (c), respectively.

$$\text{Area } A = \frac{2}{3}\text{(area of enclosing rectangle)} = \frac{2}{3}\frac{wl^2}{8}l = \frac{wl^3}{12}$$

$$\text{Area } A' = \frac{(M_A + M_B)}{2}l$$

Equating the two,

$$\frac{(M_A + M_B)}{2}l = \frac{wl^3}{12}, \quad M_A + M_B = \frac{wl^2}{6}$$

$$\text{Moment of } A \text{ about } B = \frac{wl^3}{12}\frac{l}{2} = \frac{wl^4}{24}$$

$$\text{Moment of } A' \text{ about } B = (2M_A + M_B)\frac{l^2}{6}$$

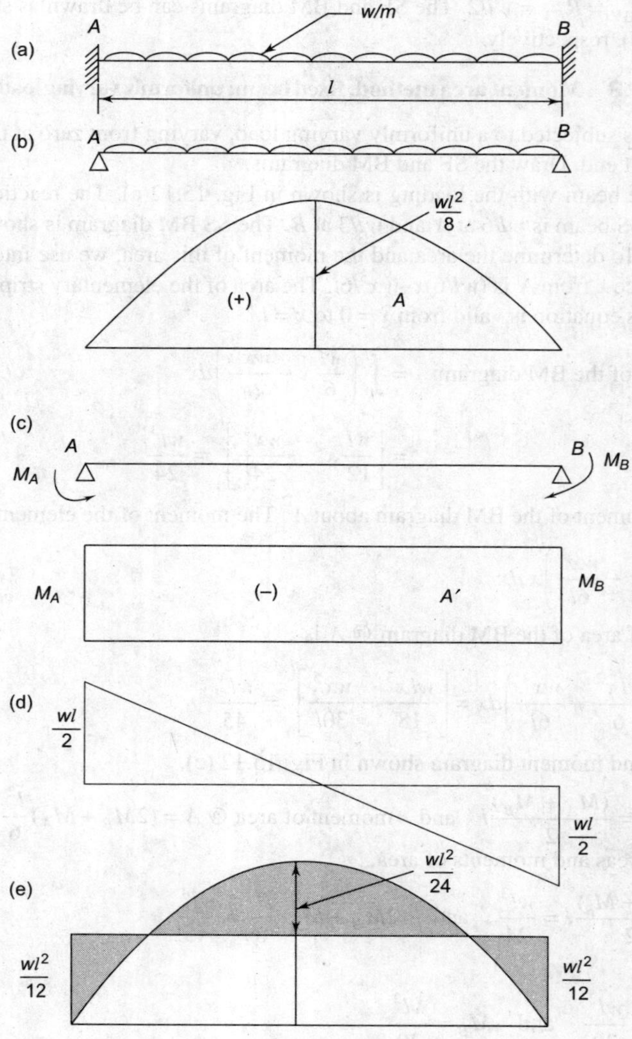

Fig. 15.11

Equating the two,

$$(2M_A + M_B)\frac{l^2}{6} = \frac{wl^4}{24}$$

$$2M_A + M_B = \frac{wl^2}{4}$$

From these equations,

$$M_A = \frac{wl^2}{12} \quad \text{and} \quad M_B = \frac{wl^2}{12}$$

Note that if you have used the symmetry of loading, you already know that $M_A = M_B$, and only one equation needs to be written. Thus, by equating areas,

$$M_A + M_B = \frac{wl^2}{6}$$

since $M_A = M_B$, $M_A = wl^2/12$, and $M_B = wl^2/12$. Since M_A and M_B are equal, the reactions are not affected: $R_{AV} = R_{BV} = wl/2$. The SF and BM diagrams can be drawn as shown in Figs 15.11(d) and (e), respectively.

Example 15.3 Moment area method, fixed beam: uniformly varying load

A fixed beam is subjected to a uniformly varying load, varying from zero at the left end to w/m at the right end. Draw the SF and BM diagrams.

Solution The beam with the loading is shown in Fig. 15.12(a). The reaction due to the loading of an SS beam is $wl/6$ at A and $wl/3$ at B. The SS BM diagram is shown in Fig. 15.12(b). To determine the area and the moment of this area, we use integration. The BM at a distance x from A is $(wl/6)x - wx^3/6l$. The area of the elementary strip is $[(wl/6)x - wx^3/6l]dx$. This equation is valid from $x = 0$ to $x = l$.

$$\text{Area of the BM diagram} = \int_0^l \left(\frac{wl}{6}x - \frac{wx^3}{6l}\right)dx$$

$$= \left[\frac{wl}{12}x^2 - \frac{wx^4}{24l}\right]_0^l = \frac{wl^3}{24}$$

We take the moment of the BM diagram about A. The moment of the elementary strip is

$$\left(\frac{wl}{6}x - \frac{wx^3}{6l}\right)x\,dx$$

The moment of area of the BM diagram @ A is

$$\int_0^l \left(\frac{wlx^2}{6} - \frac{wx^4}{6l}\right)dx = \left[\frac{wlx^3}{18} - \frac{wx^5}{30l}\right]_0^l = \frac{wl^4}{45}$$

For the fixed end moment diagram shown in Fig. 15.12 (c),

$$\text{Area} = \frac{(M_A + M_B)}{2}l \quad \text{and} \quad \text{moment of area @ } A = (2M_B + M_A)\frac{l^2}{6}$$

Equating the areas and moments of area,

$$\frac{(M_A + M_B)}{2}l = \frac{wl^3}{24} \quad \text{and} \quad (2M_B + M_A)\frac{l^2}{6} = \frac{wl^4}{45}$$

Solving,

$$M_A = \frac{wl^2}{30} \quad \text{and} \quad M_B = \frac{wl^2}{20}$$

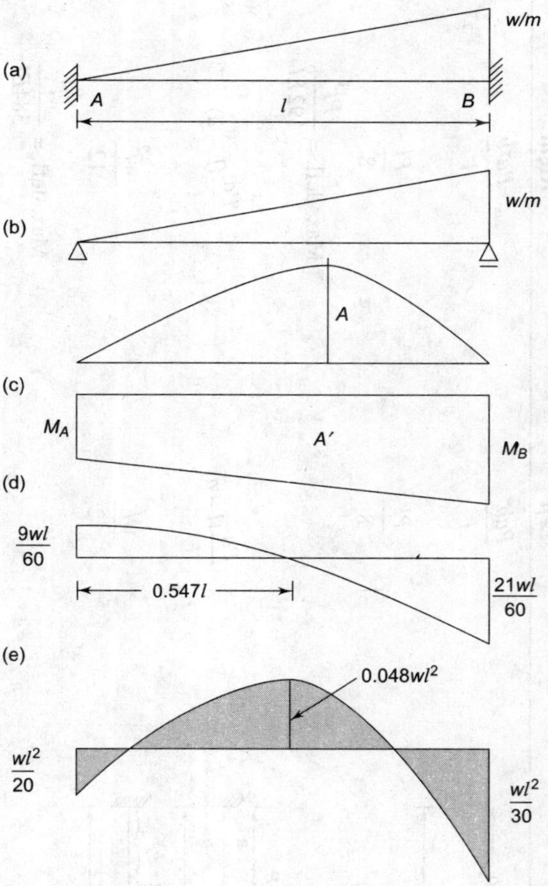

Fig. 15.12

The reactions are

$$R_{AV} = \frac{wl}{6} - \left(\frac{wl^2}{20} - \frac{wl^2}{30}\right)\frac{1}{l} = \frac{9wl}{60}$$

$$R_{BV} = \frac{wl}{3} + \left(\frac{wl^2}{20} - \frac{wl^2}{30}\right)\frac{1}{l} = \frac{21wl}{60}$$

The SF and BM diagrams are shown in Figs 15.12(d) and (e), respectively. SF becomes zero at $\sqrt{9/30}\,l$ from A. The general BM equation at a distance x from A can be written as

$$M_x = -\frac{wl^2}{30} + \frac{9wl}{60}x - \frac{wx^3}{6l}$$

The maximum positive moment at the point of zero shear is $0.048wl^2$.

Fixed end moment values are needed frequently in indeterminate structural analysis. These values for the cases of some standard loadings are given in Table 15.1.

Table 15.1 Table of fixed end moments

(For the left support, positive moment is anticlockwise; for the right support, positive moment is clockwise.)

	Loading		Fixed end moments	
			Left	*Right*
1.	(i) $a \neq b$	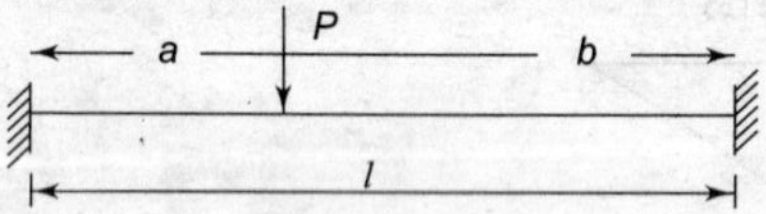	$\frac{Pab^2}{l^2}$	$\frac{Pa^2b}{l^2}$
	(ii) $a = b$		$\frac{Pl}{8}$	$\frac{Pl}{8}$
				Max. defl. = $\frac{Pl^3}{192EI}$
2.		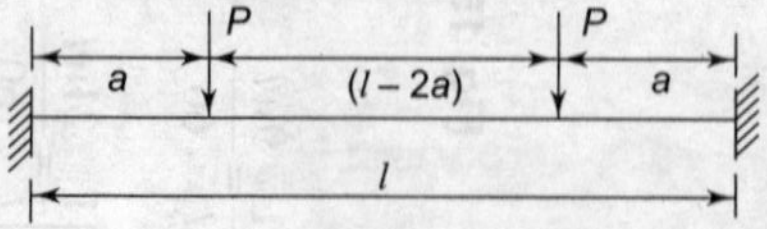	$\frac{Pa}{l}(l-a)$	$\frac{Pa}{l}(l-a)$
3.		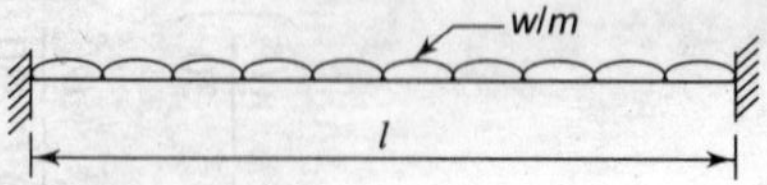	$\frac{wl^2}{12}$	$\frac{wl^2}{12}$
				Max. defl. = $\frac{wl^4}{384EI}$
4.		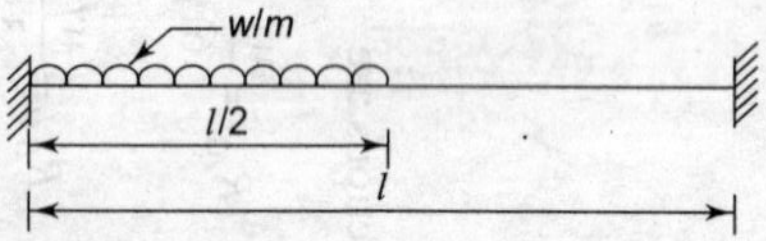	$\frac{11}{192}wl^2$	$\frac{5}{192}wl^2$
				Defl. mid-span = $\frac{wl^4}{768EI}$

(contd)

Table 15.1 *(contd)*

	Loading	Fixed end moments Left	Fixed end moments Right
5.	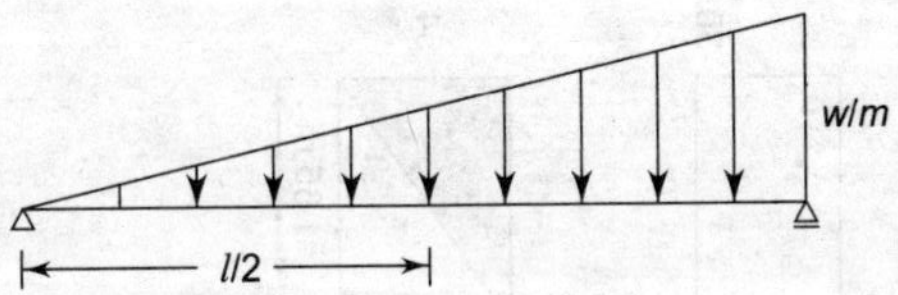	$\frac{wl^2}{30}$	$\frac{wl^2}{20}$ Defl. mid-span = $\frac{wl^4}{768EI}$
6.	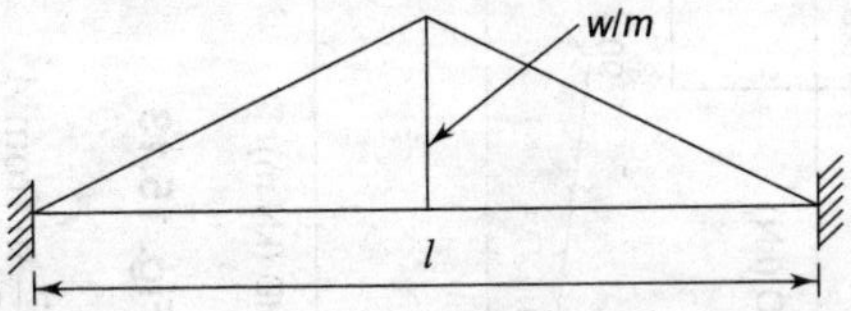	$\frac{5}{96}wl^2$	$\frac{5}{96}wl^2$ Max. defl. = $\frac{7wl^4}{3840EI}$
7.	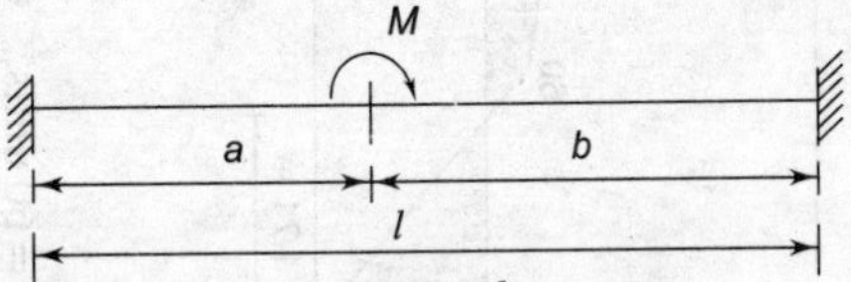	$-\frac{Mb}{l^2}(3a-l)$	$\frac{Ma}{l^2}(3b-l)$
	$a = b = \frac{l}{2}$	$-\frac{M}{4}$	$\frac{M}{4}$
8.	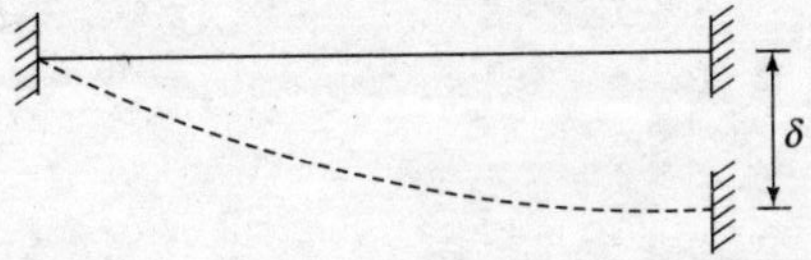	$\frac{6EI\delta}{l^2}$	$-\frac{6EI\delta}{l^2}$

Example 15.4 Moment area method, fixed beam: two point loads

A fixed beam with a span of 8 m carries two point loads, 20 kN and 40 kN, at 2 m and 5 m from the left support, respectively. Draw the SF and BM diagrams for the beam.

Solution The beam with the loads is shown in Fig. 15.13(a). The SS BM diagram due to applied loads is shown in Fig. 15.13(b). The reactions due to the loads are 30 kN each. The BM at C is 60 kN m and at D is 90 kN m. The BM diagram (A) consists of two triangles and one trapezium. The areas are as follows:

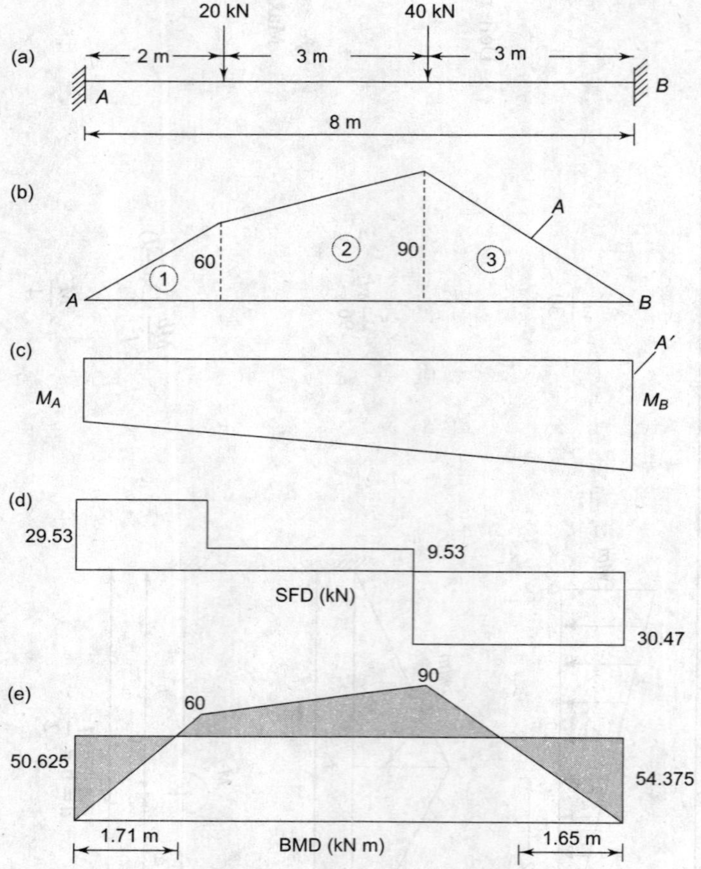

Fig. 15.13

Area of ① = 60 units, CG at $\frac{4}{3}$ m from A

Area of ② = 225 units, CG at $\frac{(60 + 2 \times 90)}{(60 + 90)} \times \frac{3}{3} = 1.6$ m from C

Area of ③ = 135 units, CG at 2 m from B

The fixed end moment diagram (A') is shown in Fig. 15.13(c). Equating the areas and moments of area,

$$\frac{(M_A + M_B)}{2} \times 8 = 60 + 225 + 135, \quad M_A + M_B = 105$$

Taking the moment @ B,

$$(2M_A + M_B) \times \frac{8^2}{6} = 60 \times \frac{20}{3} + 225 \times 4.4 + 135 \times 2, \quad 2M_A + M_B = \frac{1245}{8}$$

Solving,

$$M_A = 50.625 \text{ kN m} \quad \text{and} \quad M_B = 54.375 \text{ kN m}$$

The reactions in the fixed beam are

$$R_{AV} = 30 - \frac{54.375 - 50.625}{8} = 29.53 \text{ kN}$$

$$R_{BV} = 30 + \frac{54.375 - 50.625}{8} = 30.47 \text{ kN}$$

The SF and BM diagrams are shown in Figs 15.13(d) and (e), respectively.

Example 15.5 Moment area method, fixed beam: UD and point loads

A fixed beam has a span of 8 m and carries a uniformly distributed (UD) load of 40 kN/m over the whole length and a point load of 64 kN at 3 m from the left support. Draw the SF and BM diagrams.

Solution The beam with the loading is shown in Fig. 15.14(a). The SS BM diagram is drawn in two parts in Fig. 15.14(b), one for the point load and one for the UD load. The area of the BM diagram due to the point load is $120 \times (8/2) = 480$ units with the CG at $(8 + 5)/3$ m from B. The area of the BM diagram due to the UD load is $(2/3) \times 320 \times 8 = 5120/3$ units with the CG at mid-span. The fixed end moment diagram is shown in Fig. 15.14(c). Equating the areas and moments of area,

$$\frac{(M_A + M_B)}{2} 8 = 480 + \frac{5120}{3}, \quad M_A + M_B = \frac{1640}{3}$$

$$\frac{(2M_A + M_B)}{2} 8^2 = 480 \times \frac{13}{3} + \frac{5120}{3} \times 4, \quad 2M_A + M_B = 835$$

Solving,

$$M_A = 288.33 \text{ kN m} \quad \text{and} \quad M_B = 258.33 \text{ kN m}$$

The reactions are $R_{AV} = 40 + 160 + 3.75 = 203.75$ kN and $R_{BV} = 24 + 160 - 3.75 = 180.25$ kN. The SF diagram is shown in Fig. 15.14(d).

The general BM equation can be written as

$$M_x = -288.33 + 203.75x - \frac{40x^2}{2} - 64(x - 3)$$

the last term being valid for $x \geq 3$ only. Equating this to zero, the points of contraflexure are obtained as 1.75 m and 6.215 m from A. The BM diagram is shown in Fig. 15.14 (e).

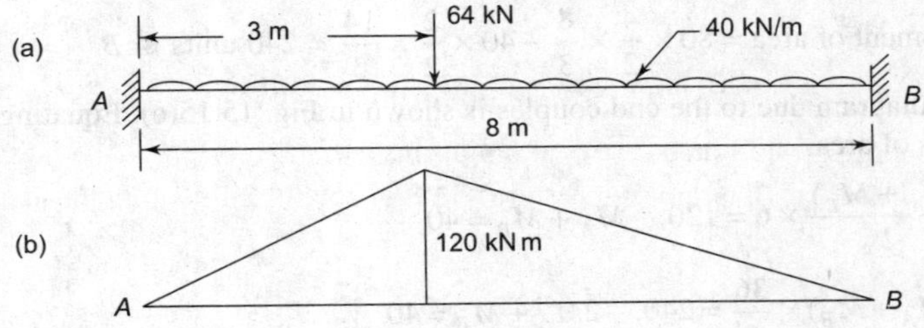

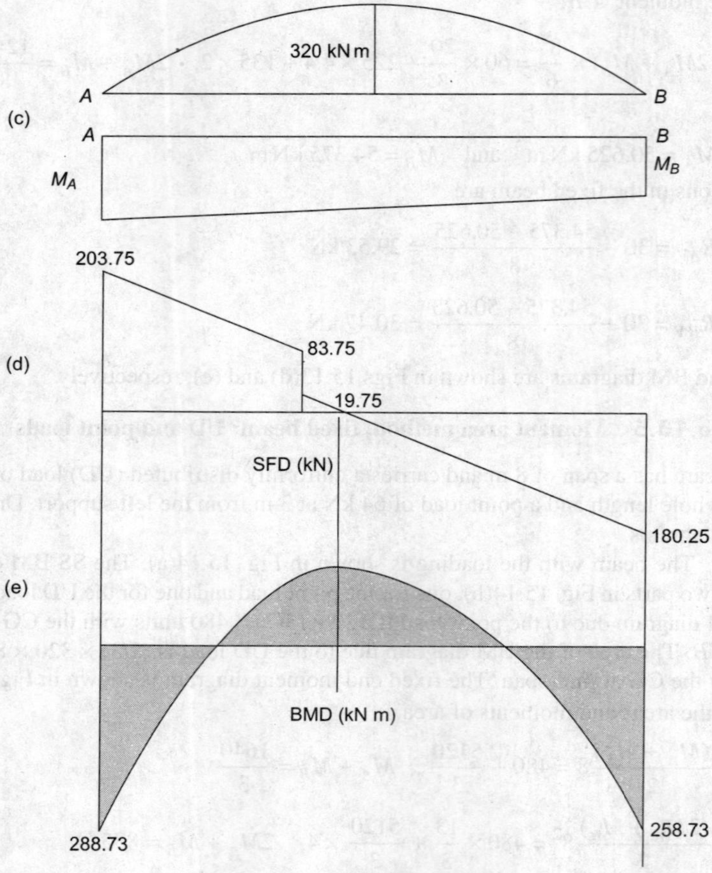

Fig. 15.14

Example 15.6 Moment area method, fixed beam: couple load

A fixed beam of 6 m span is subjected to a couple of 120 kN m at a point 2 m from the left support. Draw the SF and BM diagrams.

Solution The beam with the loading is shown in Fig. 15.15(a). The SS BM diagram due to the couple is shown in Fig. 15.15(b).

$$\text{Area} = 80 \times \frac{4}{2} - 40 \times \frac{2}{2} = 120 \text{ units}$$

$$\text{Moment of area} = 80 \times \frac{4}{2} \times \frac{8}{3} - 40 \times \frac{2}{2} \times \frac{14}{3} = 240 \text{ units @ } B$$

The SS BM diagram due to the end couples is shown in Fig. 15.15(c). Equating the areas and moments of area,

$$\frac{(M_A + M_B)}{2} \times 6 = 120, \quad M_A + M_B = 40$$

$$(2M_A + M_B) \times \frac{36}{6} = 240, \quad 2M_A + M_B = 40$$

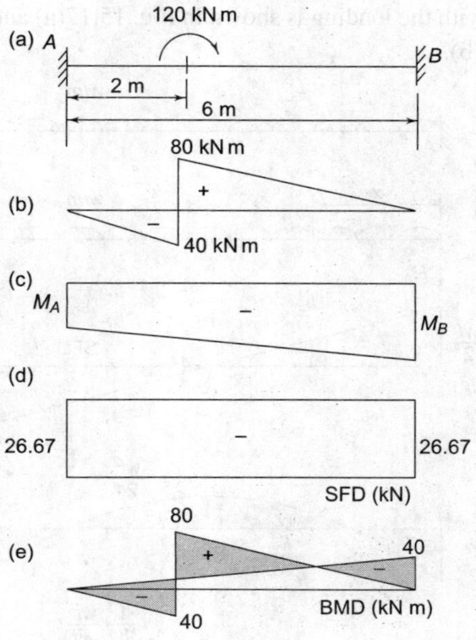

Fig. 15.15

Solving, $M_A = 0$ and $M_B = 40$. The reactions in the fixed beam are $R_{AV} = -20 - 6.67 = -26.67$ and $R_{BV} = 26.67$. The SF and BM diagrams are shown in Figs 15.15(d) and (e), respectively.

15.3.3 Double Integration Method

Considering the beam shown in Fig. 15.16, there are four conditions of deformation in a fixed beam. At $x = 0$, $\theta_A = 0$ and $y_A = 0$ and at $x = l$, $\theta_B = 0$ and $y_B = 0$. These four conditions are sufficient to evaluate the coefficients of integration, end moments, and reactions. The double integration method has the further advantage that the slopes and deflections can also be evaluated from these equations. The procedure is illustrated here with a number of examples.

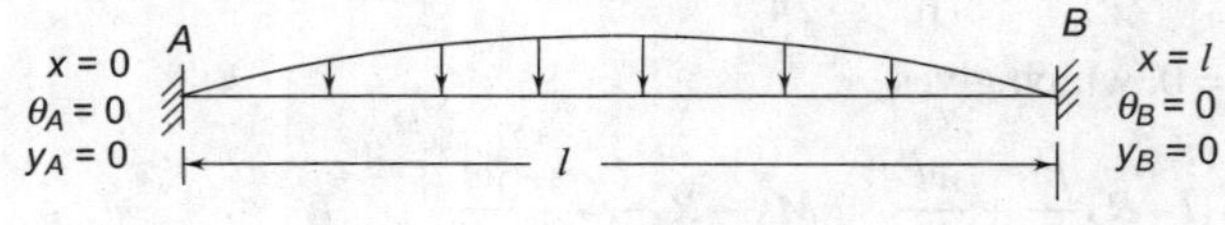

Fig. 15.16

Deformations in Fixed Beams Deformations in fixed beams can be easily determined using the double integration method. In solving fixed beams using the double integration method, we get the equations for EIy' and EIy. These can be used for finding deflections and slopes. The deflection calculation has been illustrated in some of the examples given below.

Example 15.7 Double integration, fixed beam: uniformly distributed load

A fixed beam carries a UD load of w/m throughout its length. Find the fixed end moments and draw the SF and BM diagrams.

Solution The beam with the loading is shown in Fig. 15.17(a) and the free body diagram is shown in Fig. 15.17(b).

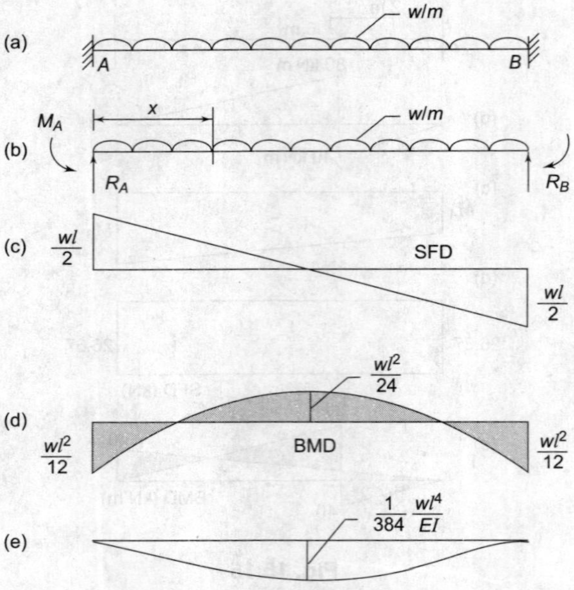

Fig. 15.17

The BM M_x at a distance x from A is

$$M_x = -M_A + R_A x - \frac{wx^2}{2}$$

$$EI\frac{d^2y}{dx^2} = -M_x = M_A - R_A x + \frac{wx^2}{2}$$

$$EI\frac{dy}{dx} = M_A x - R_A\frac{x^2}{2} + \frac{wx^3}{6} + C_1; \quad \text{at } x = 0, \quad \frac{dy}{dx} = 0 \quad \text{and} \quad C_1 = 0$$

$$EIy = M_A\frac{x^2}{2} - R_A\frac{x^3}{6} + \frac{wx^4}{24}\ C_2; \quad \text{at } x = 0, \quad y = 0, \quad \text{and} \quad \therefore C_2 = 0$$

At $x = l$, $dy/dx = 0$, which gives

$$0 = M_A l - R_A\frac{l^2}{2} + \frac{wl^3}{6}, \quad M_A - R_A\frac{l}{2} = -\frac{wl^2}{6}$$

At $x = l$, $y = 0$, which gives

$$0 = M_A\frac{l^2}{2} - R_A\frac{l^3}{6} + \frac{wl^4}{24}, \quad M_A - R_A\frac{l}{3} = \frac{wl^2}{12}$$

Solving,

$$M_A = \frac{wl^2}{12} \quad \text{and} \quad R_A = \frac{wl}{2}$$

The SF and BM diagrams are shown in Figures 15.17(c) and (d), respectively.
We have the equation for deflection as

$$EIy = (wL^2/30)x^2/2 - (3wL/20)x^3/6 + wx^5/120L$$

The`deflection at the centre is obtained by putting $x = L/2$.

$$y_C = (wL^2/30)(L/2)^2/2 - (3wL/20)(L/2)^3/6 + w(L/2)^5/120L = wL^4/768EI$$

For maximum deflection, set the equation for y' to zero.

$$EIy' = (wL^2/30)x - (3wL/20)x^2/2 + wx^4/24L = 0$$

This equation can be solved by trial and error to get $x = 0.52L$

$$EIy_{\max} = (wL^2/30)(0.52L)^2/2 - (3wL/20)(0.52L)^3/6 + w(0.52L)^5/120L$$

$$= 0.0013wL^4$$

Example 15.8 Double integration, fixed beam: uniformly varying load

A simply supported beam is subjected to a uniformly varying load, varying from zero at the left end to w/m at the right end. Draw the SF and BM diagrams for the beam.

Solution The beam with the loading is shown in Fig. 15.18(a). The general BM equation can be written from Fig. 15.18(b) as

$$M_x = -M_A + R_A x - \frac{wx^3}{6l}$$

$$EI\frac{d^2y}{dx^2} = M_A - R_A x + \frac{wx^3}{6l}$$

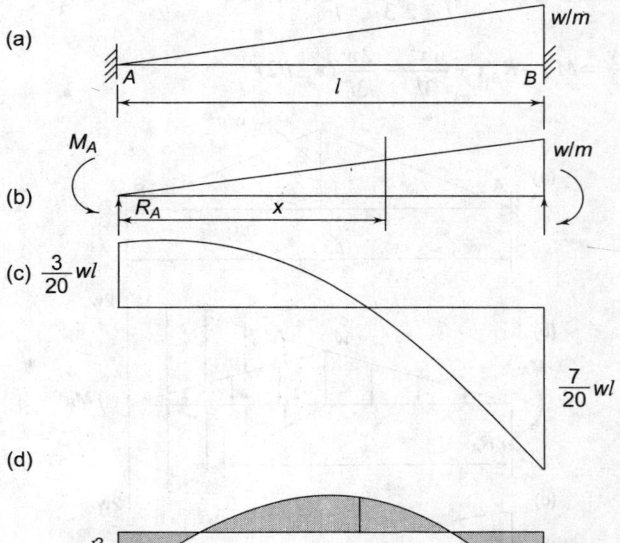

Fig. 15.18

Integrating,

$$EI\frac{dy}{dx} = M_A x - R_A\frac{x^2}{2} + \frac{wx^4}{24}C_1; \quad C_1 = 0, \text{ as at } x = 0, \frac{dy}{dx} = 0$$

$$EIy = M_A\frac{x^2}{2} - R_A\frac{x^3}{6} + \frac{wx^5}{120l} + C_2; \quad C_2 = 0, \text{ as at } x = 0, y = 0$$

$$\text{At } x = l, \quad \frac{dy}{dx} = 0, \quad M_A l - R_A \frac{l^2}{2} + \frac{wl^4}{24} = 0, \quad M_A - R_A \frac{l}{2} = -\frac{wl^2}{24}$$

$$\text{At } x = l, \quad y = 0, \quad \frac{MAl^2}{2} - \frac{R_A l^3}{6} + \frac{wl^5}{120l} = 0, \quad M_A - R_A \frac{l}{3} = -\frac{wl^2}{60}$$

Solving,

$$M_A = \frac{wl^2}{30} \quad \text{and} \quad R_A = \frac{3}{20} wl$$

Taking the moment @ B,

$$-\frac{wl^2}{30} + \frac{3wl \times l}{20} - \frac{wl}{2}\frac{l}{3} l - M_B = 0, \quad M_B = \frac{wl^2}{20}$$

The SF and BM diagrams are shown in Figures 15.18(c) and (d), respectively. ■

Example 15.9 Double integration, fixed beam: triangular load

A fixed beam carries a varying load of w/m at the centre to zero at the ends. Draw the SF and BM diagrams for the beam.

Solution The beam with loading is shown in Fig. 15.19(a). A general BM equation can be written from Fig. 15.19(b), by adding and subtracting the loads as shown, as follows:

$$M_x = -M_A + R_A x - \frac{2wx}{l}\frac{x}{2}\frac{x}{3} + \frac{2w}{l} 2\,(x - l/2)\,\frac{(x - l/2)}{2}\frac{(x - l/2)}{3}$$

$$EI\frac{d^2y}{dx^2} = M_A - R_A x + \frac{wx^3}{3l} - \frac{2w}{3l}(x - l/2)^3$$

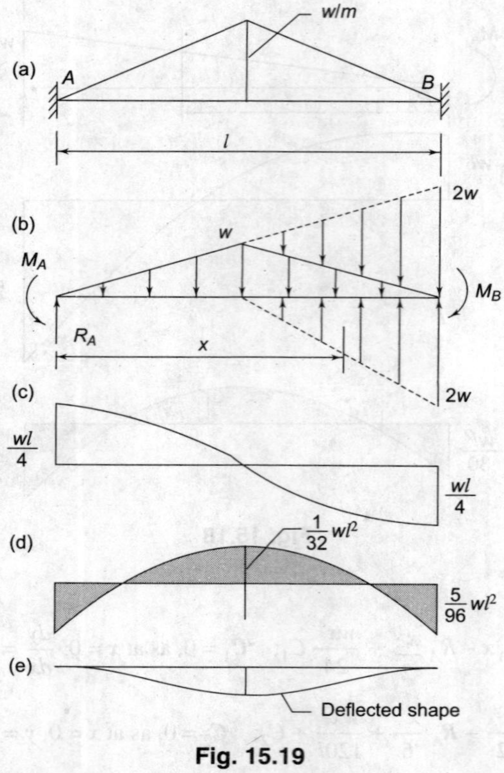

Fig. 15.19

$$EI\frac{dy}{dx}=M_Ax-R_A\frac{x^2}{2}+\frac{wx^4}{12l}-\frac{2w}{3l}\frac{(x-l/2)^4}{4}+C_1;\ C_1=0,\text{ as at } x=0, \frac{dy}{dx}=0$$

$$EIy=M_A\frac{x^2}{2}-R_A\frac{x^3}{6}+\frac{wx^5}{60l}-\frac{2w}{12l}\frac{(x-l/2)^5}{5}+C_2;\ C_2=0,\text{ as at } x=0, y=0$$

At $x = l$, $dy/dx = 0$, which gives

$$0=M_Al-R_A\frac{l^2}{2}+\frac{wl^4}{12l}-\frac{w}{6l}\frac{l^4}{16},\quad M_A-R_A\frac{l}{2}=\frac{wl^2}{96}-\frac{wl^2}{12}=\frac{7wl^2}{96}$$

At $x = l$, $y = 0$, which gives

$$0=M_A\frac{l^2}{2}-R_A\frac{l^3}{6}+\frac{wl^5}{60l}-\frac{2w}{60l}\frac{l^5}{32},\quad M_A-R_A\frac{l}{2}=\frac{wl^2}{480}-\frac{wl^2}{30}=-\frac{15wl^2}{480}$$

Solving,

$$R_A=\frac{wl}{4},\quad M_A=\frac{5}{96}wl^2,\quad \text{and } M_B=\frac{5}{96}wl^2$$

The SF and BM diagrams are shown in Figs 15.19(c) and (d), respectively. The deflection at the centre,

$$EIy=\frac{5wl^2}{96}\frac{x^2}{2}-\frac{wl}{4}\frac{x^3}{6}+\frac{wx^5}{60l}-\frac{w}{30l}(x-l/2)^5$$

At $x = l/2$,

$$y \text{ at the centre } =\left[\frac{5wl^2}{96}\frac{l^2}{8}-\frac{wl}{24}\frac{l^3}{8}+\frac{wl^5}{60\times 32\times l}\right]\frac{1}{EI}$$

$$=\frac{7wl^4}{3840EI}$$

Example 15.10 Fixed beam: part UD load

A fixed beam carries a uniformly distributed load w/m over the left half of its length. Draw the SF and BM diagrams for the beam.

Solution The beam with the load is shown in Fig. 15.20(a). From the free body diagram shown in Fig. 15.20(b), by adding and subtracting loads,

$$M_x=-M_A+R_Ax-\frac{wx^2}{2}+\frac{w(x-l/2)^2}{2}$$

$$EI\frac{d^2y}{dx^2}=M_A-R_Ax+\frac{wx^2}{2}-\frac{w(x-l/2)^2}{2}$$

$$EI\frac{dy}{dx}=M_Ax-R_A\frac{x^2}{2}+\frac{wx^3}{6}-\frac{w(x-l/2)^3}{6}+C_1;\ C_1=0,\text{ as at } x=0, \frac{dy}{dx}=0$$

$$EIy=+M_A\frac{x^2}{2}-R_A\frac{x^3}{6}+\frac{wx^4}{24}-\frac{w(x-l/2)^4}{24}+C_2;$$

C_2 is zero, as at $x = 0$, $y = 0$

At $x = l$, $dy/dx = 0$, which gives

$$0=M_Al-\frac{Rl^2}{2}+\frac{wl^3}{6}-\frac{wl^3}{48},\quad M_Al-\frac{Rl^2}{2}=\frac{wl^3}{48}-\frac{wl^3}{6}$$

At $x = l$, $y = 0$, which gives

$$0=0+\frac{M_Al^2}{2}-\frac{R_Al^3}{6}+\frac{wl^4}{24}-\frac{wl^4}{24\times 16},\quad \frac{M_Al^2}{2}-\frac{R_Al^3}{6}=\frac{wl^4}{192}-\frac{wl^4}{24}$$

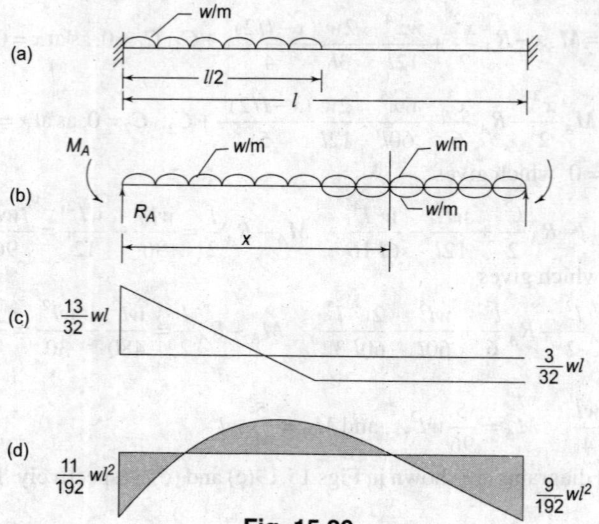

Fig. 15.20

Solving these equations,

$$R_A = \frac{13}{32}wl \quad \text{and} \quad M_A = \frac{11}{192}wl^2$$

$$R_B = \frac{3}{32}wl \quad \text{and} \quad M_B \frac{9}{192}wl^2$$

The SF and BM diagrams are shown in Figures 15.20(c) and (d), respectively.

Example 15.11 Fixed beam: UD load and couple

Draw the SF and BM diagrams for a beam loaded as shown in Fig. 15.21(a).

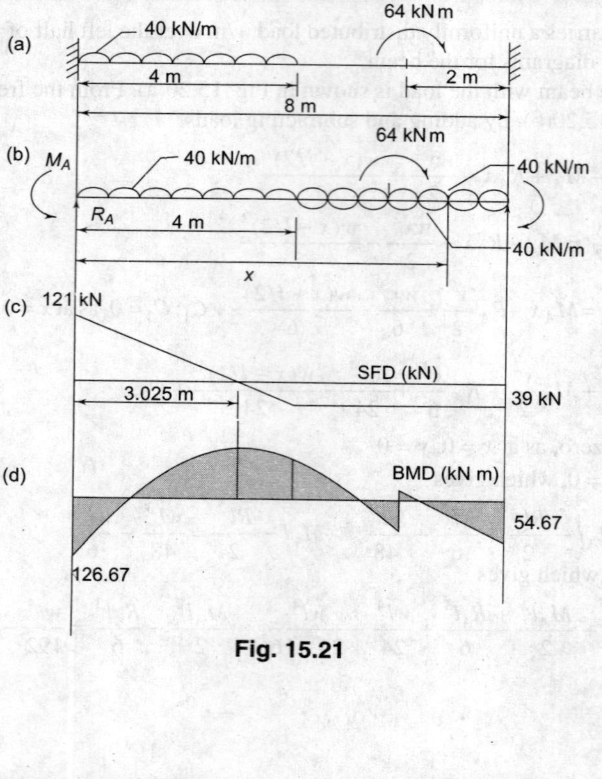

Fig. 15.21

Solution From the free body diagram shown in Fig. 15.21(b), the general expression for the BM M_x can be written as

$$M_x = -M_A + R_A x - \frac{40x^2}{2} + 64(x-6)^0 - \frac{40(x-4)^2}{2}$$

after adding and subtracting the loads shown by dotted lines. Therefore,

$$EI\frac{d^2y}{dx^2} = -M_x = M_A - R_A x + \frac{40x^2}{2} - 64(x-6)^0 - \frac{40(x-4)^2}{2}$$

$$EI\frac{dy}{dx} = M_A x - R_A\frac{x^2}{2} + \frac{40x^3}{6} - 64(x-6) - \frac{40(x-4)^3}{6} + C_1$$

$C_1 = 0$, since at $x = 0$, $dy/dx = 0$ (the last two terms are not valid when $x = 0$).

$$EIy = M_A\frac{x^2}{2} - R_A\frac{x^3}{6} + \frac{40x^4}{24} - \frac{64(x-6)^2}{2} - \frac{40(x-4)^4}{24} + C_2$$

$C_2 = 0$, since at $x = 0, y = 0$. Also, at $x = 8$, $dy/dx = 0$:

$$0 = M_A - 8 - R_A\frac{64}{2} + \frac{40\times 8^3}{6} - 64\times 2 - 40\times\frac{64}{6}, \quad M_A - 4R_A = -357.33$$

At $x = 8$, $y = 0$:

$$0 = M_A\frac{64}{2} - R_A\frac{8^3}{6} + \frac{40\times 8^4}{24} - \frac{64\times(8-6)^2}{2} - \frac{40\times(8-4)^4}{24}, \quad M_A - \frac{8}{3}R_A = 196$$

Solving these equations, we get $ssR_A = 121$ and $M_A = 126.67$. Also, the reactions at support B can be calculated as $R_B = 39$ and $M_B = 54.67$. The SF and BM diagrams are shown in Figs 15.21(c) and (d), respectively.

15.4 ADVANTAGES AND DISADVANTAGES OF FIXED BEAMS

From the examples illustrated so far, we have seen that for the same span and load, fixed beams have lower BMs than SS beams. Thus, in the case illustrated in Fig. 15.22, the SS beam will have to be designed for a BM of 80 kN m at mid-span. If the same beam were to be fixed, the maximum BM would be 53.33 kN m (hogging) at the supports and the positive BM at mid-span would be only 26.67 kN m. We have also seen that the fixed beam has only one-fifth of the deflection compared to that in the SS beam.

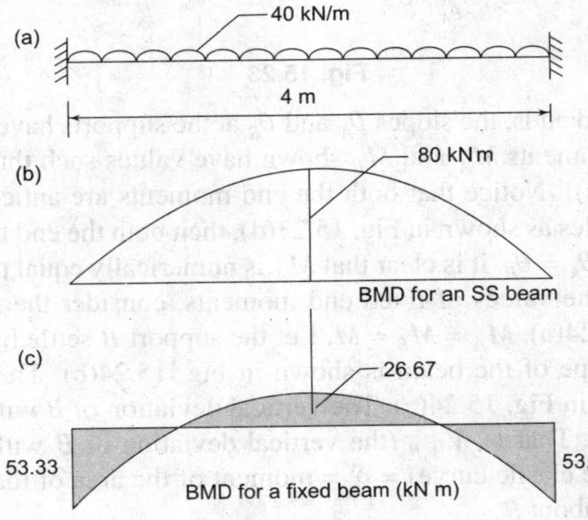

Fig. 15.22

While this shows that fixed beams have many advantages, they also have some critical disadvantages. If the condition of fixity at the ends is disturbed, as, for example, by settlement and rotational yield of supports, large BMs are induced in the beam to counter this effect. Similarly, changes in temperature also lead to large stresses. We investigate these disadvantages by taking up the case of support settlement and its effect on fixed beams.

15.5 SETTLEMENT OF SUPPORTS

Consider the beam shown in Fig. 15.23. If one support settles more than the other, leading to the differential settlement δ_v, then additional BMs are induced in the beam. To determine the effect of this settlement, consider an SS beam in which the support B settles by the amount δ_v. Since the beam can freely rotate about A, this will not create any additional effects on the SS beam. The dotted line shows the position of the beam [Fig. 15.23(b)].

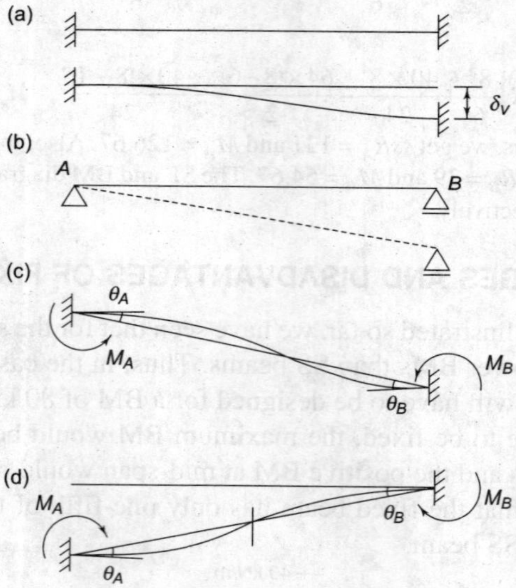

Fig. 15.23

In the case of fixed ends, the slopes θ_A and θ_B at the supports have to be zero after settlement; the moments M_A and M_B shown have values such that θ_A and θ_B are zero [Fig. 15.23(c)]. Notice that both the end moments are anticlockwise. If the left support A settles as shown in Fig. 15.23(d), then both the end moments will be clockwise. Since $\theta_A = \theta_B$, it is clear that M_A is numerically equal to M_B.

To determine the values of fixed end moments, consider the fixed end beam shown in Fig. 15.24(a). $M_A = M_B = M$. Let the support B settle by an amount δ_v. The deflected shape of the beam is shown in Fig. 15.24(b). The fixed end BM diagram is shown in Fig. 15.24(c). The vertical deviation of B with respect to the tangent at A is δ_v. That is, $V_{A,B}$ (the vertical deviation of B with respect to the tangent at A of the elastic curve) = δ_v = moment of the area of the M/EI diagram between A and B about B.

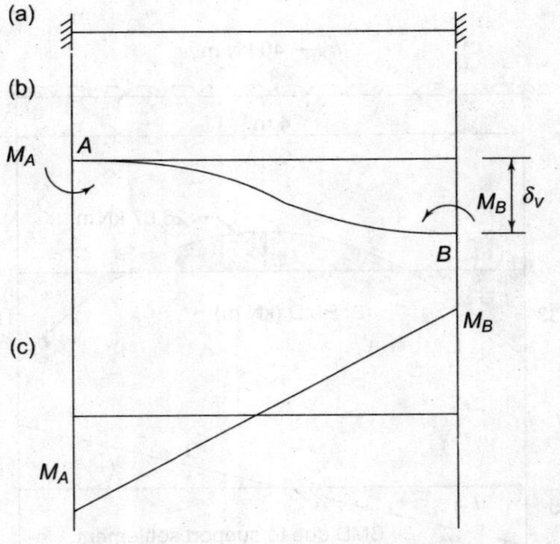

Fig. 15.24

Therefore,

$$\frac{1}{EI}\left[-M_A\frac{l}{2}\times\frac{1}{2}\left(\frac{l}{2}+\frac{2}{3}\frac{l}{2}\right)+M_B\frac{l}{2}\times\frac{1}{2}\times\frac{1}{3}\frac{l}{2}\right]=\delta_v$$

$$\frac{1}{EI}\left[-\frac{5M_A l^2}{24}+\frac{M_B l^2}{24}\right]=\delta_v$$

Since $M_A = M_B = M$,

$$M=\frac{6EI\delta_v}{l^2}$$

Thus, the settlement of supports (or the difference in it) induces equal fixing moments at the supports. Both the moments are anticlockwise if the right support is lower than the left support and clockwise if the left support is lower than the right support.

Example 15.12 Fixed beam: support settlement

A fixed beam of 4 m span carries a UD load of 40 kN/m over its whole length. Compare the maximum BM in this beam with the fixed end moments due to the right support settling by 10 mm. E = 200 GN/m^2 and I = 5000 cm^4.

Solution The beam with the loading is shown in Fig. 15.25(a). The BM diagram of this beam is shown in Fig. 15.25(b). The maximum BM in this beam is at the supports and is equal to $wl^2/12 = (40 \times 4^2)/12 = 53.33$ kN m. If the support settles by 10 mm, the end moments are as shown in Fig. 15.25(c) and are equal to $6EI\delta_v/l^2$.

$$M_A' = M_B' = \frac{6EI\delta_v}{l^2} = \frac{6\times 200\times 10^9\times 5000\times 10^{-8}\times 0.01}{4\times 4} = 37.5 \text{ kN m}$$

which is 70% of the maximum BM in the beam. The BM diagram is shown in Fig. 15.25(c).

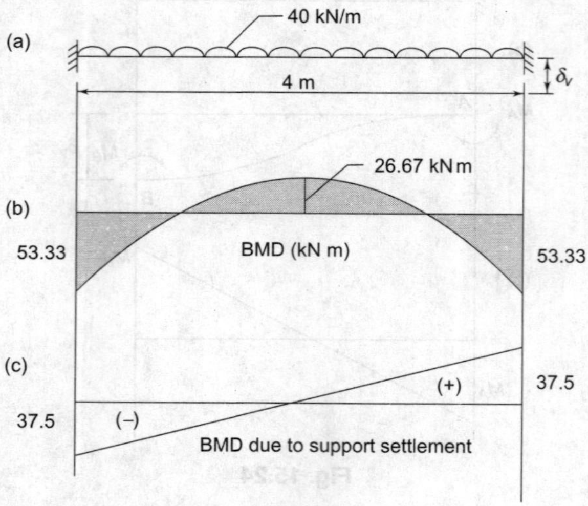

Fig. 15.25

15.6 CONTINUOUS BEAMS

A beam resting on more than two supports [Fig. 15.26(a)] is called a continuous beam. The beam ends may be overhanging, or fixed as in Fig. 15.26(b). Continuous beams are very common in buildings, bridges, etc. They are statically indeterminate structures. There are many methods for the analysis of continuous beams. One method, called *Clapeyron's theorem of three moments*, is dealt with in this section.

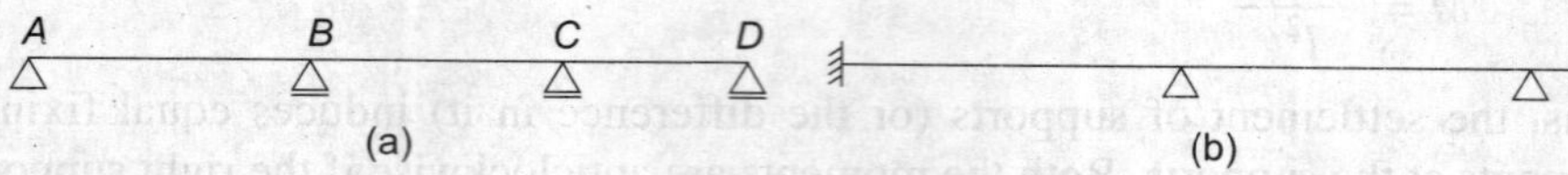

Fig. 15.26

15.6.1 Structural Action of a Continuous Beam

Due to the continuity of the beam over the supports, the loads on any span affect the behaviour of the beam on adjacent spans. Considering the beam shown in Fig. 15.27(a), the spans AB and BC can be considered to be simply supported, breaking the continuity at B. The deflected shapes of these simple spans are shown in Fig. 15.27(b). Notice the kink in the deflected shape at support B. This cannot happen in an actual structure and, due to continuity, the deflected shape may be as shown in Fig. 15.27(c). The slopes at B in the spans AB and BC have to be equal. This is achieved in effect in a continuous beam by the moments developed at support B as shown. The support moment is hogging, resulting in the final BM diagram shown in Fig. 15.27(e). The positive moments in mid-span are considerably reduced, but there is hogging moment at the supports.

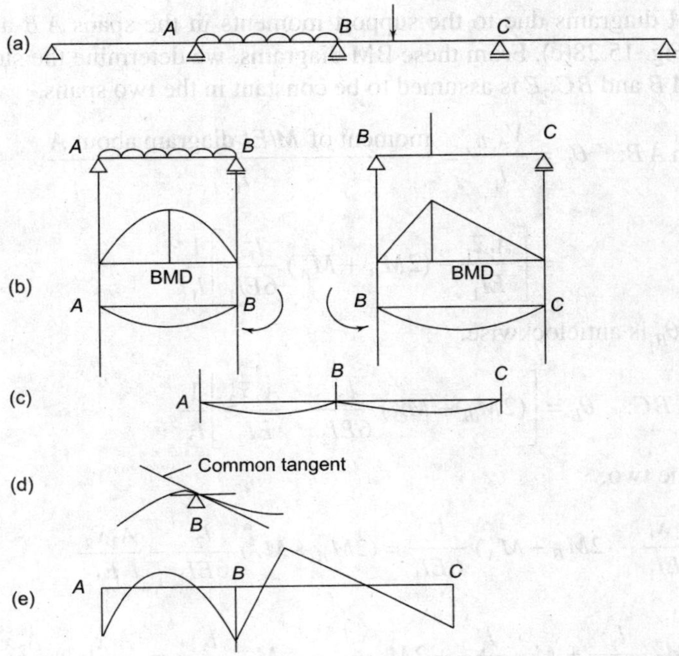

Fig. 15.27

15.6.2 Clapeyron's Theorem of Three Moments

The theorem of three moments is based on the concept of consistent deformations. This relates the support moments at any three consecutive supports to the SS moment diagram due to the loads (of consecutive spans), which when determined makes the beam statically determinate. To derive the theorem of three moments, consider any two consecutive spans AB and BC [Fig. 15.28(a)]. (The supports at A, B, and C are assumed to be at the same level for simplicity.) The two spans AB and BC are considered to be simply supported, subject to the loads. The BM diagrams are shown in Fig. 15.28(b). The support moments are M_A, M_B, and M_C, all assumed to be hogging and hence negative.

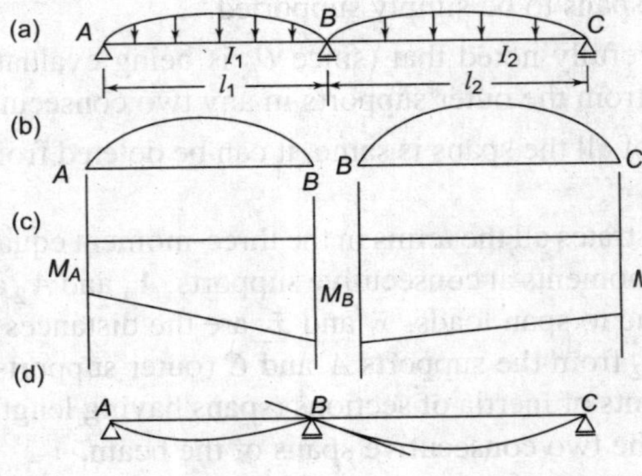

Fig. 15.28

The BM diagrams due to the support moments in the spans AB and BC are shown in Fig. 15.28(c). From these BM diagrams, we determine the slope at B in the spans AB and BC. E is assumed to be constant in the two spans.

$$\text{Span } AB: \quad \theta_B = \frac{V_{A,B}}{l_1} = \frac{\text{moment of } M/EI \text{ diagram about } A}{l_1}$$

$$= \left[\frac{A_1\bar{x}_1}{EI_1} - (2M_B + M_A)\frac{l_1^2}{6EI_1}\right]\frac{1}{l_1}$$

The slope θ_B is anticlockwise.

$$\text{Span } BC: \quad \theta_B = \left[(2M_B + M_C)\frac{l_2^2}{6EI_2} - \frac{A_2\bar{x}_2}{EI_2}\right]\frac{1}{l_2}$$

Equating the two,

$$\frac{A_1\bar{x}_1}{l_1 EI_1} - (2M_B + M_A)\frac{l_1}{6EI_1} = (2M_B + M_C)\frac{l_2}{6EI_2} - \frac{A_2\bar{x}_2}{l_2 EI_2}$$

$$2M_B\frac{l_1}{6EI_1} + M_A\frac{l_1}{6EI_1} + 2M_B\frac{l_2}{6EI_2} + M_C\frac{l_2}{6EI_2} = \frac{A_1\bar{x}_1}{EI_1 l_1} + \frac{A_2\bar{x}_2}{EI_2 l_2}$$

Since E is common in all the terms, it can be deleted. Rearranging,

$$M_A\frac{l_1}{I_1} + 2M_B\left(\frac{l_1}{I_1} + \frac{l_2}{I_2}\right) + M_C\frac{l_2}{I_2} = \frac{6A_1\bar{x}_1}{I_1 l_1} + \frac{6A_2\bar{x}_2}{I_2 l_2}$$

This is the general equation of the theorem of three moments for supports at the same level. The following points should be noted with reference to the above equation.

(i) M_A, M_B, and M_C are assumed to be hogging. In calculating their values, a positive value obtained means that they are hogging.

(ii) A_1 and A_2 are the areas of the BM diagrams obtained due to applied loads, assuming the spans to be simply supported.

(iii) It must be carefully noted that (since θ_B is being evaluated) $\bar{x}_1$ and $\bar{x}_2$ are always taken from the outer supports in any two consecutive spans.

(iv) If the I value of all the spans is same, it can be deleted from the above equation.

Figure 15.29 illustrates all the terms in the three-moment equation. M_A, M_B, and M_C are the support moments at consecutive supports. A_1 and A_2 are the areas of the SS BM diagrams due to span loads. $\bar{x}_1$ and $\bar{x}_2$ are the distances of the CGs of the BM areas A_1 and A_2 from the supports A and C (outer supports in two spans). I_1 and I_2 are the moments of inertia of sections (spans having lengths l_1 and l_2) of the beam. l_1 and l_2 are the two consecutive spans of the beam.

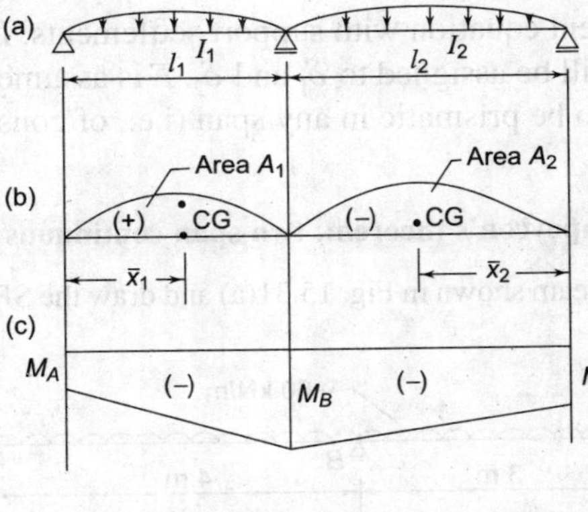

Fig. 15.29

15.6.3 Three-Moment Equation with Support Settlements

The equation of three moments derived in Section 15.6.2 assumed that the three supports are at the same level. When the supports are not at the same level, additional moments are introduced in the beam. Referring to Fig. 15.30, let δ_1 be the difference in level between the supports A and B (B is lower than A), δ_2 the difference in level between the supports B and C (B is lower than C), $V_{B,A}$ the vertical deflection of A with respect to the tangent at B, and $V_{B,C}$ the vertical deflection of C with respect to the tangent at B.

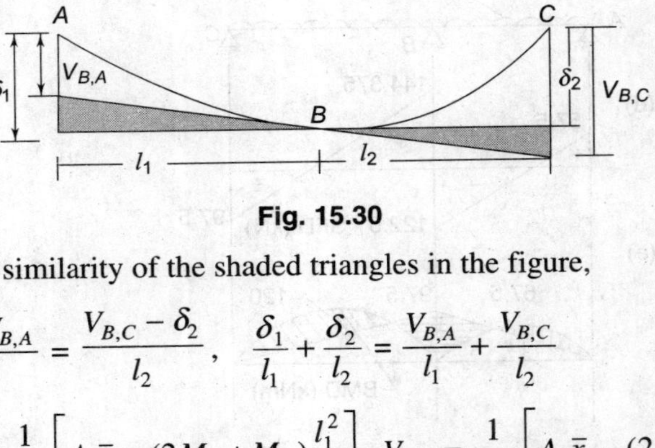

Fig. 15.30

Then, from the similarity of the shaded triangles in the figure,

$$\frac{\delta_1 - V_{B,A}}{l_1} = \frac{V_{B,C} - \delta_2}{l_2}, \quad \frac{\delta_1}{l_1} + \frac{\delta_2}{l_2} = \frac{V_{B,A}}{l_1} + \frac{V_{B,C}}{l_2}$$

$$V_{B,A} = \frac{1}{EI_1}\left[A_1\bar{x}_1 - (2M_B + M_A)\frac{l_1^2}{6}\right], \quad V_{B,C} = \frac{1}{EI_2}\left[A_2\bar{x}_2 - (2M_B + M_A)\frac{l_2^2}{6}\right]$$

$$\frac{\delta_1}{l_1} + \frac{\delta_2}{l_2} = \frac{1}{EI_1 l_1}\left[A_1\bar{x}_1 - (2M_B + M_A)\frac{l_1^2}{6}\right] + \frac{1}{EI_2 l_2}\left[A_2\bar{x}_2 - (2M_B + M_A)\frac{l_2^2}{6}\right]$$

$$(2M_B + M_A)\frac{l_1}{6EI_1} + (2M_B + M_C)\frac{l_2}{6EI_2} = \frac{A_1\bar{x}_1}{EI_1 l_1} + \frac{A_2\bar{x}_2}{EI_2 l_2} - \left[\frac{\delta_1}{l_1} + \frac{\delta_2}{l_2}\right]$$

$$M_A\frac{l_1}{I_1} + 2M_B\left(\frac{l_1}{I_1} + \frac{l_2}{I_2}\right) + M_C\frac{l_2}{I_2} = \frac{6A_1\bar{x}_1}{I_1 l_1} + \frac{6A_2\bar{x}_2}{I_2 l_2} - 6E\left[\frac{\delta_1}{l_1} + \frac{\delta_2}{l_2}\right]$$

This is the three-moment equation with support settlements. If A or C is below B, then a negative sign will be assigned to δ_1 and δ_2. E is assumed to be constant and the beam is assumed to be prismatic in any span (i.e., of constant cross section in any span).

Example 15.13 Clapeyron's theorem: two-span continuous beam

Analyse the continuous beam shown in Fig. 15.31(a) and draw the SF and BM diagrams. *EI* is constant.

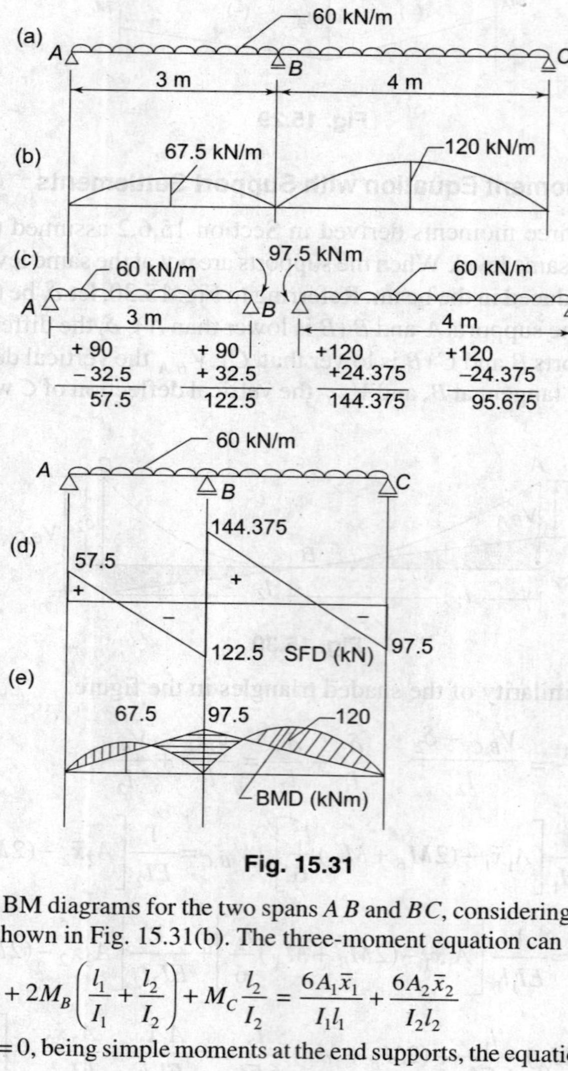

Fig. 15.31

Solution The BM diagrams for the two spans AB and BC, considering them to be simply supported, are shown in Fig. 15.31(b). The three-moment equation can be written as

$$M_A\frac{l_1}{I_1}+2M_B\left(\frac{l_1}{I_1}+\frac{l_2}{I_2}\right)+M_C\frac{l_2}{I_2}=\frac{6A_1\bar{x}_1}{I_1l_1}+\frac{6A_2\bar{x}_2}{I_2l_2}$$

Since $M_A=M_C=0$, being simple moments at the end supports, the equation reduces to

$$2M_B\left(\frac{3}{I}+\frac{4}{I}\right)=\frac{6A_1\bar{x}_1}{I\times 3}+\frac{6A_2\bar{x}_2}{I\times 4},\quad 14M_B=\frac{6A_1\bar{x}_1}{3}+\frac{6A_2\bar{x}_2}{4}$$

$$A_1=\frac{2}{3}\times 67.5\times 3=135 \text{ units} \quad \text{and} \quad \bar{x}_1=1.5 \text{ m from } A$$

$$A_2 = \frac{2}{3} \times 120 \times 4 = 320 \text{ units} \quad \text{and} \quad \bar{x}_2 = 2 \text{ m from } C$$

Therefore,

$$14M_B = \frac{6 \times 135 \times 1.5}{3} + \frac{6 \times 320 \times 2}{4}, \quad M_B = 97.5 \text{ kNm}$$

The reactions can be calculated as shown in Fig. 15.31(c). The SF and BM diagrams can be drawn as shown in Figures 15.31(d) and (e).

Example 15.14 Clapeyron's theorem: two-span beam with over-hang

Analyse the continuous beam loaded as shown in Fig. 15.32(a) and draw the SF and BM diagrams for the beam.

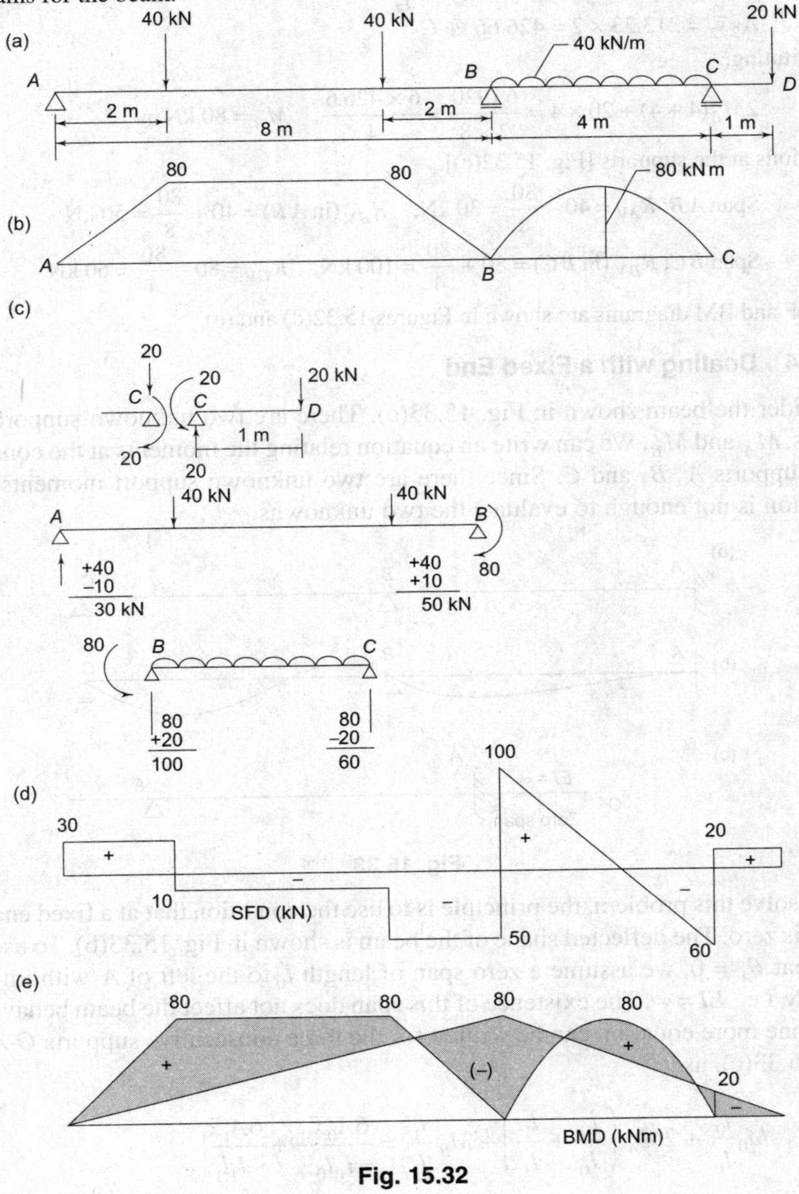

Fig. 15.32

Solution From the support conditions, $M_A = 0$ and $M_C = 20 \times 1 = -20$ kN m. The three-moment equation can be written as

$$M_A \frac{8}{2I} + 2M_B\left(\frac{8}{2I} + \frac{4}{I}\right) + M_C \frac{4}{1} = \frac{6A_1\bar{x}_1}{2I \times 8} + \frac{6A_2\bar{x}_2}{I \times 4}$$

The SS BM diagrams for the spans AB and BC are shown in Fig. 15.32(b).

Span AB: moment of area @ $A = 80 \times \frac{4}{3} + 320 \times 4 + 80 \times \frac{20}{3} = A_1\bar{x}_1$

$A_1\bar{x}_1 = 1920$

Span BC: area of BM diagram $= \frac{2}{3} \times 80 \times 4 = 213.33$

$A_2\bar{x}_2 = 213.33 \times 2 = 426.66$ @ C

Substituting,

$$2M_B(4+4) + 20 \times 4 = \frac{6.1920}{2 \times 8} + \frac{6 \times 426.6}{4}, \quad M_B = 80 \text{ kN m}$$

Reactions at the supports [Fig. 15.32(c)]:

Span AB: $R_{AV} = 40 - \frac{80}{8} = 30$ kN, R_{BV} (in AB) $= 40 + \frac{80}{8} = 50$ kN

Span BC: R_{BV} (in BC) $= 80 + \frac{80}{4} = 100$ kN, $R_{CV} = 80 - \frac{80}{4} = 60$ kN

The SF and BM diagrams are shown in Figures 15.32(d) and (e).

15.6.4 Dealing with a Fixed End

Consider the beam shown in Fig. 15.33(a). There are two unknown support moments, M_A and M_B. We can write an equation relating the moments at the consecutive supports A, B, and C. Since there are two unknown support moments, this equation is not enough to evaluate the two unknowns.

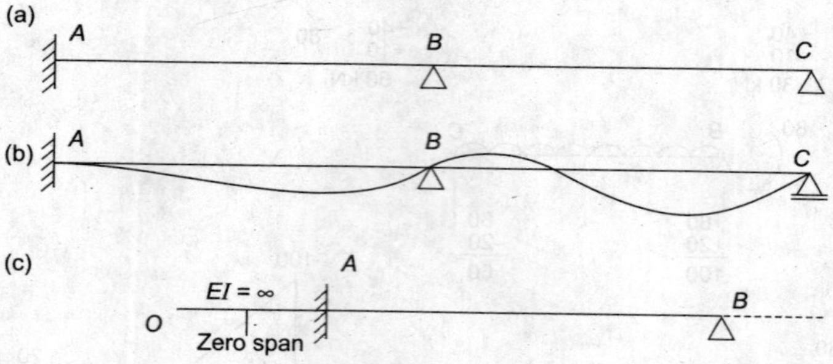

Fig. 15.33

To solve this problem, the principle is to use the condition that at a fixed end the slope is zero. The deflected shape of the beam is shown in Fig. 15.33(b). To use the fact that $\theta_A = 0$, we assume a zero span of length l_0 to the left of A with infinite rigidity, i.e., $EI = \infty$. The existence of this span does not affect the beam behaviour. Now one more equation can be written for the three consecutive supports O-A-B, Fig. 15.33(c), as

$$M_0 \frac{l_0}{I_0} + 2M_A\left(\frac{l_0}{I_0} + \frac{l_1}{I_1}\right) + M_B \frac{l_1}{I_1} = \frac{6A_0\bar{x}_0}{I_0 l_0} + \frac{6A_1\bar{x}_1}{I_1 l_1}$$

Noting that I_0 is infinity, this equation reduces to

$$2M_A\frac{l_1}{I_1}+M_B\frac{l_1}{I_1}=\frac{6A_1\bar{x}_1}{I_1l_1}$$

This is the additional equation that can be written for a continuous beam with its fixed end at A.

Example 15.15 Clapeyron's theorem: continuous beam with fixed end

Analyse the continuous beam shown in Fig. 15.34(a) and draw the SF and BM diagrams.

Solution The BM diagrams for the spans, considering them to be simply supported, are shown in Fig. 15.34(b). For the span AB, the area is 360 units. The distance of the centroid of the area is (6 + 2)/3 m from A and (6 + 4)/3 m from B. For the span BC, the area is [(20 × 4^2)/8 × 4] × (2/3) = 320/3 units. The distance of the centroid is 2 m from B and C.

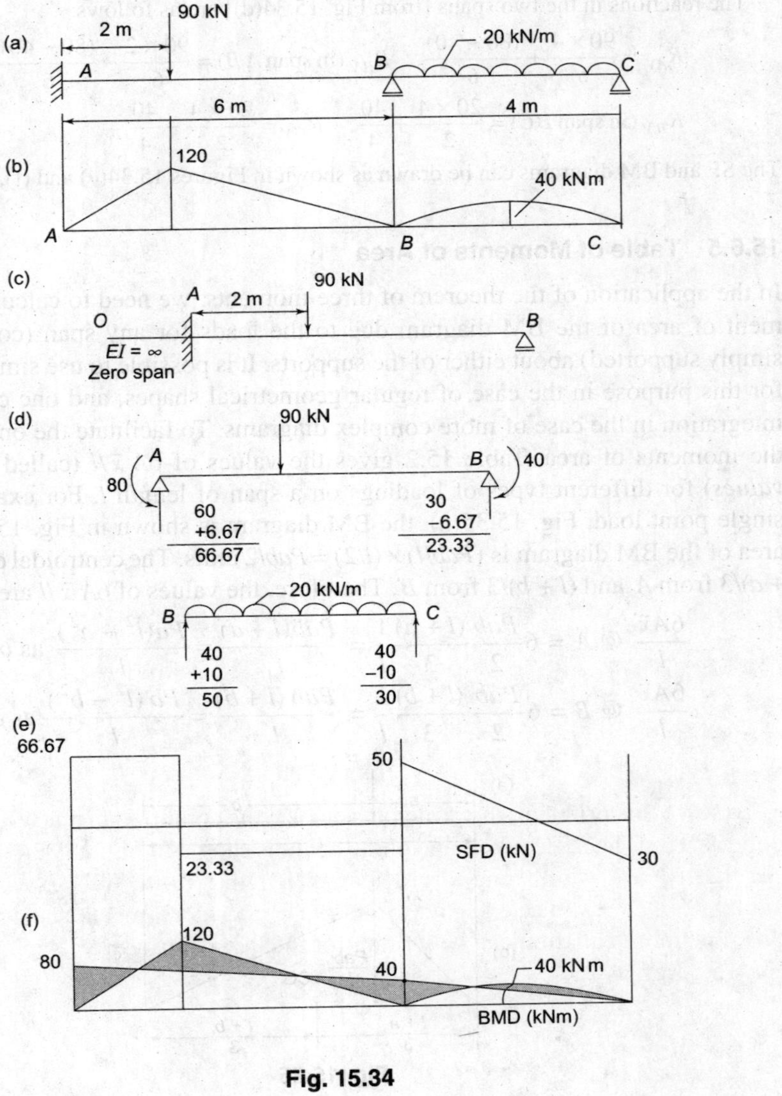

Fig. 15.34

Assuming a zero span as shown in Fig. 15.34(c), the equations for the two consecutive spans OA and AB are

$$M_0\frac{l_0}{I_0}+2M_A\left(\frac{l_0}{I_0}+\frac{l_1}{I_1}\right)+M_B\frac{l_1}{I_1}=\frac{6A_1\bar{x}_1}{l_1 I_1}$$

$$2M_A\frac{6}{1.5I}+M_B\frac{6}{1.5I}=\frac{6\times 360\times 10/3}{6\times 1.5I},\quad 2M_A+M_B=200$$

For the spans AB and BC, the three-moment equation can be written as

$$M_A\frac{6}{1.5I}+2M_B\left(\frac{6}{1.5I}+\frac{4}{I}\right)+M_C\frac{4}{I}=\frac{6\times 360\times 8}{6\times 1.5I\times 3}+\frac{6\times 320}{3}\times\frac{2}{4I}$$

$M_C=0$; therefore, $4M_A+16M_B=960$. Solving the two equations for M_A and M_B, $M_A=80$ kN m and $M_B=40$ kN m.

The reactions in the two spans [from Fig. 15.34(d)] are as follows:

$$R_{AV}=\frac{90\times 4}{6}+\frac{(80-40)}{6},\quad R_{BV}\,(\text{in span }AB)=\frac{90\times 2}{6}-\frac{(80-40)}{6}$$

$$R_{BV}\,(\text{in span }BC)=\frac{20\times 4}{2}+\frac{40}{4},\quad R_{CV}=\frac{20\times 4}{2}-\frac{40}{4}$$

The SF and BM diagrams can be drawn as shown in Figures 15.34(e) and (f), respectively.

15.6.5 Table of Moments of Area

In the application of the theorem of three moments, we need to calculate the moment of area of the BM diagram due to the loads for any span (considering it simply supported) about either of the supports. It is possible to use simple ormulae for this purpose in the case of regular geometrical shapes, and one can resort to integration in the case of more complex diagrams. To facilitate the omputation of the moments of area, Table 15.2 gives the values of $6A\bar{x}/l$ (called *load factor values*) for different types of loadings on a span of length l. For example, for a single point load, Fig. 15.35(a), the BM diagram is shown in Fig. 15.35(b). The area of the BM diagram is $(Pab/l)\times(l/2)=Pab/2$ units. The centroidal distance is $(l+a)/3$ from A and $(l+b)/3$ from B. Therefore, the values of $6A\bar{x}/l$ are as follows:

$$\frac{6A\bar{x}}{l}\;@\,A=6\frac{Pab}{2}\frac{(l+a)}{3}\frac{1}{l}=\frac{Pab(l+a)}{l}=\frac{Pa(l^2-a^2)}{l}\quad\text{as } b=l-a$$

$$\frac{6A\bar{x}}{l}\;@\,B=6\frac{Pab}{2}\frac{(l+b)}{3}\frac{1}{l}=\frac{Pab\,(l+b)}{l}=\frac{Pb\,(l^2-b^2)}{l}\quad\text{as } a=l-b$$

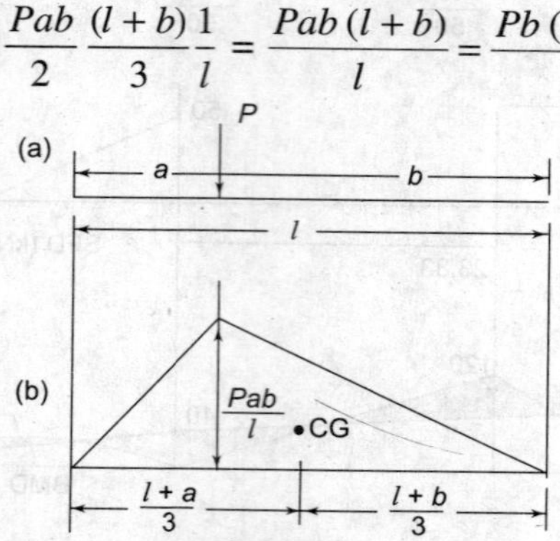

Fig. 15.35

Table 15.2 Values of $6A\bar{x}/l$ for use with Clapeyron's theorem of three moments

	Loading	$6a\bar{x}/l$	
		Left	**Right**
1.	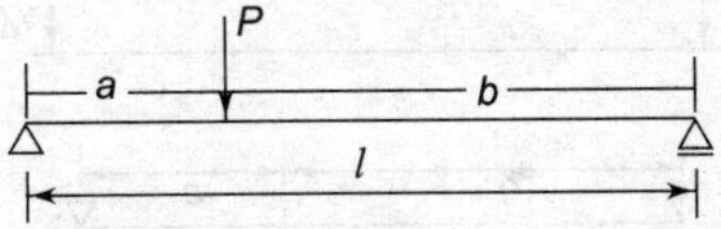	$\dfrac{Pab(l+a)}{l}$	$\dfrac{Pab(l+b)}{l}$
2.	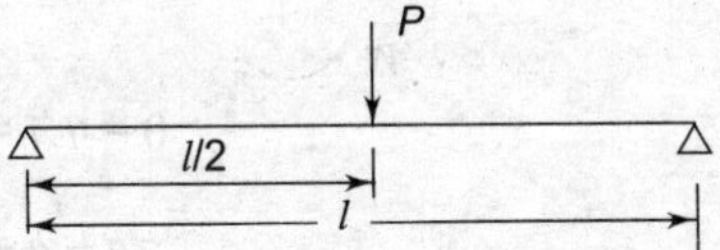	$\dfrac{3}{8}Pl^2$	$\dfrac{3}{8}Pl^2$
3.	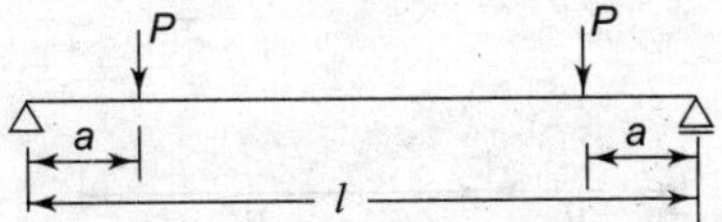	$3Pa(l-a)$	$3Pa(l-a)$
4.	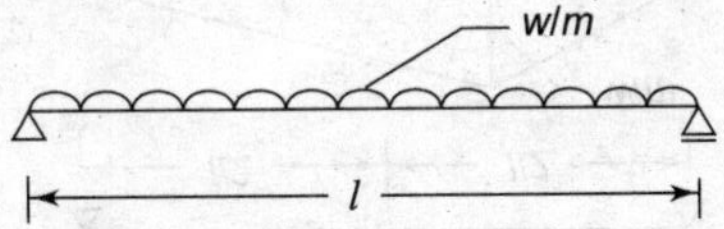	$\dfrac{wl^3}{4}$	$\dfrac{wl^3}{4}$
5.	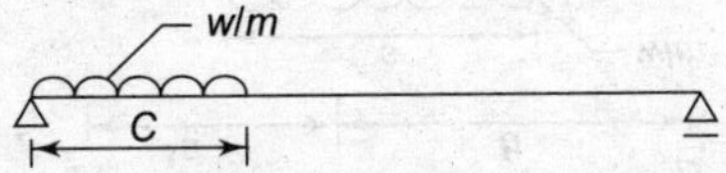	$\dfrac{wc^2}{4l}(2l^2-c^2)$	$\dfrac{wc^2}{4l}(2l-c)^2$
6.	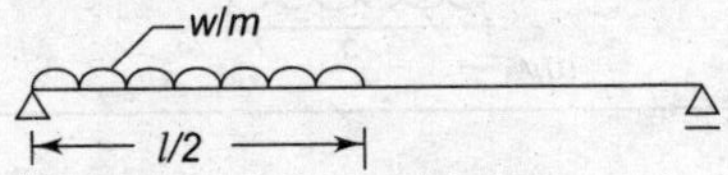	$\dfrac{7}{64}wl^3$	$\dfrac{9}{64}wl^3$

(contd)

Table 15.2 *(contd)*

	Loading	$6a\bar{x}/l$ Left	$6a\bar{x}/l$ Right
7.	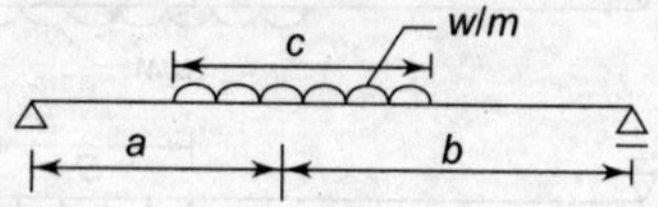	$\frac{wca}{l}\left(l^2 - a^2 - \frac{c^2}{4}\right)$	$\frac{wca}{l}\left(l^2 - a^2 - \frac{c^2}{4}\right)$
8.	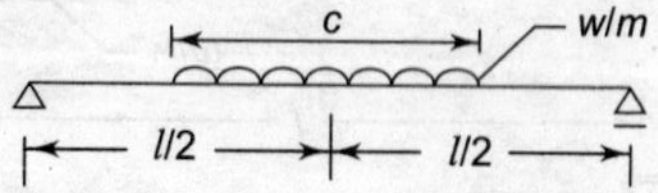	$\frac{wc}{8}(3l^2 - c^2)$	$\frac{wc}{8}(3l^2 - c^2)$
9.	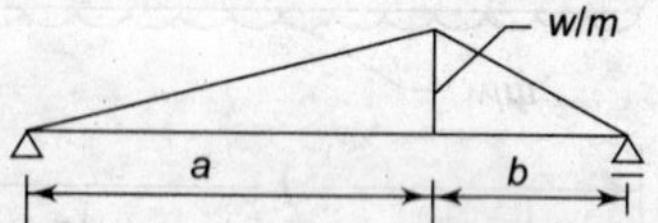		
(i)		$w\frac{(l+a)}{60}(7l^2 - 3a^2)$	$w\frac{(l+a)}{60}(7l^2 - 3b^2)$
(ii)	$a = b = l/2$	$\frac{5}{32}ql^3$	$\frac{5}{32}ql^3$
(iii)	$a = l, b = 0$	$\frac{7}{60}ql^3$	$\frac{8}{60}ql^3$
10.	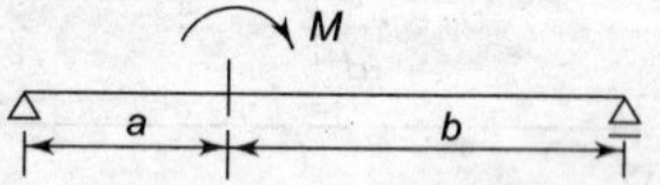	$\frac{M}{l}(l^2 - 3a^2)$	$-\frac{M}{l}(l^2 - 3b^2)$
11.	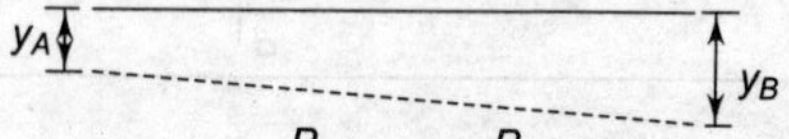	$\frac{6EI}{l}(y_B - y_A)$	$-\frac{6EI}{l}(y_B - y_A)$
12.	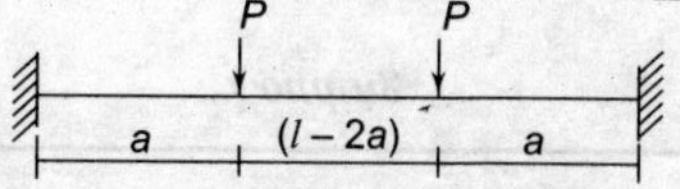	$3Pa(l - a)$	$3Pa(l - a)$

These values are listed under the 'left' and 'right' columns, respectively, of Table 15.2, since they are about the left and right supports of the span. Similarly, the values for other loading cases are listed.

Example 15.16 Clapeyron's theorem: three-span beam with overhang

Analyse the continuous beam shown in Fig. 15.36(a) and draw the SF and BM diagrams.

Solution From the end conditions of the beam, we can deduce that $M_A = 0$ and $M_D = 30 \times 1 \times 1/2 = 15$ kN m.

Spans AB and BC The simply supported BM diagrams due to loads are shown in Fig. 15.36(b).

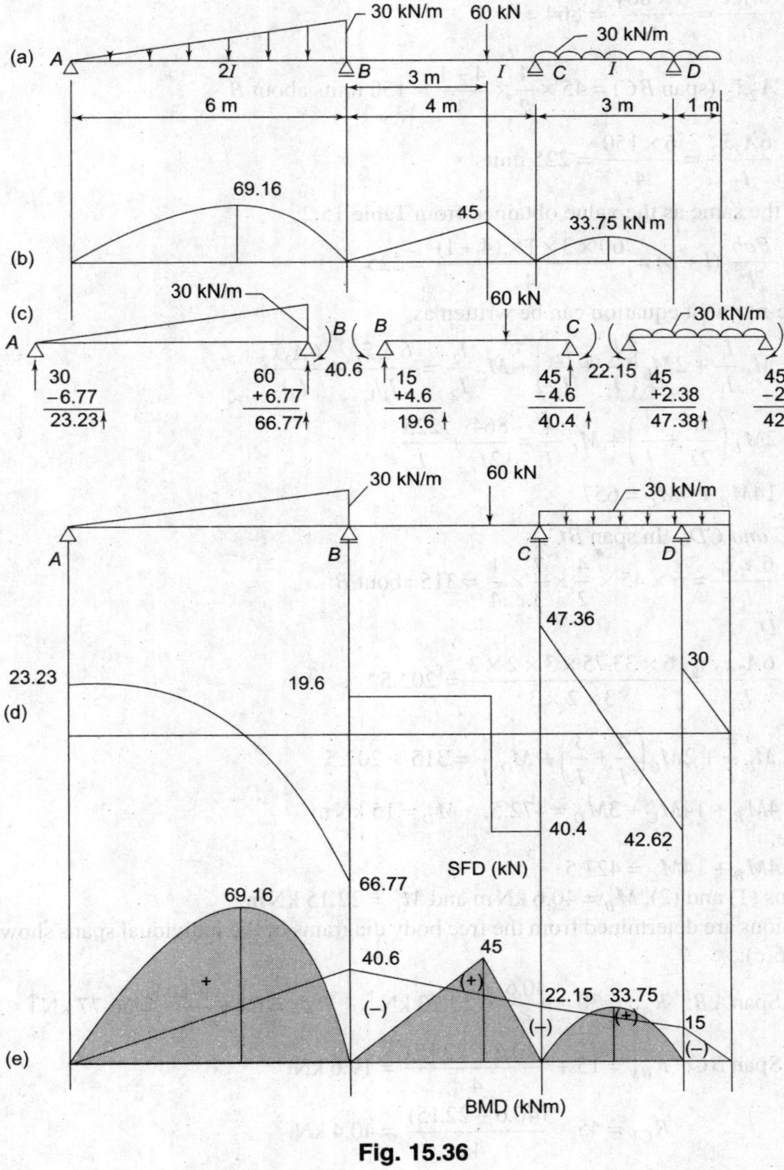

Fig. 15.36

$$A_1\bar{x}_1 \text{ (span } AB) = \int_0^b \left(30x - \frac{5x^3}{6}\right) dx\, x = \left[\frac{30x^3}{3} - \frac{5}{6}\frac{x^5}{5}\right]_0^6 = 864 \text{ units}$$

From Table 15.2, for the loading shown,

$$\frac{6A\bar{x}}{l} = \frac{8}{60} wl^3$$

$$= \frac{8}{60} \times 30 \times 6^3 = 864 \text{ units}$$

which is the same as

$$\frac{6A_1\bar{x}}{l_1} = \frac{6 \times 864}{6} = 864$$

$$A_2\bar{x}_2 \text{ (span } BC) = 45 \times \frac{4}{2} \times \frac{4+1}{3} = 150 \text{ units about } B$$

$$\frac{6A_2\bar{x}_2}{l_2} = \frac{6 \times 150}{4} = 225 \text{ units}$$

which is the same as the value obtained from Table 15.2:

$$\frac{Pab}{l}(l+b) = \frac{60 \times 3 \times 1 \times (4+1)}{4} = 225$$

The three-moment equation can be written as

$$M_A \frac{l_1}{I_1} + 2M_B\left(\frac{l_1}{I_1} + \frac{l_2}{I_2}\right) + M_C \frac{l_2}{I_2} = \frac{6A_1\bar{x}_1}{I_1 l_1} + \frac{6A_2\bar{x}_2}{I_2 l_2}$$

$$2M_B\left(\frac{6}{2I} + \frac{4}{I}\right) + M_C \frac{4}{I} = \frac{864}{2I} + \frac{225}{I}$$

$$14M_B + 4M_C = 657 \qquad (1)$$

Spans BC and CD In span BC,

$$\frac{6A_1\bar{x}_1}{l_1} = 6 \times 45 \times \frac{4}{2} \times \frac{7}{3} \times \frac{1}{4} = 315 \text{ about } B$$

In span CD,

$$\frac{6A_2\bar{x}_2}{l_2} = \frac{6 \times 33.75 \times 3 \times 2 \times 3}{3 \times 2 \times 3} = 202.5$$

$$M_B \frac{4}{I} + 2M_C\left(\frac{4}{I} + \frac{3}{I}\right) + M_D \frac{3}{I} = 315 + 202.5$$

$$4M_B + 14M_C + 3M_D = 472.5, \quad M_D = 15 \text{ kNm}$$

Therefore,

$$4M_B + 14M_C = 427.5 \qquad (2)$$

From Eqns (1) and (2), $M_B = 40.6$ kNm and $M_C = 22.15$ kNm.

The reactions are determined from the free body diagrams of the individual spans shown in Fig. 15.36(c):

Span AB: $R_{AV} = 30 - \dfrac{40.6}{6} = 23.23 \text{ kN}\uparrow$, $R_{BV} = 60 + \dfrac{40.6}{6} = 66.77 \text{ kN}\uparrow$

Span BC: $R_{BV} = 15 + \dfrac{(40.6 - 22.15)}{4} = 19.6$ kN

$R_{CV} = 45 - \dfrac{(40.6 - 22.15)}{4} = 40.4$ kN

Span CD: $R_{CV} = 45 + \dfrac{(22.15 - 15)}{3} = 47.38$ kN, $R_{DV} = 45 - 2.38 = 42.62$ kN

The total reactions are $R_{AV} = 23.23$ kN, $R_{BV} = 86.37$ kN, $R_{CV} = 87.78$, and $R_{DV} = 42.62 + 30 = 72.62$ kN.

The SF and BM diagrams can now be drawn as shown in Figs 15.36(d) and (e), respectively.

Example 15.17 Clapeyron's theorem: continuous beam with fixed end and overhang

Draw the SF and BM diagrams for a continuous beam loaded as shown in Fig. 15.37(a). I is constant.

Solution There are three unknown support moments M_A, M_B, and M_C, while $M_D = 15 \times 2 = 30$ kN m. Since A is the fixed end, an additional equation can be written for this end using the zero span shown in Fig. 15.37(b). Use values from Table 15.2.

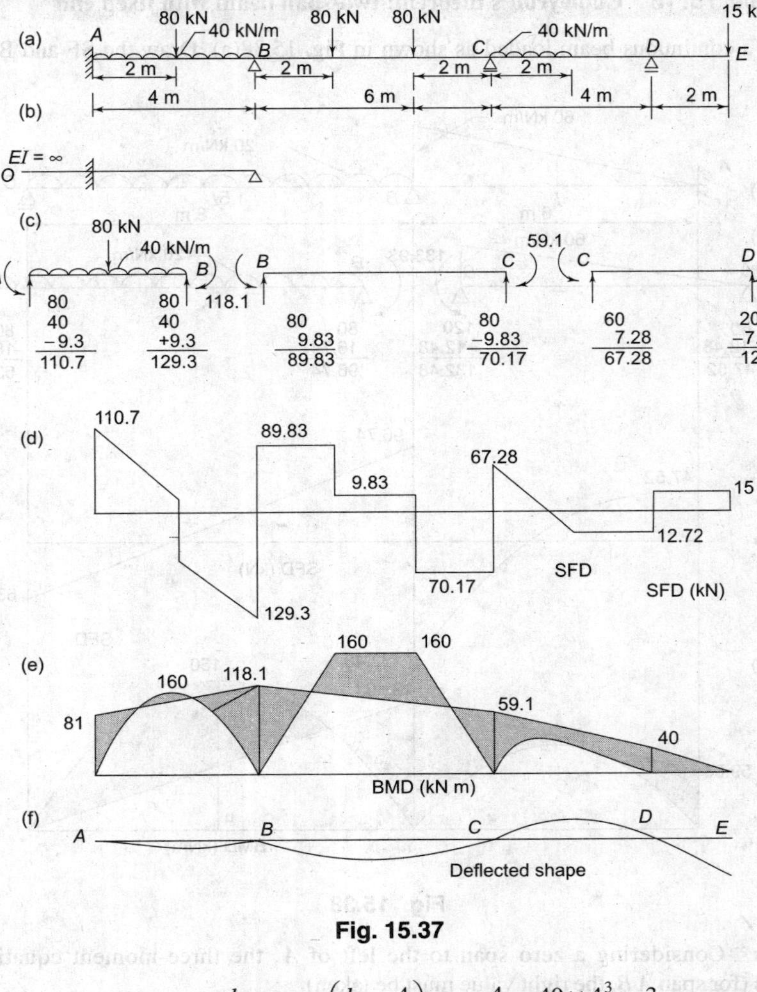

Fig. 15.37

$$\text{Spans } OA\text{-}AB\text{: } M_0 \frac{l_0}{I_0} + 2M_A\left(\frac{l_0}{I_0} + \frac{4}{I}\right) + M_B \frac{4}{I} = \frac{40 \times 4^3}{4} + \frac{3}{8} \times 80 \times 4^2$$

$$8M_A + 4M_B = 1120 \qquad (1)$$

Spans AB-BC (I is constant):

$$4M_A + 2M_B(4+6) + M_C 6 = \frac{40 \times 4^3}{4} + 80 \times \frac{3}{8} \times 4^2 + 3 \times 80 \times 2 \times 4$$

$$4M_A + 20M_B + 6M_C = 3040 \quad (2)$$

Spans BC-CD:

$$6M_B + 2M_C(6+4) + 30 \times 4 = 3 \times 80 \times 2 \times 4 + \frac{9}{64} \times 40 \times 16 = 2010$$

$$6M_B + 20M_C = 1890 \quad (3)$$

Solving for M_A, M_B, and M_C from Eqns (1)–(3), $M_A = 81$ kN m, $M_B = 118.1$ kN m, and $M_C = -59.1$ kN m. The reactions are now calculated as shown in Fig. 15.37(c): $R_{AV} = 110.7$ kN, $R_{BV} = 19.13$ kN, $R_{CV} = 137.45$ kN, and $R_{DV} = 32.72$ kN. The SF and BM diagrams are now drawn as shown in Figs 15.37(d) and (e), respectively. ■

Example 15.18 Clapeyron's theorem: two-span beam with fixed end

Analyse a continuous beam loaded as shown in Fig. 15.38(a). Draw the SF and BM diagrams.

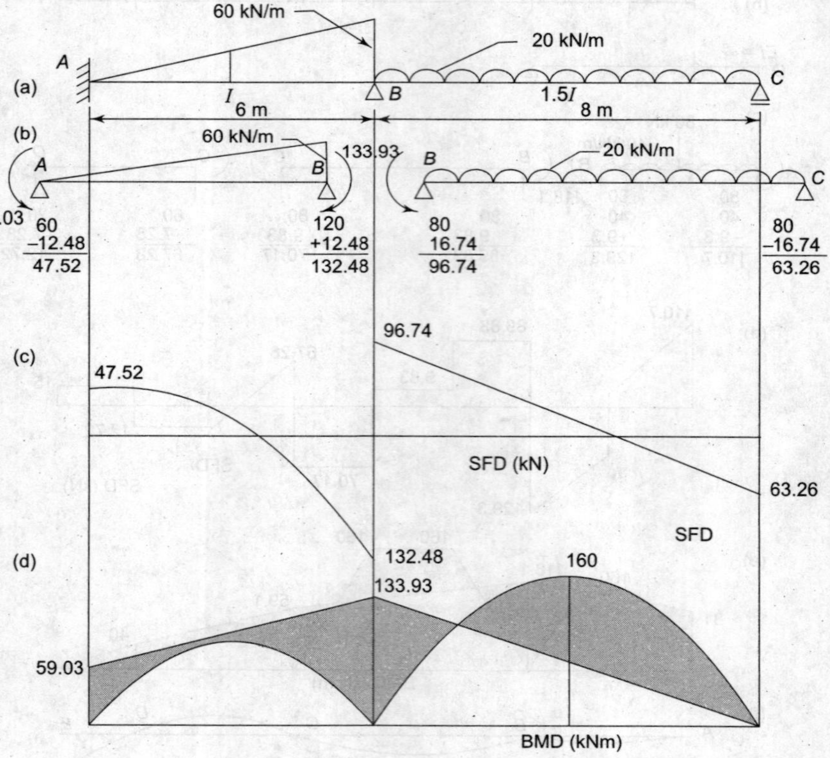

Fig. 15.38

Solution Considering a zero span to the left of A, the three-moment equation for O-A-B is (for span AB, the right value must be taken)

$$2M_A \frac{6}{I} + M_B \frac{6}{I} = \frac{7}{60} ql^3 = \frac{7}{60} \times 60 \times 6^3 = 252$$

$$12M_A + 6M_B = 252 \qquad (1)$$

For span A-B-C (for span AB, the left value must be taken),

$$M_A \frac{6}{I} + 2M_B\left(\frac{6}{I} + \frac{8}{1.5I}\right) + M_C \frac{8}{1.5I} = \frac{8}{60}\frac{ql^3}{I} + \frac{ql^3}{4\times 1.5I}$$

$$M_C = 0, \quad 6M_A + \frac{68}{3}M_B = \frac{8}{60}\times 60\times 6^3 + \frac{20\times 8^3}{4\times 1.5} = 3434.67 \qquad (2)$$

Solving for M_A and M_B from Eqns (1) and (2), $M_A = 59.03$ kN m and $M_B = 133.93$ kN m. The reactions can be calculated from the free body diagram shown in Fig. 15.38(b). The SF and BM diagrams can be drawn as shown in Figures 15.38(c) and (d), respectively. ■

Example 15.19 Clapeyron's theorem: continuous beam with support sinking

Analyse a continuous beam loaded as shown in Fig. 15.39(a). Assume that support B sinks by 10 mm. $E = 200$ GN/m^2 and $I = 78 \times 10^{-5}$ m^4.

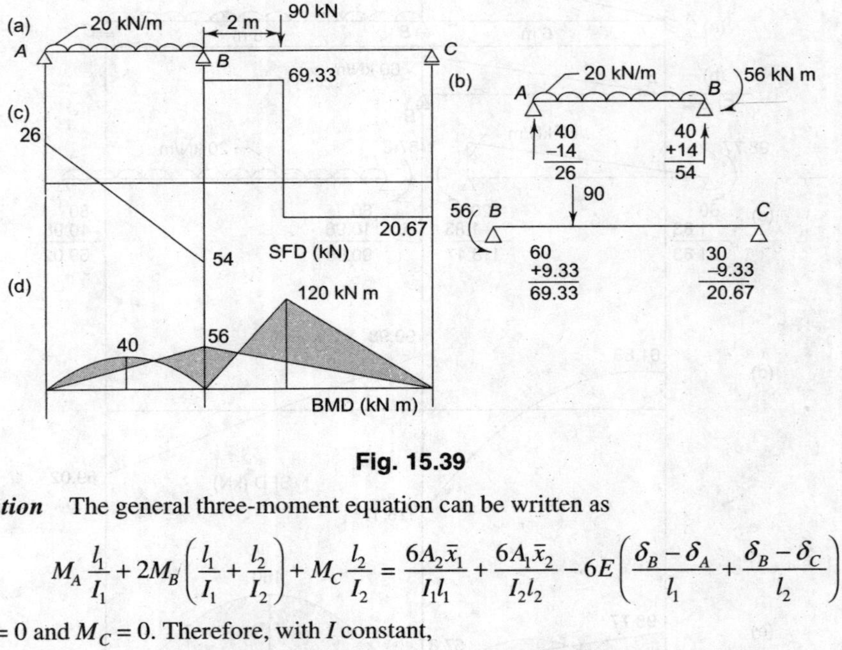

Fig. 15.39

Solution The general three-moment equation can be written as

$$M_A\frac{l_1}{I_1} + 2M_B\left(\frac{l_1}{I_1} + \frac{l_2}{I_2}\right) + M_C\frac{l_2}{I_2} = \frac{6A_2\bar{x}_1}{I_1 l_1} + \frac{6A_1\bar{x}_2}{I_2 l_2} - 6E\left(\frac{\delta_B - \delta_A}{l_1} + \frac{\delta_B - \delta_C}{l_2}\right)$$

$M_A = 0$ and $M_C = 0$. Therefore, with I constant,

$$2M_B(4+6) = \frac{20\times 4^3}{4} + \frac{90\times 2\times 4\times 10}{6} - 6EI\left(\frac{\delta_B - \delta_A}{l_1} + \frac{\delta_B - \delta_C}{l_2}\right)$$

$$20M_B = 320 + 1200 - 6\times 2\times 10^8\times 8\times 10^{-5}\left(\frac{10}{4\times 1000} + \frac{10}{6\times 1000}\right)$$

Note that $E = 200$ GN/m^2 $= 2\times 10^8$ kN/m^2 for consistency of units.

$$\delta_B - \delta_A = 10 \text{ mm} = \frac{10}{1000}\text{ m} = \delta_B - \delta_C$$

$$20M_B = 1520 - 400.32 = 1119.68, \quad M_B = 56 \text{ kN m}$$

Reactions can be calculated in the individual spans as shown in Fig. 15.39(b):

$$R_{AV} = 40 - \frac{56}{4}, \quad R_{BV}\,(\text{in span } AB) = 40 + \frac{56}{4} = 54$$

$$R_{BV}\,(\text{in span } BC) = 60 + \frac{56}{6} = 69.33$$

$$R_{CV} = 30 - \frac{56}{6} = 20.67$$

The SF and BM diagrams can be drawn as shown in Figures 15.39(c) and (d), respectively.

Example 15.20 Clapeyron's theorem: continuous beam with sinking support

Analyse the continuous beam shown in Fig. 15.40(a) if the support B sinks by 10 mm. $I = 10{,}000$ cm^4 and $E = 200{,}000$ MPa. Draw the SF and BM diagrams.

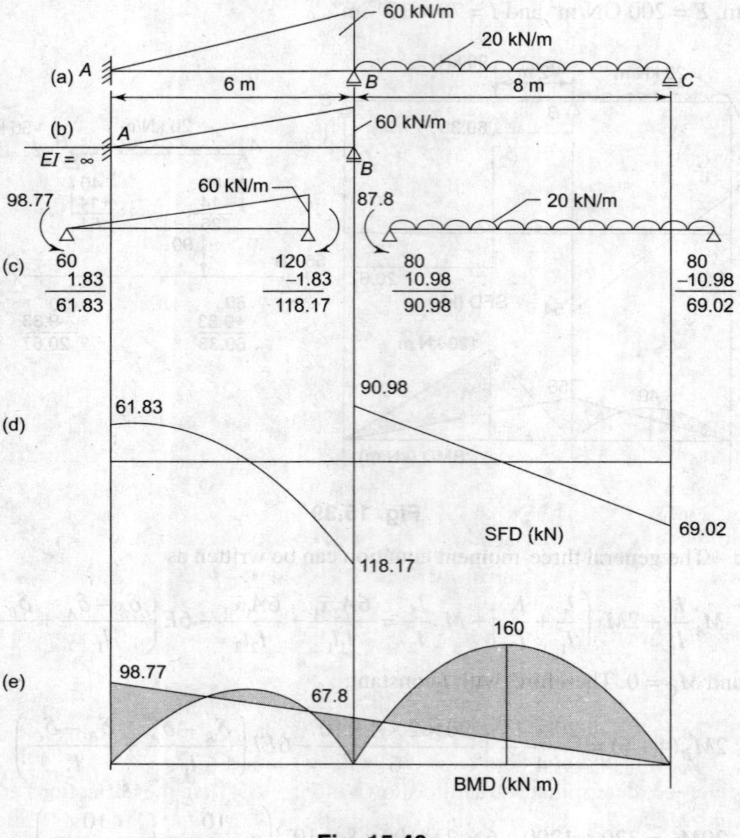

Fig. 15.40

Solution Assuming a zero span as shown in Fig. 15.40(b),

$$M_0 \frac{l_0}{I_0} + 2M_A\left(\frac{l_0}{I_0} + \frac{l_1}{I_1}\right) + M_B \frac{l_1}{I_1} = \frac{6A_1\bar{x}_1}{I_1\,l_1} - 6E\left(\frac{\delta_A - \delta_B}{l_1}\right)$$

$\delta_A = 0$, $\delta_B = 10$ mm $= 0.01$ m, $E = 200{,}000 \times 10^3$ kN/m^2 $= 2 \times 10^8$ kN/m^2, $I = 10{,}000$ cm^4 $=$ $10{,}000 \times 10^{-8}$ m^4 $= 1 \times 10^{-4}$ m^4, $I_0 = a$.

$$2M_A \times 6 + M_B \times 6 = \frac{7}{60} \times 60 \times 6^3 - 6 \times 2 \times 10^8 \times 1 \times 10^{-4}\left(\frac{0-0.01}{6}\right)$$

$$2M_A + M_B = 252 + 33.33 = 285.33 \quad (1)$$

For spans AB and BC,

$$M_A \frac{6}{I} + 2M_B\left(\frac{6}{I} + \frac{8}{1.5I}\right) + \frac{M_C \times 8}{1.5I}$$

$$= \frac{8 \times 60 \times 6^3}{60I} + \frac{20 \times 8^3}{4 \times 1.5I} - 6E\left(\frac{\delta_B - \delta_A}{b} + \frac{\delta_B - \delta_C}{8}\right)$$

$$6M_A + 2M_B\left(6 + \frac{16}{3}\right) + \frac{80 \times 8}{1.5} = 1728 + 1706.67 - 6 \times 2 \times 10^8 \times 1 \times 10^{-4}$$

$$\times\left(\frac{0.01}{6} + \frac{1.5 \times 0.01}{8}\right)$$

$$6M_A + 22.67M_B + 426.67 = 3434.67 - 425.04$$

$$6M_A + 22.67M_B = 2582.96 \quad (2)$$

$$6M_A + 3.00M_B = 856 \quad \text{[multiplying Eqn (1) by (3)]}$$

$$19.67M_B = 1726.96, \quad M_B = 87.8 \text{ kN m}$$

$$M_A = 98.77 \text{ kN m}$$

The reactions can be calculated for individual spans as shown in Fig. 15.40(c). The SF and BM diagrams are shown in Figs 15.40(d) and (e), respectively. ■

15.7 FLEXIBILITY MATRIX METHOD

We have discussed in Chapter 14 the concept of flexibility coefficients. Flexibility method, also known as force method or compatibility method, is essentially the method of consistent deformations. We illustrate the application of the flexibility method by solving the flexibility matrix form of compatibility equations.

Example 15.21 Flexibility matrix method: continuous beam

Analyse the two-span continuous beam loaded as shown in Fig. 15.41 and draw the shear force and bending moment diagrams. EI is constant.

Solution We have two ways of forming the primary released structure. We will illustrate both these cases.

(a) *When the reactive forces at B and C are taken as redundants*

The primary released structure is a cantilever of 10 m span. As the two redundants selected are forces, the corresponding deflections have to be found.

From the released, determinate structure shown in Fig. 15.41(b), the deflections at B and C can be found. Deflection at $B = \Delta_{L1}$ and deflection at $C = \Delta_{L2}$ due to the applied loads.

Deflection at C, $\Delta_{L2} = wL^4/8EI = 60 \times 10^4/8EI = 75{,}000/EI \downarrow$

To find the deflection at B, use the double integration method.

$$EIy'' = wx^2/2;\ EIy' = wx^3/6 + C_1;\ EIy = wx^4/24 + C_1x + C_2$$

At $x = 10$, $y = y' = 0$. This gives

$$60 \times 10^3/6 + C_1 = 0;\ C_1 = -10^4$$

$$60 \times 10^4/24 + (-10^4)10 + C_2 = 0;\ C_2 = 10^5 - 10^5/4 = 75{,}000$$

$$EIy = wx^4/24 - 10^4x + 75{,}000$$

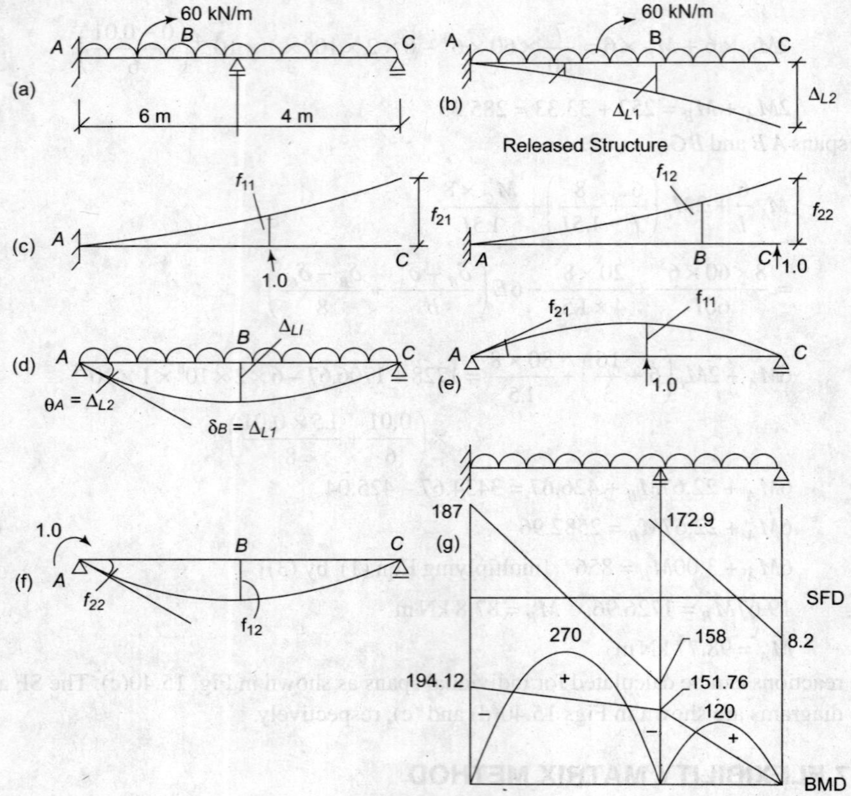

Fig. 15.41

At $x = 4$, $EI\Delta_{L1} = 60(4)^4/24 - 4 \times 10^4 + 75{,}000 = 35{,}640/EI \downarrow$

The deformation matrix due to loads is $D_L = \dfrac{1}{EI}\begin{vmatrix} 35{,}640 \\ 75{,}000 \end{vmatrix}$

The flexibility coefficients are developed from the unit loads applied as shown in Fig. 15.41(c).

When a unit load is applied at B, we have f_{11} = deflection at $B = 6^3/3EI = -72/EI \uparrow$

Due to the unit load at B, deflection at $C = f_{21}$ = deflection at B + slope at $B \times$ distance AC $= 6^3/3EI + (6^2/2EI)4 = -144/EI \uparrow$

We now apply a unit load at C as shown.

The deflection at $C = f_{22} = -10^3/3EI = -1000/3EI \uparrow$

To find the deflection at B, which gives f_{12}, use the conjugate beam shown.

Deflection at B = BM in the conjugate beam at $B = 1/EI(6 \times 4 \times 3 + 6 \times 6/2 \times 4) = 144/EI$

Therefore, $f_{12} = -144/EI$

The flexibility matrix can be formed as $(1/EI)\begin{vmatrix} -72 & -144 \\ -144 & -1000/3 \end{vmatrix}$

The compatibility condition is that the deflections at B and C are zero. This can be written as $[F]\,[R] + [\Delta_L] = 0$. Here, $[F]$ is the flexibility matrix, $[R]$ is the matrix of redundants, and $[\Delta_L]$ is the matrix of deflection due to the applied loads.

$$(1/EI)\begin{vmatrix} -72 & -144 \\ -144 & -1000/3 \end{vmatrix}\begin{vmatrix} R_B \\ R_C \end{vmatrix} + (1/EI)\begin{vmatrix} 35{,}640 \\ 75{,}000 \end{vmatrix} = 0$$

$$\begin{vmatrix} R_B \\ R_C \end{vmatrix} = \begin{vmatrix} -72 & -144 \\ -144 & -1000/3 \end{vmatrix}^{-1}\begin{vmatrix} -35{,}640 \\ -75{,}000 \end{vmatrix}$$

The inversion of the 2 × 2 matrix is given by the adjunct matrix/determinant F.
Determinant of $F = 72 \times 1000/3 - 144^2 = 3264$

$$\text{The adjunct matrix of } F = \begin{vmatrix} -1000/3 & 144 \\ 144 & -72 \end{vmatrix}$$

$$\begin{vmatrix} R_B \\ R_C \end{vmatrix} = (1/3264)\begin{vmatrix} -1000/3 & 144 \\ 144 & -72 \end{vmatrix}\begin{vmatrix} -35{,}640 \\ -75{,}000 \end{vmatrix} = \begin{vmatrix} 330.88 \\ 82 \end{vmatrix}$$

$R_B = 331$ kN and $R_C = 82$ kN

Other reactions can be calculated as

$R_A = 600 - 331 - 82 = 187$ kN

Moment at support $B = 82 \times 4 - 60 \times 4 \times 2 = -152$ kNm

which is the support moment at B.

Fixing moment at support $A = 82 \times 10 + 331 \times 6 - 60 \times 10 \times 5 = -194$ kNm

The shear force and bending moment diagrams can be drawn as shown.

(b) *When R_B and M_A are taken as redundants*

The deformations corresponding to the redundants are the slope at A and the deflection at B. First, we compute the deformations due to the loads. We have a simply supported beam.

To find the deflection due to the loads, we use the double integration method.

$EIy'' = -wLx/2 + wx^2/2; \; EIy' = -wLx^2/4 + wx^3/6 + C_1$

$EIy = -wLx^3/12 + wx^4/24 + C_1x + C_2$

As $x = 0$, $y = 0$, $C_2 = 0$. Also, $y = 0$ when $x = L$.

$0 = -wL^4/12 + wL^4/24 + C_1L; \; C_1 = wL^3/24$

General equation for slope and deflection are

$EIy' = -150x^2 + 10x^3 + wL^3/24$ (putting $w = 60$ and $L = 10$)

Slope at A is obtained when $x = 0$.

$EI\theta_A = wL^3/24 = 60 \times 10^3/24 = 2500/EI$ (clockwise)

$EIy = -50x^3 + 2.5x^4 + (2500)x$

Deflection at B $(x = 6) = -50\,6^3 + 2.5\,6^4 + 2500 \times 6 = -7440/EI \downarrow$

Now, we form the flexibility matrix. We apply a unit load at B and a unit couple at A separately.

With a unit load at B, the deflection at B and slope at A can be calculated directly from formulae.

Deflection at $B = f_{11} = a^2b^2/3EIL$ [see Table 7.2, Chapter 7]

$a = 6$, $b = 4$, and $L = 10$

$f_{11} = 36 \times 16/30EI = 19.2/EI \downarrow$

Slope at $A = f_{21} = ab(L + b)/6EIL = 6 \times 4 \times 14/60EI = -5.6/EI$ (anticlockwise)

With a unit couple applied as shown, we use the conjugate beam to find slopes and deflections. The BM diagram is positive and is a triangle. The M/EI diagram load gives the reaction at $A = 10/3EI$. This being the SF at A in the conjugate beam is the slope at A. $f_{22} = 10/3EI$. The reaction is upwards and positive. The slope is clockwise.

Deflection at $x = 6$ m is equal to the BM in the conjugate beam. This is BM at 6 m from A $= f_{12} = 5 \times 4/3 - 0.4 \times 4 \times 4/6$ (calculating from the right)

$$f_{12} = -5.6/EI \downarrow$$

The flexibility matrix $[F] = (1/EI) \begin{vmatrix} 19.2 & -5.6 \\ -5.6 & 10/3 \end{vmatrix}$

The D_L matrix is $(1/EI) \begin{vmatrix} -7440 \\ 2500 \end{vmatrix}$

The deformation compatibility equation is $[F]\,[R] = [\Delta_L]$

$$(1/EI) \begin{vmatrix} 19.2 & -5.6 \\ -5.6 & 10/3 \end{vmatrix} \begin{vmatrix} R_B \\ M_A \end{vmatrix} = (1/EI) \begin{vmatrix} -7440 \\ 2500 \end{vmatrix}$$

$$\begin{vmatrix} R_B \\ M_A \end{vmatrix} = \begin{vmatrix} 19.2 & -5.6 \\ -5.6 & 10/3 \end{vmatrix}^{-1} \begin{vmatrix} -7440 \\ 2500 \end{vmatrix} = (1/32.64) \begin{vmatrix} 10/3 & 5.6 \\ 5.6 & 19.2 \end{vmatrix} \begin{vmatrix} 330.88 \\ 194.12 \end{vmatrix}$$

$$R_B = 331 \text{ kN and } M_A = 194.12 \text{ kNm}$$

The remaining reactions can now be calculated. ■

Example 15.22 Flexibility matrix method: continuous beam

Analyse the continuous beam loaded as shown in Fig. 15.42 and draw the SF and BM diagrams.

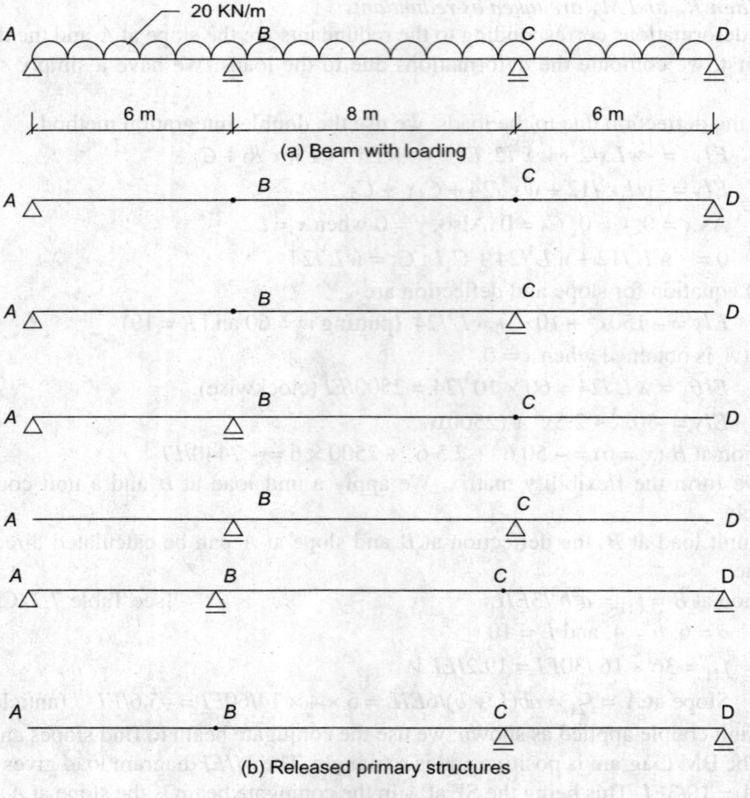

Fig. 15.42 *(contd)*

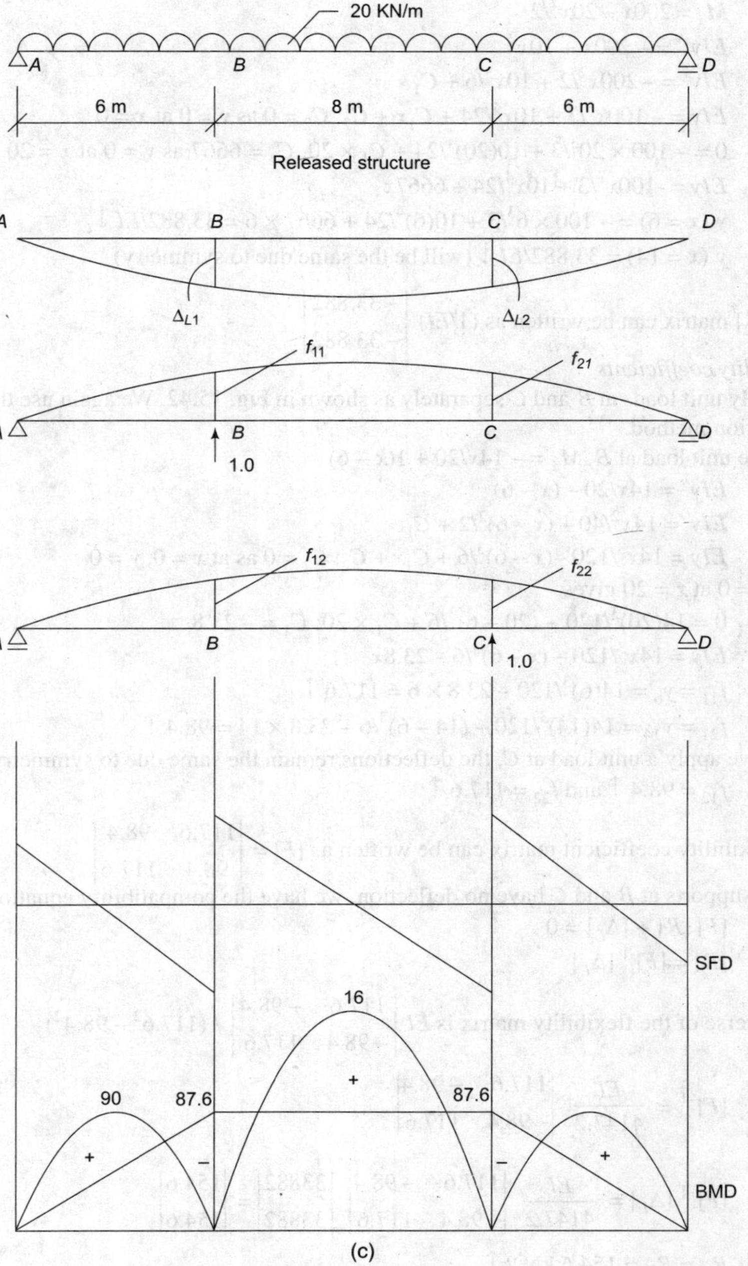

Fig. 15.42

Solution From the three span beam, we can get primary released structures in many ways. These are shown in Fig. 15.42(b). We will take the one giving a simply-supported beam. Deflections have to be found at 6 m and 14 m from the left end. We use the double integration method to find the deflections.

Deflection due to the loads

Span = 20 m; w = 20 kN/m; R_A = 200 kN

$M_x = 200x - 20x^2/2$

$EIy'' = -200x + 10x^2$

$EIy' = -200x^2/2 + 10x^3/6 + C_1$

$EIy = -100x^3/3 + 10x^4/24 + C_1x + C_2$; $C_2 = 0$ as $y = 0$ at $x = 0$

$0 = -100 \times 20^3/3 + 10(20)^4/24 + C_1 \times 20$; $C_1 = 6667$ as $y = 0$ at $x = 20$

$EIy = -100x^3/3 + 10x^4/24 + 6667x$

$y\ (x = 6) = -100 \times 6^3/3 + 10(6)^4/24 + 6667 \times 6 = 33{,}882/EI \downarrow$

$y\ (x = 14) = 33{,}882/EI \downarrow$ (will be the same due to symmetry)

The $[\Delta_L]$ matrix can be written as $(1/EI)\begin{vmatrix} -33{,}882 \\ -33{,}882 \end{vmatrix}$

Flexibility coefficients

We apply unit loads at B and C separately as shown in Fig. 15.42. We again use the double integration method.

With the unit load at B, $M_x = -14x/20 + 1(x - 6)$

$EIy'' = 14x/20 - (x - 6)$

$EIy' = 14x^2/40 - (x - 6)^2/2 + C_1$

$EIy = 14x^3/120 - (x - 6)^3/6 + C_1x + C_2$; $C_2 = 0$ as at $x = 0$, $y = 0$

Also $y = 0$ at $x = 20$ gives

$0 = 14(20)^3/120 + (20 - 6)^3/6 + C_1 \times 20$; $C_1 = -23.8$

$EIy = 14x^3/120 - (x - 6)^3/6 - 23.8x$

$f_{11} = y_6 = 14(6)^3/120 - 23.8 \times 6 = 117.6 \uparrow$

$f_{21} = y_{14} = 14(14)^3/120 - (14 - 6)^3/6 - 23.8 \times 14 = 98.4 \uparrow$

When we apply a unit load at C, the deflections remain the same due to symmetry.

$f_{12} = 98.4 \uparrow$ and $f_{22} = 117.6 \uparrow$

The flexibility coefficient matrix can be written as $[F] = \begin{vmatrix} 117.6 & 98.4 \\ 98.4 & 117.6 \end{vmatrix}$

As the supports at B and C have no deflection, we have the compatibility equation,

$[F]\,[R] + [\Delta_L] = 0$

or $[R] = [F]^{-1}\,[\Delta_L]$

The inverse of the flexibility matrix is $EI \begin{vmatrix} 117.6 & -98.4 \\ -98.4 & 117.6 \end{vmatrix} / (117.6^2 - 98.4^2)$

$$[F]^{-1} = \frac{EI}{4147.2}\begin{vmatrix} 117.6 & -98.4 \\ -98.4 & 117.6 \end{vmatrix}$$

$$[F]^{-1}\,[\Delta_L] = \frac{EI}{4147.2}\begin{vmatrix} 117.6 & -98.4 \\ -98.4 & 117.6 \end{vmatrix}\begin{vmatrix} 33882 \\ 33882 \end{vmatrix} = \begin{vmatrix} 154.6 \\ 154.6 \end{vmatrix}$$

$R_B = R_C = 154.6$ kN ? $\uparrow$

$R_A = R_B = (400 - 2 \times 154.6)/2 = 45.4$ kN

$M_B = 45.4 \times 6 - 20 \times 6 \times 3 = 87.6$ kNm $= M_C$ because of symmetry

The SF and BM diagrams can be drawn as shown.

Example 15.23 Flexibility matrix method: continuous beam

Analyse the two-span continuous beam shown in Fig. 15.43 and draw the shear force and bending moment diagrams.

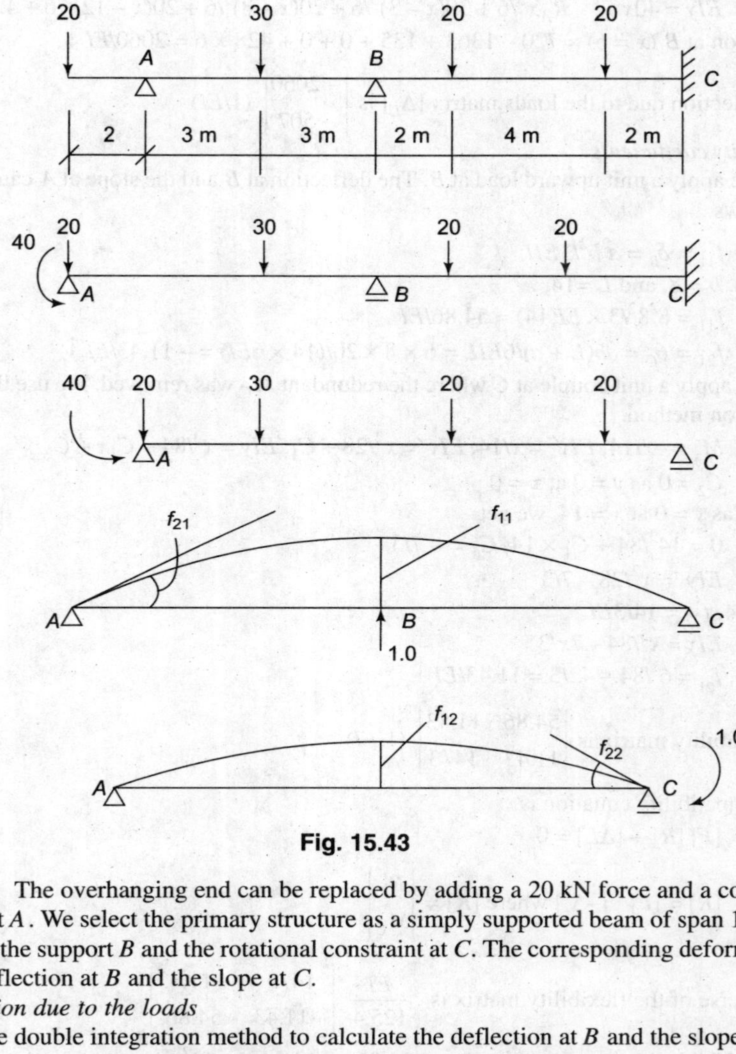

Fig. 15.43

Solution The overhanging end can be replaced by adding a 20 kN force and a couple of 40 kNm at A. We select the primary structure as a simply supported beam of span 14 m by removing the support B and the rotational constraint at C. The corresponding deformations are the deflection at B and the slope at C.

Deformation due to the loads

We use the double integration method to calculate the deflection at B and the slope at C.

Reaction at $A = (40 + 30 \times 11 + 20 \times 6 + 20 \times 2)/14 = 37.86$ kN

$M_x = -40 + R_A x - 30(x-3) - 20(x-8) - 20(x-12)$

$EIy'' = 40 - R_A x + 30(x-3) + 20(x-8) + 20(x-12)$

$EIy' = 40x - R_A x^2/2 + 30(x-3)^2/2 + 20(x-8)^2/2 + 20(x-12)^2/2 + C_1$

$EIy = 40x^2/2 - R_A x^3/6 + 30(x-3)^3/6 + 20(x-8)^3/6 + 20(x-12)^3/6 + C_1 x + C_2$

$C_2 = 0$ as at $x = 0$, $y = 0$

At $x = 14$, $y = 0$,

$0 = 40 \times 14^2/2 - 37.86 \times 14^3/6 + 30(14-3)^3/6 + 20(14-8)^3/6 + 20$
$(14-12)^3/6 + C_1 \times 14$

or $C_1 = 428$

$EIy' = 40x - R_A x^2/2 + 30(x-3)^2/2 + 20(x-8)^2/2 + 20(x-12)^2/2 + 428$

Slope at C ($x = 14$) $= 560 - 3710 + 1815 + 360 + 40 + 428$

$= -507/EI$ units (anticlockwise)

$$EIy = 40x^2/2 - R_A x^3/6 + 30(x-3)^3/6 + 20(x-8)^3/6 + 20(x-12)^3/6 + 428x$$

Deflection at B ($x = 6$) = 720 – 1363 + 135 + 0 + 0 + 428 × 6 = 2060/EI ↓

The deflection due to the loads matrix $[\Delta_L]$ is $\begin{vmatrix} -2060 \\ -507 \end{vmatrix}$ $(1/EI)$

Flexibility coefficients

First, we apply a unit upward load at B. The deflection at B and the slope at A can be found as follows.

$$f_{11} = \delta_B = a^2b^2/3EIL$$

As $a = 6$, $b = 8$, and $L = 14$,

$$f_{11} = 6^2 8^2/3 \times EI(14) = 54.86/EI$$

$$f_{21} = \theta_C = ab(L+a)/6EIL = 6 \times 8 \times 20/(14 \times 6EI) = -11.43/EI \uparrow$$

We now apply a unit couple at C where the redundant M_C was removed. We use the double integration method.

$$M_x = -x/14;\ EIy'' = x/14;\ EIy' = x^2/28 + C_1;\ EIy = x^3/84 + C_1x + C_2$$

$$C_2 = 0 \text{ as } y = 0 \text{ at } x = 0$$

Also as $y = 0$ at $x = 14$, we get

$$0 = 14^3/84 + C_1 \times 14;\ C_1 = -7/3$$

$$EIy' = x^2/28 - 7/3$$

At $x = 14$, $f_{22} = 14/3EI$

$$EIy = x^3/84 - 7x/3$$

At $x = 6$, $f_{21} = 6^3/84 - 42/3 = 11.43/EI \uparrow$

The flexibility matrix is $\begin{vmatrix} 54.86 & 11.43 \\ 11.43 & 14/3 \end{vmatrix}$ $(1/EI)$

The compatibility equation is

$$[F][R] + [\Delta L] = 0$$

or $\quad [R] = [F]^{-1}[-\Delta_L]$ where $[R] = \begin{vmatrix} R_B \\ M_C \end{vmatrix}$

The inverse of the flexibility matrix is $\dfrac{EI}{125.4}\begin{vmatrix} 14/3 & -11.43 \\ -11.43 & 54.86 \end{vmatrix}$

where 125.4 is the determinant value = 54.86s × 14/3 – 11.43^2.

$$[F]^{-1}[\Delta_L] = \frac{EI}{125.4}\begin{vmatrix} 14/3 & -11.43 \\ -11.43 & 54.86 \end{vmatrix}\begin{vmatrix} 2060 \\ 507 \end{vmatrix} = \begin{vmatrix} 30.45 \\ 34 \end{vmatrix}$$

$$R_B = 30.45 \text{ kN and } M_C = 34 \text{ kNm}$$

The other reactions can be calculated as

$$M_{AB} = -40 \text{ kNm}$$

15.8 STIFFNESS METHODS OF ANALYSIS

We have discussed in Chapter 14 the basic concepts of stiffness method of analysis. The basic procedure derived is known as slope-deflection equations. We illustrate two stiffness methods of analysis with examples of continuous beams—the slope-deflection method and the moment distribution method.

15.8.1 Slope-deflection Equations

Referring to Fig. 15.44, the slope-deflection equations are written as

$$M_{AB} = M_{FAB} + \frac{2EI}{L}\left(2\theta_A + \theta_B + \frac{3\Delta}{L}\right)$$

and

$$M_{BA} = M_{FBA} + \frac{2EI}{L}\left(\theta_A + 2\theta_B + \frac{3\Delta}{L}\right)$$

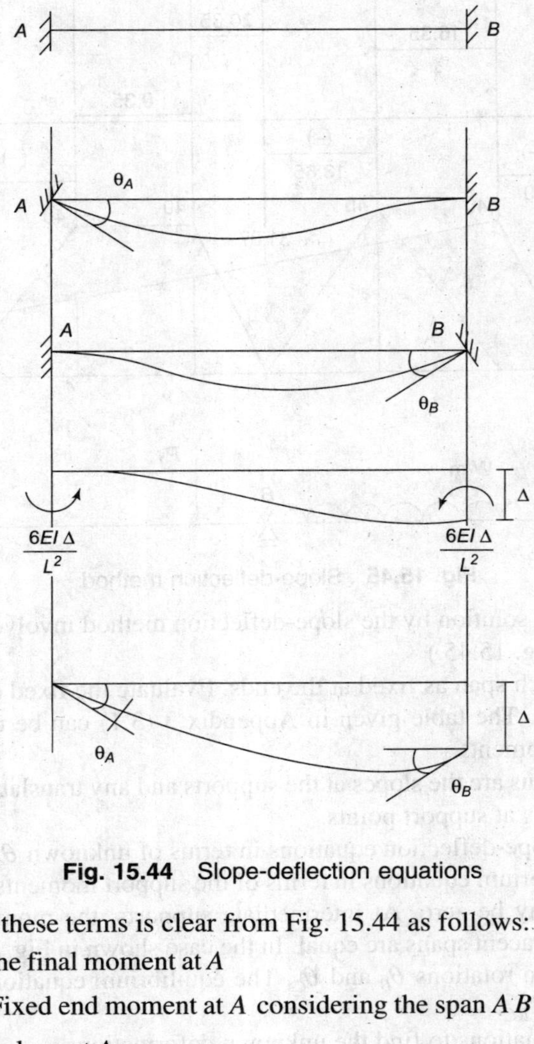

Fig. 15.44 Slope-deflection equations

The meaning of these terms is clear from Fig. 15.44 as follows:

M_{AB}: The final moment at A

M_{FAB}: Fixed end moment at A considering the span AB as a fixed beam

θ_A: The slope at A

θ_B: The slope at B

Δ: The difference in level between the supports at A and B, considered positive when the difference is due to a clockwise rotation of the axis

M_{BA}: The final moment at B

M_{FBA}: The fixed end moment at B

EI: The flexural rigidity of span AB

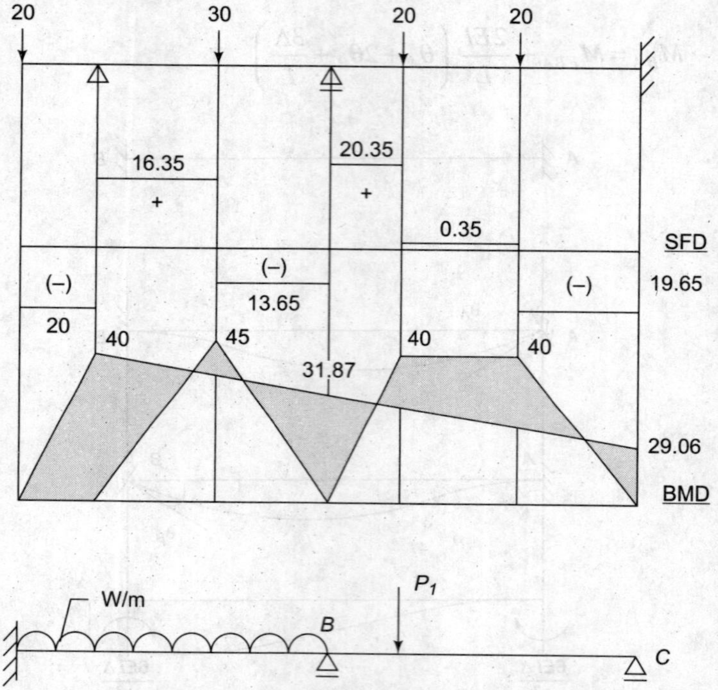

Fig. 15.45 Slope-deflection method

The procedure for solution by the slope-deflection method involves the following steps. (Refer to Fig. 15.45.)

(i) Consider each span as fixed at the ends. Evaluate the fixed end moments at the supports. The table given in Appendix 3 (3.1) can be used to find the fixed end moments.

(ii) The unknowns are the slopes at the supports and any translation perpendicular to the axis at support points.

(iii) Write the slope-deflection equations in terms of unknown θ_s and Δ_s.

(iv) Write equilibrium equations in terms of the support moments. Some support moments may be zero. At intermediate supports, the moments calculated from two adjacent spans are equal. In the case shown in Fig. 15.45, there are two unknown rotations θ_B and θ_C. The equilibrium equations are $M_{CB} = 0$ and $M_{BA} = M_{BC}$.

(v) Solve the equations to find the unknown deformations.

(vi) Substitute these deformations in the slope-deflection equations to find the support moments.

(vii) Calculate the reactions and draw SF and BM diagrams.
The following examples illustrate the procedure.

Example 15.24 Slope-deflection equations: continuous beam

Analyse the continuous beam shown in Fig. 15.46 and draw the SF and BM diagrams. *EI* is constant.

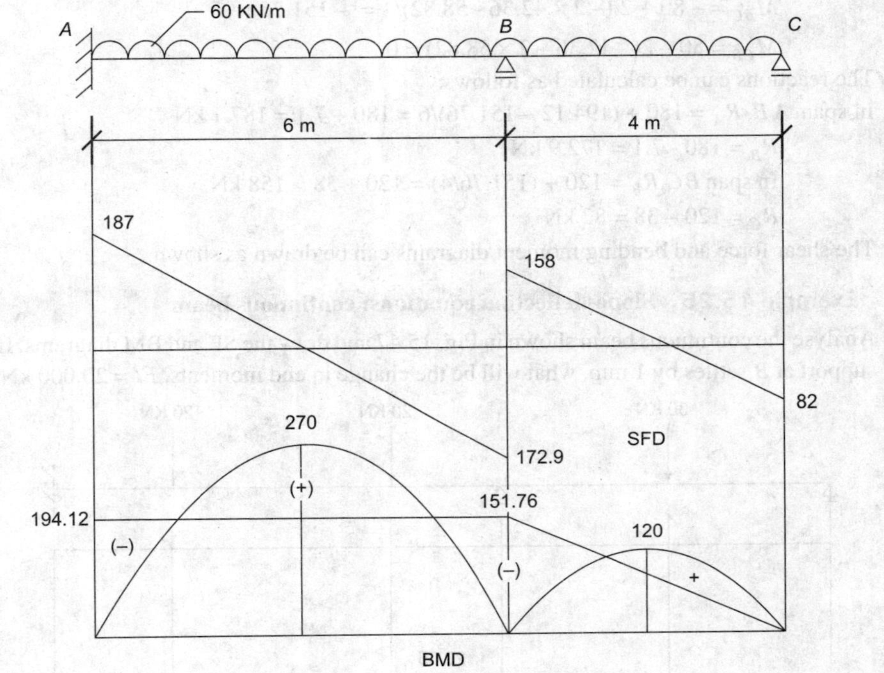

Fig. 15.46

Solution The structure has two degrees of freedom kinematically—rotations at *B* and at *C*. Also θ_B and θ_C will be taken as unknowns.

The fixed end moments in the two spans can be found out as

$M_{FAB} = -60 \times 6^2/12 = -180$ kNm $= -M_{FBA}$; $M_{FBC} = -M_{FCB} = -60 \times 4^2\ 12 = -80$ kNm

The slope-deflection equations can be written in terms of the unknown slopes θ_B and θ_C.

$$M_{AB} = -180 + 2EI(\theta_B)/6 \qquad (\theta_A = 0, \text{ being a fixed end.})$$

$$M_{BA} = 180 + 2EI(2\theta_B)/6$$

$$M_{BC} = -80 + 2EI(2\theta_B + \theta_C)/4$$

$$M_{CB} = 80 + 2EI(\theta_B + 2\theta_C)/4$$

The equilibrium conditions are

$$M_{CB} = 0;\ M_{BA} + M_{BC} = 0$$

$$M_{CB} = 0 \text{ gives } \theta_B + 2\theta_C = -160/EI$$

The second condition gives

$$180 + 2EI(2\theta_B)/3 - 80 + 2EI(2\theta_B + \theta_C)/4 = 0;\ 5EI\theta_B/3 + EI\theta_C/2 = 0$$

Solving these two equations simultaneously, we get

$\theta_B = -42.36/EI$ and $\theta_C = -58.82/EIs$

The moments can be calculated from the slope-deflection equations as

$M_{AB} = -180 - 2 \times 42.36/6 = -194.12$ kNm

$M_{BA} = 180 - 2 \times 2 \times 42.36/6 = 151.76$

$M_{BC} = -80 + 2(-2 \times 42.36 - 58.82)/4 = -151.76$

$M_{CB} = 80 - 2(-42.36 - 2 \times 58.82) = 0$

The reactions can be calculated as follows.

In span AB, $R_A = 180 + (194.12 - 151.76)/6 = 180 + 7.1 = 187.1$ kN

$R_B = 180 - 7.1 = 172.9$ kN

In span BC, $R_B = 120 + (151.76/4) = 120 + 38 = 158$ kN

$R_C = 120 - 38 = 82$ kN

The shear force and bending moment diagrams can be drawn as shown.

Example 15.25 Slope-deflection equations: continuous beam

Analyse the continuous beam shown in Fig. 15.47 and draw the SF and BM diagrams. If the support at B settles by 1 mm, what will be the change in end moments. $EI = 20{,}000$ kNm^2?

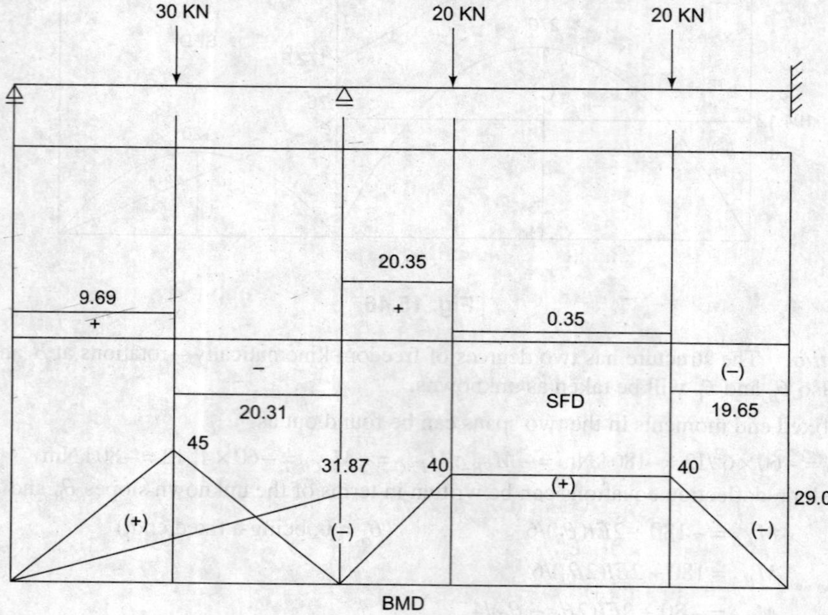

Fig. 15.47

Solution The unknowns are θ_A and θ_B. The slope-deflection equations can be framed as follows.

The fixed end moments are as follows:

$M_{FAB} = -30 \times 6/8 = -22.5$ kNm

$M_{FBA} = 22.5$ kNm

$M_{FBC} = 20 \times 2 \times 6^2/8^2 = -30$ kNm

$M_{FCB} = +30$ kNm

$M_{AB} = -22.5 + (2EI/L)(2\theta_A + \theta_B)$

$M_{BA} = 22.5 + (2EI/L)(\theta_A + 2\theta_B)$

$M_{BC} = -30 + (2EI/8)(2\theta_B)$

$M_{CB} = 30 + (2EI/8)(\theta_B)$

The conditions are

$M_{AB} = 0$ and $M_{BA} + M_{BC} = 0$

$-22.5 + (2EI/6)(2\theta_A + \theta_B) = 0; 2\theta_A + \theta_B = 67.5$

$22.5 + (2EI/6)(\theta_A + 2\theta_B) - 30 + (2EI/8)2\theta_B = 0; 2\theta_A + 7\theta_B = 45$

Solving these two simultaneous equations, $\theta_A = 35.625/EI$ and $\theta_B = -3.75/EI$

$M_{AB} = -22.5 + (2EI/6)(1/EI)(2 \times 35.625 - 3.75) = 0$

$M_{BA} = 22.5 + (2EI/6)[35.625 - 2(-3.75)] = 31.87$ kNm

$M_{BC} = -30 + (2EI/8)[2(-3.75)] = -31.875$ kNm

$M_{CB} = 30 + (2EI/8)(-3.75) = 29.06$ kNm

The reactions can be calculated as follows.

$R_A = 15 - 31.875/6 = 9.69$ kN

$R_B = 15 + 20 + 31.875/6 + (31.875 - 29.06)/8 = 40.66$ kN

$R_C = 20 - (31.875 - 27.06)/8 = 19.65$ kN

The shear force and bending moment diagrams can be drawn as shown.

When support B settles by 1 mm

The slope-deflection equations can be written as

$M_{AB} = -22.5 + (2EI/6)(2\theta_A + \theta_B - 3\Delta/L)$ and $\Delta = 1$ mm $= 0.001$ m

$6EI\Delta/L^2 = 6 \times 20{,}000 \times 0.001/36 = 3.33$

$M_{AB} = -25.83 + (2EI/6)(2\theta_A + \theta_B)$

$M_{BA} = 22.5 + (2EI/6)(\theta_A + 2\theta_B - 3 \times 0.001/6) = 18.17 + (EI/3)(\theta_A + 2\theta_B)$

$M_{BC} = -30 + (2EI/8)(2\theta_B + 3\Delta/8)$

$6EI\Delta/L^2 = 6 \times 20{,}000 \times 0.001/64 = 1.875$

$M_{BC} = -28.125 + EI\theta_B/4$

$M_{CB} = (30 + EI/4)(\theta_B) + 1.875 = 31.875 + EI\theta_B/4$

The conditions are as follows.

$M_{AB} = 0$

or $-25.83 + (EI/3)(2\theta_A + \theta_B) = 0$

or $EI(2\theta_A + \theta_B) = 77.49$ (i)

$M_{BA} + M_{BC} = 0$

or $18.17 + (EI/3)(\theta_A + 2\theta_B) - 28.125 + (EI/4)(2\theta_B) = 0$

or $EI(2\theta_A + 7\theta_B) = 59.73$ (ii)

Solving these equations simultaneously, $\theta_A = 40.225/EI$ and $\theta_B = -2.96/EI$

Substituting these back into the slope-deflection equations, we get

$$M_A = 0, M_{BA} = -M_{BC} = 29.605 \text{ kNm, and } M_C = 31.135 \text{ kNm}$$

Example 15.26 Slope-deflection equations: continuous beam

Analyse the continuous beam shown in Fig. 15.48 and draw the shear force and bending moment diagrams. *EI* is constant.

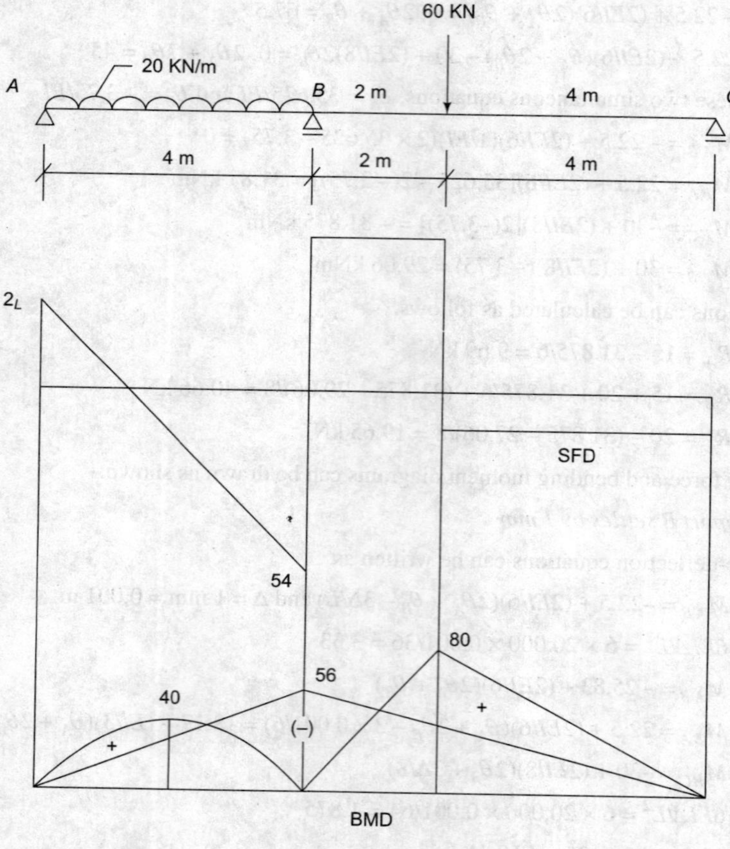

Fig. 15.48

Solution There are three unknown slopes – θ_A, θ_B, and θ_C.

The fixed end moments are

$$M_{FAB} = -20 \times 4^2/12 = -26.67 \text{ kNm}, M_{FBA} = 26.67 \text{ kNm},$$

$$M_{FBC} = -60 \times 2 \times 16/6^2 = -53.33 \text{ kNm, and } M_{FCB} = 60 \times 4 \times 4/36 = 26.67 \text{ kNm}$$

The slope-deflection equations are:

$$M_{AB} = -26.67 + (2EI/4)(2\theta_A + \theta_B)$$

$$M_{BA} = 26.67 + (2EI/4)(\theta_A + 2\theta_B)$$

$$M_{BC} = -53.33 + (2EI/6)(2v_B + \theta_C)$$

$$M_{CB} = 26.67 + (2EI/6)(\theta_B + 2\theta_C)$$

There are three conditions in terms of moments.

$$M_{AB} = 0, M_{BA} + M_{BC} = 0, \text{ and } M_{CB} = 0$$

Substituting the values derived above, we get

$$-26.67 + (2EI/4)(2\theta_A + \theta_B) = 0$$

or $\quad EI(2\theta_A + \theta_B) = 53.33 \quad$ (i)

$$26.67 + (2EI/4)(v_A + 2\theta_B) - 53.33 + (2EI/6)(2\theta_B + \theta_C) = 0$$

or $\quad EI(3\theta_A + 10\theta_B + 2\theta_C) = 160 \quad$ (ii)

Also $\quad 26.67 + (2EI/6)(2\theta_C + \theta_B) = 0 \quad$ (iii)

Solving these three equations, we get

$$\theta_A = 16/EI, \theta_B = 21.33, \text{ and } \theta_C = -50.665$$

Substituting these back into the equations, we get

$$M_{AB} = -26.67 + (1/2)(32 + 21.33) = 0$$

$$M_{BA} = 26.67 + (1/2)(16 + 42.66) = 56 \text{ kNm}$$

$$M_{BC} = -53.33 + (1/3)(42.66 - 50.665) = -56 \text{ kNm}$$

$$M_{CB} = 26.67 + (1/3)(-2 \times 50.665 + 21.33) = 0$$

The reactions can now be calculated.

$$R_A = 40 - 56/4 = 26 \text{ kN}$$

$$R_B = 40 + 60 \times (4/6) + (56/4) + (56/6) = 103.33 \text{ kN}$$

$$R_C = 60 \times (2/6) - (56/6) = 10.67 \text{ kN}$$

The SF and BM diagrams can be drawn as shown. ■

15.8.2 Moment Distribution Method

The essentials of stiffness method have been outlined in Chapter 14. The following points should be noted.

(i) If the other end is hinged, then the stiffness is only $(3/4)4EI/L$. Once this is taken into account, the joint is balanced and no more carry-over is made to this joint.

(ii) The unbalanced moment at a joint is distributed to the members according to their stiffnesses.

(iii) The carry-over factor is 1/2.

The following examples illustrate the procedure for analysis of beams.

Example 15.27 Moment distribution: two-span continuous beam

Analyse the continuous beams shown in Fig. 15.49(a) and draw the SF and BM diagrams. I is constant.

Solution Initially all the joints will be locked as in Fig. 15.49(b). The fixing moments due to the applied loads are calculated as follows:

Span AB: $M_{BC} = -\dfrac{wl^2}{12} = -\dfrac{60 \times 6^2}{12} = -180 \text{ kN m}, \quad M_{CB} = 180 \text{ kN m}$

$$M_{AB} = -\frac{60 \times 4^2}{12} = -80 \text{ kN m}, \quad M_{BA} = 80 \text{ kN m}$$

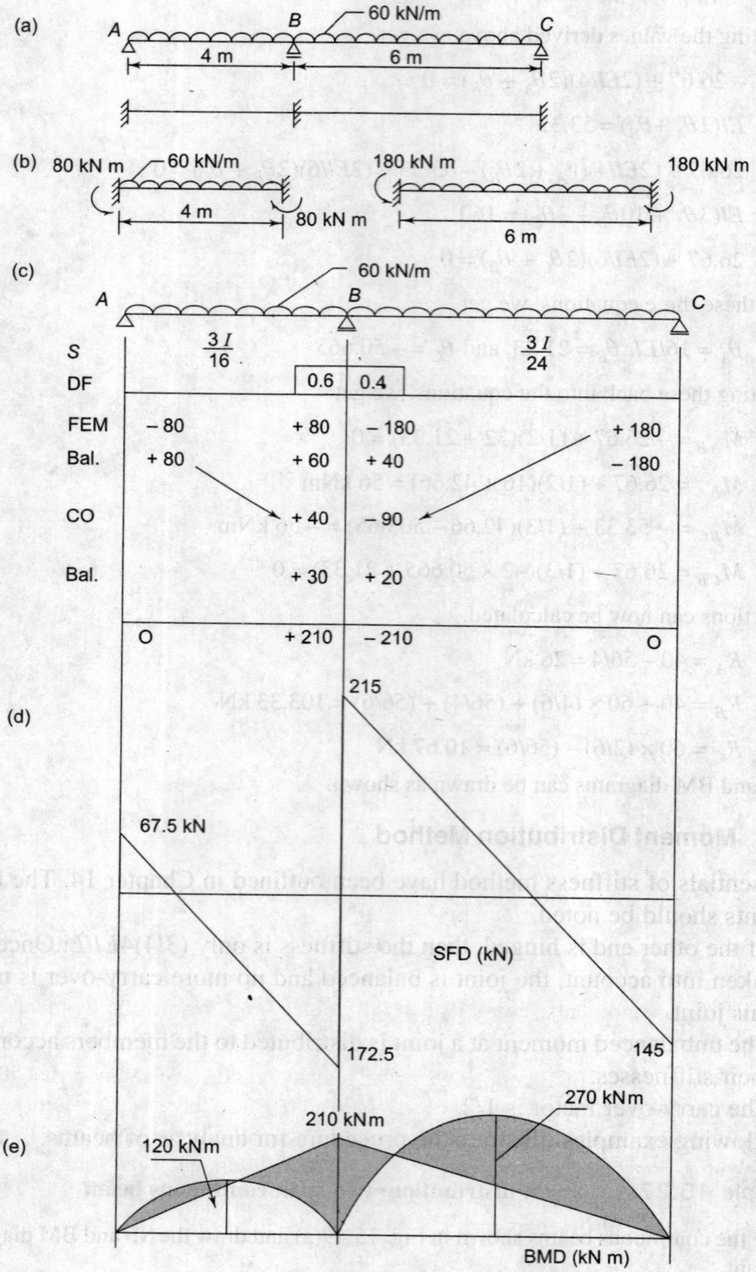

Fig. 15.49 (a–e)

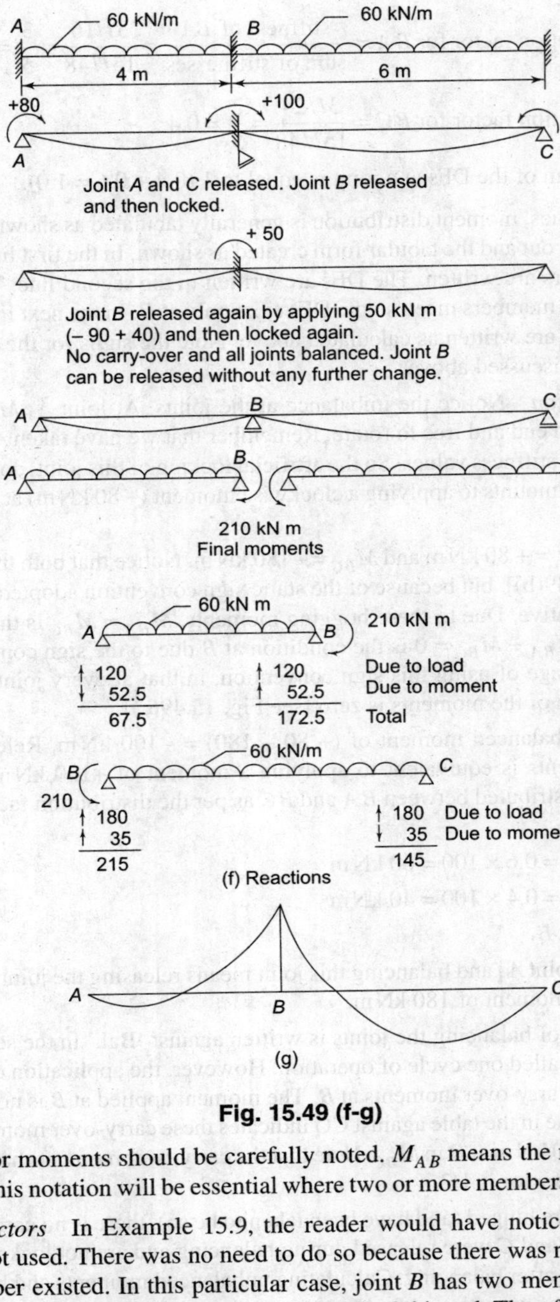

Fig. 15.49 (f-g)

The notation for moments should be carefully noted. M_{AB} means the moment at A in the member AB. This notation will be essential where two or more members meet, as at joint B.

Distribution factors In Example 15.9, the reader would have noticed that distribution factors were not used. There was no need to do so because there was no joint where more than one member existed. In this particular case, joint B has two members (BA and BC) meeting at it. For joint B, the far ends A and C are both hinged. Therefore,

$$\text{Stiffness of } BA = \frac{3}{4}\left(\frac{I}{4}\right), \quad \text{stiffness of } BC = \frac{3}{4}\left(\frac{I}{6}\right)$$

$$\text{Sum of stiffnesses} = \frac{3I}{24} + \frac{3I}{16} = \frac{15I}{48}$$

$$\text{Distribution factor for } BA = \frac{\text{stiffness of } BA}{\text{sum of stiffnesses}} = \frac{3I/16}{15I/48} = \frac{3}{5} = 0.6$$

$$\text{Distribution factor for } BC = \frac{3I/24}{15I/48} = \frac{2}{5} = 0.4$$

Check that the sum of the DFs at a joint is equal to 1 (0.4 + 0.6 = 1.0).

With these values, moment distribution is generally tabulated as shown in Fig. 15.49(c). The spans are laid out and the tabular form created as shown. In the first line, the stiffnesses (S) of the members are written. The DFs are written in the second line. There is only one joint B where two members meet, so the DFs exist only at B. In the next line, the fixing end moments (FEMs) are written as calculated above. Note the signs for the FEMs, as per the sign convention discussed above.

Moment distribution Notice the imbalance at the joints. At joint A, $M_{AB} = -80$ kN m. Joint A is a hinged end and free to rotate. Remember that we have taken A as a hinged end for calculating the stiffness values. So the artificial locking of this joint, done earlier, should be released. This amounts to applying a clockwise moment (+ 80 kN m) at A. This is known as *balancing*.

At joint B, $M_{BA} = +80$ kN m and $M_{BC} = -180$ kN m. Notice that both these moments are hogging [Fig. 15.49(b)], but because of the static sign convention adopted, M_{BA} is positive while M_{BC} is negative. Due to these hogging moments, $M_{BA} = M_{BC}$ is the condition at B. This means that $M_{BA} + M_{BC} = 0$ is the condition at B due to the sign convention adopted. This is the advantage of using this sign convention, in that at every joint one has only to check that the sum of the moments is zero [see Fig. 15.49(c)].

There is an unbalanced moment of (+ 80 − 180) = −100 kN m. Releasing joint B to balance the moments is equivalent to applying a moment of +100 kN m. However, this moment will be distributed between BA and BC as per the distribution factors determined. Therefore,

Moment in $BA = 0.6 \times 100 = 60$ kN m

Moment in $BC = 0.4 \times 100 = 40$ kN m

This balances joint B.

Joint C is like joint A, and balancing this joint means releasing the joint and applying an anticlockwise (–) moment of 180 kN m.

This procedure of balancing the joints is written against 'Bal.' in the second line of the table. This is also called one cycle of operation. However, the application of moments at A and C will lead to carry-over moments at B. The moment applied at B is not carried over to A or C. The next line in the table against CO indicates these carry-over moments: + 40 kN m at B in M_{BA} and −90 kN m at B in M_{BC}. Note that the carry-over moments have the same sign and the carry-over factor is 1/2.

Since A and C are hinged (and have been taken to be so), there is no carry-over to A and C from B. Joints A and C were released during balancing and are not locked again. Joint B is, however, locked after balancing. Calculating unbalanced moments and balancing can be taken as one cycle of operation. After carry-over, there is unbalance at joint B. Balancing joint B means applying − (−90 + 40) = 50 kN m at B, which has to be distributed between BA and BC as before. The moment in BA is (0.6 × 50) = 30 kN m and in BC is (+ 0.4 × 50) = + 20 kN m.

There is no carry-over from B to A or C and, therefore, the operation ends here as all the joints are balanced now. The net moment is obtained by summing up the moments written in all the lines against the member ends. The final moments are

$$M_{AB} = -80 + 80 = 0, \quad M_{BA} = 80 + 60 + 40 + 30 = 210 \text{ kN m}$$

$$M_{BC} = -180 + 40 - 90 + 20 = -210 \text{ kN m}, \quad M_{BC} = +180 - 180 = 0$$

The reactions are calculated as in Fig. 15.49(d).

Reactions From the free body diagram of member AB, the reaction due to the UD load of 60 kN/m is 120 kN m↑ at A and B. The reaction due to the 210 kN m clockwise moment is 210/4 kN, downwards at A and upwards at B. The net reaction at A is 67.5 kN↑ and 172.5 kN↑ at B. From the free body diagram of member BC, reactions are calculated at B and C. Due to the UD load, the reactions at B and C are equal to (60 × 6)/2 = 180 kN, upwards at B and C. The moment of 210 kN m at B induces an upward reaction at B and a downward reaction at C. The reaction is equal to 210/6 = 35 kN. The net reaction at B is (172.5 + 180 + 35) = 387.5 kN and at C is 145 kN.

Knowing the reactions at supports A, B, and C, the shear force diagram can be drawn as shown in Fig. 15.49(d). The BM diagram is shown in Fig. 15.49(e). It is generally drawn as a superposed diagram of moments due to the applied load and joint moments. The moments due to the applied loads are drawn as for the simply supported spans AB and BC. This gives maximum moments of $(60 \times 4^2)/8$ in AB and $(60 \times 6^2)/8$ in BC. The two parabolas show the variation of the moment due to the applied loads. These are positive. The joint moment at B is a hogging moment in the spans BA and BC. This is superposed on the 'applied loads' moment diagram by setting a moment of 210 kN m at B. The joint moments at A and C are zero. The net moment diagram is shown darkened and can also be drawn as shown in Fig. 15.49(g). This is the net moment diagram for the beam. ■

Example 15.28 Moment distribution: two-span beam with fixed end

Analyse the continuous beam shown in Fig. 15.50(a) and draw the SF and BM diagrams for it.

Solution The solution is shown in Fig. 15.50(c). The stepwise procedure is described below.

Fixed end moments Joint A is already fixed. Joints B and C are locked first and the FEMs are calculated. They are as follows:

$$M_{AB} = -180 \text{ kN m}, \quad M_{BA} = 180 \text{ kN m}, \quad M_{BC} = -80 \text{ kN m},$$
$$M_{CB} = 80 \text{ kN m}$$

Stiffness and distribution factors With respect to joint B, A is fixed and C is hinged. Therefore, the relative stiffnesses of the members BA and BC are the following:

$$\text{Stiffness of } BA = \frac{I}{6}, \quad \text{stiffness of } BC = \frac{3}{4}\frac{I}{4} = \frac{3}{16}I$$

$$\text{Sum of stiffnesses} = \frac{I}{6} + \frac{3}{16}I = \frac{17}{48}I$$

Distribution factors For BA, DF = $(I/6) \div (17I/48) = 8/17$, and for BC, DF = $3I/16 - 17I/48 = 9/17$. (Check: 8/17 + 9/17 = 1.0). These values are shown in Fig. 15.50(c).

Moment distribution Being a fixed end, Joint A does not need balancing. Joint B has an unbalanced moment of +100 kN m. A balancing moment of −100 kN m is distributed between BA and BC as per the DF worked out. There are the following:

$$M_{BA} = -47.06 \text{ kN m} \quad \text{and} \quad M_{BC} = -52.94 \text{ kN m}$$

Joint C is a hinged end and, hence, is fully released by applying −80 kN m. After balancing, Joint B is locked again, while joint C is left as it is. The CO moment to B from C is −40 kN m and that from B to A is −47.06 × 1/2 = −23.58. This completes one cycle of operation.

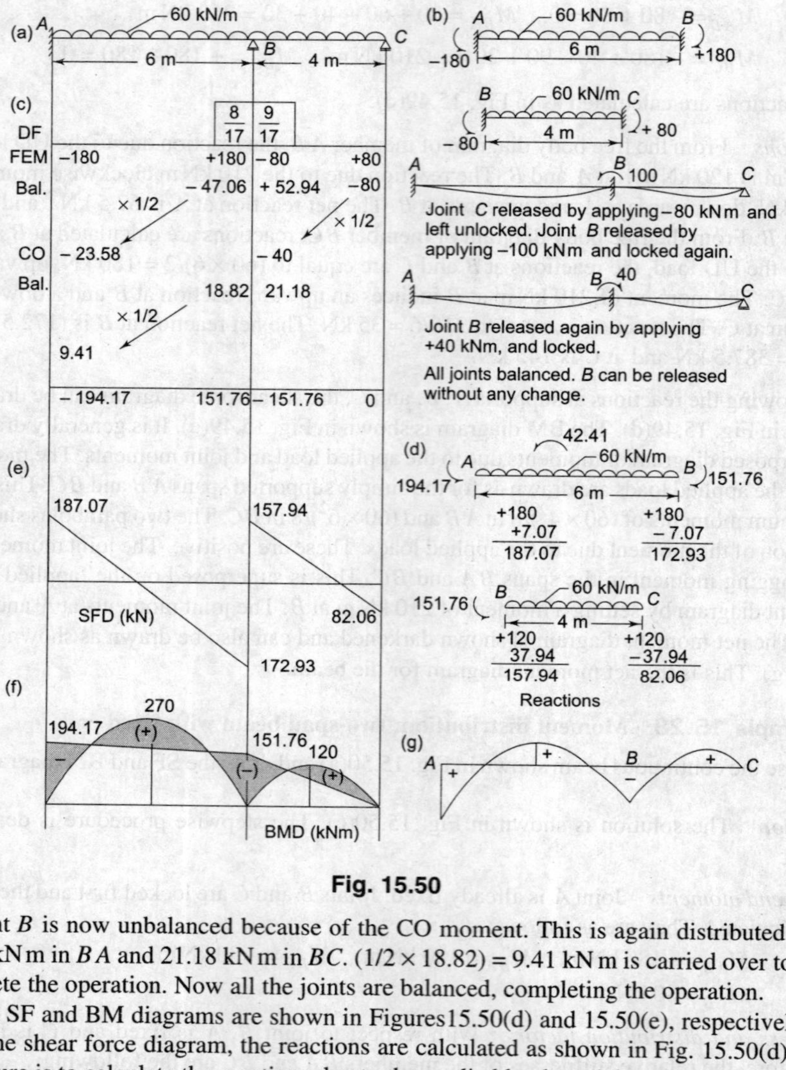

Fig. 15.50

Joint B is now unbalanced because of the CO moment. This is again distributed at B: 18.82 kN m in BA and 21.18 kN m in BC. $(1/2 \times 18.82) = 9.41$ kN m is carried over to A to complete the operation. Now all the joints are balanced, completing the operation.

The SF and BM diagrams are shown in Figures 15.50(d) and 15.50(e), respectively. To draw the shear force diagram, the reactions are calculated as shown in Fig. 15.50(d). The procedure is to calculate the reactions due to the applied loads and the end moments separately and add them. In span AB, due to the UD load, the reactions at A and B are equal to 180 kN. The joint moments act as shown. The net moment is $194.17 - 151.76 = 42.41$ kN m, inducing a reaction of 7.07 kN upwards at A and downwards at B.

The BM diagram is drawn as a superposed diagram of moments due to the applied loads and the joint moments. The moments due to the applied loads are the following:

$$\text{Span } AB\text{: } M_{\max} = \frac{60 \times 6^2}{8} = 270 \text{ kN m}$$

$$\text{Span } BC\text{: } M_{\max} = \frac{60 \times 4^2}{8} = 120 \text{ kN m}$$

The joint moments are hogging. $M_A = -194.17$ kN m and $M_B = -151.76$ kN m. The net moment diagram is shown in Fig. 15.50(g).

Example 15.29 Moment distribution: two-span beam with both ends fixed

Analyse the problem of Example 15.27 with the ends *A* and *C* fixed.

Solution The beam with the loading is shown in Fig. 15.51(a). The solution is shown in Fig. 15.51(b).

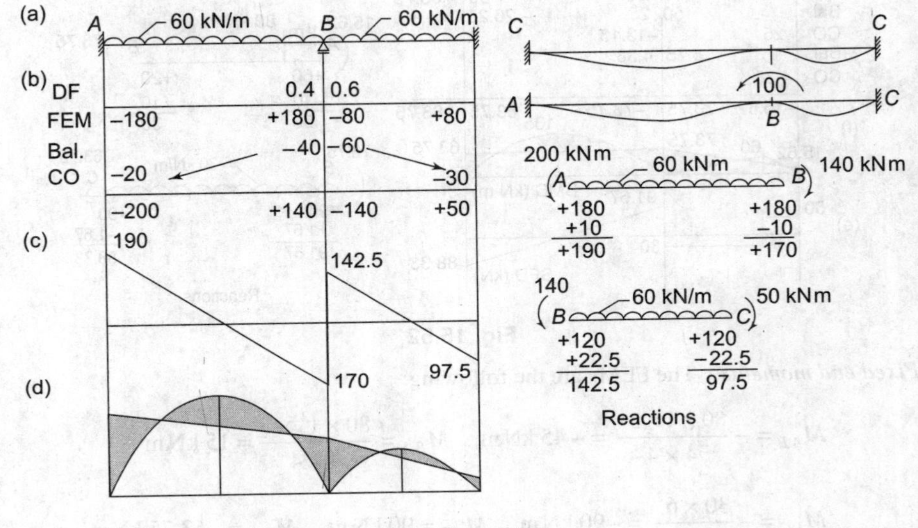

Fig. 15.51

Fixed end moments These are the same as those calculated for Example 15.27 $M_{AB} = -180$ kN m, $M_{BA} = +180$ kN m, $M_{BC} = -80$ kN m, and $M_{CB} = 80$ kN m.

Distribution factors At joint *B*, the ends *A* and *C* being fixed, the relative stiffnesses are *I*/6 for *BA* and *I*/4 for *BC*. The distribution factors are 0.4 and 0.6. These are entered in the table in Fig. 15.51(b).

Moment distribution Joints *A* and *C* are fixed and, hence, need no balancing. Joint *B* is balanced, giving the moments $M_{BA} = -40$ kN m and $M_{BC} = -60$ kN m. Moments have to be carried over from joint *B* to joints *A* and *C*, since both the joints are fixed. This completes the first cycle of moment distribution.

We now find that the joints are balanced and we sum up the moments to get $M_{AB} = -200$ kN m, $M_{BA} = -M_{BC} = 140$ kN m, and $M_{CB} = 50$ kN m. The SF and BM diagrams are shown in Figures 15.51(c) and (d), respectively.

The examples given above illustrate the basic procedure of the moment-distribution method. These problems were simple enough to allow one to obtain the final solution very fast, in just one or two cycles. The following examples further illustrate the procedure for continuous beams. ■

Example 15.30 Moment distribution: continuous beam with overhang

Analyse the continuous beam shown in Fig. 15.52(a) and draw the SF and BM diagrams.

Solution In Fig. 15.52(a), the only difference is the overhanging portion *CD*. The moment at *C* due to the load on the overhanging end is $(30 \times 1.5 \times 0.75 + 20 \times 1.5) = 63.75$ kN m. If we separate the beam at *C*, the forces at *C* will be as shown in Fig. 15.52(b). If *C* were an end support, the final moment at *C* would have been zero. Due to the overhanging end, the final moment at *C* should be − 63.75 kN m (anticlockwise) in *CD* and + 63.75 kN m in *CB*. The solution shown in Fig. 15.52(c) considers this.

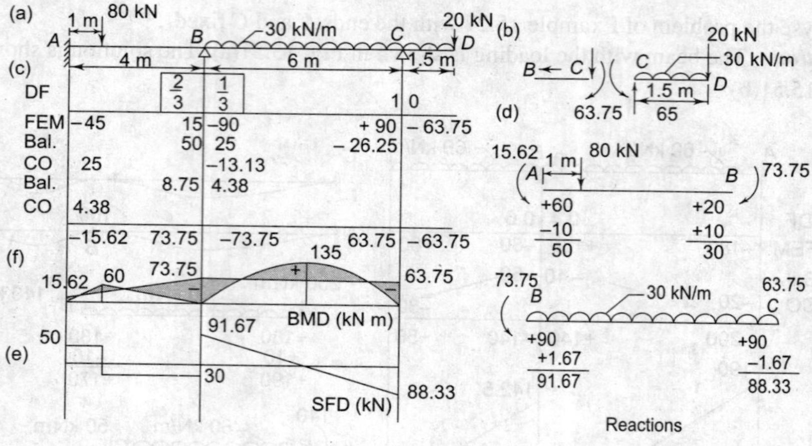

Fig. 15.52

Fixed end moments The FEMs are the following:

$$M_{AB} = -\frac{80 \times 1 \times 3^2}{4 \times 4} = -45 \text{ kN m}, \quad M_{BA} = \frac{80 \times 1^2 \times 3}{4 \times 4} = 15 \text{ kN m}$$

$$M_{BC} = -\frac{30 \times 6^2}{12} = -90 \text{ kN m}, \quad M_{CB} = 90 \text{ kN m}, \quad M_{CD} = -63.75 \text{ kN m}$$

Distribution factors At joint *B*, the stiffnesses are *I*/4 and (3/4)(*I*/6). Note that end *C* is taken as a hinged end with a known moment. The distribution factors are 2/3 and 1/3 for *BA* and *BC*, respectively. *D* being a free end, any moment applied at *C* will be applied to *CB* only. The DF is, therefore, 1.0 for *CB* and 0 for *CD*.

Moment distribution The moment-distribution procedure is shown in Fig. 15.52(c). Joint *A*, being a fixed end, needs no balancing. At joint *B*, there is an unbalanced moment of −75 kN m. The balancing moments distributed at *B* are 50 kN m in *BA* and 25 kN m in *BC*. There is an unbalanced moment of 26.25 kN m at *C*. Therefore, a balancing moment of −26.25 kN m is applied on *CB*. The first cycle of moment distribution is completed by entering the CO moments at *A* (25 kN m) and in *BC* at *B* (−13.125 kN m). The imbalance due to the CO moment at *B* is again balanced at *B*. The carry-over to *A* (4.875 kN m) completes the second cycle, which is also the end of the moment-distribution procedure, since every joint is balanced now. The final moments are

$$M_{AB} = -15.125 \text{ kN m}, \quad M_{BA} = 74.75 \text{ kN m} = -M_{BC},$$

$$M_{CB} = 63.75 \text{ kN m} = -M_{CD}$$

The SF and BM diagrams can be drawn after calculating the reactions as shown in Fig. 15.52(d). Figures 15.52(e) and (f) show the SF and BM diagrams, respectively.

Example 15.31 Moment distribution: three-span continuous beam

Analyse the three-span continuous beam shown in Fig. 15.53(a). Draw the SF and BM diagrams.

Solution Initially all the joints are locked and the FEMs calculated. The FEMs are

$$M_{AB} = -\frac{15 \times 6^2}{30} = -18 \text{ kN m}, \quad M_{BA} = \frac{15 \times 36}{20} = 27 \text{ kN m}$$

$$M_{BC} = -\frac{40\times 2\times 6^2}{8\times 8} - \frac{40\times 5\times 9}{8\times 8} = 73.13 \text{ kN m}$$

$$M_{CB} = -\frac{40\times 2^2\times 6}{64} - \frac{40\times 5^2\times 3}{64} = +61.88 \text{ kN m}$$

$$M_{CD} = -\frac{20\times 36}{12} = -60 \text{ kN m},\quad M_{DC} = 60 \text{ kN m}$$

Distribution factors At joint B, the relative stiffnesses of BA and BC are $I/6$ and $2I/8$, respectively. The distribution factors are 0.4 and 0.6. At joint C, the relative stiffnesses of CB and CD are $2I/8$ and $(3/4)(I/6)$, respectively. The distribution factors at C are 2/3 and 1/3.

Moment distribution The moment-distribution table is laid out as in Table 15.3. Joint A, being a fixed end, needs no balancing. Joint D, being a hinged end, can be unlocked and left unlocked.

Cycle I –60 kN m is applied at joint D to balance it. – 1.88 kN m is applied at C and distributed as – 0.627 kN m in CD and – 1.253 kN m in CB. 46.13 kN m is applied at B and distributed as 18.45 kN m in BA and 27.68 kN m in BC. The CO moments are 9.225 kN m at A, 0.627 kN m at B in BC, 13.84 kN m at C in CB, and – 30 kN m at C in CD.

Cycle II This causes an imbalance between B and C. These are again balanced and distributed as shown in Table 15.3. This has to be continued for a large number of cycles until the CO moments are very small. These never become zero but the cycle of operation can be stopped when the values are small. Summing up the moments, we get the final moments as $M_A = -9.82$ kN m, $M_B = 43.341$ kN m, $M_C = -84.702$ kN m, and $M_D = 0$.

The reactions can be calculated as shown in Fig. 15.53(d). The SF and BM diagrams are shown in Figures 15.53(b) and (c), respectively.

Table 15.3

Joint	A	B		C		D
Member	*AB*	*BA*	*BC*	*CB*	*CD*	*DC*
DF		0.4	0.6	2/3	1/3	
FEM	−18	27.000	−73.12	61.880	−60	60
Bal.		18.450	27.67	−1.25	−0.63	−60
CO	9.225		−0.625	13.835	−30	
Bal.		0.250	0.375	10.777	5.388	
CO	0.125		5.388	0.188		
Bal.		−0.155	−3.233	−0.125	0.063	
CO	−1.078		−0.063	−1.616		
Bal.		0.025	0.038	1.077	0.539	
CO	0.012		0.539	0.019		
Bal.		− 0.215	−0.324	−0.013	−0.006	
CO	−0.108		−0.006	−0.161		
Bal.		0.002	0.004	+ 0.108	+ 0.053	
Total	−9.824	43.357	−43.357	84.719	−84.719	0

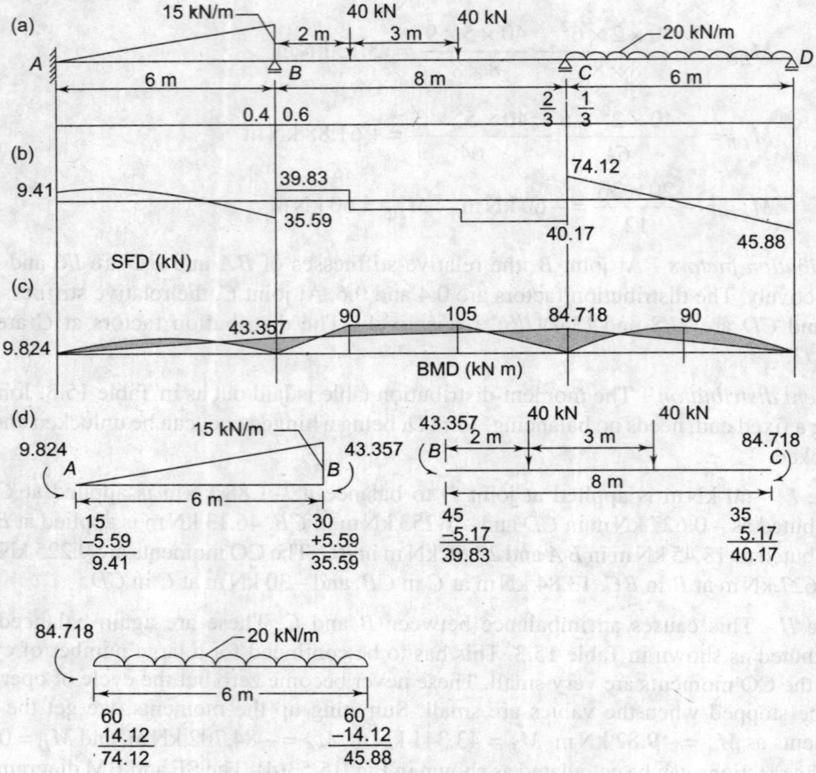

Fig. 15.53

Example 15.32 Moment distribution: two-span beam with different MI in spans

Analyse the continuous beam shown in Fig. 15.54(a). The support B settles down by 10 mm. $E = 200$ GN/m^2 and $I = 8000$ cm^4. The MI of span AB is I and that of span BC is $2I$.

Solution The FEMs are calculated due to the applied loads as well as support settlement. The effect of support settlement is shown in Fig. 15.54(b). Both the spans will have moments due to support settlements. These moments are negative (anticlockwise) in span AB and positive (clockwise) in span BC.

Fixed end moments From Fig. 15.54(b),

$$M_{AB} = -\frac{wl^2}{12} - \frac{6EI\delta}{l^2} = -\frac{15\times 36}{12} - \frac{6\times 200\times 10^6\times 8000\times 10^{-8}\times 0.01}{6\times 6}$$

$$= -45 - 26.67 = -71.67 \text{ kNm}$$

$$M_{BA} = \frac{wl^2}{12} - \frac{6EI\delta}{l^2} = +\frac{15\times 36}{12} - 26.67 = 18.33 \text{ kNm}$$

$$M_{BC} = -\frac{80\times 2\times 36}{64} - \frac{80\times 5\times 9}{64} + \frac{6\times 200\times 10^6\times (2\times 8000)\times 10^{-8}\times 01}{8\times 8}$$

$$= -116.25 \text{ kNm}$$

$$M_{CB} = \frac{80 \times 4 \times 6}{64} + \frac{80 \times 25 \times 3}{64} + 30 = 153.75 \text{ kN m}$$

Distribution factor At joint B, the stiffness of BA is $I/6$, while that of BC is (3/4) ($2I/8$), the end C being hinged. The distribution factors for BA and BC are 8/17 and 9/17, respectively.

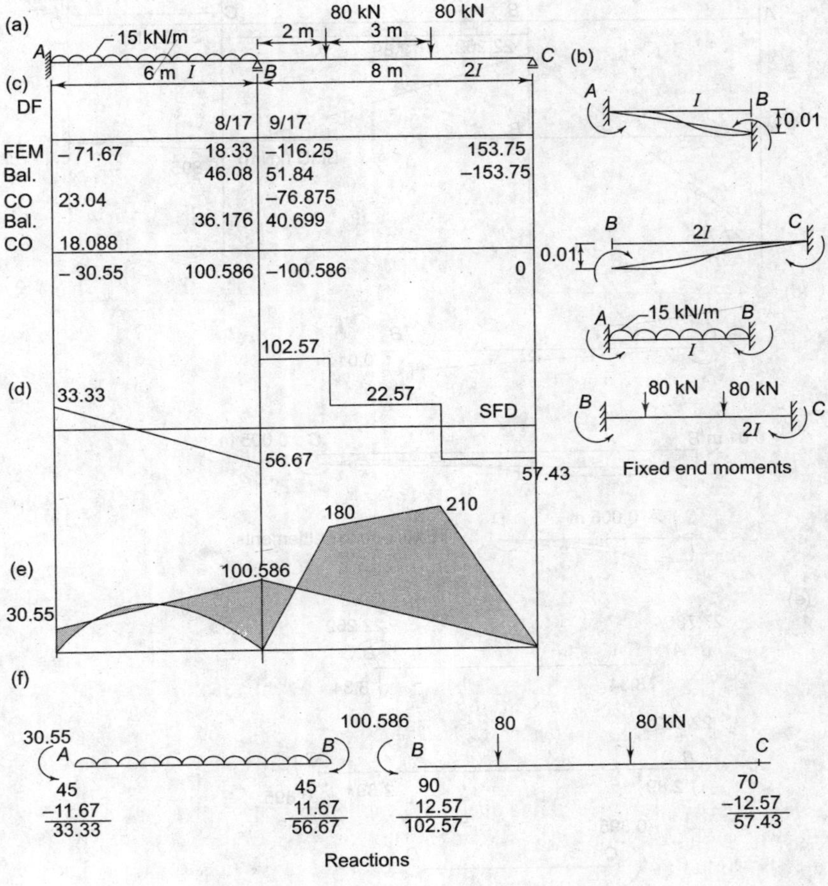

Fig. 15.54

Moment distribution This is done as shown in Fig. 15.54(c). Joints B and C are balanced first. Joint C is left in the unlocked position. The moment applied at BA is carried over to A and the moment at CB to B. The imbalance at B due to the carry-over is – 76.875 kN m. This is balanced in the second cycle and $1/2 \times 36.176$ kNm is carried over to A. The moments are now summed up. The final moments are $M_{AB} = -\,30.55$ kNm, $M_{BA} = 100.586$ kNm $= -\,M_{BC}$, and $M_{CB} = 0$. The reactions can now be calculated as shown in Fig. 16.14(f) and the SF and BM diagrams drawn as shown in Figures 15.54(d) and (e), respectively. ■

Example 15.33 Moment distribution: three-span beam with support settlement

Analyse the continuous beam shown in Fig. 15.55(a) and draw the SF and BM diagrams. Support B settles down by 10 mm, while support C has a settlement of 5 mm. $E = 200$ GN/m^2 and $I = 10{,}000$ mm^4.

Solution For consistency of units, the settlement of $B = 10$ mm $= 0.01$ m; the settlement of $C = 5$ mm $= 0.005$ m. $EI = 200 \times 10^6 \times 10{,}000 \times 10^{-?8}$ kN m^2 = 20,000 kN m^2.

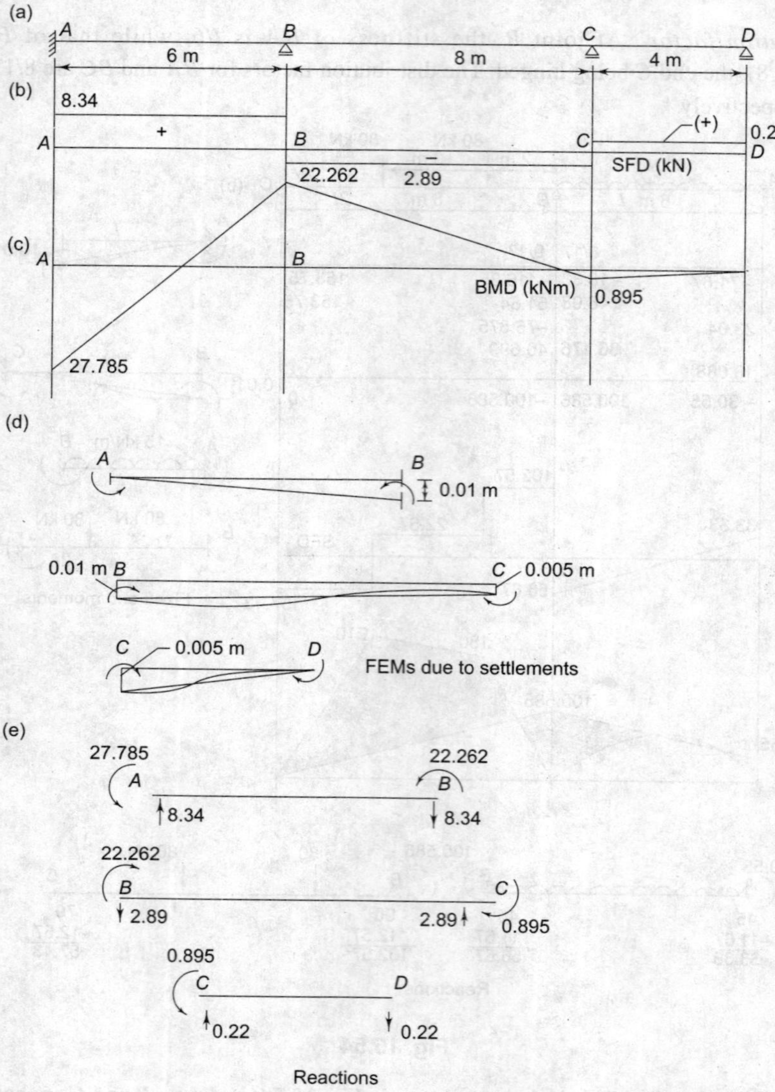

Fig. 15.55

15.9 GENERALIZATION OF STIFFNESS METHOD

Stiffness, as we have seen, is the force required to cause unit translation or rotation at a support point. Stiffness can be worked out for different types of actions as given below. (Refer to Fig. 15.56.)

Truss member A truss member is subjected to only axial forces. Due to an axial force, elongation of the member is given by $\Delta L = PL/AE$. Stiffness due to the axial load is the force required to cause unit deformation. When $\Delta L = 1$, $P = AE/L$ is the axial stiffness. This is written as $s_{11} = AE/L$.

Beam stiffness A beam is generally subjected to a shear force and bending moment. There are thus two deformations at the end of the member—a unit translation and a unit rotation. We number the translation as 1 and rotation as 2. The ends of the member are considered fixed. When a unit translation in the positive direction of the Y-axis is given to the end of this member, the actions generated at the ends of the member are the stiffness due to this deformation.

Due to the unit translation of end B, as shown in Fig. 15.56(b), the forces/couples generated are—a force of $12EI/L^3$ in the direction of 1 and a couple of $6EI/L^2$ in direction 2. The stiffness can then be written as

$$s_{11} = 12EI/L^3 \text{ and } s_{21} = -6EI/L^2$$

When a unit rotation is given at end B, the forces are $-6EI/L^2$ in direction 1 and the couple is $4EI/L$. The stiffness can be written as $s_{21} = -6EI/L^2$ and $s_{22} = 4EI/L$. The four stiffness coefficients can be written in matrix form as

$$[S] = \begin{vmatrix} 12EI/L^3 & -6EI/L^2 \\ -6EI/L^2 & 4EI/L \end{vmatrix}$$

A plane frame member The difference between a plane beam and a plane frame is the axial load that may be taken into account while considering the stiffness of the member. Referring to Fig. 15.56(c), the elements of the stiffness of the frame member can be written as

$$S = \begin{vmatrix} AE/L & 0 & 0 \\ 0 & 12EI/L^3 & -6EI/L^2 \\ 0 & -6EI/L^2 & 4EI/L \end{vmatrix}$$

A space frame member A space frame member is shown in Fig. 15.56(d). The X-axis coincides with the member axis. There are three translations and three rotations possible at the end of the member. The numbers 1, 2, and 3 are given for translations at the end of the member along the XX-, YY-, and ZZ-axes, respectively. The numbers 4, 5, and 6 are given for the rotations about the XX-, YY-, and ZZ-axes, respectively.

(i) *Unit translation along the X-axis* This will require a force of AE/L along the X-axis. It does not create any force/couple in the other five directions. $s_{11} = AE/L$.

(ii) *Translation along the Y-axis* A unit translation along the Y-axis creates a force of $12EI_{zz}/L^3$ along the Y-axis (s_{22}) and a couple of $-6EI_{zz}/L^2$ about the Z-axis (s_{62}). Note that the value of I required in this case is about the neutral axis for bending in the XY plane (I_{zz}).

(iii) *Translation along the Z-axis* The unit translation along the Z-axis is in the XZ plane. The forces generated are the same for bending about the XZ plane. The MI to be considered is about the Y-axis. The actions created are a force

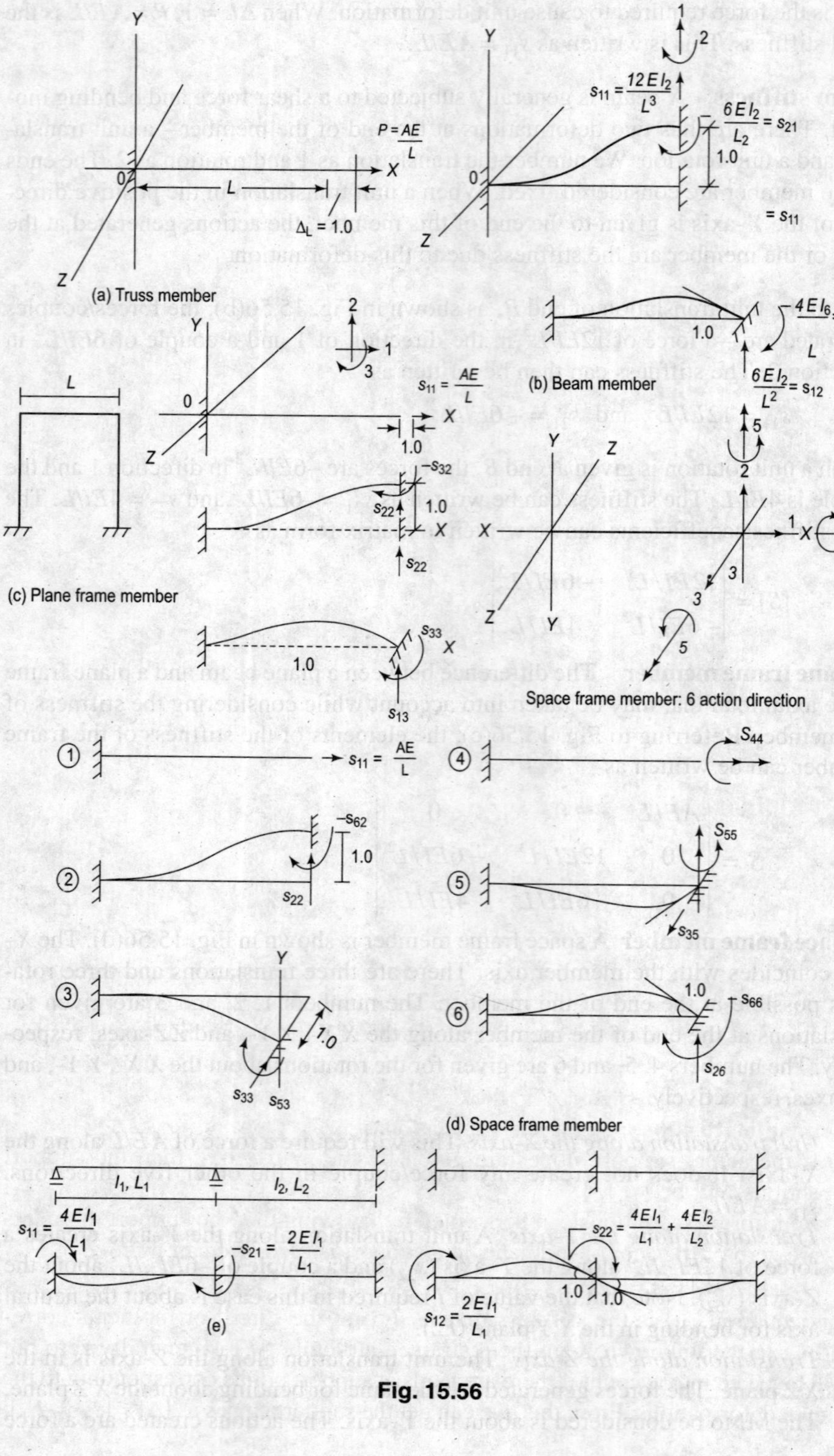
Y
$P = \frac{AE}{L}$
0
X
L
Z
$\Delta_L = 1.0$
(a) Truss member
Y
1
2
$s_{11} = \frac{12EI_2}{L^3}$
$-\frac{6EI_2}{L_2} = s_{21}$
1.0
0
X
Z
$= s_{11}$
$\frac{4EI_6}{L} = S_{22}$
1.0
(b) Beam member
$\frac{6EI_2}{L^2} = s_{12}$
L
Y
2
1
3
$s_{11} = \frac{AE}{L}$
0
X
Z
1.0
$-s_{32}$
s_{22}
1.0
X
X
s_{22}
(c) Plane frame member
s_{33}
X
1.0
s_{13}
Y
Z
5
2
4
1
X
3
3
Z
5
Space frame member: 6 action direction
1
$s_{11} = \frac{AE}{L}$
4
S_{44}
2
$-s_{62}$
1.0
s_{22}
5
S_{55}
s_{35}
3
Y
1.0
s_{33}
s_{53}
6
1.0
s_{66}
s_{26}
(d) Space frame member
I_1, L_1
I_2, L_2
$s_{11} = \frac{4EI_1}{L_1}$
$-s_{21} = \frac{2EI_1}{L_1}$
$s_{22} = \frac{4EI_1}{L_1} + \frac{4EI_2}{L_2}$
1.0
1.0
$s_{12} = \frac{2EI_1}{L_1}$
(e)

Fig. 15.56

of $12EI_{yy}/L^3$ along the Z-axis (s_{33}) and a couple of $-6EI_{yy}/L^2$ about the Y-axis (s_{53}).

(iv) *Rotation about the X-axis* As the X-axis coincides with the axis of the member, the rotation about the X-axis is torsion. Torsional stiffness is involved here, which is obtained from $T/J = G\theta/L$. Here θ is unity and T required for this is GJ/L. This value of T is the stiffness. We thus have $s_{44} = GJ/L$. It does not require any action in any other direction.

(v) *Rotation about the Y-axis* For the unit rotation about the Y-axis (bending in the XZ plane), the couple required is $4EI_{yy}/L$ (s_{55}). The reaction at the end in the direction of Z-axis is $6EI_{yy}/L^2$ (s_{35}).

(vi) *Rotation about the Z-axis* The unit rotation about the Z-axis is in the XY plane. The couple required at the end is $4EI_{zz}/L$ (s_{66}). This will also generate a reaction in the Y direction, which is equal to $-6EI_{zz}/L^2$ (s_{36}).

We have now all the stiffness values and they can be put into the stiffness matrix as given below.

$$[S] = \begin{vmatrix} AE/L & 0 & 0 & 0 & 0 & 0 \\ 0 & 12EI_{zz}/L^3 & 0 & 0 & 0 & -6EI_{zz}/L^2 \\ 0 & 0 & 12EI_{yy}/L^3 & 0 & 6EI_{yy}/L^2 & 0 \\ 0 & 0 & 0 & GJ/L & 0 & 0 \\ 0 & 0 & 6EI_{yy}/L^2 & 0 & 4EI_{yy}/L & 0 \\ 0 & -6EI_{zz}/L^2 & 0 & 0 & 0 & 4EI_{zz}/L \end{vmatrix}$$

Like the flexibility matrix, the stiffness matrix is also symmetrical about the diagonal.

15.9.1 Stiffness Matrix Method

The stiffness matrix method works in the same way as the flexibility matrix method. The procedure involves the following steps.

(i) All the supports are considered fixed in the beginning to get a restrained structure.

(ii) The kinematic indeterminacy of the structure is determined by counting the number of unknown translations or rotations. Generally, the slopes at the supports are unknowns unless there is support settlement, which will be translations.

(iii) Form the stiffness matrix. In the case of beams, these will be due to the unit rotations at supports and unit translations.

(iv) If the actual support slopes are given by the matrix [Δ], then the support actions will be [s] [Ä].

(v) Equilibrium equations can be formed by knowing that the support moments are zero or the sum of moments at a support is zero. We can write

$$[s]\,[\Delta] + [F_L] = [F]$$

In this equation, [s] is the stiffness matrix, F_L are the forces or moments corresponding to the unknown Δ*s* in the restrained structure. [F] is generally zero but should be taken into account if there are loads at support points corresponding to the unknown deformations. From this, we can get the support slopes as $[\Delta] = [s^{-1}][-F_L]$.

(vi) Knowing the slopes, the slope-deflection equations can be used to evaluate the moments.

Note that in the beam shown in Fig. 15.56(e), it is very simple to form the stiffness matrix. Since all the ends are restrained, a unit rotation at one support requires a moment of $4EI/L$.

For example, a unit rotation at support A requires a moment of $4EI/L_1$ at A. You will remember that half of this moment is carried over to support B. Also $2EI/L_1$ is the moment at support B.

Similarly, a unit rotation at B will require a unit rotation in both the spans. The moment required at support B is $(4EI/L_1 + 4EI/L_2)$. Half of $4EI/L_1$ will be carried over to A. Half of $4EI/L_2$ will be carried over to C also. But as C is a fixed end and has no unknown deformation, it is not required to be considered.

The following examples illustrate the use of stiffness matrix for solution of simple continuous beams.

Example 15.34 Stiffness matrix method: continuous beam

Analyse the continuous beam shown in Fig. 15.57 using the stiffness matrix method and find the reactions and draw the SF and BM diagrams.

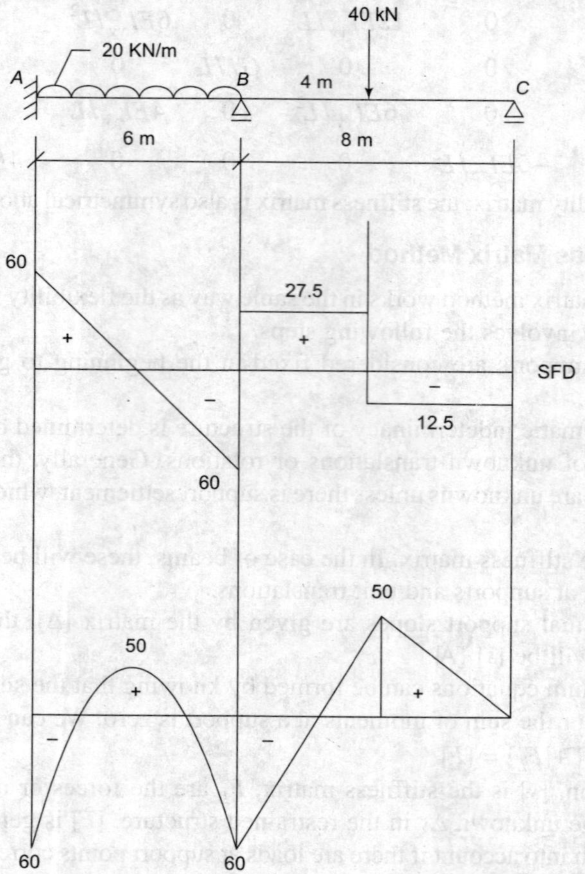

Fig. 15.57

Solution Support A, being fixed, has no deformations. Also θ_B and θ_C are the unknown rotations at supports B and C.

Fixed end moments

Considering the ends as fixed, the fixed end moments can be calculated as

$$M_{FAB} = -20 \times 6^2/12 = -M_{FBA} = -60 \text{ kNm}$$
$$M_{FBC} = -40 \times 8/8 = -M_{FCB} = -40 \text{ kNm}$$

Stiffness matrix

The rotations θ_B and θ_C are numbered 1 and 2, respectively. With restrained ends, we give a unit rotation to support B in the clockwise direction as shown. We can then write the stiffness coefficients as

$$s_{11} = (4EI/6) + (4EI/8) = 7EI/6 \text{ and } s_{21} = (1/2)(4EI/8) = EI/4$$

Note that the unit rotation at support B has to rotate both the spans BA and BC by unit angle. The moment required is the sum of the moments required to rotate the two spans. Whatever moment is applied at B in span BC, half of that moment is carried omver to the support C. This action corresponds to the second unknown θ_C.

With restrained ends, we now apply unit rotation at support C. This gives rise to the following stiffness coefficients:

$$s_{22} = 4EI/8 = EI/2 \text{ and } s_{12} = (1/2)(4EI/8) = EI/4$$

The stiffness matrix is $[S] = \begin{vmatrix} 7/6 & 1/4 \\ 1/4 & 1/2 \end{vmatrix}^{EI}$

We form the equilibrium equation as $[S]\,[\theta] + [F] = 0$. This can be written as

$$EI \begin{vmatrix} 7/6 & 1/4 \\ 1/4 & 1/2 \end{vmatrix} \begin{vmatrix} \theta_B \\ \theta_C \end{vmatrix} + \begin{vmatrix} 60 - 40 \\ 40 - 0 \end{vmatrix} = 0$$

From this, we can write $\begin{vmatrix} \theta_B \\ \theta_C \end{vmatrix} = (I/EI) \begin{vmatrix} 7/6 & 1/4 \\ 1/4 & 1/2 \end{vmatrix}^{-1} \begin{vmatrix} -20 \\ -40 \end{vmatrix}$

The inverse of the $[S]$ matrix is $\begin{vmatrix} 1/2 & -1/4 \\ -1/4 & 7/6 \end{vmatrix}$

divided by the determinant $[7/6 \times 1/2 - (1/4)^2]$.

$$\begin{vmatrix} \theta_B \\ \theta_C \end{vmatrix} = \frac{48}{25EI} \begin{vmatrix} 1/2 & -1/4 \\ -1/4 & 7/6 \end{vmatrix} \begin{vmatrix} -20 \\ -40 \end{vmatrix} = \frac{48}{25} \begin{vmatrix} 0 \\ -125/3 \end{vmatrix}$$

This gives $\theta_B = 0$ and $\theta_C = -80/EI$.

The support moments can now be calculated from the slope-deflection equations.

As $\theta_B = 0$,

$$M_{AB} = M_{FAB} = -60 \text{ kNm}$$
$$M_{BA} = M_{FBA} = 60 \text{ kNm}$$
$$M_{BC} = M_{FBC} + 2EI/L(2\theta_B + \theta_C) = -60 \text{ kNm.}$$
$$M_{CB} = M_{FCB} + 2EI/L(2\theta_C + \theta_B) = +40 - 40 = 0, \text{ which agrees with the}$$

support condition.

The reactions can be calculated as follows.

In span AB, $R_A = 60$ kN and $R_B = 60$ kN

In span BC, $R_B = 20 + (60/8) = 27.5$ kN and $R_C = 20 - (60/8) = 12.5$ kN

The shear force and bending moment diagrams can now be drawn as shown. ■

Example 15.35 Stiffness matrix method: continuous beam

Analyse the three-span continuous beam loaded as shown in Fig. 15.58. *EI* is constant.

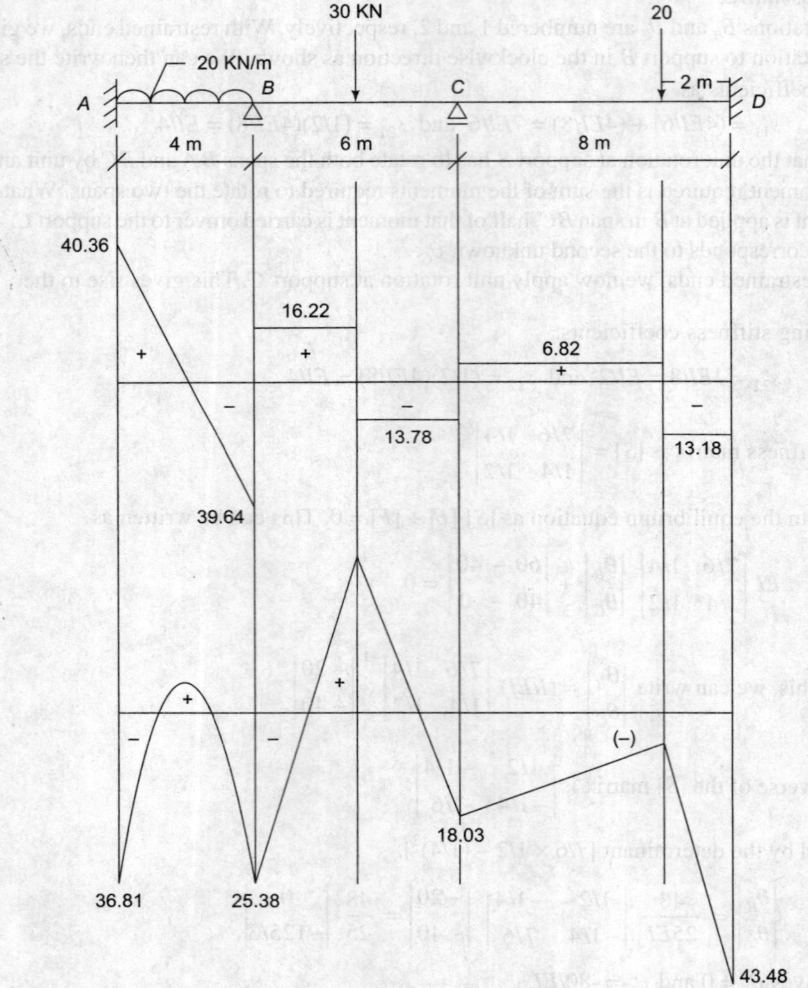

Fig. 15.58

Solution The fixed end moments are as follows:

$M_{FAB} = -wL^2/12 = 20 \times 16/12 = -26.67$ kNm

$M_{FBA} = 26.67$ kNm

$M_{FBC} = PL/8 = -30 \times 6/8 = -22.5$ kNm

$M_{FCB} = 22.5$ kNm

$M_{FCD} = -Pab^2/L^2 = -40 \times 6 \times 4/64 = -15$

$M_{FDC} = 40 \times 36 \times 2/64 = 45$ kNm

Here θ_B and θ_C are the unknown deformations. The stiffness matrix can be formed as follows.

Number 1 is given to the slope at *B* and number 2 to the slope at *C*. For a unit rotation at *B*,

$$s_{11} = (4EI/4) + (4EI/6) = 5EI/3 \quad \text{and} \quad s_{21} = (1/2)(4EI/6) = EI/3$$

For a unit rotation at *C*,

$$s_{22} = (4EI/6) + (4EI/8) = 7EI/6 \quad \text{and} \quad s_{12} = (1/2)(4EI/6) = EI/3$$

$$\text{Stiffness matrix } [S] = \begin{vmatrix} 5/3 & 1/3 \\ 1/3 & 7/6 \end{vmatrix}$$

$$\text{The matrix } [\Delta] = \begin{vmatrix} \theta_B \\ \theta_C \end{vmatrix}$$

$$\text{The matrix } [F_L] = \begin{vmatrix} 26.67 & -22.5 \\ 22.5 & -15 \end{vmatrix} = \begin{vmatrix} -4.17 \\ -7.5 \end{vmatrix}$$

The equilibrium equation is

$$[S]\,[\Delta] + [F_L] = 0 \quad \text{or} \quad [\Delta] = [S]^{-1}[-F_L]$$

$$[S]^{-1} = \frac{\begin{vmatrix} 7/6 & -1/3 \\ -1/3 & 5/3 \end{vmatrix}}{(1/EI) = \left(\dfrac{5\times7}{3\times6} - \dfrac{1\times1}{3\times3}\right)} \quad \frac{1}{1.83EI}\begin{vmatrix} 7/6 & -1/3 \\ -1/3 & 5/3 \end{vmatrix}$$

$$1.83EI \begin{vmatrix} \theta_B \\ \theta_C \end{vmatrix} \begin{vmatrix} 7/6 & -1/3 \\ -1/3 & 5/3 \end{vmatrix} \begin{vmatrix} -4.17 \\ -7.5 \end{vmatrix} \begin{vmatrix} -2.365 \\ -11.11 \end{vmatrix}$$

$$\begin{vmatrix} \theta_B \\ \theta_C \end{vmatrix} = \begin{vmatrix} 1 \\ EI \end{vmatrix} \begin{vmatrix} -1.29 \\ -6.06 \end{vmatrix}$$

Now, the moments can be calculated using the slope-deflection equations.

$$M_{AB} = -26.67 + (2EI/4) - 1.29)/EI = -26.81 \text{ kNm}$$

$$M_{BA} = 26.67 + (2EI/4)(-2\times1.29)/EI = 25.38 \text{ kNm}$$

$$M_{BC} = -22.5 + (2EI/6)(-2\times1.29 - 6.06) = -25.38 \text{ kNm}$$

$$M_{CB} = 22.5 + (2EI/6)(-1.29 - 2\times6.06) = 18.03 \text{ kNm}$$

$$M_{CD} = -15 + (2EI/8)(-2\times6.06) = -18.03 \text{ kNm}$$

$$M_{DC} = 45 + (2EI/8)(-6.06) = 43.485 \text{ kNm}$$

The reactions can be calculated as follows.

$$R_A = 40 + (26.81 - 25.38)/4 = 40.36 \text{ kN}$$

$$R_B = 40 - 0.36 = 39.64 \text{ kN}$$

$$R_B = 15 + (25.38 - 18.03)/6 = 16.22 \text{ kN}$$

$$R_C = 15 - 1.22 = 13.78 \text{ kN}$$

$$R_C = 10 - (43.485 - 18.03)/8 = 6.82 \text{ kN}$$

$$R_D = 30 + 3.18 = 33.18 \text{ kN}$$

The shear force and bending moment diagrams can be drawn as shown in the figure. ■

Example 15.36 Stiffness matrix method: continuous beam

Analyse the two-span continuous beam shown in Fig. 15.59. *EI* is constant.

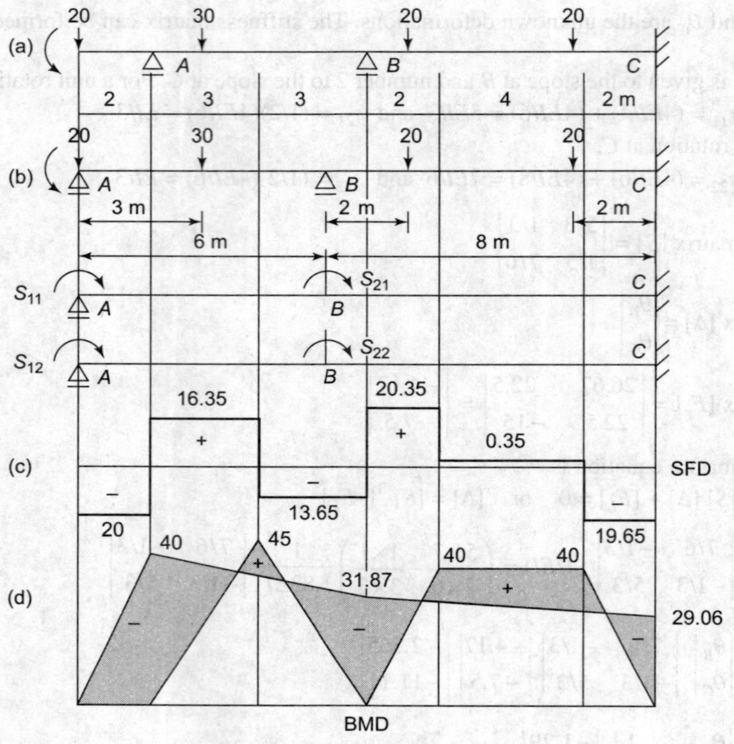

Fig. 15.59

Solution The only special feature in this case is the overhang. The overhang can be removed and the beam analysed with the loading as shown. The vertical load and an anticlockwise couple (= 40 kNm) is placed at A. Note that there is a load corresponding to θ_A at the support.

Here θ_A (1) and θ_B (2) are the unknown deformations.

We form the stiffness coefficients matrix by restraining the ends and finding the moments required to give a unit rotation at A and B separately. Referring to Fig. 15.59(b), s_{11} is the moment required to give unit clockwise rotation at A and s_{21} is the moment (corresponding to the rotation at B) at C because of the unit rotation at A.

$$s_{11} = 4EI/L = 4EI/6 = 2EI/3 \text{ and } s_{21} = (1/2)(4EI/6) = EI/3$$

The supports are restrained at A, B, and C. We give a unit clockwise rotation at B. This will require unit rotation in both the spans.

$$s_{22} = (4EI/6) + (4EI/8) = 7EI/6 \text{ and } s_{12} = (1/2)(4EI/6) = EI/3$$

$$\text{The stiffness matrix } [S] = \begin{vmatrix} 2/3 & 1/3 \\ 1/3 & 7/6 \end{vmatrix} EI$$

The fixed end moments are

$$M_{FAB} = -\,30 \times 6/8 = -\,22.5 \text{ kNm}$$

$$M_{FBA} = 22.5 \text{ kNm}$$

$$M_{FBC} = 20 \times 2 \times 6^2 + 20 \times 6 \times 2^2 = -\,30 \text{ kNm}$$

$$M_{FCB} = 30 \text{ kNm}$$

$$\text{The } [F_L] \text{ matrix is } \begin{vmatrix} -\,22.5 \\ -7.5 \end{vmatrix}$$

The $[\Delta]$ matrix is $\begin{vmatrix}\theta_B \\ \theta_C\end{vmatrix}$

The $[F]$ matrix is not zero as there is a moment of – 40 kNm at A giving $\begin{vmatrix}-40 \\ -0\end{vmatrix}$

The equilibrium equation can be formulated as

$[S]\,[\Delta] + [F_L] = [F]$

This can be written as

$$(EI)\begin{vmatrix}2/3 & 1/3 \\ 1/3 & 7/6\end{vmatrix}\begin{vmatrix}\theta_A \\ \theta_B\end{vmatrix} + \begin{vmatrix}22.5 \\ 7.5\end{vmatrix} = \begin{vmatrix}-40 \\ -0\end{vmatrix}$$

or $\quad [\Delta] = [S]^{-1}\{-[F_L] + [F]\}$

The inverse of the $[S]$ matrix is $\dfrac{3}{2EI}\begin{vmatrix}7/6 & -1/3 \\ -1/3 & 2/3\end{vmatrix}$

$$[\Delta] = \frac{3}{2EI}\begin{vmatrix}7/6 & -1/3 \\ -1/3 & 2/3\end{vmatrix}\begin{vmatrix}-17.5 \\ 7.5\end{vmatrix} = \frac{1}{EI}\begin{vmatrix}-34.375 \\ 16.245\end{vmatrix}$$

Therefore, $\theta_A = -34.375/EI$ and $\theta_B = 16.245/EI$

The slope-deflection equations can be used to find the support moments.

$$M_{AB} = -22.5 + 2EI\,\frac{2(-34.375) + (16.245)}{6EI} = -40$$

$$M_{BA} = 22.5 + 2EI(2 \times 16.245 - 34.375) = 21.87 \text{ kNm}$$

$$M_{BC} = -30 + 2EI(2 \times 16.245)/8EI = -21.87 \text{ kNm}$$

$$M_{CB} = 30 + 2EI(16.245)/8EI = 34 \text{ kNm}$$

The reactions can be calculated as follows.

In span AB,

$$R_A = 20 + 15 + (40 - 21.87)/6 = 38.02 \text{ kN} \uparrow$$

$$R_B = 15 - (40 - 21.87)/6 = 11.98 \text{ kN} \uparrow$$

In span BC,

$$R_B = 20 - (34 - 21.87)/8 = 18.49 \text{ kN} \uparrow$$

$$R_C = 20 + (34 - 21.87)/8 = 21.51 \text{ kN} \uparrow$$

15.1O ADVANTAGES AND DISADVANTAGES OF CONTINUOUS BEAMS

One can notice from the examples given above that a continuous beam has a lower BM than the case in which the spans are simply supported. Also, the maximum BM occurs at the supports, which is also an advantage. However, continuous beams have the disadvantage that any change in the assumed conditions, such as the settlement of supports, causes large changes in the BM. Continuous beams still find wide application in buildings, bridges, etc.

Summary

Fixed and continuous beams are indeterminate structures. Therefore, additional equations in terms of deformation conditions, in addition to the conditions of static equilibrium, are required to analyse such beams.

The fixed beam, while not so common in construction, is a very important element in structural analysis. The stiffness method of analysis is based on the concept of fixed beams. Fixed beams are kinematically determinate but statically indeterminate. For beams carrying vertical loads only, the static indeterminacy can be taken as 2.

For analysis, a fixed beam is considered as an SS beam with end moments at both ends to make the slopes zero. Separate moment diagrams can be drawn for applied loads and end moments, and superposed. Since the slope and deflection at the ends are zero, the angle between the tangents drawn at the ends is zero. In terms of the area-moment theorems, it means that the area of the *M*/*EI* diagram between the two supports is zero. In addition, the moment of area of the *M*/*EI* diagram between the supports about any support is zero, because the tangents drawn at the ends are collinear. In most of the common cases, the moment diagram due to the applied loads is positive, while the end moments give a hogging BM diagram. This means that

(i) the area of the *M*/*EI* diagram due to applied loads is equal to the area of the *M*/*EI* diagram due to end moments and

(ii) the moment of area of the *M*/*EI* diagram due to loads about any of the supports is equal to the moment of area of the *M*/*EI* diagram due to the end moments about the same support.

The double integration method for finding deflections can be applied to find fixed end moments. The general BM equation for the beam can be written using Macaulay's method with unknown end moments and reactions.

There are four conditions of deformation available when the slope and deflection at both the ends are zero. These four conditions can be used to evaluate the two constants of integration and the two unknown moments and reactions. The double integration method has the advantage that the slopes and deflections can also be evaluated from the equations obtained.

Fixed beams are subjected to much lower BM than the corresponding SS beams. However, they are greatly affected by support settlements and end rotations if not properly constructed.

Continuous beams are very common structural elements, finding wide application in buildings, bridges, etc. The basic characteristic of the deformation curve of a continuous beam is the continuity of the slopes over the supports. Clapeyron's theorem of three moments is used to analyse continuous beams. The general form of the equation, assuming E to be constant for all spans is

$$M_A \frac{l_1}{I_1} + 2M_B\left(\frac{l_1}{I_1} + \frac{l_2}{I_2}\right) + M_C \frac{l_2}{I_2} = \frac{6A_1\bar{x}_1}{I_1\, l_1} + \frac{6A_2\bar{x}_2}{I_2\, l_2} - 6E\left(\frac{\delta_1}{l_1} + \frac{\delta_2}{l_2}\right)$$

where M_A, M_B, and M_C are the moments at three consecutive supports, l_1, l_2 are the lengths of the spans, I_1, I_2 are the moments of inertia of the cross-sectional area in the two spans, $A_1\bar{x}_1, A_2\bar{x}_2$ are the moments of area of the BM diagrams obtained by considering the spans to be simply supported about the outer supports, and δ_1, δ_2 are the lengths by which the central support has sunk below the outer supports. If the beam has a fixed end, an equation can be written for one span adjacent to the fixed support, assuming a zero span of infinite rigidity on the other side.

The method of consistent deformation, which is a case of the flexibility method of analysis, can be generalized and put in matrix form for analysis of continuous beams using flexibility coefficients.

The slope-deflection equation is the most fundamental proposition in the stiffness method of analysis. The slope-deflection method and moment distribution method are both stiffness methods of analysis, where the joint deformations are taken as unknowns and the conditions of equilibrium are used to evaluate them. The moment distribution method is an iterative technique and is very useful in analysing continuous beams.

The generalized stiffness method of analysis using stiffness coefficients is a very powerful method of analysis. It can be put into matrix notation using stiffness coefficients and can be solved to get the unknown slopes and deflections. The slope-deflection equations can then be used to evaluate the support moments.

Continuous beams have to be designed for much less bending moments than those in the case of simply supported spans. The maximum BM is generally over the supports. However, any settlement of supports, etc. is more critical due to the large bending stresses induced in the beam.

Exercises

Review Questions

1. State the advantages and disadvantages of fixed beams.
2. A fixed beam of span l m is subjected to an anticlockwise couple. M is the moment of area at a distance a from the left end. Write the general BM equation to be solved using the double integration method. Sketch approximately the SF and BM diagrams for the beam.
3. For beams loaded as shown in Fig. 15.60, draw, without calculations, the approximate SF and BM diagrams. Also sketch the approximate deflected shape.

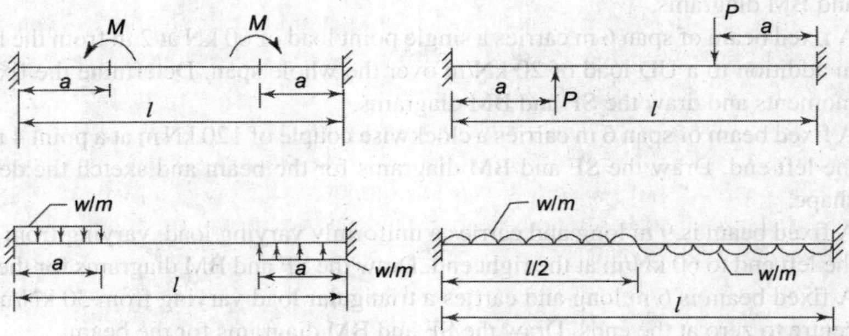

Fig. 15.60

4. A fixed beam of span 4 m has a cross section with a moment of inertia of 8000 cm^4. $E = 2 \times 10^8$ kN/m^2. If the support at the left settles by 15 mm and that on the right settles by 6 mm, calculate the fixed end moments. Draw the BM diagrams and sketch the deflected shape.
5. State the advantages and disadvantages of continuous beams.
6. Write the general three-moment equation and explain the terms used in the equation with illustrative sketches.
7. Sketch the deflected shapes for continuous beams loaded as shown in Fig. 15.61.

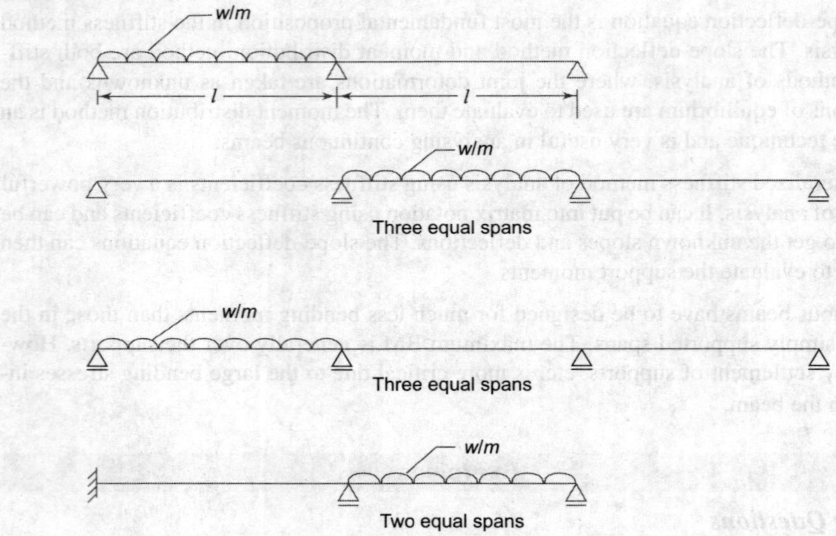

Fig. 15.61

8. Find the ratio of the total loads in two identical fixed beams, one carrying a central concentrated load and the other a UD load over the whole span, if (i) the fixed end moments are equal and (ii) the maximum deflections are equal.

Problems

1. A fixed beam of span 4 m carries two point loads of 40 kN and 60 kN at 1 m and 2 m from the left end, respectively. Determine the fixed end moments and draw the SF and BM diagrams.
2. A fixed beam of span 6 m carries a single point load of 60 kN at 2 m from the left end in addition to a UD load of 20 kN/m over the whole span. Determine the fixed end moments and draw the SF and BM diagrams.
3. A fixed beam of span 6 m carries a clockwise couple of 120 kN m at a point 4 m from the left end. Draw the SF and BM diagrams for the beam and sketch the deflected shape.
4. A fixed beam is 9 m long and carries a uniformly varying load, varying from zero at the left end to 60 kN/m at the right end. Draw the SF and BM diagrams for the beam.
5. A fixed beam is 6 m long and carries a triangular load varying from 30 kN/m at the centre to zero at the ends. Draw the SF and BM diagrams for the beam.
6. A fixed beam is 8 m long and carries a uniformly varying load, varying from zero at the centre to 20 kN/m over the supports. Find the end moments and draw the SF and BM diagrams.
7. A fixed beam of span 5 m carries a uniformly distributed load of 20 kN/m over the right half of its length in addition to a clockwise couple of 50 kN m at 1.25 m from the left support. Draw the SF and BM diagrams for the beam.
8. A fixed beam of span 6 m carries two point loads of 80 kN and 120 kN at 2 m and 4 m from the left end, respectively. The right support sinks by 8 mm. Draw the SF and BM diagrams for the beam. $E = 200$ GN/m^2 and $I = 9600$ cm^4.
9. A beam fixed at both ends carries a UD load of 20 kN/m in addition to a point load of 80 kN at 1 m from the left support. The left support sinks by 12 mm. Draw the SF and BM diagrams for the beam: $E = 2 \times 10^{11}$ N/m^2 and $I = 8000$ cm^4.

10. Compare the deflections under the load of two identical beams, one fixed and the other simply supported, if they carry a single concentrated load at one-third the span from the left support.
11. A two-span continuous beam carries a UD load of 20 kN/m over its whole length. The left span is 4 m long and the right span is 6 m long. The beam has the same section throughout. Draw the SF and BM diagrams for the beam.
12. Draw the SF and BM diagrams for a beam loaded as shown in Fig. 15.62.

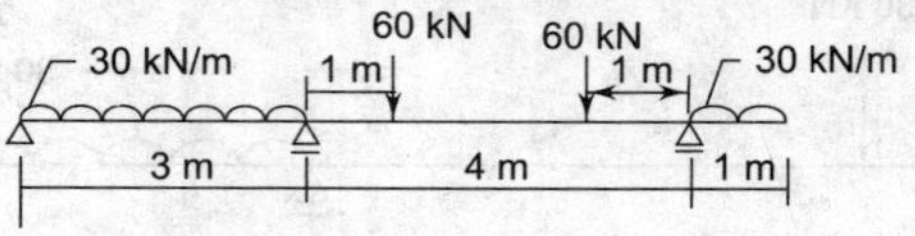

Fig. 15.62

13. Analyse the two-span continuous beam carrying the loads shown in Fig. 15.63 and draw the SF and BM diagrams.
14. A three-span continuous beam has fixed supports at both ends and carries the loads shown in Fig. 15.64. Draw the SF and BM diagrams for the beam.

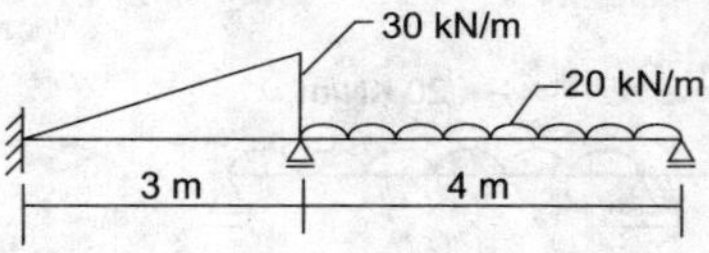

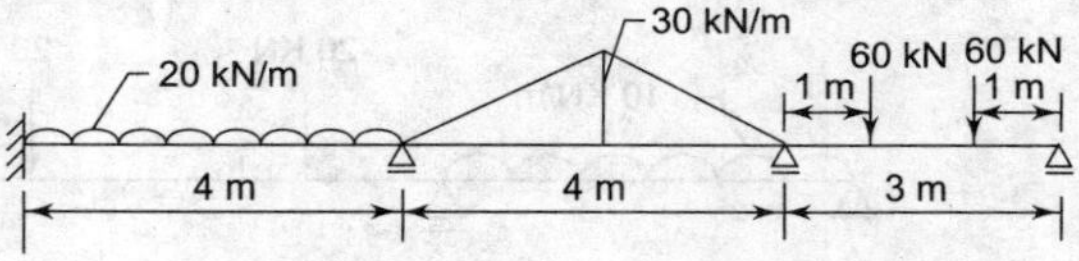

Fig. 15.63 Fig. 15.64

15. A beam 16 m long is supported at points 2 m and 10 m from the left end and is fixed at the right end. It carries the loads shown in Fig. 15.65. Draw SF and BM diagrams.

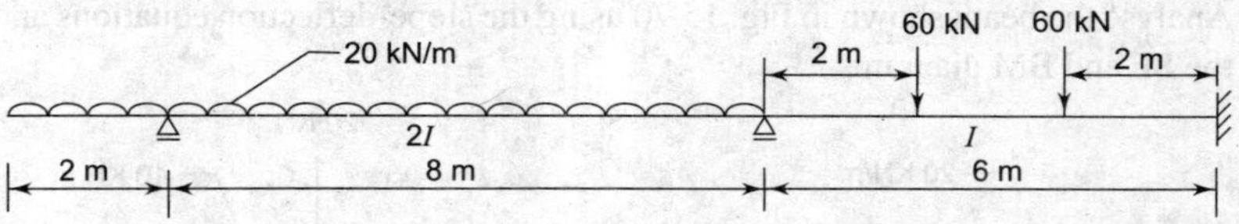

Fig. 15.65

16. Analyse the continuous beam carrying the loads shown in Fig. 15.66 and draw the SF and BM diagrams. Sketch the deflected shape.

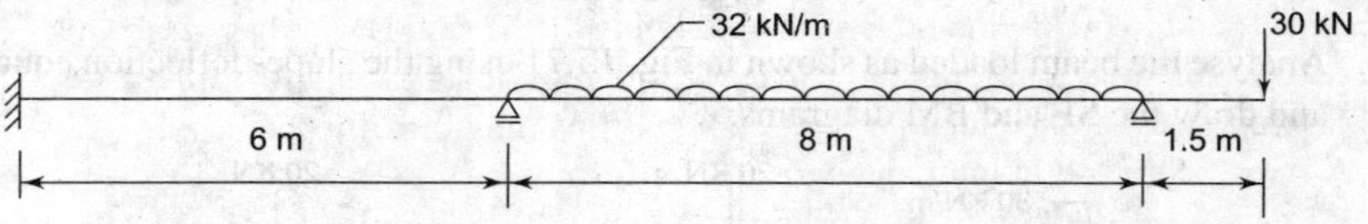

Fig. 15.66

17. Determine the end moments and draw the SF and BM diagrams for the continuous beam shown in Fig. 15.48 if the support B sinks by 15 mm. $E = 200$ GN/m^2 and $I =$ 9000 cm^4.

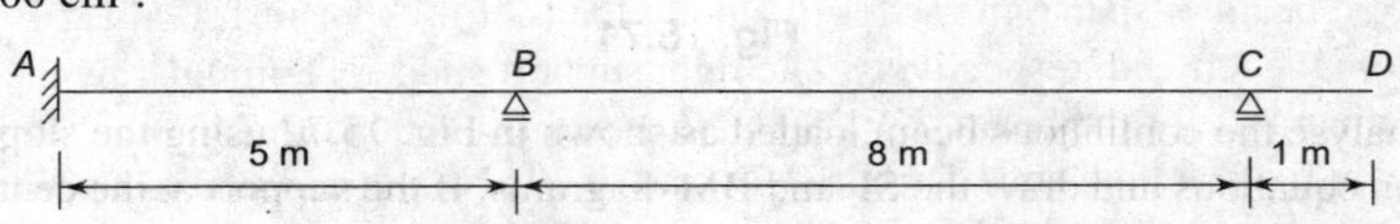

Fig. 15.67

18. A continuous beam 10 m long is supported at the ends and at a point 6 m from the left end. It carries a UD load of 30 kN/m over the whole length. The support at the middle settles by 10 mm. Draw the SF and BM diagrams for the beam if $E = 2 \times 10^8$ kN/m^2 and $I = 1 \times 10^{-4}$ m^4.

19. Analyse the two-span continuous beam shown in Fig. 15.68 by the flexibility matrix method and draw the SF and BM diagrams.

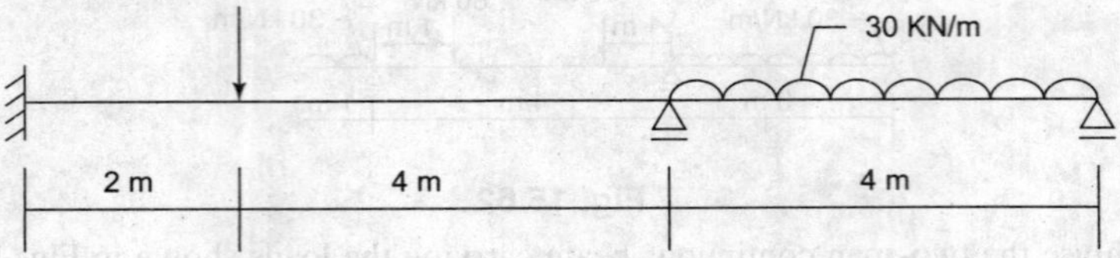

Fig. 15.68

20. Analyse the three-span continuous beam shown in Fig. 15.69 by the flexibility matrix method and draw the shear force and bending moment diagrams.

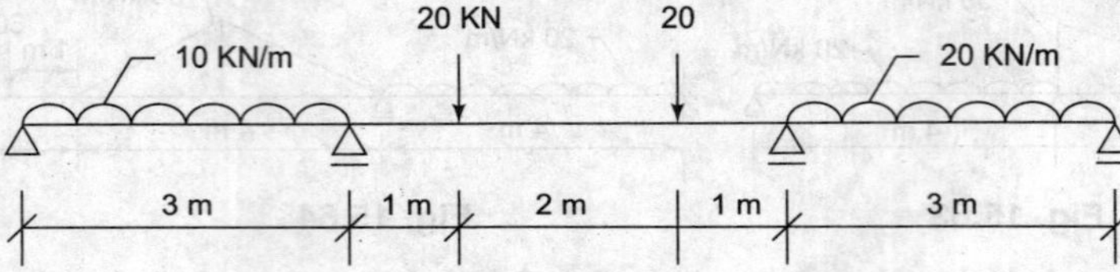

Fig. 15.69

21. Analyse the beam shown in Fig. 15.70 using the slope-deflection equations and draw the SF and BM diagrams.

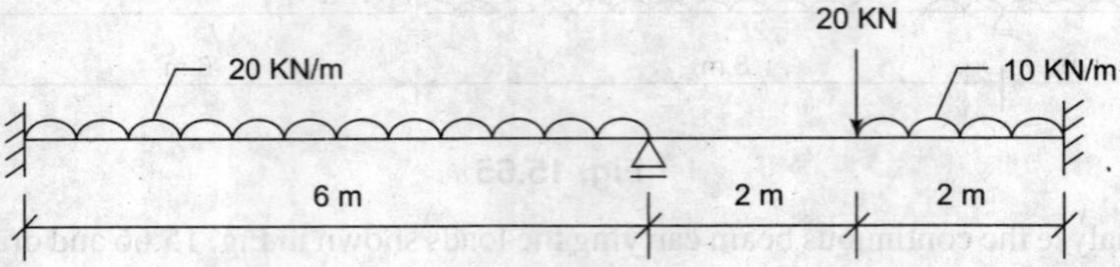

Fig. 15.70

22. Analyse the beam loaded as shown in Fig. 15.71 using the slope-deflection equations and draw the SF and BM diagrams.

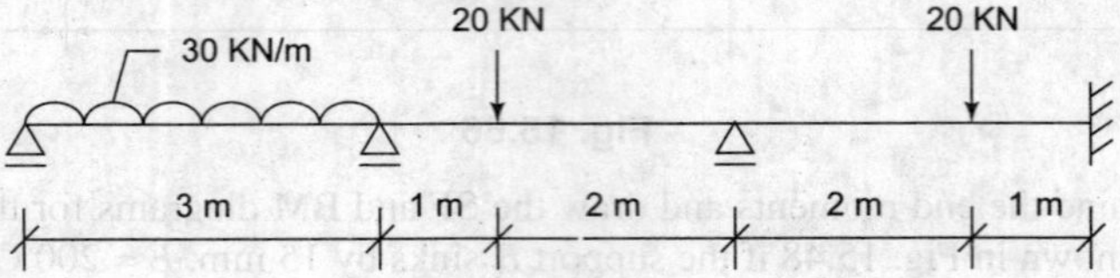

Fig. 15.71

23. Analyse the continuous beam loaded as shown in Fig. 15.72 using the slope-deflection equations and draw the SF and BM diagrams. If the support at the centre settles by 0.002 m, what will be the changes in the support moments?

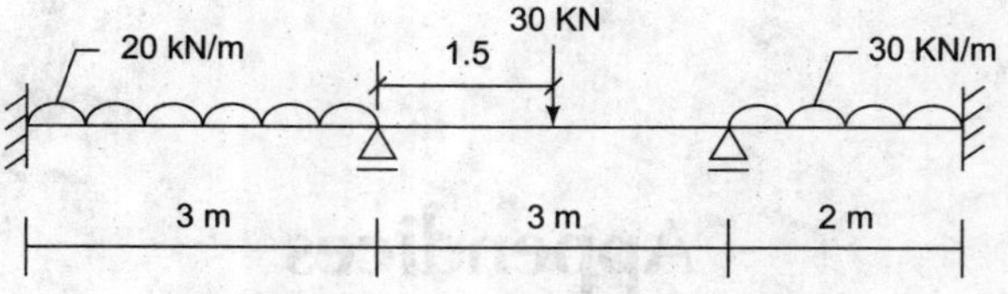

Fig. 15.72

24. Analyse the continuous beam shown in Fig. 15.73 using the stiffness matrix method and draw the SF and BM diagrams.

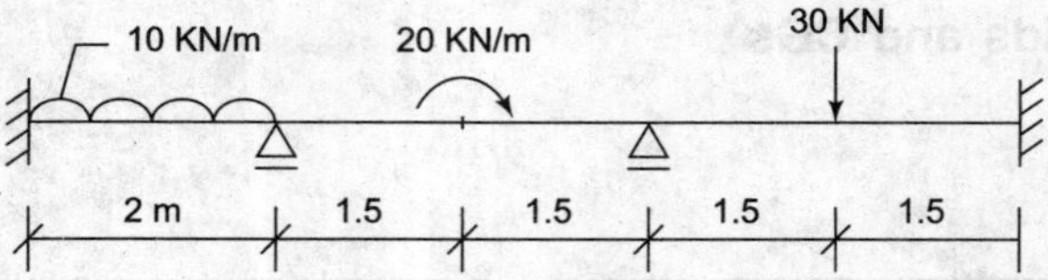

Fig. 15.73

25. Analyse the continuous beam shown in Fig. 15.74 using the stiffness matrix method and draw the SF and BM diagrams.

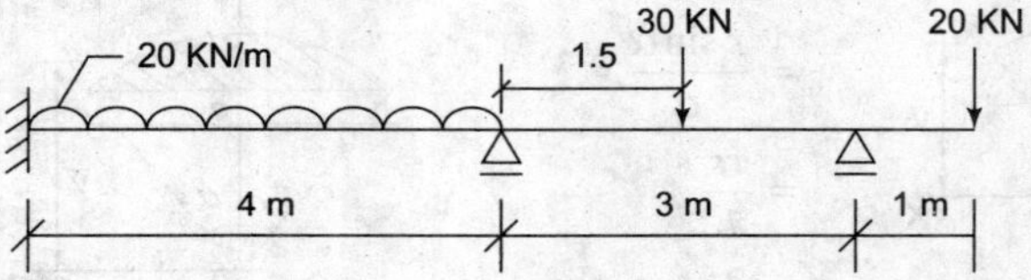

Fig. 15.74

Appendices

Appendix 1: Centroids and Moments of Inertia

1.1 Centroids and CGs

Figure	Position of centroid/CG
1. 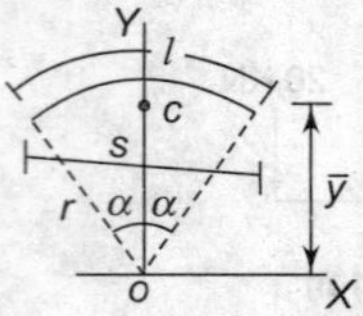	$\bar{x} = 0$ $\bar{y} = \dfrac{r \sin\alpha}{\alpha}$ $= \dfrac{rs}{b}$
2. Quarter circular arc 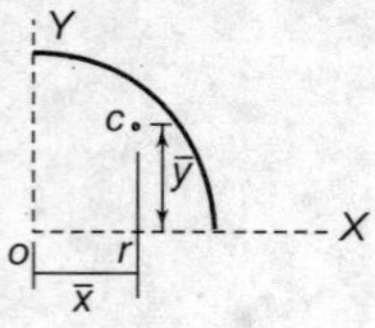	$\bar{x} = \dfrac{2r}{\pi}$ $\bar{y} = \dfrac{2r}{\pi}$
3. Semicircular arc 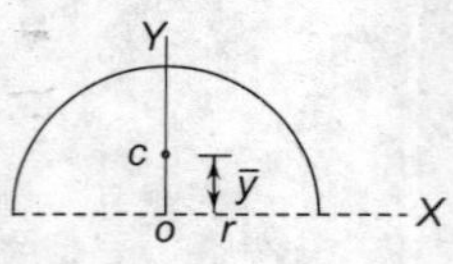	$\bar{x} = 0$ $\bar{y} = \dfrac{2r}{\pi}$
4. Sector of circle 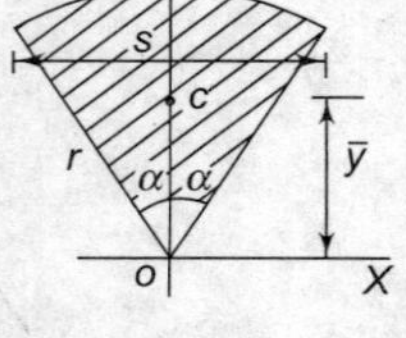	$\bar{x} = 0$ $\bar{y} = \dfrac{2r \sin\alpha}{3r}$ $= \dfrac{2rs}{3b}$
5. 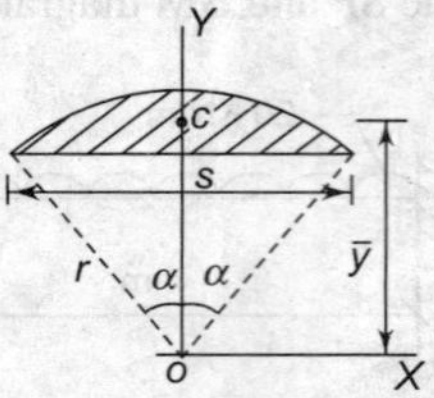	$\bar{x} = 0$ $\bar{y} = \dfrac{s^3}{12A}$ $A = \dfrac{r^2}{2}$ $(\alpha - \sin\alpha)$
6. Quarter circle 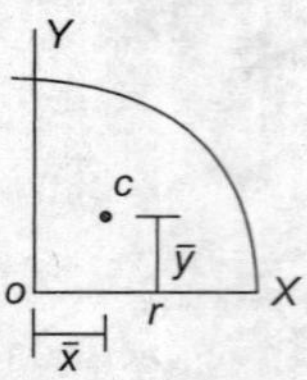	$\bar{x} = \dfrac{4r}{3\pi}$ $\bar{y} = \dfrac{4r}{3\pi}$
7. Semicircle 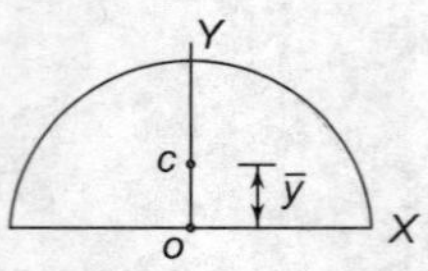	$\bar{x} = 0$ $\bar{y} = \dfrac{4r}{3\pi}$
8. Triangle 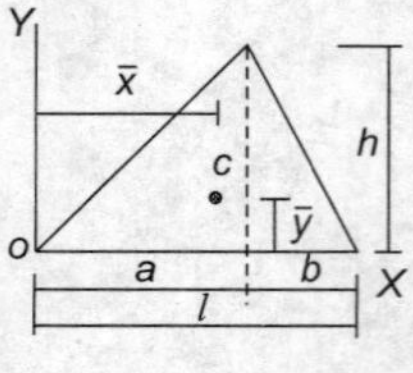	$\bar{x} = \dfrac{l+a}{3}$ $\bar{y} = \dfrac{h}{3}$

(contd)

(contd)

Figure	Position of centroid/CG

9. Trapezium

$$\bar{y} = \left(\frac{a+2b}{a+b}\right)\frac{h}{3}$$

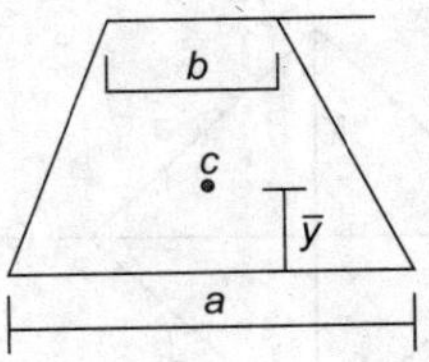

10. Parabolic area

$$\bar{x} = 0$$
$$\bar{y} = \frac{2r}{\pi}$$

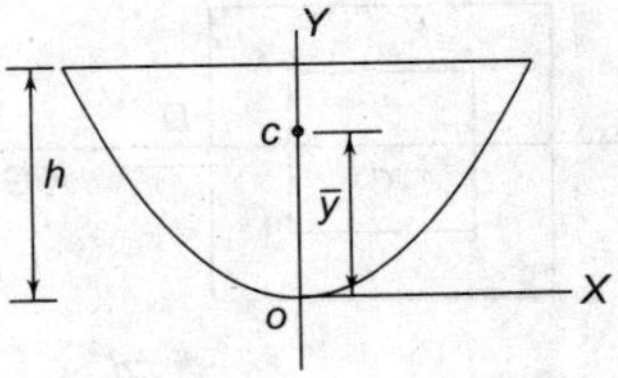

11. Semi-parabolic area

$$\bar{x} = \frac{3a}{8}$$
$$\bar{y} = \frac{3h}{5}$$

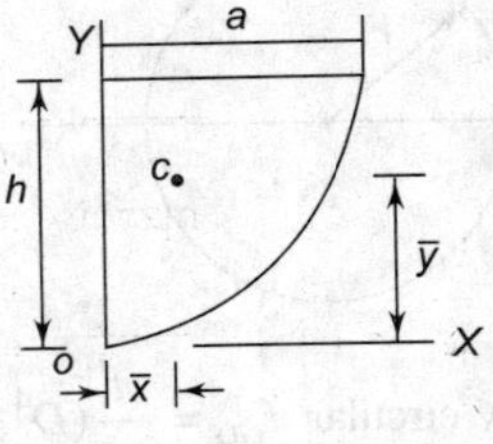

12. Spandrel

$$\bar{x} = \frac{3a}{4}$$
$$\bar{y} = \frac{3h}{10}$$

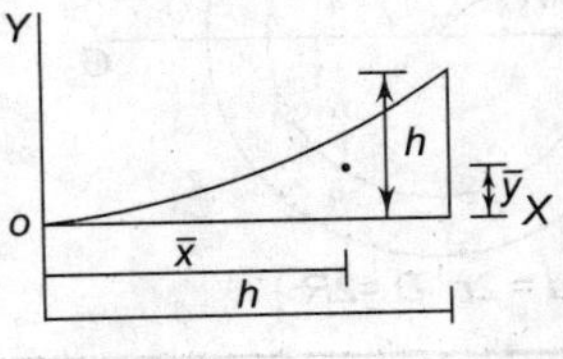

13. Circular ring

$$\bar{x} = 0$$
$$\bar{y} = \frac{2}{3}\frac{R^3 - r^3}{R^2 - r^2}\frac{\sin\alpha}{\operatorname{arc}\alpha} = \frac{2}{3}\frac{R^3 - r^3}{R^2 - r^2}\frac{s}{l}$$

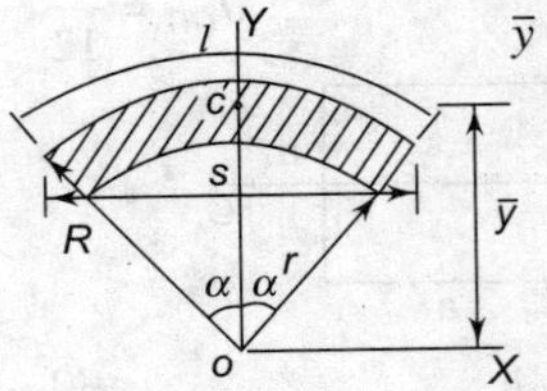

14. Hemisphere

$$\bar{x} = 0$$
$$\bar{y} = \frac{3r}{8}$$
$$\bar{z} = 0$$

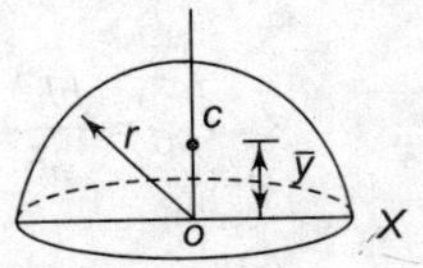

15. Spherical segment

$$\bar{x} = 0$$
$$\bar{y} = \frac{h}{2}\left[\frac{2a^2 + h^2}{3a^2 + h^2}\right]$$
$$\bar{z} = 0$$

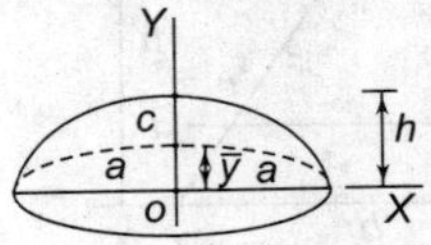

16. Cylinder

$$\bar{x} = 0$$
$$\bar{y} = \frac{h}{2}$$
$$\bar{z} = 0$$

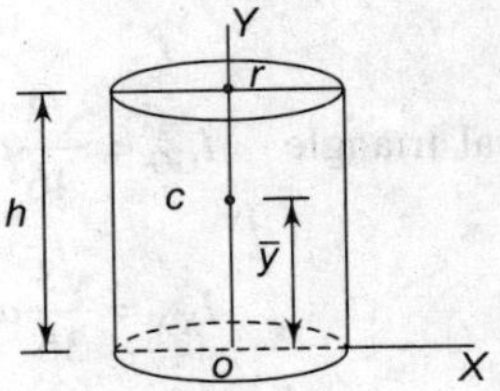

17. Right cone

$$\bar{x} = 0$$
$$\bar{y} = \frac{h}{4}$$
$$\bar{z} = 0$$

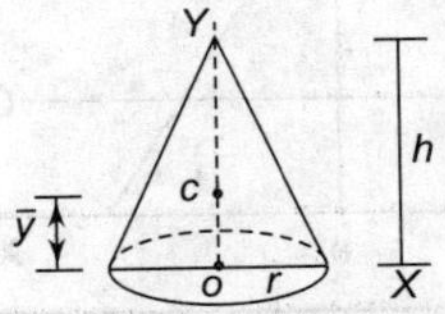

1.2 Second Moments of Area

Figure	*Second moment of area*
1. Square	$I_{GG} = \dfrac{a^4}{12}$
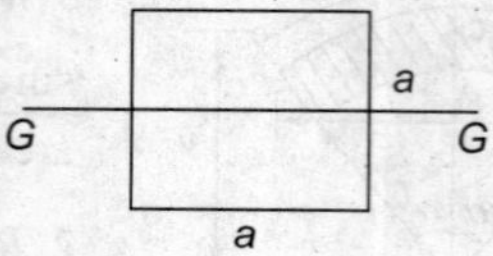	
2. Rectangle	$I_{GG} = \dfrac{ab^3}{12}$
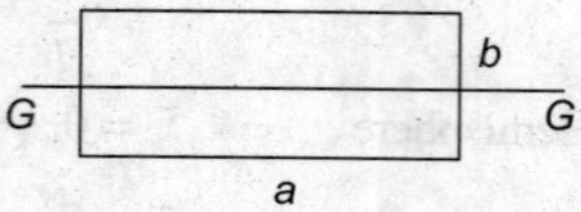	
3. Triangle	$I_{GG} = \dfrac{bh^3}{36}$ $I_{XX} = \dfrac{bh^3}{12}$
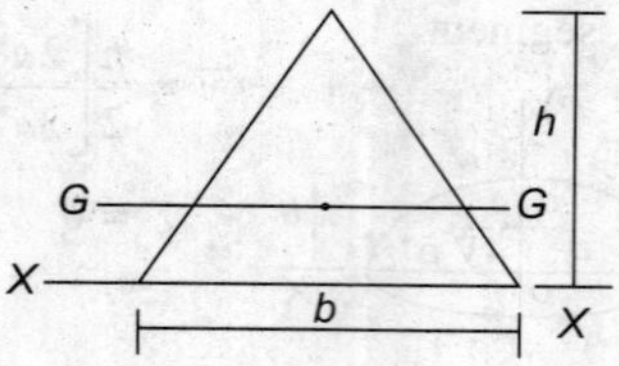	
4. Equilateral triangle	$I_{GG} = \dfrac{\sqrt{3}}{46}a^4$ $I_{XX} = \dfrac{\sqrt{3}}{36}a^4$
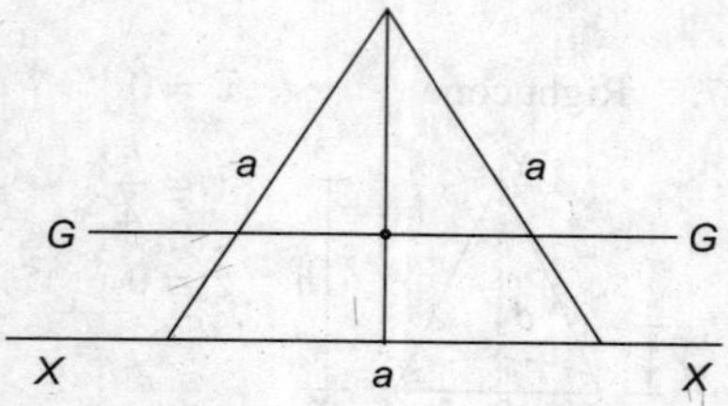	
5. Diamond	$I_{GG} = \dfrac{a^4}{12}$
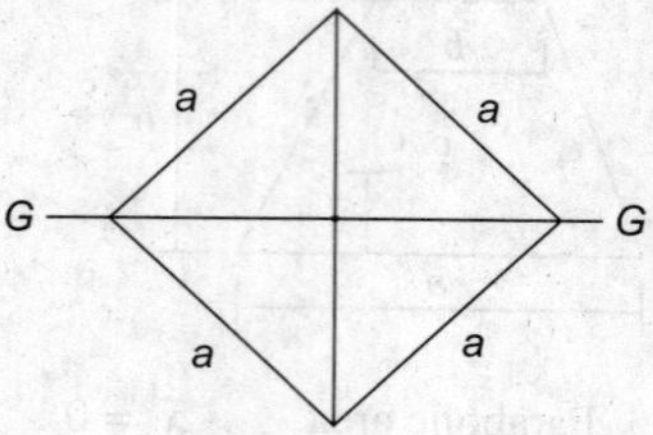	
6. Box-sector (square)	$I_{GG} = \dfrac{D^4 - d^4}{12}$
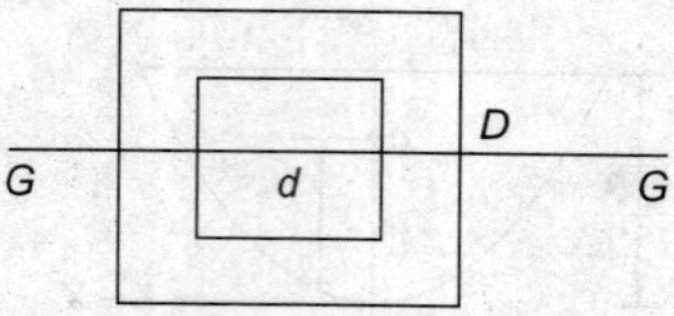	
7. Circle $d = 2r$	$I_{GG} = \dfrac{\pi d^4}{64}$ $= \dfrac{\pi r^4}{4}$
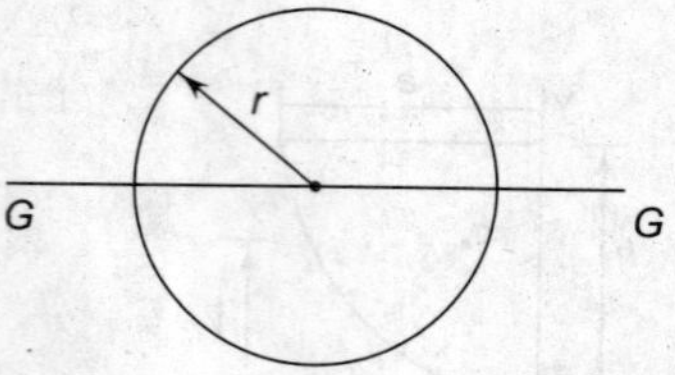	
8. Hollow-circular area	$I_{GG} = \dfrac{\pi}{64}(D^4 - d^4)$ $= \dfrac{\pi r^4}{4}\ (R^4 - r^4)$
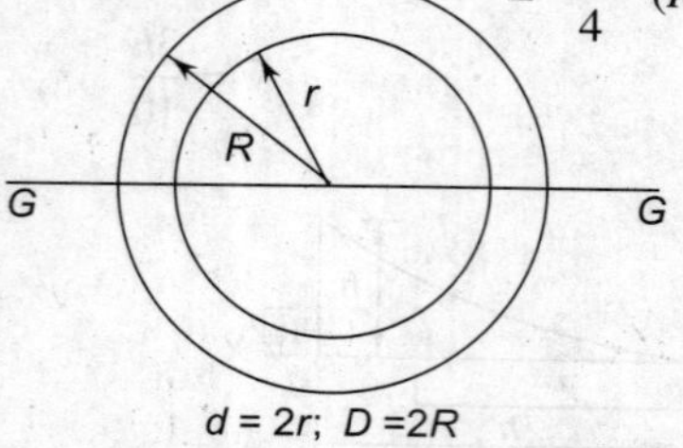	
$d = 2r;\ D = 2R$	

(contd)

(contd)

Figure	Second moment of area

9. Regular hexagon $I_{GG} = \dfrac{5\sqrt{3}}{16} a^4$

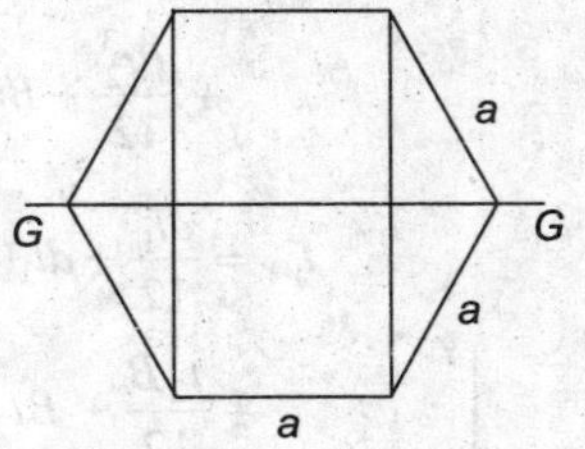

10. Ellipse $I_{GG} = \dfrac{\pi a^3 b}{4}$

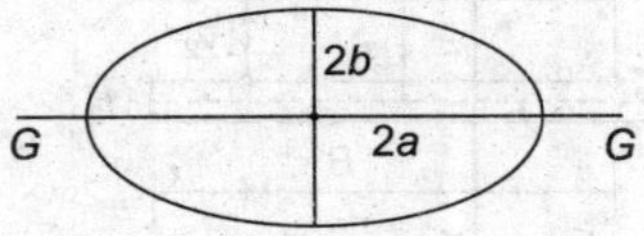

11. Semicircle $I_{XX} = \dfrac{\pi r^4}{8}$

$I_{YY} = \dfrac{\pi r^4}{8}$

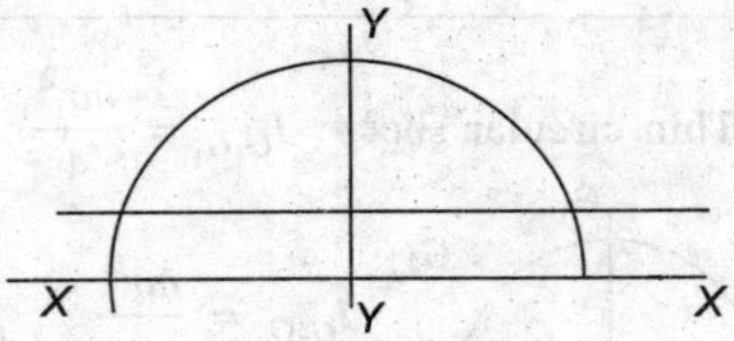

12. Quarter circle $I_{XX} = \dfrac{1}{16}\pi r^4$

$I_{YY} = \dfrac{1}{16}\pi r^4$

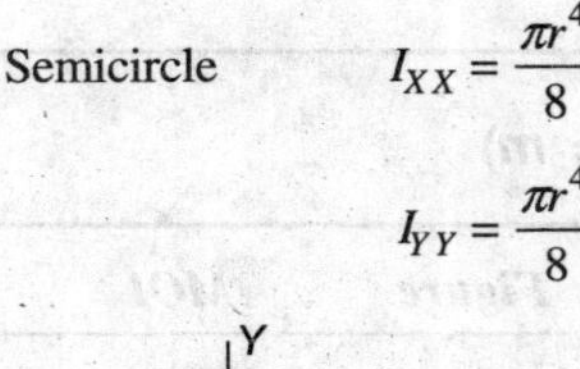

13. Parabola $I_{XX} = \dfrac{16}{105} BH^3$

$I_{GG} = \dfrac{8}{175} BH^3$

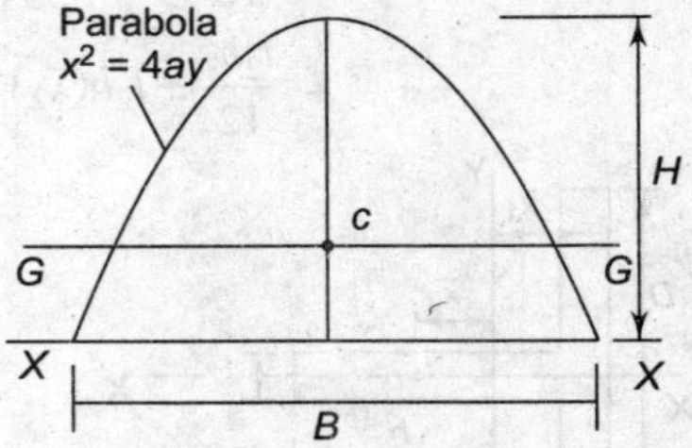

14. I-section $I_{XX} = \dfrac{BD^3}{12} - \dfrac{bd^3}{12}$

$I_{YY} = b^3\left(\dfrac{t_1}{12} + \dfrac{t_2}{12}\right) + \dfrac{dt^3}{12}$

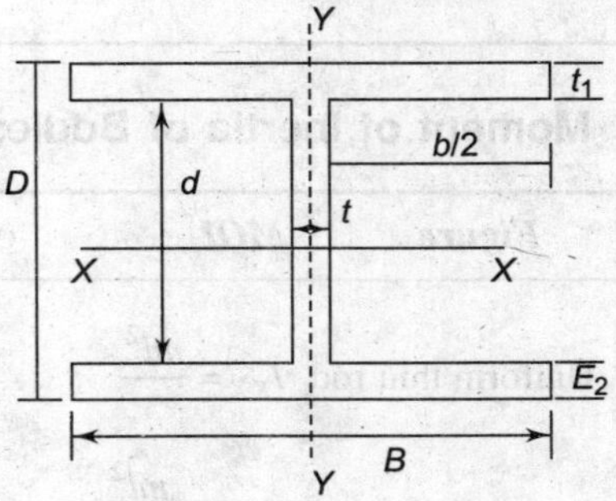

15. Channel section $I_{XX} = \dfrac{BD^3}{12} - \dfrac{bd^3}{12}$

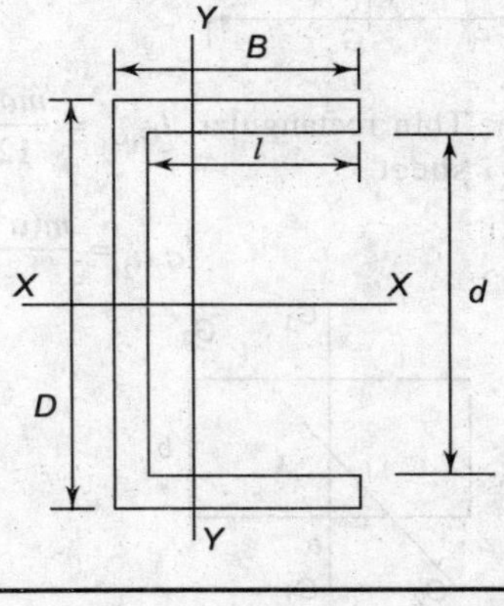

(contd)

(contd)

Figure	*Second moment of area*	*Figure*	*Second moment of area*
16. T-section	$I_{XX} = \frac{t_1 d^3}{12} + \frac{b t_2^3}{12}$ $I_{YY} = \frac{D t_1^3}{12} + D t_1 (\bar{x}_1)^2 + \frac{t_2 b_3}{12} + t_2 b (\bar{x}_2)^2$	17. Angle section	$I_{XX} = \frac{t_1 D^3}{12} + t_1 D (\bar{y}_1)^2 + \frac{B t_2^3}{12} + B t_2 (\bar{y}_2)^2$ $I_{YY} = \frac{d t_1^3}{12} + d t_1 (\bar{x}_1)^2 + \frac{t_2 B}{12} + B t_2 (\bar{x}_2)^2$

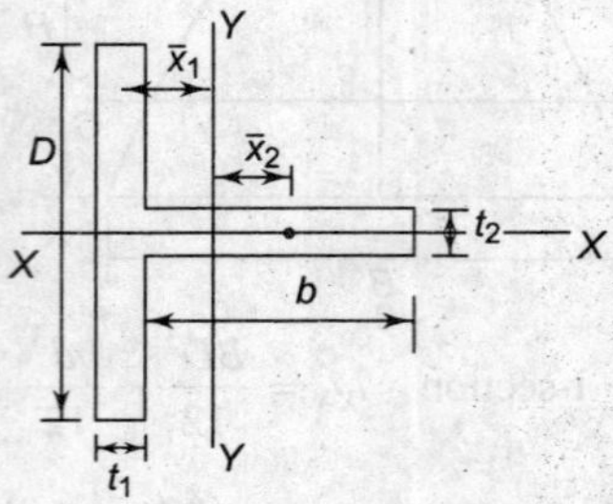

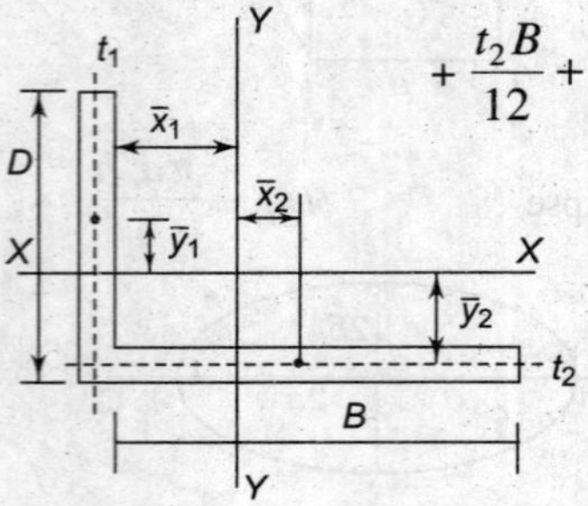

1.3 Moment of Inertia of Bodies (Mass = *m*)

Figure	*MOI*	*Figure*	*MOI*
1. Uniform thin rod	$I_{XX} = \frac{ml^2}{3}$ $I_{GG} = \frac{ml^2}{12}$	3. Thin circular sheet	$I_{G_1G_1} = \frac{mr^2}{4}$ $I_{G_2G_2} = \frac{mr^2}{2}$
2. Thin rectangular sheet	$I_{G_1G_1} = \frac{ma^2}{12}$ $I_{G_2G_2} = \frac{m(a^2+b^2)}{12}$	4. Thin circular ring	$I_{G_1G_1} = \frac{m(r_1^2+r_2^2)}{4}$ $I_{G_2G_2} = \frac{m(r_1^2+r_2^2)}{2}$

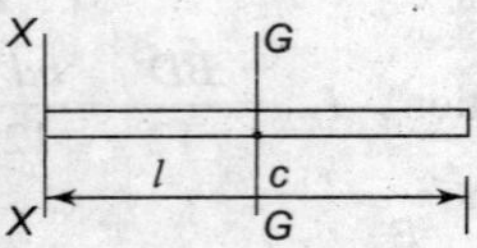

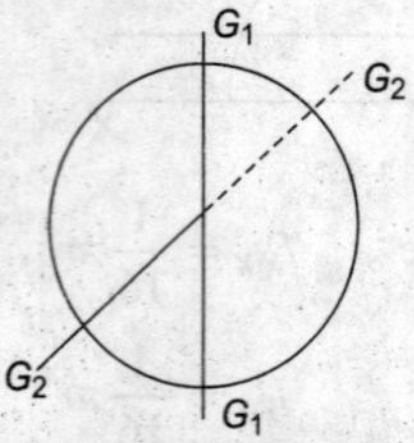

($G_2 G_2$ perpendicular to the sheet)

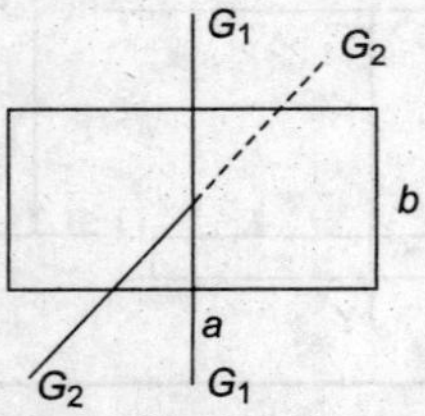

(G_2G_2 perpendicular to the sheet)

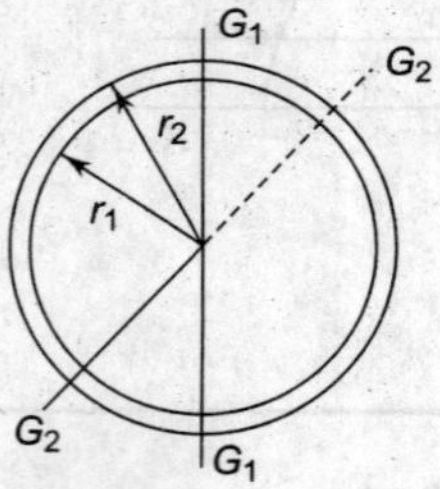

(contd)

(contd)

Figure	*MOI*

5. Rectangular parallelepiped $I_{G_1G_1} = \dfrac{m(a^2+b^2)}{12}$

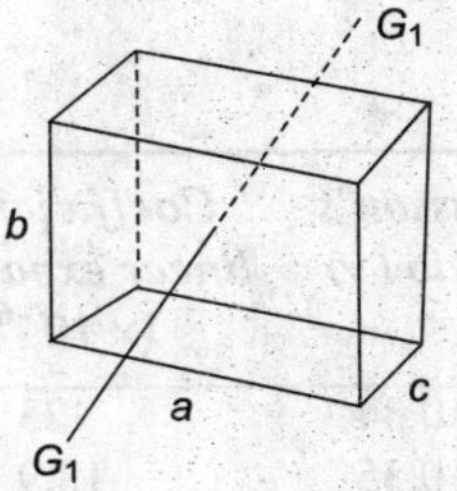

6. Sphere $I_{GG} = \dfrac{2}{5}mr^2$

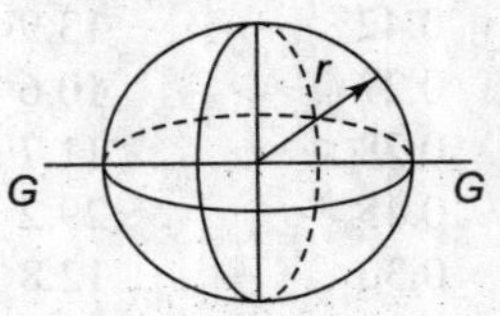

7. Right cylinder $I_{G_1G_1} = \dfrac{mr^2}{2}$

$$I_{G_2G_2} = m\left(\frac{r^2}{4} + \frac{l^2}{12}\right)$$

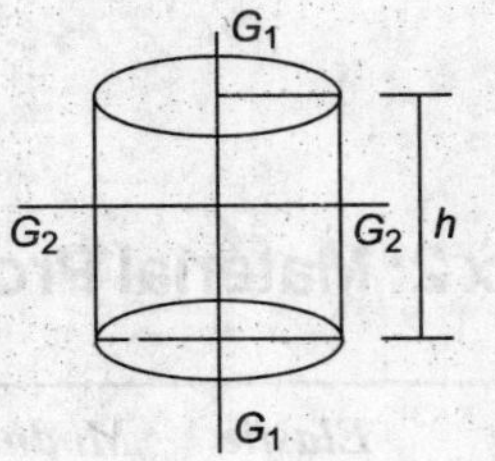

8. Right cone $I_{GG} = \dfrac{3}{10}mr^2$

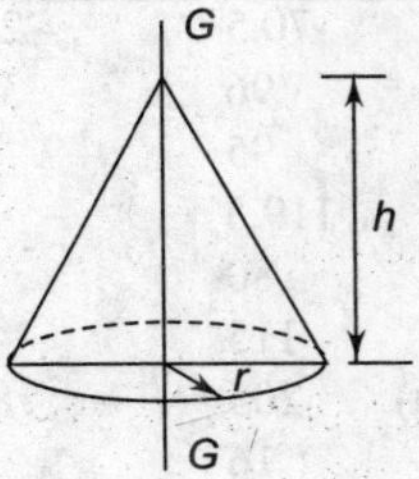

9. Ellipsoid $I_{GG} = \dfrac{m}{5}(b^2+c^2)$

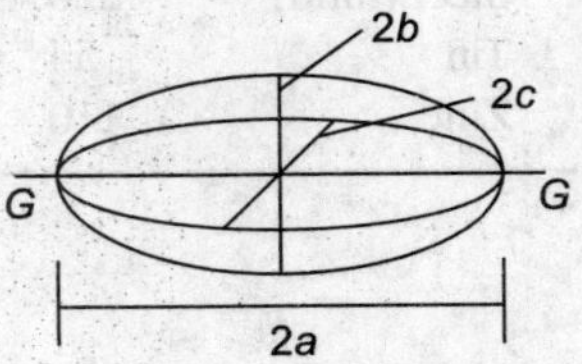

Appendix 2: Material Properties

Material	*Elastic modulus (GPa)*	*Modulus of rigidity (GPa)*	*Bulk modulus (GPa)*	*Poisson's ratio (ν)*	*Coefficient of linear expansion* ($\times 10^{-6}$)
Aluminium	70.5	25.5	75	0.34	23
Brass	96	34.5	95	0.35	18.9
Bronze	95	37	93	0.36	18
Copper	119.5	40	136.5	0.3	16.7
Gold	80	28	166	0.42	13.9
Iron (cast)	115	440	96	0.27	10.6
Iron (wrought)	205	800	153	0.28	11.7
Lead	16	6	50	0.45	29.2
Nickel	207	76	177	0.36	12.8
Silver	78	29	109	0.38	19.2
Steel (mild)	220	84.5	175	0.28	12
Tin	54	20	53	0.33	22.3
Zinc	110	43	72	0.25	31

Appendix 3: Beam Formulae

3.1 Fixed End Moments in Fixed Beams

(The positive moment is anticlockwise at the left support and clockwise at the right support.)

		Beam loading	FEM at left support A	FEM at right support B
1.	(a)	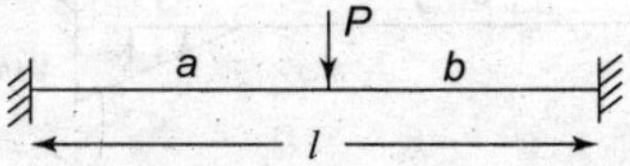	$\frac{Pab^2}{l^2}$	$\frac{Pa^2b}{l^2}$
	(b)	$a = b$	$Pl/8$	$Pl/8$
2.		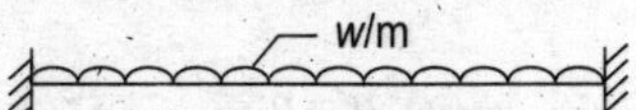	$\frac{wl^2}{12}$	$\frac{wl^2}{12}$
3.	(a)	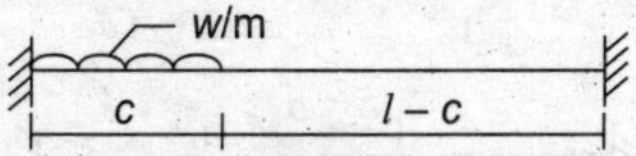	$\frac{w}{12l^2}[6l^2c^2 - 8c^3l + 3c^4]$	$\frac{w}{12l^2}[4c^3l - 3c^4]$
	(b)	$c = l/2$	$\frac{11\,wl^2}{192}$	$\frac{5}{192}wl^2$
4.		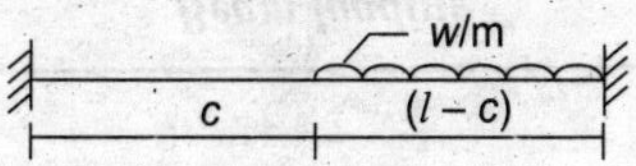	$\frac{w}{12l^2}[l^4 - 6l^2c^2 + 8lc^3 - 3c^4]$	$\frac{w}{12l^2}[l^4 - 4lc^3 + 3c^4]$

(contd)

(contd)

		Beam loading	FEM at left support A	FEM at right support B
5.	(a)	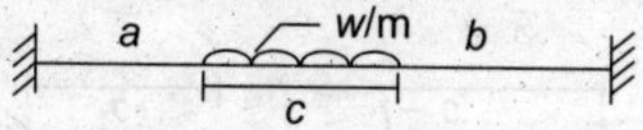	$\frac{wc}{12l^2}[12b^2a + 6b^2c + 12abc + 4ac^2 + 4bc^2 + c^3]$	$\frac{wc}{12l^2}[12a^2b + 12abc + 4bc^2 + 6a^2c + 4ac^2 + c^3]$
	(b)	$a = b$	$\frac{wc}{12l^2}[12a^3 + 18a^2c + 8ac^2 + c^3]$	$\frac{wc}{12l^2}[12a^3 + 18a^2c + 8ac^2 + c^3]$
6.		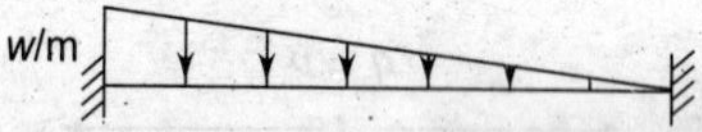	$\frac{wl^2}{20}$	$\frac{wl^2}{30}$
7.	(a)	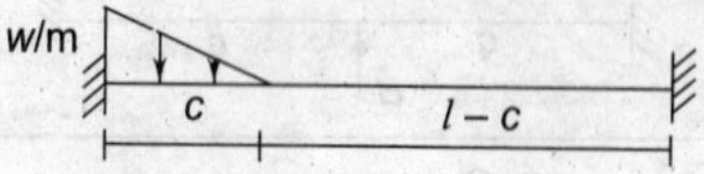	$\frac{wc^2}{60l^2}[10l^2 - 10lc + 3c^2]$	$\frac{wc^3}{60l^2}[5l - 3c]$
	(b)	$c = l/2$	$\frac{23}{960}wl^2$	$\frac{7}{960}wl^2$
8.	(a)	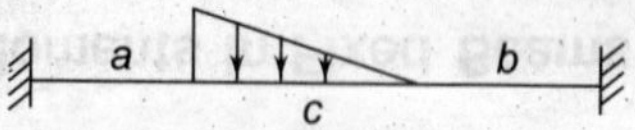	$\frac{wc}{60l^2}[30ab^2 + 10b^2c + 40abc + 10bc^2 + 15ac^2 + 3c^3]$	$\frac{wc}{60l^2}[30a^2b + 20abc + 5bc^2 + 20a^2c + 10ac^2 + 2c^3]$
	(b)	$a = b$	$\frac{w}{120l}[30a^3 + 50a^2c + 25ac^2 + 3c^3]$	$\frac{w}{120l}[30a^3 + 40a^2c + 15ac^2 + 2c^3]$

(contd)

(contd)

		Beam loading	FEM at left support A	FEM at right support B
9.	(a)	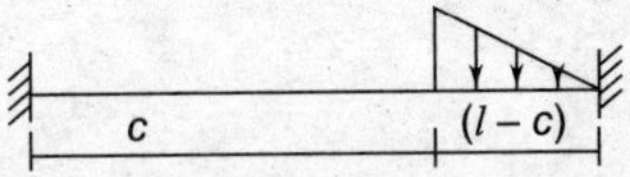	$\frac{w}{20l}[l^4 + l^3c - 9l^2c^2 + 11lc^3 - 4c^4]$	$\frac{w}{30l^2}[l^4 + l^3c + l^2c^2 - 9lc^3 + 6c^4]$
	(b)	$c = l/2$	$\frac{3}{160}wl^2$	$\frac{wl^2}{30}$
10.		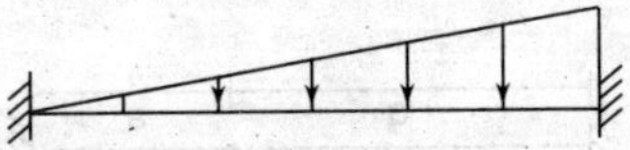	$\frac{wl^2}{30}$	$\frac{wl^2}{20}$
11.	(a)	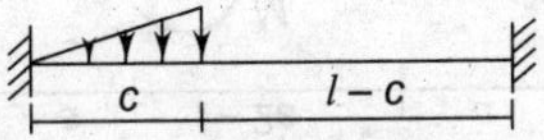	$\frac{wc^2}{30l^2}[10l^2 - 15lc + 6c^2]$	$\frac{wc^3}{20l^2}[5l - 4c]$
	(b)	$c = l/2$	$\frac{wl^2}{30}$	$\frac{3wl^2}{160}$
12.	(a)	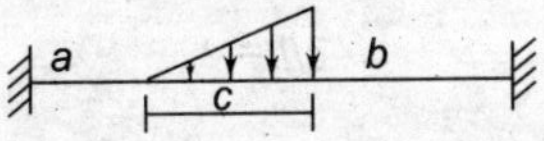	$\frac{wc}{60l^2}[30ab^2 + 5ac^2 + 20abc + 10bc^2 + 20b^2c + 2c^3]$	$\frac{wc}{60l^2}[10a^2c + 10ac^2 + 3c^3 + 30a^2b + 40abc + 15bc^2]$
	(b)	$a = b$	$\frac{w}{120l}[30a^3 + 40a^2c + 15ac^2 + 2c^3]$	$\frac{w}{120l}[30a^3 + 50a^2c + 25ac^2 + 3c^3]$

(contd)

(contd)

		Beam loading	FEM at left support A	FEM at right support B
13.	(a)	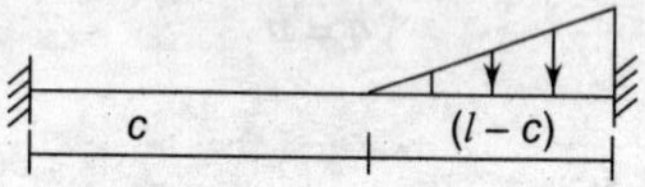	$\frac{w}{60l}[2l^4 - 3l^3c - 3l^2c^2 + 7lc^3 - 3c^4]$	$\frac{w}{60l^2}[3l^4 - 2l^3c - 2l^2c^2 - 2lc^3 + 3c^4]$
	(b)	$c = l/2$	$\frac{7}{960}wl^2$	$\frac{23}{960}wl^2$
14.		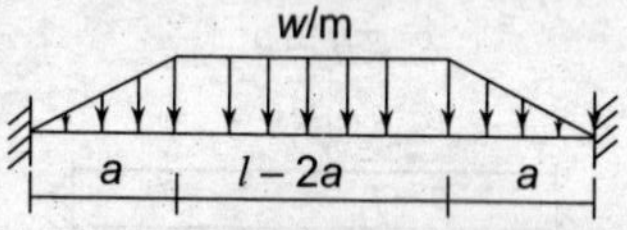	$\frac{w}{12l}[l^3 - 2a^2l + a^3]$	$\frac{w}{12l}[l^3 - 2a^2l + a^3]$
15.	(a)	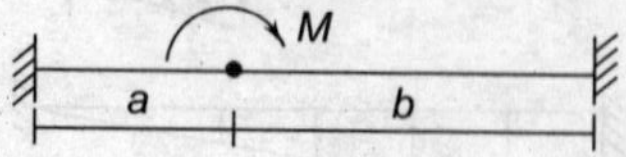	$\frac{M}{l^2}[l^2 - 4la + 3a^2]$	$\frac{Ma}{l^2}[3a - 2l]$
	(b)	$a = l/2$	$\frac{M}{4}$	$-\frac{M}{4}$

3.2 Propped Cantilevers

[l = span; w = load intensity/maximum load intensity for varying load; P = point load; A = left support; B = right support]

		Beam loading	Moment at left support	Slope at right support ($EI\theta_B$)
1.	(a)	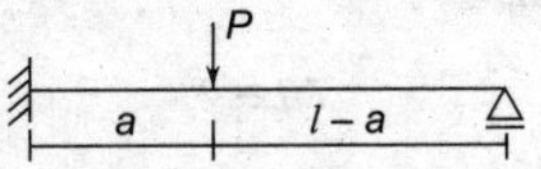	$\frac{Pa}{2l^2}[2l^2 - 3la + a^2]$	$-\frac{Pa^2}{4l}(l - a)$
	(b)	$a = l/2$	$\frac{3}{16}Pl$	$-\frac{Pl^2}{32}$
2.		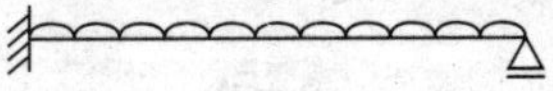	$\frac{wl^2}{8}$	$-\frac{wl^3}{8}$
3.	(a)	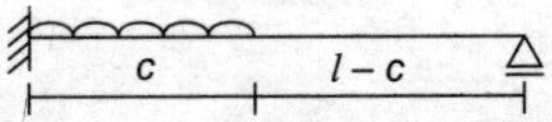	$\frac{wc^2}{8l^2}[4l^2 - 4cl + c^2]$	$\frac{-w}{48l}[4c^3l - 3c^4]$
	(b)	$c = l/2$	$\frac{9}{128}wl^2$	$\frac{-5wl^3}{768}$
4.	(a)	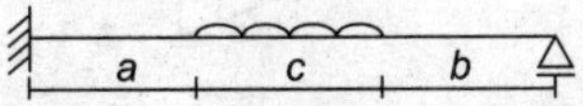	$\frac{wc}{8l^2}[8ab^2 + 4b^2c + 4a^2b + 4bc^2 + 12abc$ $+ 2a^2c + 4ac^2 + c^3]$	$\frac{-wc}{48l}[12a^2b + 12abc + 4bc^2$ $+ 6a^2c + 4ac^2 + c^3]$
	(b)	$a = b$	$\frac{wc}{8l^2}[12a^3 + 18a^2c + 8ac^2 + c^3]$	$\frac{-wc}{48l}[12a^3 + 18a^2c + 8ac^2 + c^3]$
5.	(a)	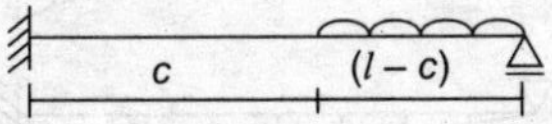	$\frac{w}{8l^2}[l^4 - 4l^2c^2 + 4c^3l - c^4]$	$\frac{-w}{48l}[l^4 - 4c^3l + 3c^4]$
	(b)	$c = l/2$	$\frac{7}{128}wl^2$	$-\frac{11}{768}wl^3$

(contd)

(contd)

		Beam loading	Moment at left support	Slope at right support ($EI\,\theta_B$)
6.		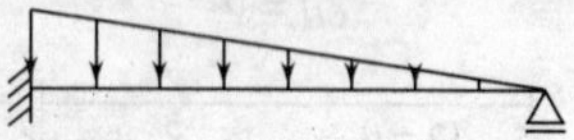	$\frac{wl^2}{15}$	$-\frac{wl^3}{120}$
7.	(a)	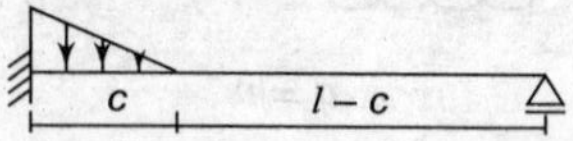	$\frac{wc^2}{120l^2}[20l^2 - 15cl + 3c^2]$	$-\frac{wc^3}{240l}[5l - 3c]$
	(b)	$c = l/2$	$\frac{53}{1920}wl^2$	$\frac{-7wl^3}{3840}$
8.	(a)	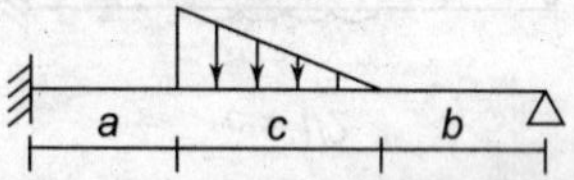	$\frac{wc}{120l^2}[60ab^2 + 30a^2b + 100abc + 20b^2c + 20a^2c + 8c^3 + 40ac^2 + 25bc^2]$	$\frac{-wc}{240l}[30a^2b + 20abc + 5bc^2 + 20a^2c + 10ac^2 + 2c^3]$
	(b)	$a = b$	$\frac{wc}{120l^2}[90a^3 + 140a^2c + 65ac^2 + 8c^3]$	$-\frac{wc}{240l}[30a^3 + 40a^2c + 15ac^2 + 2c^3]$
9.	(a)	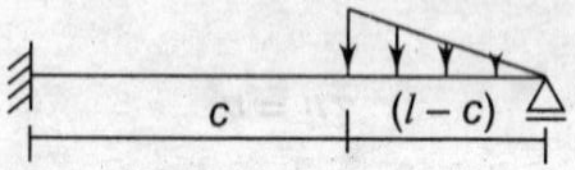	$\frac{w}{30l^2}[2l^4 + 2l^3c - 13l^2c^2 + 12lc^3 - 3c^4]$	$-\frac{w}{120l}[l^4 + l^3c + l^2c^2 - 9lc^3 + 6c^4]$
	(b)	$c = l/2$	$\frac{17wl^2}{480}$	$\frac{-wl^3}{120}$
10.		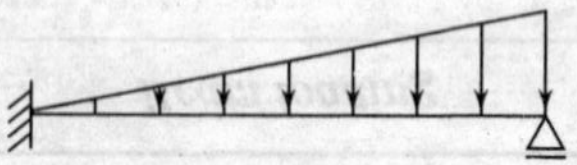	$\frac{7wl^2}{120}$	$-\frac{wl^3}{80}$

(contd)

(contd)

		Beam loading	Moment at left support	Slope at right support ($EI\theta_B$)
11.	(a)	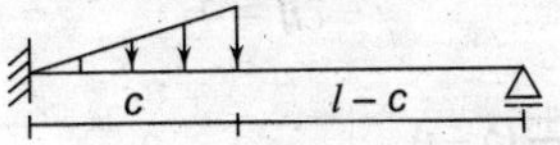	$\frac{wc^2}{120l^2}[40l^2 - 45cl + 12c^2]$	$\frac{-wc^3}{80l^2}[5l - 4c]$
	(b)	$c = l/2$	$\frac{41wl^2}{960}$	$-\frac{9}{1280}wl^3$
12.	(a)	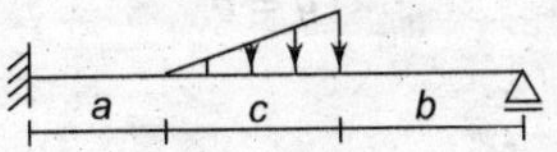	$\frac{wc}{120l^2}[60ab^2 + 20ac^2 + 30ba^2 + 10a^2c + 80abc + 35bc^2 + 40b^2c + 7c^3]$	$\frac{-wc}{240l}[10a^2c + 10ac^2 + 3c^3 + 30a^2b + 40abc + 15bc^2]$
	(b)	$a = b$	$\frac{wc}{120l^2}[90a^3 + 130a^2c + 55ac^2 + 7c^3]$	$-\frac{wc}{240l}[30a^3 + 50a^2c + 25ac^2 + 3c^3]$
13.	(a)	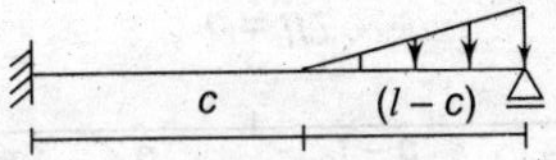	$\frac{w}{120l^2}[7l^4 - 8l^3c - 8l^2c^2 + 12lc^3 - 3c^4]$	$\frac{-w}{240l}[3l^4 - 2l^3c - 2l^2c^2 - 2lc^3 + 3c^4]$
	(b)	$c = l/2$	$\frac{37}{1920}wl^2$	$\frac{-23wl^3}{3840}$
14.	(a)	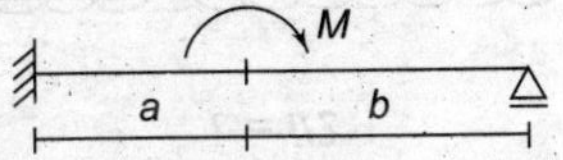	$\frac{M}{2l^2}[2l^2 - 6al + 3a^2]$	$\frac{-Ma}{4l}[3a - 2l]$
	(b)	$a = b$	$-\frac{M}{8}$	$-\frac{Ml}{16}$
	(c)	$b = 0$	$-\frac{M}{2}$	$-\frac{Ml}{4}$
15.		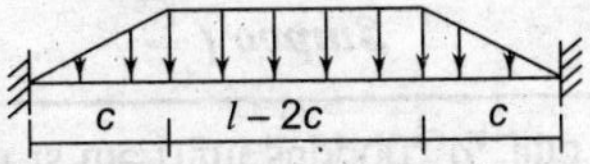	$\frac{w}{8l}(l^3 - 2c^2l + c^3)$	$-\frac{w}{48}(l^3 - 2c^2l + c^3)$

3.3 Simply Supported Beams

[A is the left support; B is the right support; θ_A and θ_B are the slopes at A and B]

		Loading	$EI\theta_A$	$EI\theta_B$
1.	(a)	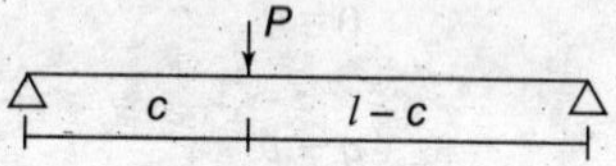	$\frac{Pc}{6l}[2l^2 - 3cl + c^2]$	$\frac{-Pc}{6l}(l^2 - c^2)$
	(b)	$c = l/2$	$\frac{Pl^2}{16}$	$\frac{-Pl^2}{16}$
2.			$\frac{wl^3}{24}$	$\frac{-wl^3}{24}$
3.	(a)	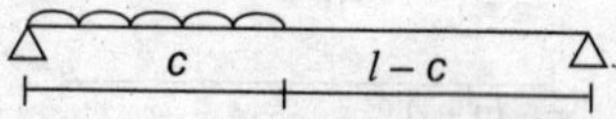	$+\frac{wc^2}{24l}[4l^2 - 4cl + c^2]$	$\frac{-wc^2}{24l}[2l^2 - c^2]$
	(b)	$c = l/2$	$\frac{3}{128}wl^3$	$\frac{-7wl^3}{384}$
4.	(a)	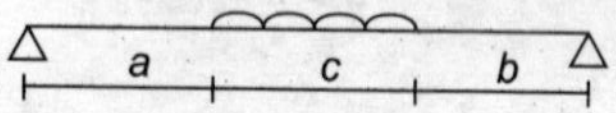	$\frac{wc}{24l}[8ab^2 + 4b^2c + 4a^2b + 4bc^2 + 12abc + 2a^2c + 4ac^2 + c^3]$	$\frac{-wc}{24l}[8a^2b + 12abc + 4bc^2 + 4a^2c + 4ac^2 + c^3 + 4ab^2 + 2b^2c]$
	(b)	$a = b$	$\frac{wc}{24l}[12a^3 + 18a^2c + 8ac^2 + c^3]$	$\frac{-wc}{24l}[12a^3 + 18a^2c + 8ac^2 + c^3]$
5.	(a)	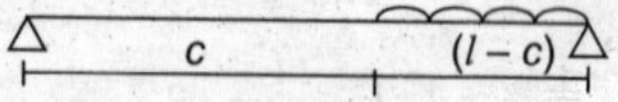	$\frac{w}{24l}[l^4 - 4l^2c^2 + 4c^3l - c^4]$	$\frac{-w}{24l}[l^2 - c^2]^2$
	(b)	$c = l/2$	$\frac{7wl^3}{384}$	$-\frac{3wl^3}{128}$

(contd)

(contd)

		Loading	$EI\theta_A$	$EI\theta_B$
6.	(a)	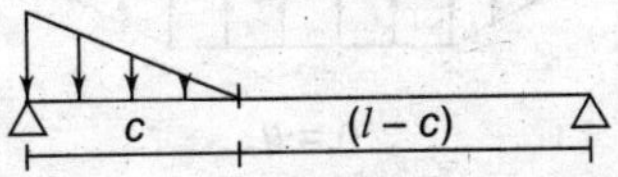	$\frac{wc^2}{360l}[20l^2 - 15lc + 3c^2]$	$\frac{-wc^2}{360l}[10l^2 - 3c^2]$
	(b)	$c = l/2$	$\frac{53wl^3}{5760}$	$\frac{-37wl^3}{5760}$
7.	(a)	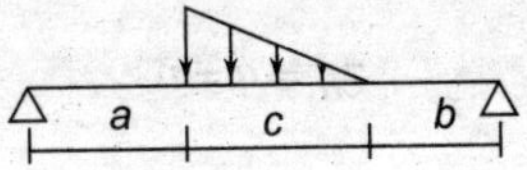	$\frac{wc}{360l}[60ab^2 + 30a^2b + 100abc + 8c^3 + 20b^2c + 20a^2c + 40ac^2 + 25bc^2]$	$\frac{-wc}{360l}[30ab^2 + 10b^2c + 60a^2b + 80abc + 20bc^2 + 40a^2c + 35ac^2 + 7c^3]$
	(b)	$a = b$	$\frac{wc}{360l}[90a^3 + 140a^2c + 65ac^2 + 8c^3]$	$\frac{-wc}{360l}[90a^3 + 130a^2c + 55ac^2 + 7c^3]$
8.	(a)	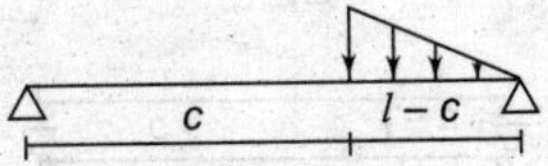	$\frac{w}{90l}[2l^4 + 2l^3c - 13l^2c^2 + 12lc^3 - 3c^4]$	$\frac{-w}{360l}[7l^4 + 7l^3c - 23l^2c^2 - 3lc^3 + 12c^4]$
	(b)	$c = l/2$	$\frac{17wl^3}{1440}$	$\frac{-41wl^3}{2880}$
9.	(a)	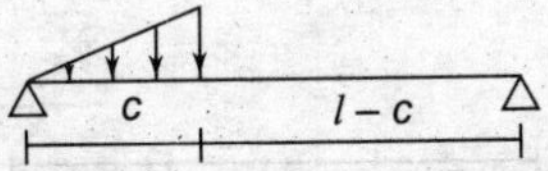	$\frac{wc^2}{360l}[40l^2 - 45lc + 12c^2]$	$\frac{-wc^2}{90l}[5l^2 - 3c^2]$
	(b)	$c = l/2$	$\frac{41wl^3}{2880}$	$\frac{-17wl^3}{1440}$

(contd)

(contd)

		Loading	$EI\theta_A$	$EI\theta_B$
10.	(a)	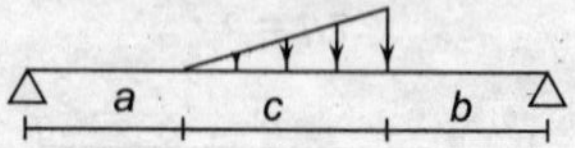	$\frac{wc}{360l}[60ab^2 + 20ac^2 + 30a^2b + 10a^2c + 80abc + 35bc^2 + 40b^2c + 7c^3]$	$\frac{-wc}{360l}[30ab^2 + 20b^2c + 20a^2c + 25ac^2 + 8c^3 + 60a^2b + 100abc + 40bc^2]$
	(b)	$a = b$	$\frac{wc}{360l}[90a^3 + 130a^2c + 55ac^2 + 7c^3]$	$\frac{-wc}{360l}[90a^3 + 140a^2c + 65ac^2 + 8c^3]$
11.	(a)	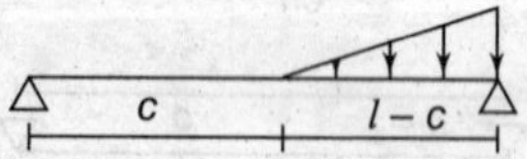	$\frac{w}{360l}[7l^4 - 8l^3c - 8l^2c^2 + 12lc^3 - 3c^4]$	$\frac{-w}{360l}[8l^4 - 7l^3c - 7l^2c^2 + 3lc^3 + 3c^4]$
	(b)	$c = l/2$	$\frac{37wl^3}{5760}$	$-\frac{53}{5760}wl^3$
12.	(a)	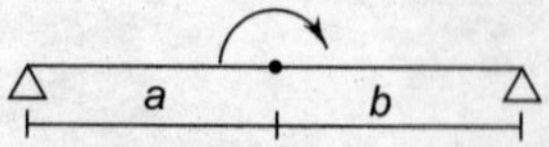	$\frac{M}{6l}[2l^2 - 3la - 3a^2]$	$\frac{-M}{6l}[l^2 - 3a^2]$
	(b)	$a = b = l/2$	$\frac{Ml}{24}$	$\frac{-Ml}{24}$
	(c)	$a = 0$	$\frac{Ml}{3}$	$\frac{-Ml}{6}$
	(d)	$b = 0$	$-\frac{Ml}{6}$	$+\frac{Ml}{3}$
13.		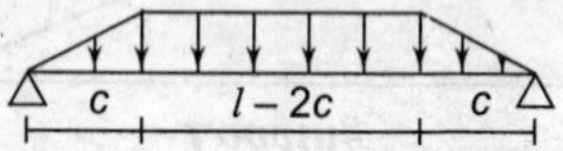	$\frac{w}{24}[l^3 - 2c^2l + c^3]$	$\frac{-w}{24}[l^3 - 2c^2l + c^3]$

Appendix 4: Answers to Problems

Chapter 1

1. Horizontal component = 89.44 N
   Vertical component = 178.88 N
2. (i) $F_B = 875$ N acting downwards
   $F_A = -375$ N acting upwards
   (ii) $F_A = 111.11$N, $F_C = 388.89$ N
3. (a) $AC = 141.4$ N
   $CD = 306.82$ N
   $CD = 280$ N
4. 732 N, 517.6 N
5. 600 N m ($R = 0$)
6. Zero
7. 1807.1 N
9. (a) 85 kN, 95 kN
   (b) 10 kN
   (c) 106.67 kN, 133.33 kN
   (d) $\dfrac{wl}{\pi}$
   (e) 108 kN; 36 kN
10. $R_A = \dfrac{+wl}{6}, x = 0.64\ l$
    $R_A = 0.14\,wl$
11. $R_{BV} = 91.25$ kN
    $R_{AH} = 20$ kN
    $R_{AV} = 58.75$ kN
12. (a) $R_{BV} = 58.56$
    $R_{AV} = 63.87$ kN
    $R_{AH} = 42.43$ kN
    (b) $R_{BV} = 72.5$ kN
    $R_{AV} = 47.5$ kN
    $R_{AH} = 30$ kN

Chapter 2

1. 1.5×10^8 mm^4
2. 0.963×10^8 mm^4
3. 2.746×10^8 mm^4
   25.5×10^8 mm^4
4. 9.98×10^6 mm^4
   1.788×10^6 mm^4
   27.67 mm
5. 19.31×10^6 mm^4
   1.91×10^6 mm^4
   100.14 mm
6. (a) 23.1×10^8 mm^4
   (b) $47{,}318.54 \times 10^6$ mm^4
   (c) 106.2×10^8 mm^4
7. $0.271a^4$
   for a hollow circular section
8. (a) 40.96×10^6 units
   (b) 624×10^8 units
   (c) 64,000 units
   (d) 60,952.4 units
9. 3×10^8 mm^4
10. 50×10^8 mm^4
11. 7.03×10^8 mm^4
    0.78125×10^8 mm^4
12. 1.733×10^8 mm^4
    0.433×10^8 mm^4
    25.1°
13. 4076×10^4
    270×10^4 mm^4
    5.59°
14. 1583.65×10^3
    130×10^3
    26.57°
17. 10318.3 mm^3
    14667 mm^3
18. 0.8687×10^6 mm^3
    0.5438×10^6 mm^3
19. 5000 cm^3
20. 1.8

Chapter 3

1. 127.324 GPa
2. 1.5 mm, 149.2 N/mm^2
3. 0.97 mm, 32.4 Mpa
4. 1.376 N/mm^2
5. 84.88 N/mm^2
   0.78 mm
6. 141.5 N/mm^2
   0.87 mm
7. 191 N/mm^2
   2.52 mm
8. 99 N/mm^2
   1.28 mm
9. 63.66 N/mm^2
   0.43 mm
10. 15.91 N/mm^2
    40.74 N/mm^2
    28.29 N/mm^2
    0.533 mm
11. 387.5 mm
12. 57.6 N/mm^2
    0.288 mm
13. 7854 N
    30 mm
14. 701.161 N
    16.49 N/mm^2
    47.77 N/mm^2
    22.97 N/mm^2
15. 0.071 mm
16. 254.29 kN
17. 0.159 mm
18. 0.4683 mm
19. 0.24 mm
20. 0.0003 mm, 125.66 N
22. 456.74 kN
23. 3.9 N/mm^2
    46.9 N/mm^2

24. 63.67 N/mm^2
127.32 N/mm^2
0.25 mm
25. 15.88 N/mm^2
215.2 N/mm^2
26. 43.12 N/mm^2
71.87 N/mm^2
27. 0.794 N/mm^2
15.87 N/mm^2
28. 91.43 N/mm^2
29. 191.6 N/mm^2
30. 207.7 N/mm^2
31. (a) 256.5 N/mm^2
64 N/mm^2
0.394 mm
0.394 mm
(b) 220.5 N/mm^2
55.13 N/mm^2
0.9187 mm
0.3063 mm
32. 8.25 N/mm^2 (tensile)
6.6 N/mm^2 (compressive)
33. 276.14 N/mm^2
99.39 N/mm^2
34. 31.56 N/mm^2, 13.15°
35. 84 N/mm^2
42 N/mm^2
36. 1508 kN
37. 202.6 N/mm^2
38. 0.00573 mm
39. 0.0056 mm,
0.1876 mm, 0.0075%
41. 0.0506 mm
0.15 mm
0.00792 mm
734.4 mm^3
42. 0.552 mm
0.442 mm
0.00947 mm
2526 mm^3
43. 0.017 mm
0.004 mm
56.5833
0.0148 mm
44. 9726a^3 m
45. 0.4896 × 10^{-3} D
47. 0.067°
48. 0.296
114.4 GPa
49. 100 GPa, 0.175,
42.55 GPa, 51.28 GPa
50. 2.51 m
51. 0.467
52. 26 N/mm^2
89 N/mm^2
53. 2.86 N/mm^2
16.33 N/mm^2
54. 7.33 N/mm^2
15.71 N/mm^2
55. 72.86 N/mm^2
44.94 N/mm^2
56. 98 N/mm^2
102.3 N/mm^2
10.8 N/mm^2, 1.16 m
57. 11.1 N/mm^2
15 N/mm^2
7.48 N/mm^2
29.9 N/mm^2
58. 283 N/mm^2
60.34 N/mm^2
59. 166.67 mm
63.66 N/mm^2
31.83 N/mm^2
60. 61.6 N/mm^2
40.42 N/mm^2
205.9 kN
61. 83.4°C
200.2 N/mm^2
62. 62°C
133.33 N/mm^2

Chapter 4

7. 4.65 m, 176.225 kNm
8. 2.19 m
95.7 kN m
11. 5.2 m, 436.5 kNm
13. 3.625 m, 157.7 kNm
16. $x = 3.29$
113.6 kN m
18. $x = 3.75$ m
100.635 kN m
0.578 m
19. $x = 1.75$ m
30.625 kN m
21. 3 m from centre
22. $x = 4.4$ m
212.52 kN m
23. $x = 0.535$, 7.465
2.485 m
25. 3.52 m

Chapter 5

1. 5 mm
2. 90.9 GPa, 33 mm
3. 394.5 N/mm^2
4. 72 kN/m,
36 kN/m, 72 kN
5. (a) 277 × 392 cm^2
(b) 85 N/mm^2
(compressive) and
186 N/mm^2 (tensile)
7. 2.05 kN m
3.08 kN m
8. ISMB 300
9. 31.775 kN m
10. 58.67 kN m
11. 101.2 × 10^6 N mm
67.49 × 10^6 N mm
12. 1.41 N/mm^2
13. 22.5 m
14. 2.02 m
15. ISNT 150 @ 224 N/m
17. 3.72 kN m
8.68 × 10^6 N mm
19. 1248 kN m
759.8 kN m
63.6 kN m
20. 3
21. 13.25 kN m
24.64 kN m
22. 33.38 × 10^6 N mm
23. 4.067 × 10^6 N mm
24. 84.15 × 10^4 N mm
25. 31.5 kN m
26. 587.4 N m
27. 351 N m
28. 8.88 N/mm^2
8.88 N/mm^2
10 N/mm^2
29. 40 N/mm^2, 45 N/mm^2,
10.6 mm
30. 16.8 N/mm^2

31. 50 N/mm^2
32. 207.733 kN
33. 20 N/mm^2
34. $\lambda_{max} = 25.3$ N/mm^2
35. $A = 0, B = 6.45$ N/mm^2, $C = 27.8$ N/mm^2, $D = 44.3$ N/mm^2, 3.86
37. ISLB 400
38. ISMB 500
    Two cover plates
    250 × 20
46. 74.4 mm
47. 37.5 mm
48. 118 mm
49. 355 mm
50. 7.5 kN
51. 155 N/mm^2 (tensile)
    273.6 N/mm^2 (compressive)
52. 4.89 kN/m
53. 8.6 kN

Chapter 6

1. 51.44 N/mm^2
2. 29 mm
3. 31.2 mm
4. 17,123 kN
6. $0.4064a$
8. 39.6 N/mm^2 (tensile)
   35.82 N/mm^2 (comp-ressive)
9. 112 kN
10. 161.9 kN/m^2
    29.19 kN/m^2
11. 3.86 m
12. 9.38 kN/m^2
    376.34 kN/m^2
13. 4.1 m
14. 179.67 kN/m^2
    114.33 kN/m^2
15. 3.896 m

Chapter 7

1. (a) $M_x = 24x - 10(x-1) - 20(x-2.5) - 20(x-3.5)$

   (b) $M_x = 11.25x - 20(x-3) - 30(x-7) + 58.75(x-8)$

   (c) $M_x = 65x - \dfrac{10x^2}{2} - 20(x-2) - 20(x-4)$

   (d) $M_x = -2x\,\dfrac{x}{2}\,\dfrac{x}{3} + 81(x-6) = \dfrac{-x^3}{3} + 81(x-6)$

   (e) $-18 + 18x - [36 - 4x]\,\dfrac{x^2}{6}$

   (f) $M_x = -170 + 50x - 10(x-1) - 20(x-3)$

   (g) $M_x = -20 + 20x - 10\dfrac{x^2}{2} + \dfrac{10(x-2)^2}{2}$

   (h) $M_x = -60 + 20x - \dfrac{10(x-2)^2}{2}$

   (i) $M_x = 33x - \dfrac{10x^2}{2} + \dfrac{10(x-4)^2}{2} - 24(x-7)$

   (j) $M_x = 20x - \dfrac{20(x-4)^2}{2}$

   (k) $M_x = 26.25x - \dfrac{20(x-3)^2}{2} + \dfrac{20(x-6)^2}{2}$

   (l) $115x - 30\dfrac{x^2}{2} + 30\dfrac{(x-8)^2}{2}\ 145(x-8)$

   (m) $45.625x - \dfrac{10x^2}{2} + \dfrac{10(x-3)^2}{2} - \dfrac{20(x-5)^2}{2}$

   (n) $M_x = -5x + 30(x-2)$

   (o) $M_x = -20 + 20(x-3)^0$

   (p) $M_x = 27.5x - \dfrac{10x^2}{2} + \dfrac{10(x-4)^2}{2} + 20(x-4)^0$

   (q) $M_x = \dfrac{M}{l}x$

   (r) $\dfrac{M}{l}x + M = M_x$

   (s) $M_x = -\dfrac{40x^2}{2} + 80(x-2) + \dfrac{40(x-2)^2}{2} + 80(x-10) - \dfrac{40(x-10)^2}{2}$

(t) $M_x = 40x - \frac{20x^2}{2} + \frac{40(x-4)^2}{2} + \frac{40(x-4)^2}{2}$

(u) $M_x = 45x - \frac{10(x-1)^3}{6} + \frac{20(x-4)^3}{6} - \frac{10(x-7)^3}{6}$

(v) $M_x = 45x - \frac{30x^2}{2} + \frac{10x^3}{6} - \frac{20(x-3)^3}{6}$

2. $\frac{86.67}{EI}$ rad

$\frac{193.33}{EI}$ units

3. $\frac{59}{EI}$

$\frac{139.2}{EI}$

4. 0.008 rad

16 mm

5. $-\frac{160}{3EI}, \frac{170.66}{EI}$

6. 0.0077 rad

0.0082 rad

16.5 mm

14.5 mm

7. $\frac{252}{EI}$

$-\frac{288}{EI}$

3.116 m

$\frac{507.2}{EI}$

8. $\frac{196.7}{EI}$

$\frac{-150}{EI}$

$\frac{413.34}{EI}$

9. $\frac{355.67}{EI}$

$\frac{-227}{EI}$

$\frac{766.68}{EI}$

$\frac{402.67}{EI}$

10. 0.0097 rad

0.0105 rads

19 mm

11. $\frac{40}{EI}$

$\frac{32}{EI}$

$\frac{64}{EI}$

12. 0.0053 rad

0.0057 rad

8.78 mm

5.4 mm

13. $\frac{240}{EI}$

$\frac{186.7}{EI}$

$\frac{+533.33}{EI}$

14. $\frac{53.33}{EI}$ (clockwise)

$\frac{26.67}{EI}$ (anticlock-wise)

$\frac{133.33}{EI}$ (downward)

$\frac{53.33}{EI}$

– 26.67 (upwards)

15. $\frac{53.33}{EI}$

$\frac{-53.33}{EI}$

$\frac{213.32}{EI}$

16. $\frac{-173.33}{EI}$

$-\frac{146.67}{EI}$

$\frac{-320}{EI}$

17. $\frac{430}{EI}$

$\frac{913.33}{EI}$

18. $\frac{130}{EI}$

$\frac{290}{EI}$

19. $\frac{240}{EI}$

$\frac{480}{EI}$

20. $\frac{33.75}{EI}$

21. $\frac{100}{EI}$

$\frac{140}{EI}$

$\frac{240}{EI}$

22. 0.0216 rad

0.0215 rad

38.8 mm

51 mm

23. $\frac{100}{EI}, \frac{120}{EI},$

$\frac{120}{EI}, \frac{346.67}{EI}$

24. $\frac{180}{EI}$

$\frac{270}{EI}$

25. $\frac{280}{EI}$

$\frac{820}{EI}$

26. $\frac{-67.5}{EI}$

$\frac{162}{EI}$

27. $\frac{53.33}{EI}$

$\frac{26.67}{EI}$

$\frac{133.32}{EI}$

$-\frac{26.67}{EI}$

28. $\frac{160}{EI}$ rad

$\frac{136.67}{EI}$

$\frac{280}{EI}$

$\frac{639.88}{EI}$

Chapter 8

4. 50.92 N/mm^2
   5.833°
5. 86.48 GPa
6. 92.6 N/mm^2
7. 1.615 kN m
   2.43°
8. 83.34 GPa
   145.5 N/mm^2
12. 88 mm
13. 76 mm
14. 82 mm
15. 50 mm
    2.17°
16. 0.424
17. 35.15 mm
18. 38.2 N/mm^2
    9.55 N/mm^2
    4.837°
19. 1.08°
    3.485°
20. 141.4 N m
    4.18°
21. 1.694
22. $L_1 = \frac{1}{2} L_2$
23. 63 mm
    53 mm
24. 1.56 mm
25. 1.272 kN m
26. 28.2 kN m
27. 8

Chapter 9

1. 79.2 N/mm^2
   55.3 N/mm^2
2. – 50.98 N/mm^2
   – 45.62 N/mm^2
3. – 260 N/mm^2
   – 40 N/mm^2
4. 30°: σ_θ = 52 N/mm^2
   (tensile)
   τ_θ = – 30 N/mm^2
   120°: σ_θ = – 52 N/mm^2
   (compressive),
   τ_θ = 30 N/mm^2
5. 204 N/mm^2
   75 N/mm^2
   ± 64.5 N/mm^2
   138 N/mm^2
   35.5°, + 80.5°
6. 82 N/mm^2
   – 44 N/mm^2
   – 80°, 10°
   σ_θ = 8 N/mm^2
   τ_θ = 62 N/mm^2
7. 114 N/mm^2
   –14 N/mm^2
   ± 62.5 N/mm^2
   50 N/mm^2
8. (a) 48 N/mm^2
   22 N/mm^2
   92 N/mm^2
   2 N/mm^2
   48 N/mm^2
   2 N/mm^2
   (b) – 107.5 N/mm^2
   – 20 N/mm^2
   – 23.75 N/mm^2
   87.5 N/mm^2
   – 16.25 N/mm^2
   – 87.5 N/mm^2
9. 164.72 N/mm^2
   75.28 N/mm^2
   ± 44.72 N/mm^2
   13.28°/103.28°
   –31.72°/121.72°
10. 220 N/mm^2
    – 120 N/mm^2
    – 14.05°, – 104.05°
11. + 20.71 N/mm^2
    – 120.71 N/mm^2
    ± 70.71 N/mm^2
    22.5°/112.5°
    – 22.5°/67.5°
12. ± 250 N/mm^2
    ± 250 N/mm^2
    – 18.43°/–108.43°
    26.67°/116.57°
13. 186.97 N/mm^2
    13.03 N/mm^2
    ± 86.97 N/mm^2
    36.64°/126.64°
    8.36°/98.36°
14. 141 N/mm^2
    30.22 N/mm^2
    55.4 N/mm^2
15. 29.9 N/mm^2
    – 144.2 N/mm^2
    87.05 N/mm^2
16. 76.82 N/mm^2
    – 57.65 N/mm^2
18. 271
    – 71.87
    179.38 N/mm^2
19. 52.5
    30.3
    60.6
    35 MPa
20. –5
    60.62 MPa
    70 MPa
21. –82.62 MPa
    68 MPa
22. $\varepsilon_x = 0.9375 \times 10^{-4}$
    $\Phi_{xy} = 0.75 \times 10^{-4}$

Chapter 10

1. 6.25 Nm
2. 94.12

3. 5093 N mm
   0.2546 mm
   0.15625 N mm
4. 7520 N mm
5. 1529 N mm
6. 127.32 N/mm^2
   0.127 mm
7. 3119 N mm
   0.0208 mm
8. 0.1 mm
   2482 N mm
9. 101.86 N/mm^2,
   2.04 mm, 204 Nm
   203.72 N/mm^2,
   4.074 mm
10. (a) 124.85 Nm
    (b) 28856 N mm
11. (a) 5595 N mm
    5596.9 N mm
    (b) 5005.64 N mm
12. 12.93 N
13. 25.92 N
14. 105 mm
15. 3.927 kg
16. 28,000 N mm
17. 165.4 N/mm^2
    82.7 N/mm^2
18. 88.52 N/mm^2
    44.26 N/mm^2
19. 578.7 Nmm, 1.16 mm
20. 8.38 kN m
21. 2.142 kN m
    0.00225 kNm
    952
22. 80 mm
23. 11.165 mm
24. 0.075 kNm
    5.625×10^{-4} kNm
25. 1.64 mm
    3.375 mm
26. 0.851 mm
27. $7wL^4/384EI$
28. 1.88 mm
29. $\theta = 0.916$
31. $\theta = 0.057°$
32. $5wL^4/384EI$
33. 500 N m
    89 N/mm^2
34. 0.355 kN m
35. 530.5 kN m
38. 28.125/*EI*
    67.5/EI
39. 0.0024 rad, 0.0023 rad, 4.2 mm, 5.5 mm
40. 53.33/*EI*
    71.87/*EI*
    110/3*EI*
    6.67/*EI*
41. 426.67/*EI*
42. 202.5/*EI*
    720/*EI*
43. 240/*EI*, 120/*EI*,
    – 180/*EI*
44. 6745/*EI*
    426/*EI*
    486/*EI*
45. 157.5/*EI*
    – 202.5/*EI*
46. 3200/*EI*
    426.67/*EI*
47. 1080/*EI*
    2160/*EI*
48. 160/*EI*
    1083/*EI*
49. 1452.5/*EI*

Chapter 11

1. (a) 46.2
   (b) 40.67
2. 526.36 kN, 58.17
3. 7185 kN, 11.2 m
4. 7.895×10^5 N
   5.03×10^5 N
6. 10730 kN
   2665 kN
7. 924 kN
   579 kN
8. 3529 kN
9. 735 kN
10. 265 kN
11. 171.93 (C), 105.27 (T)
12. 152.96 N/mm^2
13. 74.1 N/mm^2
    370 N/mm^2
    29.6 N/mm^2
14. 361.3×10^3 N
    255×10^3 N
15. 3.43 kN, 14.54 N/mm^2
16. 1.595 kN
    15.23 N/mm^2
17. Square section size
    = 105 mm, 125 mm
18. 2.96
19. 198 kN
21. 1.013 m

Chapter 12

1. 8 plates
   63.28 mm
2. 19 plates
   287.s7 N/mm^2
   51.38 mm
   2083
3. 4095 N
   45 Nm
4. 6, 12.5 mm^2
5. 8, 8 m
6. 364.5 Nmm
   6.68 mm
   7653 Nmm
7. 3.5 kN
8. 6 plates
   156.25 N/mm^2.
9. 26.65 N m
10. 43.52 N/mm^2
    35.88 mm
    143.6 N m
11. 5 springs
12. 4.04 mm
    62 mm
    49.78 mm
    1250 N mm
13. 40.53 N/mm^2
    78.71 N/mm^2
    34.7 mm
14. 98.38 N/mm^2
    13.18 N/mm^2
    100.12 N/mm^2
    50.93 N/mm^2
    0.87 mm
    0.375 rad
15. 3.4 mm, 93.5 N
16. 62.5 N mm
    468.75 N mm
17. 29.69 N, 19.8 mm
18. 2.55 mm, 6
19. 523.5 N
    0.14 mm
    0.02 rad
20. 100 N/mm^2
    50 N/mm^2
    0.000425

0.10625 mm
93226 mm^3
21. 1600 mm or 1.6 m
22. 0.21 N/mm^2
23. 0.2 mm
0.375 mm
1508 cm^3
24. 18.365 N/mm^2
80.5 N/mm^2
25. 46.67 N/mm^2
– 60 N/mm^2
26. 187 mm
27. 3.95, 15.32,
97.37, 76
28. 9.2 mm
29. 6.15 N/mm^2
30. 4.2 N/mm^2
31. 28.8 N/mm^2
100 N/mm^2
15.55 N/mm^2
32. 291.65 Nm
33. 987.24 Nm
34. 162.5 N/mm^2
35. 14.37 mm
36. 2.36 mm
37. 21.7 kN
39. 20 mm
40. 63 N/mm^2
41. 6.8
42. 12.3 mm

Chapter 13

8. 4.79 mm
9. (i) 3.35 mm
(ii) 1.38 mm
10. (i) 9.1 mm
(ii) 5.77 mm

Chapter 14

1. (a) Indeterminate, 5
(b) Indeterminate, 3
(c) Indeterminate, 3
(d) Indeterminate, 1
2. (a) Indeterminate, 5, 3
(b) Indeterminate 2, 6
(c) Indeterminate, 3, 3
(d) Indeterminate, 10, 27
(e) Indeterminate, 6, 15
3. (a) 4 (b) 6 (c) 5
(d) 4 (e) 10 (f) 6
(g) 5
4. FEM = 42.67 kNm
5. 24.44 kN
6. FEM = 33.75 kNm
8. M_A – = 23.0 kNm
MB = 34 kNm
9. M_A 104.99 kNm
M_B = 26.86 kNm
10. M_A = – 28.8 kNm
11. FEM = 86.67 kNm
12. Force in BC = 4.86 kN (compressive)
13. AO = 42.39 kN
BO = 41.14 kN
CO = 27.86 kN
14. Reaction at E = 14.3 kN
15. Force in AB = 5.52 kN
16. Force in diagonal members 30.93 kN and 49.6 kN
17. Reaction at D = 8 kN
18. Force in BC = 16.75 kN
19. Force in AC = 30.44 kN
20. Force in top chord member = 7.97 kN

Chapter 15

1. 52.5 kNm
37.5 kNm
2. 113.33 kNm
86.67 kNm
3. $M_A = -40, M_B = 0$
4. M_A = 162 kNm
M_B = 243 kNm
5. M = 56.25 kNm
6. 26.72 kNm
7. 31.78 kNm
53.68 kNm
8. 150 kNm
116.72 kNm
9. 0.25 kNm
113.6 kNm
10. $\frac{2}{9}$
11. 70 kNm
12. 48.75 kNm
13. M_A = 0.52,
M_B = 30.46
14. M_A = 28.62 kNm
M_B = 22.77 kNm
M_C = 42.78 kNm
15. M_B = 118.4 kNm
M_C = 60.8 kNm
16. M_B = 149.44 kNm
M_A = – 74.72 kNm
17. M_A – 47.06 kNm
M_B = – 29.33 kNm
18. 86 kNm
19. M_A = 15.6 kNm
M_B = 35.28 kNm
20. M_B = 12.24 kNm
M_C = 19 kNm
21. M_A = 68.25 kNm
22. M_B = 22.28 kNm
23. M_A = 15.9 kNm
24. M_A = 2.83 kNm
25. M_A = 48.3 kNm

Index